中国震例
EARTHQUAKE CASES IN CHINA
（2003～2006）

名誉主编　车　时
主　　编　蒋海昆
副 主 编　付　虹　杨马陵　马宏生

地震出版社

图书在版编目（CIP）数据

中国震例．2003～2006/蒋海昆主编．—北京：地震出版社，2014.5
ISBN 978-7-5028-4395-3
Ⅰ.①中… Ⅱ.①蒋… Ⅲ.①地震报告—中国—2003～2006 Ⅳ.①P316.2
中国版本图书馆 CIP 数据核字（2014）第 024176 号

地震版 XM2394

中国震例（2003～2006）

名誉主编：车　时
主　　编：蒋海昆
副 主 编：付　虹　杨马陵　马宏生
责任编辑：王　伟
责任校对：庞亚萍　凌　樱

出版发行：地震出版社
北京民族学院南路 9 号　　邮编：100081
发行部：68423031　68467993　　传真：88421706
门市部：68467991　　传真：68467991
总编室：68462709　68423029　　传真：68455221
专业图书事业部：68721991
E-mail：68721991@sina.com
http：//www.dzpress.com.cn

经销：全国各地新华书店
印刷：北京地大天成印务有限公司

版（印）次：2014 年 5 月第一版　2014 年 5 月第一次印刷
开本：787×1092　1/16
字数：1280 千字
印张：50
印数：1000
书号：ISBN 978-7-5028-4395-3/P（5085）
定价：200.00 元

编辑组成员

名誉主编　车　时

主　　编　蒋海昆

副 主 编　付　虹　杨马陵　马宏生

秘　　书　王　博　张　勇

内 容 提 要

《中国震例》系列丛书是研究地震和探索地震预测预报问题的重要科学资料。1988、1990、1999、2000、2002、2003 和 2008 年陆续出版了《中国震例》1 ~9 册，合计收录 1966 ~2002 年间发生的 240 次地震共 210 篇震例研究报告。本册（第 10 册）共收录 2003 ~2006 年间发生的 33 次 $M_S \geqslant 5.0$ 级、1 次 4.9 级、以及补充编写的 2002 年 1 次 4.7 级地震的震例研究共 26 篇。每个报告大体包括摘要、前言、测震台网及地震基本参数、地震地质背景、烈度分布及震害、地震序列、震源机制解和地震主破裂面、观测台网及前兆异常、前兆异常特征分析、应急响应和抗震设防工作、总结与讨论等基本内容。本书是以地震前兆异常为主的、规范化的震例研究成果，文字简明、图表清晰，便于查询、对比和分析研究。

本书可供地震预测预报、地球物理、地球化学、地质、工程地震等领域的科技人员、地震灾害管理专家学者、大专院校师生及关心地震灾害的读者使用和参考。

Synopsis

The multi-volume series book of "Earthquake Cases in China" contains important scientific data and information for seismological studies and researches on earthquake prediction and/or forecast. Volumes I to IX of this multi-volume series book were published in 1988, 1990, 1999, 2000, 2002, 2003 and 2008 with 210 case study reports on 240 earthquakes occurred from 1966 to 2002. This volume (Volume X) includes study reports on cases of 33 earthquakes of $M_S \geqslant 5.0$ and 1 earthquake with $M_S 4.9$ occurred from 2003 to 2006, as well as 1 supplementary study report on an earthquake with $M4.7$ occurred in 2002. In general, each case report includes abstract, introduction, seismic network and basic parameters of an earthquake, seismogeological background, seismic intensity distribution and earthquake damages, earthquake sequence, focal mechanism solutions and main fault plane, monitoring network and precursory anomalies, analyses on characteristics of precursory anomalies, measures of emergency response and earthquake protection, summary and discussions. This book is a collection of basic analyses and results of systematic and standardized studies on earthquake cases based mainly on the earthquake precursory anomalies. Simple and concise illustrations and distinct figures and tables are convenient for readers to get references, to make comparisons and analyses.

The book can be used and referred to by scientific and technical workers of earthquake prediction and forecast, geophysics, geochemistry, geology, engineering seismology, by earthquake disaster managers, by university and/or college teachers and students and by readers who are interested in seismic hazard reduction.

编 写 说 明

中国地震预测预报实践自1966年邢台地震开始，已走过近50年的历程，取得了显著的进展。地震预测预报是以观测为基础的科学，短临预测预报作为地震预测预报的主要目标，实现它的重要环节是获取可靠的地震前兆异常，综合分析多方面的资料，进而进行地震发生时间、地点和震级三要素的预测预报。因此，全面积累每次地震的地震地质、震害、地震参数、地震序列，尤其是地震前兆异常及预测预报和应急响应的经验教训等资料，对于地震科学研究、地震预测预报和防震减灾具有特别重要的科学价值。经过研究整理的一次或一组地震的上述系统资料，本书中称之为震例研究报告，它们是地震预测预报及其研究的基础。

1966年以来我国大陆发生了众多的5级以上地震，其中的许多地震已有不少论文和专著，但由于没有统一的规范和要求，有关资料不便进行系统的综合分析对比。为了系统地研究地震前兆和推进地震预测预报工作，国家地震局（现中国地震局，以下同）于1986～1987年安排了我国大陆1966～1985年5级以上地震的研究项目，作为该研究的成果，《中国震例》1～3册于1988年和1990年出版，含58篇震例研究报告，系统地研究总结了60次地震。1992年起国家地震局安排了1986～1991年第二批震例的研究及震例编写，共完成56篇震例研究报告，包括60次地震震例，于1999年和2000年出版了《中国震例》第4和第5册。在国家科技部社会公益性项目资助和中国地震局的联合支持下，2000～2002年完成了1992～1999年中国大陆灾害性地震震例总结研究，并在2002年和2003年出版了《中国震例》第6～8册。第9册《中国震例》于2008年出版，共收录2000～2002年发生的23次M_S≥5.0级和1次M_S4.7地震的21篇震例研究报告，同时给出了以前出版的《中国震例》1992～1999年3册中的青海震源机制的校正结果。本册（第10册）共收录2003～2006年间发生的33次M_S≥5.0级、1次4.9级地震、以及补充编写的2002年1次4.7级地震的震例研究报告共26篇。

《中国震例》的震例研究和报告编写工作基本按《震例总结规范》进行，以研究报告集的形式按地震发生日期顺序编辑成册。各总结研究报告按以下基本章节内容进行编写：

一、摘要

概述报告的主要内容。

二、前言

给出主震或重要地震的基本参数、震害、预测预报、宏观考察和研究历史等情况的概述。

三、测震台网和地震基本参数

给出地震前震中附近测震台网情况，以及主震或重要地震的基本参数。对同一地震，当不同单位给出不同参数时，则分别列出，编写人认为最合理的参数放在第一条。

四、地震地质背景

简要介绍震中附近地区的区域大地构造位置、深部构造条件、区域形变场概貌、历史地震活动及主要构造与断裂的活动习性，以及与发震构造有关的其他资料。

五、烈度分布与震害

给出烈度分布图、宏观震中的地理位置。简要介绍等震线范围、重要地表破坏现象、烈度分布特征及震害评估结果。

六、地震序列

尽可能给出全序列资料（包括直接前震和余震的有关参数）、余震震中分布图、地震序列类型、应变释放曲线或能量衰减曲线图、序列 b 值、频度衰减系数及较大余震目录等。

七、震源机制解和地震主破裂面

分别给出震源机制解图和表。对同一地震，如有不同的解，则分别列出，编写人认为最合理的解列在表中第一条。综合分析地震主破裂面与发震构造的关系。

八、观测台网及前兆异常

介绍地震前的定点前兆观测台网及其他有关观测情况。规定 $M_S \geqslant 7.0$ 级地震距震中 500km 以上，$6.0 \leqslant M_S < 7.0$ 级地震距震中 300km，$5.0 \leqslant M_S < 6.0$ 级地震距震中 200km，作为定点观测台网前兆观测资料的统计范围，给出此范围内测震台（项目）以外的其他地震前兆定点观测台站（点）或观测项目分布图，并在必要时给出前兆异常项目平面分布图。认为与此次地震孕育过程有关的全部前兆异常，包括非定点台网观测到的异常和上述规定距离以外的重要异常，均列入前兆异常登记表，并给出前兆异常图件。概述前兆异常的总体情况，以图表为主，必要时加以简要文字说明。对地震学项目以外所有定点观测台站（点）的所有观测项目或异常项目进行累加统计时，其统计学单位称为台项。对前兆异常登记表中的异常项目进行累加统计时其统计单位称为项次或条。

为保证资料的可靠性，要求所用数据的观测质量必须符合观测规范，且能够区别正常动态与异常变化。根据地震前兆观测资料清理和分析研究的结果把观测资料质量划分为三类：1 类——符合上述要求；2 类——基本符合；3 类——不符合。规定只选用 1、2 类观测资料，3 类资料不予使用，亦不进入统计。异常判定应经过全部资料和全过程的分析，经排除干扰和年变等因素后，根据一定的判据，认定与地震关系密切的变化才列入异常登记表。

规定按时间发展进程把异常分为 L、A、B、C 四个阶段类别：L——长期趋势背景异常，出现在地震前 5 年以上；A——中期趋势背景异常，出现在震前 0.5～5 年；B——为短期趋势异常，震前延续 1～6 个月；C——临震异常，震前 1 个月内。另外，对远离规定的震中距范围以外，或据现有认识水平一时无法解释，以及非常规观测的、值得研究的其他可靠和较可靠的异常现象划为 D 类，在相应的异常阶段类别前冠以 D 字样，以留下资料和记录供后续研究。对各类异常，按照其可信程度，又区别为Ⅰ、Ⅱ、Ⅲ三个等级，以下角标标示：Ⅰ——可靠；Ⅱ——较可靠；Ⅲ——参考，留作记录。D 类异常只取Ⅰ和Ⅱ两类。如：C_{II} 为较可靠的临震异常；DA_{I} 为可靠的中期 D 类异常。关于Ⅰ、Ⅱ、Ⅲ等级的确定，主要尊重研究报告作者的意见，编辑过程中仅作了个别调整，供读者参考。宏观异常在登记表中总的作为一项异常。异常登记表中各栏目，既是报告作者对异常研究的结果，亦是为了给读者提供使用、研究和参考的方便。对异常进行以上的认真审核和分类处理，既可达到去粗取精、去伪存真的目的，又可避免丢失可能有科学价值的异常记录，以利于进一步研究和资料积累。尽管如此，书中辑入的异常未必都恰当，读者可根据提供的资料和文献进一步做出判断。

全书对异常登记表中使用的观测手段和异常项目名称及图件中的常用图例作了统一规定（见异常项目名称一览表和常用图例）。

九、前兆异常特征分析

简要给出对主要异常特征的综合分析与讨论，给出要点，提出有依据的看法和待研究的问题。

十、应急响应和抗震设防工作

简要介绍（记录）预测预报、应急响应和抗震设防等方面的重要情况和工作过程，包括对强余震的监测、预测情况等。

十一、总结与讨论

从科学上讨论有技术和工作特色的经验、学术观点、教训、问题及启示。

十二、参考文献和资料

给出在震例研究和报告编写工作中研究过的主要文献和资料目录。报告中直接引用已出版文献或未出版的参考资料、图件和工作结果时均注明来源，以便读者进行核对或追踪研究。

在本系列书中，对于已发表有专著的强震，根据专著发表后的研究成果，亦按以上要求编写震例研究报告，并进行必要的资料补充，专著中发表过的异常图件一般从略，文字从简。

本书辑入的震例研究报告是前人和作者对该次震例资料整理和研究成果的集中表达，是以地震前兆异常为主的、规范化的震例科研成果。《中国震例》编辑组工作的指导思想是：经过科学整理和分析研究，给出各次地震的基本资料，既可供读者使用、参考，又可供进一步追踪研究；既具有资料性，又要反映目前研究程度；文字力求简明，避免冗长的叙述和讨论，因此尽量使用图表，便于对比。由于资料和研究程度的差异，各报告在坚持质量和科学性的前提下，根据实际情况编写和编辑，因此篇幅和章节编排不尽一致。

中国大陆地震前兆的观测与预测预报实践表明，地震孕育和发生是一个极其复杂的过程，影响因素很多，伴随这一过程有许多异常现象。我们把那些地震前出现的、与该地震孕育和发生相关联的现象称之为地震前兆，即采用了广义地震前兆的概念。本书辑录的地震前兆异常，是经过审核的、有别于正常变化背景的、可能与该地震孕育和发生相关联的异常变化，其中既可能有区域构造应力场增强引起的异常（“构造前兆异常”），又可能有来自震源的信息（“震源前兆异常”），具有不同的前兆指示意义，无疑包含着丰富的可能的前兆信息。因而震例研究报告是地震前兆研究和预测预报探索的宝贵财富，它既是进一步研究的基础资料，又可供在今后震情判定中借鉴。

震例研究报告是震后经过若干年的资料收集、发掘、整理和总结研究之后编写的，从震后总结到实现震前的科学预测预报，还要经过一段艰难的路程。结合汶川地震科学总结，根据地震形势的发展及科学认识的深入，本册在严格遵循《震例总结规范》的基础上，力图在以下方面有所加强：

（1）强调震例的史料及档案性质。要求前述第八部分在对地震学及前兆异常进行系统梳理（震后总结）的前提下，着重震前预测主要科学依据、所得结论的叙述，着重当时论证过程实事求是的还原，包括不同观点的碰撞。要求尽可能提供震前预测及震后趋势判定全面的原始证据，包括预测依据及预报凭据。对有一定预测实效的震例，更要加强预测过程、预测依据的详细辑录，详细收集当时开展科学预测的原始凭据。

（2）进一步强化震例总结的科学性。前述第九部分除已有内容的客观表达外，要求作者站在目前的角度，以当前的科学眼光，重点对当时预测过程的得失成败进行科学评述及原因分析。

（3）为突出地震预测预报这一震例研究工作的重点，并保证资料的权威性，对前述第五部分“烈度分布与震害”和第十部分“应急响应和抗震设防工作”适当简化，对应急、震害等数据直接引用相关正式资料并列出参考文献即可。

本书所辑入的震例研究报告，基于“属地为主”的原则由发生地震所在的省（自治区、直辖市）地震局负责总结研究。各报告对前人或相关的研究工作成果，特别是地震前兆研究的成果，虽尽力作了反映，但由于人员变动和资料收集的困难，以及水平限制等原因，难免仍会有疏漏，对个别异常和资料的处理亦可能会有不妥之处。

中国地震台网中心为震例工作的负责单位，承担震例研究及报告编写工作的安排、审阅、修改等工作。震例总结指导专家组由蒋海昆、刘杰、杨马陵、付虹、陈棋福、周龙泉组成。本册《中国震例》由蒋海昆、付虹、杨马陵、马宏生编辑完成，王博、张勇承担了大量事务性的工作。书中文字及图件由地震出版社王伟、庞亚萍进行了统一编辑加工和校对。《中国震例》（2003～2006）编辑组仍遵循此前制订的2～3人分别把关评审与主编审定的工作程序，确保每份报告都经历了初稿、修改稿（1次或多次）与承担单位的验收等过程。编辑组在严格遵守作者“文责自负”的前提下，在不违背原则的情况下对每份报告的体例和分析结果等进行了适当的编辑处理。编辑组虽然作了很大努力，但由于水平和条件所限，书中可能还有不周或不足之处，望予谅解并提出宝贵意见。

监测预报司预报管理处刘桂萍处长（现任中国地震台网中心副主任）对震例研究及震例编写有较深入的思考，自始至终关注并参与该项工作，编者对此表示衷心的感谢！

编　者

2013年10月　北京

About This Book

In China, practices in earthquake precursor observations and earthquake prediction and/or forecast have been carried out for more than 45 years since the Xingtai earthquake in 1966 and substantial progress has been achieved. Earthquake prediction and forecast is a science that based mainly on observations. The short term and imminent prediction or forecast of the time, magnitude and place of an earthquake is the principal goal of earthquake prediction or forecast. Successful forecast or prediction can only be achieved on the basis of acquisition of reliable data of earthquake precursory anomalies and comprehensive analyses of all data. Therefore, for earthquake research, prediction, protection and hazard mitigation, it is of particularly important scientific value to accumulate extensive data of seismogeology, earthquake disasters, earthquake parameters, earthquake sequence and especially earthquake precursor anomalies and lessons of prediction and emergency response of an earthquake. The above mentioned systematic data of an earthquake or a group of earthquakes obtained through researches and classification are treated as research reports of earthquake cases in this book. They are the foundation data for earthquake prediction or forecast and related researches.

Many earthquakes of $M_S \geqslant 5.0$ occurred in China mainland since 1966. Numerous papers and/or works related to many of them have been published and/or carried out. But due to lack of unified standards and requirements, many relevant data could not be analyzed and compared systematically and comprehensively. In order to carry out comprehensive studies on earthquake precursor anomalies and to promote earthquake prediction research, during 1986-1987 the State Seismological Bureau (named China Earthquake Administration now) launched a project for researches on earthquakes with $M_S \geqslant 5.0$ occurred in China mainland during 1966-1985. As a result, the first three volumes of "Earthquake Cases in China" were published in 1988 and 1990, containing 58 earthquake case reports of systematic studies on 60 earthquakes. Since 1992 the State Seismological Bureau had initiated the researches of the second phase on earthquake cases. Many researches were carried out and 56 research reports on 60 earthquake cases were written and published in Volume Ⅳ and Ⅴ in 1999 and 2000 respectively. With financial support for public affairs from the Ministry of Science and Technology and with support from the China Earthquake Administration, summarizations of earthquake cases occurred during 1992-1999 in China mainland were continued from 2000 to 2002, and published Volume Ⅵ, Ⅶ and Ⅷ in 2002 and 2003. The Volume Ⅸ had been published in 2008, which includes 21 research reports on 23 earthquakes with $M_S \geqslant 5.0$ and 1 earthquake with $M_S 4.7$ occurred from 2000 to 2002. And the revised focal mechanism solutions for the Qinghai earthquake cases from 1992 to 1999 in published Volume Ⅵ, Ⅶ and Ⅷ of "Earthquake Cases in China" were given. This volume (Ⅹ) includes 26 research reports on 33 earthquakes with $M_S \geqslant 5.0$ and 1 earthquake with M_S4.9 occurred from 2003 to 2006, as well as 1 supplementary study report on an earthquake with M4.7 occurred in 2002.

The book is compiled in the form of collection of reports on earthquake cases and arranged according to occurrence dates of the earthquakes. All reports of earthquake cases were written with the reference standards and requirements of "Specification for Earthquake Case Summarization". Each report contains the following basic components:

Abstract is a summary of the major contents.

Introduction gives a brief description of the occurrence time of the main shock or main earthquakes, its or their damages, the status of prediction or forecast, the macroscopic investigations and the history of earthquake studies, etc.

Seismic Network and Basic Parameters of the Earthquake gives the distribution of seismic network near the epicenter before the event (s) and the basic parameters of the main shock or main earthquakes. When the different parameters of an earthquake were given by different agencies, they are listed separately, but the first one on the list is the parameter that the authors deem most reasonable.

Seismogeological Background gives a brief description of the location of the regional geotectonic structures, deep structures, general picture of the regional deformation field, historical earthquake activity, activities of main structures and faults and other data associated with the seismogenic structures around the hypocenter.

Distribution of Seismic Intensity and Damages illustrates the distribution of seismic intensity, the geographic location of the macroseimic epicenter. The range of isoseismal lines and significant phenomena of surface destruction are described, the features of intensity distribution and the estimated earthquake damages are outlined.

Earthquake Sequence provides the whole sequence (including the relevant parameters of all direct foreshocks and aftershocks), the distribution of aftershock epicenters, the type of the sequence, the strain release curve or the energy attenuation curve, b value of the sequence, the frequency attenuation coefficient and the catalogue of major aftershocks.

Focal Mechanism Solution and Main Rupture Plane gives figures and tables of the focal mechanism solutions. When there are different solutions, they are given separately, with the most appropriate one is listed as the first one by the authors. Comprehensive analyses are made on the relation between the earthquake rupture plane and the seismogenic structure.

Monitoring Network and Precursory Anomalies describes the precursor monitoring network and other related observations. Statistical analyses are made on the precursory anomalies obtained from the networks within or more than the distance of 500km from the epicenters of the $M_S \geqslant 7.0$ earthquakes, within 300km from the epicenters of earthquakes of $6.0 \leqslant M_S < 7.0$, and within 200km from the epicenters of the earthquakes of $5.0 \leqslant M_S < 6.0$. Maps of fixed observation stations (points) or observation items (except seismic observation items) within such distances and maps of distribution of precursory anomalies (only indicating precursory items of fixed observations except seismic anomalies) are also provided. All anomalies that are assumed to be closely linked with the process of the earthquake preparation, including the important anomalies at non-fixed observation points and outside the defined distances, are listed in the summary table of precursory anomalies with corresponding figures. The overall situation of the precursory anomalies is outlined, mainly with figures and tables and with concise illustrations if necessary. The statistic unit of observation items or anomaly items of all stations (points) is called station-item.

In order to ensure the reliability of the data, the observation quality of the data must meet the observation specifications and the normal variations and anomalous changes can be distinguished. According to the result of the sorting out and analyses of the precursor observations, the quality of the observation data are classified into three classes: Type 1 — the data meet the above mentioned quality requirements; Type 2 — the data meet the quality standards in general and the normal variations and anomalies can be distinguished; Type 3 — the data don't meet the requirements. It is decided that only the first two types of data can be used, while the data of the third type will not be selected for statistical analyses. The anomalies are identified on the basis of result of analyses on all data during the whole process after eliminating contaminations, annual variations and other contamination factors. Thereafter, only anomalies identified to be closely associated with earthquakes are listed in the summary table of precursory anomalies.

The anomalies are divided into four classes L, A, B and C according to the time development of the anomalies. Class L indicates the long-term trend anomalies that appear five years or more before the earthquake; Class A is the mid-term trend anomalies which occur about six months to five years before the earthquake; Class B denotes the short-term anomalies which last for about one to six months before the earthquake; Class C means the imminent anomalies that occur within approximately one month before the impending earthquake. In addition, Class D is introduced to include certain reliable or fairly reliable anomalies that deserve further studies. They might appear at observation stations that are even further away from the epicenter than the defined distance, they could not be explained with present knowledge, or they are not obtained by conventional observations. The anomalies are further classified according to their reliabilities into degrees Ⅰ, Ⅱ and Ⅲ, with Ⅰ — reliable; Ⅱ — fairly reliable and Ⅲ — for reference. But the anomalies of Class D are only classified in degrees Ⅰ and Ⅱ. The reliability degree is marked by subscript to the bottom right of the class symbols. For example, C_{II} is a fairly reliable imminent anomaly; DA_{I} is a reliable mid-term anomaly of Class D. They are usually determined by the opinions of the authors, except a few are revised by the editors for reader's reference. The macroscopic anomalies registered in the summary table of precursory anomalies are regarded as one item of anomalies. Various items of anomalies registered in the table are the research results obtained by many authors and are provided to the readers to utilize, study and refer to with convenience. The stringent evaluation and classification of the anomalies not only serves the purpose of selecting the high quality data, but also helps to avoid the possibility of losing any scientifically valuable records of anomalies that are useful in further scientific analyses. However, the anomalies included in the book are not necessarily correct for all of them and readers should make further judgment based on the data and references provided.

The names of the observation items (precursory items) and legend in the figures are unified in the following pages.

Analyses of Features of Precursory Anomalies gives comprehensive analyses and discussions on features of the main anomalies with interpretation based on facts and opinions on problems for future study.

Measures of Emergency Response and Earthquake Prevention gives brief introduction on important situations and procedures of the work in earthquake forecast or prediction, emergency response and earthquake prevention, including the monitoring of strong aftershocks and so on.

Discussions and Concluding Remarks explores scientifically the experience, academic ideas, lessons, problems and revelations that are characteristic in technology and practical work.

References and Information lists all major references and data catalogues which have been studied during the case study and report compilation. The origins of published and unpublished data, figures and results, which are directly quoted in the reports, were given also.

Some strong earthquakes that have been studied in published monographs are also compiled with earthquake case reports, with necessary data supplemented. However, the published figures of anomalies are usually deleted and illustrations are simplified.

Each of the earthquake case reports contained in this book is the manifestation of the achievement gained by the predecessors and authors in sorting out and studying the earthquake case. They are the fruit of a systematic and standardized scientific research on earthquake cases with emphasis on precursor anomalies. The Editorial Board of Earthquake Cases in China has been working under the guide line that this book will provide readers for their use, reference and future research with basic data of each earthquake obtained through scientific sorting out and analyses. Therefore, all reports are designed to have abundant information and clearly indicate the current research level. The literal illustrations are as simple as possible without lengthy descriptions and discussions, so available figures and tables are given for comparison. Each report is compiled and written to the highest possible quality and scientific soundness. However, owing to differences in data and research extent and the actual situations, the length and format for all reports are not exactly the same.

The earthquake precursor observations and forecast or prediction practices in China mainland have shown that the preparation and occurrence of an earthquake is a rather complicated process influenced by many factors and accompanied by various anomalous phenomena. We call the anomalies appeared before an earthquake that are closely linked with the process of the preparation and occurrence of the earthquake and distinct from the normal background of variations as earthquake precursor anomalies, or as precursor anomalies in general sense. The earthquake precursory anomalies included in the book are examined to be relevant phenomena associated possibly with the process of earthquake preparation and occurrence. Among them, there may be anomalies caused by intensification of regional tectonic stress field (referred to as "tectonic precursor anomalies") and the information from a single earthquake focus (called as "focal precursor anomalies"). They have different precursory implications, undoubtedly with possible and rich precursory information. Therefore, the earthquake case reports are the valuable accumulations for studies on earthquake precursors and forecast or prediction. They provide not only basic data for further investigations, but also contribute references for future assessment of the development of earthquake activity.

However, it should be noted that those earthquake case reports have been compiled through several years of collection, analysis and exploration, and summarizing of the data after the earthquakes, and there is still a long and arduous way from the post-earthquake summarization to scientific earthquake prediction or forecast. It also should be pointed out that besides following the "Specification for Earthquake Case Summarization" strictly, some new demands have been proposed for this volume:

1. Emphasizing the historical and dossier properties of the earthquake cases. Following the systemic study on the seismogeological and precursory anomalies, a scientific and real description on evidences and conclusions are needed, especially for decision-making process before the mainshock. The original proofs for earthquake forecast or judgment of aftershock tendency are asked to be provided.

2. Emphasizing the scientific properties of the earthquake cases. In "Analyses of Features of Precursory Anomalies", besides something mentioned above, the authors also have been asked to comment on the successful or unsuccessful earthquake prediction, on the side of present scientific point of view. The analysis on reasons of successful or unsuccessful earthquake prediction are the key points of this part.

3. For projecting the earthquake forecast or prediction, which is the emphasis of the "Earthquake Case in China", as well as to ensure the reliability of the book, two parts mentioned above, "**Distribution of Seismic Intensity and Damages**" and "**Measures of Emergency Response and Earthquake Prevention**", should be simplified felicitously. It is feasible that the correlative data could be quoted directly and references be listed.

The research reports of earthquake cases collected in this book were mainly prepared by Earthquake Administrations of the provinces, autonomous regions and metropolitan cities according to the principle of the earthquake location. All efforts were made to ensure that the reports reflect the achievement of researches obtained by the predecessors or in related researches and particularly the achievement of researches on precursors to earthquakes. However, due to personnel changes and limited data accessibility, there might be inappropriate omissions or improper processing of individual anomalies and data.

China Earthquake Networks Center (CENC) is the department with responsibility for the research reports of earthquake cases. Jiang Haikun, Liu Jie, Yang Maling, Fu Hong, Chen Qifu, Zhou Longquan are members of guidance group for compilation of "Earthquake Cases in China". The editorial work of this volume was completed by Jiang Haikun, Fu Hong, Yang Maling and Ma Hongsheng. Jiang Haikun is the chief editor. Fu Hong, Yang Maling and Ma Hongsheng are the associate editors. Wang Bo and Zhang Yong are secretaries with responsibility for lots of businesslike jobs. All texts and figures in the book were edited by Wang Wei and Pang Yaping from the Seismological Press. The editorial board of "Earthquake Cases in China" (from 2003 to 2006) followed the strict working procedure of previous volumes. 2 or 3 editors were in charge of appraising the reports and the chief editor was in charge of final approving of the reports. For every report, there had to be a manuscript, a revised manuscript (revised once or several times) and the manuscript had to be examined and accepted by the institution that participated the project. Under the prerequisite that "the author is responsible for his own report or paper", the editorial board made some appropriate editing of the format and the results of analyses without violation of the principles. Though great efforts were made by the editorial board, there still might be some improper aspects in the book due to our limited scientific knowledge and work conditions. Therefore, any comments and corrections are greatly appreciated.

The Editorial Board

October 2013, Beijing

地震前兆异常项目名称一览表

学　　科	异　常　项　目　名　称
地震学	地震条带，地震空区（段），空区参数σ_H，地震活动分布（时间、空间、强度），前兆震群，震群活动，有震面积数A值，地震活动性指标（综合指标A值，地震活动熵Q^t、Q^N、Q^Σ，地震活动度γ、S（模糊地震活动度Sy）），地震强度因子M_f值，震级容量维D_0值，地震节律，应变释放，能量释放，地震频度，b值，h值，地震窗，缺震，诱发前震，前震活动，震情指数$A(b)$值，地震空间集中度C值，η值，D值；地震时间间隔，小震综合断层面解，P波初动符号矛盾比，地震应力降τ，环境应力值τ_0，介质因子Q值，波速，波速比，S波偏振，地震尾波（持续时间比τ_H/τ_V、衰减系数a、衰减速率p），振幅比，地脉动，地震波形；断层面总面积$\sum(t)$，小震调制比，地震非均匀度GL值，算法复杂性$C(n)$、AC
地形变	定点水准（短水准），流动水准；定点基线（短基线），流动基线；测距；地倾斜；断层蠕变；GPS
应力-应变	钻孔应变（体积应变，分量应变），压容应变，电感应力，伸缩应变
重　　力	定点重力，流动重力
地　　电	视（地）电阻率ρ_s；自然电位V_{SP}；地电场
地　　磁	Z变化，幅差，日变低点位移，日变畸变；总场（总强度），流动地磁；磁偏角；感应磁效应（地磁转换函数）；电磁扰动（电磁波）
地下流体	氡(水、气、土)，总硬度，水电导，气体总量，pH值，CO_2、H_2、痕量H_2、He、N_2、O_2、Ar、H_2S、CH_4、Hg（水、气）、SiO_2、Ca^{2+}、Mg^{2+}、SO_4^{2-}、HCO_3^-、Cl^-、F^-含量；地下水位，井水位；水（泉）流量，水温
气　　象	气温，气压；干旱，旱涝
其他微观动态	油气井动态；地温；长波辐射（OLR）
宏观动态	宏观现象
综　　合	前兆信息熵（H）；异常项数

The List of Earthquake Precursory Items

Subject	Precursory items
seismology	seismic band, seismic gap (segment), parameter of seismic gap σ_H, earthquake distribution (temporal, spatial, magnitude), precursory earthquake swarm, earthquake swarm activity, number of areas of earthquake occurrence (A value), index of seismic activity (comprehensive index A, seismic entropy Q^t, Q^N and Q^Σ, degree of seismic activity γ and S, fuzzy degree of seismic activity Sy), seismic intensity factor M_f value, fractal dimension of magnitude capacity D_0, earthquake rhythm, strain release, energy release, earthquake frequency, b value, h value, seismic window, earthquake deficiency, induced foreshock, foreshock activity, exponential of eathquake situation ($A(b)$ value), degree of seismic concentration C value, η value, D value of seismicity; time interval between earthquakes, composite fault plane solution of small earthquakes, sign-contradiction ratio of P-wave first motions, co-seismic stress drop τ, ambient stress τ_0, quality factor (Q value), wave velocity, wave velocity ratio, S-wave polarization, seismic coda wave (sustained time ratio τ_H/τ_V, attenuation coefficient a, attenuation rate p), amplitude ratio, microtremor, seismic waveform; total area of fault plane ($\sum(t)$), regulatory ratio of small earthquakes, degree of seismic inhomogeneity (GL value), Algorithmic Complexity ($C(n)$, AC)
deformation	fixed leveling (leveling of short route), mobile leveling; fixed baseline (short baseline), mobile baseline; ranging; tilt; fault creep; GPS
strain stress	Borehole strain (volumetric strain, 4-components strain), piezo-capacity strain, electric induction stress, extensor strain
gravity	fixed-point gravity, roving gravity
geoelectricity	apparent resistivity (ρ_s); spontaneous potential (V_{SP}); geoelectric field
geomagnetism	Z variation of geomagnetism, amplitude difference of geomagnetism, low-point drift of daily variation of geomagnetism, distortion of daily variation of geomagnetism; total intensity of geomagnetism, roving geomagnetism; magnetic declination; induced magnetic effects (geomagnetic transfer functions); electromagnetic disturbance (electromagnetic wave radiation)

continued

Subject	Precursory items
ground water	radon content in (groundwater, air, soil), total water hardness, water conductivity, total amount of gas in groundwater, pH value; CO_2, H_2, Trace hydrogen, He, N_2, O_2, Ar, H_2S, CH_4, Hg (groundwater, air), SiO_2, Ca^{2+}, Mg^{2+}, SO_4^{2-}, HCO_3^-, Cl^- and F^- content in groundwater; ground water level, well water level, (spring) water-flow quantity, water temperature
meteorology	atmospheric temperature, atmospheric pressure; drought, waterlogging
other microscopic variation	variation of oil well; ground temperature; outgoing longwave radiation (OLR)
macroscopic variation	macroscopic phenomena
comprehensive	precursor information entropy (H); anomalous item number

图件中的常用图例

Legend

微观震中
instrumental epicenter

宏观震中
macroscopic epicenter

▲ 地震台站（不分观测项目时使用）
earthquake-monitoring station

测震台
seismic station

● 初动向上
finst motion（up）

○ 初动向下
finst motion（down）

水　准
leveling

基　线
baseline

D 断层蠕变
fault creep

T 地倾斜
tilt

E 地　电
geoelectricity

ρ_s 视（地）电阻率
apparent resistivity

V_{SP} 大地电场（自然电位）
spontaneous potential

M 地　磁
geomagnetism

Mz 垂直磁场强度
vertical magnetic intensity

M_D 磁偏角
magnetic declination

G 重　力
gravity

Rn 具体标示的地球化学项目（圆内符号用相应化学组分符号标示）
marked geochemical item

Ch 不做具体标示的或一个以上的地球化学项目
unmarked geochemical items

W_E 水电导度
conductivity of groundwater

W_Q 水流量
water-flow quantity

W 水　位
water level

σ 应力应变
stress strain

验潮站
tidal gauge station

电磁扰动（电磁波）
electromagnetic disturbance

Wc 水　温
water tempeature

Gc 地　温
ground temperaure

目　　录

Contents

2002 年 9 月 3 日山西省太原 4.7 级地震

山西省地震局

张淑亮　宋美琴　吕　芳　高振强　李　丽

摘　要

2002 年 9 月 3 日在山西省太原市郝庄乡一带发生 4.7 级地震，微观震中为 37°54′N、112°48′E，宏观震中为 37°51′N、112°37′E，震中烈度为Ⅴ度。

该地震序列为孤立型，最大余震为 M_L2.3。震源机制解结果显示，主压应力轴方位为 247°，震源错动以走滑为主。推测北北西向东山山前断层为此次地震的发震断层。

震中 200km 范围内有 5 个测震台站和 13 个前兆台站或测点。震前在距震中 200km 范围内共出现异常项目 10 项，其中中期异常 4 项、短期异常 2 项、临震异常 4 项。震中 100km 范围内的异常台站和台项比分别为 33.3% 和 21.4%，101 ~ 200km 范围内的为 42.9% 和 25%。

前　言

经山西地震台网测定，2002 年 9 月 3 日 01 时 52 分山西省太原市迎泽区郝庄乡发生 4.7 级地震，微观震中为 37°54′N、112°48′E，宏观震中为 37°51′N、112°37′E，震中烈度为Ⅴ度。地震有感范围北到大同，南至临汾，西达离石，东及阳泉，波及范围较大。太原市及周边地区震感十分强烈，但没有造成房屋损坏和人员伤亡。

太原 4.7 级地震发生在汾渭断陷中部太原盆地内，盆地东、西两侧分别受太谷断裂和交城断裂控制，内部构造复杂。该区历史上发生的最大地震是 1102 年太原 6.5 级和 1614 年平遥 6.5 级地震。震中附近地震观测台站密度中等，200km 范围内有 5 个测震台、13 个前兆台点。震前出现 10 项前兆异常。

这次地震是继 1999 年大同—阳高 5.6 级地震后，山西境内发生的最大一次地震，因此，研究这次地震前各类异常变化特征，对准确掌握山西地区未来地震趋势有一定的指导意义。

本研究报告以山西省 2002 年度地震趋势研究报告为基础，收集了震后新发现的一些异常，经过重新整理资料和分析研究而完成的。

一、测震台网及地震基本参数

太原4.7级地震前，距震中200km范围内有5个测震台（图1）。其中0～100km范围1个，100～200km范围4个。山西地震台网对太原盆地的地震监测能力为$M_L \geqslant 2.5$级。4.7级地震的微观震中位于太原市迎泽区郝庄乡一带，地震基本参数见表1[1]。本地震研究所选目录来自中国地震台网中心。

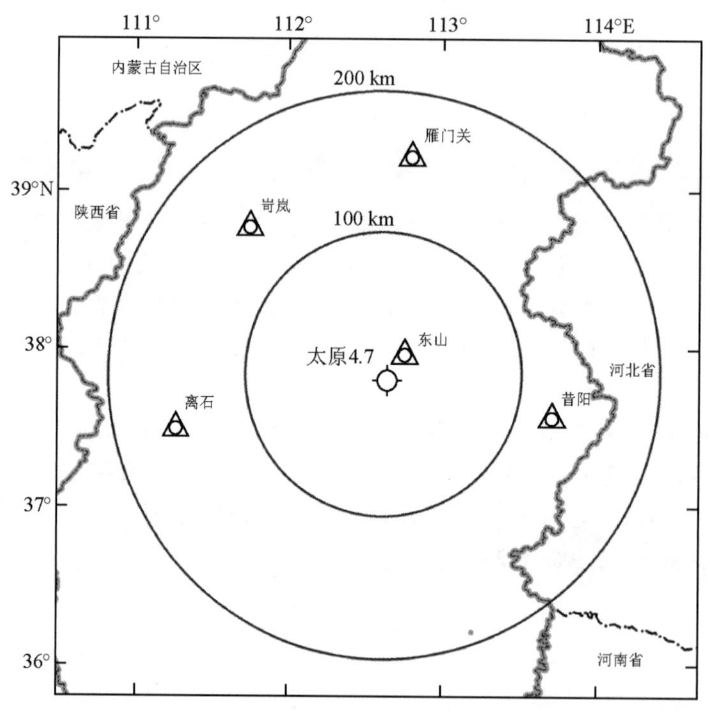

图1　太原$M4.7$地震震中附近测震台站分布

Fig. 1　Distribution of seismometric stations around the $M4.7$ Taiyuan earthquake

表1　太原4.7级地震的基本参数表

Table 1　Basic parameters of the 4.7 Taiyuan earthquake

编号	发震日期	发震时刻	震中位置		震级	震源深度	震中地名	结果来源
	年 月 日	时 分 秒	φ_N	λ_E		(km)		
1	2002 9 3	01 52 53	37°51′	112°35′	$M_L5.0$	18	太原市郝庄乡	山西省及邻近地区地震目录[1)]
2	2002 9 3	01 52	37°51′	112°37′	$M4.7$	18		中国地震台网中心
3	2002 9 3	01 52 53	37°54′	112°48′	$M4.7$		山西寿阳	国家台网大震速报目录

二、地震地质背景

太原4.7级地震发生在汾渭断陷中部太原盆地内。太原盆地除受交城断层、柴村断层、东山山前断层、太谷断层等盆地边界活动断层控制外，盆地内隐伏断层也十分发育。太原市及周边地区南北、东南、北东、北东东、北西向隐伏断层纵横交错，把该地区条块分割成大小不同的块体[1]。

太原4.7级地震发生在三给地垒断层、王家峰地垒断层、新城—亲贤断层、东山山前断层所控制的太原城区凹陷内。震中位于东西向王家峰地垒断层、南北向黄陵断层、北北西向东山山前断层三组隐伏断层的交汇部位（图2）。震中烈度为Ⅴ度，Ⅴ度区呈南北长、东西短的椭圆形，东起迎泽区张家河一带，西达万柏林区化客头乡一带，北起阳曲县泥屯镇小伽东村，南至清徐赵家堡一带（图3）。根据Ⅴ度区长轴的走向，北北西向东山山前断层为此次地震的发震断层[1]。

三、地震影响场和震害

根据地震现场调查，震区建筑物普遍为砖混结构房屋，抗震性能较好。尽管太原市及周边地区震感十分强烈，但没有发生房屋损坏和人员伤亡。调查确定地震极震区烈度为Ⅴ度。Ⅴ度区呈椭圆形，长轴为北北西向，长53km；短轴为北东东向，长34km，面积近5660km^2。其范围北起阳曲县泥屯北，南至太原市小店区北格、晋中市张庆，西达太原市万柏林区化客头，东及晋中市沛霖一带（图3）。由于地震发生在午夜，Ⅴ度区内熟睡的人们普遍从梦中惊醒，并听到地声。太原市区震感强烈，东西向晃动明显。万柏林区西铭西大街加油站墙上悬挂的镜框因地震而倾斜，小店区收费站加油桶被震倒。Ⅴ度区北部的泥屯一带，人们听到隆隆地声，从南而来；在东关口村一庙宇顶瓦被震落。Ⅴ度区南部的刘家堡、北格一带，人们普遍感到南北向晃动十分明显和强烈；晋中市北部的秋村加油站的保险柜、加油柜晃动明显，这一带人们普遍感觉到东西向晃动，晃动持续十几秒钟[1]。

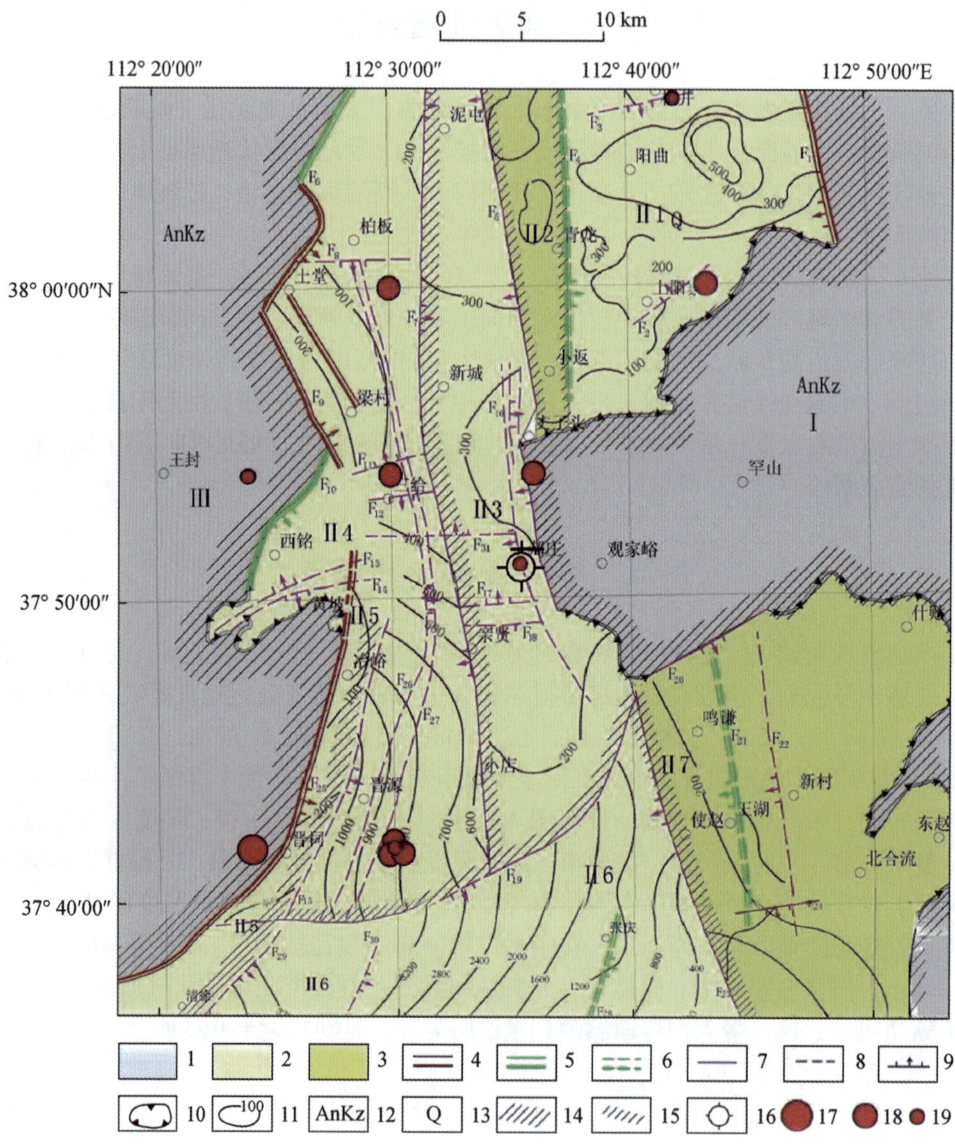

图2　太原4.7级地震区域活动构造图

Fig. 2　Map of regional fault structure and distribution of historical earthquakes around the epicenter of the 4.7 Taiyuan earthquake

1. 基岩隆起区；2. 断陷区盆地；3. 盆地内凸起；4. 全新世活动断裂；5. 晚新世活动断裂；6. 晚更新世隐伏活动断裂；7. 早、中更新世断裂；8. 早、中更新世隐伏断裂；9. 正断层；10. 盆地边界；11. 新生界等厚线；12. 前新生界；13. 第四系；14. 二级新构造单元界线；15. 三级新构造单元界线；16. 微观震中；17. 6.0～6.9级地震；18. 5.0～5.9级地震；19. 4.7～4.9级地震

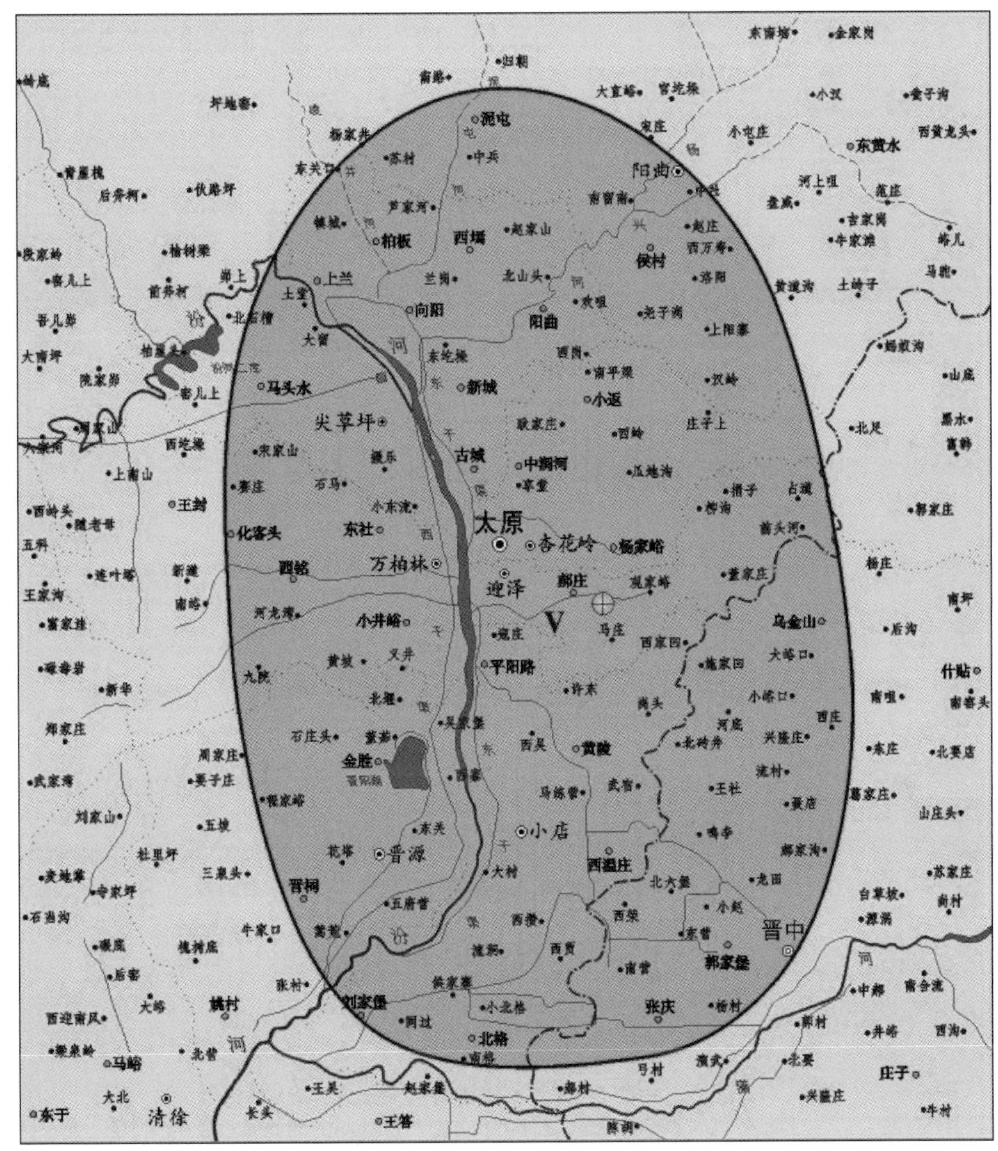

图 3 太原 4.7 级地震等震线图

Fig. 3 Isoseismal map of the 4.7 Taiyuan earthquake in 2002

四、地震序列

2002 年 9 月 3 日 01 时 52 分太原市郝庄乡（北纬 37°51′，东经 112°37′）发生 4.7 级地震，震源深度 18km。截至 10 月 21 日，共记录到 $M_L \geq 0.0$ 级地震 67 次（以近台东山台记录为准），其中：5.0 ~ 5.9 级 1 次，2.0 ~ 2.9 级 18 次、1.0 ~ 1.9 级 46 次，0.6 ~ 0.9 级 2 次，

最大余震为2.4级(图4、表2)。根据能量比99.99%和震级差2.5判断,为孤立型地震类型。

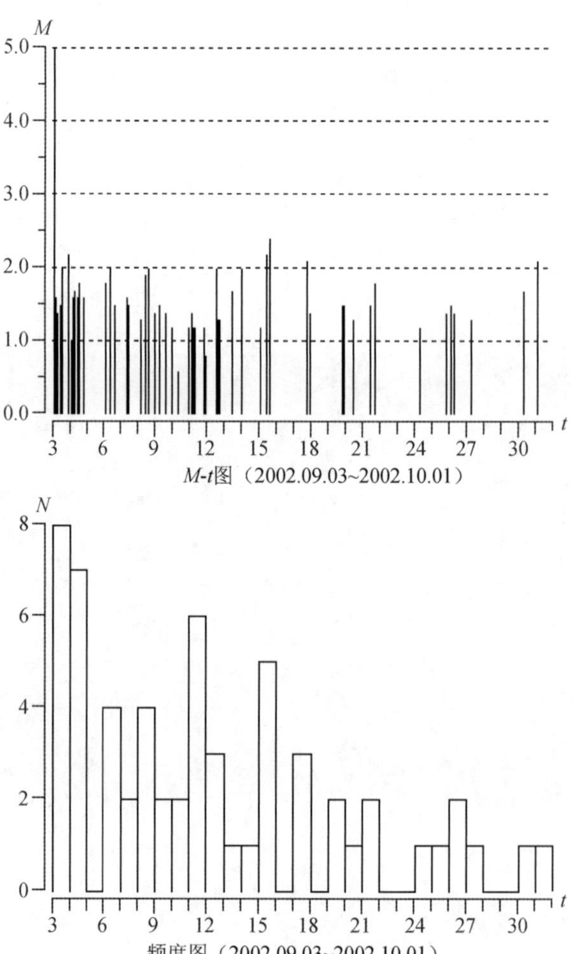

图4 太原4.7级地震时序图

Fig. 4 The sequence diagram of the 4.7 Taiyuan earthquake

表2 太原4.7级地震序列目录(M_L≥2.0级)

Table 2 Catalogue of the 4.7 Taiyuan sequence (M_L≥2.0)

编号	发震日期	发震时刻	震中位置		震级	震源深度	地名	结果来源
	年 月 日	时 分 秒	φ_N	λ_E	M_L	(km)		
1	2002 9 3	01 52 53	37°51′	112°35′	5.0	11	太原市郝庄乡	山西省及邻近地区地震目录[1)]
2	2002 9 3	02 19 00			2.1			

续表

编号	发震日期	发震时刻	震中位置		震级	震源深度	地名	结果来源
	年 月 日	时 分 秒	φ_N	λ_E	M_L	(km)		
3	2002 9 3	12 28 00			2.0			
4	2002 9 3	22 12 00	37°42′	112°34′	2.2	13	太原市郝庄乡	
5	2002 9 6	09 40 00			2.0			
6	2002 9 8	15 01 00	37°44′	112°37′	2.0		太原市郝庄乡	
7	2002 9 12	13 34 00			2.0			
8	2002 9 14	00 10 00			2.0			
9	2002 9 15	11 15 00			2.2			
10	2002 9 15	11 22 00			2.2			
11	2002 9 15	14 19 00			2.4			
17	2002 9 17	18 07 00			2.1			
18	2002 10 01	03 11 00			2.1			

五、震源参数和地震破裂面

根据山西地震台网记录，做出了太原 4.7 级地震的 P 波初动解（表 3）。图 5 为 P 波初动解的等面积投影（下半球）。结果显示，主压应力轴方向 247°，仰角 7°；主张应力轴方向 338°，仰角 2°；2 个地震节面走向分别为 23°和 292°，倾角很大，均为高角度拉张走滑正断层。其主压应力方向与山西乃至华北的主压应力方向基本一致，震源错动以走滑为主。其中节面 B 走向北西西向，与震中附近的东山山前断裂（北北西）向接近，而且断层性质一致，可以认为东山山前断裂为本次地震的发震断裂。

表 3 太原 4.7 级地震震源机制

Table 3 Focal mechanism solution of the *M*4.7 Taiyuan earthquake

节面 A			节面 B			P 轴		T 轴		N 轴		矛盾符号比	初动个数
走向 (°)	倾角 (°)	滑动角 (°)	走向 (°)	倾角 (°)	滑动角 (°)	方位 (°)	仰角 (°)	方位 (°)	仰角 (°)	方位 (°)	仰角 (°)		
23	83	−173	292	87	−3	247	7	338	2	87	83	0.18	38

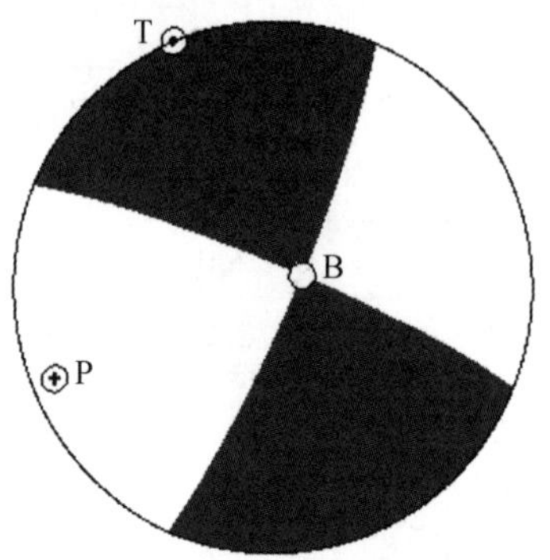

图5　太原4.7级地震震源机制解图

Fig. 5　Focal mechanism solution of the 4.7 Taiyuan earthquake

六、地震前兆观测台网和前兆异常

图6是距震中约200km范围内地震前兆观测台站（测点）分布情况，共有13个前兆台点25个测项，其中0～100km范围内有6个台点14个测项，100～200km范围内有7个台点11个测项。

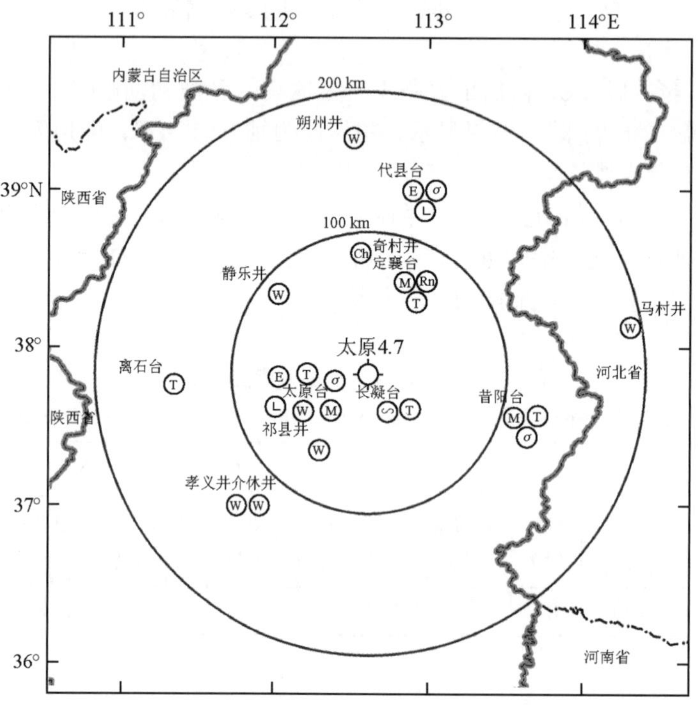

图6　太原4.7级地震前定点前兆观测台站分布图

Fig. 6　Distribution of precursory observation stations before the 4.7 Taiyuan earthquake

太原 4.7 级地震前前兆观测项目有 5 个台点 6 个测项出现异常，其中 0 ~ 100km 范围内有 2 个台点 3 个测项异常，101 ~ 200km 范围内有 1 个台点 1 个测项异常，200km 以外有 2 个台点 2 个测项异常（图 7）。

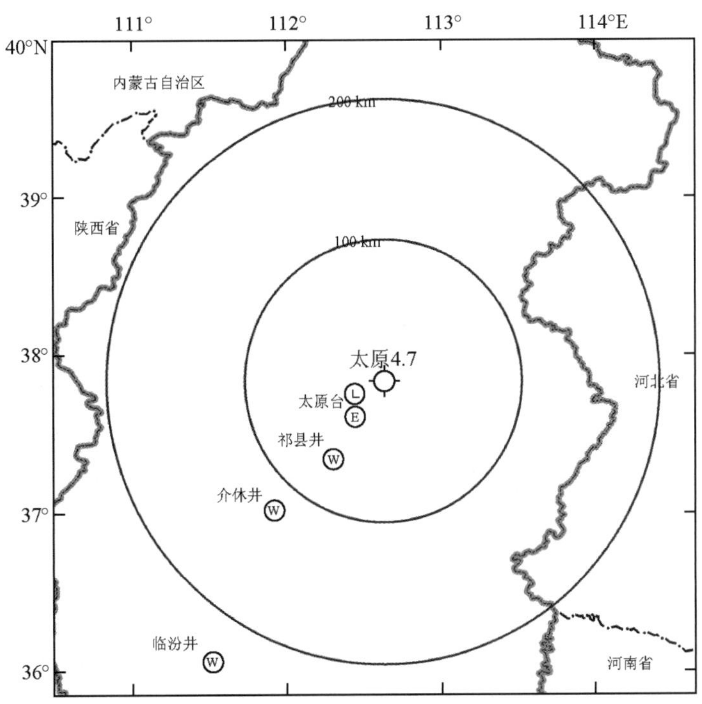

图 7　太原 4.7 级地震前定点前兆观测异常分布图

Fig. 7　Distribution of precursory anomalies on the fixed observation points before the 4.7 Taiyuan earthquake

太原 4.7 级地震前出现各类异常 10 项，其中地震活动性 4 项，定点观测异常 6 项。各异常所表现的特征见表 4 及图 8 至图 17。

表 4　异常情况登记表

Table 4　Summary table of precursory of anomalies

序号	异常项目	台站或观测区	分析方法	异常判据及观测误差	异常起止时间	震后变化	最大幅度	震中距 (km)	异常类别	图号	异常特点和备注
1	地震空段	太原盆地到忻州盆地	震中分布图	3.3 级地震围空区	1999.1 ~ 2002.9	消失			A_I	8	

续表

序号	异常项目	台站或观测区	分析方法	异常判据及观测误差	异常起止时间	震后变化	最大幅度	震中距（km）	异常类别	图号	异常特点和备注
2	地震条带	36° ~ 40°N，110° ~ 114°E	震中分布图	集中成带（M_L ≥ 2.0 级）	2002. 6. 1 ~ 7. 31	恢复正常			B_I	9	2 级地震集中在 NNW、NNE 向条带上，交汇于震中附近
3	地震平静	36° ~ 40°N，110° ~ 114°E	震中分布图	2 级地震集中活动的 2 个条带消失	2002. 8. 1 ~ 9. 3	恢复正常			C_I	10	
4	小震窗口	大同老震区	小震月频度图	月频度大于正常背景值的 1.0 倍	2002. 8. 1 ~ 8. 16	趋于平静	是背景值的 3 倍，月频次达 32 次	250	C_I	11	8 月 15 ~ 16 日地震活动频次增高，18 天后发震
5	逸出氡气	夏县	原始曲线	大于几十 ppm 背景值	2001. 8 ~ 2002. 5	恢复正常	3100ppm	330	A_I	12	异常结束发震
6	水位	介休	月均值曲线	突降	2002. 2. 27 ~ 8. 18	震后恢复	3. 34m	112	B_I	13	下降—下降加速—转折发震
7	水位	祁县	月均值曲线	破年变	2001. 12. 13 ~ 2002. 6	震后恢复	是正常年变幅的 1/2	61	A_I	14	异常恢复过程中发震
8	水位	临汾	日均值曲线	反年变	2001. 12 ~ 2002. 5	震后恢复	0. 5m	220	A_I	15	缓升—缓降—突　升—发震
9	水准	太原	日观测值	突降	2002. 9. 1 ~ 9. 2	震后恢复	1. 6mm	23	C_I	16	突降—转折—发震
10	地电	太原 EW	观测值图	突变	2002. 9. 2：20 ~ 22 时	恢复正常	0. 6Ω · m	20	C_I	17	异常结束后发震

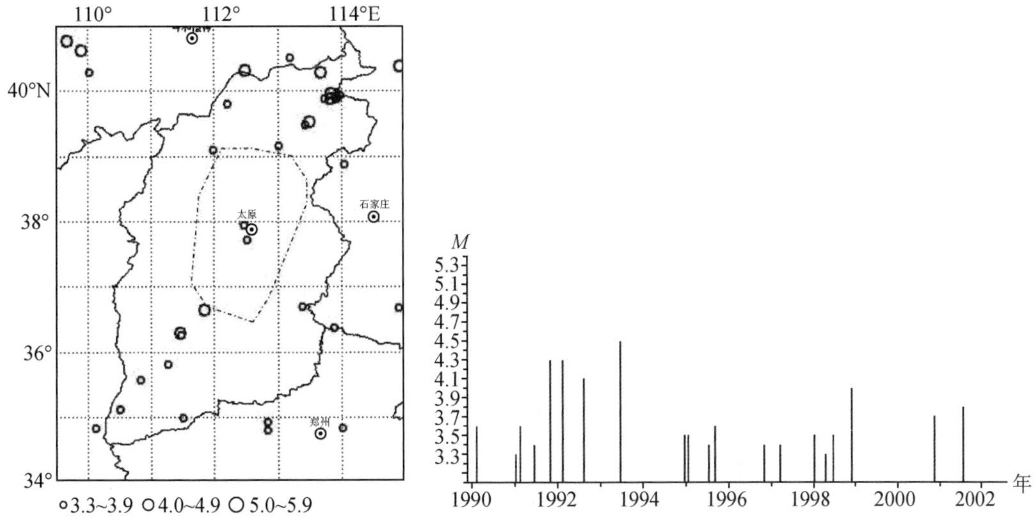

图 8 山西中北部地震空段图像及 M-t 图（1999 ~ 2001 年，$M_L \geqslant 3.3$ 级）

Fig. 8 Diagrams of the earthquake empty segment and M-t in the north of Shanxi (1999-2001, $M_L \geqslant 3.3$)

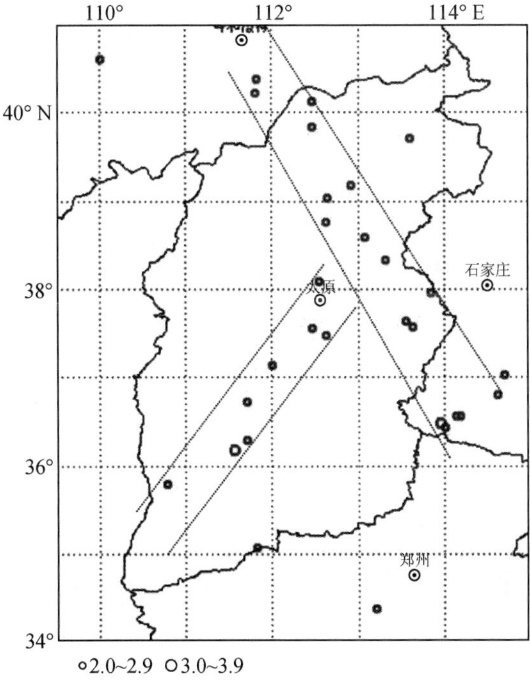

图 9 山西地区 $M_L \geqslant 2.0$ 级地震活动图像（2002. 6. 1 ~ 2002. 7. 31）

Fig. 9 Distribution of the $M_L \geqslant 2.0$ earthquake in Shanxi (2002. 6. 1-2002. 7. 31)

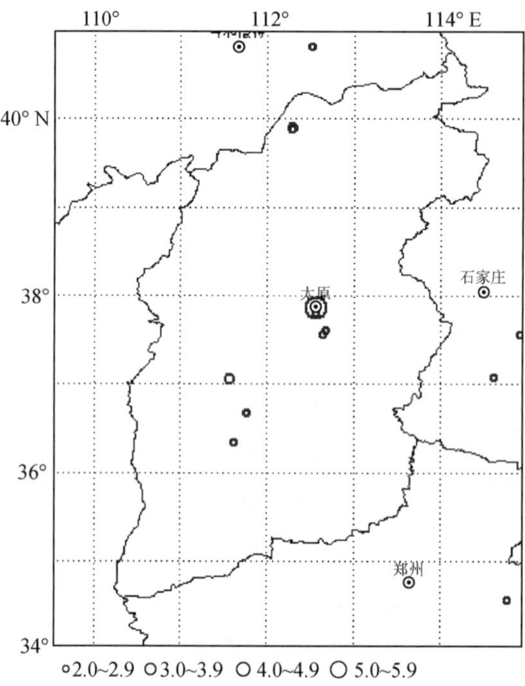

图 10　山西地区 $M_L \geqslant 2.0$ 级地震活动图像（2002. 8. 1～2002. 9. 3）

Fig. 10　Distribution of the $M_L \geqslant 2.0$ earthquake in Shanxi（2002. 8. 1-2002. 9. 31）

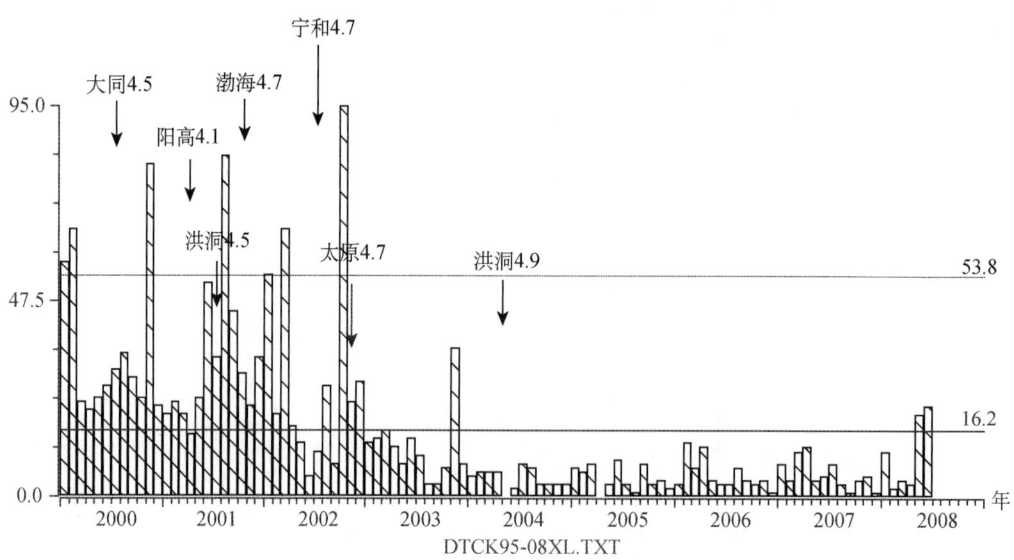

图 11　大同—阳高老震区窗口月频度图（2000～2008 年）

Fig. 11　Monthly frequency of microearthquakes in Datong-Yanggao seismic area（2000-2008）

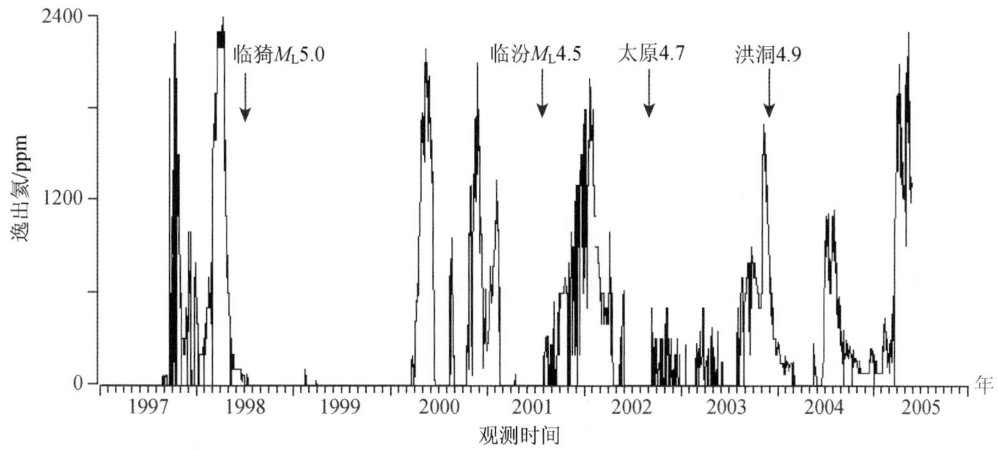

图 12　夏县逸出氦日观测值曲线（1997～2005 年）

Fig. 12　Curve of daily value of escaping helium observation at Xiaxian station（1997-2005）

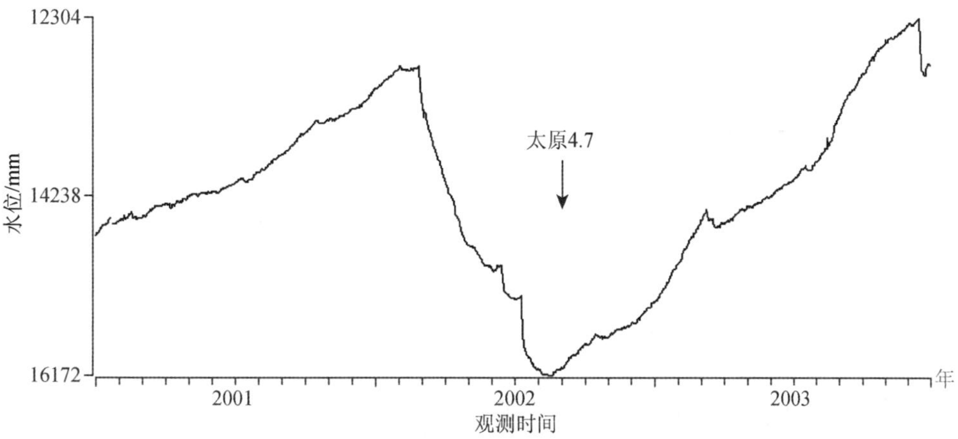

图 13　介休井水位月均值曲线（2001～2003 年）

Fig. 13　Curve of monthly mean value of well water level at Jiexiu station（2001-2003）

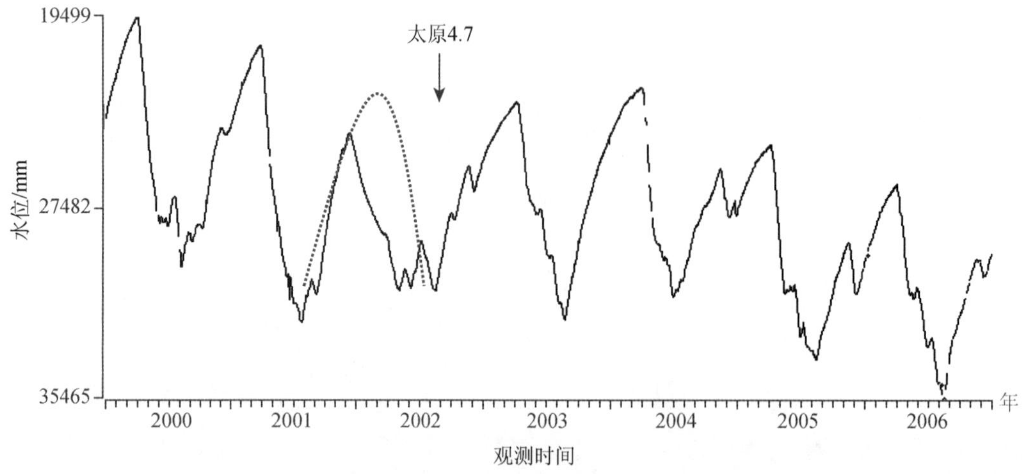

图 14　祁县井水位月均值曲线（2000 ~2006 年）

Fig. 14　Curve of monthly mean value of well water level at Qixian station（2000-2006）

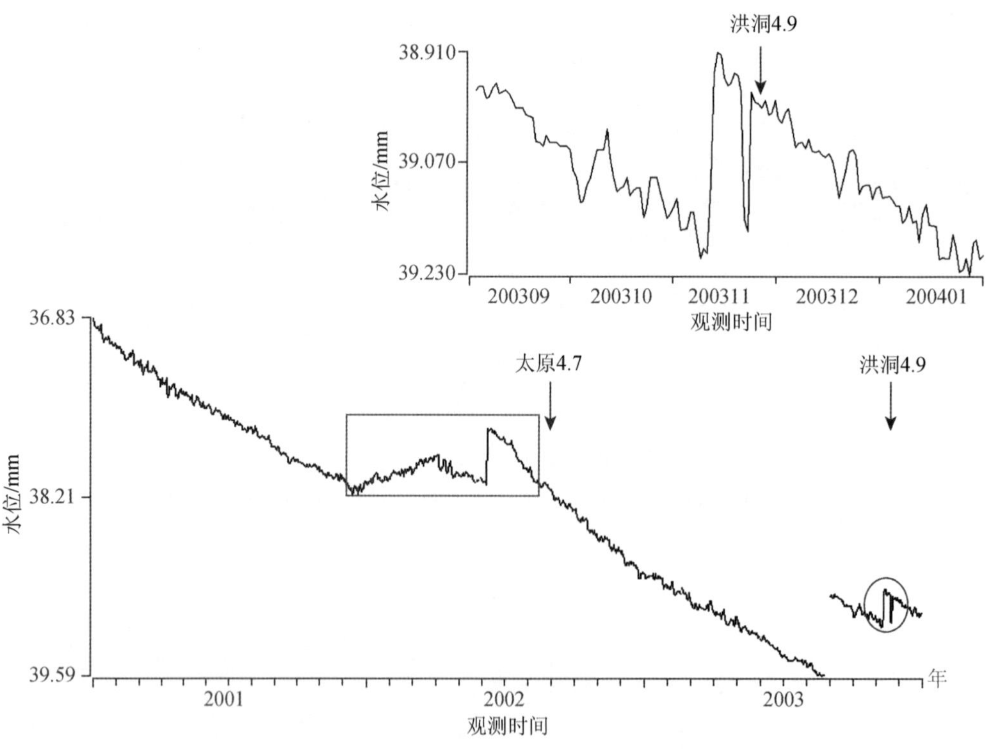

图 15　临汾行署地震局院井水位日均值图（2001 ~2003 年）

Fig. 15　Curve of daily mean value of the ground-water level in the well at Linfen Seismological Bureau（2001-2003）

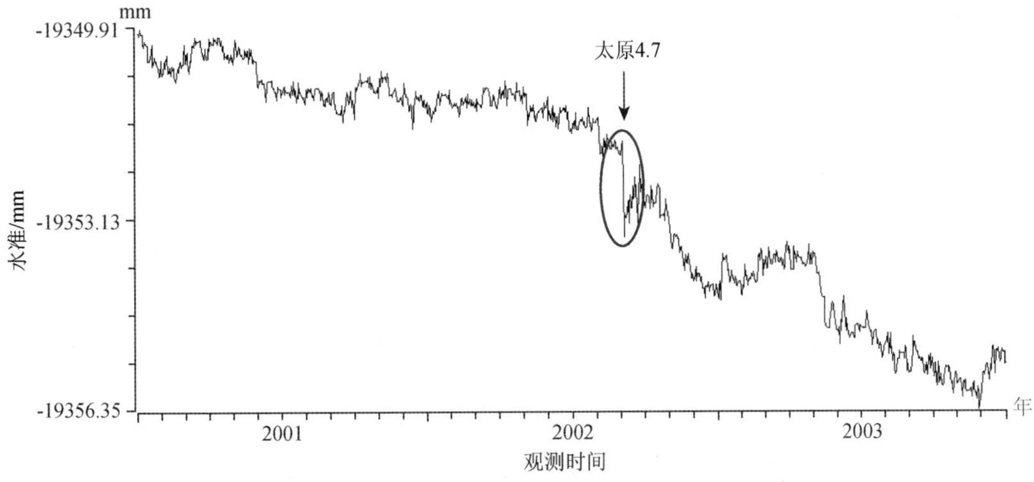

图 16　太原水准日观测值曲线

Fig. 16　Curve of daily mean observation value of leveling at Taiyuan station

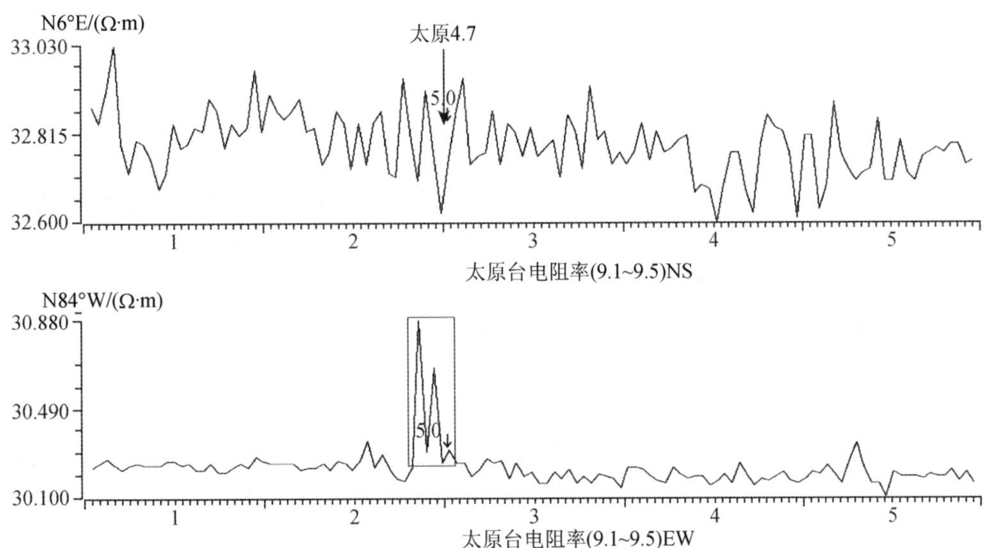

图 17　太原地电阻率整点值曲线（2002. 9. 1 ~ 2002. 9. 5）

Fig. 17　Curve of hour mean value of the apparent resistivity at Taiyuan station（2002. 9. 1-2002. 9. 5）

七、地震前兆异常特征分析

1. 时间特征

从异常测项来看，太原 4.7 级地震前地震活动性异常有 4 项，占异常总数的 40%；流体类异常 4 项，占异常总数的 40%；形变和磁电类异常各 1 项，各占总数的 10%。从异常的起始时间来看，中期异常和临震异常各 4 项，占异常总数均为 40%；短期异常 2 项，占总数的 20%，异常出现的时间以中期和临震异常为主，短期异常较少。异常出现时间最早的是地震活动性异常，大约出现在震前 3 年 8 个月，震前 1 年左右流体类异常相继出现，形变和磁电异常则出现在震前 1～2 天，甚至几个小时。4.7 级地震前各类异常随时间的演化进程是：地震活动性→流体→形变→地电阻率。

2. 空间特征

由太原 4.7 级地震前兆异常空间分布图可以看出，太原 4.7 级地震 6 项前兆异常在空间上呈 NE 向展布，其中距震中 100km 有 3 项，100～200km 范围内有 1 项，200～300km 范围内有 1 项，300km 以外有 1 项。距离震中最近的异常为太原台的水准与地电，约为 20km，最远的为夏县逸出氡气异常，约为 330km。对异常空间分布的研究表明外围区异常出现的时间早，且数量多，震中区附近异常较少，震中 100km 范围内的异常台站和台项比分别为 33.3% 和 21.4%，101～200km 范围内的为 42.9% 和 25%。各类前兆观测异常有由外围向震中推进的趋势。

八、震前预测、预防和震后响应

1. 预测情况

太原 4.7 级地震发生在山西省地震局 2002 年度地震趋势研究报告确定的年度地震值得注意地区——山西中南部地区[1)]。自 1998 年 12 月汾阳 4.0 级地震后，太原盆地的 3.3 级以上地震明显平静，仅在 2001 年 8 月 13 日发生了 1 次太原市的 3.8 级地震，形成一个从太原盆地到忻定盆地的地震空段。除此之外，在山西的中南部地区先后出现一批信度较高的前兆异常，为此，山西省地震局在 2002 年 8 月 7 日周会商会上提出"山西前兆异常相对集中于中南部，值得注意"的会商意见。尽管震前有所察觉，震前出现了几起前兆异常，并进一步强化了震情短临跟踪工作，但仍未捕捉到短期地震学前兆信息。

2. 震后趋势判定

太原 4.7 级地震发生后，山西省地震局在 2002 年 9 月 3 日紧急会商会上，根据 1989 年以来发生的 4 级以上地震均为孤立型的特点以及这次地震序列的能量比和震级差判断这次地震为孤立型地震类型。从历史地震看，1957 年太原市曾发生过 $M_S5.0$ 双震，由此判断本地震序列再发生更大地震的可能性不大。会商意见与实际地震类型一致。

3. 震后响应

地震发生后，山西省地震局立即向省委、省政府和中国地震局报告了初定的地震三要素，并启动《山西省地震局应急预案》，对震后工作做了安排部署。先后共派出 4 个考察组在震区开展了地震震害调查、地震宏观震中确定、地震烈度考察、地震构造分析等项工作。在召开的紧急会商会上，以地震的基本参数和序列给出初步判定意见。之后又连续召开 3 次紧急会商会，对震情密切跟踪、对地震序列及时分析判定，并和现场分析预报人员及时沟通，对今后的地震趋势给出了较好的判定意见。省地震局向省委、省政府和中国地震局上报了《震情通报》，及时上报震情情况、灾情情况、现场工作情况和震区的抗震救灾情况，并提出震后趋势意见和下一步的工作建议。

九、结论和讨论

2002 年 9 月 3 日太原 4.7 级地震在震中附近前兆异常与地震活动性异常较为丰富，共有 10 项信度较高的异常。最早出现的异常是震前 3 年零 8 个月在太原盆地与忻定盆地间出现地震空区，表明孕震区开始形成；震前 1 ~3 个月在孕震区范围内出现 2 级地震条带和平静，标志着震源区周围应力-应变状态的强化和地壳介质发生的显著变化；在接近震源孕区出现了太原水准和地电等临震异常。各类异常在时空域内的演化过程，在一定程度上反映了本次地震的孕震过程。

参 考 文 献

[1] 程新源等，2002 年山西太原郝庄 M_L5.0 地震宏观烈度考察报告 [J]，山西地震，2002，(4)：1 ~2

参 考 资 料

1）山西省地震局，山西省 2002 年度地震趋势研究报告，2001 年 11 月

The *M*4.7 Taiyuan Earthquake in Shanxi Province on September 3, 2002

Abstract

The *M*4.7 earthquake occurred in Haozhuang town, Taiyuan in Shanxi Province on September 3, 2002. Its instrumental epicenter was at 37°54′N, 112°48′E. The macroscopic epicenter was at 37°51′N, 112°37′E. The intensity of meizoseismal area of the earthquake was Ⅴ.

The earthquake was of isolated type, the largest magnitude of the aftershock was M_L2.3. The focal mechanism solution showed that the azimuth of principal compressive stress was 247°. It is suggested that the earthquake was the result of left lateral strike slip of the fault. The seismogenic structure was Dongshan fault that the strike was NNW.

There were 5 seismic stations and 13 precursor observational stations or measuring points within 200km from the epicenter of the earthquakes. 7 anomalies appeared before the earthquake in Shanxi province and its surrounding areas, including 4 medium term anomalies, 2 short-term anomalies and 4 imminent anomalies. Within 100km from the epicenter of the earthquake the ratios of stations with anormalies and items with anormalies were 33.3% and 21.4%, within 101-200km the ratios were 42.9% and 25%.

报告附件

附表 1　固定前兆观测台（点）与观测项目汇总表

序号	台站（点）名称	经纬度（°）		测项	资料类别	震中距 Δ（km）	备注
		φ_N	λ_E				
1	东山	37.92	112.74	测震		5.7	
2	太原	37.71	112.43	地倾斜	Ⅰ	29	
				水准	Ⅰ		
				地磁	Ⅰ		
				地电	Ⅰ		
				钻孔应变	Ⅰ		
				水位	Ⅰ		
3	长凝	37.62	112.89	电磁波	Ⅱ	32	
				地倾斜	Ⅱ		
4	定襄	38.43	113.00	地磁	Ⅰ	62	
				地倾斜	Ⅰ		
				水氡	Ⅰ		
5	祁县	37.36	112.30	水位	Ⅰ	74	
6	奇村	38.56	112.62	水氡	Ⅱ	75	
				水汞	Ⅱ		
7	静乐	38.36	112.03	水位	Ⅰ	85	
8	昔阳	37.57	113.72	测震		101	
				地倾斜	Ⅰ		
				钻孔应变	Ⅰ		
				地磁	Ⅰ		
9	介休	37.02	111.90	水位	Ⅰ	126	
10	代县	39.02	113.05	水准	Ⅰ	126	
				钻孔应变	Ⅰ		
				地电	Ⅰ		
11	离石	37.77	111.35	测震		128	
				地倾斜	Ⅰ		
12	岢岚	38.78	111.72	测震		132	
13	孝义	37.02	111.77	水位	Ⅰ	134	

续表

序号	台站（点）名称	经纬度（°）		测项	资料类别	震中距 Δ（km）	备注
		φ_N	λ_E				
14	雁门关	39.23	112.81	测震		148	
15	朔州	39.33	112.52	水位	Ⅱ	162	
16	马村	38.16	114.33	水位	Ⅰ	190	
				地热			

分类统计	$0<\Delta\leqslant100$km	$100<\Delta\leqslant200$km	总数
测项数 N	10	8	18
台项数 n	16	16	32
测震单项台数 a	1	2	3
形变单项台数 b	0	0	0
电磁单项台数 c	0	0	0
流体单项台数 d	3	4	7
综合台站数 e	3	3	6
综合台中有测震项目的台站数 f	0	2	2
测震台总数 $a+f$	1	4	5
台站总数 $a+b+c+d+e$	7	9	16
备注			

附表 2　测震以外固定前兆观测项目与异常统计表

序号	台站（点）名称	测项	资料类别	震中距 Δ（km）	按震中距 Δ 范围进行异常统计																			
					0 < Δ ≤ 100km					100 < Δ ≤ 200km					200 < Δ ≤ 300km					300 < Δ ≤ 500km				
					L	M	S	I	U	L	M	S	I	U	L	M	S	I	U	L	M	S	I	U
1	太原	水准	I	29	—	—	—	√																
		地电	I		—	—	—	√																
3	祁县	水位	I	74	—	√	—	—																
4	介休	水位	I	126						—	—	√	—											
	临汾	水位	I	220											—	√	—	—						
	夏县	氡气	I	330																—	√	—	—	
分类统计	台项	异常台项数			0	1	0	2		0	0	1	0		0	1	0	0		0	1	0	0	
		台项总数			3	3	3	3	/	1	1	1	1	/	1	1	1	1	/	1	1	1	1	/
		异常台项百分比/%			0	33	0	67	/	0	0	100	0	/	0	100	0	0	/	0	100	0	0	
	观测台站（点）	异常台站数			0	1	0	1		0	0	1	0		0	1	0	0		0	1	0	0	
		台站总数			2	2	2	2	/	1	1	1	1	/	1	1	1	1	/	1	1	1	1	/
		异常台站百分比/%			0	50	0	50	/	0	0	100	0	/	0	100	0	0	/	0	100	0	0	/
	测项总数				3					1					1					1				
	观测台站总数				2					1					1					1				
备注																								

2003年2月14日新疆维吾尔自治区石河子5.4级地震

新疆维吾尔自治区地震局

温和平　高国英　王　琼　聂晓红

摘　要

2003年2月14日新疆维吾尔自治区石河子市南部发生5.4级地震。微观震中为φ_N44°06′，λ_E85°52′。因时值冬季，加之地形和交通原因，现场考察未抵达宏观震中区，已调查区最大烈度为Ⅴ度。地震未造成人员伤亡，造成直接经济损失50万元。

此次地震为前震—主震—余震型。最大余震为2月17日M_S3.0地震。余震主要分布在主震西南侧15km范围内，呈N40°E方向线性展。地震的发震构造可能为准噶尔南缘断裂。

震中200km范围内共有地震观测台11个，34项前兆观测项目；5.4级震前，共出现地震学异常12条，主要为地震学参数异常；前兆异常20条，其中5条为中期异常，12条短期、3条临震异常，主要为定点形变和地下流体异常。2002年9月北天山中东段的流动水准和流动重力也出现了测段异常。

5.4级地震发生在新疆2003年度的5级地震危险区内。震前根据资料的异常变化，向中国地震局提出了3个月的短期预测意见，对应较好。震后趋势判定正确。

前　言

2003年2月14日01时34分20秒，新疆维吾尔自治区石河子市南部山区发生5.4级地震，微观震中为φ_N44°06′，λ_E85°52′。因时值冬季，加之地形和交通原因，现场考察未抵达宏观震中区，已调查区最大烈度为Ⅴ度。地震造成直接经济损失50万元，未造成人员伤亡[1]。

5.4级地震震中位于北天山地震带，发震构造是准噶尔南缘断裂。震中周围地区是北天山中段中强地震多发区，距1906年玛纳斯7.7级地震约为90km。距离震中200km范围内共有地震台（点）22个。震前出现地震学异常12条，有16项前兆观测项目出现20条异常。该地震发生在中期和年度确定的地震危险区内，地震前新疆地震局预报中心曾提出3个月的短期预测意见1)。

一、测震台网及地震基本参数

5.4 级地震震中区及附近是新疆地震监测能力相对较强的地区。图 1 给出震中 200km 范围内的地震台分布。震中 100km 范围内有石场、石河子、呼图壁 3 个地震台，100 ~ 200km 分别有乌鲁木齐、克拉玛依 2 个地震台。此外在震中距 200km 内还分布有乌鲁木齐遥测台网的 6 个子台，分别是硫磺沟、柳树沟、石梯子、天池、乌苏、乌什城。在研究时段内监测能力可达到 $M_S \geq 1.0$ 级地震。

表 1 列出不同作者依据不同资料给出的该地震基本参数。结合震害调查结果认为，宏观震中位置应在调查区东南或南部，新疆地震局目录和李志海[4]给出的震中位置与此更为接近，因而此次地震基本参数取表 1 中编号 1 的结果。

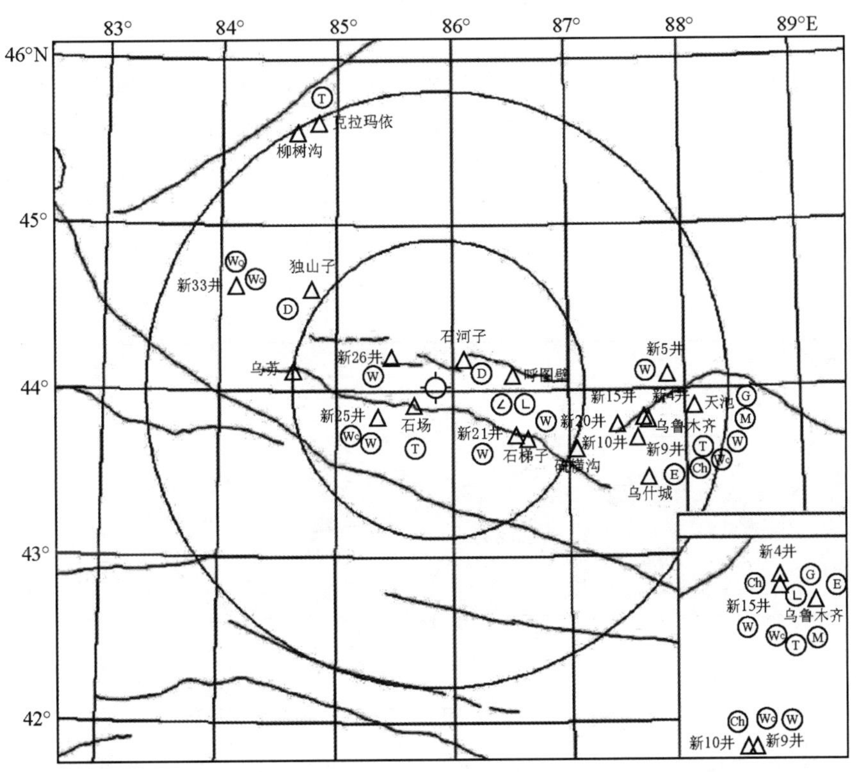

图 1　石河子 5.4 级地震前震中附近地震台站及观测项目分布图

Fig. 1　Distribution of earthquake-monitoring stations around the epicentral area before the $M5.4$ Shihezi earthquake

表 1　石河子 5.4 级地震基本参数

Table 1　Basic parameters of the *M*5.4 Shihezi earthquake

编号	发震日期	发震时刻	震中位置		震级	震源深度（km）	震中地名	结果来源
	年 月 日	时 分 秒	φ_N	λ_E				
1	2003 2 14	01 34 20	44°06′	85°52′	5.4	25	石河子	新疆地震局目录
2	2003 2 14	01 34 20	43°35′	85°41′	5.4	10	新疆石河子	新疆地震局速报
3	2003 2 14	01 34 18	43°54′	85°42′	5.4		新疆石河子	国家台网速报
4	2003 2 14	01 34 20	44°00′	85°52′	5.4	23	石河子	文献［4］
5	2003 2 14	01 34 28	44°02′	85°25′	5.3（M_W）	24	China	HRV

二、地震地质背景

北天山地震带是由多个复式褶皱组成的断褶带[1,2]。2003 年 2 月 14 日石河子 5.4 级地震发生在北天山地震带中段的石河子西南山区（图 2），属天山东西向构造带的北部地区。震中区附近的主要断裂由南向北依次是乔尔马断裂、博罗科努断裂、亚马特断裂和准噶尔南缘断裂。1900 年以来，震中 100km 范围内共发生 15 次 5 级以上地震。历史上距离该次 5.4 级地震最近的地震是 1906 年 8 月 22 日玛纳斯 7.7 级地震，二者相距 90km。

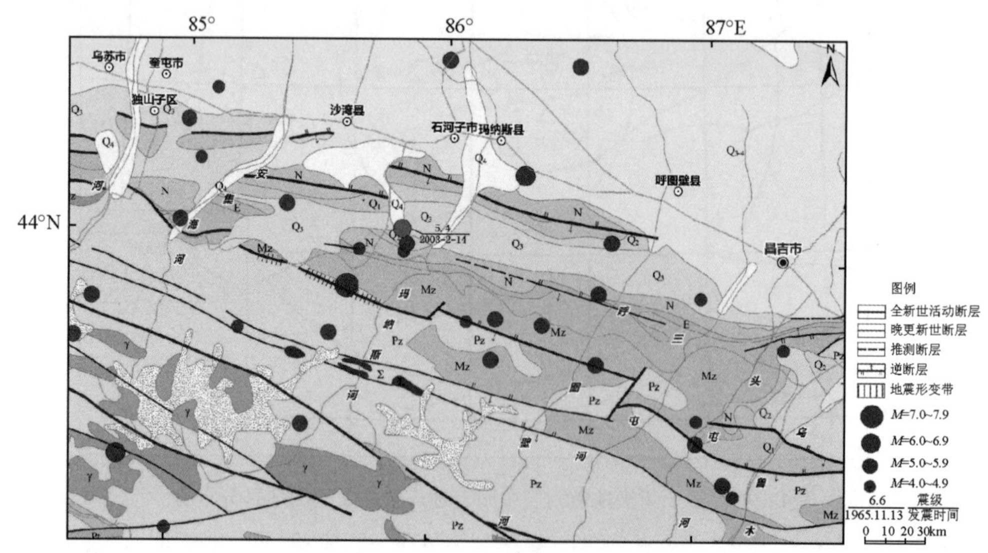

图 2　石河子 5.4 级地震附近主要构造

Fig. 2　The distribution of major seismotectonics around epicenter

地震现场考察未达宏观震中区，但依据有限资料推测此次地震的发震构造可能为依连哈比尔尕断裂[1]。该断裂是一条巨大的斜切天山的北西向断裂，主断面南倾，发育有较宽的破碎带，具逆冲和右旋走滑性质。该断裂北临的准噶尔南缘（清水河子）断裂，也是一条断层面南倾的具走滑特征的逆断层，发育有断层陡坎和切割地貌的形变带。两断裂在全新世时期有过多次显著的活动[3]，5.4 级地震的发生是断裂带最新活动的表现。

三、地震影响场和震害

这次地震石河子市区、玛纳斯县城、沙湾县城及其部分乡镇震感强烈，乌鲁木齐市普遍有感；米泉、昌吉、呼图壁、奎屯、乌苏等地部分有感。因时值隆冬，震中及附近区域大雪覆盖，加之因强烈切割的地形地貌造成道路交通困难，因此现场实地考察未抵达宏观震中，仅对可以到达的居民点进行了震害调查（图 3）。由于调查的局限性，未能勾勒出烈度等震线。

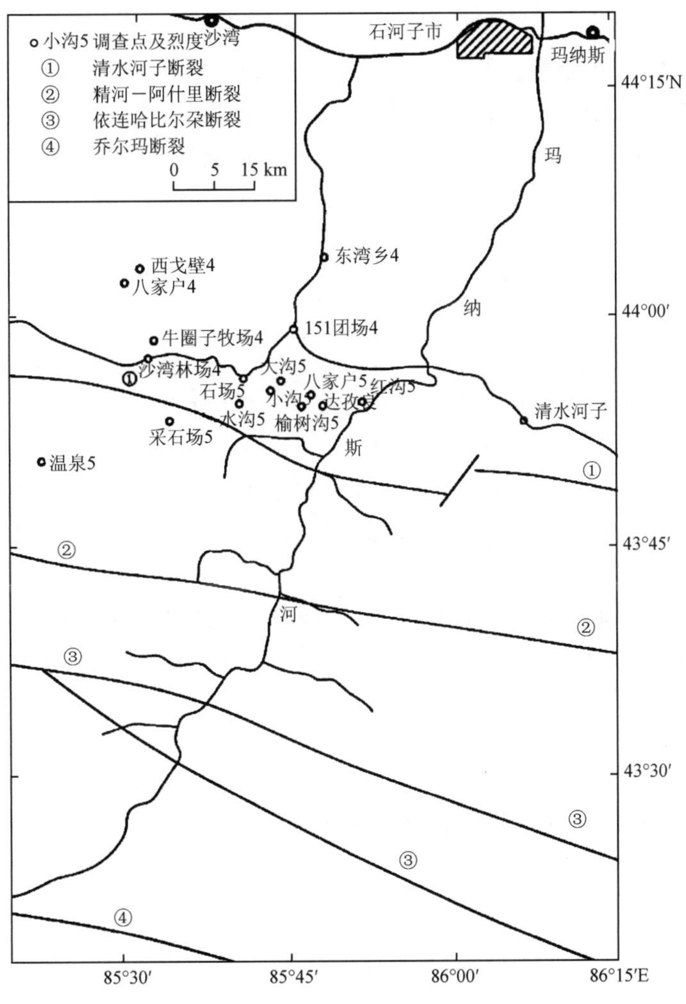

图 3 石河子 5.4 级地震现场工作图

Fig. 3 The distribution of earthquake disaster diagnoses around epicentre

Ⅴ度区：石场镇和榆树沟、小沟、水沟、大沟、达孜良、八家户等煤矿。各地居民震感强烈，分别感觉垂直和水平向振动。石场镇和榆树沟煤矿可听见轰隆隆的地声。达孜良煤矿临近山坡少量岩块滚落。本次调查表明，除20世纪50～60年代土木和砖木结构房屋在已有的陈旧破损的基础上震害有所加重外，70年代以后的砖混房屋基本保持完好[1]。例如小沟煤矿办公楼二楼纵墙墙体发育2处斜向裂缝。四层砖混结构公寓三楼几处门窗角见倒八字形裂缝，顶蓬空心板接缝处开裂，墙皮偶见脱落。沙湾县煤矿办公楼出现细小的“X”裂缝、掉灰以及个别玻璃破碎的情况，招待所和派出所楼墙体有斜裂缝，经判定均为1996年1月9日沙湾5.2级地震震害造成的陈旧性裂缝受本次地震影响有所加大。

区域内所有煤矿矿井均基本完好，甚至水沟煤矿井下工人普遍没有感觉，只有静坐的工人感觉似乎有摇动，上到地面井口后才知道系发生地震。

Ⅳ度区：石河子市区、151团部（紫泥泉镇）、沙湾县东湾乡、西戈壁乡、八家户乡等地，居民震感较强烈。建筑物主要为20世纪70年代后建设的砖木、砖混结构、钢筋混凝土结构的房屋，各类房屋没有遭受破坏。

据现场实地考察及调查情况看，Ⅴ度区范围内多数房屋基本完好，部分房屋受到轻微破坏，受损房屋原来即属老旧危房或施工质量有问题。Ⅳ度区各类房屋未遭受破坏。此次地震造成的震害损失不大，且集中分布石河子南山矿区（石场镇）和沙湾县属煤矿等地。因矿山以及部分居民的房屋遭到不同程度的破坏，造成直接经济损失50万元。

四、地震序列

2月14日石河子5.4级地震发生在区域范围内中等地震活动增强后相对平静的背景之下。震前的1时31分和32分震区分别发生了$M_S4.0$和$M_S4.9$地震。5.4级地震后截至2003年12月31日，新疆地震局台网共定位余震59次，其中5.0～5.9级1次；3.0～3.9级1次；2.0～2.9级7次；1.0～1.9级48次。据5.4级地震序列的主震及其余震呈N40°E方向展布、余震主要分布在主震西南侧15km范围内等特征（图4），推测这次地震震源破裂可能是以单侧破裂为主[4]。

由余震序列的M-t图（图5）看出，序列中最大余震为2月17日$M_S3.0$地震。余震随时间有起伏地衰减（图6），2～6月较为丰富，7月后余震明显减少，且成丛发生。12月29日后震区很少有余震发生。

在$M_S5.4$地震前2～3分钟，发生了$M_S4.0$和$M_S4.9$地震，因此可以将$M_S4.9$与5.4级地震作为双震型，而$M_S4.0$地震与最大主震震级相差1.4。因此根据地震类型的判别指标，石河子5.4级地震为前震—（双）主震—余震型。

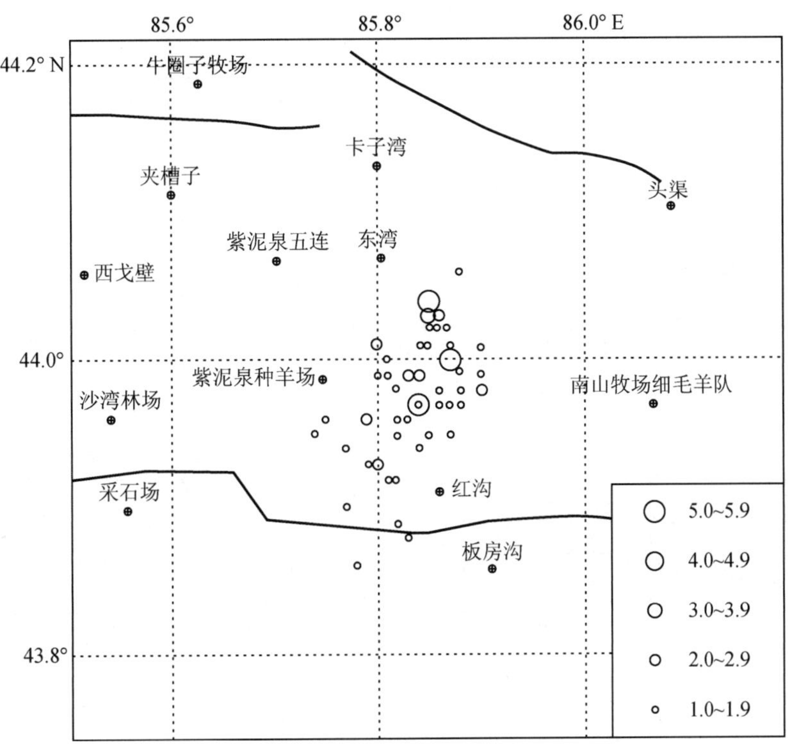

图 4 石河子 5.4 级地震序列震中分布

Fig. 4 Distribution of the sequence of *M*5.4 Shihezi earthquake

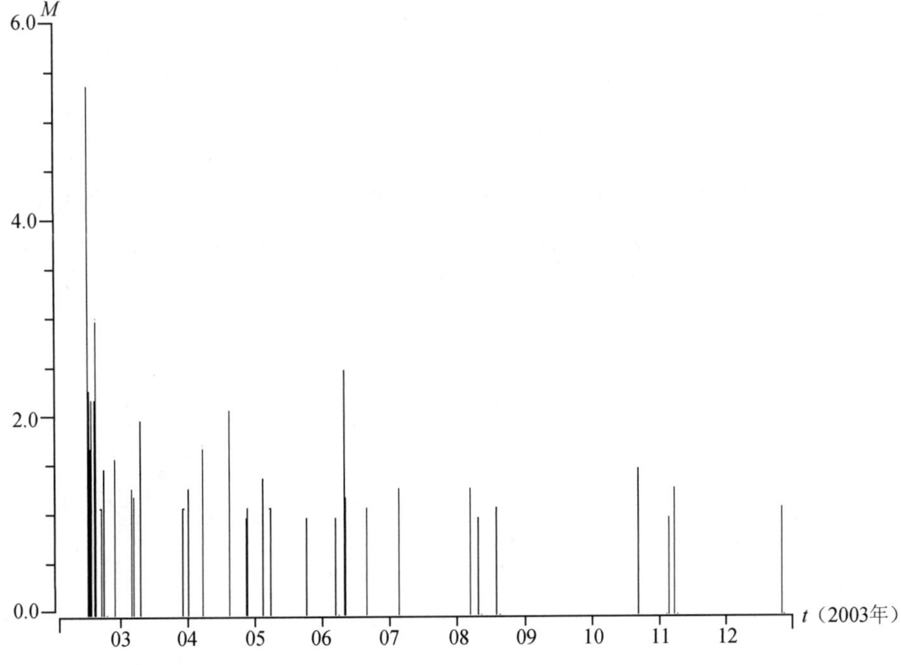

图 5 石河子 5.4 级地震序列 *M*-*t* 图

Fig. 5 *M*-*t* diagram of the *M*5.4 Shihezi earthquake sequence

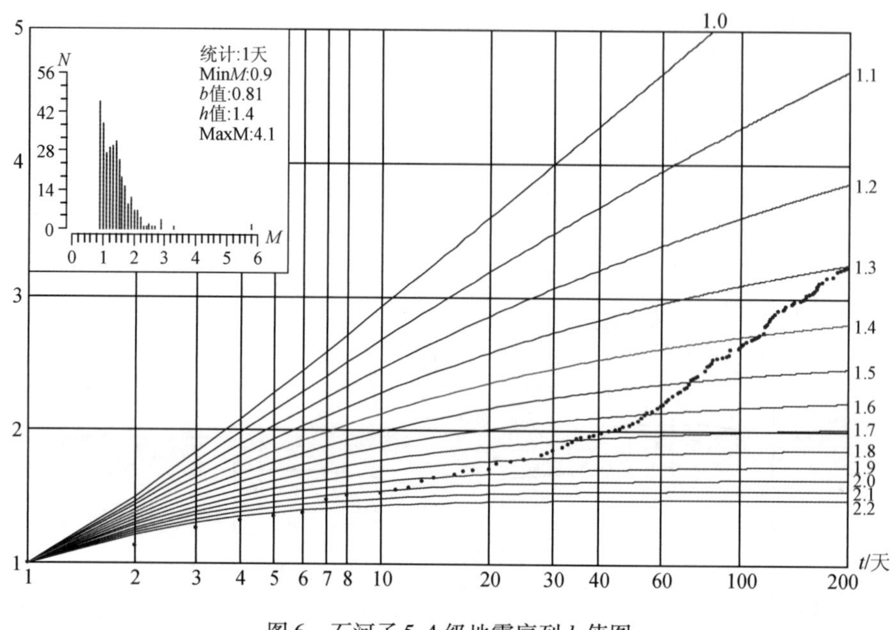

图 6　石河子 5.4 级地震序列 h 值图

Fig. 6　h diagram of the M5.4 Shihezi earthquake sequence

五、震源参数和地震破裂面

聂晓红根据新疆区域台网 14 次较清晰的 P 波初动符号，计算了这次 5.4 级地震的震源机制解（表 2、图 7），结果表明震源断错为倾滑正断性质；哈佛大学的矩张量反演的震源机制结果表明震源断错具有 NS 向主压应力下的倾滑逆冲性质。考虑到震中周围区域应力场 NS 向的压应力和中小地震震源逆断的断错性质[5]，倾滑逆断的震源断错可能更接近实际。其 N40°E 方向的震源单侧破裂方向[4]也与近 EW 向区域构造方向并不吻合。由于未能对地震宏观震中进行实际考察，因而无法取得更多的证据佐证该地震的震源破裂性质。

表 2　石河子 5.4 级地震的震源机制解

Table 2　Focal mechanism solutions of the M5.4 Shihezi earthquake

序号	节面Ⅰ			节面Ⅱ			P 轴		T 轴		N 轴		结果来源
	走向	倾角	滑动角	走向	倾角	滑动角	方位	仰角	方位	仰角	方位	仰角	
1	181	14	-153	64	83	-77	348	50	143	37	242	13	聂晓红
2	283	62	105	73	32	64	2	15	226	69	96	14	HRV

李志海利用双差地震定位法对石河子 5.4 级地震及其余震进行重新定位[4]。结果显示，余震分布在准噶尔南缘断裂（清水河子断裂）以北，主震及其余震呈 N40°E 方向线性展布，

与准噶尔南缘断裂近乎垂直；震源深度全部分布在 15～30km 范围内，深度优势分布范围为 15～25km。结合重新定位结果及震源机制解，初步认为 5.4 地震的发震构造是准噶尔南缘断裂（清水河子断裂），而非更靠南的依连哈比尔尕断裂。

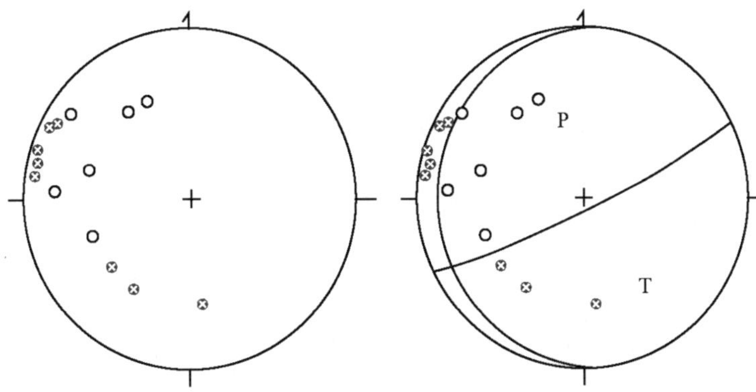

图 7　由新疆区域台网获得的石河子 5.4 级地震震源机制解

Fig. 7　Focal mechanism solutions of the *M*5.4 Shihezi earthquake from P wave records by the Xinjiang regional network

六、地震前兆异常及其特征分析[3)]

5.4 级地震发生在北天山西段，区域内台站（图 1）主要沿北天山构造走向分布，距震中最近的为石场地震台。在距震中 200km 以内共有 5 个地震台，均设有前兆观测。连同独山子、新 25 泉、新 26 泉、新 21 井、新 04 井（泉）、新 05 井、新 09 泉、新 10 井（泉）、新 15 泉、新 20 井、新 33 井等 11 个观测点，前兆观测分别有形变测 12 台项、电磁 2 台项和地下流体 20 台项。在 0～100km、101～200km 距离内地震台站的前兆观测项目分别为 7 项、27 项。此次地震前共出现 20 条前兆观测异常（表 3）。

表 3　地震前兆异常登记表

Table 3　Summary table of earthquake precursory anomalies

序号	异常项目	台站（点）或观测区	分析方法	异常判据及观测误差	震前异常起止时间	震后变化	最大幅度	震中距 (km)	异常类别及可靠性	图号	异常特点及备注
1	地震活动增强	震中 200km 范围内	$M_S \geqslant 4.0$ 级地震 *M-t* 图	4 级地震活动增强	2001.12 ～ 2002.10	恢复正常		震中周围	M_1	6	震前 1 年多 4 级地震活动增强

续表

序号	异常项目	台站（点）或观测区	分析方法	异常判据及观测误差	震前异常起止时间	震后变化	最大幅度	震中距（km）	异常类别及可靠性	图号	异常特点及备注
2	地震频度 N	乌鲁木齐区	$M_S \geqslant 2.0$ 级地震频度（1 年步长 3 个月窗长滑动）	连续高于于均值 1 年以上，之后恢复到均值线下	2001.03 ～ 2002.10	正常		震中周围	M_1	9	1995 年后首次达到异常限
3	缺震	乌鲁木齐区	$M_S \geqslant 2.0$ 级缺震分析（1 年步长 3 个月窗长滑动）	连续低于于均值 10 个月以上，后恢复到均值线上	1999.11 ～ 2002.12	正常		震中周围	M_1	9	2002 年初略有回返，其后持续至 12 月
4	b 值	乌鲁木齐区	b 值时间扫描（$M_S \geqslant$ 2.0 级，1 年步长 3 个月窗长滑动）	连续高于均值 12 个月以上，之后恢复到均值线下	1999.12 ～ 2002.12	正常		震中周围	M_1	9	高值异常，在较低值时发震
5	η 值	乌鲁木齐区	η 值时间扫描（$M_S \geqslant$ 2.0 级，1 年步长 3 个月窗长滑动）	连续低于于均值 18 个月	1999.12 ～ 2003.06	持续		震中周围	M_1	9	为 1980 年以来持续时间最长，震后未恢复
6	$A(b)$	乌鲁木齐区	$A(b)$ 值时间扫描（$M_S \geqslant 2.0$ 级，1 年步长 3 月窗滑动）	连续低于均值 12 个月，之后恢复到均值线以上	1999.10 ～ 2002.11	正常		震中周围	M_1	9	低值水平不高，但过程明显
7	YH	乌鲁木齐区	YH 值时间扫描（$M_S \geqslant 2.0$ 级，1 年步长 3 月窗滑动）	连续高于于均值 1 年以上，之后恢复到均值线下	2000.10 ～ 2002.12	正常		震中周围	M_1	9	1995 年后首次达到异常限，震前在均值线附近波动

续表

序号	异常项目	台站（点）或观测区	分析方法	异常判据及观测误差	震前异常起止时间	震后变化	最大幅度	震中距（km）	异常类别及可靠性	图号	异常特点及备注
8	b值	石场—精河区	b值时间扫描（$M_S \geqslant 2.0$级，1年步长3个月窗长滑动）	连续高于均值14个月以上	1998.11 ~ 2001.12	正常		震中周围	M_1	10	高值异常，恢复到最低值后发震
9	η值	石场—精河区	η值时间扫描（$M_S \geqslant 2.0$级，1年步长3个月窗长滑动）	连续低于于均值10个月	2000.12 ~ 2003.06	持续		震中周围	M_1	10	低值明显，过程显著
10	$A(b)$	石场—精河区	$A(b)$值时间扫描（$M_S \geqslant 2.0$级，1年步长3月窗滑动）	连续低于均值16个月	1998.10 ~ 2002.01	正常		震中周围	M_1	10	恢复到较高值时发震
11	S值	石场—精河区	S值时间扫描（$M_S \geqslant 2.0$级，1年步长3个月窗长滑动）	连续低于于均值线16个月	1998.06 ~ 2002.02	恢复正常		震中周围	M_1	10	恢复到较高值时发震
12	小震调制比R_m	石场—精河区	小震调制比（$M_S \geqslant 2.0$级，1年步长3个月窗长滑动）	连续高于自然概率12个月	2001.02 ~ 2002.11	正常	0.25	震中周围	M_1	10	形成高值过程，震前恢复至均值线及以下
13	石英水平摆	石场台	单分量时序分析	倾斜方向	2000.11 ~ 2003.11	持续		15	M_2	12	W倾明显加快

续表

序号	异常项目	台站（点）或观测区	分析方法	异常判据及观测误差	震前异常起止时间	震后变化	最大幅度	震中距（km）	异常类别及可靠性	图号	异常特点及备注
14	石英水平摆	石场台	单分量时序分析	季节转向时间及变化速率	2002.12～2003.02	恢复	0.31″	15	S_2		EW向秋季转向提前，出现大幅度速率、方向变化
15	石英水平摆	石场台	单分量时序分析	变化速率	2002.1.25	恢复	0.31″	15	I_2		EW向大幅度速率、方向变化
16	动水位	新26泉	月频次统计	脉冲频次超限	2002.2～2003.03	持续			M_2		水中油气溢出造成模拟记录出现“脉冲”
17	水准	呼图壁	时序分析	变化速率	2001.09～2002.06	正常		50	M_2		断层闭锁，震后的2003年6月再次出现
18	断层蠕变	独山子	时序分析	年变幅度	2001.01～2004.01	持续		102	M_2		断层两盘高差增大，年变幅度变大
19	石英水平摆	乌鲁木齐	单分量	倾斜方向	2000.11～2003.11	持续		140	M_1	12	由W倾转为E倾
20	井下摆倾斜仪	克拉玛依	单分量	倾斜方向	2002.10～2003.03	持续		185	S_2		NS向N倾速率减小乃至反向
21	地电阻率	乌鲁木齐	时序分析	年变畸变	2002.07～2003.11	恢复		150	S_2	13	EW向短极距夏季无峰值的年变畸变
22	水温	新04井	时序分析	趋势变化、短期突跳	2002.08～2003.02	恢复	0.15℃	145	S_1	14	缓慢上升，并伴随大幅度升降变化

续表

序号	异常项目	台站（点）或观测区	分析方法	异常判据及观测误差	震前异常起止时间	震后变化	最大幅度	震中距（km）	异常类别及可靠性	图号	异常特点及备注
23	水温	新 04 井	时序分析	短期突跳	2003. 02. 08	恢复	0. 15 ℃	145	I_1	14	大幅度升降变化
24	硫化物	新 04 井	时序分析	高值异常	2002. 10 ~ 2003. 03	持续	240 mg/L	145	S_2		逐步上升至高值后脉冲突跳
25	流量	新 10 泉	时序分析	高值异常	2002. 05 ~ 2002. 09	正常	3. 6 L/s	148	S_1	15	逐步上升至高值后缓慢降至背景值附近
26	甲烷	新 10 泉	时序分析	高值异常	2002. 05 ~ 2002. 09	持续	0. 82 %	148	S1	15	缓慢上升至高值后企稳
27	氦气	新 10 泉	时序分析	高值异常	2002. 05 ~ 2002. 09	持续	0. 056 %	148	S_1	15	上升至高值后大幅度波动
28	水汞	新 10 泉	时序分析	高值异常	2002. 05 ~ 2002. 09	正常		148	S_1	15	逐步上升至高值后缓慢降至背景值附近
29	水位	新 10 井	时序分析	突变	2002. 05 ~ 2002. 10	正常	0. 4m	148	S_1	16	快速上升至高值，快速恢复到背景值
30	水位	新 20 井	时序分析	年变畸变	2002. 11 ~ 2002. 03	持续	1. 4m	130	S_2	16	年变提前后，违年变上升
31	水位	新 21 井	时序分析	突变	2002. 01 ~ 2002. 02	恢复	0. 3m	70	S_2	16	阶跃上升，波动下降
32	水位	新 21 井	时序分析	突变	2003. 02. 14	恢复	0. 3m	70	I_2	16	阶跃上升

需要说明的是，2003 年 2 月 14 日石河子 5. 4 级地震后，2003 年 2 月 24 日在南天山西段发生了巴楚—伽师 6. 8 级地震，其后陆续有 5 级余震发生，考虑到前兆异常的场源效应，各种前兆观测资料终止时间均取为 2003 年 6 月 30 日。

1. 地震学异常

2000 年 12 月 10 日伊宁 5.0 级地震后，新疆境内 5 级地震平静 24 个月。在这种平静的背景下，2001 年 12 月至 2002 年 10 月，乌鲁木齐以西至新源的天山中段地区连续发生 8 次 4 级地震，在距震中 200km 范围内发生 6 次，且主要分布在震中以南地区[6]。但自 2002 年 11 月开始至震前，震中区及附近仅发生 3 次 3 级地震，显示出明显的增强之后的短时间平静现象（图 8、图 9）。

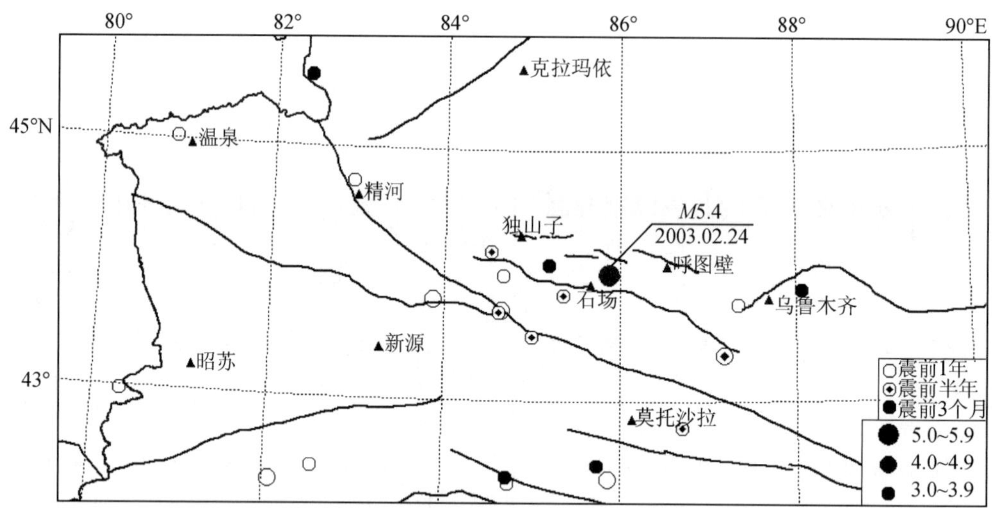

图 8　北天山地区 $M_S \geqslant 3.0$ 级地震震中分布

Fig. 8　Distribution of the $M_S \geqslant 3.0$ earthquakes around the epicenter before the $M5.4$ Shihezi earthquake

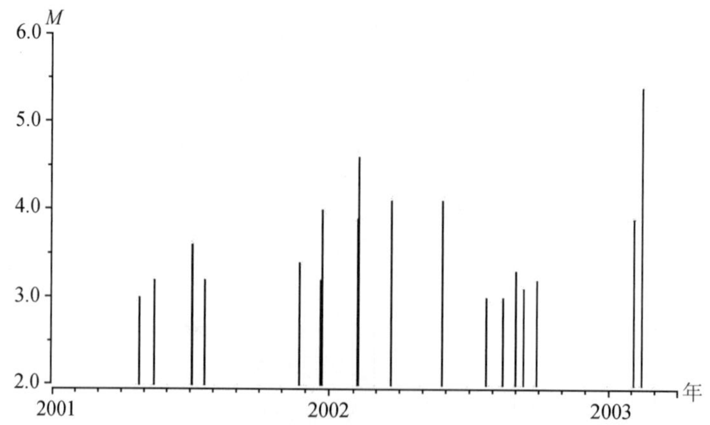

图 9　石河子 5.4 级地震震中附近 $M_S \geqslant 3.0$ 级地震 M-t 图

（北天山呼图壁至乌苏）

Fig. 9　M-t diagram for the events（$M_S \geqslant 3.0$）around the epicenter of the $M5.4$ Shihezi earthquake（from Hutubi to Wusu along North Tianshan Mountain）

新疆地震局预报中心根据地质构造和区域地震活动特点等，将新疆天山、阿勒泰、西昆仑三个地震带划分为 9 个跟踪区域。此次石河子 5.4 级地震发生在北天山的乌鲁木齐区和石场—精河区的交接部位（图 10）。以 $M_S \geqslant 2.0$ 级地震进行地震参数的时间序列跟踪分析，以 1 年为步长、3 个月为窗长进行滑动计算，结果显示乌鲁木齐区的 η 值、$A(b)$ 值、b 值、缺震具持续 3 年以上的异常，频度、YH 值异常也持续 1 年以上（图 11）；石场—精河区的 η 值、$A(b)$ 值、b 值、S 值也出现类似的变化，同时小震调制比异常也较为显著，持续 2 年以上（图 12）。

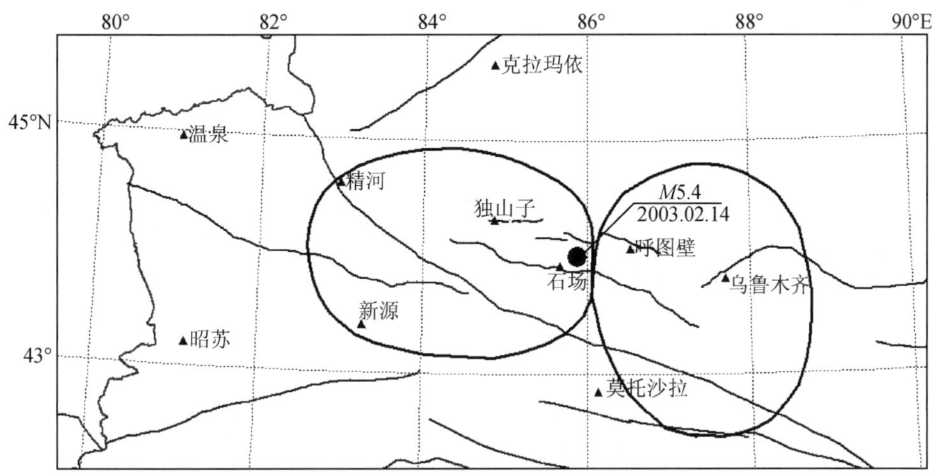

图 10 北天山地区 1980 年以来地震学参数跟踪分区示意图

Fig. 10 Distribution about track the seismology parameter area along North Tianshan Mountain

2. 前兆观测项目异常

5.4 级地震发生前，在震中 200km 范围的 5 个地震台和 8 个观测点的前兆观测项目中，北天山定点形变资料长中短期异常较为突出[7,9]，乌鲁木齐地电和地下流体资料也一定程度地显示了中短期（临）异常[8]，2002 年 9 月的流动测量显示，北天山局部地区地壳变形增强。

石场台地倾斜 EW 向自 2001 年春季转向后 W 倾加速，年变幅度增大；乌鲁木齐雅山石英摆 2000 年秋季转向后，EW 向由 W 倾转为 E 倾，年变幅减小，NS 由 N 倾速率明显减慢，年变幅减小；呼图壁跨断层水准 2001 年 9 月至 2002 年 6 月出现测值变化很小、年变畸变的断层"闭锁"现象；独山子断层蠕变观测垂直向自 2001 年起速率加快，年变幅增大。

石场地倾斜 EW 向 2002 年 12 月起连续出现大速率的方向不稳定变化（图 13）；克拉玛依地倾斜 NS 向，自 2002 年 10 月下旬起 N 倾速率明显减小，2002 年 11 月 22 日到 12 月 9 日反向 S 倾，震后恢复正常；呼图壁断层水准自 2002 年 9 月起出现解锁现象；乌鲁木齐水磨沟台地电阻率 EW 向短极距 2002 出现夏季无峰值的年变畸变现象（图 14）。

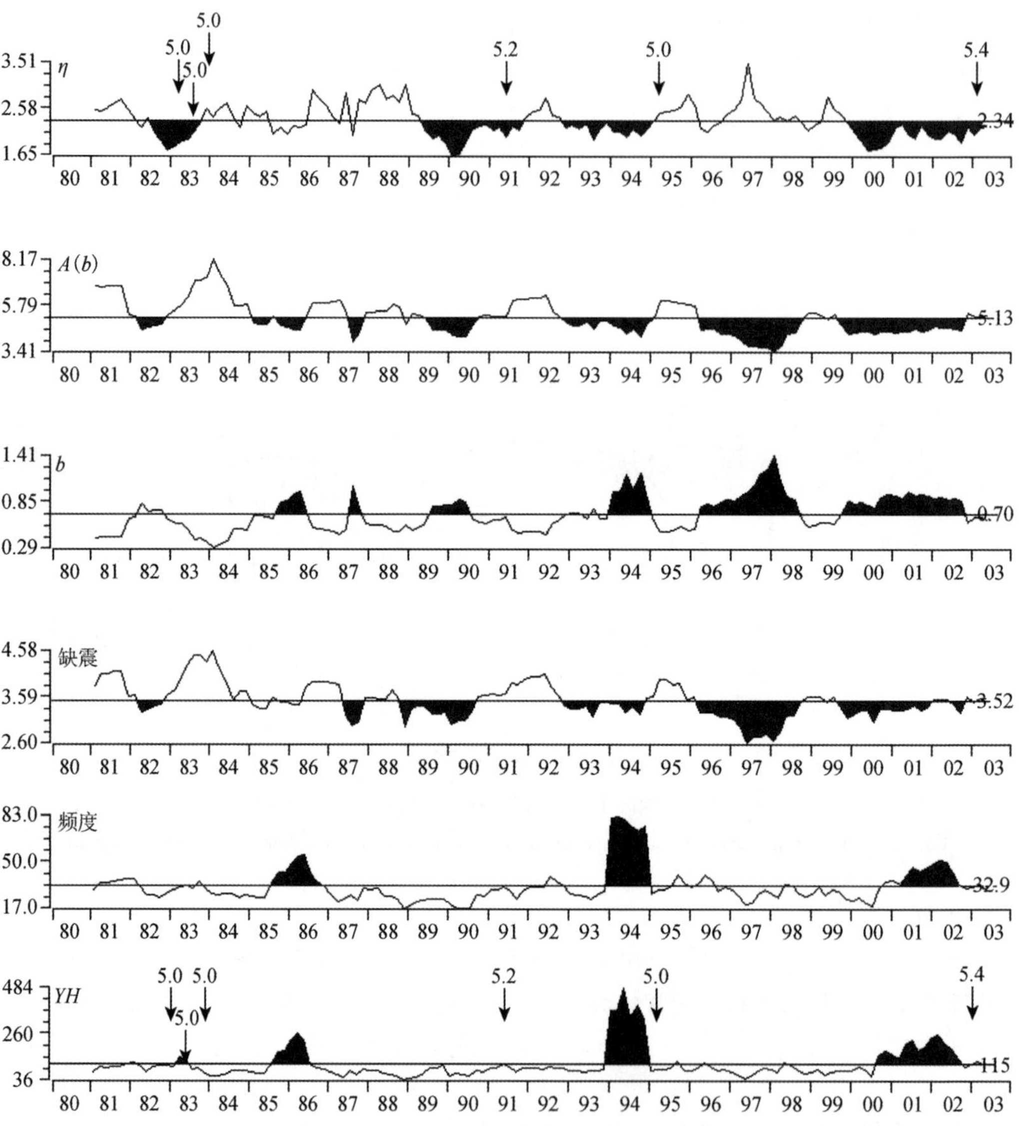

图 11　石河子 5.4 级地震前乌鲁木齐区 1980 年以来地震学参数曲线（$M_S \geqslant 2.0$ 级）

Fig. 11　Curve of seismology parameter with $M_S \geqslant 2.0$ earthquake around Urumchi area before $M5.4$ Shihezi earthquake since 1980

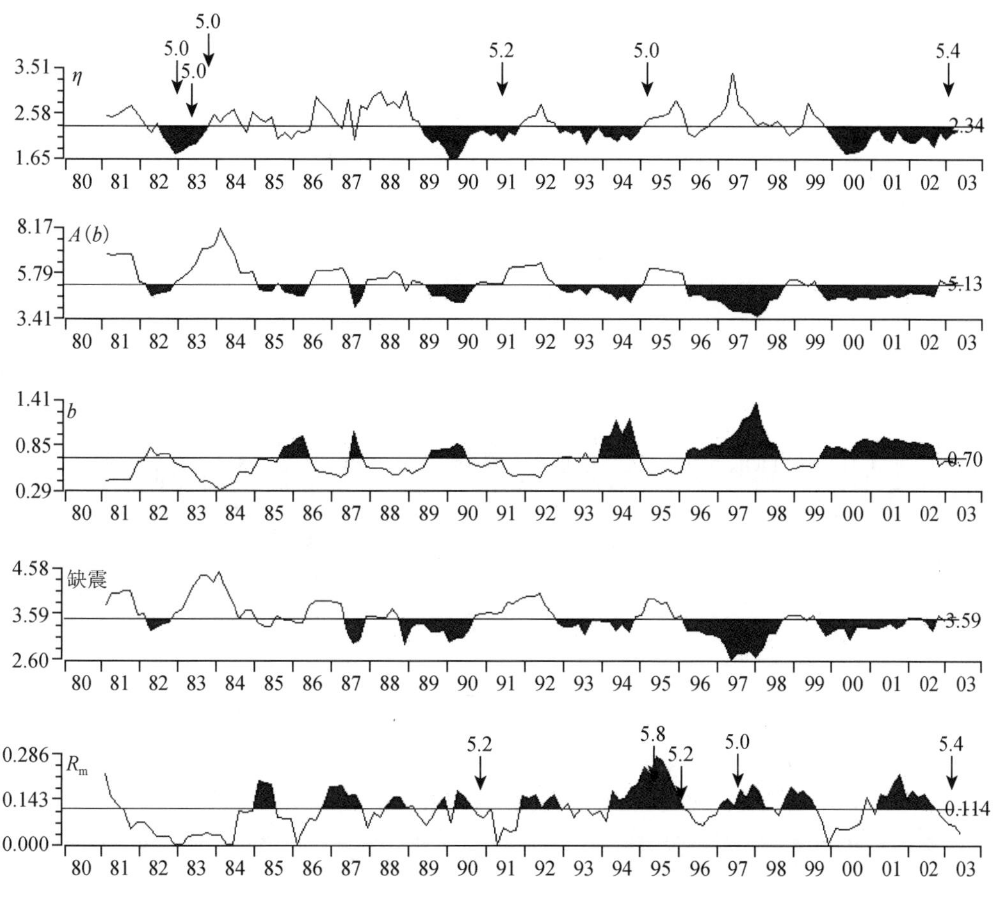

图 12　石河子 5.4 级地震前石场—精河区 1980 年以来地震学参数曲线（M_S≥2.0 级）

Fig. 12　Curve of seismology parameter with M_S≥2.0 earthquake around Shichang-Jinghe area before *M*5.4 Shihezi earthquake since 1980

乌鲁木齐新 04 井水温，自 2002 年 8 月出现趋势性上升，12 月 6、7、11 日和 2003 年 2 月 8 日出现大幅度升降变化，变幅最大达 0.15℃（图 15）；乌鲁木齐新 04 泉硫化物自 2002 年 10 月逐步上升，12 月起稳定，其后多次出现脉冲式高值，震时达到最大值；乌鲁木齐新 10 号泉的甲烷、氦气和汞，自 2002 年 5 月起几乎同步出现上升，9 月起各测项或恢复或高值企稳，直至发震（图 16）；乌鲁木齐新 10 井水位，2002 年 5 月出现快速上升 7 月 20 日达到高值后下降，10 月恢复到背景值；乌鲁木齐新 20 井，自 2002 年 11 月出现年变畸变，震后恢复；呼图壁新 21 井水位，2002 年 1 月阶跃上升后逐步下降，6 月再次阶跃上升后波浪式下降，震前 1 小时出现快速上升，震后稳定在相对高值（图 17）；新 26 井动水位自 2002 年 1 月起多次出现因井水中油气溢出而形成模拟记录的“脉冲”，震后 3 月达到高值，2003 年 4 月消失。

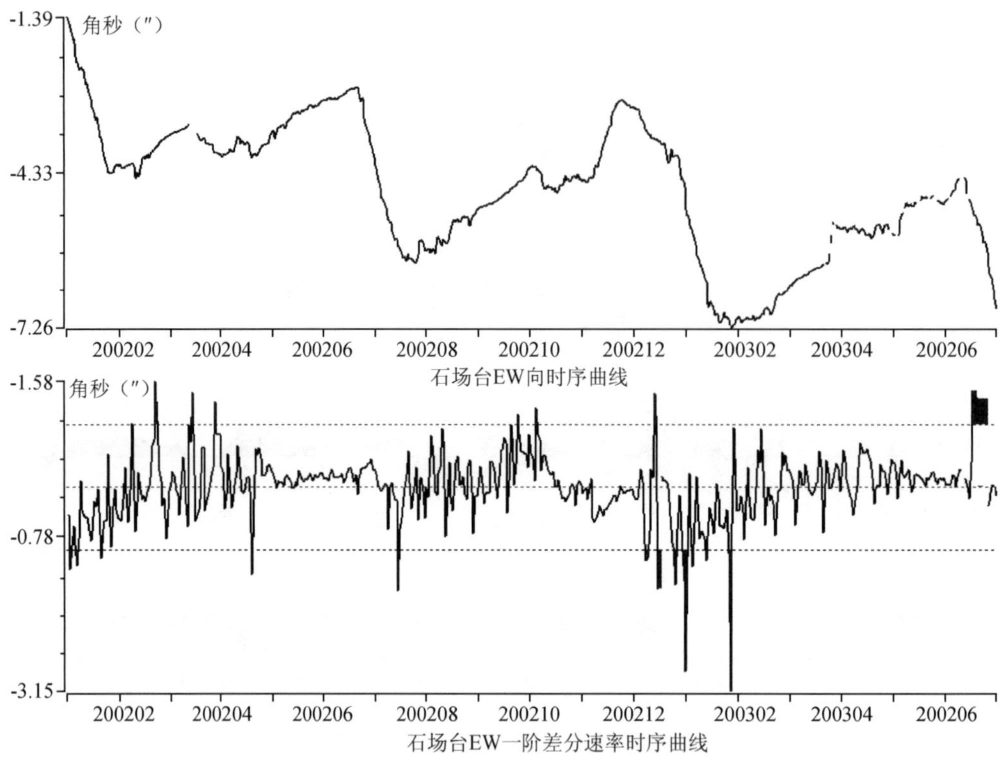

图 13　石场地倾斜 EW 向（旬均值）时序曲线图（2002.01～2003.06）

Fig. 13　Graph of 10-day mean value of Shichan tilt.（up：time sequence，bottom：rate of tilt）

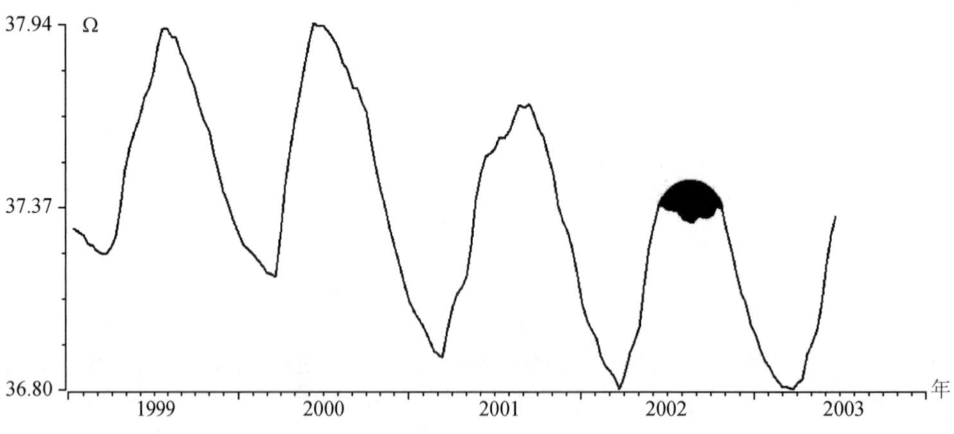

图 14　乌鲁木齐地电阻率 EW 向短极距月均值时序曲线

Fig. 14　Graph of 10-day mean value of geo- resistance of EW heading at Urumchi station

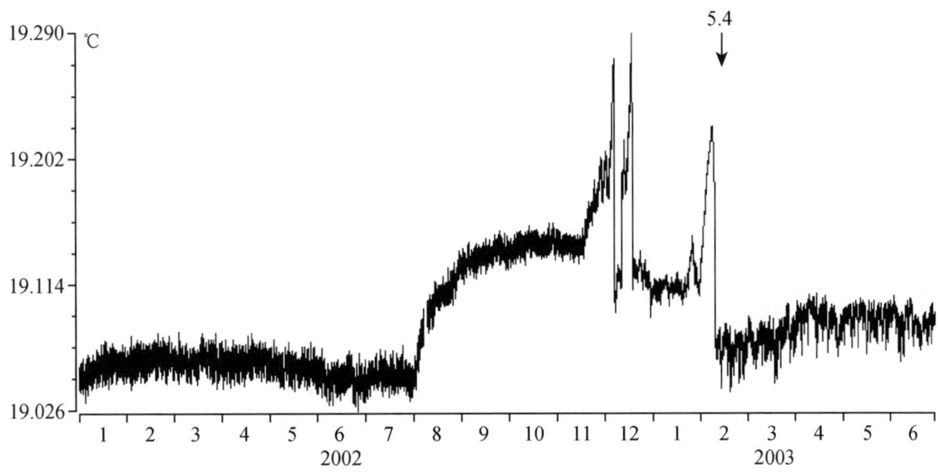

图 15　新 04 井水温时序曲线

Fig. 15　Graph of time sequence about the water temperature at Xin10 well station

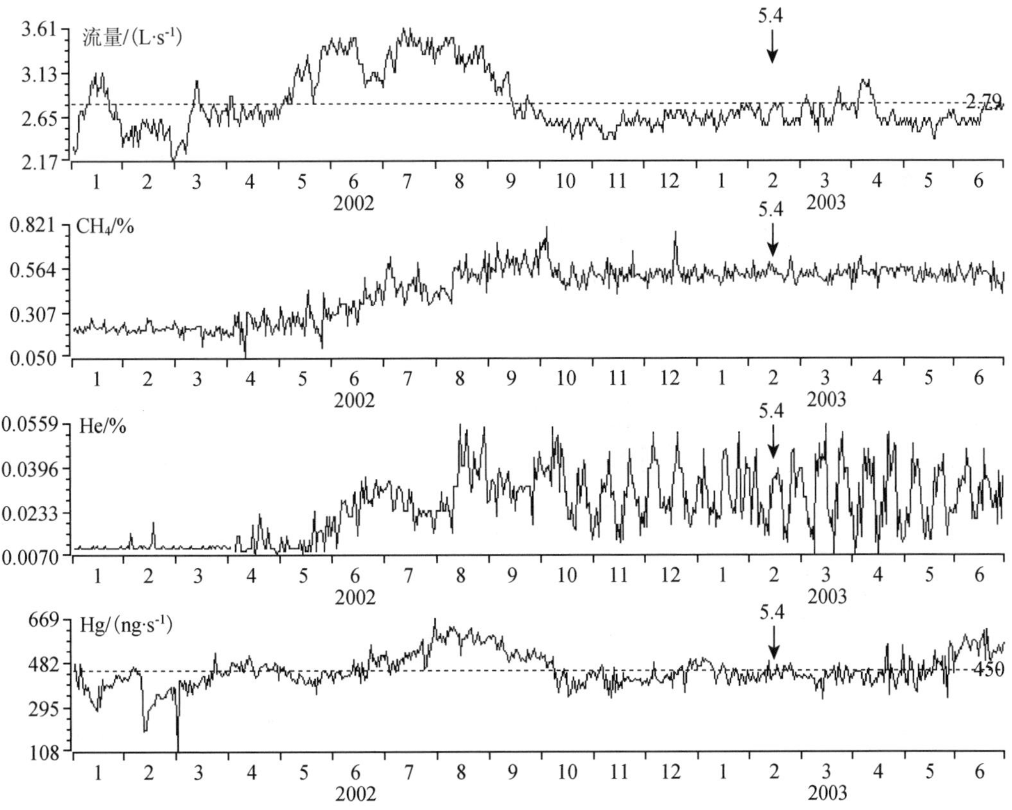

图 16　新 10 泉水化学和流量时序曲线

Fig. 16　Graph of time sequence about the water chemistry and flux at Xin10 spring station

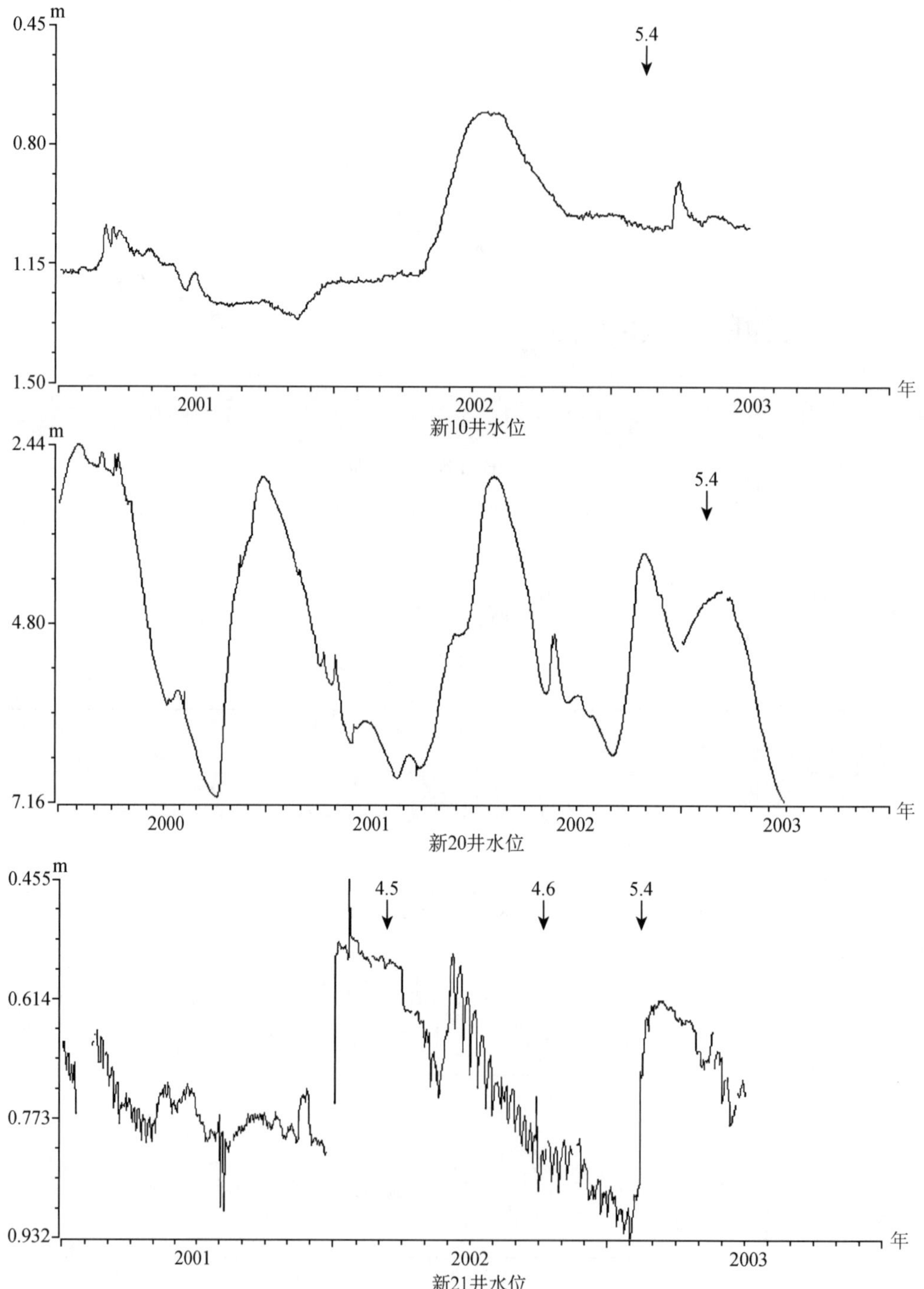

图 17 乌鲁木齐地区井水位时序曲线(自上而下依次是新 10 井、20 井、21 井)

Fig. 17 Graph of the water level time sequence in Urumchi area

(from up to bottom is Xin10 Xin20 Xin21 well)

北天山流动水准 2002 年 9 月复测时，准噶尔南缘断裂带上的庙尔沟、宁家河、巴音沟测线的速率分别为超过异常限（±1.0mm/km）的 1.7、1.07 和 -1.98mm/km，表明乌鲁木齐以西北天山中部地区断层垂直运动显著；2002 年 9 月流动重力测量结果显示，石河子南 N30-N33 测段和乌鲁木齐 N00-N14 测段的段差分别为超异常限（$\pm 300\mathrm{nm}\cdot\mathrm{s}^{-2}$）的 $-390\mathrm{nm}\cdot\mathrm{s}^{-2}$、$-470\mathrm{nm}\cdot\mathrm{s}^{-2}$，说明该区域地壳可能由于物质迁移而导致密度的变化。

七、地震前兆异常特征分析

2000 年 12 月 10 日伊宁 5.0 级地震后，新疆境内 5 级地震平静 24 个月。在这种平静的背景下，2001 年 12 月至 2002 年 10 月，乌鲁木齐以西至新源的天山中段地区中等地震活跃。与此同时北天山及周边前兆资料观测出现的异常，多数是在前期趋势异常的基础上出现的，异常特征主要表现出测值升高、阶跃或突跳、速率变化等中短期异常[7,9]。从时间特征来看基本具有同步性，在 2001 年 11 月下旬至 2002 年 3 月、2002 年 5～9 月、2002 年 12 月至 2003 年 3 月这三个时段内的涨落。

2002 年 12 月中下旬石场、克拉玛依的地倾斜先后出现速率和方向变化，可能是短期异常的突出显示，也表示异常向震中集中；至 2003 年 1 月 25 日前后石场地倾斜速率大幅变化、乌鲁木齐 04 井水温和硫化物出现高值突变[8]，则表现出明显的临震异常。

北天山地区的前兆异常过程多数持续到 2003 年 6 月前后结束。这种较为突出的异常变化本身反映了石河子 5.4 级地震前兆，但也可能与 2003 年 2 月 24 日巴楚—伽师 6.8 级地震有关。刘蒲雄研究认为，孕震过程可能对场内的某些特殊部位造成应力状态的大幅度变化，从而在这些部位可能出现远场前兆6)。因此北天山地区较为突出的群体前兆异常过程，本身显示出近场 5.4 级地震的前兆，但也可能同时反映了远场南天山西段的强震短期异常。无论是地震活动和前兆异常变化均反映出了区域应力场的增强过程。可见，震中周围地区群体异常的起伏变化，可能显示出中强震发生前因震源区应力调整所导致的敏感区（或非闭锁段）的前兆异常变化，是一种具有较高信度的地震前兆异常表现。

八、震前预测、预防和震后响应

1. 预测情况

新疆维吾尔自治区地震局在“2001～2003 年新疆地震大形势分析”中指出，“未来 1～3 年新疆地震活动将进入一个新的活跃阶段……，北天山西段有可能发生 6 级地震”2)。2003 年度地震趋势研究结果认为，“2003 年新疆地震活动水平将明显高于 2002 年，较大可能发生 6～7 级地震”，其中“乌苏—石河子地区，2003 年度有较大可能发生 5 级地震（$P=0.65$）”2)。这次 5.4 级地震就发生在上述所指出危险区内。

在未来 1～3 年和一年尺度的中短期预测基础上，新疆地震局加强了针对危险区的短临跟踪和异常落实工作，密切监视地震活动和前兆观测异常变化。在 2001 年 11 月至 2002 年 9 月，新疆境内的前兆资料异常出现了 2 次明显的起伏涨落，2002 年 12 月中旬起，北天山西段的石场、新源地倾斜和乌鲁木齐新 04 井水温等先后出现了短期变化，异常幅度明显增大。

为此新疆地震局局预报中心向中国地震局的填报 A 类预报卡，提出“2002 年 12 月 13 日至 2003 年 3 月 13 日，在 $\varphi_N 42° \sim 44°$，$\lambda_E 83° \sim 85.5°$区域内可能发生 5.5 ±0.5 级地震”[1)]。

2003 年 1 月 28 日，新疆地震局向自治区人民政府的提交了“关于近期新疆地震形势汇报”的材料，认为“2002 年 12 月 25 日乌恰 5.7 级地震和 2003 年 1 月 4 日伽师 5.4 级地震后，新疆中等地震活跃，前兆异常起伏、分布范围广……，新疆的地震形势是严峻的”，新疆地震局预报中心在 2003 年 2 月的月会商震情监视报告中明确提出新疆“未来一个月或稍长时间两个重点危险区有发生中强地震的可能”的预测意见[3)]，并于 2003 年 2 月 10 日向自治区人民政府主管副主席汇报震情时也明确提出了“近期新疆北天山有可能发生中强地震”的意见[4)]。2 月 14 日石河子 5.4 级地震发生。

2. 震后趋势判定及应急对策

地震发生后，中国地震局成立了石河子地震现场工作组，工作组由监测预报、应急救援、震害调查等相关部门领导和专家组成，并于 2 月 14 日 13 时抵达乌鲁木齐。现场工作队下午 4 时 45 分到达距震中较近的石场镇，和先期抵达的新疆地震局现场工作队汇合。在听取石河子市和新疆地震局的有关情况介绍后，现场灾害调查工作在相关区域有序展开。

震后在新疆地震局召开的紧急会商会上，新疆地震局预报中心与中国地震局的有关专家，认真分析了余震序列和主震附近地区的历史地震类型及各类前兆资料后认为，“石河子 5.4 级地震较大可能是一次双震型地震，短期内震区再次发生更大地震的可能性不大”[5)]。

石河子 5.4 级地震虽然震级不大，但波及范围包括北天山乌鲁木齐—乌苏的广大区域；由于地震发生在凌晨，石河子及其周围地区一些市民冒着零下 20℃ 左右的严寒在外徘徊，并有地震谣言产生。为应对这种混乱的局面，地震部门采取不同措施予以应对[1]：震后趋势意见未形成之前，自治区地震局和石河子市、塔城地区地震局等耐心接听各种电话询问，回答问题，安抚市民情绪，并劝返家中。震后趋势意见产生之后，新疆地震局果断决定将“短期内震区发生更大地震的可能性不大”的会商意见向社会公布。自治区及震区各级人民政府利用各种媒体正面播发震后趋势意见。这些措施对稳定群众的恐慌情绪和制止地震谣言的扩散有着重要的意义。由于措施得当，各地的地震谣言得以迅速平息。

3. 震害分析

5.4 级地震发生后，新疆维吾尔自治区地震局立即派出工作组赴灾区开展了震害评估工作，随后和中国地震局现场工作组一起，在当地政府和灾区群众的大力支持和协助下，在较短时间内完成了这次地震的震害评估工作。经调查，石场镇等Ⅴ度区范围内多数房屋基本完好，部分房屋受到轻微破坏，石河子市、沙湾部分乡场等Ⅳ度区各类房屋基本没有遭受破坏。

对本次地震未造成建筑（构）物大的破坏，初步分析认为有以下因素[2]：①地震发生在山区，地震动到达有一定距离的居民区已衰减，石场地震台数字强震仪记录次地震最大峰值加速度为 22Gal，影响烈度为Ⅴ度，不足以对建筑物造成较大破坏；②石场矿区为高烈度设防区，正规建筑物都经过抗震设防，地震对这些建筑物基本没有影响；③现场调查地区多为中山区，大部分为基岩出露区，场地条件较好。

九、结论与讨论

（1）石河子 5.4 级地震发生在北天山地震带，是新疆地震监测能力相对较强的地区，因此地震前的前兆异常较为丰富。地震前 1 年多的时间内，震中区及其附近出现了 4 级地震的增强—平静现象。乌鲁木齐区、石场—精河区的地震学参数也出现了多项异常；自 2000 年秋季至 2001 年春季，定点形变资料的趋势转折先后出现，2001 年 11 月至 2003 年 1 月，北天山的前兆群体异常三次涨落，主要为定点形变和地下流体异常，34 个前兆项目共出现 20 条异常，其中 5 条为中期异常，12 条短期、3 条临震异常；2009 年 9 月的流动测量复测结果中，北天山流动水准和流动重力的部分测段也出现了异常。在地震发生前例行召开的周、月、年度会商会上，上述所总结的各类异常，均被各有关学科专家不同程度地发现和强调。

需要指出的是，2002 年 12 月起出现的前兆异常变化，可能主要是反映了 2003 年石河子 5.4 级地震的短期前兆，但是在时间上与 12 月 25 日乌恰 5.8 级地震相吻合。同时 2003 年 2 月 14 日石河子 5.4 及 2 月 24 日巴楚—伽师 6.8 地震后，北天山前兆异常依然活跃，此后中俄边境和中哈边境先后发生了 9 月 27 日 7.9 和 12 月 1 日 6.1 级地震。因此前兆群体异常的涨落可能是 2002 年后的一组强震成丛活动时前兆异常的场动态演化。

（2）石河子 5.4 级地震发生在新疆维吾尔自治区地震局中期预测和年度地震危险区内，震前新疆地震局向中国地震局提出了 3 个月的短期预测意见，对应较好。新疆地震局在地震发生后向当地人民政府提出了正确的震后趋势判定意见，为稳定社会局面起到了积极作用，取得了一定的社会效果。

（3）石河子 5.4 级地震震级较小，引起的关注程度不够，相关研究文献不多。大震速报的震中位置与最终确定的震中相距较远，不同作者的震源机制结果差异较大，加之地震现场工作受天气和地形的局限未能到达震中区域，因此此次地震发震构造和震源断错性质的认识有较大差异。虽然前文根据双差定位的主余震分布情况，确定发震构造为准噶尔南缘断裂的倾滑性质的单侧破裂，但是该问题仍有待深入研究。

（4）石河子 5.4 级地震的短期预报的实现，北天山等地的前兆资料异常的分析判定起到了非常重要的作用，部分文献中也作了一些总结。但由于震例总结要求前兆异常主要局限于 200 公里范围内，因而地震前显示出明显中短期（临）异常的库尔勒、精河、新源地倾斜资料未被统计在内。此外通过仔细甄别，对一些文献中列出的异常予以舍弃。如乌鲁木齐地磁资料在震前多次出现低点时间提前、石场台 CO_2 的短期变化、温泉地倾斜的异常变化、石河子跨断层的阶跃现象等。这些“异常”经过认真排查后发现，部分异常信度不高，有些则存在明显的环境干扰。

参 考 文 献

[1] 新疆地震局地震现场工作队，新疆石河子南 2003 年 2 月 14 日 5.0 和 5.4 级地震现场工作报告

[2] 罗福忠等，2003 年 2 月 14 日新疆石河子南 5.0 和 5.4 级地震灾害减轻因素的分析，内陆地震，19（1），86～89，2005

[3] 宋和平，论新疆深大断裂特征与地震的关系（3），内陆地震，20（3），198～210，2006

[4] 李志海等，2003 年新疆石河子 5.4 级地震序列重新定位及发震断层与震源机制分析，地震研究，29（2），109～113，2006

[5] 龙海英、高国英等，北天山中东段中小地震震源机制解及应力场反演，地震，28（1），93～99，2008

[6] 李莹甄、赵翠萍，2003 年新疆石河子 5.4 级地震前地震活动异常分析，内陆地震，17（4），309～316，2003

[7] 杨又陵、张翼，2003 年 2 月 14 日石河子 *M*5.4 地震地形变异常及其演化分析，内陆地震，19（3），241～248，2005

[8] 高小其、许秋龙，2003 年 2 月 14 日新疆石河子 5.0、5.4 级地震地下流体前兆异常特征的分析，内陆地震，18（1），64～71，2004

[9] 杨又陵、高国英，北天山西段中强地震前地倾斜异常的演化分析，大地测量与地球动力学，25（1），98～101，2005

参 考 资 料

1）新疆维吾尔自治区地震局，2003 年新疆地震局报中国地震局中短临地震预报意见卡，2003.02.13

2）新疆维吾尔自治区地震局，2003 年度新疆地震趋势研究报告，2002.11

3）新疆维吾尔自治区地震局，2003 年度新疆震情监视报告，2003.01

4）新疆维吾尔自治区地震局，2003 年 2 月 10 日向自治区政府汇报材料及有关批示，2003.02

5）新疆维吾尔自治区地震局，2003 年度新疆震情监视报告，2003.02.14，2003.02.16

6）刘蒲雄，地震现场震情分析方法指南，见：中国地震局监测预报司，地震监测预报骨干培训教材（下册），110～143，2000

Shihezi Earthquake of *M*5.4 on February 14, 2003 in Xinjiang Uygur Autonomous Region

Abstract

An earthquake of *M*5.4 occurred south to Shihezi in Xinjiang Uygur Autonomous Region on February 14 2003. Its epicenter was $\varphi_N 44°06'$, $\lambda_E 85°52'$. Because of the hypsography and traffic hardness, The macroscopic epicenter not be investigated, The maximal intensity at the investigated area was Ⅴ. There was no loss of people, the direct economic loss was 500 thousand Yuan.

The earthquake sequence belonged to preshock-mainshock-aftershock type and the magnitude of the largest aftershock was $M_S 3.0$. Aftershocks distributed southwest to the main shock, and distributing along the direction of N40°E. The seismological structure was the South brim of Zhungaer basin fault.

Within the distance of 200km from the epicenter, there were 11 earthquake monitoring stations, and has 34 observation items for earthquake precursor. There were 12 seismology anomalies before the earthquake, there were 20 earthquake-precursor anomalies, 5 of them were medium term anomalies, 12 of them were short-term anomalies and 3 of them were imminent, and the anomalies also be found in the result of ambulatory survey along the north Tianshan mountain at Sep. 2002.

The *M*5.4 earthquake occurred in the seismically dangerous area in Xinjiang as specified in 2003. Based on abnormal variations before the earthquake, a short-term forecasting for 3 months was submitted to CSB and the earthquake occurred as expected. After the earthquake, the judgement on the seismic activity was correct.

报告附件

附表1　固定前兆观测台（点）与观测项目汇总表

序号	台站（点）名称	经纬度（°）		测项	资料类别	震中距Δ（km）	备注
		φ_N	λ_E				
1	石场	43.91	85.69	测震△	Ⅰ	15	
				地倾斜（摆式仪）	Ⅱ		
2	沙湾新25泉	44.20	85.50	水温	Ⅲ	30	
				硫化物	Ⅱ		
3	沙湾新26泉	43.84	85.38	水位	Ⅱ	45	
4	石河子	44.20	86.11	测震△	Ⅱ	25	
				断层蠕变	Ⅱ		1996年故障，2001年改造
5	呼图壁	44.08	86.53	测震△	Ⅱ	50	
				水准	Ⅱ		
6	呼图壁新21井	43.73	86.56	水位	Ⅱ	70	
7	独山子	44.61	84.80	断层蠕变	Ⅱ	102	
8	乌鲁木齐基准台	43.81	87.69	测震△	Ⅰ	150	
				钻孔应变	Ⅱ		
				地磁	Ⅰ		
				地电阻率	Ⅱ		
		43.80	87.58	地倾斜（摆式仪）	Ⅰ	140	
				地倾斜（水管仪）	Ⅱ		2000年底，"九五"项目更新仪器
				地应变	Ⅱ		
		43.80	87.60	重力	Ⅱ	143	
		43.83	87.64	水准	Ⅲ	145	

续表

序号	台站（点）名称	经纬度（°）		测项	资料类别	震中距Δ（km）	备注
		φ_N	λ_E				
9	乌鲁木齐新 04 井（泉）	43.83	87.65	水质	Ⅱ	145	
				气体	Ⅱ		
				水温	Ⅰ		
				硫化物	Ⅱ		
10	乌鲁木齐新 05 井	44.09	87.86	水位	Ⅱ	160	
11	乌鲁木齐新 09 泉	43.70	87.62	水质	Ⅱ	148	
				水氡	Ⅱ		
12	乌苏新 33 井	44.63	84.15	流量	Ⅱ	150	
13	乌鲁木齐新 10 井（泉）	43.70	87.61	水位	Ⅰ	148	
				水质	Ⅱ		
				气体	Ⅱ		
				流量	Ⅰ		
				水氡	Ⅱ		
				水汞	Ⅰ		
14	乌鲁木齐新 15 泉	43.82	87.65	水质	Ⅱ	145	
				水氡	Ⅱ		
15	乌鲁木齐新 20 井	43.79	87.42	水位	Ⅱ	130	
16	克拉玛依	45.61	84.85	测震△	Ⅰ	185	
				地倾斜（垂直摆）	Ⅱ		
17	乌苏	44.12	84.64	测震	Ⅰ	55	
18	石梯子	43.71	86.65	测震	Ⅰ	75	
19	硫磺沟	43.64	87.06	测震	Ⅰ	108	
20	乌什城	43.46	87.69	测震	Ⅰ	162	
21	天池	43.89	88.09	测震	Ⅰ	178	
22	柳树沟	45.56	84.67	测震	Ⅰ	187	

续表

序号	台站（点）名称	经纬度（°）		测项	资料类别	震中距Δ（km）	备注
		φ_N	λ_E				

分类统计	$0<\Delta\leqslant100$km	$100<\Delta\leqslant200$km	总数
测项数 N	12	34	46
台项数 n	8	14	22
测震单项台数 a	2	4	6
形变单项台数 b	0	1	1
电磁单项台数 c	0	0	0
流体单项台数 d	3	7	10
综合台站数 e	3	2	5
综合台中有测震项目的台站数 f	3	2	5
测震台总数 $a+f$	5	6	11
台站总数 $a+b+c+d+e$	8	14	22
备注			

附表 2 测震以外固定前兆观测项目与异常统计表

序号	台站（点）名称	测项	资料类别	震中距Δ（km）	按震中距Δ范围进行异常统计																			
					$0<\Delta\leqslant100$km					$100<\Delta\leqslant200$km					$200<\Delta\leqslant300$km					$300<\Delta\leqslant500$km				
					L	M	S	I	U	L	M	S	I	U	L	M	S	I	U	L	M	S	I	U
1	石场	地倾斜	Ⅱ	15	—	√	√	√																
2	石河子	断层蠕变	Ⅱ	25	—	—	—	—																
3	沙湾新 25 泉	水温	Ⅲ	30	—	—	—	—																
		硫化物	Ⅱ		—	—	—	—																
4	沙湾新 26 泉	水温	Ⅱ	45	—	√	—	—																
5	呼图壁	水准	Ⅱ	50	—	√	—	—																
6	呼图壁新 21 井	水位	Ⅱ	70	—	—	√	√																
7	独山子	断层蠕变	Ⅱ	102						—	√	—	—											
8	乌鲁木齐新 20 井	水位	Ⅱ	130						—	—	√	—											
9	乌鲁木齐基准台	地倾斜	Ⅰ	140						—	√	—	—											
		水管仪	Ⅱ							—	—	—	—											
		地应变	Ⅱ							—	—	—	—											
		重力	Ⅱ	143						—	—	—	—											
		水准	Ⅲ	145						—	—	—	—											

续表

序号	台站（点）名称	测项	资料类别	震中距Δ（km）	按震中距Δ范围进行异常统计																			
					0＜Δ≤100km					100＜Δ≤200km					200＜Δ≤300km					300＜Δ≤500km				
					L	M	S	I	U	L	M	S	I	U	L	M	S	I	U	L	M	S	I	U
9	乌鲁木齐基准台	钻孔应变	Ⅱ	150						—	—	—	—											
		地磁	Ⅰ							—	—	—	—											
		地电阻率	Ⅱ							—	—	√	—											
10	乌鲁木齐新04井（泉）	水质	Ⅱ	145						—	—	—	—											
		气体	Ⅱ							—	—	—	—											
		水温	Ⅰ							—	—	√	√											
		硫化物	Ⅱ							—	—	√	—											
11	乌鲁木齐新15泉	水质	Ⅱ	145						—	—	—	—											
		水氡	Ⅱ							—	—	—	—											
12	乌鲁木齐新09泉	水质	Ⅱ	148						—	—	—	—											
		水氡	Ⅱ							—	—	—	—											
13	乌鲁木齐新10井（泉）	水位	Ⅰ	148						—	—	√	—											
		水质	Ⅱ							—	—	—	—											
		气体	Ⅱ							—	—	√	—											
		流量	Ⅰ							—	—	√	—											
		水氡	Ⅱ							—	—	√	—											
		水汞	Ⅰ							—	—	√	—											

续表

序号	台站（点）名称	测项	资料类别	震中距 Δ（km）	按震中距 Δ 范围进行异常统计																			
					$0<\Delta\leqslant100$km					$100<\Delta\leqslant200$km					$200<\Delta\leqslant300$km					$300<\Delta\leqslant500$km				
					L	M	S	I	U	L	M	S	I	U	L	M	S	I	U	L	M	S	I	U
14	乌苏 新 33 井	流量	Ⅱ	150						—	—	—	—											
15	乌鲁木齐 新 05 井	水位	Ⅱ	160						—	—	—	—											
16	克拉玛依	地倾斜	Ⅱ	185						—	—	√	—											
分类统计	台项	异常台项数			0	3	2	2	/	0	2	10	1	/										
		台项总数			7	7	7	7	/	27	27	27	27	/										
		异常台项百分比/%			0	43	29	29	/	0	7	37	4	/										
	观测台站（点）	异常台站数			0	3	2	2	/	0	2	5	1	/										
		台站总数			6	6	6	6	/	10	10	10	10	/										
		异常台站百分比/%			0	50	33	33	/	0	20	50	10	/										
	测项总数				7					27														
	观测台站总数				6					10														
备注																								

2003年2月24日新疆维吾尔自治区巴楚6.8级地震

新疆维吾尔自治区地震局

王　琼　孙甲宁　高国英　曲延军

摘　要

2003年2月24日在新疆维吾尔自治区巴楚—伽师县交界发生6.8级地震，宏观震中为39°20′N，77°38′E，震中烈度为Ⅸ度。地震造成268人死亡，4853余人受伤，其中重伤2058人，直接经济损失13.98亿元。

巴楚6.8级地震发生在柯坪塔格推覆构造系西段前缘部位，发震构造为NWW走向的北倾盲逆断层巴什托普隐伏断裂。6.8级地震序列类型为主震—余震型，最大余震为M_S5.8。余震分布在主震两侧约50km范围内，呈NWW和NE向共扼分布，其中NWW向余震区为主体活动区。

6.8级地震震源断错类型为逆冲型，近EW走向的节面Ⅰ为断层面。主压应力P轴方位339°，倾角水平，张应力T轴近于垂直。推测6.8级地震是在区域构造应力场NW向的挤压应力作用下，产生的逆冲倾滑破裂过程。

6.8级地震震中300km以内共有8个地震台，均设有测震观测。其中6个台站有前兆观测，包括乌什、阿合奇及阿图什地倾斜，乌什钻孔应变，阿克苏跨断层形变，喀什地磁，喀什、乌恰地电等观测项目。此次地震前测震学存在中强地震平静—增强、中等地震平静、地震学参数时空扫描和震情窗四类异常；定点前兆存在3项中期异常、4项短期（临）异常。

巴楚6.8级地震前，新疆地震局根据地震活动、前兆异常做出了不同程度的中短期预测，强度和时间判定较为准确，但地点判定存在偏差。震后对4次5级强余震做出了较为准确的预测。

前　言

2003年2月24日10点03分，新疆维吾尔自治区巴楚县、伽师县交界发生6.8级地震，微观震中为39°37′N，77°16′E。经考察宏观震中为39°20′N，77°38′E，震中烈度达Ⅸ度。这次地震有感范围大，喀什市、疏勒县、疏附县、英吉沙县、叶城县、阿克苏等地强烈有感。地震造成268人死亡，4853余人受伤，其中重伤2058人，直接经济损失13.98亿元。

巴楚 6.8 级地震位于天山南麓山前冲积平原与塔里木盆地西北部交接地带。由于受印度板块和欧亚板块碰撞作用，该区所在的南天山地震带西段与西昆仑地震带交汇区域构造运动强烈，是中国大陆强震活动最为频繁的地区。震区周围曾发生多次 6 级以上地震，其中包括 1902 年阿图什 8.2 级大震，1961 年巴楚 6 级震群、1997 ~ 1998 年伽师强震群。此次地震前测震学存在中强地震平静—增强、中等地震平静、地震学参数时空扫描和震情窗等四类异常；定点前兆存在 3 项中期异常、4 项短期（临）异常。

巴楚 6.8 级地震发生在 1997 ~ 1998 年伽师强震群区东南约 20 ~ 30km 处，位于 1998 年 8 月 27 日伽师 6.4 级地震后 3.5 级以上余震活动区。6.8 级地震前，新疆地区 6 级地震平静长达 4 年、5 级地震呈现出“平静—短期增强”的活动特征，震源区附近中等以上地震呈现“活跃—平静”的起伏活动特征，尤其是震前 1 年，震区范围内无 3.5 级以上地震发生。巴楚 6.8 级地震可能与 1997 ~ 1998 年伽师强震群活动存在一定程度的内在联系，可能缘于 1997 ~ 1998 年伽师强震群的向外扩展[1]。

巴楚 6.8 级地震发生在 2003 年度确定的地震危险区东侧，地震前新疆地震局曾提出 1 个月的短期预测意见，特别是 6.8 级地震前对新疆地震形势发展趋势把握基本正确，但地点存在偏差。巴楚 6.8 级地震后，对地震类型和强余震作出了准确的判定和预报，提出过 4 次较为准确的强余震短临预报意见，其中 3 月 12 日伽师 5.8 级地震余震前提出了 15 天的临震预报。

在有关文献和资料的基础上[1 ~ 21；1 ~ 7)]，经过对资料的重新整理和分析研究，最终完成此研究报告。

一、测震台网及地震基本参数

图 1 给出震中 300km 范围内的地震台站分布，所有 8 个地震台均有测震观测，在研究时段内基本可达到 $M_S \geqslant 2.5$ 级地震不遗漏。100km 内仅有阿图什测震台；100 ~ 200km 分别有喀什、乌恰、巴楚、阿合奇 4 个测震台；200 ~ 300km 有阿克苏、乌什和塔什库尔干测震台。震中附近区域当时属于新疆监测能力较为薄弱的地区，包括伽师—巴楚震区在内的喀什以东地区，地震监测能力可达 $M_L \geqslant 2.5$ 级，而喀什以西至乌恰、塔什库尔干地区地震监测能力为 $M_L \geqslant 3.0$ 级，乌恰—塔什库尔干以西的边境地区仅能达到 $M_L \geqslant 3.5$ 级。

表 1 列出不同来源给出的这次地震的基本参数，其中编号（1）结果为《巴楚—伽师强震序列的深入研究》课题采用双差定位得到的震中位置（双差定位目录的震级采用新疆地震局目录），本次地震基本参数取表 1 中编号 1.1 结果[1)]。

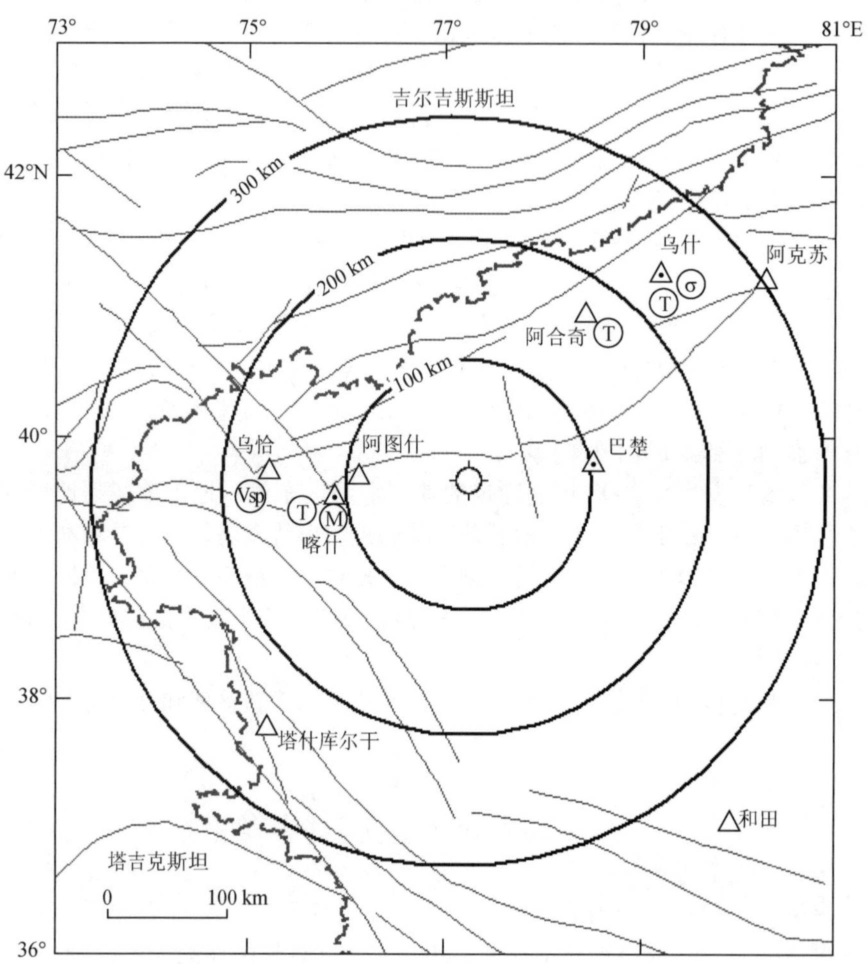

图 1　巴楚 6.8 级地震前震中附近地震台站及观测项目分布图

Fig. 1　Distribution of earthquake-monitoring stations and observational items around the epicentral area before the *M*5.4 Jiashi and *M*6.8 Bachu earthquake

表 1　巴楚 6.8 级地震基本参数

Table 1　Basic parameters of the *M*5.4 Jiashi earthquake and *M*6.8 Bachu earthquake

编号	发震日期	发震时刻	震中位置		震级	震源深度	震中地名	结果来源
	年 月 日	时 分 秒	φ_N	λ_E	*M*	(km)		
1.1	2003 2 24	10 03 42	39°34′	77°16′	6.8	20	巴楚	资料1)
1.2	2003 2 24	10 03 42	39°37′	77°16′	6.6	27	巴楚	新疆地震局2)
1.3	2003 2 24	10 03 41	39°35′	77°20′	6.8	8	巴楚	中国地震台网3)
1.4	2003 2 24	10 03 43	39°31′	77°13′	6.3	26	巴楚	ISC

二、地震地质背景

2003 年 2 月 24 日巴楚 6.8 级地震位于 1997 ~ 1998 年伽师强震群东南约 20 ~ 30km。深部地球物理探测资料表明[2]，巴楚 6.8 级地震偏向柯坪塔格推覆构造系西段前缘部位，主体发震构造为柯坪塔格推覆构造系西段南缘隐伏的近 EW 向多重推覆盲逆断层。石油物探结果显示[3]，塔里木盆地西部的断裂构造十分发育，巴楚 6.8 级地震震区东侧存在一组 NW 向的由多条隐伏逆断层和走滑断层组成的断裂构造，它将巴楚隆起与喀什坳陷分割开来。巴楚 6.8 级地震就发生在上述 NW 向断裂带的一个分支——NWW 走向的北倾盲逆断层巴什托普隐伏断裂上。震区附近地区主要断裂分布见图 2，北部为托特拱拜孜断裂和柯坪断裂，西部为卡兹克阿尔特断裂，东部有普昌断裂。

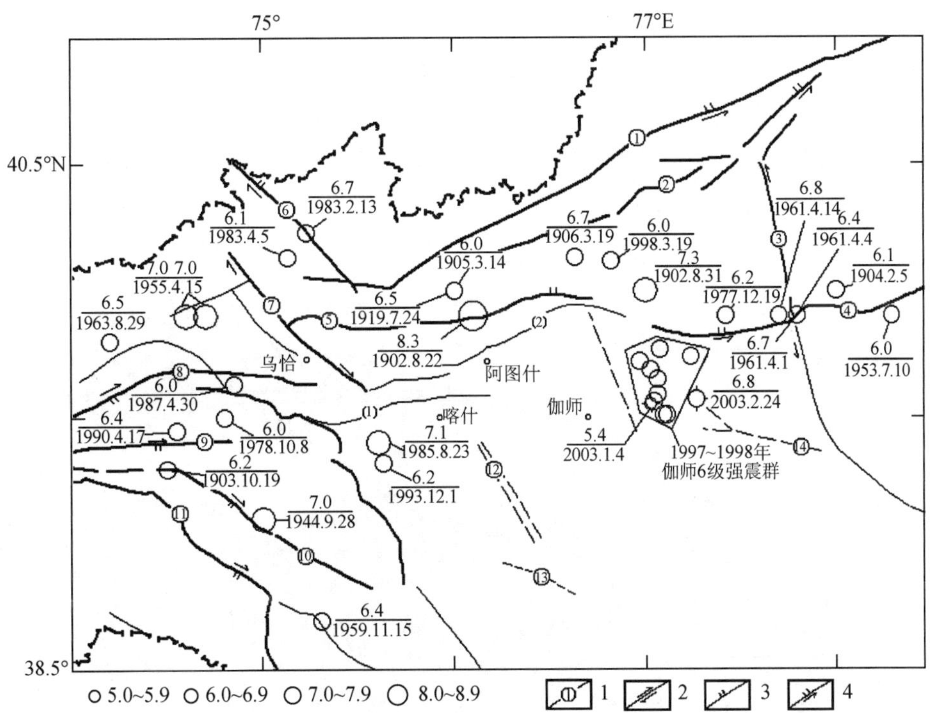

图 2 伽师—巴楚附近地区主要断裂及历史地震震中分布图

Fig. 2 Major faults and distribution of historical earthquakes around Jiashi and Bachu area

图例说明：1. 断层编号；2. 走滑断层；3. 逆断层、逆掩断层；4. 逆走滑断层

断裂名称：①麦丹塔格断层；②卡拉铁克壳断层；③普昌断层；④柯坪断层；⑤托特拱拜孜—阿尔帕勒克断裂；⑥塔拉斯—费尔干纳断层；⑦库孜贡断层；⑧卡兹喀尔阿尔特断裂；⑨肯恩别尔特断层；⑩马尔坎苏—奥依塔克断层；⑪塔什库尔干断层；⑫羊达曼断层；⑬英吉沙断层；⑭巴什托普断层

活动背斜名称：（1）喀什；（2）阿图什

柯坪塔格推覆构造系晚第四纪以来活动强烈，发育数排活动褶皱—逆断裂带，现今构造

运动强烈，地震活动频度高、强度大，是新疆 6 级以上地震活动最频繁的断裂，强震基本上呈近东西向展布。曾发生多次 6 级以上地震，其中包括 1902 年阿图什 8.2 级大震，1961 年巴楚 6 级震群、1997 ~ 1998 年伽师强震群。巴楚 6.8 级地震位于柯坪塔格推覆构造系西段南缘，距其最近的历史地震，是 1997 ~ 1998 年伽师强震群，二者相距约 20 ~ 30km 范围（图 2）。

在 1°×1°布格重力异常图上，巴楚 6.8 级地震发生在围绕巴楚隆起显示高重力值展布的近南北向斜坡带，重力梯度为 $-1.04 \times 10^{-5} m/s^2/km$。布格重力异常值为 $-250 \times 10^{-5} m/s^2$[4]。

1955 ~ 1988 年大地水准测量[5]表明，该区域受印度洋板块向欧亚板块挤压的作用，以 2mm/a 的速率隆起。GPS 观测结果显示[6,7]，柯坪塔格推覆构造系现今地壳缩短速率为 19.0 ~ 20.0mm/a。

分析认为，巴楚 6.8 级地震发震构造为柯坪塔格推覆构造系西段南缘的北倾盲逆断层巴什托普隐伏断裂。

三、地震影响场和震害[8]

据现场实地考察，巴楚 6.8 级地震宏观震中为 39°20′N，77°38′E，震中区烈度为Ⅸ，等震线呈东西向椭圆形分布（图 3）。

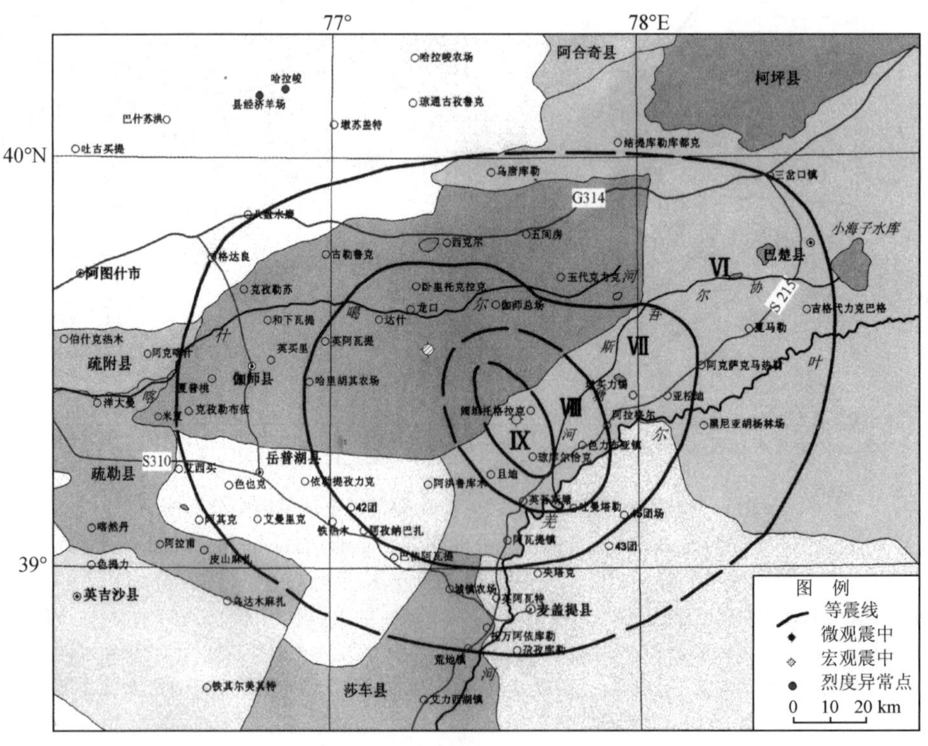

图 3　巴楚 6.8 级地震烈度等震线图

Fig. 3　Isoseismal map of the *M*6.8 Bachu earthquake

Ⅸ度区：南部边界位于巴楚县琼库尔恰克乡政府南，北部边界至巴楚—伽师交界的沙漠地带（位于琼五井以北）。长轴38km，走向北西方向，短轴14km，面积421km^{2}左右。在Ⅸ度区内土木结构房屋90%以上倒塌，其余的局部倒塌、屋顶坍塌或墙体普遍出现裂缝。倒塌、局部倒塌、屋盖塌落等的砖木结构平房约占70%～80%。未经抗震设防的砖混结构房屋毁坏、局部倒塌、严重破坏占一定比例，但仍有基本完好的砖混结构房屋。砖柱厂房、粮库遭到严重破坏。大部分水塔遭受破坏，塔筒身出现斜裂或环裂，部分水塔倒塌。大面积出现喷砂冒水，最大喷砂孔直径达3m。柏油路面多处裂缝，松软地面出现不规则张裂缝，河岸张裂、崩塌。在伽师—巴楚交界的沙漠地带发现有北北西向排列的地裂缝带。

Ⅷ度区：包括巴楚县的琼库尔恰克乡南部，色布里亚镇、阿拉格尔乡和英吾斯塘乡的部分村庄。长轴59km，走向北西方向，短轴42km，面积1573km^2。Ⅷ度区内土木结构房屋40%倒塌，50%严重破坏或中等破坏。砖木结构房屋普遍出现裂缝，50%严重破坏。未经严格抗震设防的砖混结构房屋遭到严重破坏和中等破坏。砖柱厂房遭到严重破坏。部分水塔遭受破坏，塔筒身出现斜裂或环裂，近一半水塔基本完好。喷砂冒水现象多分布在地势低洼，地下水位较高的地区（如叶尔羌河两岸）。

Ⅶ度区包括巴楚县西南部、岳普湖县东部、伽师县南部、麦盖提县北部、农三师42团等。长轴95km，方向北北东，短轴92km，面积4999km^2。Ⅶ度区长轴方向与高烈度区不同。Ⅶ度区土木结构房屋个别倒塌或严重破坏，40%～50%为中等破坏。墙体出现较大交叉裂缝、斜裂缝、女儿墙塌落等的砖木结构平房约占50%～60%。部分砖混结构房屋遭到中等破坏或普遍出现轻微裂缝。岳普湖县城内有个别砖混结构房屋遭到中等破坏。县城自来水管线多处震裂。单层（单跨）钢筋混凝土柱厂房有轻微破坏。Ⅶ度区内仍然存在砂土液化现象，伽师县卧里托克拉克乡龙口村克孜河岸边分布有小面积喷砂冒水，喷砂孔直径20～30cm。在龙口以南约6km，克孜河南岸沼泽地带见到规模较大的喷砂冒水现象，并伴随有地裂缝。

Ⅵ度区：北边到达伽师县西克尔以北山区，西边到达岳普湖县、伽师县西部和阿图什市东部的格达良乡，东边到达巴楚县城，南边到达麦盖提县尕孜库勒村。包括巴楚县、岳普湖县、伽师县、麦盖提县北部、莎车县局部和阿图什市局部，以及兵团农三师部分团场。Ⅵ度区长轴180km，走向北北东，短轴131km，面积11573km^2。Ⅵ度区土木结构房屋墙体出现裂缝、外闪，个别有严重破坏。砖木结构平房部分墙体出现裂缝，大部分基本完好。砖混结构房屋大部分基本完好，个别墙体出现轻微裂缝，女儿墙断裂。

Ⅴ度区：包括距离震区120km的阿图什市哈拉峻乡，其房屋的破坏程度相当于Ⅵ度。该乡部分土木结构房屋墙体普遍裂缝，个别倒塌。新建的砖木结构学校教室承重横梁下、窗角、门框处的墙体普遍裂缝。

1997年伽师强震群发生后在伽师县东部灾区由“安居工程”重建的大量板墙房经受住了本次地震的考验，此次地震极震区的板墙房也表现出良好的抗震性能。

巴楚6.8级地震是半个多世纪以来新疆境内发生的最强烈的地震之一，也是新中国成立以来新疆发生的损失最大的一次地震，灾区群众蒙受了巨大生命和财产损失。地震波及喀什地区及克州，共计5县1市37个乡镇931个村（场），灾区面积达21498km^2，极震区烈度达Ⅸ。巴楚县的琼库尔恰克乡、阿拉格尔乡、色力布亚镇等6个乡镇遭到了毁灭性破坏。

震区受灾人口达 51 万多人，共造成 268 人死亡，4853 人受伤，其中重伤 2058 人；19899 户民房倒塌；死亡牲畜 70958 头（只）。根据国家地震灾害评估委员会的评估，地震造成直接经济损失 13.71 亿元。

四、地震序列[9,10]

巴楚 6.8 级地震序列空间特征分析采用精确定位结果[1)]，序列时间特征分析采用巴楚单台序列目录。

从巴楚 6.8 级地震余震空间分布来看（图 4），余震主要分布在主震两侧约 50km 范围内，呈 NWW 和 NE 向共扼分布，其中 NWW 向余震活动区与近东西走向的等震线方向一致。6.8 级地震的余震沿主震向两侧扩展，其中 NWW 向余震区为主体活动区，发生 20 次 4 级以上地震，其中包含 3 次 5 级地震；NE 向余震分布区发生 9 次 4 级以上地震，其中包含 2 次 5

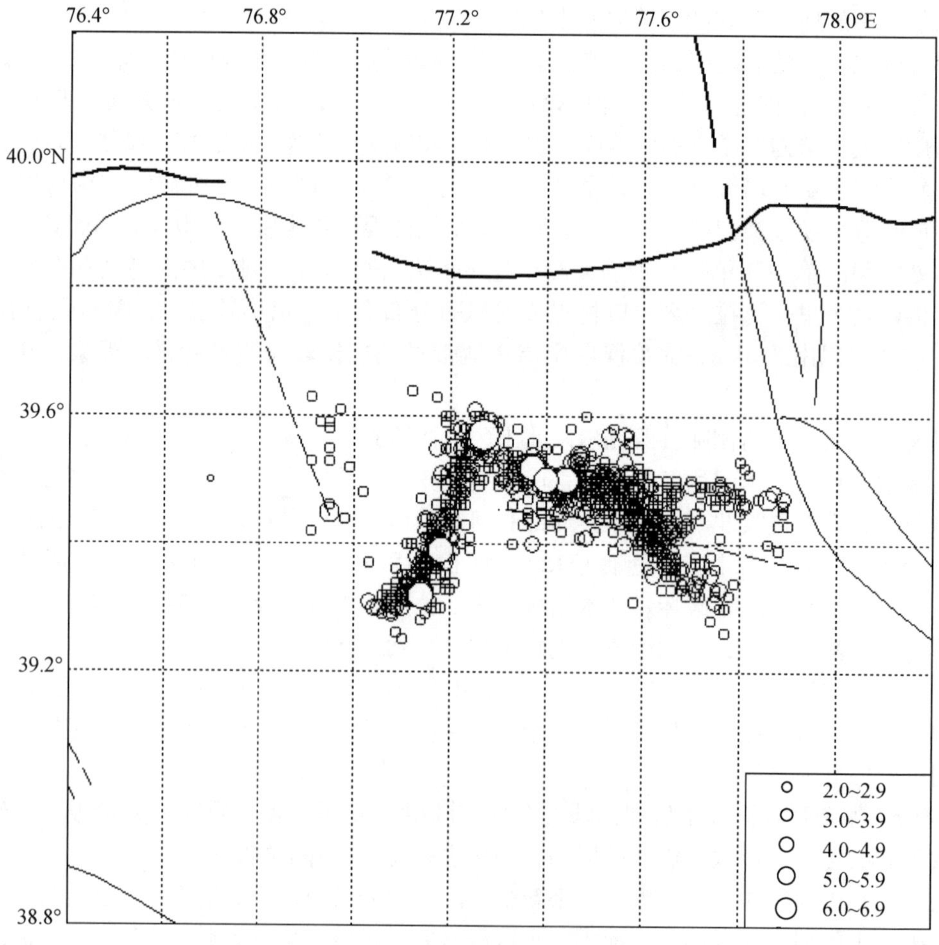

图 4　巴楚 6.8 级地震序列分布图

Fig. 4　Distribution of the M6.8 Bachu earthquake sequence

级地震，5 月 4 日岳普湖 5.8 级地震后，该方向的余震扩展至主震西南约 20km 之外的区域。

6.8 级地震后，2003 年 3 月 7 日至 7 月 31 日，距 6.8 级地震约 30km 处的伽师震源区附近出现 15 次 2～4 级地震活动，这组小震活动偏离 6.8 级地震余震主体活动区，可能缘于 6.8 级地震的发生，触发了伽师老震区的地震活动。周云好[1)]的研究结果表明，本次地震的破裂面走向为近东西向，震源破裂为双侧破裂，这与 6.8 级的余震空间分布是一致的。

表 2 巴楚 6.8 级地震序列目录[1)]（$M_S \geqslant 5.0$ 级）

Table 2 Catalogue of the *M*6.8 Bachu earthquake sequence（$M_S \geqslant 5.0$）

编号	发震日期	发震时刻	震中位置		震级	震源深度	震中地名	结果来源
	年 月 日	时 分 秒	φ_N	λ_E	M_S	(km)		
1	2003 02 24	10 03 41	39°34′	77°16′	6.8	20	巴楚	资料 1)
2	2003 02 25	11 52 42	39°31′	77°22′	5.6	28		
3	2003 03 12	12 47 52	39°30′	77°26′	5.8	26		
4	2003 03 31	07 15 47	39°30′	77°24′	5.3	25		
5	2003 05 04	23 44 36	39°23′	77°10′	5.8	25		
6	2003 05 16	22 30 01	39°19′	77°08′	5.4	26		

根据巴楚单台序列定位结果，2003 年 2 月 24 日巴楚 6.8 级地震后，截至 2004 年 1 月 30 日，共定出 2.0 级以上地震 3707 次，其中 5.0～5.9 级 7 次；4.0～4.9 级 50 次；3.0～3.9 级 547 次；2.0～2.9 级 3103 次。

6.8 级地震序列的 *M-t* 图（图 5），6.8 级地震序列开始衰减较为缓慢，至 3 月 16 日 M_L5.1 余震发生后衰减速率加快。5 月 4 日 M_L5.2 余震至 6 月 5 日 M_L5.1 余震，序列再次出现起伏增强活动，之后序列开始逐渐衰减，但强度衰减较慢。

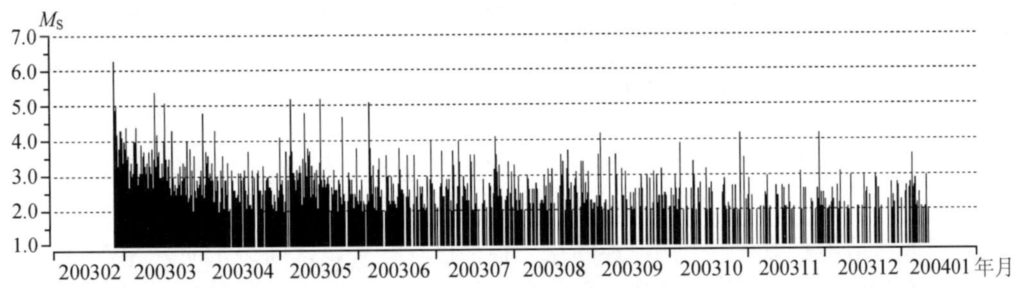

图 5 巴楚 6.8 级地震序列 *M-t* 图

Fig. 5 *M-t* diagram of the *M*6.8 Bachu earthquake sequence

地震序列频度分布（图 6）表明，6.8 级地震序列丰富，2 月 25 日 M_L5.0 余震后，序列频度逐步衰减；5 月序列频度明显增加，之后序列开始缓慢衰减。

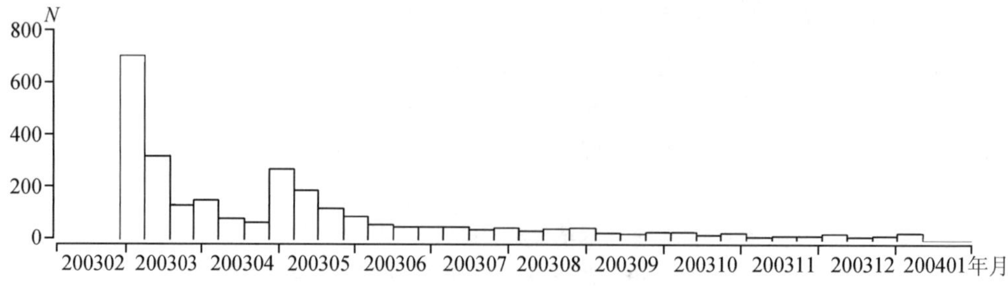

图6 巴楚6.8级地震序列频次分布

Fig. 6 Variation of earthquake frequency of the *M*6.8 Bachu earthquake sequence

由巴楚单台记录的巴楚6.8级余震序列目录，计算得到序列 b 值为0.99，高于南天山地震带平均0.70的背景 b 值，序列 p 值为0.91，h 值为1.2，主震释放能量占全序列能量的93.6%，主震与最大余震震级相差1.0。根据地震类型的判别指标，巴楚6.8级地震序列类型为主震—余震型。

五、震源参数和地震破裂面

根据新疆区域台网26个清晰的P波初动符号，测定了这次6.8级地震的震源机制解（表3、图7），矛盾符号比为0.19。

表3 巴楚6.8级地震的震源机制解

Table 3 Focal mechanism solutions of the *M*6.8 Bachu earthquake

序号	节面Ⅰ			节面Ⅱ			P轴		T轴		N(B)轴		X轴		Y轴		结果来源
	走向	倾角	滑动角	走向	倾角	滑动角	方位	仰角	方位	仰角	方位	仰角	方位	仰角	方位	仰角	
1	272	50	134	47	50	48	339	0	69	65	250	26					文献［11］
2	74	53	53	306	50	129	190	1	282	61	99	29					许力生
3	239	33	62	92	61	107	169	15	37	69	263	15					HRV
4	273	6	99	84	84	89	175	39	353	51	84	1					USGS

巴楚6.8级地震震源断错为倾滑逆断层。根据巴什托普隐伏断裂走向（NWW向）、极震区烈度分布和余震分布特征分析认为，近EW走向的节面Ⅰ为断层面。主压应力P轴方位339°，仰角水平，张应力T轴近于垂直。由震源机制解结果分析，此次6.8级地震是在区域构造应力场NW向的挤压应力作用下，产生的逆冲倾滑破裂过程[11]。巴楚6.8级地震震源断错性质与柯坪块体周围多次6级地震倾滑逆断层性质基本一致，也反映出柯坪块体构造运动特性。

周云好1) 使用IRIS全球地震台网宽频带地震仪记录的垂直向P波波形资料，利用波形拟合和有限断层模型反演方法研究了此次巴楚 6.8 地震的震源破裂时空过程。结合地质构造、余震分布及地震宏观考察资料，综合理论地震图和观测地震图拟合结果，认为此次地震是一次具有逆倾和一定右旋走滑分量的、由一点向双侧和深处扩展的三侧破裂事件，震源持续时间为 20s。6.8 级地震余震在 NWW 向和 NE 向呈现共轭分布，其中余震主要分布在 NWW 方向，5 月 4 日岳普湖 5.8 级强余震发生后 NE 向余震区扩展，意味着 6.8 级地震是一次双侧破裂事件。6.8 级地震后续 5 次强余震的深度大于主震表明，本次 6.8 级地震破裂具有向深处扩展的特征。

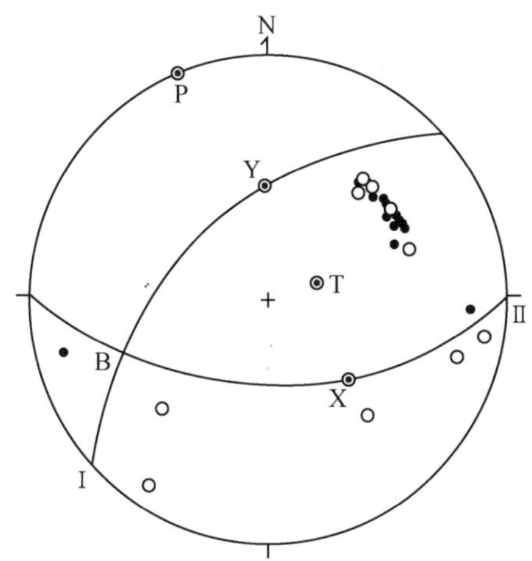

图 7　由新疆区域台网巴楚 6.8 级地震震源机制解

Fig. 7　Focal mechanism solutions of the M_S6.8 Bachu earthquake from Pwave records by the Xinjiang regional network

六、地震前兆观测台网及前兆异常

震中附近地震台站及观测项目分布见图 1。地震发生在塔里木盆地的西北缘，位于台网内偏西部位，台站主要分布在震中东、西两侧。区内台站主要沿南天山西段构造走向分布，南北两个方向分别是天山和塔里木盆地，无地震观测台站，震区属于地震监测能力较弱的地区。

在震中 300km 范围以内共有 8 个地震台，均设有测震观测。此外，6 个台站有其他前兆观测，包括地倾斜（乌什、阿合奇、阿图什台）、钻孔应变（乌什台）、跨断层形变（阿克苏台）、地磁（喀什台）和地电（喀什、乌恰台）等观测项目。在 0 ~ 100、101 ~ 200、201 ~ 300km 范围内分别有地震台站 1、4 和 3 个，其中测震以外前兆观测台站 1、3 和 2 个，观测项目分别为 1 项、5 项和 3 项。此次地震前共出现 10 个异常项目 12 条异常（表4），除此之外，2003 年度新疆地区趋势会商会提出南天山西段至喀什—乌恰交汇区存在 10 条测震学异常和 2 条定点前兆异常（附件8、附件9）。另外，喀什、莎车和巴楚地办各有土层应力观测点 1 个。

中国地壳运动观测网络在新疆地区布设有 3 个基准站和 5 个基本站，其中在伽师及其邻区有 2 个基准站、1 个基本站和 5 个过渡点（图 8）[12]。

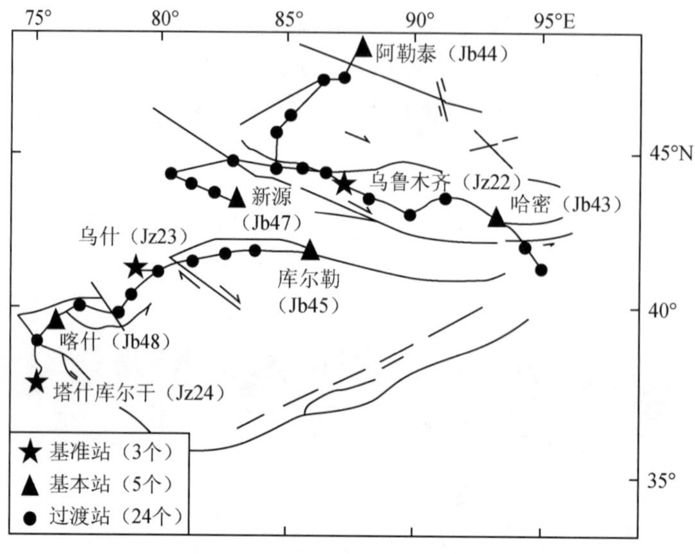

图 8　新疆地区重力联测路线及构造示意图

Fig. 8　Sketch for distribution of the gravity survey routes and tectonics in Xinjiang

在伽师及其邻区的天山南麓和帕米尔东北侧布设了以喀什（基本站）、伽师为中心，共 43 个 GPS 观测点及原攀登计划中的若干土层点共同构成的监测网。这些 GPS 点均匀分布在 φ_N34°50′～41°50′，λ_E73°15′～80°50′的范围内，平均点间距离为 60～100km，基本覆盖了南疆西部地区。

表 4　地震前兆异常登记表

Table 4　Summary table of earthquake precursory anomalies

序号	异常项目	台站（点）或观测区	分析方法	异常判据及观测误差	震前异常起止时间	震后变化	最大幅度	震中距（km）	异常类别及可靠性	图号	异常特点及备注	备注
1	中强地震平静—短期增强	新疆地区	中期地震平静、增强	6 级地震平静、5 级地震平静—短期增强	1998. 8～2003. 2 2000. 3～2003. 2				M_1	9	6 级地震平静 4 年、5 级地震平静 2 年和短期增强	震前

续表

序号	异常项目	台站（点）或观测区	分析方法	异常判据及观测误差	震前异常起止时间	震后变化	最大幅度	震中距（km）	异常类别及可靠性	图号	异常特点及备注	备注
2	地震平静	震中 50km 范围内	M_S≥3.5 级地震空间分布	M_S ≥ 3.5 级地震平静	2002.1 ~ 2002.12	地震发生在平静区		震中周围	M_1	15	震前震中附近约 50km 内 M_S ≥ 3.5 级地震平静	
3	地震频度	震中 250km 范围内	M_S≥3.0 级地震频度（1 年步长 2 个月窗长滑动）	持续 7 个月高值	2001.12 ~ 2002.6	恢复正常	最低值	震中周围	M_1	16	强震多发生在低值异常结束后出现的高值异常恢复后	震前/见新疆 2003 年度地震趋势研究报告附件 9
4	缺震	震中 250km 范围内	M_S≥2.5 级缺震分析（1 年步长 3 个月窗长滑动）	持续 2 年低于均值	2000.5 ~ 2003.7	恢复正常	最低值	震中周围	M_1	17	6 级地震都发生在低值异常恢复过程中	
5	*b* 值	震中 250km 范围内	*b* 值时间扫描（M_S ≥ 2.5 级，1 年步长 1 个月窗长滑动）	*b* ≥ 0.83 持续近 2 年	2001.2 ~ 2002.12	恢复正常	最高值	震中周围	M_1	18	1997 ~ 1998 年伽师强震群发生在低 *b* 值异常过程中，其他强震发生在低值后出现的高值异常变化过程	
6	小震调制比 R_m	震中 250km 范围内	小震调制比（M_S ≥ 2.5 级，1 年步长 3 个月窗长滑动）	R_m >0.12	1999.2 ~ 2001.3	恢复正常	0.15	震中周围	M_1	19	中强震多发生在调制比异常过程中或恢复后	

续表

序号	异常项目	台站(点)或观测区	分析方法	异常判据及观测误差	震前异常起止时间	震后变化	最大幅度	震中距(km)	异常类别及可靠性	图号	异常特点及备注	备注
7	地倾斜(石英摆周记)	乌什	单分量日均值	加速变化	2002.10～2003.1	基本恢复	0.4″	240	M_3	20	打破年变、异常方向背向未来震中	震前/见新疆2003年度地震趋势研究报告附件8
8	钻孔应变(压容)	乌什	N52°E元件日均值	打破正常年变，年变幅度增大	2001.8～2002.7	基本恢复	3×10^{-6}	240	M_2	21	持续时间长、异常幅度大，年变相位逆变	
9	地倾斜(竖直摆)	喀什	单分量整点值	速率突变	2003.1.9～1.11	恢复正常	0.12″	120	S_2	22	速率快，持续时间度	震前/见附件1、附件2
10	地倾斜(石英摆)	阿合奇	日均值矢量	打破正常年变	2003.1.5～3.31	基本恢复		170	S_2	23	矢量方向打破正常年变，频繁拐弯和打结	震前/见附件1
11	地倾斜(石英摆日记)	乌什	日均值矢量	矢量模明显增大	2003.1.5～2.24	基本恢复		240	S_2	24	矢量模明显增大，临震方向不稳定	震前/见附件1
12	钻孔应变(压容)	乌什	N52°E元件旬均值	压应变持续增大	2003.1.1～2.24	基本恢复		240	S_2	25	持续时间长、异常幅度大，临震压—张反复变化	震前/见附件2

1. 地震学异常

1996～1998年新疆西南部塔里木盆地西北缘的阿图什—伽师地区和西南缘的喀喇昆仑

—和田地区出现一组强震活动，共发生 12 次 6 级地震和 1 次 7 级地震。之后的 1999 ~ 2002 年新疆强震活动逐步减弱，尤其是 2001 ~ 2002 年新疆中强地震进入了显著的平静状态，2 年仅发生 1 次 5 级地震。2 月 24 日巴楚 6.8 级地震前，新疆地区 6 级以上地震平静长达 4 年 6 个月、5 级地震平静 2 年的状态被 2002 年 12 月 25 日乌恰 5.7 级地震打破。

1）震前异常情况

1998 年 8 月 27 日伽师 6.0 级地震后，新疆地区处于 6 级地震平静 4 余年、5 级地震平静 2 年的异常弱活动状态（图 9）。2002 年 12 月之后，新疆地区中强地震呈现短期增强趋势，接连发生乌恰 5.7 级、2003 年 1 月 4 日伽师 5.4 级和 2 月 14 日石河子 5.4 级地震，2 月 24 日巴楚 6.8 级地震发生在新疆地区中强地震短期增强的过程中。

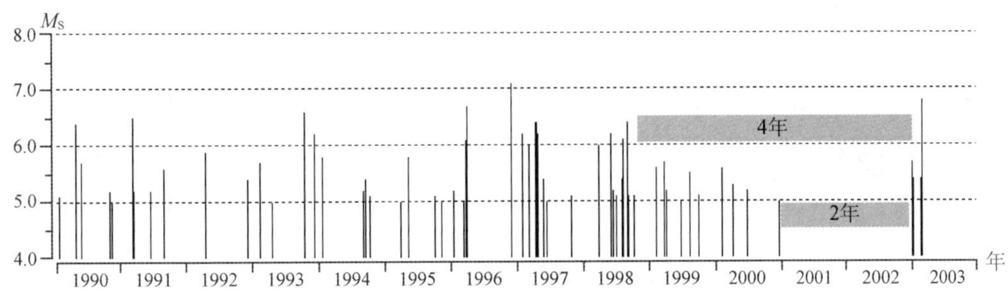

图 9　1990 年以来新疆地区 5 级以上地震 M-t 图

Fig. 9　M-t diagram of $M \geqslant 5.0$ earthquakes since 1900 in Xinjiang

巴楚 6.8 级地震前，2002 年 11 月召开的 2003 年度地震趋势会商会上，新疆地震局提出的南天山西段及其周围的测震学中期异常主要有（附件 9）：3 级地震频度偏低（2001.10 ~ 2002.9，图 10a）、b 值高值异常（2000.10 ~ 2002.9，图 10b）、缺震（1995.8，图 11a）、η 值低值（2000.1 ~ 2001.8，图 11b）、$A(b)$ 值低值（2001.1 ~ ，图 11c）、震情窗（2002.3 ~ 2002.10）等。

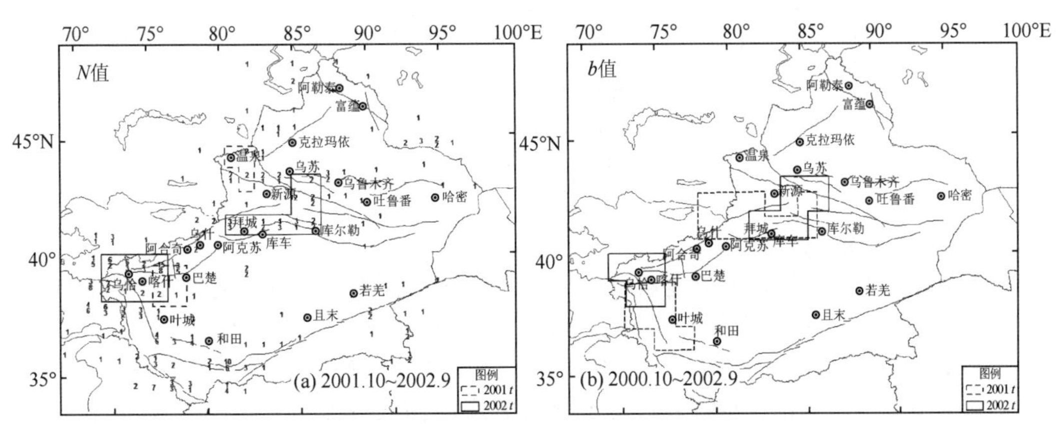

图 10　新疆地区地震学参数空间异常分布图

Fig. 10　Distribution for spatial anormaly of seismic parameters in Xinjiang

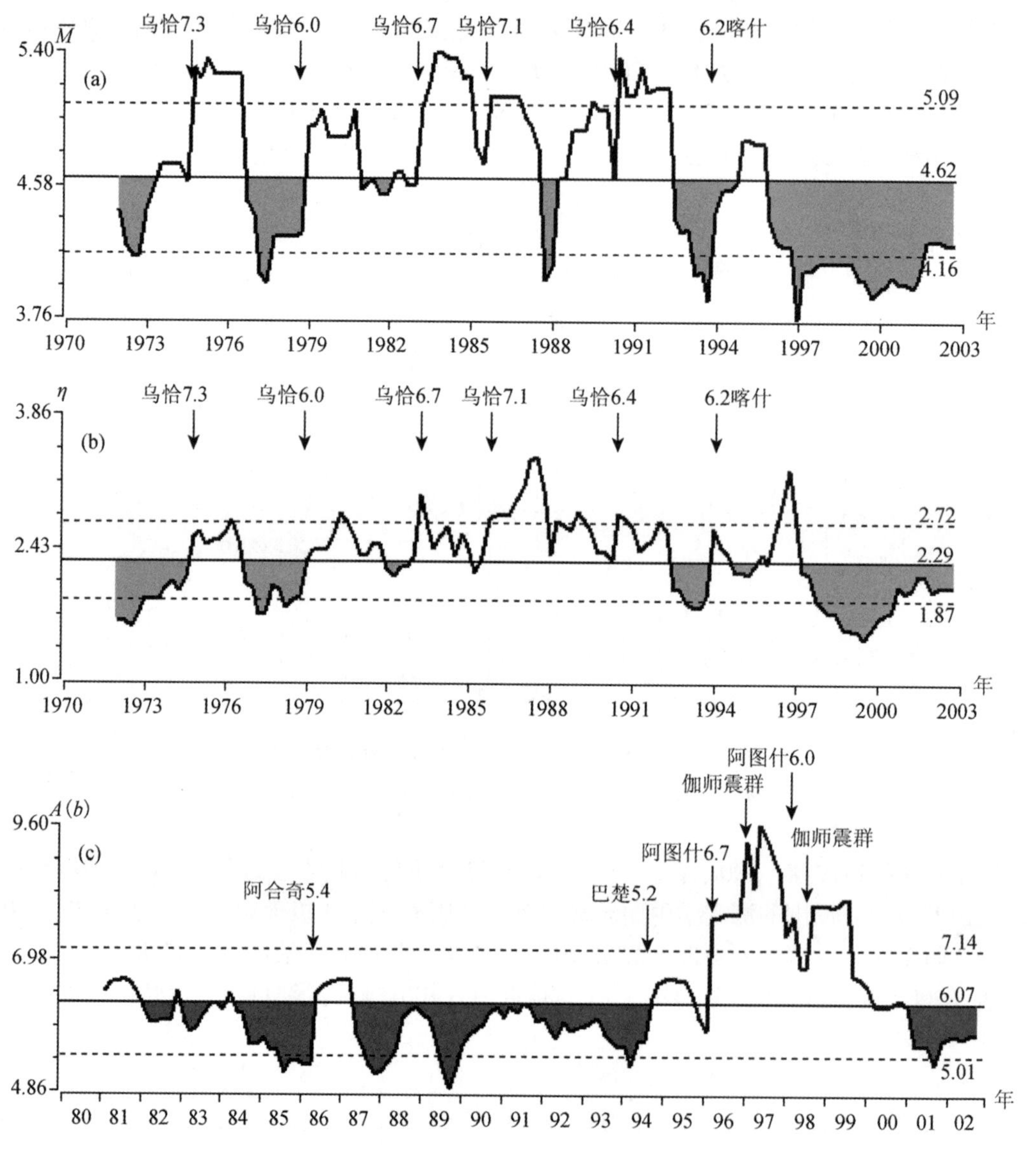

图 11　地震学参数时序异常曲线

Fig. 11　Curve for time anormaly of seismic parameters in the western segment of south Tianshan mountain

2）震后异常总结分析

2003 年 1 月 4 日伽师 5.4 级地震与巴楚 6.8 级地震相距约 20km，前者发生在 1997～1998 年伽师强震群震源区，而后者发生在老震区东南约 20km 处（图 12）。

巴楚 6.8 级地震发生在 1997～1998 年 M_S≥3.5 级余震向 NE 方向扩展的区域。1998 年 8 月 27 日伽师 6.4 级地震前，1997～1998 年伽师强震群 M_S≥3.5 级地震呈 NNE 向分布（图 13 空心三角），8 月 27 日伽师 6.4 级地震后 M_S≥3.5 级余震向 NE 扩展（图 13 空心五角星），巴楚 6.8 级地震发生该 NE 向扩展区。

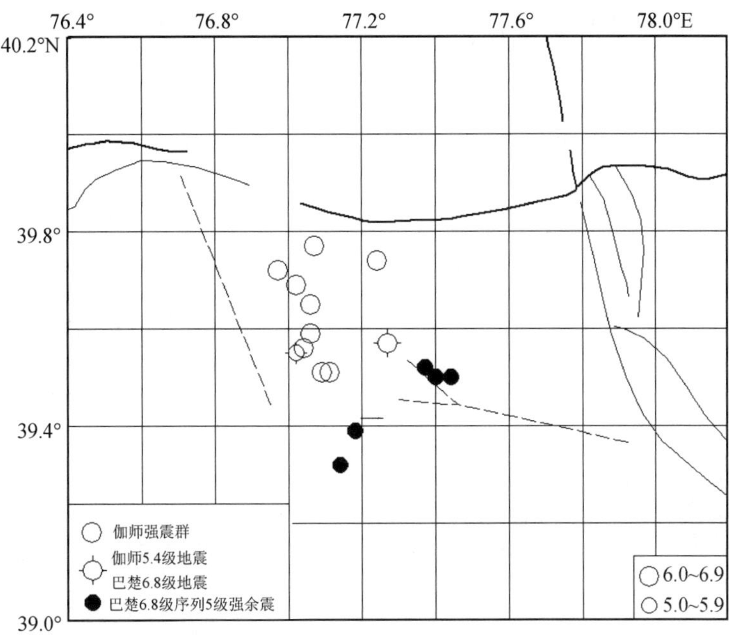

图 12　1997 ~ 1998 年伽师强震群和 2003 年伽师 5.4 级和巴楚 6.8 级地震震中分布图

Fig. 12　Distribution of 1997-1998 Jiashi strong earthquake Swarm and 2003 Jiashi M_S5.4 and Bachu M_S6.8 earthquake

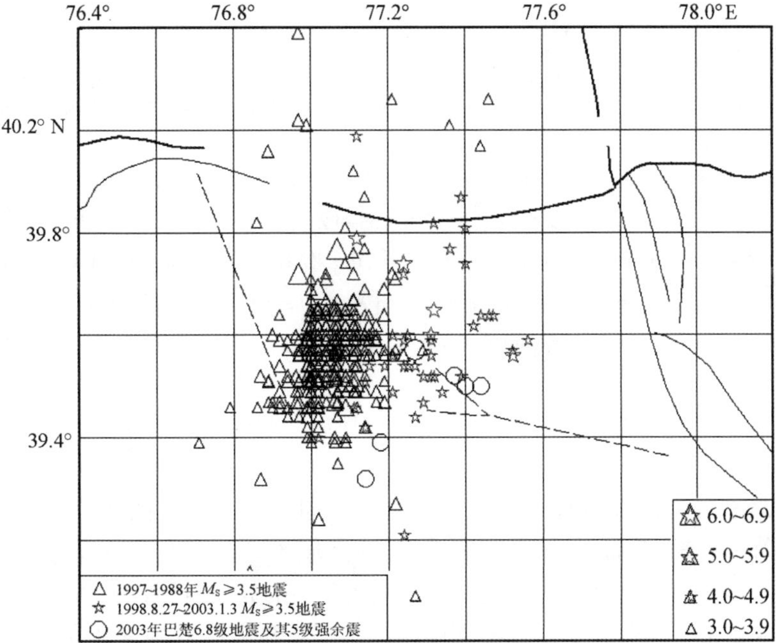

图 13　1997 ~ 2003 年伽师—巴楚地区 3.5 级以上地震震中分布图

Fig. 13　Distribution of 1997-2003 $M \geqslant 3.5$ aftershocks in Jiashi-Bachu region

巴楚 6.8 级地震震中周围 50km 范围内 3.5 级以上地震呈现出活跃—平静特征（图 14a）。1998 年 8 月 27 日伽师 6.4 级地震后至 2000 年 12 月，震区 3.5 级以上地震活跃（空圈），集中分布在巴楚 6.8 级地震震源区附近；2001～2002 年震源区附近 3.5 级以上地震呈现平静状态，仅于 2001 年发生 2 次 3.5 级以上地震（图 14b）。

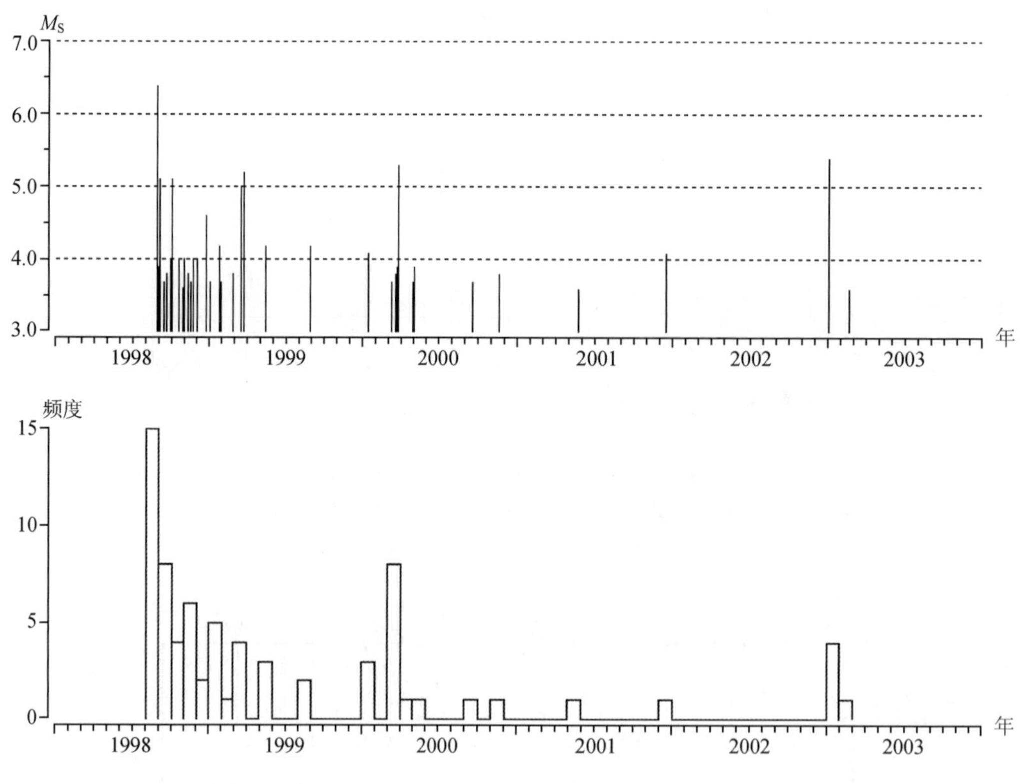

图 14　巴楚 6.8 级地震前（1998.8.27～2003.2.23）震中附近 50km 范围内的 3.5 级以上地震 *M-t* 图和频度图

Fig. 14　*M-t* diagram and frequency distribution of the $M_S \geqslant 3.5$ earthquakes within 50km of the $M_S 6.8$ earthquake source region (from Aug. 27, 1998 to Feb. 23, 2003) before the $M_S 6.8$ earthquake

对震中周围 250km 范围内 $M_S \geqslant 2.0$ 级地震进行地震参数时间进程曲线可知，6.8 级地震前地震频次出现持续近 1.5 年高值异常（图 15）；缺震曲线分析表明（图 16），该区域 3 组 6 级以上地震前缺震异常较为显著，6.8 级地震前 3 年即 2000 年开始出现明显的缺震现象，震前 1 年达最低值；*b* 值时序曲线显示（图 17），该区域 3 组 6 级以上地震前 2 组 *b* 值存在明显的低值—高值异常的变化过程，震前 2 年存在较明显的高 *b* 值异常，高值异常恢复过程中发震，震后恢复正常；震前 2 年该区小震调制比出现较显著高值异常变化过程（图 18），高值异常恢复后发震。

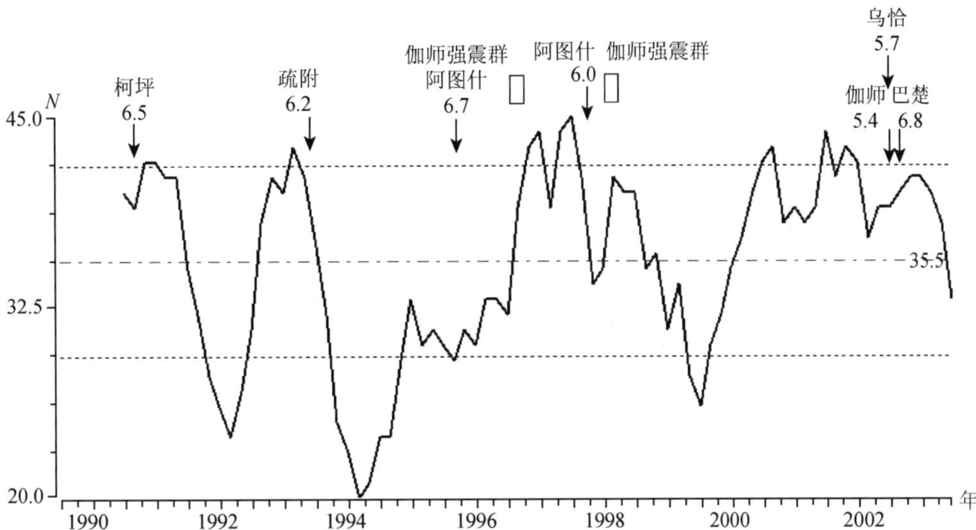

图15　巴楚6.8级地震震中周围地区1990年以来$M_S \geqslant 3.0$级地震频次变化曲线

(□表示1997～1998年伽师强震群)

Fig. 15　Variation of $M_S \geqslant 3.0$ earthquake frequency around the epicenter of $M_S 6.8$ Bachu earthquake since 1990

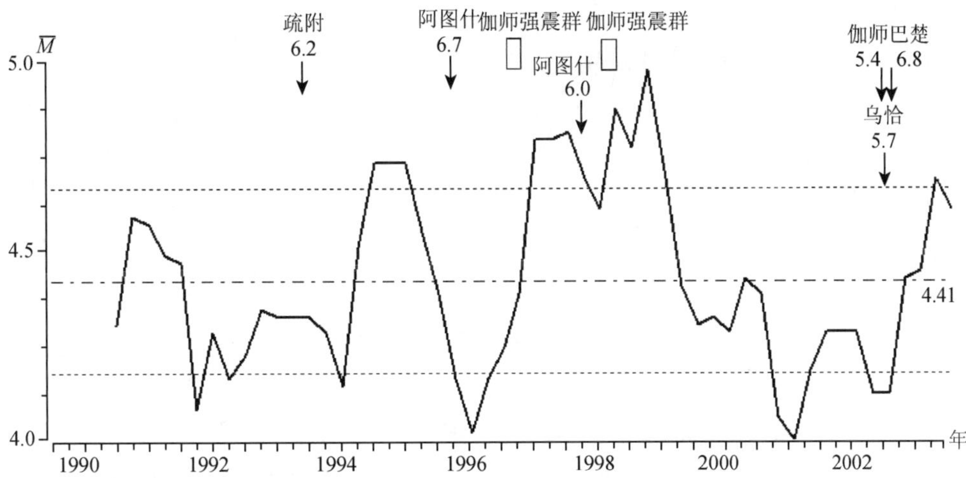

图16　巴楚6.8级地震震中周围地区1990年以来缺震曲线（$M_S \geqslant 2.5$级）

Fig. 16　Curve of $M_S \geqslant 2.5$ earthquake deficiency around the epicenter of $M_S 6.8$ Bachu earthquake since 1990

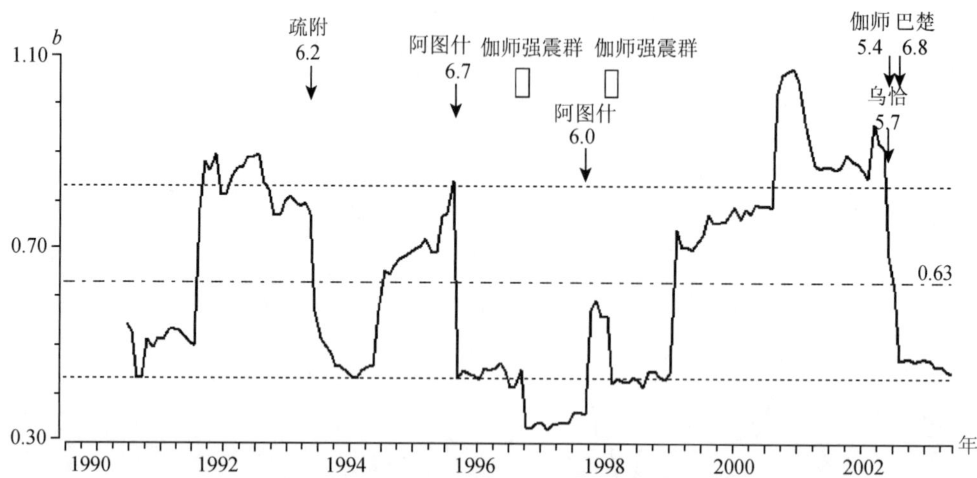

图 17　巴楚 6.8 级地震震中周围地区 1990 年以来 b 值时序曲线（$M_S \geqslant 2.5$ 级）

Fig. 17　b-value curve of $M_S \geqslant 2.5$ earthquakes around the epicenter of $M_S 6.8$ Bachu earthquake since 1990

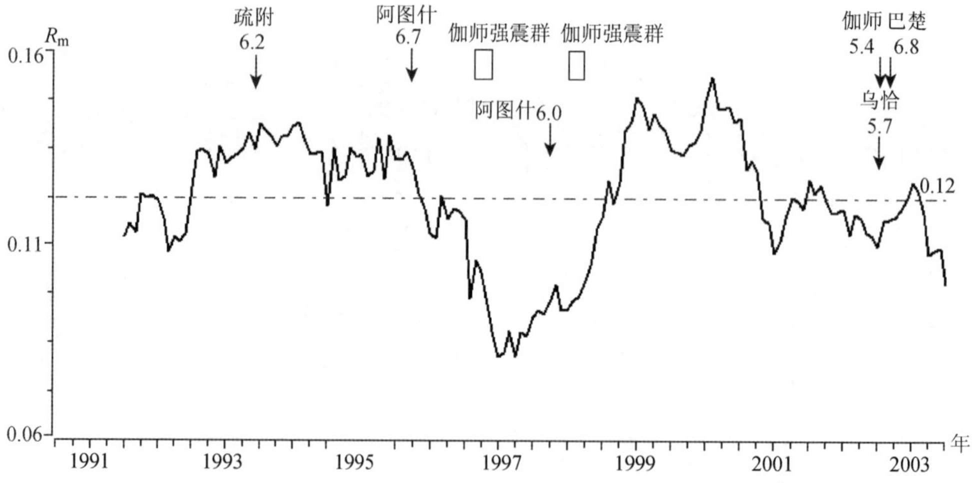

图 18　巴楚 6.8 级地震震中周围地区 1990 年以来小震调制比 R_m 时序曲线

Fig. 18　Curve of regulatory ratio（R_m）of small earthquakes for earth-tide around the epicenter of $M_S 6.8$ Bachu earthquake since 1990

2. 其他前兆观测项目异常

1）定点前兆观测项目异常

在 300km 范围的 8 台项前兆观测项目中，阿图什地倾斜、阿克苏跨断层形变震前无明显异常变化；乌什地倾斜（周记）、乌什钻孔应变和喀什地磁震前存在趋势异常，其他项前兆观测项目在震前存在不同程度的短期和临震异常。

（1）震前中期异常（附件 8）* 。

南天山地区定点前兆观测项目少，巴楚 6.8 级地震前出现中期异常的台项也相对较少。乌什地倾斜自 1999 年春季转向后，其 NS 向由 S 倾转为 N 倾，2001 ~2002 年 N 倾速率稳定，2002 年 10 月起 N 倾速率明显加快，约为往年同期的 2 倍左右，2003 年 1 月初趋势转折后发生了 2003 年 2 月 24 日巴楚—伽师 6.8 级地震，3 月初该资料北南向速率恢复到往年同期水平（图 19）；喀什地磁总强度 2000 年速率出现明显的加速转折现象，2002 年则出现减速变化。

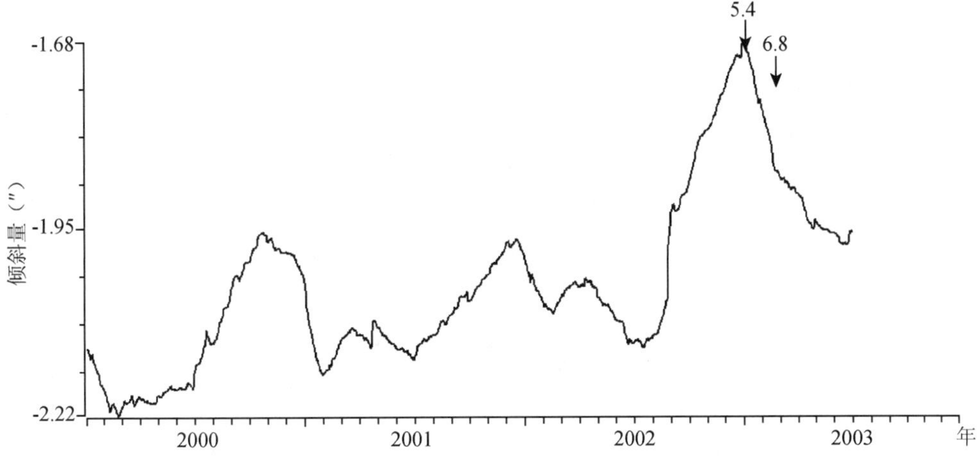

图 19 乌什地倾斜（周记）北南向日均值时序曲线[13]

Fig. 19 Curve of daily mean value of Tilt in the NS direction at Wushi station

乌什台钻孔应变有比较好的年变[14]。2001 年 8 月出现破年变变化，11 月测值比正常年变增大 1.5 倍。2002 年 3 ~6 月，年变相位逆反，测值为正常年变幅度的 2 倍多。2002 年 7 月初，应变性质由压应变转为张应变；7 月中旬之后基本恢复压应变的年变趋势，但压应变幅度比正常大得多。地震前 4 天，幅度达正常年变幅的 12 倍以上（图 20）。2001 年下半年以来，乌什台钻孔应变测值波动幅度较大，即地应力出现张、压交替变化，一定程度表明应变场处于不稳定状态。

（2）震前短期和短临异常。

① 喀什台地倾斜 EW 向 1 月 9 ~11 日出现大速率 E 倾突变（变化量达 0.12″），倾斜方向朝向本次地震震中，变化量相当于正常时期近 1 个月的变化量，固体潮畸变为一条直线（图 21），突变结束后固体潮形态基本恢复正常[15]（附件 1）。

---

* 《新疆维吾尔自治区 2003 年度地震趋势研究报告》附表 1。

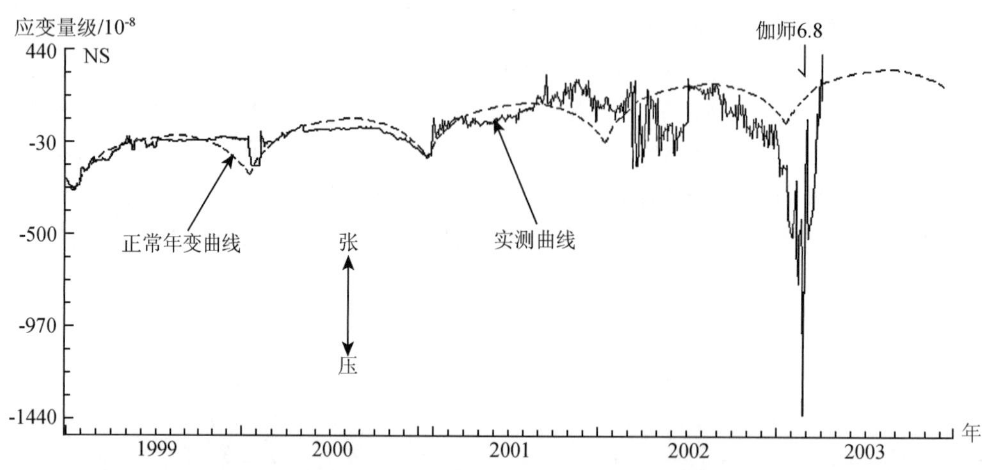

图 20　巴楚 6.8 级地震前乌什台钻孔应变日均值曲线

Fig. 20　Curve of daily mean value of borehole strain at Wushi station

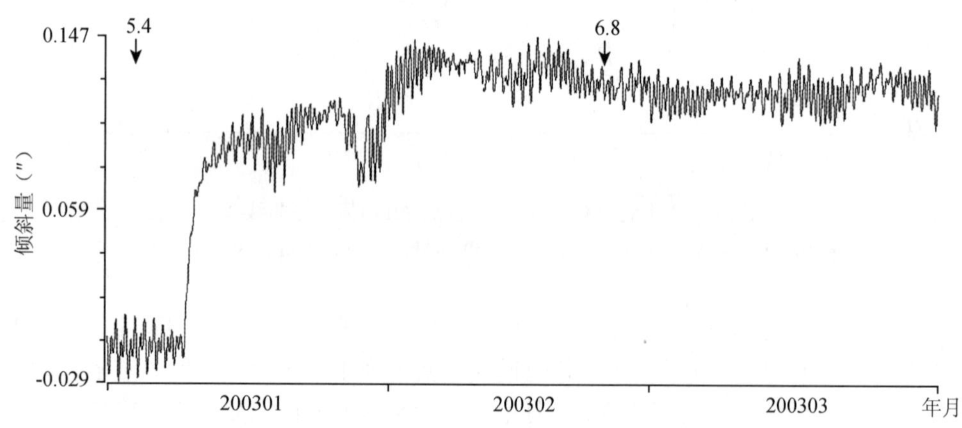

图 21　喀什台地倾斜东西分量整点值时序曲线（2003.1～3）

Fig. 21　Curve hourly value of tilt in the EW direction at Kashi station (from Jan. to Mar., 2003)

② 2003 年 1 月 4 日伽师 5.4 级地震后 1 周之内，阿合奇和乌什地倾斜相继出现加速变化[13,16,17]（附件 1）。阿合奇地倾斜异常多表现为转折和打结，2003 年 1 月 20 日矢量转向北东方向，2 月 6～20 日矢量年变转为北西西向，并且出现打结，2 月 21 日恢复到正常西南的方向。6.8 级地震发生当天矢量方向受地震干扰，2 天后恢复正常。3 月 1 日后矢量再次转为南东东方向（图 22），3 月 12 日发生 5.8 级强余震，至 3 月 14 日后恢复正常。此后矢量方向及单分量速率基本正常。自 2003 年 1 月 9 日乌什地倾斜（日记）两分量出现加速变化，持续 20 天左右后恢复正常变化速率，该变化期间矢量模长约为 2002 年同期矢量模长的 1.5 倍，但方向相对稳定；2 月 6 日后矢量方向不稳定，存在多处转折和拐弯现象，震后资料变化基本转为正常年变化（图 23）。

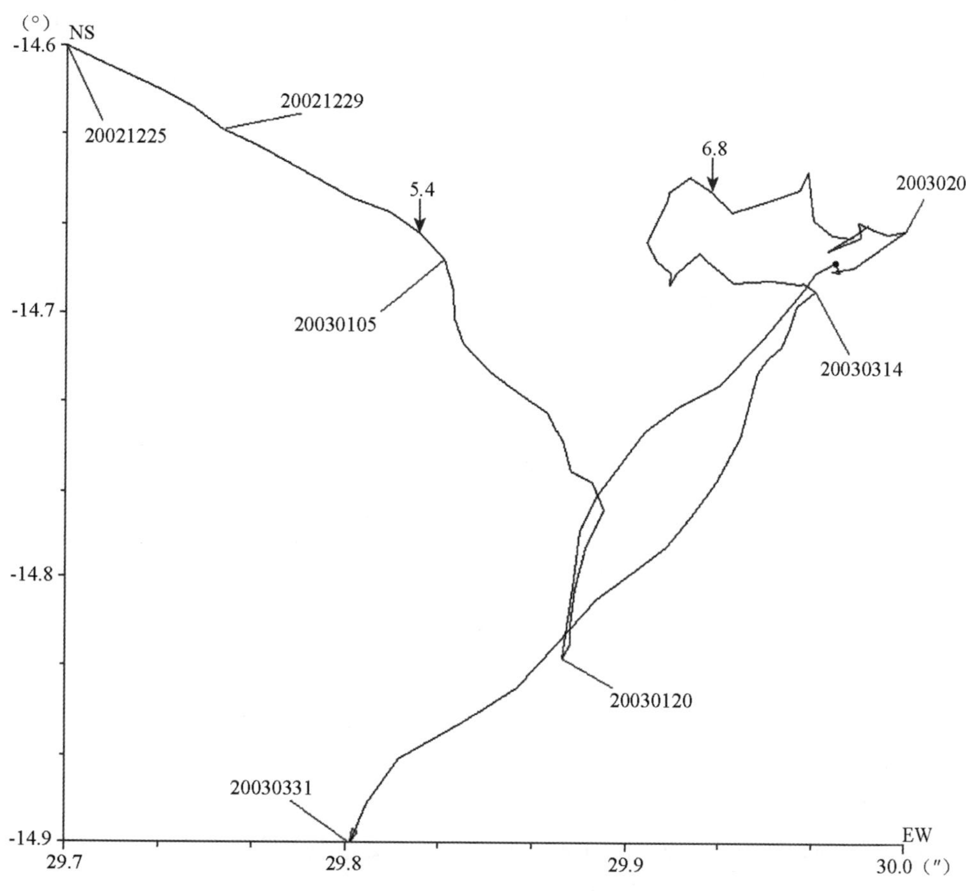

图 22　阿合奇地倾斜 2002 年 12 月至 2003 年 3 月矢量合成图

Fig. 22　Composite vector map of daily mean value of tilt at Aheqi station from Dec., 2002 to Mar., 2003

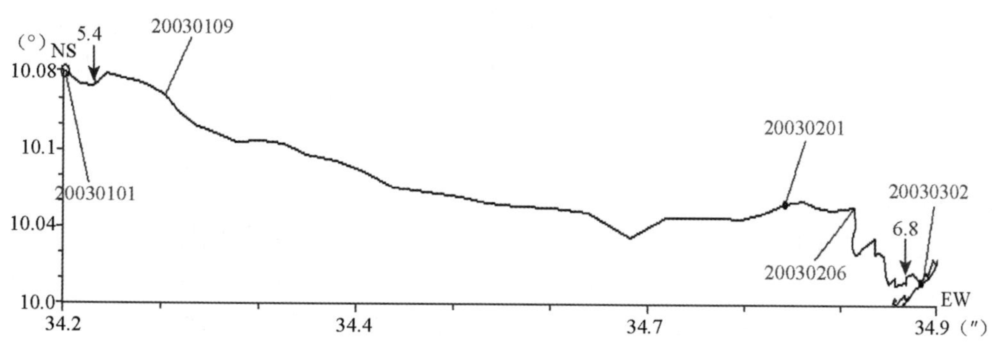

图 23　乌什地倾斜（日记）2003 年 1 ~ 3 月矢量合成图

Fig. 23　Composite tilt vector map of daily mean value at Wushi station from Jan. to Mar., 2003

③ 6.8 级地震前，乌什台钻孔应变出现压应变明显增大的短期异常变化[14]，从 1 月 1 日到 1 月 20 日的 20 天内，压应变增大 1.45×10^{-5}，是正常年变幅度的 10 倍（图 24a），其间发生 1 月 4 日伽师 5.4 级地震，震后压应变继续增大，1 月 20 日达到最大值（附件 2）。2 月 20 ~ 21 日，乌什台钻孔应变出现快速张压变化，应变量级 1.3×10^{-5}（图 24b）。这种反复张压变化一定程度表明区域应力场处于不稳定的状态，可能是临震前地壳应变失稳的表现。

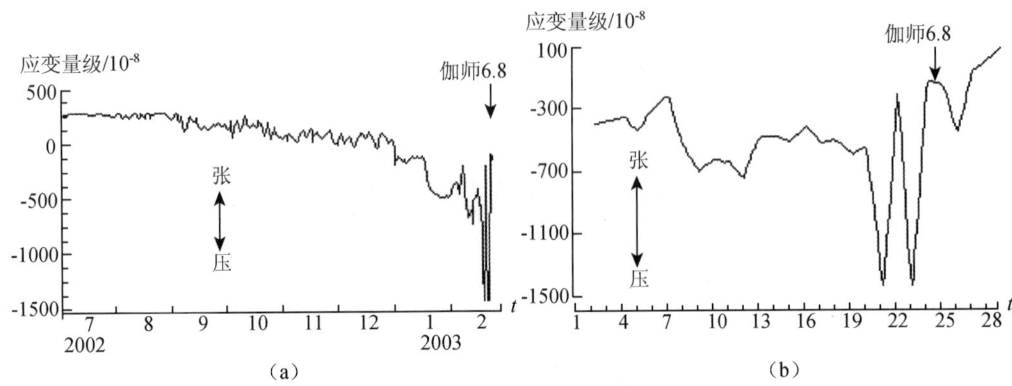

图 24　巴楚 6.8 级地震前乌什台钻孔应变短期（a）和临震（b）异常图像

Fig. 24　The short term and impending anomaly of borehole strain at Wushi station

2）*流动观测异常*

流动重力和 GPS 观测结果表明，6.8 级地震前，震源区及其附近流动重力测值出现较大幅度的正负值交替变化和剪应变高值集中区。

（1）流动重力异常[12]。

2000 ~ 2002 年整个新疆地区重力出现了与 1998 ~ 2000 年反向的变化：新疆南部重力呈负值变化，异常等值线最小值为 $-40\times10\text{m}\cdot\text{s}^{-2}$；喀什—乌什—乌鲁木齐地区处于重力变化正负值交替变化的等值线密集区（图 25）。

震区周围的塔什库尔干—喀什—伽师—阿克苏和乌什—阿克苏—库车—库尔勒重力剖面 2000 ~ 2002 年（图 26 虚线）测值与 1998 ~ 2000（图 27 实线）相比出现明显变化，由上期急剧增加的重力正值变化变为重力负值变化，6.8 级地震震区附近区域重力测值出现较明显的异常变化。

① 塔什库尔干—阿克苏剖面 1998 ~ 2000 年塔什库尔干—阿克苏剖面中的塔什库尔干—盖孜地区重力测值由正值逐步变为负值；盖孜—伽师—巴楚测段重力测值出现急剧增加，重力测值变化量达 $60\times10^{-8}\text{m}\cdot\text{s}^{-2}$；2000 ~ 2002 年，在盖孜—伽师—巴楚测段，重力变化由上期急剧增加的重力正值变化转为重力负值变化。正负重力异常变化梯度带的零线附近是物质增减差异剧烈的地区，能量易于积累，2003 年 2 月 6.8 级地震就发生在 2 个测期重力测值变化最大的伽师—巴楚测段。

② 乌什—阿克苏—库车—库尔勒剖面 1998 ~ 2000 年乌什—阿克苏地区重力测值急剧增

加，重力变化量达 $50\times10^{-8}m\cdot s^{-2}$；2000～2002 年，乌什—阿克苏地区重力测值转为与上期相反的负值。

（2）GPS 观测异常[18]。

GPS 观测结果表明，2001 年 8 月至 2003 年 1 月测期喀什—乌帕尔—乌恰地区和巴楚—伽师地区出现两个剪应变高值集中区，剪应量由 2001 年前的 10×10^{-8}～20×10^{-8}增至 30×10^{-8}～50×10^{-8}，区域最大剪应变值为 58×10^{-8}，集中分布在伽师—巴楚地区。巴楚 6.8 级地震发生在剪应变高值集中区附近。

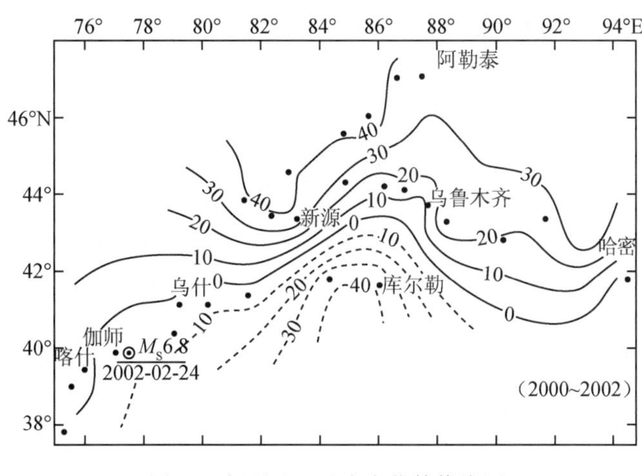

图 25　新疆地区重力变化等值线图

Fig. 25　Isoline map of gravity variation in Xinjiang

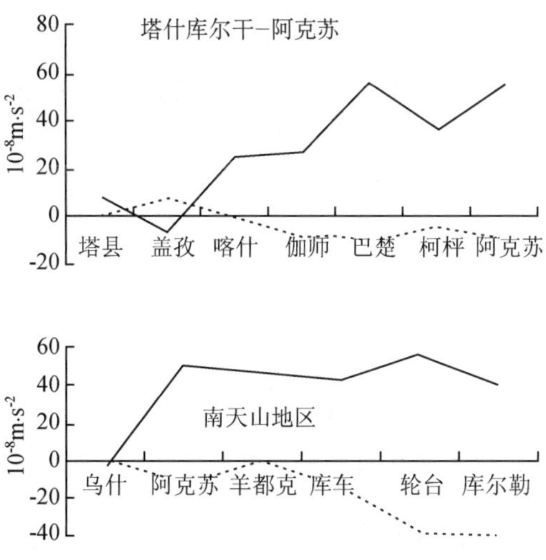

图 26　新疆地区重力变化剖面图

Fig. 26　Profiles of gravity variation in Xinjiang

七、地震前兆异常特征分析

地震学异常：巴楚 6.8 级地震发生在新疆地区 4 年多无 6 级地震、5 级地震平静 2 年的背景下，但震前 2 个月中强地震活动显著增强，先后发生 2001 年 12 月 25 日乌恰 5.8 级、1 月 4 日伽师 5.4 级和 2 月 14 日石河子 5.4 级地震。2003 年 1 月 4 日伽师 5.4 级地震与巴楚 6.8 级地震相距约 20km，前者发生在 1997～1998 年伽师强震群震源区，而后者发生在老震区东南约 20km 处。空间上看巴楚 6.8 级地震位于 1997～1998 年伽师强震群老震区东南约 20km 处，1998 年 8 月 27 日伽师 6.4 级地震后伽师强震群 3.5 以上余震具有向 NE 方向扩展的趋势，巴楚 6.8 级地震发生在该 NE 向扩展区范围内。巴楚 6.8 级震中周围 50km 范围内 3.5 级以上地震震前呈现活跃—平静特征，尤其是震前 1 年震区无 3.5 级以上地震发生。部分地震活动性参数的时间扫描分析，在震前出现较明显的持续 1～3 年的异常变化，其中缺震和频度异常明显。

其他前兆观测资料异常：在 300km 范围的 8 台项前兆观测项目中，除了阿图什地倾斜、阿克苏跨断层形变震前无明显异常变化外，其余 6 个台项前兆观测资料均出现不同程度的中期、短期或短临异常。

（1）异常时间演化。前兆观测异常趋势异常和短期异常数目相当，临震异常稀少，缺少宏观异常。2001 年 8 月乌什钻孔应变出现年变形态异常；2002 年 10 月乌什周记地倾斜（7 天更换 1 次模拟记录图纸）NS 向出现加速 N 倾变化；2000～2002 年流动重力测量结果也表明在盖孜—伽师—巴楚地区存在一定的异常变化。在中期异常背景下，乌什台钻孔应变仪于 2003 年 1 月 1 日记录到压应变明显增大的短期异常变化；1 月 5 日之后，乌什日记倾斜仪（每日更换 1 次模拟记录图纸）和阿合奇石英摆倾斜仪矢量合成曲线出现多次拐弯和打结等异常变化；1 月 9～11 日喀什竖直摆倾斜仪 EW 向出现突然加速 E 倾变化；震前 4 天乌什台钻孔应变仪记录到了大幅度的压—张反复变化短临异常现象。

（2）异常空间分布。巴楚 6.8 级地震前，定点前兆测项中期异常主要集中在离震中较远的乌什台，而流动重力和 GPS 测量异常区更靠近震中地区；短期异常变化最早出现在离震中较远的乌什台，依次是阿合奇和喀什台，而离震中较近的喀什台相对较晚，在空间上具有逐渐向震中发展的态势；临震异常变化仍主要出现在离震中较远的乌什台。

（3）流动重力和 GPS 观测。6.8 级地震前，震源区及其附近流动重力测值出现明显的较大幅度的正负值交替变化和剪应变高值集中区。

总体来说，6.8 级地震前南天山西段出现的短期前兆异常与北天山地区的多项异常在时间上具有同步性，但南天山西段异常持续时间和异常变化量级不及北天山地区突出。北天山地区的多数异常过程持续到 2003 年 6 月前后结束，而且呈现出较为明显的群体前兆异常过程。这可能与近场区 2003 年 2 月 14 日北天山石河子 5.4 级地震有关，根据目前的研究，也难以排除是南天山西段 2 月 24 日巴楚 6.8 级强震的远场短期异常反映[16]。

八、震前预测、预防和震后响应

1. 预测情况[9]

新疆地震局对2003年度新疆地区地震形势的预测意见为“2003年度新疆较大可能发生6~7级地震，优势发震地区为南天山西段与西昆仑交汇的喀什—乌恰地区”，巴楚6.8级地震距该危险区约10km，总体来说，2003年度的新疆地震趋势预测基本正确，但地点略有偏差*。

在中期和一年尺度的中短期预测基础上，新疆维吾尔自治区地震局对危险区及其周围各类异常加强了跟踪分析，密切监视地震活动和前兆观测异常变化。6.8级地震短期震情判定的主要依据为：① 地震活动异常平静的背景下，近期地震活动的显著增强预示着新疆发生强震的短期危险性增大；② 南天山西段的喀什—乌恰周围地区长达8年的6级地震平静仍在持续；③ 前兆观测资料显示出明显的短期异常特征，主要集中在北天山和南天山西段；④ 2003年1月4日伽师5.4级地震后，南天山西段仅有的5台项前兆观测项目都出现不同程度的短期异常变化，预示着异常区域周围存在发生中强地震的可能。

2002年12月25日乌恰5.7级地震后，新疆的地震形势和前兆观测资料发生了转折性变化。依据震情的发展，新疆地震局提出短期预测意见：① 12月26日新疆地震局临时会商会意见为：“此次5.7级地震的发生，打破了新疆24个月无5级地震和35个月无5.5级地震的平静，这次地震的发生对新疆地震形势会产生一定的影响。经过几年地震平静之后，新疆地震活动可能会进入一个新的活跃时段”（附件3）；② 2002年12月25日乌恰5.7级地震和2003年1月伽师5.4级地震后，依据地震活动和前兆资料的异常变化，在2003年1月28日新疆地震局月会商会上，综合结论为“2月份南、北天山西段2个地震重点危险区有发生中强地震的可能”（附件2）；③ 2月14日石河子5.4级地震后，特别是2月17日在北天山西段的伊宁北部又发生4.7级地震，根据短期内地震形势的突变，分析认为新疆近期发生6级以上地震的可能性很大，最危险区域为南天山西段；④ 在震情的跟踪判定过程中，及时向自治区人民政府主管领导和中国地震局监测预报司通报震情：在1月28日给自治区人民政府汇报材料中明确提出“新疆随时都有发生6级以上地震的可能”（附件7）；2月14日石河子5.4级地震后给政府的汇报中又一次明确提出“近期新疆有发生6级地震的可能，危险区域为南天山西段”。

2002年12月中、下旬，全疆地震活动出现转折的同时，前兆资料也显示出较显著的短期异常变化过程。在北天山地震危险区周围地倾斜、跨断层观测和地下流体等测项先后出现一系列较为同步的异常变化；2003年1月10日前后，南天山西段地倾斜和钻孔应变也显示出一定程度的短期或短临异常特征。

巴楚6.8级地震的追踪预测，从中期到短期均有据可依，虽未能实现临震预测，但仍可从中得到不少宝贵的经验和教训。1月4日伽师5.4级地震后，多数专家分析认为该地震可能是一次晚期强余震。地震活动性方面，喀什—乌恰交汇区6级地震持续平静8年，5级地

---

* 新疆维吾尔自治区2004年度地震趋势研究报告。

震平静被2002年12月25日乌恰5.7级地震打破。前兆方面，6.8级地震前全疆定点前兆观测资料存在一定数量的趋势和较明显的短临异常，趋势异常多数出现在北天山西段，南天山西段的前兆测项少、异常持续时间短。

但伽师5.4级地震后，南天山西段5台项前兆观测项目都出现了不同程度的短期异常变化。综合分析认为，未来6级以上地震的最危险区域可能是喀什—乌恰周围地区。震后总结看来，当时对乌什、阿合奇台和喀什台前兆资料临震前的一些资料的突变现象和地点指示意义认识不足，是我们未能作出较准确短临预报的因素之一。实际上，前兆异常分布区域与未来强震发生位置的关系一直以来都是地震预测研究的难题，由此，强震危险地点判定也是地震预测三要素中最难的。

2. 震后趋势判定

地震发生后，中国地震局组织专家与新疆维吾尔自治区地震局业务人员赶赴大震现场，进行现场地震监测预报工作。现场工作组与新疆维吾尔自治区地震局分析预报人员密切跟踪序列的发展，对巴楚、喀什地震台余震序列进行分析，结合该区历史地震活动情况，分别于3月20日和3月21日对震后地震趋势进行初步判定，认为该地震属主震—余震型；并于2003年5月15日的专题研讨会上提出，6.8地震序列类型可能属于主—余震型。巴楚6.8级地震后，新疆维吾尔自治区地震局对地震类型和强余震作出了准确的判别和预报7)（附件4、附件5）：提出过4次较为准确的强余震短临预报，其中3月12日伽师5.8级地震余震前提出了15日天的临震预报（附件6），同时也有3次5级强余震漏报。

3. 震害分析[19,20]

2003年2月24日6.8级地震发生当日，中国地震局组成以新疆维吾尔自治区地震局、中国地震局机关、工程力学研究所、分析预报中心、综合观测中心、地震工程中心、地球物理研究所、喀什地区地震局、克孜勒苏柯尔克孜自治州、阿克苏中心地震台等单位参加的联合现场工作队赴灾区开展地震灾害损失评估工作。

此次6.8级地震受灾面积较大，Ⅵ度以上（含Ⅵ度）地区面积达18566km^2，包括喀什地区巴楚、伽师、岳普湖、麦盖提、莎车县及克州阿图什市，共计5县1市37个乡镇931个村（场）。重灾区位于巴楚县的琼库尔恰克乡、阿拉格尔乡、色力布亚镇等6个乡镇。

震区位于天山南麓山前冲积平原与塔里木盆地西北部交接地带，震区北部为南天山山脉，西南部为昆仑山脉。与快速隆起的天山和昆仑山相反，震区为快速沉降地区，地势平坦，第四纪沉积层厚度较大，地表以下100m范围内为第四纪晚期以来堆积的粉土和粉细砂层。由于喀什噶尔河穿流其中，地下水位较高，地表盐碱化现象严重，场地土条件较差，易发生砂土液化。

灾区地震地质灾害主要分布于Ⅵ、Ⅸ度区的色力布亚镇、琼库尔恰克乡、英吾斯塘乡和阿拉格尔乡等地区，其类型主要为场地砂土液化、地裂缝与河岸滑塌等，造成农田毁坏、道路裂缝、建筑物破坏和河流、沟渠岸崩塌等灾害。

（1）砂土液化广泛分布，主要表现为喷砂冒水。除极震区外，在Ⅶ度以上地区沿水位较高的漫滩和低阶地、极震区以东广大地区，河道纵横，地下水位高的地区均有分布。砂土液化与震区为粉砂和细砂土层沉积层和地下水埋深较浅等有关。

（2）地震地表破裂分布较为局限，在伽师县及巴楚县交界附近的塔西南石油天然气开发公司柯柯亚作业区的琼五井附近，发现三条长约 1~1.5km 的地表破裂带。地裂缝带由多条裂缝斜列组成，推测为受断裂控制形成的次级地表破裂。

（3）建（构）筑物遭受严重破坏，其中巴楚、伽师、岳普湖等县的基础工程与设施灾情特别严重，离震中较远的阿图什哈拉峻乡的房屋建筑也遭遇了严重灾害。

在 6.8 级地震极震区内土木结构房屋 90% 以上倒塌，其余的局部倒塌、屋顶坍塌或墙体普遍出现裂缝。个别房屋墙体出现轻微裂缝；倒塌、局部倒塌、屋盖塌落等的砖木结构平房约占 70%~80%。未经抗震设防的砖混结构房屋毁坏、局部倒塌、严重破坏占一定比例，但仍有基本完好的砖混结构房屋；砖柱厂房、粮库遭到严重破坏；大部分水塔遭受破坏，塔筒身出现斜裂或环裂，部分水塔倒塌。

灾区房屋普遍为土木结构、地基不良，抗震能力差；砖结构房屋由于使用砂土砌砖，造成墙体抗剪能力差；工程场地的地质条件差、地下水位高；1997~1998 年伽师强震群活动对灾区已造成了不同程度破坏，本次地震的发生加重了对灾区房屋的破坏。

本次地震造成的严重震害，与震区地质构造环境、场地条件和房屋结构有关。

九、结论与讨论

（1）6.8 级地震前 1 个月距其 20km 的 1997~1998 年伽师强震群震源区发生了 1 月 4 日伽师 5.4 级地震，5.4 级地震震源断错类型为走滑型。伽师强震群中强地震震源断错类型以走滑和正断层为主、破裂无明显的方向性。5.4 级地震震源断错性质与 1997~1998 年伽师强震群的强震基本一致，显示出 1997~1998 年伽师强震群活动的继承性活动特征。

巴楚 6.8 级主震发生在 1997~1998 年老震区东南 20km，该区在 1998 年 8 月 27 日伽师 6.4 级地震后曾突破老震区余震活动范围，出现 3.5 级以上地震的集中活动，表明巴楚 6.8 级地震可能与 1997~1998 年伽师强震群向外扩展有关，其震源断错性质和破裂方式不同于 1997~1998 年伽师强震群中的强震。巴楚 6.8 级地震震源破裂方式为双侧破裂，破裂由 6.8 级震中向 NWW 向和 NE 两侧扩展。其构造成因可能是，1997~1998 年伽师强震群 NNW 向发震断层的错动，触发了近 EW 向的柯坪塔格推覆构造系前缘盲逆断层巴什托普隐伏断层的活动。综上分析表明，伽师 5.4 级地震可能属于 1997~1998 年伽师强震群晚期强余震活动，而巴楚 6.8 级可能是一次新的破裂过程。上述认识还有待于进一步的深入探讨。

（2）巴楚 6.8 级地震发生在老震区东南 20km 处，但序列特征却差别较大，巴楚 6.8 级地震序列类型单一，具有明显的主余型特征。而 1997~1998 年伽师震区中强地震序列类型复杂，是由双震型、孤立型、前—主—余震型、主—余震型等多种地震类型组合成的强震群[21]。

（3）2003 年 2 月 24 日巴楚 6.8 级地震发生在 1997~1998 年伽师强震群老震区东南约 20km 处。巴楚 6.8 级前震源区 3.5 级以上地震呈现活跃—平静特征，尤其是震前 1 年震区 3.5 级以上地震呈现平静状态。部分地震活动性参数震前存在较为明显的中期异常。研究区域范围内前兆观测资料以地倾斜为主的中短（临）期异常较为明显，其主要异常特征为倾斜速率出现加速变化、矢量方向和模长打破正常年变；从观测到的 3 个台地倾斜的临震异常形态分析，其特征以突变为主。从异常的空间分布特征来看，本次地震前定点前兆中期异常主要集中在离震中较远的乌什台，流动重力和 GPS 测量异常区则更靠近震中地区；短期异常变化最早出现在离震中较远的乌什台，依次是阿合奇和喀什台，而离震中较近的喀什台相对较晚，在空间上具有逐渐向震中发展的态势；临震异常变化主要出现在离震中较远的乌什台。在巴楚 6.8 级地震前，研究区域范围内中短期异常显著，其空间演化过程对未来震中具有一定的指示意义。

（4）在年度趋势预测的基础上，依据震情发展和阶段性异常特征，新疆地震局在震前提出了短期预测意见，对于巴楚 6.8 级地震强度和时间判定较为准确，但地点判定存在偏差。1 月 4 日伽师 5.4 级地震后，多数专家分析认为该次地震可能是一次晚期强余震，结合地震活动和前兆异常特征，认为发生 6 级以上地震的最危险区域应该是喀什—乌恰交汇区周围。但预期的强震仍然发生在伽师老震区，这是我们没有认识到的问题，也进一步体现出地震预报的难度之所在。

总结反思认为，巴楚 6.8 级地震未能作出较准确短临预报主要是对几点认识不足：① 1997 ~ 1998 年伽师震区已连续发生了 7 次 6 级地震，且当时仅看到了伽师 5.4 级地震晚期强余震的一面；② 喀什—乌恰周围地区存在发生中强地震的显著背景，即 1993 ~ 2002 年该区 5 级以上地震一直处于显著平静状态，9 年时间仅发生 2 次 5 级地震，无 6 级以上地震发生；③ 统计分析表明，喀什—乌恰交汇区与柯坪块体中强地震活动具有交替活动特征。当时在震情判定过程中，认为伽师强震群活动基本结束，之后喀什—乌恰周围发生强震的可能性更大。④ 6.8 级地震前，南天山西段有 4 项前兆观测项目出现突变，其中 3 项还出现了临震异常。由于南天山西段前兆测项目少、异常持续时间短，当时对这些临震前的突变现象认识也不足。

巴楚 6.8 级地震后，新疆地震局对地震类型和强余震作出了较准确的判别和预报[7)]：提出过 4 次较为准确的强余震短临预报，其中 3 月 12 日伽师 5.8 级地震余震前提出了 15 日天的临震预报，并在其中 1 次强余震发生前向当地人民政府提出了较好的临震预测意见，取得了一定的社会效益和减灾实效。

参 考 文 献

[1] 苏逎秦 . 2004. 从 1997 ~ 1998 年伽师强震群到 2003 年巴楚—伽师—岳普湖地震—震群演化特征及地震趋势分析 . 内陆地震，18（1）：29 ~ 38

[2] 徐锡伟，张先康，冉勇康 . 2006. 南天山地区巴楚—伽师地震（M_S6.8）发震构造初步研究 . 地震地质，28（2）：161 ~ 178

[3] 沈军，陈建波，王翠等 . 2006. 2003 年 2 月 24 日新疆巴楚—伽师 6.8 级地震发震构造 . 地震地质，28（2）：205 ~ 212

[4] 苏逎秦，王海涛 . 2003. 中国震例（1997 ~ 1999）［M］. 地震出版社，4 ~ 7

[5] 彭树森 . 1993. 大地形变测量所反映的天山最新构造运动 . 内陆地震，7（2），136 ~ 141

[6] 王琪，丁国瑜，乔学军，王晓强，游新兆 . 2000. 天山现今地壳快速缩短与南北地块的相对运动［J］. 科学通报，40（14）：1543 ~ 1547

[7] 王晓强，李杰，D Alexander Zubovich，王琪 . 2007. 利用 GPS 形变资料研究天山及邻近地区地壳水平位移与应变特征 . 地震学报，29（1）：31 ~ 37

[8] 新疆维吾尔自治区地震局，2005. 中国新疆巴楚—伽师 6.8 级地震图集［M］. 新疆科学技术出版社，

35 ~ 37
[9] 高国英，杨又陵，温和平 . 2005. 2003 年巴楚—伽师 6. 8 级地震中短期异常与预测 . 地震地磁观测与研究，26（6）：22 ~ 28
[10] 曲延军，聂晓红 . 2005. 2003 年新疆巴楚—伽师 6. 8 级地震序列特征分析 . 内陆地震，19（1）：65 ~ 73
[11] 高国英，聂晓红，夏爱国 . 2004. 2003 年伽师 6. 8 级地震序列特征和震源机制的初步研究 . 中国地震，20（2）：179 ~ 186
[12] 祝意青，胡斌，李辉等 . 2003. 新疆地区重力变化与伽师 6. 8 级地震 . 大地测量和地球动力学，23（3）：66 ~ 69
[13] 温和平，牛安福，张翼 . 2003. 新疆巴楚—伽师 6. 8 级地震的短期形变前兆特征 . 内陆地震，17（2）：176 ~ 181
[14] 蒋靖祥，尹光华，王在华等 . 2004. 新疆乌什钻孔应变异常特征与强震关系初步研究——以伽师 M_S6. 8 地震为例 . 地震学报，26（增刊）：64 ~ 70
[15] 蒋靖祥 . 2004. 2003 年 2 月 24 日巴楚—伽师 6. 8 级地震前喀什台地倾斜异常 . 内陆地震，18（1）：94
[16] 高国英，温和平，杨又陵 . 2005. 2003 年新疆巴楚—伽师 6. 8 级地震异常的阶段性特征 . 地震地磁观测与研究，26（6）：22 ~ 28
[17] 潘振生，刘辉，李士柱，赖爱京 . 2004. 乌什、阿合奇地倾斜在伽师地震前临震异常特征分析 . 内陆地震，18（2），187 ~ 192
[18] 李杰，王晓强，王琪等 . 2004. 乌恰伽师地区 GPS 地壳运动监测网研究 . 内陆地震，18（3）：281 ~ 288
[19] 罗福忠，胡伟华，赵纯青等 . 2006. 巴楚—伽师 6. 8 级地震地质灾害及未来地震地质灾害 . 内陆地震，20（1）：33 ~ 39
[20] 尹力峰，唐丽华 . 2003. 2003 年 2 月 24 日新疆巴楚—伽师 6. 8 级地震灾区房屋建筑震害现象综述 . 17（3）：209 ~ 215
[21] 朱令人，苏乃秦，杨马陵 . 1998. 1997 年新疆伽师强震群及三次成功的临震预报［J］. 中国地震，14（2）：101 ~ 115

参 考 资 料

1）王海涛等，《巴楚—伽师强震序列的深入研究》研究报告，2005
2）新疆维吾尔自治区地震局，新疆地震目录库（1996）
3）中国地震台网地震目录
4）张先康、赵金仁、张成科等，深地震宽角反射/折射剖面深部结构的探测和研究，“九五”国家科技攻关计划（96-913-07-02-01）专题研究报告，2000. 11
5）新疆维吾尔自治区地震局，新疆伽师 2003 年 1 月 24 日 5. 4 级地震震害损失评估、烈度考察报告，新疆维吾尔自治区地震局科技档案，流水号：6777
6）朱令人、周仕勇、单新建等 . 伽师强震群成因综合研究，“九五”国家科技攻关计划（96-913-07-04）专题研究报告，2000. 11
7）新疆维吾尔自治区地震局，2004 年度地震趋势研究报告，2003. 12

Bachu M_S 6. 8 Earthquake on February 24, 2003 in Xinjiang Uygur Autonomous Region

Abstract

An earthquake occurred on February 24, 2003 in the boundary of Bachu-Jiashi county Xinjiang Uygur Autonomous Region with the macroscopic epicenter at 39°20′N, 77°38′E, and the epicenter intensity was Ⅸ. The earthquake caused 268 people death, and 4853 people injured and 2058 people injured heavily. Direct economic loss was about RMB 1. 398 billion yuan.

Bachu M_S6. 8 earthquake locates in the front edge part of the western segment in Kalpintag structure system, and seismogenic structure is the northward blind reverse fault of the Bashituopu fault in the NWW direction. Bachu M_S6. 8 earthquake is characteristic of mainshock-aftershock type, and the maximum aftershock is M_S5. 8. The aftershocks distributes in the range of 50km of the two sides of the mainshock, and appear the conjugative distribution in the NWW and NE direction, and the aftershock area in the NWW direction are main part.

Rupture characteristics of M_S6. 8 earthquake is reverse fault, and the nodal section I in the EW direction is fault plane. The P axes azimuth of principal compressive stress is 339°, and the plunge angle is horizontal, and T axes of tensile stress is nearly vertical. The M_S6. 8 earthquake resulted from thrust rupture under the action of the NW compressive stress of the regional tectonic stress field.

There was 8 seismic stations in the range of 300km from the M_S6. 8 earthquake, the 8 stations have all seismometric observation, and 6 stations haves precursory observation, including the observation items of Wushi station tilt and borehole strain, Aheqi and Atushi station tilt, Akesu Fault-crossing, Kashi geomagnetism and geoelectricity, and Wuqia geoelectricity. There are quiescence—activity of mid-strong earthquake, quiescence of moderate earthquake, spatial and temporal scanning of seismic parameters and seismic window before the earthquake. There are 3 mid-term anomaly and 4 short-term (impending) anomaly of fixed-point precursor.

Earthquake administration of Xinjiang Uygur Autonomous Region have made middle-short term prediction according to seismicity and precursory anomaly before the M_S6. 8 earthquake (strength and time are nearly correct, but location is out of the source), and made correctly the short-term predictions for 4 strong aftershocks.

报告附件

附件一：

秘密★三个月

震情监视报告

单位	新疆地震局预报中心		会商会类型	临时会商会
期数	（2003）第5期		会商会地点	局会商室
	（总字）第130期		会商会时间	2003年1月11日18时
主持人	王季达		发送时间	1月10日20时00分
签发人	王季达		收到时间	月　日　时
Apnet网络编码		AP65	发送人	龙海英

2003年1月11日18时，新疆地震局对近几天喀什及附近地区前兆资料出现的异常变化，进一步分析讨论如下：

喀什台地倾斜1月9日开始出现大幅度东倾变化，至10日23时变化幅度约为0.08角秒，目前有减缓趋势。阿合奇地倾斜东西向1月9日出现畸变。乌什台地倾斜南北向1月10日速率增大。

乌恰5.7级和伽师5.4级地震序列无明显异常变化。

综合分析认为维持第4期会商意见。

附件二：

秘密★三个月

震情监视报告

单位	新疆地震局预报中心		会商会类型	月会商会
期数	（2003）第8期		会商会地点	局会商室
	（总字）第133期		会商会时间	2003年1月28日12时
主持人	高国英		发送时间	1月28日16时00分
签发人	王海涛		收到时间	月　日　时
Apnet网络编码		AP65	发送人	王琼

1. 本月震情实况

2002年12月23日至2003年1月25日，全疆共交出地震308次，其中$M_L<2.0$级95

次；2.0～2.9 级 161 次；3.0～3.9 级 45 次；4.0～4.9 级 5 次；5.0～5.9 级 1 次；6.0～6.9 级 1 次；最大地震为 12 月 25 日乌恰 $M_L6.0$。另外，边境地区共交出 46 次地震。分区统计情况如下：

乌鲁木齐地区 50 次（$M_{Lmax}=3.7$ 级）；北天山西段 31 次（$M_{Lmax}=3.5$ 级）；南天山东段 36 次（$M_{Lmax}=4.4$ 级）；柯坪块区 133 次（$M_{Lmax}=5.7$ 级）；乌恰地区 20 次（$M_{Lmax}=6.0$ 级）；西昆仑地区 12 次（$M_{Lmax}=3.8$ 级）；富蕴地区 7 次（$M_{Lmax}=3.1$ 级）；其他地区 19 次（$M_{Lmax}=3.8$ 级）。

2. 地震活动及前兆异常情况

1 月 4 日伽师 5.4 级地震后，除震区外，全疆其他地区的地震活动相对平静，特别是北天山地区只有 2 次 3 级地震。2 级以上地震相对集中在库尔勒附近地区和柯坪地区，且库尔勒附近地区形成北东和北西两个小震条带，条带的交汇区在和静一带。利用数字地震资料分析 1 月 22 日和静 $M_L4.4$ 地震，应力降相对较高，计算 4.4 级震中附近的小震地震波 Q 值，认为这一带地区有应力逐渐增强趋势。柯坪区本月 2 级以上地震有明显增强；本月阿合奇、和田、阿图什、塔什库尔干地震台均记录到小震群事件；和田、塔什库尔干地震窗出现超限异常。

前兆异常 12 月以来此起彼伏，2003 年 1 月以来先后出现异常的有：喀什地倾斜 1 月 10～12 日加速东倾；新源、石场、温泉、精河地倾斜变化不稳定。1 月 25～26 日，精河、石场、温泉地倾斜又出现新的不稳定变化，但时间较短；新源台地倾斜前期异常已恢复；乌什地应力近 10 天出现较大变化；04 井水位、水温 1 月 25 日出现同步异常变化，水位 2 天上升 20cm，27 日有恢复迹象；10 泉甲烷、氦气仍持续高值；库尔勒跨断层测量异常仍持续。

3. 综合分析

上月乌恰 5.7 级地震和本月伽师 5.4 级地震后，全疆 3 级以上地震活动主要集中在柯坪块体。自去年 10 月以来乌鲁木齐西南——库尔勒北 3～4 级地震较为活跃。近期全疆多次出现小震群活动。前兆异常此起彼伏，分布范围广，多为突发异常。分析认为，新疆震情是严峻的，存在发生中强以上地震的可能，特别值得注意的是两个重点危险区，但目前临震异常不显著。需继续跟踪震情的发展，密切注意地震活动及前兆资料的变化。

4. 地震趋势预测

（1）未来一月新疆地震活动水平可能与上月持平。

（2）乌鲁木齐地区未来一周发生 $M_S5.0$ 以上地震的可能性不大。

（3）未来一个月或稍长时间两个重点危险区有发生中强地震的可能。

附件三：

秘密★三个月

震情监视报告

<table>
<tr><td>单位</td><td colspan="2">新疆地震局预报中心</td><td>会商会类型</td><td>临时会商会</td></tr>
<tr><td rowspan="2">期数</td><td colspan="2">（2002）第 66 期</td><td>会商会地点</td><td>局会商室</td></tr>
<tr><td colspan="2">（总字）第 122 期</td><td>会商会时间</td><td>2002 年 12 月 26 日 11 时</td></tr>
<tr><td>主持人</td><td colspan="2">高国英</td><td>发送时间</td><td>12 月 26 日 19 时 00 分</td></tr>
<tr><td>签发人</td><td colspan="2">王海涛</td><td>收到时间</td><td>月　日　时</td></tr>
<tr><td colspan="2">Apnet 网络编码</td><td>AP65</td><td>发送人</td><td>王琼</td></tr>
</table>

2002 年 12 月 26 日上午 11 时，新疆地震局再次召开临时会商会，对 25 日乌恰 5.7 级地震后地震形势进行了进一步的分析和判定。

乌恰 5.7 级地震后至 26 日 17 时，喀什仅记录到 2 次 2 级余震，分别为 26 日 08 时乌恰 M_L2.3 和 08 时 44 分 M_L2.2 地震。一次 5.7 级地震后无余震发生，在该区 5 级地震序列中属少见，还待继续关注序列发展。分析认为北天山地区部分较为突出的前兆资料的异常变化，不足以对应这次 5.7 级地震。

此次 5.7 级地震的发生打破了新疆 24 个月无 5 级地震和 35 个月无 5.5 级地震的平静，这次地震的发生对新疆地震形势会产生一定的影响。经过几年地震平静之后，新疆地震活动可能会进入一个新的活跃时段。特别要关注北天山地区前兆资料的异常变化。

综合分析认为：近几天存在发生 5 级强余震的可能，也不排除发生相当震级双震的可能。

附件四：

秘密★三个月

震情监视报告

<table>
<tr><td>单位</td><td colspan="2">新疆地震局预报中心</td><td>会商会类型</td><td>临时会商会</td></tr>
<tr><td rowspan="2">期数</td><td colspan="2">（2003）第 17 期</td><td>会商会地点</td><td>局会商室</td></tr>
<tr><td colspan="2">（总字）第 142 期</td><td>会商会时间</td><td>2003 年 2 月 24 日 12 时</td></tr>
<tr><td>主持人</td><td colspan="2">高国英</td><td>发送时间</td><td>2 月 24 日 14 时 30 分</td></tr>
<tr><td>签发人</td><td colspan="2">王海涛</td><td>收到时间</td><td>月　日　时</td></tr>
<tr><td colspan="2">Apnet 网络编码</td><td>AP65</td><td>发送人</td><td>王琼</td></tr>
</table>

2003 年 2 月 24 日新疆伽师发生 6.8 级地震。新疆地震局 12 时召开临时会商会，对震后趋势进行了专题讨论：

(1) 截至 24 日 10 时 58 分，巴楚台记录余震 24 次，最大余震 4.6。其中 2.0～2.9 级地震 1 次；3.0～3.9 级地震 20 次；4.0～4.9 级地震 3 次；

(2) 6.8 级地震位于 1997～1998 年伽师强震群区的东南端，余震活动有继续朝东南方向扩展的迹象；伽师震区 6 级地震有成组活动特征；

(3) 据此次 6.8 级地震的震源机制解的初步结果分析，这次 6.8 级地震与 1997 年和 1998 年伽师震群的结果明显不同，主压应力方向为近东西向；

另外，24 日 06 时乌鲁木齐发生一次 M_S3.7 有感地震。

综合分析认为：近期伽师震区及附近地区仍有发生 6 级左右强震的可能。

附件五：

秘密★三个月

震情监视报告

单位	新疆地震局预报中心		会商会类型	临时会商会
期数	(2003) 第 18 期		会商会地点	局会商室
	(总字) 第 143 期		会商会时间	2003 年 2 月 24 日 22 时
主持人	高国英		发送时间	2 月 25 日 01 时 30 分
签发人	王海涛		收到时间	月　日　时
Apnet 网络编码		AP65	发送人	王琼

2003 年 2 月 24 日 22 时新疆地震局的有关专家和中国地震局分析预报中心刘杰、陈荣华研究员一起对伽师 6.8 级地震震后趋势进行了认真而热烈地分析讨论，分析意见如下：

(1) 截至 24 日 23 时 49 分，喀什台记录余震 138 次，最大余震 M_L5.1。其中 3.0～3.9 级地震 84 次；4.0～4.9 级地震 33 次；5.0～5.9 级地震 3 次。

(2) 目前余震分布在 1997～1998 年伽师震群区的东南端，并有继续向东南方向扩展的趋势。余震丰富、分布较为分散，序列衰减不太正常。此次 6.8 级地震的发生已突破该区历史强震的最大震级，可能是伽师强震群在塔里木盆地内的一次新的破裂活动，具备发生成组强震的孕震环境。分析认为，震区仍有可能发生较大的调整性地震。

(3) 本次 6.8 级强震与 1997～1998 年的伽师震群及 2003 年 1 月 4 日伽师 5.4 级地震的震源机制结果明显不同。另外 GPS 测量结果表明，2002 年 12 月 25 日乌恰 5.7 级地震后，乌恰西部高剪应变区消失，在喀什西南地区形成新的相对高剪应变区，伽师地区剪应变增强也比较明显。

综合分析认为：近期伽师震区及附近地区仍有可能发生 6 级左右强震。目前正密切跟踪余震序列的变化，特别注意余震活动的突然平静和增强现象，随时对震情作出判定。

附件六：

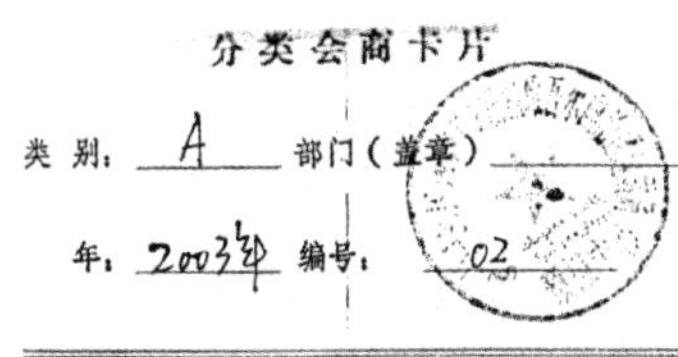

分类会商卡片

类 别：A 部门（盖章）

年：2003年 编号：02

预报效果评价：

会商时间：2003.2.25 地点：新疆地震局会商室

主 持 人：王国英

预报意见：

时 间：2003.2.25～2003.3.27

区 域：39°-40°N，76.8°-78.0°E

震 级：M_S5.5±

上报时间和部门：2003年2月25日 新疆地震局

说　明

1.类别：指分析预报工作管理条例第21条中的A、B、C、D四类。
2.预报意见：
A类必须明确填报地震可能发生的时间段（三个月内），哪些地县区域以及震级范围。
B类必须明确填报地震可能发生的若干地县区域范围和震级范围；时间至少几个月以上，甚至二、三年。
C类只指出未来一定时间内可能发生中强以上或较大地震。区域范围可较大。
D类必须明确指出未来一定时间某区域范围内没有破坏性地震或可能引起较强社会反应的地震发生。
3.预报效果评价包括地震活动实况、三要素预报的正确程度、预报依据的科学性以及决策能力评估。

※※※※※※※※※※※※※※※※※

（附 图）

依据：

1. 巴楚—伽师6.8级地震后，余震密集，衰减缓慢，与1997—1998年伽师强震群活动状态类似。
2. 6.8级地震区在1997—1998年伽师震群区的东南端，其主压应力方向为北西向，与1997—1998年伽师6级地震主压应力北北东方向明显不同，可能是一次新的破裂活动。地下构造特征又可能与在伽师震区类似，故可能出现与伽师震群类似的地震活动。
3. 余震活动范围在扩大，余震主要沿南东和北北西两个方向扩展。余震初动方向一致。

附件七：

新疆维吾尔自治区地震局

机密★1年

关于近期新疆地震形势的汇报材料

新疆维吾尔自治区人民政府：

自1996－1998年阿图什－伽师强震群活跃之后，新疆地震活动进入了一个相对平静时段，特别是2000年－2002年12月连续2年没有发生5级地震，4年5个月没有6级地震发生，5级和6级地震平静分别达历史最长时间。2002年12月25日乌恰发生一次5.7级地震，2003年1月4日在伽师老震区又发生一次5.4级地震，打破了长时间无5级地震平静。有关分析表明新疆地震活动有可能进入一个新的活跃时段。

2002年11月召开的2003年度新疆地震趋势会商会上，在对全疆各种观测资料进行深入分析、科学论证的基础上，对2003年新疆地震趋势进行了预测。研究后认为2003年度新疆地震活动水平将高于2002年度，较大可能发生6－7级地震，并划定了地震危险区。其中乌恰－喀什地区有较大可能发生6－7级地震；乌苏－石河子和乌什－柯坪地区有可能发生5级地震。

2003年1月全国地震趋势会商会分别把新疆乌恰－喀什和乌苏－精河地区划为6－7级和6级左右全国重点危险区。新疆的震情发展已引起中国地震局高度重视。

5 级地震平静打破后，新疆多项前兆观测出现异常突变，在一定程度上表明我区震情较为严峻的状态，随时都有发生 6 级左右地震的可能。面临着较为严峻的地震形势，新疆地震局高度重视震情发展，目前已在乌恰县乌合沙鲁乡、膘尔托阔依乡架设了 2 个临时数字地震台，并派出 9 名工作人员组成 3 个 GPS 观测组对南至塔什库尔干，北至乌什，西至乌鲁克恰提，东至巴楚一带的伽师 GPS 网进行复测（预计 2003 年 2 月 5 日结束），以严密跟踪南天山西段的地震活动。近期，我局已接连向专业地震台站和地方地震局下发了两个文件，就进一步加强震情监视工作提出了具体要求。我们将密切注意地震活动情况，并随时将震情发展情况上报政府。

新疆维吾尔自治区地震局

二〇〇三年一月二十八日

附件八：

附表 1

2003 年度会商前兆异常项目表

填表单位：新疆地震局　　　　填表人：朱　燕　　　　2002 年 11 月 23 日

前兆项目	异常台站	台站位置 纬度（°）	经度（°）	异常特征 起始时间	幅度	加速时间	转折时间	持续时间	结束时间	异常可信度	预报三要素	分析方法及备注	图件编号
地倾斜	阿勒泰 CZB	47.9	88.1	2002.05	两分向速率变化较大	2002.05	2002.06	5 个月	持续	0.6	阿尔泰地震带存在发生 5 级以上地震的趋势异常背景	形态法、时序曲线、卡尔曼滤波、多项式拟合等	4.24
	富蕴 JB	47.0	89.5	2001.9	NS 向年变幅增大，掉格频繁，秋季转向推迟	2001.07	2001.10	15 个月	持续	0.6			4.25
	精河 SQ	44.6	83.0	2001.03	2001 年~2002 年 NS 向年变形态异常，年变幅减小	2001.03	2001.03	19 个月	持续	0.6	北天山存在发生 5-6 级地震背景		4.34 4.35
	新源 CZB	43.4	83.3	2002.5	2000 年以来 NS 向南倾速率减缓	2002.05	2000.03	21 个月	2002.10	0.7			4.36
	温泉 SQ-70	45.0	81.0	2002.6	6 月开始年变消失，矢量曲线连续打结		2002.06	5 个月	持续	0.7			4.37
	石场 SQ-70	43.9	89.7	2002.10	EW 向秋季转向时间提前近 2 个月		2002.10	1 个月	持续	0.6			
	乌什 SQ 周记	41.2	79.2	1999.04	2000 年—2002 年春季向北速率逐渐减慢，2001 年 NS 向年变形态异常，		2001.03	47 个月	持续	0.6	柯坪块存在发生 5-6 级地震的背景		4.54
断层测量	克孜尔	41.8	82.4	2001.03	2001~2002 年年变发生畸变，3~9 月测值打破年变规律		2001.03	17 个月	持续	0.7	库轮拜地区存在发生 5-6 级左右地震的可能	形态法、时序曲线	4.54
地磁	乌鲁木齐	43.8	87.7	2000	2000 年总强度年变化速率转折后连续两年保持同一速率		2000	两年	持续	0.6	北天山存在发生 5-6 级地震背景	差分速率分析	
	喀什	39.6	75.9	2001	2000 年~2002 年已出现两次年变率转折		2000 2001	两年	持续	0.7	喀什--乌恰地区存在 6 级地震背景		4.55
地电	乌鲁木齐	43.8	87.7	2001.10	乌鲁木齐地电视电阻率测值，2001 年开始缓慢趋势性下降		2001.10	12 个月	持续	0.6	北天山存在发生 5-6 级地震背景	时序曲线分析	
地下流体	新 10 井水位	43.7	87.6	2002.05	5 月趋势性升高，8 月开始趋势性下降	2002.05	2002.07	6 个月	持续	0.6	北天山存在发生 5-6 级地震背景	时序曲线分析	4.41
	新 10 泉流量	43.7	87.6	2002.05	5 月趋势性升高，8 月开始趋势性下降	2002.05	2002.08	5 个月	2002.10	0.6			4.42a
	新 10 泉甲烷	43.7	87.6	2002.05	5 月开始趋势性上升	2002.05	2002.09	6 个月	持续	0.7			4.42b
	新 10 泉氦气	43.7	87.6	2002.05	5 月开始趋势性上升	2002.05	2002.09	6 个月	持续	0.7			4.42c
	新 21 井水位	44.1	86.5	2002.01	元月出现不稳定的突跳。6 月持续性上升后呈波浪形缓慢下降趋势	2002.01	2002.05	10 个月	持续	0.6			4.43

附件九：

附表 2

2003 年度会商测震异常项目表

填表单位：新疆地震局　　　　填表人：李莹甄　　　　2002 年 11 月 23 日

异常项目	研究区域	起始震级 (Ms)	异常范围 纬度(°)	异常范围 经度(°)	异常开始时间	异常特征量	异常结束时间	异常可信度	预报三要素	分析方法及备注	图件编号
空区（段）	新疆	4.0	38.2-39.6 42.0-44.0	73.0-76.0 82.5-85.0	1996 2001	长轴 204km，短轴 130km 长轴 236km，短轴 163km		0.7	异常区附近 6.3±0.5 级以上 异常区附近 5.6±0.5 级以上	由背景空区长轴计算 由背景空区长轴计算	4.3
地震频度	新疆	3.0	41.5-44.5 38.5-40.5	80.5-86.5 73.5-78.5	2001.10 2001.10	增高 降低	2002.09	0.68	异常区附近 5-6 级	空间扫描	4.7
		2.5 2.0	45.5-49.0 42.5-44.5	87.0-92.5 86.0-88.5	2000.10 2001.02	低值 高值	异常持续 2002.08	0.7 0.68	异常区 5 级 异常区 5 级	时间进程分析	4.21 4.26
b 值	新疆	2.5 2.5	42.0-43.0 38.0-39.0	81.5-85.0 74.0-77.0	2000.10 2000.10	超出背景值 20%	2002.09	0.65	异常区附近 5-6 级左右 异常区附近 6 级	b 值空间扫描	4.8
		2.5 3.0	43.0-45.0 34.5-36.7	82.5-86.0 77.5-82.0	2001.02 2002.02	低值	2002.02 异常持续	0.60	异常区 5 级 异常区 5-6 级	b 值时间进程分析	4.33 4.58
小震调制比 Rm	新疆	2.0	44.0-46.0 42.0-44.0 38.0-41.0 35.0-37.0	80.5-82.5 82.5-85.5 73.5-76.5 75.5-82.5	2001.10	≥0.27		0.62	异常区附近 5 级 异常区附近 5-6 级 异常区附近 6-7 级 异常区附近 6-7 级	空间扫描	4.8
			38.0-40.5	73.0-77.0	2000.10		异常持续	0.63	异常区 6 级	时间进程分析	4.52

续前表

异常项目	研究区域	起始震级 (Ms)	异常范围 纬度(°)	异常范围 经度(°)	异常开始时间	异常特征量	异常结束时间	异常可信度	预报三要素	分析方法及备注	图件编号
地震非均匀度 GL	新疆	2.0	42.0-45.0 38.0-42.0 35.0-38.0	80.5-82.5 73.0-79.5 76.5-82.5	2001.10 2001.10 2001.10	≥3.5		0.68	异常区附近 5-6 级 异常区附近 6-7 级 异常区附近 6 级	空间扫描	4.9
震情窗	新疆		42.0-44.0 38.5-42.0	84.5-87.0 73.0-77.5	2002.03 2002.03	异常窗口网反映地震能力范围的重叠	2002.10	0.52 0.56	异常区附近 5.5 级 异常区 6.3 级	空间图像	4.11
A 值	新疆	2.0	42.5-45.0 40.5-42.5 38.0-40.5 35.0-37.0	83.0-87.5 77.5-82.0 73.5-74.4 76.5-82.0	2001.04	高值	2002.09	0.60	异常区附近 5-6 级 异常区附近 5-6 级 异常区附近 6 级 异常区附近 5 级	空间扫描	4.12
地震增强-平静区	新疆	2.5	42.0-44.0 39.0-42.0 34.0-37.0	84.0-87.0 73.0-80.0 78.0-82.0	2001.09	增强平静	2002.0		异常区 5 级以上	空间扫描	4.1
地震波振幅比	新疆	1.0	43.2-44.2	83.0-85.2	2002.09.15	持续低值	2002.11.18		异常区 5 级	空间扫描 时间进程	4.14 4.13
应力降	新疆	1.0	42.0-44.0	83.0-85.0	2002.01	相对高值	2002.10.07		异常区 5 级	空间扫描	4.5
缺震	新疆	2.5 2.0 3.0 2.0	45.5-49.0 40.0-42.0 34.5-36.7 38.0-40.0	87.0-92.5 78.0-81.0 77.5-82.0 73.0-77.0	2000.10 2000.02 2001.07 1995.08	低值	异常持续 2002.02 异常持续 异常持续	0.58	异常区 5-6 级 异常区 5-6 级 异常区 5-6 级 异常区 6 级	时间进程分析	4.23 4.46 4.57 4.50
η 值	新疆	2.0 2.5 3.0 3.0	42.5-44.5 43.0-45.0 34.5-36.7 38.0-40.5	86.0-88.5 82.5-86.0 77.5-82.0 73.0-77.0	2000.01 2001.04 2001.06 2000.01	低值	异常持续 2002.02 异常持续 2001.08	0.62 0.58 0.58 0.60	异常区 5 级 异常区 5 级 异常区 5-6 级 异常区 6 级	时间进程分析	4.28 4.30 4.56 4.51

续前表

A(b)值	新疆	2.5	43.0-45.0	82.5-86.0	2001.01	低值	2002.02	0.60	异常区5级	时间进程分析	4.31
		2.0	41.0-43.0	80.5-84.0	2001.02		异常持续	0.58	异常区5级		4.45
		2.0	40.0-42.0	78.0-81.0	1999.10		异常持续	0.58	异常区5-6级		4.47
		2.0	39.0-41.0	76.0-78.5	2001.01		异常持续	0.58	异常区5-6级		4.49
模糊地震活动度S	新疆	2.5	45.5-49.0	87.0-92.5	2000.10	低值	异常持续	0.70	异常区5-6级	时间进程分析	4.22
		2.0	42.5-44.5	86.0-88.5	1999.06		2002.04	0.58	异常区5级		4.27
		2.5	43.0-45.0	82.5-86.0	1998.06		2002.02	0.62	异常区5级		4.32
		2.0	41.5-43.0	80.5-84.0	2000.08		2002.04	0.58	异常区5级		4.44
		2.0	40.0-42.0	78.0-81.0	1999.08		异常持续	0.60	异常区5-6级		4.48
		2.5	38.0-40.5	73.0-77.0	1995-10		异常持续	0.58	异常区6-7级		4.53
		3.0	34.5-36.7	77.5-82.0	2001.04		异常持续	0.68	异常区5-6级		4.59
尾波持续时间比	新疆	1.0	38.5-40.5	74.5-76.5	2001.01	低值	2002.09	0.64	异常区6级	时间进程分析	
地震活动演化指数YH值	新疆	2.0	42.5-44.5	86.0-88.5	2000.06	高值	异常持续	0.58	异常区5级	时间进程分析	4.29

2003 年 4 月 17 日青海省德令哈 6.6 级地震和 2004 年 2～5 月 5 次 5～5.9 级强余震

青海省地震局

陈玉华　马玉虎　王培玲　刘文邦　张晓清

摘　要

2003 年 4 月 17 日青海省德令哈市西北发生 6.6 级地震，宏观震中位于德令哈西查伊沟上游，震中烈度为Ⅷ度，地震未造成人员伤亡，直接经济损失 4345 万元。

此次地震序列为主震—余震型，最大余震 5.9 级，余震分布在主震西侧，与极震区长轴走向一致。节面Ⅱ为主破裂面，主压应力 P 轴方位北北东向，发震构造为大柴旦—宗务隆山—青海南山断裂。

震中 300km 范围内共有地震观测台 7 个，其中测震单项观测台 3 个，测震和前兆综合观测台 5 个。7 个前兆观测项目，震前 3 个测项出现 8 条异常，其中 6 条为中期异常，2 条短期异常，地震学异常明显，其他主要为水氡、气氡、地电异常。

德令哈 6.6 级地震发生在 2003 年度青海省 5～6 级地震重点注意区内。

前　言

2003 年 4 月 17 日 08 时 48 分，青海德令哈发生 6.6 级地震。中国地震台网测定的微观震中为 N37.50°，E96.80°，现场考察判定宏观震中为 N37°33.5′，E96°27′，位于德令哈西查伊沟上游，震中烈度为Ⅷ度。地震未造成人员伤亡，直接经济损失 4345 万元。

德令哈地震区是青海主要强震活动区，历史上多次发生 6 级强震。6.6 级地震前发现有地震频度、地震非均匀度、中等地震活动平静等 3 个地震学项目 4 条异常。前兆观测格尔木水氡趋势性异常。该震中区是 2003 年度青海省地震趋势会商重点应注意地区[1)]。2004 年 2～5 月该区再次发生 5 次 5 级中强震。

地震发生后，中国地震局、青海省政府、青海省地震局、海西州政府立即启动大震应急预案，进入紧急状态，多次向震区派出考察队，开展震害评估、地震地质考察和余震监测等工作。极震区震害以山体崩塌、滚石及房屋倒塌为主。

一、测震台网及地震基本参数

德令哈6.6级地震发生前，震中周围300km范围共有测震观测项目的台站7个（图1），其中0～100km范围仅德令哈1个台站；100～200km范围没有台站；200～300km范围6个台站，分别为格尔木、都兰、镜铁山、金佛寺、嘉峪关、肃南地震台。该区域地震台网分布极不均匀，除德令哈、都兰、格尔木3个台属青海地震台网外，其余4个台隶属甘肃省地震台网。该区台网可监控M_L2.5以上地震。本次地震基本参数如表1。

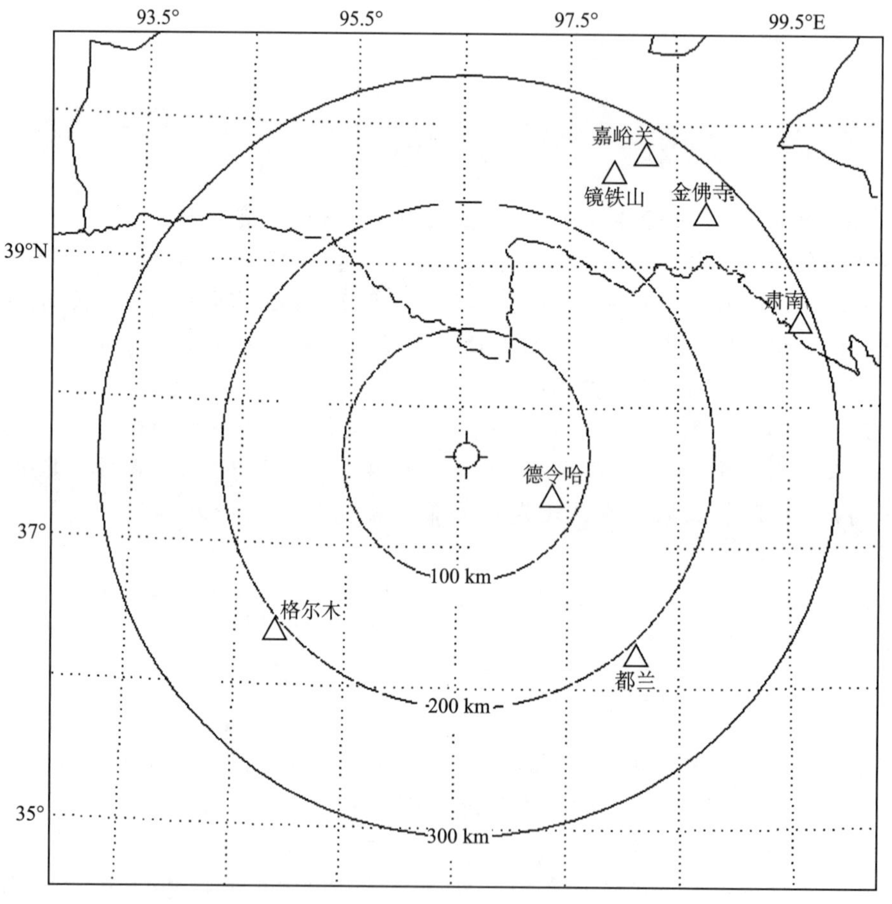

图1　德令哈6.6级地震前距震中300km范围内测震台网分布图

Fig. 1　Distribution of seismometric stations around the epicentral area within 300km before the M_S6.6 Delingha earthquake

表 1　地震基本参数

Table 1　Basic parameters of the earthquake

编号	发震日期	发震时刻	震中位置（°）		震级		震源深度	震中地名	结果来源
	年 月 日	时 分 秒	φ_N	λ_E	M_S	M_W	（km）		
1	2003 04 17	08 48 41.1	37.50	96.80	6.6		15	德令哈	CNS
3	2003 04 17	08 48 33.2	37.67	96.57	6.6			德令哈	青海台网[2)]
4	2003 04 17	08 48 38.5	37.529	96.476	6.3		14	德令哈	NEIC
5	2003 04 17	08 48 38.58	37.529	96.476		6.2	10	德令哈	GSN
6	2003 04 17	08 48 46.3	37.53	96.45		6.3	16	德令哈	HRV

二、地震地质背景

德令哈地震区位于青藏高原北部的柴达木盆地东北边缘，区域内地质构造错综复杂，断裂构造纵横交错，历史上发生的中强地震较多，且 6 级以上地震沿大型活动断裂分布特征明显（图 2）。

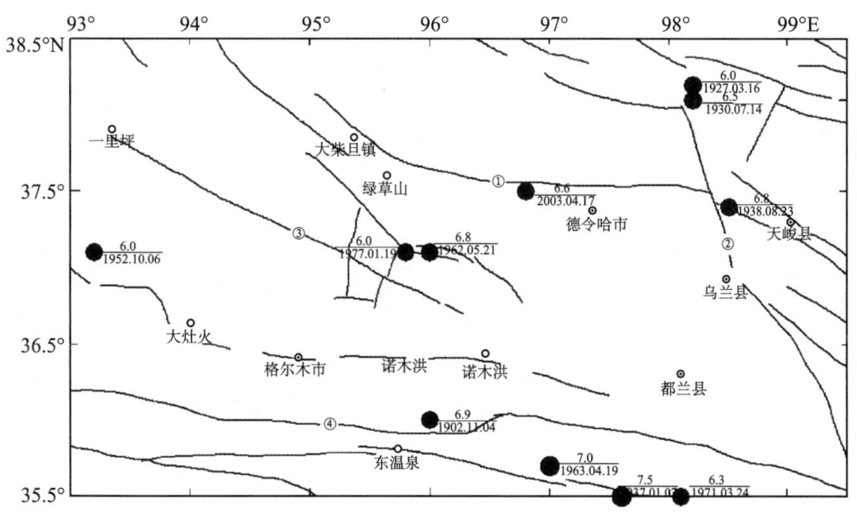

图 2　大柴旦 6.3 级地震震区附近地震地质构造及 6 级以上地震分布

Fig. 2　Map of geological structure and earthquake distribution of $M_S \geq 6$ around the M_S6.6 Delingha area

①大柴旦—宗务隆山—青海南山断裂带；②鄂拉山断裂带；
③柴达木中央断裂带；④东昆仑断裂带

德令哈 6.6 级地震发震构造为大柴旦—宗务隆山活动断裂带，该断裂带是柴达木盆地的主要断裂构造。断裂带在青海北部地壳断裂系统中占有重要位置，它不但规模大延伸长，且控制着不同时代地层的分布以及中酸性岩浆岩的产出。亦是柴达木断陷区与祁连断隆带两个新构造区的分区边界[1]。大柴旦—宗务隆山断裂西起柴达木山南的鱼卡北，经大柴旦，东延进入宗务隆山南坡和巴音以东，由一组北西—北西西—近东西向断裂组成，总体呈NWW—SEE 向展布[2]。资料显示，在 NNE—SSW 向压应力作用下，该断层主要以逆冲活动为主。

德令哈 6.6 级地震宏观震中位于该断裂的西段（大柴旦—宗务隆山南坡）。该段现代地震活动剧烈，1938 年曾发生 6.0 级地震，2003 年德令哈 6.6 级地震后，2004 年 2～5 月发生 5 次 5 级以上强余震、2008 年 11 月 10 日发生 6.3 级地震、2009 年 8 月 28 日又发生 6.4 级地震。

三、地震影响场和震害

据震区现场考察资料3)，德令哈 6.6 级地震宏观震中位于德令哈西查伊沟上游，极震区烈度为Ⅷ度。等震线形状呈椭圆形，长轴走向北西（图 3），与宗务隆山—青海南山断裂带总体方向一致。Ⅵ度区以上受灾总面积约 5360km²。

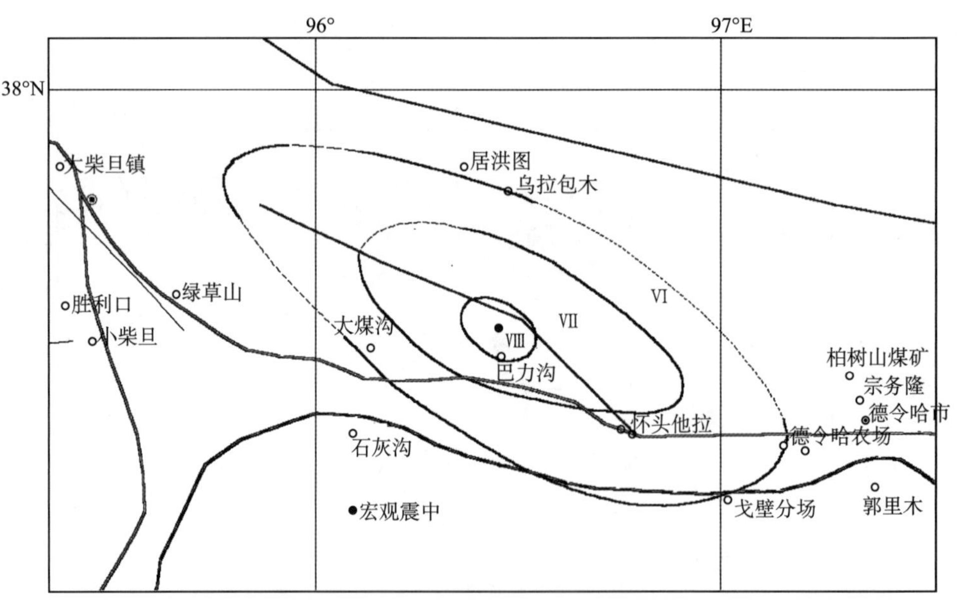

图 3　德令哈 6.6 级地震烈度等震线图2)

Fig. 3　Isoseismal map of the M_S6.6 Delingha earthquake

Ⅷ度区：烈度等震线为北西向分布的椭圆，长轴沿 306°方向展布，长轴约 17.4km、短轴约 11.3km，面积约 154km²。

震区出现大量山坡滚石，滚石直径最大 0.8m。多处沟壁崩塌，崩塌体最大宽 7m。大面积的山体崩塌，局部发育 NE 走向地表裂缝。

震区内人员和房屋稀少、调查中仅见到 6 户牧民 7 间土木结构房屋。房屋部分墙体倒塌，墙体裂缝较多。地震时河流阶地跨塌压死羊 12 只。

Ⅶ度区：该烈度区长轴方向为 296°，长轴约 76.3km、短轴约 30.5km，区域面积 1673km^2（不含Ⅷ度区面积）。该烈度区多为无人区，震害以地震地质灾害为主。表现为地表裂缝、崩塌、滑塌、喷砂冒水等。地表裂缝均出现于第四系冲洪积滩中。

Ⅵ度区：该烈度区长轴方向 297°，长轴约 134.7km，短轴约 50.7km。区域面积 3533km^2（不含Ⅷ度和Ⅶ度区面积）。该烈度区沿长轴方向以北仅有 2 户居住帐篷的牧民，无房屋建筑。震害以地表裂缝为主，有喷砂冒水、沟壁崩塌现象。地表裂缝均出现于戈壁滩中和河流阶地上，受地形、河流控制方向较混杂。以南有数个居民点，包括：怀头他拉镇、怀头他拉农场、戈壁乡、大煤沟等地。区内房屋建筑以中度和严重破坏较多。区内大部分人震感强烈，个别人感到头晕，有桌椅摇晃、器皿晃动、门窗作响的感觉。

对震区 2004 年 2 月 25 日 5.0 级，3 月 17 日 5.2 级，5 月 4 日 5.5、5.1 级及 5 月 11 日 5.9 级强余震，仅做了震害损失评估工作，未做烈度区划工作。

6.6 级地震发生在人员稀少地区，未造成人员伤亡，直接经济损失 4345 万元。

四、地 震 序 列

2003 年 4 月 17 日至 2004 年 12 月 31 日德令哈 6.6 级地震序列共记录到 M_L2.0 以上地震 665 次，其中 2.0≤M_L≤2.9 级 534 次，3.0≤M_L≤3.9 级 112 次，4.0≤M_L≤5.3 级 13 次，5.0≤M_S≤5.9 级 5 次，M_S6.6 级 1 次。由图 4 的 3 项序列参数曲线可看出该序列呈现出两个明显的活动段，前期（2003 年 4 月 17 日至 2003 年 11 月上旬）序列呈指数衰减，符合主余震序列特征。2003 年 11 月下旬后小震活动出现增强趋势，2004 年 2 月 25 日至 5 月底序列出现两次起伏活动发生 5 次 5 级以上强余震，最大震级地震为 5 月 11 日 5.9 级，该序列类型显示主余型或多震型特征。

表 2　德令哈 6.6 级地震序列目录（M_L≥4.0 级）

Table 2　Catalogue of the M_S6.6 Delingha earthquake sequence（M_L≥4.0）

编号	发震日期	发震时刻	震中位置		震级		震中地名	结果来源
	年 月 日	时 分 秒	φ_N	λ_E	M_L	M_S		
1	2003 04 17	08 48 41.1	37°30′	96°48′		6.6	德令哈	CNS
2	2003 04 17	19 29 52.8	37°22′	96°26′		4.7	德令哈	青海台网[1)]
3	2003 04 18	01 11 45.7	37°37′	96°27′		4.9	德令哈	青海台网[1)]
4	2003 04 20	03 41 30.7	37°21′	96°25′		4.9	德令哈	青海台网[1)]
5	2003 04 21	13 51 26.2	37°32′	96°49′		4.9	德令哈	青海台网[1)]
6	2003 05 03	14 12 50.7	37°29′	96°37′		4.7	德令哈	青海台网[1)]
7	2004 02 19	00 29 20.9	37°42′	96°46′	4.0		德令哈	青海台网[1)]

续表

编号	发震日期	发震时刻	震中位置		震级		震中地名	结果来源
	年 月 日	时 分 秒	φ_N	λ_E	M_L	M_S		
8	2004 02 25	04 21 43.8	37°30′	96°41′		5.0	德令哈	青海台网[1)]
9	2004 02 26	00 07 49.3	37°28′	96°38′	4.0		德令哈	青海台网[1)]
10	2004 03 02	15 13 40.4	37°38′	96°38′		4.4	德令哈	青海台网[1)]
11	2004 03 02	15 17 20.8	37°42′	96°40′	4.0		德令哈	青海台网[1)]
12	2004 03 02	20 30 21.7	37°36′	96°39′		4.7	德令哈	青海台网[1)]
13	2004 03 02	20 37 36.5	37°28′	96°44′	4.2		德令哈	青海台网[1)]
14	2004 03 17	05 23 12.8	37°29′	96°37′		5.2	德令哈	青海台网[1)]
15	2004 05 04	13 04 56.6	37°26′	96°34′		5.5	德令哈	青海台网[1)]
16	2004 05 04	19 36 01.1	37°31′	96°34′		5.1	德令哈	青海台网[1)]
17	2004 05 11	07 27 19.5	37°43′	96°24′		5.9	德令哈	青海台网[1)]
18	2004 5 14	20 37 52.2	37°34′	96°37′	4.0		德令哈	青海台网[1)]
19	2004 8 26	02 45 26.6	37°28′	96°32′		4.7	德令哈	青海台网[1)]

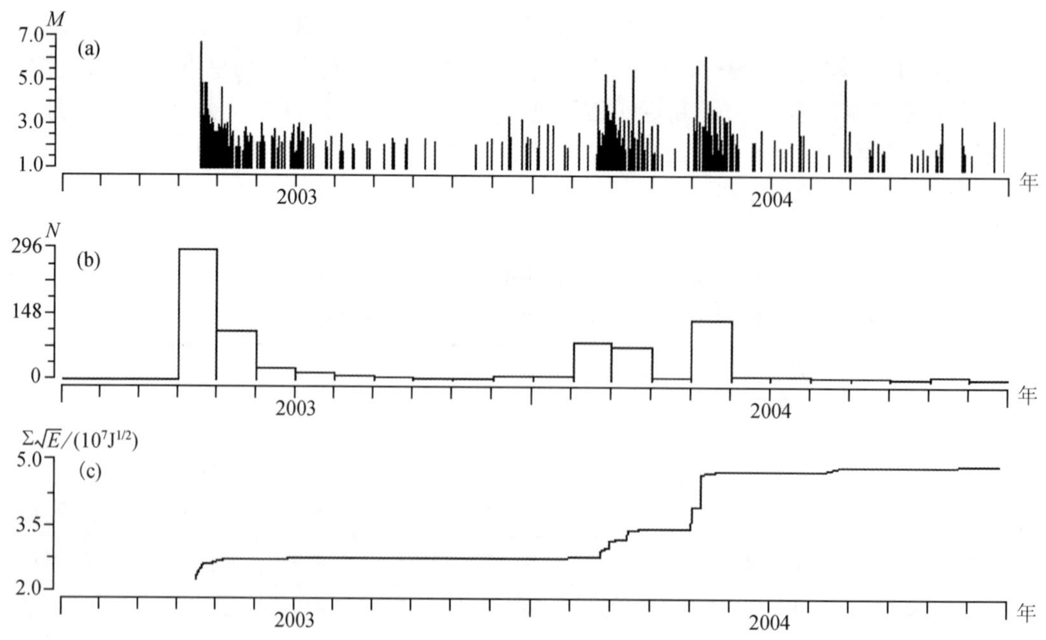

图4　德令哈6.6级地震序列 M-t（a）、N-t（b）、应变释放（c）曲线

Fig. 4　M-t（a）、N-t（b） and curve of strain release of the M_S6.6 Delingha earthquake sequence

图 5 为青海省地震台网记录的德令哈 6.6 级地震序列震中分布图，其中空心圆为序列前期（2003 年 4 月 17 日至 2003 年 10 月）震中分布，实心圆为后期（2003 年 11 月至 2004 年 12 月）震中分布。可见前期余震密集分布在主震以西区域，余震区以发震构造为中心呈近南北向分布；后期该序列余震活动往东南方向延展明显。形成一北西向的余震分布区。整个序列 M_L4 以上地震基本集中分布在主震以西。

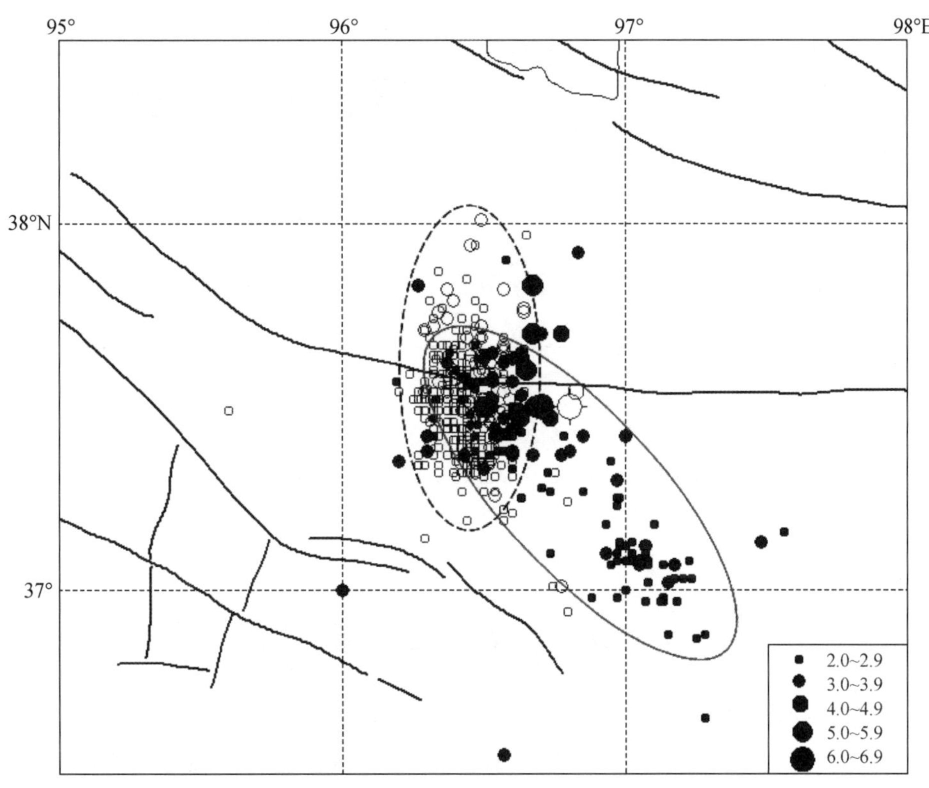

图 5　青海地震台网交出的德令哈 6.6 级地震序列震中分布图

Fig. 5　Epicentral distribution of the M_S6.6 Delingha earthquake sequence by Qinghai seismic network

北京地质大学孙长虹等[3]利用祁连山周边 8 套 REFTEK 三分量宽屏带地震仪组成的流动数字地震观测台网记录到的德令哈（2003 年 4 月 17 日至 10 月 25 日）序列 $M_L \geq 1.0$ 级 117 次地震事件进行了精确定位。定位结果序列余震空间分布与发震构造（大柴旦—宗务隆山活动断裂带）吻合较好（图 6）。

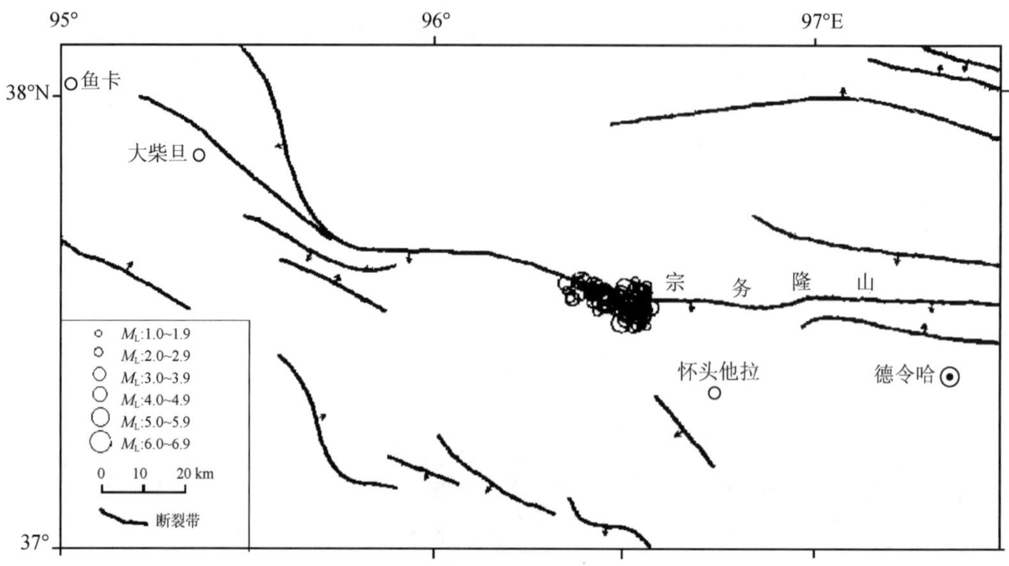

图6　精定位的德令哈地震序列（2003 年 4 月 17 日至 10 月 25 日）沿大柴旦—宗务隆山断裂带分布[3]

Fig. 6　Relocated earthquake distribution of the Delingha earthquake sequence along the Dachaidan-Zongwulong mountain fault

五、震源参数和地震破裂面

青海省地震局张晓清利用 13 个全球数字化地震台网观测的波形数据，应用矩张量反演方法求得 2003 年 4 月 17 日德令哈 6.6 级地震的地震矩张量解，结果显示：走向为 114°节面（即节面Ⅱ）为该次地震的主破裂面，该结果与哈佛大学给出结果基本吻合。该次地震为逆冲兼有少量走滑的破裂模式。

表3　德令哈 6.6 级地震地震矩张量解

Table 3　Moment tensor solutions of the M_S6.6 Delingha earthquake

节面Ⅰ			节面Ⅱ			矩张量						M_W	结果来源
走向	倾角	滑动角	走向	倾角	滑动角	M_{xx}	M_{yy}	M_{zz}	M_{xy}	M_{yz}	M_{zx}		
294	29	88	116	61	91	3.430	−2.870	−0.565	1.990	−1.010	1.370		哈佛大学
286	49	85	114	41	96	−2.563	−0.7492	3.315	0.7745	−0.4939	1.370	6.3	张晓清

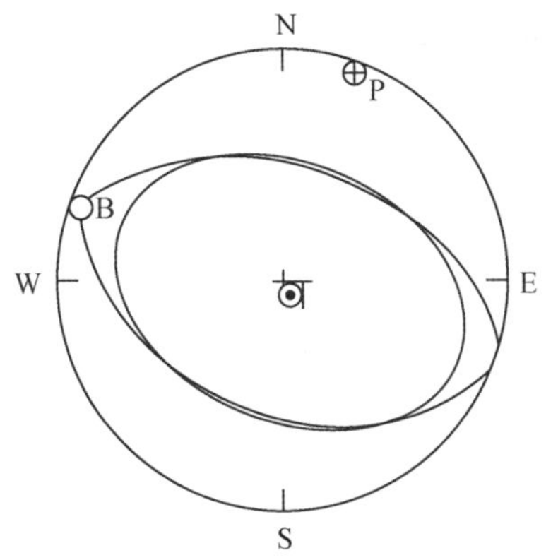

图 7－1 德令哈 6.6 级地震的矩张量解（青海省地震局张晓清）

Fig. 7－1 Moment tensor solutions of the M_S6.6 Delingha earthquake

表 4 德令哈 6.6 级地震震源机制解

Table 4 Focal mechanism solutions of the M_S6.6 Delingha earthquake

节面Ⅰ			节面Ⅱ			P 轴		T 轴		N 轴		结果来源
走向	倾角	滑动角	走向	倾角	滑动角	方位	仰角	方位	仰角	方位	仰角	
262	40	66	112	54	109	8	189	73	73	15	281	GS
294	29	88	116	61	91	16	205	74	29	1	296	HRV

哈佛大学给出的主震震源机制解结果与重新定位后的德令哈地震序列的空间展布优势方位、与烈度分布图优势方向基本一致，表明德令哈地震序列的发震构造为近 NWW—SEE 向的断层。大柴旦—宗务隆山断裂带总体呈 NWW 向展布，断裂的西段（大柴旦—宗务隆山南坡）穿过重新定位后的德令哈地震序列区。主震紧邻该断裂为其北侧，大部分余震紧邻该断裂为其南侧。

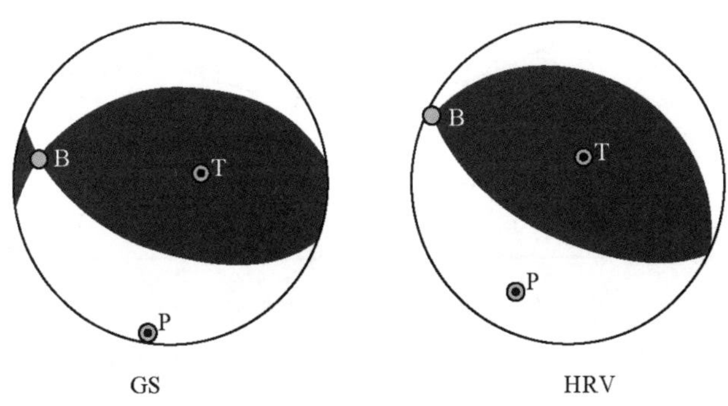

图 7－2 其他机构给出的德令哈 6.6 级地震的震源机制解

Fig. 7－2 Focal mechanism solutions of the M_S6.6 Delingha earthquake by other agenicies

六、地震前兆观测台网及前兆异常

德令哈6.6级地震前震中300km范围内共有测震以外的前兆观测台站5个，有地磁、水（气）氡、电阻率、自然电位和地倾斜等7个观测项目（图8）。其中0 ~ 100km仅有1台1测项；100 ~ 200km无台无测项；200 ~ 300km 4台6测项。该区台站分布稀疏，观测项目少。

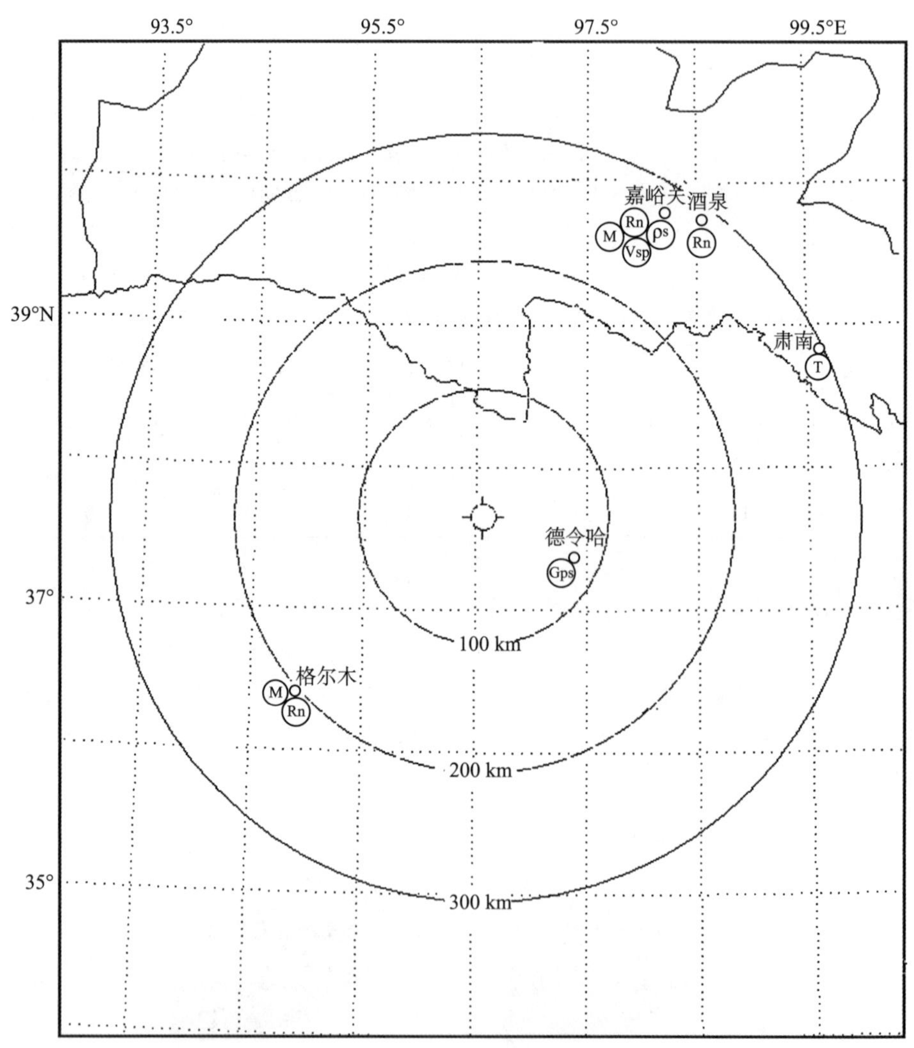

图8　德令哈6.6级地震前定点前兆观测台站分布图

Fig. 8　Distribution of the precursory monitoring stations before the M_S6.6 Delingha earthquake

此次地震前300km范围的5个地震台站7个观测项目记录到4条前兆异常（表5，图9至图14）。表5中给出的地震学异常是在日常监视预报中使用的常规方法得到的异常。2002

年 6 月 1 日至 2003 年 4 月 16 日大陆西部地区 $M_L \geqslant 4$ 级地震在相对活跃的态势下，围绕青海北部及甘肃南部呈现一明显的 $M_L \geqslant 4$ 级地震大面积平静区[4]（图 9）；三湖（N36°～38°，E94°～98°）地区 $M_L \geqslant 3$ 级地震 12 个月窗长 1 个月步长地震活动性时间滑动扫描显示出该区的多项异常1)（图 10）。表 5 中列出了震中距 300km 范围内的 4 项测震以外的前兆异常（图 11 至图 14），除嘉峪关地电阻率外，其余均为地下流体水（气）氡异常。据文献［5～7］，距震中 400～1000km 还分析出一些其他前兆异常。

表 5　前兆异常情况登记表

Table 5　Summary table of precursory anomalies

序号	异常项目	台站（点）或观测区	分析方法	异常判据及观测误差	震前异常起止时间	震后变化	最大幅度	震中距 Δ（km）	异常类别及可靠性	图号	异常特点及备注
1	地震平静	中国大陆西部	$M_L \geqslant 4$ 级地震分布		2002.6～2003.4.16	6.6 主震与余震均发生在平静区内	平静区面积约 60 万平方公里		M_1	9	2002 年 6 月至震前 $M_L \geqslant 4$ 级地震大范围平静[4]
2	地震频度	36°～38° 94°～98°	N-t 曲线 $M_L \geqslant 3.0$ 级	频次超出 2 倍均方差后快速回返	2001 年底至 2003 年初	恢复正常			M_1	10a	2001 年底 $M_L \geqslant 3.0$ 级地震频度出现增强至 2002 年 10 月达到最高后快速下降，震前形成完整的异常形态1)
3	地震非均匀度	36°～38° 94°～98°	GL-t 曲线 $M_L \geqslant 3.0$ 级	≥1.0	2001 年底至 2003 年初	恢复正常			M_1	10b	2001 年底至 2002 年底形成高值过程1)

续表

序号	异常项目	台站(点)或观测区	分析方法	异常判据及观测误差	震前异常起止时间	震后变化	最大幅度	震中距Δ(km)	异常类别及可靠性	图号	异常特点及备注
4	地震平静	36° ~ 38° 94° ~ 98°	$M_L \geqslant 3.5$ 级	地震平静	1998 年底至 2001 年底	活跃			M_1	10c	$M_L \geqslant 3.5$ 级地震持续平静近 4 年显现活跃1)
5	水氡	格尔木	日均值	趋势下降—快速下降—转平	2002. 04 ~ 2003. 03	正常	5.2 Bq/L	208	M_2	11a	2002 年 4 月至 2003 年 1 月形成趋势性下降过程。2003 年 2 月出现快速下降,3 月回返转平[5]
			日均值傅立叶周期分析			正常			U_3	11b	
6	地电	嘉峪关(NW 道)	月均值	破年变	2002. 3 ~ 2003. 1	恢复正常年变	1.95 %	286	M_1	12	2002 年 3 ~ 8 月出现快速上升,破年变[6]
7	气氡	嘉峪关	日均值	破年变、高值突跳	2003. 2 ~ 2003. 6		35 Bq/L	286	S_2	13	2003 年 2 月后测值出现多次突跳,形态破年变突出[6]
8	水氡	酒泉	日均值	突跳下降后转折上升	2003. 3. 20 ~ 3 月底	震后一段时间正常,2003 年 8 月后出现起伏变化		290	S_2	14	2003 年 3 月 20 日测值突跳下降至 3 月底达低值转折上升[6]

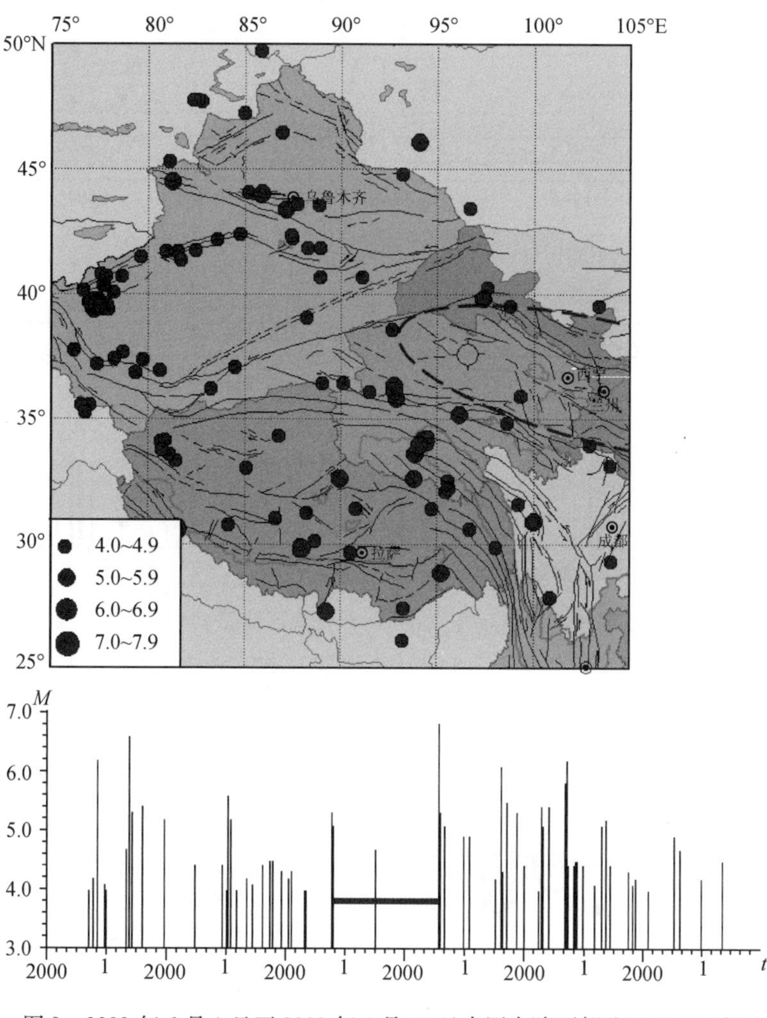

图 9　2002 年 6 月 1 日至 2003 年 4 月 17 日中国大陆西部地区 M_L≥4 级地震分布及平静区 M-t 图

Fig. 9　Earthquake distribution of M_L≥4 in the west china and calm zone M-t diagram from June 1, 2002 to April 17, 2003

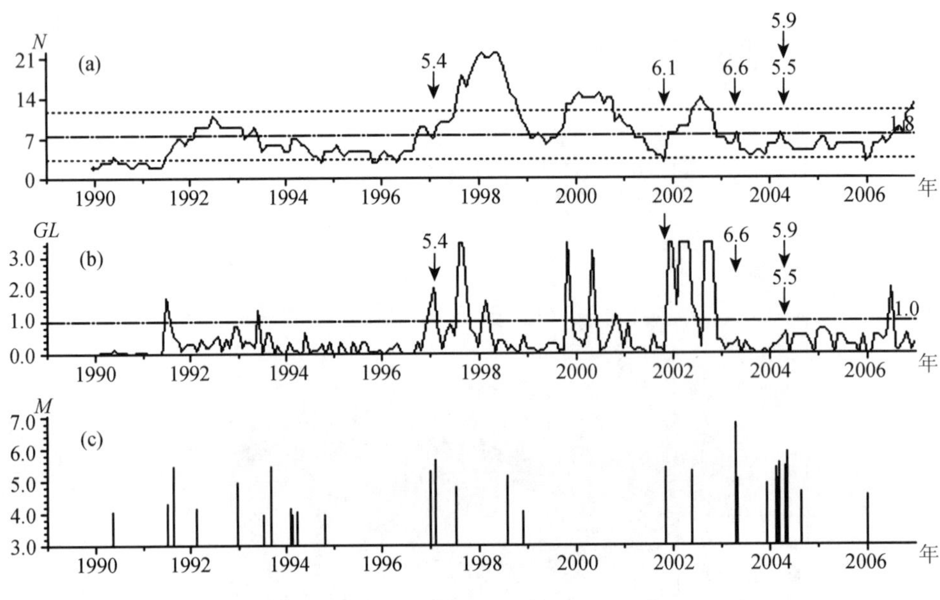

图 10　三湖（N36°～38°，E94°～98°）地区地震学时间扫描曲线

Fig. 10　Seismology time scanning curve in sanhu（N36°-38°，E94°-98°）region

（a）*N-t*；（b）*GL-t*；（c）*M-t*

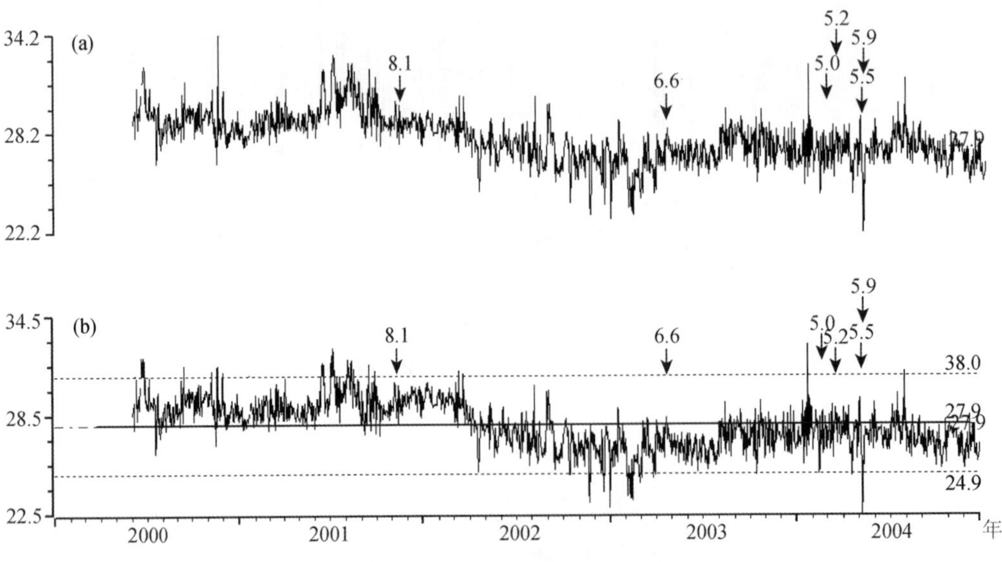

图 11　格尔木水氡日均值曲线（a）及周期分析（b）

Fig. 11　Daily average value of Geermu water radon（a）and cycle analysis（b）

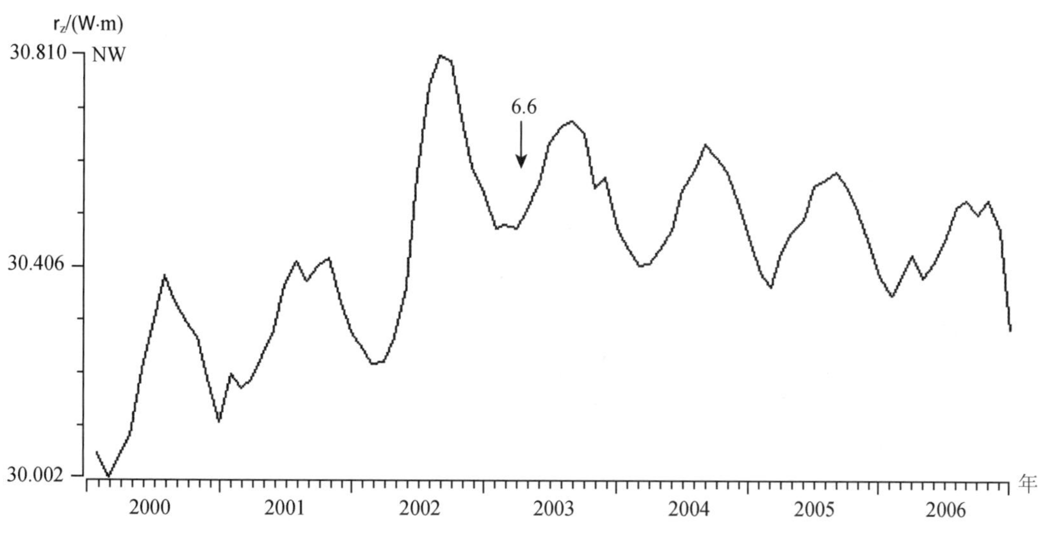

图 12 嘉峪关地电阻率 NW 道月均值

Fig. 12 Month average value of Jiayuguan apparent resistivity in NW component

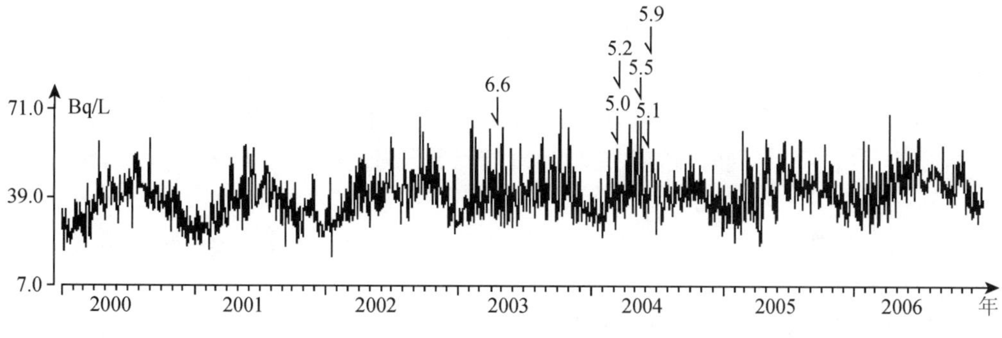

图 13 嘉峪关气氡日均值曲线

Fig. 13 Daily average value of Jiayuguan gas radon

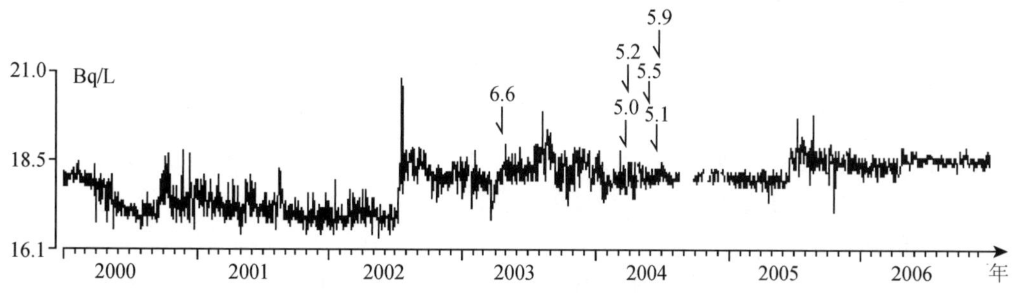

图 14 酒泉水氡日均值曲线

Fig. 14 Daily average value of Jiuquan water radon

七、地震前兆异常特征分析

1. 地震学异常

德令哈6.6级地震孕育中期阶段震源区地震活动突出的异常是M_L4以上地震长达近1年的大面积平静。震例研究表明[8]，中国大陆西部M_L4以上地震大面积、长时间平静是该区6.5级以上地震前的主要中期异常指标。震前1年多震区$M_L \geqslant 3$级地震增长显著，同时GL值出现高值异常过程，地震活动性的显著异常为2003年度地震危险区判定提供了依据。

2. 前兆观测资料异常

德令哈6.6级地震发生在监测能力较低的地区，震中300km范围内仅有5个定点前兆观测台站，震前3个前兆观测台记录到4条前兆异常（表5，图11至图14）。主要异常特征是快速上升、突跳和破年变，以中短期异常为主。

八、震前预测、预防和震后响应

1. 震前预测

2003年度会商将震中（三湖）地区列为2003年度地震重点注意区1)。提出多项地震活动性异常以及前兆观测趋势异常。主要有：该区历史上$M_L \geqslant 5$级地震具有较好的活跃—平静特征，平静打破后多有中强以上地震对应，本次出现的4年平静被2002年5月22日、30日的M_L5.3、M_L5.1地震打破，显示该区可能进入新一轮活动阶段；该区蠕变曲线出现加速，外推震级6级左右；同时还存在多项地震活动参数时间扫描异常及格尔木水氡低值趋势异常等。

震中地区前兆观测台站稀少，没有记录到明显的短临异常变化，无短临预测意见。

2. 趋势判定

青海省地震局根据该区历史地震及区域地震地质背景研究，在德令哈6.6级地震发生2小时做出了早期趋势判断意见，认为该地震序列为主余型可能性较大。地震第3天（4月19日）中国地震局现场专家组根据序列的衰减及参数判定该序列为主余震序列。

2004年2月25日至5月11日德令哈震区发生系列M5以上强余震，最大震级为M5.9。此前，2003年11月起青海东北部地区（距6.6级地震400km以外）记录到多期前兆异常4)，青海地震局预报中心多次召开临时会商会讨论地震趋势，因为异常集中在青海东北部，认为近期青海库玛断裂带中段、南北地震带中段有发生中强震的危险（意见均反映在震情会商报告中）。此后德令哈震区发生（2月25日5.0级，3月17日5.2级，5月4日5.5级、5.1级和5月11日5.9级）一系列中强震。

3. 地震应急

地震发生后，中国地震局和青海省政府领导十分重视。地震发生当日中午14时20分，中国地震局局长宋瑞祥率中国地震局第一批工作组抵达西宁，与在机场等候的青海省副省长马培华等省政府有关部门汇合后，立即乘车赶赴震区。按照省委书记苏荣、省长赵乐际的指示精神，在震区开展现场指导、抗震救灾工作。青海省地震局、海西州政府立即启动大震应急预案，进入紧急状态，在第一时间向震区派出了考察队。中国地震局第二批考察队 16 人携带先进的仪器设备，迅速从北京出发，17 日下午飞往兰州，转乘汽车赶赴灾区，连续行车十几个小时，18 日早晨到达德令哈。在德令哈市设立了地震现场工作指挥部，指挥开展地震灾害调查和损失评估、震情监测、科学考察、建筑物安全鉴定等工作。

九、结论与讨论

（1）此次地震发生在 2003 年度青海省地震重点注意区内。具有一定的中期地震预测效应。

（2）震前大区域地震活动图像及震区地震活动性参数扫描显示出的异常对未来发震地点有一定的指示意义。尤其是大面积的 M_L4 以上地震平静不仅反映出该区域应力处于闭锁状态，还对于未来主震震级有一定的提示作用。平建军等（2001）研究认为 M_L4 以上地震平静是华北地区强震前的一个重要震兆特征，中国大陆西部震例研究认为[8]，大面积 M_L4 以上地震平静后该空区及边缘发生 6.5 级以上的概率极高。在青藏高原地震活动相对活跃，而前兆观测台网相对稀疏的地区，地震活动图像变化规律的研究，可以做为强震的中期预测指标。

（3）震前虽记录到一些较清晰的前兆异常，但由于地震发生在监测台网稀疏的青藏高原东北隅，区内前兆观测台站少，布局不均匀。而且震中距 300km 范围内一些台站和测项集中在甘肃境内。尽管震前发现一些异常变化，但关注的地区很难和德令哈地区的地震相联系，更多的可能是祁连地震带或青海东北部区域。

青海东北部（震中距大于 400km）的一些台站，从 2003 年 1 月底起记录到一些突出的前兆短临变化[4)]：如平安地温（$\Delta=500$km）1 月 28 日 14 时、20 时，29 日 11 时，2 月 2 日 23 时出现单点突跳，最大幅度 0.3℃；3 月 4、5、7 日该测项再次记录到大幅度上升异常，整点值最大达 14.99℃。平安电磁波（$\Delta=500$km）EW 道同步出现异常，5 日 14 时达 13984 脉冲数，为异常以来的最大值；乐都气氡（$\Delta=520$km）在 2 月 16～18 日、28 日出现两次幅度较大的突跳，最大变化幅度 8.0Bq/L；同时，西宁水氡（$\Delta=439$km）2 月 18～20 日出现 4.5Bq/L 的上升。这几项前兆观测异常在以往青海东北部中强震例中多有较好的映震效果。如 1994 年 1 月 3 日、2 月 16 日共和 6.0、5.8 级地震[9]前，1995 年 12 月 18 日玛多 6.2 级地震[10]前，2000 年 9 月 12 日兴海 6.6 级地震[11]前，这些测项多出现类似异常。青海省地震局预报中心针对这些异常多次紧急会商，由于这些短临异常时间和地点较为集中，以前又有较好的震例对应，当时认为这些异常可能仍反映青海东北部地区以及南北地震带北段有中强震危险[4)]。由于受前兆台网的布局限制，震前很难将这些异常归于德令哈地区，震后总结认为这些突出的短临变化应该是德令哈 6.6 级地震的前兆短临异常。

（4）野外考察推测发震断裂与序列精定位结果以及哈佛大学给出的主震震源机制解均与大柴旦-宗务隆山-青海南山断裂带十分吻合。

（5）青海省台网建设在“九五”、“十五”、“十一五”期间有了很大提高，但距全国水平还相距甚远，全国地震台点密度平均为每万平方公里 0.88 个，青海省仅为每万平方公里 0.18 个。而且地震观测台网分布极不均衡，主要集中在青海东北部区，西部及南部区台网稀疏，甚至存在观测“盲区”。前兆观测台网的分布密度还很低，即使在地震重点监视防御区，监测系统仍不能很好地满足震情监测和科学研究的需求。此次地震总结，再一次反映出台网建设的重要性，尤其是对未来重点地震危险区，没有合理的台网布局，震前短临跟踪以及地震序列跟踪和地震趋势判断工作将难以有效开展。

参 考 文 献

[1] 青海省地质矿产局．中华人民共和国地质矿产部抵制专报——区域地质（第24号）：青海省区域地质志［M］．北京：地质出版社，1991：544

[2] 曾秋生．青海地震综合研究．北京：地震出版社，1999，103

[3] 孙长虹，钱荣毅，肖国林，孟小红．2003年青海德令哈地震序列的重新定位和发震构造．物探与化探，2006年第1期，79～82

[4] 陈玉华，马文静，崔国信．德令哈6.6级地震前地震活动及地震序列特征．高原地震，2003年第3期，1～8

[5] 王培玲，都昌庭，马文静．德令哈6.6级地震前青海省前兆异常特征分析．高原地震，2003年第3期，9～14

[6] 燕明芝，李晓峰，张昱，郑卫平，张苏平，张小美．德令哈6.6级地震前甘肃地区前兆异常的初步分析．高原地震，2003年第3期，15～26

[7] 燕明芝，白亚平，梅秀苹，姚军，李春燕．震（前）兆远场异常与德令哈6.6级地震临震特征分析．地震地磁观测与研究，2004年第3期，15～23

[8] 陈玉华，孙宏斌．中国大陆西部中等地震平静作为6.5级以上强震标志的研究．高原地震，2004，16（4），8～12

[9] 马文静，陈玉华．1994年1月3日和2月16日青海省共和6.0和5.8级地震，中国震例（1992～1994）．北京：地震出版社，282～306

[10] 陈玉华，张敏，马玉虎．1995年12月18日青海省玛多6.2级地震，中国震例（1995～1996）．北京：地震出版社，247～257

[11] 马玉虎，王培玲，李永强．2000年9月12日青海省兴海6.6级地震，中国震例（2000～2002）．北京：地震出版社，106～140

[12] 姜枚，许志琴，钱荣毅，王亚军，张立树．从德令哈地震分析青藏高原北缘东段的深部构造活动．中国地质，2006年第2期，268～273

[13] 夏玉胜，李文巧，张铁军．德令哈6.6级地震震害评估．高原地震，2003年第3期，45～51

[14] 夏玉胜，吴红岩，庞希华．“震害预测系统”在德令哈6.6级地震应急救灾决策中发挥功效．高原地震，2003年第3期，52～56

[15] 孙洪斌，张铁军，都昌庭．德令哈6.6级地震的地震地质灾害及发震构造分析．高原地震，2003年第3期，34～41

[16] 秦松涛，孙洪斌．德令哈6.6级地震前卫星红外长波辐射OLR的分析．高原地震，2003年第3期，43～51

参 考 资 料

1）2003年青海省地震趋势研究报告

2）青海省地震台网观测报告

3）德令哈6.6级地震综合科学考察报告

4）2003年周、月、临时会商报告

Delingha M_S 6.6 Earthquake on April 17, 2003 in Qinghai Province and 5 Times M_S 5-M_S 5.9 Strong Aftershocks from February to May in 2004

Abstract

An earthquake of M_S6.6 occurred in northwest of delingha in Qinghai Province, on April 17, 2003. Its Macroscopic epicenter was located in Delingha xichayi ditch upstream, the epicenter intensity was Ⅷ in the meizoseismal area. No one was injured or killed, the direct economic loss was 43.45 million yuan RMB.

The earthquake sequence was of main shock-aftershock type, The largest aftershock was M_S5.9, the aftershock distribution on the west of main shock, the long axis towards consistent in the meizoseismal area, Section II was main rupture surface, the azimuth of P axis of the principal compressive stress was in NNE direction , the seismogenic structure was Dachaidan-Zongwulong mountain-Qinghai Nanshan mountain fault.

Within the distance of 300km from the epicenter of the earthquake there were 7 seismic observation stations, 3 of them were seismometric stations, 5 of them were comprehensive monitoring stations. Within 7 precursory observation items, before the earthquake there were 3 anomalies items including 8 anomalies, 6 medium term anomalies and 2 short term anomalies, seismological anomaly obvious, other major were water radon, geoelectric anomaly.

Delingha M_S6.6 earthquake occurred in the Qinghai M_S5-M_S6 earthquake focus area in 2003.

2003 年 7 月 21 日、10 月 16 日云南省大姚 6.2、6.1 级地震

云南省地震局

李永莉　王世芹　毛慧玲　毛　燕　付　虹

摘　要

2003 年 7 月 21 日、10 月 16 日云南省大姚先后发生 6.2、6.1 级地震，宏观震中分别位于大姚县昙华、石羊，六苴镇海古簸、外期地一带，极震区烈度Ⅷ度，呈北西向椭圆形，两次地震造成 19 人死亡，重伤 87 人、轻伤 557 人，经济损失约合 100750 万元。

6.2、6.1 级地震序列为震群型，6.2 级地震最大余震为 M_L4.9，6.1 级地震最大余震为 M_S4.7，序列余震分布的优势方向为北西向。据两次地震的震源机制解，地震的发震构造为北西向隐伏断裂。节面Ⅰ为 6.2 级地震主破裂面，节面Ⅱ为 6.1 级地震主破裂面，主压应力 P 轴方位为 NNW 向，综合推测 6.2、6.1 级地震均为北西西向右旋走滑错动的结果。

震中附近地区 300km 内共有地震台 71 个均有测震观测项目，其中测震观测台 9 个，测震和前兆综合观测台 52 个，定点前兆观测台 62 个。地震前共有 19 项异常项，63 条前兆异常，其中地震学 4 项异常项目 8 条异常，定点前兆 11 项异常项目 46 条异常，中期异常 30 条，短临异常 24 条。此外，震前 9 条宏观异常。6.2 级地震前 3、4 级地震在震源区和近场区平静是中期阶段最显著的地震事件，短临期阶段震源区及附近地区的小地震活跃为地震时间、地点提供了信息；云南省内 3 级地震平静 18 天是 6.1 级地震前短期阶段的最显著地震事件，6.2 级地震序列于 8、9 月的增频为 6.1 级地震的临震标志。震中周围 100km 内前兆异常台站和台项比为 70% 和 36%，101 ~ 200km 为 74% 和 32%，201 ~ 300km 为 24% 和 8%。文中最后从预报的角度对地震学和前兆异常的协调性在强震孕育不同阶段的特征与作用等问题作了讨论。

云南省地震局对大姚 6.2 级地震作了较好的长期、中短期预测，6.1 级短临阶段预报中心作出了较准确的预测，于 2003 年 8 月 22 日以震情反映（200305）上报中国地震局和云南省人民政府，提出了明确的短期预测预报意见，取得了一定的减灾实效。

前　言

2003 年 7 月 21 日 23 时 16 分 30.1 秒云南大姚发生了 6.2 级地震，据云南测震台网测定，微观震中为 25°57′N，101°14′E，震源深度为 6km；宏观震中位于大姚县昙华、石羊一带，极震区烈度Ⅷ度。2003 年 10 月 16 日 20 时 28 分 03.9 秒云南大姚再次发生 6.1 级地震，据云南测震台网测定，微观震中为 25.55°N，101.18°E，震源深度为 5km；宏观震中位于大姚县六苴镇海古簸、外期地一带，极震区烈度Ⅷ度，两次地震共造成 19 人死亡，重伤 87 人、轻伤 557 人。直接经济损失约 100750 万元[1,2)]。

震区地处川滇菱块中南部，是云南主要强震活动区，在此次地震震中附近曾发生过 2000 年姚安 6.5 级地震。震中附近 300km 范围内共有地震台 71 个，震前出现了 15 项异常项目 45 条前兆异常，有 9 条宏观异常。

云南省地震局对大姚两次地震作了较准确的长期和中期预报，对大姚 6.2 级地震的预测中期较成功，短临有察觉，在震前一个月向中国地震局和云南省人民政府，提出了明确的滇西北至滇中地区存在 6 级地震短期预测意见，为此云南省人民政府对省地震局通报表彰。

这次研究报告是在对大姚 6.2、6.1 级地震已有的总结基础上，按《震例研究和报告编写规范》要求又重新作了一次清理研究而完成的，有些结果和认识与原来的工作稍有出入，报告中给出的结果和认识可能更多代表了作者的观点和认识，力求全面和客观。

一、测震台网及基本参数

1. 区域台网

大姚 6.2、6.1 级地震震中（以 6.2 级地震震中为主）周围 300km 范围内有 41 个区域测震台（包括数字记录和模拟记录），其中 100km 范围内有 6 个，101～200km 内有 17 个，201～300km 内有 18 个（图 1）。云南省测震台网对这次地震发震地区的 M_L2.0 以上的地震基本能控制，本次地震的基本参数如表 1 所示。此次地震基本参数取云南台网的结果，但震级则按要求取中国台网中心的结果。

2. 数字流动台网

大姚 6.2 级地震后，于 7 月 22 日在地震现场围绕地震震中，布设了昙华乡电信所、石羊镇、达么、新街、马茨、六苴与三台等 7 个流动数字化地震记录台，构成了近震源临时监测台网。自 2003 年 7 月 22 日 08 时 56 分至 8 月 21 日 13 时 40 分共记录余震 2173 次，流动台站分布位置如图 2 所示。

大姚 6.1 级地震后，于 10 月 17 日在地震现场围绕地震震中，布设了大姚县城、新街、六苴、昙华与马茨等 5 个流动数字化地震记录台，构成了近震源临时监测台网。自 2003 年 10 月 17 日 05 时 13 分至 11 月 17 日 15 时 00 分共记录余震 3793 次，流动台站分布位置如图 2 所示。

表 1　地震基本参数

Table 1　Basic parameters of the earthquake

编号	发震日期	发震时刻	震中位置		震级				震源深度 (km)	震中地名	结果来源
	年 月 日	时 分 秒	φ_N	λ_E	$M(M_S)$	M_L	M_b	M_W			
1	2003 07 21	23 16 30. 1	25°57′	101°14′	6. 1				6	大姚	云南台网
		23 16 32. 3	26. 0°	101. 2°	6. 2				6		中国台网
		23 16 31. 93	25. 975°	101. 290°				6. 0	14		美国 USGS
		23 16 31. 0	25. 975°	101. 290°	6. 0		5. 4	5. 9	10		美国 NEIC
		23 16 38. 3	25. 98°	101. 32°	6. 0		5. 4	6. 0	15		美国哈佛大学
2	2003 10 16	20 28 03. 9	25. 55°	101. 18°	5. 9				5	大姚	云南台网
		20 28 04. 5	26. 0°	101. 3°	6. 1				5		中国台网
		20 28 09. 07	25. 954°	101. 254°				5. 5	25		美国 USGS
		20 28 09. 0	25. 954°	101. 254°	5. 6		5. 2	5. 6	33		美国 NEIC
		20 28 11. 5	25. 89°	101. 52°	5. 6		5. 2	5. 6	33		美国哈佛大学

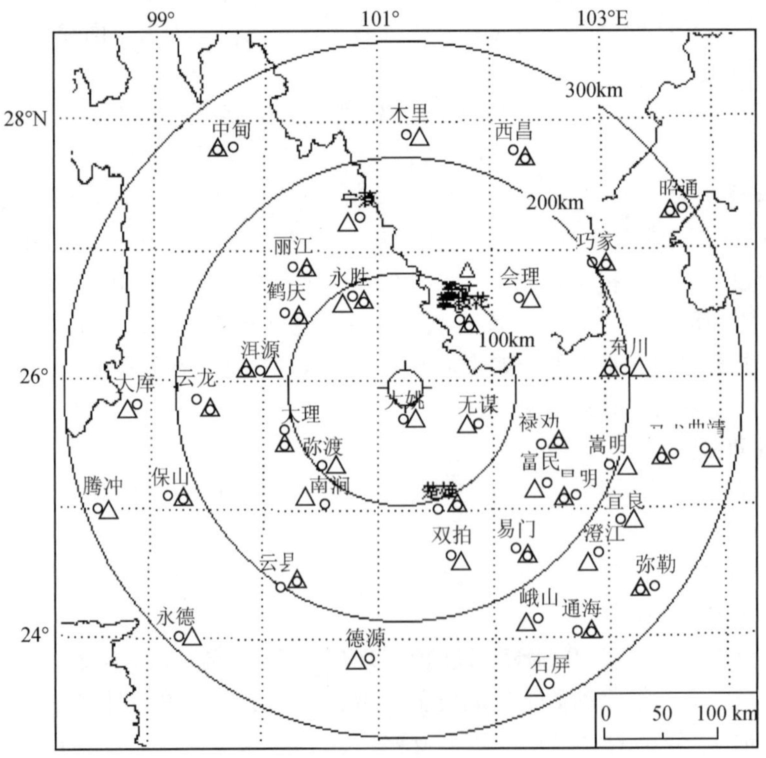

图 1　大姚 6. 2、6. 1 级地震震中附近测震台站分布图

Fig. 1　Distribution of seismometric stations around the epicentral area of the M_S6. 2, 6. 1 Dayao earthquakes

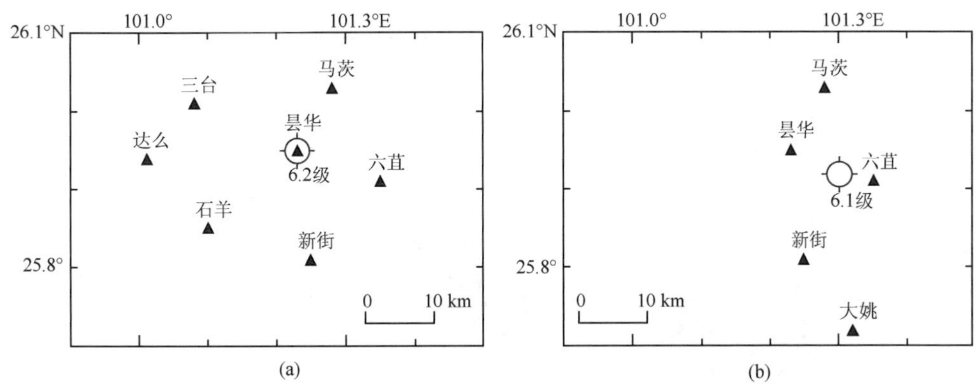

图 2　大姚 6.2、6.1 级地震流动数字地震台站分布图

Fig. 2　Distribution of mobile digital seismic stations around the epicentral area of the M_S6.2, 6.1 Dayao earthquakes

（a）6.2 级地震数字台站；（b）6.1 级地震数字台站

二、地震地质背景

震区地处川滇菱形块体中南部，属扬子准地台川滇台背斜内的滇中中台陷。震区位于由程海断裂、红河断裂和元谋断裂围限而成的滇中块体内部，区域内主要分布有一系列近南北向的走滑活动断裂带，南部发育以楚雄—南华活动断裂带为代表的一组北西向断裂及褶皱构造，但极震区地表没发现明显的断裂构造，仅见一系列白垩系红层构成的宽缓褶皱[3)]。

在震区及其附近发育有三组不同走向的断裂构造，按其发育的成熟度，依次是北西向，近南北向和北东向，其中北东向断裂是在地表未造成两盘地层缺失的一组小规模断裂。由航片解释发现，其走向从大姚盆地东南边缘起向南西方向经由七街至姚安盆地的北西边缘。震区构造变形较简单，北西向褶皱紧密线性排列，构成“S”形转折。褶皱构造多以北西 300°~330°方向延伸，向北撒开之势，褶皱为不对称的背、向斜组成，向斜东缓西陡，岩层倾角一般为 40°~60°。

震区历史上中强地震较为活跃，据该区最早地震记录公元 1488 年至今统计，震区及其附近地区共发生 $M_S \geq 5$ 级地震 15 次（扣除余震），其中 $M_S \geq 6$ 级地震 2 次。最大地震为 6.5 级。

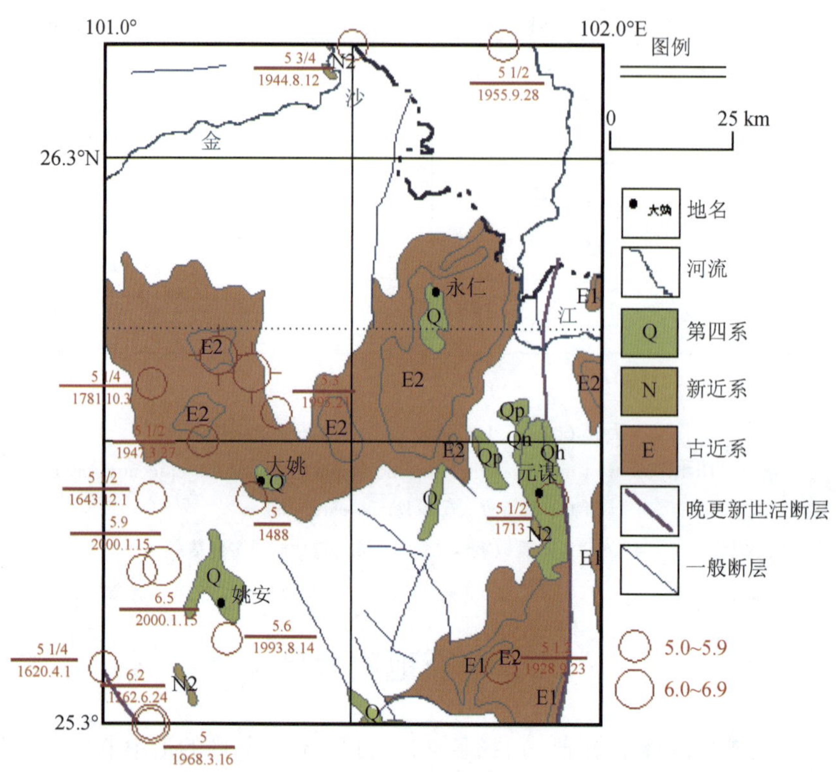

图 3　大姚 6.2、6.1 级地震震区地质构造及历史地震分布图

Fig 3　Map of geological structure and distribution of historical earthquakes around the M_S6.2，6.1 Dayao earthquakes

表 2　大姚 6.2、6.1 级地震震中附近历史地震目录（$M_S \geqslant 5.0$ 级）

Table 2　Catalogue of historical earthquakes around the M_S6.2，6.1 Dayao epicentral area（$M_S \geqslant 5.0$）

编号	发震日期	震中位置（°）		震级 M_S	震中地名	结果来源
	年 月 日	φ_N	λ_E			
1	1488	25.7	101.3	5.0	大姚	文献［2］
2	1620 04	25.4	101.0	$5\frac{1}{4}$	南华西北	文献［2］
3	1643 12	25.7	101.1	$5\frac{1}{2}$	姚安、盐丰间	文献［2］
4	1713	25.7	101.9	$5\frac{1}{2}$	元谋	文献［2］
5	1781 10 03	25.9	101.1	$5\frac{1}{4}$	大姚、盐丰	文献［2］
6	1928 09 23	25.4	101.8	$5\frac{1}{2}$	禄丰黑井	文献［2］

续表

编号	发震日期	震中位置（°）		震级 M_S	震中地名	结果来源
	年 月 日	φ_N	λ_E			
7	1944 08 12	26.5	101.5	$5\frac{3}{4}$	华坪东南	文献［2］
8	1947 03 27	25.8	101.4	$5\frac{1}{2}$	大姚东北	文献［2］
9	1955 09 28	26.5	101.5	$5\frac{1}{2}$	永仁	文献［2］
10	1962 06 24	25.3	101.0	6.2	南华普朋	文献［2］
11	1968 03 16	25.3	101.1	5	南华西北	文献［2］
12	1993 02 01	25.8	101.3	5.3	大姚	资料 5）
13	1993 08 14	25.4	101.1	5.6	姚安	资料 5）
14	2000 01 15	25.57	101.08	5.9	姚安	资料 5）
15	2000 01 15	25.58	101.12	6.5	姚安	资料 5）

三、地震影响场和震害

1. 烈度分布和震害

据震区的考察资料[1)]，大姚 6.2 级地震宏观震中位于大姚县昙华、石羊一带，极震区烈度Ⅷ度，等震线呈北西向椭圆形分布。

Ⅷ度区分布在大姚县境内，北起昙华乡莱西拉，南至石羊镇郭家，东自昙华乡平风山，西到三台乡干河，面积约 185km^2。

Ⅶ度区分布在大姚县境内，北起过拉地，南至新街，东自六苴，西到阿白米，面积约 692km^2。

Ⅵ度区范围：北起大姚县三台乡施拉，南达大姚县金碧镇七街，西起宾川县平川，东至大姚县赵家店、江头，面积约 2292km^2。

根据调查震害特征为震害分布范围广、地震地质灾害突出：各类房屋建筑遭受不同程度的破坏，极灾区内各类房屋毁坏及严重破坏比率高，土木结构 37.87%，砖木结构 20.84%，多层砌体 13.52%，并有Ⅳ度破坏现象。震区大部为山区，地形起伏且正值雨季，地震诱发山体滑坡、崩塌、地裂在极灾区分布广泛。这次地震使大姚、宾川、永仁等县的生命线工程、水利设施遭受不同程度的损失。

此次地震主要涉及楚雄州的大姚、永仁和大理州的宾川等县，包括 19 个乡镇、159 个行政村，灾区面积约 3169km^2（Ⅵ度区以上区域）；受灾人口 322962 人，涉及 80880 户。地震造成 16 人死亡，重伤 72 人、轻伤 515 人。直接经济损失约 59190 万元[1)]。

据震区的考察资料[2)]，大姚 6.1 级地震宏观震中位于大姚县六苴镇海古簸、外期地一带，极震区烈度Ⅷ度，等震线呈北西西向分布。

Ⅷ度区分布在大姚县境内，北起昙华乡迷窝簸，南至新街乡碧么、六苴镇大坡阱，东自六苴，西到昙华乡板房，面积约 83km²。

Ⅶ度区北近大姚县桂花乡马茨，南至大姚县新街，东起永仁县宜就乡外普拉，西到大姚县石羊镇永丰，面积约 561km²。

Ⅵ度区范围：北近大姚县桂花乡大村，南至大姚县仓街镇，东起元谋县虎溪，西至宾川县东升，面积约 2715km²。

根据调查震害特征为新旧震害叠加，震害分布范围广。在本次地震极震区，海古簸村北出现延续约 2000m 长的地表裂缝，走向 280°～290°。海古簸、外期地一带出现边坡崩塌现象。这次地震使大姚、永仁、元谋等县的生命线工程、水利设施遭受不同程度的破坏。

此次地震灾区主要涉及楚雄州的大姚、永仁和元谋县，包括 21 个乡镇，161 个行政村，灾区面积约 3359km²；受灾人口 283169 人，涉及 68442 户。地震造成大姚县 3 人死亡，重伤 15 人、轻伤 42 人。经济损失约 41560 万元[2)]。

大姚 6.2、6.1 级地震的分度震害特征如下：

Ⅷ度区的主要震害特征为：土木及砖木结构个别房屋倒毁，部分墙体大部倒塌，大多数墙体开裂或局部倒塌，梭瓦、掉瓦现象普遍。砖混结构房屋部分墙体开裂严重，多数墙体开裂明显。

Ⅶ度区的主要震害特征为：土木及砖木结构房屋个别墙倒架歪，部分墙体局部倒塌，多数墙体开裂、梭瓦。砖混结构房屋个别墙体开裂严重，少数墙体普遍开裂，部分墙体开裂明显。框架结构个别承重构件轻微裂缝，部分填充墙开裂、抹灰层脱落。

Ⅵ度区的主要震害特征为：土木和砖木结构房屋除个别年久失修者倒塌外，主要以轻微破坏和基本完好为主，震害现象主要为墙体开裂，少量梭瓦。砖混结构房屋，个别墙体出现贯通裂缝，少数墙体出现显见裂纹。框架结构房屋个别承重梁可见细微裂纹，部分填充墙开裂。

大姚 6.2、6.1 级两次地震共造成 19 人死亡，重伤 87 人、轻伤 557 人，直接经济损失约达 100750 万元。

2. 强震动观测和地震前后增设的流动观测[8]

2003 年 7 月 21 日 23 时 16 分云南大姚昙华乡发生 M_S6.2 地震，架设在昆明 38 层楼和 7 层隔震结构建筑以及宾川地震局办公室地面的数字强震动仪都记录了这次地震。地震发生后，云南省地震局立即派出强震动观测应急组人员携两台 SMA.1、两台数字强震动仪及中国地震局增派的三台强震动仪器一起奔赴震区，开展强震动流动观测工作，当晚在震中和极震区附近的大姚县城、昙华、碧么、石羊、纳家庄、三台 6 个点布设 7 台强震仪。23 日为了尽可能不丢失随时可能发生的大余震，临时选择了小井河点架设仪器。在石羊和昙华同时架设了 SMA.1 和数字强震动仪，实际有 5 个观测点，完成台站布设。2003 年 8 月 5 日，震区再次发生 M_S4.6 强余震，强震动观测组人员于 8 月 8 日进入震区，检查并回收获取的强震动记录。截至 8 月 10 日止，布设于震区的 7 台强震动仪记录到了本次地震的 M_S4.5、M_S4.6 两次强烈余震及多次余震。

2003 年 10 月 16 日大姚县 M_S6.1 地震后，云南省地震局立即派出由 3 位观测人员携带两台 GDQJ 数字强震仪到震区开展强震动观测工作，于次日完成完成架设大姚法院、昙华乡中学、新街乡政府、六苴乡政府以及碧么村小学 5 个观测点、6 台仪器的工作。强震观测截至 2003 年 12 月 6 日，共获取强震记录 13 次 39 条记录，记录到最大余震 3.7 级 2 次。其中，由于现场应对正确，捕获了 2003 年 11 月 26 日 21 时 38 分 54.7 秒发生的 5.0 级主震记录，最大加速度记录为 210Gal。

四、地震序列

大姚 6.2 级、6.1 级地震前没有观测到直接前震，地震序列至 2004 年 4 月 30 日止，共记录到 2664 次地震，其中 1.0～1.9 级 1763 次，2.0～2.9 级 777 次，3.0～3.9 级 105 次，4.0～4.9 级 17 次。6.2、6.1 级地震的 M_L≥4 级地震序列目录见表 5。

表 3　大姚 6.2、6.1 级地震序列目录（M_L≥4.0 级）

Table 3　Catalogue of the M_S6.2, 6.1 Dayao earthquake sequence（M_L≥4.0）

编号	发震日期	发震时刻	震中位置（°）		震级		震源深度	震中地名	结果来源
	年 月 日	时 分 秒	φ_N	λ_E	M_L	M_S	(km)		
1	2003 7 21	23 16 30	25.95	101.23		6.2	6	大姚	资料 1）
2	2003 7 22	08 09 38	25.93	101.20	4.3		4	大姚	资料 1）
3	2003 7 23	12 42 30	26.00	101.05	4.3		8	大姚	资料 1）
4	2003 7 24	07 52 41	25.93	101.27	4.7		10	大姚	资料 1）
5	2003 8 5	20 49 05	25.93	101.23	4.9	4.6	0	大姚	资料 1）
6	2003 8 12	23 59 41	26.00	101.03		4.5	9	大姚	资料 1）
7	2003 8 18	18 00 17	25.92	101.22	4.8	4.7	8	大姚	资料 1）
8	2003 9 27	09 06 30	25.93	101.22	4.2		10	大姚	资料 1）
9	2003 10 16	20 28 04	25.92	101.3		6.1	5	大姚	资料 1）
10	2003 10 16	21 07 35	25.95	101.27	4.0		0	大姚	资料 1）
11	2003 10 17	06 41 33	25.97	101.27		4.7	7	大姚	资料 1）
12	2003 10 20	05 01 12	25.90	101.17	4.0		0	大姚	资料 1）
13	2003 10 27	10 16 53	25.95	101.32	4.1		0	大姚	资料 1）
14	2003 11 01	02 09 22	25.93	101.22		4.7	3	大姚	资料 1）
15	2004 1 23	15 52 54	25.95	101.30	4.0		10	大姚	资料 1）
16	2004 3 24	23 58 31	25.92	101.30	4.4	4.8	12	大姚	资料 1）
17	2004 3 25	00 39 26	25.90	101.28	4.2	4.5	10	大姚	资料 1）
18	2004 3 26	05 14 55	25.90	101.27	4.6	4.9	12	大姚	资料 1）
19	2004 4 16	18 38 09	25.92	101.32	4.3	4.0	10	大姚	资料 1）

根据云南台网资料，大姚6.2、6.1级地震序列具有以下特征：

1. 地震频度起伏大，余震强度低

姚安6.2级地震序列无前震，地震后余震没有5级，最大余震只有M_L4.9。从序列M-t图可见，24小时内最大余震为7月22日08时09分38秒发生的M_L4.3地震，同时7月21至23日4级余震出现增频，之后7月26至8月5日4级地震出现平静同时3级地震增频，8月5日发生M_S4.6余震，之后3级地震无明显的衰减，接着8月18日发生了最大余震M_L4.7，8月19日至8月25日3级地震无明显的起伏，从9月1日至9月27日出现3、4级地震的缓慢增频（震后40 ~50天左右），之后出现长达6天的2级以上地震平静；10月16日发生显著的M_L3.9地震（与大姚6.1级地震相距2km），时间间隔仅9小时后发生第二主震——大姚6.1级地震，之后从10月16日M_S4.7余震后至10月20日3、4级地震快速衰减（图4）。

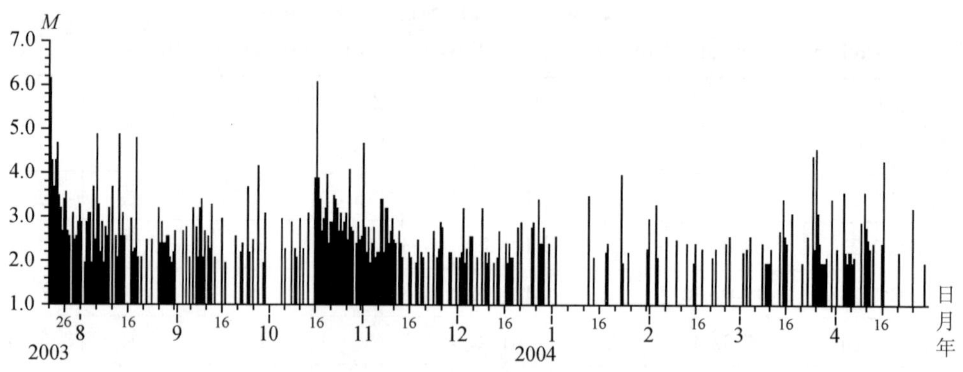

图4 大姚6.2、6.1级地震序列M-t图

Fig. 4 M-t diagram of the M_S6.2，6.1 Dayao earthquake sequence

6.2级地震后序列出现3、4级地震显著的密集—平静—增频现象，特别是9月27日的4.2级地震标志了序列活动的增强。由于6.2级与6.1级地震活动的重叠，从M-t图上无法分辨大姚6.2级地震的余震与6.1级地震的前震。

2. b值、h值、p值和应变释放

大姚6.2、6.1级地震序列云南台网记录到的序列完整震级下限震级是2.0，取2级以上的地震计算系列参数。大姚6.2、6.1级地震释放的总能量为8.80×10^{13}J，其中6.2地震能量为4.91×10^{13}J约占整个序列总能量的55.7%，6.1级地震能量为3.31×10^{13}J。应变释放曲线显示6.1级地震后余震迅速衰减（图5）。分别统计6.2、6.1级地震序列的能量比为：6.2地震约占6.2地震序列总能量的96%，6.1地震约占6.1地震序列总能量的96.5%。

6.2、6.1级序列的b值为0.81接近震区周围的背景b值0.8。6.2级地震后序列的日频度衰减基本正常：h值为1.1大于1，b值为0.82，p值为0.80。显示了该序列的地震类型趋于主震—余震型。主震后50天（即9月8日）左右出现明显的上翘，偏离了余震序列的衰减，出现了增频。

6.1级地震后序列的日频度相对于6.2级地震衰减较慢，序列的b值为0.94高于震区周围的背景b值0.8，h值为1，p值为0.55，序列衰减基本正常（图6至图9）。

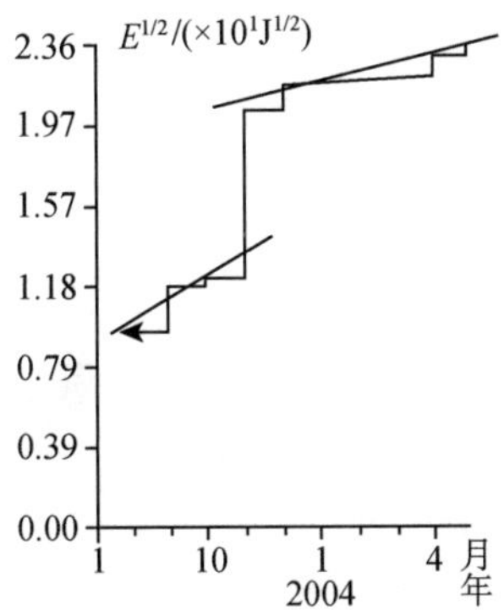

图5 大姚6.2、6.1级地震序列应变释放曲线

Fig. 5 Strain release of the M_S6.2, 6.1 Dayao earthquake sequence

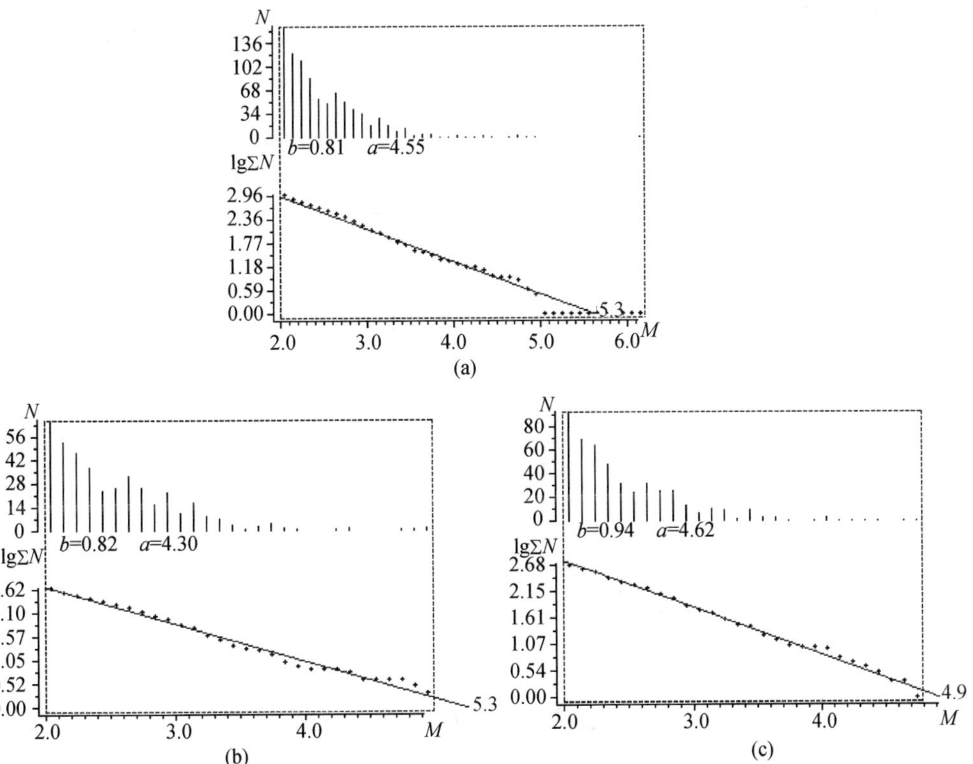

图6 大姚6.2、6.1级地震序列b值曲线

Fig. 6 b-value curve of the M_S6.2, 6.1 Dayao earthquake sequence

(a) 大姚6.2、6.1级地震；(b) 大姚6.2级地震；(c) 大姚6.1级地震

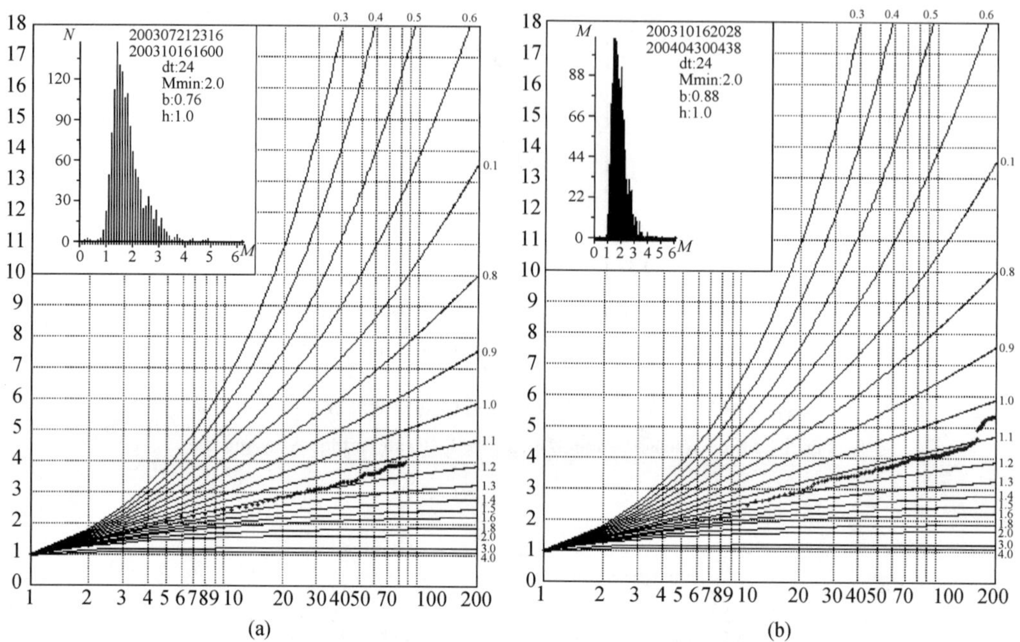

图 7　大姚 6.2、6.1 级地震序列 h 值曲线

Fig. 7　h-value curve of the M_S6.2，6.1 Dayao earthquake sequence

（a）大姚 6.2 级地震；（b）大姚 6.1 级地震

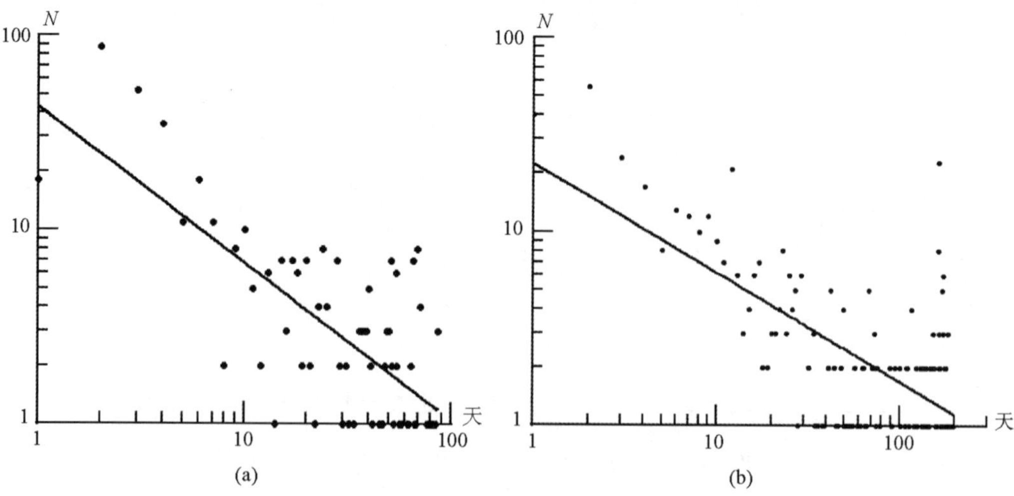

图 8　大姚 6.2、6.1 级地震序列 p 值曲线

Fig. 8　p-value curve of the M_S6.2，6.1 Dayao earthquake sequence

（a）大姚 6.2 级地震；（b）大姚 6.1 级地震

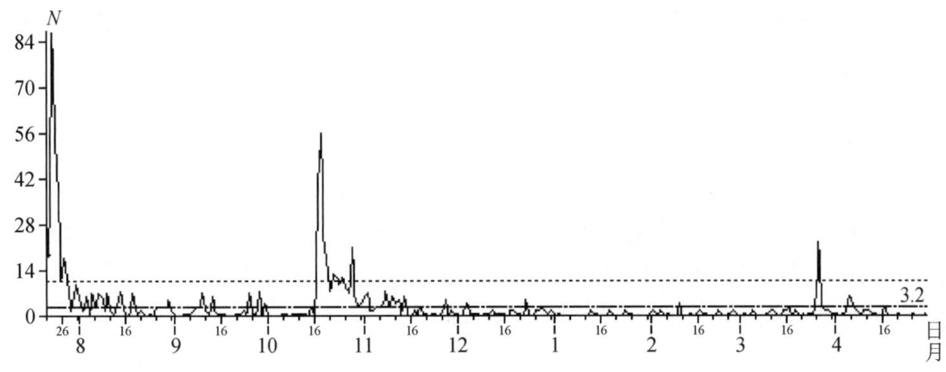

图9　大姚6.2、6.1级地震序列日频度曲线

Fig. 9　Daily N-value curve of the M_S6.2, 6.1 Dayao earthquake sequence

3. 累积频度 N

累积频度 N 可反映频度 N 随时间的变化的变化关系。N-t 曲线若为幂函数关系，则序列可能为主余震型；若为指数函数关系，则序列可能为震群型；若为线性函数关系，则序列可能为主余震型或少数为震群型。

以序列4级地震为统计下限，6.2级地震后3天内的 N-t 曲线呈指数形式，依此判定该序列趋向于震群型（图10）。

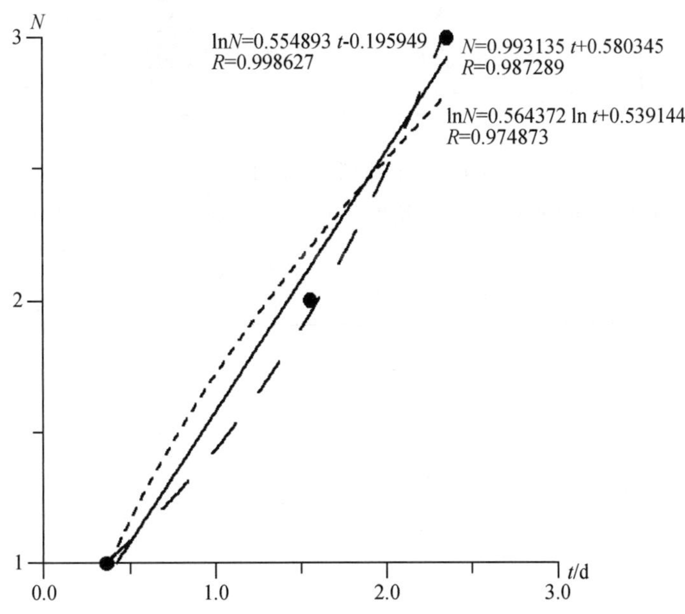

图10　大姚6.2级地震序列4级地震日频度曲线

Fig. 10　Daily N-value curve of the M_S6.2 Dayao earthquake sequence with $M\geqslant4.0$

4. 余震分布

如图 11 整个序列分布为北西向，长轴长约为 42km，6.1 级地震位于 6.2 级地震的东南，相距约 8km。全序列的余震主要分布于 6.1 级地震的西北方，6.1 级地震的余震主要分布于两次主震之间，有对第一主震余震震中填空的现象。

图 12 为大姚 6.2、6.1 级地震序列的余震震源深度剖面随时间分布图，可见大姚 6.2 级地震与 6.1 级地震的余震深度，除 7～9 月份部分余震深度为 20～30km 外，大多数分布于 0～16km，大姚 6.1 级后地震的余震深度则集中于 5～10km 左右，呈现 6.1 级地震后的余震分布浅部的现象。图 15 为 6.1 级地震后的余震分布，相对于 6.2 级地震较为集中。图 16 为全序列 $M \geqslant 4$ 级地震震中分布图，其优势分布方向为北西向。

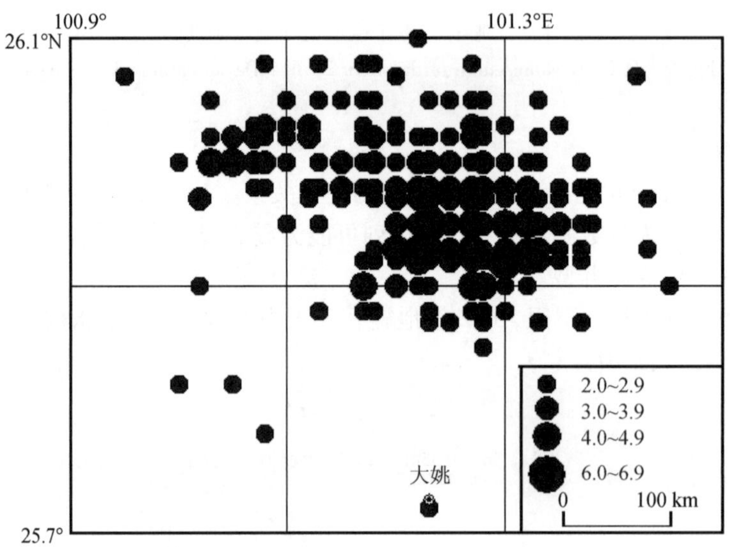

图 11　大姚 6.2、6.1 级地震序列余震震中分布图（$M_L \geqslant 2.0$ 级）

Fig. 11　Distribution of the M_S6.2, 6.1 Dayao sequence ($M_L \geqslant 2.0$)

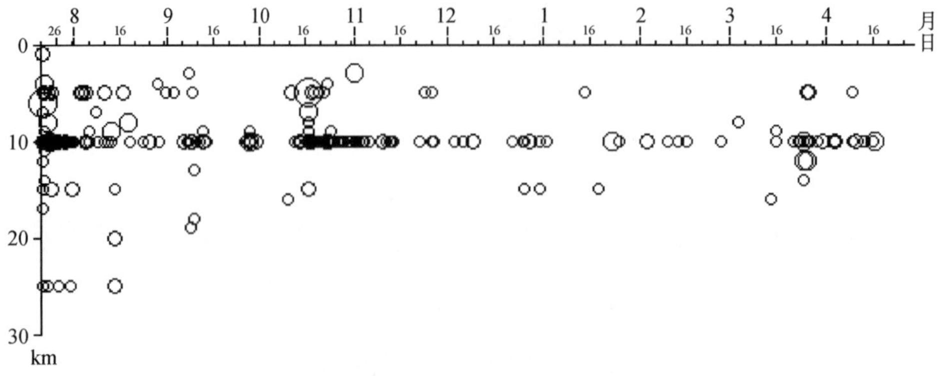

图 12　余震深度分布图

Fig. 12　Depths distribution of the sequence

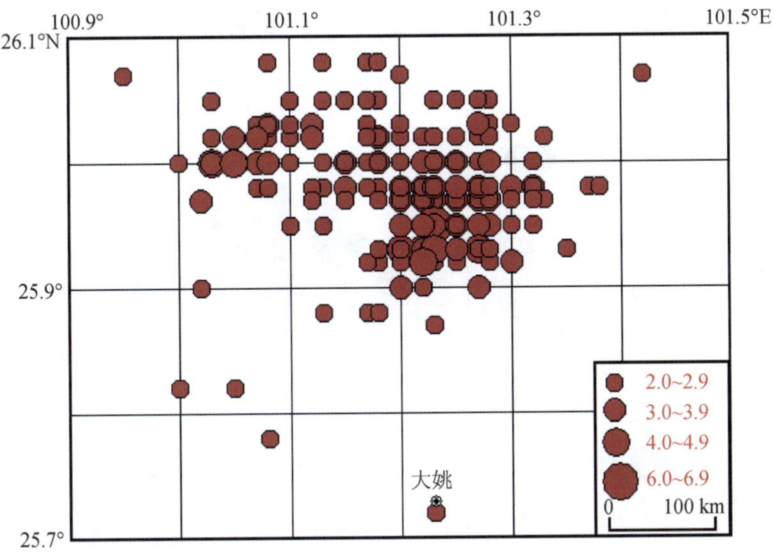

图 13　大姚 6.2 级地震震中分布

Fig. 13　Epicentral distribution of the M_S6.2 Dayao earthquake

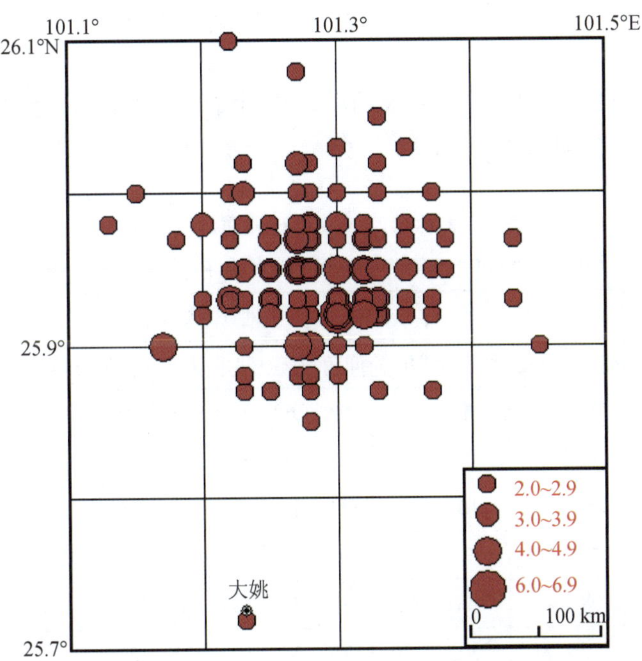

图 14　大姚 6.1 级地震震中分布

Fig. 14　Epicentral distribution of the M_S6.1 Dayao earthquake

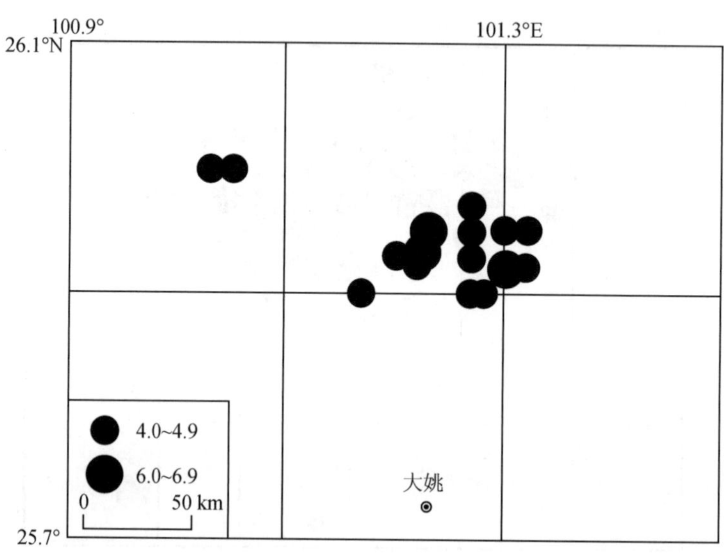

图 15　全序列 $M \geqslant 4$ 级地震分布

Fig. 15　Distribution of Dayao earthquake sequence with $M \geqslant 4.0$

五、震源参数和地震破裂面

根据云南地震遥测台网记录，用 P 波初动符号图解断层面法求得大姚 6.2、6.1 级地震的震源机制解见表 4 和图 16。

表 4　大姚 6.2、6.1 级地震震源机制解

Table 4　Focal mechanism solutions of the M_S6.2，6.1 Dayao earthquakes

序号	节面 I			节面 Ⅱ			P 轴		T 轴		N 轴		X 轴		Y 轴		结果来源
	走向	倾角	滑动角	走向	倾角	滑动角	方位	仰角	方位	仰角	方位	仰角	方位	仰角	方位	仰角	
1.1	208	86	175	38	85	4	1	353	6	263	83	95					王绍晋
	200	80	−6	291	84	−170	11	155	3	65	78	321					HRV
	109	80	175	200	85	10	4	334	11	65	78	226					USGS
2.1	17	88	10	287	80	178	6	151	8	242	80	26					王绍晋
	18	88	−1	108	89	−178	2	333	88	135	1	243					HRV
	97	60	178	189	88	30	20	319	60	192	22	57					USGS

根据震源机制解结果（王绍晋），6.2 级（1.1）地震主压应力 P 轴方位 1°，节面 I（断层面）为北西向。6.1 级（2.1）地震主压应力 P 轴方位 6°，节面 Ⅱ（断层面）为北西西

表5　地震矩张量解

Table 5　Moment tensor solution of the earthquake

编号	节面Ⅰ			节面Ⅱ			矩张量（×10^{18}）						地震矩 M_0（N·m）	矩震级 M_W	结果来源
	走向	倾角	滑动角	走向	倾角	滑动角	M_{xx}	M_{yy}	M_{zz}	M_{xy}	M_{yz}	M_{zx}			
1.1	109	80	175	200	85	10	0.16	0.72	0.56	0	0.79	0.18	10e+18	6.0	USGS
	200	80	-6	291	84	-170	-0.013	-0.537	0.55	0.175	-0.64	0.55	9.2e+17	5.9	HRV
2.1	97	60	178	189	88	30							2.1e+17	5.5	USGS
	18	88	-1	108	89	-178	-0.383	-1.53	19.1	-0.102	-2.31	-0.001	2.9e+17	5.6	HRV

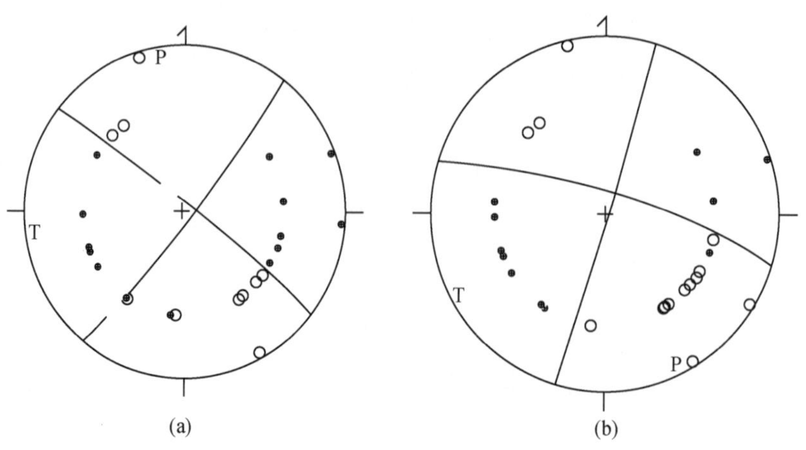

图16　大姚6.2、6.1级地震震源机制解

Fig. 16　Focal mechanism solution of the M_S6.2, 6.1 Dayao earthquake

（a）大姚6.2级地震；（b）大姚6.1级地震

向。这两次地震的等震线极震区长轴方向与余震分布为北西向，与震区褶皱紧密线性北西向排列的构造方向基本吻合，为此，认为6.2、6.1级地震是在近南北向主压应力作用下，北西向线性构造发生右旋走滑错动的结果，推测北西向线性构造可能就是大姚6.2、6.1级地震的发震构造。

六、地震前兆观测台网及前兆异常

大姚6.2、6.1级地震震中300km范围内共有71个地震台站，其中测震台站41个，其他前兆观测台站62个（表6），测震、水氡、水位、水温、水质、地电、形变、地磁、跨断层短水准、短基线等25个观测项目。其中0～100km、101～200km和201～300km分别有地震台10、25和36个，测震观测台6、17和18个，前兆台10、19和33个，测震学以外的观测项目16、16和23个，台项28、57和110个。这些前兆定点台站，震前都有连续观测的资料。图1和图17分别是距震中300km范围内的测震台站和定点前兆观测台站分布图。

此次地震前共出现15个项目，54条异常（表7）。其中4个项目8条地震学异常，11个项目46条前兆观测异常（图17），9条宏观异常。各类异常的具体情况详见表7和图18至图58。

表6　大姚6.2、6.1级地震前定点前兆观测项目登记表

Table6　Summary table of precursory monitoring items on the fixed observation points before the M_S6.2, 6.1 Dayao earthquake

编号	前兆观测台站	观测项目	编号	前兆观测台站	观测项目
1	大姚	水位、水温	2	元谋	视电祖率、自然电位
3	宾川	水位	4	华坪	水氡

续表

编号	前兆观测台站	观测项目	编号	前兆观测台站	观测项目
5	攀枝花	水氡、水位、重力、形变	6	南华	水温、水位
7	攀矿	倾斜	8	永胜	水位、地磁、形变、短水准、短基线、伸缩
9	弥渡	倾斜、水温、水氡、水汞、二氧化碳、伸缩	10	楚雄	水位、短水准、短基线、倾斜、地磁、伸缩
11	大理	水氡、水位、水汞、流量、二氧化碳、短水准、短基线	12	南涧	水氡
13	罗茨	水位、水氡	14	武定	水温、水位
15	鹤庆	水氡、二氧化碳	16	洱源	水氡、水位、水温、水汞、二氧化碳
17	丽江	水温、水位、水氡、地磁、倾斜、短水准、短基线	18	剑川	水位、短水准、短基线
19	盐源	水氡	20	易门	水位、水氡、水温
21	景东	水位、水氡	22	昆明	倾斜、水位
23	云龙	倾斜	24	小哨	水位
25	东川	倾斜、水氡	26	巧家	地磁、水氡、碳酸氢根离子、镁离子、氟离子
27	会泽	水位、水氡	28	云县	倾斜、伸缩
29	昌宁	水氡、流量、电导率、pH、CL^-	30	寻甸	水氡、短水准、短基线
31	玉溪	水位、水氡、水温、水汞	32	宜良	短水准、短基线、倾斜
33	澄江	水温、水位	34	保山	倾斜、水位、水氡、水温、水汞、pH、钙粒子、镁离子、氟离子、硫酸根离子、碳酸氢根离子
35	峨山	水温、短水准、短基线	36	西昌	倾斜、水位、水氡、应力、重力
37	镇源	水温、水位	38	怒江	水温、水氡、水位
39	江川	水位、水温	40	六库	水位、水氡、水温
41	施甸	水氡、水位、钙离子、镁离子、碳酸氢根离子	42	通海	地磁、倾斜、短水准、短基线

续表

编号	前兆观测台站	观测项目	编号	前兆观测台站	观测项目
43	临沧	水位、水氡、水温、氟离子	44	中甸	水位、水温
45	华宁	水温	46	曲靖	水位
47	高大	水温、、水汞	48	鲁甸	水温、氟离子、硫酸根离子、pH
49	元江	水氡	50	曲江	水位、水氡、水温、水汞
51	弥勒	水位、水氡、水汞	52	景谷	水位、水温
53	鱼洞	水温、水位	54	石屏	倾斜、短水准、短基线
55	宣威	水温、水位	56	腾冲	视电祖率、自然电位、地磁、水位、水氡、钙离子、镁离子、硫酸根离子
57	昭通	倾斜、水温	58	龙陵	水氡、流量、水温、电导率、硫酸根离子、氟离子、碳酸氢根离子

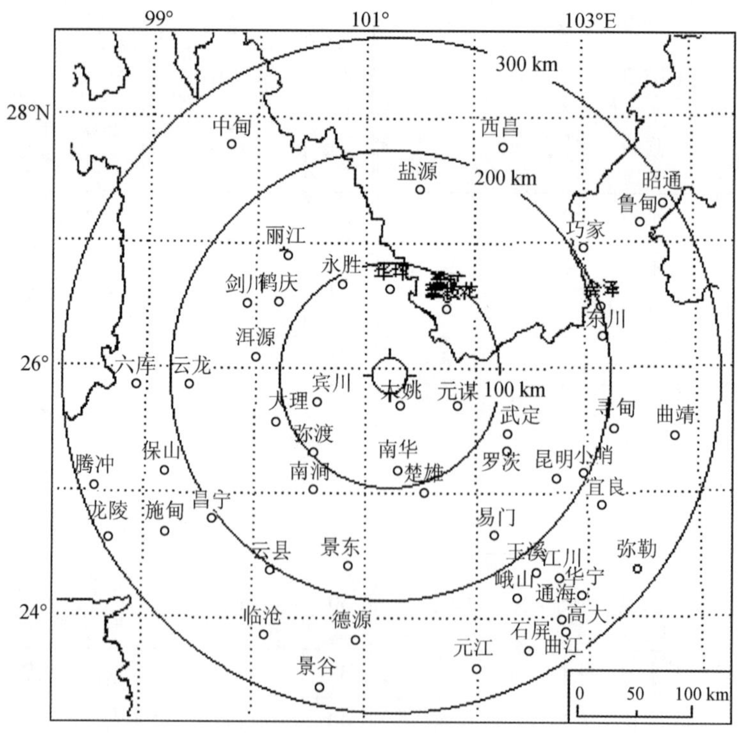

图17　大姚6.2、6.1级地震定点前兆观测台站的分布图

Fig. 17　Distribution of the precursor observation stations before the M_S6.2, 6.1 Dayao earthquake

表 7　大姚震前的异常情况登记表

Table 7　Summary table of precursory anomalies before the Dayao earthquake swarm

序号	异常项目	台站（点）或观测区	分析方法	异常判据及观测误差	震前异常起止时间	震后变化	最大幅度	震中距 Δ/km	异常类别及可靠性	图号	异常特点及备注	震前提出/震后总结
1	地震频度	21°～29°N 97°～106°E	$M_L \geqslant 4$ 级地震 0.5 平方度为单元扫描	$N \geqslant 3$	2001.11～2002.10	正常	4 次		M_1	18[7)]	云南 70% 以上的 $M_S \geqslant 6$ 级地震前震中附近地区出现 N 值异常	震前提出
2		21°～29.5°N 97°～106°E	$M_L \geqslant 4$ 级地震 3 月累计 1 月滑动	2 个月以上 $\geqslant -2\sigma$ 或 $\leqslant 2\sigma$	2002.11～2003.3	正常	0 次		M_1	19a[1]	连续 2 个月低值后回升，为云南地区强震前的异常特征	震前提出
3		21°～29.5°N 97°～106°E	$M_L \geqslant 3$ 级地震 5 月累计 1 月滑动	3 个月以上 $\geqslant 2\sigma$ 或 $\leqslant -2\sigma$	2003.11～2003.6	正常	45 次		M_1	19b[1]	连续 3～6 月低值后回升	震前提出
4	b 值	23.8°～28°N 97°～101.5°E	M_L：2.5～4.5 级地震以 6 个月累计 1 月滑动	$\geqslant 2\sigma$ 或 $\leqslant -2\sigma$	2002.6～2003.4	正常	1.12		M_1	20	连续 3 月高值异常后恢复，为滇西地区 6 级地震前的异常特征	震前提出
5	地震平静	23.8°～28°N 97°～101.5°E	$M_L \geqslant 4$ 级地震间隔时间	$\Delta t \geqslant 130$ 天	2002.7.25～2003.6.13	正常	315 天		M_1	21[1]	中短期阶段滇西地区出现 4 级地震平静	震前提出

续表

序号	异常项目	台站（点）或观测区	分析方法	异常判据及观测误差	震前异常起止时间	震后变化	最大幅度	震中距 Δ/km	异常类别及可靠性	图号	异常特点及备注	震前提出/震后总结
6	地震平静	云南省内	$M_L \geqslant 3$ 级地震间隔时间	$\Delta t \geqslant 6$ 天	2003.6.21 ~ 2003.8.9	正常	19 天		S_1	表 8	1996 年来中强地震后云南省内 3 级地震平静为中强震连发指标	6.1 级震前提出
7	地震窗	27.2° ~ 28.2°N 99° ~ 101°E	$M_L \geqslant 3$ 级中甸窗地震月频度	$N \geqslant 2$	2003.4 与 2003.6	正常	3 次		S1	22a[1]	小震窗活动	震前提出
8	地震窗	25.6° ~ 26.2°N 100.1° ~ 100.4°E	$M_L \geqslant 2$ 级南涧窗地震月频度	$N \geqslant 1$	2003.5	正常	4 次		S1	22b[1]	0.6° × 0.3° 小震活动	震前提出
9	水位	大姚	五日均值	大于阈值	2002.6 ~ 2003.5	异常	2.1m	26	M_1	23[1]	大姚高水位多次较好的对应附近地区的 6 级地震。6.2 级后地震水位再次上升	震前提出
10	水位	大姚	五日均值	大于阈值	2003.7 ~ 2003.10	正常	0.3m	26	S_1	23[1]	大姚高水位多次较好的对应附近地区的 6 级地震。6.2 级后地震水位再次上升	震前提出
11	水位	宾川	日均值	破坏正常年动态	2003.2.27 ~ 2003.7.21	异常	150cm	68	S_1	24[1]	年变幅度增大	震前提出
12	水位	南华	日均值	破坏正常年动态	2002.10 ~ 2003.4	正常	趋势	83	M_1	25[1]	年变幅度增大	震前提出

续表

序号	异常项目	台站（点）或观测区	分析方法	异常判据及观测误差	震前异常起止时间	震后变化	最大幅度	震中距Δ/km	异常类别及可靠性	图号	异常特点及备注	震前提出/震后总结
13	水位	永胜	五日均值	破坏正常年动态	2002.11～	异常	54cm	96	M_1	26[1]	年变幅度减小。2004年3月结束	震前提出
14		楚雄	旬均值	破坏正常动态	2002.6～2002.8	正常	1.85m	106	M_1	27[1]	该水位出现破坏正常动态，短期水位再次上升异常	震前提出
15		楚雄	旬均值	破坏正常动态	2003.3～2003.10	正常	0.5m	106	S_1			
16		武定	月均值	破坏正常年动态	2002.4～2002.11	正常	趋势	120	M_1	28[1]	高水位异常	震前提出
17		月溪	月均值	破坏正常年动态	2001～	异常	71cm	124	M_1	29[1]	大理水位2次年变幅度增大，均对应了附近6级以上地震。2004年3月结束	震前提出
18		施甸	旬均值	破坏正常年动态	2002.2～	异常	25cm	248	M_1	30[1]	2003年12月结束	震前提出
19	高精度水温	丽江	日均值	变化≥0.004℃	2002.4.11～2003.6.5	正常	0.004℃	143	M_2	31	中期出现升温异常	震后总结

续表

序号	异常项目	台站（点）或观测区	分析方法	异常判据及观测误差	震前异常起止时间	震后变化	最大幅度	震中距 Δ/km	异常类别及可靠性	图号	异常特点及备注	震前提出/震后总结
20	水氡	弥渡	旬均值	$R_n \geq 29.5$Bg/L	2001.11～2002.12	正常	31.8 Bq/L	100	M_2	32[1]	中期出现高值异常。6.1级地震前有所波动	震前提出
21		南涧	日均值	变化≥70%	2003.5.22与2003.8.12	正常	46.8 Bq/L	113	S_2	33[1]	单点突降	6.1级震前提出
22		鹤庆	日均值	变化≥46%	2003.6.28 2003.8.25与2003.9.22	正常	10.6 Bq/L	128	I_2	34[1]	临震单点突降	6.1级震前提出
23		易门	日均值	变化≥50%	2003.8.9	正常	50 Bq/L	170	S_2	35[1]	单点突降，的异常	6.1级震前提出
24		巧家	旬均值	$R_n \geq 2\sigma$ 或 $R_n \leq -2\sigma$	2003.4～2003.6	异常	15.2 Bq/L	199	S_2	36[1]	6.2级地震前高值异常，之后6.1级地震前转为低值异常	震前提出
25		巧家	旬均值	$R_n \geq 2\sigma$ 或 $R_n \leq -2\sigma$	2003.7～2003.10	正常	128 Bq/L	199	S_2			
26		昌宁	五日均值	$R_n \leq 64$Bg/L	2002.11～2003.10	正常	14 Bq/L	205	M_2	37	低值异常一致持续至第二个地震前	震后总结
27		施甸	月均值	≤11Bg/L	2003.4～2003.10	正常	10.6 Bq/L	248	S_2	38	低值异常在7、8月时有所恢复，9月又再次下降	震后总结

续表

序号	异常项目	台站（点）或观测区	分析方法	异常判据及观测误差	震前异常起止时间	震后变化	最大幅度	震中距 Δ/km	异常类别及可靠性	图号	异常特点及备注	震前提出/震后总结
28	水氡	临沧	旬均值	破坏正常动态	2002.10 ~ 2003.6	正常	5.2 Bq/L	260	M_2	39	趋势下降，于2003年1~4短期有所回升	震后总结
29		元江	五日均值	破坏正常动态	2003.1 ~	异常	227.5 Bq/L	280	S_2	40	异常表现为氡上升，尤其是临震波动幅度增大。2003年12月结束	震后总结
30		龙陵	旬均值	$R_n \geqslant 200$Bg/L 或 $R_n \leqslant 80$Bg/L	2002.12 ~ 2003.4	正常	220 Bq/L	300	M_2	41	震前水氡测值出现升高，达异常指标	震后总结
31	汞	下关	日均值	≥175ng/L	2003.5.20 ~ 2003.10.15	正常	280 ng/L	100	S_2	42[1]	震前水汞含量增加，6、9、10月临震突升	震后总结
32		下关（气）	日均值	≥2.5ng/L	2003.5.1 ~ 2003.6.10	正常	19.6 ng/L	100	S_2	43[1]	震前气汞含量增加	震前提出
33		下关（气）			2003.8.10 ~ 2003.10.5	正常	2.9 ng/L	100	S_2			

续表

序号	异常项目	台站（点）或观测区	分析方法	异常判据及观测误差	震前异常起止时间	震后变化	最大幅度	震中距 Δ/km	异常类别及可靠性	图号	异常特点及备注	震前提出/震后总结
34	汞	洱源	五日均值	≥57ng/L	2003. 5 ~ 2003. 6	正常	61 ng/L	130	S_2	44[1]	震前水汞含量增加	震后总结
35		玉溪（气）	日均值	≥88ng/L	2003. 5 ~	异常	350 ng/L	221	S_2	45[1]	震前气汞含量增加，6. 2 级地震后汞含量更高。2003 年 12 月结束	震后总结
36		高大（气）	日均值	≥23ng/L	2003. 5. 20 ~ 2003. 6. 20	正常	40. 3 ng/L	264	S_2	46[1]	震前气汞含量增加	震后总结
37		高大（气）			2003. 8. 10 ~	异常	67 ng/L				震前气汞含量大幅度增加。2003 年 11 月 5 日结束	震后总结
38	氟离子	巧家	旬均值	破坏正常动态	2002. 11. 1 ~ 2003. 5. 28	正常	0. 08 mg/L	199	M_2	47[1]	震前地下水中氟离子含量增加	震后总结
39		龙陵	五日均值	破坏正常动态	2002. 8 ~ 2003. 6	正常	1. 6 mg/L	300	M_2	48[1]	震前地下水中氟离子含量减少，恢复时发震	震后总结

续表

序号	异常项目	台站（点）或观测区	分析方法	异常判据及观测误差	震前异常起止时间	震后变化	最大幅度	震中距 Δ/km	异常类别及可靠性	图号	异常特点及备注	震前提出/震后总结
40	镁离子	巧家	旬均值	破坏正常动态下降	2002.11.1 ~ 2003.7.20	正常	3 mg/L	199	M_2	49[1]	震前地下水中镁离子含量减少，恢复时发震	震后总结
41	碳酸氢根	巧家	旬均值		2002.11.1 ~ 2003.6.20	异常	26 mg/L			50[1]	震前地下水中碳酸氢根离子含量减少，恢复时发震	震后总结
42					2003.7 ~ 2003.10	正常	11 mg/L			51[1]	6.2级震后地下水中碳酸氢根离子含量再次减少	震后总结
43	地磁 Z	永胜	日均值	≥7nT	2003.6.12 ~	异常	16nT	96	I_2	52a[1,9]	异常形态为变幅增大，6.1级地震前变幅更大。2003年12月9日结束	震前提出
44		丽江	日均值	≥9nT	2003.7.10 ~	异常	18.8nT	143	I_2	52b[1,9]	异常形态为变幅增大，6.1级地震前变幅更大。2003年12月5日结束	震前提出

续表

序号	异常项目	台站（点）或观测区	分析方法	异常判据及观测误差	震前异常起止时间	震后变化	最大幅度	震中距 Δ/km	异常类别及可靠性	图号	异常特点及备注	震前提出/震后总结
45	伸缩	永胜 73	东西向月均值	破坏正常年动态	2002. 7 ~	异常	2. 8 $\times10^{-6}$	96	M_2	53a[1]	异常年变幅变小，2004 年 7 月结束	震后总结
46		永胜 73	南北向月均值		2001. 11 ~		2. 7 $\times10^{-6}$		M_2	53b[1]	异常年变幅变大趋势下降，2004 年 7 月结束	震后总结
47		弥渡 73	东西向日均值	破坏正常年动态	2003. 3 ~	异常	2. 1 $\times10^{-6}$	100	S_3	54a[1]	异常年变幅变小，临震为突降，2004 年 3 月结束。有抽水干扰	震后总结
48		弥渡 73	南北向日均值		2003. 3 ~		0. 17 $\times10^{-6}$		S_3	54b[1]	异常年变幅变小，临震为突升，2004 年 3 月结束。有抽水干扰	震后总结

续表

序号	异常项目	台站（点）或观测区	分析方法	异常判据及观测误差	震前异常起止时间	震后变化	最大幅度	震中距 Δ/km	异常类别及可靠性	图号	异常特点及备注	震前提出/震后总结
49	地倾斜	弥渡62	东西向日均值	破坏正常年动态	2002.10～	异常	2.1角秒	100	M_3	55a[1]	异常年变幅变大加速上升，2004年7月结束	震后总结
50		弥渡62	南北向日均值		2002.10～	异常	1.4角秒		M_3	55b[1]	异常年变幅变大加速下降，2004年7月结束	震后总结
51		楚雄83	东西向旬均值	破坏正常年动态	2002.10～2003.10	正常	0.3角秒	106	M_2	56[1]	异常年变幅变小，临震为突降	震后总结
52		云龙83	东西向五日均值	破坏正常年动态	2002.3～2003.10	正常	40角秒	186	M_2	57a[1]	异常年变幅变大趋势上升	震前提出
53		云龙83	南北向五日均值		2003.3～2003.10	正常	1.6角秒		M_2	57b[1]	异常年变幅变大趋势下降	震前提出
54	短水准	楚雄	1—2边旬均值	破坏正常年动态	2002.10～	异常	2.1mm	106	M_2	58[1]	异常年变幅变大趋势上升。2004年7月结束	震前提出

续表

序号	异常项目	台站（点）或观测区	分析方法	异常判据及观测误差	震前异常起止时间	震后变化	最大幅度	震中距 Δ/km	异常类别及可靠性	图号	异常特点及备注	震前提出/震后总结
55	宏观	姚安胡家山水库	破坏正常动态	黄褐色、浑浊明显	2002.11.5～2003.10	正常		震中附近	M_2		4个水库发浑，为时间预测提供了依据	震前提出的异常[7]
56		姚安洋派水库	破坏正常动态	灰黄色、浑浊程度较轻								
57		姚安马游水库	破坏正常动态	老库区浑浊明显，黄褐色								
58		大姚妙峰水库	破坏正常动态	黄褐色、浑浊明显								
59		弥渡石碑水库	破坏正常动态	浑浊	2003.7.8～2003.10	正常		震中附近	I_2			震前提出的异常[7]
60		华宁县地震局院内的一口深机井	破坏正常动态	水位下降	2003.4.10～	正常	90cm	震中附近	S_2		结束时间不详	6.1级震前提出的异常[8]
61		鹤庆辛屯镇连义村80眼民用井水	破坏正常动态	发浑、变黑冒泡	2003.7.26～2003.8.4	正常		震中附近	S_2		水色变化	6.1级震前提出的异常[9]
62		龙陵14号泉	破坏正常动态	涌出乳白色水	2003.8.14～	正常		震中附近	S_2		水色变化	6.1级震前提出的异常[9]
63		嵩明小街镇龙塘	破坏正常动态	冒泡水发浑，呈土红色	2003.8.18～2003.8.20	正常		震中附近	S_2		水色变化	6.1级震前提出的异常[9]

在震中周围 100km 内有前兆观测台站 10 个台项 28 个，出现异常台站 7 个异常台项 10 个，异常台站和异常台项百分比分别为 70% 和 36%；在 201 ~ 200km 范围内有前兆台站 19 个台项 57 个，出现异常台站 14 个异常台项 18 个，异常台站和异常台项百分比分别为 74% 和 32%；在 201 ~ 200km 范围内有前兆台站 33 个台项 110 个，出现异常台站 8 个异常台项 9 个，异常台站和异常台项百分比分别为 24% 和 8%。

表 7 中给出的地震学异常，基本上是在日常监视预报中使用的常规方法得到的异常，3、4 级地震频度的连续几个月低值异常后显著增加，曾多次对应过云南地区的 6 级以上地震，根据 4 级地震活动的密集区，在云南省 2003 年度地震趋势研究报告中对该地震作了中期预测。追踪地震活动异常区 4 级地震平静的间隔时间、b 值变化、2、3 级小震窗的活动则确定了 6 级地震进入了短期时段。

表 7 中给出的前兆观测异常，大多数的异常判别标准是经验的，但多数异常是震前就看到并给出的，因此对指导未来的地震预报有一定的参考价值。按照规定前兆曲线时间应延长至震后 2 年，但下关、玉溪气汞观测为“九五”所建，于 2001 年中始观测，因 2004、2005 年受印度尼西亚 8 级巨震影响较大，时间扩展后则看不清 2003 年异常，因而时间只至 2003 年。规定地倾斜 1 个台站只算 1 个观测台项，实际观测中同一台站同一仪器、不同的方向曾出现多项异常，作为统计只计算了 1 项，但作为异常还是在表 7 和图 17 中给出。

表 8　1997 年云南 5 级地震后省内 3 级平静 6 天以上与后续中强震关系

Table 8　Relationships of strong earthquake and Anomalous quiescence of M_L3.0 after the M_S5.0 earthquake in the Yunnan

编号	5 级地震	3 级平静时间	后续中强震	间隔时间	备注
1	1998.10.27　5.2　宁蒗	17 天	1998.11.19　6.2　宁蒗	22 天	前震
2	1999.11.25　5.2　澄江	25 天	2000.1.15　6.5　姚安	50 天	
3	2003.7.21　6.2　大姚	19 天	2003.10.16　6.1　大姚	75 天	双震

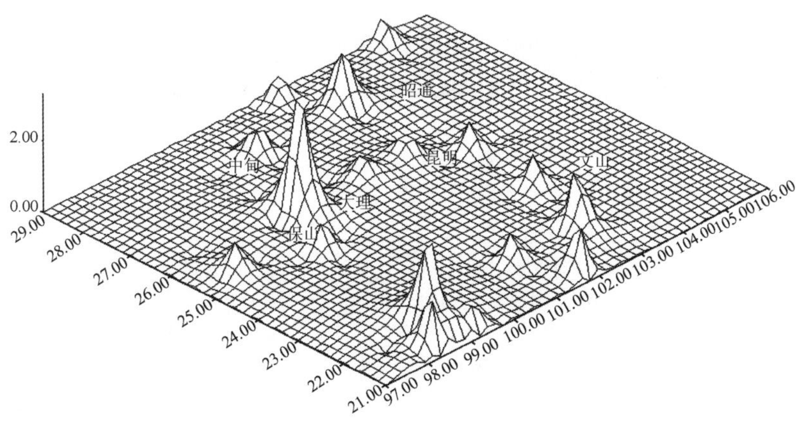

图 18　云南地区 4 级地震频度空间分布图

Fig. 18　Distribution of $M_L \geqslant 4.0$ frequency in Yunnan

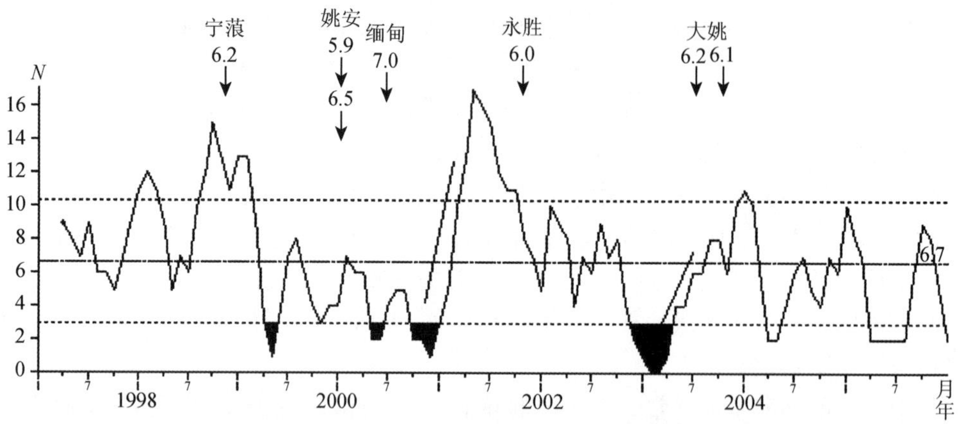

图 19a　云南地区 4 级地震月频度

Fig. 19a　Monthly frequency of $M_L \geqslant 4.0$ earthquakes in Yunnan

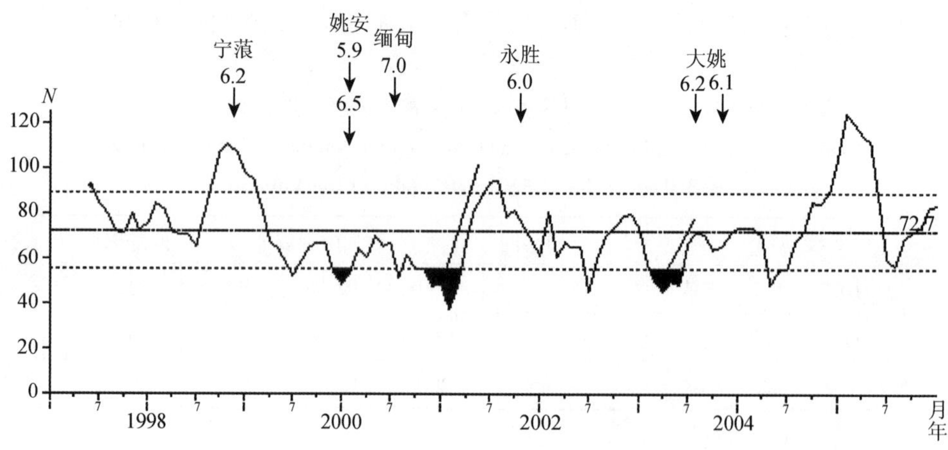

图 19b　云南地区 3 级地震月频度

Fig. 19b　Monthly frequency of $M_L \geqslant 3.0$ earthquakes in Yunnan

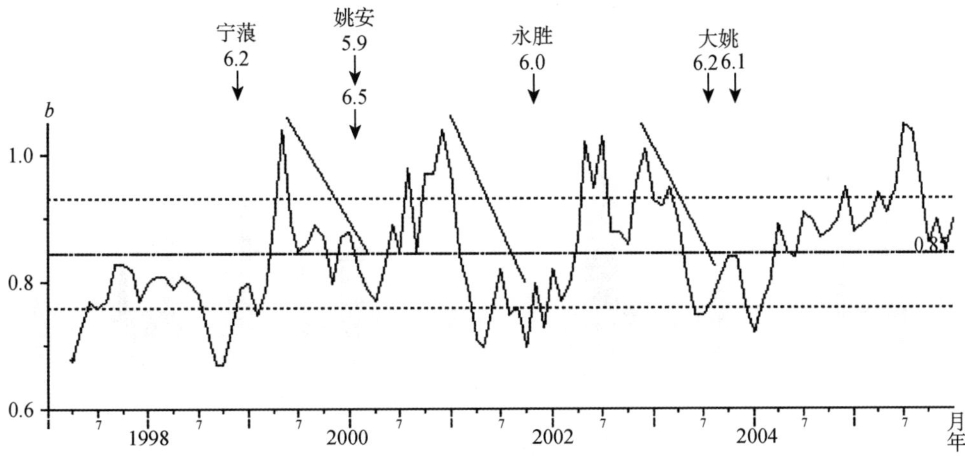

图 20 云南地区 M_L：2.5～4.5 级地震月 b 值

Fig. 20 The b-t curve of M_L (2.5-4.9) earthquakes in the Yunnan area

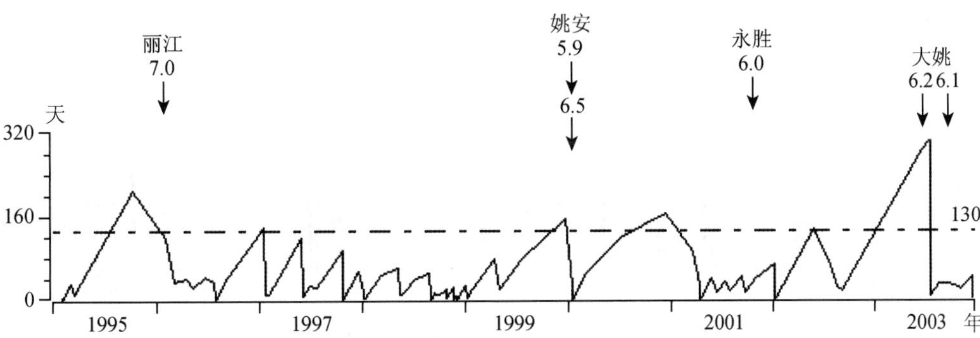

图 21 滇西地区 $M_L \geq 4$ 级地震时间间隔曲线

Fig. 21 The ΔT-t curve of $M_L \geq 4.0$ earthquakes in the Yunnan area

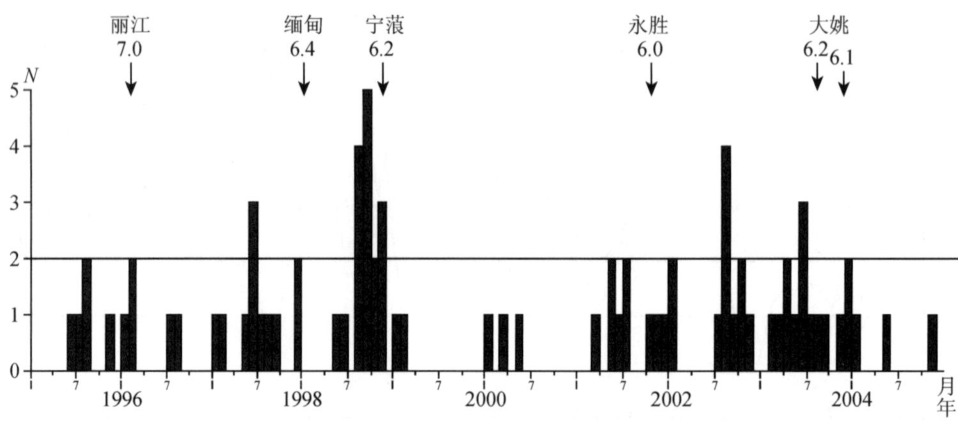

图 22a　中甸窗 $M_L \geqslant 3$ 级地震月频度

Fig. 22a　Monthly frequency of $M_L \geqslant 3.0$ earthquakes in Zhongdian seismic window

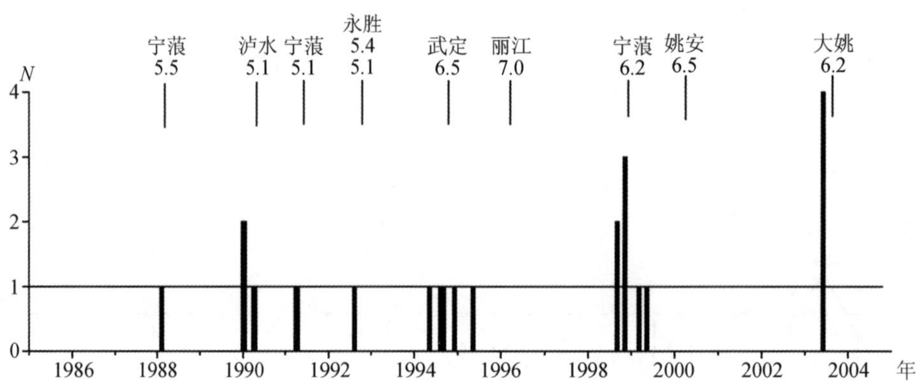

图 22b　南涧窗 $M_L \geqslant 2$ 级地震月频度

Fig. 22b　Monthly frequency of $M_L \geqslant 2.0$ earthquakes in Nanjian seismic window

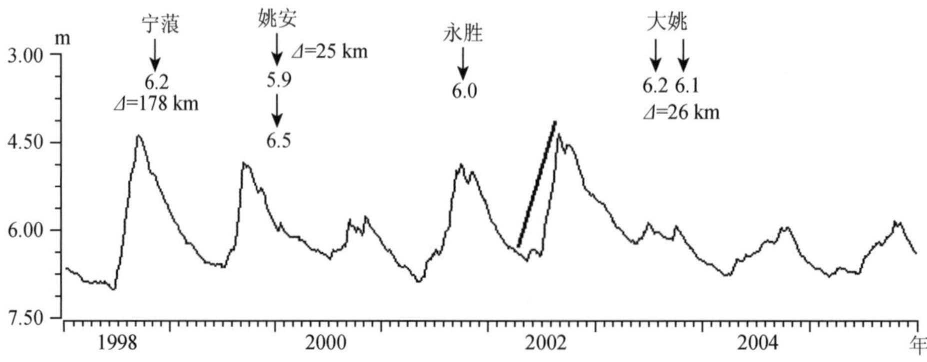

图 23　大姚水位五日均值曲线

Fig. 23　5-day mean value of water level at Dayao station

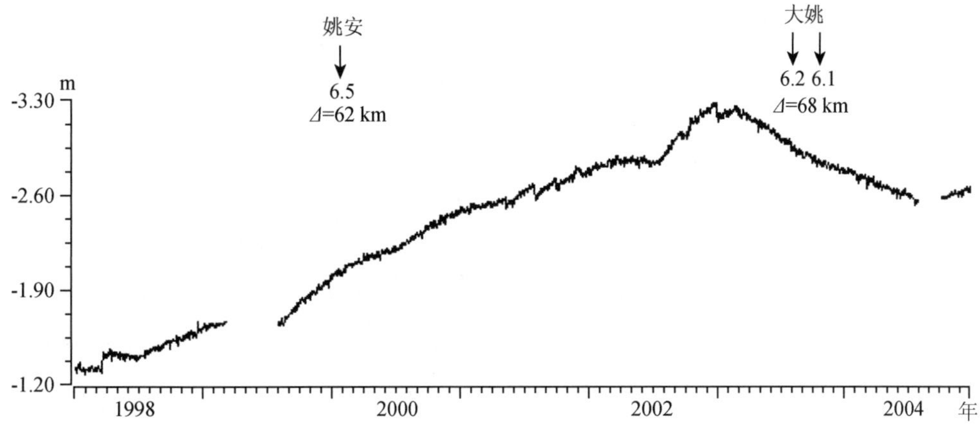

图 24　宾川水位日均值曲线

Fig. 24　Daily mean value of water level at Bingchuan station

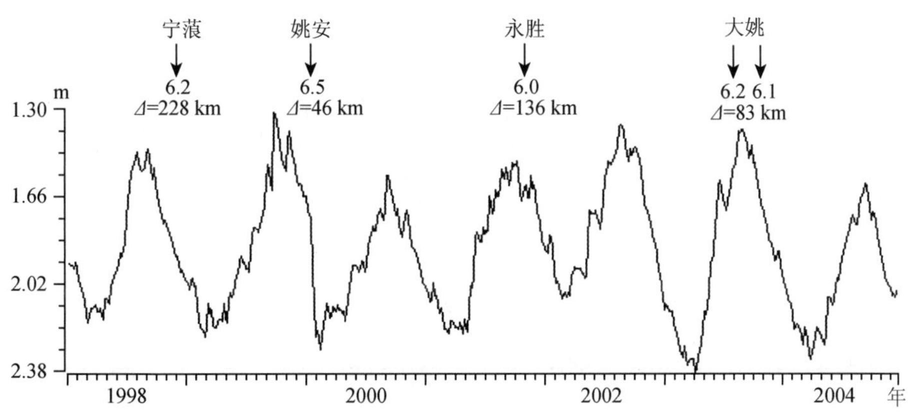

图 25　南华水位日均值曲线

Fig. 25　Daily mean value of water level at Nanhua station

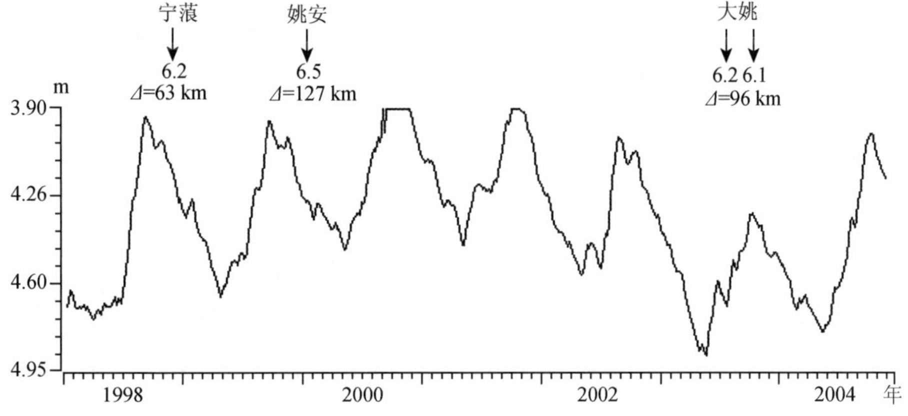

图 26　永胜水位五日均值曲线

Fig. 26　5-day mean value of water level at Bingchuan station

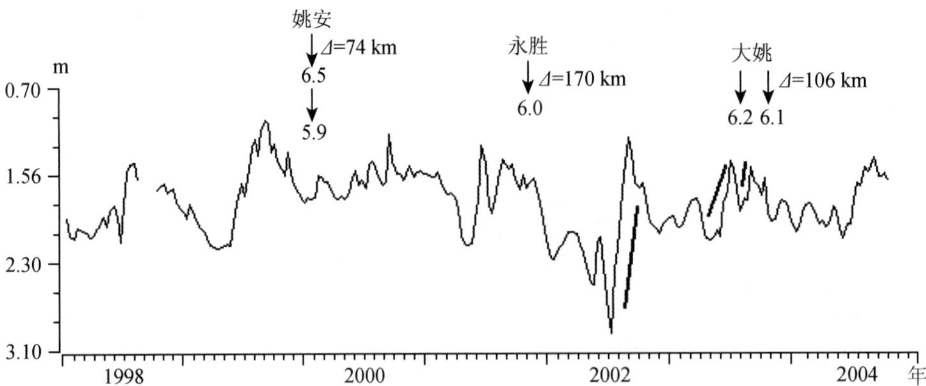

图 27　楚雄水位旬均值曲线

Fig. 27　10-day mean value of water level at Chuxion station

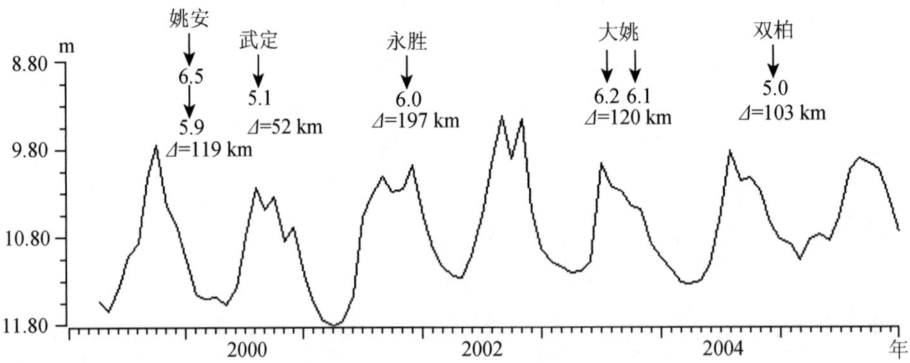

图 28　武定水位月均值曲线

Fig. 28　Monthly mean value of water level at Wuding station

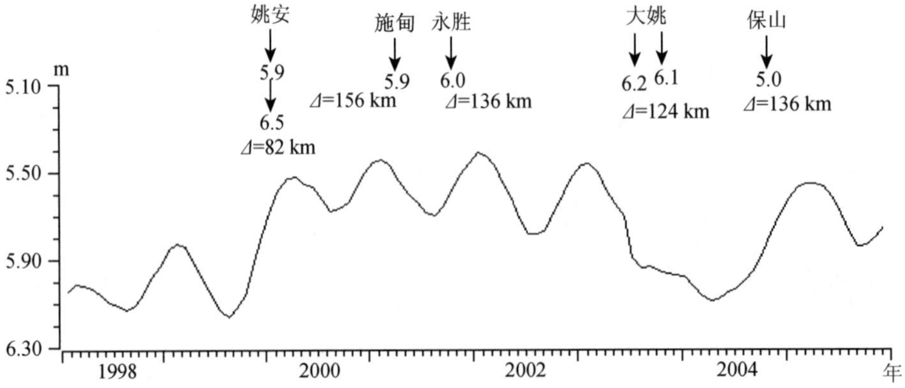

图 29　大理月溪水位月均值曲线

Fig. 29　Monthly mean value of water level at Yuxi station

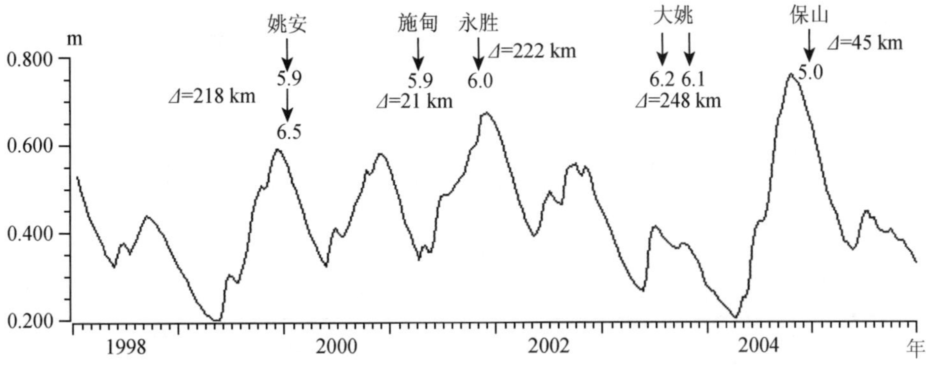

图30 施甸水位旬均值曲线

Fig. 30 10-day mean value of water level at Shidian station

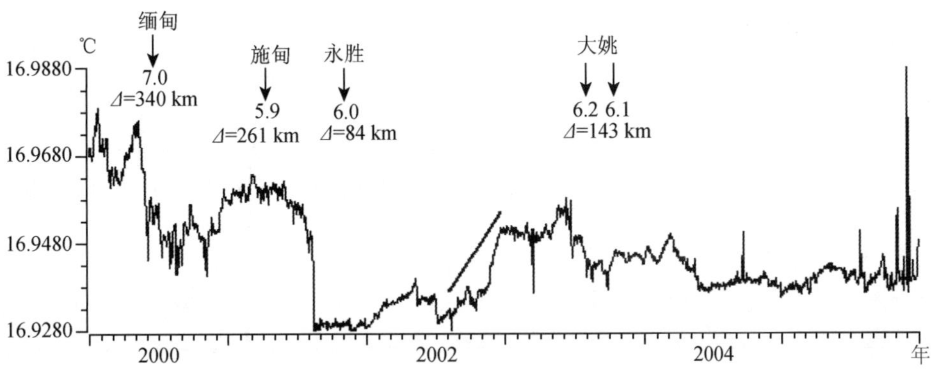

图31 丽江水温日均值曲线

Fig. 31 Daily mean value of water temperature at Lijiang station

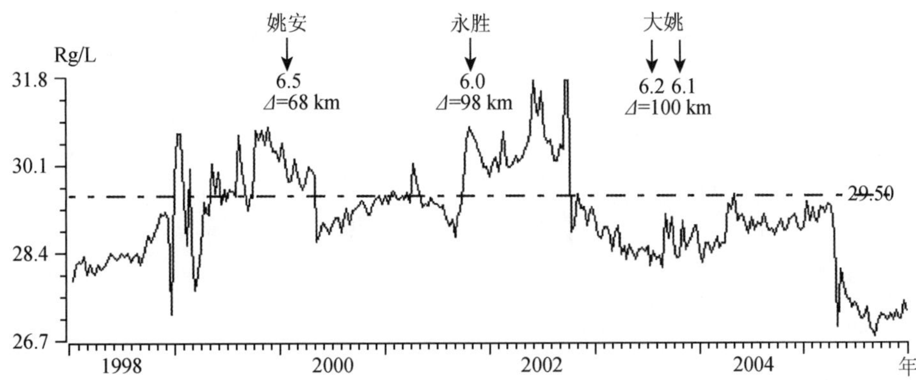

图32 弥渡水氡旬均值曲线

Fig. 32 10-day mean value of radon content in groundwater at Midu station

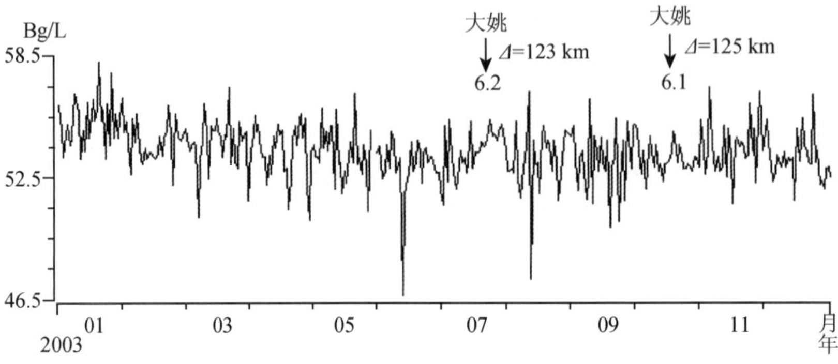

图 33　南涧水氡日均值曲线

Fig. 33　Daily mean value of radon content in groundwater at Nanjian station

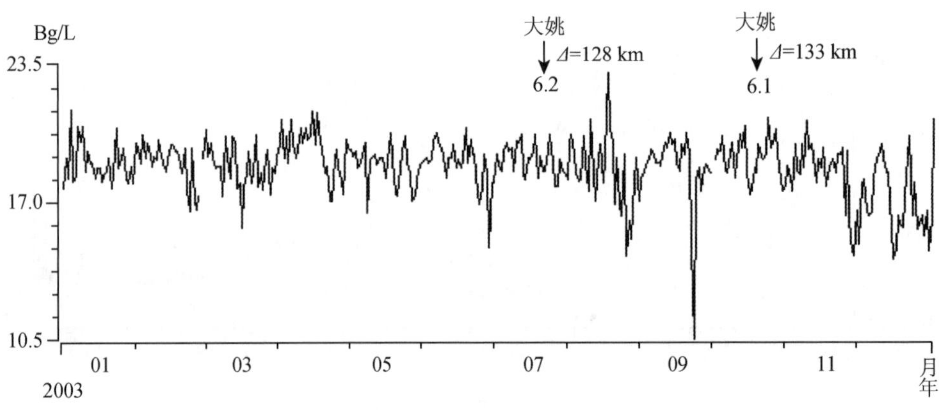

图 34　鹤庆水氡日均值曲线

Fig. 34　Daily mean value of radon content in groundwater at Heqing station

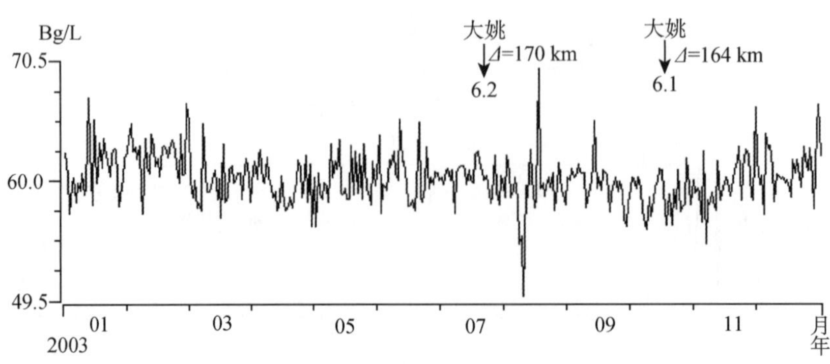

图 35　易门水氡日均值曲线

Fig. 35　Daily mean value of radon content in groundwater at Yimen station

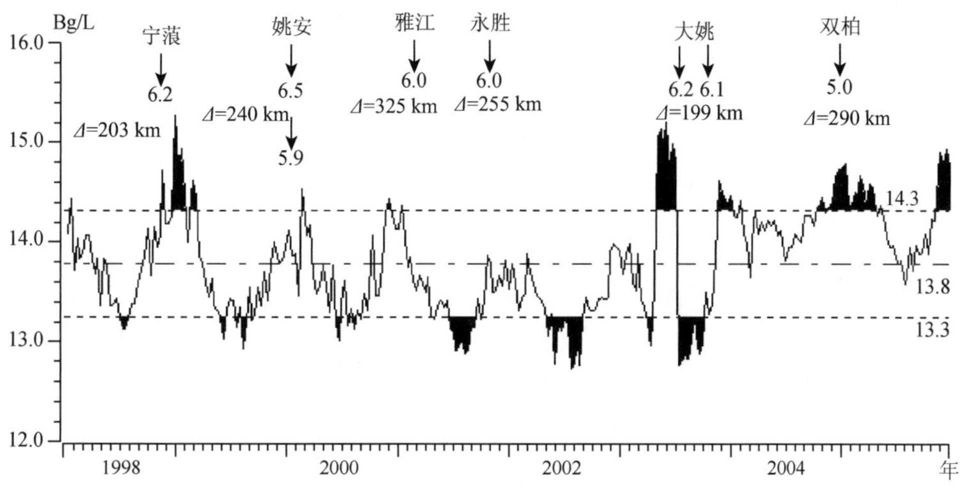

图 36　巧家水氡旬均值曲线

Fig. 36　10-day mean value of radon content in groundwater at Qiaojia station

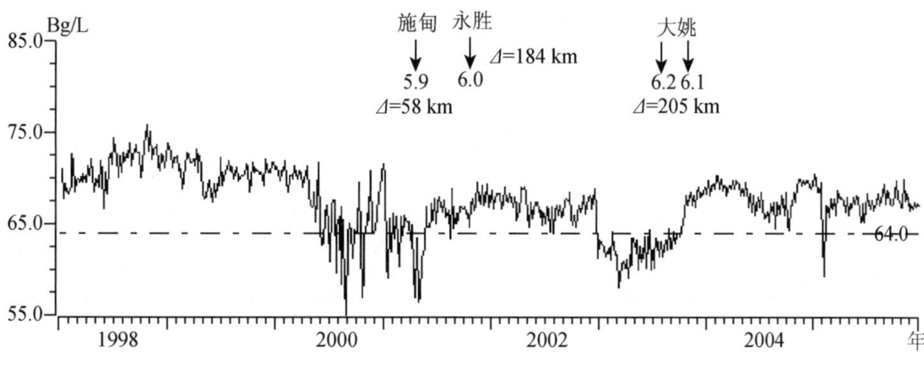

图 37　昌宁水氡五日均值曲线

Fig. 37　5-day mean value of radon content in groundwater at Changning station

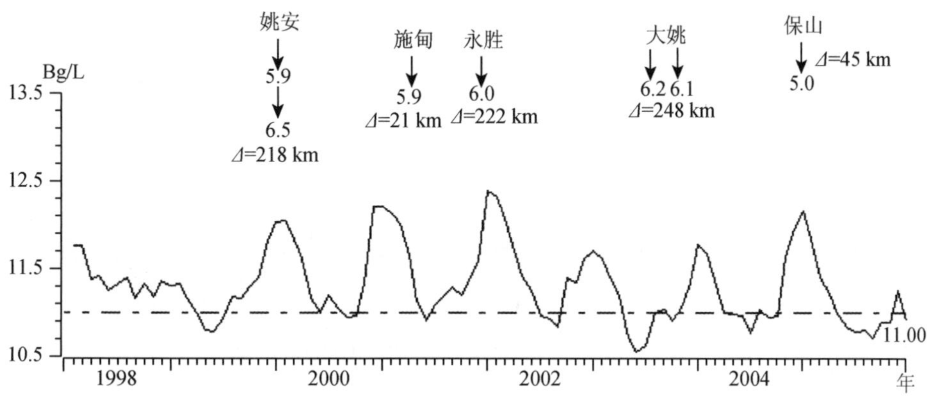

图 38　施甸水氡月均值曲线

Fig. 38　Monthly mean value of radon content in groundwater at Shidianstation

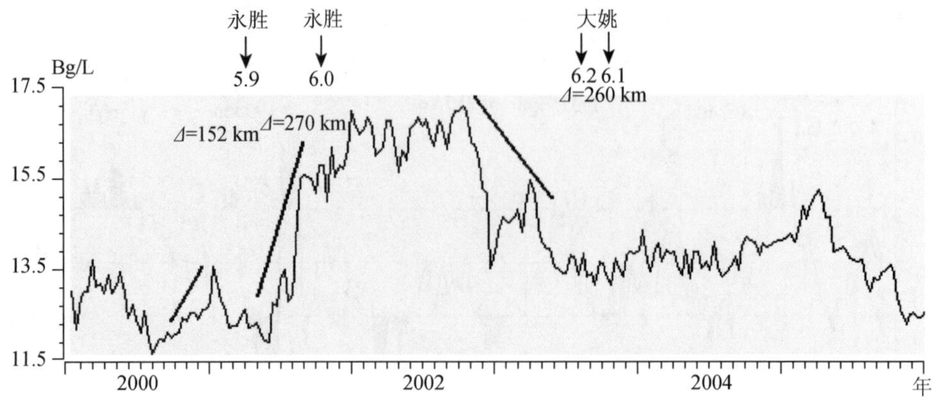

图 39　临沧水氡旬均值曲线

Fig. 39　10-day mean value of radon content in groundwater at Lingcang station

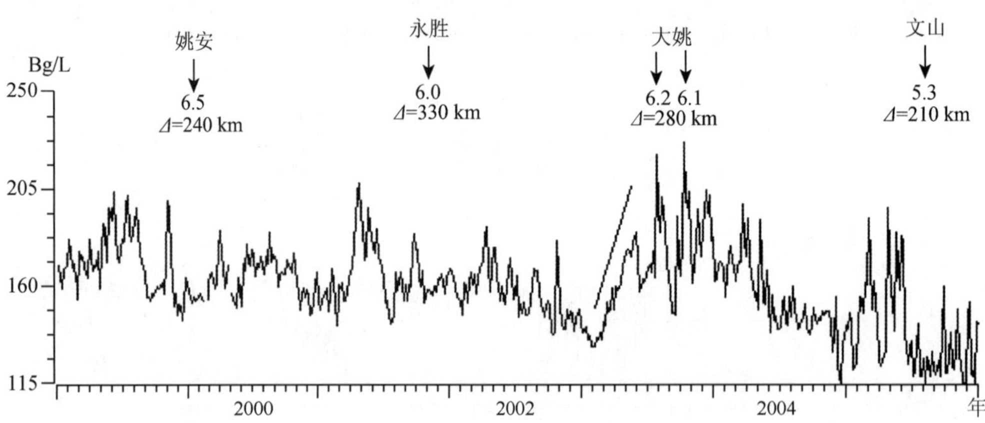

图 40　元江水氡五日均值曲线

Fig. 40　5-day mean value of radon content in groundwater atYuanjiang station

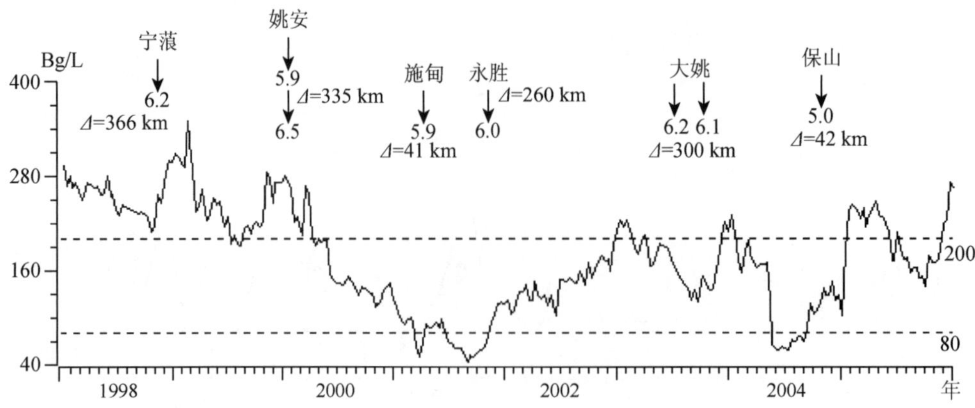

图 41　龙陵水氡旬均值曲线

Fig. 41　10-day mean value of radon content in groundwater at Longling station

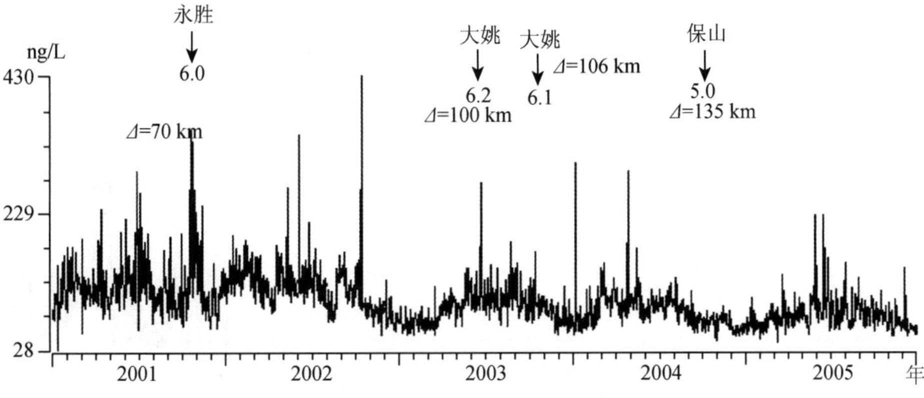

图 42 下关水汞日均值曲线

Fig. 42 Daily mean value of mercury content in groundwater at Xiaguan station

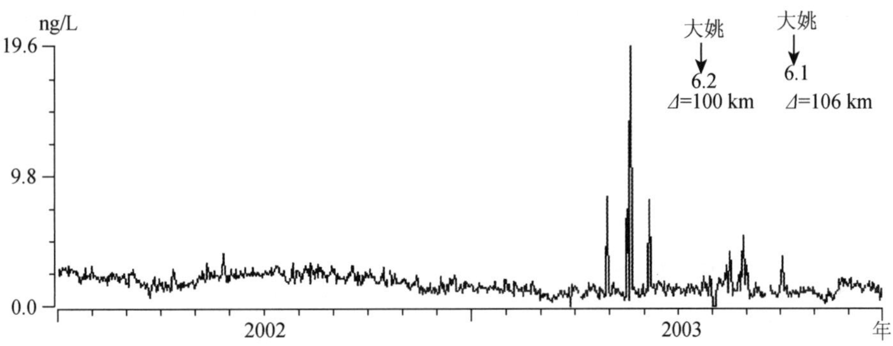

图 43 下关气汞日均值曲线

Fig. 43 Daily mean value of mercury content at Xiaguan station

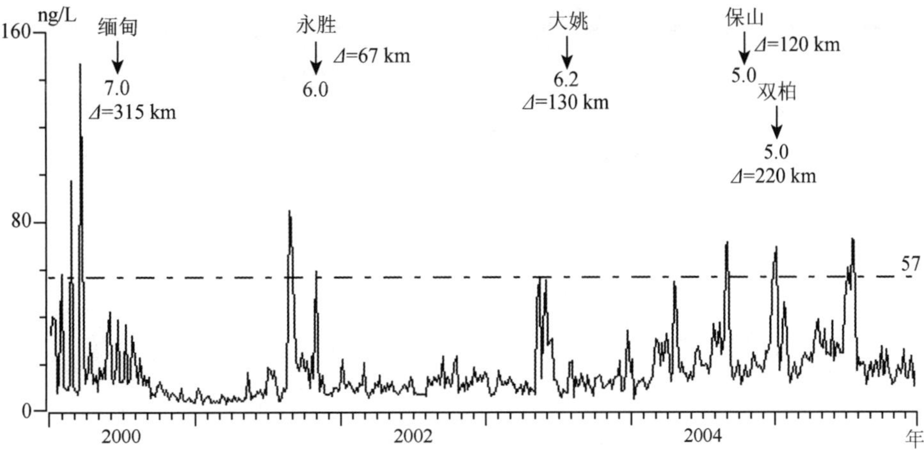

图 44 洱源水汞五日均值曲线

Fig. 44 5-day mean value of mercury content in groundwater at Eryuan station

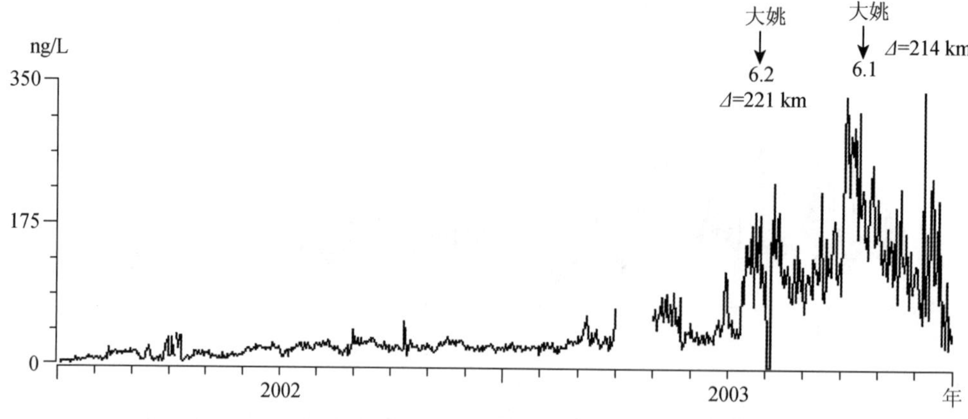

图 45　玉溪气汞日均值曲线

Fig. 45　Daily mean value of mercury content at Yuexi station

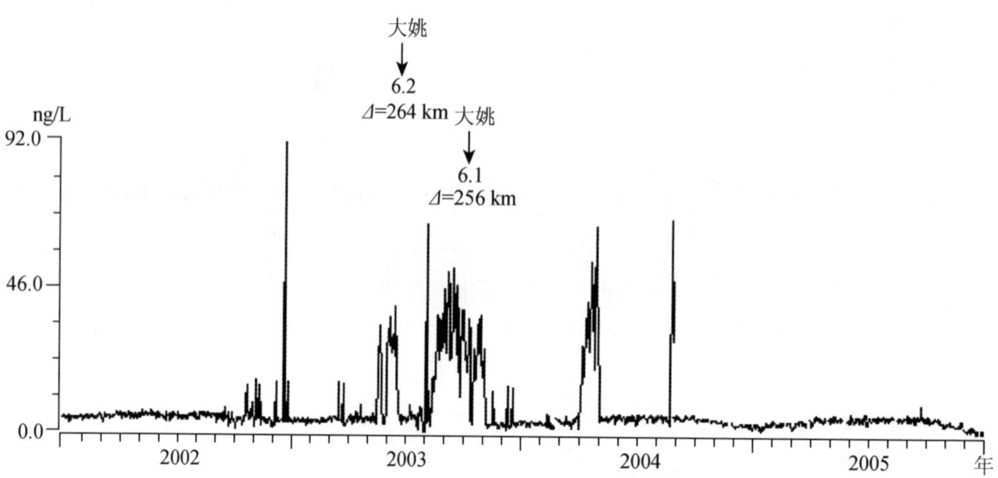

图 46　高大气汞日均值曲线

Fig. 46　Daily mean value of mercury content at Gaoda station

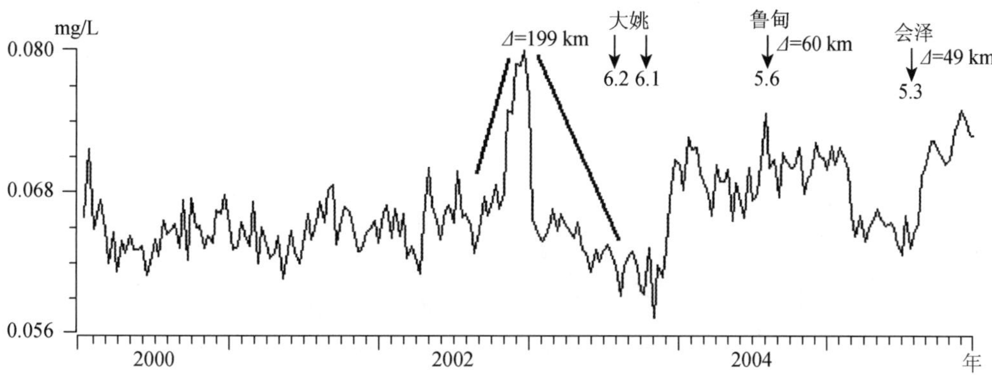

图 47 巧家氟离子旬均值曲线

Fig. 47 10-day mean value of F^- in groundwater at Qiaojia station

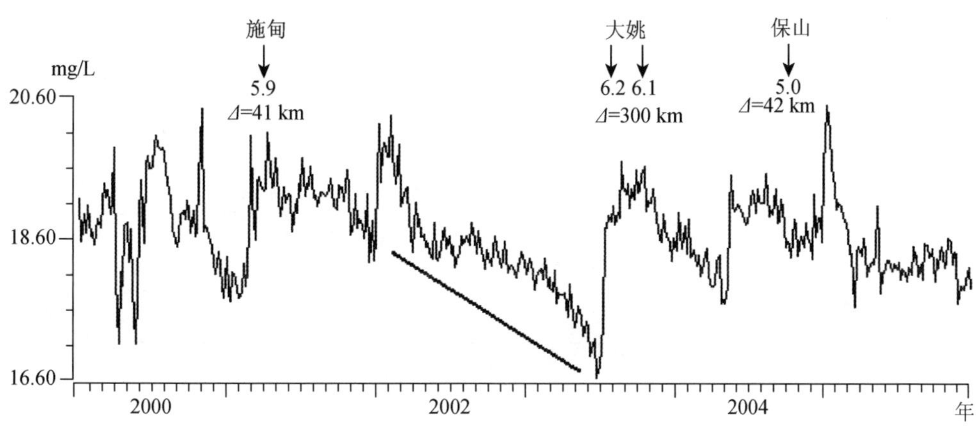

图 48 龙陵氟离子五日均值曲线

Fig. 48 5-day mean value of F^- in groundwater at Longling statio

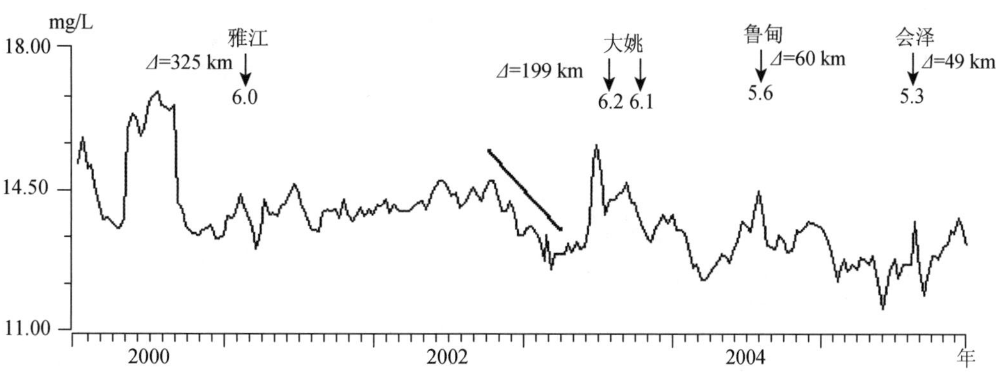

图 49 巧家镁离子旬均值曲线

Fig. 49 10-day mean value of Mg^{2+} in groundwater at Qiaojia statio

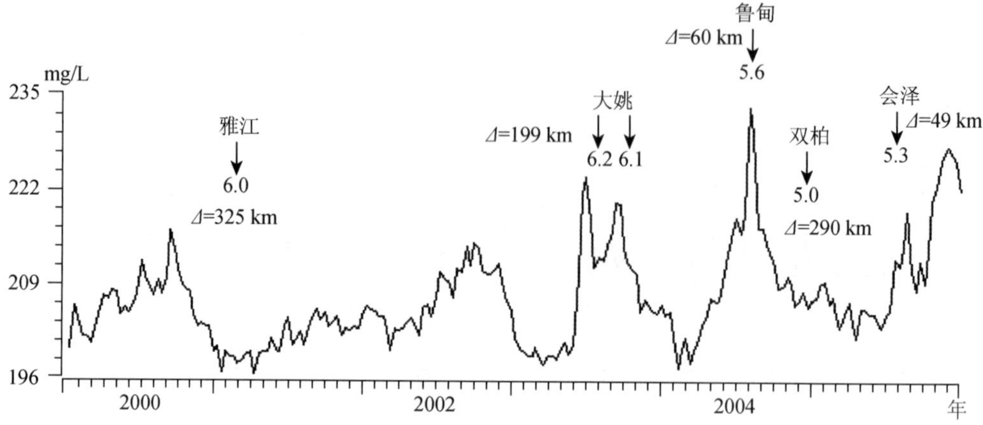

图 50　巧家碳酸氢根离子旬均值曲线

Fig. 50　10-day mean value of HCO_3^- in groundwater at Qiaojia statio

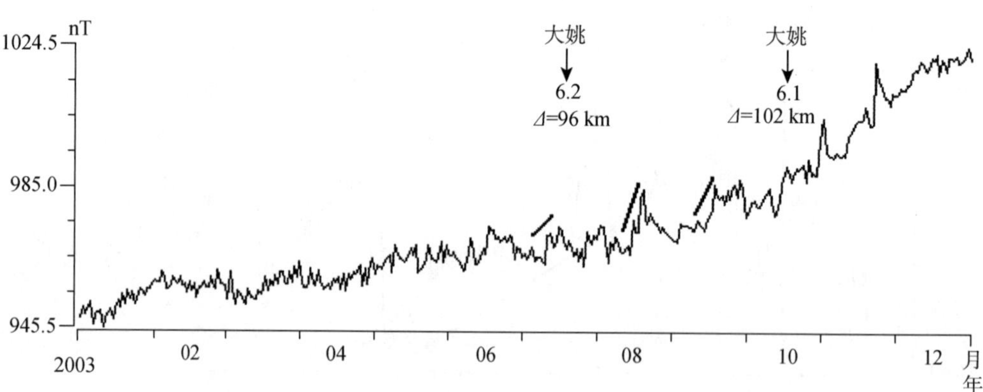

图 51　永胜地磁 *Z* 分量日均值曲线

Fig. 51　Daily mean value of vertical magnetic intensityt at Yongsheng station

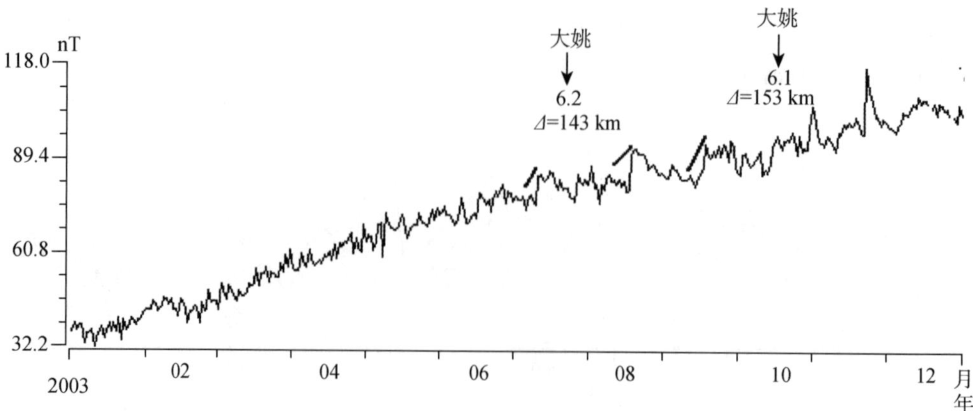

图 52　丽江地磁场 *Z* 分量日均值曲线

Fig. 52　Daily mean value of vertical magnetic intensityt at Lijiang station

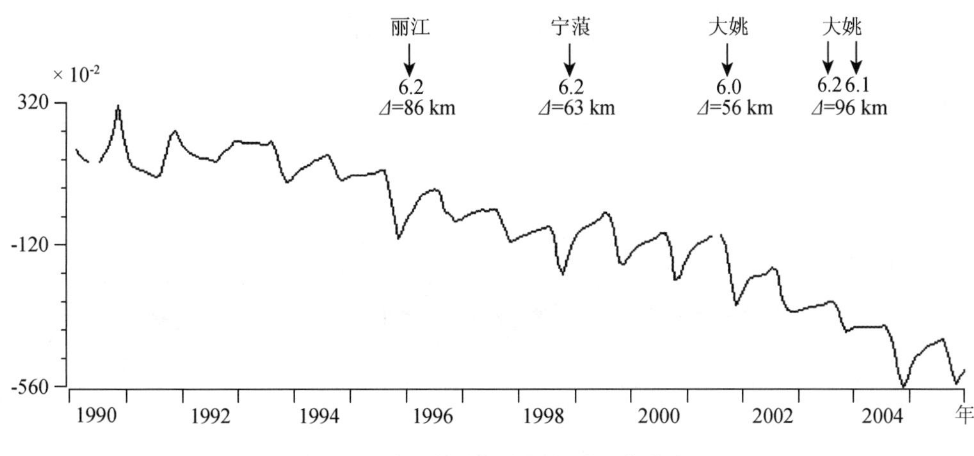

图 53a　永胜伸缩仪东西向月均值曲线

Fig. 53a　Monthly mean value of EW-extensometert at Yongsheng station

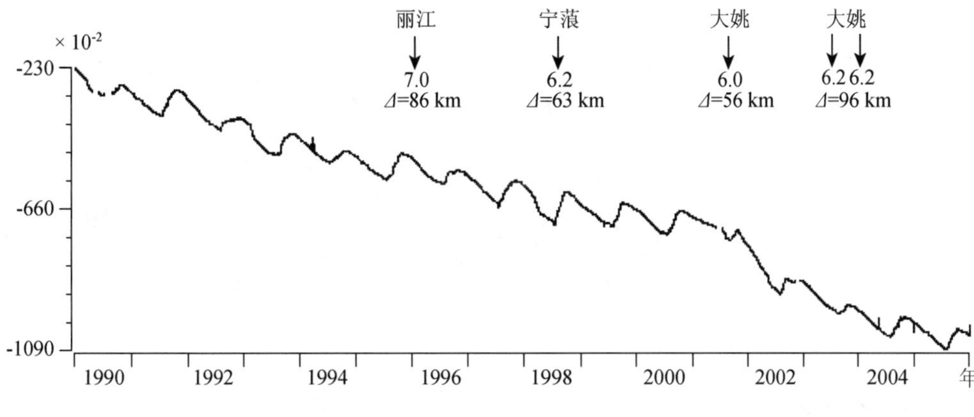

图 53b　永胜伸缩仪南北向月均值曲线

Fig. 53b　Monthly mean value of NS-extensometert at Yongsheng station

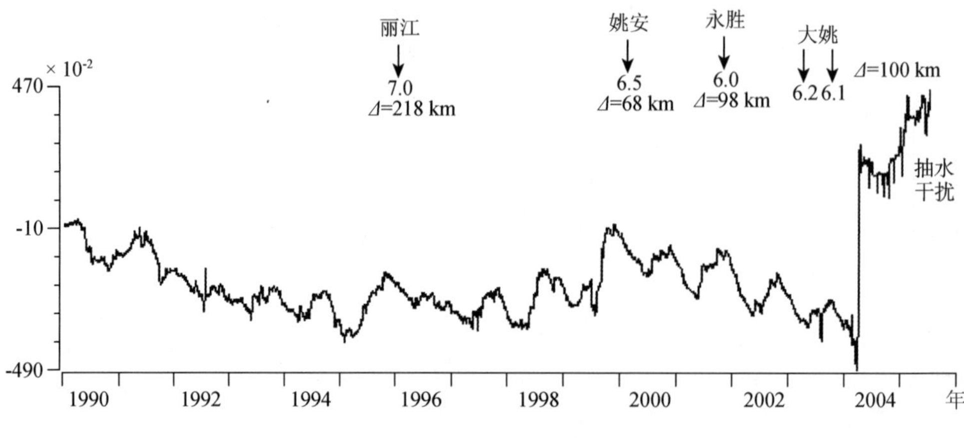

图 54a　弥渡伸缩仪东西向日均值曲线

Fig. 54a　Daily mean valueo of EW-extensometert at Midu station

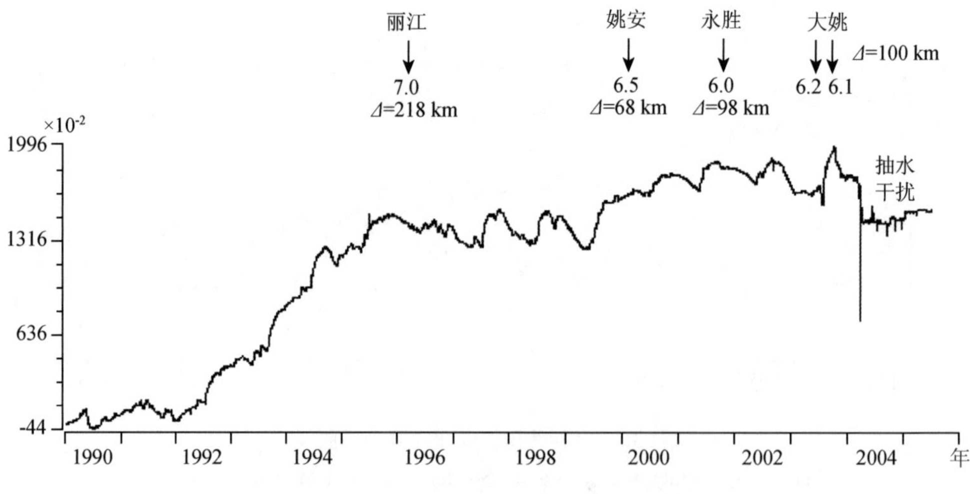

图 54b　弥渡伸缩仪南北向日均值曲线

Fig. 54b　Daily mean value of NS-extensometert at Midu station

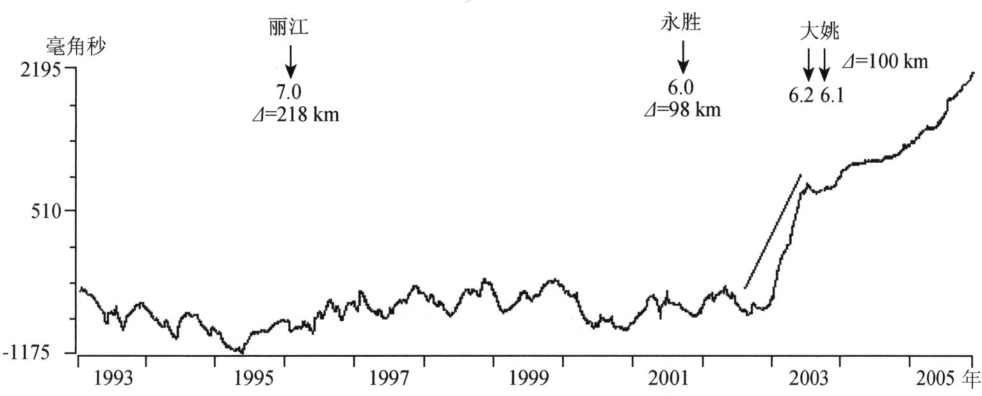

图 55a　弥渡石英管倾斜东西向日均值曲线

Fig. 55a　Daily mean value of EW-quartz tiltat Midu station

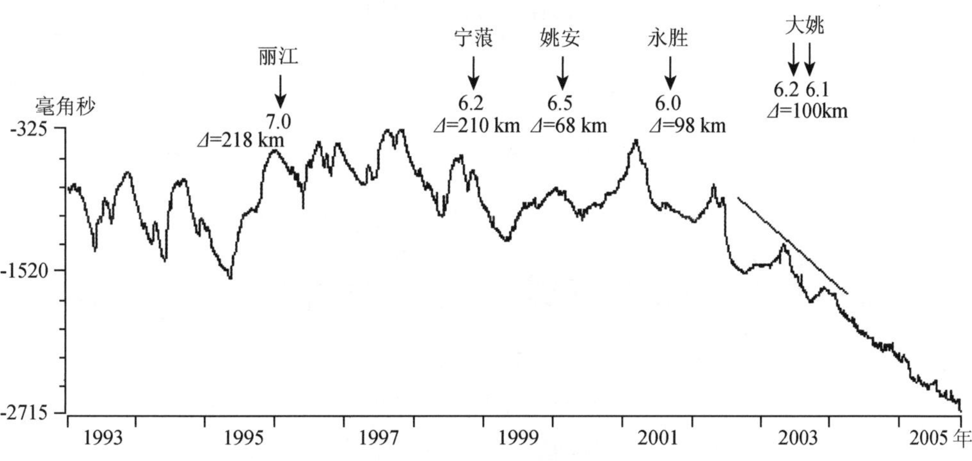

图 55b　弥渡石英管倾斜南北向日均值曲线

Fig. 55b　Daily mean value of NS-quartz tiltat Midu station

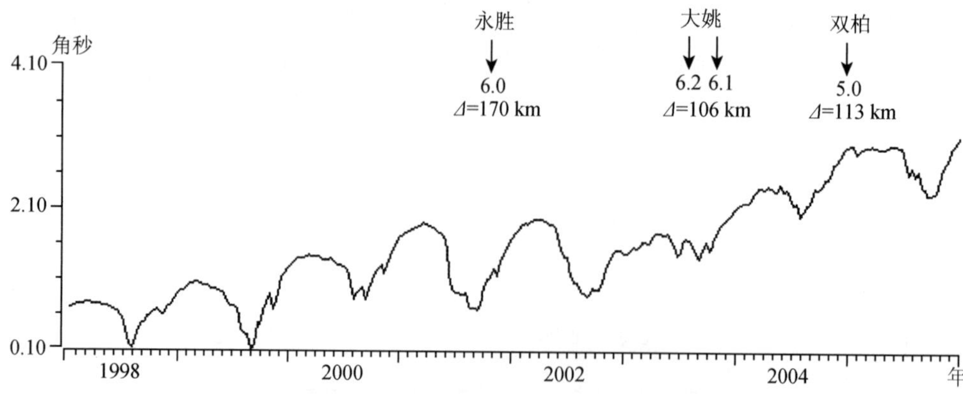

图 56　楚雄石英摆倾斜东西向旬均值曲线

Fig. 56　10-day mean value of EW -quartz tiltat Chuxiong station

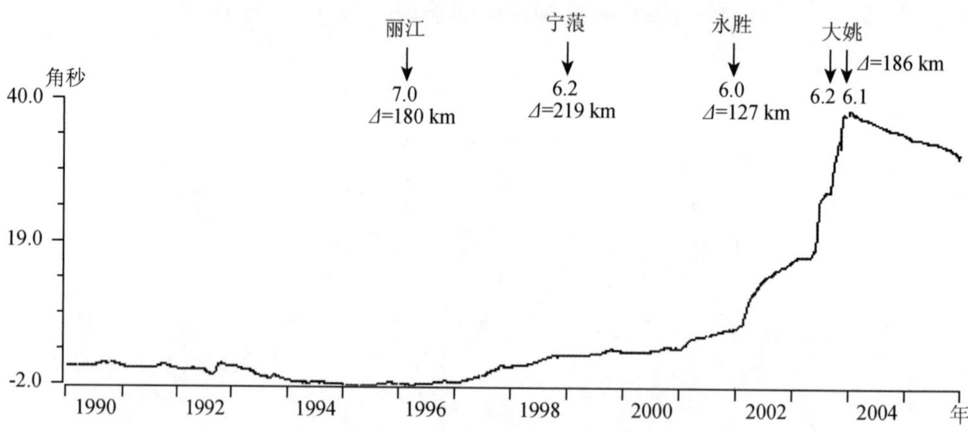

图 57a　云龙石英摆倾斜东西向五日均值曲线

Fig. 57a　5-day mean value of EW-quartz tiltat Yunlong station

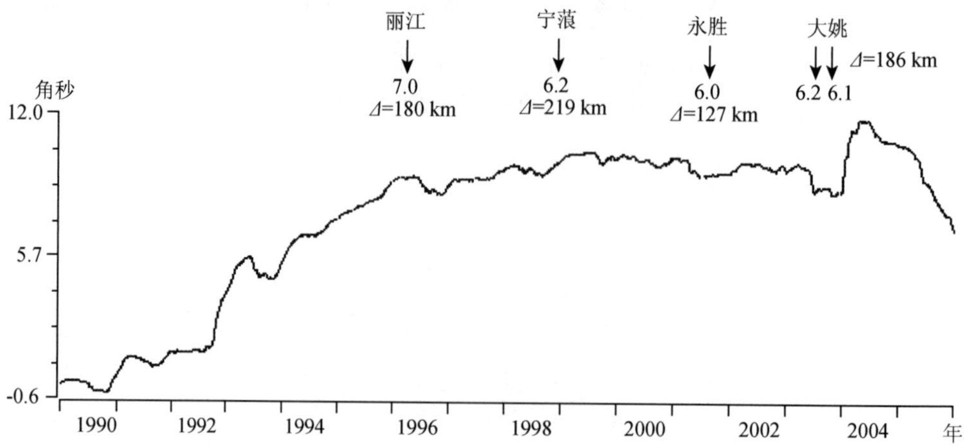

图 57b　云龙石英摆倾斜南北向五日均值曲线

Fig. 57b　10-day mean value of NS -quartz tiltat Yunlong station

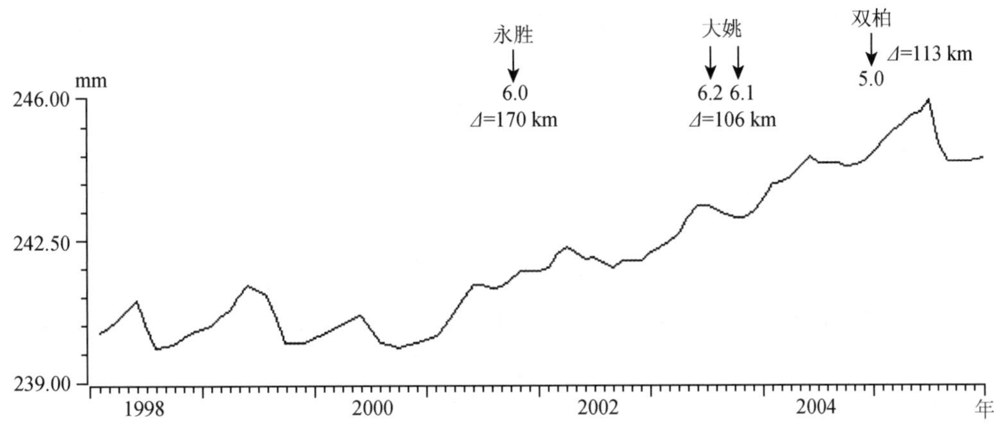

图 58 楚雄短水准 1—2 边旬均值

Fig. 58 10-day mean value of 1-2 side of short leveling Chuxiong station

七、前兆异常特征分析

1. 地震学异常

震前 1 年近场区的 4 级地震增长比较显著，为年度判定特别是发震地点的判断提供了依据。地震孕育中期阶段震源区和近场区 3、4 级地震活动处于平静，短期 2003 年 4、5 月开始震源区小震活动增加，尤其是 6 月 13 日中甸 4.1 级地震后至临震，滇西地区 3 级地震活动显著增强，以 2、3 级地震震群为主。

平静是个能量积累的过程，也是云南比较好的预报指标：中短期阶段云南省内 $M \geqslant 6$ 级地震平静，区域中等地震平静时间间隔达异常指标。2001 年 11 月 27 日永胜 6.0 级地震后出现了长达 18 个多月（632 天）的中强地震平静，滇西地区出现了达 315 天（2002 年 7 月 25 日姚安 4.0 级地震至 2003 年 6 月 13 日中甸 4.1 级地震）的 4 级地震平静异常，从 1996 年丽江 7 级地震后，滇西地区 4 级地震平静大于 130 天后均有 6 级以上地震发生。2003 年 6 月 21 日 6.2 级地震后云南省再次出现 3 级地震平静达 19 天，出现云南省内 1997 年以来中强地震连发指标。

2. 前兆异常

前兆异常以中短期异常为主，没有 5 年以上的长期异常。主要是 2001 年 10 月 27 日永胜 6 级地震与该地震同处一个地震活动区域，因此大姚 6.2 级地震前没有 5 年以上的长趋势异常，给出的基本上是中短期和短临异常。

从 2002 年始永胜、弥渡、云龙与楚雄的形变仪以及短水准的准同步大幅度破年变化（图 53 至图 58），为我们判断年度有 6 级地震提供了依据。前兆对于地点的预测有一定的困难，从云南的震例看，能对地点有指示意义的主要是流体物理量观测中的水位和水温。近年来云南的 6 级地震都发生在高水位地区，水位的升高表示孔隙压力增大，它预测的物理意义也较为清楚，2002 年大姚、宾川、南华、楚雄、大理、武定等水位都出现了年变化幅度加大，水位比往年升高的现象，即大姚震中周围出现了高水位异常。同时丽江的水温也出现增温现象，最大变幅为0.004℃，为中强地震的地点提供了参考依据。2002年11月处于断裂附近地区的大姚、姚安出现的4个水库库水发浑现象。短期较为突出的异常为水氡、水汞于2003年5月出现了多台项准同步、多年未见的突跳变化，临震前几天永胜、丽江的 Z 分量出现单点突跳。

7月21日大姚6.2级地震后，形变异常、水位、水氡、汞等前兆持续性异常未结束、又出现新的异常，地点主要集中在红河断裂中北段，宏观异常在大姚地震后空间范围有所扩展，分布在弥渡、鹤庆、龙陵、嵩明等地。据云南地区多年中强地震震例总结的经验，从时间、地点上确认云南省滇西地区还存在中强地震的危险，而永胜、丽江的 Z 分量于2003年8、9月再次出现更大幅度的突跳，为6.1级地震的临震预测提供了时间依据。

表9是给出M、S、I类异常在不同震中距的异常台站比例，从表可见震中周围100～200km范围内出现异常台站的比例最高，M类异常数量最多，而临震异常未出现于震源区，这与云南其他地震的前兆异常有一定的差异[6,7]。但从表9分析异常出现的时间早晚与震中距没有明显的关系。

表9　不同震中距的各类异常统计表

Table 9　Summary table of precursory anomalies with different the epicentral

震中距离/km	$0<\Delta\leqslant100$			$100<\Delta\leqslant200$			$200<\Delta\leqslant300$		
异常类别	M	S	I	M	S	I	M	S	I
异常百分比	40%	30%		32%	32%	11%	12%	12%	

八、震前预测、预防和震后响应

1. 地震预测情况

云南省地震预报中心于2002年11月17日至2003年9月填报预报卡4份，中期与短临预报卡各半，虚报一次，并以震情反映一份与会议纪要两份向云南省政府通报震情。楚雄州、大姚县地震局各报一份短临卡分别对两次大姚地震作出预报，一虚一准确。具体如下：

中期预报为云南省地震预报中心分别于2002年11月17日与2003年6月17～19日的分类会商卡B（2003-1）、B（2003-2），两次预报意见①中均指出滇西华坪—剑川—云龙—永平—下关—永仁一带，震级6级左右，2003年大姚的6.2、6.1级地震发生在我们圈定给出的滇西地震危险预测区内，做了时、空、强的准确中期预报。

短临预报为：2002年12月25日云南省地震预报研究中心，在联合楚雄州地震局、姚安县地震局、大姚县地震局等，调查了妙峰、胡家山、马游、梨园、洋派水库，确认水库发浑真实、可靠是一起宏观异常事件后，向中国地震局填写A类卡指出2003年1月1日至2003年3月31日（1）滇西北—滇中的丽江、大理、楚雄及川滇交界；存在5～6级地震。地点上再次指出了滇西北—滇中一带的危险性，但时间偏早属于虚报。同时2003年1月6

日，楚雄州地震局向中国地震局和云南省地震局填报临震卡［200301］，预报“楚雄的南华、姚安、大姚、永仁与丽江、攀枝花、大理交界附近 1 月底前，存在发生 5.5～6.4 级地震的危险，”该预报地点准确、时间偏早属于虚报。

2003 年 2 月 26 日在云南省监测预报处在楚雄主持召开的短临预报研讨会，会商结论意见为：“2003 年上半年（尤其注意 3、4 月份），省内楚雄、丽江、大理地区存在发生 5～6 级地震的危险。”2003 年 7 月 14 日云南省地震局确定 7 月 20 日在中甸召开滇西地区震情会，遗憾的是全体会议代表赶到中甸的第 2 天 7 月 21 日发生了大姚 6.2 级地震。对“7.21”大姚 6.2 级地震的预测中期较成功，短临有察觉，但未作出临震预报。

2003 年 8 月 22 日云南省地震局鉴于大姚 6.2 级地震后仍未缓解的震情形势以及新出现的突出异常情况，以“云南省近期震情分析”《震情反映（2003 05）》上报中国地震局和云南省人民政府，明确提出滇西北至滇中地区存在 6 级左右地震的危险。同时 2003 年 8 月 27～30 日和 9 月 8～10 日中国地震局组织在兰州、成都召开“中国大陆 7 级地震研讨会”的会议与“南北地震带紧急会商”会议，指出：“南北地震带近期存在发生 6 级地震的危险性。”“滇西至滇西北地区存在发生 6 级左右地震的可能。”而 2003 年 9 月 6 日[17)]大姚县地震局依据大姚水位、地温等部分前兆手段及大姚 6.2 级地震余震活动向云南省地震局填报了短临预报卡，预报：“2003 年 9 月 7 日至 12 月 7 日震级 5.0～5.9，大姚、永仁、元谋、宁蒗、永胜及以北的川、滇交界一带”，作出了较正确的短临预报。

2003 年 10 月 11 日，云南省政府召开了全省防震减灾工作会议，云政办发［2003］211 号，“云南省人民政府办公厅关于印发李汉柏副省长在全省防震减灾工作会议上讲话的通知”。皇甫岗局长首先就全国和云南近期地震形势做了专题报告，明确了近期工作的重点，提出需要采取的紧急措施。李汉柏副省长在会上作了重要指示，提出“地震部门不坚持 24 小时值班不放过；应急队伍不落实不放过；防震减灾机构、人员不落实不放过；物资储备不落实不放过；没有应急预案不放过”，对加强各项措施的落实起到重要作用。2003 年 12 月 4 日，云南省政府发文云政发［2003］175 号，“云南省人民政府关于表彰省地震局的决定”，指出“省地震局对大姚“7.21”和“10.16”地震作出了较为准确的中期、短期预报”。对大姚两次地震的预报给予了高度评价。

2. 地震应急

2003 年 7 月 21 日 23 时 16 分 30.1 秒大姚 6.2 级地震发生后，云南省地震局立即启动《云南省地震应急预案》，震后 20 分钟在云南省地震局皇甫岗局长的主持下，召开紧急地震应急会议，同时召开紧急会商会，研究震情动态，判定地震趋势。应急队由乔森副局长率领，在震后 1 小时，分批开赴大姚地震灾区。中国地震局也随即派出了 19 人的现场工作组到灾区指导工作。各小组于 22 日凌晨 5 时左右全部抵达大姚灾区，开展工作于 2003 年 7 月 29 日圆满完成任务。

10 月 16 日大姚 6.1 级地震发生后 5 分钟内，皇甫岗局长、胡永龙和乔森副局长，陈勤局长助理赶到地震应急指挥室，立即启动《云南省破坏性地震应急预案》二级对策及《云南省地震局破坏性地震应急预案》，成立前后方指挥部，指挥地震应急工作。地震发生后应急队由乔森副局长率队于 17 日凌晨 4 时左右全部抵达大姚灾区。中国地震局也随即派出了 15 人的现场工作组到灾区指导工作。地震发生 7 小时后，现场工作队各小组连夜开展灾害损失评估、强震观测、震情监视等工作。于2003年10月23日圆满完成了各项任务后，应急工作得到了中国地震局党组与领导的肯定，同时还受到了云南省委、省政府领导的表扬。

3. 震后趋势判定

云南省地震局数字遥测台网共23个台，大姚地区的地震监控能力下限达到1.5级，在对大姚6.2、6.1级地震序列跟踪监视和震后趋势判定中，使用了云南省地震遥测中心记录的序列目录；同时云南地区也具有丰富的前兆监测手段，在地震序列的跟踪监视过程中也发挥了作用。对大姚地震序列早期判定中，预报研究中心及现场工作组分析了该地震发生区域的历史资料认为：该地震距离最近的是1993年2月1日的大姚5.3级地震，仅相距16km；该区地震类型较为复杂，既有震群型也有主—余震型。大姚历史上4次5级以上地震均为主—余震型地震，如1488年大姚5.0级地震，1781年大姚5.5级地震，1947年大姚东北5.5级地震，1964年大姚5.4级地震和1993年大姚5.5级地震。姚安历史上有地震记载以来共发生3次5级以上地震，均为主—余震型地震，如1643年姚安5.5级地震，1962年姚安5.0级地震和1993年姚安5.6级地震。且该地震序列的频度衰减基本正常，但*M-t*图显示，主震后余震震级有一个逐渐增大的现象；另根据王绍晋震源机制解结果，大姚6.2级地震的主破裂方向约为N50°W，几次较大余震的震源机制解与主震机制结果相当一致，主震均为纯走滑断裂；该次地震的余震序列震中分布走向较为清晰，约为N60°W。目前最大余震只有3.8级，震级偏小。同时大姚6.2级地震前云南有部分前兆异常较为突出，如建水的应力异常和2001年永胜6.0级地震前相当，宾川的水位异常持续时间4个月可以与6级地震匹配，洱源的水汞异常似已结束，初步认为震区发生更大地震的可能性不大。初步判定为：①大姚6.2级地震类型为主—余震型；余震活动将持续一段时间，近期尚应注意4级余震的活动，对大姚地震序列的动态加强跟踪。②短期内云南地区警惕发生5～6级地震的危险，密切监视地震活动动态发展趋势，重点加强对川滇菱块西边界及滇西南地区震情的动态跟踪。

云南省地震预报研究中心连续七天加密会商，现场工作组每天实时对云南省地震局遥测台网资料、大姚县地震局的模拟测震资料和现场流动数字化台网的资料分析处理，到宏观震中区及周围调查收集震前及震后的宏观异常现象；并对大姚震区及周边的前兆观测资料进行实时的收集、处理分析进行会商讨论，对震情发展动态做出估计。现场工作组每天与省地震局监测预报处和地震分析预报研究中心汇报、沟通和了解有关情况，跟踪监视大姚地震序列和全省地震情况提交《震情简报》。这些工作对灾区取到了安定人心、稳定社会秩序的作用，为震后救援及恢复重建提供了良好的氛围。

九、总结与讨论

(1) 震中100km范围内70%的前兆观测台站出现异常，101～200km范围内有74%的前兆观测台站出现异常，201～300km范围内有24%的前兆观测台站出现异常，101～200km范围内出现异常的台站最多，但200km内的异常台站比例似乎与震中距离无关，这与以往的很多震例存在明显的差异[3]，造成前兆观测台站异常比例分布特异，同时也为用前兆异常分布预报地震地点带来了困难。

可能有如下两个原因：第一云南地区2001年始全面进行了“九五”数字化前兆台网改

造，大姚地震前部分前兆仪器记录受到了更替影响；第二从图 17 前兆异常分布可见，大姚地震前，前兆分布主要沿红河断裂带从滇西的丽江、永胜至滇南的玉溪、通海一带，小滇西的龙陵、施甸与昌宁一带，巧家也出现前兆密集。而 2003 年除大姚两次 6 级地震外，还发生了 2003 年 9 月 22 日小滇西境外的缅甸（19.8°N，95.2°E）7.2 级地震，2003 年 11 月 15、26 日鲁甸的 5.1、5.0 级地震（与巧家台震中距为 65km），由于云南地震多源场的存在及多条构造同时活动，其他的前兆异常密集区域可能与这些地震有一定的关系。

云南地区利用多水井观测站的优势，提出了利用“水位后效”[4]——水位、水温震后效应，确定后续地震的地点问题。水位震后效应是地震波传播后，震后应力调整，在某一地块或断裂上井孔含水层在区域应力场和地震波传播形成附加应力场迭加作用引起的变化，这种变化包含着前兆信息。该方法为云南地区中短期预测预报后续中强地震的地点的指标之一。2001 年昆仑山口西 8.1 级地震水位后效井位，形成滇西北东条带及滇东北东条带两条带分布，在大姚一带呈条带交汇的密集区；大姚 6.2 级地震水位、水温后效井位相对较集中于大姚、下关、保山及楚雄一带（图 59）。用水位、水温的震后效应及高水位集中区确定中强地震的地点，是大姚地震前成功应用前兆观测预测预报地点的有效方法。

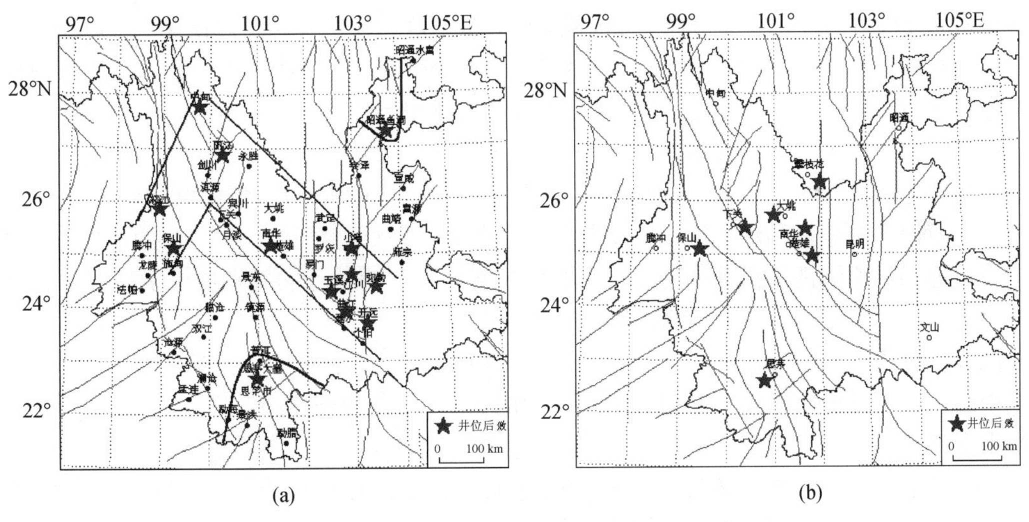

图 59　2001、2003 年云南地区水位后效台站分布图

Fig. 59　Distribution of stations of water level reffecting in 2001，2003 before eartquakes

（a）2001 年昆仑山口西 8.1 级地震水位后效；（b）2003 年大姚 6.2 级地震水位后效

（2）对大姚 6.2、6.1 级地震序列的判定过程中，常规的序列参数计算大多数趋向于主余震型，仅 4 级地震累积频度曲线对 6.2 级地震序列的判定为震群，因此与大姚实际发生的地震类别存在误判的问题。但动态追踪 3、4 级地震变化，发现 6.2 级地震后序列 *M-T* 曲线出现起伏大、余震强度偏低，有显著的密集—平静—增频现象，特别是 9 月 27 日的 4.2 级地震标志了序列地震活动的增强。同时对大姚地震附近地区的历史震例分析认为：该地震距离最近的是 1993 年 2 月 1 日的大姚 5.3 级地震，仅相距 16km；该区地震类型较为复杂，既有以 2000 年 1 月 15 日姚安 5.9、6.5 级地震的震群型，也有类似 1993 年 8 月 14 日姚安 5.6

级地震的主震余震型。因此，综合地震附近地区历史震例、动态分析序列的强度、频次变化为减少误差，准确判定地震序列类型的方法之一。

（3）根据云南省6级地震平静—活动的规律，4级地震密集区为准确判断年度地震的强度、地点提供了中期预测依据，结合前兆异常等特征对大姚6级地震作了较准确地中长期预报，实现了短临预测，尤其是6.1级地震前的成功短临预报。但短临阶段因地震以2、3级地震活动为主，对于云南地震较活跃的区域，该小震活动现象较为普遍而造成地震活动增长不显著；同时这次地震给出的37项定点前兆异常中，1个月内出现的临震异常仅有2项，大多数为中期和短期异常。回顾整个预报过程，发现目前我们没有能判定地震孕育已进入临震，在几天内要发震的判据和指标，同时鉴于目前前兆观测数字化资料时间较短，需重新认识正常、异常动态及逐步建立异常预测指标体系，环境变更影响一些台站正常工作，甚至台站报废、停测，使部分地区前兆观测台网更加稀疏、空白等，更增加了临震预报的难度。

（4）宏观异常实际和前兆异常一样[5]，也都存在长、中、短临不同尺度的异常[4]。从时间上分析两次大姚6级地震前，2002年11月姚安县胡家山、马由、洋派水库库水发浑，大姚县妙峰水库发浑等显著宏观异常，并非是短临信息。由于认识的局限性，云南省地震预报研究中心于2002年12月25日预测未来3个月内滇西地区有6级左右地震，但时间偏早属于虚报。大姚地震的经验再一次表明宏观异常和前兆异常一样，亦存在中、短、临异常；宏观异常是前兆异常的补充，前兆微观异常是基础，不能单凭宏观异常显著而预报地震。在实际预测预报中应考虑到宏观异常信息的复杂性，也说明了目前地震预报仍然存在很多困难。

（5）中期阶段近场区附近地震活动增强、中短期阶段前兆观测异常台项非线性增长，短期阶段部分趋势异常结束，台项出现减少是云南大部分$M \geqslant 6$级地震的共性特征[4]。大姚地震前云南地区6地震活动规律与形变趋势异常的结合较好地预测了这次地震的震级。中期震中附近近场区4级地震活动增强，出现高水位异常等，为地点的判别提供了判据。震源区2、3级地震活动增强、前兆异常数量从非线性增长到出现减少呈转折状态、地磁的突变异常是地震孕育进入短临阶段的显著特征。中期阶段地震学异常为震级和地点判定提供主要依据，中短期特别是短临阶段，前兆异常对时间的指示优于地震学指标，二者结合可以提高地震预报准确率。

（6）能否做到有效的防震减灾，实现明显的社会效益和减轻地震灾害损失，科学预测与政府减灾行动的有机结合是最关键的工作。

2003年2月云南省地震局监测处在楚雄组织召开的短临预报研讨会，促使各地州地震部门高度重视震情的发展，充分调动云南省各级地震部门的积极性，部分地震部门并向其相关政府部门做了震情汇报，取得了显著的社会经济效益。从思想和措施上对将发生的强震有预防准备，实际取到了地震预报的效果。

2003年7月16日大姚6.2级地震后，鉴于仍未缓解的震情形势以及新出现的突出异常情况，云南省地震局于2003年8月22日以“云南省近期震情分析”《震情反映（2003 05）》上报中国地震局和云南省人民政府，明确提出滇西北至滇中地区存在6级左右地震的危险。2003年8月28日，云南省地震局制定《云南省2003年下半年及近期地震短临跟踪实施方案》（云震发测〔2003〕71号）。要求有关地、州、市地震部门立即进入地震短临应震状态，并在全省范围内开展地震短临跟踪工作。2003年10月11日，云南省政府召开了全省防震减灾工作会议（云政办发［2003］211 号），皇甫岗局长的专题报告明确了近期工作的重点，提出需要采取的紧急措施。李汉柏副省长在会上作了重要指示，对加强各项措施的落实起到重要作用。从措施上对将发生 6.1 级地震取到了较好预报效果。

参 考 文 献

［1］苏有锦，2003 年 7 月 21 日、10 月 16 日云南大姚 6.2 级和 6.1 级地震预测预报回顾与讨论，国际地震动态，2004（1），18～21

［2］云南省地震局等，西南地震简目，成都：四川科技出版社，1988

［3］中国地震局监测预报司，西南地区强地震短期前兆特征和预测方法研究，北京：地震出版社，2005

［4］陈立德、付虹，地震预报基础与实践，北京：地震出版社，2003

［5］付虹、万登堡、张立，云南地区地震宏观异常研究，地震研究，2003，26（3），209～216

［6］陈棋福主编，中国震例（1992～1994），北京：地震出版社，2002

［7］陈棋福主编，中国震例（1995～1996），北京：地震出版社，2002

［8］崔建文、任增云、赵永庆等，大姚 6.2 级地震的强地震动观测研究，地震研究，2004：27（2），133～139

［9］李树华，云南大姚 6.2 级、6.1 级地震前电磁异常特征分析，地震地磁观测与研究，2004：25（5），1～10

［10］张俊伟、李永莉、叶建庆、王琼伟，成组连发过程中的大姚 6.2 和 6.1 级地震，地震研究，2004：27（增刊），23～28

［11］李永莉、苏有锦、田秀美、钱晓东，云南地区 3 次 6 级间歇性双震第二主震的前震特征分析，地震研究，2005：28（4），114～118

参 考 资 料

1）云南省地震局，2003 年 7 月 21 日大姚 6.2 级地震灾害损失评估报告

2）云南省地震局，2003 年 10 月 16 日大姚 6.1 级地震灾害损失评估报告

3）云南省地质局二区测大队，1∶20 万大姚幅区域地质调查报告，1969

4）云南省地震局预报监测处，滇西地区短临预报研讨会会议纪要，2003 年 2 月

5）云南省地震局，云南地震目录（区域台网）

6）云南省地震局，2005 年前强震危险性预测研究，1994 年 10 月

7）云南省地震局，云南省 2003 年度地震趋势研究报告，2002 年 11 月

8）云南省地震局，云南省 2003 年年中地震趋势研究报告，2003 年 6 月

9）云南省地震局，云南省 2004 年度地震趋势研究报告，2003 年 11 月

10）中国地震局，震例研究和报告编写规范（试行本）

11）中国地震局，中国强地震目录（公元前 23 世纪至公元 1999）

12）云南省地震局，云南省近期震情分析《震情反映（2003.05）》，2003 年 8 月

13）云南省地震局，云南省 2003 年下半年及近期地震短临跟踪实施方案（云震发测［2003］71 号），2003 年 8 月

14）中国地震局，关于印发《中国大陆 7 级地震研讨会纪要》的通知，2003 年 9 月

15）中国地震局，关于印发南北地震带紧急会商会纪要的通知，2003 年底

16）楚雄州地震局，申报“2003 年大姚 6.2、6.1 级地震监视预报与减灾对策”的减灾防震奖文件，2004 年

17）大姚县地震局，申报“‘7.21’、‘10.16’大姚 6.2、6.1 级地震监测预报、防灾对策及减灾实效”的减灾防震奖文件，2004 年

Dayao Earthquakes of M_S 6.2 on July 21, 2003 and M_S 6.1 on October 16, 2003 in Yunnan Province

Abstract

An earthquake of M_S6.2 and an earthquake of M_S6.1 took place seperately on July 21, 2003 and on Oct. 16 in Dayao county, Yunnan Province. The macroseismic epicenter was located around Tanhua, Shiyang and around Haigubo, Waiqidi of Liuju township in Dayao county. The meizoseismal area in ellipse went along North-West and the intensity of it was Ⅷ. In the both earthquakes, 19 persons were killed, 87 persons were severely wounded and 557 persons were slightly injured. The economic loss was 1007.5 million Yuan.

The two earthquakes were of swarm type. The largest aftershock of the M_S6.2 earthquake was of M_L4.9, and the largest aftershock of M_S6.1 was of M_S4.7. The aftershocks sequence mainly distribute along North-Weat. The focal mechanism solutions show that the eismogenic structure s are the North-West hidden fault The nodal plane I was the main rupture of this M_S6.2 event. The nodal plane Ⅱ was the main rupture of this M_S6.1 event. It is suggested that the earthquakes were caused by the right lateral displacement of the fault under in the NWW direction the action of the principle compressive stress in NNW direction.

There were 71 seismic stations within 300km from the epicenter, among which there being 9 stations with seismomentric observations, 52 stations with seismometric and precursory observations and 62 stations with precursory observations. Before both mainshocks, there were 19 anomaly items including 63 precursory anomalies, among which there were 4 anomaly items including 8 anomalies in seismic activity and 11 items including 37 anomalies in fixed point precursor. In addition, 9 macroscopic anomalies, 29 mid-terms anomalies and 24 short impending anomalies were acquires before the two earthquakes. Anomalous quiescence of M_L3.0-4.0 surrounding the epicentral region before the earthquake were significant features with mid-term anomalies. In the short-term, acticities of small shocks around the epicenter provided information about time and place of earthquake. Quiescence of M_L3.0 lasting 18 days is the significant feature in short-term stage before M_S6.2 earthquake in Yunnan. At the same time, that quiescence of M_S6.2 increased from August to September was imminent anomalies of M_S6.1 event. Within the distance of 100km from the epicenter, the ratios of stations and items with anomalies were 70% and 36% respectively. Within the distance of 101～200km from the epicenter, the ratios were 74% and 32% respectively and Within the distance of 201～300km from the epicenter, the ratios were 24% and 8% respectively. Many proplems such as harmonly of seismometric and precursor anomalies and prediction of earthquake are discussed in this paper.

The long-term, mid-term and short-term prediction of the earthquake made by Seismogical Bureau of Yunnan province, and imminet prediction of the earthquake before the M_S6. 1 earthquake was very exact. On Aug. 25, 2003, the Seismological Bureau of Yunnan province reported the forecasting to the government of the province and the Seismological Bureau of china. They mitigated the loss to the lowest level.

报告附件

大姚地震序列目录：

2003	721231630	25.95	101.236.20	6	1
2003	7212320 3	25.90	101.203.70	0	1
2003	721232211	25.95	101.223.10	0	1
2003	721232211	0.00	0.003.30	0	1
2003	721232229	26.00	101.253.60	0	1
2003	721232933	25.98	101.223.20	0	1
2003	7212330 2	26.02	101.252.80	0	1
2003	721233048	0.00	0.002.00	0	1
2003	721233144	25.98	101.232.20	0	1
2003	721233153	0.00	0.002.20	0	1
2003	721233712	25.97	101.182.60	0	1
2003	721233926	25.92	101.252.90	15	1
2003	721234322	26.05	101.032.40	12	1
2003	721234444	25.92	101.182.60	0	1
2003	721234552	25.98	101.223.20	0	1
2003	721235022	0.00	0.002.20	0	1
2003	721235716	26.02	101.053.80	1	0
2003	721235951	26.00	101.172.50	0	1
2003	722 0 140	0.00	0.002.30	0	1
2003	722 0 324	26.03	101.082.20	10	1
2003	722 0 340	26.00	101.202.60	25	1
2003	722 0 432	26.00	101.132.10	0	1
2003	722 0 546	0.00	0.002.00	0	1
2003	722 0 613	0.00	0.002.10	0	1
2003	722 0 643	25.97	101.222.70	0	1
2003	722 0 925	0.00	0.002.00	0	1
2003	722 014 7	25.97	101.232.70	0	1
2003	722 01649	25.97	101.272.00	5	1
2003	722 02021	25.98	101.072.50	5	1
2003	722 02845	25.87	101.232.80	0	1
2003	722 03523	26.03	101.123.40	10	1
2003	722 04552	26.00	101.252.10	17	1
2003	722 05039	0.00	0.002.00	0	1
2003	722 05523	25.97	101.223.10	0	1

2003	722 05653	0. 00	0. 002. 30	0	1
2003	722 1 915	25. 95	101. 232. 30	0	1
2003	722 11722	25. 97	101. 202. 10	10	1
2003	722 12458	25. 97	101. 233. 10	0	1
2003	722 126 3	0. 00	0. 002. 00	0	1
2003	722 12713	25. 97	101. 222. 60	0	1
2003	722 13335	26. 00	101. 152. 70	0	1
2003	722 13438	25. 98	101. 182. 60	10	1
2003	722 15124	0. 00	0. 002. 30	0	1
2003	722 15325	25. 97	101. 232. 70	0	1
2003	722 15912	25. 90	101. 222. 60	0	1
2003	722 2 910	26. 00	101. 082. 50	0	1
2003	722 21713	26. 02	101. 102. 00	10	1
2003	722 22714	26. 00	101. 252. 00	7	1
2003	722 22758	25. 98	101. 172. 30	0	1
2003	722 23755	26. 05	101. 232. 80	10	1
2003	722 24119	25. 98	101. 072. 50	0	1
2003	722 24535	25. 98	101. 232. 80	10	1
2003	722 246 2	0. 00	0. 002. 20	0	1
2003	722 31111	0. 00	0. 002. 00	0	1
2003	722 32258	25. 97	101. 222. 70	0	1
2003	722 33543	25. 95	101. 252. 60	0	1
2003	722 420 8	25. 97	101. 322. 40	0	1
2003	722 422 4	26. 00	101. 102. 40	0	1
2003	722 42321	25. 88	101. 172. 40	0	1
2003	722 42630	25. 98	101. 203. 30	0	1
2003	722 42813	26. 02	101. 232. 00	10	1
2003	722 44752	26. 00	101. 233. 00	0	1
2003	722 51535	25. 97	101. 232. 70	0	1
2003	722 52428	26. 00	101. 222. 20	10	1
2003	722 55329	0. 00	0. 002. 00	0	1
2003	722 6 3 7	25. 97	101. 182. 10	9	1
2003	722 65158	26. 02	101. 272. 20	11	1
2003	722 65456	26. 07	101. 202. 50	10	1
2003	722 659 0	25. 95	101. 102. 30	0	1
2003	722 72810	25. 98	101. 232. 10	5	1
2003	722 8 938	25. 93	101. 204. 30	4	1
2003	722 81244	0. 00	0. 002. 10	0	1

2003	722 85723	26. 05	101. 182. 70	0	1
2003	722 85933	26. 08	101. 172. 50	10	1
2003	722 91318	0. 00	0. 002. 00	0	1
2003	722 92315	26. 02	101. 202. 80	10	1
2003	722 92729	0. 00	0. 002. 10	0	1
2003	722 95651	26. 02	101. 183. 10	5	1
2003	7221015 3	25. 98	101. 272. 70	10	1
2003	7221057 2	0. 00	0. 002. 10	0	1
2003	72211 8 4	25. 98	101. 282. 20	10	1
2003	7221136 9	0. 00	0. 002. 30	0	1
2003	722112549	26. 07	100. 952. 20	10	1
2003	7221148 1	26. 02	101. 102. 90	10	1
2003	7221254 3	26. 05	101. 272. 80	0	1
2003	7221335 8	0. 00	0. 002. 10	0	1
2003	722145236	26. 03	101. 282. 20	14	1
2003	722153038	0. 00	0. 002. 00	0	1
2003	722163136	26. 00	101. 252. 90	0	1
2003	72217 717	0. 00	0. 002. 30	0	1
2003	7221723 8	0. 00	0. 002. 10	0	1
2003	72218 2 7	25. 95	101. 272. 60	0	1
2003	72218 840	25. 98	101. 303. 40	0	1
2003	722181513	25. 98	101. 252. 90	0	1
2003	722182130	0. 00	0. 002. 00	0	1
2003	722184836	0. 00	0. 002. 00	0	1
2003	722185712	25. 97	101. 152. 20	0	1
2003	72219 0 6	26. 00	101. 223. 70	10	1
2003	722193333	0. 00	0. 002. 10	0	1
2003	722204452	0. 00	0. 002. 10	0	1
2003	722205445	25. 98	101. 232. 60	0	1
2003	722221256	25. 92	101. 172. 40	0	1
2003	7222234 5	25. 98	101. 232. 60	0	1
2003	722225242	26. 00	101. 282. 70	0	1
2003	722232550	0. 00	0. 002. 00	0	1
2003	723 0 048	25. 97	101. 252. 90	0	1
2003	723 0 948	0. 00	0. 002. 10	0	1
2003	723 040 4	25. 97	101. 282. 80	0	1
2003	723 04140	25. 95	101. 222. 60	0	1
2003	723 047 4	25. 98	101. 282. 70	0	1

2003	723 1 854	0.00	0.002.20	0	1
2003	723 22250	0.00	0.002.00	0	1
2003	723 35314	0.00	0.002.10	0	1
2003	723 35510	0.00	0.002.30	0	1
2003	723 4 631	25.97	101.222.30	0	1
2003	723 4 711	25.98	101.132.20	0	1
2003	723 42426	25.98	101.282.60	0	1
2003	723 42832	26.00	101.272.40	10	1
2003	723 54419	26.02	101.102.50	10	1
2003	723 63046	25.88	101.132.20	0	1
2003	723 63636	25.98	101.382.60	0	1
2003	723 74218	26.00	101.182.90	0	1
2003	723 8 329	25.97	101.272.80	0	1
2003	723 8 623	0.00	0.002.00	0	1
2003	723 81613	25.92	101.202.30	0	1
2003	723 84727	0.00	0.002.20	0	1
2003	723 9 5 8	25.82	101.052.30	25	1
2003	723 93750	0.00	0.002.00	0	1
2003	723114134	0.00	0.002.00	0	1
2003	72312 120	0.00	0.002.10	0	1
2003	7231212 1	25.95	101.273.10	0	1
2003	723124230	26.00	101.054.30	8	1
2003	723124846	26.02	101.082.60	10	1
2003	723125645	25.95	101.252.40	10	1
2003	7231311 2	25.93	101.252.30	0	1
2003	723131827	25.88	101.172.40	0	1
2003	7231342 2	25.97	101.273.00	0	1
2003	723134333	0.00	0.002.00	0	1
2003	723142150	0.00	0.002.20	0	1
2003	723144628	25.95	101.233.60	10	1
2003	72315 2 2	25.97	101.152.70	0	1
2003	72315 816	25.82	101.002.40	25	1
2003	723154948	0.00	0.002.30	0	1
2003	723161010	0.00	0.002.00	0	1
2003	723162358	0.00	0.002.10	0	1
2003	723164329	0.00	0.002.30	0	1
2003	72319 116	0.00	0.002.20	0	1
2003	72320 2 9	0.00	0.002.10	0	1

2003	723204727	26. 02	101. 332. 80	0	1
2003	723221730	25. 97	101. 253. 20	0	1
2003	723223110	25. 98	101. 223. 80	0	1
2003	723224135	0. 00	0. 002. 00	0	1
2003	723224222	0. 00	0. 002. 00	0	1
2003	723224656	25. 97	101. 252. 40	0	1
2003	723225630	25. 93	101. 232. 40	0	1
2003	72323 158	0. 00	0. 002. 20	0	1
2003	72323 827	25. 97	101. 222. 90	0	1
2003	724 055 1	0. 00	0. 002. 00	0	1
2003	724 1 351	25. 93	101. 202. 90	10	1
2003	724 11821	0. 00	0. 002. 10	0	1
2003	724 15324	26. 08	101. 182. 90	0	1
2003	724 22454	25. 98	101. 122. 20	10	1
2003	724 23120	0. 00	0. 002. 30	0	1
2003	724 342 0	26. 08	101. 082. 60	0	1
2003	724 35124	25. 97	101. 202. 50	0	1
2003	724 43626	0. 00	0. 002. 30	0	1
2003	724 44632	25. 97	101. 252. 50	5	1
2003	724 61839	25. 98	101. 232. 90	0	1
2003	724 73941	0. 00	0. 002. 20	0	1
2003	724 75157	0. 00	0. 002. 00	0	1
2003	724 75241	25. 93	101. 274. 70	10	1
2003	724 8 813	25. 97	101. 272. 80	10	1
2003	724 915 0	0. 00	0. 002. 30	0	1
2003	724 92122	0. 00	0. 002. 30	0	1
2003	724 92734	0. 00	0. 002. 20	0	1
2003	724 94822	25. 92	101. 172. 40	0	1
2003	724113033	26. 00	101. 232. 90	10	1
2003	7241138 5	0. 00	0. 002. 30	0	1
2003	724123719	25. 95	101. 223. 50	10	1
2003	72413 143	25. 95	101. 233. 10	5	1
2003	724133311	0. 00	0. 002. 10	0	1
2003	72414 210	25. 97	101. 202. 80	10	1
2003	724151850	25. 90	101. 273. 20	15	1
2003	7241536 0	25. 95	101. 203. 00	10	1
2003	72416 611	25. 97	101. 252. 70	0	1
2003	72417 624	0. 00	0. 002. 10	0	1

2003	724171337	26.00	101.182.90	10	1
2003	724211725	25.98	101.252.50	0	1
2003	724222637	25.97	101.122.50	0	1
2003	724222715	25.93	101.223.20	10	1
2003	72423 818	25.98	101.082.20	0	1
2003	724231110	26.00	101.083.10	10	1
2003	725 03054	26.02	101.122.30	5	1
2003	725 14315	25.98	101.232.60	5	1
2003	725 521 5	25.98	101.232.70	0	1
2003	725 72257	25.95	101.272.20	0	1
2003	725 91948	0.00	0.002.30	0	1
2003	72511 140	26.00	101.132.30	10	1
2003	7251415 1	26.00	101.252.70	10	1
2003	725143127	0.00	0.002.00	0	1
2003	7251433 6	0.00	0.002.10	0	1
2003	725144837	0.00	0.002.20	0	1
2003	725233046	25.95	101.322.60	10	1
2003	726 315 9	26.03	101.083.40	10	1
2003	726 418 1	26.05	101.172.10	10	1
2003	726 72512	26.03	101.202.70	10	1
2003	726 74445	0.00	0.002.10	0	1
2003	726 856 6	25.97	101.283.10	10	1
2003	726 9 254	26.00	101.002.20	0	1
2003	726 92947	25.98	101.252.30	10	1
2003	726 94136	0.00	0.002.30	0	1
2003	726125225	25.97	101.253.30	10	1
2003	726132434	25.98	101.253.60	10	1
2003	7261328 9	0.00	0.002.00	0	1
2003	7261425 9	26.03	101.102.90	10	1
2003	72615 534	25.78	101.082.00	25	1
2003	726155012	26.05	101.282.40	10	1
2003	7261645 2	25.98	101.182.10	10	1
2003	7262058 0	26.05	101.152.80	10	1
2003	726205951	26.05	101.172.50	10	1
2003	727 71644	26.03	101.072.50	10	1
2003	727 73516	25.92	101.252.20	0	1
2003	727 74138	26.03	101.172.20	0	1
2003	727 84320	25.97	101.272.70	10	1

2003	727 952 9	0. 00	0. 002. 00	0	1
2003	727 959 1	26. 02	101. 282. 40	10	1
2003	72710 533	26. 03	101. 302. 50	10	1
2003	727102528	0. 00	0. 002. 10	0	1
2003	727114333	25. 95	101. 272. 40	0	1
2003	727124834	26. 02	101. 222. 60	10	1
2003	727212639	26. 02	101. 122. 60	10	1
2003	728 45220	26. 00	101. 172. 20	0	1
2003	728201410	25. 97	101. 232. 20	0	1
2003	729 3 812	25. 98	101. 222. 80	10	1
2003	729 42740	25. 97	101. 023. 10	10	1
2003	729 42953	26. 00	101. 072. 50	10	1
2003	729 73132	26. 02	101. 032. 50	10	1
2003	7291342 0	25. 97	101. 252. 50	10	1
2003	729181655	25. 92	101. 232. 20	0	1
2003	729184717	0. 00	0. 002. 20	0	1
2003	7292248 5	25. 97	101. 272. 60	10	1
2003	730 212 1	26. 00	101. 252. 10	0	1
2003	730 85820	25. 92	101. 252. 10	0	1
2003	730125527	25. 98	101. 252. 70	10	1
2003	730153132	25. 95	101. 232. 10	0	1
2003	730165726	0. 00	0. 002. 00	0	1
2003	73017 728	0. 00	0. 002. 20	0	1
2003	730204636	0. 00	0. 002. 00	0	1
2003	730205026	25. 90	101. 022. 30	25	1
2003	7302131 2	26. 02	101. 172. 90	10	1
2003	730214154	25. 97	101. 302. 10	0	1
2003	731 140 4	25. 95	101. 233. 00	15	1
2003	731 234 9	25. 98	101. 153. 30	15	1
2003	731124550	25. 97	101. 252. 60	10	1
2003	731182125	0. 00	0. 002. 00	0	1
2003	731212848	25. 98	101. 222. 90	10	1
2003	8 1 14040	0. 00	0. 002. 00	0	1
2003	8 1202610	26. 00	101. 102. 00	0	1
2003	8 2 84618	25. 95	101. 132. 90	5	1
2003	8 2113747	0. 00	0. 002. 20	0	1
2003	8 2141211	0. 00	0. 002. 00	0	1
2003	8 215 837	0. 00	0. 002. 10	0	1

2003	8 2154810	25.98	101.252.00	0	1
2003	8 2231755	26.02	101.073.10	5	1
2003	8 3 93640	26.00	101.183.10	5	1
2003	8 4 93240	0.00	0.002.00	0	1
2003	8 4153935	0.00	0.002.10	0	1
2003	8 4163328	26.02	101.123.70	10	1
2003	8 4201918	0.00	0.002.30	0	1
2003	8 4211140	0.00	0.002.40	0	1
2003	8 4212111	26.00	101.153.00	5	1
2003	8 4214214	25.97	101.252.70	10	1
2003	8 5 35429	25.98	101.132.50	0	1
2003	8 52049 5	25.93	101.234.90	0	1
2003	8 5205454	26.00	101.282.30	10	1
2003	8 6 05057	26.02	101.182.60	10	1
2003	8 6 137 8	25.98	101.252.10	9	1
2003	8 6 73459	0.00	0.002.00	0	1
2003	8 6 92326	0.00	0.002.20	0	1
2003	8 6103219	25.98	101.233.30	0	1
2003	8 6105136	26.00	101.322.70	0	1
2003	8 622 748	0.00	0.002.20	0	1
2003	8 7 11238	0.00	0.002.00	0	1
2003	8 7 64526	0.00	0.002.00	0	1
2003	8 7 840 8	0.00	0.002.00	0	1
2003	8 711 419	25.97	101.322.70	0	1
2003	8 7164255	26.05	101.252.90	0	1
2003	8 7192613	25.98	101.202.00	7	1
2003	8 8 813 9	26.05	101.102.80	10	1
2003	8 8173154	0.00	0.002.00	0	1
2003	8 9 827 3	25.93	101.272.60	0	1
2003	8 9104737	25.93	101.352.30	0	1
2003	8 9134621	25.97	101.332.40	0	1
2003	8 9154821	0.00	0.002.00	0	1
2003	8 9205229	25.98	101.232.60	0	1
2003	8 9212522	0.00	0.002.30	0	1
2003	8 9222748	26.03	101.273.20	0	1
2003	81017 6 7	25.97	101.233.70	5	1
2003	8101945 3	26.00	101.232.50	10	1
2003	811163920	0.00	0.002.10	0	1

2003	812 42744	25. 98	101. 282. 60	10	1
2003	812183416	0. 00	0. 002. 10	0	1
2003	812212140	25. 97	101. 202. 50	10	1
2003	812235941	26. 00	101. 034. 90	9	1
2003	813 14842	26. 03	101. 082. 30	10	1
2003	813 15711	25. 98	101. 273. 00	0	1
2003	813 84748	25. 98	101. 272. 50	0	1
2003	813 95636	0. 00	0. 002. 00	0	1
2003	813114229	25. 95	101. 252. 60	0	1
2003	813203913	0. 00	0. 002. 00	0	1
2003	813224248	26. 00	101. 073. 00	25	1
2003	813224433	26. 02	101. 073. 10	20	1
2003	814 05629	26. 05	101. 132. 50	15	1
2003	814205627	25. 95	101. 272. 60	0	1
2003	814221631	0. 00	0. 002. 10	0	1
2003	81423 549	0. 00	0. 002. 00	0	1
2003	817 02018	0. 00	0. 002. 20	0	1
2003	817 023 4	25. 72	101. 232. 10	0	1
2003	817 05713	26. 00	101. 033. 00	5	1
2003	817 126 0	25. 92	101. 282. 90	0	1
2003	817 64147	0. 00	0. 002. 20	0	1
2003	8171141 2	0. 00	0. 002. 10	0	1
2003	817191038	0. 00	0. 002. 30	0	1
2003	81818 017	25. 92	101. 224. 80	8	1
2003	81818 558	0. 00	0. 002. 30	0	1
2003	819 54429	25. 97	101. 272. 10	10	1
2003	820 64222	0. 00	0. 002. 00	0	1
2003	820 85738	25. 92	101. 222. 10	0	1
2003	82121 7 3	25. 97	101. 272. 50	0	1
2003	823193745	25. 95	101. 252. 50	10	1
2003	8251525 3	25. 93	101. 253. 20	10	1
2003	8251655 2	25. 95	101. 252. 70	0	1
2003	825222642	25. 93	101. 282. 90	0	1
2003	826 74935	0. 00	0. 002. 10	0	1
2003	826 92827	26. 05	101. 172. 40	0	1
2003	82619 356	25. 92	101. 172. 90	0	1
2003	827131550	0. 00	0. 002. 30	0	1
2003	827132221	26. 00	101. 082. 40	0	1

2003	82719 219	25.92	101.222.40	0	1
2003	828 92128	25.95	101.252.50	0	1
2003	828 92622	25.95	101.252.00	4	1
2003	828201258	25.95	101.252.60	0	1
2003	829 33828	0.00	0.002.10	0	1
2003	829 5 040	25.97	101.182.20	10	1
2003	829 75053	25.95	101.272.60	10	1
2003	829 829 5	25.92	101.222.20	0	1
2003	82923 621	25.95	101.222.10	0	1
2003	830 81110	0.00	0.002.00	0	1
2003	83014 2 6	25.95	101.282.20	0	1
2003	831 44433	26.00	101.232.70	5	1
2003	9 3 5 130	26.08	101.132.70	5	1
2003	9 4 6 144	25.98	101.282.80	0	1
2003	9 5105513	0.00	0.002.10	0	1
2003	9 6133138	25.97	101.273.20	10	1
2003	9 6133352	25.97	101.232.90	0	1
2003	9 712 646	25.98	101.232.80	10	1
2003	9 71948 0	0.00	0.002.00	0	1
2003	9 7214630	26.02	101.232.10	3	1
2003	9 8145451	25.95	101.273.20	10	1
2003	9 8145451	25.98	101.252.20	19	1
2003	9 8145546	25.98	101.323.10	10	1
2003	9 91256 4	0.00	0.002.00	0	1
2003	9 9125738	25.97	101.322.90	5	1
2003	9 9125752	25.90	101.273.40	10	1
2003	9 9131047	0.00	0.002.00	0	1
2003	9 9152741	26.07	101.422.00	18	1
2003	9 91638 8	0.00	0.002.10	0	1
2003	9 9195130	25.98	101.372.00	13	1
2003	910 4 344	25.98	101.322.70	10	1
2003	910 523 8	25.97	101.302.40	10	1
2003	911152859	25.93	101.282.60	10	1
2003	912 34025	25.72	101.232.30	0	1
2003	912 73933	0.00	0.002.20	0	1
2003	9121946 7	0.00	0.002.00	0	1
2003	912205318	25.92	101.303.30	10	1
2003	912205536	25.95	101.322.00	0	1

2003	912205952	25.98	101.322.00	9	1
2003	913155459	0.00	0.002.00	0	1
2003	913163615	25.95	101.302.10	10	1
2003	915183449	25.95	101.283.00	0	1
2003	916203656	0.00	0.002.00	0	1
2003	920 82726	25.98	101.322.60	0	1
2003	921195054	0.00	0.002.20	0	1
2003	922122224	0.00	0.002.40	0	1
2003	9221253 7	0.00	0.002.40	0	1
2003	924 65954	25.97	101.223.70	0	1
2003	924 7 145	25.92	101.232.60	0	1
2003	924 71525	26.00	101.283.10	0	1
2003	924 74249	0.00	0.002.10	0	1
2003	924 83915	0.00	0.002.00	0	1
2003	9242239 2	25.88	101.182.20	0	1
2003	924231631	0.00	0.002.10	0	1
2003	92516 133	25.97	101.252.50	10	1
2003	926 44115	0.00	0.002.10	0	1
2003	927 9 630	25.93	101.224.20	10	1
2003	927 9 939	0.00	0.002.10	0	1
2003	927101719	25.95	101.273.30	10	1
2003	927103422	25.97	101.272.70	10	1
2003	927114655	0.00	0.002.00	0	1
2003	927115314	25.95	101.252.20	9	1
2003	92716 710	26.02	101.172.70	10	1
2003	927202949	25.95	101.322.40	10	1
2003	929 1 248	0.00	0.002.00	0	1
2003	929132519	26.02	101.053.10	10	1
2003	929175150	0.00	0.002.00	0	1
2003	929231259	0.00	0.002.20	0	1
2003	10 5 33832	25.95	101.253.00	0	1
2003	10 6142320	0.00	0.002.30	0	1
2003	10 8185550	25.98	101.282.90	0	1
2003	10 9204111	0.00	0.002.30	0	1
2003	1010 6 5 1	25.93	101.182.10	16	1
2003	1011171120	25.97	101.203.00	5	1
2003	1012141914	25.97	101.282.30	10	1
2003	1014 01847	25.98	101.253.10	10	1

2003	1014 84947	25.97	101.222.60	10	1
2003	1014102548	25.98	101.282.10	10	1
2003	1016115630	25.95	101.223.90	0	1
2003	101613 744	0.00	0.002.20	0	1
2003	10162028 4	25.92	101.306.10	5	2
2003	1016203343	0.00	0.002.40	0	1
2003	10162037 5	0.00	0.002.20	0	1
2003	1016203757	25.72	101.232.20	0	1
2003	1016203919	25.72	101.232.60	0	1
2003	1016203932	25.95	101.273.10	15	1
2003	1016204055	0.00	0.002.60	0	1
2003	1016204150	0.00	0.002.00	0	1
2003	1016204232	0.00	0.002.70	0	1
2003	1016204414	0.00	0.002.20	0	1
2003	1016204843	0.00	0.002.10	0	1
2003	1016205119	0.00	0.002.70	0	1
2003	1016205246	25.72	101.232.20	0	1
2003	10162054 3	25.72	101.232.00	0	1
2003	1016205455	0.00	0.002.00	0	1
2003	10162055 1	0.00	0.002.10	0	1
2003	1016205614	0.00	0.002.60	0	1
2003	101621 530	25.72	101.232.10	0	1
2003	101621 735	25.95	101.274.00	0	1
2003	1016211050	25.72	101.232.40	0	1
2003	1016211319	25.97	101.432.20	9	1
2003	1016211719	0.00	0.002.20	0	1
2003	1016212719	25.97	101.302.20	10	1
2003	1016212814	25.95	101.323.70	0	1
2003	1016213452	25.93	101.432.70	0	1
2003	1016214132	25.95	101.322.70	0	1
2003	1016214417	0.00	0.002.00	0	1
2003	1016215517	0.00	0.002.10	0	1
2003	1016215531	25.90	101.322.30	10	1
2003	101622 359	0.00	0.002.10	0	1
2003	101622 459	0.00	0.002.10	0	1
2003	101622 546	25.93	101.322.70	0	1
2003	1016221544	0.00	0.002.10	0	1
2003	1016222256	0.00	0.002.30	0	1

2003	1016223024	25. 95	101. 332. 60	0	1
2003	1016223713	25. 97	101. 252. 60	10	1
2003	1016224337	25. 93	101. 272. 80	10	1
2003	1016234429	25. 72	101. 232. 10	0	1
2003	1016235310	25. 97	101. 272. 80	10	1
2003	1016235632	25. 98	101. 303. 60	0	1
2003	1017 01341	0. 00	0. 002. 30	0	1
2003	1017 014 0	25. 97	101. 302. 60	10	1
2003	1017 028 3	25. 97	101. 253. 20	0	1
2003	1017 05326	25. 95	101. 302. 90	10	1
2003	1017 1 611	25. 97	101. 302. 80	0	1
2003	1017 1 914	25. 97	101. 272. 60	0	1
2003	1017 13942	25. 97	101. 322. 80	10	1
2003	1017 22422	25. 97	101. 322. 00	10	1
2003	1017 3 242	25. 93	101. 272. 10	10	1
2003	1017 3 910	25. 93	101. 282. 80	10	1
2003	1017 321 5	25. 72	101. 232. 00	0	1
2003	1017 34158	0. 00	0. 002. 00	0	1
2003	1017 419 7	25. 95	101. 382. 00	8	1
2003	1017 54839	0. 00	0. 002. 10	0	1
2003	1017 621 6	25. 93	101. 302. 30	0	1
2003	1017 62218	25. 95	101. 302. 20	10	1
2003	1017 632 1	25. 92	101. 322. 80	10	1
2003	1017 64133	25. 97	101. 274. 70	7	2
2003	1017 64621	0. 00	0. 002. 00	0	1
2003	1017 64943	0. 00	0. 002. 00	0	1
2003	1017 65123	0. 00	0. 002. 30	0	1
2003	1017 65514	25. 95	101. 272. 70	10	1
2003	1017 65558	25. 97	101. 282. 80	10	1
2003	1017 72746	25. 92	101. 303. 00	10	1
2003	1017 73345	0. 00	0. 002. 00	0	1
2003	1017 73627	0. 00	0. 002. 00	0	1
2003	1017 73652	25. 72	101. 232. 30	0	1
2003	1017 75715	0. 00	0. 002. 20	0	1
2003	1017 75850	0. 00	0. 002. 20	0	1
2003	1017 75936	25. 97	101. 302. 80	10	1
2003	1017 81025	0. 00	0. 002. 00	0	1
2003	1017 81440	25. 93	101. 253. 90	0	1

2003	1017 923 4	25.97	101.182.10	10	1
2003	1017101236	0.00	0.002.00	0	1
2003	1017103228	25.95	101.282.60	10	1
2003	1017111142	25.92	101.352.50	10	1
2003	1017114454	25.95	101.283.10	0	1
2003	1017115346	25.98	101.283.50	0	1
2003	10171233 4	25.97	101.272.30	10	1
2003	1017131843	25.95	101.323.10	0	1
2003	1017133937	0.00	0.002.00	0	1
2003	1017135251	0.00	0.002.30	0	1
2003	1017151311	25.92	101.352.60	10	1
2003	1017152943	25.95	101.302.80	10	1
2003	1017153339	25.93	101.272.20	0	1
2003	1017153958	25.95	101.302.20	10	1
2003	1017155222	0.00	0.002.00	0	1
2003	1017155525	0.00	0.002.00	0	1
2003	101718 6 2	0.00	0.002.00	0	1
2003	1017193725	25.72	101.232.00	0	1
2003	1017223128	25.95	101.273.40	0	1
2003	1017224256	25.93	101.322.70	10	1
2003	1017225828	25.95	101.283.20	0	1
2003	101723 033	0.00	0.002.20	0	1
2003	1017231131	0.00	0.002.20	0	1
2003	1017231838	25.95	101.283.00	0	1
2003	1018 115 2	0.00	0.002.00	0	1
2003	1018 23952	25.92	101.303.40	10	1
2003	1018 24522	25.92	101.333.30	5	1
2003	1018 34815	25.95	101.283.10	0	1
2003	1018 42933	25.93	101.372.60	0	1
2003	1018 51446	25.93	101.302.70	10	1
2003	1018 51543	26.00	101.372.60	0	1
2003	1018 52923	0.00	0.002.00	0	1
2003	1018 7 211	25.93	101.322.80	10	1
2003	1018 82935	25.93	101.282.50	10	1
2003	10181050 5	25.72	101.232.00	0	1
2003	1018111233	25.95	101.322.00	10	1
2003	10181159 7	25.97	101.332.40	10	1
2003	101813 130	25.95	101.322.60	0	1

2003	1018132831	25. 93	101. 302. 20	10	1
2003	101816 919	0. 00	0. 002. 10	0	1
2003	1018171850	25. 98	101. 252. 30	10	1
2003	101818 913	25. 95	101. 282. 60	10	1
2003	10181851 2	25. 97	101. 282. 00	5	1
2003	1018193445	0. 00	0. 002. 00	0	1
2003	1018205350	25. 92	101. 282. 50	0	1
2003	101823 545	0. 00	0. 002. 00	0	1
2003	1018231147	0. 00	0. 002. 00	0	1
2003	1018234734	25. 93	101. 332. 70	10	1
2003	1019 02739	0. 00	0. 002. 00	0	1
2003	1019 1 4 1	25. 72	101. 232. 00	0	1
2003	1019 12655	0. 00	0. 002. 20	0	1
2003	1019 21256	25. 93	101. 303. 00	0	1
2003	1019 32645	25. 95	101. 322. 20	10	1
2003	1019 51822	0. 00	0. 002. 20	0	1
2003	1019 63415	0. 00	0. 002. 80	0	1
2003	1019 65429	25. 97	101. 182. 20	10	1
2003	1019 917 8	0. 00	0. 002. 10	0	1
2003	10191039 1	0. 00	0. 002. 00	0	1
2003	1019114531	0. 00	0. 002. 30	0	1
2003	1019134123	25. 95	101. 253. 20	0	1
2003	1019165631	25. 92	101. 322. 30	0	1
2003	10191832 3	25. 97	101. 302. 90	0	1
2003	1019185720	25. 93	101. 323. 00	0	1
2003	1019214430	25. 92	101. 303. 10	0	1
2003	101922 256	25. 93	101. 272. 80	10	1
2003	1020 0 1 7	0. 00	0. 002. 00	0	1
2003	1020 414 4	25. 98	101. 232. 10	10	1
2003	1020 5 112	25. 90	101. 174. 00	0	1
2003	1020 65540	25. 72	101. 232. 20	0	1
2003	1020143334	0. 00	0. 002. 10	0	1
2003	102015 940	25. 95	101. 282. 70	5	1
2003	102016 039	0. 00	0. 002. 00	0	1
2003	10201846 2	25. 72	101. 232. 00	0	1
2003	1021 03143	25. 92	101. 332. 40	10	1
2003	1021 559 5	25. 95	101. 322. 40	0	1
2003	1021 64816	25. 88	101. 232. 30	0	1

2003	1021 7 154	26.00	101.272.90	0	1
2003	10211226 5	26.00	101.302.80	10	1
2003	102113 610	25.95	101.232.50	10	1
2003	1021131346	25.98	101.352.60	0	1
2003	1021144933	0.00	0.002.20	0	1
2003	102116 6 2	25.72	101.232.10	0	1
2003	1021161028	0.00	0.002.00	0	1
2003	1021174643	0.00	0.002.30	0	1
2003	10212229 5	0.00	0.002.30	0	1
2003	1021233648	25.93	101.282.90	5	1
2003	1022 01557	0.00	0.002.10	0	1
2003	1022 34022	25.93	101.282.90	10	1
2003	1022 759 7	25.92	101.282.10	0	1
2003	1022 935 1	25.97	101.302.40	0	1
2003	102210 715	25.95	101.233.50	10	1
2003	10221218 9	0.00	0.002.10	0	1
2003	1022133622	25.95	101.302.20	0	1
2003	1022133754	25.93	101.272.20	0	1
2003	1022135715	25.97	101.302.20	0	1
2003	10221419 0	25.93	101.202.10	0	1
2003	1022211732	0.00	0.002.10	0	1
2003	1022225119	25.92	101.302.90	10	1
2003	1023 24415	25.95	101.303.00	10	1
2003	1023 45334	25.92	101.322.10	0	1
2003	1023 72317	25.72	101.232.00	0	1
2003	1023 73332	25.95	101.283.40	10	1
2003	1023 84116	25.72	101.232.00	0	1
2003	1023 951 0	26.00	101.152.60	10	1
2003	1023 95348	25.93	101.282.00	4	1
2003	10231059 4	25.98	101.283.20	10	1
2003	1023155619	25.95	101.282.90	10	1
2003	1023175228	0.00	0.002.00	0	1
2003	1024 04029	0.00	0.002.20	0	1
2003	1024 05744	25.98	101.282.50	0	1
2003	1024 05830	25.93	101.322.70	0	1
2003	1024 14719	25.95	101.372.00	9	1
2003	1024 75558	26.00	101.302.50	0	1
2003	102412 342	25.95	101.332.60	0	1

2003	1024134339	25. 93	101. 333. 10	0	1
2003	1024142748	0. 00	0. 002. 00	0	1
2003	1024162740	25. 72	101. 232. 00	0	1
2003	1024191748	25. 97	101. 282. 30	0	1
2003	1024204331	25. 98	101. 352. 30	0	1
2003	102421 821	25. 92	101. 352. 50	0	1
2003	1025 05311	25. 88	101. 272. 00	0	1
2003	1025 05548	25. 93	101. 352. 50	10	1
2003	1025 22015	25. 72	101. 232. 20	0	1
2003	1025 93212	25. 92	101. 282. 60	0	1
2003	1025123225	0. 00	0. 002. 20	0	1
2003	10251346 3	25. 87	101. 282. 20	0	1
2003	1025155834	25. 93	101. 322. 70	0	1
2003	1025211655	25. 90	101. 272. 50	0	1
2003	1025221556	25. 97	101. 302. 90	0	1
2003	1026 715 2	25. 95	101. 282. 50	10	1
2003	1026 71555	25. 95	101. 282. 30	10	1
2003	1026 92428	0. 00	0. 002. 30	0	1
2003	1026113637	25. 95	101. 323. 10	0	1
2003	1026115554	25. 88	101. 282. 00	0	1
2003	1026194014	25. 95	101. 272. 40	10	1
2003	1026235522	0. 00	0. 002. 30	0	1
2003	1027 14757	25. 72	101. 232. 00	0	1
2003	1027 151 5	25. 93	101. 272. 00	0	1
2003	1027 71250	25. 95	101. 282. 50	0	1
2003	1027 73031	25. 95	101. 332. 50	0	1
2003	1027 83751	25. 95	101. 372. 10	10	1
2003	1027 94544	0. 00	0. 002. 00	0	1
2003	1027101653	25. 95	101. 324. 10	0	1
2003	10271027 9	25. 93	101. 323. 40	0	1
2003	10271326 6	25. 90	101. 272. 10	0	1
2003	102714 647	25. 95	101. 222. 20	0	1
2003	1027145622	25. 92	101. 252. 30	0	1
2003	1027162526	25. 97	101. 222. 30	10	1
2003	1027163751	0. 00	0. 002. 20	0	1
2003	102717 0 4	25. 90	101. 282. 00	0	1
2003	1027192946	25. 92	101. 282. 50	0	1
2003	1027195158	25. 92	101. 272. 30	0	1

2003	102720 050	25.72	101.232.10	0	1
2003	10272118 6	25.72	101.232.10	0	1
2003	102722 148	25.95	101.272.30	0	1
2003	10272213 0	25.93	101.282.80	10	1
2003	1027224527	25.90	101.302.30	0	1
2003	1028 238 8	25.93	101.252.00	0	1
2003	1028 74410	25.93	101.302.30	10	1
2003	1028 930 9	25.97	101.322.30	0	1
2003	1028 94034	25.90	101.322.30	0	1
2003	1028113334	25.95	101.252.00	0	1
2003	1028133426	25.95	101.332.70	10	1
2003	1029143437	25.88	101.272.00	0	1
2003	1029184331	25.72	101.232.00	0	1
2003	1029212758	25.92	101.322.40	10	1
2003	1030 9 523	25.98	101.322.90	0	1
2003	10301157 2	25.98	101.322.70	10	1
2003	10302026 2	25.95	101.252.10	0	1
2003	103021 422	25.90	101.272.20	0	1
2003	1031 03618	25.92	101.252.00	0	1
2003	1031 3 152	26.00	101.282.50	10	1
2003	1031 81646	25.95	101.322.60	0	1
2003	1031115312	25.72	101.232.20	0	1
2003	1031135032	25.72	101.232.20	0	1
2003	103116 722	0.00	0.002.00	0	1
2003	11 1 2 922	25.93	101.224.70	3	2
2003	11 1 34214	25.98	101.272.80	10	1
2003	11 1 5 650	25.93	101.272.00	10	1
2003	11 1 75856	0.00	0.002.00	0	1
2003	11 113 746	0.00	0.002.10	0	1
2003	11 119 6 1	26.05	101.332.80	10	1
2003	11 1214759	0.00	0.002.20	0	1
2003	11 2124628	25.92	101.352.80	10	1
2003	11 2203516	25.97	101.282.20	0	1
2003	11 3 54943	0.00	0.002.00	0	1
2003	11 3164851	0.00	0.002.10	0	1
2003	11 4 04150	25.93	101.282.40	10	1
2003	11 410 9 9	26.02	101.282.80	10	1
2003	11 4173022	25.72	101.232.10	0	1

2003	11 5 5 322	25.72	101.232.00	0	1
2003	11 516 911	25.72	101.232.20	0	1
2003	11 52242 5	25.97	101.272.20	10	1
2003	11 6 932 8	25.72	101.232.00	0	1
2003	11 615 814	25.87	101.232.30	0	1
2003	11 6193918	25.93	101.333.40	0	1
2003	11 6202850	25.92	101.352.20	0	1
2003	11 7 247 8	25.98	101.322.60	0	1
2003	11 7 44732	25.95	101.303.40	0	1
2003	11 7 45942	25.95	101.302.20	0	1
2003	11 7 5 759	25.72	101.232.00	0	1
2003	11 7 55819	25.85	101.282.10	0	1
2003	11 7102626	25.93	101.322.10	0	1
2003	11 7152512	25.93	101.232.10	0	1
2003	11 7162155	25.87	101.372.20	0	1
2003	11 8 31543	25.97	101.283.20	0	1
2003	11 82049 2	25.95	101.303.20	0	1
2003	11 82052 3	0.00	0.002.10	0	1
2003	11 9152721	25.92	101.322.40	0	1
2003	11 9164016	25.92	101.282.60	0	1
2003	11 9164750	25.92	101.202.60	0	1
2003	11 917 247	25.98	101.322.70	0	1
2003	11 9184453	25.97	101.332.60	0	1
2003	11 9192510	25.90	101.232.30	0	1
2003	1110112327	25.93	101.222.30	0	1
2003	1110152853	25.98	101.203.00	10	1
2003	1110153354	0.00	0.002.10	0	1
2003	1110193813	25.97	101.322.50	0	1
2003	1111 919 6	25.95	101.322.40	10	1
2003	1111 957 2	25.72	101.232.20	0	1
2003	1111 95711	25.72	101.232.00	0	1
2003	111110 5 1	25.72	101.232.10	0	1
2003	1111165612	0.00	0.002.00	0	1
2003	11121851 2	26.02	101.232.70	10	1
2003	1113 013 1	25.97	101.332.30	0	1
2003	1113 14140	25.95	101.302.30	0	1
2003	1113 24810	25.95	101.282.40	10	1
2003	1113184618	25.72	101.232.10	0	1

2003	1113191347	26.00	101.272.10	10	1
2003	111321 414	0.00	0.002.00	0	1
2003	1116 24753	0.00	0.002.20	0	1
2003	1116115255	25.92	101.302.10	0	1
2003	1118 22035	25.72	101.232.00	0	1
2003	1118171156	25.88	101.232.30	0	1
2003	1118214012	25.95	101.272.50	0	1
2003	1119191648	25.93	101.322.20	0	1
2003	112018 031	25.72	101.232.10	0	1
2003	1122 82022	26.02	101.332.20	10	1
2003	1123152951	25.98	101.322.70	5	1
2003	1125 04210	25.72	101.232.10	0	1
2003	1125165439	0.00	0.002.30	0	1
2003	1126 438 6	26.00	101.282.90	10	1
2003	1126 44252	26.02	101.272.20	5	1
2003	1126 63820	0.00	0.002.50	0	1
2003	1126141858	26.00	101.272.70	10	1
2003	11261545 5	26.00	101.332.80	10	1
2003	1128204335	25.97	101.282.10	0	1
2003	1128205058	25.98	101.322.20	0	1
2003	1129133536	25.72	101.232.20	0	1
2003	113023 620	0.00	0.002.10	0	1
2003	12 2 62849	25.95	101.282.10	0	1
2003	12 21535 7	25.97	101.272.20	0	1
2003	12 3 04241	25.87	101.282.10	0	1
2003	12 3 75036	25.95	101.323.20	0	1
2003	12 3114157	25.93	101.252.10	10	1
2003	12 319 756	25.88	101.302.00	0	1
2003	12 4 65616	0.00	0.002.30	0	1
2003	12 5111330	25.97	101.322.60	0	1
2003	12 6 25818	26.10	101.222.60	10	1
2003	12 7162147	25.88	101.302.10	0	1
2003	12 9105841	25.97	101.323.20	10	1
2003	1210 92732	25.72	101.232.20	0	1
2003	1211 64238	25.95	101.372.00	0	1
2003	121118 2 8	25.72	101.232.20	0	1
2003	1213 915 9	26.00	101.332.00	0	1
2003	1214125751	0.00	0.002.10	0	1

2003	1215 24625	25.95	101.332.70	0	1
2003	1217 75328	25.87	101.252.40	0	1
2003	12171310 7	25.72	101.232.00	0	1
2003	1218 33959	26.00	101.222.40	0	1
2003	1218161253	25.87	101.332.10	0	1
2003	1219132215	25.72	101.232.10	0	1
2003	1220193250	25.95	101.302.80	0	1
2003	1222 1 838	25.97	101.282.90	0	1
2003	1222 25212	25.97	101.302.40	0	1
2003	1222 34726	0.00	0.002.20	0	1
2003	1222 517 4	25.98	101.282.80	0	1
2003	1222 517 8	25.90	101.302.80	10	1
2003	1225 92012	25.97	101.282.80	10	1
2003	1225192318	25.93	101.302.50	15	1
2003	1226 2 221	25.93	101.302.90	10	1
2003	1226 44956	25.95	101.282.70	15	1
2003	12271631 9	0.00	0.002.10	0	1
2003	1227181546	25.72	101.232.10	0	1
2003	1227233154	25.93	101.303.40	10	1
2003	1228 14834	26.08	101.272.40	0	1
2003	1228173118	25.95	101.332.40	0	1
2003	1229 62457	25.95	101.332.80	0	1
2003	1230231259	25.92	101.372.40	15	1
2003	1230234353	25.93	101.332.40	10	1
2004	1 2 43134	25.98	101.132.60	10	1
2004	112132410	25.97	101.382.10	0	1
2004	112153734	25.93	101.323.50	0	1
2004	11415 811	25.95	101.302.10	5	1
2004	11812 438	0.00	0.002.20	0	1
2004	118173417	26.03	101.352.40	15	1
2004	123155254	25.95	101.304.00	10	1
2004	123215920	0.00	0.002.00	0	1
2004	125153232	25.93	101.332.20	10	1
2004	131181942	25.72	101.232.30	0	1
2004	2 1 73415	26.02	101.273.00	0	1
2004	2 1 73741	25.97	101.302.60	0	1
2004	2 3162448	25.92	101.253.30	10	1
2004	2 3184630	0.00	0.002.10	0	1

2004	2 6164024	25.95	101.272.60	0	1
2004	2 9223316	25.95	101.282.50	10	1
2004	2 9223441	0.00	0.002.20	0	1
2004	2 9224131	0.00	0.002.00	0	1
2004	2 9225648	0.00	0.002.30	0	1
2004	213111147	25.95	101.282.40	10	1
2004	215231541	0.00	0.002.00	0	1
2004	216 04545	25.95	101.282.20	10	1
2004	216114954	25.72	101.232.40	0	1
2004	218161825	25.97	101.282.30	0	1
2004	222 124 3	25.98	101.282.10	0	1
2004	222182128	0.00	0.002.30	0	1
2004	2261031 7	0.00	0.002.40	0	1
2004	227132239	25.72	101.232.00	0	1
2004	2271527 8	25.98	101.322.60	10	1
2004	3 2223856	0.00	0.002.20	0	1
2004	3 4 03527	25.98	101.232.30	8	1
2004	3 4185939	25.72	101.232.10	0	1
2004	3 5 42243	25.93	101.302.60	0	1
2004	3 9 016 2	0.00	0.002.40	0	1
2004	310 45641	0.00	0.002.00	0	1
2004	310155324	25.72	101.232.00	0	1
2004	311 42223	0.00	0.002.00	0	1
2004	311163031	0.00	0.002.30	0	1
2004	314174752	25.98	101.372.70	0	1
2004	314174755	26.00	101.302.40	16	1
2004	31521 440	25.95	101.333.40	0	1
2004	315214133	25.93	101.352.60	10	1
2004	316 02716	25.97	101.382.20	9	1
2004	316 44655	25.97	101.352.00	10	1
2004	316201817	25.72	101.232.40	0	1
2004	318 85527	26.00	101.233.10	0	1
2004	318144823	0.00	0.002.30	0	1
2004	321131314	26.03	101.302.00	10	1
2004	323 73318	25.98	101.252.60	10	1
2004	324235831	25.92	101.304.40	12	1
2004	325 0 351	25.93	101.322.70	10	1
2004	325 02116	25.95	101.282.30	14	1

2004	325 03926	25.90	101.284.20	10	1
2004	325 12040	25.93	101.282.50	10	1
2004	325 124 7	25.92	101.282.80	10	1
2004	325 15048	25.97	101.352.00	10	1
2004	325 823 0	25.93	101.322.30	0	1
2004	325212036	25.95	101.302.10	0	1
2004	326 32613	25.88	101.282.10	0	1
2004	326 51455	25.90	101.274.60	12	1
2004	326 51829	25.93	101.322.70	10	1
2004	326 52131	25.97	101.332.00	10	1
2004	326 52253	25.92	101.282.60	10	1
2004	326 524 7	25.72	101.232.00	0	1
2004	326 52738	25.72	101.232.40	0	1
2004	326 54458	25.72	101.232.10	0	1
2004	326 54848	25.95	101.282.20	10	1
2004	326 642 2	25.93	101.302.50	10	1
2004	326 657 9	25.72	101.232.50	0	1
2004	326 75553	25.72	101.232.30	0	1
2004	326 91356	25.72	101.232.20	0	1
2004	326103328	25.92	101.303.10	5	1
2004	326103724	25.93	101.272.10	0	1
2004	326103814	25.72	101.232.10	0	1
2004	32611 6 0	25.97	101.332.50	10	1
2004	3261145 7	25.92	101.273.10	5	1
2004	326121911	25.72	101.232.10	0	1
2004	326124932	25.92	101.322.10	10	1
2004	32613 322	25.72	101.232.10	0	1
2004	32613 826	25.72	101.232.00	0	1
2004	326142817	25.72	101.232.00	0	1
2004	327 037 0	25.93	101.322.10	10	1
2004	327 652 7	0.00	0.002.00	0	0
2004	327111522	0.00	0.002.40	0	1
2004	327134349	0.00	0.002.00	0	1
2004	328153539	25.72	101.232.00	0	1
2004	32821 135	25.72	101.232.00	0	1
2004	329 95621	25.72	101.232.10	0	1
2004	329121621	25.95	101.352.10	10	1
2004	331 72548	25.97	101.273.40	10	1

2004	4 1204935	0.00	0.002.30	0	1
2004	4 3235756	25.92	101.273.10	10	1
2004	4 4 04356	25.97	101.253.60	10	1
2004	4 4 12859	25.95	101.302.40	10	1
2004	4 4 45343	25.93	101.272.40	10	1
2004	4 41310 8	0.00	0.002.00	0	1
2004	4 4152454	25.93	101.372.20	0	1
2004	4 5 83251	25.72	101.232.00	0	1
2004	4 5115631	0.00	0.002.20	0	1
2004	4 51222 4	25.72	101.232.00	0	1
2004	4 5221035	25.90	101.452.20	0	1
2004	4 5225939	0.00	0.002.00	0	1
2004	4 5235618	25.72	101.232.00	0	1
2004	4 6 740 4	0.00	0.002.20	0	1
2004	4 61130 6	0.00	0.002.00	0	1
2004	4 6154252	25.72	101.232.00	0	1
2004	4 7 6 5 0	0.00	0.002.10	0	1
2004	4 9121317	25.97	101.332.90	5	1
2004	4 9145346	25.97	101.322.70	10	1
2004	410125832	25.93	101.302.30	0	1
2004	410142524	25.92	101.303.60	10	1
2004	411 23924	25.95	101.302.80	0	1
2004	411173113	25.72	101.232.50	0	1
2004	412 740 2	25.72	101.232.30	0	1
2004	41316 733	25.97	101.332.40	10	1
2004	416 122 6	25.97	101.382.40	10	1
2004	4161838 9	25.92	101.324.30	10	1
2004	416231621	0.00	0.002.00	0	1
2004	42117 3 1	25.98	101.272.20	0	1
2004	426 92429	25.95	101.353.20	0	1
2004	430 42946	25.72	101.232.00	0	1

2003年8月16日内蒙古自治区巴林左旗、阿鲁科尔沁旗交界5.9级地震

1）内蒙古自治区地震局；2）中国地震台网中心

高立新[1]　胡　博[1]　张永仙[2]

摘　要

2003年8月16日内蒙古自治区赤峰市巴林左旗、阿鲁科尔沁旗交界发生5.9级地震，宏观震中位于巴林左旗白音沟乡和阿鲁科尔沁旗乌兰哈达乡，震中烈度为Ⅷ度，极震区呈东西向椭圆形分布，地震造成4人死亡，60人重伤，1004人轻伤，直接经济损失8.06亿元。

此次地震序列为主震—余震型，最大余震M_L4.7。余震分布在主震西侧，呈北西西方向展布，余震分布与近东西向分布的等震线方向一致，最大余震距离主震约24km。5.9级地震的震源机制解，节面Ⅱ为主破裂面，主压应力P轴方位为NEE方向。本次地震的发震构造是下水泉沟断裂。

这次地震周围台站较少，震中200km范围内有测震台站、前兆台站8个。震前共出现7个项目14条异常，其中2条为长期异常，10条为中期异常，2条为短期异常，未出现短临异常。地震活动性图像主要以长期和中期背景性异常为主，地震学指标异常主要呈现中期异常性质。

5.9级地震发生在年度值得注意地区内，内蒙古自治区地震局曾有过明确的中期预测意见，取得了一定的社会效益。

前　言

2003年8月16日18时58分，内蒙古自治区赤峰市巴林左旗、阿鲁科尔沁旗交界发生5.9级地震。内蒙古自治区地震台网测定的微观震中为43°55′N、119°52′E，宏观震中为44°03′N、119°42′E，位于微观震中北部的巴林左旗白音沟乡和阿鲁科尔沁旗乌兰哈达乡，震中烈度达Ⅷ度。这次地震震感范围很大，赤峰市的巴林左旗、阿鲁科尔沁旗、巴林右旗、翁牛特旗、林西县均强烈有感，另外，辽宁、吉林、北京、河北等部分地区都有震感。地震造成4人死亡，60人重伤，1004人轻伤，直接经济损失8.06亿元。

5.9级地震所在的内蒙古东部地区是内蒙古主要的中强地震活动区。在震中西南约

188km 处，曾发生过 1940 年 1 月 19 日库伦旗 6.0 级地震。震中附近地震观测台站较少，200km 范围内有地震台 8 个。震前出现 7 个观测项目的 14 条前兆异常。

2003 年 8 月 16 日巴林左旗、阿鲁科尔沁旗交界 5.9 级地震发生在区域地震活动相对平静的背景下，2002 年 10 月 20 日，位于 5.9 级地震震中西北约 210km 处的西乌珠穆沁旗曾发生 4.8 级地震，2004 年 3 月 24 日，距离 5.9 级地震震中西北约 190km 处的东乌珠穆沁旗再次发生 5.9 级地震，几次中强地震是否存在内在联系，还有待进一步研究。

该次地震发生前，内蒙古地震局曾有明确中期预报意见，2003 年度地震趋势研究报告，将巴林左旗和阿鲁科尔沁旗所在的辽蒙交界地区列为值得注意地区，认为该地区存在发生中强地震的可能性[1)]。2003 年年中内蒙古自治区地震趋势研究报告提出："下半年该区发生 5～6 级地震的危险性进一步增大"。尽管震前有所察觉，但仍未捕捉到短期地震学前兆信息。

本研究报告是在有关资料和文献的基础上经过重新整理资料和分析研究而完成的[1～6];[1,2)]。

一、测震台网及地震基本参数

5.9 级地震发生在内蒙古地震监测能力较弱的区域，特别是震中西北没有测震台站分布，图 1 给出了震中范围 200km 范围内的测震台站和前兆测项分布，8 个台站中天山、鲁北、林西、新惠为测震台站，赤峰为测震和前兆综合观测台站，翁牛特、大甸子、开鲁为前兆观测台站。100km 范围内仅有天山台，其余台站均分布在 100～200km 之间。除以上 5 个测震台站，在测定震中时，还利用了距离震中约 240km 宁城台的数据。在研究时段内基本可达到 $M \geqslant 2.5$ 级地震不遗漏。

表 1 给出了不同来源给出的这次地震的基本参数，经对比分析，认为内蒙古自治区地震局经过修订后的震中位置更为精确，因此，这次地震基本参数采用表 1 中编号 1 的结果。

表 1　巴林左旗、阿鲁科尔沁旗交界 5.9 级地震基本参数

Table 1　Basic parameters of the *M*5.9 earthquake in boundary area between Balinzuoqi and Alukeerqinqi

编号	发震日期	发震时刻	震中位置		震级	震源深度	震中地名	结果来源
	年 月 日	时 分 秒	ψ_N	λ_E		(km)		
1	2003 8 16	18 58 42	43°55′	119°52′	M_S5.9	20	巴林左旗和阿鲁科尔沁旗间	内蒙古地震局修订[1)]
2	2003 8 16	18 58 42	43°54′	119°57′	M_S6.1	20		内蒙古地震局速报目录
3	2003 8 16	18 58 45	43°55′	119°52′	M_L6.2	20		中国地震台网中心
4	2003 8 16	18 58 43	43.90°	119.70°	M_S5.9	15		国家台网大震速报目录
5	2003 8 16	18 58 43	43.77°	119.64°	M_b5.5	24		美国国家地震信息中心
6	2003 8 16	18 58 46	43.80°	119.47°	M_W5.4 M_b5.5 M_S5.1	30		美国哈佛大学

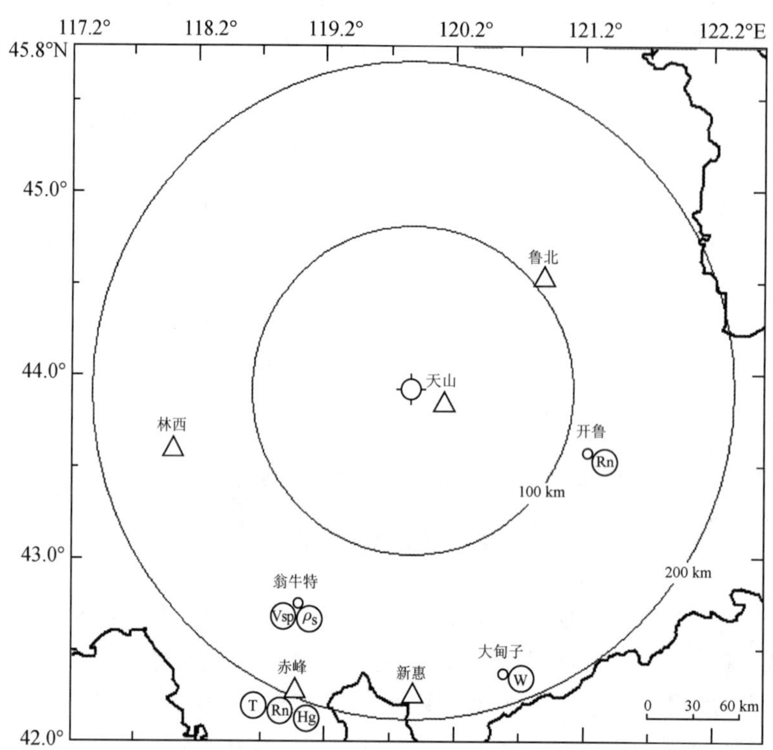

图 1　巴林左旗、阿鲁科尔沁旗交界 5.9 级地震前震中附近地震台站及观测项目分布图

Fig. 1　Distribution of earthquake-monitoring stations around the epicentral area before *M*5.9 earthquake in boundary area between Balinzuoqi and Alukeerqinqi

二、地震地质背景

本次地震发生的大地构造位置是天山—兴蒙地槽系的二级构造单元兴安地槽褶皱带和内蒙地槽褶皱带的交接地区，兴蒙地槽褶皱带的南部。在新构造分区中位于东北断块区的大兴安岭断隆的东边缘，与松辽拗陷交接的部位，也是大兴安岭山脉南部东坡边缘向松辽平原过渡的区域，新构造运动以大面积的缓慢隆升为特征。

布格重力异常图上位于纵贯我国南北的北东向重力异常梯度上，异常值在 -60 毫伽左右，莫霍面等深线图上处于大兴安岭隆起东坡地壳厚度变异带，地壳厚度 36 ~ 37km。

区域地震活动属于东北地震区大兴安岭地震亚区。震区附近历史上无破坏性地震记载，在中国地震烈度区划图（1990 版）中被列为小于Ⅵ度的不设防区。破坏性地震主要发生在东部的松辽（开鲁）盆地中。

从图 2 分析，震区共发育有三组断裂，分别为北东（包括北北东）、近东西和北西向断裂。其中北东向断裂和近东西向断裂最为显著，前者主要分布在大兴安岭隆起带，是隆起带主体构造线的一部分，沿断裂曾发生一定数量的地震，震级一般在 5 级左右，未有 6 级以上地震记录。后者主要展布在松辽盆地及边缘，一般长度都较大，第四纪以来仍有活动，沿断裂往往成为地貌分界线。沿养畜牧断裂和西辽河断裂分别于 1940 年和 1942 年有 6 级地震发生。北西向断裂数量最少，长度最短，卫星照片解译显示有线性构造，一般沿河谷发育。

由图 2 可见，水泉子沟断裂是最靠近震中的北西向断裂，断裂性质不明，两侧地层分别为上白垩统和上侏罗统，长度约 25km。断裂沿水泉子沟—西山湾洼地发育，在地貌上表现为线性沟谷。宏观震中（极震区）正好位于该断裂的位置，本次地震等烈度线长轴方向也与该断裂走向一致，现场考察认为该断裂是发震断裂。另据震源机制解结果显示主压应力 P

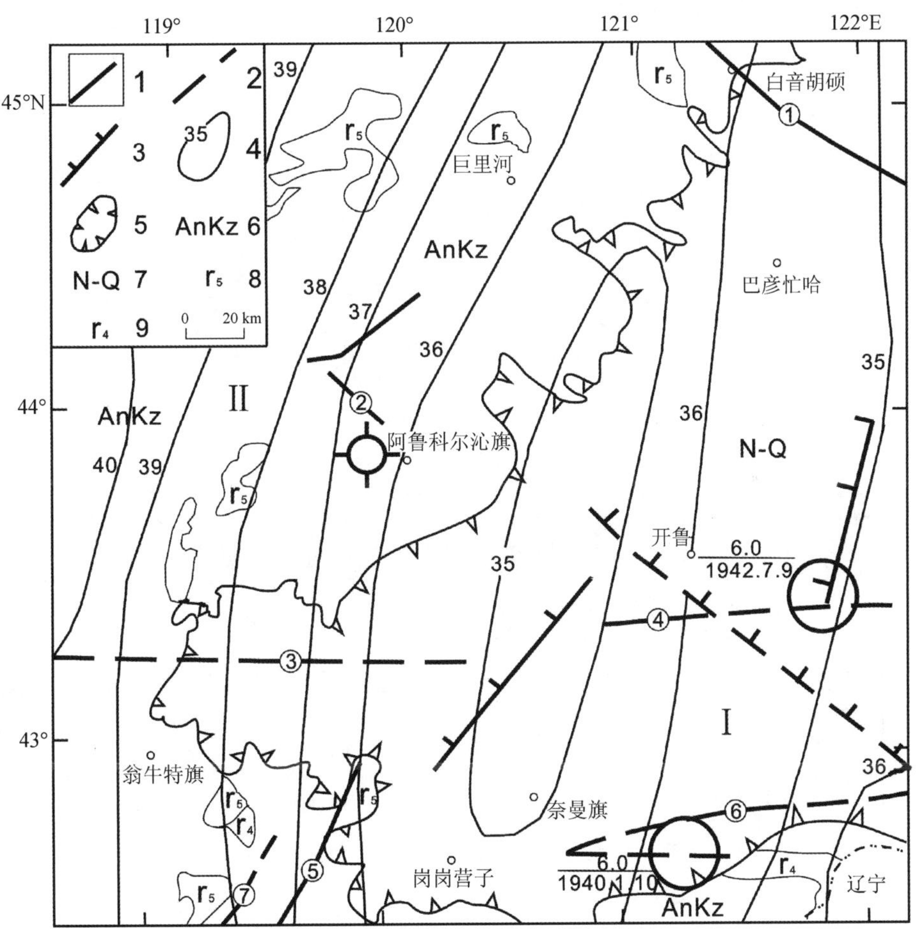

图 2 巴林左旗、阿鲁科尔沁旗交界 5.9 级地震震区断裂构造及历史震中分布图

Fig. 2 Major faults and distribution of historical earthquakes around the epicenter of the *M*5.9 earthquake in boundary area between Balinzuoqi and Alukeerqinqi

图例说明：1. 实测断层；2. 隐伏断层；3. 正断层；4. 地壳等厚度（km）；5. 盆地边界；6. 前新生界隆起区；7. 上第三纪至第四纪盆地；8. γ_5 燕山期花岗岩；9. γ_4 海西期花岗岩

构造名称：①霍林河断裂；②水泉沟断裂；③西拉木伦河断裂；④西辽河断裂；⑤老哈河断裂；⑥养畜牧断裂；⑦哈拉道口断裂；Ⅰ. 开鲁盆地；Ⅱ. 大兴安岭隆起

轴方向为 NEE，发震断层为左旋走滑逆冲断错，符合北西向断层在 NEE 向主压应力下的错动方式。

综合以上分析认为，北西向的水泉子沟断裂为本次地震的发震构造。其他断裂远离震中，与本次地震应没有构造关系。

三、地震影响场和震害

据现场实地考察及调查资料，本次地震的宏观震中为 44°03′N，119°42′E，震中区烈度为Ⅷ度，等震线呈东西向椭圆形分布（图 3）。

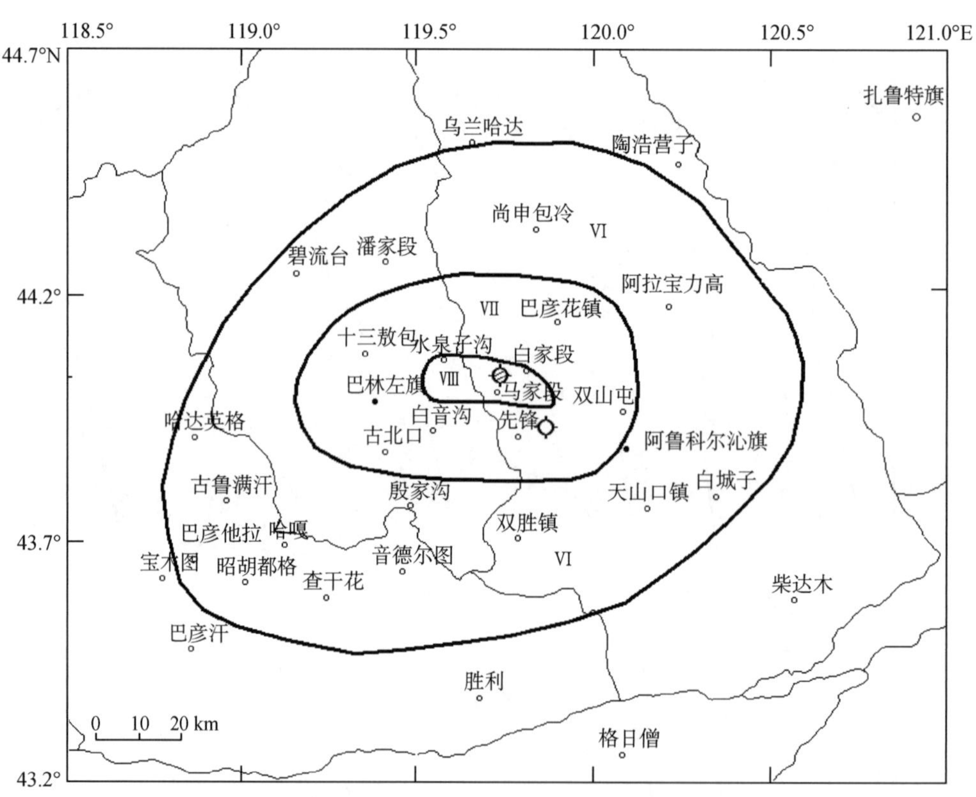

图 3　巴林左旗、阿鲁科尔沁旗交界 5. 9 级地震烈度等震线图

Fig. 3　Isoseismal map of *M*5. 9 earthquake in boundary area between Balinzuoqi and Alukeerqinqi

1. Ⅷ度区破坏特征

极震区（Ⅷ度区）：位于巴林左旗白音沟乡、阿鲁科尔沁旗乌兰哈达乡、先峰乡、东沙布台乡境内。西起白音沟乡水泉沟村，东至乌兰哈达乡凤凰山村，南起先峰乡孟家段村，北至东沙布台乡宝山村，面积 263km^2。其基本形态呈椭圆形，长轴为 30km，呈东西方向，短轴约 10km。

人们普遍反映地震时听到轰隆一声巨响，感到心惊胆颤，上下颠簸，坐立不稳，行走十分困难。骑自行车和摩拖车的人感到忽高忽低车不由人。山区岩石飞滚，沟谷滑坡塌方，房屋倒塌，烟雾滚滚，尘土飞扬。西山湾村、宏发村居民述说震动和烟雾由西北方向而来南东方向而去。

极震区属底山丘陵区，基岩裸露，低洼地带第四系厚度约10m，主要为冲洪积、坡积层。农民房屋建筑在Ⅱ类场地条件上，房屋以土木、砖木和砖土混合墙体结构的为主。砖木结构房屋地震中40%左右的房屋遭到严重毁坏或毁坏（含局部倒塌）。住宅区民房烟筒70%左右倒塌毁坏，房顶脊瓦震动错位，盖瓦波浪起伏承重墙出现水平和剪切裂缝，少数墙体震酥，个别非承重构件严重破坏，房顶震裂，门窗变形，玻璃震碎。只有5%左右的房屋轻微损坏或基本完好，个别承重构件轻微裂缝。土木房屋有60%左右的毁杯倒塌。综合考虑房屋结构、工程质量、建筑年代、场地条件等因素，房屋的平均震害指数为0.31~0.5。

水泉沟村、西山湾村等土木结构房屋普遍严重毁坏，前、后墙体倒塌。砖木结构房屋局部倒塌或严重破坏，房屋墙体开裂外闪，裂缝宽度达5~10cm，剪切裂缝主要发生在后墙和前墙。个别屋盖前后坡塌落。滚石和沟谷滑坡塌方主要分布于西山村、宝山村带。

2. Ⅶ度区破坏特征

Ⅶ度区：西起罕吐柏下朝阳营子村，东到扎嘎斯台苏木乌兰哈达村，南起隆昌镇，北至东沙布台乡姚家段村，面积3009km^2，其基本形态呈椭圆形，长轴方向近东西向，长约77km，短轴方向为近南北向，长约47km，长短轴之比为1.5。

地震时人们普遍听到放炮或闷雷似的响声，感到强烈颠簸晃动，室内的人惊慌逃出室外。有30%左右的土木、砖木房屋受到严重毁坏或倒塌，有40%左右的房屋为中等破坏，仅有30%左右的房屋轻微损坏或基本完好。土木结构房屋普遍发生裂缝，少数房屋受到严重破坏，或局部倒塌。砖木房屋也有轻微破坏，表现为墙体、窗角、门框裂缝，砖砌围墙有部分倒塌受到中等损坏。砖混结构楼房有个别受到轻微破坏，主要是槽型板之间裂缝。据调查资料统计，在Ⅶ度区一般房屋的平均震害指数为0.15~0.3，震害比为严重毁坏4%，中等损坏8%，轻微损坏20%。居民住宅烟筒25%左右的倒塌。

3. Ⅵ度区破坏特征

Ⅵ度区：西起白音勿拉苏木的哈日白其村，东至扎嘎斯台苏木，南起巴拉奇如德苏木的萨如拉塔拉村，北至乌兰哈达苏木，面积10362km^2，长轴方向略呈北东向，长约150km，短轴方向为北西向，长约100km。

大多数土木、砖木结构的房屋承重墙体出现裂缝受到轻微破坏，个别房屋局部塌落，非承重墙体明显破裂，少数年久失修的老旧土房受到严重毁坏或倒塌。据调查资料统计约有5%的老旧土房严重破坏，15%的土木或砖木平房中等破坏，30%的轻微损坏，50%的基本完好。

天山口镇仁义村，土木结构房屋墙体通裂、外倾，个别墙倒。砖木旧房墙角开裂、墙体裂缝。柴达木苏木乌那嘎查，土木房屋普遍裂缝、掉土块，个别局部倒塌。巴拉奇如德苏木阿日诺嘎查，土木结构民房纵横墙交接处裂缝较大，新建砖木房屋细微裂缝，基本完好或轻微破坏。

4. 震害

本次地震受灾范围包括赤峰市的巴林左旗、阿鲁科尔沁旗、巴林右旗、翁牛特旗、林西县和锡林郭勒盟的西乌珠穆沁旗，其中巴林左旗、阿鲁科尔沁旗受灾最为严重。克什克腾旗、宁城县、敖汉旗、通辽市、奈曼旗、扎鲁特旗、开鲁县、锡林浩特市等地强烈有感。本次地震共有 4 人死亡，60 人重伤，1004 人轻伤。灾区面积为 11979km^2（按评估区域统计），受灾范围涉及巴林左旗、阿鲁科尔沁旗、巴林右旗 3 个旗，37 个乡镇苏木，共 408 个嘎查村，灾区人口共计 480869 人，13739 户。根据地震现场灾害评估工作和地震灾害评估委员会确认，本次地震直接经济损失 8.06 亿元。

四、地震序列

为了保证序列的完整性，主要利用距离 5.9 级地震 24km 的天山地震台单台测定的结果以及可以定位的地震组成 5.9 级地震序列，截至 2003 年 11 月 12 日，序列基本结束，共记录到 $M_L \geqslant 0.0$ 级地震 200 次，其中 $0.0 \leqslant M_L \leqslant 0.9$ 级 17 次，$1.0 \leqslant M_L \leqslant 1.9$ 级 122 次，$2.0 \leqslant M_L \leqslant 2.9$ 级 49 次，$3.0 \leqslant M_L \leqslant 39$ 级 9 次，$4.0 \leqslant M_L \leqslant 4.9$ 级 2 次，最大余震是 2003 年 10 月 17 日的 $M_L 4.7$ 地震。200 次余震中 62 次地震进行了定位，表 2 给出了 $M_L \geqslant 3.0$ 级的余震序列目录。

从 5.9 级地震序列分布可以看出（图 4），余震活动主要分布在主震的西侧 25km 范围内，呈北西西方向展布，余震分布与近东西向分布的等震线方向一致，最大余震距离主震约 24km。

表 2　巴林左旗、阿鲁科尔沁旗交界 5.9 级地震序列目录（$M_L \geqslant 3.0$ 级）

Table 2　Catalogue of the M5.9 earthquake sequence in boundary area between Balinzuoqi and Alukeerqinqi（$M_L \geqslant 3.0$）

编号	发震日期	发震时刻	震中位置		震级	震源深度	震中地名	结果来源
	年 月 日	时 分 秒	ψ_N	λ_E	M_L	(km)		
1	2003 8 16	19 59 48	43.90°	119.78°	3.5	5	巴林左旗、阿鲁科尔沁旗交界	资料 1)
2	2003 8 16	20 52 45	43.88°	119.75°	3	5		
3	2003 8 16	21 17 49	43.92°	119.77°	3.6	5		
4	2003 8 17	13 24 19	43.92°	119.72°	4	5		
5	2003 8 18	7 51 5	43.93°	119.77°	3.4	5		
6	2003 8 19	22 38 42	43.88°	119.73°	3.2	5		
7	2003 9 1	22 6 28	43.88°	119.77°	3	5		
8	2003 9 17	12 42 6	43.88°	119.73°	3.1	5		
9	2003 10 8	10 52 36	43.88°	119.77°	3.2	5		
10	2003 10 17	9 38 25	43.92°	119.57°	4.7	8		
11	2003 10 20	18 33 40	43.98°	119.57°	3.7	9		

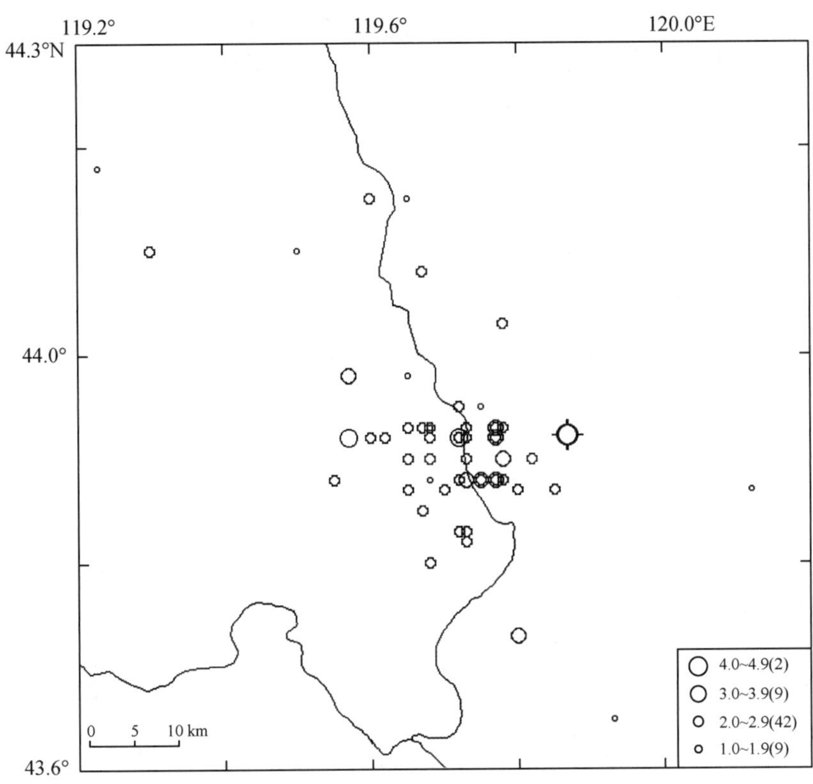

图4 巴林左旗、阿鲁科尔沁旗交界5.9级地震序列分布

Fig. 4 Distribution of the *M*5.9 earthquake sequence in boundary area between Balinzuoqi and Alukeerqinqi

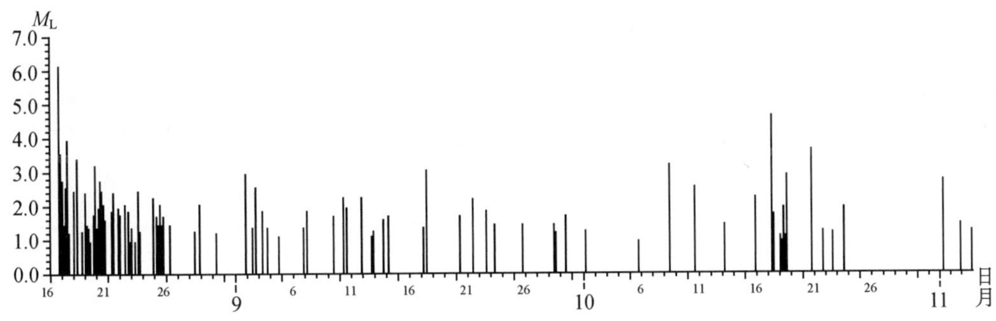

图5 巴林左旗、阿鲁科尔沁旗交界5.9级地震序列 *M-t* 图

Fig. 5 *M-t* diagram of the *M*5.9 earthquake sequence in boundary area between Balinzuoqi and Alukeerqinqi

由余震序列的 *M-t* 图可以看出（图5），余震主要集中发生在8月16~25日间，且多数为 $M_L \leqslant 3.0$ 级地震，之后余震逐渐衰减，11月12日后，余震活动基本结束。最大余震 M_L4.7 地震发生在10月17日。

地震序列频次分布图可以看出（图6），余震活动主要集中在前10天内，除10月17、18日出现小的起伏外，序列衰减基本正常。序列的蠕变曲线也反映了序列衰减的基本特征，除10月17日发生 M_L4.7最大余震曲线出现阶变外，整个序列蠕变基本平稳（图7）。

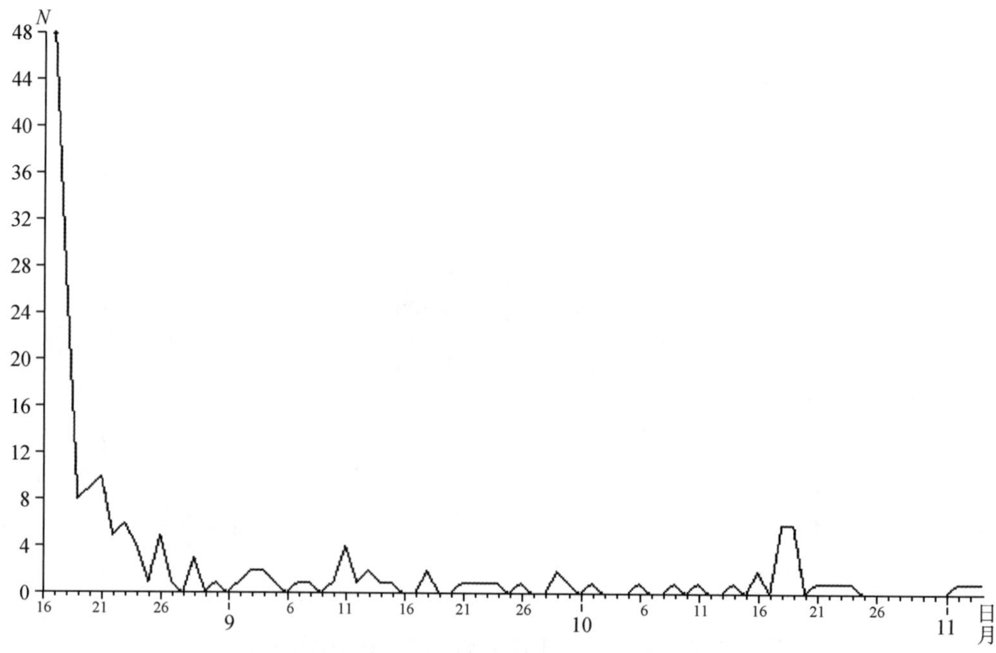

图6　巴林左旗、阿鲁科尔沁旗交界5.9级地震序列频次分布

Fig. 6　Variation of daily frequency of the M5.9 earthquake sequence in boundary area between Balinzuoqi and Alukeerqinqi（M_L）

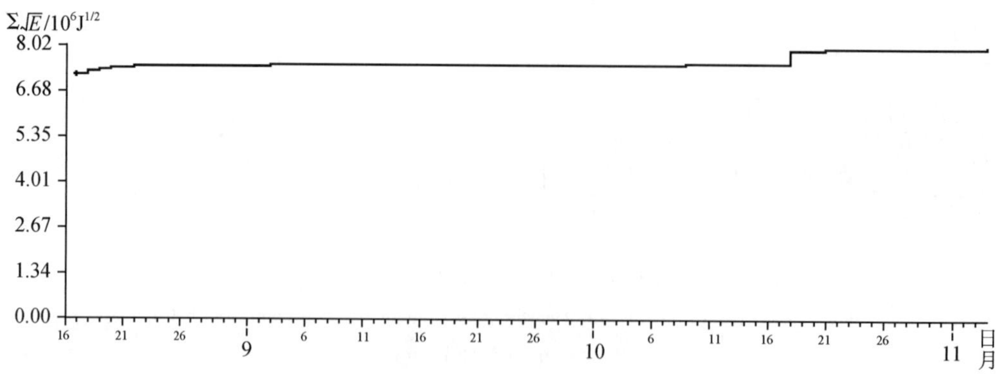

图7　巴林左旗、阿鲁科尔沁旗交界5.9级地震序列蠕变曲线

Fig. 7　Strain release curve of the M5.9 earthquake sequence in boundary area between Balinzuoqi and Alukeerqinqi（M_L）

5.9 级地震余震序列的 b 值为 0.68（图 8），p 值为 0.65，h 值为 1.5（图 9），主震释放能量占全序列能量的 99.3%，序列参数正常，主震与最大余震震级差为 1.5。根据地震序列判别指标，巴林左旗和阿鲁科尔沁旗间 5.9 级地震为主震—余震型。

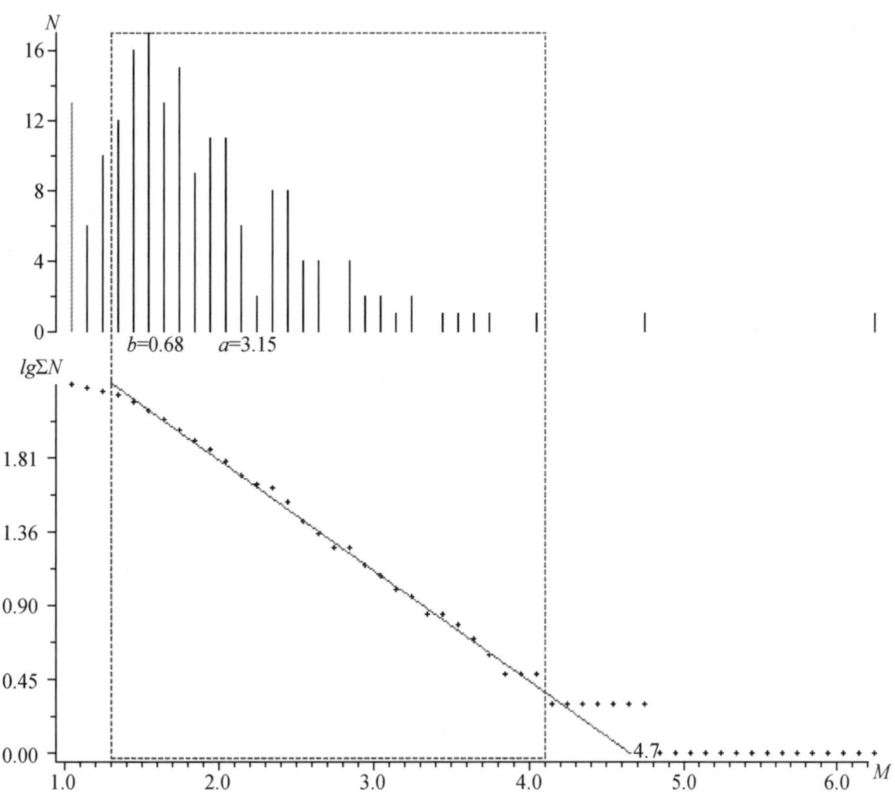

图 8 巴林左旗、阿鲁科尔沁旗交界 5.9 级地震序列 b 值拟合曲线

Fig. 8 b-value fitting curve for the $M5.9$ earthquake sequence in boundary area between Balinzuoqi and Alukeerqinqi（M_L）

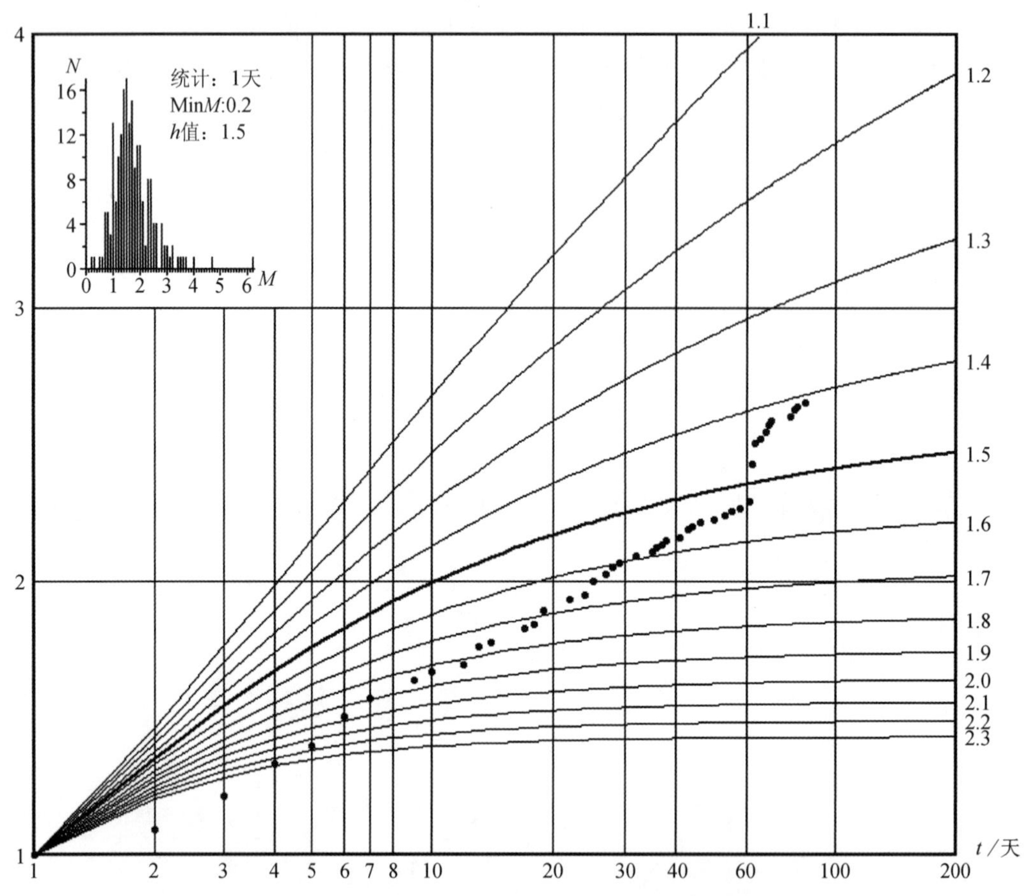

图9　巴林左旗、阿鲁科尔沁旗交界 5.9 级地震序列 h 值拟合曲线

Fig. 9　h-value diagrams the $M5.9$ earthquake sequence in boundary area between Balinzuoqi and Alukeerqinqi（M_L）

五、震源参数和地震破裂面

根据收集的内蒙古及辽宁省台网 17 个台站的记录清晰的初动符号，采用乌尔夫网做图法测定了这次 5.9 级地震的震源机制解（表 3、图 10a），矛盾符号比为 0.059。主压应力为 NEE 方向，P 轴走向为 268°，节面Ⅰ走向 200°，倾角 65°，滑动角 154°，节面Ⅱ走向 319°，倾角 70°，滑动角 30°。为增加震源机制解的可信度，再次收集辽宁省、黑龙江省、吉林省、河北省部分测震台的 P 波初动符号，测定结果与图 10a 所得结果基本接近。从震源机制解结果来看，此次地震的震源断错为倾滑逆断层，且滑动分量较大，5.9 级地震是在 NEE 向主压应力作用下，截面Ⅱ左旋错动的结果。

结合余震序列分布优势方向以及地质构造等资料，截面Ⅱ是这次 5.9 级地震的主破裂面。

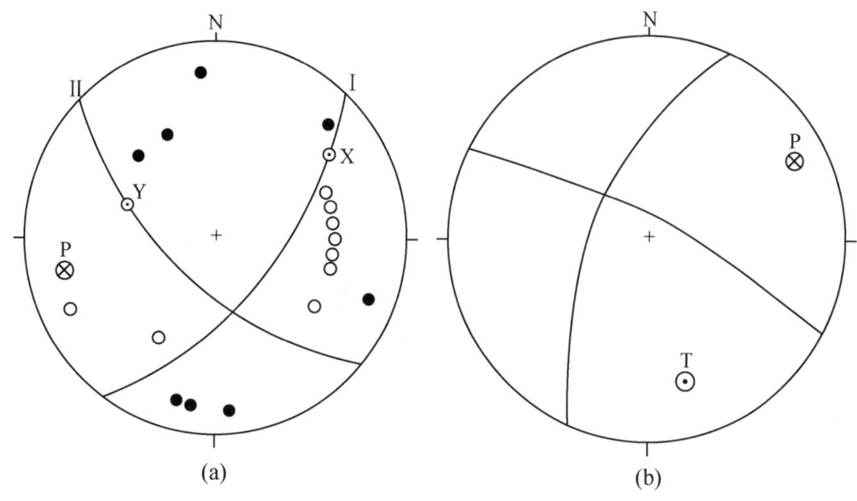

图 10 巴林左旗、阿鲁科尔沁旗交界 5.9 级地震震源机制解

Fig. 10 Focal mechanism of the *M*5.9 earthquake in boundary area between Balinzuoqi and Alukeerqinqi from P wave records by the Inner Mongolia regional network (a) and USGS (b)

(a) 内蒙古区域台网；(b) USGS

表 3 巴林左旗、阿鲁科尔沁旗交界 5.9 级地震震源机制解参数

Table 3 Focal mechanism solutions of the *M*5.9 earthquake in boundary area between Balinzuoqi and Alukeerqinqi

编号	节面 I			节面 II			P 轴		T 轴		B 轴		X 轴		Y 轴		结果来源
	走向	倾角	滑动角	走向	倾角	滑动角	方位	仰角	方位	仰角	方位	仰角	方位	仰角	方位	仰角	
1	200	65	154	319	70	30	268	17	8	52	177	68	55	25	315	47	资料 2)
2	202	56	174	296	85	34	64	20	165	27	303	55					USGS
3	37	83	113	142	24	16	108	34	332	46	215	24					CENC
4	317	57	10	222	82	146											Harvard University

六、地震前兆观测台网和前兆异常

震中附近地区的地震台及观测项目分布见图 1。地震发生在大兴安岭隆起带南部与松辽盆地过渡地带，测震和前兆监测能力均较薄弱。

在震中 200km 范围内有 5 个测震台站，4 个前兆台站。在 0～100、100～200km 范围内分别有地震台站 1、7 个，4 个前兆台站为赤峰台、翁牛特台、大甸子台、开鲁台，包括赤峰台地倾斜、赤峰水氡、赤峰水汞、翁牛特地电阻率、翁牛特自然电位、大甸子井地下水位、开鲁水氡共计 7 个观测项目，开鲁水氡、大甸子井水位出现异常现象（表 4）。

表 4 地震前兆异常登记表

Table 4 Summary table of earthquake precursory anomalies

序号	异常项目	台站或观测区	分析方法	异常判据及观测误差	震前异常起止时间	震后变化	最大幅度	震中距（km）	异常类别	图号	异常特点及备注	异常发现时间
1	中等地震外围活跃与内部平静	震中 200km 外围	活动图像	M_L4.0 以上地震丛集分布	1990.1～2003.6	正常	$N \geqslant 3$		L_1	11a、b	M_L4.0 以上地震在 5.9 级地震震中西北集中分布	震前
2	震中附近中等地震活跃与平静	震中 200km 范围	活动图像	M_L3.5 以上地震活跃与平静	1970～1995	正常	$N \geqslant 38$、$\Delta t \geqslant 1009$ 天	震中周围	L_1	12	震前震中 200km 范围出现 M_L3.5 以上地震活跃预平静图像	震前
3	地震空区	41.9°～46.9°N，115.5°～122.0°E	M_L2.5 以上地震震中分布	围空地震数 $N \geqslant 8$，持续 12 个月	2000.5～2003.1	正常		震中附近	M_1	14a	空区长轴 186km、短轴 128km，5.9 级地震发生在空区西南边缘	震前
4	地震条带	41.9°～46.9°N，115.5°～122.0°E	M_L3.0 以上地震震中分布	地震成带状分布	2000.8～2003.4	正常		震中附近	M_1	14b	条带长 364km，宽 113km，5.9 级地震发生在条带东南约 120km 处	震前
5	η 值	5.9 级地震震中周围 200km	M_L2.5～5.0，12 个月窗长，1 个月步长滑动	1 倍均方差，结合曲线形态	2000.1～2002.8	正常	大于 1 倍均方差	震中周围	M_1	15a	震前 2 年出现低值之后回升，震前 1 年再次低值	震后

续表

序号	异常项目	台站或观测区	分析方法	异常判据及观测误差	震前异常起止时间	震后变化	最大幅度	震中距（km）	异常类别	图号	异常特点及备注	异常发现时间
6	缺震	5.9级地震震中周围200km	M_L2.5～5.0，12个月窗长，1个月步长滑动	1倍均方差，结合曲线形态	2000.1～2002.9	正常	小于1倍均方差	震中周围	M_1	15a	震前2～1年持续低值变化	震后
7	地震频度	5.9级地震震中周围200km	M_L2.5～5.0，12个月窗长，1个月步长滑动	1倍均方差，结合曲线形态	1999.12～2002.9	正常	小于1倍均方差	震中周围	M_1	15a	震前2～1年持续低值变化	震后
8	*D*值	5.9级地震震中周围200km	M_L2.5～5.0，12个月窗长，1个月步长滑动	1倍均方差，结合曲线形态	2001.7～2003.1	正常	大于1倍均方差	震中周围	M_1	15a	震前2～1年持续高值变化	震后
9	小震调制比	5.9级地震震中周围200km	M_L2.5～5.0，12个月窗长，1个月步长滑动	1倍均方差，结合曲线形态	2002.7～2002.9	正常	大于1倍均方差	震中周围	M_1	15b	震前1年出现高值变化	震后
10	地震非均匀度*GL*	5.9级地震震中周围200km	M_L2.5～5.0，12个月窗长，1个月步长滑动	1倍均方差，结合曲线形态	2001.8～2003.7	正常	大于1倍均方差	震中周围	M_1	15b	震前2年出现高值，之后恢复，震前1年再次出现高值	震后

续表

序号	异常项目	台站或观测区	分析方法	异常判据及观测误差	震前异常起止时间	震后变化	最大幅度	震中距(km)	异常类别	图号	异常特点及备注	异常发现时间
12	加卸载响应比	5.9级地震震中周围200km	M_L2.5 ~ 5.0,12个月窗长,1个月步长滑动	1倍均方差,结合曲线形态	2001.9 ~ 2003.7	正常	大于1倍均方差	震中周围	M_1	15b	震前2年出现高值,之后恢复,震前1年再次出现高值	震后
13	水氡	开鲁	日均值	突跳下降	2002.12 ~ 2003.5	恢复	57ng/L	110	S_2	16	大幅度突跳下降	震前
14	水位	大甸子	日均值	下降后回升	2002.9 ~ 2003.8	恢复	0.2m	185	S_2	17	大幅度下降后回升	震前

1. 地震学异常

1）中等地震外围活跃与内部平静

1990 年以来震中 200km 范围内未发生 $M_L \geq 5.0$ 级地震，但在震中西北约 350km 范围内，分别发生 1992 年 6 月锡林浩特 $M_L5.3$、1999 年 1 月锡林浩特 $M_L5.5$、2002 年 10 月西乌珠穆沁旗 $M_L5.2$ 地震，震中东南约 210km 处 1996 年 7 月发生敖汉 $M_L5.5$ 地震，中等地震内部平静、外围活跃，5.9 级地震西北形成了明显的中等地震增强活动区。特别是 2002 年 10 月西乌珠穆沁旗 $M_L5.2$ 地震，距离 5.9 级地震震中约 220km，该次地震发生后 10 个月即发生 5.9 级地震，具有显著地震的性质[3]。1998 年 4 月阿鲁科尔沁旗 $M_L5.0$ 地震后，5.9 级地震震中 200km 范围内未发生 $M_L \geq 5.0$ 级地震，$M_L \geq 5.0$ 级地震持续平静 15 年（图 11a）。1900 年以来震中 200km 范围内未发生 $M_L \geq 4.0$ 级地震，其分布特点，仍然具有外围活跃、内部平静的特征，集中活跃区位于震中西北约 210～350km 的西乌珠穆沁旗附近地区，震中 100～200km 范围内仅发生 2 次 $M_L \geq 4.0$ 级地震（图 11b）。因此，中等地震外围活跃与内部平静是 5.9 级地震前地震活动图像的一个显著特征。

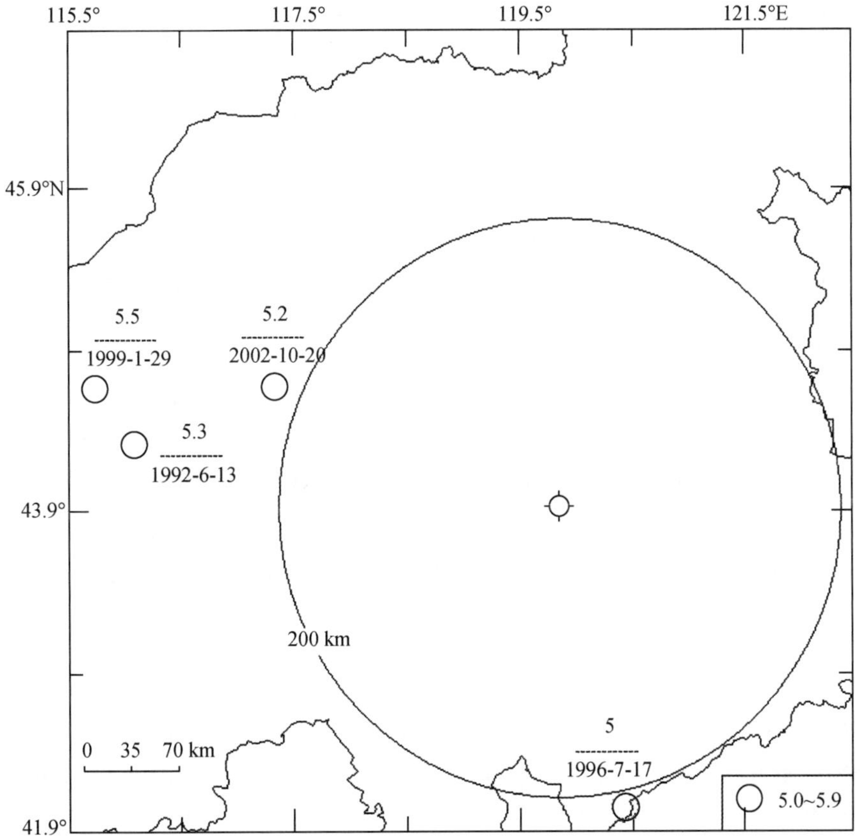

图 11a 1990 年以来巴林左旗、阿鲁科尔沁旗交界 5.9 级地震震中周围地区 $M_L \geq 5.0$ 级地震分布

Fig. 11a Distribution of the $M_L \geq 5.0$ earthquakes around the epicenter of the $M5.9$ earthquake in boundary area between Balinzuoqi and Alukeerqinqi since 1990

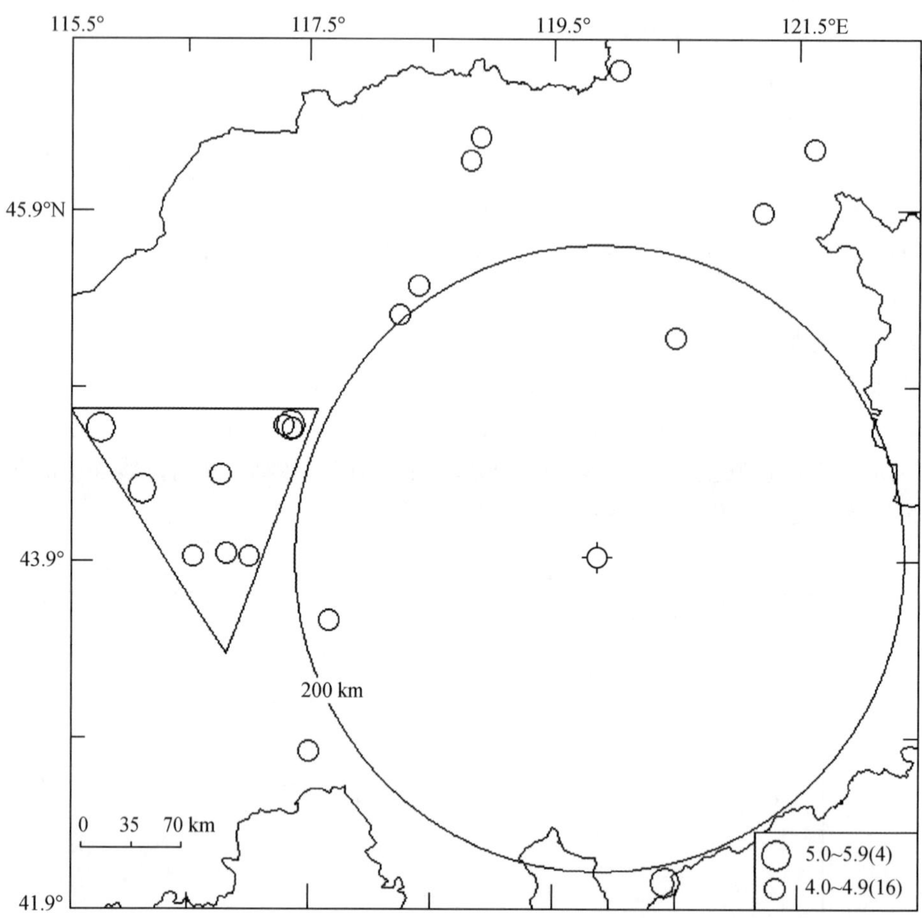

图 11b　1990 年以来巴林左旗、阿鲁科尔沁旗交界 5.9 级地震震中周围地区 M_L≥4.0 级地震分布

Fig. 11b　Distribution of the M_L≥4.0 earthquakes around the epicenter of the *M*5.9 earthquake in boundary area between Balinzuoqi and Alukeerqinqi since 1990

2）震中附近中等地震活跃与平静

1970 年 1 月至 2003 年 6 月，5.9 级地震震中 200km 范围发生 M_L≥3.5 级地震 38 次，其中 M_L≥4.0 级地震 8 次，最大地震是 1971 年 11 月 8 日距离 5.9 级地震震中 70km 的阿鲁科尔沁旗 M_L5.1 地震，次大地震是 1988 年 4 月 16 日距离 5.9 级地震震中 30km 的阿鲁科尔沁旗 M_L5.0 地震。从空间分布来看，8 次 M_L≥4.0 级地震中有 5 次分布在震中 100km 范围内，可见，从 1970 年 1 月至 1995 年 12 月震源区附近中等地震较为活跃（图 12a、b）。

从时间分布来看，8 次 M_L≥4.0 级地震，平均时间间隔为 335 天，以 1998 年 12 月 19 日克什克腾旗 M_L4.4 地震为标志，震源区附近 M_L≥3.5 级地震开始出现平静，直至 2002 年 9 月 28 日西乌珠穆沁旗 M_L3.6 地震发生，震源区附近 M_L≥3.5 级地震持续平静 1009 天，是 1970 年以来震中 200km 范围内 M_L≥3.5 级地震最大平静时间间隔（图 12b）。

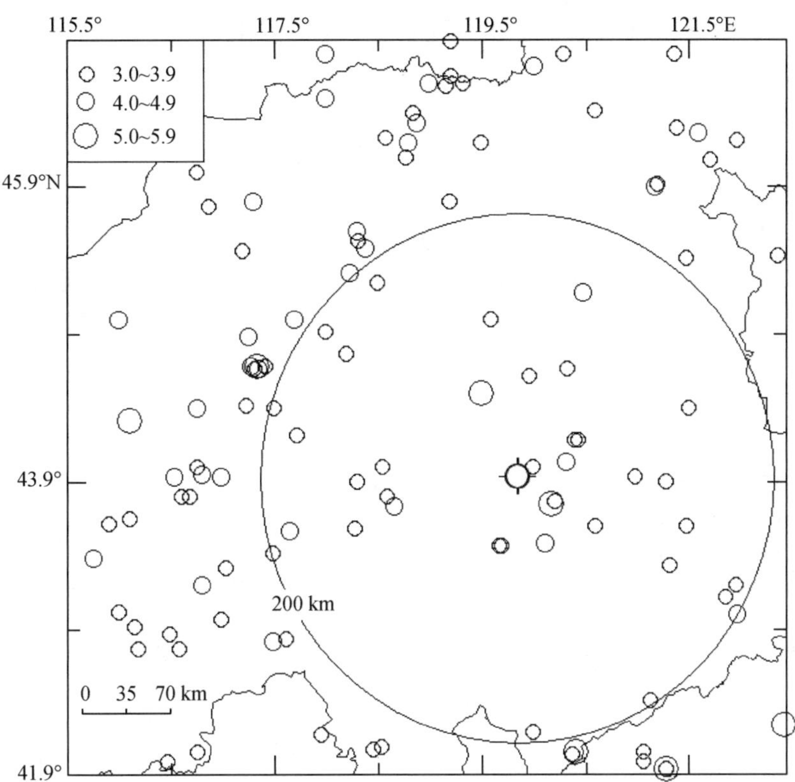

图12a 巴林左旗、阿鲁科尔沁旗交界5.9级地震震中周围地区 $M_L \geqslant 3.5$ 级地震分布（1970.1～2003.6）

Fig. 12a Distribution of the $M_L \geqslant 3.5$ earthquakes around the epicenter of the $M5.9$ earthquake in boundary area between Balinzuoqi and Alukeerqinqi (1970.1-2003.6)

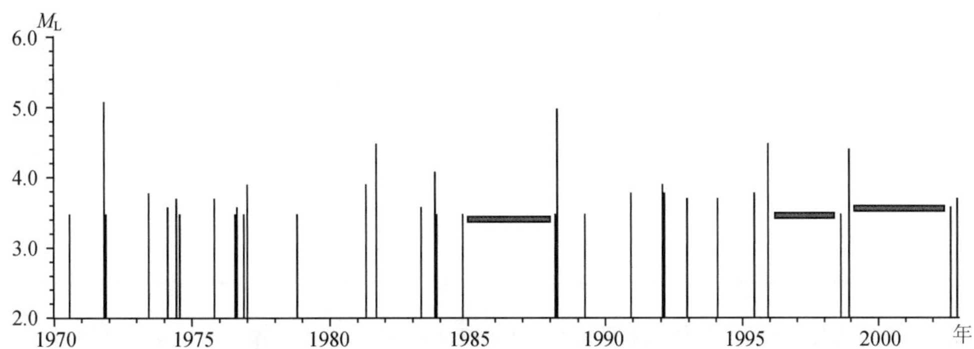

图12b 巴林左旗、阿鲁科尔沁旗交界5.9级地震震中200km范围 $M_L \geqslant 3.5$ 级地震 M-t 图（1970.1～2003.6）

Fig. 12b M-t diagram for the $M_L \geqslant 3.5$ earthquakes within 200km of the $M5.9$ earthquake in boundary area between Balinzuoqi and Alukeerqinqi (1970.1-2003.6)

3）地震空区和条带

研究结果发现，震前1年震源区附近存在地震空区（图13）。空区形成于2000年5月至2003年1月，持续12个月，空区长轴186km，呈北西向展布，短轴128km，围空震级下限 $M_L \geq 2.5$ 级。空区形成后其边缘附近发生显著地震（2002年10月20日西乌珠穆沁旗 $M_L5.2$ 地震），2003年8月16日空区边缘发生巴林左旗与阿鲁科尔沁旗间5.9级地震。

2002年10月20日西乌珠穆沁旗 $M_L5.2$ 地震后，在西乌珠穆沁旗—阿尔山一带形成 $M_L3.0$ 地震条带（图13b）。条带形成时间为2000年8月至2003年4月，持续10个月，条带长364km，呈NNE向展布，宽113km，震级下限 $M_L \geq 3.0$ 级。5.9级地震发生在条带东南120km处。

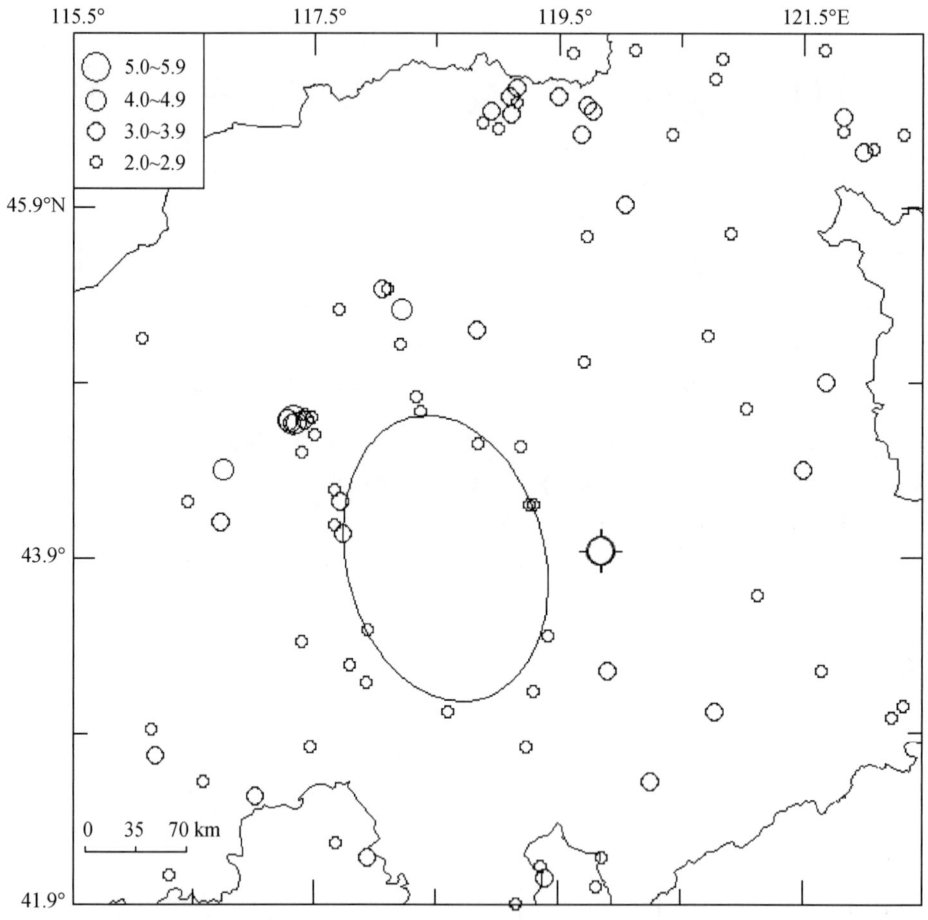

图13a 巴林左旗、阿鲁科尔沁旗交界5.9级地震前的地震空区图像（2000.5 ~ 2003.1）

Fig. 13a Image of earthquake gap before the *M*5.9 earthquake in boundary area between Balinzuoqi and Alukeerqinqi (2000.5-2003.1)

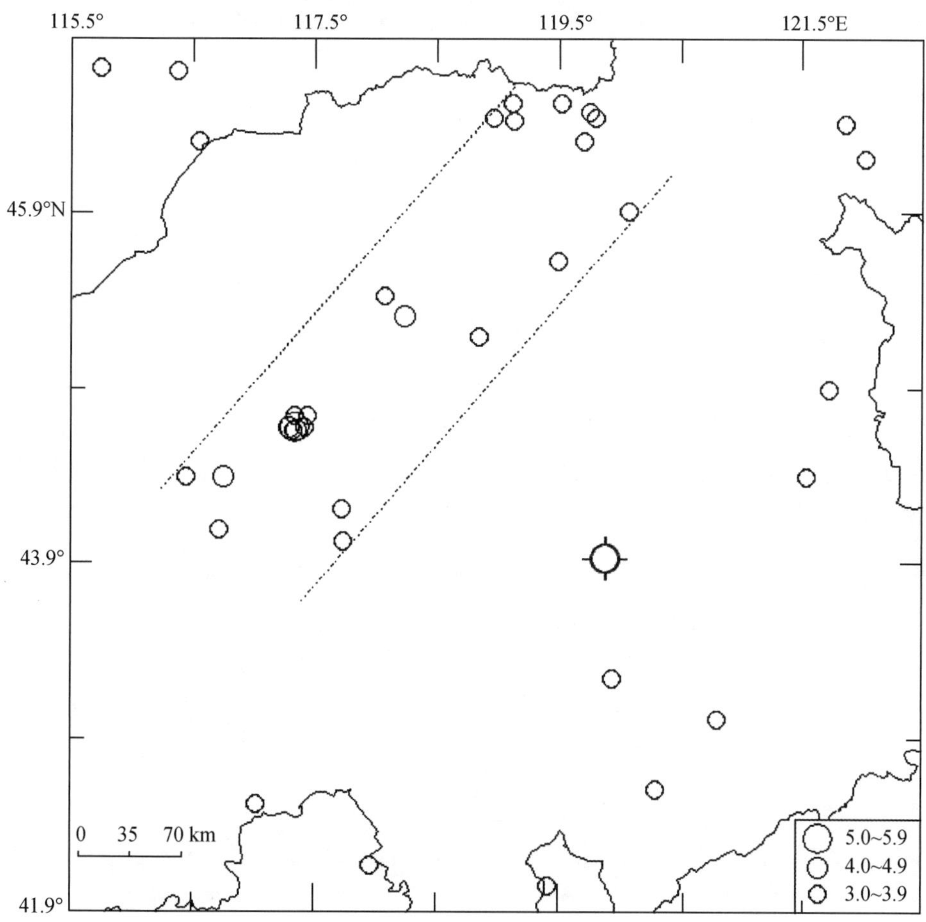

图 13b 巴林左旗、阿鲁科尔沁旗交界 5.9 级地震前的地震条带图像（2000.8～2003.4）

Fig. 13b Image of earthquake band before the M5.9 earthquake in boundary area between Balinzuoqi and Alukeerqinqi（2000.8-2003.4）

4）地震学指标异常

以巴林左旗、阿鲁科尔沁旗交界 5.9 级地震震中为圆心，半径 $r=200$km 为研究区，选取 b 值、η 值、强度因子 M_f、集中度 C 值、危险度 D 值、非均匀度 GL、频度 N、缺震、调制比 R_m 和加卸载响应比 Y 等 10 种地震学参数，按时间扫描窗长 12 个月，扫描步长 1 个月，对 1988 年 1 月至 2003 年 6 月，$M_L \geqslant 2.5$ 级地震资料进行时间扫描分析，以 1 倍均方差作为异常判定标准，进行异常提取。

表 5　震前 2.0 ~ 0.5 年震源区附近 (r = 200km) 地震学异常统计表

Table 5　Statistic table of seismology abnormity around epicenter before two years to half year (r = 200km)

震前/年	b	η	M_f	C	D	GL	N	缺震	调制比 R_m	响应比 Y
2.0 ~ 1.0		√			√	√	√	√		√
1.0 ~ 0.5		√			√		√	√	√	√
0.5 ~ 0						√				√

从图 14a、b 和表 5 可以看出，震前 2 年左右震源区附近 10 种地震学参数有 6 种存在异常，占 60%。震前 1.0 ~ 0.5 年震源区附近 10 种地震学参数有 6 种存在异常，占 60%，震前 0.5 年震源区附近 10 种地震学参数仅有 2 地震非均匀度 GL 值和加卸载响应比 2 种参数存在异常，可见，地震学指标异常主要出现在震前 2.0 ~ 1.0 年，震前随着时间推移，5.9 级地震中短期阶段震源区附近地震学异常项次逐渐减少（图 14a、b）。

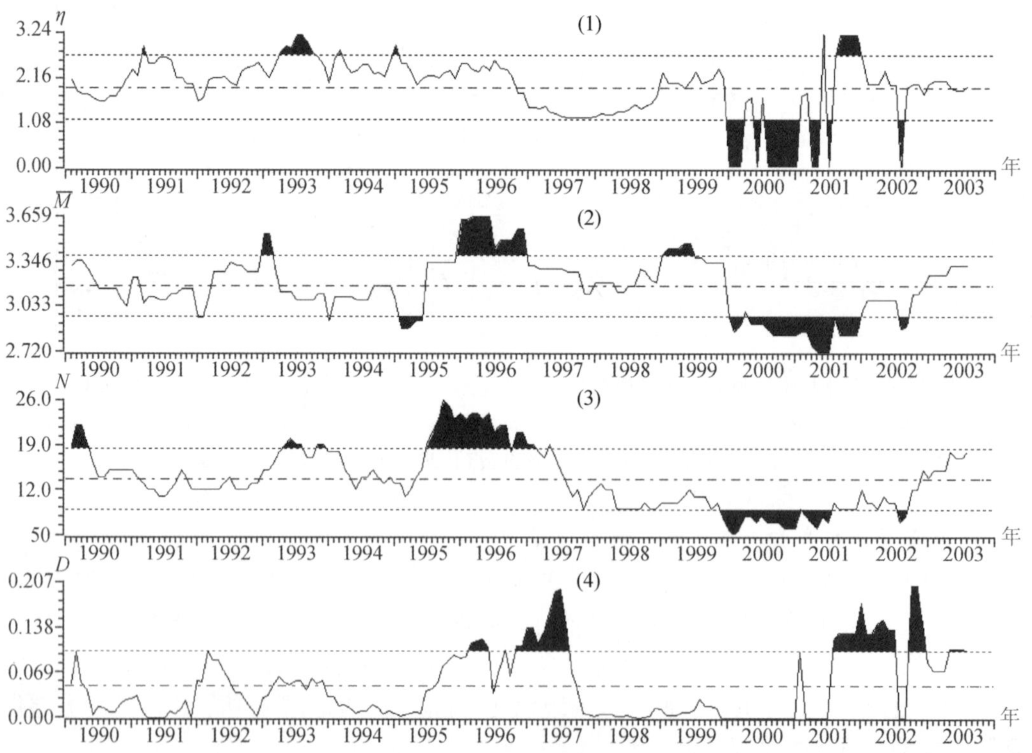

图 14a　巴林左旗、阿鲁科尔沁旗交界 5.9 级地震前地震学参量时序曲线

Fig. 14a　Curve of seismology parameter before the M5.9 earthquake in boundary area between Balinzuoqi and Alukeerqinqi

(1) η 值；(2) 缺震；(3) 地震频度；(4) D 值

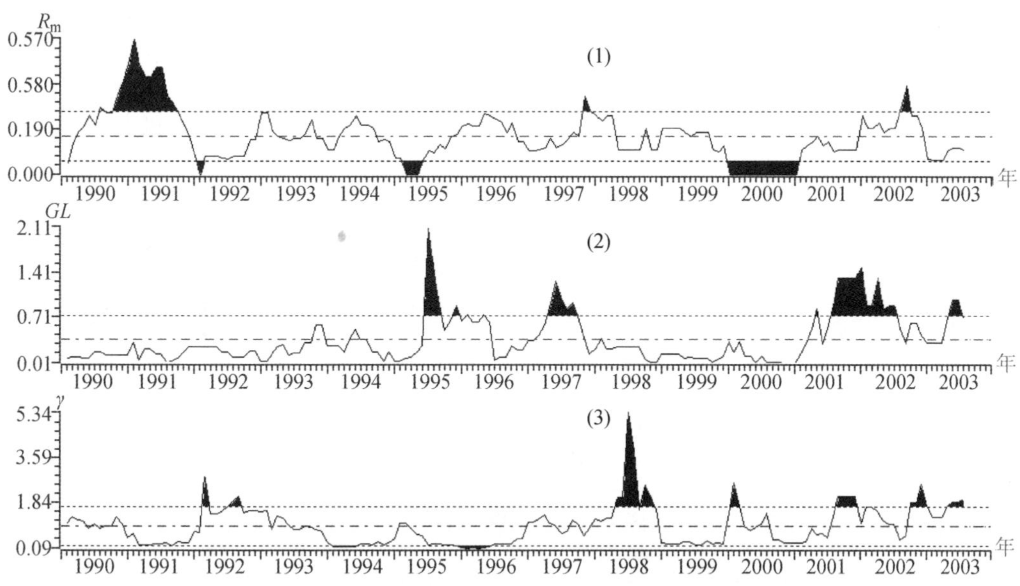

图 14b 巴林左旗、阿鲁科尔沁旗交界 5.9 级地震前地震学参量时序曲线

Fig. 14b Curve of seismology parameter before the *M*5.9 earthquake in boundary area between Balinzuoqi and Alukeerqinqi

（1）小震调制比；（2）地震非均匀度 *GL* 值；（3）加卸载响应比

2. 前兆异常

1）5.9 级地震 200km 范围内前兆异常分析

5.9 级地震震中周围 100km 范围内没有前兆台站，200km 范围内有 4 个前兆台站共计 7 个观测项目，其中赤峰地倾斜、赤峰水氡、赤峰氺汞、翁牛特自然电位未出现异常变化，翁牛特地电阻率、大甸子井地下水位、开鲁水氡分别不同程度出现中段期异常变化。距离 5.9 级地震震中 220km 左右的内蒙古库伦井地下水位和通辽地热也曾出现趋势性上升变化，但经过异常核实和进一步分析，认为这两项异常信度较低，不作为 5.9 级地震前前兆异常进行分析[4)]。

（1）开鲁水氡。

开鲁水氡距离震中约 110km，该测点 1996 年 6 月 7～30 日有一个小的下降过程，1996 年 7 月 17 日发生敖汉 4.7 级地震，地震后，观测数据变化一直较平稳，2002 年 12 月 21～26 日观测数据出现突然下降，而且幅度很大，是过去没有过的，这次异常可能是 2002 年 10 月 20 西乌珠穆沁旗 5.0 级的震后效应。2003 年 4 月 18 日至 5 月 19 日又出现和上次一样的突降，对该异常进行核实后[5)]，并未发现干扰因素。异常持续一个多月后恢复正常变化，直到 3 个月后的 2003 年 8 月 16 日发生巴林左旗 5.9 级地震（图 15）。

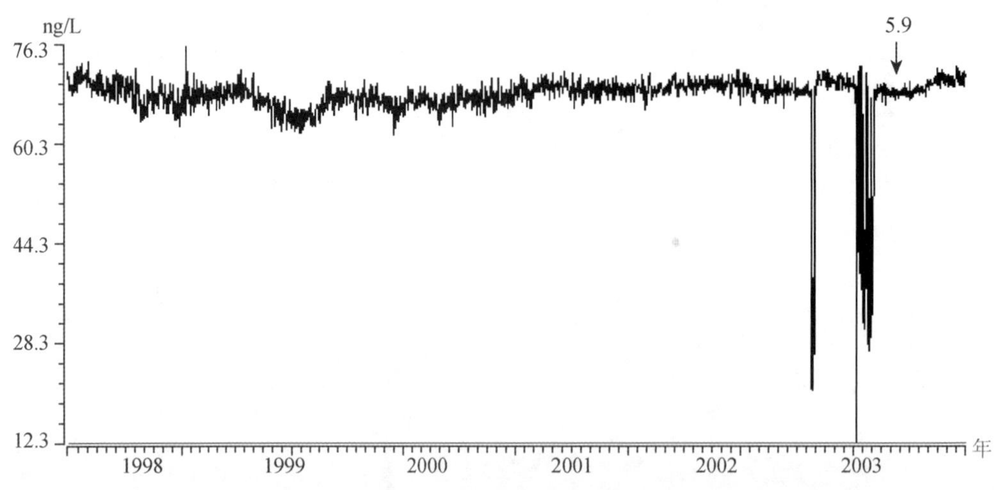

图 15　开鲁水氡日均值图

Fig. 15　Curve of daily mean value of radon content in groundwater at Kailu station

（2）大甸子井地下水位。

大甸子井距离震中约 185km，该井点 1994 年初水位出现加速上升，1995 年 7 月达到最高值后转折下降，1996 年 7 月 17 日发生敖汉 4.7 级地震，震后继续下降，1997 年初恢复正常年变化。2002 年 11 月打破正常年动态，再次出现下降趋势，下降幅度是 1997 年以来最大的，2003 年 3 月下降到最低值，随后出现上升趋势，2003 年 8 月 16 日发生 5.9 级地震，震后异常仍未出现结束的迹象（图 16），异常出现分析人员到现场进行了核实[6)]。

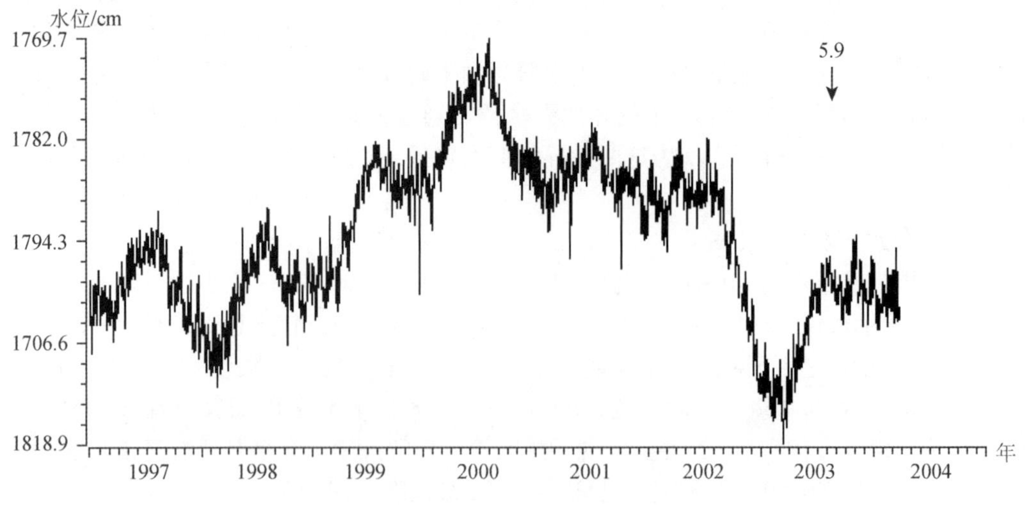

图 16　大甸子井地下水位日均值图

Fig. 16　Curve of daily mean value of well water level at Dadianzi station

2）5.9 级地震 200km 范围外围前兆分析

5.9 级地震 200km 外围主要涉及辽宁省和吉林省部分前兆测项。

收集 5.9 级地震震中周边辽宁省前兆资料，发现，5.9 级地震前辽宁朝阳—北票地区出现前兆异常，主要以形变异常为主[3]，包括桃花吐跨断层水准、北票跨断层水准、哈达户梢水氡、盘锦地热、朝阳地倾斜在 5.9 级地震前不同程度出现异常变化。

（1）桃花吐水准 1—3 测段、3—4 测段及水位异常。

该测点距离 5.9 级地震 251km，桃花吐水准 1—3 测段高差，1990 年以来，观测资料一直有较好的年变规律，2000、2001 年年变规律开始变得模糊。2003 年 5 月出现大幅度下降异常，幅度达 1mm，3—4 测段高差 2000 年以来其年变幅度逐渐变小，显示出一定的趋势异常。其辅助测项水位观测 2003 年 1 月出现下降趋势，下降幅度高达 2.4m，6 月 25 日至 8 月 10 日反向上升 1.6m（图 17）。

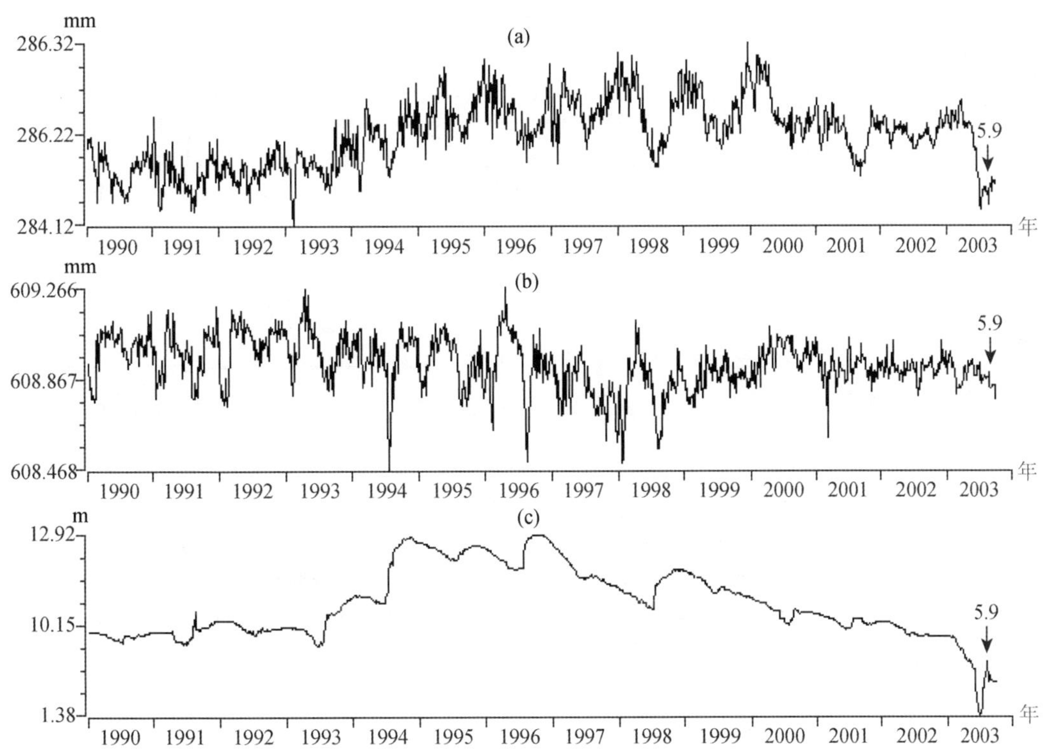

图 17 桃花吐测点水准、地下水位日均值曲线（据辽宁省地震局）

Fig. 17 Curve of daily mean value of leveling and water level at Taohuatu measuring point (from the Seismological Bureau of Liaoning province)

（a）水准 1—3 测段；（b）水准 3—4 测段；（c）地下水位

该异常表明，桃花吐断层运动方式是西盘相对东盘下降，表明了该断层在 5.9 级地震前的剧烈活动。

（2）北票跨断层水准 7—E、N1—W、W—7 测段高差异常。

该测点距离 5.9 级地震 239km，北票跨断层水准 7—E 测段，2001 年初出现趋势上升，2002 年中上升速率加快，至巴林左旗—阿鲁科尔沁旗间 5.9 级地震累计上升幅度达 3mm；N1—W 测段从 2001 年开始出现趋势下降，2003 年初下降进一步加快，累计下降达 1.6mm；W—7 测段从 2001 年 2 月开始下降，2002 年 4 月出现加速下降，至 2003 年 8 月累计下降幅度 4mm（图 18）。

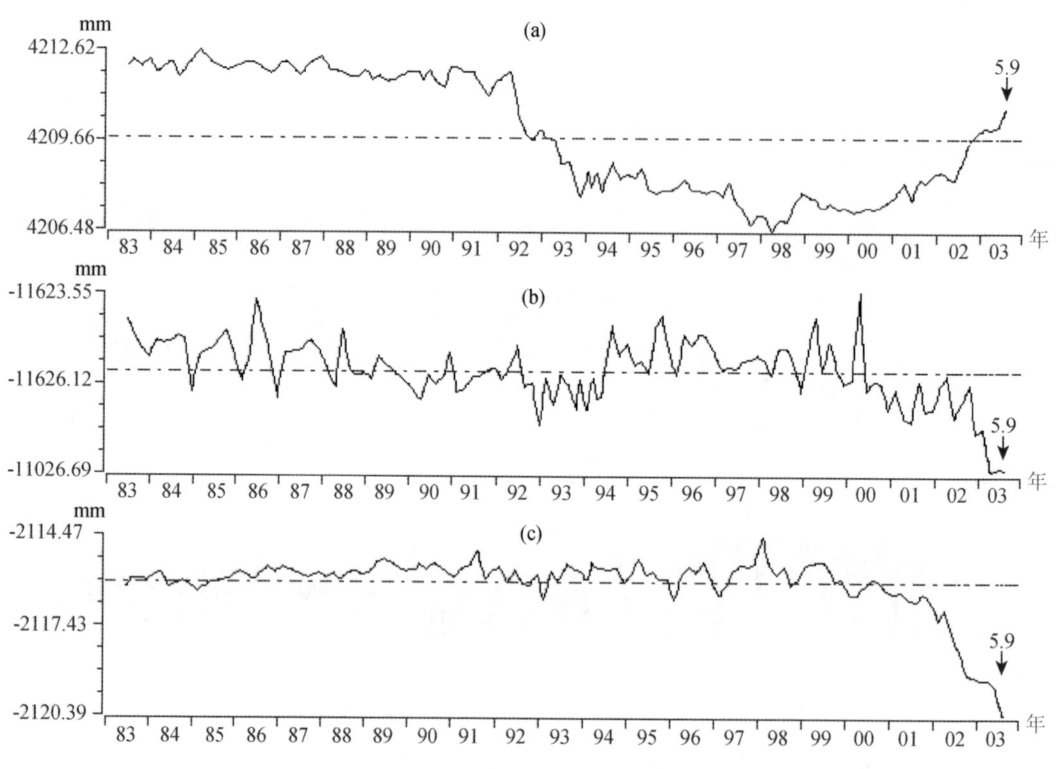

图 18　北票测点水准日均值图（据辽宁省地震局）

Fig. 18　Curve of daily mean value of leveling at Beipiao measuring point (from the Seismological Bureau of Liaoning province)

（a）7—E 测段；（b）N1—W 测段；（c）W—7 测段

（3）盘锦高七井地热异常。

该测点距离 5.9 级地震 350km，该井地热在趋势性上升的背景下，2003 年 8 月 12 日出现突跳上升，幅度达 4‰℃，以后持续上升最大幅度约 8‰℃，辽宁地震局对此异常落实后，未发现明显干扰因素，认为此异常具有一定的可信度（图 19）。

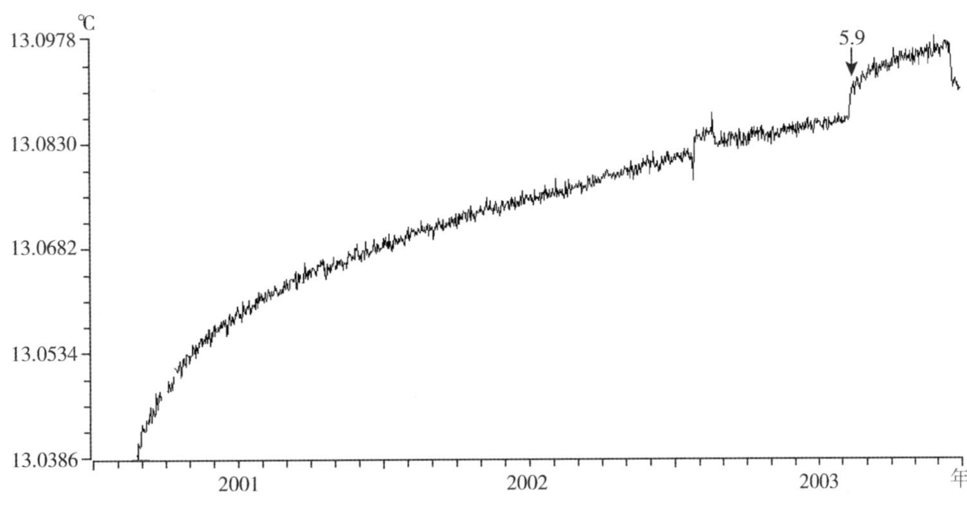

图 19　盘锦地热日均值图（据辽宁省地震局）

Fig. 19　Curve of daily mean value of ground temperature at PanJin measuring point (from the Seismological Bureau of Liaoning province)

（4）阜新哈达户梢水氡异常。

该测点距离 5.9 级地震 255km，该井水氡 2003 年 7 月 24 日突降 6Bq/L，8 月 26 日再次突降 4Bq/L，9 月 26 日又下降 1.5Bq/L，异常出现后辽宁地震局分析人员到台站对此异常进行了落实，未发现明显干扰，认为此异常具有一定的可信度（图 20）。

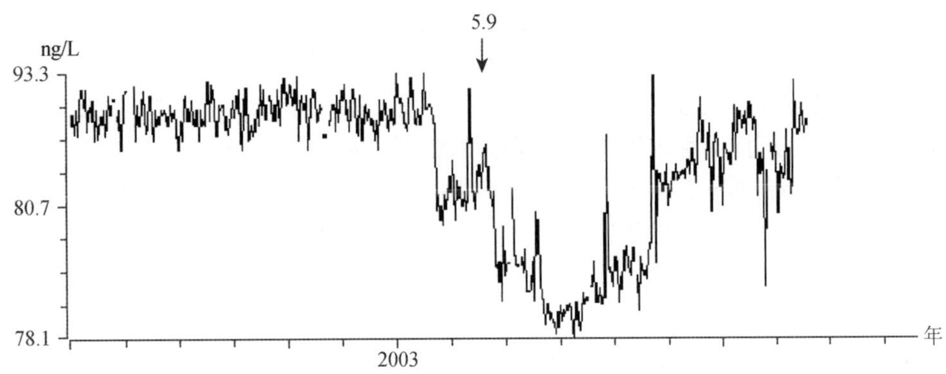

图 20　哈达户梢水氡异常图（据辽宁省地震局）

Fig. 20　Curve of daily mean value of radon content in groundwater at Hadahushao (from the Seismological Bureau of Liaoning province)

（5）朝阳台倾斜仪异常。

该测点距离 5.9 级地震 248km，2002 年开始，NS 向和 EW 向同步出现转折变化，NS 向由趋势性的南倾转折为北倾，北倾幅度达 80ms；EW 向有趋势东倾转折为西倾，西倾幅度达 60ms。特别是 2003 年 7 月后这种向北、向西倾斜的趋势异常显示更加明显，认为此异常的信度较高（图 21）。

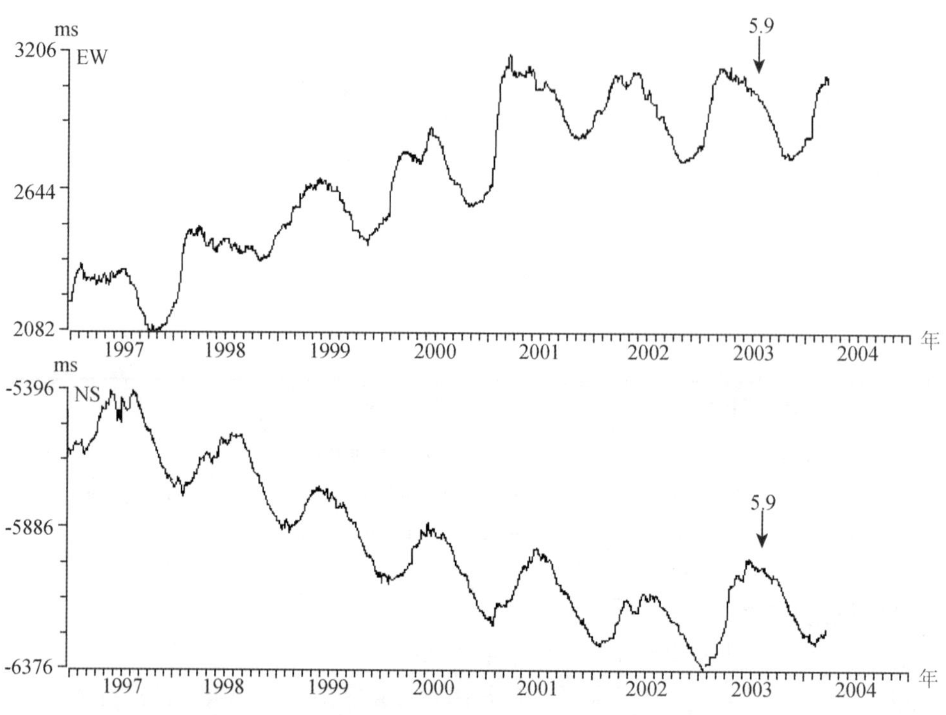

图 21　朝阳地倾斜日均值图（据辽宁省地震局）
Fig. 21　Curve of daily mean value of tilt at Chaoyang
(from the Seismological Bureau of Liaoning province)

收集 5.9 级地震震中周边吉林省前兆资料和相关文献，位于吉林省中部的长春、双阳、丰满、磐石台地区，震前 1 年内出现了不同程度的中短期及临震异常，异常也主要以形变异常为主[4]，包括双阳台水管倾斜仪、双阳台伸缩仪、丰满台水管倾斜仪、磐石台石英倾斜仪、长春台体应变。由于以上这些异常出现在距离 5.9 级地震震中 250～500km 之间，经过分析认为以上测项信度较低，故不进行统计和分析。

3. 宏观异常

大量震例研究表明，地震前动物行为异常现象表现得非常突出，如能及时捕捉到这些异常现象，就能为地震部门预报地震、减轻地震灾害提供依据。

震前，与动物生存、生活环境有密切相关的地球物理场与地球化学场的变化，有可能使某些动物因不能适应上述条件的变化，出现异常现象。震源区应力积累，造成地壳深部流体（水、二氧化硫等）沿断层和岩石裂隙上溢，可能使两栖动物穴居动物出现行为异常；震前岩层微破裂引起的振动和电磁辐射，空气中的阳离子浓度增加，促使动物神经激素增加，有可能引起鸟类和哺乳类动物出现行为异常反应[7]。

地震现场工作结束后，阿鲁科尔沁旗地震局，对地震前后的宏观异常现象进行了考察7)。通过对12个苏木乡镇的调查核实，据不完全统计，震前出现了250多起宏观异常现象，其中动物近200余起，地下水6起，人感、地声、地光等多起。经调查分析，宏观异常基本在极震区及其附近区域，异常区走向西北。其范围北至西沙布台乡，南到天山口镇，东西展布长，南北展布短，形似靴形。强震前动物的鸣叫与平日不同。人们普遍反映其声音发颤，透着悲哀，像嚎哭一样，无人惊吓而突然发疯似的乱跳、乱蹿，表现惊恐状，烦躁不安。具有突发性、群体性特征。

越接近震中异常种类越多，异常量越高，行为异常越强烈，异常出现的时间也越早。在震前1~3天动物异常占70%左右，震前几小时占20%左右。

最早出现行为异常的是穴居动物和水生动物，而后是家禽和牲畜（如鸡、鸭、狗、猪、马、驴等）行为异常最晚，大约在震前几分钟才出现异常。

8月16日5.9级地震前地下水水位普遍上升，伴有发浑、发苦、翻花。由于大部分井是封闭状态，无法观测其变化全貌。

极震区地声无明显方向性，听到地声与发震时间间隔约几分钟左右。

七、地震前兆异常特征分析

1. 地震学异常

中等地震外围活跃与内部平静，1900年以来震中200km范围内未发生$M_L \geq 4.0$级地震，其分布特点具有外围活跃、内部平静的特征，集中活跃区位于震中西北约210~350km的西乌珠穆沁旗附近地区，震中100~200km范围内仅发生2次$M_L \geq 4.0$级地震。

震中附近中等地震活跃与平静，以1998年12月19日克什克腾旗$M_L4.4$地震为标志，震源区附近$M_L \geq 3.5$级地震开始出现平静，直至2002年9月28日西乌珠穆沁旗$M_L3.6$地震发生，震源区附近$M_L \geq 3.5$级地震持续平静1009天，是1970年以来震中200km范围内$M_L \geq 3.5$级地震最大平静时间间隔。

地震空区和条带，震前1年震源区附近存在地震空区，2002年10月20日西乌珠穆沁旗$M_L5.2$地震具有显著地震性质，2003年8月16日空区边缘发生巴林左旗与阿鲁科尔沁旗间5.9级地震。

震前2年左右震源区附近10种地震学参数有6种存在异常，占60%，震前1.0~0.5年震源区附近10种地震学参数有6种存在异常，占60%，震前0.5年震源区附近10种地震学参数仅有地震非均匀度GL值和加卸载响应比2种参数存在异常，可见，地震学指标异常主要出现在震前2.0~1.0年，震前随着时间推移，5.9级地震中短期阶段震源区附近地震学异常项次逐渐减少。

2. 前兆异常

由于震中附近前兆监测能力较低，5.9级地震200km范围内仅开鲁水氡和大甸子水位两个测项震前出现短期异常[5,6]，均为短期异常，开鲁水氡震前出现大幅度突跳下降，大甸子水位则是大幅度下降转折回升后发震。

5.9级地震200km范围外围的辽宁省的桃花吐水准、北票水准、阜新哈达呼梢水氡、盘锦地热出现异常变化。吉林省几项前兆异常变化信度较低。

3. 宏观异常

狗异常行为表现为昼夜狂吠、乱咬、惊吓外逃；猫异常行为表现为惊逃；鸡异常行为表现为夜不进窝、不停悲啼、惊飞；鸭子异常行为表现为夜不进窝、不停悲啼、惊飞；蛇异常行为表现为惊逃出洞，聚集在一起。鱼异常行为表现为死亡、撞鱼缸、跃出水面；鼠异常行为表现为惊慌外逃；驴异常行为表现为嚎叫；猪异常行为表现为跳圈、嚎叫；地下水异常表现为井水水位上升、井水变浑、变苦、喷出、翻花；地光、地声异常表现为刺眼的地光、地声、烟雾。

4. 异常的时空演化特征

车用太研究华北大同、包头、张北等3次地震前地下流体异常数量随时间变化时，发现震前2～1个月或稍长时间，异常数量明显增加[10)]，本次地震也大致具有这样的特点，震前异常主要以中长期异常为主，兼有中期异常出现，震前5～3个月短期异常集中出现。尽管该地区前兆异常台站稀疏，异常的空间演化存在向震中、向外围迁移的趋势。

在内蒙古的赤峰—通辽地区，震前未出现中长期异常，而出现了两项短期异常。在异常空间分布上，由于近场没有前兆台站分布，异常主要分布在中远场，随着时间的推移，异常开始出现向震中逼近、向中远场扩散的迹象。主要表现在，震前4个月，距离震中最近的开鲁水氡出现异常，震前2个月处于中远场的辽宁一侧的前兆测项集中出现了短临异常，震前异常向震中逼近和中远场扩散，出现丰富的短临异常是该次地震前兆异常的主要特征[5]。

八、震前预测、预防和震后响应

1. 预测情况

这次5.9级地震发生在内蒙古自治区地震局2003年初确定的年度地震值得注意地区—辽蒙交界附近[8)]。多年来，根据地震活动的起伏变化，我们一直关注辽蒙交界地区地震形势的发展，特别是近几年该区严重缺震，引起地震部门高度重视。2002年10月20日西乌珠穆沁旗发生M_L5.0地震，打破了内蒙东部中强地震平静格局。针对当时严峻而复杂的地震形势，2003年年中内蒙古自治区地震趋势研究时我们提出："下半年该区发生5～6级地震的危险性进一步增大"[9)]。尽管震前有所察觉，震前出现了几起前兆异常，并进一步强化了震情短临跟踪工作，但仍未捕捉到短期地震学前兆信息。

2. 震后趋势判定

5.9级地震发生后，根据序列衰减的各项指标，认为序列发展正常，认为短时间内发生更大地震的可能性较小。根据5.9级地震序列发展趋势和历史地震序列类比，认为这次5.9级地震为主—余震型地震。历史上5.9级地震周围曾发生过几次中强地震，1976年8月29日阿巴嘎旗5.4级为孤立型地震，1985年6月21日苏尼特右旗5.3级为孤立型地震，1991年9月30日苏尼特右旗5.4级为孤立型地震，1999年1月29日锡林浩特5.2级地震为孤立型地震，1986年9月9日阿巴嘎旗5.6级地震为主—余型地震。根据历史上没有发生过双震型或震群型地震特点，认为短时间内震区地震活动升级的可能性不大[2)]。

1986年3、8月黑龙江德都发生5.4和5.9级地震；1941年5月、1942年9月黑龙江绥化发生2次6.0级地震；1940年1月、1942年7月库仑、通辽发生2次6.0级地震；1980

年2月、1981年4月博克图发生2次5.6级地震。根据中强地震具有成对活动的特点，分析认为震区附近地区未来2年存在发生5~6级地震的危险[2)]。

2002年6月29日东北7.2级深震后，内蒙古东北地区地震活跃，超过内蒙古中西部地区，先后发生2002年10月西乌旗4.8、2002年12月鄂温克4.7、2003年8月巴林左旗—阿鲁科尔沁旗间5.9级地震。6月26日牙克石发生 $M_L3.1$ 响应地震。内蒙古自治区东北地区地震活跃。分析认为：未来1~2年，东北地区的地震监测预报应予重视[2)]。

3. 震后响应

地震发生后，内蒙古自治区地震局第一时间向自治区党委、政府和中国地震局报告了初定的地震三要素，所有应急人员迅速赶到机关，启动《内蒙古自治区地震局应急预案》，对救灾工作做了安排部署。地震当晚20时，自治区地震局召开了第一次紧急会商会，初步介绍了地震的基本参数和地震序列，给出初步判定意见。之后10几天内，连续召开14次紧急会商会，对震情密切跟踪、对地震序列及时分析判定，并和现场分析预报及时沟通，判定本次地震为主—余震型，对几次4级以上余震及震区今后的地震趋势给出了较好的判定意见。自治区地震局向党委、政府和中国地震局上报了《震情通报》12期，及时上报震情情况、灾情情况、现场工作情况和震区的抗震救灾情况，并提出震后趋势意见和下一步的工作建议。

九、结论和讨论

1. 主要结论

巴林左旗、阿鲁科尔沁旗交界5.9级地震震中烈度Ⅷ度，等震线呈东西向椭圆形分布。地震共造成4人死亡，60人重伤，1004人轻伤。灾区面积为11979km^2（按评估区域统计），受灾范围涉及巴林左旗、阿鲁科尔沁旗、巴林右旗3个旗，37个乡镇苏木，共408个嘎查村，灾区人口共计480869人，13739户。根据地震现场灾害评估工作和地震灾害评估委员会确认，本次地震直接经济损失8.06亿元。

巴林左旗与阿鲁科尔沁旗间5.9级地震共记录到 $M_L \geqslant 0.0$ 级地震200次，其中 $3.0 \leqslant M_L \leqslant 3.9$ 级9次，$4.0 \leqslant M_L \leqslant 4.9$ 级2次，最大余震是2003年10月17日的4.7级地震。序列参数正常，为主震—余震型地震序列，震后趋势判定较正确。

2. 讨论

（1）5.9级地震后，开鲁水氡仍在较为平静的背景下发展，未出现大的脉冲式跳跃，说明5.9级地震前该测项大幅度的脉冲式跳跃是5.9级地震前兆异常的可信度较高，大甸子井水位在5.9级地震后再未出现过震前如此大幅度的下降变化，也证明了震前该测项的大幅度下降变化可能就是5.9级地震的前兆异常变化。

由于地震孕育和前兆的复杂性，确认某个测项出现的前兆变化是否是某个中强地震的地震前兆，目前的研究手段和分析方法还无法进行定量的分析和确定，只能通过经验和统计关系进行定性分析，一般来说，测项位于中强地震的孕震范围内，其作为中强地震的前兆异常的可能性更大，但是由于中强以上地震具有远场效应，位于孕震区外围的前兆测项出现的异常变化，也可能反映某次中强地震的前兆变化，比如5.9级地震前辽宁出现的几项前兆异常变化，可能就是该次5.9级地震的中远场地震效应。

（2）巴林左旗、阿鲁科尔沁旗为抗震设防标准较低的地区，却发生极震区烈度Ⅷ度的破坏性地震，类似现象其他地方也有。如何科学、合理制定适当抗震设防标准，为我们工程地震工作提出了更高的要求，应予认真研究。

根据烈度调查结果，本次地震的高烈度区长轴方向北西西，与北西向发震构造方向一致，受到发震构造控制。同时极震区位于微观震中的北东侧，反映出地震的破裂过程可能是自北西向东南破裂。从余震展布方向和分布特点、地震烈度衰减及震源机制解结果分析，发震断层倾向南西。极灾区的分布于发震断层走向倾向及松散的粉砂、细砂质场地土有关。在未来的建筑规划设计中应当考虑场地条件作用2)。

（3）对东北地区的地震活动性做进一步研究表明，从1999年开始，东北地区的浅源地震进入第5个活跃期，其可能的活跃时段为1999年至2010年前后，类比前4个活跃时段，第5个活跃期可能要发生7～8次中强以上地震，截至2003年8月16日5.9级地震，东北地区已发生3次浅源地震，未来的几年仍存在发生4～5次中强以上浅源地震的可能[5]。

另外，该地区中强地震活动存在“成对活动”特征，5.9级地震的发生，未来1～3年该地震附近区域仍然存在发生相当水平地震的危险性。为此，系统和深入地研究该地区的前兆异常特征和演化过程，对于下一次中强震的预测具有重要意义。东北地区中强地震“成对出现”，以及目前出现的前兆异常，未来一段时间，东北地区仍然存在发生中强地震的危险性[5]。

参 考 文 献

［1］刘芳，薛丁，曹井泉等.2003年8月16日巴林左旗与阿鲁科尔沁旗间M_S5.9地震参数及序列特征．华北地震科学，2004，22（3）：44～46

［2］刘芳，曹井泉．巴林左旗—阿鲁科尔沁旗5.9级地震余震序列精确定位及分布特征．东北地震研究，2004，20（4）：22～27

［3］王永江，王辉，李子涛等．朝阳—北票地区的形变异常与巴林左旗5.9级地震关系的探讨．东北地震研究，2006，22（2）：34～39

［4］张兴科，焦伟，孙旭丽等．内蒙巴林左旗5.9级地震前吉林定点形变异常特征.2004，20（4）：28～32

［5］高立新．巴林左旗—阿鲁科尔沁旗5.9级地震前前兆异常特征分析［J］．东北地震研究，2004，20（4）：8～15

［6］高立新，索亚峰．巴林左旗—阿鲁科尔沁旗5.9级地震前地下流体异常特征分析［J］．地震，2005，25（1）：103～110

［7］陈鑫连主编．动物奇观［M］．北京：地震出版社，1988

［8］高孟潭，许力生，郭文生等.2003年8月16日内蒙古M_S5.9地震震害分布特征及其成因［J］．地震学报，2005，27（2）：205～212

参 考 资 料

1）内蒙古自治区地震局，内蒙古地震台网观测报告（2003年8～11月）

2）内蒙古自治区地震局，2003年8月16日巴林左旗和阿鲁科尔沁旗间5.9级地震现场与科考工作报告，

2003 年 12 月
3）内蒙古自治区地震局，内蒙古地震台网观测报告（巴林左旗—阿鲁科尔沁旗间 5.9 级地震序列），2004 年 4 月
4）内蒙古自治区 2006 年度地震趋势研究报告，库伦 CK3 地震观测井水位趋势上升异常核实报告，2006 年
5）内蒙古自治区 2004 年度地震趋势研究报告，2003 年开鲁水氡异常核实报告，2004 年
6）内蒙古自治区 2004 年度地震趋势研究报告，赤峰市敖汉旗大甸子 CK14 井水位异常核实报告，2004 年
7）阿鲁科尔沁旗地震局，巴林左旗和阿鲁科尔沁旗间 5.9 级地震宏观考察报告，2003 年
8）内蒙古自治区 2003 年度地震趋势研究报告，内蒙古自治区地震局，2002 年 12 月
9）内蒙古自治区 2003 年年中地震趋势研究报告，内蒙古自治区地震局，2003 年 6 月
10）车用太，1999，拉张型构造区强震中段期前兆异常机理研究专题报告

*M*5.9 Earthquake in boundary area between Balinzuoqi and Alukeerqinqi on August 16, 2003 in Inner Mongolia Autonomous Region

Abstract

An earthquake of *M*5.9 occurred in boundary area between Balinzuoqi and Alukeerqinqi of Chifeng city in the Inner Mongolia Autonomous Region on August 16, 2003. Its macroscopic epicenter was located between Baiyingou village of Balinzuoqi and Wulanhala village of Alukeerqinqi. The intensity at the meizoseismal area was Ⅷ and its shape was about oval-shaped with major axis in EW direction. The earthquake caused 4 death, 60 serious injured and 1004 people commonly injured. The direct economic loss was 806 million Yuan.

This earthquake sequence belongs to mainshock-aftershock type, the magnitude of the largest aftershock was M_L4.7 o. Aftershocks distributed west to the main shock and along NWW direction almost as the same direction as the major axis of isoseismal line, The distance between the largest aftershock and main shock was about 24km. The focal mechanism solution showed that nodal plane Ⅱ was the main rupture surface. The P axis of main compressive stress was NEE direction. Xiashuiquangou fault was confirmed as the rupture fault.

There were fewer stations around the *M*5.9 epicenter. There were 8 seismic station and precursory stations within 0-200km from the epicenter, There were 7 observation terms include Chifeng tilt etc. There were 14 anomalies before the earthquake, 2 of them were long-term anomalies, 10 of them were medium term anomalies, and 2 of them were short-term anomalies There were not any imminent anomalies. The image of seismic activity was main long-term background anomalies), and the anomalies of seismic parameters are mainly medium-term anomalies.

There was specific medium-term forecasting opinion for the Seismological Bureau of Inner Mongolia Autonomous Region, and has gained certain social benefit.

2003 年 8 月 21 日四川省盐源 5.0 级地震

四川省地震局

吴小平　阮　祥

摘　要

2003 年 8 月 21 日四川省凉山州盐源县发生 5.0 级地震。微观震中位于 N27°26′，E101°18′，宏观震中在盐源县博大乡、大草乡、盐塘镇一带。震中烈度为Ⅵ度。极震区呈北北西向的椭圆形。地震造成 9 人受伤，其中 2 人重伤。直接经济损失约 2000 万元。

此次地震序列为主震—余震型，最大余震 4.5 级。余震呈散状分布，推测震源机制节面Ⅱ为主破裂面，主压应力 P 轴方位北北西向，地震发震构造为盐源弧形断裂。

震中附近地区 200km 内共有地震观测台 48 个，其中测震观测台 30 个，定点前兆观测台 18 个。地震前出现的异常共 12 条。地震前出现的异常多为中期和短期异常，其中 8 条为中期异常，3 条为短期异常，1 条为临震异常。

前　言

2003 年 8 月 21 日 10 时 17 分，四川省凉山州盐源县境内发生 5.0 级地震。四川地震台网测定的微观震中为 N27.42°、E101.27°，宏观震中为 N27°28′、E101°10′，位于盐源县博大乡、大草乡、盐塘镇一带，极震区烈度为Ⅵ度。这次地震有感范围较大：盐源、木里、攀枝花等地都有不同程度的震感。重灾区主要在盐源县博大乡红岩村、甘胜村、几坡村、大草乡水坝村、盐塘乡郑家田村等地，地震造成 9 人受伤（其中 2 人重伤），直接经济损失 2086 万元。5.0 级地震发生在川滇菱形块体的中部，该区历史上多次发生中强地震，1976 年曾发生 6.7、6.4 级强震。震区共有地震台 48 个，震前出现 9 个项目的 12 条前兆异常。

一、测震台网及地震基本参数

盐源 5.0 级地震震中 100km 内有测震台 7 个，101 ~ 200km 内有测震台 23 个（有人值守和遥测台网）。它们于 1970 ~ 1985 年间陆续建立。1985 年以后西昌遥测地震台网投测，使其震区小震监测能力达 $M_L2.0$[1]。101 ~ 200km 范围内的西部地区达到 $M_L2.5$ 地震监测能力。盐源 5.0 级地震发生在川滇两省交界的四川一侧，测震台主要分布在震中东面和南面的四川地区，分布见图 1。根据四川地震台网[4)]和中国地震台网测定[3)]的结果，5.0 级地震的基本参数见表 1。仪器测定的微观震中与宏观震中有一定的距离，相差 20km。

表 1　地震基本参数表

Table 1　Basic parameters of the earthquake

编号	发震日期	发震时刻	震中位置		震级		震源深度	震中地名	资料来源
	年 月 日	时 分 秒	λ_E	φ_N	M_L	M_S	(km)		
1	2003 08 21	10：17：48.8	101.27°	27.42°	5.1		5	四川盐源	资料 4)
2	2003 08 21	10：17：51.2	101.3°	27.4°		5.0	10	四川盐源县	资料 3)
3	2003 08 21	10：17：47.7	101.27°	27.52°	4.6		3	四川盐源	资料 10)
4	2003 08 21	10：17：54.6	101.400°	27.353°		4.4	33	中国四川省	NEIC
5	2003 08 21	10：17：51.2	101°10′	27°28′		5.0		盐源县博大乡至大草间	资料 1)

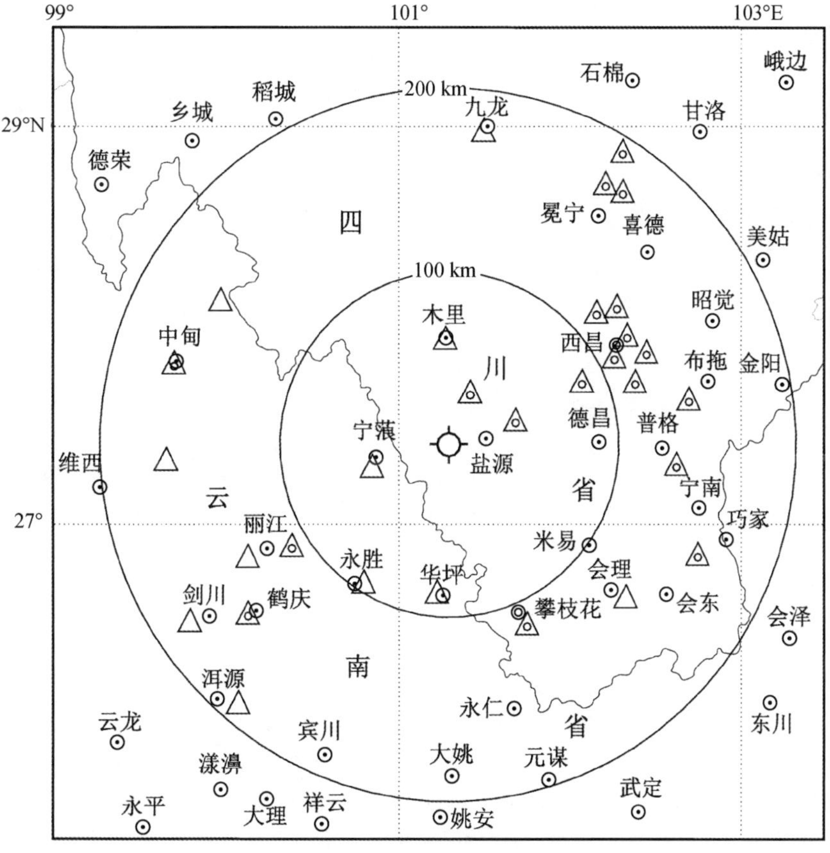

图 1　盐源 5.0 级地震前测震台站分布图

Fig. 1　Distribution of seismic stations around the M_S5.0 Yanyuan earthquake

二、地震地质背景

盐源 5.0 级地震发生在由鲜水河断裂带、安宁河断裂带、小江断裂带和金沙江、红河断裂带所围限的川滇菱形块体的中部。区内发震构造带主要由盐源弧形断裂带和辣子弧形断裂带组成。震区及其附近地区先后发生过 1467 年盐源 6.5 级地震和 1478 年盐源 6 级地震、1976 年盐源下甲米 6.7 级地震[5]、1976 年盐源辣子乡 6.4 级地震、1998 年宁蒗、盐源间 6.2 级地震及 4 次 5.0 ~ 5.3 级地震、2001 年盐源泸沽湖 5.8 级地震[2,3]（图 2）。最大为 1976 年 11 月 7 日盐源下甲米 6.7 级地震1)。2003 年 8 月 21 日盐源 M_S5.0 地震就位于 1478 年盐源 6 级地震震区的东南隅，该区域集中有记载以来的 4 次中强震。

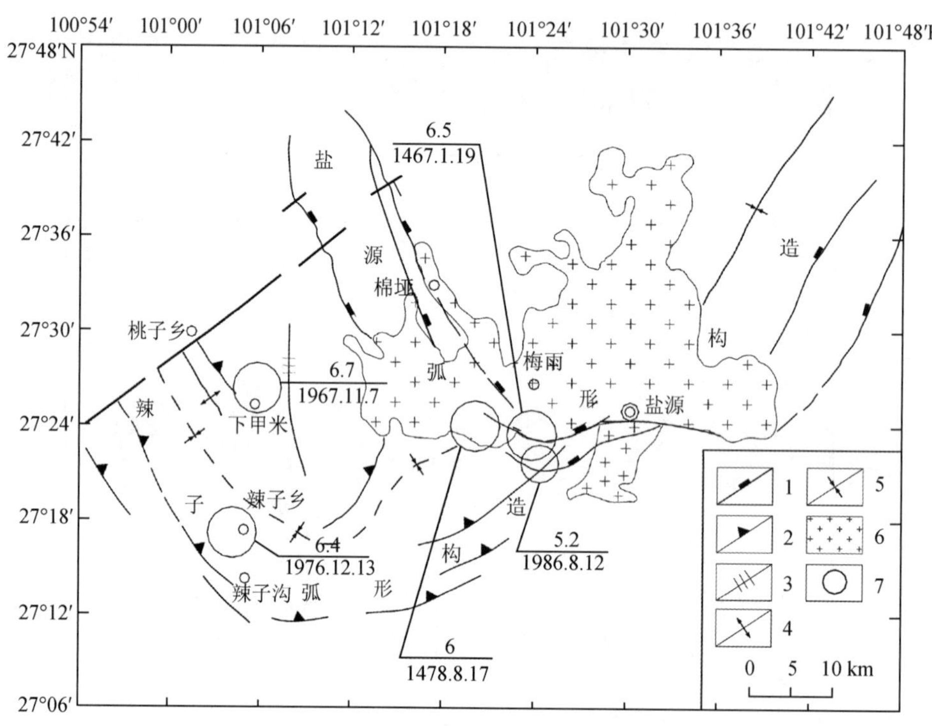

图 2　盐源 M_S5.0 地震震区附近地震地质构造及历史地震分布图

Fig. 2　Seismogeological tectonic and distribution of historic earthquakes in the area around the M_S5.0 Yanyuan earthquake

1. 盐源弧形构造成分；2. 辣子弧形构造成分；3. 挤压带；4. 背斜；5. 向斜；6. 新生代盆地；7. 地震震中

震区内的盐源弧形断裂带控制了盐源新生代断陷盆地的沉积[3]，盐源盆地南侧受断裂控制。1467 年 1 月 19 日盐源 6.5 级地震和 1478 年 8 月 17 日盐源 6 级地震震中在梅雨附近，推测其发震构造部位为盐源弧形断裂带的弧顶及西翼。盐源弧形构造带的新活动性，其西翼较东翼为注目。盐源以西的新、老第三纪地层中断裂、褶皱发育。跨西翼弧形断层的短水准形变观测结果表明，北东盘平均每年下降 0.1mm。

另外，辣子弧形构造带现今活动较强烈，尤以西翼明显，1976 年 11 月 7 日盐源下甲米 6.7 级地震和 1976 年 12 月 13 日盐源辣子乡 6.4 级地震都发生在该弧形断裂带西翼。其震害分布特征，各烈度区长轴方向均近南北。6.7 级地震的地裂缝分布在桃子乡至大厦村一带。6.4 级地震的地裂缝分布在辣子乡的潘家湾与辣子沟一带。

盐源、辣子弧形构造带展布于区域性布格重力异常梯级带和地壳厚度变异带和现代地壳垂直形变的上升区，其深部重力异常值为 $2.0\times10^{-3}m/s^2$；由重力反演地壳厚度为 55km，居里温度深度为 30km。该区域构造应力场的优势方向为北北西向[3]。

三、地震影响场和震害

据现场实地考察及调查资料，本次地震的宏观震中为 27°28′N，101°10′E，位于盐源县博大乡，大草乡、盐塘镇一带1)。重灾区主要在博大乡红岩村、甘生村、几波村、大革乡水草坝村、盐塘乡郑家田村等地，极震区烈度为Ⅵ度。等烈度线呈长轴方向为北北西向的椭圆形。烈度沿北北西、南南东方向衰减较慢，沿北东东、南西西方向较快。

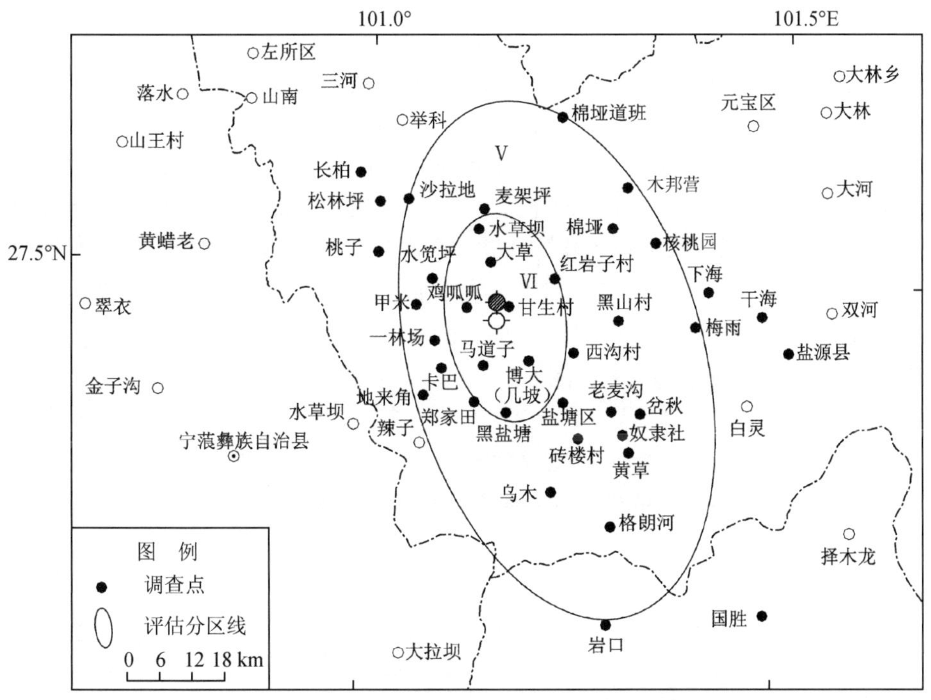

图 3 盐源 5.0 地震烈度分布及调查点分布图1)

Fig. 3 Intensity isoline map of the Yanyuan $M_S5.0$ earthquake

Ⅵ度区范围：东起红岩子，西至鸡呱呱，南达黑盐塘，北抵水草坝。呈一北北西向的近似椭圆形，面积约为 $50km^2$。民居墙体开裂，墙体局部垮塌，屋面梭瓦、脊瓦震倒、震裂和杉板房(当地居民称“黄板房”)屋顶压石震落等现象。个别地方地表可见到边坡地裂缝或小规模山体崩塌以及地震滚石等地质灾害现象。

Ⅴ度区范围:东起木邦营,核桃园、梅雨,西至甲米、地来角,南达格朗河以南,北抵棉垭道班附近。呈一北北西向的椭圆形,面积约350km²。

地震涉及盐源县10个乡(镇)和一个县属牧场、盐边县3个乡,其重灾区(破坏相对较重的地区)包括博大乡、大草乡所辖部分地区,以及盐塘乡郑家田村、梅雨镇黑山村,面积约50km²;轻灾区涉及盐源县黄草镇、梅雨镇、盐塘乡、棉植乡、下海乡、巫木乡、长柏乡、桃子乡、县属艳牛山牧场,以及盐边县岩口乡、洼洛乡和国胜乡少部分地区,总面积约400km²。

这次地震造成9人受伤,其中2人重伤,直接经济损失2000万元。

主要震害特征为:

(1) Ⅵ区内严重破坏房屋为14200m²,占总房屋面积的5.8%。中等破坏房屋为69350m²,占总面积的28.37%。境内的5个小型水库堤坝出现水平向裂缝,现场可见最大裂缝宽度约为2cm,有渗漏的现象。

(2) Ⅴ区内中等破坏房屋为810m²,占总面积的0.5%。部分沟坎渠道有下陷、局部垮塌现象。

四、地震序列

据四川地震台网测定,2003年8月21日至2003年10月24日共记录到盐源5.0级地震序列827次,其中1.0～1.9级地震331次,2.0～2.9级地震29次,3.0～3.9级地震5次,4.0～4.9级地震3次,5.0级地震1次[4)],最大余震为2003年8月21日M_L4.5地震。主震与余震的级差为0.5。M_S5.0地震释放能量占总释放能量的50%,当日地震释放能量占总释放能量的98.8%。该序列为主—余震型。取1.0级作为震级下限,序列衰减系数$p=0.97$,衰减正常(图5)。序列b值为0.6(图6)。

表2 盐源5.0级地震序列目录($M_L \geqslant 2.5$级)[4)]

Table 2 Catalogue of the Yanyuan M_S5.0 earthquake sequence ($M_L \geqslant 2.5$)

编号	发震日期	发震时刻	震中位置		震级		震源深度	震中地名
	年 月 日	时 分 秒	北纬(°)	东经(°)	M_L	M_S	(km)	
1	2008 08 21	10 17 51	27.42	101.27	5.1	5.0	5	盐源
2	2003 08 21	10 21 42	27.43	101.17	3.0			盐源
3	2003 08 21	11 35 54	27.43	101.17	2.5			盐源
4	2003 08 21	13 38 55	27.43	101.22	4.1			盐源
5	2003 08 21	18 58 14	27.43	101.32	4.5			盐源
6	2003 08 21	23 20 27	27.47	101.23	4.3			盐源

续表

编号	发震日期	发震时刻	震中位置		震级		震源深度	震中地名
	年 月 日	时 分 秒	北纬（°）	东经（°）	M_L	M_S	(km)	
7	2003 08 21	23 40 53	27.43	101.13	2.6			盐源
8	2003 08 23	03 32 21	27.33	101.23	2.6			盐源
9	2003 08 23	16 55 52	27.37	101.13	2.5			盐源
10	2003 08 23	17 25 24	27.43	101.17	2.6			盐源
11	2003 08 23	19 28 56	27.43	101.13	3.6			盐源
12	2003 08 23	21 08 01	27.35	101.18	2.7			盐源
13	2003 08 26	11 16 15	27.45	101.13	3.0			盐源
14	2003 08 31	17 49 05	27.42	101.12	2.5			盐源
15	2003 10 03	00 57 20	27.40	101.20	3.8			盐源

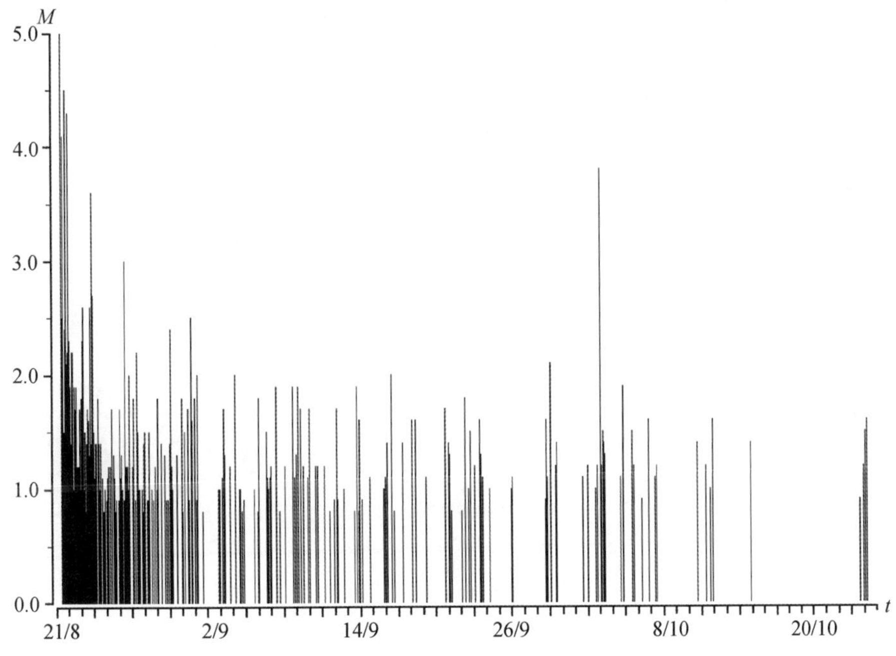

图 4　盐源 5.0 级地震序列 M-t 图

Figure 4　M-t diagram of the M_S5.0 Yanyuan earthquake sequence

图 8 为 2003 年 8 月 21 日至 10 月 3 日 $M_L \geqslant 1.7$ 级地震震中分布图。主震位于盐源弧形断裂弧顶处，4 级以上余震活动主要分布在盐源弧形断裂带西翼至弧顶一带，小震级余震则向主震以西延伸在甲米断裂段东侧呈现密集分布，总体上余震主要分布盐源弧形构造内，呈近东西向展布。

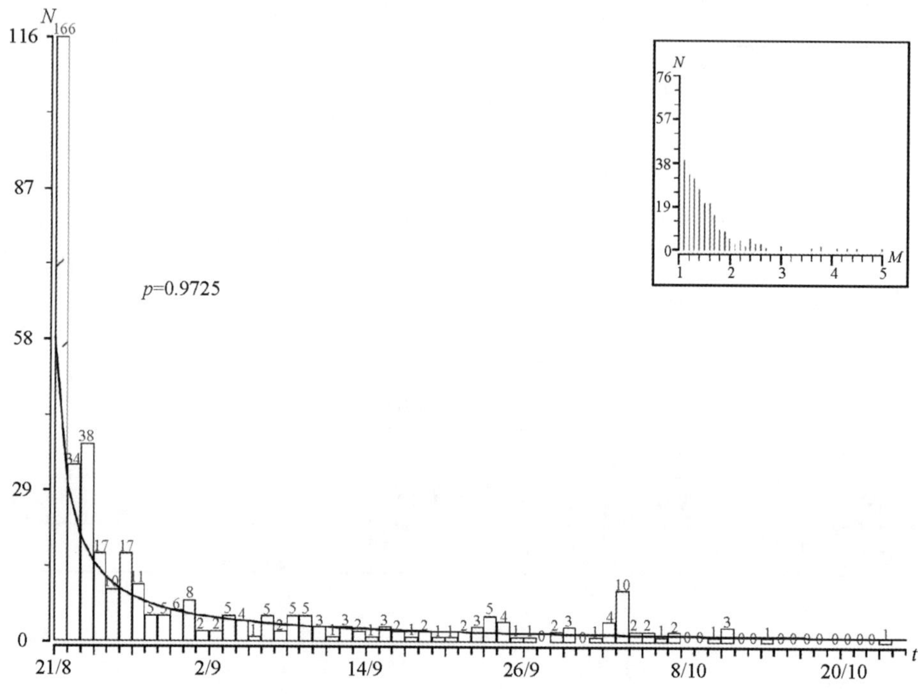

图5　盐源5.0级地震序列衰减p值曲线

Figure 5　p-value of attenuation of the M_S5.0 Yanyuan earthquake sequence

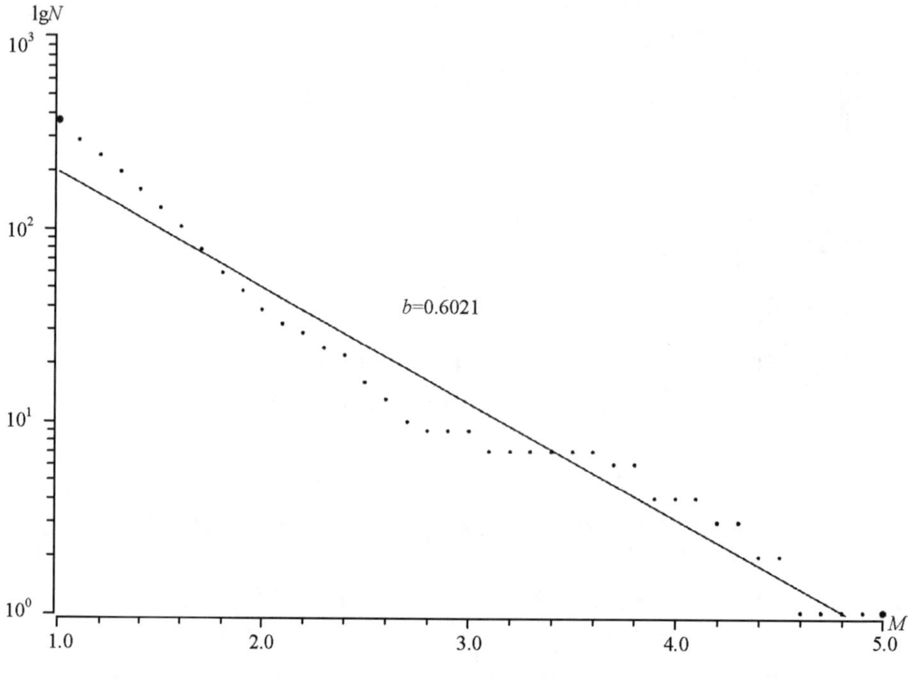

图6　盐源5.0级地震序列震级频度关系

Figure 6　lgN-M curve of the Yanyuan M_S5.0 earthquake sequence

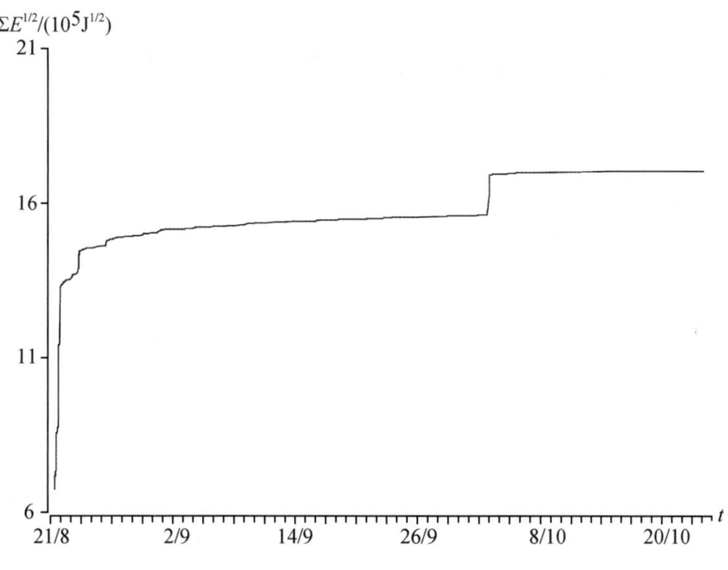

图 7 盐源 5.0 级地震序列蠕变曲线

Figure 7 Curve of strain release of the M_S5.0 Yanyuan earthquake sequence

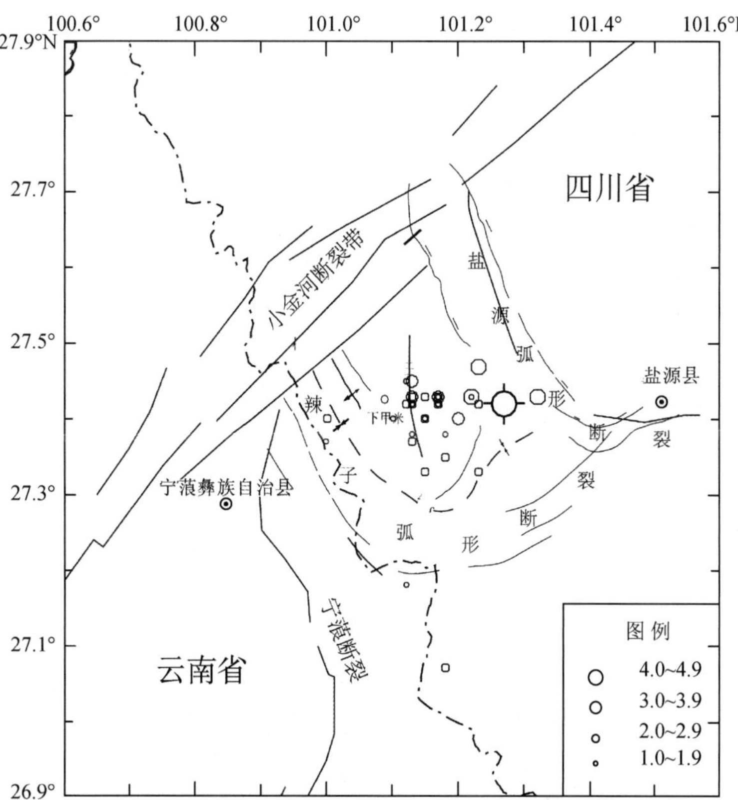

图 8 盐源 5.0 级地震余震震中分布图

Figure 8 Distribution of the M_S5.0 Yanyuan earthquake sequence

五、震源机制解和地震主破裂面

根据四川台网和西昌台网，以及部分云南台网 P 波初动资料，并采用震源上半球投影方法得到的 2 次 $M_L \geqslant 4.5$ 级地震的震源机制解，其结果列于表 3 中（由杜方提供）。结合余震活动和等震线分布，推断 $M_S5.0$ 地震 P 波初动节面Ⅱ为断层面，为带逆冲分量的右旋走滑型断层，$M_L4.5$ 地震为带正倾分量的走滑型断层。见图 9。

表 3　2003 年 8 月 21 日 $M_S5.0$ 和 $M_L4.5$ 地震震源机制解

Table 3　Focal mechanism solutions of the $M_S5.0$ and $M_L4.5$ Yanyuan earthquakes

时间	震级	节面Ⅰ			节面Ⅱ			X 轴		Y 轴		P 轴		T 轴		N 轴	
		走向	倾向	倾角	走向	倾向	倾角	方位	仰角	方位	仰角	方位	仰角	方位	仰角	方位	仰角
8.21	5.0	47	NW	66	122	SW	60	212	30	317	24	353	4	267	40	79	50
8.21	4.5	23	NW	83	117	NE	56	27	34	293	7	334	29	75	18	193	55

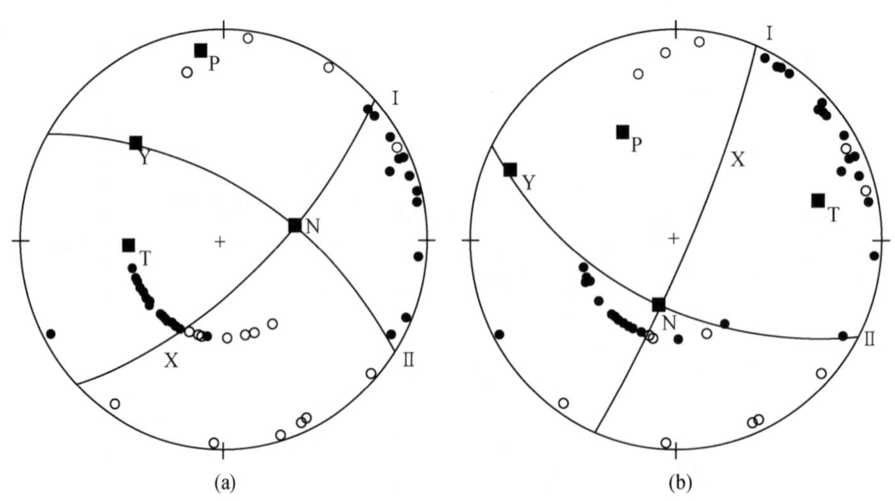

图 9　盐源 $M_S5.0$（a）和 $M_L4.5$（b）地震震源机制

Figure 9　Focal mechanism solutions of the $M_S5.0$（a） and $M_L4.5$（b） Yanyuan earthquakes

六、地震前兆观测台网及前兆异常

图 10 为距盐源 5.0 级地震震中 200km 范围内定点前兆观测站分布图。该范围内共有定点前兆观测台 18 个。有水位、高精度水温、水氡、水质、地磁、地电、重力、应力、地倾斜、短水准、短基线等 21 个前兆观测项目 98 个台项。这些前兆观测台项多数有 5 年以上连

续可靠的观测资料。前兆观测台主要分布在地震东南部地区，西北部地区观测台站比较稀疏。

此次地震前共出现 12 条前兆异常。测震学异常 1 条，即盐源—滇西北地震带 2003 年上半年 $M_L \geqslant 2.5$ 级出现低 b 值异常（表 4、图 12）。定点观测异常有 11 条，即西昌水氡、水位、地磁、地电、钻孔应变；攀枝花水、氡、水位、水温、地倾斜；云南洱源水氡、丽江地磁[5)]（表 4，图 13 至图 23）。

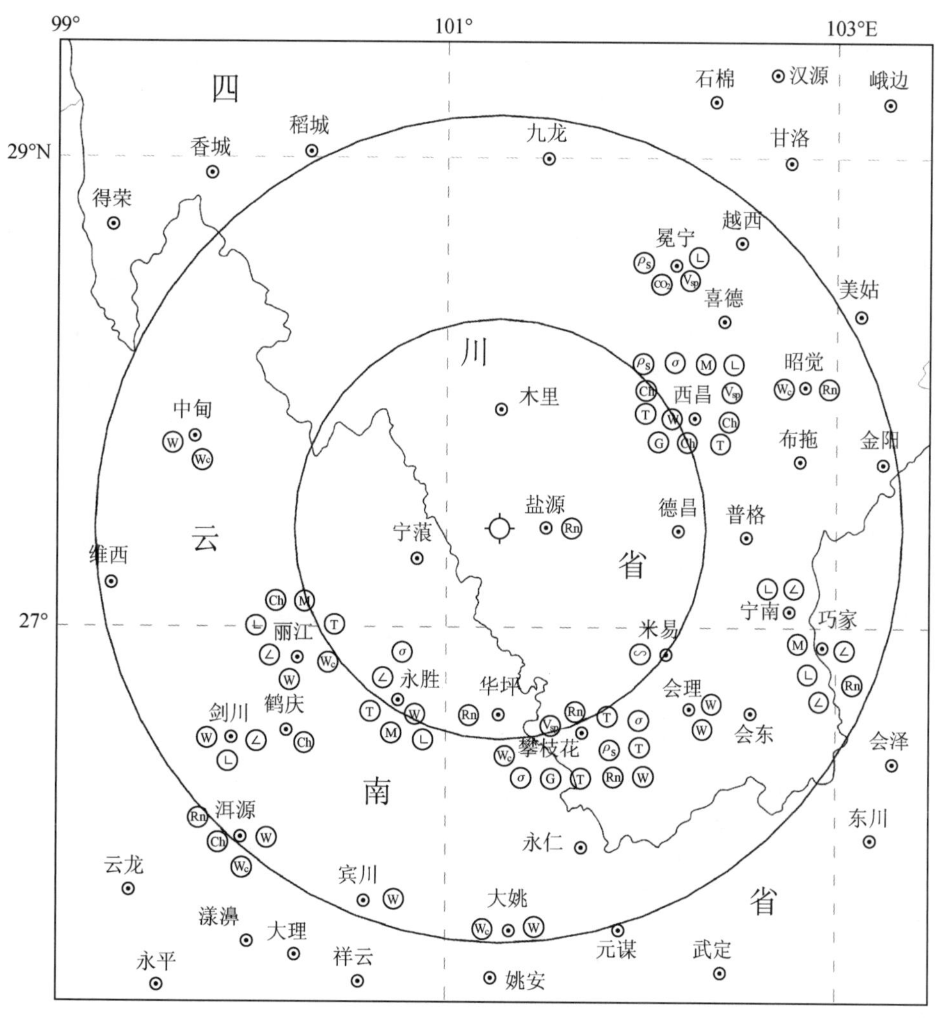

图 10　盐源 5.0 级地震前定点前兆观测台站分布图

Figure 10　Distribution of precursor stations around the M_S5.0 Yanyuan earthquake

盐源地震前，四川地震台网已经具备了较高的地震监控能力，所以地震活动性的资料信度高。这里给出的异常是在日常震情跟踪监视与预报工作中常用的、震前发现的异常。表 4 所列出的 12 条异常，其信度及预报效能简述如下：

表 4　异常情况登记表

Table 4　Summmary table of precursory anomalies

序号	异常项目	台站或观测区	分析方法	异常判据及观测误差	震前异常起止时间	震后变化	最大幅度	震中距 Δ/km	异常类别及可靠性	图号	异常特点及备注
1	地震 *b* 值	盐源—滇西北地震带	用最大拟然法按时间滑动计算	1 倍均方差低 *b* 值	2003.1～	异常继续存在			M_3	12	四川短临跟踪工作中常用的指标，震前出现明显的低 *b* 值异常
2	水氡	攀枝花沙沟	日测值	趋势下降变化，累计下降幅度接近 10%	2001 年 9 月 26 日至 2002 年 8 月底	逐步恢复正常	10%	105	M_1	13	震前出现趋势下降，为震前发现的异常
3	水氡	西昌新村	日测值	趋势下降变化	2002.9～2002.10	逐步恢复正常	10%	105	M_1	14	震前出现趋势下降，为震前发现的异常
4	水氡	云南洱源	旬均值	趋势下降	2002 年 5 月中旬	异常仍未恢复	10%	120	M_1	15	震前出现趋势下降，为震前发现的异常
5	水位	攀枝花川 05 井	日均值	与 2001 年同期相比较	2002 年 11 月至 2003 年 5 月	继续下降	53mm	105	M_1	16	为震前发现的异常
6	水温	攀枝花仁和	日均值	长时间波动	2002 年以来	逐步恢复正常	0.5℃	105	M_1	17	为震前发现的异常
7	钻孔应变 EW	西昌小庙	EW 日均值	持续上升	2002 年 1 月至 2003 年 8 月	逐步恢复正常	6×10^{-6}	105	M_1	18	为震前发现的异常 2003 年 8 月转平

续表

序号	异常项目	台站或观测区	分析方法	异常判据及观测误差	震前异常起止时间	震后变化	最大幅度	震中距 Δ/km	异常类别及可靠性	图号	异常特点及备注
8	磁偏角	西昌	月均值	趋势下降	从 2002 年起	逐步恢复正常	10%	105	M_1	19	为震前发现的异常
9	水位	西昌川 03 井	日均值	与 2002 年同期相对比	2003 年 3～5 月	逐步恢复正常	98mm	105	S_1	20	为震前发现的异常
10	倾斜 JB 仪	攀枝花南山	NS 向日均值	持续上升	2003 年 5 月	逐步恢复正常	5800ms	100	S_1	21	震前出现明显的上升变化
11	地电阻率	西昌小庙	EW 向五日均值	下降异常	2003 年 5 月	出现新的下降	2%	105	S_1	22	为震前发现的异常
12	地磁 Z 分量	丽江	日均值	出现形态异常	2003 年 8 月以来	出现新的下降	2%	120	S_1	23	为震前发现的异常

（1）地震活动性异常项目：资料信度高，异常清晰，具有较好的对应率，预报效能好。

（2）定点前兆观测项目：所有观测项目均有 10 年以上连续、完整的资料，并具有较好的地震重现性和较好的地震对应率，预报效能好。

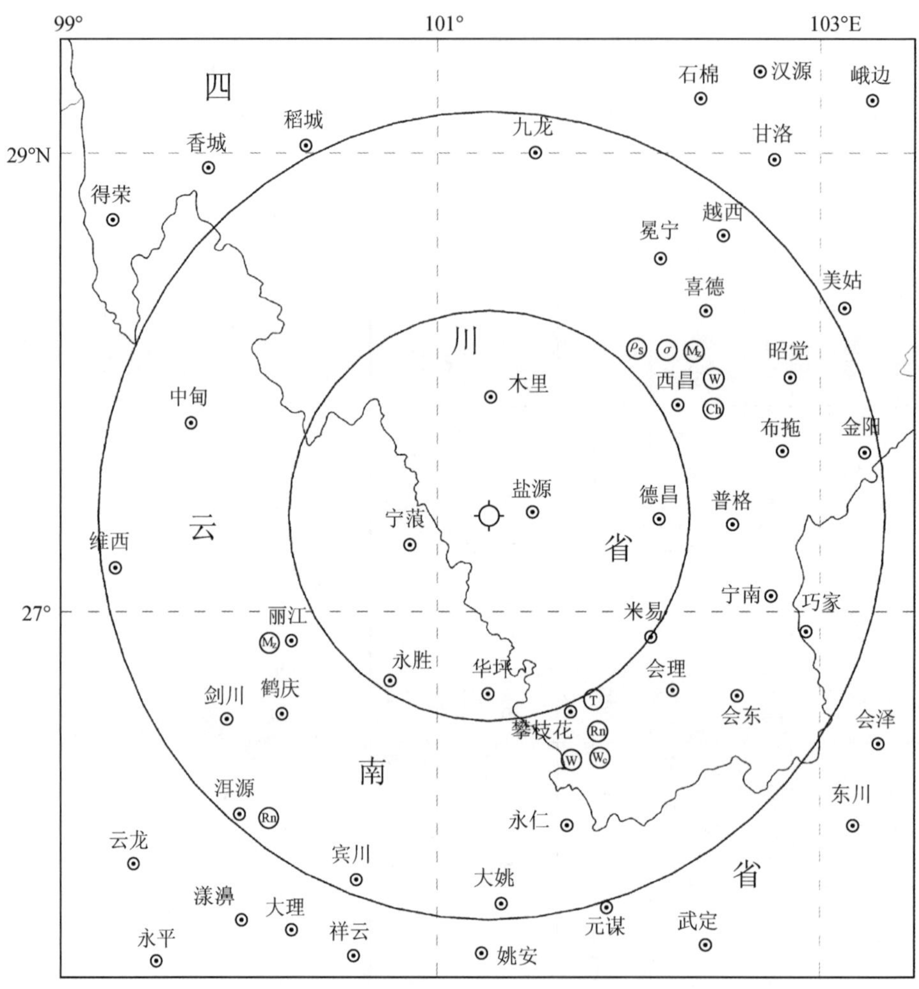

图 11　盐源 5.0 级地震前定点前兆观测异常分布图

Figure 11　Distribution of precursory anomalies at the stations before the yanyuan $M_S5.0$ earthquake

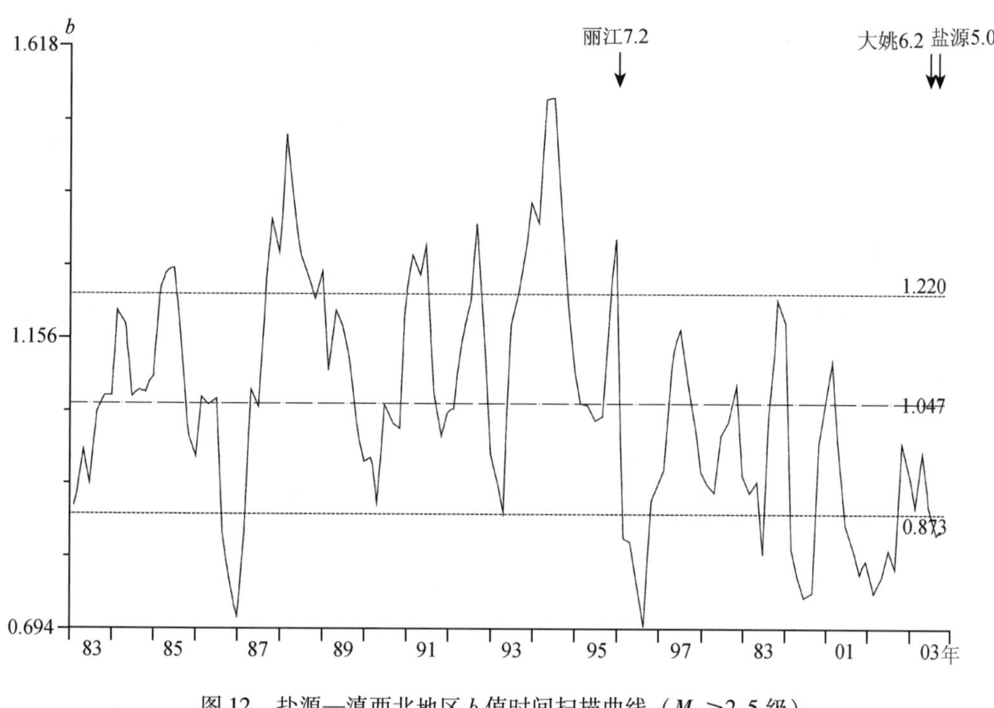

图 12　盐源—滇西北地区 b 值时间扫描曲线（$M_L \geqslant 2.5$ 级）

Figure 12　Time variation of the b-value in the region of Yanyuan-northwest Yunnan（$M_L \geqslant 2.5$）

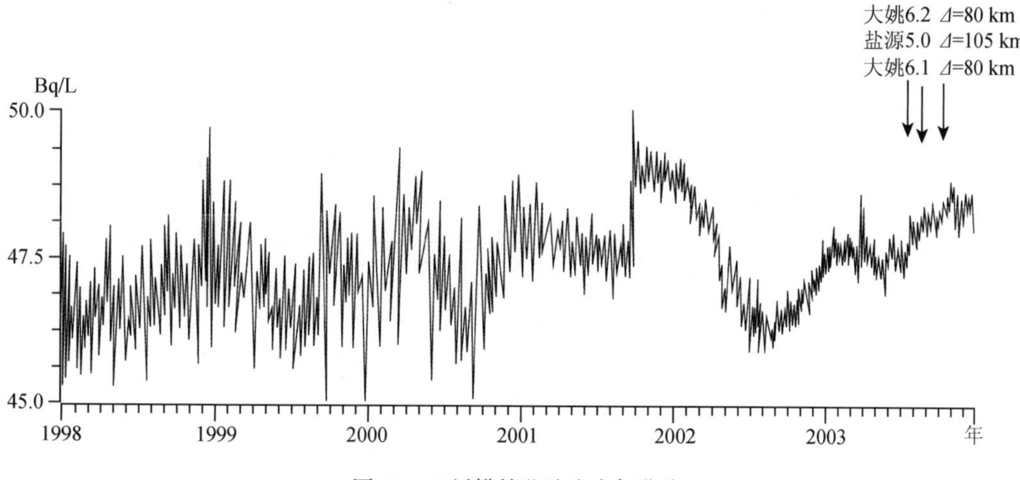

图 13　四川攀枝花沙沟水氡曲线

Figure 13　Time variation of water radon at Shagou station, Panzhihua city, Sichuan Province

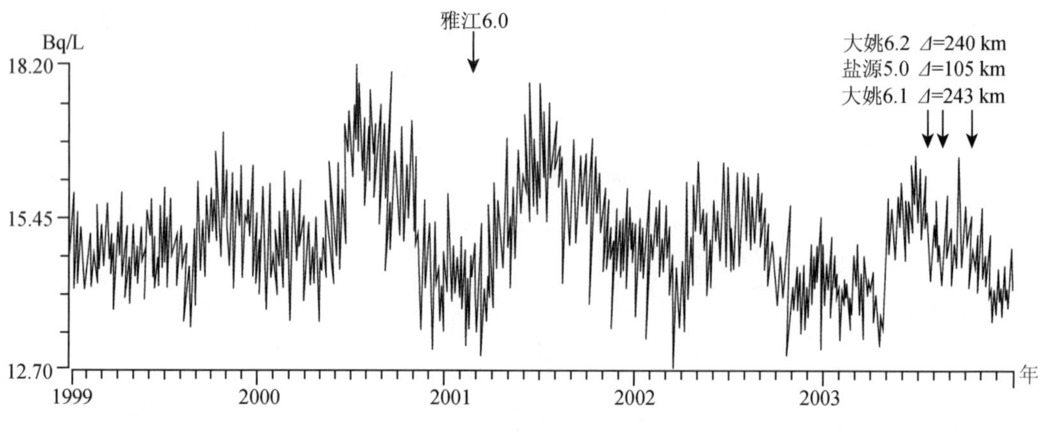

图 14　四川西昌新村水氡曲线

Figure 14　Time variation of water radon of Xincun station, Xichang city, Sichuan Province

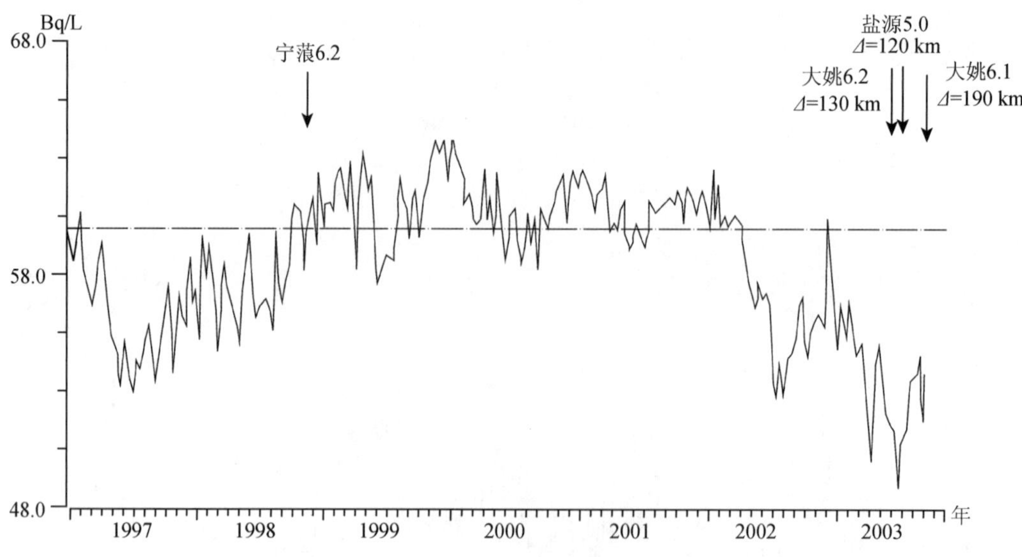

图 15　云南洱源江干水氡旬均值

Figure 15　Time variation of water radon mean value in every ten days of Jianggan, Eryuan, Yunnan Province

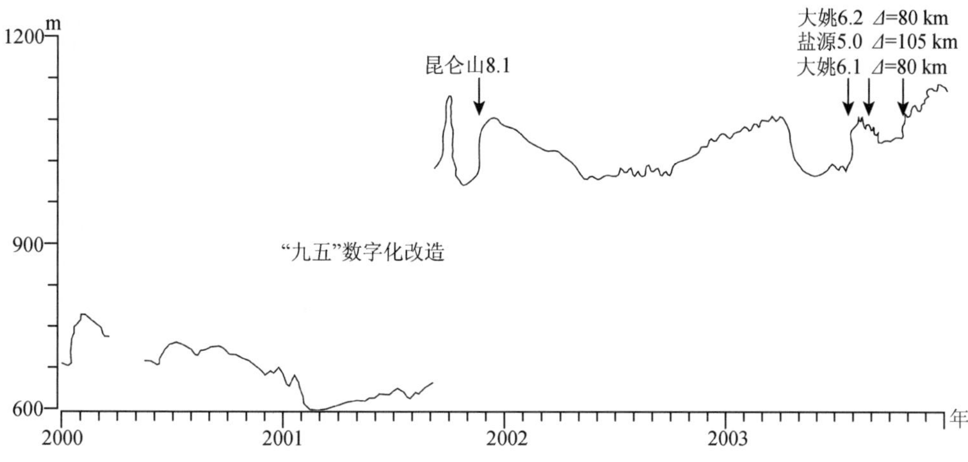

图 16 攀枝花川 05 井水位曲线

Figure 16 Time variation of water level of 05 well in Panzhihua, Sichuan Province

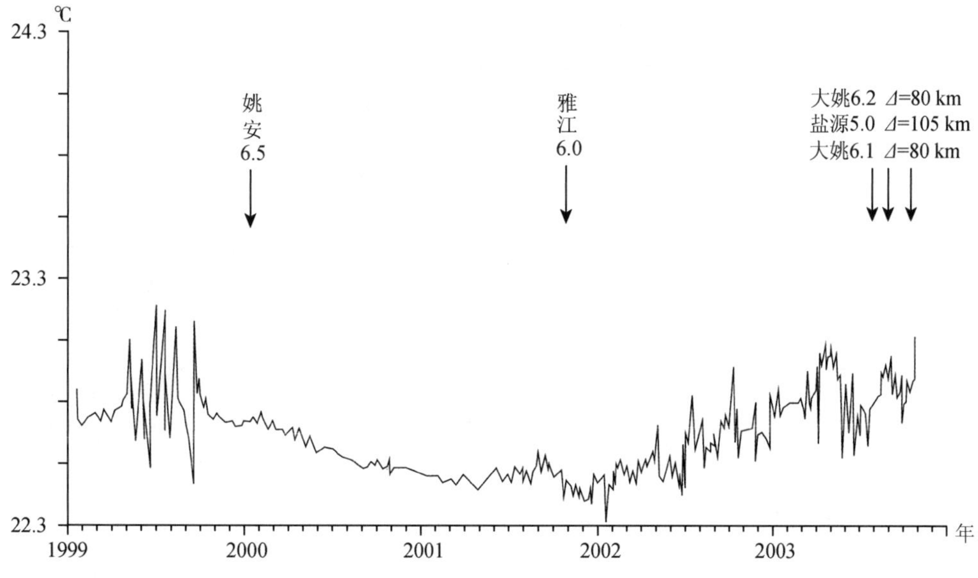

图 17 攀枝花仁和台钻孔水温日均值曲线

Figure 17 Time variation of daily mean value of borehole water temperature of Renhe station, Panzhihua

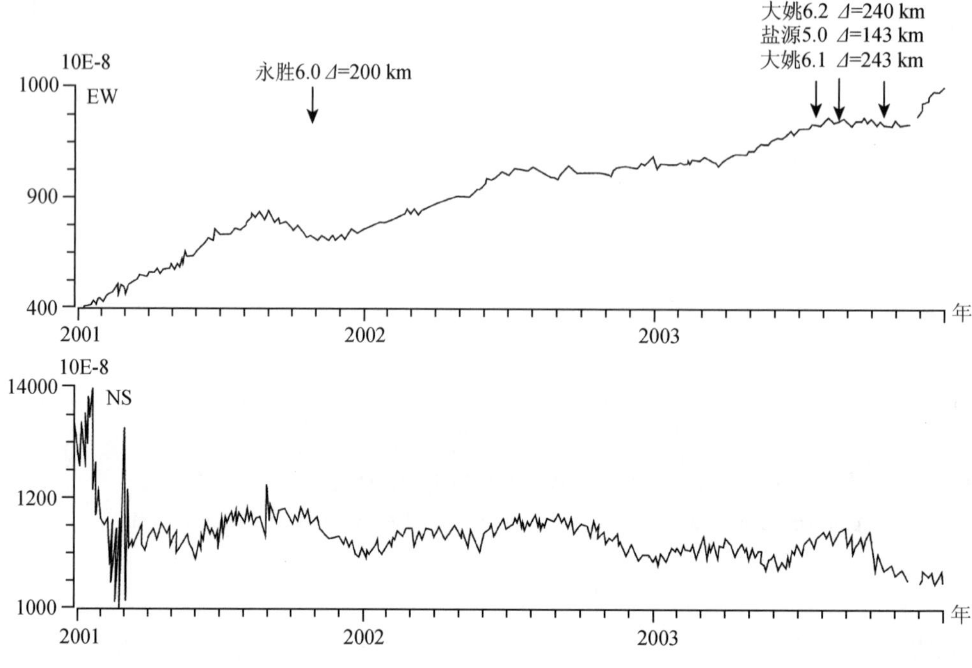

图 18　西昌小庙台钻孔应变 EW 向日均值曲线

Figure 18　Time variation of daily mean value of borehole deformation of Xiaomiao station, Xichang

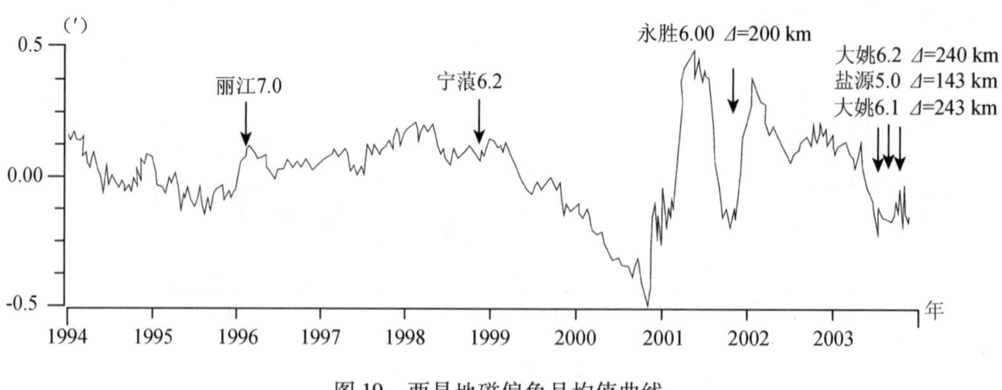

图 19　西昌地磁偏角月均值曲线

Figure 19　Time variation of monthly mean value of geomagnetic declination of Xichang station

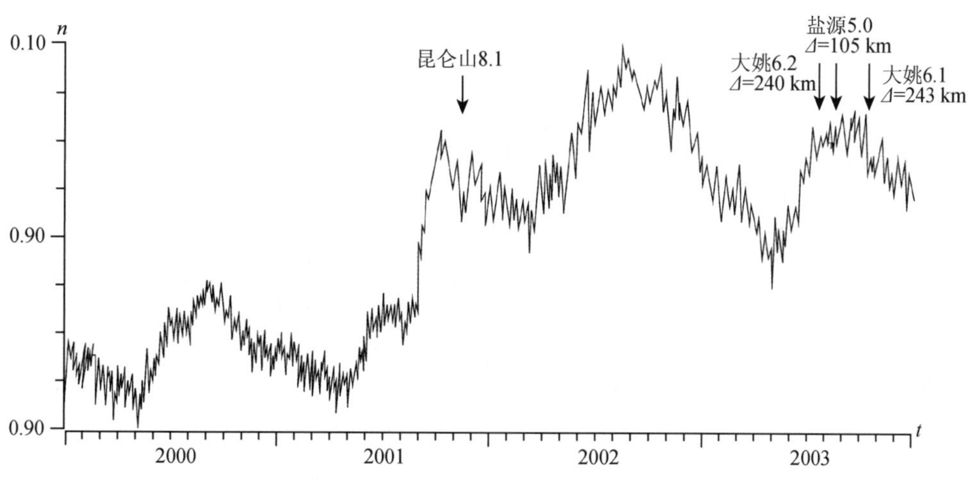

图 20　西昌川 03 井水位曲线

Figure 20　Time variation of water level of 03 well in Xichang, Sichuan Province

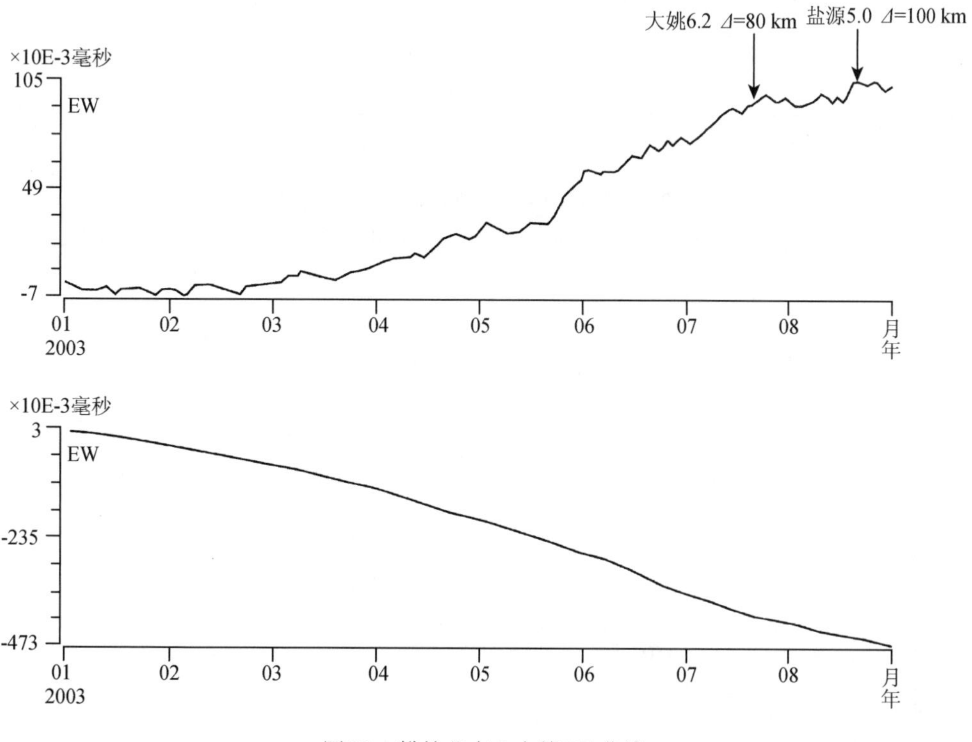

图 21　攀枝花南山水管 NS 曲线

Figure 21　Time variation of water-pipe tiltmeter in NS direction of Nanshan station, Panzhihua

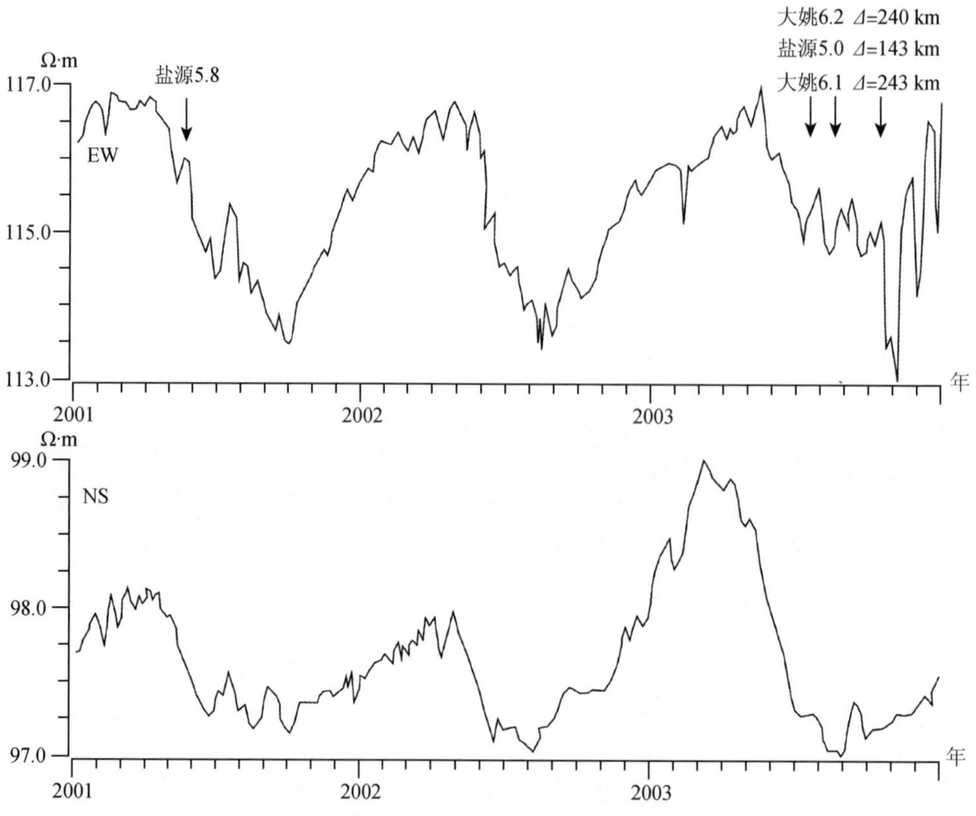

图 22　西昌台地电阻率 EW 向五日均值曲线

Figure 22　Time variation of earth resistivity in EW direction of every 5 days mean value of Xichang station

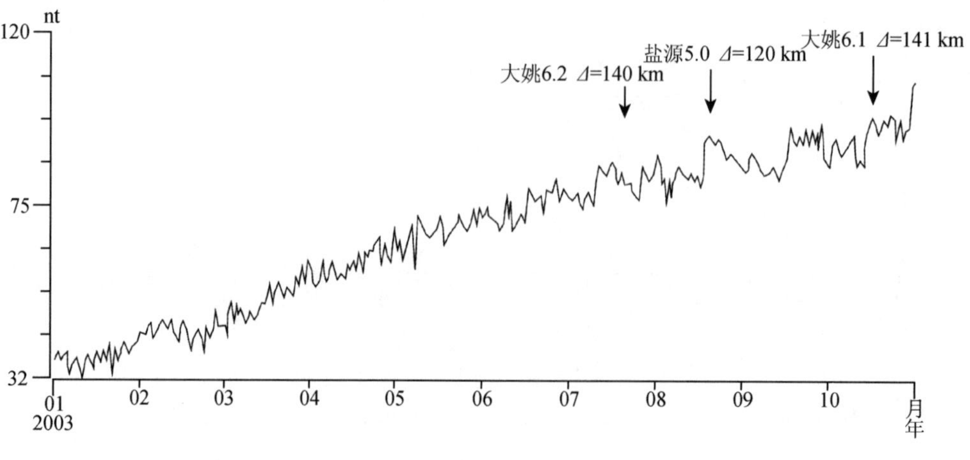

图 23　丽江地磁垂直分量日均值

Figure 23　Time variation of daily mean value of geomagnetic vertical component of Lijiang station

七、前兆异常特征分析

1. 震前盐源—滇西北地震带低 *b* 值中期异常显著

闻学泽、易桂喜[4]在“十五”短临预报攻关课题研究中提出川滇地区地震活动统计单元应充分考虑地震构造背景的地震区、带划分，将川滇地区划分为若干地震区（带），该方案具有较好的短期预报意义，该预报指标可操作性强，在近年四川省地震预测预报实际工作中得到了应用[4]。盐源—滇西北地震带就是其中之一，该地震带 2003 年上半年 $M_L \geqslant 2.5$ 级出现低 b 值异常，短时间恢复正常又出现低 b 值异常（图 12），预示该带及附近未来一年内有可能发生 5 级以上地震。

2. 震前前兆异常数量不多，仅 8 项，短临异常 4 项

攀枝花、西昌、云南洱源水氡（图 13 至图 15）均在 2001 年底至震前相继出现趋势下降变化，累计下降幅度接近 10%。

攀枝花水位、水温（图 17、图 18）多年观测资料年变动态规律明显，2002 年 11 月至 2003 年 5 月水位打破了正常年变动态规律，经现场落实，无环境和人为干扰，确认为是异常。水温 2002 年初以来出现长时间 0.5℃的波动。

西昌小庙钻孔应变（图 19）EW 向 2002 年 1 月至 2003 年持续上升，最大幅度达 6×10^{-6}，为明显异常。

西昌地磁偏角（图 21）2002 年初出现幅度接近 10% 的下降异常。

短临异常 4 项是：西昌 03 井水位、西昌小庙地电阻率 EW 项、攀枝花南山倾斜 JB 仪、于 2003 年 5 月相继出现水位偏低、持续下降等不同程度的异常。云南丽江地磁 Z 分量 2003 年 8 月出现形态下降异常。

3. 前兆异常特征

总体上看，盐源及其附近区域在震前异常项次不多。清理作为 2003 年 8 月 21 日盐源 5.0 级地震的 12 项前兆异常按时间分类，其中震前半年至 5 年的中期异常 8 项，震前 1 ~ 6 个月短期异常 3 项，震前 30 天内的临震异常 1 项。清理出的 12 项异常中 11 项出现在 2003 年 7 月 21 日云南大姚 6.2 级地震前。

异常空间特征看，除盐源—滇西北地震带低 b 值异常外，其余的 11 项前兆异常主要分布在震中的东南面，震中距离分析，100 km内无异常显示，所有异常都出现在 101 ~ 200 km 范围，且集中在攀枝花、西昌，这与台网测项的空间分布有关。

八、震前预测、预防和震后响应

1. 震前预测

四川省地震局攀枝花基准台7)、攀枝花市防震减灾工作领导小组办公室8)、凉山州地震局9)分别在他们的《2003 年度地震趋势研究报告》中明确指出：盐源—木里—宁蒗一带 2003 年底前有可能发生 5 ~ 6 级地震，信度 0.6。预测时间、地点、震级均正确。

2. 震后响应

地震发生后，四川省地震局立即启动“破坏性地震应急预案”，立即派出了以王力副局长带队的大震现场工作队奔赴震区开展地震灾害调查及评估工作。中国地震局也排出了侯建盛和陈荣华研究员从北京直赴灾区指导工作。大震现场工作组由中国地震局、四川省地震局、凉山州地震局和盐源县地震办公室共35人组成。

3. 序列判定

对盐源5.0级地震序列，四川省地震局后方工作组在规定的时间内给出了趋势判定意见，序列类型判定为主震—余震型，结果证明序列判断是正确的。8月22～24日地震现场震情监视工作组向当地政府和有关部门发出《震情简报》，对序列做出较准确的震后趋势判定。

4. 灾害评估

现场调查组于8月21～29日历时9天在现场开展了灾情调查和评估工作。19人组成的现场工作组分别对博大乡的几坡村—红岩子村一线，梅雨乡—棉垭乡一线，博大乡、盐塘乡、桃子乡、长柏乡一线的震害情况进行了调查，攀枝花防震减灾办公室现场工作组也对盐边县岩口乡、国胜乡，盐源县黄草镇的地震灾害进行了调查。2003年9月5日四川省灾害损失评定委员会对盐源5.0级地震现场工作组提交的《2003年8月21日四川盐源5.0级地震灾害损失评估报告》进行了评审。

九、总结与讨论

历史上盐源及附近地区地震类型比较复杂，有主震—余震型（2001年5月24日盐源5.8级地震），双震型（1976年11月7日、12月13日盐源—宁蒗6.7、6.4级地震）。本次地震类型属主震—余震型。

震前的定点前兆观测异常不多，持续时间不长，震兆和观测前兆异常不配套。所分析的一次项的异常幅度、异常发展时间变化、空间展布，对该次地震的强度、发震时间、危险地点的预测难以提供确切信息。本次地震前，即2003年7月21日云南大姚发生6.2级地震，该地震距四川攀枝花观测台75km，距四川西昌观测台230km，距云南丽江观测台153km。盐源5.0级地震距上述台站均在100km左右范围，分析认为上述所列异常项也可能是云南大姚6.2级地震的前兆异常显示，或认为这些异常项是对该区域地震的叠加显示效应，不是仅对盐源5.0级地震的指示意义。尤其，本次地震短临阶段，除了部分异常出现转折或结束，多数异常仍未结束，2003年10月16日云南大姚再次发生6.1级地震。

参 考 文 献

[1] 杜方、吴江，四川盐源—云南宁蒗间5.8级地震发震特征，四川地震，2008

[2] 唐荣昌、韩渭滨主编，四川活动断裂与地震，北京：地震出版社，1993

[3] 程万正，1986年8月12日四川省盐源5.2级地震，北京：地震出版社，1986

[4] 闻学泽、易桂喜，川滇地区地震活动统计单元的新划分，地震研究，2003年6月

[5] 1976年11月7日四川省盐源6.7级地震，中国震例（1976～1980），北京：地震出版社

参 考 资 料

1）四川省地震局盐源 5.0 级地震现场工作组，2003 年 8 月 21 日四川盐源 5.0 级地震灾害损失评估报告，2003 年 8 月

2）四川省地震局，情况简报，18 ~ 22 期，2003 年

3）中国地震简目汇编组，中国地震简目（$M_L \geqslant 4.5$），2003

4）四川省地震局，四川地震台网目录（2003）

5）四川省地震局预报研究所，2003 年四川省地震趋势研究报告（打印稿），四川省地震局，2002. 11

6）四川省地震局预报研究所，2004 年四川省地震趋势研究报告（打印稿），四川省地震局，2003. 11

7）四川省地震局攀枝花基准台，2003 年地震趋势研究报告（打印稿），2003. 11，卷宗 3192

8）攀枝花市防震减灾工作领导小组办公室，2003 年地震趋势研究报告（打印稿），2003. 11，卷宗 3193

9）凉山州地震局，2003 年度地震趋势研究报告（打印稿），2003. 11，卷宗 3194

10）云南省地震局，云南省地震目录（2003），铅印本

Yanyuan Earthquake of M_S 5.0 on 21 August 2003 in Sichuan Province

Abstract

The M_S5.0 earthquake occurred on 21 Aug 2003 in Yanyuan County, Liangshan Prefecture, Sichuan Province. The macro-seismic epicenter was located in the area of junction of Boda, Dacao and Yantang three townships in Yanyuan County, Sichuan Province. The seismic intensity at the epicenter area was VI. The isoseismic map shape of meizoseismal area is an elliptic with major axis in NNW direction. The M_S5.0 earthquake took place on the Luomizui fault with EW trending which is on the top of Yanyuan arc tectonic. During the earthquake no person was killed and 9 persons were injured including 2 heavy injured. Area of collapsed houses was 736920m^2 and 5 discharge culverts of reservoir were leakage. The total economic loss was about 20.86 million Yuan (RMB).

The sequence character of this M_S5.0 Yanyuan earthquake is mainshock-aftershock type. The biggest aftershock is M_L4.5 on 21 Aug 2003. The predominance distribution for aftershock is about NW direction. According to the predominance for aftershock and macro-seismic intensity we infer that the nodal plane Ⅰ with right lateral strike slipping from P-wave onset solution is the fault plane. While M_L4.5 earthquake has same result as M_S5.0 earthquake.

There are 48 observation stations in the range of 200km near epicenter, in which 30 are seismic stations and 18 are fixed precursory observatories. 12 anomalies appeared before M_S5.0 earthquake and no anomaly were observed within the range of 100km from the epicenter. These anomalies are mostly middle-term and short-term anomalies. Whereas two earthquakes with magnitude larger than 6 occurred in Dayao county, Yunnan Province before and after the M_S5.0 event respectively, we conclude that these anomalies are the reflection to this area for a certain time not only for M_S5.0 earthquake.

2003 年 9 月 2 日新疆维吾尔自治区阿克陶 5.9 级地震

新疆维吾尔自治区地震局

李志海　高国英

摘　要

2003 年 9 月 2 日新疆维吾尔自治区阿克陶县西南发生 5.9 级地震，中国地震台网测定微观震中 38°50′N，75°10′E。根据科考结果，此次地震震中烈度Ⅶ度，等震线长轴为 NW 向。由于此次地震发生在山区，未造成人员伤亡。

阿克陶 5.9 级地震序列为主震—余震型，最大余震为 4.1 级，余震分布在主震两侧，沿 NWW 方向展布。震源性质为正断走滑型，近南北走向的节面Ⅱ为断层面。主压应力 P 轴方位 46°，仰角 312°。此次地震可能是青藏块体向北挤压作用下，塔什库尔干断裂发生走滑断错所造成。

阿克陶 5.9 级地震震中位于边境地区，震中 300km 范围内有 4 个地震台，前兆观测项目 2 个，3 台（套）观测仪器。地震前出现 2 项测震异常，1 项前兆异常。

前　言

2003 年 9 月 2 日 7 时 16 分，新疆维吾尔自治区阿克陶县西南发生 5.9 级地震，据中国地震台网测定微观震中为 38°50′N，75°10′E。中国地震局与新疆地震局现场工作组联合科考的宏观震中为 38°41′N，75°20′E。这次地震震中位于阿克陶县西南公格尔峰北坡，震中烈度Ⅶ度，地震未造成人员伤亡，直接经济损失 847.37 万元[1)]。

阿克陶 5.9 级地震发生在西昆仑地震带西段，该区是新疆强震主要活动区。1900 年以来震中 200km 范围内共发生 7 级地震 6 组（7 次），其中 8 级地震 1 次。阿克陶 5.9 级地震震中位于中国新疆边境，靠近塔吉克斯坦，海拔 4600m。震中附近地震观测台站较少，300km 范围内仅有 4 个地震台，2 个前兆观测项目，3 台（套）观测仪器。震前只有 1 套仪器出现异常。

阿克陶 5.9 级地震发生在新疆中强震较为活跃的背景之下。阿克陶 5.9 级地震前，距阿克陶 5.9 级地震震中约 110km 和 220km 分别发生 2002 年 12 月 25 日乌恰 5.7 级地震和 2003 年 2 月 24 日伽师—巴楚 6.8 级地震，另外北天山 2003 年 2 月 14 日石河子 5.4 级，2003 年 9 月 27 日中、俄、蒙交界区 7.9 级地震，2003 年 12 月 1 日昭苏 6.0 级地震。

阿克陶 5.9 级地震发生在 2003 年度划定的地震危险区内2)。

在有关文献和资料的基础上[1];1~5)，经过对资料的重新整理和分析研究，该研究报告得以完成。

一、测震台网及地震基本参数

震中 300km 范围内共有 4 个地震台，这 4 个地震台都分布在震中 200km 范围内（图 1），其中 100km 范围内仅有塔什库尔干地震台，100~200km 范围内有乌恰、喀什、阿图什三个地震台。喀什台是数字地震台，其余为模拟地震台。阿克陶 5.9 级地震位于西昆仑地震带西段边境地区，监测能力较弱，但研究时段内可以保证 $M_S \geqslant 2.5$ 级地震不遗漏。

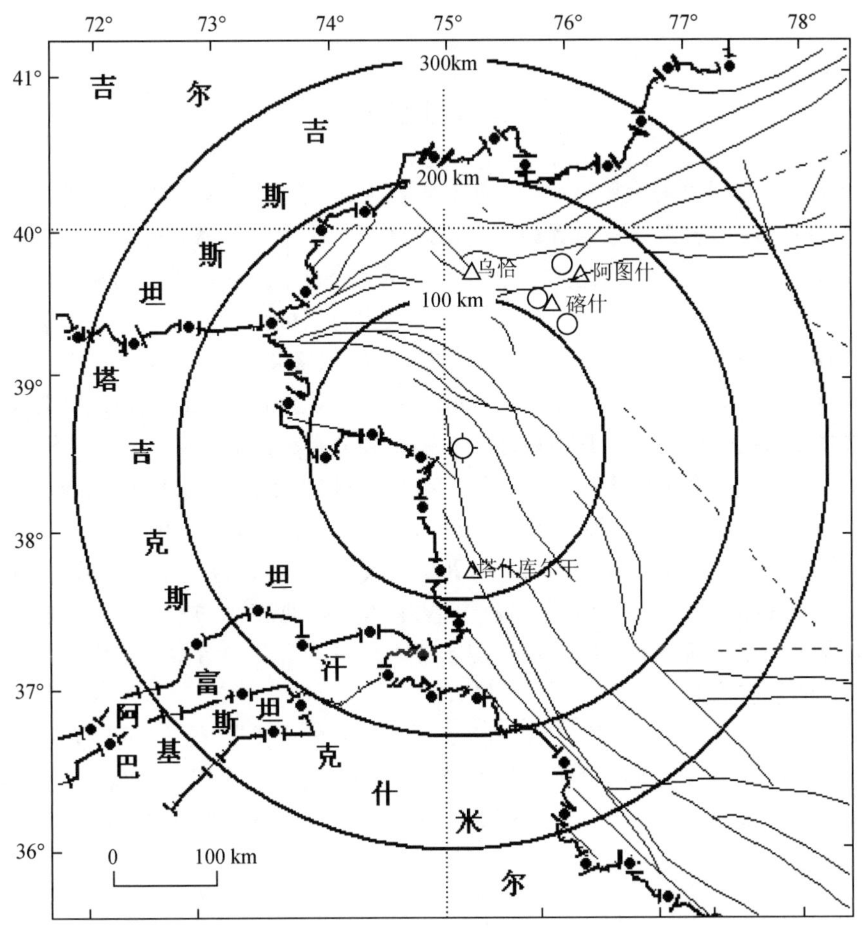

图 1　阿克陶 5.9 级地震前震中附近地震台站及观测项目分布图

Fig. 1　Distribution of earthquake-monitoring stations around the epicentral area before the M5.9 Aketao earthquake

表 1 列出阿克陶 5.9 级地震的基本参数，结果来源于不同地震部门。

表 1　阿克陶 5.9 级地震基本参数（哈佛大学和 UGSG 结果以格林威治时间为准）

Table 1　Basic parameters of the *M*5.9 Aketao earthquake

编号	发震日期	发震时刻	震中位置		震级		震源深度 (km)	震中地名	结果来源
	年 月 日	时 分 秒	φ_N	λ_E	M_S	M_W			
1	2003 9 2	07 16 32	38°34′	75°08′	5.6		28	公格尔山	新疆地震局
2	2003 9 2	07 16 33	38°50′	75°10′	5.9		15	阿克陶	CENC
3	2003 9 1	23 16 41.20	38°42′	75°19′	5.6	5.7	15	新疆西部	美国哈佛大学
4	2003 9 1	23 16 35.05	38°36′	75°20′		5.6	7	新疆西部	USGS

二、地震地质背景

2003 年 9 月 2 日阿克陶 5.9 级地震发生在西昆仑海西期褶皱系。西昆仑地区地层出露齐全，元古界为绿色变质岩系，古生界为浅变质碎硝岩，中生代为海相碎屑岩及碳酸岩等，第三系以浅海陆相湖泊碎屑岩堆积为主，第四系主要是分布于山前地带、河流两岸的冲积、轰积、风积和冰川堆积。震区位于帕米尔地区西昆仑山海拔最高的区域，这里山势陡峭，公格尔峰海拔 7719m，是西昆仑山脉中国境内最高峰，峰顶终年积雪，沿其山坡广布冰川，有 20 余条冰舌向下散射，北坡冰舌最长达 23km，冰川面积 300km^2，此次地震宏观震中就位于公格尔峰的北坡雪线附近[1];1)。其附近地区主要断裂分布见图 2。

从图 2 来看，宏观震中位于布伦口断裂，布伦口断裂位于西昆仑帕米尔弧构造带南缘，长约 100km，走向 NNW，倾角 60° ~ 80°，倾向 270°，正断性质。台网确定的微观震中位于塔什库尔干断裂，塔什库尔干断裂长度约 120km，倾向 275°，倾角 50° ~ 80°，具有右旋走滑正断性质。塔什库尔干断裂上曾于 1895 年发生 7.5 级地震5)，震中位于该断裂南段，此次阿克陶地震发生在该断裂的北段。根据现场考察组对宏观震中考察，认为塔什库尔干断裂是阿克陶地震的发震构造。

从地震带划分来看，阿克陶 5.9 级地震发生于西昆仑地震带北段，西昆仑地震带地震活动频繁，是新疆境内地震活动频度最高的地震带，自有文字记录以来曾发生过 1944 年乌恰南、1974 年乌孜别里山口和 1985 年乌恰南 3 次 7 级以上地震，发生 6 ~ 6.9 级地震 12 次，并发现有古地震遗迹5)。

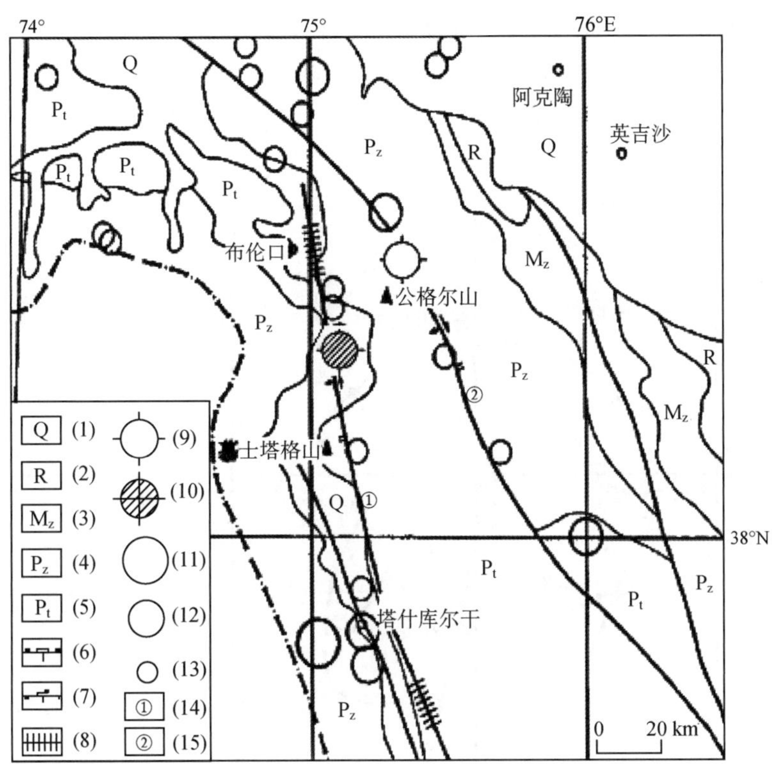

图 2 阿克陶 5.9 级地震附近地区主要断裂及历史地震震中分布图

Fig. 2 Major faults and distribution of historical earthquakes around *M*5.9 Aketao earthquake area

（1）第四系；（2）第三系；（3）中生界；（4）古生界；（5）元古界；（6）逆走滑断层；（7）正走滑断层；（8）古地震地表破裂带；（9）宏观震中；（10）微观震中；（11）7.0～7.9 级地震；（12）6.0～6.9 级地震；（13）5.0～5.9 级地震；（14）①为塔什库尔干断裂；（15）②为布伦口断裂

三、地震影响场和震害

阿克陶 5.9 级地震震区为地广人稀的高山区，交通条件差，地震现场科学考察和烈度调查有一定难度和局限性。根据《中国地震烈度表》，现场共调查了 20 个点，其中Ⅵ度区 10 个，Ⅴ度区 7 个，Ⅳ度区 2 个，Ⅶ度点上滚石、雪崩等现象依据地方政府提供的录像资料确定，由此圈定出Ⅶ度区、Ⅵ度区和Ⅴ度区范围，烈度等震线长轴 NNW 走向，与构造走向基本一致。据现场实地考察及资料分析，本次地震宏观震中为 38°41′N，75°20′E[1)]（图 3）。

Ⅶ度点：分布在阔什喀尔其西南，区内地震造成山体崩塌、滚石、雪崩。

Ⅵ度区：包括盖子村、布伦口乡、丘鲁克敦巴舍村、苏巴什村、卡拉库勒湖、盖兹检查站等地。区内土木结构房屋墙体裂缝，不稳定器物翻倒，砖混结构房屋出现裂缝，公路两边陡坡基岩风化层碎石散落，崩塌在公路上形成石堆。

Ⅴ度区：调查点有皮拉勒、托喀依、木吉、琼让、恰克尔艾格勒、乌依塔格、塔什米力克等，区内人感到先上下震动，后左右晃动。旧老土木届欧房屋出现裂缝。

Ⅳ度区：调查点有阿克陶、乌帕尔等，区内有震感，门窗作响。

震中烈度达Ⅶ度，地震未造成人员伤亡，直接经济损失达 847.37 万元[1)]。

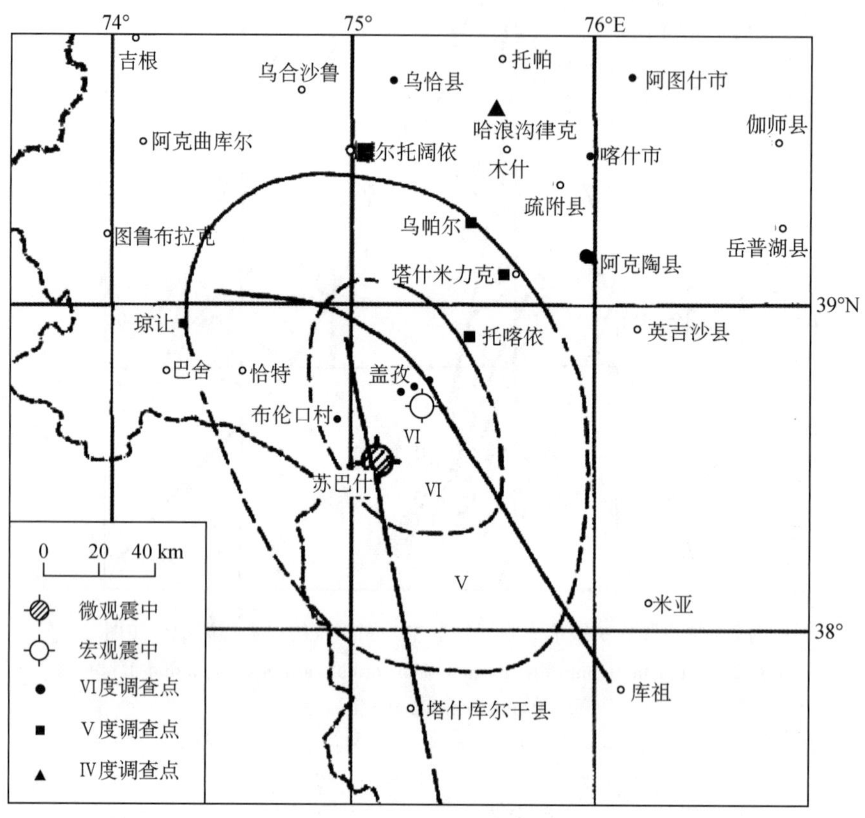

图 3　阿克陶 5.9 级地震烈度等震线图

Fig. 3　Isoseismal map of the M5.9 Aketao earthquake

四、地 震 序 列

阿克陶 5.9 级地震发生在区域地震活动相对活跃的背景之下，震前 2002 年 12 月 25 日距离 114km 处发生乌恰 M_S5.7 地震，2003 年 2 月 24 日距离 217km 处发生巴楚—伽师 M_S6.8 地震。震前 1 年震中附近 60km 范围内未发生 M_S3.0 以上地震，主震后截至 2003 年 12 月 31 日，新疆地震局台网共定位 $M_S \geqslant 2.0$ 级地震 62 次，其中 5.0 ~ 5.9 级 1 次；4.0 ~ 4.9 级 1 次；3.0 ~ 3.9 级 11 次；2.0 ~ 2.9 级 49 次。表 2 给出新疆地震台网确定的序列 $M_S \geqslant 3.5$ 级地震目录。

表 2 阿克陶 5.9 级地震序列目录（M_S≥3.5 级）

Table 2 Catalogue of the *M*5.9 Aketao earthquake sequence（M_S≥3.5）

编号	发震日期	发震时刻	震中位置		震级	震源深度	震中地名	结果来源
	年 月 日	时 分 秒	φ_N	λ_E	M_S	(km)		
1	2003 09 02	07 16 32	38°34′	75°08′	5.6	28	阿克陶	资料 1）
2	2003 09 02	07 46 27	38°37′	74°56′	3.6	17		
3	2003 09 02	08 04 34	38°35′	75°12′	4.1	23		
4	2003 09 02	15 13 27	38°35′	75°06′	3.8	22		
5	2003 09 03	09 27 03	38°34′	75°09′	3.7	29		

从 5.9 级地震序列空间分布图可以看出（图 4），余震分布为 NWW 方向，余震分布在主震两侧，与 NNW 方向的等震线方向有一定夹角。

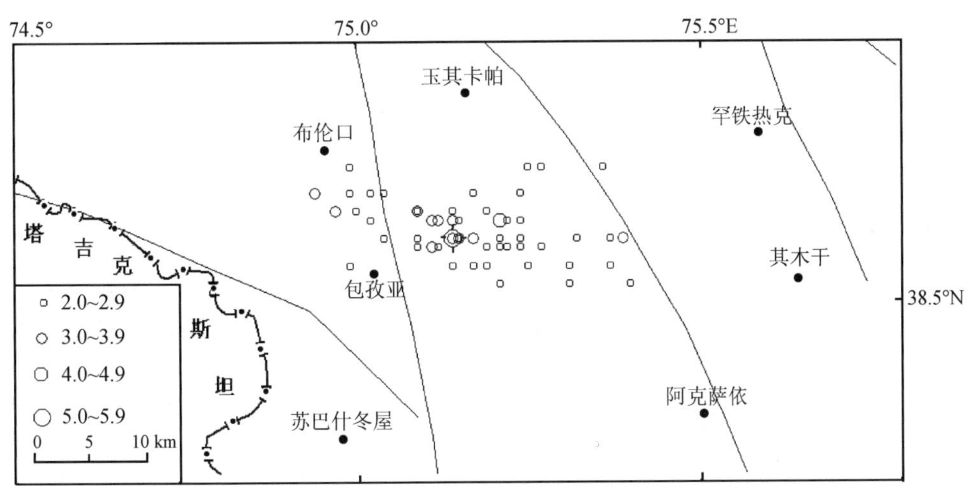

图 4 阿克陶 5.9 级地震序列分布

Fig. 4 Distribution of the *M*5.9 Aketao earthquake sequence

截至 2003 年 11 月 30 日，喀什单台共记录到 229 次余震，其中 M_L1.0~1.9 地震 39 次，M_L2.0~2.9 地震 144 次，M_L3.0~3.9 地震 39 次，M_L4.0~4.9 地震 7 次。由序列的 *M-t* 图（图 5）看出，余震频度呈起伏衰减，强度衰减较慢。

截至 2003 年 11 月 30 日，余震序列 *b* 值为 0.77（图 6），接近南天山地震带平均 0.70 的背景 *b* 值。序列 *p* 值为 0.4，*h* 值为 1.1。最大余震为 M_L4.6 地震，主震释放能量占全序列能量的 95.1%，主震与最大余震震级相差 1.3，根据地震类型的判别指标，此次地震类型为主震—余震型。

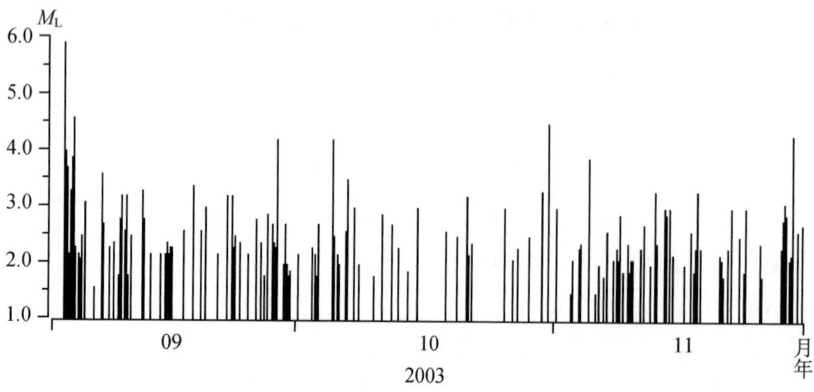

图5　阿克陶 5.9 级地震序列 *M*-*t* 图

Fig. 5　*M*-*t* diagram of the *M*5.9 Aketao earthquake sequence

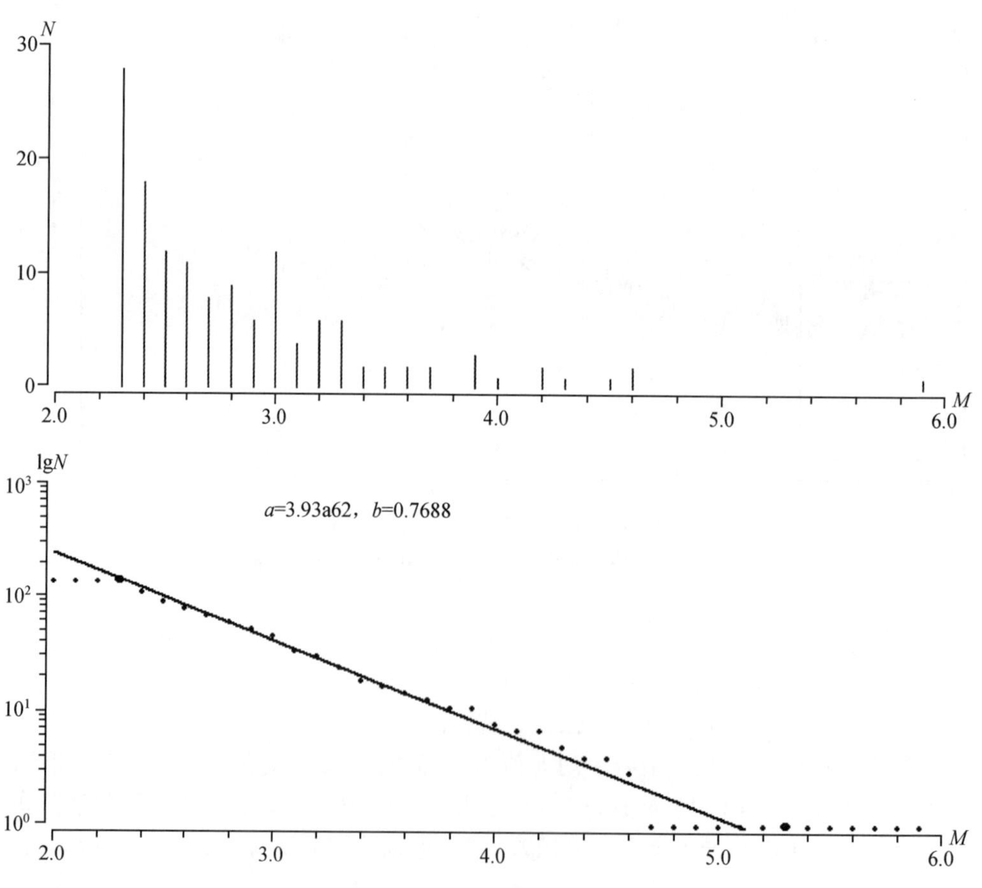

图6　阿克陶 5.9 级地震序列 *b* 值拟合曲线

Fig. 6　*b*-value fitting curve for the *M*5.9 Aketao earthquake sequence

五、震源参数和地震破裂面

根据新疆区域台网清晰 P 波初动符号测定了这次 5.9 级地震的震源机制解。另外，哈佛大学和美国地质调查局也根据全球台网测定了这次地震的震源机制解，结果见表 3。鉴于此次地震发生在边境，新疆台网包围不好，本文将引用哈佛大学震源机制解，见图 7。

表 3　阿克陶 5.9 级地震的震源机制解

Table 3　Focal mechanism solutions of the *M*5.9 Aketao earthquake

序号	节面 I			节面 II			P 轴		T 轴		N（B）轴		结果来源
	走向	倾角	滑动角	走向	倾角	滑动角	方位	仰角	方位	仰角	方位	仰角	
1	181	W	62	293	NNE	55	145	48	238	5	332	42	聂晓红
2	107	47	-153	358	71	-46	46	312	14	57	40	160	HRV
3	148	45	-111	357	49	-70	75	334	2	73	15	163	USGS

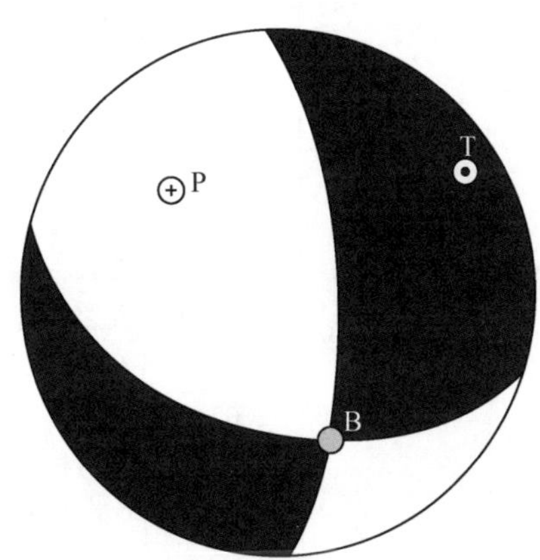

图 7　阿克陶 5.9 级地震震源机制解（据哈佛大学）

Fig. 7　Focal mechanism solutions of the *M*5.9 Aketao earthquake by Harvard

根据哈佛大学结果分析认为此次地震震源断错为走滑正断层。综合极震区烈度分布和震区断裂性质分析认为，近南北走向的节面 II 为断层面。主压应力 P 轴方位 46°，仰角 312°，张应力 T 轴方位 14°。此次地震可能是青藏块体在向北的挤压作用下，塔什库尔干断裂发生走滑断错造成的。

根据阿克陶 5.9 级地震等震线长轴方向、微观震中位置、余震展布综合分析认为，塔什库尔干断裂是这次地震的发震构造是塔什库尔干断裂。

六、地震前兆观测台网及前兆异常

震中附近区域地震台站及观测项目分布如图 1 所示。地震发生在塔什库尔干断裂西端，测震台站主要分布在震中东北和南侧，在震中北部、西北部和东南部无台站分布。震中周围是新疆地震监测能力较弱的区域。

震中 200km 范围内有 4 个地震台，均有测震观测项目；其中，2 个地震台有前兆观测项目，包括地倾斜、地磁观测项目。震中 0 ~ 100km 范围内有 1 个测震台站，无前兆观测项目；震中 101 ~ 200km 范围内有 3 个测震台站，2 台（3 套）前兆观测项目。阿克陶地震前共出现 2 个异常项目 3 条异常（表 4）。

表 4　地震前兆异常登记表

Table 4　Summary table of earthquake precursory anomalies

序号	异常项目	台站（点）或观测区	分析方法	异常判据及观测误差	震前异常起止时间	震后变化	最大幅度	震中距（km）	异常类别及可靠性	图号	异常特点及备注
1	地震空区[2)]	38° ~ 40°N，73° ~ 76°E	$M_S \geqslant 4.0$ 级地震空间分布	$M_S \geqslant 4.0$ 级地震围空	1996.9 ~ 2002.9	地震发生在空区边缘		震中周围	M_1	8	4 级地震空区于 2002 年 12 月 16 日被打破后，先在空区边缘发生了 2002 年 12 月 25 日乌恰 5.7 级地震，后发生了此次 5.9 级地震
2	η 值[2)]	38° ~ 40.5°N，72.5° ~ 76.5°E	η 值时序扫描（$M_S \geqslant 3.0$，9 个月窗长 3 个月步长滑动）	持续 4 个月低值	2003.6 ~ 2003.10	逐渐恢复正常		震中周围	S_1	9	2003 年 9 月 2 日阿克陶 5.9 级地震后 1 个月异常恢复

续表

序号	异常项目	台站（点）或观测区	分析方法	异常判据及观测误差	震前异常起止时间	震后变化	最大幅度	震中距（km）	异常类别及可靠性	图号	异常特点及备注
3	地倾斜（竖直摆）	喀什	单分量 5 日均值	速率变化不稳	2003.5 ~ 2003.8	数据断记	0.5″	120	S_2	10	2003 年 8 月 27 日仪器被雷击断记，直至 2004 年 2 月才恢复观测

1. 地震学异常

2003 年 9 月 2 日阿克陶 5.9 级地震前后，新疆地区 5 级以上地震处于活跃状态，震中附近 200km 范围内发生了 3 次 5 级地震，分别为 2002 年 12 月 25 日乌恰 5.7 级地震，2003 年 9 月 2 日阿克陶 5.9 级地震，2004 年 3 月 21 日中国与塔吉克斯坦边境 5.0 级地震；这一时期，2003 年 2 月 14 日北天山石河子发生 5.4 级地震，2003 年 2 月 24 日距离阿克陶地震约 220km 的伽师发生 6.8 级地震，2003 年 9 月 27 日中、俄、蒙交界发生 7.9 级地震。

1996 年 9 月至 2002 年 9 月，震区附近形成 4 级地震空区[2)]，空区长轴约 200km，短轴约 120km。2002 年 12 月 16 日空区内部发生 4.5 级地震，空区被打破，其后于 2002 年 12 月 25 日在空区边缘发生乌恰 5.7 级地震，2003 年 9 月 2 日在空区边缘的空段内发生此次阿克陶 5.9 级地震（图 8）。

以 9 个月为窗长、3 个月为滑动步长，对震中周围地区（38°~40.5°N，72.5°~76.5°E）$M_S \geqslant 3.0$ 级地震进行地震学参数时间扫描计算，结果显示以均值线为阙值线，η 值低值回返过程中或回返后，扫描区内都有 $M_S \geqslant 5.5$ 级地震发生，对应率为 100%。2003 年 6~10 月间 η 值出现 4 个月的低值异常，阿克陶 5.9 级地震发生在低值过程中，震后 1 个月恢复正常（图 9）。

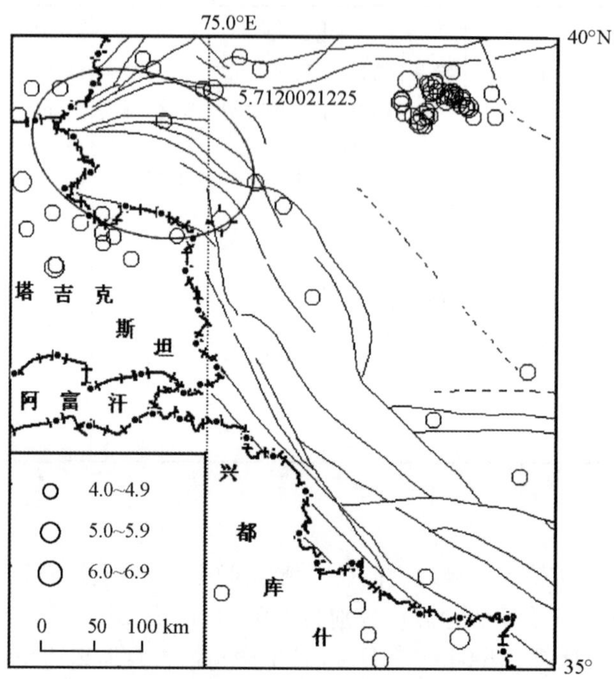

图 8　阿克陶 5.9 级地震震前震中周围地区 $M_S \geqslant 4.0$ 级地震震中分布

Fig. 8　Distribution of the $M_S \geqslant 3.0$ earthquakes around the epicenter before the *M*5.9 Aketao earthquake

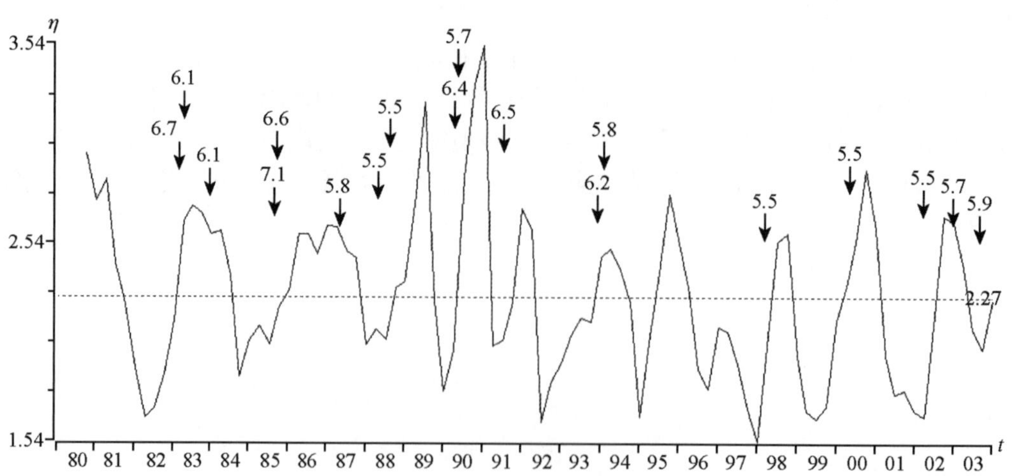

图 9　阿克陶 5.9 级地震震中周围地区 1980 年以来 $M_S \geqslant 3.0$ 级地震 η 值变化曲线

Fig. 9　Variation of $M_S \geqslant 2.5$ earthquake frequency around the *M*5.9 Aketao earthquake since 1980

2. 其他前兆观测项目异常

震中 300km 范围内的 2 台项（3 套）前兆观测项目中，喀什台竖直摆地倾斜仪自 1998 年 11 月开始观测，喀什台地磁观测自 1984 年开始观测，阿图什台竖直摆倾斜仪自 1994 年开始观测。阿克陶地震之前，喀什台竖直摆地倾斜南北向 2003 年转向时间比往年明显推后 40 天左右，5 月中旬转向后北倾速率明显加快；该台地倾斜东西向 6 月也出现类似的加速东倾变化，到 8 月 27 日该仪器因被雷击而停止观测[2)]。之前两分量的倾斜量变化分别为往年同期的 2 倍左右，矢量合成曲线向背离震中方向加速倾斜，仪器停记 5 天后在台站西南 120km 处发生阿克陶 5.9 级地震（图 10）。而喀什台地磁观测仪和阿图什竖直摆地倾斜仪未发现异常。

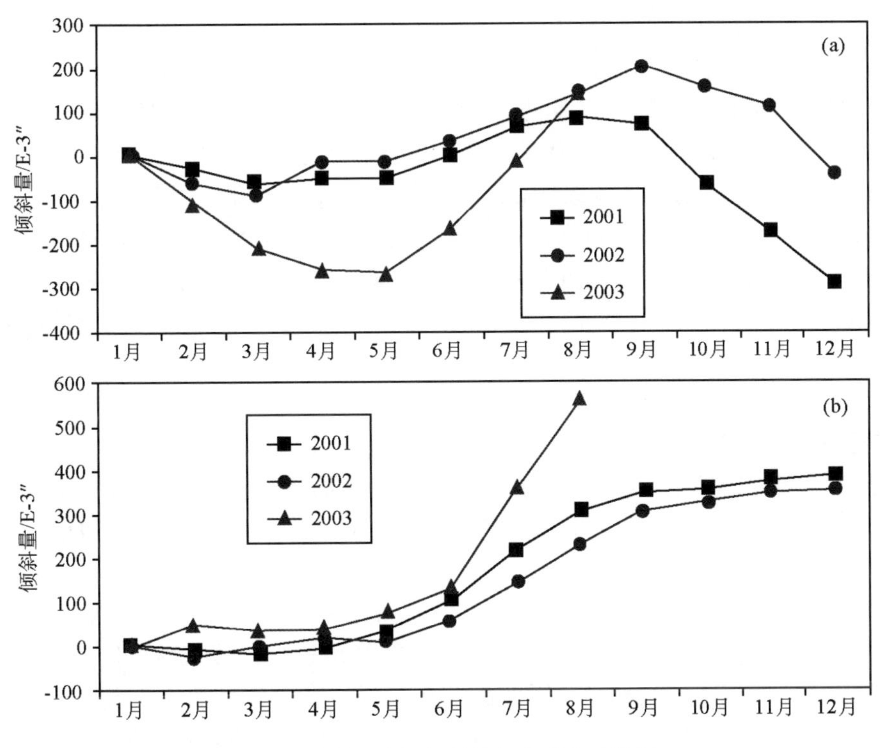

图 10　喀什竖直摆倾斜仪年变形态比较图

Fig 10　Annual curve of vertical pendulum tiltmeter

（a）北南向；（b）东西向

七、地震前兆异常特征分析

地震学异常：1996 年 9 月至 2002 年 9 月，喀什—乌恰区周围形成 4 级地震围空，2002 年 12 月 16 日空区被打破，同年 12 月 25 日在空区边缘发生乌恰 5.7 级地震，而后 2003 年 9 月 2 日在空区边缘的空段内发生此次阿克陶 5.9 级地震。对震区附近 3 级以上地震的时序扫描来看，η 值在 2003 年 6 ~ 10 月间出现 4 个月的低值异常，阿克陶 5.9 级地震发生在低值过程中，震后1个月恢复正常。

前兆观测资料异常：阿克陶5.9级地震发生在监测能力较弱的西昆仑西段，震中300km范围内2台项（3套）前兆观测项目中，喀什台竖直摆地倾斜仪东西和南北向的倾斜分量是往年同期的2倍，但震前1周该仪器由于雷击而断记。

八、震前预测、预防和震后响应

1. 预测情况

2003年9月2日阿克陶5.9级地震发生在新疆地震局划定的2003年度喀什—乌恰6～7级地震危险区内[2)]。震级也在误差范围内，年度预测对应情况较好。

在中期和一年尺度中短期预测基础上，新疆地震局对危险区及其周围各类异常加强了跟踪分析，密切监视地震活动和前兆观测的异常变化。2003年7月25日，新疆地震局预报中心对喀什—乌恰地区提出短临预报意见，预测2003年7月28日至8月28日喀什—乌恰地区有发生5～5.5级地震的可能，具体见附件。主要依据是①2003年7月16～24日伽师震区3级余震平静，此平静时长为6.8级地震序列中最长时段；②2003年7月以来，5月4日伽师5.8级余震区内余震强度明显增强；③2003年6月5日伽师5级余震后，余震序列强度衰减缓慢，但7月24日后其强度有较明显增强；④2003年7月15日喀什地磁低点时间提前8小时，16日其地磁总强敌差之异常，18日其地磁响应比也出现异常变化；⑤2003年7月23～24日，乌什台钻孔出现压性—回张—压性的不稳定变化过程。结果于2003年9月2日距离预测区56km的阿克陶发生5.9级地震[3,4)]。

2. 震害分析

阿克陶5.9级地震发生后，中国地震局和新疆地震局派出联合工作组赴灾区开展震害评估。在当地政府的大力支持下，工作组与有关部门协同作战，在较短的时间内完成了这次地震的震害评估工作[1)]。

阿克陶5.9级地震发生在高山地区，人口密度低，因此地震造成的灾害损失不是很大。由于受环境地理条件限制，工作组仅实地考察了20个调查点，其中震中附近的阔什喀尔其村震感最为强烈，震区土木结构房屋毁坏的占0.37%，严重破坏的占8.63%，中等破坏的占11.18%，轻微破坏的占17.90%，基本完好的占61.92%，破坏的房屋主要是布伦口乡，吉木乡和奥依塔克乡。震中最高烈度为Ⅶ度，范围较小主要位于阔什喀尔其村西南侧的山上；其余地区烈度都绝大部分小于Ⅵ度（图3）。这次地震破坏主要是农村房屋高、跨度大、房顶房泥过厚，房屋抗震能力差，这也是房屋破坏严重和掉顶多的原因之一[1)]。

九、结论与讨论

阿克陶5.9级地震发生在西昆仑地震带北段，由于震中位于中塔边境，监测能力差，300km范围内只有4个测震台，2台（3套）前兆观测仪器。根据哈佛大学结果分析认为此次地震震源断错为走滑正断层。综合极震区烈度分布和震区断裂性质分析认为，近南北走向的节面Ⅱ为断层面。主压应力P轴方位46°，仰角312°，张应力T轴方位14°。此次地震可能是青藏块体在向北的挤压作用下，塔什库尔干断裂发生走滑断错造成的。此次地震序列余震较为丰富，根据主震占全部地震序列能量比和震级差分析认为，此次地震类型可能为主震—余震型。

阿克陶 5.9 级地震震区为地广人稀的高山区，交通条件差，地震现场科学考察和烈度调查有一定难度和局限性。中国地震局和新疆地震局组成联合地震现场灾害评估组，通过现场工作发现，此次地震发生在山区，未造成人员伤亡。根据《中国地震烈度表》，现场共调查了 20 个点，其中Ⅵ度区 10 个，Ⅴ度区 7 个，Ⅳ度区 2 个，Ⅶ度点上滚石、雪崩等现象依据地方政府提供的录像资料确定，由此圈定出Ⅶ度区、Ⅵ度区和Ⅴ度区范围，烈度等震线长轴 NNW 走向，与构造走向基本一致。通过评估，此次地震共造成直接经济损失 847.37 万元。

震前，阿图什台地倾斜仪和喀什台地磁仪震前未发现异常，只有喀什台地倾斜 NS 向和 EW 出现了异常变化，但 8 月份该观测仪器因为雷击而停记。新疆地震局预报中心填报了一张预报卡，2003 年 7 月 28 日至 8 月 28 日以 39.5°N，76.5°E 为中心，半径 100km 有可能发生 5.0～5.5 级。这次预报主要针对 2003 年 2 月 28 日巴楚 6.8 级地震强余震的预报[3,4)]，结果于 2003 年 9 月 2 日在距离预报区 56km 的阿克陶发生了 5.9 级地震。阿克陶 5.9 级地震发生在新疆及周边中强震活跃时段，2002 年 12 月 25 日乌恰 5.7 级地震开始，新疆及周边 1 年内先后发生了 2003 年 2 月 14 日石河子 5.4 级地震，2003 年 2 月 24 日伽师 6.8 级地震，2003 年 9 月 2 日阿克陶 5.9 级地震，2003 年 9 月 27 日中俄蒙交界 7.9 级地震，2003 年 12 月 1 日昭苏 6.0 级地震，由此可将 2003 年前后，新疆中强震很活跃，阿克陶 5.9 级地震就在此状态下发生的。

参 考 文 献

[1] 杨纪林、宋立军、吐尼亚孜 · 沙吾提等，2003 年 9 月 2 日阿克陶 5.9 地震宏观烈度考察和发震构造探讨，内陆地震，18 (4)，353～358，2004

参 考 资 料

1) 新疆维吾尔自治区地震局，2003 年 9 月 2 日新疆阿克陶 5.9 级地震烈度考察及灾害损失评估现场工作报告，卷宗 6840

2) 新疆维吾尔自治区地震局，2003 年度新疆地震趋势研究报告，2002.11

3) 新疆维吾尔自治区地震局，2003 年新疆地震局报中国地震局中短临地震预报意见卡，卷宗 6904

4) 新疆维吾尔自治区地震局，2003 年新疆地震局中短临地震预报意见卡，卷宗 6905

5) 新疆维吾尔自治区地震局，2002，新疆通志 · 地震志，新疆：新疆人民出版社

Aketao M_S 5.9 Earthquake on September 2, 2003 in Xinjiang Uygur Autonomous Region

Abstract

An Earthquake of M_S5.9 occurred in the southwest of Aketao country, Xinjiang Uygur Autonomous Region on September 2, 2003. The micro-epicenter is 38°50′N, 75°10′E measured by seismic net center, CEA. From the result of scientific expedition, the epicenter intensity is Ⅶ, and isoseismal long axis direction is NW, the earthquake occurred in the mountainous area, no person was injured.

This earthquake sequence is main-after type, the most aftershock is M_S4.1, the aftershock distribute two sides of main earthquake, and the direction is NWW direction. The epicenter type is slip-normal fault, the plane Ⅱ is the fault plane. The main stress P axis is 46°, elevation is 312°. The Qinghai-Xizang block move in North direction, the Tashikuergan fault slip-offset make the earthquake occurred.

Aketao M_S5.9 earthquake lie in the border between China and Tajikistan, there is four seismic stations and two precursor stations in the 300km region of epicenter. There is two abnormality of seismology and one abnormity of precursor.

报告附件

附件二：预报卡证明材料

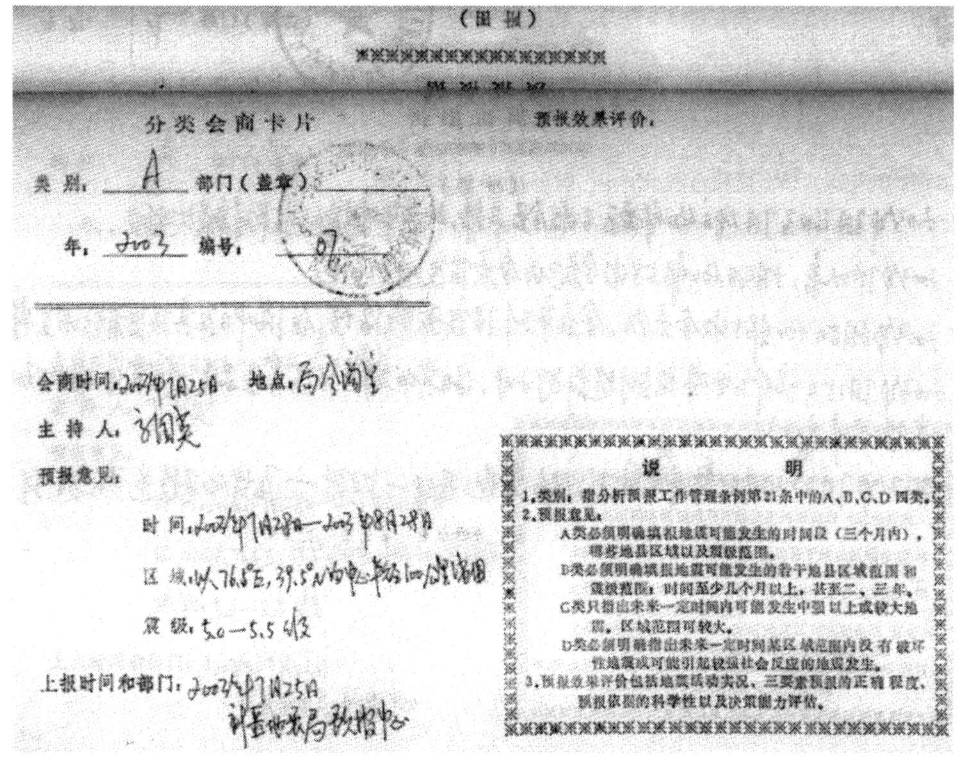

分类会商卡片

预报效果评价：

类别：A　部门（盖章）

年：2003　编号：07

会商时间：2003年7月25日　地点：局会商室

主持人：刘国英

预报意见：

时间：2003年7月28日—2003年8月28日

区域：以76.5°E，39.5°N为中心半径100公里范围

震级：5.0—5.5级

上报时间和部门：2003年7月25日

新疆地震局预报中心

说　明

1.类别：指分析预报工作管理条例第21条中的A、B、C、D四类。

2.预报意见：

A类必须明确填报地震可能发生的时间段（三个月内），哪些地县区域以及震级范围。

B类必须明确填报地震可能发生的若干地县区域范围和震级范围，时间至少几个月以上，甚至二、三年。

C类只指出未来一定时间内可能发生中强以上或较大地震，区域范围可较大。

D类必须明确指出未来一定时间某区域范围内没有破坏性地震或可能引起较强社会反应的地震发生。

3.预报效果评价包括地震活动实况、三要素预报的正确程度、预报依据的科学性以及决策能力评估。

预报根据

（附图）

1. 2003年7月16日至7月24日伽师震区3级余震平静，此平静时长为6.8级后到目前最长时段。
2. 2003年7月以来，7月16日伽师5.8级余震区内余震强度明显增强。
3. 2003年6月5日伽师5级余震后，余震序列强度衰减缓慢，但7月24日后其强度有较明显增强。
4. 2003年7月15日喀什台地磁低点时提前约5小时，16日其地磁总强度差值异常，18日其地磁[illegible]也出现异常变化。
5. 2003年7月23—24日乌什台钻孔应力出现压性—拉张—压性的不稳定变化过程。

2003年9月27日中、俄、蒙交界7.9级地震

新疆维吾尔自治区地震局

龙海英　高国英　孙甲宁

摘　要

2003年9月27日在中国、俄罗斯、蒙古交界发生7.9级地震，震中位于新疆阿勒泰地区以北，距我国国界约110km。此次地震震级大，地震波及范围广、面积大，但由于震中在境外，离国内人员聚集区较远，因而未造成国内人员伤亡，国内房屋受损也不多，倒塌更少。在中国境内距离震中最近的区域属于交通不便、通讯不畅的高寒、高海拔山区，工作队无法到达进行实地考察。地震造成直接经济损失7635.6万元。

此次地震序列类型为双震型，最大余震7.3级。余震分布在主震的东南—北西两侧。此次地震为右旋走滑断错，节面Ⅰ为主破裂面，主压应力P轴方位近NS向。破裂相对简单，破裂持续时间约37s，最大位错3.6m。断层面上显示出两个显著的、滑动量超过2.0m的破裂区。发震构造为俄罗斯境内的乌列盖NW向断裂带，与余震分布相吻合。

震中附近属地震监测能力较弱地区，500km范围内共有地震台2个，遥测台1个。2个台有地倾斜前兆观测项目。震前共出现3个异常项目7条异常，除震前新疆阿勒泰地区出现3.5级地震增强活动外，其他均为地倾斜异常。

7.9级地震发生在新疆阿勒泰地区的我国国境之外，震前未做出预测、预报。

前　言

2003年9月27日19时33分，中、俄、蒙交界发生7.9级地震。新疆维吾尔自治区地震局台网测定的微观震中为50°15′N，88°02′E。这次地震震中位于俄罗斯阿尔泰南部山区楚亚盆地北部、楚亚和库赖山间盆地以及查干—乌尊（Chagan-Uzun）地块之间[1]，距我国新疆边境约110km。地震有感范围大，包括俄罗斯的阿尔泰、哈卡斯自治州、图瓦自治共和国、克拉斯诺亚尔斯克边疆区、新西伯利亚和科麦罗沃州，蒙古共和国西北部和哈萨克斯坦共和国的东部[1]。地震也波及到中国新疆北疆大部分地区，阿勒泰地区强烈有感，克拉玛依、乌鲁木齐、石河子等地普遍有感。地震在新疆境内未造成人员伤亡，但造成直接经济损

失7635.6万元。

7.9级地震发生在阿尔泰地区，位于阿勒泰—戈壁阿尔泰褶皱带与蒙古地块交接的乌列盖深大断裂带北侧。阿尔泰地区曾发生过多次强震，如1905年杭爱8级地震，1931年新疆富蕴8级地震，1957年戈壁阿尔泰8.3级地震，1991年阿尔泰山西部斋桑7.3级地震，以及多次5~6级地震发生，表明阿尔泰山是亚洲中部一个非常活跃的地震带[2]。新疆境内阿尔泰地区最近一次6级地震为1996年3月13日阿尔泰6.1级地震，距7.9级地震约为195km。

震中附近新疆境内地震观测台站较少，500km范围内仅有2个地震台和1个遥测台。震前出现3个异常项目共计7条异常。

由于此次7.9级地震发生在中国、俄罗斯、蒙古交界，现场工作考察队无法考察境外区域震害，收集了俄罗斯科学院地球物理调查所（GSRAS）关于宏观震中、震中烈度等的调查结果。新疆境内最大烈度点为Ⅵ。

一、测震台网及地震基本参数

图1给出震中500km范围内的地震台站分布，仅有2个测震观测台，1个遥测观测台。其中0~200km范围内无观测台，201~300km范围内有阿勒泰地震台，301~400km范围内有富蕴地震台，401~500km范围内有布和图遥测台。其中阿勒泰和富蕴台还有地倾斜前兆观测项目。7.9级地震发生在中、俄、蒙交界地区，新疆境内该区域监测能力相对较弱，阿勒泰地区地震的监测能力基本可达到M_L2.5地震。这次地震的基本参数见表1。由于新疆台网和国家台网速报给出此次地震为7.9级，本文取此次地震震级为7.9级。

表1 中、俄、蒙交界7.9级地震基本参数

Table 1 Basic parameters of M_S7.9 earthquake

编号	发震日期	发震时刻	震中位置		震级		震源深度	震中地名	结果来源
	年 月 日	时 分 秒	φ_N	λ_E	M_S	M_W	(km)		
1	2003 09 27	19 33 23	50.15°	88.02°	7.9		35	友谊峰东北	新疆地震局
2	2003 09 27	19 33 25	50.02°	87.87°	7.7*		16	俄罗斯西伯利亚西南部	中国地震台网
3	2003 09 27	19 33 26	50.06°	87.73°		7.3	16		USGS
4	2003 09 27	19 33 36	50.02°	87.86°	7.5	7.3	15		HRVD

注：国家台网速报目录震级为7.9。

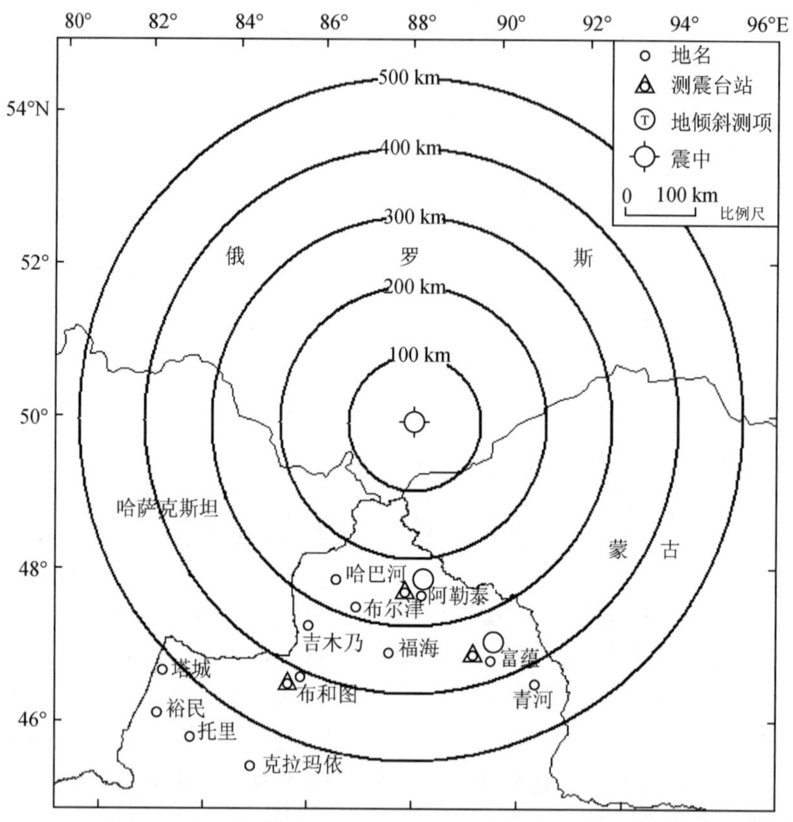

图1　中、俄、蒙交界 7.9 级地震前震中附近新疆境内地震台站及观测项目分布图

Fig. 1　Distribution of seismic stations surrounding the epicenter area in Xinjiang before M_S7.9 earthquake in the boundary of China, Russia and Mongolia

二、地震地质背景

阿尔泰山位于蒙古高原的西南翼，是欧亚大陆内部远离板块碰撞边界的内陆活动造山带。跨越中国新疆、蒙古西部、哈萨克斯坦东北部和俄罗斯西伯利亚[3]。古生代以来，阿尔泰山曾是蒙古—西伯利亚板块与准噶尔—哈萨克斯坦板块的接合带；中生代由于西伯利亚板块向南运动而继续活动，形成侏罗纪山间盆地和燕山期火山岩；燕山运动之后又经历多期构造活动，但构造活动强度逐渐减弱；渐新世以来，阿尔泰地区构造活动明显增强，大体上发育 3 种类型的活动断裂：NNW 向的大型右旋走滑断裂（如富蕴断裂）、NW 向的大型右旋走滑逆冲断裂（如额尔齐斯断裂）、以及近 EW 向的大型左旋走滑断裂（如杭爱断裂）。作为天山挤压构造区与贝加尔拉张构造区的过渡带，阿尔泰地区是距离板块边界最远的内陆第四纪构造显著活动和强震多发的区域[2]。

2003 年 9 月 27 日中、俄、蒙交界 7.9 级地震发生在阿尔泰—蒙古西部地区，在大地构造单元划分上属于阿尔泰—萨彦褶皱区。阿尔泰—萨彦褶皱区西与哈萨克斯坦褶皱区为邻，南与准噶尔地块和天山褶皱区相毗邻，东北部是西伯利亚地块（图 2）。阿尔泰—蒙古西部地区地层发育较完整：元古界为一套火山岩、碎屑岩和碳酸盐建造，厚达数千米；古生界为海相、海陆交互相碎屑岩，中性、中酸性火山岩以及含煤岩系；而新生界为巨厚的河湖相碎屑岩和冰、冲洪积堆积物[4]。该地区由于构造演化、地壳运动的差异，形成不同展布方向、不同构造运动形式、而又互相交叉切割影响的错综复杂的构造环境。阿尔泰—蒙古西部地区大致可划分为 3 个不同性质的构造单元，由北向南分别为唐努山褶皱带、蒙古中间地块和阿尔泰—戈壁阿尔泰褶皱带（图 2）[4]。

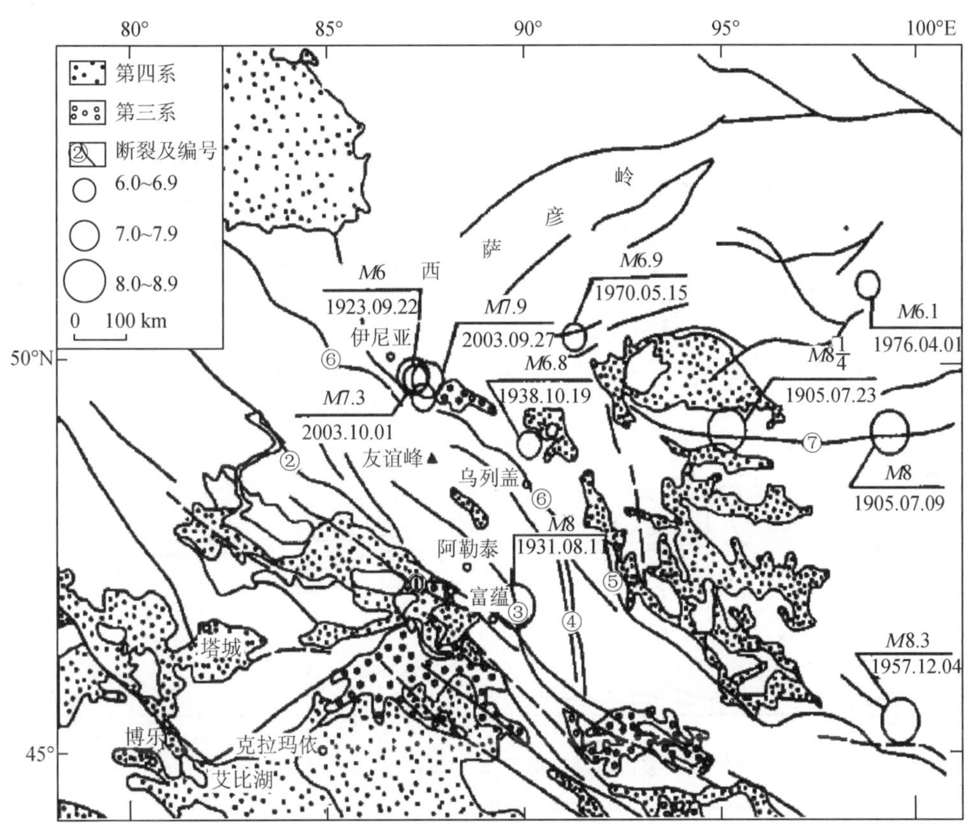

图 2 区域地震构造环境[4]

Fig. 2 Map of geological structure in the epicenter and its adjacent area

①额尔齐斯河谷断裂；②额尔齐斯断裂；③二台断裂；④布尔根断裂；⑤额尔格朗图断裂；⑥乌列盖断裂；⑦杭爱断裂

阿尔泰—蒙古西部地区深部地质结构较为复杂，它主要表现在莫霍界面起伏变化和地幔异常两方面。

1. 区域地壳厚度变化[8]

在阿尔泰—蒙古西部及毗邻地区，由卫星测量得到的 1°×1°布格重力异常图正、负异常区形态多样，重力场较为复杂（图 3）。在天山以北的准噶尔盆地，重力等值线稀疏，中部为近东西向分布的椭圆形重力高，反映准噶尔盆地的莫霍界面及上地幔为一个形态不规则的宽缓隆起区。以重力值推算，该区地壳厚度为 42～45km。阿尔泰山的重力场为北西向分布的重力值很低的负异常带，西北端宽，往东向变窄，但负异常中心并不在阿尔泰山主峰。按重力值推算，阿尔泰山地壳厚度均在 47km 以上，而富蕴以东的阿尔泰山区最大地壳厚度可达 54km。在阿尔泰山西南缘有一条规模较大的、北西向延伸的重力梯度带。这条梯度带大致沿哈巴河、二台至蒙古的达兰扎兰加德一线分布，重力梯度值高达 1.2mGal/km，反映这一地带的莫霍界面和上地幔呈一向北东方向倾斜的大陆坡。

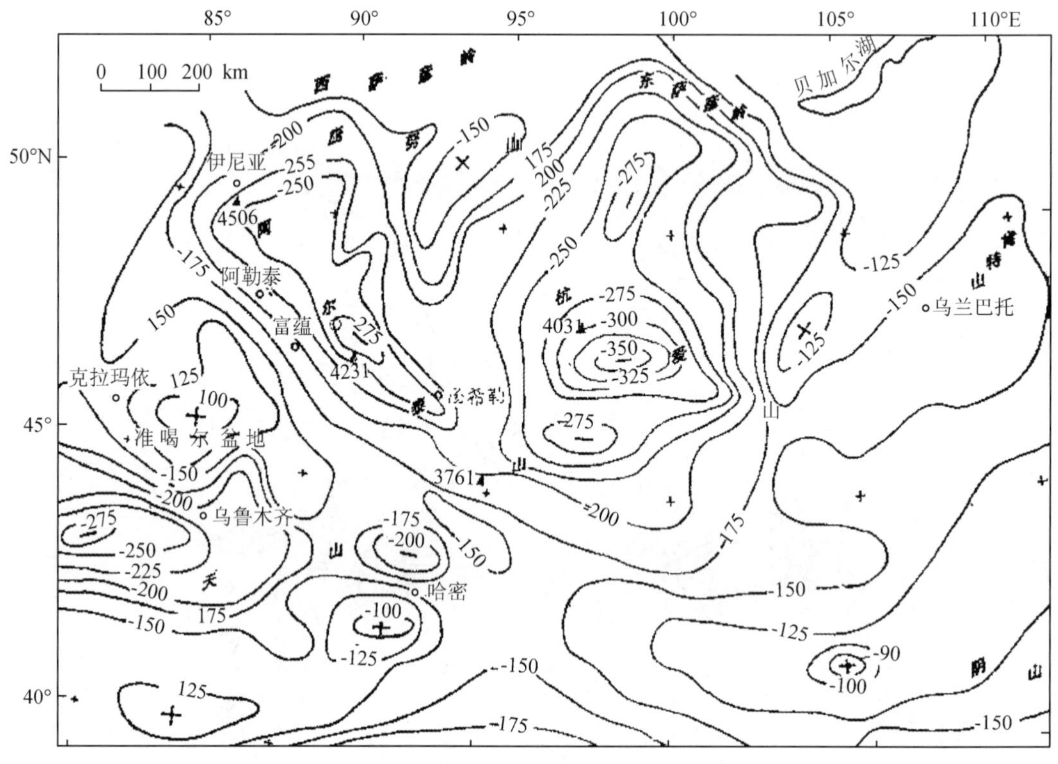

图 3　阿尔泰—蒙古西部地区 1°×1°布格重力（mGal）异常图[8]

Fig. 3　The anomaly map of 1°×1° Bouguer gravity (mGal) in the western region of Altay-Mongolia

蒙古西部地区的重力异常场，在东萨彦岭以南为三个不同方向延伸的重力低，成近南北向的重力负异常区。其中心在杭爱山，重力值低于－350mGal，呈近东西向延伸，是亚洲大陆腹地重力值最低的地方。它反映这一带莫霍界面和上地幔形成了很深的拗陷，拗陷中心在杭爱山区，是地壳厚度最大的地方。据推算其厚度超过 57km。唐努山一带的重力场，为伸入阿尔泰山和杭爱山重力值之间的一个三角形重力高值区域，其中心在乌布苏湖以东。－150mGal重力等值线呈北东向展布，它反映这里存在着一个北东向分布的长条状上地幔隆起。

根据重力资料推测，准噶尔盆地地壳厚度为 42～45km，向东北伸入阿尔泰山区地壳厚度剧增到 47～54km，杭爱山地壳厚度更大，最深处逾 57km；贝加尔裂谷地带莫霍界面上隆，地壳厚度减小为 35 ~ 37km，北面的西伯利亚地块地壳厚度为 38 ~ 40km 之间。

2. 地幔异常[8]

根据地震波速度资料，阿尔泰—蒙古西部一带深部存在着一个地幔低速非均质异常区（图 4）。异常区北部边界在贝加尔湖附近，在那里局部熔融体已达到地壳底部，基性和超基性岩浆上冲至地壳形成“反向山根”，热流值达到 2 ~ 2.2HFU。它造成地幔异常的低速非均质体呈舌形，从贝加尔湖一带自北往南倾斜，其西南边界大致与阿尔泰山—戈壁阿尔泰山一带相当，低速非均质体顶面往西南倾斜，深至莫霍界面以下，逾 400 ~ 500km。在蒙古西部大湖区一带，低速非均质体顶面往深部倾斜梯度剧增，至二台断裂附近已深入莫霍界面以下超过 600km，剖面上呈铲状。这个低速非均质体往东、东南方向延伸变化不大，其边界埋深于莫霍界面以下 300 ~ 400km。

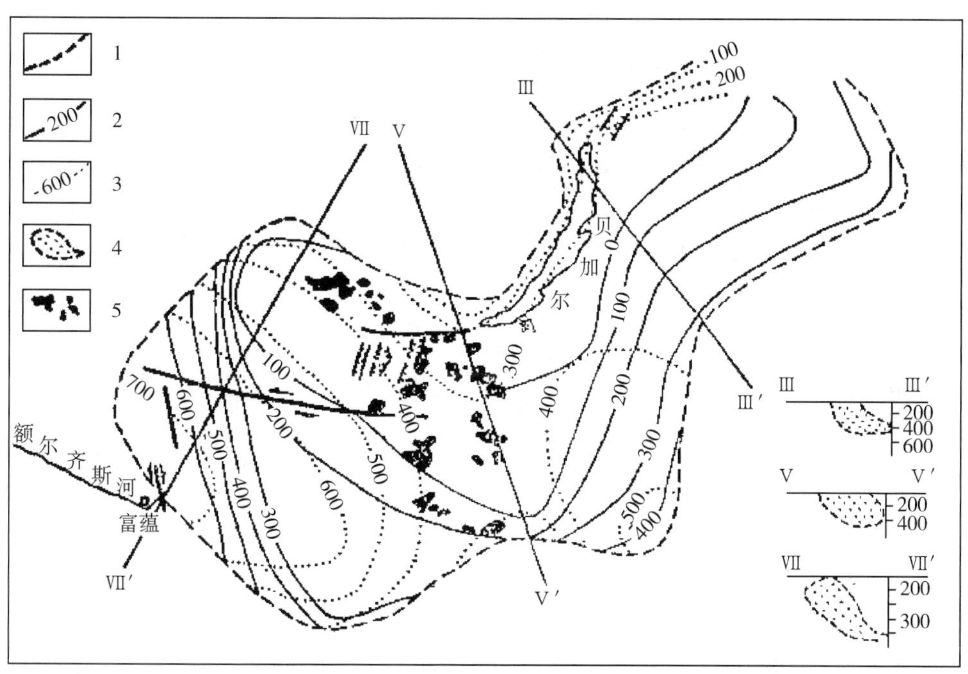

图 4 蒙古—阿尔泰一带地幔异常区[8]

Fig. 4 The anomaly map of the mantle in Mongolia-Altay area

1. 低速非均质体范围；2. 由莫霍面计算得到的非均质体上界面等深线；3. 由莫霍面计算得到的非均质体下界面等深线；4. 剖面上的低速非均质体；5. 新生代玄武岩

这个地幔异常分布区自西伯利亚块体南部开始，往南到杭爱山及其以南地区，从始新世至全新世曾有过六次岩浆喷发，溢出大量玄武岩，呈近南北向的楔形分布，北宽南窄，宽度为 500 ~ 600km。

这些深部构造控制着地壳浅部种种地质构造现象，而阿尔泰山西南一带正处在这个非常重要的区域深部构造变化的边界线上。该区域地质构造作用为这种特有的深部地质结构所控制。

中、俄、蒙交界7.9级地震发生在阿尔泰—戈壁阿尔泰褶皱带与蒙古地块交接的乌列盖深大断裂带北侧（图2）。该断裂是分割阿尔泰、西萨彦、蒙古3个块体的构造，总体呈北西展布，断面北倾，倾角陡，延伸长度大于500km。受南部印度—欧亚板块碰撞效应以及北部蒙古—西伯利亚块体的作用，该区域北北西走向的断裂主要为右旋走滑形式，而北西西向则主要为右旋走滑逆冲断裂，同时沿构造线发育了一系列由走滑断裂控制的小型山间盆地(洼地)[3]。

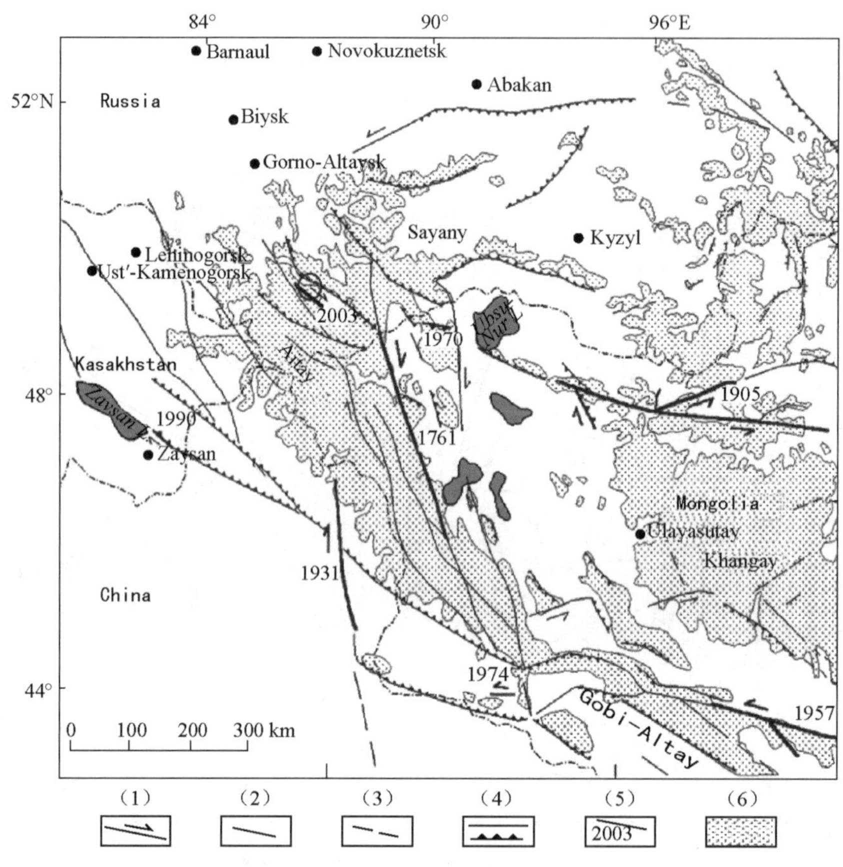

图5 蒙古和西伯利亚地区的地震构造图[1]

Fig. 5 The seismotectonic map of Mongolia and Siberia area

(1) 走滑断裂；(2) 逆断裂；(3) 正断裂；(4) 地震活动的区域；

(5) 2003年9月27日地震的断裂（圆圈为地震位置）；(6) 比海拔2000m更高的区域

中、俄、蒙交界7.9级地震是位于蒙古西部断裂往西北方的继续延伸和破裂，与蒙古—西伯利亚地区的现代构造模式有一定的关系（图3），那里重复发生了好几个7级以上的大地震。在现代构造应力场中，走向NW和近南北向的断裂是一个右旋走滑的逆断裂，NE走向和近东西向的断裂则为有正断层成分的左旋断裂[1]。

综合判断认为，中、俄、蒙交界7.9级地震发震构造为乌列盖活动断裂带[4]。乌列盖断裂带主体延伸方向为北西向，该断裂规模较大，破裂带较宽，在局部区域方向略有变化。而卫星影像解译的 7.9 级地震发震断裂方向偏北北西，且线状平直，表明走滑运动明显。

三、地震影响场和震害1)

此次地震发生后，新疆地震局立即启动破坏性地震应急预案，及时了解灾情，并派出地震现场工作队于次日赶赴阿勒泰震区开展地震灾害评估和科学考察工作。由于震中在中国、俄罗斯、蒙古交界，在中国境内距离震中最近的区域属于交通极为不便、通讯不畅的高寒、高海拔山区，工作队无法到达进行实地考察，只能进行抽样核实，因此未能给出新疆境内烈度分布，只能给出一些点的烈度调查值。这次地震震害波及我国的范围为阿勒泰地区的六县一市，面积达 11 万平方公里，在调查过程中发现一些Ⅵ度点，如：

距边境最近的喀纳斯（Ⅵ度）：多数人站立不稳，少数人惊逃户外，木结构房屋出现细微裂缝，少数有倾斜，部分木结构房屋砖混基础出现裂缝；公路护坡出现裂缝。

喇嘛昭乡和阿勒泰市金鑫铅锌矿（Ⅵ度）：金鑫铅锌矿尾坝垮塌。

卧依莫克乡（Ⅵ度）：多数新建砖木结构房屋出现裂缝。

富蕴县铁买克乡（Ⅵ度）：乡政府办公室、乡招待砖木结构房屋出现裂缝，窗台瓷砖断裂，乡小学土木结构房屋酥裂，少数居民土木结构房屋出现裂缝。

阿勒泰市（Ⅵ度弱）：多数人站立不稳，少数人惊逃户外，大多数房屋基本完好，各类结构房屋大多数基本完好，个别房屋（含多层砖混房屋）墙体出现裂缝；少数桥梁出现细微裂缝。

布尔津县（Ⅵ度弱）：多数人站立不稳，少数人惊逃户外，部分房屋女儿墙裂缝，少数砖混结构房屋（如县公安局、交警大队办公楼、山区林场家属院等）出现裂缝。

富蕴县可可托海镇（Ⅵ度弱）：镇锅炉房铁烟囱倾倒，粮管所及少数居民的房屋出现裂缝。

但由于无法深入到离震中较近的高海拔山区进行调查，同时根据上报资料和实地抽样核实发现Ⅵ度点稀少而且分散，故无法圈定Ⅵ度圈。

Ⅴ度影响较为普遍，如抽样的阿勒泰市汗德尕特乡、布尔津县的卧依莫克乡、也格孜托别乡、也格孜托别乡尔格尔胡木村、哈巴河县城及哈巴河加依勒马乡、齐巴尔乡、齐巴尔乡萨雅铁烈克村、哈巴河县库尔拜乡、富蕴县城等，房屋受损情况主要是少数房屋出现细微裂缝，人们普遍有感。

由于地震发生在中、俄、蒙交界，新疆地震局现场工作考察队无法考察境外区域震害。根据俄罗斯科学院地球物理调查所（GSRAS）的调查[5]，震中地区包括戈尔诺—阿尔泰自治州的 Bel’tir、库赖和阿克塔什镇，Bel’tir 地区的烈度为Ⅷ度，阿克塔什镇的烈度是Ⅵ ~ Ⅶ。该地震造成国外的震害主要有：火炉被破坏，烟囱垮塌，在窗户之间的墙壁上形成典型“X”裂缝，矿渣混凝土建筑物墙壁和角落张裂歪斜；由于木结构房屋具有较好的抗震性能，其震害主要表现为檩木移位和部分房屋屋顶被破坏。其他地区的烈度分布分别是：克麦罗沃州塔什塔戈尔市烈度是Ⅵ，普罗科皮耶夫斯克市烈度是Ⅴ ~ Ⅵ，新西伯利亚、乌斯季卡缅诺戈尔斯克和塞米巴拉金斯克的烈度是Ⅳ，哈卡斯自治州阿巴坎市的烈度是Ⅲ ~ Ⅳ，克拉斯诺亚尔斯克、东哈萨克州的斋桑市和克麦罗沃市的烈度是Ⅲ，阿尔泰边疆区巴尔瑙尔市、阿拉木图市、塔尔迪库尔干市和阿斯塔那市（哈萨克首都）的烈度是Ⅱ ~ Ⅲ[1]。

此次地震震中在境外，因而尽管震级大，地震波及范围广、面积大，但由于离国内人员聚集区较远，未造成国内人员伤亡，国内房屋受损也不多，倒塌更少。

四、地 震 序 列

中、俄、蒙交界 7.9 级地震震前无直接前震，但震前 8 月中旬至 9 月中旬近 20 天时间内，富蕴—阿勒泰地区接连发生 4 次 $M_S \geqslant 3.5$ 级地震，这种增强活动在该区域历史上非常少见。7.9 级地震后截至 2004 年 1 月 27 日，新疆维吾尔自治区地震局编目确定 $M_S \geqslant 4.0$ 级以上地震 73 次，其中 7.0 ~ 7.9 级 2 次，6.0 ~ 6.9 级 1 次，5.0 ~ 5.9 级 14 次，4.0 ~ 4.9 级 56 次。表 2 给出俄、蒙、中交界 7.9 级地震序列 $M_S5.0$ 以上地震序列目录。

表 2　中、俄、蒙交界 7.9 级地震序列目录（$M_S \geqslant 5.0$）

Table 2　Catalogue of $M_S7.9$ earthquake sequence (more than $M_S5.0$)

编号	发震日期	发震时刻	震中位置		震级	震源深度	震中地名	结果来源
	年 月 日	时 分 秒	φ_N	λ_E	M_S	(km)		
1	2003 09 27	19 33 23	50°09′	88°01′	7.9	35	友谊峰东北	资料 2）
2	2003 09 27	19 58 04	50°03′	88°01′	5.8		友谊峰东北	
3	2003 09 27	21 03 03	50°22′	88°13′	5.2		友谊峰东北	
4	2003 09 27	21 16 41	50°09′	88°01′	5.6		友谊峰东北	
5	2003 09 27	21 38 04	49°50′	88°00′	5.0		友谊峰东北	
6	2003 09 27	23 31 19	50°16′	87°52′	5.0		友谊峰北	
7	2003 09 28	02 52 48	50°07′	87°48′	6.4		友谊峰北	
8	2003 09 29	23 27 00	50°07′	88°01′	5.0		友谊峰东北	
9	2003 09 30	16 27 07	50°03′	87°58′	5.4		友谊峰北	
10	2003 10 01	09 03 23	50°16′	88°02′	7.3		友谊峰东北	
11	2003 10 01	11 03 09	50°12′	88°15′	5.0		友谊峰东北	
12	2003 10 01	11 58 44	50°06′	88°04′	5.2		友谊峰东北	
13	2003 10 10	00 06 01	50°12′	87°48′	5.4		友谊峰北	
14	2003 10 13	13 26 39	50°12′	87°45′	5.9		友谊峰北	
15	2003 10 17	13 30 20	50°18′	87°45′	5.7		友谊峰北	
16	2003 10 23	08 25 49	49°36′	88°06′	5.4		友谊峰东北	
17	2003 11 17	09 35 48	50°09′	87°55′	5.6		友谊峰北	

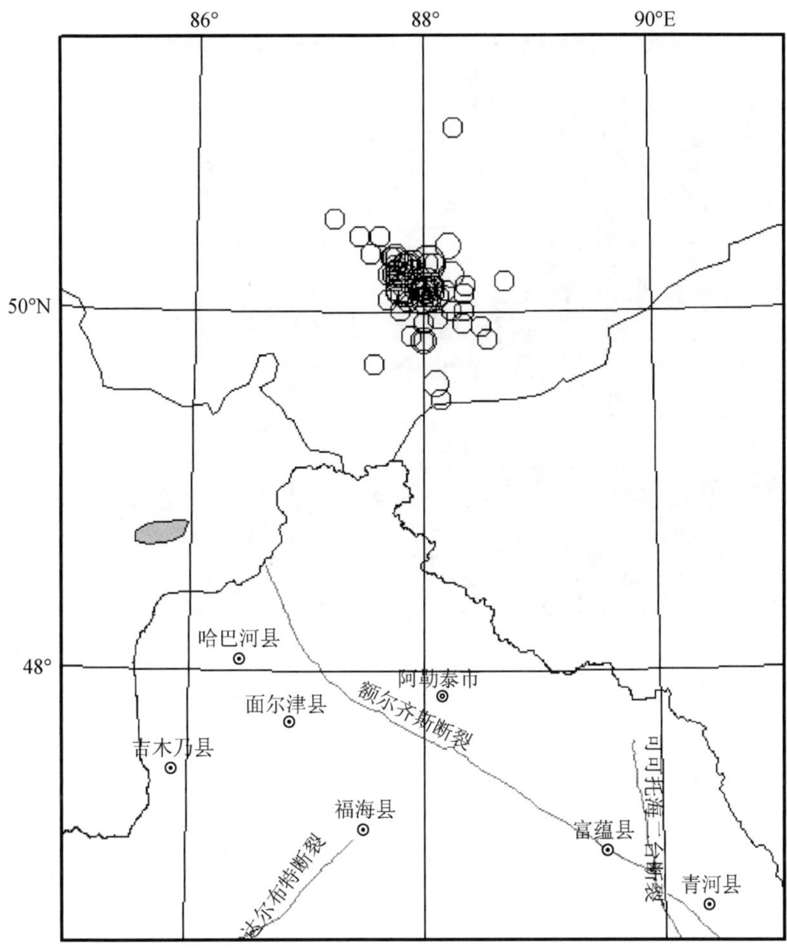

图 6 中、俄、蒙交界 7.9 级地震序列 $M_S \geqslant 4.0$ 级余震分布

Fig. 6 The distribution of M_S7.9 earthquake sequences with more than M_S4.0 in the boundary of China, Russia and Mongolia

从余震分布来看（图 6），余震呈东南—北西向分布，主要分布于主震东南和北西两侧约 50km 范围内。与俄罗斯科学院西伯利亚分院地球物理研究所记录的 2 级以上余震分布[1]（图 7）基本一致。

根据 NEIC 全球台网定位结果，截至 2004 年 1 月底共发生 146 次 M_b3.5 以上地震，由此得到 7.9 级地震序列分布（图 8）。图 8 同样显示，余震分布的优势走向为北西向，余震区长轴约 110km。2003 年 10 月 1 日 7.3 级强余震发生前，余震主要发生在 7.9 级地震的东南侧。7.3 级强余震发生在主震西北约 25km 处，此后多数余震分布在 7.3 级余震周围。

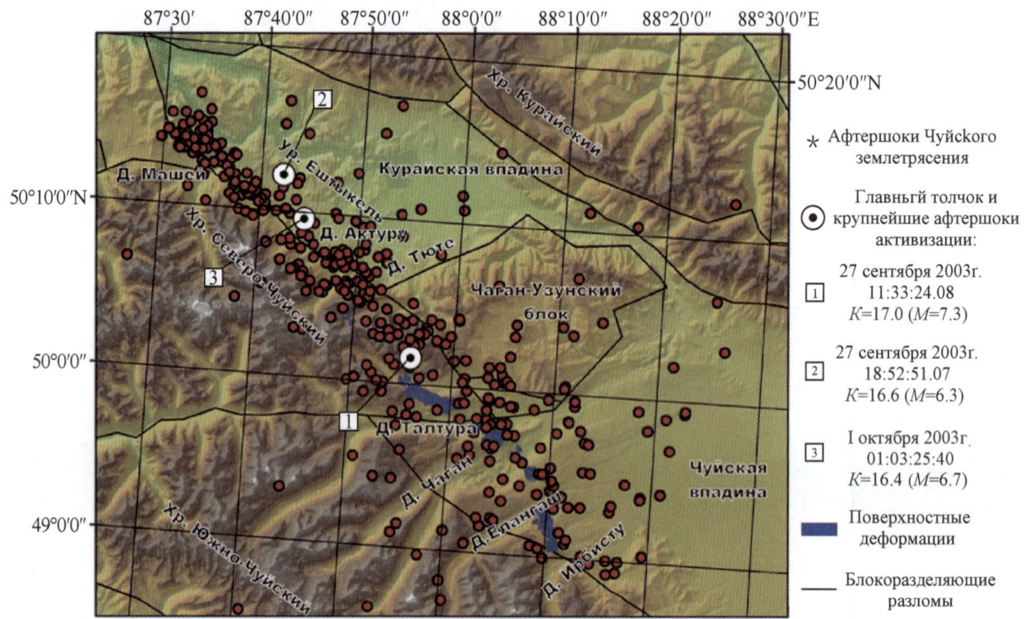

图 7　中、俄、蒙交界 7.9 级地震序列分布（俄罗斯科学院西伯利亚分院地球物理研究所提供）

Fig. 7　The distribution of China, Russia and Mongolia M_S7.9 earthquake sequence

(Provided by Sciences Institute of Geophysics of Siberian branch of Russian Academy)

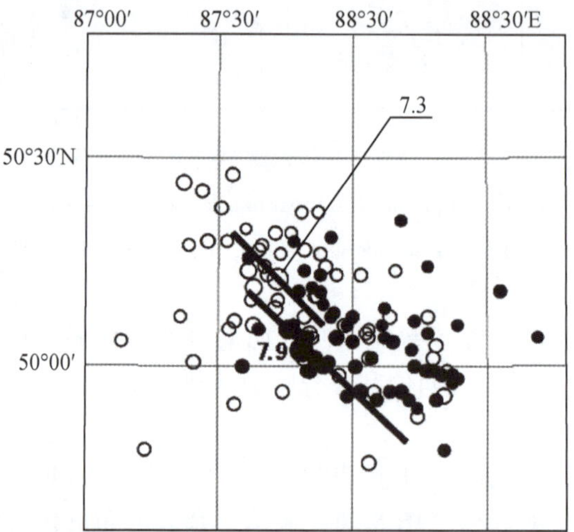

图 8　中、俄、蒙交界 7.9 级地震序列 $M_b \geqslant 4.0$ 级余震分布（据 NEIC 目录）

Fig. 8　The distribution of M_S7.9 earthquake sequences with more than M_b4.0 (NEIC catalogue)

（实心圆代表 2003 年 9 月 27 日至 2003 年 10 月 1 日 01 时发生的地震；空心圆代表 2003 年 10 月 1 日 01 时至 2004 年 1 月 20 日发生的地震；直线为反演得到断层面上的破裂长度及方向示意）

由余震序列 M-t（图 9a）和日频次（图 9b）可以看出，余震主要集中发生在主震后前 10 天内，多数为 4 级余震（由于观测环境限制，4 级以下地震新疆地震目录不完整）。序列最大余震为 10 月 1 日 7.3 级强余震，10 月中旬至 11 月初，M_S4.0 以上余震迅速衰减，11 月中旬序列出现起伏增强，11 月中旬后至 2004 年 1 月中旬，4 级地震活动明显减弱，最后一次 4 级余震发生在 2004 年 1 月 25 日，震级 M_S4.7。

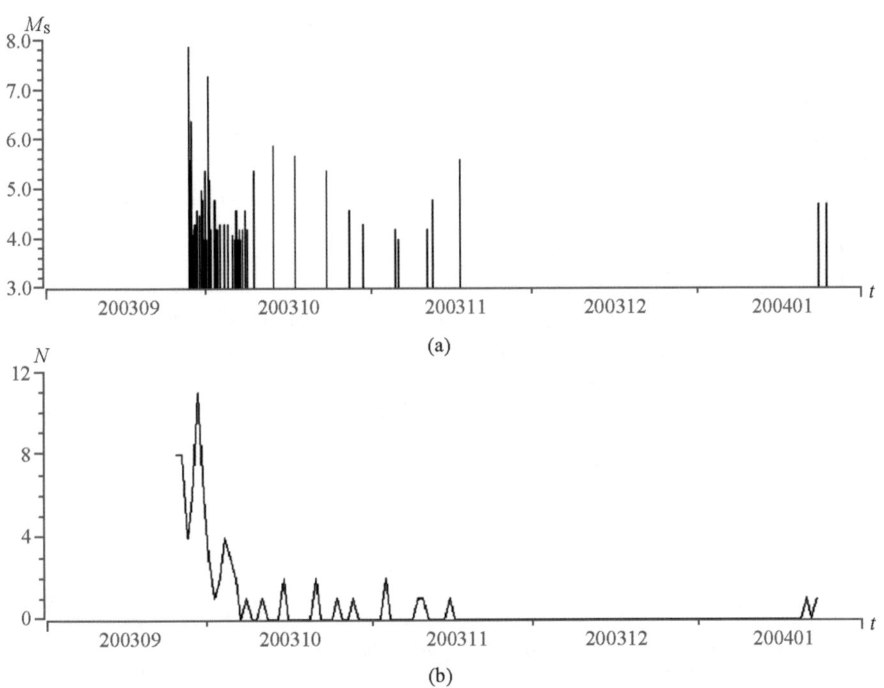

图 9 中、俄、蒙交界 7.9 级地震序列 M-t 图（a）和日频次图（b）

Fig. 9 The M-t map (a) and the daily frequency curve (b) of M_S7.9 earthquake sequences in the boundary of China, Russia and Mongolia

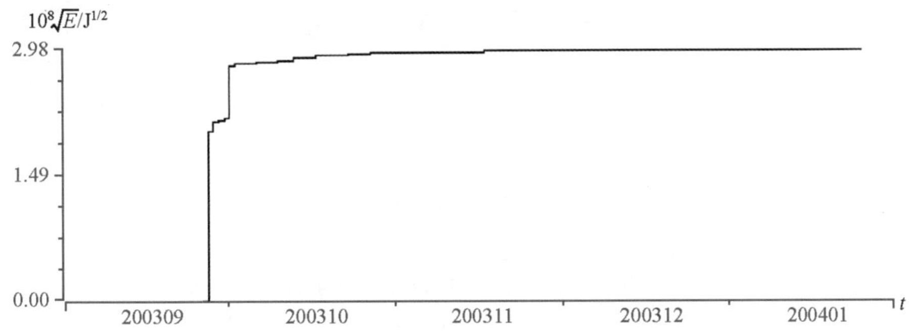

图 10 中、俄、蒙交界 7.9 级地震序列蠕变曲线

Fig. 10 The curve of strain release of M_S7.9 earthquake sequences in the boundary of China, Russia and Mongolia

依据7.9级地震序列蠕变曲线（图10），主震能量与总序列能量之比为67.5%，最大余震为10月1日 M_S7.3 地震，次大余震为9月28日 M_S6.4 地震，主震与最大余震的级差为0.6，根据有关地震序列判别标准，属双震型地震序列。G-R 关系 b 值为0.71，修改的大森公式衰减系数 p 值为0.68。此次地震序列余震初期衰减很快，其后序列虽有起伏，但总体处于衰减过程中，且持续时间较短。

五、震源参数和地震破裂面

1. 震源机制

此次地震发生在新疆境外地区，由于该区域附近新疆地震台稀疏，新疆地震局未能给出震源机制，以下是其他机构及个人利用远震资料给出的最佳双力偶震源机制解结果（表3、图11）。

表3　中、俄、蒙交界7.9级地震最佳双力偶震源机制解

Table 3　Focal mechanism solutions of M_S7.9 earthquake

序号	节面Ⅰ			节面Ⅱ			P轴		T轴		N(B)轴		标量地震距 M_0 ($\times10^{20}$N·m)	矩震级 M_W	结果来源
	走向	倾角	滑动角	走向	倾角	滑动角	方位	仰角	方位	仰角	方位	仰角			
1	129	59	164	228	76	32	355	12	93	32	248	55	1.0	7.3	HRVD
2	130	85	-160	38	70	-5	356	18	262	10	143	69	1.0	7.3	USGS
3	127	79	171	219	81	11	353	2	83	14	257	76	0.97	7.3	文献［7］
4	136	73	120	253	34	32	203	22	81	53	306	29	15	8.0	文献［6］

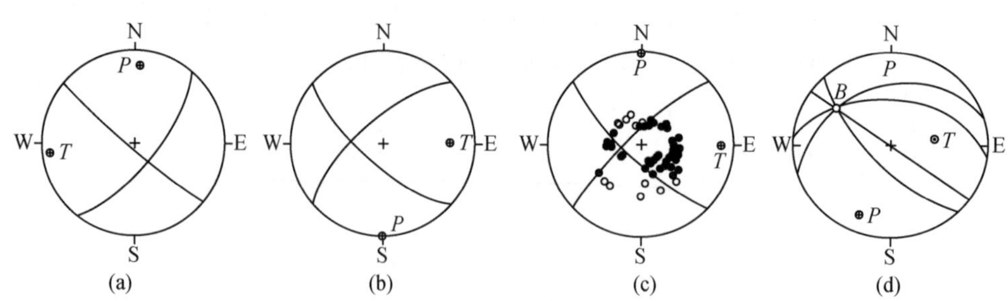

图11　中、俄、蒙交界7.9级地震最佳双力偶震源机制解

Fig. 11　The focal mechanism solution of the best double power couple of M_S7.9 earthquake in the boundary of China, Russia and Mongolia

实心圆代表P波初动为正，空心圆代表P波初动为负

（a）Harvd；（b）USGS；（c）文献［7］；（d）文献［6］

从表3结果分析，此次地震为走滑型破裂。据7.9级地震发生后的余震分布，认为节面

Ⅰ为主破裂面，它与附近一条规模较小的北西向断层走向基本一致。压应力 P 轴近南北，接近水平；张应力 T 轴走向近东西。地震发生时，接近直立的断层两盘发生右旋走滑错动，并带有一定的逆冲分量[7]。此外，根据地震宏观破裂推断的 1931 年 8 月 11 日富蕴 8.0 级地震断错性质为走滑型，主压应力轴走向为 194°N，近 NS 向，倾角为 12°，即为 NNE 向的水平压应力[8]。据此可以认为，阿尔泰地区构造应力场的主压应力 P 轴方位为 NNE—SN，近水平作用方式。在此构造应力场作用下，高倾角断层发生了右旋走滑性质的错动[7]。

震后根据美国地质调查局（USGS）给出的 CMT 资料得到 7.9 级地震及其余震 CMT 解的空间分布（图 12），俄罗斯科学院西伯利亚分院地球物理研究所利用在震区架设的 5 个临时地震台和阿尔泰地区运行的 12 个固定数字地震台的资料也给出了 7.9 级地震及其余震的震源机制解的空间分布（图 13）。由图可见，除主震外，序列余震也多以走滑型地震为主。

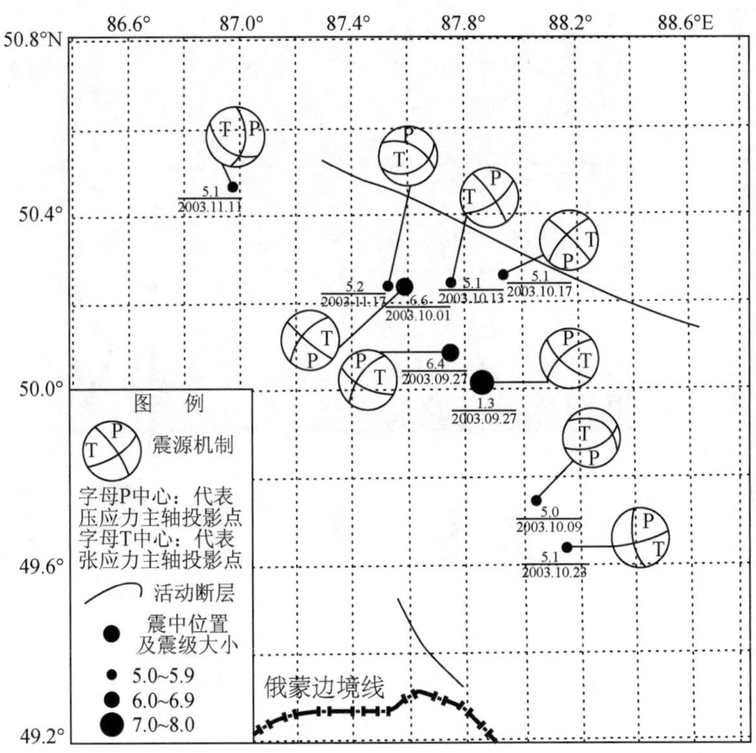

图 12　中、俄、蒙交界 7.9 级地震及其余震 CMT 解的空间分布（据 USGS）

Fig. 12　The spatial distribution of CMT solution about M_S7.9 earthquake and its aftershocks in the boundary of China, Russia and Mongolia (According to USGS)

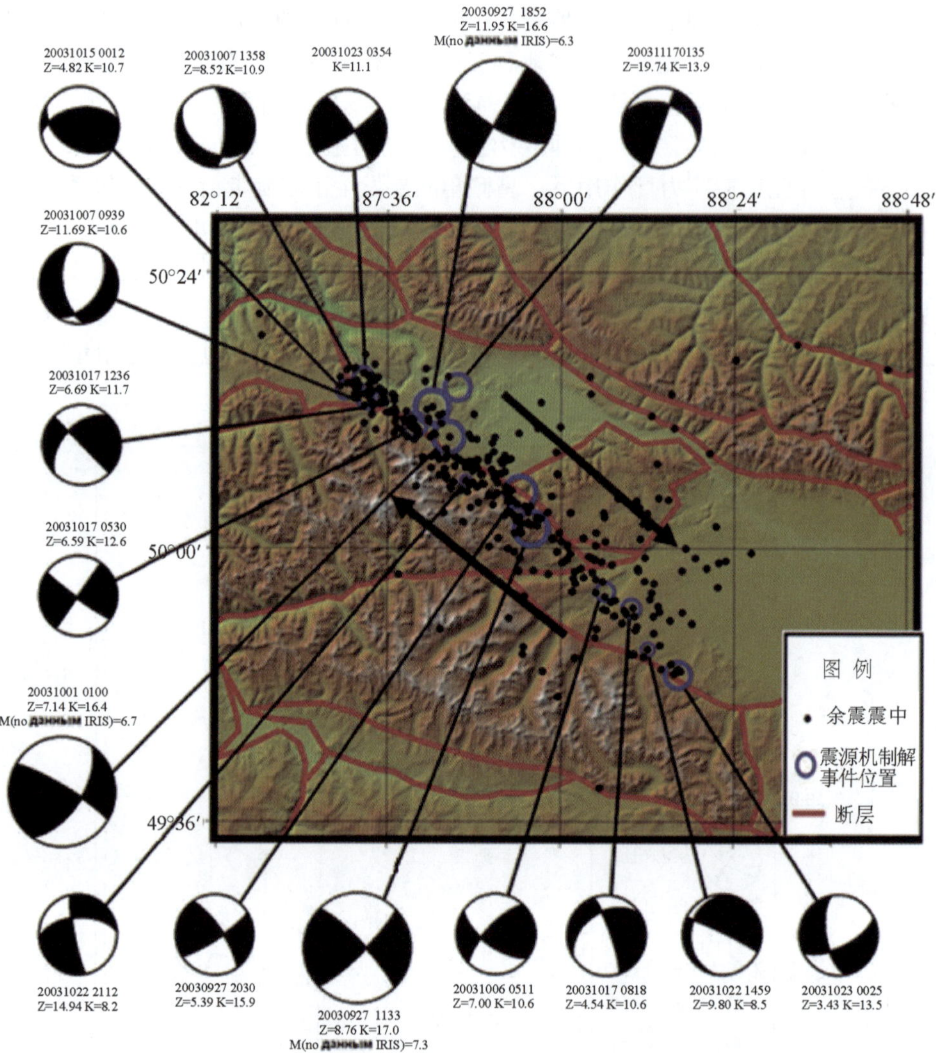

图 13　中、俄、蒙交界 7.9 级地震及其余震的震源机制解的空间分布
(据俄罗斯科学院西伯利亚分院地球物理研究所)

Fig. 13　The spatial distribution of focal mechanism solution about M_S7.9 earthquake and its aftershocks
(Provided by Sciences Institute of Geophysics of Siberian branch of Russian Academy)

2. 断层破裂特征[7,8]

图 14 为赵翠萍等[7]反演得到的 M_S7.9 地震的震源时间函数，可以看出这是一个相对简单的破裂过程，破裂持续约 37s，但上升时间较快，能量主要在前 20s 以内释放。图 15 为断层面上最终滑动量的静态分布，可见破裂主要发生在长 110km、宽 30km 的浅部。在起始破裂点以上的浅部，以走滑错动为主；而在深部及起始破裂点以西，则显示一定的逆冲成分。起始破裂处不是滑动量最大的地方。断层面上有两个显著的、滑动量超过 2.0m 的破裂区；

一个是震中东南的破裂区，最大滑动量约 3.6m，分布比震源深度浅，长约 60km、深 0 ~ 15km；另一个位于震源以西，深 20 ~ 30km，长近 20km，最大滑动量为 2.1m。

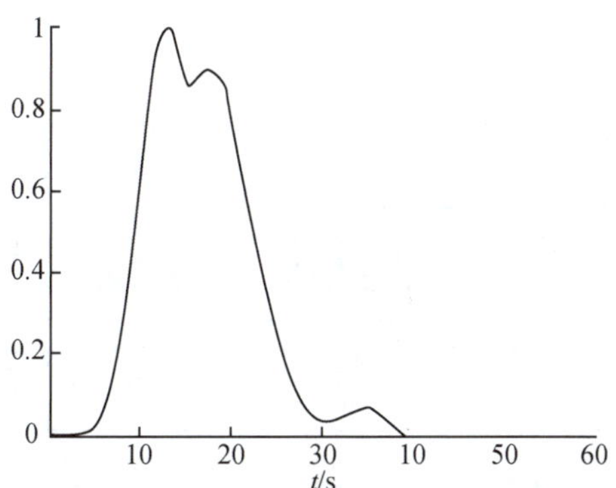

图 14　中、俄、蒙交界 M_S7.9 地震归一化的震源时间函数[7]

Fig. 14　The unitary time function of the epicenter about M_S7.9 earthquake in the boundary of China, Russia and Mongolia

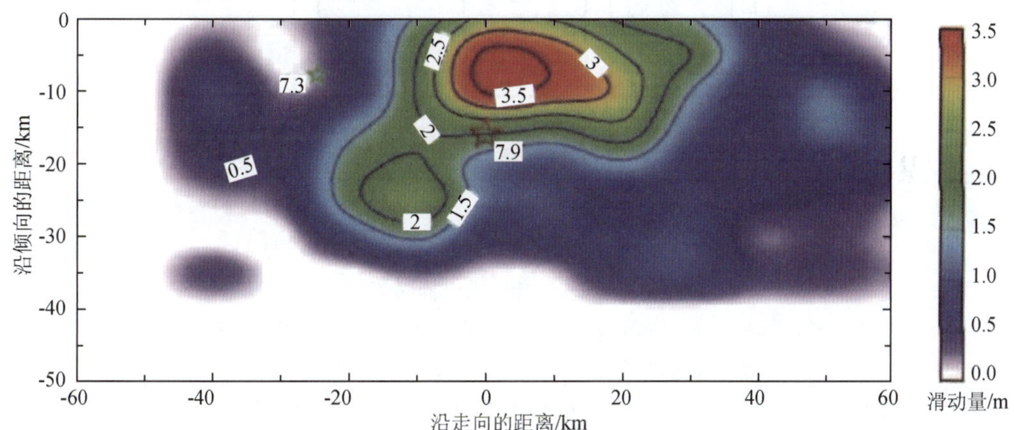

图 15　中、俄、蒙交界 M_S7.9 地震断层面上最终位错的静态分布[7]

Fig. 15　The static state distribution of displacement in the fault plane about M_S7.9 earthquake in the boundary of China, Russia and Mongolia

图 16 为断层面上每秒间隔的累积破裂过程快照[7]，整个破裂过程主要发生在震源东南的浅部及西北的较深部位，破裂自起始破裂点开始，前 7s 首先在其上方浅部发生，形成一个破裂区；此后在其西北 20 ~ 30km 处逐渐形成另一破裂区。前 15s 内破裂主要集中发生在这两个区域，并且浅部区域的破裂更为迅速，滑动量更大，持续时间更长，面积也较大，为主要破裂区域。该破裂在 M_S7.3 强余震发生的部位迅速减小，第 17s 后，在 M_S7.3 强余震位置的西部出现破裂，而在 M_S7.3 地震发生的部位留下一个与主破裂区之间的空白区（箭头所指区域），显示出由于不均匀障碍体的存在，破裂传播出现不连续和停止。第 25s 时，

第一个破裂区的滑动位移已达 3. 6m，滑动量大的区域位于震中东南较浅部位，而震中附近滑动量相对较小。滑动量超过 0. 5m 的区域均位于地下 35km 以内，即震源破裂基本上发生在中上地壳。断层面上的最大应力降达 42MPa，平均应力降为 1. 3MPa。

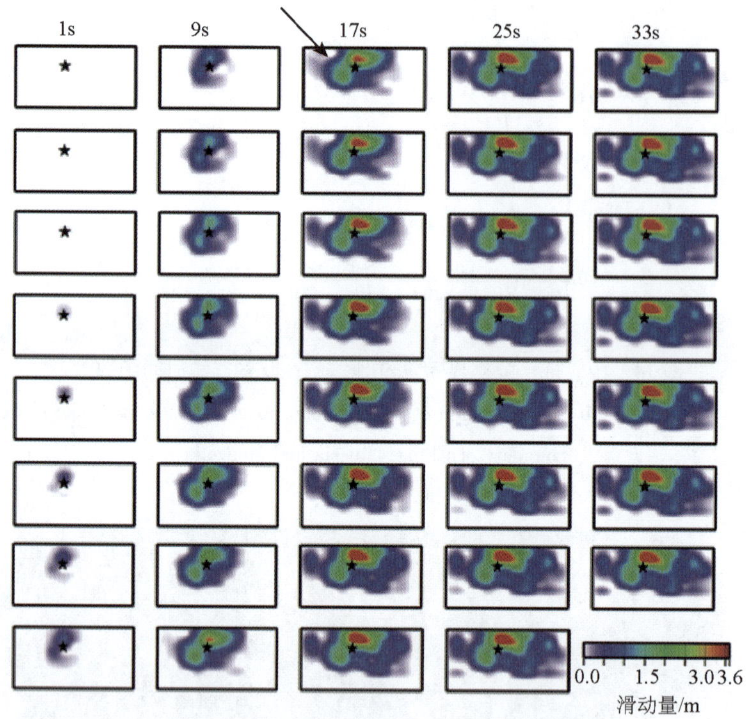

图 16　中、俄、蒙交界 M_S7. 9 地震破裂过程快照

（1 ~ 35s，每秒的断层面坐标同图 8）[7]

Fig. 16　The snapshot of rupture process about the M_S7. 9 earthquake in the boundary of China, Russia and Mongolia

图中每列上方的数字表示每列第一个快照的时刻，箭头所指为空白区的位置

此外，这次 M_S7. 9 地震断层面长度与余震区长度大体相同，但与 1931 年 8 月 11 日富蕴 8. 0 级地震长达 176km 的地表破裂带[8]相比要小得多。

六、地震前兆观测台网及前兆异常

震中附近地区地震台站及观测项目分布如图 1 所示。地震发生在新疆境外，区内台站主要分布在阿勒泰—富蕴地区，均在震中以南地区。

震中 500km 以内仅有 2 个地震台和 1 个遥测台，设有测震和地倾斜前兆观测。其中 201 ~ 300、301 ~ 400、401 ~ 500km 距离内各有 1 个台站，分别为阿勒泰地震台、富蕴地震台和布和图遥测台，其中前两个台设有地倾斜前兆观测。此次地震前共出现 3 个异常项目 7 条异常（表 4）。

表 4 地震前兆异常登记表

Table 4 Summary table of precursory anomalies

序号	异常项目	台站（点）或观测区	分析方法	异常判据及观测误差	震前异常起止时间	震后变化	最大幅度	震中距（km）	异常类别及可靠性	图号	异常特点及备注
1	地震增强活动	阿勒泰地震带	$M_S \geqslant 3.5$ 级		2003.08 ~ 09.16	恢复正常	4 次	震中南部	I_2	17、18	3.5 级以上地震出现明显增强活动
2	地倾斜（竖直摆）	阿勒泰	单分量速率	大速率不稳定变化	2002.05 ~ 2003.07	基本稳定	EW: 8.9″ NS: −29.2″	255	M_2	20	两分量加速形变
3	地倾斜（竖直摆）	阿勒泰	单分量速率	短期速率不稳定	2003.09	逐渐稳定		255	I_2	20	EW 向速率不稳定变化
4	地倾斜（金属摆）	富蕴	单分量速率	加速变化	2000.10 ~ 2002.12	恢复正常	NS: 21.3″ EW: −0.78″	373	M_1	21	NS 向加速南倾 EW 向由 E 倾转为缓慢 W 倾
5	地倾斜（金属摆）	富蕴	合成矢量	合成矢量偏转	1999.01 ~ 2002.12	恢复	模长达 3 ~4 倍	373	M_1	21	矢量方向不稳定，矢量模长达到正常时的 3 ~ 4 倍
6	地倾斜（金属摆）	富蕴	记录曲线	掉格频次多记录格值不稳定	2001.04 ~ 2003.08	基本恢复	格值: EW: 8.7% NS: 9.3%	373	M_2 S_2	22	掉格持续时间长，掉格频次多，掉格幅度大；记录格值不稳定变化

续表

序号	异常项目	台站（点）或观测区	分析方法	异常判据及观测误差	震前异常起止时间	震后变化	最大幅度	震中距（km）	异常类别及可靠性	图号	异常特点及备注
7	地倾斜（金属摆）	富蕴	单分量速率	速率变化不稳定	2003. 07. 25～	仍不正常		373	I_2	22	两分向速率变化不稳定
8	地倾斜（竖直摆）	克拉玛依	单分量速率	停滞变化	2003. 4. 25～	基本恢复	NS：-2. 5″	554	I_1	23	NS 向出现停滞变化，持续了 154 天
9	地倾斜（竖直摆）	克拉玛依	单分量速率	日变形态消失	2003. 8. 24～2003. 9. 4	基本恢复	EW：0. 6″	554	I_1	24	EW 向日变形态消失，观测值出现大幅度的反复突变，最大幅度达 0. 6″

1. 地震学异常

阿勒泰地震带历史上 M_S3. 5 以上地震活动相对较弱，但 2000 年 10 月至 2001 年该区域 3. 5 级以上地震明显增多；2002 年相对平静；7. 9 级地震发生前的 5 月发生 1 次 5. 8 级地震，8 月底至 9 月中旬阿勒泰地震带出现 3. 5 级以上地震活动明显增强，属该地震带的少有现象（图 17 和图 18）。且 3. 5 级地震集中活动是在阿勒泰地震带出现大面积的 4. 2 级以上平静的

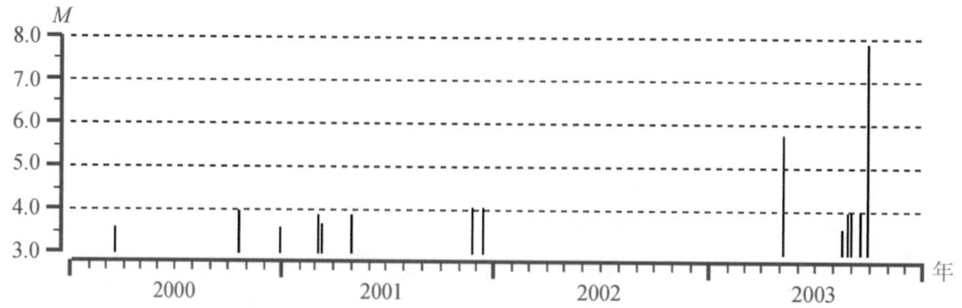

图 17　中、俄、蒙交界 7. 9 级地震前阿勒泰—富蕴地区 3. 5 级以上地震活动 M-t 图

Fig. 17　The M-t map of seismic activity about earthquakes with more than M_S3. 5 in Altay-Fuyun area before M_S7. 9 earthquake in the boundary of China, Russia and Mongolia

背景下发生的，这种平静时间长达近 5 年（图 19）。地震发生之后，该区域 3.5 级以上地震又恢复平静。

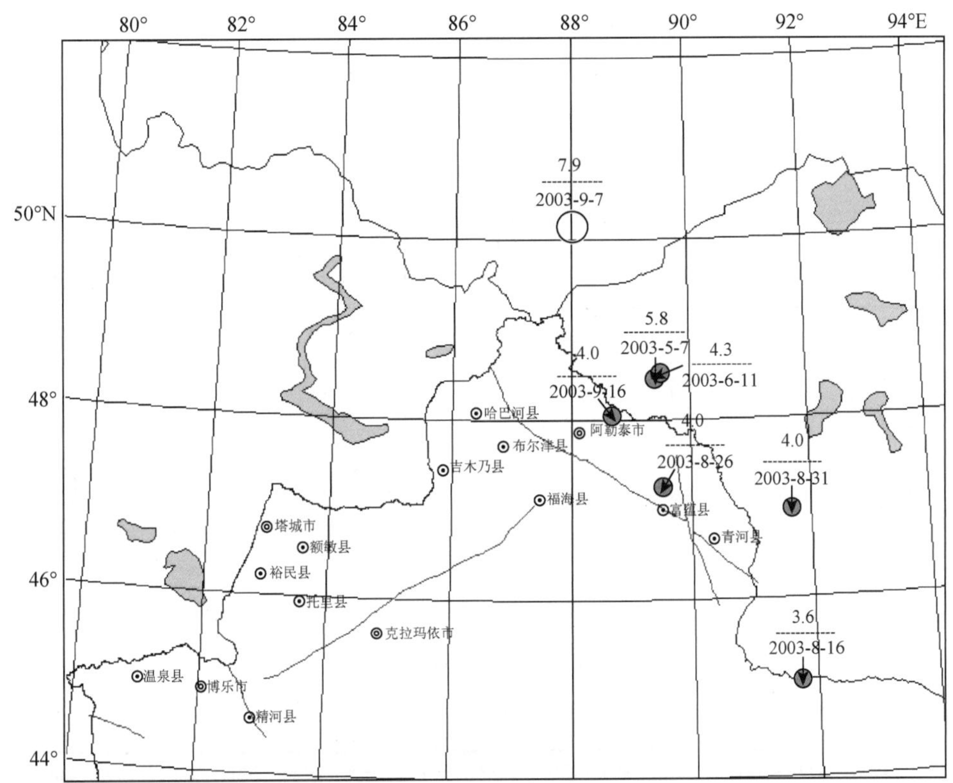

图 18 中、俄、蒙交界 7.9 级地震前阿勒泰—富蕴地区 3.5 级地震活动分布

Fig. 18 The distribution of seismic activity about earthquakes with more than M_S3.5 in Altay-Fuyun area before M_S7.9 earthquake in the boundary of China, Russia and Mongolia

2. 前兆异常[9]

震中 500km 范围内的 2 台项前兆观测项目均出现异常。

阿勒泰台（Δ =255km）从 1993 年开始使用 CZB 竖直摆倾斜仪进行观测，仪器安装在纵深约 100m 的山洞内。该洞顶部覆盖大于 40m，洞内较潮湿，年温差小于 0.5℃。由于近年仪器更换较为频繁，资料连续时间仅为 2～3 年。1996 年 3 月 13 日阿勒泰 M_S6.1 地震（Δ =60km）前曾记录到明显的短期前兆异常。2000 年至 2002 年 12 月 26 日使用同一台仪器观测，资料连续。2002 年 12 月 26 日观测值出现突变，观测人员分析认为可能是仪器故障而停测，2003 年更换了一台精度较低的同类仪器进行观测。

7.9 级地震前阿勒泰台地倾斜异常可分为两个时段来认识[9]。

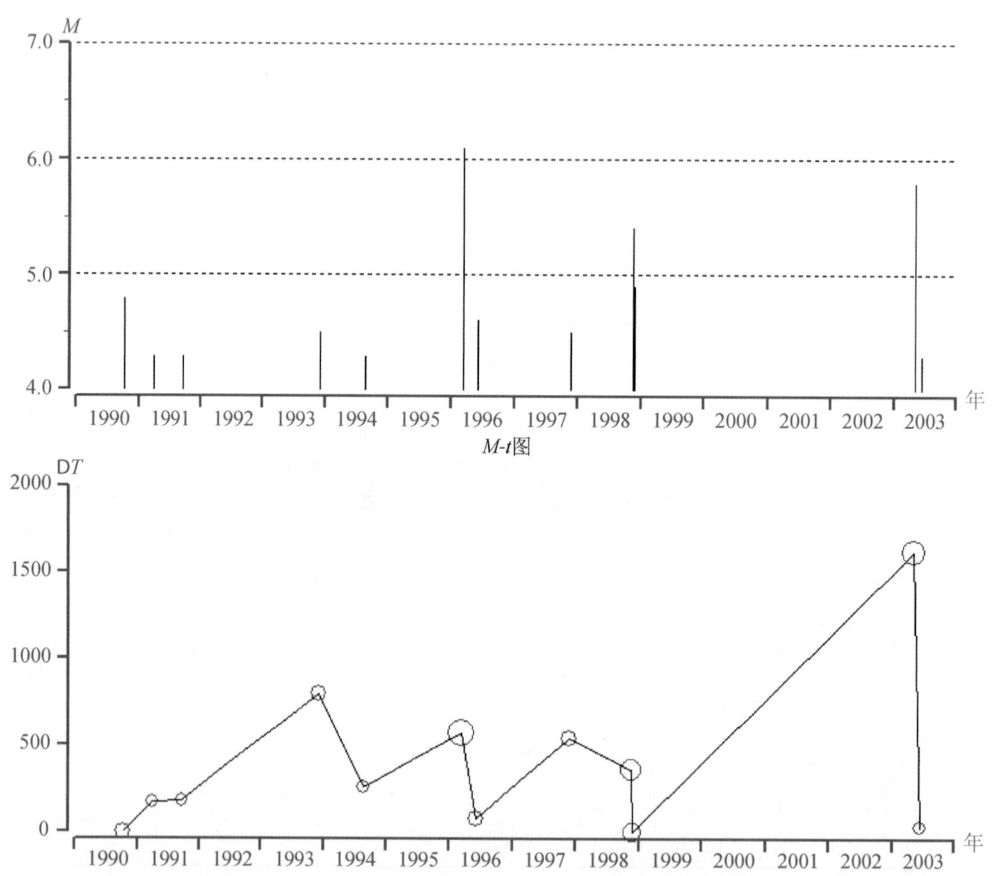

图 19　中、俄、蒙交界 7.9 级地震前阿勒泰地震带 4.2 级以上地震 M-t 和时间间隔图

Fig. 19　The curves of M-t and time interval about earthquakes with more than M_S4.2 in Altay earthquake belt before M_S7.9 earthquake in the boundary of China, Russia and Mongolia

中期异常：加速变形阶段。

2002 年 5 月底两个方向观测值在平稳变化的背景上出现快速的不稳定变化，一直持续到仪器停测。EW 向在 2002 年 5 月 8 ~ 22 日先出现短时间的反向西倾之后突转快速东倾，最大速率达 0.54″/d；NS 向在 2002 年 5 月 22 日由北倾转为 S 倾，速率忽快忽慢，最大速率达到 -0.66″/d。截至 12 月 26 日，EW 向累积变化量为 8.9″，NS 向累积量高达 -29.2″。2003 年更换仪器后，1 ~ 3 月速率平稳，但 4 ~ 7 月又出现加速变化，EW 向变化累积量达 12.3″，NS 向累积量达 13.5″，8 月份以后基本稳定（图 20）。

短期或短临异常：短期速率变化阶段

震前 10 天，EW 向速率在减小的背景下出现短期的不稳定变化，持续到 10 月中旬以后逐渐稳定。

富蕴台（Δ = 373km）使用 JB 金属摆倾斜仪，1976 年开始观测，仪器安装在纵深 18m 的山洞内，顶部覆盖层约 10m，洞内潮湿，年温差约 1℃。1993 年对观测仪器进行改造，光

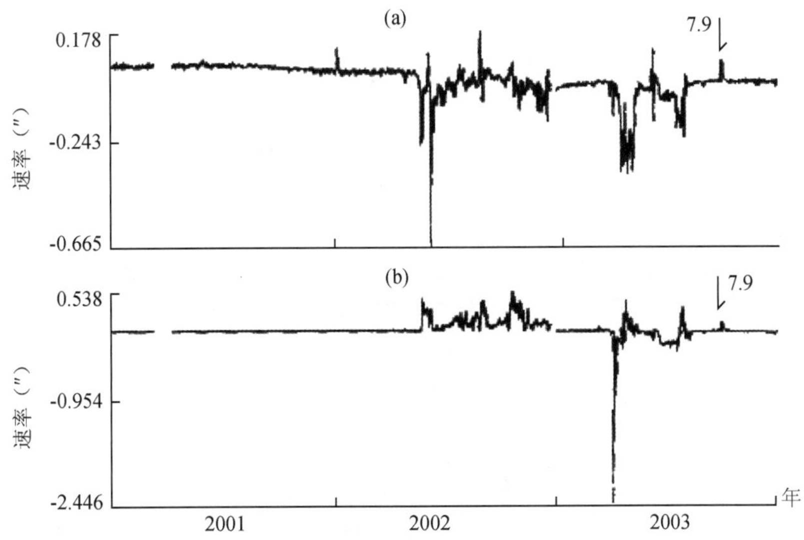

图 20　2000 ~ 2003 年阿勒泰台地倾斜速率变化曲线[9]

Fig. 20　The curve of speed ratio change about the tilt in the Altay station from 2000 to 2003

（a）NS 向；（b）EW 向

杠杆由 1m 延长到 4m，工作周期上升到 20s 左右，提高了观测精度。目前观测资料年变清晰，固体潮日变形态基本规则。在 1986 年 4 月 23 ~ 24 日富蕴 2 次 5 级地震（$\Delta = 50$km）、1990 年 6 月 14 日斋桑 M_S7. 3（$\Delta = 360$km）和 1996 年 3 月 13 日阿勒泰 M_S6. 1（$\Delta = 188$km）地震前曾记录到不同程度的形变异常信息。

7. 9 级地震前富蕴台倾斜仪异常主要表现为以下特征[9]：

中期异常。

（1）加速南倾。

富蕴台地倾斜 NS 向的正常趋势呈南倾，速率为 −3. 54″/a。2000 年 10 月至 2002 年 12 月出现大幅度的加速变化并持续 27 个月，平均速率达 −8. 99″/a，是正常时段的 2. 5 倍，累计变化量达 21. 3″。2003 年初恢复正常，8 个月后发生 M_S7. 9 地震（图 21a）。EW 向的正常趋势为缓慢东倾。M_S7. 9 地震前，EW 向异常变化主要表现为两个阶段：第 1 阶段在 1996 年 3 月 13 日阿勒泰 6. 1 级地震后，EW 向出现大幅度的东倾变化并持续 33 个月（1996 年 7 月至 1999 年 3 月），累计量达 10. 2″；第 2 个阶段从 2001 年 3 月至 2002 年 5 月，EW 向出现反向变化，由东倾转为缓慢西倾并持续 15 个月，累积量 −0. 78″，与 NS 向加速南倾在时间上基本一致。分析认为 EW 向的大幅变化可能与 7. 9 级地震的中期孕震过程有关（图 21b）。

（2）合成矢量出现异常。

富蕴台地倾斜正常合成矢量方向在 S25° ~ 35°E 之间，1999 ~ 2002 年向西偏转了 20°，2001 ~ 2002 年为 S2°E，2003 年又转回 S20°E。M_S7. 9 地震位于富蕴台的 N20°W 方向，矢量方向在震前 4 年出现连续的向 WS 方向的偏转，震前恢复到远离震中的方位。此外，震前 3

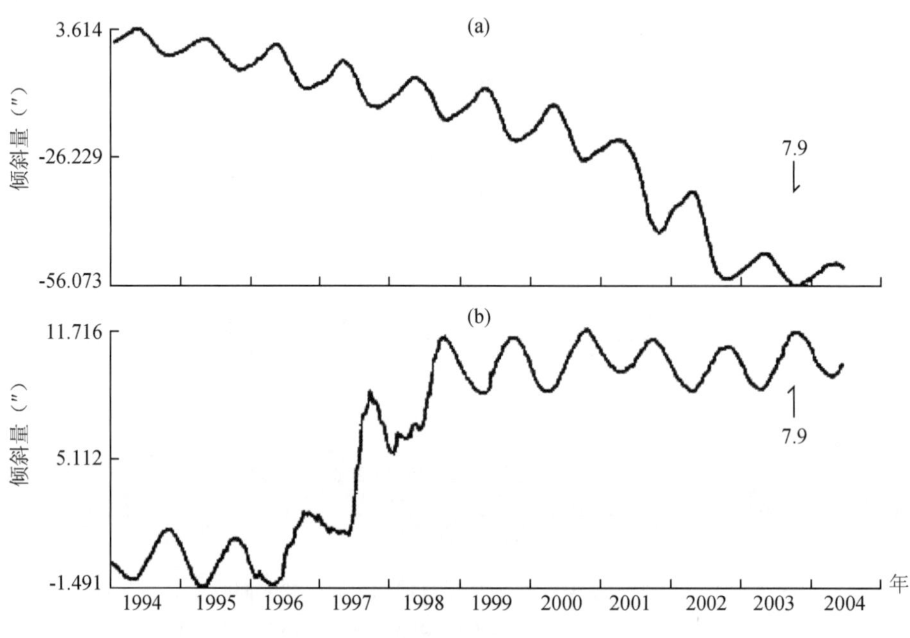

图 21　富蕴台地倾斜时序曲线[9]

Fig. 21　The time sequence curve of tilt in Fuyun station

（a）NS 向；（b）EW 向

年矢量模长达到正常时期的 3～4 倍。

（3）记录曲线出现大幅度掉格图像。

富蕴台地倾斜在 M_S7.9 地震前（2001～2003 年）NS 向记录到大量的掉格图像，而 EW 向却记录不多。1999～2004 年 NS 向地倾斜日掉格频次、日累计幅度和日最大幅度表 5 和图 13 所示。7.9 级地震前掉格的特征是：①掉格持续时间长达 3 年，高峰期为每年的 5～10 月，2001 年持续时间较长（从 4 月到次年的 1 月），直到 M_S7.9 地震后恢复正常（图 22a）；②掉格年频次是正常年份的 8～9 倍，2001、2002 年累计掉格频次在 1350 次左右，2003 年减少至 997 次。分析日频次发现，2001、2002 年的最高日频次分别为 36 次/日和 37 次/日，而 2003 年的 7～8 月有 3 天超过 40 次，其中 8 月 9 日达到 60 次，表现为震前 2 个月掉格频次增强；③掉格幅度明显增大，以每年累计掉格 450mm 为正常值，则 2001 年增大了 13 倍，2002 年增大 9 倍，而 2003 年增大 3 倍，震后均快速恢复（图 22b）。由日掉格的最大幅度分析可见，2001～2003 年掉格明显减弱，即临近地震前掉格幅度减小（图 22c）；④掉格日频次与幅度无关，有时 1 天掉格几十次，每次 1～2mm，有时仅 1 次，幅度达几十毫米，如 2001 年 7 月 31 日单次掉格幅度达 52mm；⑤以向南掉格为主，极少数向北，与同期地倾斜的大幅度南倾相一致。

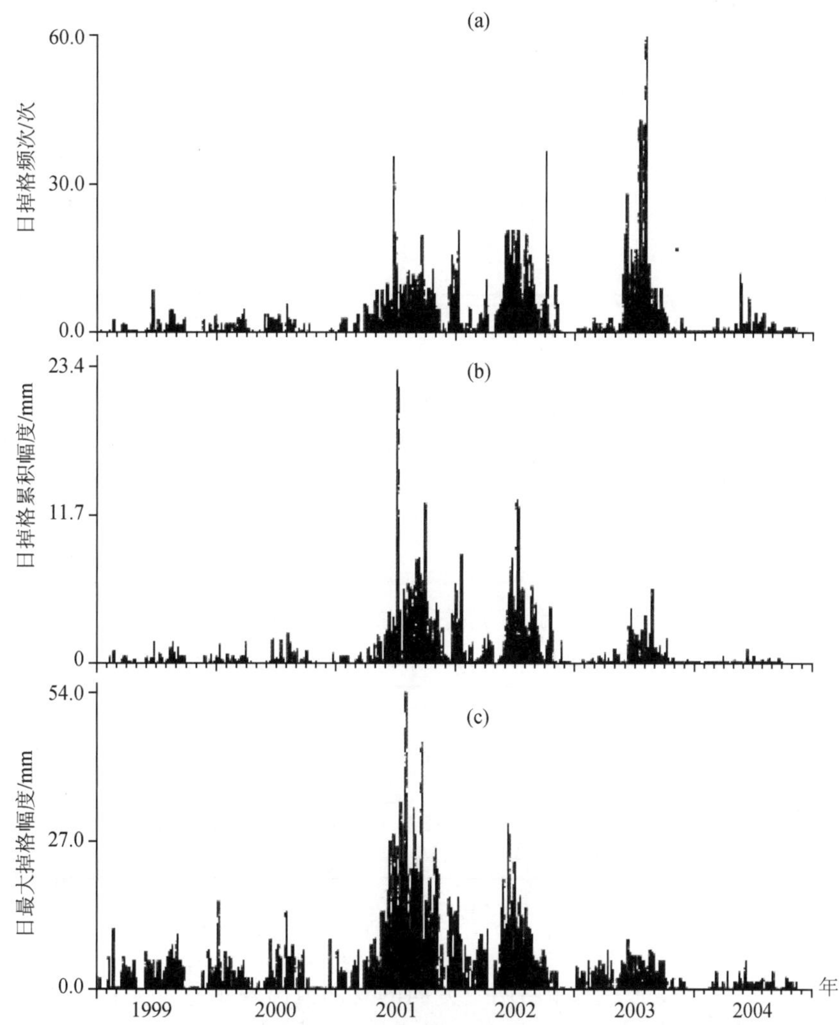

图 22 富蕴台地倾斜 NS 向日掉格频次、最大掉格幅度统计曲线[9]

Fig. 22 The statistic curves of day-frequency and the biggest scope about lose-grid value in NS of tilt about Fuyun station

（a）频次；（b）累积幅度；（c）最大掉格幅度

表 5 富蕴台 NS 向记录曲线掉格统计

Table 5 The statistics of lose-grid recode in NS of tilt about Fuyun station

年份	平均日频次（次/日）	日最大幅度（mm）	最高日频次（次/日）	年累积次数（次/年）	年累积幅度（mm）
1999	1.9	11	9	165	444
2000	2.0	16	6	173	448

续表

年份	平均日频次 (次/日)	日最大幅度 (mm)	最高日频次 (次/日)	年累积次数 (次/年)	年累积幅度 (mm)
2001	5.9	54	36	1322	5817
2002	6.8	30	37	1395	4079
2003	5.2	9	60	997	1298
2004	2.1	5	12	132	137

(4) 记录格值不稳定。

2001～2003 年伴随着大幅度的南倾和大量掉格现象，富蕴台地倾斜的记录格值出现不稳定变化。表 6 列出 2000～2003 年格值的年均值和中误差。可以看到 2002 年的误差最大，2003 年明显减小。地震前后格值的变化更加显著，EW 向和 NS 向分别相差了 8.7% 和 9.3%。

表 6　富蕴台格值统计表

Table 6　The statistics of lose-grid value about Fuyun station

分量	2001 年 (″/mm)	误差 (%)	2002 年 (″/mm)	误差 (%)	2003 年 (″/mm)	误差 (%)
EW 向	4.035	7.58	3.994	9.17	3.990	5.31
NS 向	4.067	8.16	4.236	7.61	4.117	2.94

短期或短临异常：短期不稳定变化。

2003 年 7 月 25 日至震前，地倾斜两个方向的观测值出现不稳定变化，速率时大时小，在矢量曲线上表现为小的打结过程。仪器调零后光点很难稳定，这种现象一直持续到地震之后。

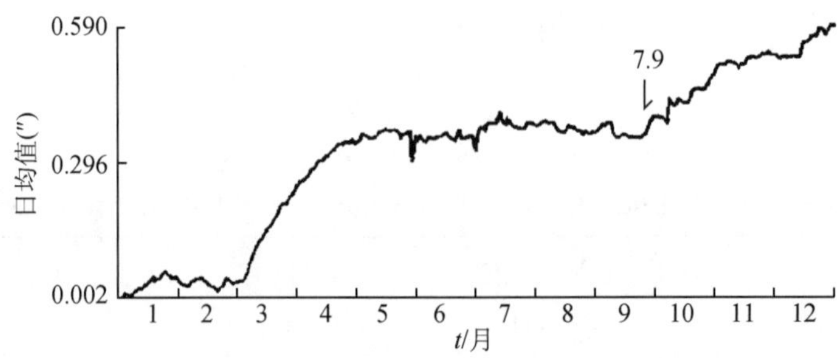

图 23　2003 年克拉玛依台地倾斜 NS 向日均值曲线[9]

Fig. 23　The curve of day-mean value in NS of tilt about Kelamayi station in 2003

此外，距 7.9 级地震 500km 之外的克拉玛依台地倾斜也出现明显的异常显示。克拉玛依台（Δ=554km）从 2002 年开始地倾斜观测，使用的是 CZB 井下竖直摆倾斜仪。观测井深 80m，两分向观测值无年变，呈线性上升趋势[9]。

自 2003 年 4 月 25 日开始，NS 向出现停滞变化，持续 154 天，M_S7.9 地震后恢复正常，累计异常量 -2.5″（图 23），具有中短期异常特征。

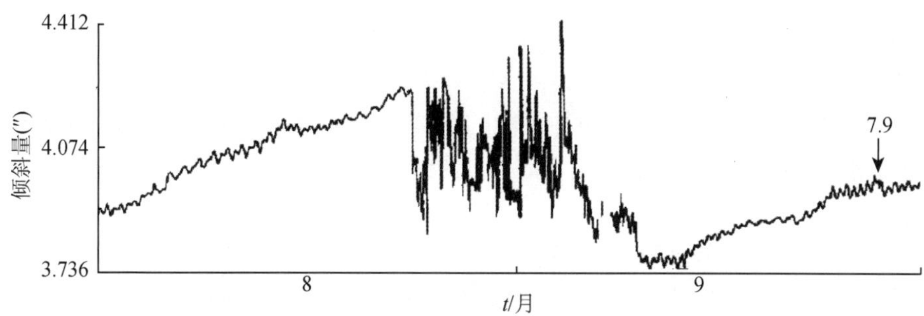

图 24 2003 年 8～9 月克拉玛依台地倾斜 EW 向整点值曲线[9]

Fig. 24 The curve of hour value in EW of tilt about Kelamayi station from Aug. to sept. in 2003

2003 年 8 月 24 日开始，EW 向日变形态消失，观测值出现大幅度的反复突变，最大幅度达 0.6″，直到 9 月 4 日以后减缓，9 月 9 日仪器调零后恢复正常（图 24）。因为同期 EW 向的输出电压在 -0.16～-0.43V 之间，属正常观测范围，NS 向记录也正常，因而判断仪器工作正常，上述突变具有短临异常特征。

除以上 3 个台外，震前 10 天前后 850km 范围的其他 5 个台站的地倾斜也记录到打结、加速等短临异常变化现象。其中石场台（Δ=715km）在 9 月 14～26 日地倾斜的矢量曲线连续打结，震后恢复正常；精河台（Δ=722km）2 套倾斜仪（FSQ、SQ-70）的 EW 向在 9 月 24～26 日速率加大；温泉台（Δ=780km）9 月 18～22 日矢量曲线打结；新源台（Δ=832km）9 月 5～25 日 NS 向速率停滞。

七、地震前兆异常特征分析[9]

（1）中、俄、蒙交界 M_S7.9 地震前多台地倾斜记录到明显的渐进式形变异常过程。趋势异常从震前 3 年开始，首先表现在富蕴台 NS 向出现大幅度的南倾、曲线掉格和记录格值的变化；阿勒泰台地倾斜在震前 16 个月出现大幅度速率变化，震前 6 个月又发生一组大速率变化；之后是富蕴台的速率不稳定变化和克拉玛依台 NS 向的速率停滞异常；震前 1 个月克拉玛依台的 EW 向出现日变畸变；震前 10 天 850km 范围内有 6 个台出现了程度不同的短临异常，其中最显著的是石场台倾斜仪连续的矢量打结现象。

（2）就新疆境内有限的资料来看，中期异常主要集中表现在位于阿勒泰地震带上的两个台站的地倾斜上；中短期异常扩展到 600km 范围的克拉玛依台，850km 范围内的 6 台倾斜仪也记录到较明显的临震异常。

(3) 异常具有方向性。趋势异常主要表现在 NS 向，位于阿勒泰地震带的 2 台倾斜仪均为大幅度南倾，富蕴台 EW 向变化很小，阿勒泰台 NS 向变化量是 EW 向的约 3.3 倍，但短期、短临异常在 EW 向表现得较为突出。

(4) 富蕴台地倾斜持续 3 年的记录曲线掉格现象是本次地震观测到的最特殊的形变异常，也是新疆地倾斜从未观测到的、持续时间最长的记录图像异常。

(5) 中、俄、蒙交界 M_S7.9 地震前 500km 范围的 2 台地倾斜的 NS 向分别记录到 29″和 21″的异常量。尽管是两台不同的观测仪器，但异常的量级与震中距呈现近台大远台小的特征。地震之后资料均恢复正常，分析认为异常量级是可靠的。

(6) 富蕴台地倾斜掉格异常有两种可能的解释。一是可能与观测仪器的性能有关。5～9 月是 NS 向南倾加速时段，每年的这一时段记录曲线上都会叠加一些掉格图像，可能是倾斜仪吊丝对快速倾斜变化的不连续响应，因此当南倾速率成倍增大时，这种不连续响应也随之加剧；二是在多次强震前倾斜仪的记录曲线出现畸变、掉格等图像，一般是震前几天，最长出现在震前 5 个月。文献 [9] 中，笔者认为富蕴台地倾斜的掉格异常是第二种解释的可能性较大。因为这种现象可解释为是主破裂前震源区微破裂的扩展、预滑等结果。以此思路分析认为，掉格现象可能与震源区外围的某些薄弱部位的缓慢破裂过程有关，随着局部区域应力的释放，破裂逐渐减弱。强大的主压应力在震源区集中，直至主破裂发生后，外围区缓慢破裂过程趋于消失。

八、震前预测、预防和震后响应

1. 震前预测和震后趋势判定

中、俄、蒙交界 7.9 级地震发生在新疆监测能力弱的区域，未做出明确的长、中、短、临预测，仅在 2003 年度新疆地震趋势预测中，将阿勒泰—富蕴地区列为中强震值得注意地区[3)]。

地震发生后，新疆维吾尔自治区地震局立即派现场工作组赶赴现场，同时进行现场地震监测预报工作。后经分析并结合该区历史地震活动情况，判定中、俄、蒙交界 7.9 级地震序列为双震型地震。

2. 震害

中、俄、蒙交界 7.9 级地震发生后，新疆维吾尔自治区地震局派出工作组赴灾区开展震害评估工作。此次地震发生在中国、俄罗斯、蒙古交界，距我国边境约 110km，但由于地震震级大，震害波及我国的范围为阿勒泰地区六县一市，面积达 11 万平方公里，在调查过程中发现一些Ⅵ度点，但无法圈定Ⅵ度圈。

7.9 级地震造成直接经济损失 7635.6 万元。

本次地震仅造成少量房屋倒塌。有些地方多层砖混结构房屋比平房震害严重，可能是由于震中较远，长周期地震波作用所致。在砂土、地下水位高的盐碱地区，房屋破坏较严重。本次地震发生时，牧民冬季转场已到山前，因而未造成人员伤亡。但是山区牧民的学校、牧办、牧道、桥梁等受到一定程度的破坏。

九、结论与讨论

（1）中、俄、蒙交界7.9级地震发生在新疆阿勒泰地区北缘的境外地区，距新疆边境大约110km。地震位于俄罗斯NW向乌列盖断裂构造带北侧。研究区域范围内前兆异常主要是以地倾斜为主的长、中、短、临异常，中期异常主要是位于阿勒泰地震带上的两个台站的地倾斜异常，中短期异常扩展到600km范围的克拉玛依台，850km范围内的6台倾斜仪也记录到较明显的临震异常。趋势异常主要表现NS向变化，位于阿勒泰地震带的2台倾斜仪均为大幅度南倾，富蕴台EW向变化很小，阿勒泰台NS向变化量是EW向的3.3倍。短期、短临异常在EW向表现得较为突出。富蕴台地倾斜持续3年的记录曲线掉格现象是本次地震观测到的最特殊的形变异常，也是新疆地倾斜从未观测到的、持续时间最长的记录图像异常。尽管是两台不同的观测仪器，但异常的量级与震中距呈现近台大远台小的特征[9]。

（2）据有关研究[7]，此次地震是一次个相对简单的震源破裂过程，整个破裂过程持续37s，但上升时间较快，能量主要在前20s以内释放，释放标量地震矩0.97×10^{20}N·m。破裂主要发生在长110km，宽30km的地壳范围内，最大位错3.6m。断层面上显示出两个显著的、滑动量超过2.0m的破裂区。

（3）由于发生在境外，国内针对研究区的监测能力较弱。尽管2002年6月阿勒泰倾斜仪出现大幅度变化后新疆地震局即开始对阿勒泰地区地震活动给予关注，但并未对该地震做出任何预报。

参考文献

[1] 尹光华、蒋靖祥、沈军，2003年阿尔泰强烈地震综述，内陆地震，20（2），183～192，2006

[2] 洪顺英、申旭辉、赖木收等，阿尔泰山东缘主要活动断裂影像特征分析，地震地质，28（1），119～128，2006

[3] 沈军、李莹甄、汪一鹏、宋方敏，阿尔泰山活动断裂，地学前缘，10（特刊），131～140，2003

[4] 柔洁、宋和平、娄少平、刘志坚，俄、蒙、中交界7.9级地震，国际地震动态，304（4），16～20，2004

[5] Rogozhin E A, Ovsyuchenko A N, Geodakov A R, et al., A strong earthquake of 2003 in Gomyi Altai, Russian Journal of Earth Sciences, 2003, 5 (6), 439-454

[6] 张洪由、许力生，中、俄边界连续发生强烈地震，国际地震动态，10，37～38，2003

[7] 赵翠萍、陈章立、郑斯华、刘杰，2003年9月27日中、俄、蒙边界$M_S7.9$地震震源机制及破裂过程研究，地震学报，27（3），237～249，2005

[8] 新疆维吾尔自治区地震局，富蕴地震断裂带，北京：地震出版社，1～200，1985

[9] 杨又陵、刘建民、于克滋、杜新民，2003年俄、中、蒙交界$M7.9$地震前地倾斜异常，内陆地震，19（2），97～104，2005

[10] 布赖特，岩石破坏前倾斜和地震活动性异常的实验研究，见：国家地震局科技监测司编，地震理论与实验译文集，北京：地震出版社，215～219，1979

[11] 许昭永、杨润海、赵晋明等，岩石破坏前的短临应变前兆研究，地震研究，24（3），191～196，2001

[12] Bonafede M, Boschi E, Dragoni M，缓慢震源过程的力学模式，［美］金森博、［意］博斯基主编，地

震：观测、理论和解释，柳百琪、周冉等译，北京：地震出版社，1992

参 考 资 料

1）新疆维吾尔自治区地震局，2003 年 9 月 27 日新疆阿勒泰北俄罗斯境内 7.9 级地震现场科学考察报告，卷宗 6889，2003. 11

2）新疆维吾尔自治区地震局，新疆地震目录

3）新疆维吾尔自治区地震局，2003 年度地震趋势研究报告，2002. 11

M_S 7. 9 Earthquake on September 27, 2003 in the Boundary of China, Russia and Mongolia

Abstract

An earthquake of M_S7. 9 occurred in the realm of Russia on Sept. 27, 2003. And the epicenter of the earthquake located at the north of Altay area in Xinjiang Uygur Autonomous Region. There was about 110 km away from the bound of Xinjiang Uygur Autonomous Region. Because the epicenter of the earthquake located in the periphery of Xinjiang Uygur Autonomous Region, it was far from collective habitation in China. So no person was injured and no person was killed. Because the earthquake was strong, it spread to very much area. However, the injured building was few. And the downfall building was better few and so on. In the latest area in China away from the epicenter, the traffic was very discommodious and the communication was block. And it belonged to very chill and height area. The work team could not reach there. And they could not investigate in that area. The economic loss caused by the earthquake was 76. 356 million Yuan RMB.

The earthquake sequence was double-event type. The biggest aftershock was M_S7. 3. The distribution of aftershocks located the two sides of ES—NW in main earthquake. The earthquake was strike fault with dextrorotation. The strike of node plane I was the main rupture plane of the earthquake. The azimuth of P axis was near NS. The rupture was simple. The last time of rupture was 37 s. The biggest displacement was 3. 6 m. The rupture plane appeared two distinct rupture areas with more than 2. 0 m of displacement. The seismogenic structure was NW fault of Uliegai in Russa. It was accordant with the distribution of aftershocks.

The observation ability of the epicenter and its adjacent area was very weak. There were 2 seismic stations and 1 remote sensing station within 500 km from the epicenter. There were 2 stations with tilt precursory observation. There were 7 items of 3 anomalies before the earthquake. And the earthquakes with more than M_S3. 5 appeared concentrated activity in Altay area. Others belonged to tilt anomalies.

M_S7. 9 earthquake occurred in outside of Altay area in Xinjiang Uygur Autonomous Region. Before the earthquake, the Earthquake Administration of Xinjiang Uygur Autonomous Region did not make a prediction.

2003年10月25日甘肃省民乐—山丹6.1级地震

甘肃省地震局

梅秀苹　郑卫平　刘小凤　杨立明

摘　要

2003年10月25日甘肃省民乐县和山丹县交界处相继发生6.1和5.8级地震。宏观震中位于民乐县永固镇姚寨子、元圈子和山丹县霍城镇上河西寨子、刘庄等地，极震区烈度为Ⅷ度，地震烈度呈不规则椭圆形分布，总体走向北西向。地震造成10人死亡（其中直接死亡8人），受伤46人，其中重伤14人，轻伤32人，直接经济损失逾5亿元。

此次地震序列为双主震—余震型。余震全部分布在主震西侧约35km范围内，与极震区长轴方向较为一致，节面Ⅱ为主破裂面，主压应力轴方位北东向。地震的发震构造是民乐—永昌隐伏断裂与童子坝隐伏断裂。

震中300km范围内共有17个地震观测台，有7个测震观测项目，2个综合观测台，7个前兆观测项目。震前共出现17项异常，其中7项前兆观测异常，10项测震学异常。地震学异常较为明显。

民乐—山丹6.1、5.8级地震发生在甘肃地震局2003年度地震危险区边缘，地震强度预测较为准确，震前甘肃地震局向中国地震局提出了3个月的短期预测意见，短期预测较为准确。

前　言

2003年10月25日20时41分和48分，甘肃省民乐县和山丹县交界处相继发生6.1和5.8级地震。中国地震台网测定的微观震中为38.4°N、101.2°E和38.4°N、101.1°E，甘肃省地震局台网测定的微观震中为38.3°N、101.0°E和38.32°N、100.97°E，宏观震中位于民乐县永固镇姚寨子、元圈子和山丹县霍城镇上河西寨子、刘庄等地，极震区烈度为Ⅷ度。这次地震的波及面和有感范围很大，张掖市、武威市、酒泉市、金昌市、青海省祁连县、门源县等地强烈有感，兰州市、西宁市、银川市普遍有感。地震造成10人死亡（其中直接死亡8人），受伤46人（其中重伤14人，轻伤32人），直接经济损失逾5亿元。

民乐—山丹6.1、5.8级地震发生在河西走廊过渡带中部地区的民乐盆地与大黄山隆起的交汇部位，是甘肃主要强震活动区。在震中以北约67km处，曾发生过1954年山丹$7\frac{1}{4}$级地震，在震中东南约120km处，曾发生过1927年古浪8级大震。震中300 km范围内共有17个地震台，震前出现13个异常项目，22条异常。民乐—山丹6.1级地震发生在甘肃省地震局确定的2003年度地震重点危险区的边缘，震前甘肃省地震局向中国地震局和甘肃省政府提出了3个月的短期预测意见。

在震前预报意见的指导下，在河西地区的安西、玉门、张掖、民乐和河东地区的通渭、礼县、漳县架设了7台数字流动强震仪器，其中在极震区民乐县架设的流动强震台捕捉到了主震三分向记录。在祁连山中西段的金佛寺、黄泥沟、下河清、玉门市架设了4台数字微震流动仪，进行加密流动观测，完整的记录到了本次地震及余震。震后在现场布设了ZD9A大地电场观测仪，为强余震跟踪和预测预报提供技术支持。

地震发生后，甘肃省地震局立即启动了地震应急预案，成立了应急指挥部，于21时30分派出第一批地震现场工作队赶赴灾区，开展灾害损失评估、地震科学考察、现场震情分析、震情监测、强震观测等工作。中国地震局迅速派出了28人的地震现场工作队，携带通讯设备、流动台网等到达地震现场，与甘肃省地震局现场工作组汇合，成立了现场工作指挥部，开展了地震监测预报、灾害调查、震灾评估、科学考察、强震观测、震后趋势判定等工作，共发送“现场工作简讯”18期，“震情简报”3期，为政府组织开展抗震救灾工作提供了科学、详实的信息和依据，为灾区社会稳定发挥了积极作用，取得了显著的社会效益。

本研究报告是在有关文献和资料的基础上[1～11];1～5)经过重新整理和分析研究完成的。

一、测震台网及地震基本参数

图1给出了震中300km范围内的地震台分布，在研究时段内基本可达到$M\geqslant2.5$级不遗漏。100km内有山丹、河西堡和祁连测震台；100～200km分别有肃南、红崖山、石岗和湟源测震台，200～300km分别有景泰和永登测震台。震前甘肃省地震局架设了多台强震仪，其中的一台就架设在极震区民乐县永固乡，从而获得了极震区6.1级和5.8级三分向的强震记录，为地震工程研究及工程抗震设防，积累了资料。流动台网架设情况如图4。

表1列出了不同来源给出的民乐—山丹地震基本参数。因中国地震台网中心编制的正式地震目录中以中国地震台网测定的结果为准，因此把此结果列为第一位。但从等震线（图3）和余震序列的定位结果来看（图5），不管是宏观震中还是微观震中都与甘肃地震台网所测定的震中位置更为接近。

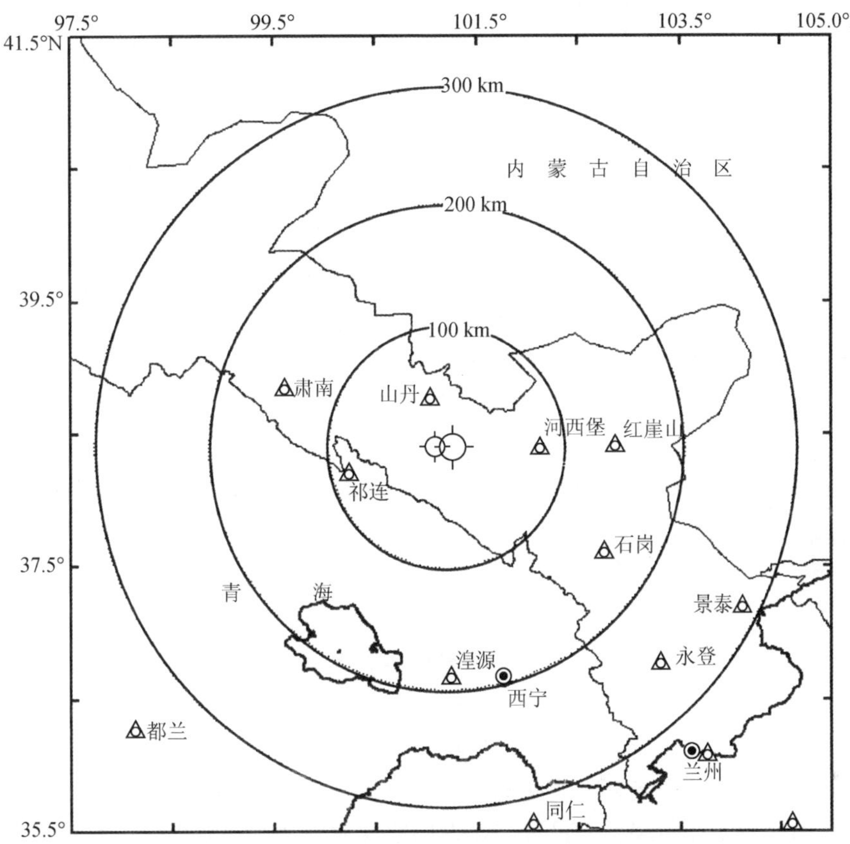

图 1 民乐—山丹 6.1 级地震前震中附近地震台站

Fig. 1 Distribution of earthquake-monitoring stations around the epicentral area before the M6.1 Minle-shandan earthquake

表 1 民乐—山丹 6.1 级地震基本参数

Table 1 Basic psrameters of the M6.1 Minle-shandan earthquake

编号	发震日期	发震时刻	震中位置 (°)		震级				震源深度 h/km	震中地名	结果来源
	年 月 日	时 分 秒	φ_N	λ_E	$M(M_S)$	M_L	M_b	M_W			
1	2003 10 25	20 41 36	38.4	101.2	6.1				33	甘肃省民乐、山丹	中国地震台网
	2003 10 25	20 48 03	38.4	101.1	5.8				15	甘肃省民乐、山丹	中国地震台网

续表

编号	发震日期	发震时刻	震中位置(°)		震级				震源深度 h/km	震中地名	结果来源
	年 月 日	时 分 秒	φ_N	λ_E	$M(M_S)$	M_L	M_b	M_W			
2	2003 10 25	20 41 36	38.3	101.0	6.0				12	甘肃省民乐、山丹	甘肃地震台网
	2003 10 25	20 47 59	38.32	100.97	5.8				10	甘肃省民乐、山丹	甘肃地震台网
3	2003 10 25	12 41 35	38.4	100.95	5.8		5.7	5.8	10	GANSU, CHINA	美国哈佛大学(HRV)*
	2003 10 25	12 47 59	38.38	100.97	5.7		5.5	5.8	10	GANSU, CHINA	美国哈佛大学(HRV)*

注:* 协调世界时(UTC)。

二、地震地质背景

民乐—山丹双震极震区所在的大地构造位置处在祁连山褶皱系之北的河西走廊坳陷盆地带内，在区域构造上被河西走廊南北两侧的深大断裂所围限(图 2)[1]。河西走廊盆地坳陷带南界为祁连山北缘逆冲断裂系，具有向盆地内部逆冲的特点；北界为以龙首山南缘断裂带为主的走廊北缘断裂系，也多具向盆地内部挤压逆冲的特点[2]。祁连山—河西走廊地区的新构造活动是在青藏高原大幅度整体抬升的背景下进行的，其运动方式以断块运动和拱曲运动为主，表现出继承性、新生性和间歇性活动的特点[1]。现代垂直形变测量表明，区内主要山系仍以上升运动为主，盆地仍以下降运动为主，说明了新构造运动的继承性。然而该区的新构造运动还表现出一定的新生性。在祁连山—河西走廊地区，许多规模较大的断裂晚第三纪—第四纪早期主要为逆倾滑性质，但自中更新世以后运动性质发生了改变，区内北西西、北北西和北东东向大断裂多被水平走滑所代替，形成多条大型走滑断裂带，如祁连山北缘的海原—祁连断裂带、昌马断裂带等。极震区还发育有两条隐伏活动断裂带，分别为民乐—永昌隐伏断裂和童子坝河隐伏断裂(图 2)[2]。

本次地震发生在河西走廊过渡带中部地区的民乐盆地与大黄山隆起的交汇部位，同时也是走廊过渡带与北北西隆起区的构造复合部位[3,4]，主要断裂是走廊过渡带南北两侧的边界断裂、发育在过渡带内的近东西向断裂和大黄山隆起区西端的一些边界断裂，这些断裂部分在晚更新世有过活动，部分为全新世活动断裂。综合等震线展布特征、震源机制解、余震分布特征和现场震害考察的结果，分析认为，民乐—永昌隐伏断裂与童子坝隐伏断裂是本次民乐—山丹 6.1、5.8 级地震的孕震和发震构造1)。

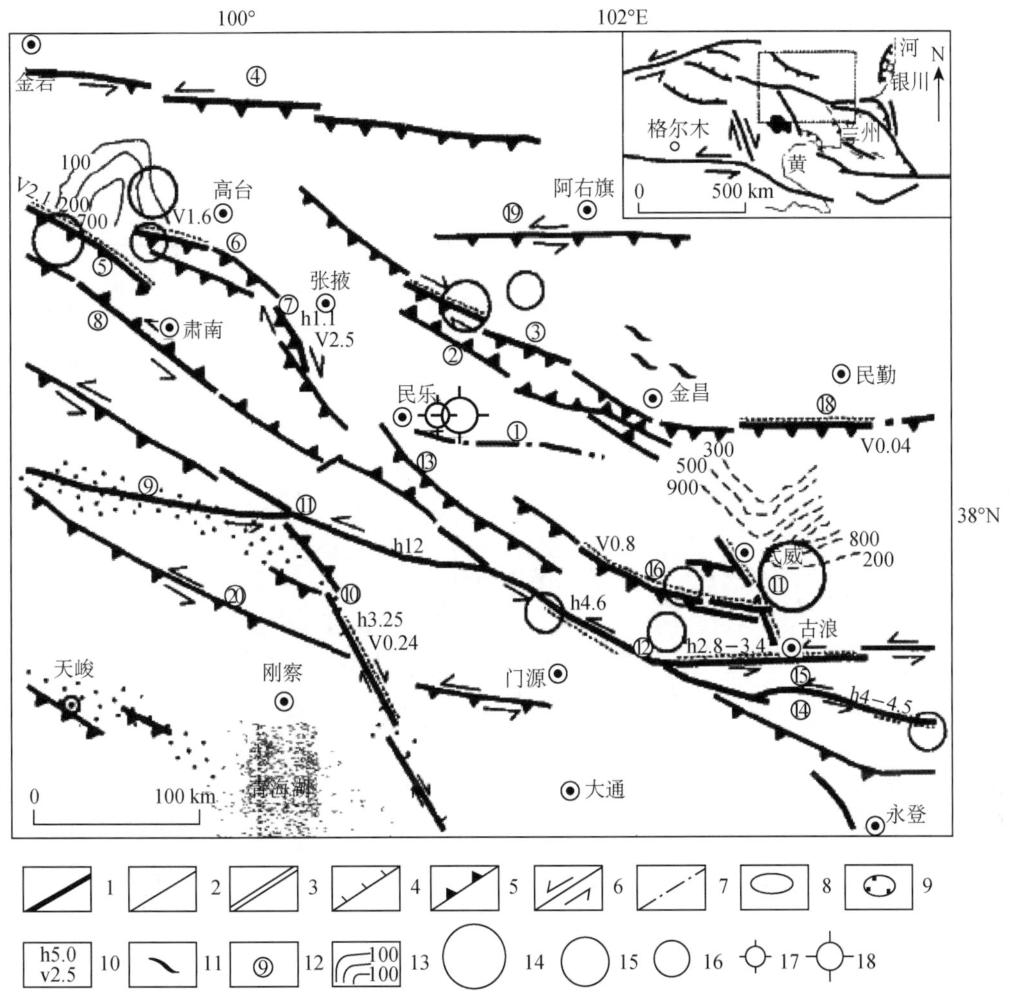

图 2 民乐—山丹附近地区主要断裂及历史地震震中分布图[1]

Fig. 2 Major faults and distribution of historical earthquakes around Minle-shandan area

图例说明：1. 全新世断裂；2. 晚更新世断裂；3. 地震破裂带；4. 正断层；5. 逆断层；6. 走滑断层；7. 隐伏断层；8. 第三纪盆地；9. 第四纪盆地；10. 水平和垂直滑动速率；11. 活动褶皱；12. 断层编号；13. 第四纪等厚线；14. $M\geqslant8.0$；15. $7.0\leqslant M\leqslant7.9$；16. $6.0\leqslant M\leqslant6.9$；17. 5.8 级地震震中位置；18. 6.1 级地震震中位置

①民乐—永昌隐伏断裂；②龙首山南缘断裂；③龙首山北缘断裂；④北大山断裂；⑤佛洞庙—红崖子断裂；⑥榆木山北缘断裂；⑦榆木山东缘断裂；⑧肃南—祁连断裂；⑨木里—江仓断裂；⑩热水—日月河断裂；⑪托莱山断裂；⑫冷龙岭断裂；⑬祁连山北缘断裂；⑭老虎山—毛毛山断裂；⑮天桥沟—黄羊川断裂；⑯皇城—双塔断裂；⑰武威—天祝隐伏断裂；⑱河西堡—四道山断裂；⑲桃花拉山—阿右旗断裂；⑳大通山—大坂山断裂

三、地震影响场和震害

现场考察表明，宏观震中大致在永固镇、霍城镇一带。震中区烈度Ⅷ度，地震烈度呈不规则椭圆形分布，总体走向北西（图3）。

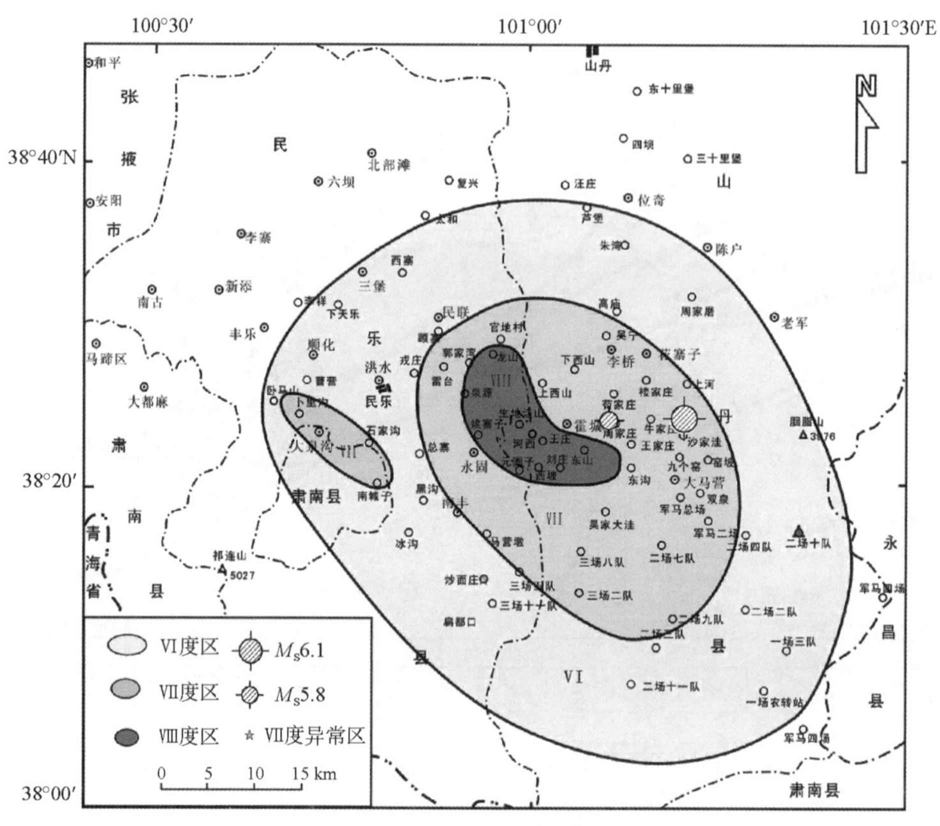

图3　甘肃民乐—山丹6.1、5.8级地震等震线图1)

Fig. 3　Isoseismal map of the *M*6.1 Minle-shandan earthquake1)

Ⅷ度区：长轴总体走向300°，长轴长18km，短轴长6.8km，呈弯豆状，北起龙山村，向南经翟寨子、姚寨子转向东，过生地洼山、河西村、王庄至刘庄以东，面积约96km²。Ⅷ度区内以土木结构房屋为主，这类房屋抗震性能差，震后毁坏、开裂现象十分普遍；简易棚圈、土坯围墙和砖柱大门毁坏；砖木结构房屋较少，80%左右出现墙体裂缝，部分房角塌落，出现结构性破坏。地面出现裂缝、塌陷和山体滑塌。

Ⅶ度区：长轴总体走向303°，长轴长39km，短轴长25km，西起民乐县民联乡的郭家湾村，东到兰州军区军马场一场四队附近，面积约765km²。Ⅶ度区内多数土木结构房屋出现裂缝，简易棚圈和土坯围墙毁坏，砖木及砖混结构房屋出现门、窗洞口上的斜裂缝、八字裂缝和垂直裂缝且一般都裂通，承重墙体的剪切斜裂缝和部分剪刀形裂缝。檐瓦掉落，屋顶烟囱掉落。地面出现重力裂缝。

Ⅵ度区：长轴总体走向 290°，长轴长 69km，短轴长 49km，西起民乐县民三堡乡和顺化乡一带，东到兰州军区军马场四场附近，面积约 2654km^2。Ⅵ度区内土木结构房屋出现裂缝墙皮脱落，个别土坯围墙毁坏，砖木及砖混结构房屋出现小裂缝，外贴瓷砖开裂和脱落，梁、柱与填充墙脱开。

这次地震的总体特点是，震害和有感面积大，烈度偏高、破坏较重。地震造成 10 人死亡（直接死亡 8 人），伤数百人。张掖市民乐县、山丹县、肃南县，14 个乡镇 4.7 万多户、近 20 万人口受灾。倒塌房屋 23 万多间，严重破坏 5 万多间。本次地震所造成的直接经济损失总额为 5 亿多元。

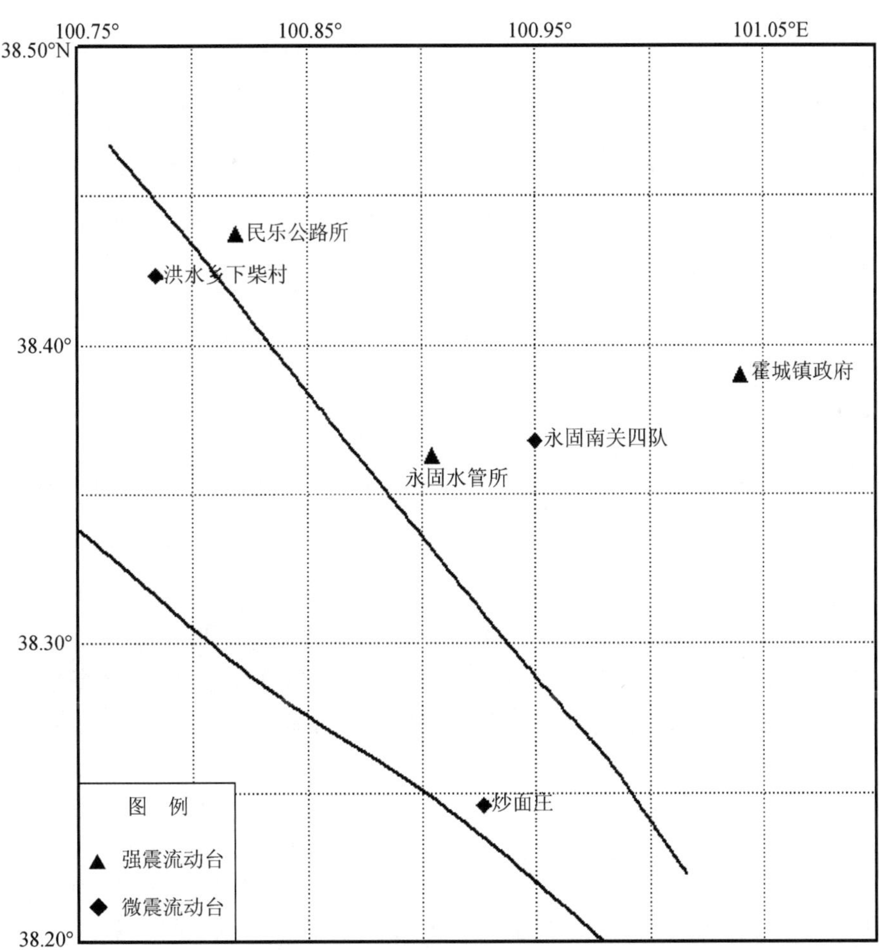

图 4 民乐—山丹 6.1 级地震前后布设的流动测震台网

Fig. 4 Distribution of mobile earthquake-monitoring network before and after the *M*6.1 Minle-shandan earthquake

四、地震序列

据甘肃省地震台网目录，2003 年 10 月 25 日民乐—山丹 6.1、5.8 级地震后，余震活动丰富。主震后截至 2004 年 12 月 31 日，定出 $M\geqslant2.0$ 级地震 87 次，其中无 5.0～5.9 级地震，4.0～4.9 级地震 5 次，3.0～3.9 级地震 14 次，2.0～2.9 级地震 68 次。最大余震为主震后约两个小时发生的 $M4.5$。最大余震与两个主震的震级差分别为 $M_S1.6$、$M_S1.3$，序列震级衰减相对稳定。

表 2　民乐—山丹 6.1、5.8 级地震序列目录（$M\geqslant4.0$ 级）

Table 2　Catalogue of the $M6.1$ Minle-shandan earthquake sequence（$M\geqslant4.0$）

编号	发震日期	发震时刻	震中位置（°）		震级 M	震源深度 h/km	震中地名	结果来源
	年 月 日	时 分 秒	φ_N	λ_E				
1	2003 10 25	20 41 36	38.4	101.2	6.1	33	民乐—山丹	中国地震台网
2	2003 10 25	20 48 03	38.4	101.1	5.8	15	民乐—山丹	中国地震台网
3	2003 10 25	21 25 24	38.35	100.95	4.3	12		甘肃地震台网
4	2003 10 25	21 29 38	38.33	100.97	4	11		
5	2003 10 25	21 30 26	38.33	100.93	4.1	11		
6	2003 10 25	22 30 33	38.35	100.95	4.5	10		
7	2003 10 25	22 34 41	38.35	100.95	4.1	10		

从 6.1、5.8 级地震序列分布图（图 5）可以看出，余震全部分布在主震西侧约 35km 范围内，余震分布与等震线方向较为一致，也与民乐—永昌隐伏断裂与童子坝河隐伏断裂所组成的大黄山隆起区前缘逆冲断裂的走向一致。深度定位结果也说明了这一点[11]。主震、余震精定位结果显示，民乐—山丹地震主要发生于民乐—永昌隐伏断裂上，该断裂具有逆冲性质，走向 NWW，倾向 NE，倾角约为 60°。余震主要沿着民乐—永昌隐伏断裂 NWW 向分布[11]。

由余震序列的 M-t 图（图 6）看出，余震主要集中发生在前 4 天内，多数为 2.5 级左右地震。之后，序列衰减迅速，地震序列频次分布（图 7）也说明了这一点。

第一主震与第二主震的震级相差 0.3，小于 0.6，为双主震型。余震序列 b 值为 0.58，序列 p 值为 0.77，h 值为 1.5，两次主震释放能量占全序列能量的 90%。根据地震类型的判别指标，民乐—山丹地震 6.1、5.8 级地震为双主震—余震型。

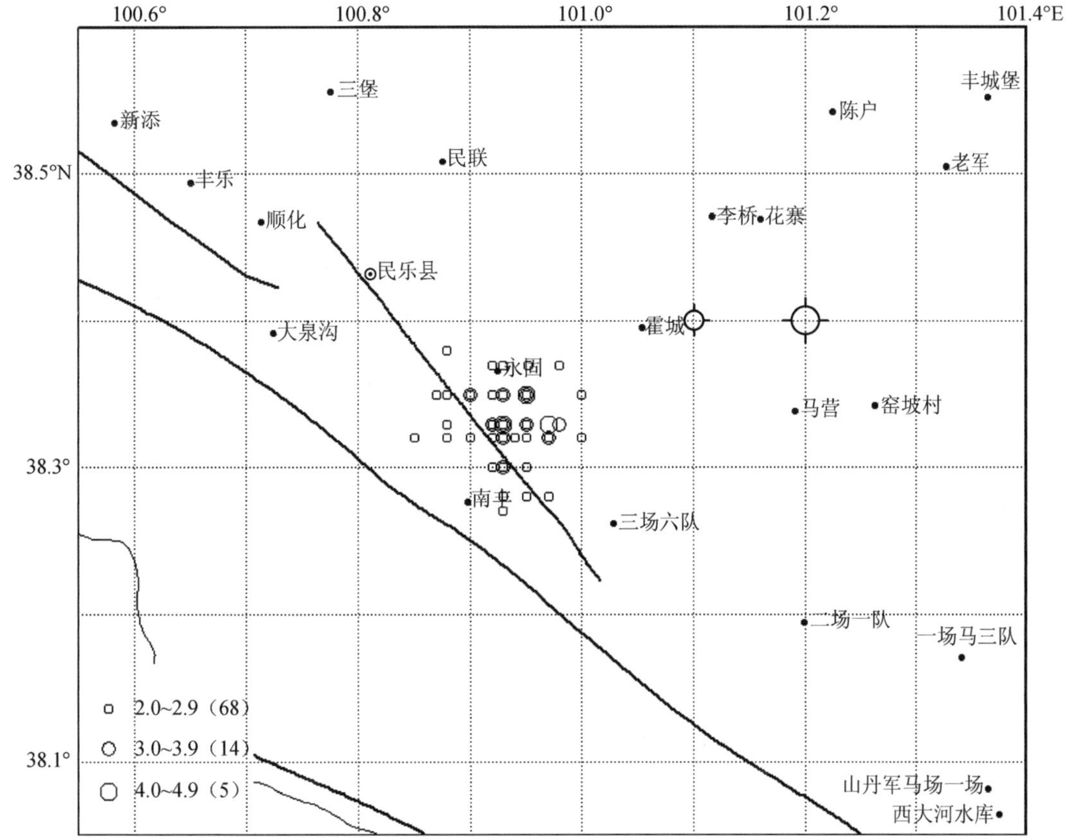

图 5　民乐—山丹 6.1、5.8 级地震序列分布

Fig. 5　Distribution of the M6.1 Minle-shandan earthquake sequence

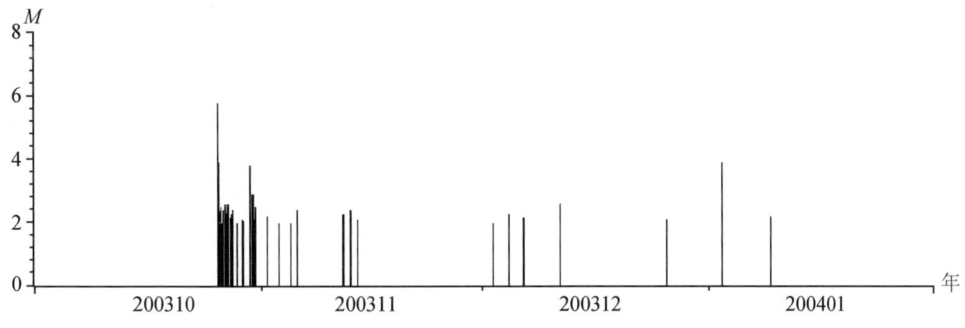

图 6　民乐—山丹 6.1、5.8 级地震序列 M-t 图

Fig. 6　M-t diagram of the M6.1 Minle-shandan earthquake sequence

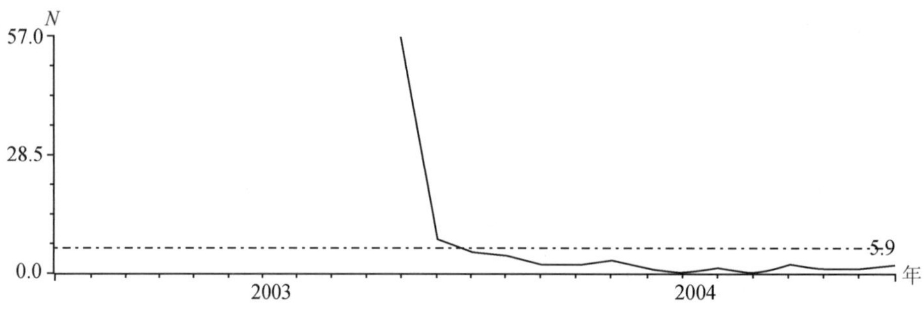

图 7　民乐—山丹 6.1、5.8 级地震序列频度

Fig. 7　Variation of earthquake frequency of the M6.1 Minle-shandan earthquake sequence

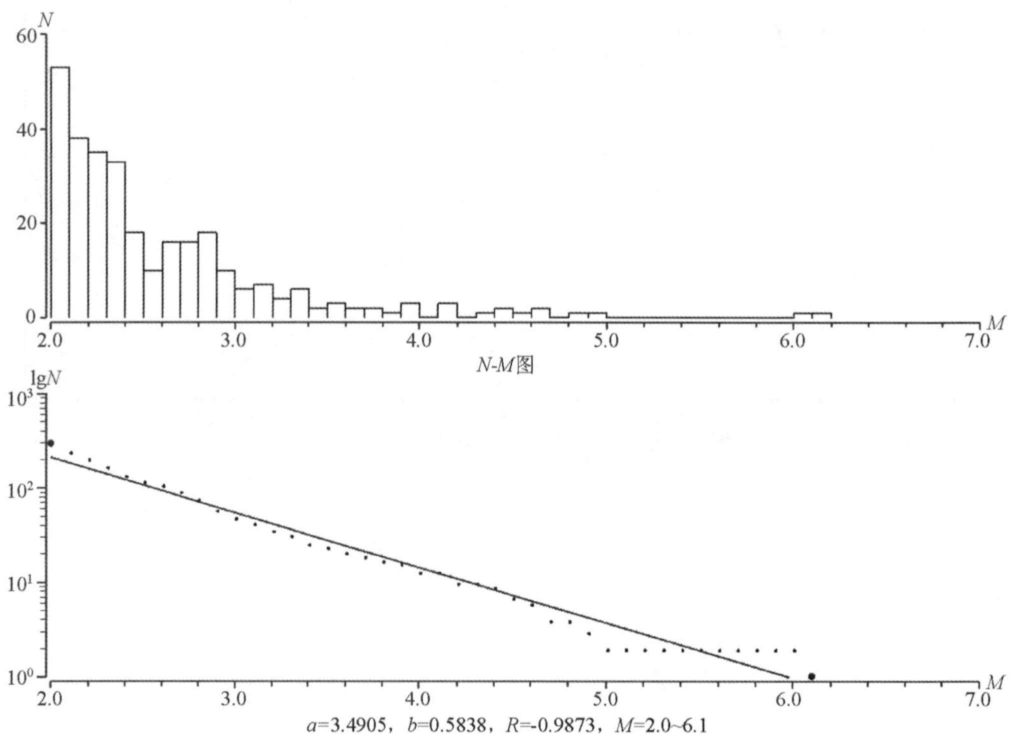

图 8　民乐—山丹 6.1、5.8 级地震序列 b 值拟合曲线

Fig. 8　b-value fitting curve of the M6.1 Minle-shandan earthquake sequence

五、震源参数和地震破裂面

中国地震局兰州地震研究所荣代潞、刘旭宙[6]据甘肃、青海、宁夏地震台网记录的 30 多个地震初动资料，做出的震源机制解见表 3、图 9。美国哈佛大学给出的震源机制解

（HRV）（http：//www. globalcmt. org/CMTfiles. html）也列于表 3。地震矩张量解列于表 4。

表 3　民乐—山丹 6.1 级地震震源机制解

Table 3　Focal mechanism solutions of the *M*6.1 Minle-shandan earthquake

编号	节面Ⅰ			节面Ⅱ			P 轴		T 轴		N（B）轴		X 轴		Y 轴		结果来源
	走向	倾角	滑动角	走向	倾角	滑动角	方位	仰角	方位	仰角	方位	仰角	方位	仰角	方位	仰角	
1	191	76	15	285	75	166	238	1	148	21	330	69					荣代潞 刘旭宙[6]
2	7	77	134	110	45	19	65	20	317	41	172	42					张洪由 许力生[7]
3	108	40	55	331	58	116	290	66	136	22	42	10					HRV
	85	54	21	343	73	142	298	38	142	49	38	12					

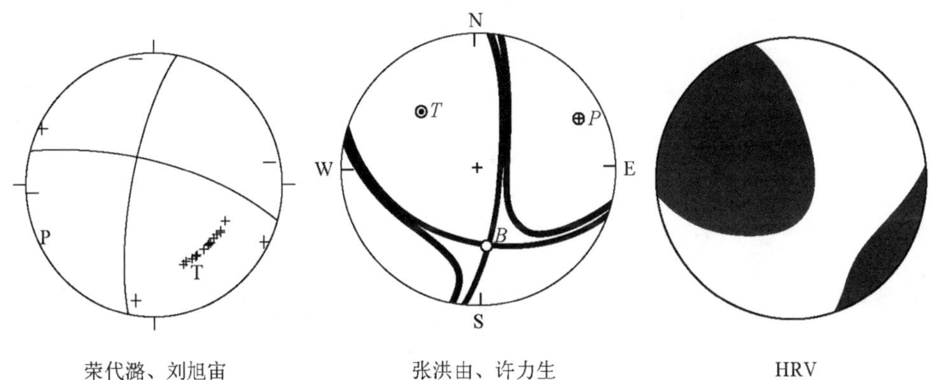

荣代潞、刘旭宙　　张洪由、许力生　　HRV

图 9　民乐—山丹 6.1 级地震震源机制解

Fig. 9　Focal mechanism solution of the *M*6.1 Minle-shandan earthquake

表 4　民乐—山丹 6.1 级地震矩张量解

Table 4　Moment tensor solution of the *M*6.1 Minle-shandan earthquake

编号	节面Ⅰ			节面Ⅱ			矩张量						地震矩 M_o	矩震级 M_w	结果来源
	走向	倾角	滑动角	走向	倾角	滑动角	M_{xx}	M_{yy}	M_{zz}	M_{xy}	M_{yz}	M_{zx}			
1	108	40	55	331	58	116	3.49	−2.24	−1.25	−1.26	1.31	5.27	24E+24	5.8	HRV
	85	54	21	343	73	142	1.75	−2.72	0.96	0.94	3.76	4.08	24E+24	5.8	
2	7	77	134	110	45	19							1.27E+18	6.0	张洪由 许力生

本次地震属于中等强度地震，没有形成地表破裂带，地面变形也不是太多见，经调查，仅在极震区部分地方有规模不等的地裂缝。经调查，在极震区范围内的低山丘陵地区，由于为第四纪粉砂土、黄土状粉土、松散冲—洪积砾石层分布，因此地震时沿山脊多发育滑坡、边坡崩塌等地震次生灾害现象，同时部分地段形成了轻微的地面变形，产生了地裂缝，其分布与发震断层走向基本一致1)。

六、地震前兆观测台网及前兆异常

震中附近地区的地震台站及观测项目分布见图 10。震中 300km 范围内共有 17 个地震台，其中 7 个测震观测单项台，2 个综合观测台，7 个前兆单项观测台；总计有 7 个定点观测项目，有测震、水氡、水位、电阻率、形变、地磁等。6.1 级地震前共出现异常 17 项次，其中测震学异常 10 项次。在震中 300km 以内有水氡、水位、形变和地电阻率共 7 项次异常（表 5）。震前甘肃省地震局根据地震活动性及应变、水氡等项目异常作了不同程度的中、短期和短临预测。

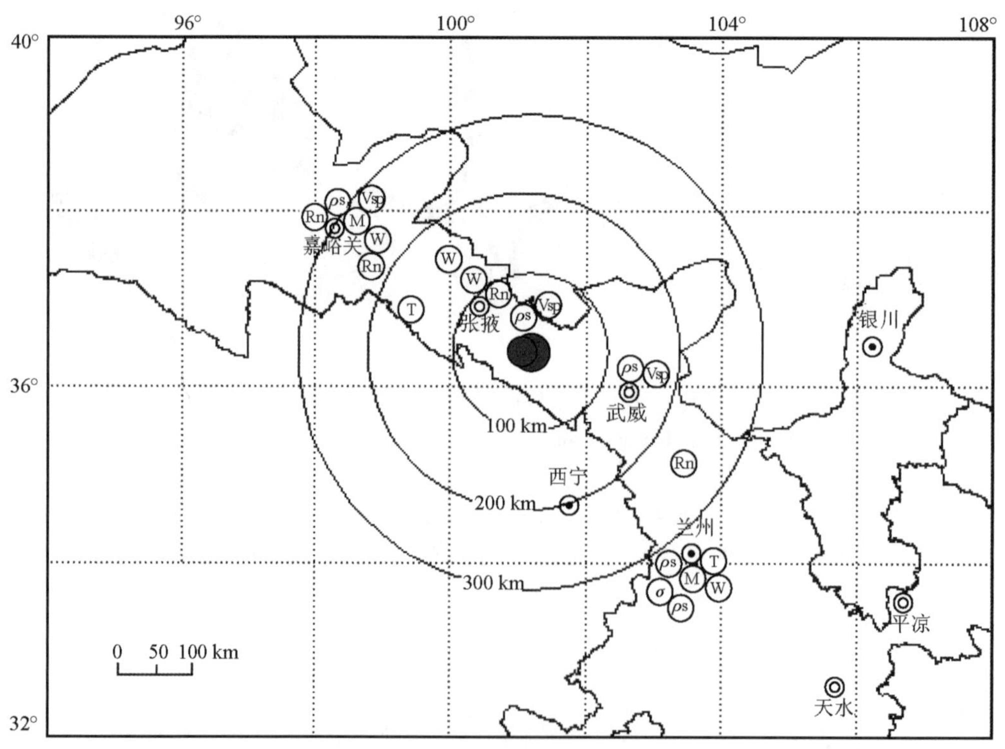

图 10　民乐—山丹 6.1、5.8 级地震附近地区的地震台站及观测项目分布

Fig. 10　Distribution of precursory monitoring stations before the *M*6.1 Minle-shandan earthquake

表5 民乐—山丹6.1级地震前兆异常登记表

Table 5 Table of precursory anomaly of the M 6.1 Minle-shandan earthquake

序号	异常项目	台站（点）或观测区	分析方法	异常判据及观测误差	震前异常起止时间	震后变化	最大幅度	震中距Δ（km）	异常类别及可靠性	图号	异常特点及备注
1	地震空区	38°~41°N，96°~102°E	$M\geqslant3.0$ 级地震空间分布	$M\geqslant3.0$ 级地震平静区段	1996.7.28 ~ 2003.6.14	地震发生在空区内	长轴160km，围空时间7年	震中周围	L_1		
		38°~41°N 96°~102°E	$M\geqslant2.0$ 级地震空间分布	$M\geqslant2.0$ 级地震平静区段	2002.4.29 ~ 2003.6.14	地震发生在空区内	长轴120km，围空时间14.5月	震中周围	M_1		
2	地震活动增强	38°~41°N，96°~103°E	$M\geqslant3.0$ 级地震空间分布	连续发生6次 $M\geqslant3.5$ 级地震	2002.11.17 ~ 2003.10.1	恢复正常		震中周围	M_1		
3	显著地震有序迁移	甘东南—祁连山地震带	$M\geqslant4.0$ 级地震空间分布	短时间内显著事件有序发生	2003.6.19 ~7.18			甘东南—祁连山地震带	S_1	11	礼县震群—景泰—肃南由东往西有序迁移
		甘东南—祁连山地震带	显著地震空间分布	短时间内显著事件有序发生	2003.10.5 ~10.10			甘东南—祁连山地震带	S_1		甘谷—肃南—肃南由东往西有序迁移
4	响应地震	甘肃张掖东北 39.3°N，100.82°E 39.27°N，100.8°E	显著地震快速响应	中强震后短时间内显著事件的响应地区	2003.4.18 4.19		$M3.3$ $M3.0$	110km 160km	M_1		2003年4月17日青海德令哈发生6.6级地震，震后这两次地震快速响应发生

续表

序号	异常项目	台站（点）或观测区	分析方法	异常判据及观测误差	震前异常起止时间	震后变化	最大幅度	震中距Δ（km）	异常类别及可靠性	图号	异常特点及备注
5	蠕变	祁连山中西段	蠕变时间曲线	蠕变加速	2001.9～2003.10.10	恢复正常		震中周围及震中以西地区	M_1	12(b)	
6	小震调制比	祁连山中西段	小震调制比曲线	$R_m \geqslant 0.27$	2003.6～2003.10.24	恢复正常	0.38	震中周围	S_1		
7	b 值	祁连山中西段	$M \geqslant 1.2$ 级地震 b 值时间扫描曲线	$\geqslant 2\sigma$	2003.7～2003.10.24	逐渐恢复正常	0.3	震中周围及震中以西地区	S_1	12(a)	正常背景时段 b 值为 0.78，2003 年 7 月出现 $\leqslant 2\sigma$ 的低值异常
8	η 值	祁连山中西段	$M \geqslant 1.2$ 地震 η 值时间扫描曲线	$\geqslant 2\sigma$	2003.9～2003.10.24	逐渐恢复正常		震中周围及震中以西地区	I_1		震后总结的异常
9	C 值	祁连山中西段	$M \geqslant 1.2$ 级地震 C 值时间扫描曲线	$\geqslant 2\sigma$	2002.10～2003.10.24	逐渐恢复正常		震中周围及震中以西地区	M_1	13(a)	震后总结的异常
10	D 值	祁连山中西段	$M \geqslant 1.2$ 级地震 D 值时间扫描曲线	$\geqslant 2\sigma$	2002.10～2003.10.24	逐渐恢复正常		震中周围及震中以西地区	M_1	13(b)	震后总结的异常
11	电阻率	山丹	日均值	破年变	EW：2003.02～	恢复正常	1.50%	50	M_1	14	EW、NW 道为高值，NS 道为低值，震前有预测

续表

序号	异常项目	台站（点）或观测区	分析方法	异常判据及观测误差	震前异常起止时间	震后变化	最大幅度	震中距Δ（km）	异常类别及可靠性	图号	异常特点及备注
11	电阻率	山丹	日均值	破年变	NS：2003.05 ~	恢复正常	0.80%		S_1		
					NW：2003.04 ~	恢复正常	0.80%		S_1		
12	水位	临泽	日均值	破年变	2003.06 ~	恢复正常	0.4m	120	S_2	15(a)	破年变性低值
13	水位	高台	日均值	破年变	2003.03 ~	恢复正常	4.5m	150	M_2	15(b)	破年变性低值
14	地倾斜	肃南	日均值	趋势改变	NS：2002.09 ~	恢复持续	2650ms	150	M_2	16	相对北西倾斜，存在仪器干扰，震前有预测
					EW：2002.04 ~	恢复正常	500ms				
15	电阻率	武威	日均值	趋势改变	NS：2002.05 ~	恢复正常	11.50%	160	M_1	17	低值异常，震前有预测
				阶变	EW：2003.08	恢复正常	3.40%		S_1		
16	水氡	酒泉	日均值	趋势改变	2003.05 ~	恢复正常	1Bq/L	270	S_1	18	尖峰形高值异常，震前有预测
17	水位	酒泉	日均值	破年变	2003.08 ~ 2003.10	恢复正常	0.8m	270	S_2	19	破年变性高值

1. 地震学异常[8,9]

（1）地震空区。

地震前，祁连山中东段地区存在2级和3级地震空区，其中2级地震空区形成时间为2002年4月29日至2003年6月14日，围空时间为14.5个月，空区长轴NWW向，长120km；3级地震空区形成时间为1996年7月28日至2003年6月14日，围空时间为7年，空区长轴NW向，长160km。民乐—山丹地震就发生在空区内部。

（2）响应地震。

2003年4月17日青海德令哈发生6.6级地震，4月18、19日在甘肃张掖市东北发生了*M*3.3、3.0地震，表现为受德令哈地震影响的快速呼应特征，这两次响应地震分别与民乐—山丹6.1级主震相距约110、160km。

（3）显著地震的有序分布和迁移。

2003年6月19日至7月18日期间，甘肃及边邻地区地震活动表现出了十分显著的活动异常。甘东南地区在多年地震活动水平低下的基础上，6月19日发生了礼县震群，先后累计发生中小地震60余次，经分析该震群属于前兆性震群。随后，7月2日在景泰发生一次*M*4.4地震、7月18日在肃南发生一次*M*4.6地震，整体上表现出由东往西有序迁移的特征（图11）。

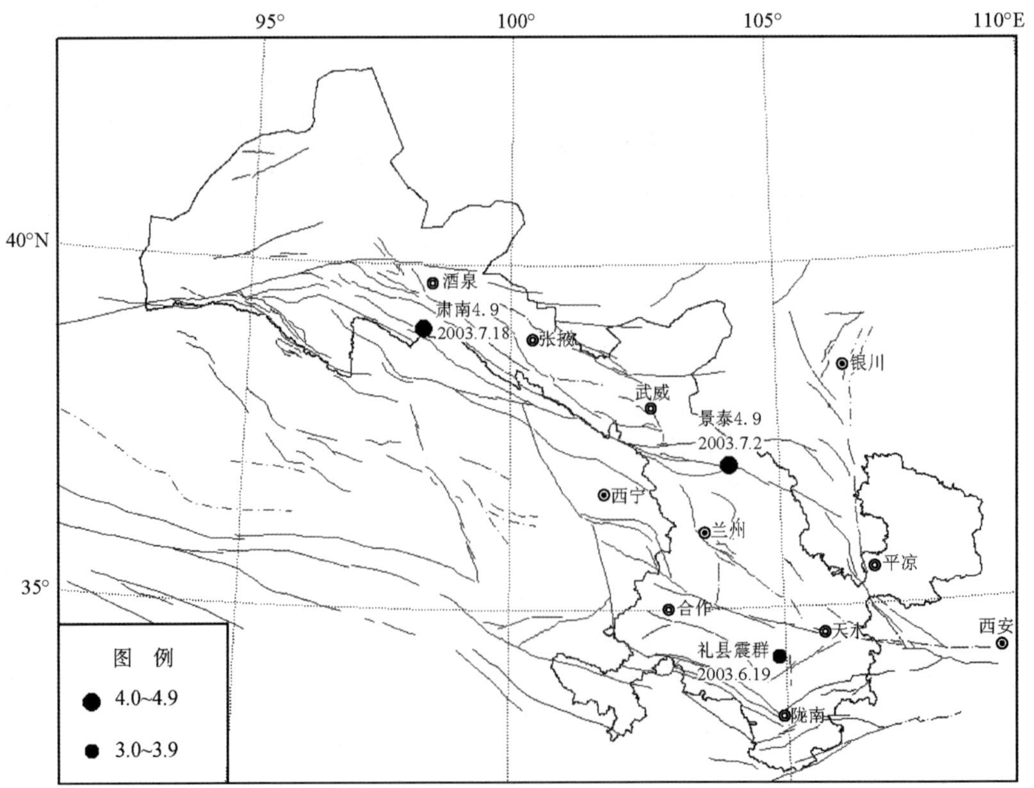

图11　2003年6月19日至7月18日中等地震的有序迁移

Fig. 11　Distribution of the middle earthquakes occurred from June 19 to July 18, 2003

10 月 5 ~ 10 日，类似的地震活动图像再次出现：在 10 月 5 日 6 点 56 分至 7 点 31 分的不到一个小时内，距离礼县较近的甘谷县，连续发生三次地震，强度依次为 *M*1.7、1.4、1.6；随后，10 月 6 日在肃南发生一次 *M*3.9 地震（38.85°N，100°E）、10 月 10 日在肃南再次发生一次 *M*4.6 地震（39.52°N，98.25°E），表现出由东往西有序迁移的特征。意味着青藏块体东北缘地区地震活动的有序性、整体性均在不断加强。

（4）另据震后总结，民乐—山丹地震前还出现过蠕变加速、*C*、*D*、*b*、η 值和小震调制比等地震学异常[10]（图 12、图 13）。

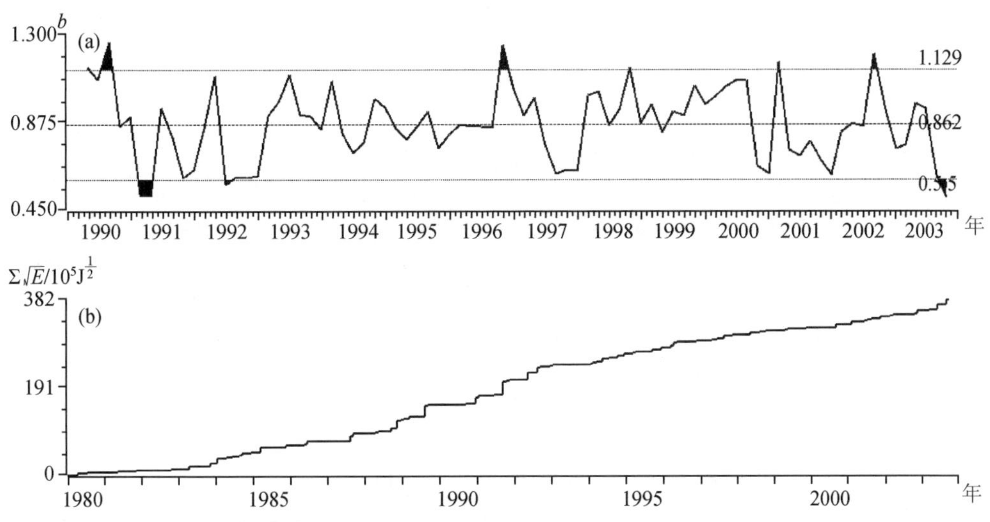

图 12　民乐—山丹地震前祁连山中西段地区 1990 年以来 $M\geqslant1.2$ 级地震 *b* 值时序曲线（a）和 1980 年以来蠕变曲线（b）

Fig. 12　Strain curve（b）and b-value curve（a）of $M\geqslant1.2$ earthquakes in the middle and western Qilian Mountain before Minle-shandan earthquake

2. 其他前兆观测项目异常

2003 年 10 月 25 日民乐—山丹 6.1、5.8 级地震发生在甘肃省中西部地区，300km 范围内定点前兆观测台站不多，受地理环境的限制，大多分布在震中的北西—南东方向。震中 300km 范围内定点前兆有 7 个测项、6 个台项的异常（图 14 至图 19），0 ~ 100、101 ~ 200 和 201 ~ 300km 范围内各有 1、4 和 1 个台项的异常，最近的为山丹地电，距震中约 50km；300km 内最远的为酒泉水氡和水位，距震中 270km。

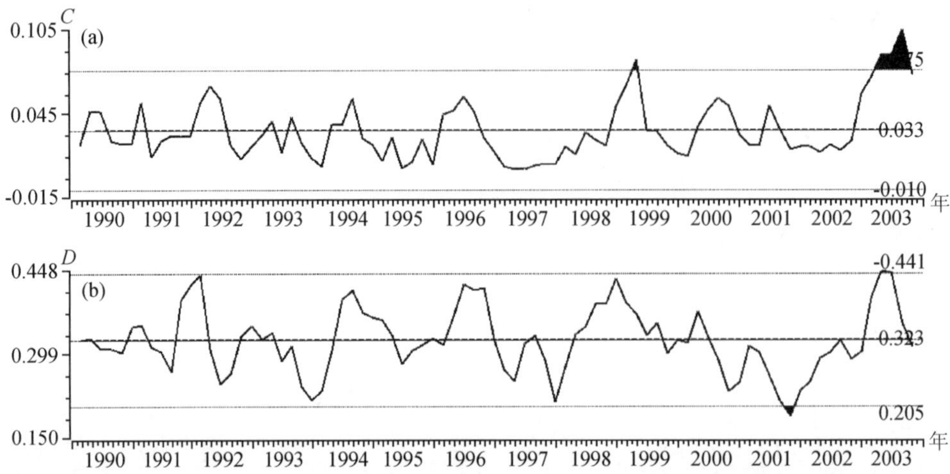

图 13　民乐—山丹地震前祁连山中西段地区 1990 年以来 $M \geqslant 1.2$ 级地震 C 值（a）和 D 值时序曲线（b）

Fig. 13　C-value（a）and D-value curve（b）of $M \geqslant 1.2$ earthquakes in the middle and western Qilian Mountain before Minle-shandan earthquake

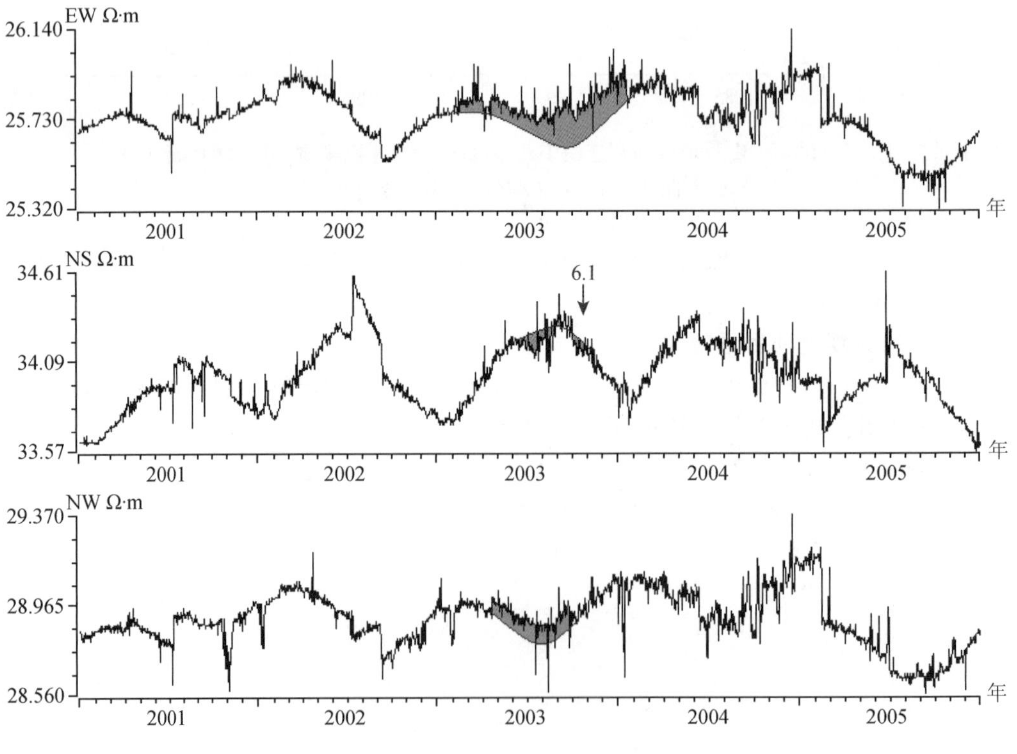

图 14　山丹地电阻率日均值曲线

Fig. 14　Curve of daily mean value of apparent resistivity at Shandan station

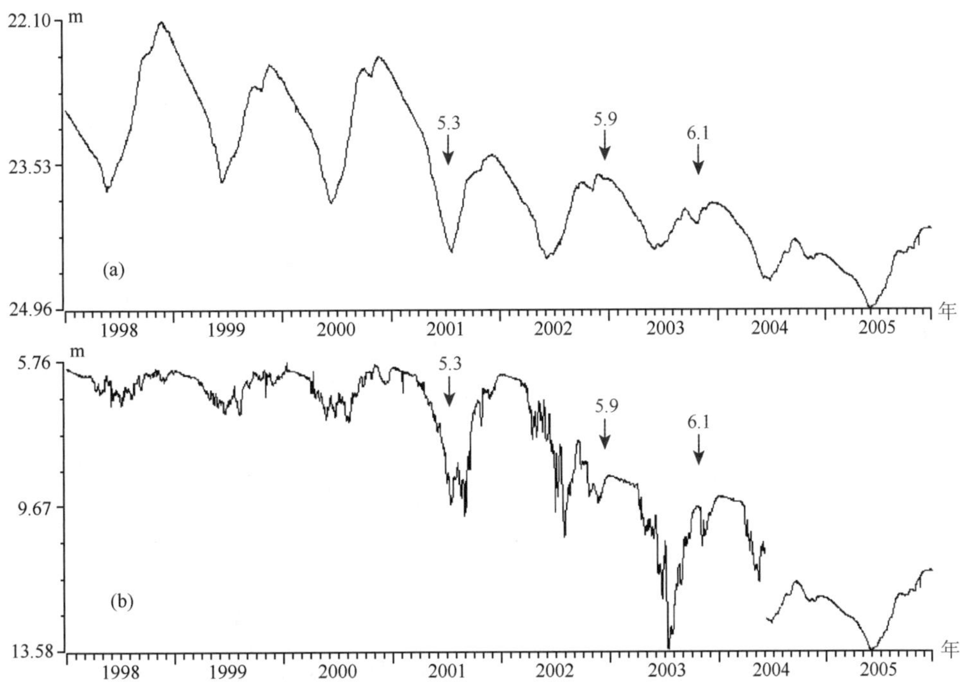

图 15　临泽（a）和高台（b）水位日均值曲线

Fig. 15　Daily value curve of water level at Linze（a）and Gaotai（b）station

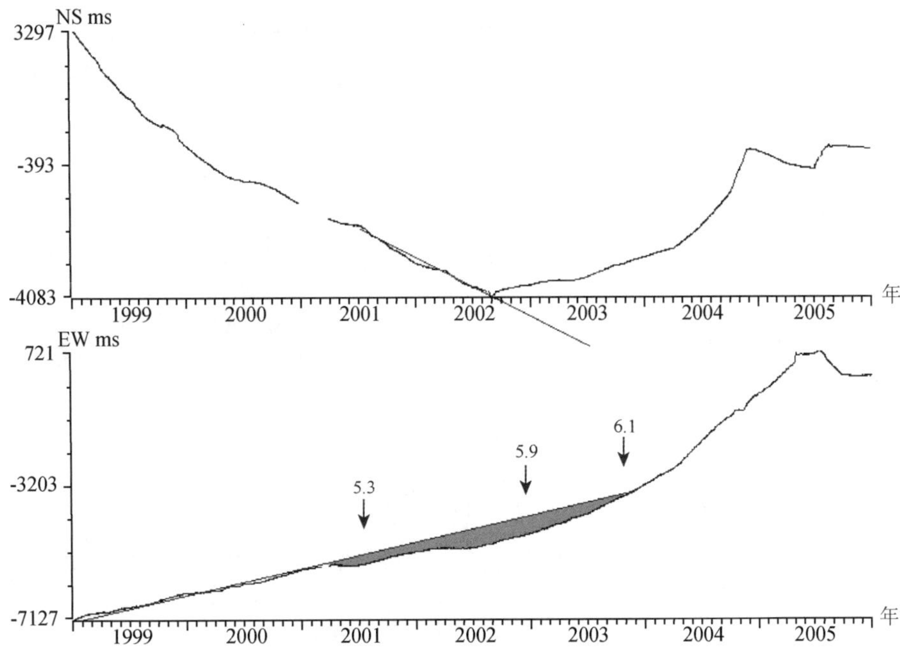

图 16　肃南倾斜日均值曲线

Fig. 16　Curve of daily mean value of tilt at Sunan station

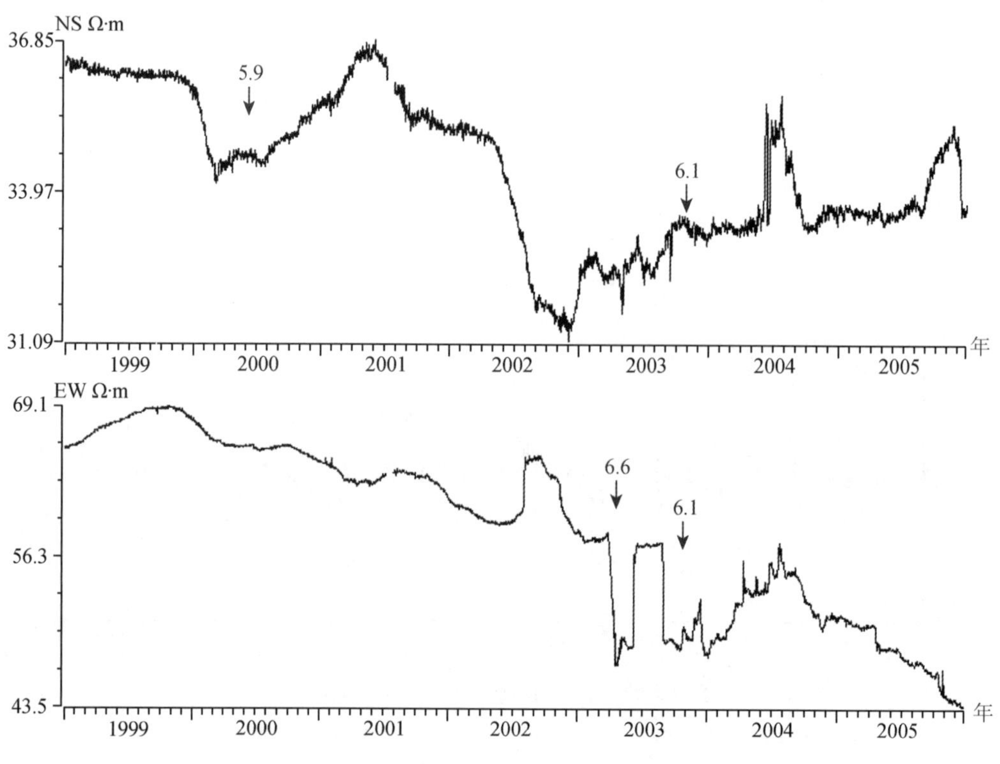

图 17 武威地电阻率日均值曲线

Fig. 17 Curve of daily mean value of apparent resistivity at Wuwei station

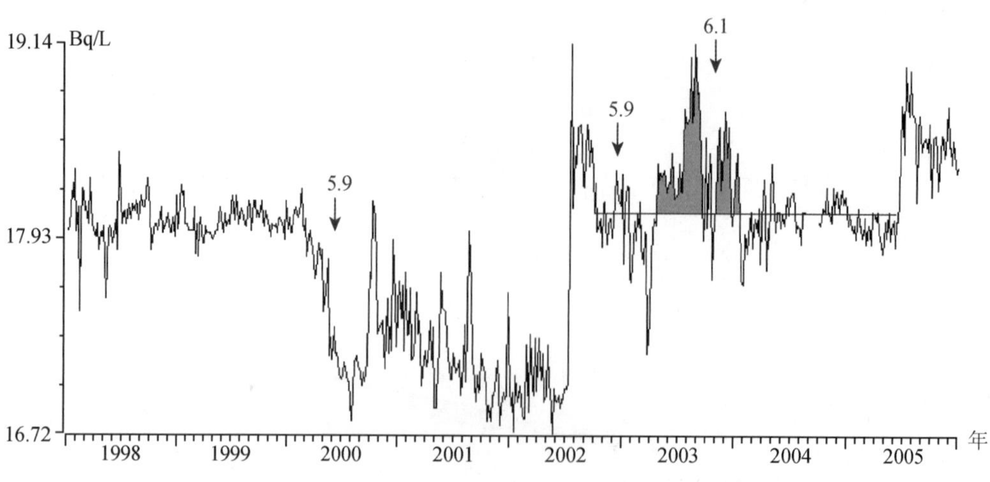

图 18 酒泉水氡五日均值曲线

Fig. 18 Curve of 5-day mean value of radon content in groundwater at Jiuquan

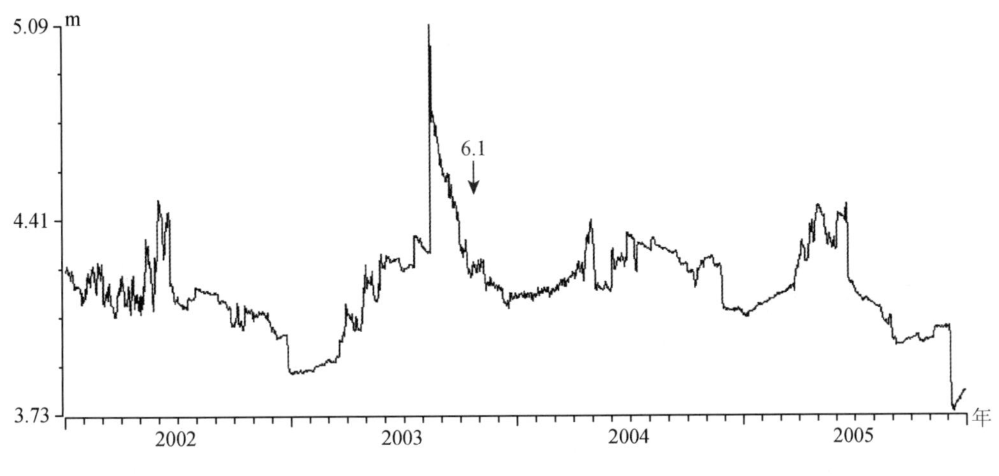

图 19　酒泉水位日均值曲线

Fig. 19　Curve of daily mean value of water level at Jiuquan station

另据杨立明等的总结[8,9]，民乐—山丹地震前，甘肃省前兆观测出现了共计 18 个台项的异常变化，呈现出前兆异常成组同步活动的态势。主要特征表现为异常测项种类齐全，包括地电、应力、形变、水位、水氡等观测项目；空间上集中在祁连山中东段附近（6 台项）、甘东南地区（10 台项）及祁连西段地区（2 台项）；时间上，7 月份出现了 4 台项的异常、8 月份 9 台项、9 月份 5 台项；主要异常点位包括兰州地电、平凉水位、宕昌水管仪、武都应变、酒泉水氡、兰州大滩水位、清水流量、武山 1 号泉水氡、22 号井水氡、武威地电、刘家峡应力、山丹地电、临夏地电、礼县流量、礼县水氡、肃南水管仪、武都应变、清水水氡等。主要图件见图 20 至图 23。

1）8 月 19 ~ 31 日：具有较高映震效能前兆异常的同步出现阶段

武威地电、刘家峡应力、清水流量等是甘肃地区具有较高预报效能的前兆观测。8 月 19 ~ 31 日期间，这几个前兆观测相对同步出现了显著的异常变化（图 20）：

（1）武威地电：8 月 28 日至 9 月 2 日期间，观测曲线出现了类似于德令哈地震前的变化，8 月 28 日测值快速下降，至 9 月 2 日累计下降幅度达到 16.7%（同图 17）。

（2）刘家峡应力：8 月 18 ~ 27 日期间，刘家峡应力两道观测曲线出现了类似于玉门地震前的各向异性变化。其中 EW 向 8 月 17 日开始转折上升，8 月 27 日加速上升，9 月 29 日发生转折回返；NS 向 8 月 21 日开始转折下降，10 月 2 日转折回返（图 21）。

（3）清水流量：8 月初以来，清水流量在多年趋势下降的基础上，出现了显著的上升变化，8 月 29 日起出现加速上升变化，异常变化十分显著（图 22）。

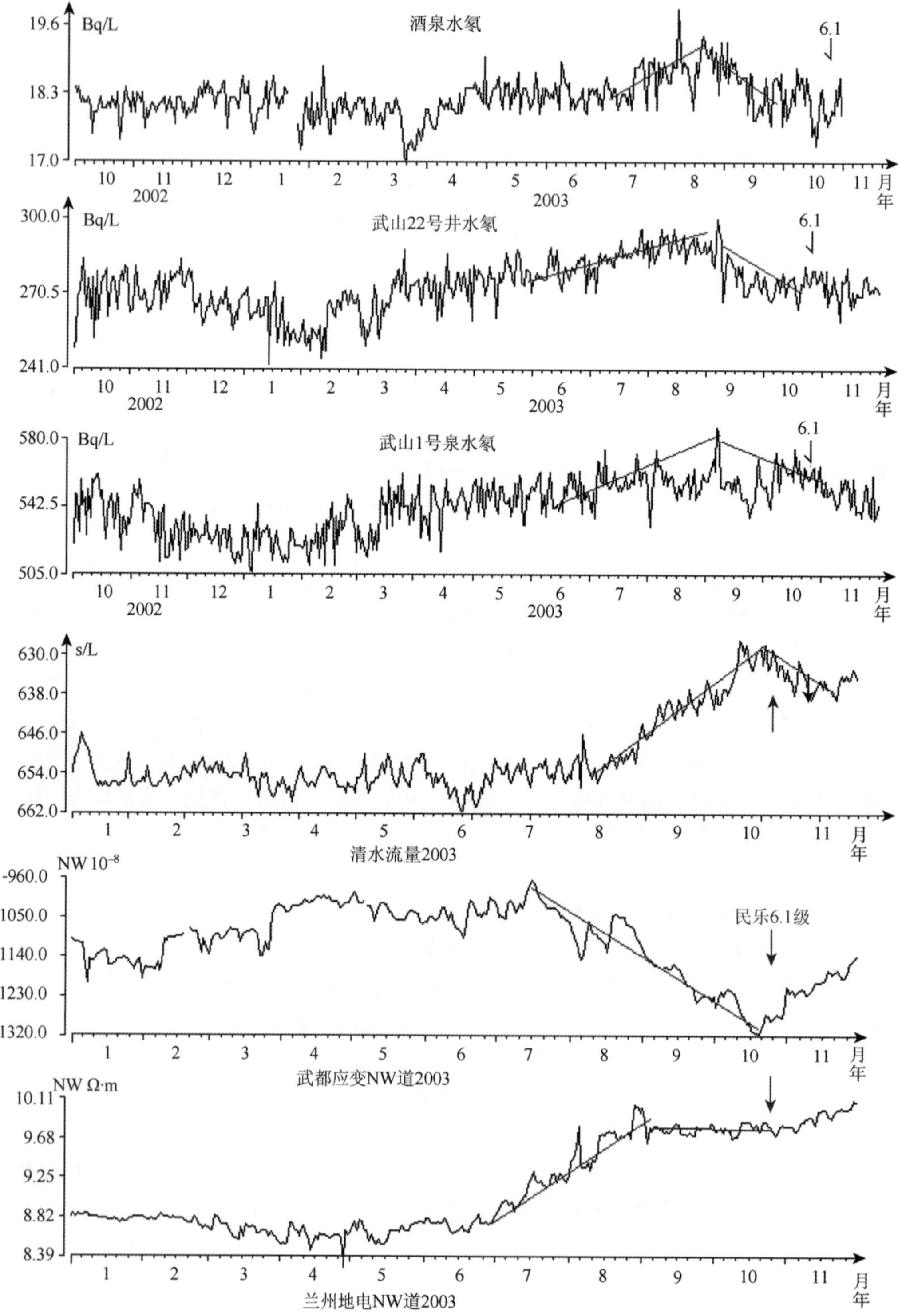

图 20　2003 年 7～9 月甘肃地区群体前兆异常成组活动特征

Fig. 20　Synchronized variation of precursory anomalies in Gansu province from July to September in 2003

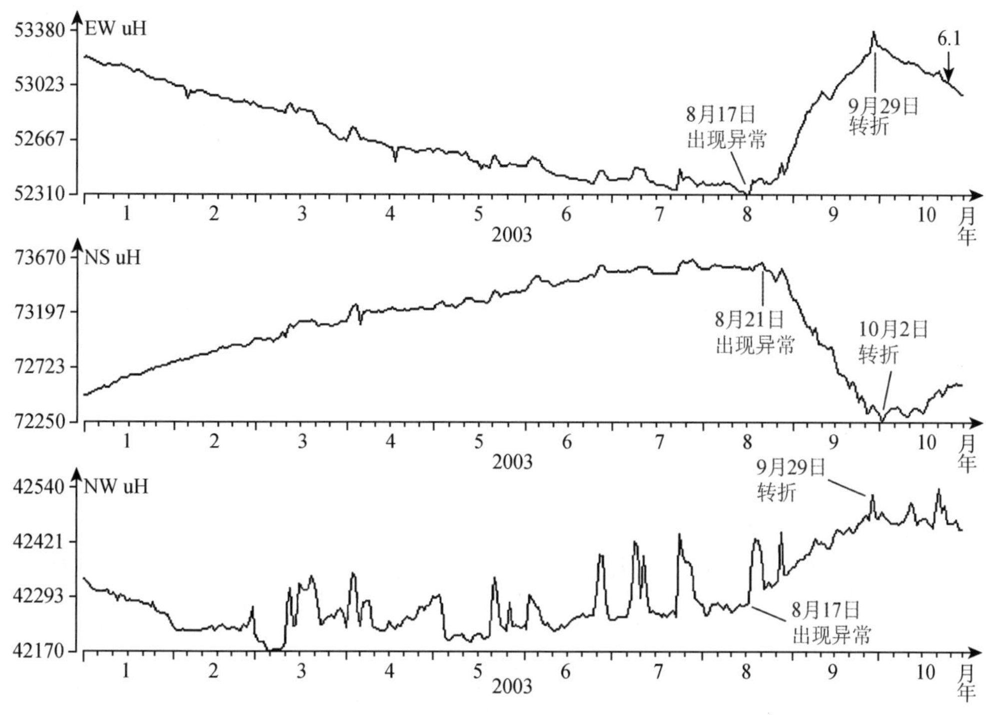

图 21　刘家峡应力异常变化

Fig. 21　Curve of daily mean stress value at Liujiaxia station

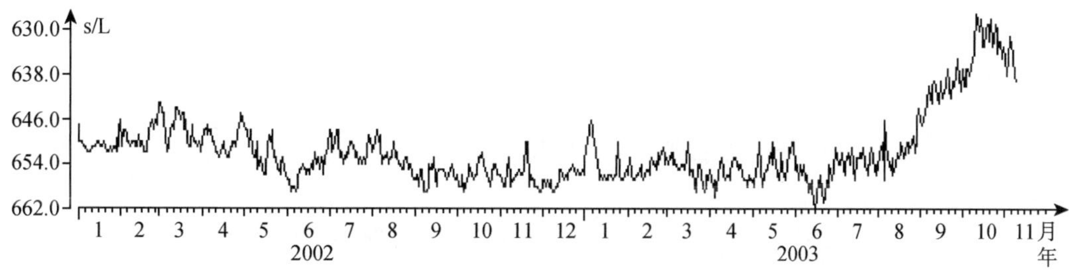

图 22　清水流量观测曲线

Fig. 22　Curve of daily mean value of water level at Qingshui station

2）10 月初—地震发生：转折—突跳及部分宏观异常出现阶段

在此阶段，出现了多项前兆异常的转折变化及显著的地震活动。

主要包括武威地电、刘家峡应力、清水水氡、灵台水氡、临夏地电、礼县流量、礼县水氡、嘉峪关气氡等。以临夏地电为例予以说明，临夏地电 10 月 4 日快速下降，7 日达到最低点，然后回返，10 日前后基本恢复（图 23）。

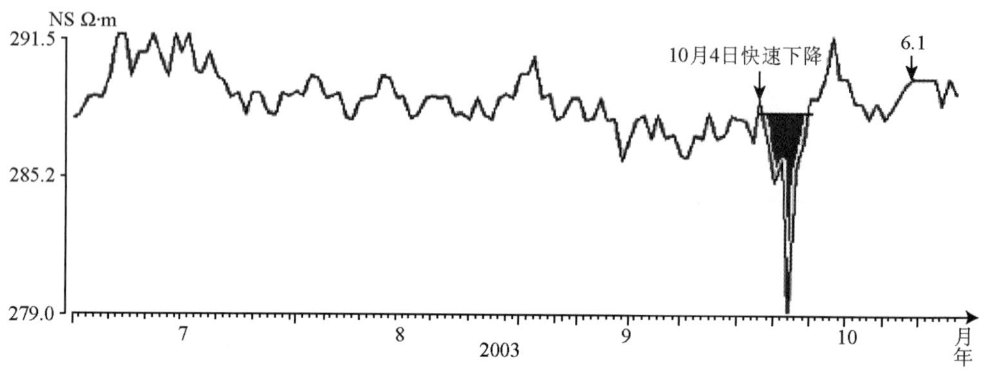

图 23　临夏地电观测曲线

Fig. 23　Curve of daily mean value of apparent resistivity at Linxia station

在此阶段也出现了部分宏观异常现象。

刘家峡水库水位 4 个测点于 7 月 20 日出现陡降；经落实，刘家峡水库水位的异常变化及应力观测站的异常，观测仪器正常，环境也无大的变化，很可能反映了该地区局部的应力场变化，其异常变化是可信的[5)]。

礼县地震局 10 月 22 日报告，礼县自来水公司的水井突然上涨约 3m，原因不明。甘肃省地震局及时派出现场异常落实组开展落实工作，民乐地震发生时该工作组正在现场工作。

10 月 23 日下午接兰州市地震局报告，兰化厂地震测报站电磁辐射观测于 10 月 19 日开始出现高值异常。为此，由甘肃省地震局派出考察组赶赴观测现场进行异常落实。分析认为该异常具有一定的信度[5)]。

七、地震前兆异常特征分析

综上所述，民乐—山丹地震前兆异常主要特征如下：

1. 震兆异常特征

震前祁连山中东段地区 3 级地震空区为长期异常，2 级地震空区为中短期异常。震中周围地震活动增强、震中附近出现响应地震、震中周围及震中以西地区蠕变加速、C 值和 D 值为中期异常。甘东南—祁连山地震带显著地震有序活动和震中周围及震中以西地区 b 值为短期异常。震中周围及震中以西地区 η 值为临震异常。

2. 定点前兆异常待征

前兆异常流体为短期异常、形变为中期异常。电阻率存在中期和短期异常。前兆异常的时空分布特征分析如下：

1）空间分布特征

震中 300km 范围内定点前兆异常主要分布在祁连山中西段地区。另外，位于甘东南地区的震中距为 330km 的刘家峡应力、370km 的临夏电阻率、600km 的清水流量等都在震前有明显的异常，但同年 11 月 13 日岷县 5.2 级地震与该地震发震时间接近、距离这几项异常台站也较近，在异常形态上难以区分，可能为这两个地震的共同反映。

2）时间进程特征

6.1 级地震前，地震活动性以中期异常为主要特征，10 项异常中，有 6 条为半年以上的中期异常，占地震活动性异常项次的 60%。而 300km 范围内的定点前兆 7 项、11 个测道的异常中，半年以上的中期异常为 5 个测项，占异常测项总数的 45%；1～6 个月的短期异常为 6 个测项，占异常总数的 55%；没有 1 个月以内的临震异常。可见定点前兆以短期异常为主要特征。

3）异常的数量特征

震前 300km 范围内共 9 个前兆观测台站 12 个观测项目（图 9），有 6 个台站的 7 个测项、11 个测道出现前兆异常，其中 100km 范围内共 2 个前兆观测台站，有 1 个台站的 1 个测项、2 个测道出现前兆异常，分别占异常项目和测道总次数的 50% 和 50%；101～200km 范围内共 4 个前兆观测台站，有 4 个台站的 4 个测项、6 个测道出现前兆异常，占异常项目和测道总次数的比例均为 100%；201～300km 范围内共 3 个前兆观测台站，有 1 个台站的 2 个测项、2 个测道出现前兆异常，分别占异常项目和测道总次数的 33% 和 22%。可见 101～200km 范围内异常测项比和测道比均较高。

3. 宏观异常特征

民乐—山丹地震前宏观异常为地下水和电磁波异常，异常出现时间至发震时间相隔 2 天至 5 个月，属短临异常。短临宏观异常主要分布于地震以东、以南地区，且距震中较远。但 11 月 13 日岷县 5.2 级地震与这几项宏观异常点距离较近，难以区分这些宏观异常是哪一次地震的，可能为这两个地震的共同反映。

八、震前预测、预防和震后响应

1. 预测情况

研究[2)]指出，祁连山地震带的地震活动受青藏高原北部地区地震活动的制约和控制。1900 年以来，青藏高原北部地区 $M_S \geqslant 5.0$、5.5、6.0 级等不同强度层次地震活动存在明显的活跃—平静交替的特征，可以分为 7 个活跃期和 6 个平静期。该区目前处于 2000 年景泰 5.9 级、兴海 6.6 级等地震开始的活跃期内；2001 年昆仑山口西 8.1 级、2002 年玉门 5.9 级、2003 年德令哈 6.6 级等多次 6 级左右或以上地震的发生是这种活跃状态发展的结果。因而，近几年青藏高原北部地区及祁连山地震带中强地震的活跃状态，为民乐—山丹 6.1 级地震的发生提供了基本环境和背景。

青藏高原北部地区地震活动既存在活跃期—平静期交替轮回的特征，也存在活跃期内低一个层次的密集丛集特征，绝大部分地震分布在不同的密集丛集活动段内。经分析，1950 年以来，活跃期内独立的 5.5 级以上地震的时间间隔统计分布特征是：76% 的 5.5 级以上地震的时间间隔不超过 6 个月，处于不同的密集丛集段内。2002 年 9 月以来，本区中强以上地震活动的密集丛集现象突出，先后发生了 2002 年 12 月 14 日玉门 5.9 级、2003 年 4 月 17 日德令哈 6.6 级及 3 次 5.0～5.4 级地震。这种特征及其进一步发展有利于该区继续发生中强以上地震，从而为民乐—山丹 6.1 级地震的预报提供了中短期依据。

1900 年以来，青海西、南部发生 6 级以上地震后，甘肃地区均有 5 级以上迁移地震发生，迁移的时间基本集中在 1 年以内，优势时段为震后 5～10 个月；空间以祁连山中东段占优势，同时祁连西段、甘东南等区域均有呼应地震发生2)。2003 年 4 月 17 日德令哈 6.6 级地震的发生为甘肃地区发生 5 级以上地震提供了中短期预报依据。

在中短期预测基础上，甘肃省地震局对民乐—山丹 6.1 级地震前短期阶段地震活动和前兆观测异常变化加强了跟踪分析，并密切监视各种资料的发展和变化。2003 年 6 月 19 日至 7 月 18 日，甘肃及边邻地区地震活动表现出了十分显著的活动异常。甘东南地区在多年地震活动水平低下的基础上，6 月 19 日发生了礼县震群，先后累计发生中小地震 60 余次，经分析该震群属于前兆性震群。随后，7 月 2 日在景泰发生一次 M_S4.4 地震、7 月 18 日在肃南发生一次 M_S4.6 地震，整体上表现出由东往西有序迁移的特征。

在此阶段，前兆观测出现了多个台项的异常变化，呈现出前兆异常成组同步活动的态势。主要特征表现为异常测项齐全，包括地电、应力、形变、水位、水氡等观测项目。空间上在祁连山中东段附近、甘东南地区及祁连西段地区均有分布。异常时间以中短期异常为主。主要异常点包括兰州地电、平凉水位、宕昌水管仪、武都应变、酒泉水氡、兰州大滩水位、清水流量、武山 1 号泉水氡、22 号井水氡、武威地电、刘家峡应力、山丹地电、临夏地电、礼县流量、礼县水氡、肃南水管仪、武都应变、清水水氡等。

2003 年 8 月 19～31 日，具有较高映震效能前兆异常的同步出现阶段。武威地电、刘家峡应力、清水流量等是甘肃地区具有较高预报效能的前兆观测。这期间，这几个前兆观测相对同步出现了显著的异常变化。武威地电于 8 月 28 日至 9 月 2 日，观测曲线出现了类似于德令哈地震前的变化，8 月 28 日测值快速下降，至 9 月 2 日累计下降幅度达到 16.7%。刘家峡应力于 8 月 18～27 日两道观测曲线出现了类似于玉门地震前的各向异性变化。其中 EW 向 8 月 17 日开始转折上升，8 月 27 日加速上升，9 月 29 日发生转折回返；NS 向 8 月 21 日开始转折下降，10 月 2 日转折回返。清水流量自 8 月初以来在多年趋势下降的基础上，出现了显著的上升变化，8 月 29 日起出现加速上升变化，异常变化十分显著。

2003 年 10 月初至地震发生前，出现了多项前兆异常的转折变化及显著的地震活动，且出现了部分宏观异常现象。主要包括武威地电、刘家峡应力、清水水氡、灵台水氡、临夏地电、礼县流量、礼县水氡、嘉峪关气氡等。如临夏地电 10 月 4 日快速下降，7 日达到最低点，然后回返，10 日前后基本恢复。

10 月 5～10 日，出现了类似于第一阶段的地震活动的有序迁移，在 10 月 5 日 6 点 56 分至 7 点 31 分的不到一个小时内，距离礼县较近的甘谷县，连续发生三次地震，强度依次为 2.5、2.2、2.4；随后，10 月 6 日在肃南发生一次 M_S3.9 地震（38.85°N，100°E）、10 月 10 日在肃南再次发生一次 M_S4.6 地震（39.52°N，98.25°E），表现出由东往西有序迁移的特征。

宏观现象异常的出现：2003 年 3 月份始，甘肃东部地区出现大面积的前兆异常。3 月 12～16 日甘肃省地震局派出专业人员到平凉市、天水市、定西市等地方地震局和平凉、天水、清水、武山、通渭、定西等地震台，对所涉及的观测资料和主要异常做了详细的现场落实工作。经分析研究，平凉水位、天水应力、天水水位、天水地磁、武山水氡、清水流量和清水水氡等前兆异常可信度较高。

武威地电 EW 道地电阻率日均值分别于 2003 年 3 月 27 日、8 月 28 日出现大幅度快速下降。甘肃省地震局组织专家和预报人员分别于 4 月 17 日、8 月 31 日赴现场进行异常落实。经落实，排除了干扰因素，认为异常情况真实可靠，确属电阻率的前兆异常变化。

经甘肃省电力实验研究所报告，刘家峡水库水位 4 个测点于 7 月 20 日出现陡降；刘家峡应力于 8 月中旬出现异常。为此，甘肃省地震局预报人员和甘肃省电力实研所专家于 9 月 9 日赶往刘家峡。经落实，刘家峡水库水位的异常变化及应力观测站的异常，观测仪器正常，环境也无大的变化，很可能反映了该地区局部的应力场变化，其异常变化是可信的。

据礼县地震局 10 月 22 日报告，礼县自来水公司的水井突然上涨约 3m，原因不明。甘肃省地震局及时派出现场异常落实组开展落实工作，民乐地震发生时该工作组正在现场工作。2003 年 10 月 23 日下午接兰州市地震局报告，兰化厂地震测报站电磁辐射观测于 10 月 19 日开始出现高值异常。为此，由预报人员组成考察组赶赴观测现场进行异常落实。分析认为该异常具有一定的信度。

通渭温泉水汞自 2002 年 12 月以来出现大幅度高值异常，至今已持续一年时间，不论是幅度还是持续时间，都属罕见。经落实，异常已排除。

7 月 30 日、8 月 27 日、9 月 28 日甘肃省地震局月会商结论认为，祁连山地震带和甘东南地区存在发生 5 ~6 级地震的可能。10 月 6 日临时会商结论认为，短期内祁连山地震带和甘东南地区存在发生 6 级左右地震的可能，10 月 10 日临时会商结论认为，未来 2 个月内祁连山地震带，尤其是祁连山西段、中东段等地区存在发生 6 级左右地震的可能。

2003 年 9 月 20 日，甘肃省地震局根据地震活动、前兆异常特征向中国地震局填报了地震短期预报卡。在此基础上，9 月 28 日，甘肃省地震局向省政府进行了专门汇报，并提交了书面报告，其中结论意见指出，“未来 3 个月内甘肃及边邻地区，尤其是祁连山中东段、甘东南等地震重点危险区存在发生 6 级左右地震的可能”，并于 9 月 30 日向省政府提交了“甘肃及边邻地区近期震情分析报告”。短期预报意见提出后，10 月 8 日省政府召开会议，专题研究甘肃省地震灾害紧急救援队组建、震情短临跟踪、地震安全性评价和抗震设防要求规范管理等工作。10 月 10 日甘肃省地震局紧急震情会商会进一步缩短了预报期限，认为未来 2 月内发震的可能性较大，并将有关意见报送省政府。在地震预报意见的指导下，震前甘肃省地震局架设了多台强震仪，其中的一台就架设在极震区民乐县永固乡，从而获得了极震区的 6.1 级和 5.8 级三分向的强震记录，为地震工程研究及工程抗震设防，积累了资料。

2. 震后趋势判定

民乐—山丹 6.1、5.8 级地震后，中国地震局组织专家与甘肃省地震局地震现场应急队员赶赴地震现场，进行现场地震监测预报工作，甘肃省地震局及时召开紧急震情会商会。根据 6.1 和 5.8 两次地震相隔不到 7 分钟的时间，震级差仅为 0.3 的特征，现场工作组和甘肃省地震局分析预报人员判定该地震属双主震—余震型（与祁连山地震带以往表现为主震—余震型的地震类型不完全相一致）。根据现场抗震救灾工作的实际需要，分阶段给出了震后 24 小时、未来 3 天、一周等时段的原震区震后趋势判定结果，为政府组织抗震救灾工作提供了科学、客观、明确的依据，为稳定震区人心发挥了积极的作用，具有显著的社会效益。

3. 震害分析

地震发生后，中国地震局与甘肃省地震局在地震现场开展了震害评估工作，10 月 26 ~

29 日在现场工作期间，每天有 50 余名灾评人员分乘 15 辆车对 5000km^2 范围内的民乐、山丹、肃南三县共 11 个乡镇以及山丹军马总场、二场、三场内的 74 个抽样点，179 个调查点进行了现场调查，累计行程约 4 万余公里，获得了珍贵的第一手资料[12];1)。宏观震害考察结果显示本次地震极震区范围大，灾害重；经济损失大，人员伤亡小。民乐—山丹 6.1、5.8 级地震发震时间间隔 7 分钟，震中位置距离较近，存在明显的震害累积和叠加效应，是本次地震损失惨重的一个重要原因。同时，由于震区农村房屋以生土建筑为主，结构设计不合理，场地选址不当，地基处理不善，导致民房震害非常严重。地震造成 10 人死亡（直接死亡 8 人），数百人受伤；张掖市民乐县、山丹县、肃南县 14 个乡镇受灾住户为 4.7 万多户，受灾人数近 20 万人；共有 23 万多间房屋倒塌，其中 5 万多间遭到严重破坏；地震造成的直接经济损失总额为 5 亿多元。

九、结论与讨论

（1）民乐—山丹 6.1、5.8 级地震发生在祁连山地震带、河西走廊过渡带中部地区的民乐盆地与大黄山隆起的交汇部位。2002 年 9 月以来，本区中强以上地震活动的密集丛集现象突出，先后发生了 2002 年 12 月 14 日玉门 5.9 级、2003 年 4 月 17 日德令哈 6.6 级及 3 次 5.0 ~5.4 级地震。这种特征及其进一步发展有利于该区继续发生中强以上地震，从而为民乐—山丹 6.1 级地震的预报提供了中短期依据。

在中短期预测基础上，民乐—山丹 6.1 级地震前短期阶段甘肃及边邻地区地震活动表现出了十分显著的活动异常，整体上表现出由东往西有序迁移的特征。在此阶段，前兆观测呈现出异常成组同步活动的态势，异常测项种类齐全，包括地电、应力、形变、水位、水氡等观测项目。短期阶段，2003 年 8 月 19 ~31 日，具有较高映震效能前兆异常的同步出现阶段。2003 年 10 月初至地震发生前，出现了多项前兆异常的转折变化及显著的地震活动，且出现了部分宏观异常现象。

（2）此次地震发生在甘肃地震局 2003 年度地震危险区的边缘，地震强度预测较为准确，震前甘肃地震局向中国地震局提出了 3 个月的短期预测意见，短期预测较为准确。震前向甘肃省政府进行了专题汇报，省政府研究并部署了甘肃省地震灾害紧急救援队组建、震情短临跟踪、地震安全性评价和抗震设防要求规范管理等工作，产生了重大社会效益。在地震预报意见的指导下，震前甘肃省地震局架设了 7 台强震仪，其中的一台架设在极震区民乐县永固乡，从而获得了极震区非常珍贵的三分向强震记录。本次地震的成功预报取得了一定的减灾实效，得到了中国地震局和甘肃省政府的表彰和奖励。

（3）本次地震属于中等强度地震，没有形成地表破裂带，地面变形也不是太多见，经调查，仅在极震区部分地方有规模不等的地裂缝1)。因此给发震断层的确定带来了一定的困难。但极震区地震时沿山脊发育的滑坡、边坡崩塌等地震次生灾害现象和部分地段轻微的地面变形、地裂缝等的分布有助于发震断层的推断。

参 考 文 献

[1] 袁道阳，2003，青藏高原东北缘晚新生代以来的构造变形与时空演化［D］，北京：中国地震局地质研究所博士学位论文

[2] 国家地震局地质研究所、国家地震局兰州地震研究所，1993，祁连山—河西走廊活动断裂带［M］，北京：地震出版社

[3] 何文贵、郑文俊，2004，2003 年 10 月 25 日民乐—山丹 6.1、5.8 级地震烈度和发震构造特征［J］，西北地震学报，26（3），240 ~ 245

[4] 郑文俊、何文贵，2005，2003 年甘肃民乐—山丹 6.1、5.8 级地震发震构造及发震机制探讨［J］，地震研究，28（2），133 ~ 140

[5] 闵祥仪、姚凯、何新社，2003，2003 年 10 月 25 日甘肃省民乐—山丹 M_S6.1 地震强震近场记录和分析［J］，西北地震学报，25（4），289 ~ 292

[6] 刘旭宙，2004，甘肃近期几次中强地震震源机制解［J］，西北地震学报，26（1），94 ~ 95

[7] 张洪由、许力生，2003，2003 年 10 月 25 日甘肃省西部发生 6.1 级和 5.8 级地震［J］，国际地震动态，（11），39

[8] 杨立明、王兰民，2004，民乐—山丹 6.1 级地震短期预报的科学总结［J］，西北地震学报，26（1），1 ~ 9

[9] 杨立明，2004，甘肃省民乐 6.1 级地震短期预报的简要回顾与启示［J］，国际地震动态，（1），15 ~ 17

[10] 肖丽珠、刘小凤、张小美，2005，2003 年民乐—山丹 6.1 和 5.8 级地震序列类型及地震学参数异常特征［J］，地震研究，28（1），28 ~ 33

[11] 莘海亮、张元生、郭晓、李稳，2008，2003 年民乐—山丹 6.1、5.8 级地震序列精确定位［J］，地震研究，31（2），129 ~ 133

[12] 石玉成、马尔曼、陈永明、付长华，2005，2003 年民乐—山丹 6.1、5.8 级地震房屋震害分析，西北地震学报，27（3），260 ~ 266

参 考 资 料

1）甘肃省地震局、中国地震局兰州地震研究所，2003 年 10 月 25 日甘肃省民乐—山丹 6.1、5.8 级地震现场科学考察报告，2004.12

2）甘肃省地震局，2001 年度甘肃省震情趋势研究报告，2000.11

3）甘肃省地震局，2002 年度甘肃省震情趋势研究报告，2001.11

4）甘肃省地震局，2003 年度甘肃省震情趋势研究报告，2002.11

5）甘肃省地震局，2004 年度甘肃省震情趋势研究报告，2003.11

Minle-Shandan Earthquakes of *M*6. 1 and 5. 8 on October 25, 2003 in Gansu Province

Abstract

Earthquakes of *M*6. 1 and 5. 8 occurred in Minle and Shandan county of Gansu province on October 25, 2003. Macroseismic epicenter was located in Yaozhaizi and Yuanquanzi Village of Yonggu town in Minle county and Shanghexizhaizi and Liuzhuang village of Huocheng town in Shandan county. The intensity at the epicenter was Ⅷ and meizoseismal area was irregular oval-shaped with major axis in NW direction. 10 people were killed, 46 people were injured and the direct economic loss was 50 million Yuan.

The earthquake sequence belonged to double mainshock-aftershock type. Aftershocks distributed 35 km far away from and west to the main shock as the almost same direction of the major axis of the meizoseismal area. The seismogenic structure was Minle-Yongchang and Tongziba hidden faults.

Within the distance of 300 km from the epicenter, there were 17 seismic stations, 7 of them with seismometric observations, 2 with integrated observations, and 7 with other precursory observations. Several strong-motion earthquake observation instruments were set up in the epicenter area before the earthquakes and strong motion records of mainshocks were obtained. There were 17 anomalies before the earthquake. 7 of them were precursory anomalies, 10 of them were seismic anomalies.

The *M*6. 1 and 5. 8 earthquakes occurred in the seismically dangerous area in Gansu province as specified in 2003. Based on abnormal variations before the earthquakes, a short-term forecasting for 3 months was submitted to China Earthguake Adminstration and the earthquakes occurred as expected.

报 告 附 件

附件一：预测预报证明材料

1. 甘肃省地震局，短期预报卡片，2003 年 9 月 19 日（附图 1）。

2. 甘肃省地震局，民乐公路所强震台记录的主震强震观测记录图（附图 2）。

3. 甘肃省政府，关于商请表彰甘肃省地震局的函，2003 年 11 月 17 日（附图 3）。

4. 张掖市人民政府，张掖市人民政府对省地震局民乐—山丹 6. 1 级地震预报工作的评价，2004 年 3 月 25 日（附图 4）。

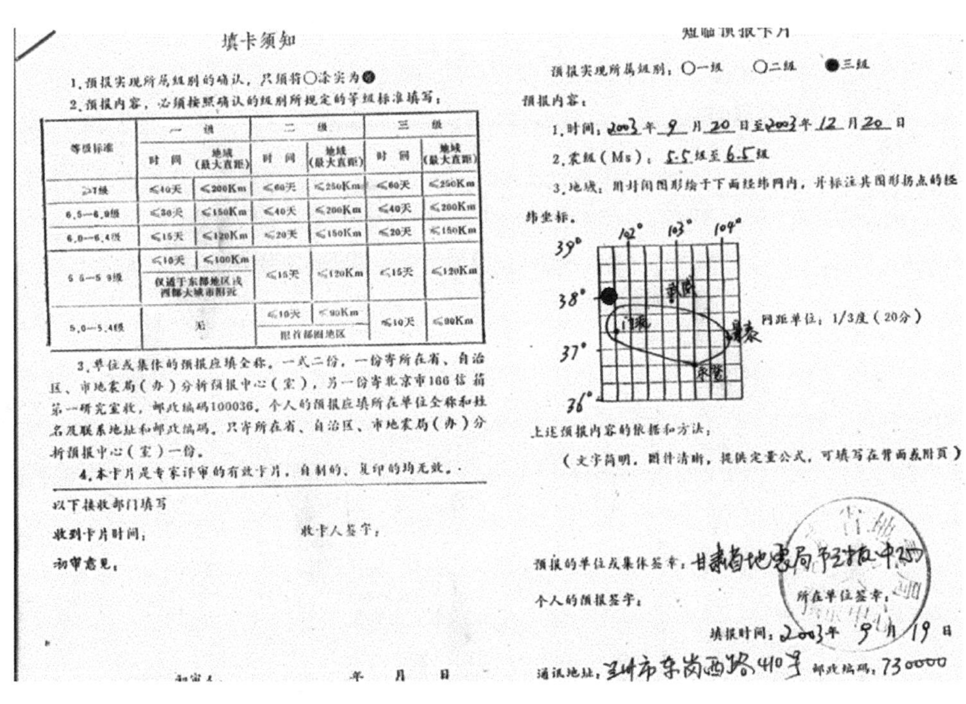

填卡须知

1. 预报实现所属级别的确认，只须将○涂实为●

2. 预报内容，必须按照确认的级别所规定的等级标准填写：

等级标准	一级		二级		三级	
	时间	地域（最大直距）	时间	地域（最大直距）	时间	地域（最大直距）
≥7级	≤40天	≤200Km	≤60天	≤250Km	≤60天	≤250Km
6.5—6.9级	≤30天	≤150Km	≤40天	≤200Km	≤40天	≤200Km
6.0—6.4级	≤15天	≤120Km	≤20天	≤150Km	≤20天	≤150Km
5.5—5.9级	≤10天	≤100Km	≤15天	≤120Km	≤15天	≤120Km
	仅适于东部地区及西部大城市附近					
5.0—5.4级	无		≤10天	≤90Km	≤10天	≤90Km
			限首都圈地区			

3. 单位或集体的预报应填全称。一式二份，一份寄所在省、自治区、市地震局（办）分析预报中心（室），另一份寄北京市166信箱第一研究室收，邮政编码100036。个人的预报应填所在单位全称和姓名及联系地址和邮政编码。只寄所在省、自治区、市地震局（办）分析预报中心（室）一份。

4. 本卡片是专家评审的有效卡片，自制的、复印的均无效。

以下接收部门填写

收到卡片时间： 收卡人签字：

初审意见：

初审人： 年 月 日

短期预报卡片

预报实现所属级别：○一级 ○二级 ●三级

预报内容：

1. 时间：2003 年 9 月 20 日至 2003 年 12 月 20 日

2. 震级（Ms）：5.5 级至 6.5 级

3. 地域：用封闭图形绘于下面经纬网内，并标注其图形拐点的经纬坐标。

上述预报内容的依据和方法：

（文字简明，图件清晰，提供定量公式，可填写在背面或附页）

预报的单位或集体签章：甘肃省地震局预报中心

个人的预报签字：

所在单位签章：

填报时间：2003 年 9 月 19 日

通讯地址：兰州市东岗西路410号 邮政编码：730000

附图 1 甘肃省地震局预报中心 9 月 20 日填报的短期预报卡片

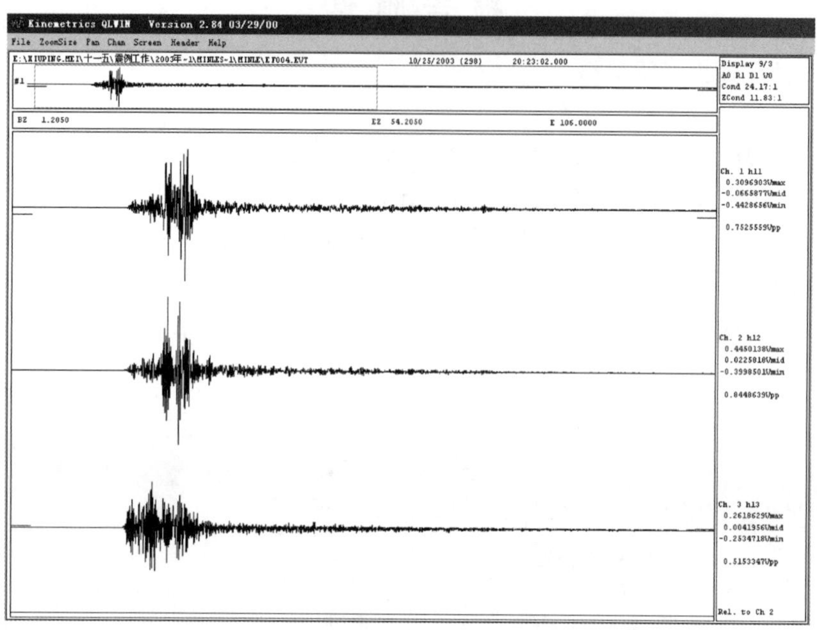

附图 2　民乐公路所记录的主震强震观测记录图

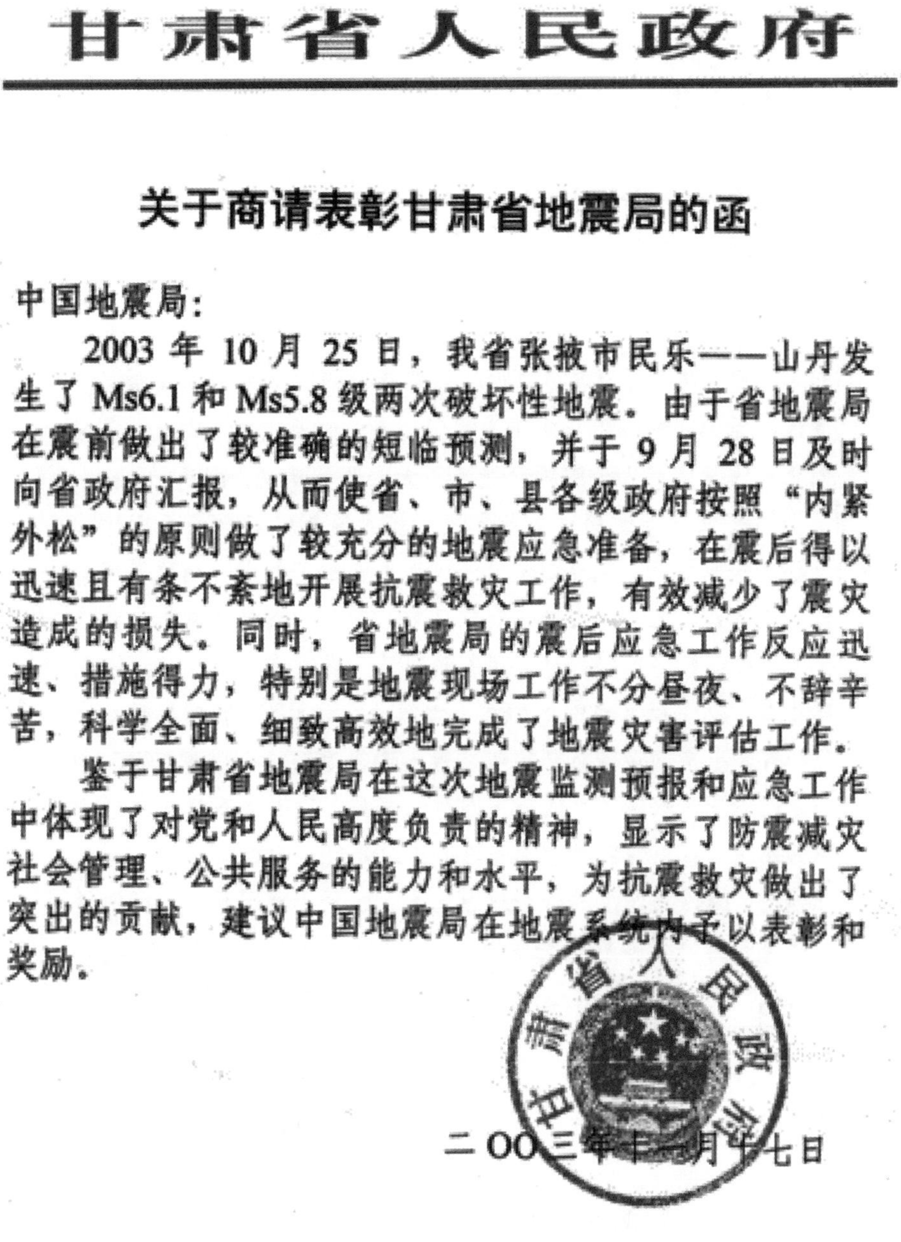

甘肃省人民政府

关于商请表彰甘肃省地震局的函

中国地震局:

2003 年 10 月 25 日，我省张掖市民乐——山丹发生了 Ms6.1 和 Ms5.8 级两次破坏性地震。由于省地震局在震前做出了较准确的短临预测，并于 9 月 28 日及时向省政府汇报，从而使省、市、县各级政府按照“内紧外松”的原则做了较充分的地震应急准备，在震后得以迅速且有条不紊地开展抗震救灾工作，有效减少了震灾造成的损失。同时，省地震局的震后应急工作反应迅速、措施得力，特别是地震现场工作不分昼夜、不辞辛苦，科学全面、细致高效地完成了地震灾害评估工作。

鉴于甘肃省地震局在这次地震监测预报和应急工作中体现了对党和人民高度负责的精神，显示了防震减灾社会管理、公共服务的能力和水平，为抗震救灾做出了突出的贡献，建议中国地震局在地震系统内予以表彰和奖励。

甘肃省人民政府

二〇〇三年十一月十七日

附图 3　甘肃省政府关于商请表彰甘肃省地震局的函

张掖市人民政府对省地震局民乐—山丹 6.1 级地震预报工作的评价

2003 年 10 月 25 日民乐－山丹强烈地震发生前，根据省地震局的预报意见，本着内紧外松的原则，我市地震系统启动了应急预案，全体人员进入紧急状态，加密了资料的报送，强化了震前短临跟踪、应急等工作，市政府对市地震局的应急工作高度重视，进行了具体的安排部署，启动了短临跟踪实施方案，加强了宏观监测、群测群防等工作。于 9 月 24 日启动我市辖五县一区宏观观测点，对应急车辆不能越野上山的问题，震前调拨三菱越野车一辆，用于地震应急工作。地震发生后，全市城乡居民人心慌慌，居民在中心广场或街头露天过夜，省局又发出了 24 小时内老震区发生更大地震的可能性不大的预报意见，据此，市政府主要领导要求地震部门领导带领工作人员到广场和街头现场动员群众回家过夜，并通过电话对城乡居民进行解答，劝其不要在露天过夜，以免冻坏身体，取得了好的效果，解除了震慌，稳定了人心，维护了社会秩序。接着省局又分别发布了三天内、一周内老震区发生更大地震可能性不大的预报意见。我们根据预报意见，有条不紊的按照张掖市破坏性地震应急预案，分解任务，全力抢险救灾，取得了抗震救灾阶段性成果。我们认为省局的地震预报意见与实际情况吻合，对我市抗震救灾工作起到了不可替代的作用，为最大限度的减轻地震灾害，起到了重要作用，并取得了明显的经济效果和社会效益，我们对省局的工作表示感谢。

张掖市人民政府办公室

二〇〇四年三月二十五日

附图 4　张掖市人民政府对省地震局民乐—山丹 6.1 级地震预报工作的评价

附录：民乐—山丹地震预报大事记

2002 年	
11 月	甘肃省地震局年度会商会
2003 年	
1 月	全国会商会，确定地震重点危险区
6 月	甘肃省地震局年中会商会
6 月 19 日	礼县前兆震群，甘肃省地震局紧急震情会商会
7 月 2 日	景泰发生 M_S4.4 地震，甘肃省地震局召开紧急震情会商会
7 月 18 日	肃南发生 M_S4.6 地震，甘肃省地震局召开紧急震情会商会
8 月 25 日	中国地震局 7 级地震研讨会
9 月 8～11 日	中国地震局南北地震带紧急震情会商会
9 月 11～18 日	甘肃省地震局对张掖市和武威市的地震应急工作进行了检查
9 月 17 日	甘肃省地震局修订印发了《甘肃省地震灾情速报实施细则（试行）》
9 月 20 日	甘肃省地震局填报了地震短期预报卡
9 月 20 日至 10 月 6 日	进行了祁连山中东段地区大地电磁测深的加密复测；启动了中法合作流动地震台网的加密数据处理工作
9 月 21 日	中国地震局第二形变测量中心开展形变加密观测工作
9 月 24 日起	加密了有关台站的观测次数，缩短数据报送周期
9 月 24 日	启动了第一批宏观观测点
9 月 26 日	中国地震局重点地区震情通报和工作部署会议
9 月 28 日	甘肃省地震局局长向省政府副省长进行了专门汇报
9 月 30 日	甘肃省地震局向省政府提交了专题报告
10 月 8 日	李膺副省长主持召开会议，专题研究甘肃省地震灾害紧急救援队组建、震情短临跟踪、地震安全性评价和抗震设防要求规范管理等工作
10 月 6 日	肃南发生 M_S3.9 地震，甘肃省地震局召开紧急震情会商会
10 月 10 日	肃南发生 M_S4.2 地震，甘肃省地震局召开紧急震情会商会
10 月 11 日	张掖市地震局针对三次有感地震应急中存在的主要问题向市政府做了专门汇报
10 月 15 日	省政府办公厅下发了《关于组建甘肃省地震灾害紧急救援队的通知》
10 月 16 日	启动了第二批宏观观测点。两批共计启动了 25 个县 280 个宏观观测点
10 月 23 日下午	兰化厂地震测报站电磁辐射异常落实
10 月 24 日	下午张掖市政府给市地震局调拨了一辆越野车用于地震应急
10 月 25 日	民乐—山丹交界 6.1、5.8 级地震

2003 年 11 月 13 日甘肃省岷县 5.2 级地震

甘肃省地震局

郑卫平　代　炜　刘小凤

摘　要

2003 年 11 月 13 日甘肃省岷县发生了 5.2 级地震，宏观震中位于岷县北部维新乡兹那村，震中烈度Ⅷ度，极震区为近东西向的椭圆形。地震造成 1 人死亡，128 人受伤，直接经济损失 1 亿元。

此次地震序列为主震—余震型，最大余震为 3.2 级。余震分布大致呈北西向，与极震区长轴走向基本一致。节面Ⅰ为主破裂面，主压应力 P 轴方向近 EW。发震构造为北西向临潭—宕昌断裂。

震中附近属地震监测能力较强地区，200km 范围内共有 16 个地震观测台，有测震等 7 个观测项目。震前共出现 5 个异常项目 19 条异常。震中周围 100km 范围内前兆观测异常台站和台项比均为 100% 和 100%，101～200km 范围内为 55% 和 28%。

甘肃省地震局对岷县 5.2 级地震做出了较好的中期预测，该地震发生在甘肃省甘东南地震重点危险区边缘。震前未提出短临预测意见，震后趋势判定基本正确。

前　言

2003 年 11 月 13 日 10 时 35 分，在甘肃省岷县、临潭、卓尼三县交界地区发生了一次 5.2 级地震。甘肃省地震局台网测定的微观震中为 34°46′N，103°56′E。宏观震中位于岷县维新乡兹那村漆家湾社、中寨乡北部山区，极震区烈度Ⅷ度。地震造成 1 人死亡，128 人受伤，直接经济损失总额约为 1 亿元人民币。

岷县 5.2 级地震前存在地震学异常 8 条，分别为 2 级地震空区、前兆震群、显著地震活动、频度、b 值、C 值、D 值和蠕变加速等。前兆观测有形变、应力、水氡、电阻率等 11 个异常项目。

地震发生后，甘肃省地震局和定西市地震局按照“甘肃省破坏性地震应急预案”，分别派出应急考察队、地震现场工作组赶赴地震灾区，在当地政府的协助下，开展地震灾害损失评估、地震地质考察、震情动态跟踪监视等工作，并在震区布设了流动地震台网。

甘肃省地震局对岷县 5.2 级地震做出了较好的中期预测，该地震发生在甘肃省地震局确定的 2003 年度地震重点危险区边缘，震前没有提出短临预报意见，震后地震趋势判定基本正确。

本报告是在有关文献[1~12]和资料1~3)的基础上经过重新整理和分析研究而完成的。

一、测震台网及地震参数

岷县 5.2 级地震震中 200km 范围内分布有 7 个测震台，100km 范围内有合作台和岷县台 2 个台，100～200km 分别有安宁台、定西台、天水台、武都台、迭部台等 5 个台（图 1）。甘肃省测震台网基本可监测到 $M_L \geq 2.0$ 级地震。

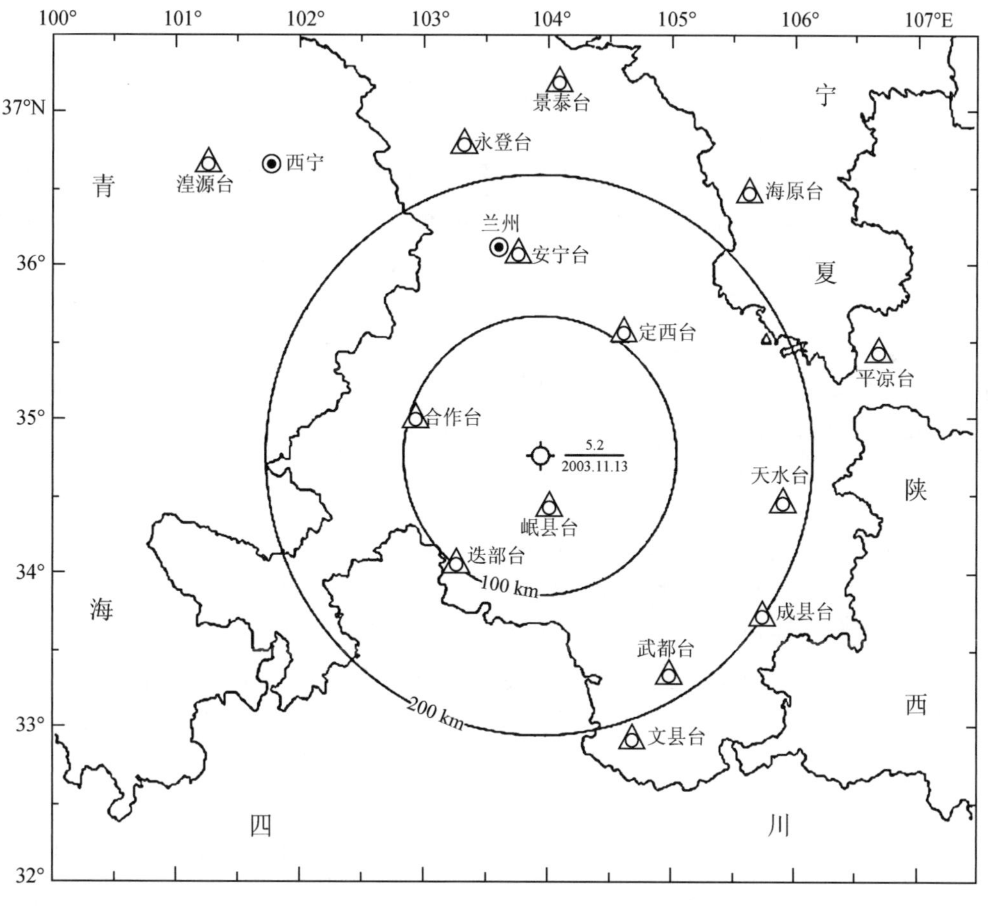

图 1　岷县 5.2 级地震前震中附近地震台站分布图

Fig. 1　Distribution of earthquake-monitoring station around the epicentral of *M*5.2 Minxian earthquake

表 1　岷县 5.2 级地震基本参数

Table 1　Basic parameters of the *M*5.2 Minxian earthquake

编号	发震日期	发震时刻	震中位置 (°)		震级				震源深度	震中地名	结果来源
	年 月 日	时 分 秒	φ_N	λ_E	M_S	M_L	M_b	M_W	h/km		
1	2004 11 13	10 35 00	34.7	103.9	5.2					甘肃岷县，临潭、卓尼间	中国地震台网
2	2003 11 13	10 35 12	34.7	103.9	5.2				22	甘肃岷县	甘肃地震台网
3	2005 11 13	10 35 14	34.65	103.89				4.6			CCDSN

CCDSN：China Center of Digital Seismic Net2work。

二、地震地质背景

岷县 5.2 级地震发生在青藏高原东缘的甘东南地区，是南北地震带与东昆仑断裂带的构造交汇部位，也是北西西向东昆仑断裂带与西秦岭北缘断裂带之间的构造转换区，区域范围内发育北西—北西西向和北东向两组断裂，本次地震就发生在北西向临潭—宕昌断裂带中段的北支。临潭—宕昌断裂带从西端的合作以西开始，向东经过临潭、岷县至宕昌南东与礼县—罗家堡断裂[2]交汇，构成一 V 字形构造。根据前人的初步调查[3]，断裂带全长大于 250km，由数条规模不等、相互平行或斜列的断裂组合而成。在合作—岷县间，该断裂分为南、北两支，在岷县东南一带又归并为一体，延伸到宕昌以南。断裂总体呈 NWW—NW 向展布，为一向 NE 方向突出的弧形，倾向 NE，倾角 50° ~ 70°，具左旋兼逆断性质。该断裂控制了合作、临潭、宕昌等第三纪盆地的形成、演化及构造变形，其新活动导致断裂沿线山脊、水系、洪积扇被断错，形成断崖、断层垭口、断坎、断陷槽地等。

从历史记载和地表地貌现象分析，该断裂新活动性强，地震活动水平较高，曾发生 1573 年岷县 6¾级地震和 1837 年临洮—岷县间 6 级地震[1,4]（图 2）。

本次地震没有形成地表破裂带，据此次地震的震源机制解，节面 I 走向与临潭—宕昌断裂带中段的走向基本吻合，且 P 轴方向近 EW，在此挤压力作用下断裂发生左旋走滑兼逆冲运动，这与临潭—宕昌断裂带中段的运动特征是一致的；从宏观调查圈定的等震线图来看，极震区由于受地形影响较大，与临潭—宕昌断裂的走向差异较大，但Ⅶ、Ⅵ的长轴方向与该断裂走向基本一致，宏观震中位于该断裂中段北支的上盘，与断裂距离较近，而且震害特征也明显具有方向性，垂直断裂方向的震害衰减明显比平行于断裂方向的快。因此，临潭—宕昌断裂带的中段北支前缘的次级断裂应是岷县 5.2 级地震的孕震和发震构造。

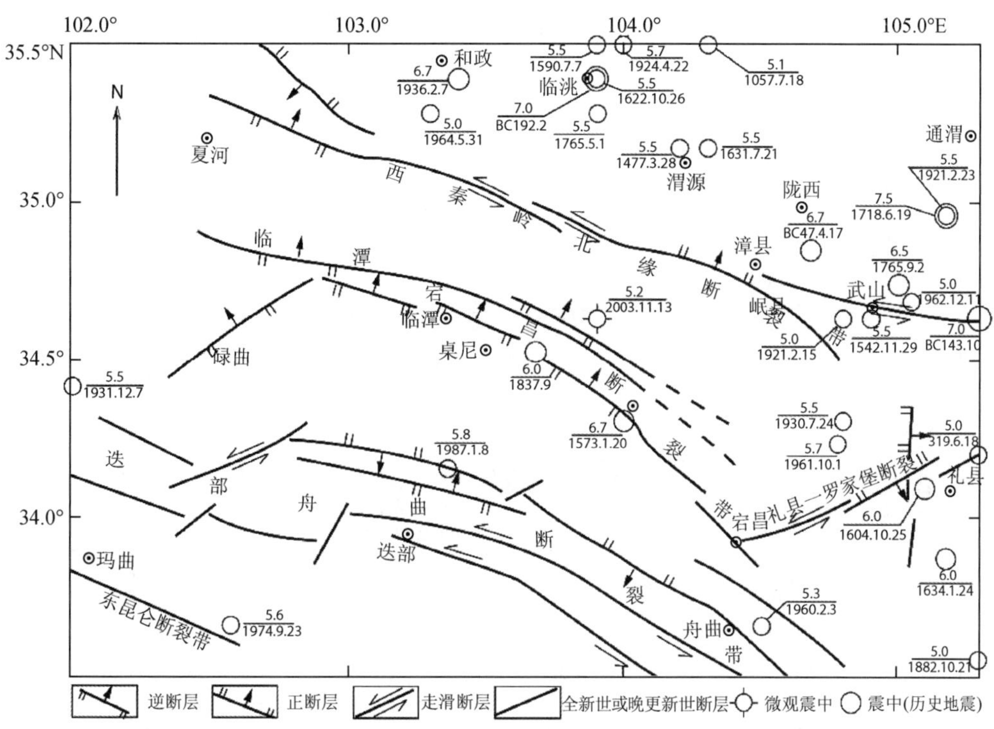

图 2 岷县 5.2 级地震附近主要断裂和历史地震分布图

Fig. 2 Distribution of main faults and historical earthquakes around Minxian area

三、地震影响场和震害

根据现场考察，岷县 5.2 级地震的宏观震中位于甘肃省岷县堡子乡兹那村和中寨乡扎拉村的交汇部位，极震区烈度为Ⅷ度，呈近椭圆形分布（图 3）。

Ⅷ度区：包括岷县堡子乡兹那村和明泉村、中寨镇扎拉村和大哈村。房屋类型主要是砖木结构和简易棚圈两类，及少量的砖混结构（主要是几所学校）和砖柱土坯结构或土坯结构类房屋。房屋破坏特征主要是木架结构整体顺山坡方向发生移位，部分倒塌或局部倒塌，房屋木柱与横梁的移位达 5cm 左右，部分达 10cm，房屋结构发生整体的变形，多数已变为危房。例如兹那村漆家湾社，70% 以上此类房屋发生不均匀移位，已基本变为危房；砖柱土坯结构或土坯结构的房屋多已墙体开裂，裂缝宽度多在 5 ~ 15cm，部分墙体倒塌或屋顶垮塌；棚圈大多数已倒塌；学校等的砖混结构房屋破坏相对较轻。另外，山体及路边坡发生崩塌和裂缝，裂缝宽度最宽可达 20 ~ 30cm，松动岩体及石块顺山坡发生滚动等现象也较为常见。

Ⅶ度区：主要分布在岷县堡子乡和中寨镇的绝大部分村庄及临潭县陈旗乡和卓尼县洮砚乡的部分村庄，房屋类型与Ⅷ度区基本相同，只是砖混结构房屋略有增加。房屋破坏特征主要是：墙体开裂，裂缝宽度在 3 ~5cm；公用住房的烟囱倒塌；民用住房顶部木架结构变形；

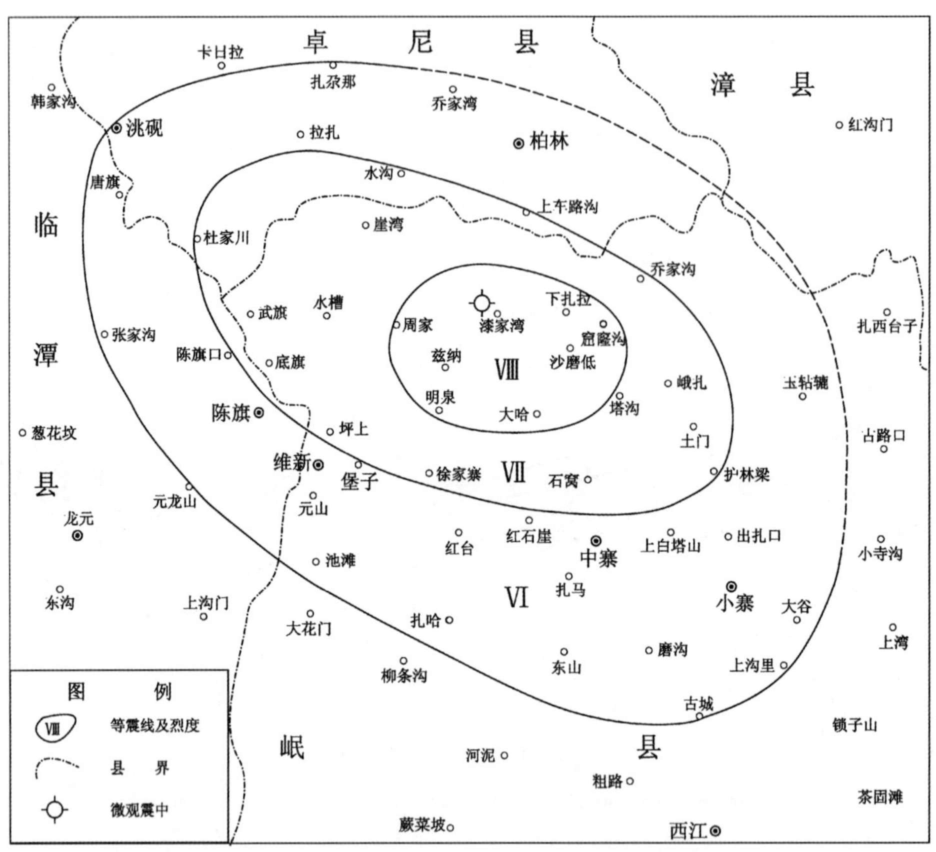

图3　岷县5.2级地震震害烈度等震线图

Fig. 3　Isoseismal map of the M5.2 Minxian earthquake

土坯房和棚圈等部分倒塌；但砖混结构房屋基本没多大破坏。另外部分桥梁有轻微的变形，少量地方地面有轻微裂缝，如中寨镇塔沟村房屋多已出现不同程度的裂缝，木架结构变形幅度为2~3cm。

Ⅵ度区：包括岷县维新乡、小寨乡、西江乡、临潭陈旗乡、卓尼柏林乡、洮砚乡、柏林乡等半数以上靠近震中的村庄。该区房屋类型与Ⅶ度区相同。房屋破坏类型主要是：少数房屋墙体轻微开裂、裂缝宽度1~5cm不等；部分房屋木架结构轻微变形；少量土坯房和棚圈倒塌；但结构相对较好的砖混结构类房屋几乎没有破坏；砖木结构类房屋破坏也较轻，出现轻微裂缝。

尽管本次地震震级不高，但由于震区特殊的地形地貌条件，造成极震区烈度达到了Ⅷ度，远高于同等震级地震的灾害。地震造成10余个乡村房屋不同程度的破坏，死亡1人，伤128人，直接经济损失总额约1亿元人民币。

四、地震序列

根据甘肃省地震台网的测定，岷县 5.2 级地震发生后余震不是很发育。截至 2004 年 8 月 31 日，序列共记录到余震 12 次，其中 0 ~ 0.9 级地震 1 次，1.0 ~ 1.9 级地震 6 次，2.0 ~ 2.9 级地震 2 次，3.0 ~ 3.9 级地震 3 次，最大余震为 2003 年 11 月 13 日的 3.2 级，最大余震与主震的震级差为 2.0（表 2，图 4）。依据地震类型判定指标，本次地震序列类型为主震—余震型，序列衰减基本正常。余震主要分布在主震附近，且随着时间的推移，余震逐渐向外围扩散（图 5），但由于余震次数较少，其优势方向不甚明显，大致呈北西向，无法据此判定其发震构造。岷县 5.2 级地震发生在区域地震活动水平相对平静的背景下，震前无明显前震活动。

表 2　岷县 5.2 级地震序列目录

Table 2　Catalogue of the *M*5.2 Minxian earthquake sequence

编号	发震日期	发震时刻	震中位置（°）		震级 M_S	震源深度 h/km	震中地名	结果来源
	年 月 日	时 分 秒	φ_N	λ_E				
1	2003 11 13	10 35 12	34.75	103.93	5.2	10	岷县	资料 1）
2	2003 11 13	11 08 51	34.75	103.90	3.2	10		
3	2003 11 14	00 04 51	34.77	103.87	1.7	10		
4	2003 11 15	01 05 42	34.75	103.88	3.2	5		
5	2003 11 15	01 09 33	34.75	103.82	1.5	10		
6	2003 11 27	04 38 48	34.75	103.85	1.2	10		
7	2003 12 26	04 11 05	34.98	103.83	2.2	0		
8	2004 02 12	12 08 08	34.82	103.83	0.8	0		
9	2004 02 22	09 11 14	34.75	103.93	3.0	15		
10	2004 03 03	22 44 38	35.00	103.93	1.1	0		
11	2004 06 19	10 17 11	34.70	103.62	2.2	0		
12	2004 07 03	02 10 15	34.72	103.43	1.2	0		
13	2004 08 24	10 44 05	34.70	104.03	1.3	7		

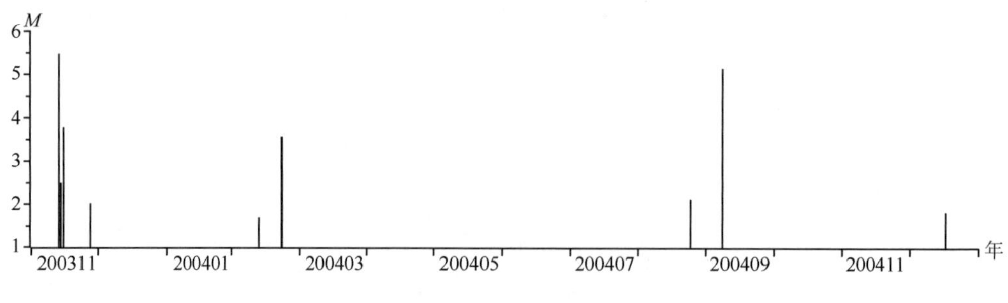

图 4　岷县 5.2 级地震序列 *M-t* 图

Fig. 4　*M-t* diagram of the *M*5.2 Minxian earthquake sequence

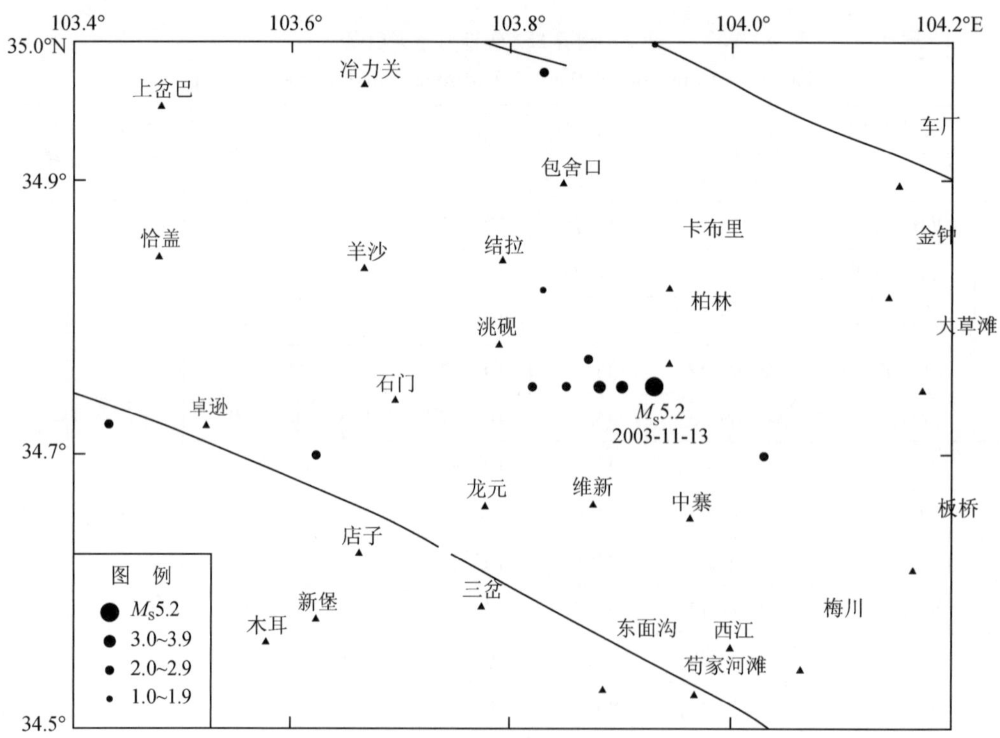

图 5　岷县 5.2 级地震余震震中分布图

Fig. 5　Distribution of the *M*5.2 Minxian earthquake sequence

五、震源参数和地震破裂面

岷县 5.2 级地震后，收集甘肃台网的地震波形资料，读取 P 波初动，求得震源机制，符号矛盾比 0.095，该地震震源机制为逆冲型（表 3、表 4 和图 6）。

表 3 岷县 $M5.2$ 级地震的震源机制解

Table 3 Focal mechanism solutions of the $M5.2$ Minxian earthquake

编号	节面 I			节面 II			P 轴		T 轴		N(B)轴		X 轴		Y 轴		结果来源
	走向	倾角	滑动角	走向	倾角	滑动角	方位	仰角	方位	仰角	方位	仰角	方位	仰角	方位	仰角	
1	330	27	67	175	65	101	256	20	106	68	350	10					荣代潞[4]
2	330	27	67	175	65	101	256	20	106	68	350	10					CCDSN
3	338	44	67	188	50	110	264	3	163	74	355	15					刘旭宙[5]
4	325	28	96	138	62	87	231	17	41	73	140	3					中国数字台网中心

表 4 岷县 $M5.2$ 级地震的矩张量解

Table 4 Moment tensor solutions of the $M5.2$ Minxian earthquake

编号	节面 I			节面 II			矩张量						地震矩	矩震级	结果来源
	走向	倾角	滑动角	走向	倾角	滑动角	M_{xx}	M_{yy}	M_{zz}	M_{xy}	M_{yz}	M_{zx}	M_0	M_w	
1	160	72	115	284	30	38									荣代潞[4]

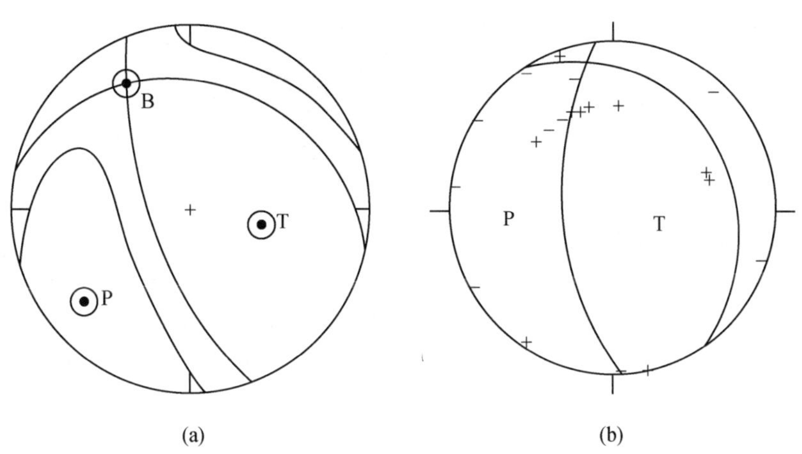

图 6 岷县 5.2 级地震的矩张量解和相应的最佳双力偶解（a）及其震源机制解（b）

Fig 6 Solution of moment tensor and the best even- solution (a) and Focal mechanism solutions (b) of the $M5.2$ Minxian earthquake from P wave records

震源机制解中的节面 I 走向 330°，与临潭—宕昌断裂带中段的走向基本吻合，且 P 轴方向近 EW，在此挤压力作用下断裂发生左旋走滑兼逆冲运动，这与临潭—宕昌断裂带中段的运动特征是一致的，同时也反映了该地区的构造应力特征。

六、地震前兆观测台网及其前兆异常

岷县 5.2 级地震震中 200km 范围内共有 16 个地震台，其中 4 个测震观测单项台，6 个综合观测台，6 个前兆单项观测台（图 7，表 5），总计有 7 个定点观测项目，包括测震、水氡、水位和流量、电阻率、地倾斜、地磁、应力和应变等。岷县 5.2 级地震前共出现异常 19 项次，其中测震学异常 8 项次。在震中 200km 以内有水氡、地倾斜、应变和地电阻率共 11 项次异常[7~12]。震前甘肃省地震局根据地震活动性及应变、水氡等项目异常作了一定程度的中期预测。

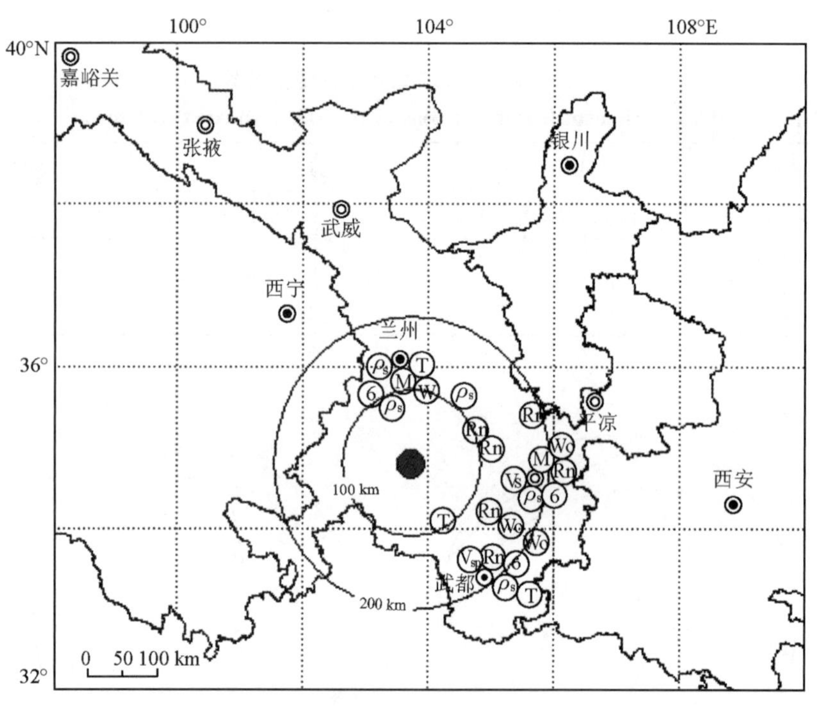

图 7　岷县 5.2 级地震前定点前兆观测台站分布图

Fig. 7　Distribution of precursory monitoring stations before the M_S5.2 Minxian earthquake

表 5 地震异常登记表

Table 5 Anomalies catalogue of the earthquake

序号	异常项目	台站（点）或观测区	分析方法	震前异常起止时间	最大幅度	震中距 Δ（km）	图号	异常特点及备注
1	地震空区	震中 200km 范围内	$M_L \geqslant 2.0$ 级空间分布	2002.9～2003.11			8	空区走向 NW，长轴 195km，持续时间为 14 个月。震后总结的异常
2	前兆震群	震中 100km 范围内	活动分布	2003.6.19		100		礼县震群是一条震前提出的重要的短期异常。$M_{max}=3.9$ 级，ρ 值 =0.3830，K 值 =0.8025，h 值 =0.8，b 值 =0.6959，为前兆震群
3	显著地震	震中 100km 范围内	活动分布	2003.7.10～13		100		7 月 10 日、13 日礼县 3.3 级地震，与主震震级差为 M_S1.9，空间距离 100km，时间间隔 4 个月。为震后总结的异常
4	频度	甘东南	M1.2～4.9 两月累计滑动	2002.3～12	57	震中周围		区域地震活动增强，频度高于 1 倍均方差。震后总结的异常
5	b 值	甘东南	$M \geqslant 1.2$ 级地震 b 值时间扫描	2002.1～2002.9	0.26	震中周围		正常背景时段 b 值为 0.84，2002 年 1 月出现 $\leqslant 1\sigma$ 的低值异常
6	C 值	甘东南	$M \geqslant 1.2$ 级地震 C 值时间扫描	2002.4～10；2003.7～	0.39	震中周围		高于 2 倍均方差。震后总结的异常
7	D 值	甘东南	$M \geqslant 1.2$ 级地震 D 值时间扫描	2002.5～2002.9	0.3	震中周围		高于 1 倍均方差。震后总结的异常

续表

序号	异常项目	台站（点）或观测区	分析方法	震前异常起止时间	最大幅度	震中距Δ（km）	图号	异常特点及备注
8	蠕变	甘东南	蠕变时间曲线	2001.9～2003.10		震中周围		蠕变曲线加速，震后总结的异常
9	地倾斜	宕昌	日均值	NS：2002.07～	490ms	80	9	相对西南倾斜
				EW：2002.05～	600ms			
10	电阻率	临夏	日均值付氏分析	NS：2002.02～	4.1%	101	10	低值异常
				EW：2002.05～	3.5%			
11	水氡	武山	日值	1#：2002.06～2003.07	80Bq/L	104	11	破年变低值异常
				22#：2003.02～09	22Bq/L			破年变高值异常
12	应力	刘家峡	日均值	NS：2003.09～	1373uH	131	12	各向异性异常
				EW：2003.09～	989uH			
				NW：2003.09～	433uH			
13	电阻率	通渭	日均值	NS：2003.08～	1.1%	144	13	
14	水氡	礼县	五日均值	2003.05～	5Bq/L	149	14	破年变低值异常
15	水氡	西和	日值	2003.09～	3.5Bq/L	167	15	破年变高值异常
16	电阻率	天水	日均值	NS：2003.03～	0.7%	181	16	各向异性异常
			日均值付氏分析	EW：2002.06～2003.08	0.8%			

续表

序号	异常项目	台站（点）或观测区	分析方法	震前异常起止时间	最大幅度	震中距Δ（km）	图号	异常特点及备注
17	电阻率	武都	日均值	EW：2003.03 ~	1.6%	191	17	各向异性异常
				NW：2003.02 ~09	2.5%			
				NW1：2003.06 ~	1.3%			
18	应变	武都	日均值	NW：2003.07 ~	3.2×10^{-8}	191	18	各向异性异常
				NS：2003.09 ~	3.8×10^{-8}			
				NE：2003.07 ~	3.0×10^{-8}			
				EW：2003.07 ~	4.0×10^{-8}			
19	水氡	静宁	日均值	2003.03 ~	3.5Bq/L	197	19	低值异常

1. 地震学异常

地震空区：震前在距震中 200km 范围内出现 2 级以上地震活动平静过程，持续时间为 14 个月（图 8）。空区走向 NW，长轴 195km。2003 年岷县地震发生在空区内部，之后该空区逐渐解体。

前兆震群：岷县 5.2 级地震前，在距离主震震中约 100km 处发生 2003 年 6 月 19 日礼县震群。根据序列类型判定结果，$M_{max}=3.9$，$\Delta M\leqslant0.5$ 级的地震有 4 次，判定该序列为震群型。序列参数分别为，$U=0.1905$，$U\leqslant0.50$ 为非前兆性震群；$F=0.3655$，$F\leqslant0.70$ 为非前兆性震群；$\rho=0.3830$，$\rho<0.55$ 为前兆震群；$K=0.8025$，$K>0.70$ 为前兆震群；$h=0.8000$，$h<1.00$ 为前兆震群；$b=0.6959$，$b>0.65$ 为前兆震群，综合判定该震群为前兆震群。

显著地震：岷县 5.2 级地震前，在距离主震震中约 100km 处发生 2003 年 7 月 10 日和 13 日礼县 3.3 级地震。这 2 次地震与主震震级差为 $M_S1.9$，空间距离 100km，与主震之间的时间间隔 4 个月。

根据地震活动性参数时间扫描分析结果，震前出低 b 值异常，震后总结发现，频度、C 值、D 值和蠕变加速等地震学参数也存在异常（表 5）。

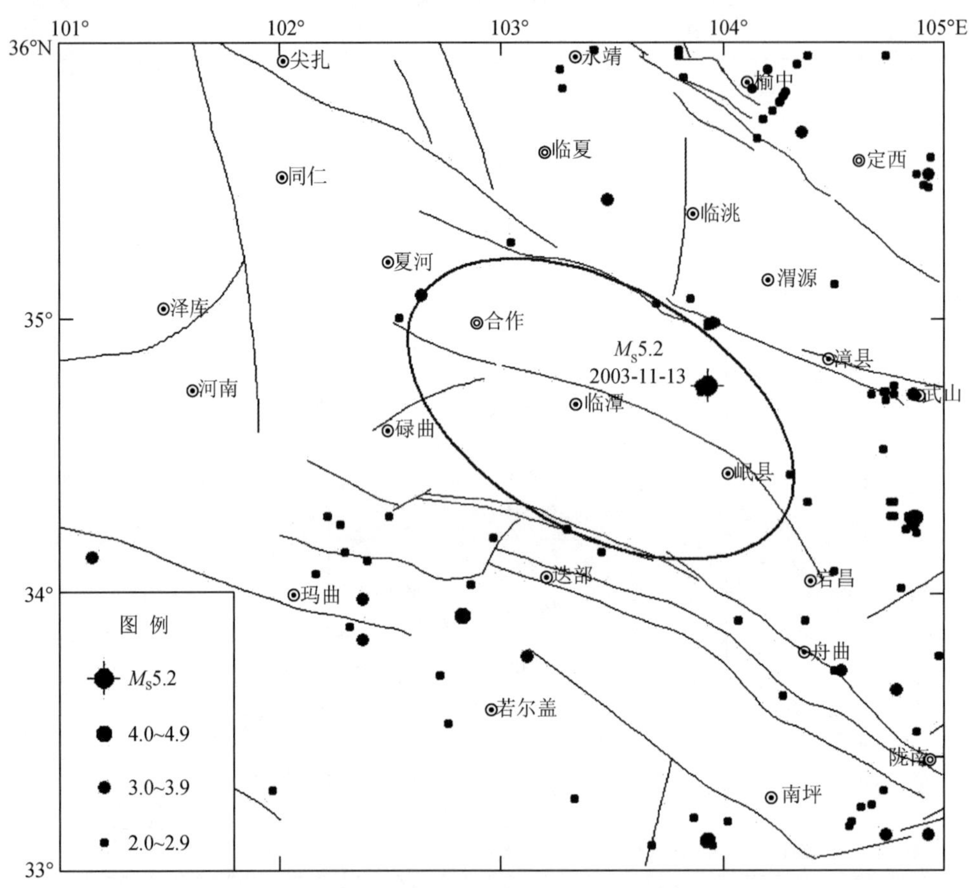

图 8　震中周围 $M_L \geqslant 2.0$ 级地震空间分布图（2002.9～2003.11）

Fig. 8　Distribution of $M_L \geqslant 2.0$ earthquakes around the Minxian（2002.9-2003.11）

2. 前兆异常

中期异常：宕昌水管仪 NS、EW 道分别于 2002 年 7、5 月出现转折性异常，呈相对南西异常倾斜（图 9）；经付氏分析处理，临夏电阻率 NS、EW 道分别于 2002 年 2、5 月出现明显低值异常（图 10）；武山 1 号井、22 号泉水氡分别于 2002 年 6 月、2003 年 2 月出现破年变性异常，并在震前结束（图 11）；礼县水氡于 2003 年 5 月出现破年变性低值异常（图 14）；天水电阻率 EW 道于 2002 年 6 月出现破年变性低值异常（图 16）；武都电阻率 3 道于震前出现各向异性的破年变性异常（图 17）；静宁水氡于 2003 年 3 月出现破年变性低值异常（图 19）。

短期异常：武山 22 号井水氡于 2003 年 9 月出现快速下降性异常恢复（图 11）；刘家峡应力 3 道于 2003 年 9 月出现各向异性异常，呈 NS 向受压、EW 向舒张的受力状态（图 12）；通渭电阻率 NS 道于 2003 年 8 月出现高频振荡（图 13）；西和水氡 2003 年 9 月出现破年变性高值异常（图 15）；天水电阻率 NS 道于 2003 年 3 月出现高值异常，震前恢复（图 16）；武都电阻率 3 道异常于震前基本结束，均出现快速变化过程（图 17）；武都应变 4 道于 2003 年 7～9 月出现各向异性异常（图 18）。

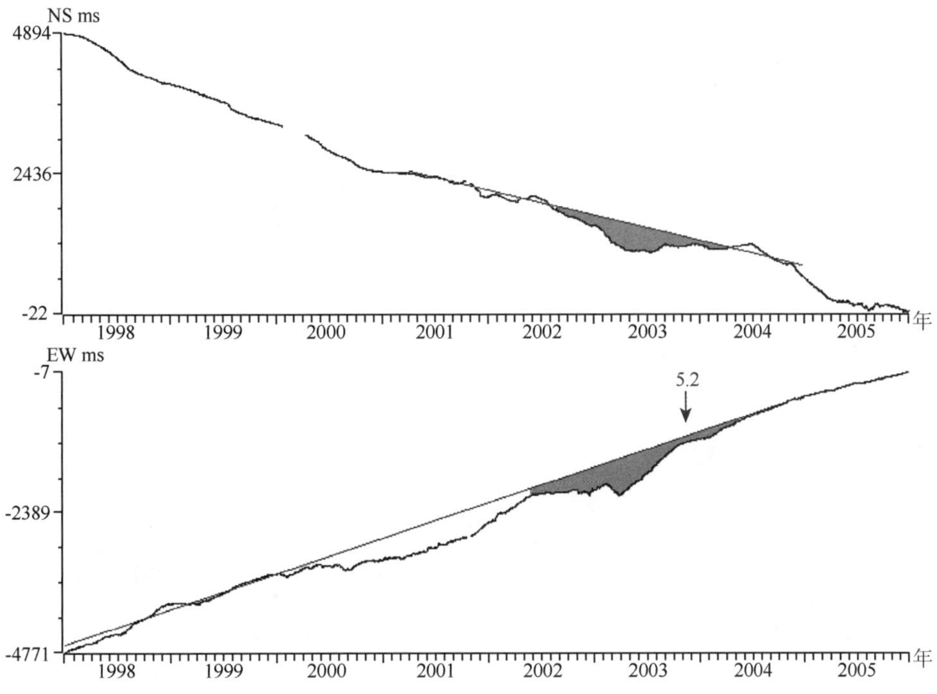

图 9 宕昌倾斜日均值曲线

Fig. 9 Curve of daily mean value of tilt at Dangchang station

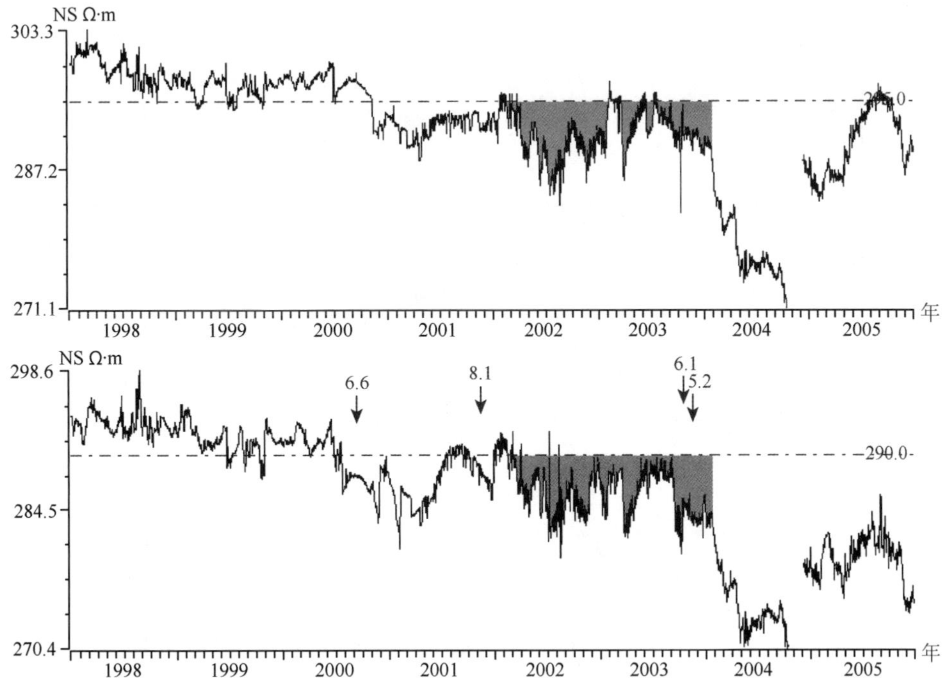

图 10 临夏地电阻率付氏分析残差曲线

Fig. 10 Curve of daily mean value of apparent resistivity at Linxia station

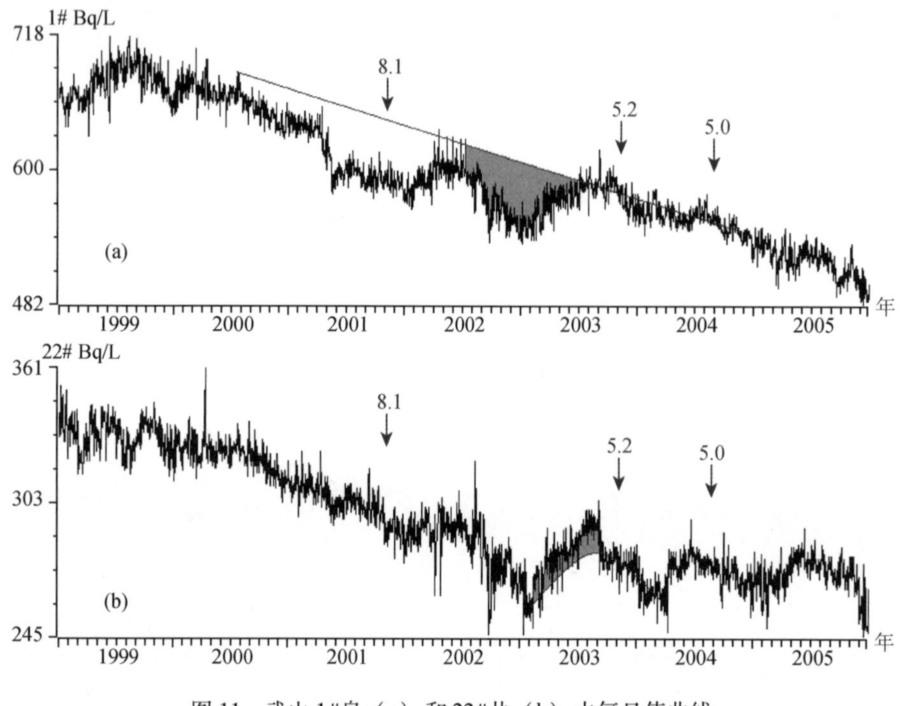

图 11　武山 1#泉（a）和 22#井（b）水氡日值曲线

Fig. 11　Daily value curve of radon content in groundwater at Wushang 1# and 22# station

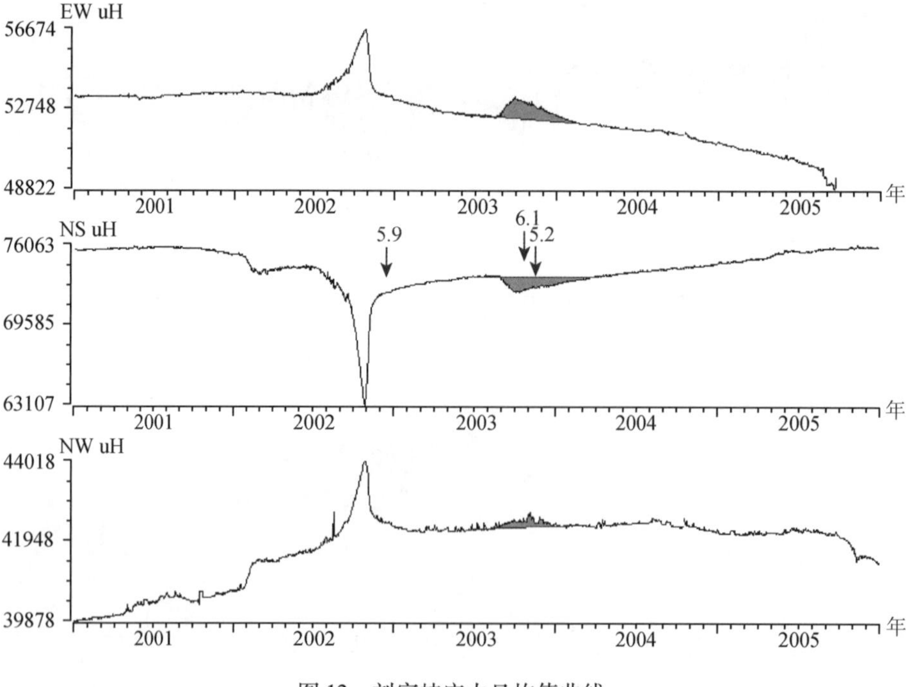

图 12　刘家峡应力日均值曲线

Fig. 12　Curve of daily mean value of electric induction stress at Liujiaxia station

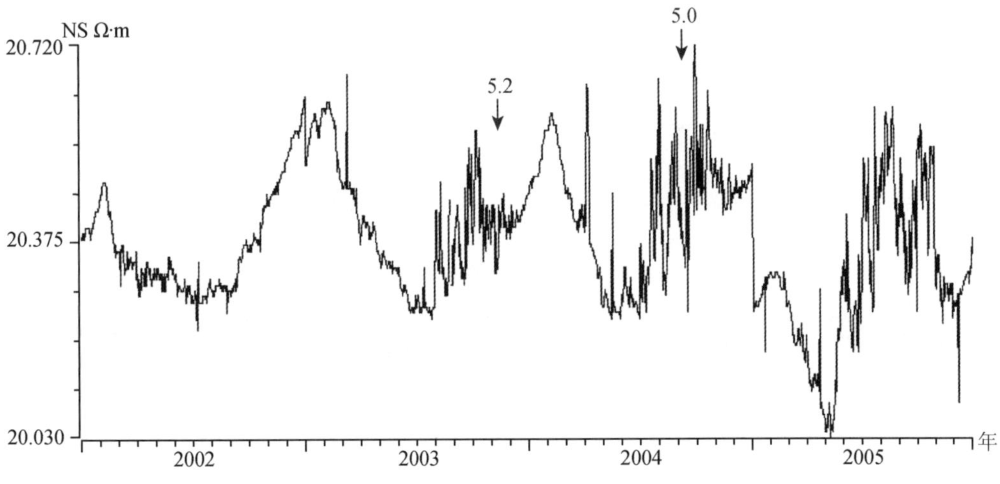

图 13　通渭地电阻率日均值曲线

Fig. 13　Curve of daily mean value of apparent resistivity at Tongwei station

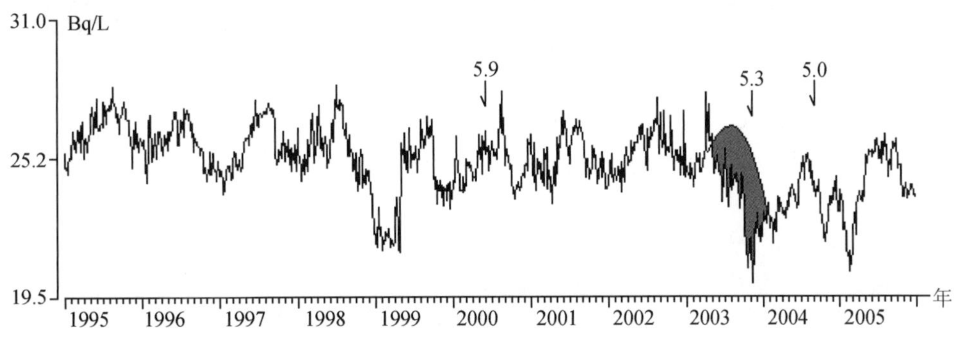

图 14　礼县水氡五日均值曲线

Fig. 14　Curve of 5-day mean value of radon content in groundwater at Lixian station

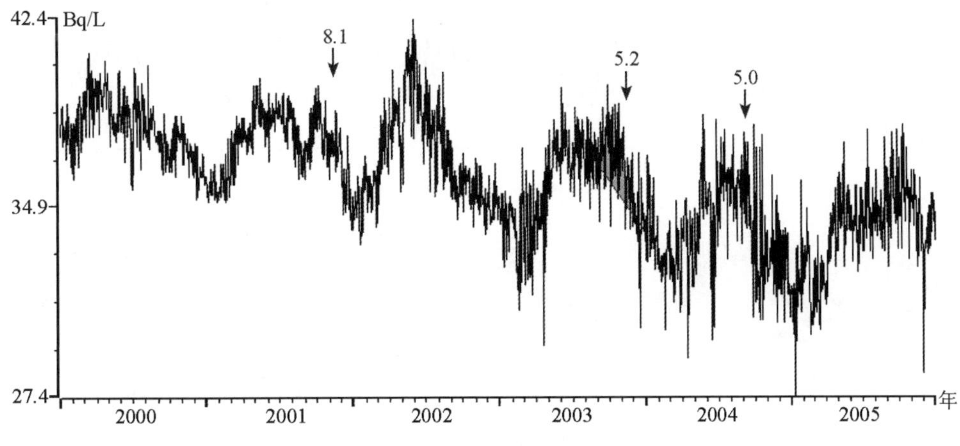

图 15　西和水氡日值曲线

Fig. 15　Daily value curve of radon content in groundwater at Xihe station

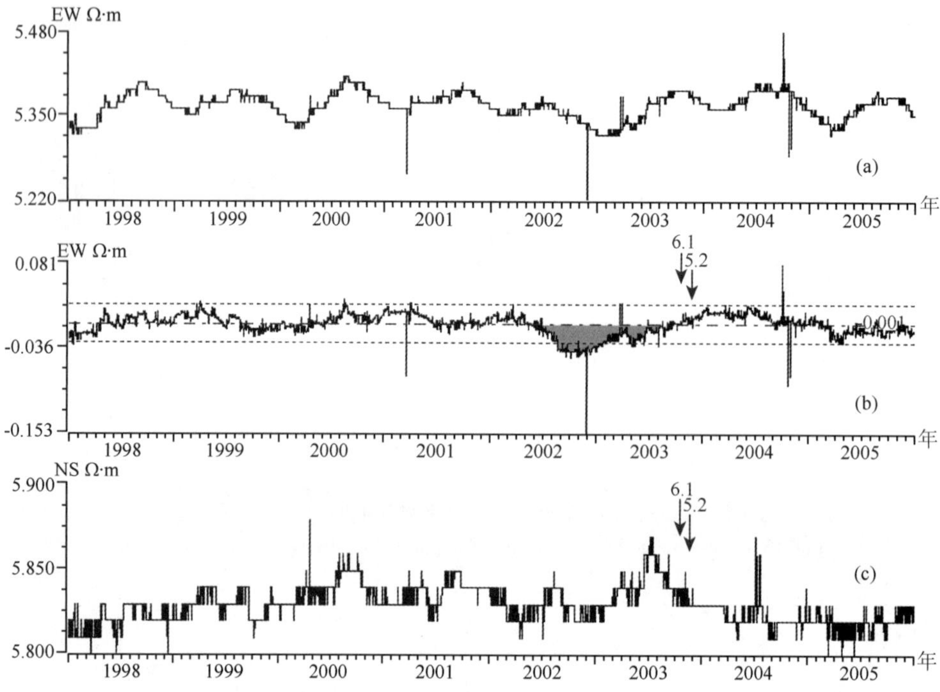

图 16　天水地电阻率日均值（a、c）和矩平残差（b）曲线

Fig. 16　Curve of daily mean value of apparent resistivity at Tianshui station

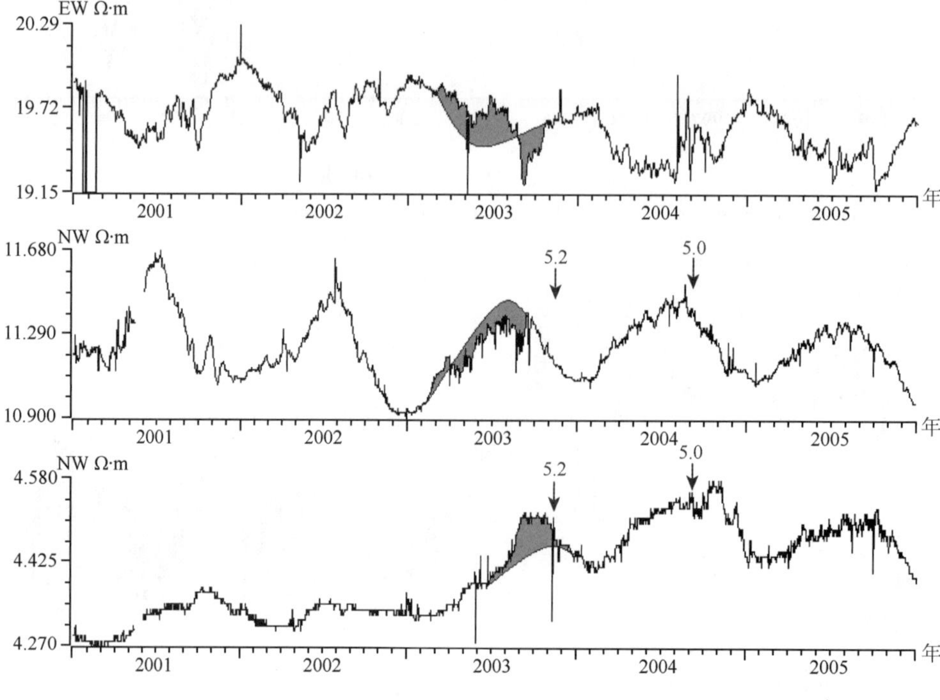

图 17　武都地电阻率日均值曲线

Fig. 17　Curve of daily mean value of apparent resistivity at Wudu station

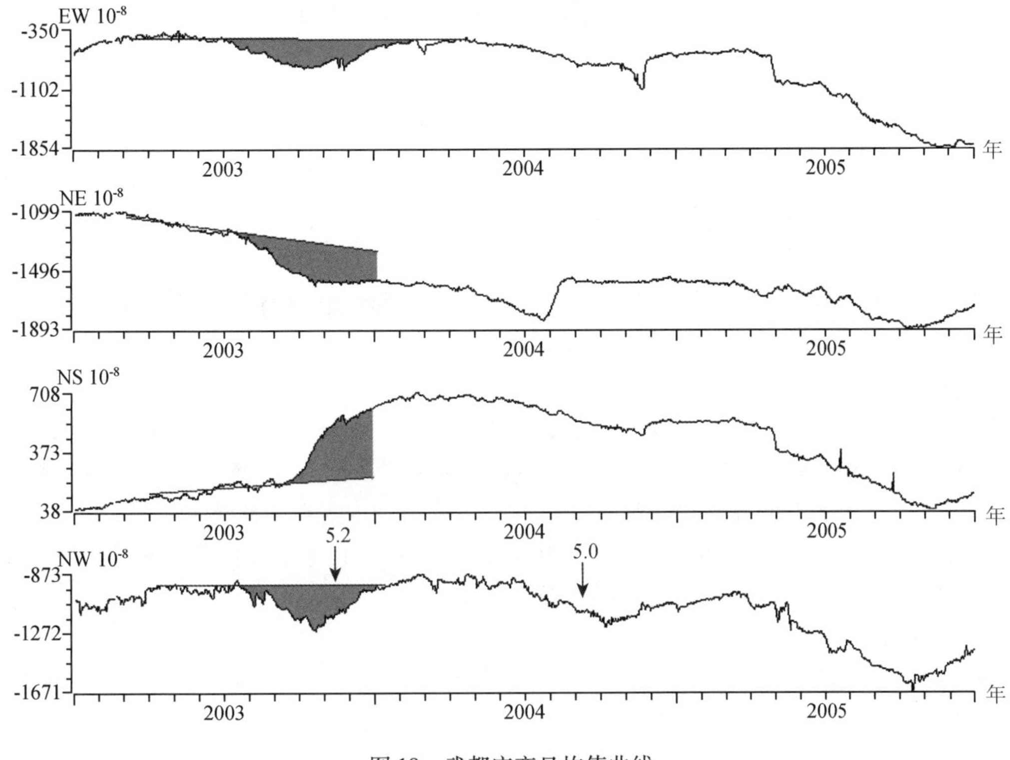

图 18 武都应变日均值曲线

Fig. 18 Curve of daily mean value of borehole strain at Wudu station

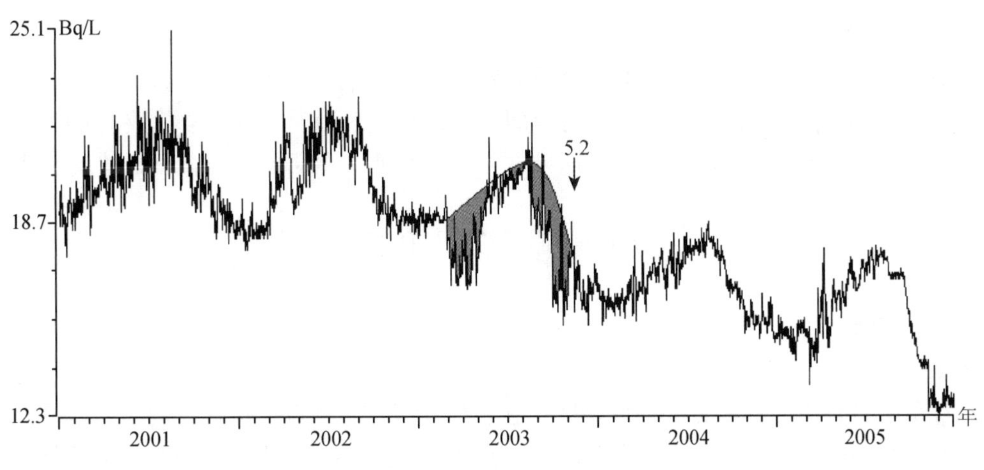

图 19 静宁水氡日值曲线

Fig. 19 Daily value curve of radon content in groundwater at Jingning station

七、地震前兆异常特征分析

1. 地震学异常特征

岷县 5.2 级地震前，地震活动性异常相对较少，仅出现 1 项中期异常，即持续 1 年左右的 2 级地震空区，还有 2 项短期异常，分别为 6 月 19 日礼县震群和 7 月 10～13 日显著地震活动。在 2001 年 11 月 14 日昆仑山口西 8.1 级地震后，位于东昆仑断裂带东端的甘东南—甘青川交界地区 3、4 级地震活动明显增强，从 2001 年 11 月 23 日至 2002 年 12 月 21 日发生 3.0～4.0 级地震 10 次，2003 年 1～10 月仅发生 1 次 3 级地震，存在较明显的增强—平静—发震的特征。

2. 前兆观测异常特征

2003 年 11 月 13 日岷县 5.2 级地震发生在甘肃省南部临潭—宕昌断裂带的中段北支前缘的次级断裂上，200km 范围内定点前兆观测台站较多，受台网布局的限制，大多分布在震中的北、东方向，西南方向没有观测台站。震中 200km 范围内定点前兆有 11 个测项、10 个台项的异常，0～100km 和 101～200km 范围内各有 1 个和 9 个台项的异常，最近的为宕昌倾斜，距震中约 80km；200km 内最远的为静宁水氡，距震中 197km。

1）空间分布特征

岷县 5.2 级地震前 200km 范围内共 12 个前兆观测台站 6 个观测项目（图 1），有 10 个台站的 11 个测项、22 个测道出现前兆异常，其中 100km 范围内共 1 个前兆观测台站，有 1 个台站的 1 个测项、2 个测道出现前兆异常，均占观测项目和测道总数的 100% 和 100%；101～200km 范围内共 11 个前兆观测台站，有 9 个台站的 10 个测项、20 个测道出现前兆异常，分别占观测项目和测道总数的 40% 和 39%；可见 100km 范围内异常测项比和测道比均较高。

2）时间进程特征

在 200km 范围内的定点前兆 11 个测项的异常中，半年以上的中期异常为 6 个台项，占异常测项总数的 55%；出现 1～6 个月的短期异常的有 7 个测项，占异常总数的 64%；没有 1 个月以内的临震异常。定点前兆以中期异常为主要特征，短期异常具有一定的成组性。

八、震前预测、预防和震后响应

1. 震前预测情况

在 2003 年度甘肃省地震趋势研究报告中，研究确定了三个地震危险区，其中祁连山中东段和甘东南甘青川交界地震重点危险区存在发生 6 级左右地震的可能，河西西段注意监视区存在发生 5～6 级地震的可能。2 个地震重点地震危险区被全国地震趋势会商会论证为 2003 年度全国地震重点危险区。岷县 5.2 级地震发生在甘东南甘青川交界地震重点地震危险区东部边界附近，空间距离约 50km。

在 2003 年 9 月 24 日月会商意见中，认为祁连山地震带存在发生 5～6 级地震的可能，同时应密切关注甘东南地区的震情发展。2003 年 10 月 29 日月会商、11 月 5 日周会商结论意见认为甘东南、祁连山西段存在发生 5～6 级地震的可能。该结论一直持续至 2003 年 11 月 13 日发生岷县 5.2 级地震。

2. 震后趋势判定

岷县 5.2 级地震发生后，甘肃省地震局立即组织地震预报评审委员会专家和全体分析预报人员召开紧急会商会，及时开展地震序列类型和震后趋势判定工作。11 月 13～17 日共召开了 4 次会商会，结论意见认为“近期原震区及其邻区仍有发生强震的可能”或“近期原震区及其邻区仍有发生 5 级左右地震的可能”，并认为“余震序列不发育”，但均未提出本次地震序列类型特征。2003 年 11 月 19 日周会商意见和 11 月 26 日月会商意见中也未提及地震序列类型。

3. 震后响应

地震发生后，甘肃省地震局组织专家和应急人员赶赴现场，进行现场的地震监测预报工作。现场工作组与甘肃省地震局分析预报人员密切跟踪地震序列的发展，密切关注前兆资料和异常的发展变化，结合该地区历史地震活动情况，快速作出震后趋势的初步判定，为震后应急救援和现场考察工作提供了准确、有力的支持。

该地震发生后，甘肃省地震局在震区增上了流动地震台网，对地震序列类型判定、震后趋势判定及现场震情监测工作起到了积极的作用。

4. 震害特征

尽管岷县 5.2 级地震震级不高，但由于震区特殊的地形地貌条件，造成极震区烈度达到了Ⅷ度，远高于同等震级地震的灾害。通过现场考察，认为造成震区烈度明显偏高和震害较为严重的原因有以下几点：

（1）地基条件的稳定性差。本次地震极震区的几个村庄，建筑在陡峭的地形条件上，高角度的斜坡多为松散的黄土堆积，坡体本身与基岩自由面之间就结合不稳定，当在瞬间受到外力作用时，上部的松散土体很容易顺坡向下滑动，因此造成其上建筑物的移位和变形等。

（2）房屋结构不合理。极震区的房屋多为砖木结构，前后墙之间的结构差异本身就造成了在受力过程中的应力局部集中，后墙体容易与木架结构脱离，同时又使木架发生整体变形综上所述，该地区的民用住房结构不合理及地基条件的不稳定是造成震害加大的主要因素，同时也为震后重建工作和相似地区民用住房的改造敲响了警钟，提高农村民用房屋结构的抗震性能已是当务之急。

九、结论与讨论

（1）2003 年 11 月 13 日甘肃省岷县 5.2 级地震发生在东昆仑断裂带与西秦岭北缘断裂带的构造转换区。本次地震宏观震中位置位于甘肃省岷县堡子乡兹那村和中寨乡扎拉村的交汇部位，极震区烈度为Ⅷ度。极震区为一近椭圆形，其震害及烈度均明显高于同等强度的其他地震，其原因主要是因为该地区陡峭的地形地貌条件加重了震害。这次地震的震害特征也为今后的农村民用房屋的抗震性能的提高以及在农村普及防震知识提供了很好的素材。

（2）岷县 5.2 级地震前没有做出明确的短临预测意见，仅发现一些较突出的异常现象。

在2003年10月25日民乐—山丹6.1、5.8级地震震情跟踪过程中，认为甘东南地区出现的突出地震活动性和前兆异常现象尚不能完全对应该地震，如礼县震群、武都应变、清水流量、礼县水氡和流量、灵台水氡等异常在震后仍然持续；另一方面考虑到祁连山地震带发生6级以上地震后短期内有可能向甘东南地区迁移，类似震例为1986年门源6.4级地震后间隔4.5月发生了1987年迭部5.8级地震。因此，虽然该地震前没有提出明确的短临预测意见，但在震前的震情跟踪判定过程中仍获得了有意义的经验和教训。

（3）岷县5.2级地震的前兆异常共计19项，其中8项地震学异常，11项定点前兆异常，主要以中期异常为主，短期异常具有一定的成组性。在震情跟踪过程中，有一些异常被认为与10月25日民乐—山丹6.1、5.8级地震有关，因而该地震前只在中期尺度作出了较好的预测，而没有作出短临预报。此外，该地震与2003年山丹6.1、5.8级地震、2004年岷县5.0级地震，无论是时间上的成组性，还是空间上的相关性均存在密切的联系，在前兆异常的发展阶段和空间分布上也是密不可分，这种关联性在以往震例中也比较少见，还有待进一步的研究和探索。

（4）基于宏观灾害调查认为，本次地震的孕震和发震构造为临潭—宕昌断裂带的中段北支前缘的次级断裂，临潭—宕昌断裂带在历史上曾多次发生过中强地震。该地震发生在区域地址活动水平相对平静的背景下，震前无明显前震活动。据有关文献[4~6]研究，此次地震的震源机制解中的结果，与临潭—宕昌断裂带中段的走向和运动特征基本吻合，也与宏观调查结果是一致的。

参 考 文 献

[1] 郑文俊、雷中生、袁道阳等，1573年甘肃岷县地震史料考证与发震构造探讨，中国地震，23（1），75～82，2007

[2] 郑文俊、雷中生、袁道阳等，1837年甘肃岷县北6级地震考证，地震，27（1），120～130，2007

[3] 郑文俊、刘小凤等，2003年11月13日甘肃岷县M_S5.2地震基本特征，西北地震学报，27（1），61～65，2005

[4] 荣代潞、李亚荣，甘肃岷县两次中强地震震源机制研究，地震研究，29（4），344～348，2006

[5] 刘旭宙，甘肃近期几次中强地震震源机制解，26（1），94～95，2004

[6] 李亚荣、荣代潞、刘旭宙等，2003年11月13日岷县M_L5.5地震震源机制的矩张量反演，华南地震，26（1），126～132，2006

[7] 李亚荣、荣代潞等，2003年岷县5.5级地震地震学前兆特征及预报意义，西北地震学报，29（2），150～155，2007

[8] 肖丽珠等，礼县震群与岷县—临潭—卓尼县5.2级地震，华南地震，24（3），11～17，2004

[9] 张昱等，2003～2004年岷县—卓尼地震与甘肃东部地区水化异常的关系探讨，高原地震，19（3），17～20，2007

[10] 燕明芝等，岷县—卓尼5.2级地震与地电前兆异常的初步研究与发震构造分析，地震地磁观测与研究，26（4），50～56，2005

[11] 高曙德、苏永刚等，武都地电阻率在岷县5.2级地震前的异常变化，地震，25（2），115～121，2005

[12] 蒲小武、高原等，武都形变对岷县M_S5.2地震的前兆异常反映，高原地震，16（2），66～71，2004

参 考 资 料

1）甘肃地震局，甘肃地震目录（2002～2003）
2）甘肃地震局，2003 年度甘肃省震情趋势研究报告
3）甘肃地震局，2004 年度甘肃省震情趋势研究报告
4）甘肃地震局，甘肃省地震局震情档案（2002）
5）甘肃地震局，甘肃省地震局震情档案（2003）

Minxian Earthquake of M_S 5.2 on November 13，2003 in Gansu Province

Abstract

In Minxian，Gansu province occurred an earthquake of M_S5.2 on Nov. 13，2003. Its macroseismic epicenter was located in the north mountainous area of Minxian. The intensity at the epicenter was Ⅷ. The meizoseismal area was oval-shaped with major axis in EW direction. 1 people was killed，128 people were injured and the direct economic loss was 100 million Yuan.

The sequence of the earthquake was of mainshock-aftershock type and the magnitude of the largest aftershock was M3.2. Aftershocks distributed In northwestern direction as the same direction of the major axis of the meizoseismal area. Nodal plane I was the main rupture surface and the P axis of main compressive stress was in EW direction. The seismogenic structure was the Lintan-Dangchang fault.

The epicenter and its adjacent area were located in the area with power monitoring capability in Gansu. Within the distance of 200km from the epicenter，there were 16 seismic stations. There were 19 anomalies before the earthquake. Within the distance of 100km from the epicenter，the proportion of anomal stations and anomal items in stations is 100% and 100%. Within the distance of 101-200km from the epicenter，the proportion is 55% and 28%.

The medium term forcast by earthquake administration of Gansu province was good. The earthquake occurred by the seismically dangerous area of magnitudes 6 in Gansu as specified. After the earthquake，the judgement on the seismic activity was correct.

2003年11月15、26日及2004年8月10日云南省鲁甸5.1、5.0和5.6级地震

云南省地震局

刘　强　王世芹

摘　要

2003年11月15、26日和2004年8月10日云南省鲁甸东南相继发生5.1、5.0和5.6级地震。宏观震中位于鲁甸县桃源一带。前两次地震的极震区烈度为Ⅶ度，2004年M_S5.6地震极震区烈度为Ⅷ度，等震线形状呈北北东向椭圆形。三次地震共造成8人死亡，重伤217人，轻伤498人，直接经济损失总计6.048亿元。

综合野外现场地质考察和余震分布，推测三次地震均是在北北西向的主压应力作用下，左旋走滑的结果。震源机制解节面Ⅱ为主破裂面，三次地震主压应力轴P轴方位为北北西287°，发震构造为NNW向的新活动断裂。

震中周围200km范围内共有地震台站35个，其中测震台14个，有其他前兆观测台站21个。震前共出现5个异常项目12条异常，定点前兆震前共出现3个异常项目10条前兆异常，主要为短临异常。

本次中强震群发生在云南省地震局年度预测区内。三次地震序列为丛集性震群型，最大余震4.3级，余震呈北西向分布。

本文对异常特征、异常判别等问题进行了分析和讨论，对本区域破坏性地震预测预报有一定借鉴意义。

前　言

2003年11月15日02时49分、11月26日21时38分和2004年8月10日18时26分云南省鲁甸县相继发生5.1、5.0和5.6级地震。据云南地震台网测定，微观震中分别为27°10′N、103°37′E，27°12′N、103°38′E和27°10′N、103°36′E。宏观震中位于鲁甸县桃源一带。前两次地震的极震区烈度为Ⅶ度，5.6级地震极震区烈度为Ⅷ度。三次地震有感范围较大，昭通市巧家、永善、大关、彝良等县和曲靖市会泽县、宣威市及贵州省威宁县等地有震感。三次地震共造成8人死亡，重伤217人，轻伤498人，直接经济损失60480万元。

三次5级地震发生在云南强震活动区，该区域历史上$M_S \geq 5$级地震活动频繁，最大地

震为 1974 年大关—永善 7.1 级地震。震中 200km 范围共有地震台 35 个，震前出现 5 个异常项目 12 条前兆异常。

每次地震发生后，云南省地震局均立即启动破坏性地震应急预案，当即派出工作组奔赴地震现场。中国地震局、云南省地震局、昭通市地震局组成地震现场工作指挥部，下设秘书、震情监测预报、地震灾评及科考、强震观测、通讯及后勤保障等 6 个组，三次地震分别有 38、27、38 人开展地震现场工作。现场工作组对震后的余震序列和震情动态进行了跟踪和监视，提交了三份《地震灾害损失评估报告》[1~3]。

一、测震台网及地震基本参数

鲁甸三次地震前，震中周围 200km 范围内共有测震台 14 个，其中 100km 范围内有 2 个，101 ~200km 范围内有 12 个（图 1），云南地震台网对该地区的 M_L1.0 以上地震完全能够控制。三次地震的基本参数列于表 1。

位于鲁甸茨院乡的数字强震台获得 5.0 级地震记录，震中距 4km，南北方向最大加速度 209.8cm/s^2，东西方向最大加速度 157.2cm/s^2，垂直方向最大加速度 141.5cm/s^2。

表 1 地震基本参数

Table 1 Basic parameters of the earthquake

编号	发震日期	发震时刻	震中位置		震级	震源深度	震中地名	结果来源
	年 月 日	时 分 秒	φ_N	λ_E	M_S	(km)		
1	2003 11 15	02 49 42.7	27°10′	103°37′	5.0		鲁甸	资料 4)
		02 49 43	27.20°	103.60°	5.1	10	鲁甸	资料 5)
		02 49 46.5	27.37°	103.97°	5.6	33	鲁甸	USGS
2	2003 11 26	21 38 54.7	27°12′	103°38′	4.6		鲁甸	资料 4)
		21 38 59	27.30°	103.60°	5.0	8	鲁甸	资料 5)
		21 38 57.8	27.28°	103.75°	4.7	33	鲁甸	USGS
3	2004 08 10	18 26 13.1	27°10′	103°36′	5.6		鲁甸	资料 4)
		18 26 14	27.20°	103.60°	5.6	10	鲁甸	资料 5)
		18 26 14.7	27.27°	103.87°	5.4	6	鲁甸	USGS

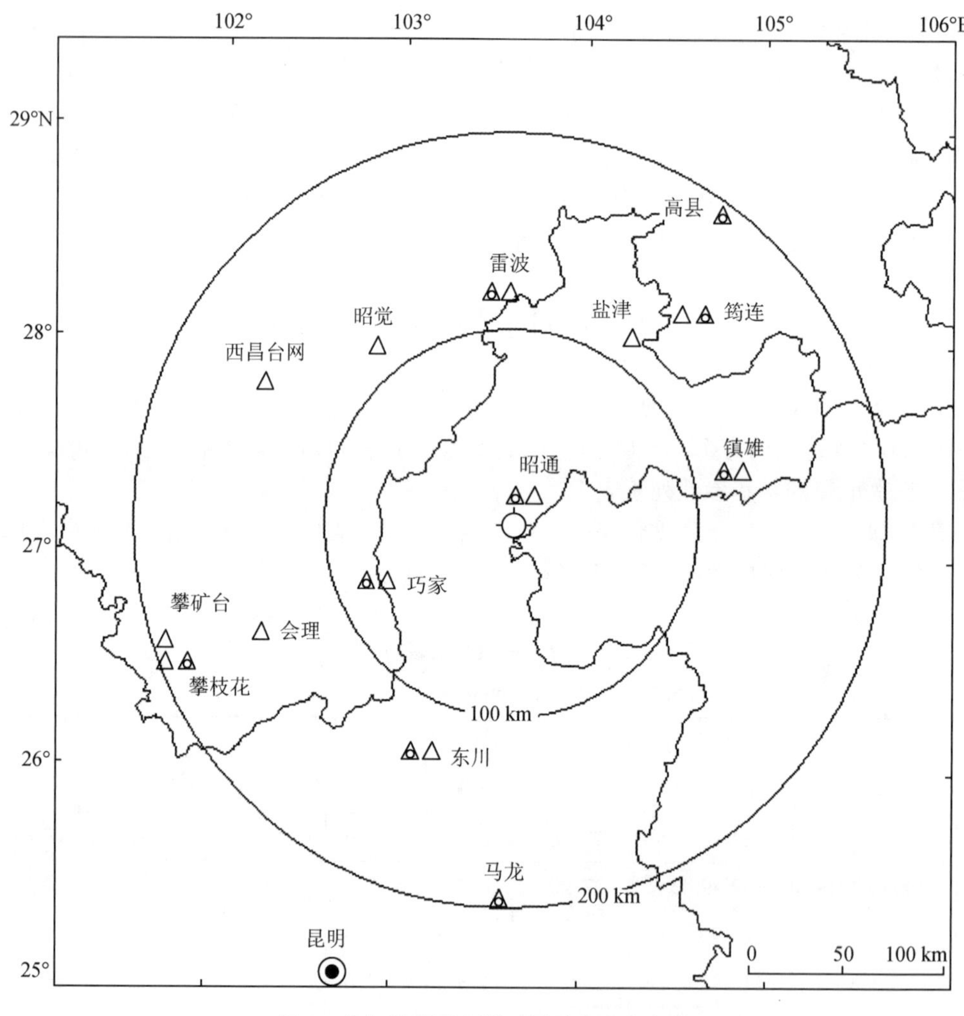

图 1　鲁甸地震震中附近测震台网分布图

Fig. 1　Distribution of seismometric stations around Ludian earthquakes area

二、地震地质背景

震区地处川滇菱形块体东侧，属于扬子准地台滇东台褶带的滇东北台褶束。北东向断裂、褶皱发育。震区地质构造纵横交错，三条北东向构造的巧家—莲峰断裂、洒渔河断裂与近南北向小江断裂与北西向威宁—大关—马边断裂斜插相交。鲁甸三次中强地震均发生在洒渔河断裂上。北东走向的洒渔河断裂为走滑性质断裂，南起巧家以南的双河经鲁甸、昭通、彝良后隐伏于盐津一带，断裂全长约 240km，该断裂历史上以中强地震活动为主，与其相交的威宁—大关—马边断裂和小江断裂则是 6 ~ 7 级强震带（图 2）。

地震震中区附近 100km 范围内，1965 年以来曾经发生过 5.0 ~ 5.9 级地震 18 次，6.0 ~

6.9 级地震 4 次，7.0 ~ 7.9 级地震 1 次，最大地震为 1974 年 5 月 11 日大关-永善 7.1 级地震，与该三次鲁甸地震震中相距约 124km（表 2）。

该三次地震等震线形状呈椭圆形，极震区烈度线的长轴走向为北北东，与洒渔河断裂走向一致。

表 2 鲁甸地震震区附近历史地震目录（$M_S \geq 5.0$ 级）

Table 2 Catalogue of historical earthquakes around the $M_S5.0$ Ludian earthquakes area（$M_S \geq 5.0$）

编号	发震日期	发震时刻	震中位置		震级	震源深度	震中地名	结果来源
	年 月 日	时 分 秒	φ_N	λ_E	M_S	(km)		
1	1966 02 05	23 12 27	26.10°	103.10°	5.0		云南东川	资料 5）
2	1966 02 13	18 44 36	26.10°	103.10°	6.5		云南东川	资料 5）
3	1966 10 11	18 06 20	27.90°	103.80°	5.1		云南永善	资料 5）
4	1970 07 31	21 10 49	28.60°	103.70°	5.7	44	四川雷波	资料 5）
5	1971 08 16	12 57 59	28.90°	103.80°	5.0	24	四川马边	资料 5）
6	1971 08 17	02 53 51	28.90°	103.80°	5.4	7	四川马边	资料 5）
7	1971 08 17	06 37 32	28.80°	103.60°	6.3	19	四川马边	资料 5）
8	1971 08 17	17 36 13	28.70°	103.70°	5.2	35	四川沐川	资料 5）
9	1971 08 18	01 07 39	28.80°	103.60°	7.1	17	四川马边	资料 5）
10	1971 08 23	13 36 04	28.70°	103.70°	5.2	5	四川马边	资料 5）
11	1973 04 22	13 46 21	27.70°	104.00°	5.2		云南彝良	资料 5）
12	1973 06 29	05 59 27	28.90°	103.70°	5.4	10	四川马边	资料 5）
13	1973 06 30	01 52 01	28.90°	103.80°	5.0		四川马边	资料 5）
14	1973 08 02	16 58 16	27.90°	104.70°	5.4		云南彝良	资料 5）
15	1974 05 11	03 25 18	28.20°	104.10°	5.0	14	云南大关	资料 5）
16	1975 03 08	01 36 09	28.40°	104.20°	5.2	30	云南盐津	资料 5）
17	1985 04 18	13 52 52	25.80°	102.80°	5.4	5	云南禄劝	资料 5）
18	1988 04 15	18 58 15	26.20°	102.50°	5.7	9	四川会东	资料 5）
19	1993 08 07	16 36 04	29.00°	103.70°	5.9		四川沐川	资料 5）
20	1994 12 30	02 58 23	28.90°	103.50°	5.4		四川马边	资料 5）
21	1995 04 26	11 46 30	25.90°	103.70°	5.1		四川沐川	资料 5）
22	1995 10 24	06 46 49	29.00°	102.20°	6.2		云南武定	资料 5）
23	1998 12 01	15 37 56	26.40°	104.10°	6.5		云南宣威	资料 5）

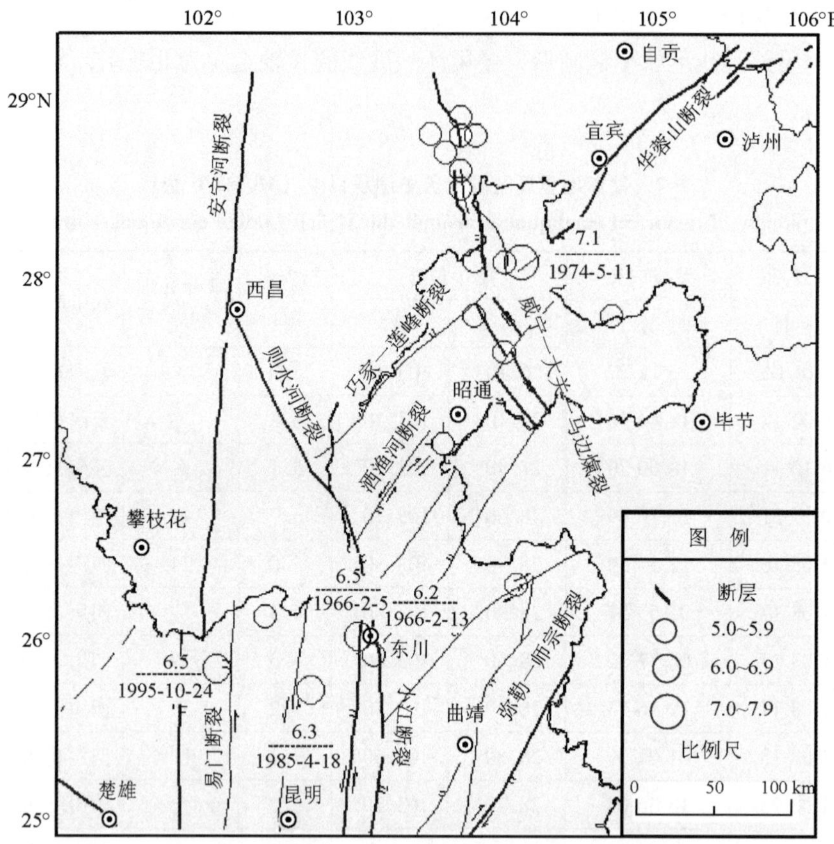

图 2　鲁甸地震震区地质构造及历史地震分布图

Fig. 2　Map of geological structure and distribution of historical earthquakes around Ludian earthquakes area

三、地震影响场和震害

(1) 据现场调查，5.1 和 5.0 级地震宏观震中位于鲁甸县桃源乡铁家湾和昭阳区永丰镇迎水一带。极震区烈度为Ⅶ度，等震线呈椭圆形，长轴走向为北东向（图 3a）。通过估算两次地震发震断层长度，建立适当的弹性位错模型，计算得到破裂长度 5.0～5.6km，破裂宽度 1.2～1.3km，平均位错量为 0.14～0.16m[1]。

两次地震造成Ⅵ度、Ⅶ度两个破坏区。Ⅶ度区北自昭阳区永丰镇，南至昭阳区布嘎乡新街以南，东起昭阳区布嘎乡赵家院子，西到鲁甸县桃源乡普芝噜以西，面积约 138km^2；Ⅵ度区北自昭阳区永丰镇凤凰，南至鲁甸县大水井乡政府所在地，东起昭阳区守望乡袁家包包和贵州省威宁县新寨，西至鲁甸县小寨乡政府所在地，面积约 462km^2。两次地震灾区总面积约 600km^2。

两次地震受灾地区主要涉及昭通市鲁甸县和昭阳区，包括 11 个乡（镇），47 个行政村，波及曲靖市会泽县和贵州省威宁县，昭通市巧家、永善、大关、彝良普遍有感。受灾人口 246732 人，涉及 59105 户。5.1 级地震死亡 4 人，重伤 24 人，轻伤 70 人，直接经济损失 1.919 亿元1)；5.0 级地震由于震前采取有效的防震避险措施，无人员死亡，重伤 2 人，轻伤 22 人，直接经济损失 0.93 亿元2)。

两次地震形成较大范围的Ⅶ度破坏区和少数调查点有Ⅷ度破坏现象的主要原因[1]：震源浅，地表振动强烈；盆地区广泛分布有煤层、膨胀土、淤泥土和松散的粉砂、粉土等不良影响地层，且埋藏浅；震区建筑结构简易，抗震能力极差；两次地震震害叠加。

两次地震破坏区涉及贵州省威宁县部分乡村，贵州省Ⅵ度区面积约 $57km^2$。由于贵州省灾区的建筑面积、建筑单价等基本情况不清楚，因此，没有对贵州省灾区的经济损失进行评估。

（2）5.6 级地震宏观震中位于鲁甸县桃源乡一带。极震区烈度为Ⅷ度，等震线基本呈椭圆形，长轴走向为北东向（图 3b）。估算该地震发震断层长度，建立适当的弹性位错模型，计算得到破裂长度 9.8km，破裂宽度 2.2km，平均位错量为 0.28m[2]。

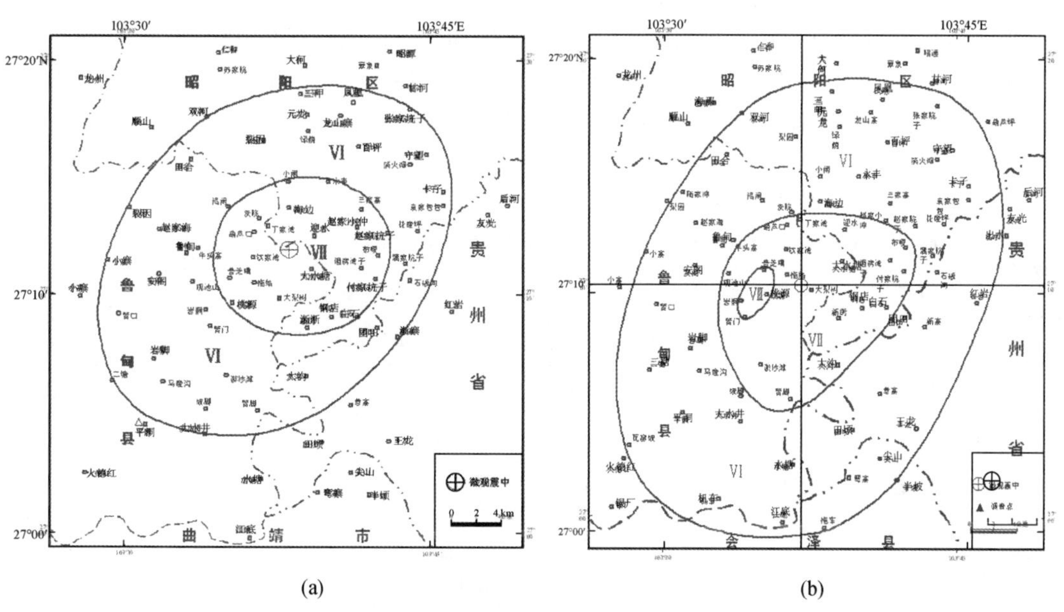

图 3　鲁甸 $M_S5.1$、$M_S5.0$（a）和 $M_S5.6$（b）地震等震线图

Fig. 3　Isoseismal map of the $M_S5.1$, $M_S5.0$ and $M_S5.6$ Ludian earthquake

地震造成Ⅵ度、Ⅶ度、Ⅷ度三个破坏区。Ⅷ度区主要分布在鲁甸县桃源乡，北起普芝噜，南至箐门，东自桃源村东，西到岩洞，面积约 $8km^2$；Ⅶ度区北近昭阳区永丰镇赵家小冲，南至鲁甸县大水井乡箐脚，东起昭阳区布嘎乡布嘎村，西到鲁甸县文屏镇砚池山，面积约 $182km^2$；Ⅵ度区北自昭通市昭阳区蒙泉乡凤凰北，南至曲靖市梨园乡拖车村，东近贵州省威宁县友光村，西近鲁甸县小寨乡小寨村，面积约 $697km^2$。灾区总面积约 $887km^2$。

受灾地区主要涉及昭通市鲁甸县、昭阳区和曲靖市会泽县，包括 12 个乡（镇），68 个行政村，波及贵州省威宁县，昭通市巧家、永善、大关、彝良和曲靖市宣威市有感。受灾人口 313556 人，涉及 75158 户。死亡 4 人，重伤 191 人，轻伤 406 人，直接经济损失 3.199 亿元[3)]。

该次地震比鲁甸 5.1 和 5.0 级地震破坏更严重，形成了Ⅷ度破坏区及较大范围的Ⅶ度和Ⅵ度破坏区，盆地区震害比基岩区重。其主要原因是[3]：①盆地区广泛分布有软弱地层；②震区建筑结构简易，抗震能力极差；③三次地震震区基本重叠，震害累积效应表现突出。

本次地震破坏区涉及贵州省威宁县部分乡村，贵州省Ⅵ度区面积约 146km^2，其经济损失未进行统计。

（3）三次地震极震区最大烈度为Ⅷ度，等震线基本呈椭圆形，长轴走向为北东向。三次地震共造成 8 人死亡，重伤 217 人，轻伤 498 人，直接经济损失总计 6.048 亿元。

四、地 震 序 列

昭通测震台距三次鲁甸地震震中约 18km，对 M_L1.0 以上的地震完全能够控制。

据云南区域台网目录（月报目录）[4)]，鲁甸 5.1 和 5.0 级地震截至 2004 年 2 月，共有 M_L≥1.0 级余震 51 次，其中，1.0～1.9 级 24 次，2.0～2.9 级 18 次，3.0～3.9 级 7 次，5.0～5.9 级 2 次，最大余震 M_L3.8，3 级以上地震目录见表 3。

表 3　鲁甸 5.1 和 5.0 级地震序列目录（M_L≥3.0 级）

Table 3　Catalogue of the M_S5.1 and M_S5.0 Ludian earthquake sequence（M_L≥3.0）

编号	发震日期	发震时刻	震中位置		震级		震源深度	震中地名	结果来源
	年 月 日	时 分 秒	φ_N	λ_E	M_L	M_S	(km)		
1	2003 11 15	02 49 43	27.20°	103.60°		5.1	10	鲁甸	资料 5)
2	2003 11 16	02 50 09	27°11′	103°33′	3.8			鲁甸	资料 4)
3	2003 11 18	05 50 02	27°10′	103°35′	3.8		10	鲁甸	资料 4)
4	2003 11 21	00 56 35	27°05′	103°38′	3.0		10	鲁甸	资料 4)
5	2003 11 21	17 24 31	27°11′	103°37′	3.1		10	鲁甸	资料 4)
6	2003 11 23	05 19 05	27°07′	103°42′	3.6		10	鲁甸	资料 4)
7	2003 11 26	21 38 59	27.10°	103.60°		5.0	8	鲁甸	资料 4)
8	2003 11 30	06 29 04	27°12′	103°40′	3.2			昭通	资料 4)
9	2003 12 05	15 55 59	27°07′	103°39′	3.2			鲁甸	资料 4)

鲁甸 5.6 级地震截至 2004 年 12 月，共有 M_L≥1.0 级余震 82 次，其中，1.0～1.9 级 52 次，2.0～2.9 级 17 次，3.0～3.9 级 9 次，4.0～4.9 级 3 次，5.0～5.9 级 1 次最大余震 M_L4.3，3 级以上地震目录见表 4。

表 4 鲁甸 5.6 级地震序列目录（$M_L \geqslant 3.0$ 级）

Table 4 Catalogue of the M_S5.6 Ludian earthquake sequence（$M_L \geqslant 3.0$）

编号	发震日期	发震时刻	震中位置		震级		震源深度（km）	震中地名	结果来源
	年 月 日	时 分 秒	φ_N	λ_E	M_L	M_S			
1	2004 08 10	18 26 13	27.20°	103.60°		5.6	10	鲁甸	资料 5）
2	2004 08 10	18 31 02	27°11′	103°34′	3.9		10	鲁甸	资料 4）
3	2004 08 10	19 49 20	27°10′	103°36′	4.0			鲁甸	资料 4）
4	2004 08 10	20 48 12	27°12′	103°33′	4.3		4	鲁甸	资料 4）
5	2004 08 10	20 54 59	27°13′	103°29′	3.2		10	鲁甸	资料 4）
6	2004 08 10	21 27 33	27°09′	103°36′	4.1		4	鲁甸	资料 4）
7	2004 08 11	16 12 15	27°08′	103°33′	3.8			鲁甸	资料 4）
8	2004 08 12	02 56 39	27°06′	103°33′	3.7			鲁甸	资料 4）
9	2004 08 14	02 22 17	27°03′	103°38′	3.0		10	鲁甸	资料 4）
10	2004 09 01	02 19 06	27°10′	103°39′	3.6			鲁甸	资料 4）
11	2004 10 11	02 54 36	27°14′	103°38′	3.2		16	昭通	资料 4）
12	2004 10 24	20 57 02	27°13′	103°38′	3.1		8	昭通	资料 4）
13	2004 11 12	00 33 07	27°10′	103°36′	3.1		4	鲁甸	资料 4）

把鲁甸 5.1 和 5.0 级地震作为一个序列分析，5.1 级释放的能量占全序列能量的 57.57%，5.0 级释放的能量占全序列能量的 34.54%，为双震型序列。5.0 序列的 b 值为 0.50（图 4），h 值为 0.7（图 5），p 值为 0.07。图 6 至图 8 分别为全序列 M-t、N-t 及应变释放图。5.1 级主震后至 5.0 级主震前 12 天内无 2 级以下地震发生，序列衰减较慢；5.0 级地震后序列衰减不正常。图 9 为全序列 1 级以上地震的震中分布图，从余震分布来看近似 NW 向，与等震线分布存在差异。

鲁甸 5.6 级地震序列主震释放的能量占全序列能量的 95.52%，为主余震型序列。序列的 b 值为 0.43（图 10），h 值为 1.0（图 11），p 值为 0.43。图 12 至图 14 分别为该序列 M-t、N-t 及应变释放图。序列有起伏，衰减基本正常。图 15 为全序列 1 级以上地震的震中分布图，从余震分布来看为 NW 向，与等震线分布并不完全一致。

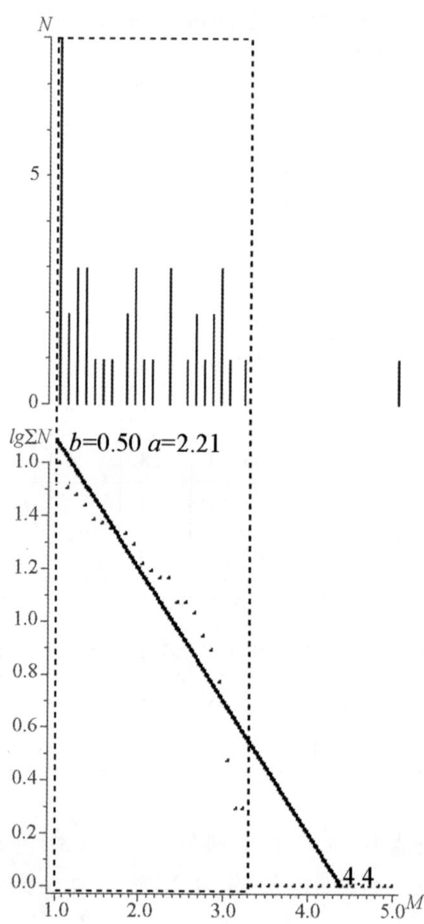

图 4　鲁甸 5.0 级地震序列 b 值曲线

Fig. 4　b-value curve of the M_S5.0 Ludian earthquake sequence

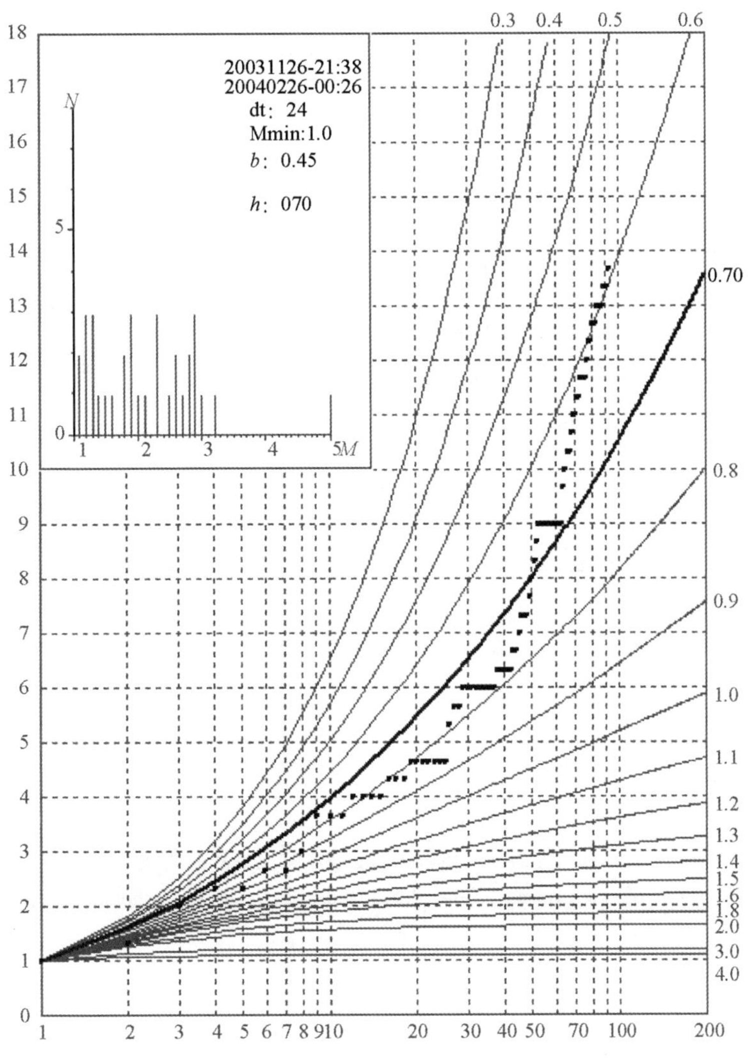

图 5 鲁甸 5.0 级地震序列 h 值曲线

Fig. 5 h-value curve of the M_S5.0 Ludian earthquake sequence

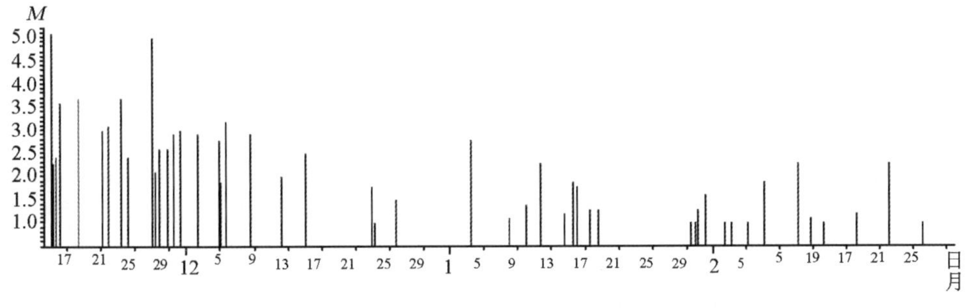

图 6 鲁甸 5.1 和 5.0 级地震 M-t 图

Fig. 6 M-t diagram of the M_S5.1 and M_S5.0 Ludian earthquakes sequence

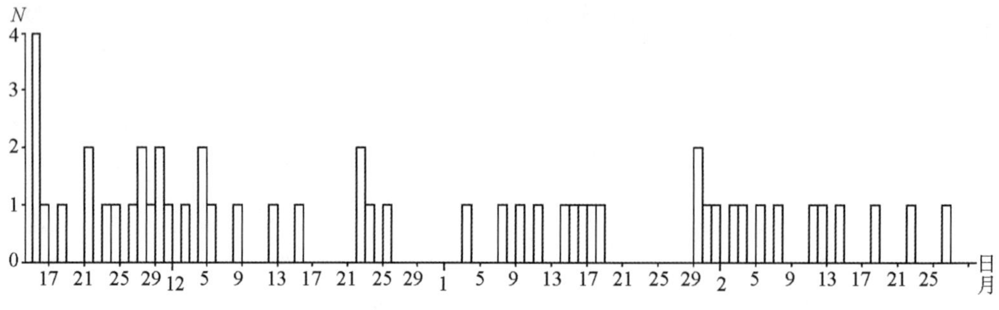

图7　鲁甸5.1和5.0级地震日频度曲线

Fig. 7　Curve of daily frequency of the M_S5.1 and M_S5.0 Ludian earthquakes sequence

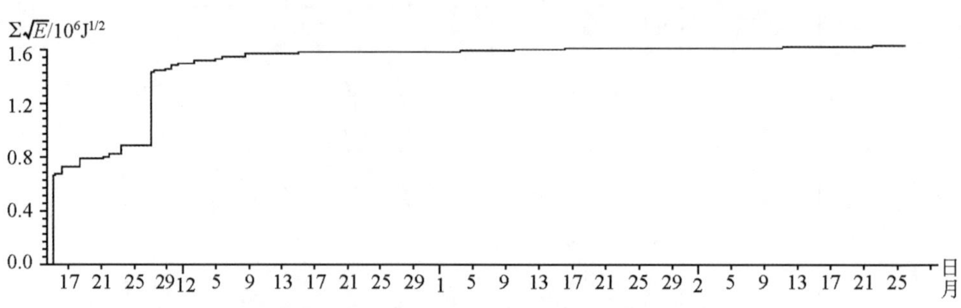

图8　鲁甸5.1和5.0级地震序列应变释放曲线

Fig. 8　Curve of strain release of the M_S5.1 and M_S5.0 Ludian earthquakes sequence

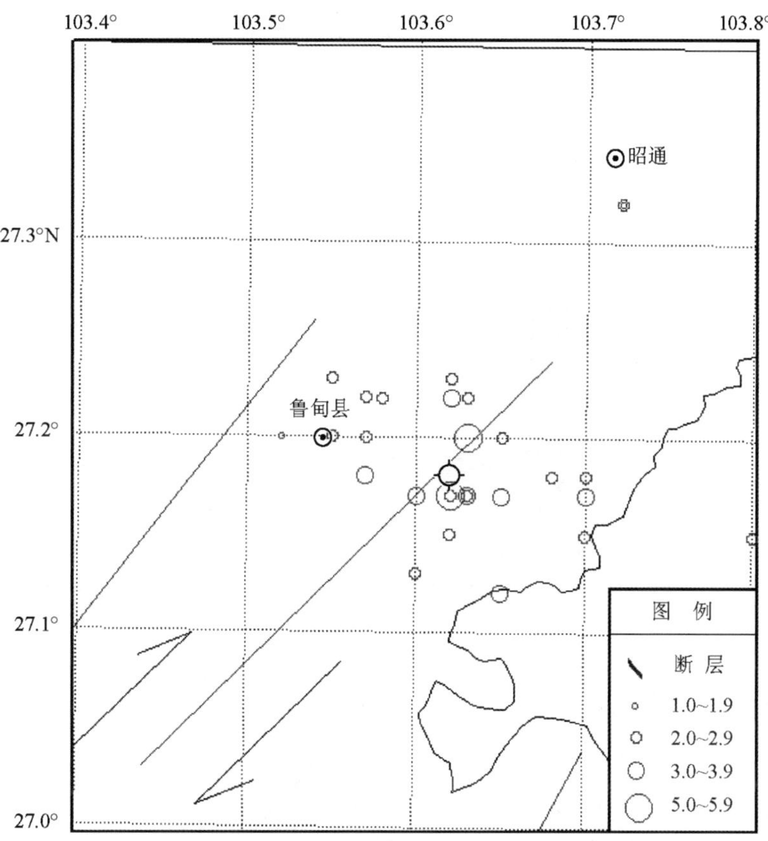

图 9　鲁甸 5.1 和 5.0 级地震序列震中分布图（$M_L \geqslant 1.0$ 级）

Fig. 9　Epicenter distribution of the $M_S 5.1$ and $M_S 5.0$ Ludian earthquakes sequence（$M_L \geqslant 1.0$）

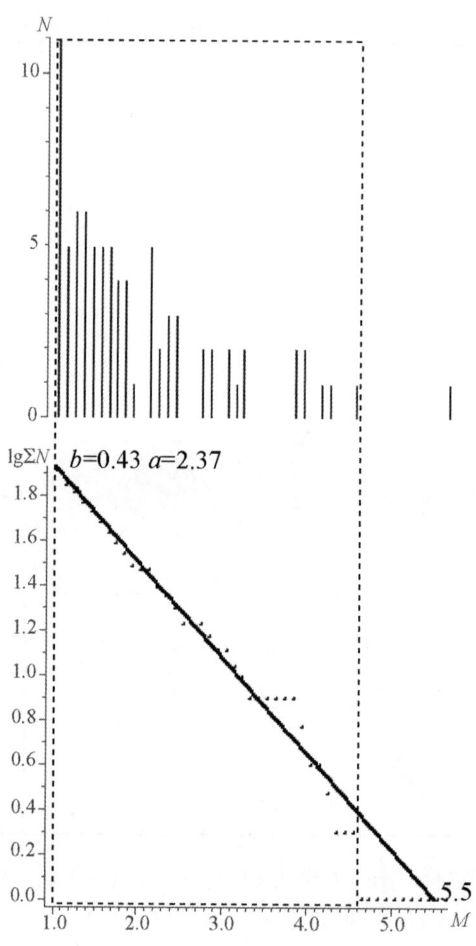

图 10 鲁甸 5.6 级地震序列 b 值曲线

Fig. 10 b-value curve of the M_S5.6 Ludian earthquake sequence

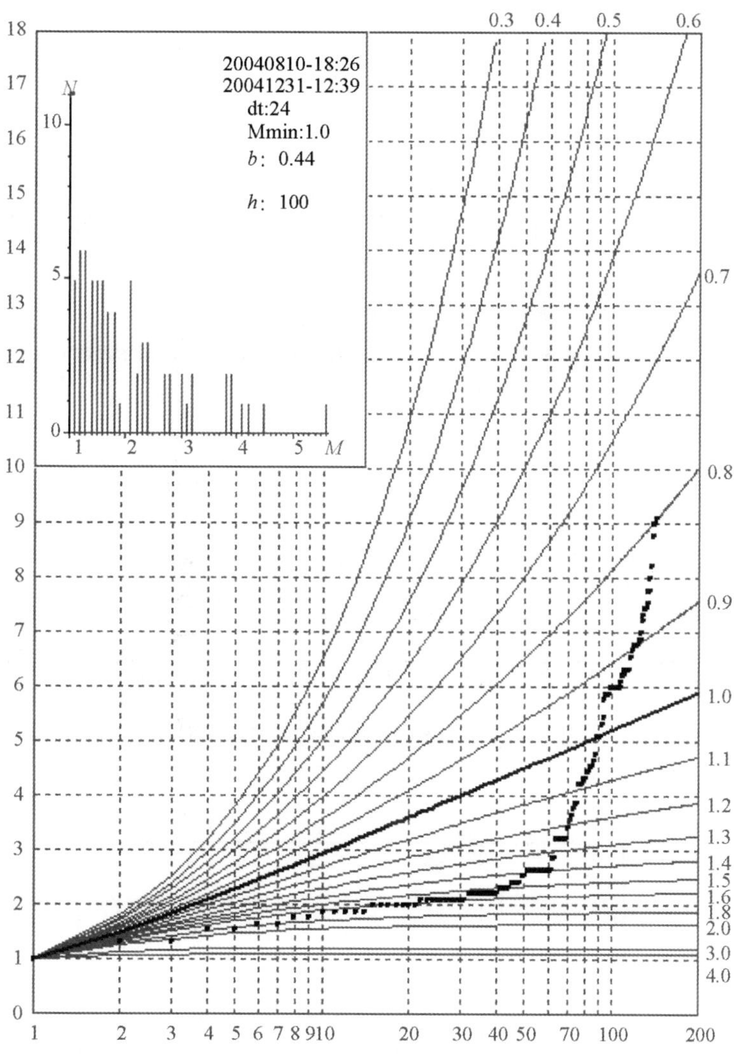

图 11　鲁甸 5.6 级地震序列 h 值曲线

Fig. 11　h-value curve of the M_S5.6 Ludian earthquake sequence

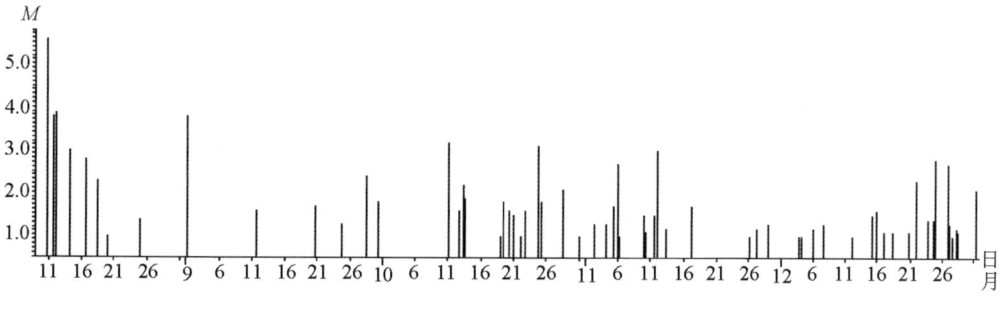

图 12　鲁甸 5.6 级地震 M-t 图

Fig. 12　M-t diagram of the M_S5.6 Ludian earthquake sequence

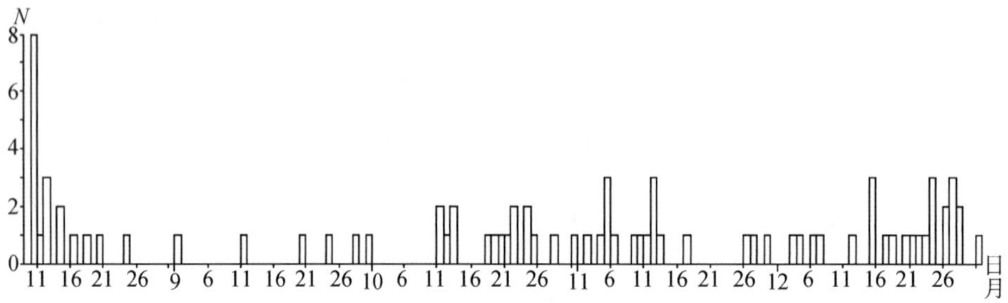

图 13　鲁甸 5.6 级地震日频度曲线

Fig. 13　Curve of daily frequency of the M_S5.6 Ludian earthquake sequence

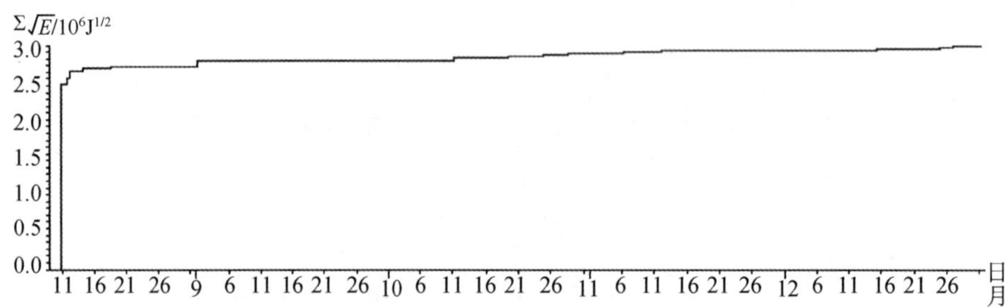

图 14　鲁甸 5.6 级地震序列应变释放曲线

Fig. 14　Curve of strain release of the M_S5.6 Ludian earthquake sequence

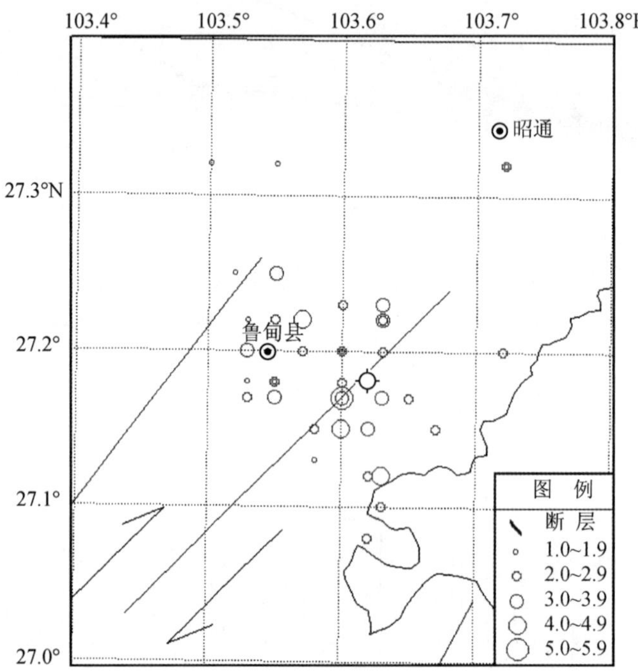

图 15　鲁甸 5.6 级地震序列震中分布图（$M_L \geqslant 1.0$ 级）

Fig. 15　Epicenter distribution of the M_S5.6 Ludian earthquake sequence（$M_L \geqslant 1.0$）

五、震源参数和地震破裂面

分别收集三次鲁甸地震序列云南地震台网 P 波初动符号 26、26、32 个，用 P 波初动符号图解断层面法求得三次地震的震源机制解见表 5 和图 16（Schmidt 网下半球投影）。从三次地震序列震中分布图来看，余震分布优势方向为 NNW 向。

根据波谱参数求得的破裂方向，结合震源机制解和地震矩张量解判定鲁甸三次地震主破裂倾向于沿 NNW 方向，自 SW 向 NNW 或近 SN 向破裂[4]。

节面Ⅱ与洒渔河断裂走向近似垂直，走向为 331°，倾向 NE，倾角 71°较陡，分析认为，节面Ⅱ可能是鲁甸三次地震的主破裂面，三次地震发震构造为一条 NNW 向的近似垂直于洒渔河断裂的新活动断裂，本区域可能存在一条 NNW 向的新活动构造。

鲁甸三次地震等效释放应力场主压应力轴 P 轴方位约 287°，仰角约 10°，仰角较小，说明以水平作用为主。三次地震均是在北北西向的主压应力作用下，左旋走滑的结果。

表 5 鲁甸三次地震震源机制解

Table 5 Focal mechanism solution of the M_S5.1, M_S5.0 and M_S5.6 Ludian earthquakes

地震	节面Ⅰ			节面Ⅱ			P 轴		T 轴		B 轴		X 轴		Y 轴		结果来源
	走向	倾向	倾角	走向	倾向	倾角	方位	仰角	方位	仰角	方位	仰角	方位	仰角	方位	仰角	
*M*5.1	239	NW	89	329	NE	64	287	18	191	19							王绍晋
*M*5.0	241	NW	86	333	NE	73	288	9	196	15							王绍晋
*M*5.6	239	NW	80	332	NE	76	286	3	195	17							王绍晋

注：M_S5.1、M_S5.0 矛盾符号比 0.04；M_S5.6 矛盾符号比 0.09。

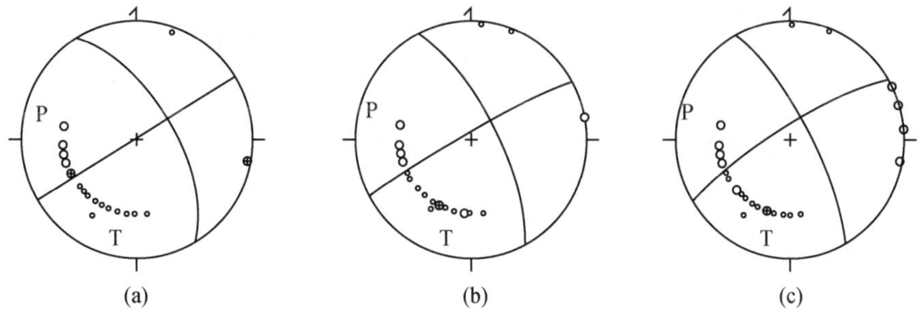

图 16 鲁甸地震震源机制解

Fig. 16 Focal mechanism solution of the M_S5.1 (a), M_S5.0 (b) and M_S5.6 (c) Ludian earthquakes

六、地震前兆观测台网及前兆异常

图 1 和图 17 分别是震中周围 200km 范围测震台站和前兆观测台站分布图，100km 范围内地震台站相对较少，大部分台站主要分布在 101～200km 范围内。

由图可见，震中 200km 范围内共有 35 个地震台站，其中开展测震观测的台站 14 个，有其他前兆观测的台站 21 个，包括水位、水温、水氡、水质、磁电、地倾斜、地应变及跨断层短基线、短水准等 14 个观测项目共 64 个测项。在震中距 0～100km 及 101～200km 范围内分别有前兆观测台站 6 个和 15 个。

取震中周边地区的洒渔河断裂及其相邻断裂带上 2000～2004 年 3 级以上地震作为地震活动研究对象（图 18），研究发现，该丛集性震群发生前本地区出现 3 级平静异常（图 19），5.6 级地震前出现 3 级地震增频异常（3 个月为窗长，1 月为步长，图 20）。

除了地震活动性出现的 2 项异常外，该丛集性震群发生前还存在 10 个台项的定点及流动前兆测项异常（图 21），包括水氡、水温、重碳酸氢根离子、氟离子、钙镁离子等。各项异常情况详见表 6 和图 22 至图 31。

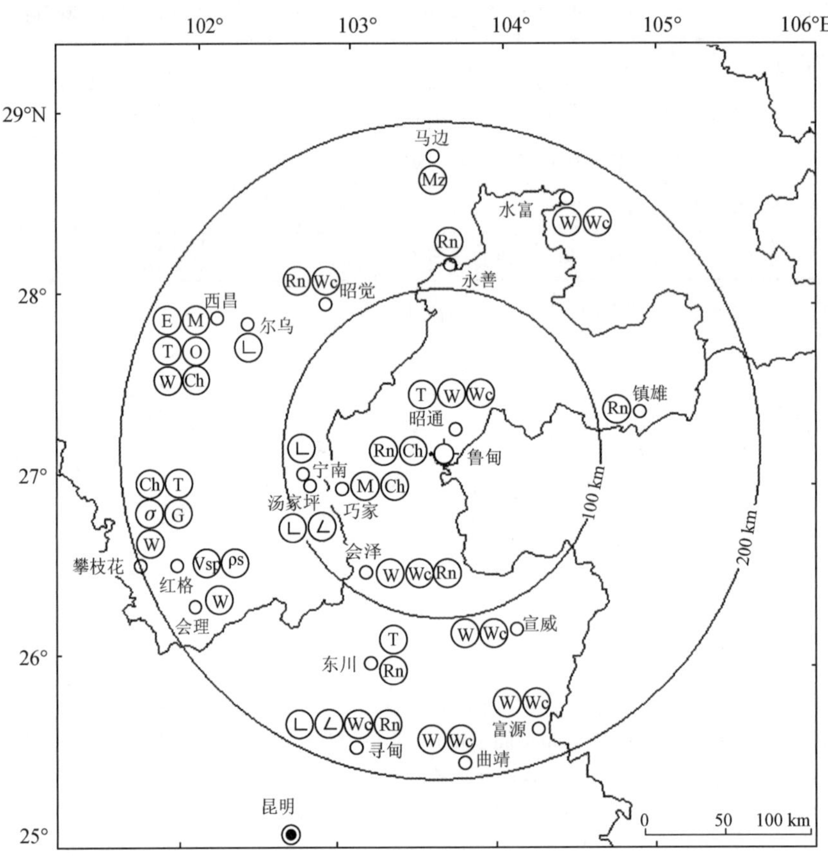

图 17　鲁甸地震前前兆观测台站分布图

Fig. 17　Distribution of the precursory monitoring stations before the Ludian earthquakes

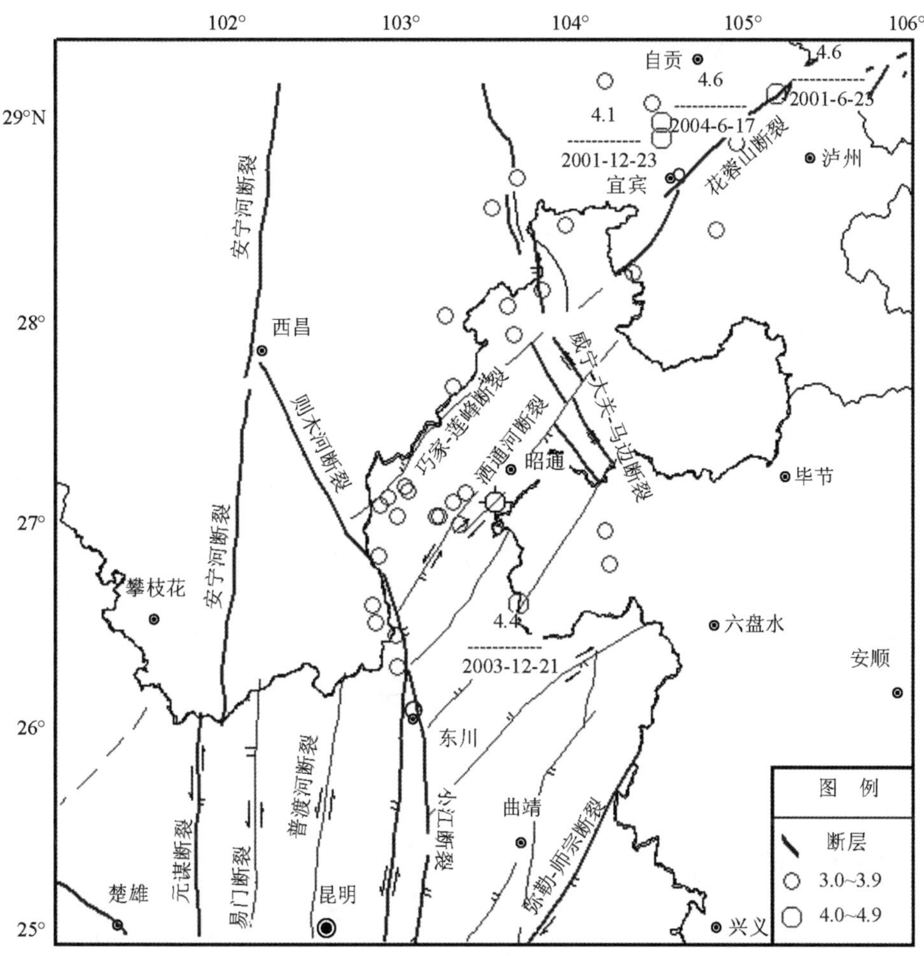

图 18　鲁甸地震震区 $M_L \geqslant 3.0$ 级地震分布图（2000～2004 年）

Fig. 18　Map of distribution of $M_L \geqslant 3.0$ earthquake around Ludian area (2000-2004)

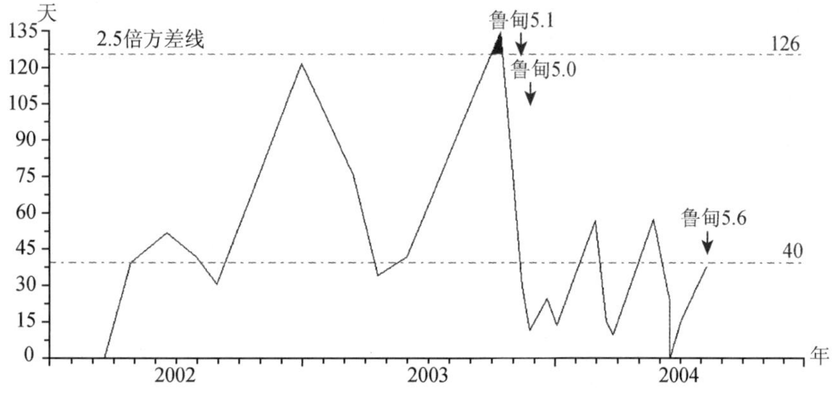

图 19　鲁甸地震前震中周围 ΔT-t 图（$M_L \geqslant 3.0$ 级）

Fig. 19　ΔT-t diagram around the epicentral area before the Ludian earthquakes ($M_L \geqslant 3.0$)

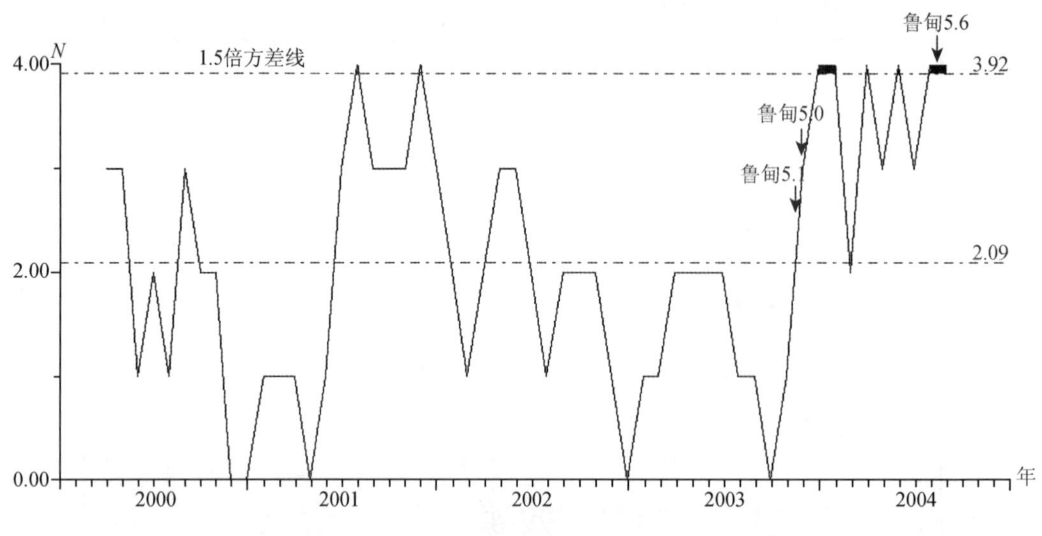

图 20　鲁甸地震前震中周围 N-t 图（$M_L \geqslant 3.0$ 级）

Fig. 20　N-t diagram around the epicentral area before the Ludian earthquakes（$M_L \geqslant 3.0$）

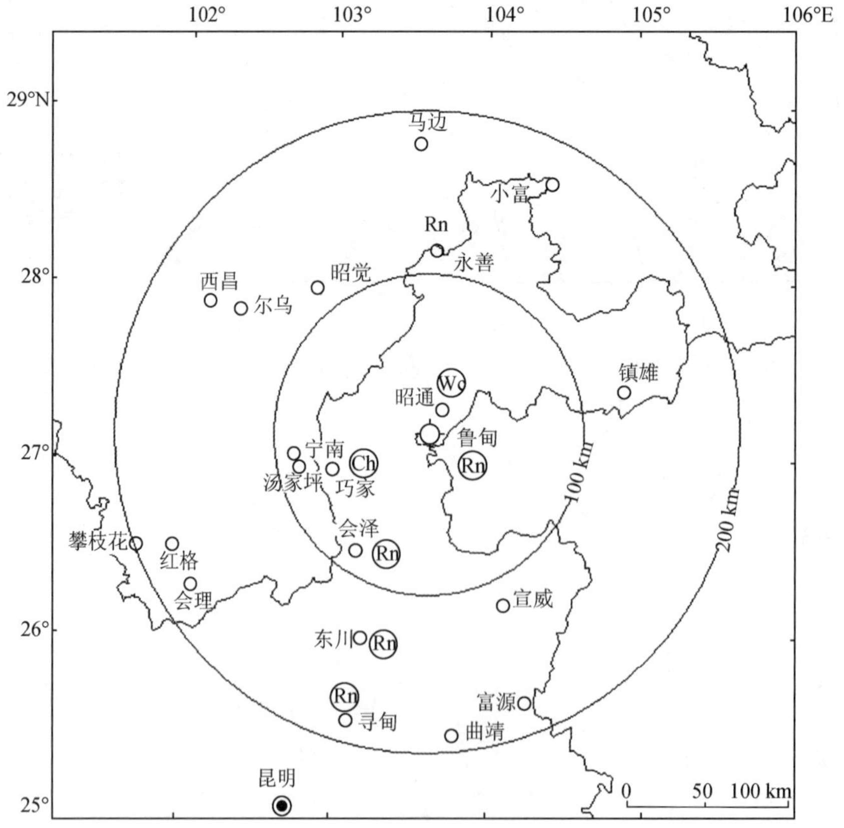

图 21　鲁甸地震前前兆异常分布图

Fig. 21　Distribution of the precursory anomalies before the Ludian earthquakes

表 6 异常情况登记表

Table 6 Summmary table of precursory anomalies

序号	异常项目	台站（点）或观测区	分析方法	异常判据及观测误差	震前异常起止时间	震后变化	最大幅度	震中距 Δ/km	异常类别及可靠性	图号	异常特点及备注	震前提出/震后总结
1	地震时间间隔	发震断层附近地区	$M_L \geqslant 3.0$ 级	$\Delta T \geqslant 2.5$ 倍方差	2003.6.3 ~ 2003.10.16	正常	135 天	≤300	S_1	19	发震断层附近地区 $M_L \geqslant 3.0$ 级平静	震后总结
2	地震频度	发震断层附近地区	$M_L \geqslant 3.0$ 级月频次	$N \geqslant 1.5$ 倍方差	2004.1 ~ 2004.7	正常	4 次	≤300	M_2	20	发震断层附近地区 $M_L \geqslant 3.0$ 级增频	震前提出
3	水氡	寻甸	五日均值	破正常动态	2003.9.15 ~ 2003.10.25 2004.2.25 ~ 2004.8.10	持续	26.8Bq/L	187	S_2	22	持续高值	震前提出
4	水氡	东川	五日均值	破年变	2004.4.30	结束	31.2Bq/L	137	S_1	23	加速上升破年变后转折发震	震后总结
5	水氡	鲁甸	五日均值	低值超 2 倍方差	2003.5.31 ~ 2003.6.15 2004.2.29 ~ 2004.4.30	结束	46.8Bq/L 43.8Bq/L	5	S_1	24	持续下降后正常时段发震	震后总结
6	水氡	永善	五日均值	破正常动态，超 2 倍方差	2004.6.30 ~ 7.25	结束	2.8Bq/L	116	S_1	25	大幅振荡结束后转折上升 15 天发震	震后总结
7	水氡	会泽	五日均值	低值超 2 倍方差	2003.8.15 ~ 2003.9.20 2004.3.5 ~ 2004.3.25 2004.7.15 ~ 2004.8.10	异常	1.4Bq/L 1.6Bq/L 1.2Bq/L	86	S_2 S_2 I_2	26	低值异常结束后发震	震后总结

续表

序号	异常项目	台站（点）或观测区	分析方法	异常判据及观测误差	震前异常起止时间	震后变化	最大幅度	震中距 Δ/km	异常类别及可靠性	图号	异常特点及备注	震前提出/震后总结
8	水温	昭通	日均值	低值破正常动态	2003. 9. 25 ~ 2003. 9. 27 2004. 7. 28 ~ 2004. 8. 8	结束	0. 0013℃ 0. 0039℃	18	S_1 I_1	27	回升后发震	震前提出
9	重碳酸氢根	巧家	五日均值	破年变超 2 倍方差	2003. 6. 20 ~ 2003. 9. 5 2004. 7. 25 ~ 2004. 8. 10	结束	225. 0mg/L 235. 0mg/L	67	S_1 I_1	28	高值异常结束后发震	震前提出
10	氟离子	巧家	五日均值	破正常动态超 2 倍方差	2004. 6. 10 ~ 2004. 7. 31	结束	0. 018mg/L	67	S_1	29	持续上升后转折发震	震前提出
11	钙离子	巧家	五日均值	破正常动态超 1. 8 倍方差	2003. 5. 25 ~ 2003. 9. 5 2004. 3. 31 ~ 2004. 7. 25	结束	5. 5mg/L 8. 8mg/L	67	S_1	30	持续上升后转折发震	震前提出
12	镁离子	巧家	五日均值	破正常动态超 1. 8 倍方差	2003. 5. 20 ~ 2003. 9. 5 2004. 3. 5 ~ 2004. 7. 31	结束	2. 6mg/L 2. 7mg/L	67	S_1	31	持续上升后转折发震	震前提出

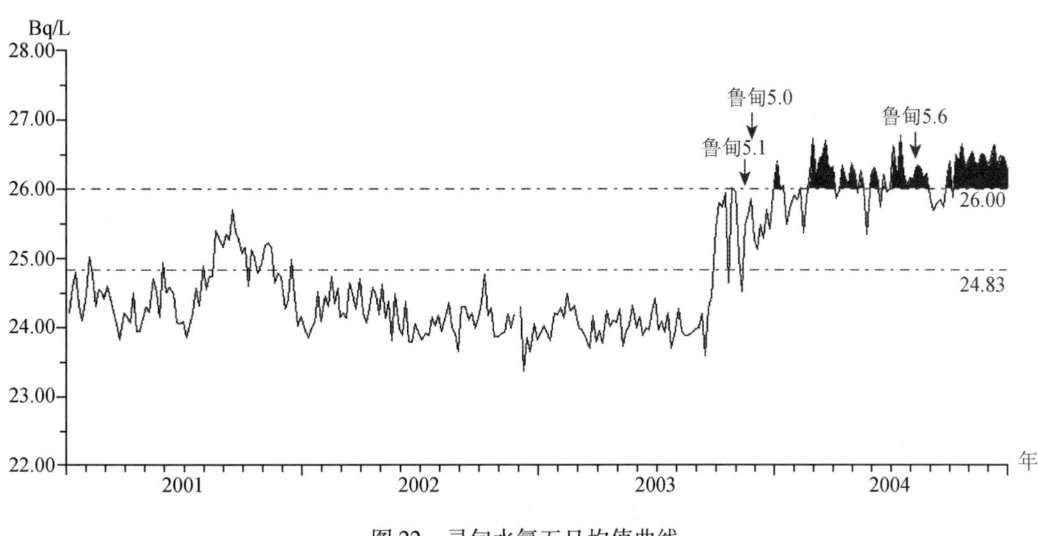

图 22 寻甸水氡五日均值曲线

Fig. 22 Curve of 5-day mean value of radon content in groundwater at Xundian station

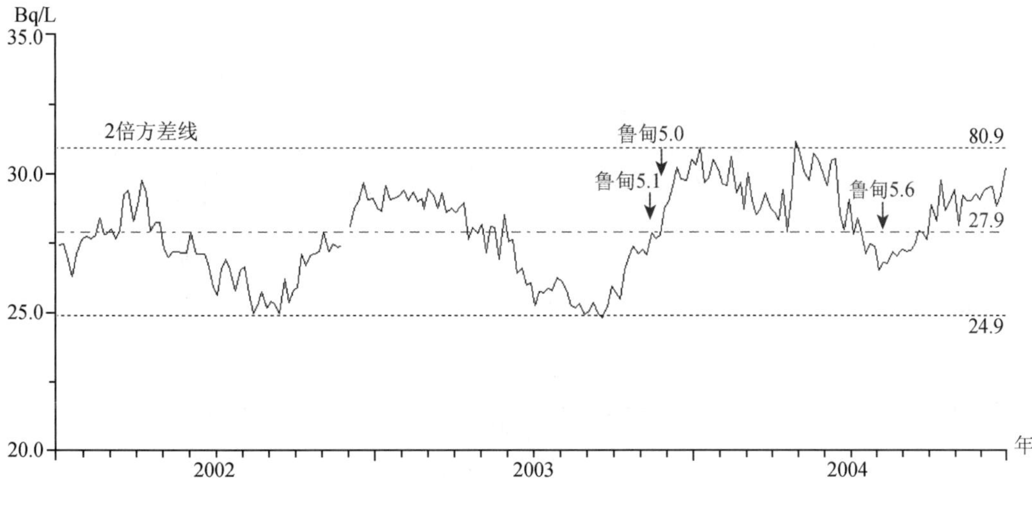

图 23 东川水氡五日均值曲线

Fig. 23 Curve of 5-day mean value of radon content in groundwater at Dongchuan station

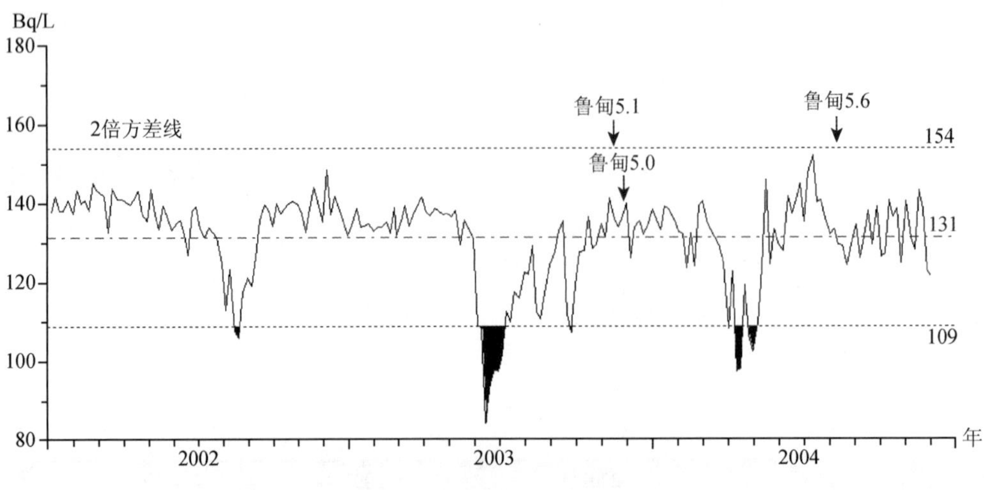

图 24　鲁甸水氡五日均值曲线

Fig. 24　Curve of 5-day mean value of radon content in groundwater at Ludian station

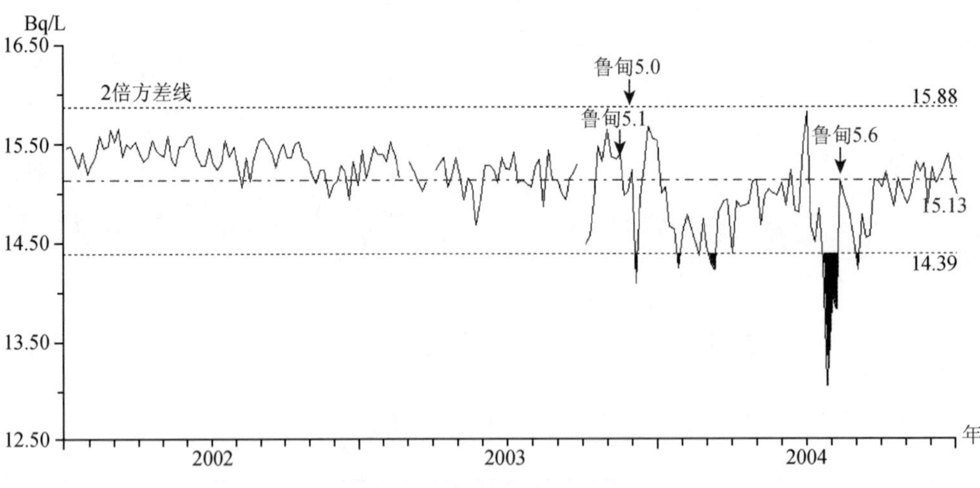

图 25　永善水氡五日均值曲线

Fig. 25　Curve of 5-day mean value of radon content in groundwater at Yongshan station

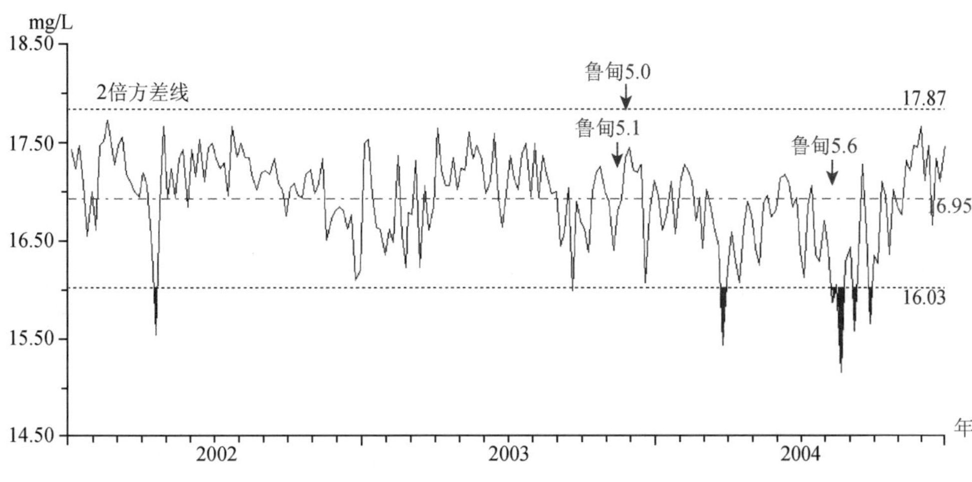

图 26　会泽水氡五日均值曲线

Fig. 26　Curve of 5-day mean value of radon content in groundwater at Huize station

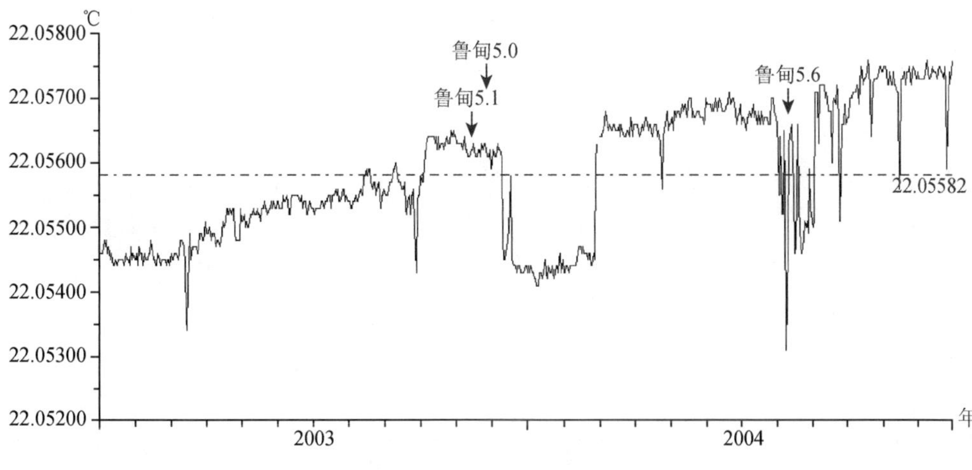

图 27　昭通深井水温日均值曲线

Fig. 27　Curve of daily mean value of deep well tempareture at Zhaotong station

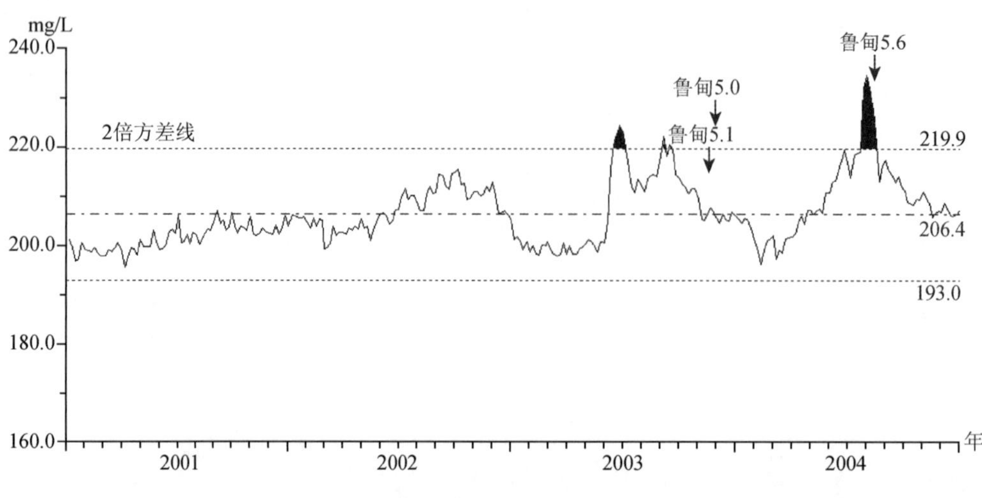

图 28　巧家重碳酸氢根离子五日均值曲线

Fig. 28　Curve of 5-day mean value of HCO_3^- content in groundwater at Qiaojia station

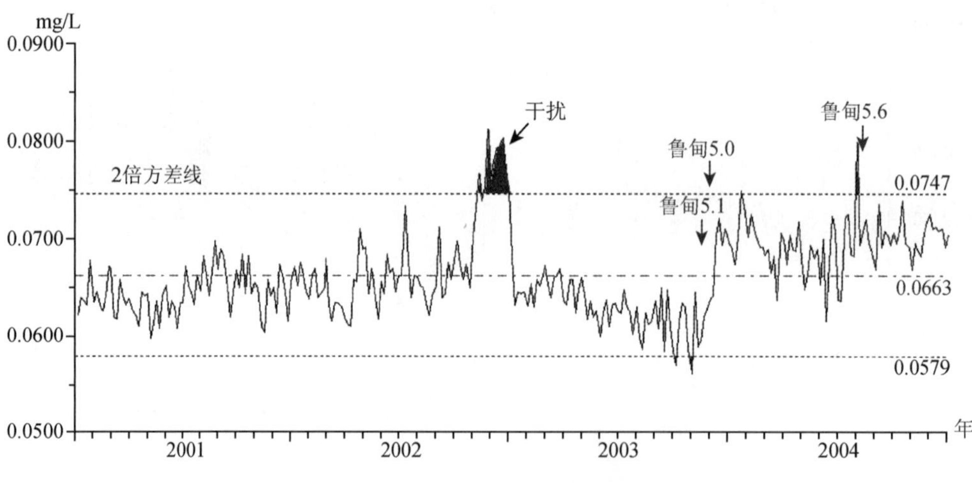

图 29　巧家氟离子五日均值曲线

Fig. 29　Curve of 5-day mean value of F^- content in groundwater at Qiaojia station

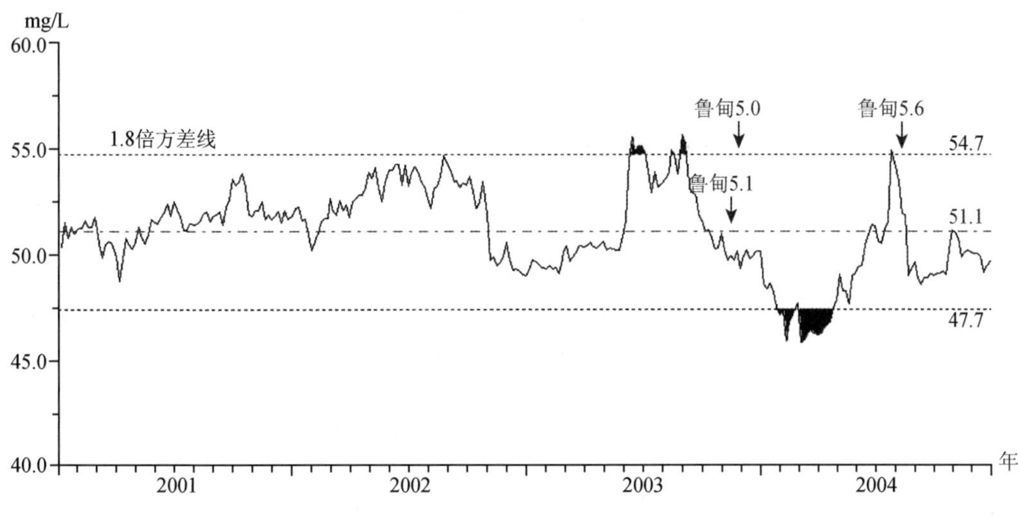

图 30　巧家钙离子五日均值曲线

Fig. 30　Curve of 5-day mean value of Ca^{2+} content in groundwater at Qiaojia station

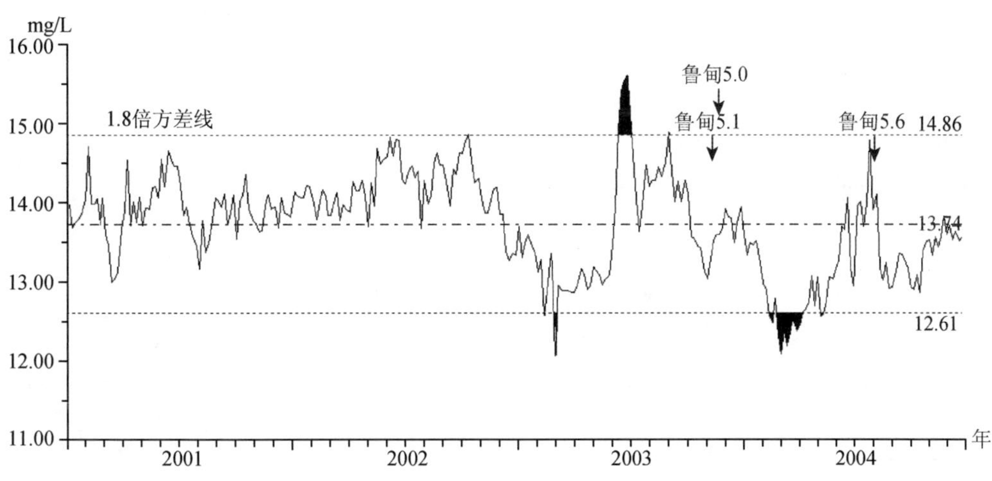

图 31　巧家镁离子五日均值曲线

Fig. 31　Curve of 5-day mean value of Mg^{2+} content in groundwater at Qiaojia station

七、地震前兆异常特征分析

综合分析后认为，鲁甸震群具有如下前兆异常特征：

1. 地震活动性参数中、短期异常显著

震区所在的洒渔河断裂及构造上紧密联系的巧家—莲峰断裂、威宁—大关—马边断裂及小江断裂北段等滇东北地区，2003 年 6 月 3 日至 10 月 16 日出现长达 135 天 3 级地震平静异常，趋过 2.5 倍方差阀值；3 级地震月频次（$w = 3\text{m}$）在鲁甸 5.6 级地震前的 2003 年 12 月和 2004 年 1、3、5、7 ~ 8 月持续出现超 1.5 倍方差高频次异常，说明该次中强震群的发生是滇东北地区进入中强以上地震高活动期的标志，2005 年的会泽 5.3 和 2006 年盐津两次 5.1 级地震更突显了这一趋势，而 2001 年 7 月、11 月出现的高频次现象，可能与 2001 年 10 月 27 日永胜 6.0 级地震有关，这也反映了滇东北地区构造活动与川滇菱块存在一定的关联性。

2. 前兆测项异常以水化学为主，异常沿北东向展布特征明显

鲁甸丛集性中强震群发生之前，前兆异常以水化学测项为主，异常有水氡、水温、重碳酸氢根离子、氟离子、钙镁离子等 6 个项目 10 台项异常，异常时序特征表现为测值破坏正常年变和超差变化；空间上穿越震中区呈北东向分布，与洒渔河断裂的北东走向一致。

由图 21 可知，异常均出现在云南区域，这可能与北东走向的洒渔河断裂参与活动有关。前兆异常以水化学为主，主要原因有三：一是震区次级断裂较发育，2003 年以来构造活动持续增强，浅部与深部的水体交换较迅速，集中反映为震前短期内水中氡含量的增加；二是本区域水化学测项数较多，占总测项的 44%；三是其他部分测项受环境干扰较大，如昭通石英摆、东川金属摆及巧家地磁等受建设施工影响。

表 7　鲁甸地震前前兆异常百分比统计表

Table 7　Anomalous ratio of observation stations and items before the Ludian earthquakes

$\Delta \leqslant 100\text{km}$		$101 < \Delta \leqslant 200\text{km}$	
异常台站/%	异常台项/%	异常台站/%	异常台项/%
67	37	20	7

表 7 为鲁甸丛集性中强震群前前兆异常百分比统计表，表中显示震中周围 100km 内 6 个台站有 4 个台出现异常，占 67%，19 个测项中有 7 个测项异常，占 37%，异常量较多；而 101 ~ 200km 内 15 个台中有 3 个台出现异常，占 20%，45 个测项中有 3 个测项异常，占 7%，异常的相对量较小。前兆异常主要集中在震中 100km 范围内，孕震区范围较小，与该次震群地震震级不大、震源深度较浅（约 10km）有关。

表 8 给出了三次鲁甸地震前的中、短、临异常数量分布情况。5.1 和 5.0 级地震前出现 1 ~ 6 个月的短期异常 8 项，无中期和临震异常出现；5.6 级地震前则分别出现了中、短、临异常 1 项、9 项、3 项，异常在时间上较为清晰。

表 8 鲁甸地震前异常数统计表

Table 8 Anomalies of items before the Ludian earthquakes

M_S5.1 和 M_S5.0			M_S5.6		
中期异常数（M）	短期异常数（S）	临震异常数（I）	中期异常数（M）	短期异常数（S）	临震异常数（I）
0	8 项	0	1 项	9 项	3 项

八、震前预测、预防和震后响应

1. 震前预测

地震前云南省地震局曾作出过一定程度的中期预报。云南省地震局分别在 2003 年度和 2004 年度地震趋势研究报告[6,7]中明确指出：2003 年度和 2004 年度滇东双柏—易门—建水—开远—华宁—宜良—寻甸—巧家一带，存在发生 6 级左右地震危险。

2. 震后响应

鲁甸三次 5 级以上地震发生后，云南省地震局均立即启动了破坏性地震应急预案，震后 1 ~ 3 小时内派出地震现场工作队奔赴震区；中国地震局派出由地震应急救援司组织的专家组（共有 10 人次）均于次日抵达震区；中国地震局和云南省地震局以及昭通市地震局领导和专家组成了地震现场工作指挥部，下设秘书、震情监测预报、地震灾评及科考、强震观测、通讯及后勤保障等 6 个组，共有 103 人次开展地震现场工作。经过 2003 年 11 月 15 ~ 19 日、11 月 27 ~ 29 日以及 2004 年 8 月 11 ~ 15 日共 13 天的工作，出动灾评人员 40 余人次，车辆 29 车次，对云南省鲁甸县、昭阳区、会泽县和贵州省威宁县进行灾情调查，行程 4.7 万余公里，完成 260 个调查点、159 个抽样点的调查工作，提交了《鲁甸 5.1 级地震灾害损失评估报告》[1)]、《鲁甸 5.0 级地震灾害损失评估报告》[2)]和《鲁甸 5.6 级地震灾害损失评估报告》[3)]。

3. 震后趋势判定

鲁甸 5.1 级地震震型判定错误，5.0 级地震余震强度判定偏大，5.6 级地震震型、震后趋势判定正确。

（1）2003 年 11 月 15 日鲁甸 5.1 级地震，11 月 21 日云南省地震局会商意见：该地震序列衰减基本正常，为主震—余震型，震区近期仍要注意 3-4 级地震。

（2）2003 年 11 月 26 日鲁甸 5.0 级地震，12 月 3 日云南省地震局会商意见：该地震序列衰减不正常，震区近期仍要注意 4 级左右地震的发生。

（3）2004 年 8 月 10 日鲁甸 5.6 级地震，8 月 16 日云南省地震局会商意见：序列衰减基本正常，初步判定鲁甸 5.6 级地震序列为主震—余震型，近几天老震区可能会有 4 级左右余震的起伏。

4. 震害分析

鲁甸 5.1 级、5.0 级两次地震形成了较大范围的Ⅶ度破坏区，甚至有Ⅷ度的破坏现象。

主要原因大致如下：①震源深度浅、地表振动强烈。②从工程地质条件来看，灾区地处昭阳—鲁甸盆地，广泛分布有煤层、膨胀土、淤泥和松散的粉砂、粉土等不良地层，且埋藏较浅。③从建筑结构来看，当地绝大多数土木结构房屋是土搁梁房屋，没有木屋架，主要由夯土墙承重；砖木结构和砖混结构房屋的墙体灰浆标号低。

鲁甸 5.6 级地震造成Ⅵ度、Ⅶ度、Ⅷ度区三个破坏区。个别砖混结构房屋有倒塌现象，部分砖混结构房屋底层承重墙体开裂，呈“X”形或斜向或横向贯穿裂缝，多数砖混结构墙体见裂纹或裂缝；砖混结构房屋的震害程度超过砖木和土木结构；地震灾害较 5.1、5.0 级地震严重。可能原因是：本次地震震中紧邻 5.1、5.0 级地震，相距仅 2 ~ 3km，地震灾区重叠，发震时间间隔仅 9 个月，新老震害叠加。

九、结论与讨论

（1）鲁甸震群发生前，震区出现 135 天 3 级地震平静异常，震群发生前 1 个月，3 级平静被 2003 年 10 月 16 日巧家 3.0 级地震打破，该地震距鲁甸未来震群的震中仅 50km。鲁甸地震震中与 2003 年大姚 6.2 和 6.1 级地震震中相距约 270km，其孕震区存在交叉，所以说，鲁甸震群发生前的 3 级平静可能与大姚两次地震后的区域应力场减弱有一定关系；同时，3 级地震长时间平静被打破对预测本地区破坏性地震有一定短临指示意义。

（2）鲁甸 5.1 和 5.0 级地震后，5.6 级地震之前，2004 年 1 ~ 7 月滇东北地区出现 3 级地震的显著增频，3 级活动水平明显高于鲁甸 5.1 和 5.0 级地震前，地震加速活动趋势明显，这是本地区震群活动的一大特点。

（3）鲁甸 5.1 级主震后至 5.0 级主震前 12 天内无 2 级以下地震发生，序列衰减较慢；5.0 级地震后序列衰减不正常；5.6 级地震余震序列稍有起伏，衰减基本正常。前两次地震序列衰减不正常有利于对后期 5.6 级地震的判断；序列 b 值偏小，均未超过 0.5，与本地区地质构造特点有关。

（4）对于 6 级以上强震，目前一般是根据震源机制解结合余震总体分布走向、野外地震地表破裂带展布或等震线长轴方向推测震时的主破裂面的破裂方向；而对于中强地震，主破裂面的确定却是一难点。鲁甸三次中强地震的余震分布呈 NNW 向，我们结合地震矩张量解结果把主破裂面定为 NNW 向，并提出可能存在一条 NNW 向的新活动构造，这与等震线长轴走向不一致，还有待进一步研究。

（5）前兆异常主要集中在震中 100km 范围内，震中区 67% 的台站出现异常。该次鲁甸震群孕震区范围较小，可能与三次中强地震震级不大、震源深度较浅（约 10km）有关。

（6）鲁甸 5.1 和 5.0 级地震前出现短期异常 8 项，无中期和临震异常出现，说明小震级地震发生有更大的随机性；而 5.6 级地震前则分别出现了中、短、临异常，异常对地震孕育过程的指示较为清晰，有利于震前作出有减灾实效的预报。

参 考 文 献

[1] 非明伦、付正新、谢英情等，云南鲁甸 5.1、5.0 级地震震害分析 [J]，防灾减灾工程学报，24 (4)：432 ~ 439，2004

[2] 刘强、倪四道、刘杰，云南滇东北地区丛集式中强地震应力触发特征 [J]，云南大学学报（自然科学版），29 (S1)：478 ~ 487，2007

[3] 非明伦、余庆坤、谢英情等，鲁甸 5.6 级地震震害分析 [J]，地震研究，29 (1)：87 ~ 91，2006

[4] 阮祥、程万正，云南鲁甸 3 次 $M_S \geqslant 5.0$ 级地震破裂参数的研究 [J]，地震学报，30 (1)：97 ~ 102，2008

参 考 资 料

1）云南省地震局，2003 年 11 月 15 日鲁甸 5.1 级地震灾害损失评估报告，2003.11

2）云南省地震局，2003 年 11 月 26 日鲁甸 5.0 级地震灾害损失评估报告，2003.11

3）云南省地震局，2004 年 8 月 10 日鲁甸 5.6 级地震灾害损失评估报告，2004.08

4）云南省地震局，云南地震目录（2003，2004）

5）中国地震局分析预报中心，中国地震目录（2003，2004）

6）云南省地震预报中心，云南省 2003 年度地震趋势研究报告，云南省地震局，2002.11

7）云南省地震预报中心，云南省 2004 年度地震趋势研究报告，云南省地震局，2003.11

M_S 5.1 (November 15, 2003), M_S 5.0 (November 26, 2003) and M_S 5.6 (August 10, 2004) Ludian Earthquakes in Yunnan Province

Abstract

On Nov. 15, 2003, Nov. 26, 2003 and Aug. 10, 2004, three intermediate-strength earthquakes occurred in Southeast of Ludian, Yunnan provice. The macroseismic epicentre was located Taoyuan township, Ludian county. The meizoseismal area was elliptic with major axis in NNE direction and intensity Ⅶ for M_S5.1 and M_S5.0, and intensity Ⅷ for M_S5.6. For three earthquakes, 8 people were killed, but 217 injured and 498 of them were wounded severely, and the direct economic loss of events was about 604.8 million Yuan.

According to on-the-spot geological investigations and distribution of aftershocks, it is suggested that three mainshocks were the result of the left-lateral slip of the nodal plane Ⅱ under the action of the principal stress of compression in the NNW direction. The focal mechanism solutions showed that the azimuth of the principal axis of compression was 287° (NNW). The nodal plane Ⅱ was the main rupture plane and the seismogenic structure was a new active fault with NNW direction.

There were 35 seismic stations in the region within 200km around the epicenter area. Among them there were 14 stations for seismometric observations, 21 stations with precursor monitoring observations. Before earthquakes, 12 items of anomalies appeared. Among them there were 10 items of precursory anomalies which are mainly imminent and short-term on the fixed observation points.

Before Ludian earthquakes, a mid-term forecast was put forward by the Seismological Administration of Yunnan province. We defined them clustering earthquakes type. Aftershocks were distributed in NW direction and the magnitude of the largest aftershock is M_L4.3.

The characteristics of anomaly, anomaly distribution and identification of anomalies are analyzed and discussed in the paper. The research report of earthquake cases is valuable for destructive earthquake prediction in this region.

报告附件

附表 1 固定前兆观测台（点）与观测项目汇总表

序号	台站（点）名称	经纬度（°）		测项	资料类别	震中距Δ（km）	备注
		φ_N	λ_E				
1	鲁甸	27.17	103.57	水氡	Ⅰ	5	
				pH、HCO_3^-、F^-	Ⅱ		
2	昭通地震台	27.32	103.72	测震△◬		18	
				地倾斜（石英摆）	Ⅱ		
3	昭通渔洞	27.40	103.57	水位	Ⅰ	25	
				水温	Ⅰ		
4	巧家	26.98	102.98	地磁 Z、D、H	Ⅲ	67	
				水氡	Ⅰ		
				HCO_3^-、Ca^{2+}、Mg^{2+}、F^-	Ⅰ		
5	会泽	26.52	103.15	水位	Ⅱ	87	
				水温	Ⅱ		
				水氡	Ⅰ		
6	盐津	28.07	104.25	测震△		115	
7	永善	28.23	103.65	水氡	Ⅰ	117	
8	宣威	26.22	104.10	水位	Ⅱ	117	
				水温	Ⅱ		
9	昭觉	28.00	102.85	测震△		119	
				水氡	Ⅱ		
				水温	Ⅱ		
10	雷波	28.27	103.57	测震△◬		120	
11	镇雄	27.45	104.87	测震△◬		126	
				水氡	Ⅱ		
12	东川	26.11	103.20	测震△◬		127	
				地倾斜（摆式）	Ⅱ		
				水氡	Ⅰ		
13	筠连	28.18	104.52	测震△◬		141	
14	会理	26.65	102.25	测震△		149	

续表

序号	台站（点）名称	经纬度（°）		测项	资料类别	震中距Δ（km）	备注
		φ_N	λ_E				
15	西昌	27.83	102.23	测震台网△		156	
				地磁 F、Z、D、H	Ⅱ		
				电场、电阻率	Ⅱ		
				地倾斜（摆式）	Ⅱ		
				地应力	Ⅱ		
				水位	Ⅱ		
				水氡	Ⅱ		
				水汞	Ⅱ		
16	水富	28.60	104.40	水位	Ⅱ	177	
				水温	Ⅱ		
17	富源	25.67	104.25	水位	Ⅰ	179	
				水温	Ⅰ		
18	红格	26.53	101.93	电场	Ⅱ	183	
				电阻率	Ⅱ		
19	会理川06	26.31	102.06	水位	Ⅱ	183	
20	马边	28.83	103.53	地磁Z	Ⅱ	184	
21	寻甸	25.53	103.25	水温	Ⅰ	187	
				水氡	Ⅰ		
22	曲靖	25.48	103.78	水位	Ⅰ	189	
				水温	Ⅰ		
23	高县	28.64	104.74	测震⛛		195	
24	马龙	25.43	103.58	测震⛛		196	
25	攀矿台	26.60	101.73	测震△		199	
				地倾斜（摆式）	Ⅱ		
				水氡	Ⅱ		

续表

序号	台站（点）名称	经纬度（°）		测项	资料类别	震中距Δ（km）	备注
		φ_N	λ_E				
26	攀枝花	26.50	101.74	测震△◬		200	
				重力	Ⅱ		
				地倾斜（摆式、连通管）	Ⅱ		
				地应力	Ⅱ		
				水位	Ⅱ		
				水氡	Ⅱ		
				水汞	Ⅱ		

分类统计	0 < Δ≤100km	100 < Δ≤200km			总数
测项数 N	13	15			28
台项数 n	19	52			71
测震单项台数 a	0	6			6
形变单项台数 b	0	0			0
电磁单项台数 c	0	1			1
流体单项台数 d	0	2			2
综合台站数 e	5	12			17
综合台中有测震项目的台站数 f	1	6			7
测震台总数 $a+f$	1	12			13
台站总数 $a+b+c+d+e$	5	21			26
备注					

附表 2　测震以外固定前兆观测项目与异常统计表

序号	台站（点）名称	测项	资料类别	震中距 Δ（km）	按震中距 Δ 范围进行异常统计																			
					0 < Δ ≤ 100km					100 < Δ ≤ 200km					200 < Δ ≤ 300km					300 < Δ ≤ 500km				
					L	M	S	I	U	L	M	S	I	U	L	M	S	I	U	L	M	S	I	U
1	鲁甸	水氡	Ⅰ	5	—	—	√	—																
		PH、HCO_3^-、F^-	Ⅱ		—	—	—	—																
2	昭通地震台	地倾斜（石英摆）	Ⅱ	18	—	—	—	—																
3	昭通渔洞	水位	Ⅰ	25	—	—	—	—																
		水温	Ⅰ		—	—	√	√																
4	巧家	地磁 Z、D、H	Ⅲ	67	—	—	—	—																
		水氡	Ⅰ		—	—	—	—																
		HCO_3^-	Ⅰ		—	—	√	√																
		Ca^{2+}	Ⅰ		—	—	√	—																
		Mg^{2+}	Ⅰ		—	—	√	—																
		F^-	Ⅰ		—	—	√	—																
	会泽	水位	Ⅱ	87	—	—	—	—																
		水温	Ⅱ		—	—	—	—																
		水氡	Ⅰ		—	—	√	√																
	永善	水氡	Ⅰ	117						—	—	√	—											
	宣威	水位	Ⅱ	117						—	—	—	—											
		水温	Ⅱ							—	—	—	—											

续表

序号	台站（点）名称	测项	资料类别	震中距Δ（km）	按震中距Δ范围进行异常统计																			
					0 < Δ ≤ 100km					100 < Δ ≤ 200km					200 < Δ ≤ 300km					300 < Δ ≤ 500km				
					L	M	S	I	U	L	M	S	I	U	L	M	S	I	U	L	M	S	I	U
	昭觉	水氡	Ⅱ	119						—	—	—	—											
		水温	Ⅱ							—	—	—	—											
	镇雄	水氡	Ⅱ	126						—	—	—	—											
	东川	地倾斜（摆式）	Ⅱ	127						—	—	—	—											
		水氡	Ⅰ							—	—	√	—											
	西昌	地磁 F、Z、D、H	Ⅱ	156						—	—	—	—											
		电场、电阻率	Ⅱ							—	—	—	—											
		地倾斜（摆式）	Ⅱ							—	—	—	—											
		地应力	Ⅱ							—	—	—	—											
		水位	Ⅱ							—	—	—	—											
		水氡	Ⅱ							—	—	—	—											
		水汞	Ⅱ							—	—	—	—											
	水富	水位	Ⅱ	177						—	—	—	—											
		水温	Ⅱ							—	—	—	—											
	富源	水位	Ⅰ	179						—	—	—	—											
		水温	Ⅰ							—	—	—	—											
	红格	电场	Ⅱ	183						—	—	—	—											
		电阻率	Ⅱ							—	—	—	—											

续表

序号	台站（点）名称	测项	资料类别	震中距Δ（km）	按震中距Δ范围进行异常统计																			
					0＜Δ≤100km					100＜Δ≤200km					200＜Δ≤300km					300＜Δ≤500km				
					L	M	S	I	U	L	M	S	I	U	L	M	S	I	U	L	M	S	I	U
	会理川06	水位	Ⅱ	183						—	—	—	—											
	马边	地磁 *Z*	Ⅱ	184						—	—	—	—											
	寻甸	水温	Ⅰ	187						—	—	—	—											
		水氡	Ⅰ							—	—	√	—											
	曲靖	水位	Ⅰ	189						—	—	—	—											
		水温	Ⅰ							—	—	—	—											
	攀矿台	地倾斜（摆式）	Ⅱ	199						—	—	—	—											
		水氡	Ⅱ							—	—	—	—											
	攀枝花	重力	Ⅱ	200						—	—	—	—											
		地倾斜（摆式、连通管）	Ⅱ							—	—	—	—											
		地应力	Ⅱ							—	—	—	—											
		水位	Ⅱ							—	—	—	—											
		水氡	Ⅱ							—	—	—	—											
		水汞	Ⅱ							—	—	—	—											

续表

序号	台站（点）名称	测项	资料类别	震中距Δ（km）	按震中距Δ范围进行异常统计																			
					0 < Δ ≤ 100km					100 < Δ ≤ 200km					200 < Δ ≤ 300km					300 < Δ ≤ 500km				
					L	M	S	I	U	L	M	S	I	U	L	M	S	I	U	L	M	S	I	U
分类统计	台项	异常台项数			0	0	7	3		0	0	3	0											
		台项总数			18	18	18	18		40	40	40	40											
		异常台项百分比/%			0	0	39	17		0	0	8	0											
	观测台站（点）	异常台站数			0	0	4	3		0	0	3	0											
		台站总数			5	5	5	5		15	15	15	15											
		异常台站百分比/%			0	0	80	60		0	0	20	0											
	测项总数				12					14														
	观测台站总数				5					15														
备注																								

2003 年 11 月 25 日山西省洪洞 4.9 级地震

山西省地震局

张淑亮　宋美琴　吕　芳　高振强　李　丽

摘　要

2003 年 11 月 25 日在山西省洪洞县甘亭镇发生 4.9 级地震，地震微观震中 36°10′N，111°37′E，宏观震中位于洪洞县甘亭镇一带。这次地震极震区烈度为Ⅵ度。地震造成 11 人受伤，直接经济损失 54.94 万元。

该地震序列为主余型，最大余震为 M_L2.8。根据地震现场考察结果、序列小震分布和烈度等震线长轴等，推测洪洞 4.9 级地震的发震断裂可能为罗云山断裂，但该结果与震源机制解推断的发震断层具有逆断性质的结果相悖。

洪洞 4.9 级地震前距震中 200km 范围内分布有 7 个测震台和 11 个前兆台站或测点。震前共出现异常项目 10 项，其中中期异常 2 项、短期异常 6 项、临震异常 2 项。前兆异常以短临异常为主，中期异常较少。震中 100km 范围内的异常台站和台项比分别为 60% 和 55.6%，101 ~ 200km 范围内的为 16.7% 和 10%。

前　言

2003 年 11 月 25 日 13 时 40 分，山西省洪洞县甘亭镇发生 M4.9 地震。山西地震台网测定的微观震中为 36°10′N，111°37′E，宏观震中位于洪洞县甘亭镇南羊獬、北羊獬一带，与微观震中一致。这次地震没有造成直接人员伤亡，但距震中 30km 的山西浮山县城关示范小学学生在避震时因拥挤造成挤伤、踩伤 11 人。地震直接经济损失 54.94 万元。

洪洞 4.9 级地震位于临汾断陷盆地中部临汾凹陷内的山西省洪洞县甘亭镇。临汾断陷盆地内历史上曾发生过 1303 年洪洞 8 级和 1695 年临汾 7¾级地震，是山西地震带中地震活动频度高、强度大的一个构造单元。

洪洞 4.9 级地震位于临汾断陷盆地中部的临汾凹陷内。临汾断陷盆地内历史上曾发生过 1303 年洪洞 8 级和 1695 年临汾 7¾级地震，是山西地震带中地震活动频度高、强度大的一个构造单元。

这次地震的震中 200km 范围内有 7 个测震台、11 个前兆台点。震前出现 10 条异常。其中地震活动异常 4 项，前兆异常 6 项。前兆异常以短临异常为主，中期异常较少。

本研究报告是以山西省 2003 年度地震趋势研究报告为基础，并收集震后新发现的一些异常，经过重新整理资料和分析研究而完成的。

一、测震台网及地震基本参数

1. 测震台网

距震中 200km 范围内分布有 7 个测震台（图 1），测震台网密度较低。其中 0～100km 范围 3 个，100～200km 范围 4 个。4.9 级地震发生的临汾盆地内，监测能力为 $M_L \geqslant 2.0$ 级，两边山区监测能力 $M_L \geqslant 2.5$ 级。

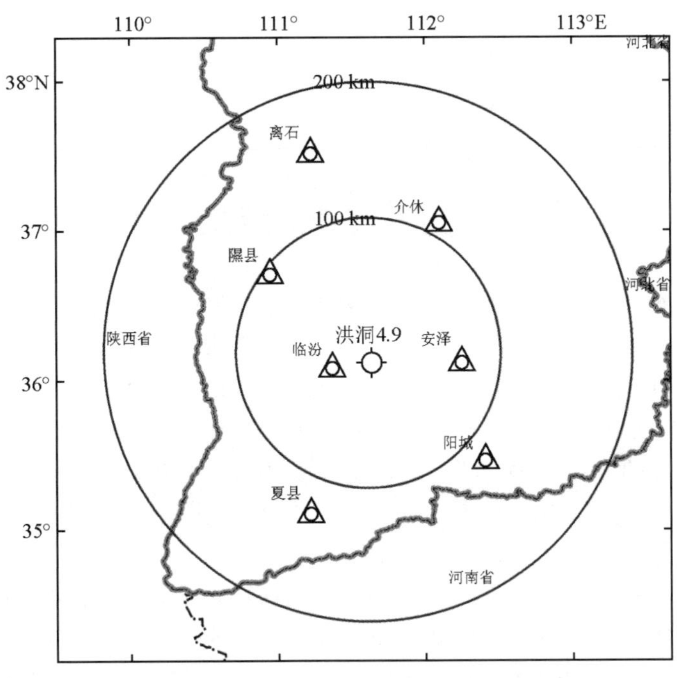

图 1 洪洞 M4.9 地震震中附近测震台站分布

Fig. 1 Distribution of seismometric stations around the M4.9 Hongtong earthquake

2. 地震基本参数

表 1 列出了不同来源的洪洞 4.9 级地震基本参数。本震例研究所选参数来源于国家大震速报目录。

表1　洪洞 $M4.9$ 级地震的基本参数表

Table 1　Basic parameters of the $M4.9$ Hongtong earthquake

编号	发震日期	发震时刻	震中位置		震级	震源深度	震中地名	结果来源
	年 月 日	时 分 秒	ψ_N	λ_E		(km)		
1	2003 11 25	13 40 34.7	36°10′	111°37′	$M_L4.9$	19	洪洞甘亭镇	山西省地震局速报目录
2	2003 11 25	13 40 34.8	36°10′	111°38′	$M_L5.0$	19	洪洞甘亭镇	山西省及邻近地区地震目录[1)]
3	2003 11 25	13 40	36°10′	111°37′	$M4.9$	20	洪洞甘亭镇	中国地震台网中心
4	2003 11 25	13 40 32	36°12′	111°36′	$M4.9$	15	山西洪洞	国家台网大震速报目录

二、地震地质背景

洪洞4.9级地震震中区位于临汾断陷盆地的中部，涉及洪洞凹陷和临汾凹陷。

临汾盆地东侧受NNE向霍山山前断裂、浮山断裂控制，西侧受NNE向罗云山山前断裂控制，南侧受NEE向峨嵋断裂北缘断裂控制，形成临汾、洪洞、侯马3个凹陷和辛置、浮山、襄汾3个凸起。临汾盆地是新构造运动最强烈的地区之一，为强震多发盆地，历史上发生过15次 $M\geqslant5.0$ 级地震，其中5.0～5.9级地震10次，6.0～6.9级地震3次，7.0～7.9级1次，8.0～8.9级1次，最大地震为1303年9月25日洪洞8.0级地震。1970年以来，本区共发生过 $M\geqslant2.0$ 级地震167次（余震除外），其中 $M2.0$～2.9地震153次，$M3.0$～3.9级地震10次，$M4.0$～4.9地震4次。最大地震为1989年12月25日侯马 $M4.9$ 地震[1]。

洪洞4.9级地震震中距罗云山断裂12km，位于该断裂20km深度的投影上。罗云山断裂是临汾凹陷的主控断裂，走向北北东，是全新世活动断裂。深部地球物理勘探研究表明，该断裂深度大于20km，控制临汾盆地发育，断裂活动以正断为主兼有走滑。断裂的土门—峪里段，是活动最强烈的地段。盆地中的汾东、汾西断裂与其平行，规模小于罗云山断裂，向下延伸小于10km，是罗云山断裂的次级构造。断裂东部的魏村2001年7月25日曾发生过 $M_L4.5$ 地震，震源深度为12km。魏村地震震中位于洪洞4.9级地震震中北西10km，表明该断裂断面的埋深由魏村至甘亭增加了8km。根据烈度等震线走向与罗云山断裂走向一致判断，洪洞4.9地震的发震断裂为罗云山断裂[1]。

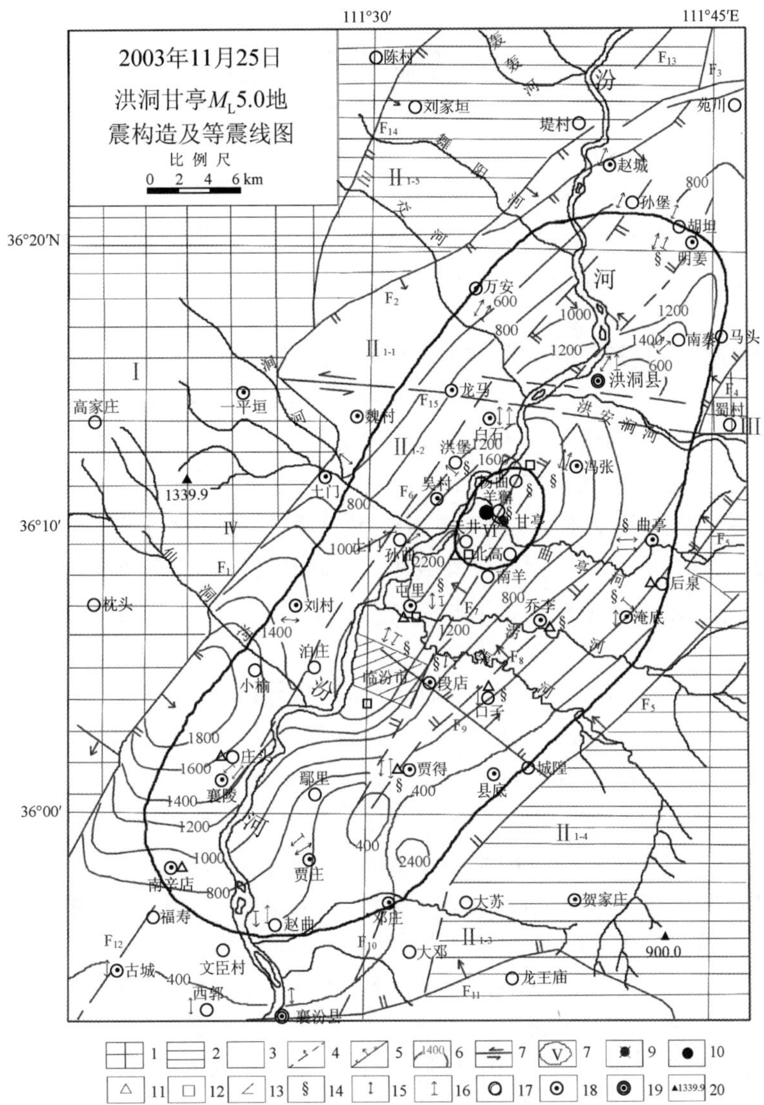

图 2　洪洞 4.9 级地震区域活动构造及等震线图

Fig. 2　Map of regional fault structure and isoseismal map related to the *M*4.9 Hongtong earthquake

1. 强烈隆起区；2. 盆地中凸起；3. 断陷盆地；4. 隐伏活动断裂；5. 活动断裂；6. 新生代断陷幅度等值线（m）；7. 走滑断层；8. 烈度等值线及烈度值；9. 微观震中；10. 宏观震中；11. 屋脊倒塌；12. 墙壁裂缝；13. 墙倒塌；14. 垂直晃动；15. 水平晃动方向；16. 地声传播方向；17. 村庄；18. 乡镇；19. 县城；20. 海拔高程（m）

新构造分区：Ⅰ. 罗云山断块隆起区；Ⅱ. 汾渭断陷带；Ⅱ$_{1}$. 临汾断陷盆地；Ⅱ$_{1-1}$. 洪洞凹陷；Ⅱ$_{1-2}$. 临汾凹陷；Ⅱ$_{1-3}$襄汾凸起；Ⅱ$_{1-4}$. 浮山凸起；Ⅱ$_{1-5}$. 辛置凸起；Ⅲ. 霍山隆起区

主要断裂编号：F$_1$. 罗云山断裂；F$_2$. 万安断裂；F$_3$. 辛东断裂；F$_4$. 霍山断裂；F$_5$. 大阳断裂；F$_6$. 汾西断裂；F$_7$. 汾东断裂；F$_8$. 乔李断裂；F$_9$. 郭家庄断裂；F$_{10}$. 塔儿山西断裂；F$_{11}$. 襄汾凸起北缘断裂；F$_{12}$古城断裂；F$_{13}$赤峪断裂；F$_{14}$下团堡断裂；F$_{15}$苏堡断裂

三、地震影响场和震害

洪洞4.9地震发生在山西中部经济相对发达的临汾盆地内，房屋建筑相对较好。根据《地震现场调查规范》要求，按照GB 17742—1999中国地震烈度表的标准，结合区内房屋的破坏程度及地震时人的感觉制定了该区烈度划分的标志，Ⅵ度区以房屋成片的轻微破坏为标志，Ⅴ度区、Ⅳ度区以人的感觉的强弱为标志。共划分出Ⅵ、Ⅴ、Ⅳ三个烈度区（图2）[2]。

Ⅵ度区呈椭圆形，长轴7km，走向北北东向，短轴4.6km，面积25km^2，范围涉及山西洪洞县甘亭镇甘亭村、北羊獬、南羊獬、天井村、上桥村、杨曲村、侯建村、北高等16个村庄。地震对年久的砖木结构、土木砖结构房屋造成了轻微破坏。主要表现为烟囱破坏、屋脊倒塌、墙体裂缝，屋脊、烟囱普遍向北倒，屋脊倒塌长度6～10m，墙体上部裂缝宽2～15mm，个别房屋开裂为中等破坏。Ⅵ度区中以南羊獬、北羊獬村破坏最重，南羊獬有170间房屋受到轻微破坏，有4间房屋属中等破坏。地震时Ⅵ度区几乎所有的人都听到了像打雷一样的地声，感觉到地面上下颠簸，南北晃动，家中家具、门窗、房梁作响，多数人惊惶失措，仓惶逃出[2]。

Ⅴ度区也呈走向北北东的椭圆形，长轴55km，走向北北东，短轴22km。西起山西临汾市尧都区刘村，东至大阳村，北起山西洪洞县明姜镇，南至山西襄汾县赵曲村，面积1210km^2。地震对土窑、砖木结构的房屋仅有零星的破坏，主要表现为屋脊倒塌、窑洞倒塌、烟囱倒塌、墙体轻微裂缝等。烟囱屋脊倒塌多为年久的砖木、土、砖、木结构房屋，遭破坏的窑洞为年久失修的无人居住的砖窑或土窑，灾害点零散分布，所占的灾害比例为Ⅴ度区现有房屋的0.1%左右。Ⅴ度区范围内，部分人听到地声、感到上下颠簸，大多数人普遍感觉到水平晃动，多数午休的人被惊醒而跑出室外，部分商店、药店中货架上摆放的不稳定商品翻倒，门窗屋顶、屋架颤动作响，灰土掉落。距极震区较近的临汾市尧都区屯里镇办公楼抹灰墙出现细小裂缝及掉墙灰现象[2]。

在Ⅴ度区外，距震中30km的浮山县城关示范小学学生从教学楼三楼向下跑时因拥挤造成挤伤、踩伤11人，其中轻伤7人，4人住院治疗。

四、地 震 序 列

洪洞4.9级地震始于2003年11月25日，历时10天，结束于12月04日。安泽台和隰县台较完整地记录了地震序列，共记录$M_L \geqslant 0.0$级地震49次，其中1.0～1.9级31次，2.0～2.9级17次，5.0～5.9级1次，最大地震M_L5.0，次大地震M_L2.8级。根据次大地震与主震震级差以及能量比判断，这次地震为主余型。洪洞4.9级地震的余震呈北北东向展布（表2、图3），同Ⅴ度、Ⅵ度烈度区形状相似，且与临汾盆地的走向一致。说明发震断裂走向北北东。

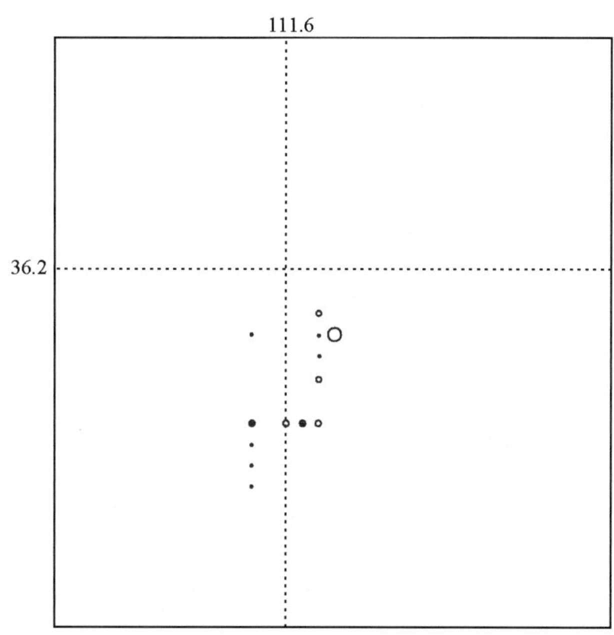

图 3　洪洞 4.9 级地震余震震中分布图

Fig. 3　Epicentral distribution of the $M4.9$ Hongtong earthquake sequence

表 2　洪洞 $M4.9$ 级地震序列目录（$M_L \geqslant 2$ 级）

Table 2　Catalogue of the $M4.9$ Hongtong earthquake sequence（$M_L \geqslant 2.0$）

编号	发震日期	发震时刻	震中位置		震级	震源深度	震中地名	结果来源
	年 月 日	时 分 秒	φ_N	λ_E	M_L	(km)		
1	2003 11 25	13 40 35	36°10′	111°38′	5.0	18	洪洞县甘亭镇	山西省及邻近地区地震目录[1)]
2	2003 11 25	13 48 50	36°09′	111°37′	2.6	19	洪洞县甘亭镇	
3	2003 11 25	13 51 33			2.1			
4	2003 11 25	13 55 34			2.3			
5	2003 11 25	14 08 57			2.1			
6	2003 11 25	14 11 17			2.1			
7	2003 11 25	14 16 59			2.3			
8	2003 11 25	14 26 25	36°09′	111°37′	2.8	18	洪洞县甘亭镇	
9	2003 11 25	14 55 22			2.0			
10	2003 11 25	15 11 48	36°11′	111°37′	2.3	12	洪洞县甘亭镇	
11	2003 11 25	17 29 23			2.0			

续表

编号	发震日期	发震时刻	震中位置		震级	震源深度	震中地名	结果来源
	年 月 日	时 分 秒	φ_N	λ_E	M_L	(km)		
12	2003 11 25	20 39 41			2.0			
13	2003 11 25	22 19 08	36°09′	111°36′	2.0	19	洪洞县甘亭镇	
14	2003 11 25	23 30 39	36°09′	111°37′	2.5	21	洪洞县甘亭镇	
15	2003 11 26	22 45 25			2.4			
16	2002 12 01	04 26 00			2.4			
17	2002 12 01	16 18 09			2.1			
18	2002 12 03	03 08 52	36°08′	111°34′	2.0	19	临汾市东张乡	

这次地震共释放应变能约 $7.7\times10^5J^{1/2}$，其中最大地震释放的应变能占总应变能的84%。由安泽台走时差 t_{S-P}在6.2～6.9s估计，地震分布在5km范围内。

由图4可以看出：这次地震的特点是总频次少于50次，但主震震级较大，最高日频次8次，延续时间较短，序列高潮时段很短。

第1阶段：序列活动的高潮阶段（11月25～26日），5.0级地震后强度和频度快速衰减。

第2阶段：序列活动的衰减阶段（11月27至12月4日），不论强度还是频度，均很低，余震断断续续地发生。

五、震源参数和地震破裂面

根据山西地震台网记录，采用P波初动符号法和波形反演CAP方法分别做了洪洞4.9级地震的震源机制解（表3），图5为CAP方法震源机制解结果的等面积投影（下半球）。结果显示，洪洞4.9级地震发震断层具有逆断层性质，而余震分布、地震现场考察及烈度等震线推测的发震断层为罗云山断裂。深部地球物理勘探研究表明，该断裂活动以正断为主兼有走滑。而震源机制解节面Ⅱ走向与罗云山断裂接近，但断层错动性质不一样，因此，罗云山断裂不是本次地震的发震断裂。现有地质考察在震中区地表未发现逆冲型断裂，因此，该地震的发震断层还需考察震区隐伏断裂，开展野外深部探测工作来确定。

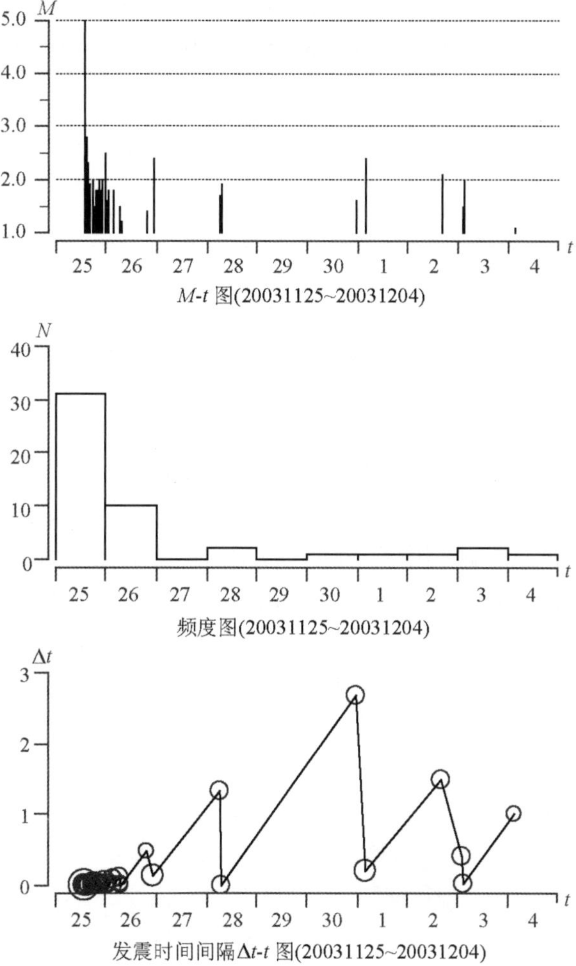

图 4　2003 年 11 月 25 日洪洞 *M*4.9 级地震序列时序图

Fig. 4　The sequence diagram of *M*4.9 Hongtong earthquake on November, 25, 2003

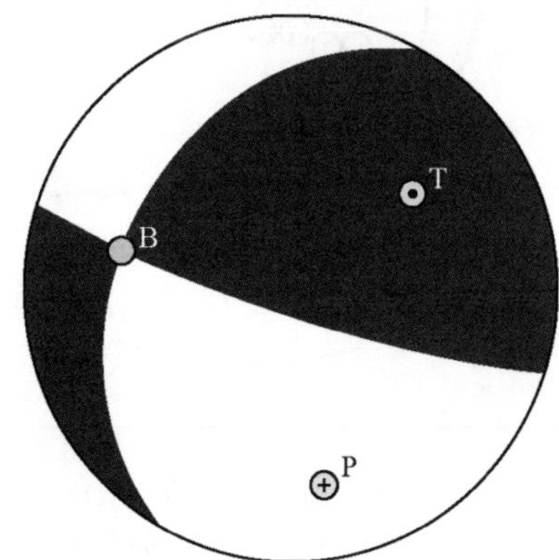

图 5　洪洞 *M*4.9 级地震震源机制解图

Fig. 5　Focal mechanism solution of the *M*4.9 Hongtong earthquake

表 3　洪洞 4.9 级地震震源机制解

Table 3　Focal mechanism solution of the *M*4.9 Hongtong earthquake

节面 A			节面 B			P 轴		T 轴		N 轴		使用方法/矛盾比	使用台站个数（符号数）	结果来源
走向（°）	倾角（°）	滑动角（°）	走向（°）	倾角（°）	滑动角（°）	方位（°）	仰角（°）	方位（°）	仰角（°）	方位（°）	仰角（°）			
108	82	125	201	35	3	170	28	51	42	282	24	CAP 法	6	资料 3）
108	88	124	200	34	3	170	34	48	37	287	34	P 波初动法/0.31	63	文献［3］

六、前兆观测台网及前兆异常

图 6 是洪洞 4.9 级地震前距震中约 200km 范围内地震前兆观测台站（测点）分布情况，共有 11 个前兆观测台点，其中 0 ~100km 范围内有 5 个台点，101 ~200km 范围内有 6 个台点。图 7 是洪洞 4.9 级地震前兆异常分布图，6 项前兆异常均分布在 200km 以内，其中 0 ~100km 范围内有 5 个测项 3 个台点异常，101 ~200km 范围内有 1 个台点 1 个测项的异常。

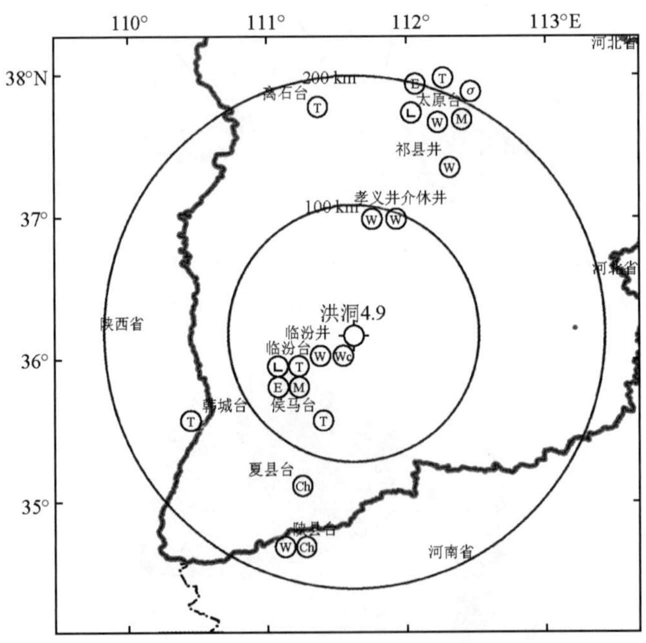

图 6　洪洞 4.9 级地震前定点前兆观测台站分布图

Fig. 6　Distribution of precursory observation stations before the *M*4.9 earthquake

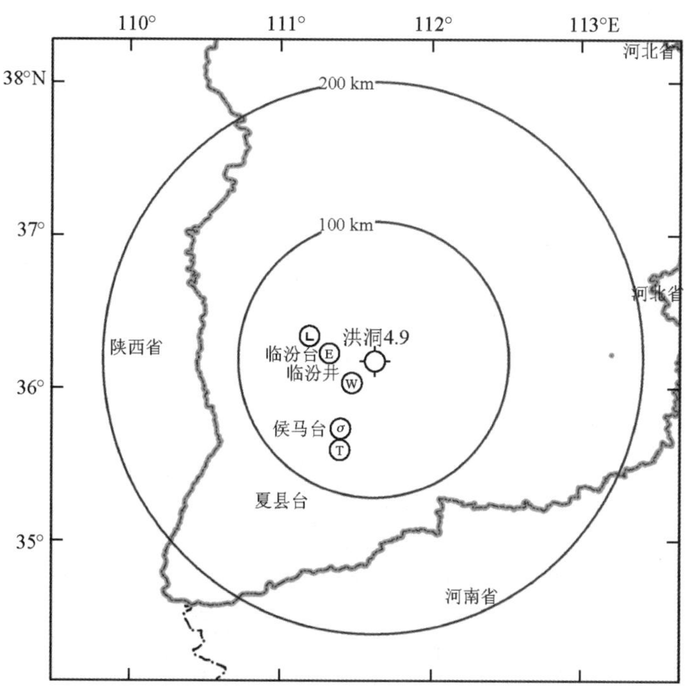

图 7　洪洞 4.9 级地震前定点前兆异常分布图

Fig. 7　Distribution of precursory anomalies on the fixed observation points before the M4.9 Hongtong earthquake

洪洞 4.9 级地震前山西及邻区共出现异常项目 10 项，其中地震活动性 4 项，定点观测项目异常 6 项，各类异常特征见表 4 及图 8 至图 17。

表 4　异常情况登记表

Table 4　Summary of precursory of anomalies

序号	异常项目	台站或观测区	分析方法	异常判据及观测误差	异常起止时间	震后变化	最大幅度	震中距（km）	异常类别	图号	异常特点及备注	备注
1	地震条带	36°～42°N，110°～114°E	震中分布图	集中成带（M_L≥2.0 级）	2003.10.03～10.23	恢复正常			B_I	8	2 级地震在 NNE 向条带上集中，地震发生在条带南端	震后发现的异常
2	地震空区	36°～38°N，113°～116°E	震中分布图	围空	2003.10.03～10.23	空区消失			B_I	8	震前空区打破，地震发生在空区边缘	震后发现的异常

续表

序号	异常项目	台站或观测区	分析方法	异常判据及观测误差	异常起止时间	震后变化	最大幅度	震中距（km）	异常类别	图号	异常特点及备注	备注
3	地震活动因子	31°～43°N，108°～124°E	5项因子空间扫描	异常小区比值较高	2002.11～2003.9	异常消失	>0.43		A_I	9	异常集中在山西中南部—收缩至山西南部—震前消失	震前发现的异常
4	小震窗口	大同老震区	小震月频度图	月频度大于背景值的1倍	2003.11.1～11.22	趋于平静	>3.5倍背景值	460	C_I	10	异常结束后22天后发震	震前发现的异常
5	地电	临汾	观测值曲线	突变	EW：2003.8～2003.10 NS：2003.11.4～11.7	恢复正常	EW：1.8Ω·m NS：1.2Ω·m	25	B_I	11	突降—缓降—突升—发震	震前认为异常属干扰
6	水位	临汾	日均值曲线	突降	2003.11.10～11.23	震后恢复	0.3m	15	C_I	12	突升—突降—突升—发震	震前发现的异常
7	逸出氡气	夏县	原始曲线法	大于几十ppm背景值	2003.8～2004.5	恢复正常	1600ppm	120	B_I	13	异常结束后发震	震前发现的异常
8	水准	临汾	短水准	上升加速	2002.7～2003.11	恢复正常		25	A_I	14	加速—转平—加速—发震	震前发现的异常
9	水管	侯马	日均值	NS：南倾加速 EW：加速东倾	2003.6.24～11.25 2003.9～2003.11	继续加速 恢复正常	780ms	68	B_I B_I	15	南倾加速—发震 东倾—西倾—东倾—转平—发震	震后发现的异常

续表

序号	异常项目	台站或观测区	分析方法	异常判据及观测误差	异常起止时间	震后变化	最大幅度	震中距（km）	异常类别	图号	异常特点及备注	备注
10	伸缩	侯马	日均值	NS：反向加速 EW：西倾加速	2003.9.5~2003.11 2003.8~11.18	基本恢复 恢复后又西倾	5750×10^{-10}	68	B_I B_I	16	北倾—南倾—发震 加速西倾—加速东倾—转平—发震	震后发现的异常

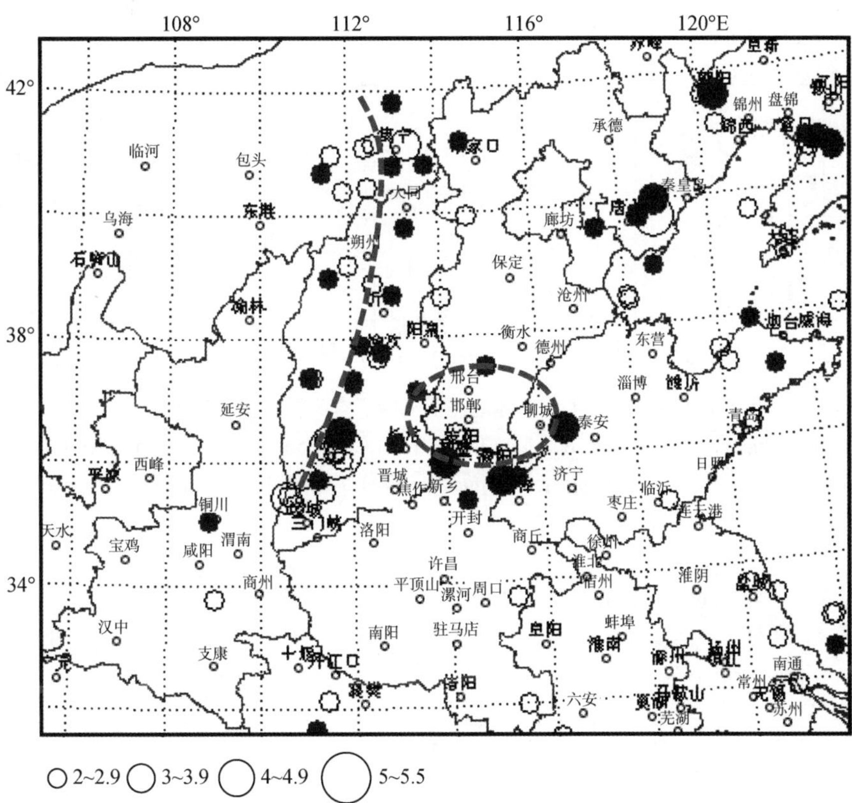

图 8　洪洞 4.9 级地震前 $M_L\geqslant2$ 级地震条带与空区图像（2003.10.3~2003.10.23）

Fig. 8　Distribution of $M_L\geqslant2.0$ events before the $M4.9$ Hongtong earthquake（2003.10.3-2003.10.23）

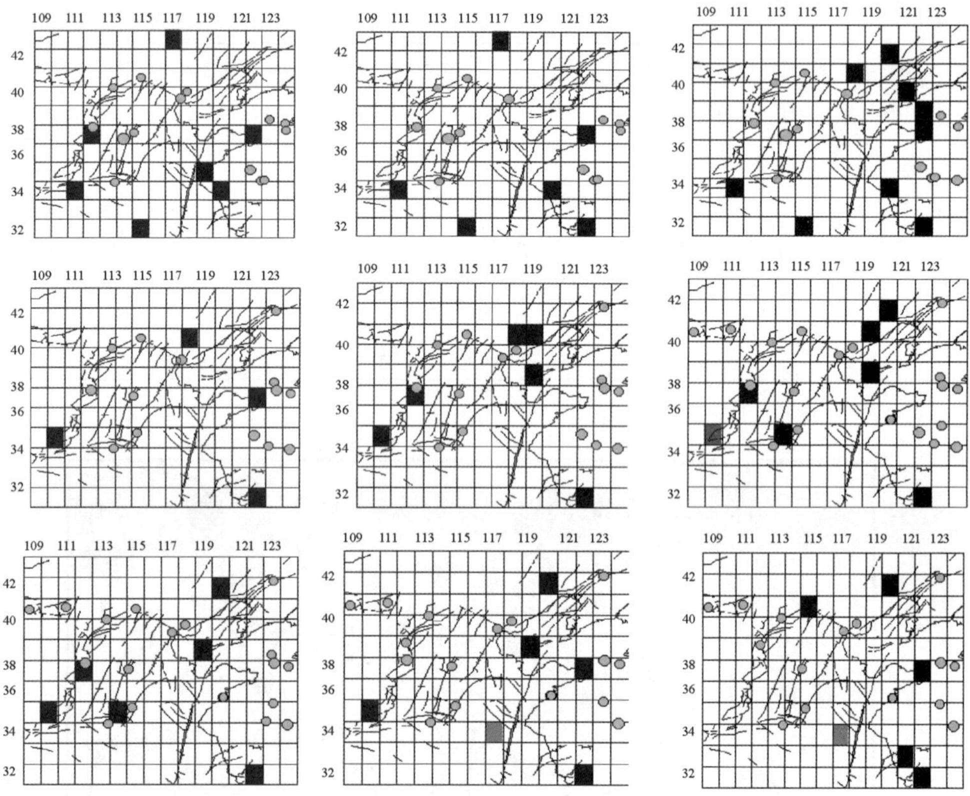

图 9　洪洞 4.9 级地震前华北地区地震活动因子空间扫描图（2003.2 ~ 2003.10）

Fig. 9　Spatial scanning of seismicity factor A value in Northern China before the *M*4.9 Hongtong earthquake (2003.2-2003.10)

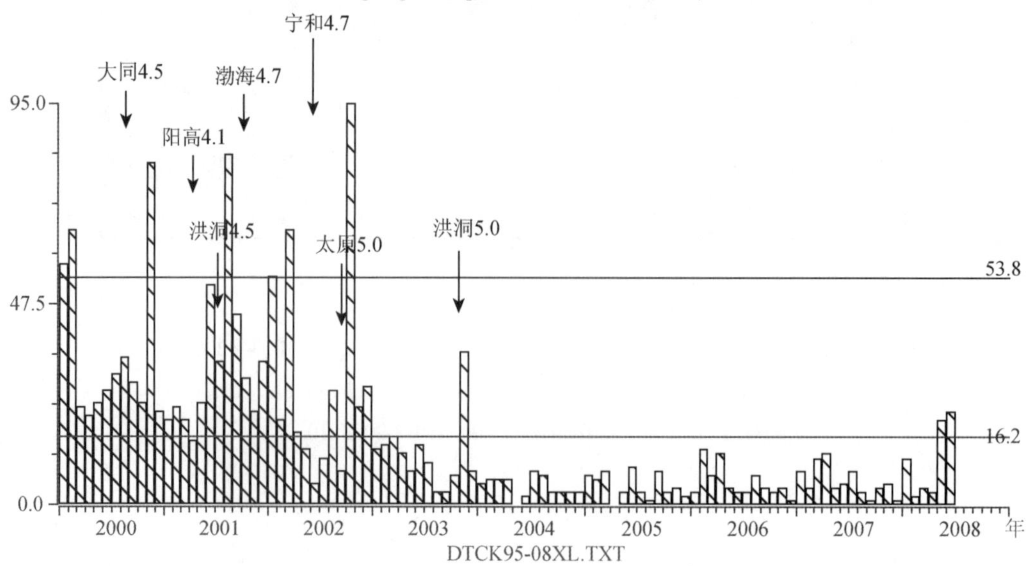

图 10　大同-阳高老震区窗口月频度图（2000 ~ 2008 年）

Fig. 10　Monthly frequency of microearthquakes in Datong-Yanggao seismic area (2000-2008)

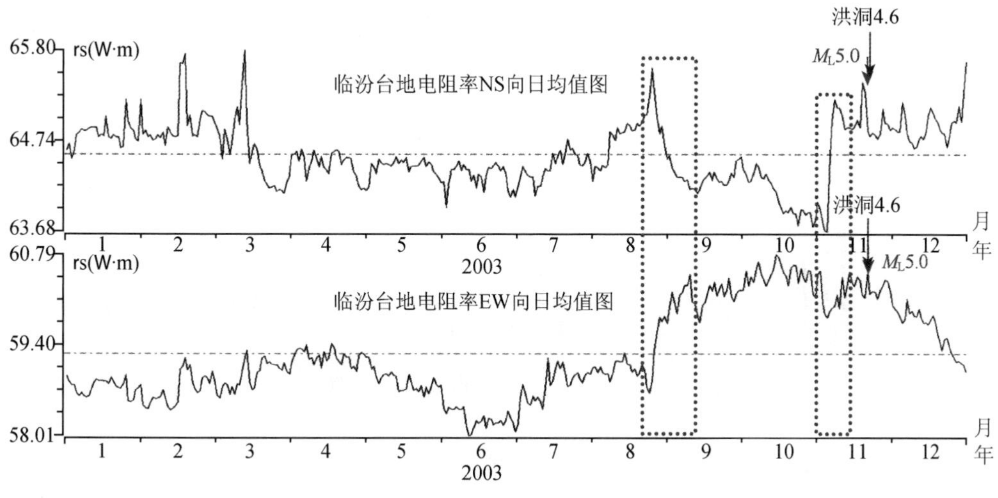

图 11 临汾地电日均值图（2003 年）

Fig. 11 Daily mean value of the apparent resistirity at Linfen station (2003)

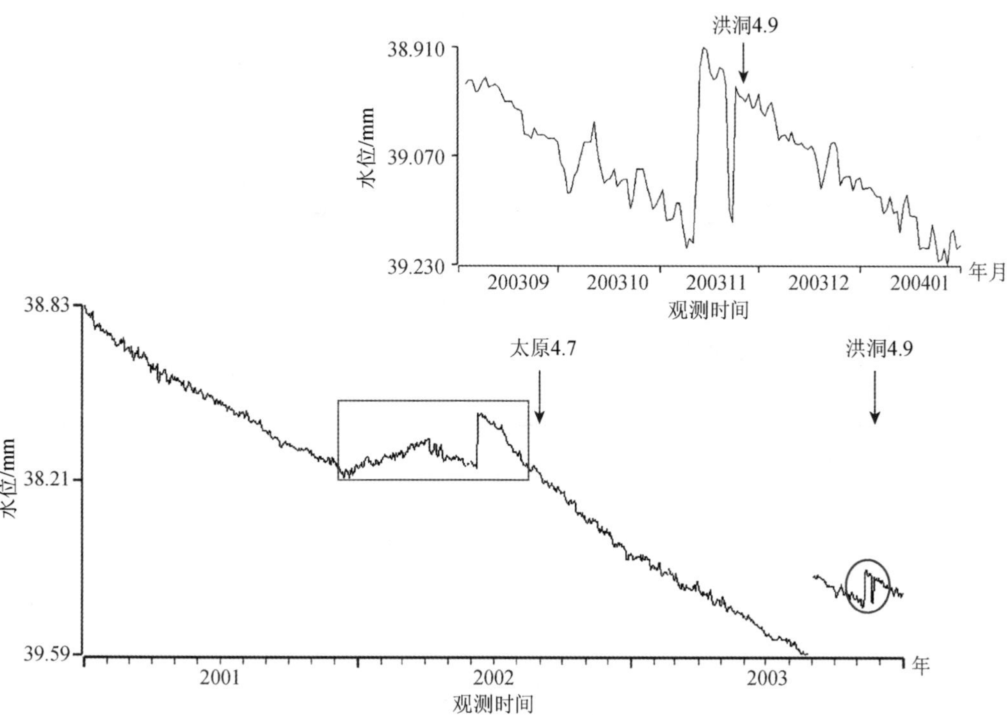

图 12 临汾行署地震局院井水位日均值图（2001 ~ 2003 年）

Fig. 12 Curve of daily mean value of well water level at Linfen Seismological Bureau (2001-2003)

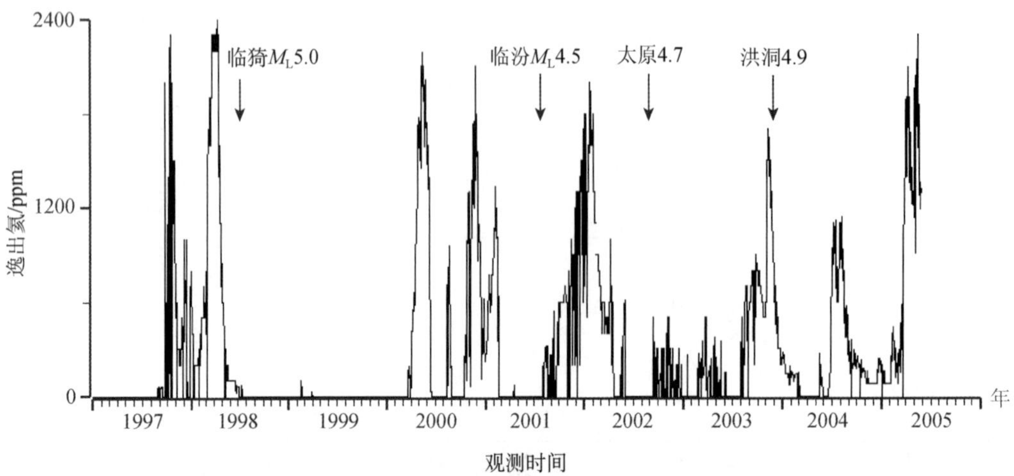

图 13　夏县逸出氦日观测值曲线（1997～2005 年）

Fig. 13　Curve of daily mean value of escaping helium observation at Xiaxian station（1997-2005）

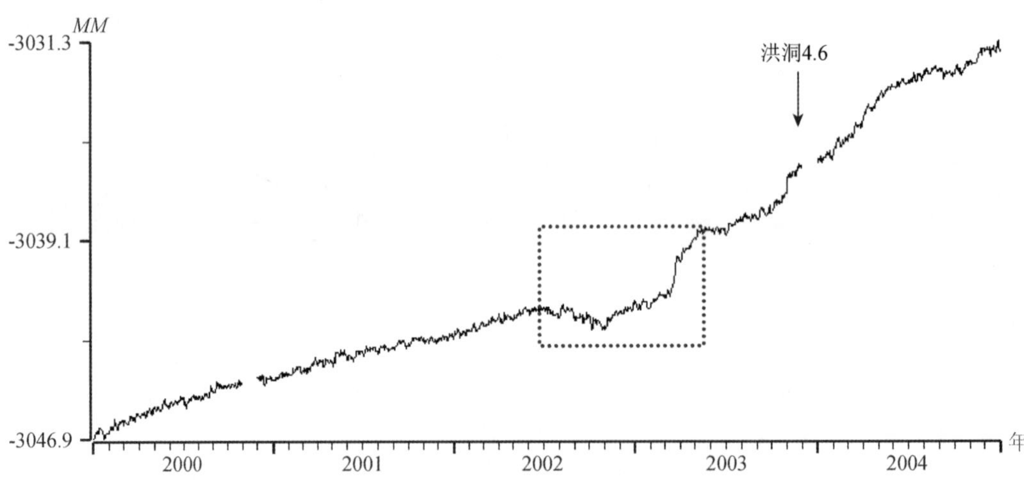

图 14　临汾水准日观测值曲线（2000～2004 年）

Fig. 14　Curve of daily mean value of leveling measurement at Linfen station（2000-2004）

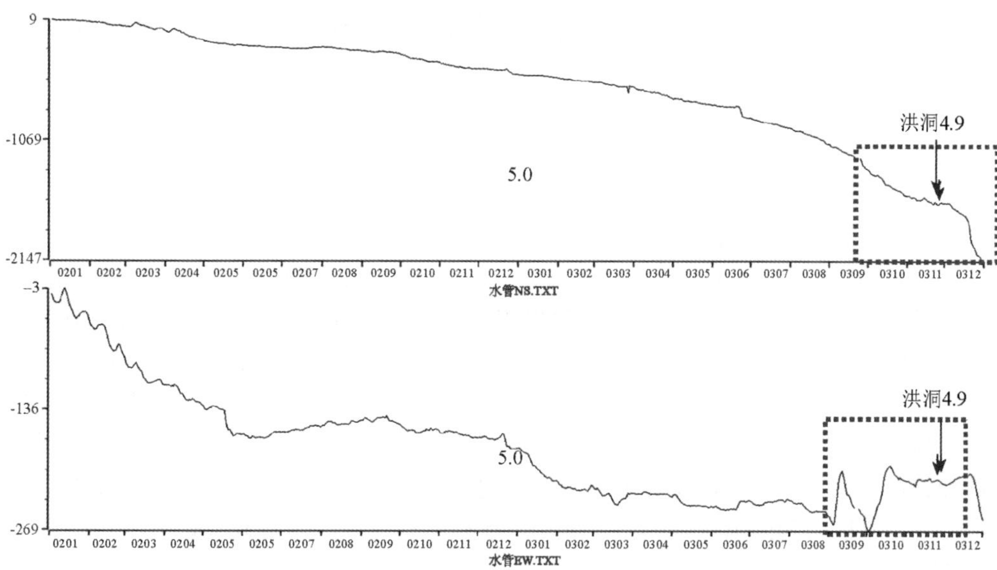

图15 洪洞4.9级地震前侯马水管日均值变化曲线

Fig. 15 Curve of daily mean value of tilt using watertube tilmeter at Houma station before the *M*4.9 Hongtong earthquake

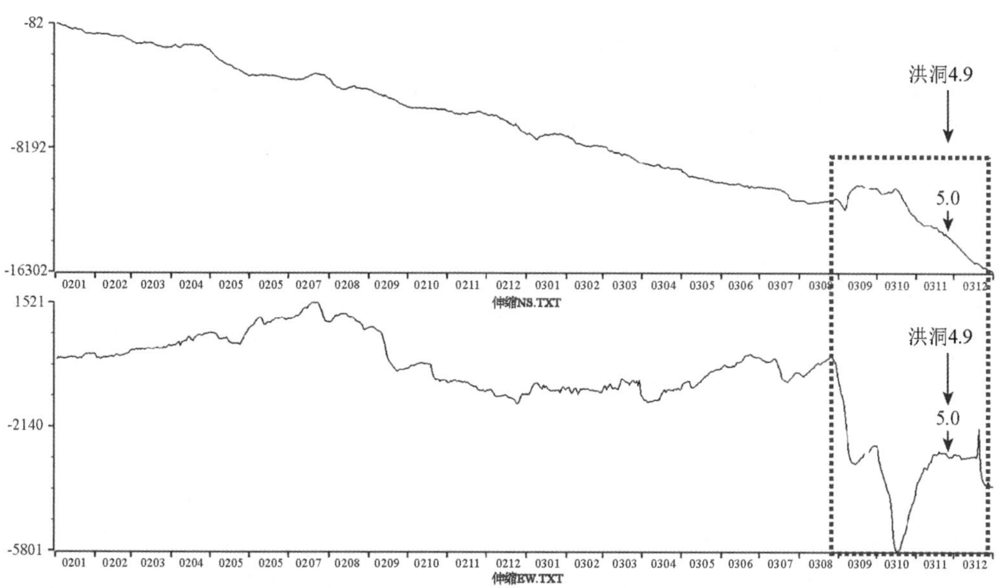

图16 洪洞4.9级地震前侯马伸缩仪日均值变化曲线

Fig. 16 Curve of daily mean value of extensometer at Houma station before the *M*4.9 Hongtong earthquake

七、地震前兆异常特征分析

1. 时间特征

从异常测项来看，洪洞 4.9 级地震前地震活动性异常有 4 项，占异常总数的 40%；流体类异常 2 项，占异常总数的 20%；形变类异常 3 项，占总数的 30%。磁电异常 1 项，占总数的 10%。从异常的起始时间来看，起始时间大于 1 年的中期异常 2 项，占异常总数的 20%；起始时间在震前 3~6 个月的短期异常 6 项，占总数的 60%，起始时间在震前 1 个月内的临震异常 2 项，占总数的 20%。异常出现的时间以短期异常为主，中期和临震异常较少，占总数的 40%。异常出现时间最早的是临汾短水准异常，大约在震前 17 个月，其次是地震活动 5 项因子异常，出现在震前 1 年左右。2003 年 6 月下旬至 11 月初又相继出现了侯马水管、夏县逸出氡气、临汾地电、侯马伸缩、2 级地震条带和空区等 6 项异常，大同老震区窗口和临汾行署地震局院水位观测井的异常则出现在震前不足 1 个月的时间段内。4.9 级地震前各类异常随时间的演化进程是：形变→地震活动性→流体→地电阻率。

2. 空间特征

由洪洞 4.9 级地震前兆异常空间分布图 11 及表 3 可以看出，洪洞 4.9 级地震 6 项前兆异常分布在 120km 以内，其中距震中 0~100km 有 5 项，100~200km 范围内有 1 项。距离震中最近的异常为临汾行署地震局院内的水位观测井约 15km，最远的为夏县逸出氡气异常，距离震中约为 120km。异常主要集中在 100km 以内，且以形变类异常和流体类异常为主，这种空间分布格局，可能与前兆台网布局有一定的关系，2003 年 11 月以前，山西前兆测项主要集中在中部的太原盆地及其相邻地区，南北两端测项少，且比较分散。震中 100km 范围内的异常台站和台项比分别为 60% 和 55.6%，101~200km 范围内的为 16.7% 和 10%。各类前兆观测异常有由外围向震中推进的趋势。

八、震前预测、预防和震后响应

1. 预测情况

山西省 2003 年度地震趋势研究报告将山西中南部列为值得注意的地区，有发生 5 级左右地震的可能[2)]。洪洞 4.9 级地震就发生在年度会商报告所确定的值得注意地区之内。震前未出现显著的地震活动性图像异常和可靠的前兆异常，未能作出中短期预测。

2. 震后趋势判定

洪洞 4.9 级地震发生后，预报人员根据该地区 1967 年 12 月 18、20 日蒲县连续发生的 2 次 5.0 级地震和 2001 年 7 月 25 日临汾市连续发生 2 次 4.5、3.6 级地震具有双震型的特点，结合山西南部地震活动因子、大同小震窗口频度异常、夏县逸出氡气、临汾地电等异常，综合分析认为，该区应注意同等地震的发生，即地震序列为双震型。11 月 27 日根据地震序列的发展特点和地质构造背景资料，及时将双震型地震序列更正为主余型地震序列。

3. 震后响应

洪洞 4.9 级地震发生后，山西省地震局迅速启动应急预案，20 分钟内派出地震现场工作队。山西省地震局迅速将震情、灾情报告中国地震局、中共山西省委、山西省人民政府及山西省有关厅局，并向媒体公布震情、灾情信息，解答群众来人、来电提出的问题，平定了群众的恐慌心理。山西省临汾市地震局、运城市地震局也启动震时工作程序，加强震情监视及震情值班和震害调查。山西省地震局在紧急会商会后，提出初步的震后趋势判断意见。在之后 10 几天内，连续召开 9 次紧急会商会，密切跟踪序列的发展及各项前兆异常新变化，并和现场分析预报人员及时沟通，及时对震后趋势判断意见进行修订。

九、总结与讨论

1. 总结

洪洞 4.9 级地震时间演化特征是：异常出现时间最早的是临汾短水准异常，大约在震前 17 个月，其次是地震活动 5 项因子异常，出现在震前 1 年左右。2003 年 6 月下旬至 11 月初又相继出现了侯马水管、夏县逸出氡气、临汾地电、侯马伸缩、2 级地震条带和空区等 6 项异常，大同老震区窗口和临汾行署地震局院水位观测井的异常则出现在震前不足 1 个月的时间段内。4.9 级地震前各类异常随时间的演化进程是：形变→地震活动性→流体→地电阻率。

空间演化特征是：距离震中最近的异常为临汾行署地震局院内的水位观测井约 15km，最远的为夏县逸出氡气异常，距离震中约为 120km。异常主要集中在 100km 以内，且以形变类异常和流体类异常为主。各类前兆观测异常有由外围向震中推进的趋势。

2. 讨论

如前所述，洪洞 4.9 级地震震源机制解结果与现场考察结果相悖，正需开展近场深部地球物理勘探等工作来进一步佐证。

根据历史地震活动特点对洪洞 4.9 级地震震后趋势进行判断的结果与实际情况相差较大。其原因是：在进行趋势判断时仅依据地震活动特点没有结合震中区地质构造特点。因此在进行震后趋势判断时必须要两者兼顾，才有可能对震后趋势做出较准确的判定结果。

参 考 文 献

[1] 山西洪洞甘亭 M_L5.0 地震现场工作队. 2003 年 11 月 25 日山西洪洞甘亭 M_L5.0 地震考察报告 [J]. 山西地震，2004，(1)：1 ~ 9

[2] 王跃杰等. 2003 年 11 月 25 日山西洪洞甘亭 M_L5.0 地震灾害损失评估 [J]. 山西地震，2004，(1)：10 ~ 14

[3] 宋美琴，郑勇，葛粲，李斌. 山西地震带中小震精确位置及其显示的山西地震构造特征. 地球物理学报，2012，55 (2)：513 ~ 525

参 考 资 料

1) 山西省及邻近地区地震目录，2003 年 11 月

2) 山西省地震局，山西省 2003 年度地震趋势研究报告，2002 年 11 月

3) 山西省地震局，山西地区历史中强地震和现代中小地震（序列）类型及区域特征研究报告，2009 年 7 月

*M*4. 9 Hongtong Earthquake in Shanxi Province on November 25, 2003

Abstract

On Nov. 25, 2003, an earthquake of *M*4. 9 occurred in Ganting town, Hongdong county in Shanxi Province. Its instrumental epicenter was at 36°10′N, 111°37′E. The macroscopic epicenter was at Ganting town in Hongtong country. The intensity of meizoseismal area of the earthquake was Ⅵ. 11 students were injured, Direct economic losses was about 549 400 yuan RMB.

The earthquake was of mainshock-aftershock type, and the largest magnitude of aftershock distributed within 15km was M_L2. 8. Field investigation, Epicentral distribution of the earthquake sequence and isoseismal line indicated that the seismogenic structure of the *M*4. 9 earthquake may be Luoyunshan fault, contrary to the focal mechanism solution indicating the reverse fault.

Within 200km from the epicenter of the earthquake, there were 7 seismic stations and 11 precursory observational stations or measuring points 10 anomalies appeared before the *M*4. 9 earthquake in Shanxi province and its surrounding areas, including 2 medium term anomalies, 6 short-term anomalies and 2 imminent anomalies. The short-imminent anomalies were more than medium-term ones. Within 100km from the epicenter of the earthquake the ratios of stations with anormalies and items with anormalies were 60% and 55. 6%, within 101 ~ 200km the ratios were 16. 7% and 10%.

报告附件

附表 1 固定前兆观测台（点）与观测项目汇总表

序号	台站（点）名称	经纬度（°）		测项	资料类别	震中距 Δ（km）	备注
		φ_N	λ_E				
1	临汾	36.07	111.51	水位	Ⅱ	15	
				水温	Ⅱ		
2	临汾	36.08	111.36	测震		25	
				地倾斜	Ⅰ		
				水准	Ⅰ		
				地电	Ⅰ		
				地磁	Ⅰ		
3	安泽	36.16	112.25	测震		58	
4	侯马	35.59	111.40	地倾斜	Ⅰ	68	
5	隰县	36.70	110.95	测震		85	
6	孝义	37.02	111.77	水位	Ⅰ	96	
7	介休	37.02	111.90	水位	Ⅰ	98	
8	介休	37.02	112.07	测震		103	
9	阳城	35.44	112.41	测震		109	
10	夏县	35.12	111.24	测震		120	
				水氡	Ⅰ		
				水汞	Ⅰ		
				氦气	Ⅰ		
				氢气	Ⅰ		
11	韩城	35.57	110.46	地倾斜	Ⅰ	120	
12	祁县	37.36	112.30	水位	Ⅰ	146	
13	离石	37.77	111.35	测震		155	
				地倾斜	Ⅰ		
14	陕县	34.7	111.1	水位		170	
				水氡			
				pH 值			

续表

序号	台站（点）名称	经纬度（°）		测项	资料类别	震中距Δ（km）	备注
		φ_N	λ_E				
15	太原	37.71	112.43	地倾斜	Ⅰ	186	
				水准	Ⅰ		
				地磁	Ⅰ		
				地电	Ⅰ		
				钻孔应变	Ⅰ		
				水位	Ⅰ		

分类统计	0＜Δ≤100km	100＜Δ≤200km	总数
测项数 N	7	11	12
台项数 n	12	20	32
测震单项台数 a	2	2	4
形变单项台数 b	0	0	0
电磁单项台数 c	0	0	0
流体单项台数 d	3	2	5
综合台站数 e	1	3	4
综合台中有测震项目的台站数 f	1	2	3
测震台总数 $a+f$	3	4	7
台站总数 $a+b+c+d+e$	6	7	13
备注			

附表 2　测震以外固定前兆观测项目与异常统计表

序号	台站（点）名称	测项	资料类别	震中距 Δ（km）	按震中距 Δ 范围进行异常统计									
					0 < Δ ≤ 100km					100 < Δ ≤ 200km				
					L	M	S	I	U	L	M	S	I	U
1	临汾	水位	Ⅰ	15	—	—	—	√						
2	临汾	水准	Ⅰ	25	—	√	—	—						
		地电	Ⅰ		—	—	√	—						
3	侯马	水管	Ⅰ	68	—	—	√	—						
4		伸缩	Ⅰ		—	—	√	—						
5	夏县	氡气	Ⅰ	120						—	—	√	—	
分类统计	台项	异常台项数			0	1	3	1		0	0	1	0	
		台项总数			5	5	5	5	/	1	1	1	1	/
		异常台项百分比/%			0	20	60	20	/	0	0	100	0	/
	观测台站（点）	异常台站数			0	1	2	1		0	0	1	0	
		台站总数			3	3	3	3	/	1	1	1	1	/
		异常台站百分比/%			0	33	67	33	/	0	0	100	0	/
	测项总数				5					1				
	观测台站总数				3					1				
备注														

2004 年 3 月 24 日内蒙古自治区东乌珠穆沁旗 5.9 级地震

1) 内蒙古自治区地震局；2) 中国地震台网中心

高立新[1]　胡　博[1]　张永仙[2]

摘　要

2004 年 3 月 24 日 09 时 53 分在内蒙古自治区东乌珠穆沁旗发生 5.9 级地震，宏观震中位于东乌珠穆沁旗翁根苏木格日乐图嘎查，等震线呈东西向椭圆形分布，极震区烈度达Ⅶ度呈东西走向近似椭圆形，地震造成 1 人死亡，3 人重伤，2 人轻伤，直接经济损失 2.38 亿元。

截至 2005 年 5 月 28 日，序列基本结束，共记录到 $M_L \geqslant 0.0$ 级地震 56 次，最大余震是 2004 年 5 月 28 日的 $M_L 3.8$ 级地震。余震活动主要分布在主震的西南侧 17km 范围内，呈北东方向展布，余震分布与近东西向分布的等震线方向大体一致，最大余震距离主震约 6km。东乌珠穆沁旗 5.9 级地震为孤立型地震序列。震源机制解结果显示，这次 5.9 级地震的震源断错为倾滑逆断层，且滑动分量较大，是在 NWW 向主压应力作用下右旋错动的结果。发震构造可能为二连—宝日格斯台牧场断裂。

这次地震周围台站较少，震中 200km 范围内共有林西、林东 2 个测震台站，没有前兆台站。地震活动性图像主要以长期背景性异常为主，地震学指标异常主要呈现中期异常性质。

前　言

2004 年 3 月 24 日 9 时 53 分，内蒙古自治区锡林郭勒盟东乌珠穆沁旗发生 5.9 级地震。内蒙古自治区地震台网测定的微观震中为 45°21′N、118°22′E，宏观震中为 45°19′N、118°18′E，位于东乌珠穆沁旗翁根苏木格日乐图嘎查。震中烈度达Ⅶ度。地震有感范围东至吉林省白城市，南至北京市，西到集宁市。锡林浩特市、赤峰市、通辽市等地强烈有感，另外，辽宁省、吉林省等部分地区都有震感。锡林郭勒盟的东乌珠穆沁旗、西乌珠穆沁旗受灾最为严重，牧区的房屋遭到严重破坏，部分倒塌。地震造成 1 人死亡，3 人重伤，2 人轻伤，直接经济损失 2.38 亿元。

5.9 级地震震中附近地震观测台站较少，监测能力较弱。200km 范围内有地震观测台站

2 个，没有前兆观测台站分布。震前震中附近出现地震活动图像和地震学指标 2 个项目共计 10 条异常。

2004 年 3 月 24 日东乌珠穆沁旗 5.9 级地震发生在区域地震活动相对平静的背景下，2002 年 10 月 20 日，位于 5.9 级地震震中西南约 110km 处的西乌珠穆沁旗曾发生 4.8 级地震，2003 年 8 月 16 日，距离 5.9 级地震震中东南约 200km 处的巴林左旗和阿鲁科尔沁旗间发生 5.9 级地震[1~4]，几次中强地震是否存在内在联系，还有待进一步研究。

本研究报告是在有关文献和资料的基础上[1~5];1~5)经过重新整理资料和分析研究而完成的。

一、测震台网及地震基本参数

5.9 级地震发生在内蒙古地震监测能力较弱的区域，特别是震中西北没有测震台站分布，图 1 给出了震中一定范围内的测震台站和前兆测项分布，震中 100km 范围内没有测震

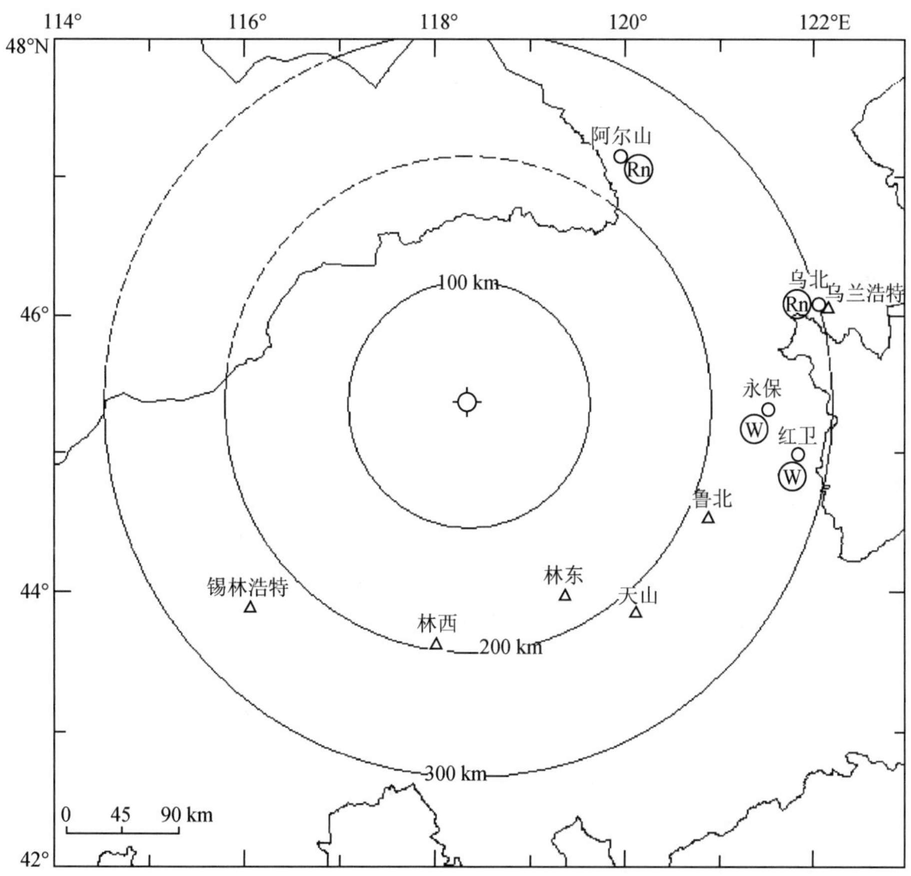

图 1 东乌珠穆沁旗 5.9 级地震前震中附近地震台站及观测项目分布图

Fig. 1 Distribution of earthquake-monitoring stations around the epicentral area before the *M*5.9 Dongwuzhumuqinqi earthquake

和前兆台站分布，200km 范围内只有林西、林东台 2 个测震台站，201km～300km 间有锡林浩特、天山、鲁北、乌兰浩特 4 个测震台站。

表 1 给出了不同来源给出的这次地震的基本参数，经对比分析，认为内蒙古自治区地震局经过修订后的震中位置更为精确，因此，这次地震基本参数采用表 1 中编号 1 的结果。

表 1　东乌珠穆沁旗 5.9 级地震基本参数

Table 1　Basic parameters of the *M*5.9 Dongwuzhumuqinqi earthquake

编号	发震日期	发震时刻	震中位置		震级	震源深度（km）	震中地名	结果来源
	年 月 日	时 分 秒	φ_N	λ_E				
1	2004 3 24	09 53 39	45°21′	118°22′	M_L6.2	5	东乌珠穆沁旗	内蒙古地震局修订1)
2	2004 3 24	09 53 39	45°19′	118°22′	M_S5.9	20		内蒙古地震局速报目录
3	2004 3 24	09 53 49	45°23′	118°15′	M_L6.4	18		中国地震台网中心
4	2004 3 24	09 53 49	45.45°	118.17°	M_w5.5 M_b5.7			美国 USGS
5	2004 3 24	09 53 51	45.38°	118.15°	M_b5.3 M_w5.6 M_S5.2	13		美国哈佛大学
6	2004 3 24	09 53 49	45.38°	118.26°	M_b5.6	18		美国国家地震信息中心 NEIC

二、地震地质背景

1. 区域地质构造环境

根据文献［5］的划分，本次地震所处的大地构造部位是天山—兴蒙地槽系的二级构造单元兴安地槽褶皱带。该带始于早古生代，经历了加里东和海西两个阶段，后者是其主要发展阶段。早二叠世末的晚期海西运动使地槽褶皱封闭，燕山期主要活动表现为强烈的断裂作用和大规模的岩浆喷出，断裂活动产生许多北北东向的断陷盆地（图 2）。

区域地震活动属于东北地震区的大兴安岭地震亚区。震区历史上地震活动相对较弱，截至 2000 年之前一直无破坏性地震记载，被列为Ⅴ度以下的不设防区。但本世纪以来先后发生 2002 年 10 月 20 日西乌珠穆沁旗 4.8 级、2003 年 8 月 16 日巴林左旗 5.9 级地震和本次 5.9 级地震，表明该区地震活动有增强的趋势。

2. 震区的地质构造特征

震区总的构造方向为北北东向和近东西向，发育有多条北东或北北东向的新生代活动断裂，以及东西走向的西拉木伦河第四纪活动断裂，而北西向的断裂不仅数量少，规模也很小。其中震中附近的断裂有两条（图 2），分别为：

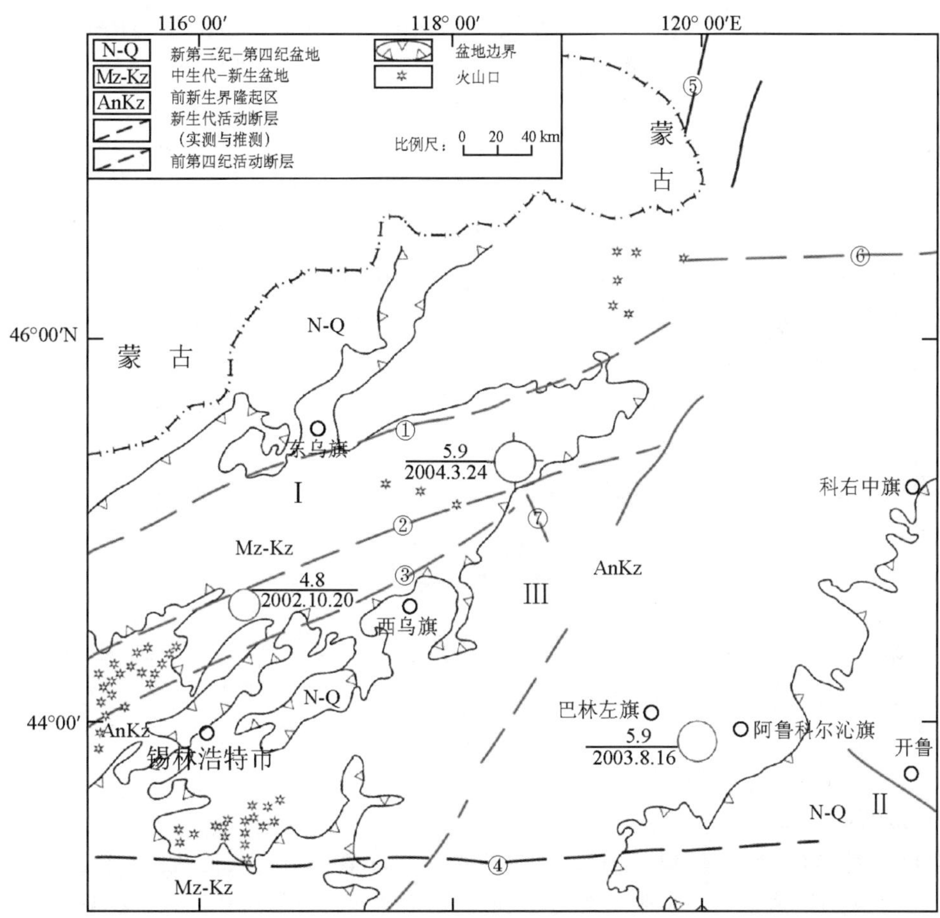

图2 东乌珠穆沁旗5.9级地震附近主要断裂及历史地震震中分布图[2]

Fig. 2 Major faults and distribution of historical earthquakes around Dongwuzhumuqinqi area

图例说明：①那仁宝力格—军马场断裂；②二连—宝日格斯台断裂；
③白音温都尔—巴彦花断裂；④西拉木伦河断裂；⑤阿尔山断裂；
⑥满都呼宝力格—图木吉断裂；⑦哈拉道口断裂；
Ⅰ. 巴音和硕盆地；Ⅱ. 开鲁盆地；Ⅲ. 大兴安岭隆起

（1）二连—宝日格斯台牧场断裂。是距离震中最近的一条断裂，以NE50°走向西起二连浩特市北，经阿拉善宝力格东延至霍林格勒市南，全长超过600km。断裂在地貌上控制了巴音和硕中新生代盆地南边界，沿断裂有新生代玄武岩喷溢，为新生代活动断裂。它是该地区的主要控震构造，2002年10月20日西乌珠穆沁旗北4.8级地震就发生在该断裂带上。

（2）德林敖包北推测断层。位于高力罕牧场东15km的巴格翁牛特北大沟，沿断裂为平直宽阔大沟，走向315°，长约13km。此断裂北东盘为二叠纪硬砂岩夹板岩，南西盘为印支期花岗岩，沿断裂有北西走向的流纹岩（J3）溢出，其走向为北西与区域产状直交，显然受断裂控制。推测北东盘相对北移，属燕山早期新华夏系左旋张扭性断裂。

3. 发震构造判定

本次地震未在地表产生具有"地震断层"意义的地表破裂，这给发震构造的判定带来了一定的困难。但根据现场震害调查的结果，位于震中以东地区建筑物的破坏和人的感觉基本以右旋错动为主。震区所在的我国东北地区基本处于北东东—南西西向的水平主压应力为主的现代构造应力场中，表现为以北东向断裂右旋或右旋逆断、北西向断裂左旋或左旋逆断的剪切—挤压构造环境[3]。

根据本次地震发生后两天内强余震震中的精定位结果，余震分布比较集中，主要呈北东向展布，所有余震均分布在主震西南部，与主震的连线为北东向。此外本次的地震微、宏观震中连线也为北东向。

值得注意的是，现场调查得到的极震区Ⅶ度和Ⅵ度等震线的长轴方向为近东西和北西西向，这与以上分析得到的发震构造走向不一致，原因可能是圈定等震线时受自然条件的影响（如人口与建筑物的分布比较稀少），存在一定的误差。

综合以上分析认为本次地震的发震构造为 NE50°走向的二连—宝日格斯台牧场断裂。

三、地震影响场和震害[2)]

据现场实地考察及调查资料，本次地震的宏观震中为 44.32°N，118.30°E，震中区烈度为Ⅶ度，等震线呈东西向椭圆形分布（图 3）。地震共造成 1 人死亡，3 人重伤，2 人轻伤。灾区面积为 1.87 万平方公里，范围涉及东乌珠穆沁旗、西乌珠穆沁旗，16 个苏木乡镇，灾区人口共计 5.3 万人。本次地震总经济损失 2.38 亿元。

1. Ⅶ度区破坏特征

Ⅶ度区主要涉及到西乌珠穆沁旗的巴棋苏木、高日罕牧场及东乌珠穆沁旗的翁根苏木，包括了 11 个嘎查。人们普遍反映地震时听到轰隆声如同重型车辆开过来，感到强烈颠簸晃动，坐立不稳，惊慌逃出室外。室外行走十分困难，骑摩托车的人感到后面有人拉车，前行不便。区内土木结构房屋普遍发生裂缝，30% 严重破坏，45% 达到中等以上破坏，或局部倒塌，没有人员死亡。砖木结构房屋也有 20% 以上达到中等破坏，表现为前后墙体开裂或局部倒塌，窗角及门框部位墙体普遍产生裂缝。仅有 10% 左右的房屋轻微损坏。据调查资料统计，一般房屋的平均震害指数为 0.25 ~ 0.35。

巴棋苏木达布斯图：土木结构房屋普遍发生裂缝，前后墙体严重开裂，局部倒塌，砖木结构房屋重墙体出现宽 3 ~ 5cm 裂缝。

翁根苏木：格日勒图嘎查及哈日根图等 5 个嘎查，土木结构房屋后墙及墙角部分倒塌，承重墙严重裂缝宽约 3 ~ 5cm，包日塔拉嘎查土木结构房屋局部倒塌。砖木结构房屋，墙体严重开裂，横墙和纵墙裂缝比较严重。

2. Ⅵ度区破坏特征

Ⅵ度区包括了东、西乌珠穆沁旗的 15 个苏木 62 个嘎查。据调查资料统计，土木结构房屋（年久失修的老旧房屋）受严重破坏、毁坏或倒塌的占 20%，中等破坏的占 30%，轻微损坏的占 20%。砖木结构房屋严重破坏的占 10%，中等破坏的占 30%，轻微损坏的占 20%。高日罕牧场巴彦德勒分场、翁根苏木哈日根图、乌里雅斯太镇巴彦高毕破坏较为严

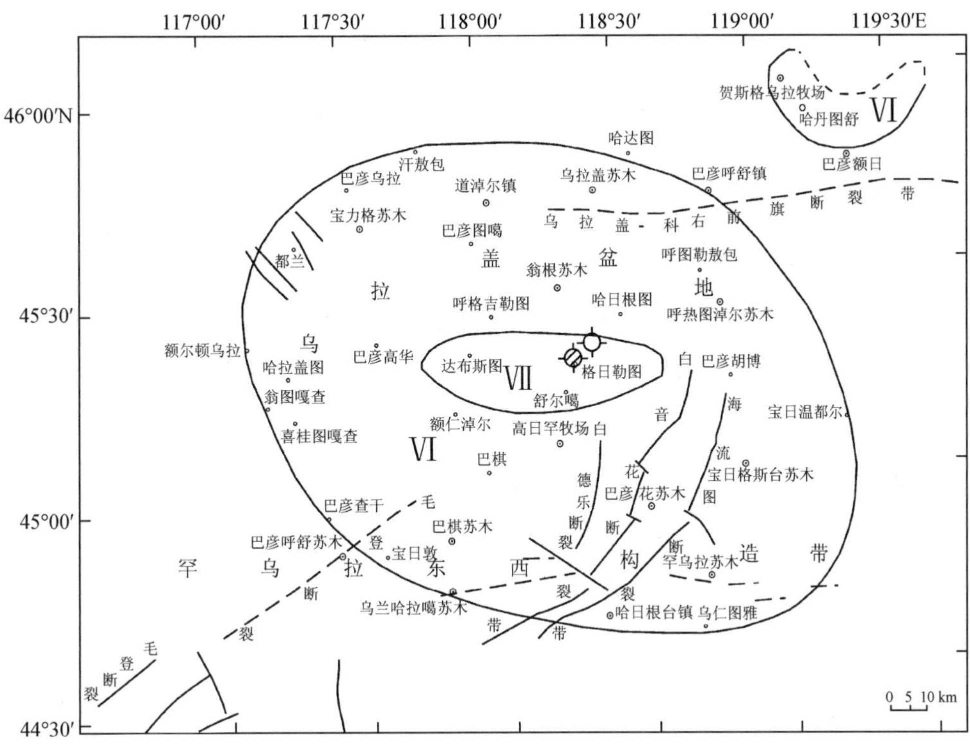

图 3　东乌珠穆沁旗 5.9 级地震烈度等震线图

Fig. 3　Isoseismal map of M5.9 Dongwuzhumuqinqi earthquake

重，大多数土木、砖木结构的房屋承重墙体出现裂缝受到严重破坏局部塌落，少数房屋非承重墙体明显破裂。宝日格斯台苏木巴彦胡舒、巴彦温都尔，巴彦花苏木的伊和花、唐斯格、巴彦都日格，巴棋苏木的额仁淖尔、巴棋宝拉格等嘎查，大部分土木、砖木结构的房屋墙体出现裂缝受到中等破坏，个别房屋局部塌落，非承重墙体多处明显裂缝。

3. 地震地面运动

在调查中了解到，多数人感到震时大地在向南北或北西方向抖动，近南北向行驶的摩托车有前拉后拽的感觉难以驾驶。震害调查表明，多数房屋的前后墙体严重向外闪或倒塌，居民住宅烟筒多数向南北或北西—南东倒塌。有的住宅烟筒顺向动，东西向墙具北西扭动特征。乌拉盖苏木海拉庙白塔顶部饰物明显南移。上述种种现象表明，这次地震时地面运动特征为南北—北西向运动为主并兼反向扭动2)。

4. 地震烈度异常分析

这次地震震中区的地震烈度达到Ⅶ度，随着远离震中区地震烈度一般逐渐降低，但在部分地区出现了地震烈度异常。通过现场实际调查，认为造成地震烈度异常的主要原因是场地差异。说明地震烈度不仅受距离震中的距离控制，而且还受场地因素的控制2)。

四、地 震 序 列

主震前未发生前震。为了保证序列的完整性，主要利用距离5.9级地震240km的锡林浩特地震台单台测定的结果以及可以定位的地震组成5.9级地震序列，截至2004年5月28日，序列基本结束，共记录到$M_L \geqslant 0.0$级地震56次，其中$0.0 \leqslant M_L \leqslant 0.9$级4次，$1.0 \leqslant M_L \leqslant 1.9$级33次，$2.0 \leqslant M_L \leqslant 2.9$级13次，$3.0 \leqslant M_L \leqslant 39$级5次，最大余震是2004年5月28日9时的$M_L 3.8$级地震，次大余震为2004年3月24日9时的$M_L 3.5$级地震。56次余震中17次地震进行了定位，表2给出了$M_L \geqslant 2.5$级余震序列目录。

从5.9级地震序列分布可以看出（图4），余震活动主要分布在主震的西侧，整体成北东方向展布，余震分布与近东西向分布的等震线方向一致，最大余震距离主震约2.3km。

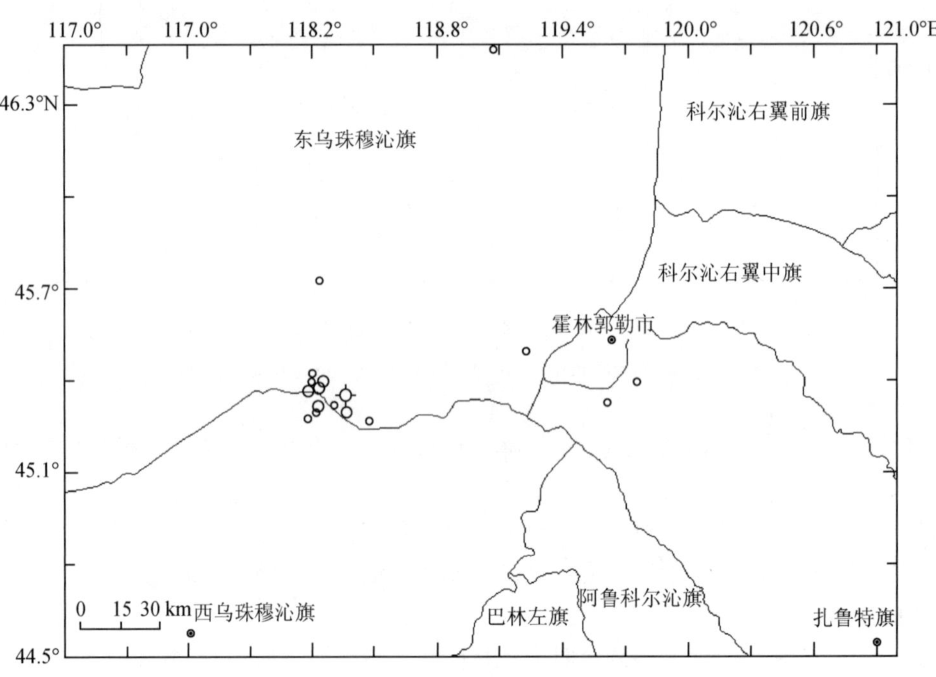

图4　东乌珠穆沁旗5.9级地震序列分布

Fig. 4　Distribution of the *M*5.9 Dongwuzhumuqinqi earthquake sequence

表 2 东乌珠穆沁旗 5.9 级地震序列目录（$M_L \geqslant 2.5$ 级）

Table 2 Catalogue of the *M*5.9 Dongwuzhumuqinqi earthquake sequence（$M_L \geqslant 2.5$）

编号	发震日期	发震时刻	震中位置		震级	震源深度	震中地名	结果来源
	年 月 日	时 分 秒	φ_N	λ_E	M_L	(km)		
1	2004 03 24	09 53 39	45.35°	118.37°	6.2	5	东乌珠穆沁旗	资料 1)
2	2004 03 24	10 01 59	45.4°	118.25°	3.5	12		
3	2004 03 24	10 07 39	45.38°	118.23°	3.1	18		
4	2004 03 24	11 20 27	45.3°	118.22°	2.5	8		
5	2004 03 24	16 20 34	45.32°	118.23°	3.0	19		
6	2004 03 25	16 20 26	45.37°	118.18°	3.0	15		
7	2004 04 11	09 48 02	45.4°	119.75°	2.6	10		
8	2004 04 15	14 32 48	45.33°	119.62°	2.5	15		
9	2004 05 20	00 03 58	45.43°	118.2°	2.5	8		
10	2004 05 28	09 46 00	45.3°	118.37°	3.8	8		

由余震序列的 *M-t* 图可以看出（图 5），余震主要集中发生在 3 月 24～26 日间，且多数为 $M_L \leqslant 3.0$ 级地震，之后余震逐渐衰减，5 月 28 日后，余震活动基本结束。最大余震 $M_L3.8$ 地震发生在 5 月 28 日。

地震序列频次分布图可以看出（图 6），余震活动主要集中在前 5 天内，序列衰减基本正常。5.9 级地震余震序列的 *b* 值为 0.60（图 7），*p* 值为 0.75，*h* 值为 1.6（图 8），主震释放能量占全序列能量的 99.99%，序列参数正常，主震与最大余震震级差为 2.4。根据地震序列判别指标，东乌珠穆沁旗 5.9 级地震类型为孤立型地震。

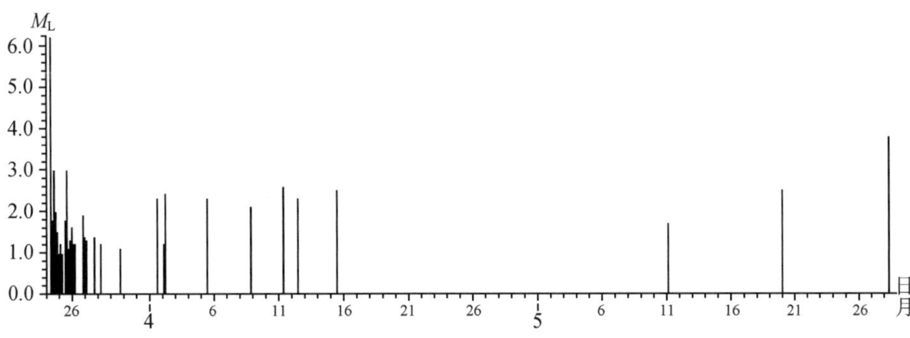

图 5 东乌珠穆沁旗 5.9 级地震序列 *M-t* 图

Fig. 5 *M-t* diagram of the *M*5.9 Dongwuzhumuqinqi earthquake sequence

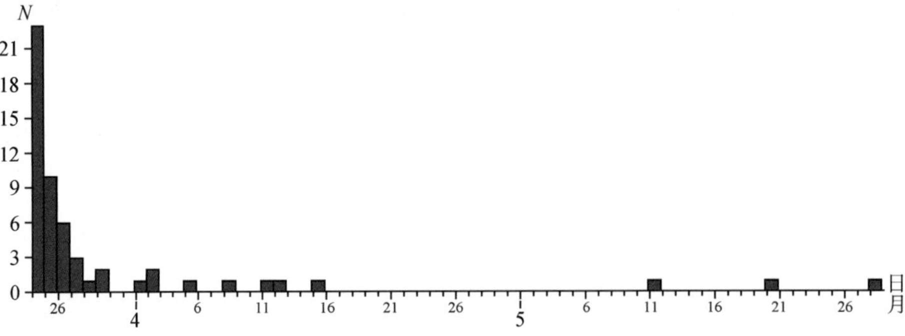

图6 东乌珠穆沁旗5.9级地震序列频次分布

Fig. 6 Variation of daily frequency of the *M*5.9 Dongwuzhumuqinqi earthquake sequence

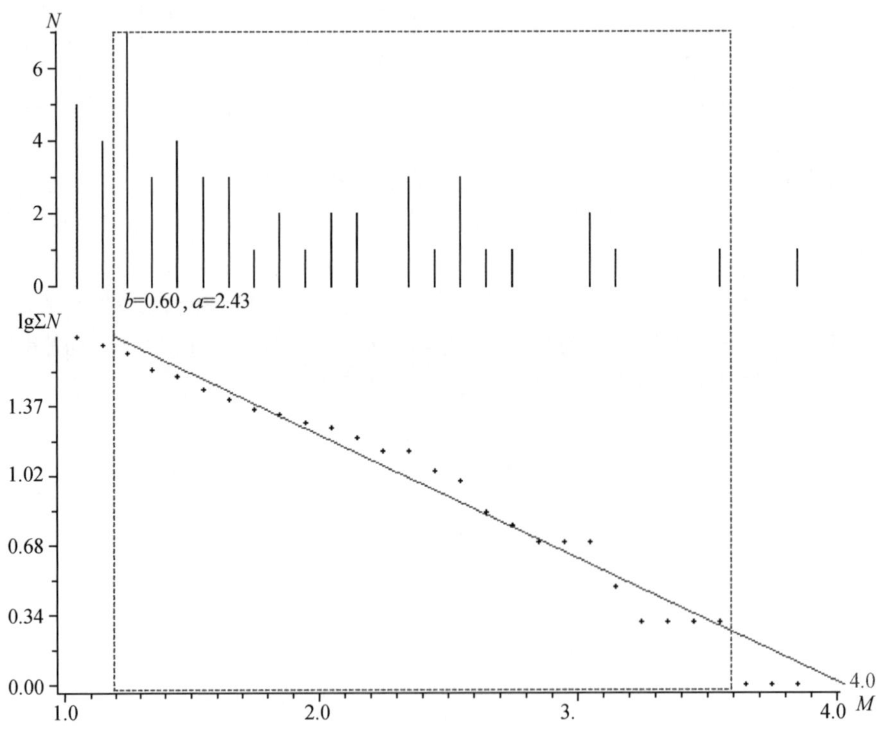

图7 东乌珠穆沁旗5.9级地震序列 *b* 值拟合曲线

Fig. 7 *b*-value fitting curve for the *M*5.9 Dongwuzhumuqinqi earthquake sequence

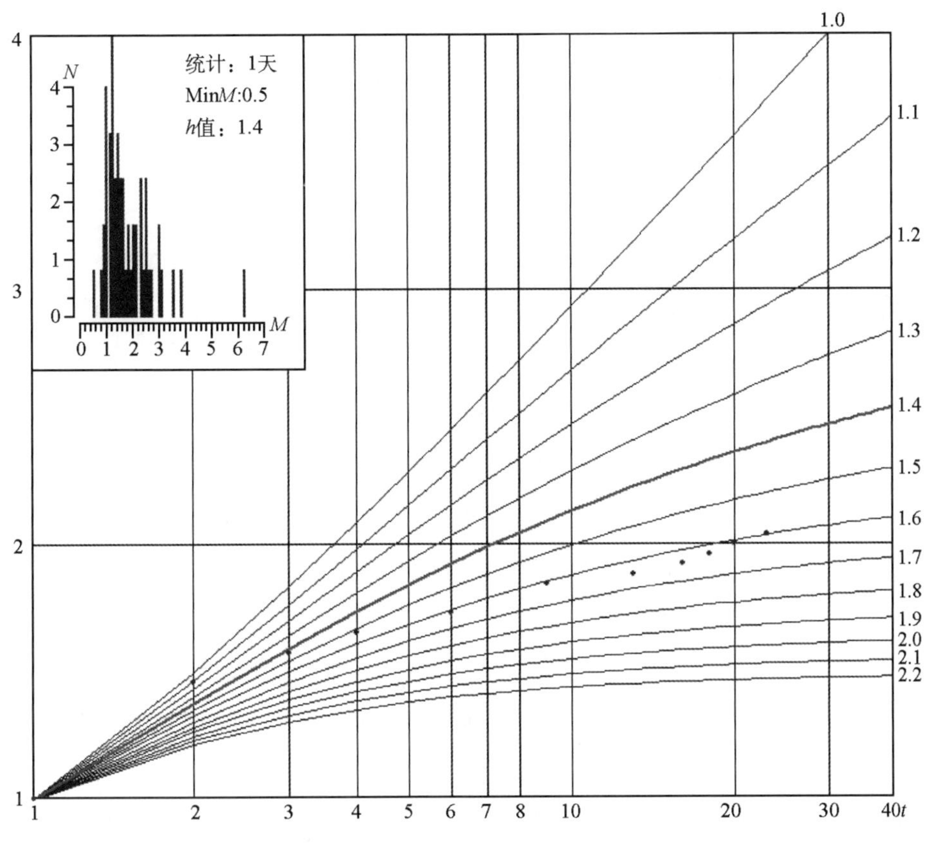

图 8 东乌珠穆沁旗 5. 9 级地震序列 h 值拟合曲线

Fig. 8 h-value diagrams the M5. 9 Dongwuzhumuqinqi earthquake sequence

五、震源参数和地震破裂面

5. 9 级地震发生后，内蒙古呼和浩特数字地震台网收集了内蒙古自治区、辽宁省、黑龙江省、吉林省、河北省测震台网的的 31 个测震台站 31 个 P 波初动符号，用 Mapsis 软件提供的震源机制解程序作 5. 9 级地震的震源机制解，在此基础上，采用手工乌尔夫网做图法求解 5. 9 级地震震源机制解（图 9）。

所收集的 31 个测震台站主要分布在 5. 9 级地震的西南、东南、东北方向，地震的西北方向为蒙古境内，没有初动符号。这些台站大多分布在距离 5. 9 级地震 300 ~ 700km 之间，所用初动符号大多数为首波 Pn 符号，也有个别直达波 Pg 符号。由于台站分布的位置所限，初动符号分布的四象限结果较差，因此求得唯一的震源机制解结果变得较为困难，震源机制解的多解性在所难免。通过手工乌尔夫网作图，基本上可以求得 2 个结果，下面列出内蒙古地震局求解的 3 个震源机制解结果（表 3）。以上结果的滑动角是利用两个节面的走向和倾角计算的。

表3也列出美国地质调查局USGS和中国地震局数字地震台网CCDSN的结果，以便今后做进一步分析和考证。

结合余震序列分布优势方向、现场宏观考察的烈度等震线分布方向以及地质构造等资料，结果Ⅰ、Ⅱ中的节面Ⅰ是这次5.9级地震的主破裂面。

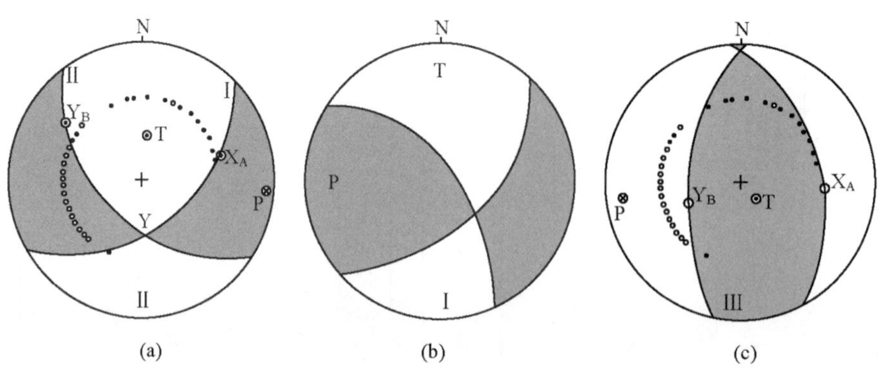

图9　东乌珠穆沁旗5.9级地震震源机制解

Fig. 9　Focal mechanism of the *M*5.9 Dongwuzhumuqinqi earthquake from P wave records

表3　东乌珠穆沁旗5.9级地震震源机制解参数

Table 3　Focal mechanism solutions of the *M*5.9 Dongwuzhumuqinqi earthquake

编号	节面Ⅰ			节面Ⅱ			P轴		T轴		B轴		X轴		Y轴		结果来源
	走向	倾角	滑动角	走向	倾角	滑动角	走向	仰角	走向	仰角	走向	仰角	走向	仰角	走向	仰角	
1	39	60	25	332	55	163	98	5	4	41	181	40	63	35	310	30	内蒙古地震局
2	53	64	16	316	75	153	258	25	1	7	112	60	227	13	323	25	
3	356	21	79	182	59	92	263	14	124	72	359	2	92	21	250	58	
4	348	54	126	117	49	51											CCDSN
5	158	29	80	349	62	96											USGS

六、地震前兆观测台网和前兆异常特征分析

震中附近地区的地震台及观测项目分布见图1。地震发生在北东向和东西向构造控制的乌拉盖断陷盆地内，测震和前兆监测能力均较为薄弱。在震中100km范围内没有台站分布，200km范围内共有2个测震台站，没有综合观测台站和前兆台站分布。201～300km范围内有4个测震台站和5个前兆观测台站，没有综合观测台站分布。

表 4 地震前兆异常登记表

Table 4 List of earthquake precursory anomalies

序号	异常项目	台站或观测区	分析方法	异常判据及观测误差	震前异常起止时间	震后变化	最大幅度	震中距（km）	异常类别	图号	异常特点及备注	异常发现时间
1	中强地震持续活跃	41°～47°N，113°～123°E	地震活动图像	图像分析	1986.9～2003.11	正常			L_1	10	5.9 级地震前区域中强地震活跃	震前
2	地震条带	41°～47°N，113°～123°E	M_L3.0 以上地震震中分布	地震成带状分布	2003.9～2004.4	正常		震中附近	M_1	11	条带长 150km，5.9 级地震发生在条带西南约 45km 处	震后
3	b 值	5.9 级地震震中周围 200km	M_L2.5～5.0，12 个月窗长，1 个月步长滑动	1 倍均方差，结合曲线形态	2003.7～2003.12	正常	大于 1 倍均方差	震中周围	M_1	12a	低 b 值异常，震后回升	震后
4	η 值	5.9 级地震震中周围 200km	M_L2.5～5.0，12 个月窗长，1 个月步长滑动	1 倍均方差，结合曲线形态	2002.6～2002.8	正常	大于 1 倍均方差	震中周围	M_1	12a	低值异常，震后平稳发展	震后
5	M_f 值	5.9 级地震震中周围 200km	M_L2.5～5.0，12 个月窗长，1 个月步长滑动	1 倍均方差，结合曲线形态	2002.5～2002.9，2003.10～2004.2	正常	大于 1 倍均方差	震中周围	M_1、S_1	12a	2 次异常持续时间均较短	震后

续表

序号	异常项目	台站或观测区	分析方法	异常判据及观测误差	震前异常起止时间	震后变化	最大幅度	震中距(km)	异常类别	图号	异常特点及备注	异常发现时间
6	C 值	5.9级地震震中周围200km	M_L2.5～5.0，12个月窗长，1个月步长滑动	1倍均方差，结合曲线形态	2002.3～2002.12	正常	大于1倍均方差	震中周围	M_1、S_1	12a	震前1年多出现异常	震后
7	D 值	5.9级地震震中周围200km	M_L2.5～5.0，12个月窗长，1个月步长滑动	1倍均方差，结合曲线形态	2003.4～2004.2	正常	大于1倍均方差	震中周围	M_1、S_1	12b	异常时段集中，震后恢复	震后
8	频度	5.9级地震震中周围200km	M_L2.5～5.0，12个月窗长，1个月步长滑动	1倍均方差，结合曲线形态	2003.3～2003.9	正常	大于1倍均方差	震中周围	M_1	12b	高值异常，震后恢复	震后
9	缺震	5.9级地震震中周围200km	M_L2.5～5.0，12个月窗长，1个月步长滑动	1倍均方差，结合曲线形态	2002.12～2003.11	正常	小于1倍均方差	震中周围	M_1、S_1	12b	异常时段较集中，震后恢复	震后
10	小震调制比	5.9级地震震中周围200km	M_L2.5～5.0，12个月窗长，1个月步长滑动	1倍均方差，结合曲线形态	2003.10～2003.12	正常	大于1倍均方差	震中周围	S_1	12b	异常出现在震前半年到震前3个月，震后恢复	震后

1. 地震学异常

1）中强地震持续活跃

有记载以来，5.9级地震震中100km范围内未发生过M_S≥4.8级以上地震，101～200km范围内仅发生过两次M_S≥4.8级以上地震，分别是2002年10月20日内蒙古西乌珠穆沁旗 4.8 和 2003 年 8 月 16 日内蒙古巴林左旗和阿鲁科尔沁旗间 5.9 级地震。可见震中区附近历史中强地震活动较弱。但进入 20 世纪 80 年代后，震中附近区域中强地震连续发生，5.9 级地震震中西 280km 处 1986 年 9 月 9 日发生阿巴嘎旗 5.6 级地震，5.9 级地震震中西 220km 处 1999 年 1 月 29 日发生锡林浩特 5.2 级地震，之后再次连续发生 2002 年西乌珠穆沁旗 4.8、2003 年巴林左旗和阿鲁科尔沁旗间 5.9 级地震、2003 年 11 月 3 日巴林右旗再次发生 4.3 级地震，显示了区域中强地震持续活跃状态。

从几次中强地震的空间迁移来看，向东迁移的特征较为显著，1986 年 9 月阿巴嘎旗 5.6 级→1999 年 1 月锡林浩特 5.2 级→2002 年西乌珠穆沁旗 4.8 级→2003 年 8 月巴林左旗和阿鲁科尔沁旗间 5.9 级→2004 年 3 月东乌珠穆沁旗 5.9 级地震，几次中强地震活动不但显示了向东迁移的空间特征，而且也具有等间隔发震的性质，主要体现在阿巴嘎旗 5.6 级→1999 年 1 月锡林浩特 5.2 级→2002 年西乌珠穆沁旗 4.8 级→2004 年 3 月东乌珠穆沁旗 5.9 级，几次地震的间隔均约为 110km 左右。

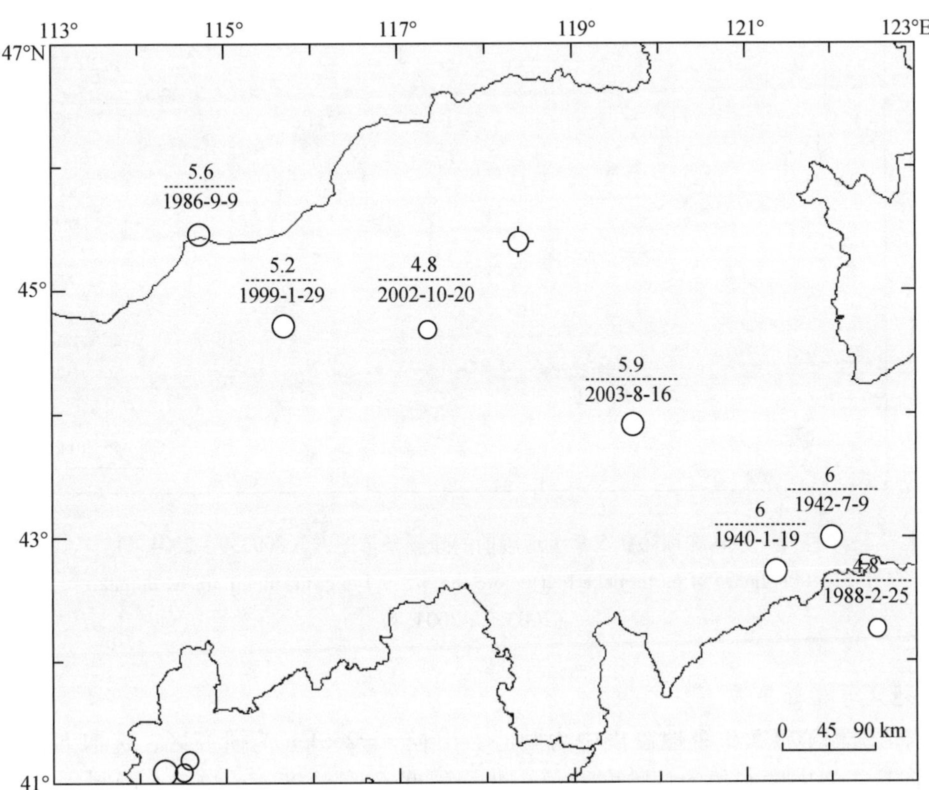

图 10　东乌珠穆沁旗 5.9 级地震震中周围地区 $M_S \geq 4.8$ 级地震分布

Fig. 10　Distribution of the $M_S \geq 4.8$ earthquakes around the epicenter of the $M5.9$ Donhwuzhumuqinqi earthquake

2）地震条带

2003 年 8 月 16 日巴林左旗和阿鲁科尔沁旗间 5.9 级地震后，在 5.9 级地震东北一带形成 M_L2.0 地震条带（图 11）。条带形成时间为 2003 年 9 月至 2004 年 4 月，持续 20 个月，条带长 150km，呈 NW 向展布，震级下限 $M_L \geqslant 2.0$ 级。5.9 级地震发生在条带西南约 45km 处。

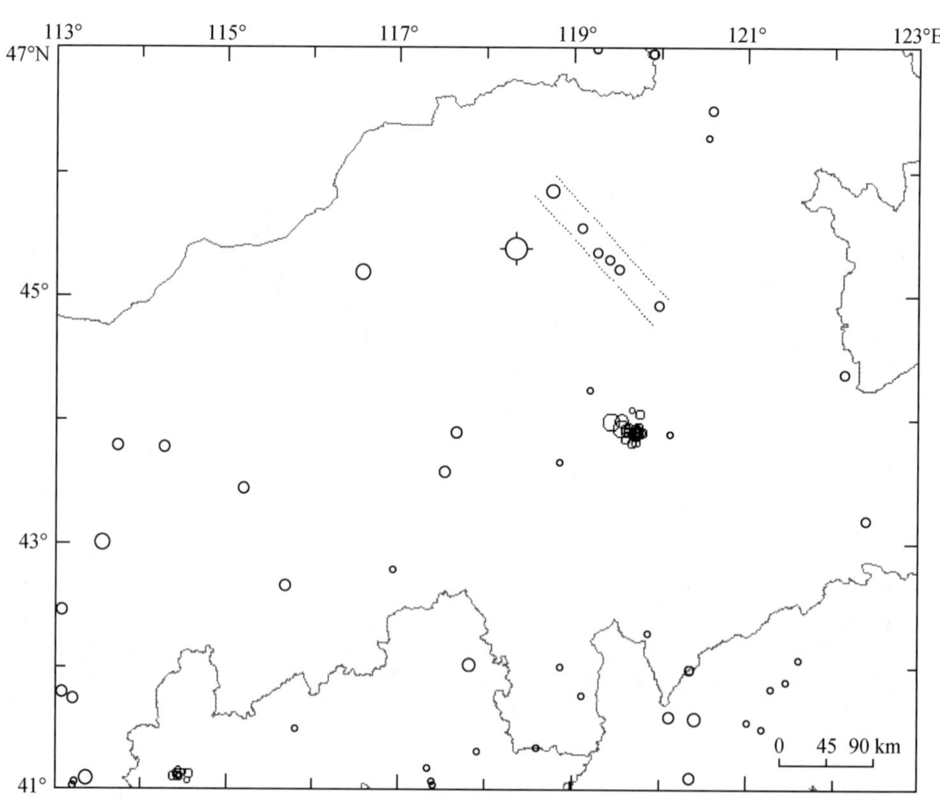

图 11　东乌珠穆沁旗 5.9 级地震前的地震条带图像（2003.9 ~ 2004.4）

Fig. 11　Image of earthquake belt before the *M*5.9 Dongwuzhumuqinqi earthquake (2003.9 ~ 2004.4)

3）地震学指标异常

以东乌珠穆沁旗 5.9 级地震震中为圆心，半径 $r = 200$km 为研究区，选取 b 值、η 值、强度因子 M_f、集中度 C 值、危险度 D 值、非均匀度 GL、频度 N、缺震、调制比 R_m 和加卸载响应比 Y 等 10 种地震学参数，按时间扫描窗长 12 个月，扫描步长 1 个月，对 1994 年 1 月至 2007 年 12 月，$M_L \geqslant 2.5$ 级地震资料进行时间扫描分析，以 1 倍均方差作为异常判定标准，进行异常提取。

表 5 震前 2.0～0.5 年震源区附近（r=200km）地震学异常统计表

Table 5 statistics table of seismology abnormity around hypocenter before two years to half year（r=200km）

震前/年	b	η	M_f	C	D	GL	N	缺震	调制比 R_m	响应比 Y
2.0～1.0		√	√					√		
1.0～0.5	√			√	√		√	√		
0.5～0			√		√				√	

从图 12a 和图 12b 和表 5 可以看出，震前 2 年左右震源区附近 10 种地震学参数有 8 种存在异常，占 80%，其中，震前 2.0～1.0 年有 3 种存在异常，震前 1.0～0.5 年有 5 种参数存在异常，震前 0.5 年到发震仅有 3 种参数存在异常，可见，地震学指标异常主要出现在震前 1 年到半年（图 12a、图 12b）。

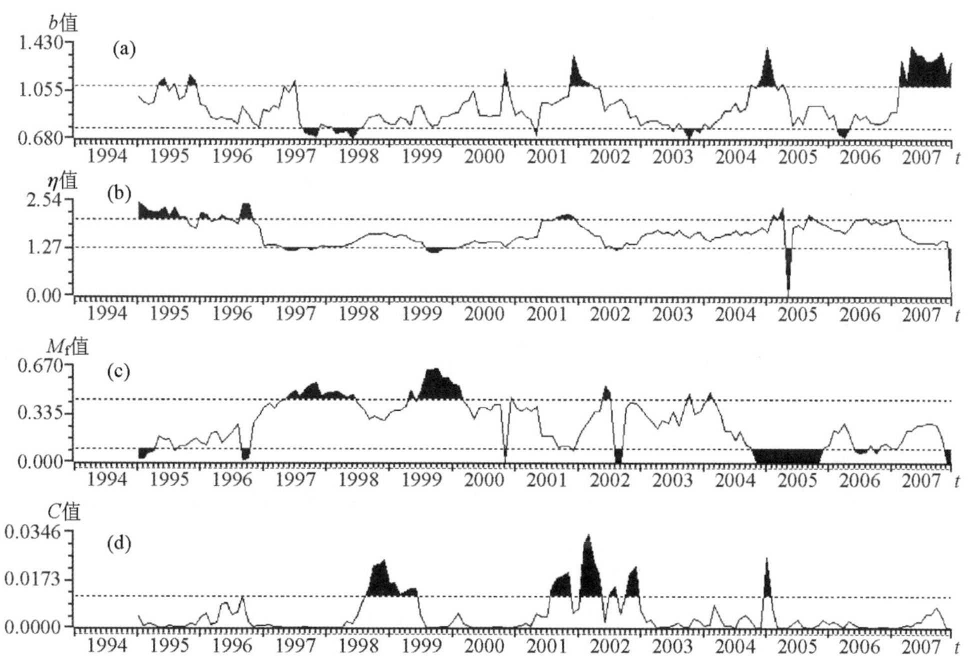

图 12a 东乌珠穆沁旗 5.9 级地震前地震学参量时序曲线

Fig. 12a Curves of seismic parameters before the M5.9 Dongwuzhumuqinqi earthquake

(a) b 值；(b) η 值；(c) M_f 值；(d) C 值

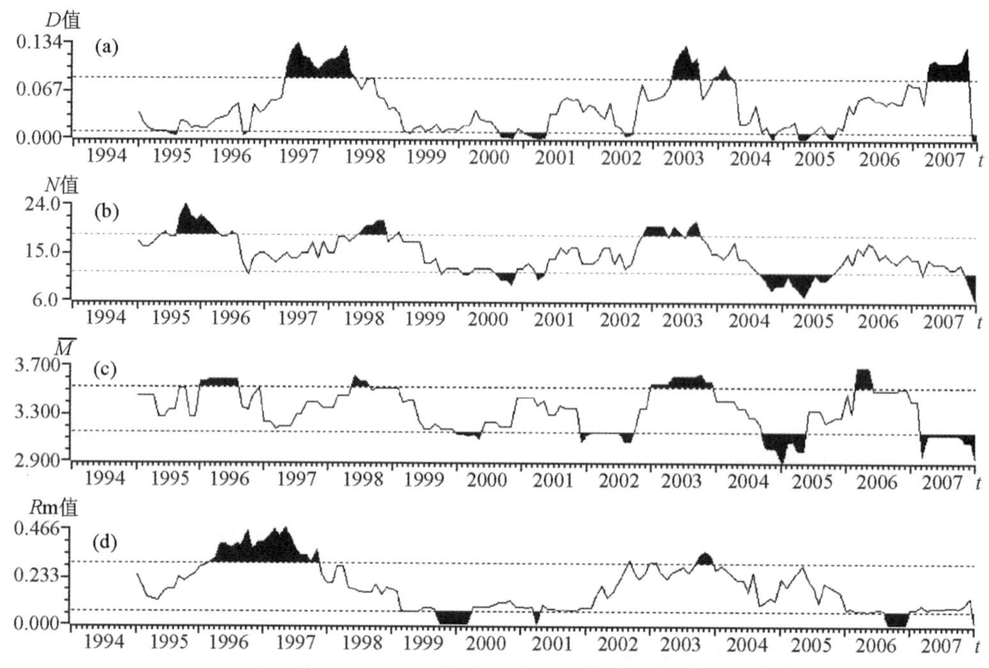

图 12b　东乌珠穆沁旗 5.9 级地震前地震学参量时序曲线

Fig. 12b　Curves of seismic parameters before the *M*5.9 Dongwuzhumuqinqi earthquake

(a) *D* 值；(b) 地震频度；(c) 缺震；(d) 小震调制比

2. 前兆异常

5.9 级地震震中周围 200km 范围内没有前兆台站，201～300km 范围内有 5 个前兆台站共计 8 个观测项目，均未出现异常。

3. 宏观异常

由于震中区为蒙古族聚居区，极震区人口稀少，5.9 级地震调查到的宏观异常极少且可靠性较低，不进行分析。

七、震前预测、预防和震后响应

1. 预测情况

2004 年度内蒙古自治区地震局年度会商，将赤峰—通辽地区列为 2004 年度全区重点危险区，5.9 级地震未发生在确定的危险区之内。震前未出现显著的地震活动性图像异常，由于监测能力限制，震前未有可靠的前兆异常记录，震前未作出中短期预测。

2. 震后趋势判定

通过对东北研究区历史地震和该研究区最近几次地震活动特征和相关的统计分析，该区域中强地震活动类型主要为孤立型、主—余型地震，历史上该区域没有发生过双震型或震群型地震。5.9 级地震发生后，大震现场震情判定组与内蒙古地震局分析预报中心对该序列进行了分析判定，认为该地震序列属于孤立型地震的可能性较大，会商结论、分析结果与实际地震序列类型一致。

3. 震后响应

5.9 级地震发生在 2004 年 3 月 24 日 9 时 53 分，内蒙古自治区地震局测震台网 15 分钟准确测定了地震三要素，10 时 20 分，向自治区党委、政府和中国地震局报告了初定的地震三要素和初步的灾情情况，10 时 30 分，内蒙古自治区人民政府领导赶到自治区地震局，了解地震情况，指挥地震应急工作，11 时，按照《内蒙古自治区地震局应急预案》召开紧急会议，对救灾工作做了安排部署，组成 13 人现场工作队，12 时，现场工作队赶赴震区。同时，内蒙古地震局分析预报人员及有关职能部门负责人，按职责有条不紊地开展地震应急各项工作。12 时，现场工作队从呼和浩特市向震区出发，连续奔袭近 900km，次日凌晨 1 时到达震区，与先期到达的中国地震局现场应急队会合密切跟踪地震序列发展，并召开地震现场紧急会议，成立地震现场指挥部，全面开展抗震救灾工作。

自治区地震局向内蒙古自治区党委、政府和中国地震局上报了《震情通报》6 期，及时上报震情情况、灾情情况、现场工作情况和震区的抗震救灾情况，并提出震后趋势意见和下一步的工作建议。

八、结论和讨论

1. 主要结论

5.9 级地震震中烈度Ⅶ度，等震线呈近东西向椭圆形分布。地震共造成 1 人死亡，3 人重伤，2 人轻伤。灾区面积为 1.87 万平方公里，范围涉及东乌珠穆沁旗、西乌珠穆沁旗，16 个苏木乡镇，灾区人口共计 5.3 万人。本次地震总经济损失 2.38 亿元。

5.9 级地震共记录到 $M_L \geqslant 0.0$ 级地震 56 次，其中 $3.0 \leqslant M_L \leqslant 3.9$ 级 5 次，最大余震是 2004 年 5 月 28 日的 $M_L3.8$ 地震。序列为孤立型地震序列，震后趋势判定正确。

2. 讨论

2004 年内蒙古自治区地震趋势会商结论认为“2004 年度，内蒙古自治区中强地震活动将进入新的活跃时段，发生 6 级左右地震的危险性将进一步增大”，同时将赤峰—通辽地区列为全区 5 ~ 6 级地震危险区，主要预测依据是东北地区大形势分析结果、区域地震活动图像演化和前兆异常变化特征。5.9 级地震偏离预测的危险区约 200km，年度会商结论基本把握了该地区的地震趋势[3)]。5.9 级地震发生在内蒙古自治区地震监测能力较低、地震活动较弱、没有前兆监测能力的地区，震前未发生前震序列，如何准确判定类似区域的地震活动趋势，进一步提高地震监测能力、提高前兆监测能力、加强地震地质基础研究、加强地震预测探索实践，可能是地震科学工作者的必然选择，也是 5.9 级地震给我们的最大启示。

东乌珠穆沁旗是抗震设防标准较低的地区，却发生极震区烈度Ⅷ度的破坏性地震，类似现象其他地方也有。如何科学、合理制定适当抗震设防标准，为工程地震工作者提出了更高的要求，应予认真研究。

5.9 级地震震中区的地震烈度达到Ⅶ度，随着远离震中区地震烈度一般应该逐渐降低，但在部分地区出现了地震烈度异常，通过现场实际调查，认为造成地震烈度异常的主要原因是场地差异。东乌珠穆沁旗在未来的城市规划设计中应考虑场地条件作用，选择潜在地震危险小的地区、选择场地地震反应较小的地段、选择工程结构地震反应较小的地段、选择地震地质灾害较小的地段，应作为东乌珠穆沁旗建设工程选址抗震设防的一般原则。

参 考 文 献

[1] 高立新，巴林左旗—阿鲁科尔沁旗 5.9 级地震前地下流体异常特征分析 [J]，地震，2005，25（1）：103 ~ 110

[2] 内蒙古自治区地震局，内蒙古自治区 2003 年和 2004 年两次 5.9 级地震，北京：地震出版社，2005

[3] 高立新、阎海滨、丁风和等，东乌珠穆沁旗 5.9 级地震参数及序列特征，东北地震研究，2005，21（2），16 ~ 23

[4] 高立新，中国松辽盆地构造环境及东北地区地震活动特征分析，地震，2008，28（4）：59 ~ 67

[5] 谢富仁、崔效锋、赵建涛等，中国大陆及邻区现代构造应力场分区，地球物理学报，2004，47（4）：654 ~ 662

参 考 资 料

1）内蒙古自治区地震局，内蒙古地震台网观测报告（2004 年 3 ~ 5 月）

2）内蒙古自治区地震局，2004 年 3 月 24 日东乌珠穆沁旗 5.9 级地震现场与科考工作报告，2004 年 4 月

3）内蒙古自治区地震局，内蒙古自治区 2004 年度地震趋势研究报告，2003 年 12 月

4）内蒙古自治区地质研究队，内蒙古自治区地质图说明书（1∶100 万），1981 年

5）内蒙古自治区地震局，内蒙古自治区地震构造图（内部资料），1982 年

Dongwuzhumuqinqi *M*5.9 Earthquake on March 24, 2004 in Inner Mongolia Autonomous Region

Abstract

An earthquake of *M*5.9 occurred in Dongwuzhumuqinqi of Inner Mongolia Autonomous Region on March 24, 2004. Its macroscopic epicenter was located in Wengensumugeriletugacha of Dongwuzhumuqinqi. Its isoseismal was oval-shaped with major axis in EW direction. The intensity at the meizoseismal area was Ⅶ and its shape was about oval-shaped with major axis in EW direction. The earthquake has caused 1 death, 3 people were serious injured, 2 people were commonly injured and the direct economic loss was about 238 million Yuan.

The earthquake sequence ended up to May 28 2005, 56 earthquakes of $M_L \geqslant 0.0$ were recorded, The magnitude of the largest aftershock was $M_L3.8$. The aftershock distributed southwestern from the main shock within 170km and along northeastern) direction as the same direction of isoseismal line, the distance between the largest aftershock and main shock was about 6km. This earthquake sequence belonged to isolation type. The focal mechanism solution showed that the hypocenter displacement of the *M*5.9 earthquake was dip-slip thrust fault and there was larger slip component, was the solution of nodal plane Ⅰ right-handed displacement under the main compressive stress of NWW direction. The causative structure was Erlian-baorigesitai pasture fault.

There is none station near the *M*5.9 epicenter and there were only 2 seismic stations named Lindong and Linxi within 200km from the epicenter. The image of seismic activity was main long-term abnormity, and the seismic parameter abnormity was mainly medium-term anomaly.

2004 年 9 月 17 日广东省阳江 4.9 级地震

广东省地震局

叶东华　万永芳　杨马陵　刘　锦

摘　　要

2004 年 9 月 17 日广东省阳江市发生 4.9 级地震。地震有感范围北到广州、肇庆，东到深圳、澳门，西到湛江、茂名等地。极震区烈度Ⅵ度，等震线呈椭圆形，长轴走向 NE 至 NEE，面积 214.8km^2。地震造成直接经济损失约 2208 万元人民币，无人员伤亡。

阳江 4.9 级地震为主震—余震型，序列衰减较快，最大余震 3.6 级。根据震源机制解、余震分布、烈度等震线及当地构造分布等综合判定，NEE 向的平冈断裂为该次地震的发震断层。

震中 200km 范围内有 12 个地震台站，其中测震台 6 个，定点前兆观测台 9 个，含地倾斜、水氡、水位、水温、水质、逸出气、地磁等 13 个观测项目共计 27 个台项。4.9 级地震前出现 8 个异常测项共 9 条异常，主要为中短期异常。

该地震是 10 多年来在广东省陆地发生的最大地震，广东省地震局首次尝试在网站上向社会公布震后趋势意见，对稳定社会，安定民心起了积极作用。地震当天震区就恢复了正常的社会生活秩序。

前　　言

2004 年 9 月 17 日 02 时 31 分 21 秒，广东省阳江市发生 4.9 级地震。据广东地震台网测定，微观震中为 111.87°E，21.77°N。极震区烈度Ⅵ度，长轴走向 NE 至 NEE，面积约 214.8km^2。此次地震有感范围较大，北到广州、肇庆，东到深圳、澳门，西到湛江、茂名等地，有感面积 71530km^2。地震造成直接经济损失约 2208 万元人民币，但无人员伤亡。

震后 1.5 小时，广东省地震局做出震后趋势判定，提出“近期内发生更大地震的可能性很小”的判定意见。同时分两批派出 12 人组成的工作队赶赴地震现场，与阳江市地震部门现场工作队汇合，开展震害评估、监测预测和地震科考。当日，中国地震局也派专家到现场指导工作。

阳江震区位于东南沿海地震带西段，周围 100km 范围内历史上发生过 5 级以上地震 4

次，是现代广东陆域地震活动水平较高的地区。1969 年 7 月 26 日曾发生 6.4 级地震，其后余震持续不断。本次 4.9 级地震是该区 1987 年以来发生的最大地震。2003 年底完成的《2004 年度广东省地震趋势研究》判定，2004 年阳江地区存在发生 4 级左右地震的可能性，并填写了分类会商（预报）卡（见附件 1）。此次 4.9 级地震发生在预测区内，除震级稍低以外，地点和时间的中期预测正确，但未能作出短期和临震的预测。

一、测震台网及地震基本参数

震中 300km 范围内有测震台 15 个，其中 100km 内有 1 个，即阳江测震台，101～200km 有湛江、信宜、台山、珠海、肇庆 5 个测震台（图 1），监测能力可达 $M_L \geqslant 1.0$ 级。表 1 列出不同来源的地震基本参数，其结果基本一致，本文取表 1 中编号 2 的测定结果。

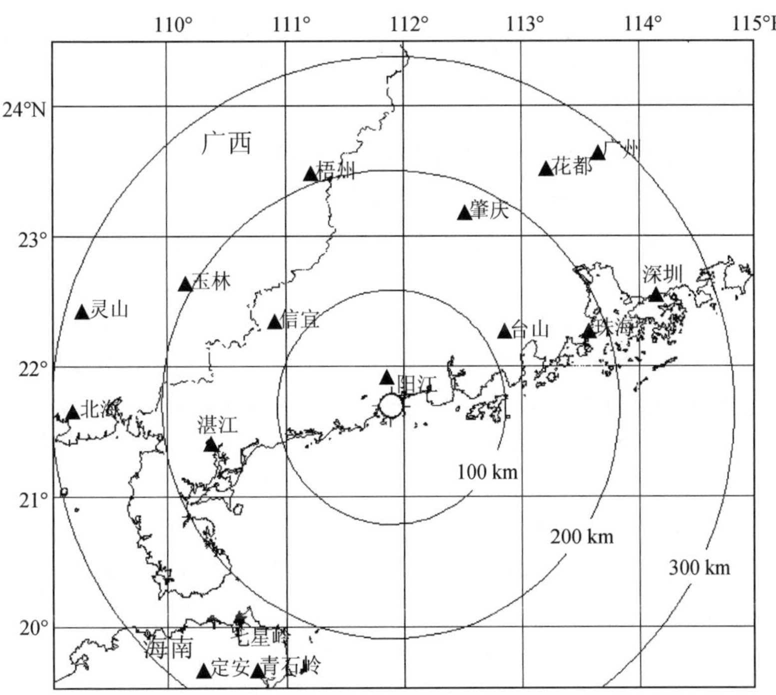

图 1　阳江 4.9 级地震前的地震台站分布图

Fig. 1　Distribution of earthquake-monitoring stations around the epicentral area before the *M*4.9 Yangjiang earthquake

表1 阳江4.9级地震基本参数

Table 1 Basic parameters of the *M*4.9 Yangjiang earthquake

编号	发震日期	发震时刻	震中位置(°)		震级		震源深度	震中地名	结果来源
	年 月 日	时 分 秒	φ_N	λ_E	M	M_L	(km)		
1	2004 9 17	02 31 19	21.7	111.9	4.9		10	阳江	国家数字地震台网
2	2004 9 17	02 31 21	21.77	111.87		5.2	12	阳江	广东省数字地震台网

二、地震地质背景

阳江地区位于丰头河以及漠阳江下游冲积平原，属华南断块粤西间歇性的断块隆起区。其西部以北东向四会—吴川断裂为界，东达西江断裂，北界是近东西向的高要—长宁断裂。该地区构造差异运动较强，地震活动较多[1]。

阳江震区重力和航磁异常等值线均呈北东东方向分布，与平冈断裂延伸方向相同。地壳厚度34km，其所在的区域范围存在层状的地壳结构，阳江以东是以广州为中心的隆起，以西是以信宜为中心的拗陷；4.9级震中位于隆起和拗陷的转折部位[1]。

震中区及附近的主要断裂有北东走向的恩平—新丰断裂、吴川—四会断裂、平岗断裂，北西走向的茶山—闸坡断裂、洋边海断裂、白沙—双捷断裂等，北东向断裂和北西向断裂是晚更新世至今的全新世活动断裂，两组共同组成一个覆盖全区的正交破裂网格。本次4.9级、1969年7月26日6.4级地震均发生在平冈断裂与洋边海断裂交汇区域[1];1)(图2)。

平冈断裂主要出露于平冈镇东西青年农场一带，向东延伸至埠场后被第四纪覆盖，继续沿那龙河延伸至丹载，陆地长度约25km，向西进入洋边海。走向50°N～60°E，SE倾，倾角近直立。断裂由硅化岩、构造角砾岩和石英脉组成，构造带宽度沿走向变化较大，宽者近百米，窄者仅几米。据槽样揭示，断层横切洪积扇；经断层物质石英形貌电镜扫描得知，断裂有过多期次正反向活动，主要表现为逆冲粘滑运动。第四纪地质调查、地震活动等资料证实，该断裂属晚更新世至今仍在活动的全新世活动断裂。洋边海断裂(又称丰头河断裂)，走向310°N～320°W，倾向北东或南西，倾角65°～75°。北段主要沿洋边海延伸，大部分被海水淹没，仅在公石岛露出水面，南段位于海陵岛沙角村、麻汀村、广陵村一带，主要由硅化构造角砾岩，石英脉组成，宽6～30m不等，强烈挤压破碎，断层上的擦痕显示断层具左旋走滑的性质1)。

阳江4.9级地震烈度和震害分布表明，沿平冈断裂走向破坏最重。地震微观、宏观震中均位于洋边海的平冈—溪头镇一带，也表明本次地震与北东向的平冈断裂活动有关。

该区是现代广东省陆地地震活动水平较高的地区，震中100km范围内(东经110.9°～112.9°、北纬21.0°～22.5°范围)历史上曾发生过4次5级以上地震(图2)，最大为1969年7月26日阳江6.4级地震。阳江6.4级地震后，又相继发生了1986年5.0级、1987年4.8级等多次5级左右地震。

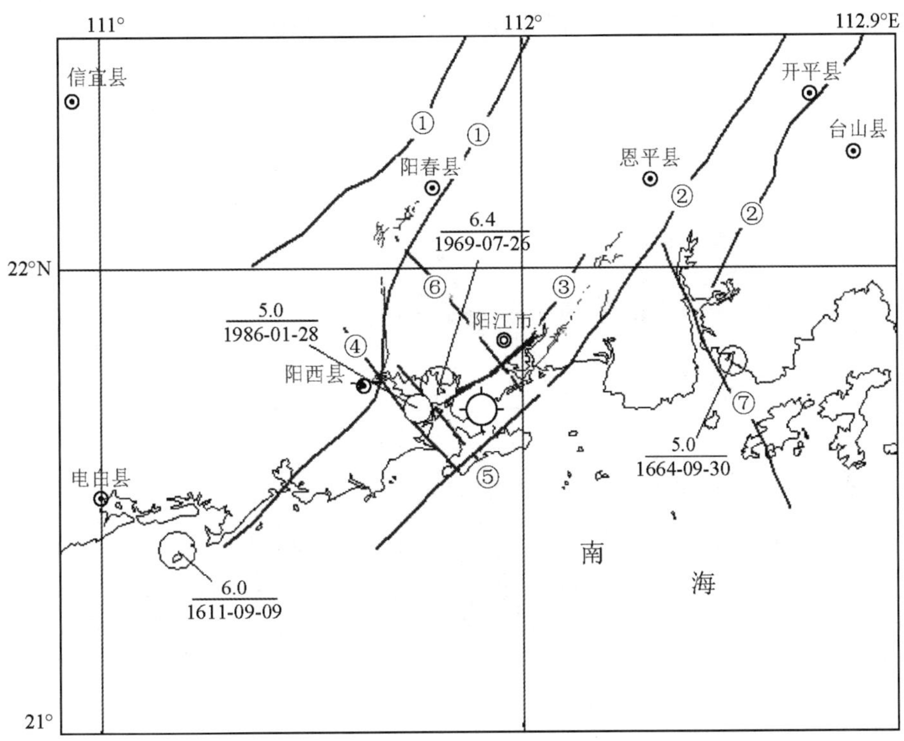

图2　震中附近主要断裂及历史地震分布图[1]

Fig. 2　Major faults and distribution of historical earthquakes around Yangjiang area

断裂名称：①吴川-四会断裂；②恩平-新丰断裂；③平岗断裂；④茶山-闸坡断裂；⑤洋边海断裂；⑥白沙-双捷断裂；⑦镇海湾断裂

三、地震影响场和震害

据现场宏观考察资料1)，4.9级地震宏观震中位于阳江市平冈镇、阳西县溪头镇之间（111.8°E，21.7°N），极震区烈度Ⅵ度，烈度等震线呈椭圆形，长轴走向NE至NEE，与1969年6.4级、1986年5.0级和1987年4.8级等地震的烈度等震线长轴方向大致相近。Ⅵ度区长轴约24km，短轴约14.4km，面积约214.8km^2（图3）。地震造成的有感范围北到广州、肇庆，东到深圳、香港、澳门，西到湛江、茂名，有感面积约为71530km^2。

Ⅵ度区主要表现为多数旧砖瓦房旧裂缝扩大，部分出现新裂缝，常见瓦片滑移并漏光、碎裂、掉落；多数人惊慌外逃，器皿翻倒；Ⅴ度区主要表现为部分旧砖瓦房旧裂缝扩大，个别出现新裂缝，部分瓦片滑移，个别掉落，多数人惊醒，个别外逃。Ⅵ度和Ⅴ度区砖混结构和框架结构楼房只有个别出现墙体裂缝现象。未发现喷砂冒水、地裂缝、砂土液化、滑坡等地质灾害现象。

地震造成14个镇的57个行政村共1620多间房屋（主要为老旧民房）遭受不同程度的

瓦片滑移、碎裂、掉落、墙体轻微裂缝、旧裂缝扩大等轻微破坏，受影响户数达 21195 户，受损房屋面积约 2.6 万平方米。地震影响范围大，但对建筑物破坏较轻。

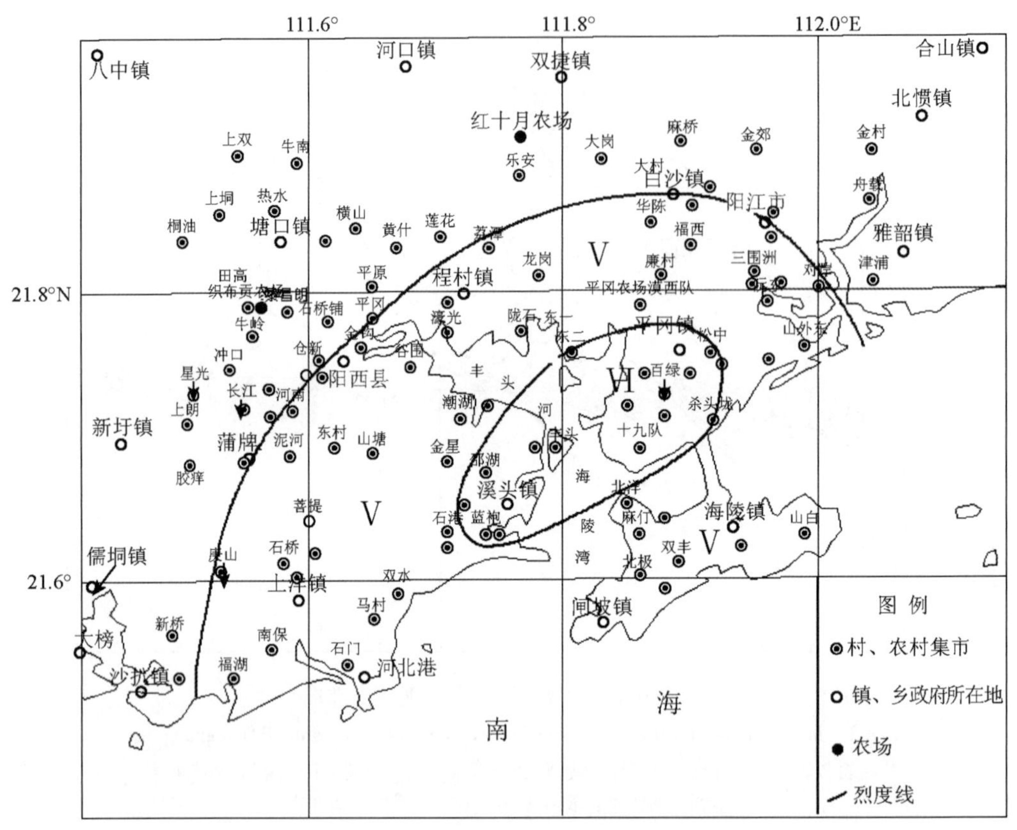

图 3 阳江 4.9 级地震等震线图

Fig. 3 Isoseismal map of the *M*4.9 Yangjiang earthquake

四、地 震 序 列

4.9 级地震发生在区域地震活动较为平静的背景下，震前无明显的前震活动。主震后截至 2004 年 12 月 16 日，共记录 $M_L \geqslant 1.0$ 级余震 80 次，其中 $M_L 3 \sim 3.9$ 余震 1 次，$M_L 2.0 \sim 2.9$ 余震 9 次，$M_L 1.0 \sim 1.9$ 余震 70 次。最大余震是 11 月 11 日的 $M_L 4.0$（$M3.4$）地震（图 4）。$M_L \geqslant 2.0$ 级地震序列目录见表 2。

表 2 阳江 4.9 级地震序列目录（$M_L \geqslant 2.0$ 级）

Table 2 Catalogue of the *M*4.9 Yangjiang earthquake sequence（$M_L \geqslant 2.0$）

编号	发震日期	发震时刻	震中位置（°）		震级		震源深度	震中地名	结果来源
	年 月 日	时 分 秒	φ_N	λ_E	M_L	M	（km）		
1	2004 09 17	02 31 21	21.77	111.87	5.2	4.9	12	阳江	广东数字地震台网
2	2004 09 17	03 07 52	21.85	111.82	2.0		8		
3	2004 09 17	12 10 44	21.75	111.83	2.1		8		
4	2004 09 25	04 28 49	21.8	111.83	2.2		12		
5	2004 09 26	13 47 30	21.75	111.82	2.1		8		
6	2004 10 11	08 03 45	21.82	111.85	2.1		7		
7	2004 10 14	06 10 26	21.77	111.82	2.2		11		
8	2004 11 11	16 08 32	21.78	111.85	2.3		7		
9	2004 11 11	17 42 05	21.78	111.83	4.0	3.4	6		
10	2004 11 13	22 27 50	21.75	111.78	2.1		10		
11	2004 11 24	10 44 46	21.82	111.88	3.1	2.4	9		
12	2004 11 24	11 09 36	21.8	111.88	2.0		8		

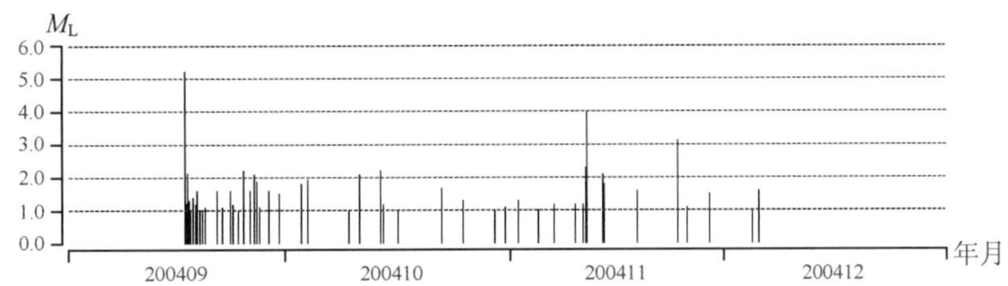

图 4 阳江 4.9 级地震序列 *M-t* 图

Fig. 4 *M-t* diagram of the *M*4.9 Yangjiang earthquake sequence

地震全序列释放能量 9.96×10^{11} J，4.9 级主震释放能量占全序列的 99%，主震与最大余震震级差为 1.5，属于主震—余震型序列（图 6）。取起算震级 M_L0.8，计算地震序列的 b 值为 0.72，用截距法推算最大余震 M_L3.6（图 7）；序列 h 值为 1.4（图 8）。从地震序列 *M-t*、*N-t* 和蠕变曲线图（图 4 至图 6）可见，震后前 7 天序列衰减较快，h 值接近 2.0（图 8）。余震活动所表现出的前期频度、强度衰减快的特点，与阳江地区此前发生的多次 4 ~ 5 级地震后的衰减特征类似。地震活动频度相对较高时段集中在震后 10 天（2004 年 9 月 17 ~ 27 日），共发生 $M_L \geqslant 1.0$ 级 49 次，其中 $M_L \geqslant 2.0$ 级地震 4 次。此后，余震活动迅速衰减至每日 2 ~ 3 次 M_L1 小震的水平。2004 年 11 月 11 日 M_L4.0（*M*3.4）最大余震发生前，未出现小震增多现象。

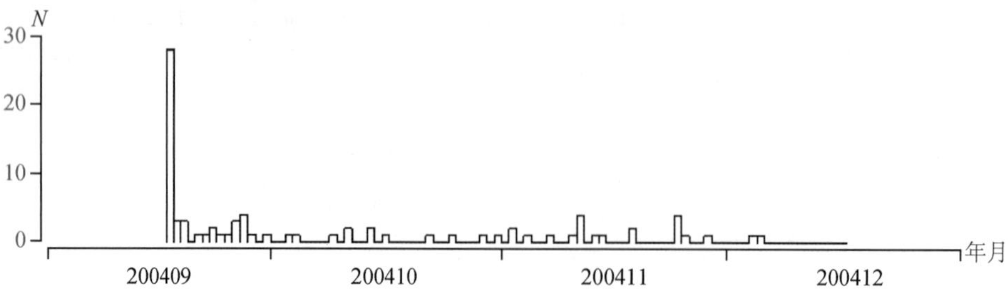

图 5　阳江 4. 9 级地震序列日频度图

Fig. 5　Variation of earthquake frequency of the *M*4. 9 Yangjiang earthquake sequence

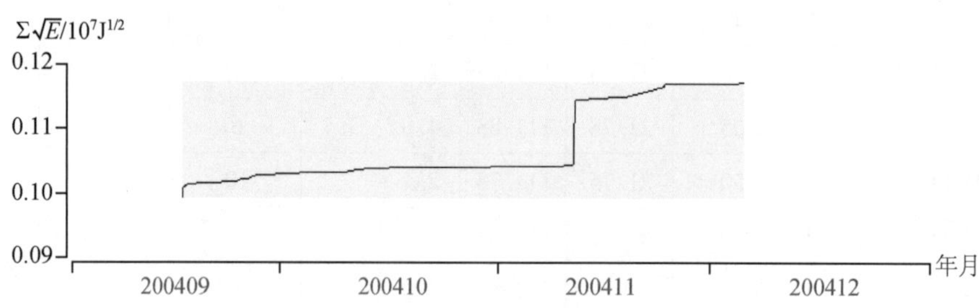

图 6　阳江 4. 9 级地震序列蠕变曲线图

Fig. 6　Squirm diagram of the *M*4. 9 Yangjiang earthquake sequence

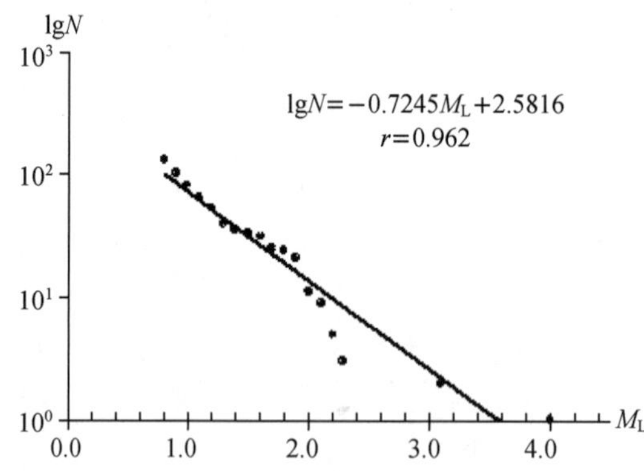

图 7　阳江 4. 9 级地震余震序列 *b* 值曲线图（$M_L \geqslant 0.8$ 级）

Fig. 7　*b*-value fitting curve for the *M*4. 9 Yangjiang earthquake sequence（$M_L \geqslant 0.8$）

由阳江 4. 9 地震序列空间分布可见（图 9），余震主要分布在平冈断裂西北侧，集中分布在 NE、NW 两组断裂交汇区域内，呈北东向展布。

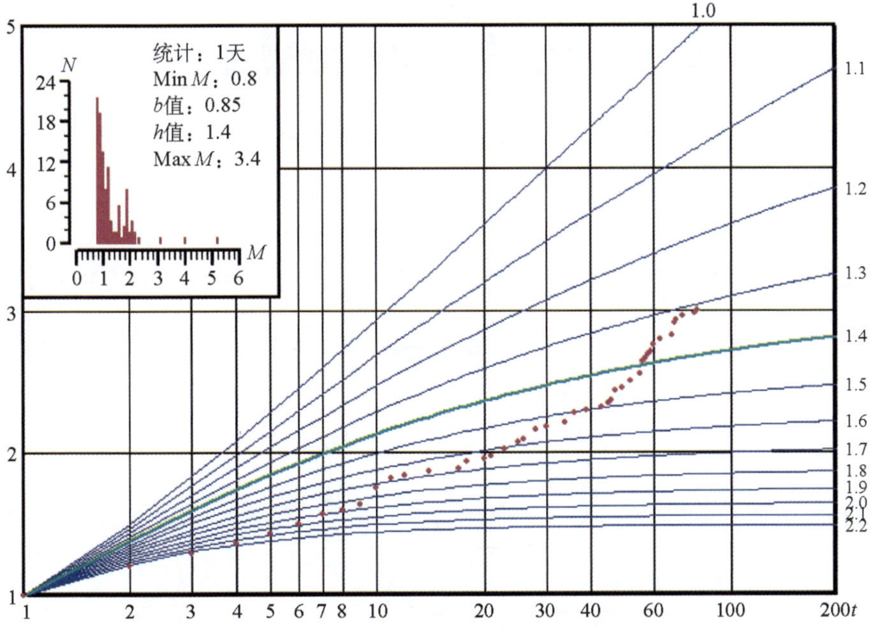

图8 阳江4.9级地震序列 h 值图

Fig. 8 h-value fitting curve for the M4.9 Yangjiang earthquake sequence

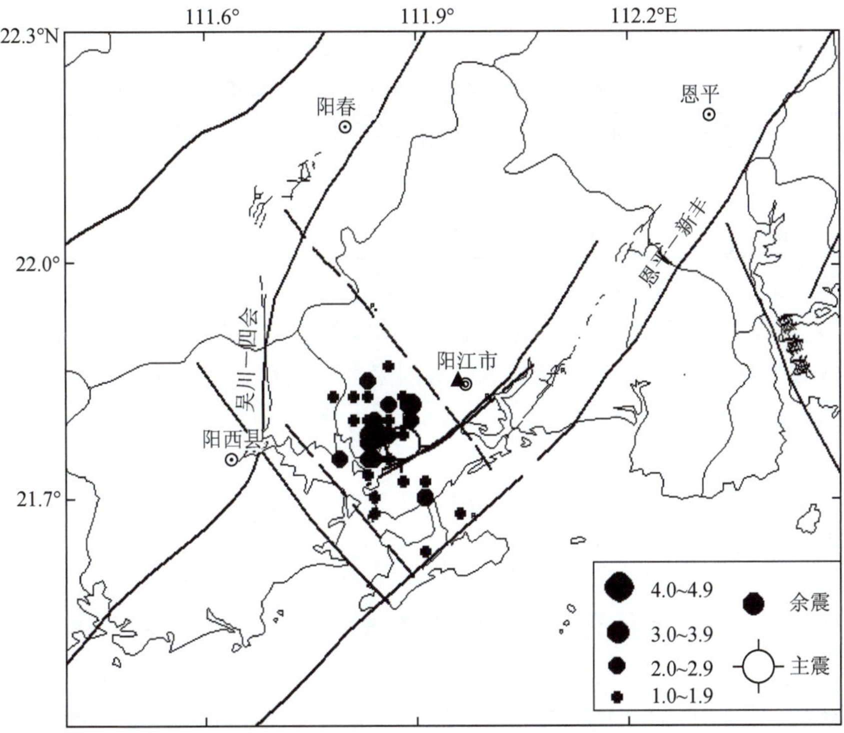

图9 阳江4.9级地震序列空间分布（$M_L \geq 1.0$ 级）

Fig. 9 Distribution of the M4.9 Yangjiang earthquake sequence（$M_L \geq 1.0$）

五、震源参数和地震破裂面

利用广东、海南和广西台网30个台的45个初动，获得2004年9月17日阳江4.9级地震的震源机制解，矛盾符号比为0.09[2]，具体结果见表3和图10。

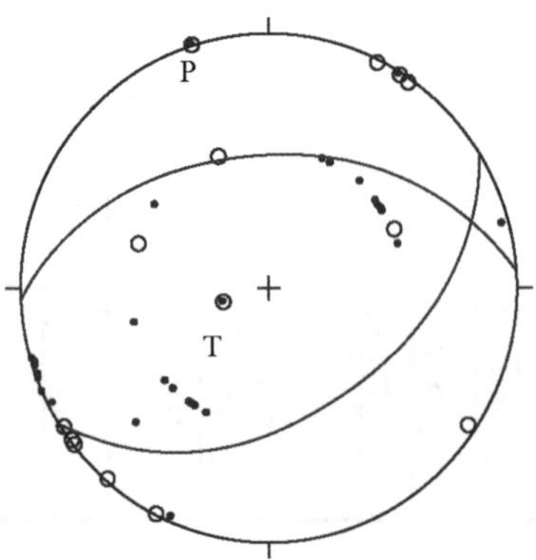

图10　阳江4.9级地震的震源机制（下半球投影）
Fig. 10　Focal mechanism solutions of the *M*4.9 Yangjiang earthquake
圆圈代表初动向上，黑点代表初动向下

表3　阳江4.9级地震的震源机制解

Table 3　Focal mechanism solutions of the *M*4.9 Yangjiang earthquake

节面Ⅰ（°）			节面Ⅱ（°）			P轴（°）		T轴（°）		N轴（°）		结果来源
走向	倾角	滑动角	走向	倾角	滑动角	走向	倾角	走向	倾角	走向	倾角	
57	47	69	267	47	111	342	0	252	75	72	15	康英等

据现场宏观考察1)，极震区（Ⅵ度）走向为NE60°，与节面Ⅰ的走向基本一致。结合余震分布、烈度线走向、震中附近的断裂走向等因素，综合判定节面Ⅰ为本次地震的破裂面。4.9级地震是在近NS向主压应力作用下沿NEE向发生的破裂，破裂方向和平冈断裂吻合，属逆断层性质的破裂。

利用广东台网资料计算此次4.9地震的震源参数为：地震矩1.69×10^{15}N·m，应力降1.7MPa，震源半径约876m[2]。

六、地震前兆观测台网以及前兆异常

震中200km范围内有12个地震台站，其中测震台6个，定点前兆观测台9个（图1、图11），包括地倾斜、水氡、井水位、水温、水质、逸出气、地磁、电磁波等13个观测项目的25个观测台项，观测资料类别均属Ⅰ、Ⅱ类资料。此外，阳江—雷州半岛地区每年进行两期流动重力观测。在震中 0 ~ 100km、101 ~ 200km 范围内分别有测震台站 1 个和 5 个，定点前兆台站 3 个和 6 个，前兆观测台项 5 个和 20 个。在震中 200km 范围内，阳江 4.9 级地震前共出现 8 个异常项目的 9 条异常（表 4）。

1. 地震测震学异常情况

本次地震前，地震学异常不显著。2004 年 1 月至 2004 年 9 月 16 日，阳江地区共发生 $M_L \geqslant 1.0$ 级地震 91 次，其中 $M_L \geqslant 2.0$ 级 21 次，最大为 $M_L 3.1$ 级地震。震前近 9 个月以及临震前几天的小震活动水平均没有显著增强。震前 1 年对华南地区进行的地震学空间扫描显示，阳江地区仅响应比、演化指数 YH 值两项指标出现短时间异常2)。此外，2002 年 10 月到 2004 年 1 月阳江地区小震调制比出现持续 1 年多的高值异常[3]（图 12）；算法复杂性 $C(n)$ 从 2003 年 10 月至 2004 年 4 月期间出现中期异常[4]（图 13）。

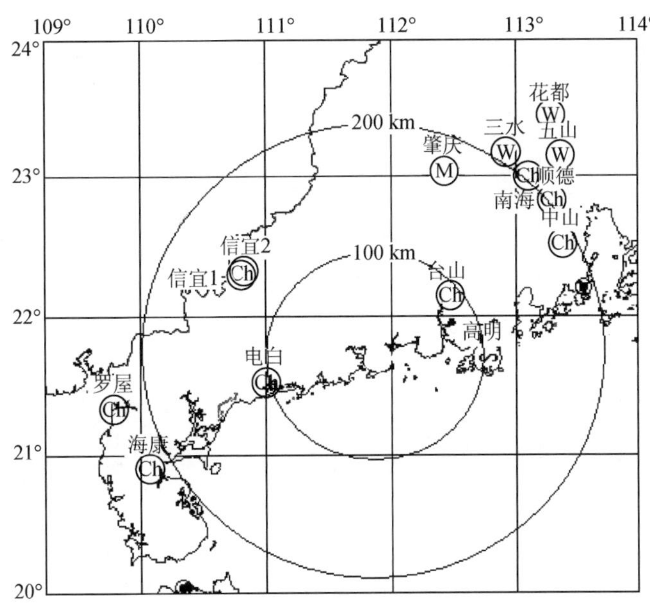

图 11 阳江 4.9 级地震震中附近前兆站点分布图

Fig. 11 Distribution of earthquake-monitoring stations around the epicentral area before the *M*4.9 Yangjiang earthquake

表 4 前兆异常登记表

Table 4 Summary table of earthquake precursory anomalies

序号	异常项目	台站或观测区	分析方法	异常判据及观测误差	震前异常起止时间	震后变化	最大幅度	震中距(km)	异常类别及可靠性	图号	异常特点及备注	备注
1	小震调制比 R_m	震中周围 21.4°～22.1°, 111.5°～112°	调制比($M_L \geq 1.0$级,窗长1年、步长3个月)	$R_m \geq 0.27$	2002.10～2004.1	基本恢复	0.4	震中附近	M_1	13	1986、1987年阳江中强地震前出现异常	震前发现异常
2	算法复杂性 $C(n)$		算法复杂性($M_L \geq 1.0$级)	$C(n)$ 超方差	2003.10～2004.4	恢复		震中附近	M_1	14	高值异常	震后发现异常
3	气体总量	信宜1号井	日值	超差、高值异常	2004.4～12	持续	1.87 ml/l	130	S_1	15	高值异常	
4	CO_2	信宜1号井	日值	超差、高值异常	2004.4.12～12 2004.7.7～8.13	基本恢复	0.027	130	S_1	15	高值异常	
5	水氡	信宜2号井	日值	超差、高值异常	2003.7～2004.5 2004.7～9	基本恢复	24 Bq/L	130	S_1	15	高值异常	
6	水温	台山	整点值	突降	2004.9.5	9.19～22突变	0.015	80	I_2	16	低值异常	
7	水温	电白	整点值	先升后降	2004.9.9～14	正常	0.008 ℃	100	I_2	16	高值异常	
8	地磁 Z 幅差	肇庆	幅相法(整点值)	高值异常	2004.8.20～9.15	恢复		160	I_2	17	高值异常	震前发现异常

续表

序号	异常项目	台站或观测区	分析方法	异常判据及观测误差	震前异常起止时间	震后变化	最大幅度	震中距（km）	异常类别及可靠性	图号	异常特点及备注	备注
9	流动重力	阳江—雷州半岛测区	重力场期变化	震前变化量高于往年	2003～2004	没有恢复			M_1	18		震前发现异常
		岗美—阳江—闸坡测点	重力测线点值	同步转折反向上升						19		

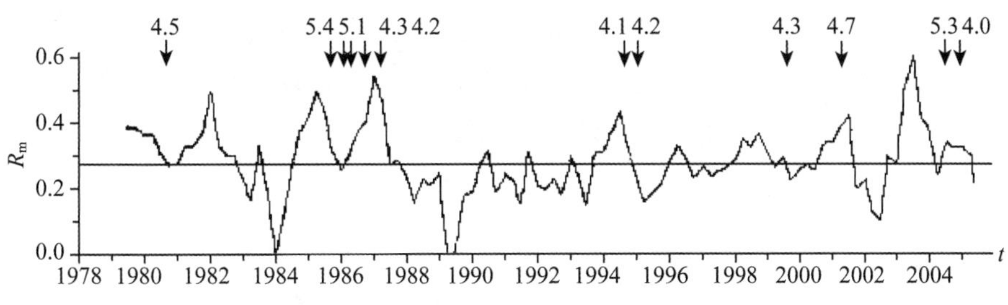

图 12　阳江地区小震调制比曲线（$M_L \geqslant 1.0$ 级）

Fig. 12　Curve of regulatory ratio（R_m） of small earthquakes in Yangjiang region（$M_L \geqslant 1.0$）

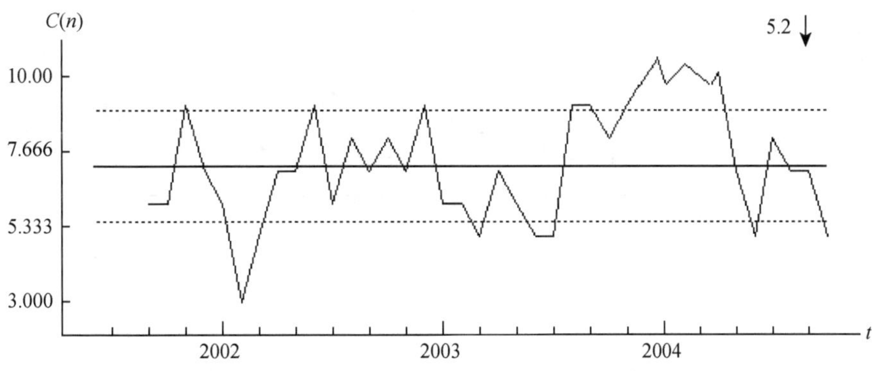

图 13　阳江地区小震算法复杂性 $C(n)$ 扫描曲线（$M_L \geqslant 1.0$ 级）

Fig. 13　Curve of algorithmic complexity（$C(n)$） of small earthquakes in Yangjiang region（$M_L \geqslant 1.0$）

2. 其他前兆观测项目异常情况[3];2,3)

（1）信宜地下流体异常（图 14）。

信宜 1 号井气体总量和 CO_2 高值异常。信宜 1 号井位于震中北西方向约 130km，其气体总量 2003 年 9～12 月曾出现显著的高值异常时段，但之后邻近地区并未出现相应的地震活动。2004 年 4 月气体总量再度大幅上升，8 月 31 日达到最高值，此后转折下降，至 2004 年底逐渐回降至正常值范围。CO_2 于 2004 年 4 月开始大幅突跳，震前出现两个突跳丛集时段，分别是 2004 年 4 月 12 日至 5 月 21 日和 7 月 7 日至 8 月 13 日。

信宜 2 号井水氡高值异常。信宜 2 号井位于震中北西方向约 130km，其水氡值自 2003 年 7 月起超异常上限，高值异常持续了约 10 个月，到 2004 年 5 月中旬回落至正常范围，2004 年 7 月再度上升，震前 1 天出现单日突跳，达 407Bq/L（正常背景值约 375Bq/L）。

（2）电白台和台山台水温异常（图 15）。

电白台水温先升后降。电白台位于震中以西方向约 100km，电白水温 2004 年 9 月 9 日开始呈趋势上升，到 9 月 13 日累积上升约 0.008℃，9 月 13 日出现突降，震前恢复至正常水平。

台山台水温突降。台山台位于震中北东方向约 80km，台山水温 2004 年 9 月 5 日出现突降，幅度约为 0.015℃，17 日 14 时至 17 时上升 0.017℃。震后在 9 月 19～22 日再次出现幅度达 0.142℃的大幅突降—回升过程，22 日后恢复至正常动态。

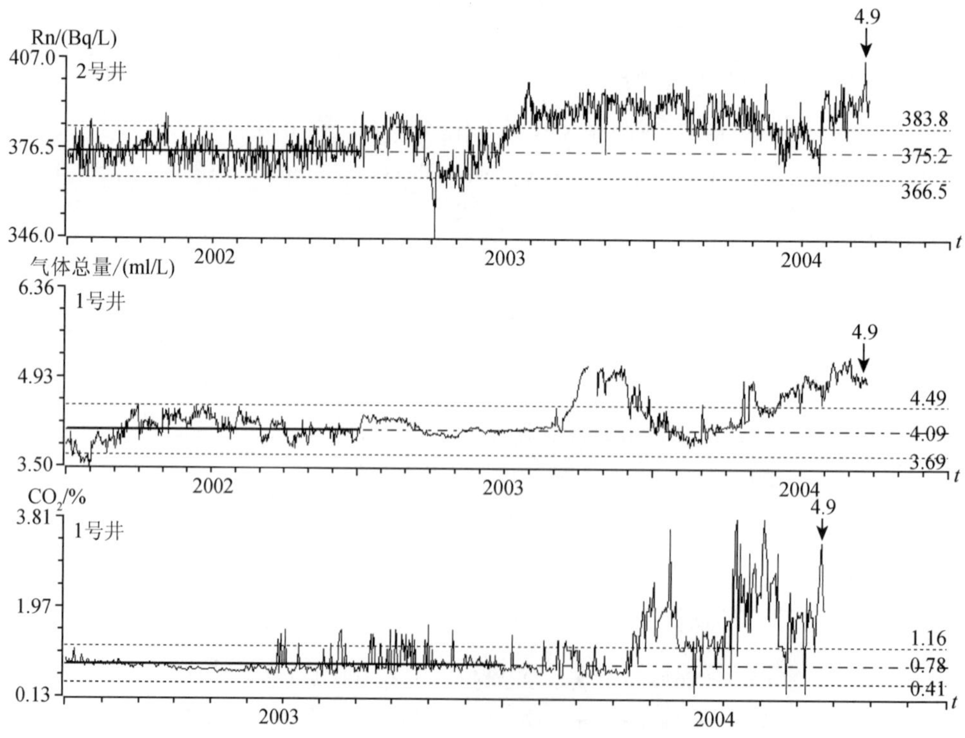

图 14　信宜 2 号井水氡，信宜 1 号井气体总量和 CO_2 日值曲线

Fig. 14　Curves of the radon in groundwater of Xinyi No. 2 well and the Daily value curves of gas gross and CO_2 of Xinyi No. 1 well

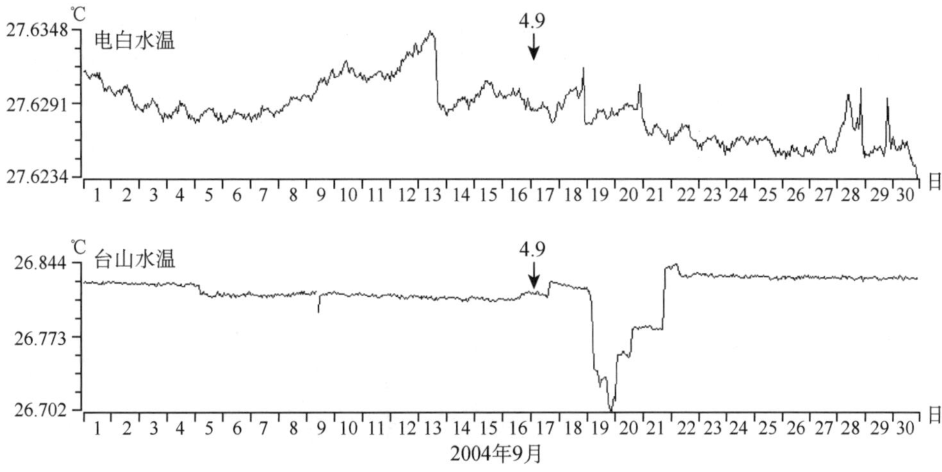

图 15 电白台、台山台水温小时值曲线

Fig. 15 Hourly value curves of water temperature in Dianbai station and Taishan station

（3）肇庆台地磁异常。肇庆台位于震中以北约 160km。对肇庆、琼中、邕宁三个地磁台分别进行两两地磁台 Z 分量的幅相法分析，其瞬时差幅度曲线可见（图 16），肇庆—邕宁、肇庆—琼中地磁幅相差在 2004 年 8 月底至 9 月 12 日出现异常段，由此判断肇庆台 Z 分量出现短期异常。

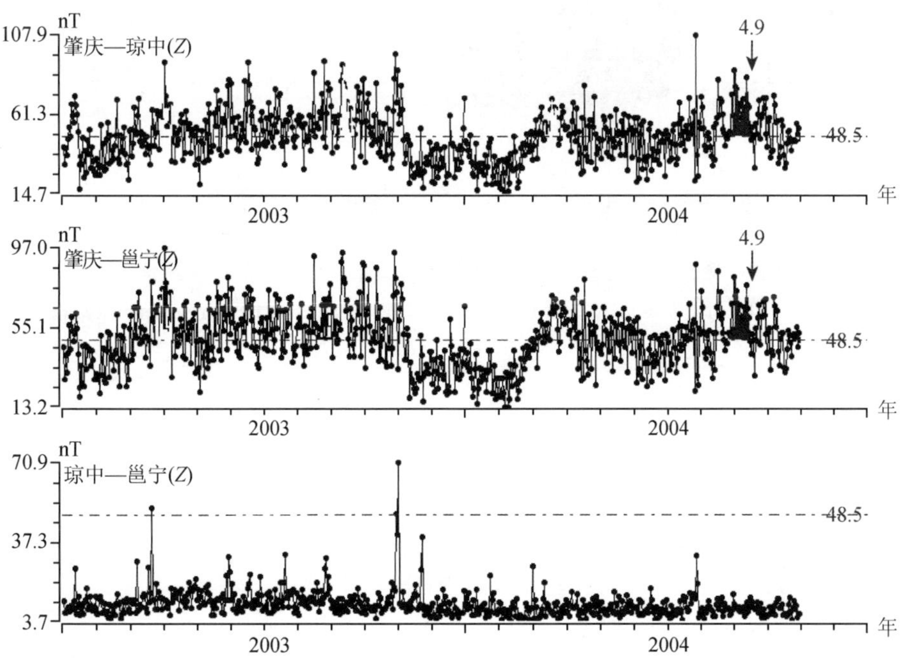

图 16 地磁垂直分量幅相法分析曲线

Fig. 16 Curves of the geomagnetism amplitude and phase analysis method

（4）阳江—雷州流动重力震前变化量高于往年。2004 年 3 月阳江—雷州半岛测区的流动重力复测显示，本期重力场变化量明显高于 2003 年（图 17），出现南北正负异常区。该异常区以东西向遂溪断裂为界，以南是海康—客路—河头为中心的负异常区，中心点期变化 $-41\times10^{-8}m/s^2$；以北是以观珠—高州—那霍—电城为中心的正变化区，中心点最大期变化 $35\times10^{-8}m/s^2$。岗美—阳江—闸坡测线点值的变化显示（图 18），从 2003 年开始各测点重力出现同步的转折反向上升，2004 年变化速率明显加强，其中闸坡、海陵测点变化最大。

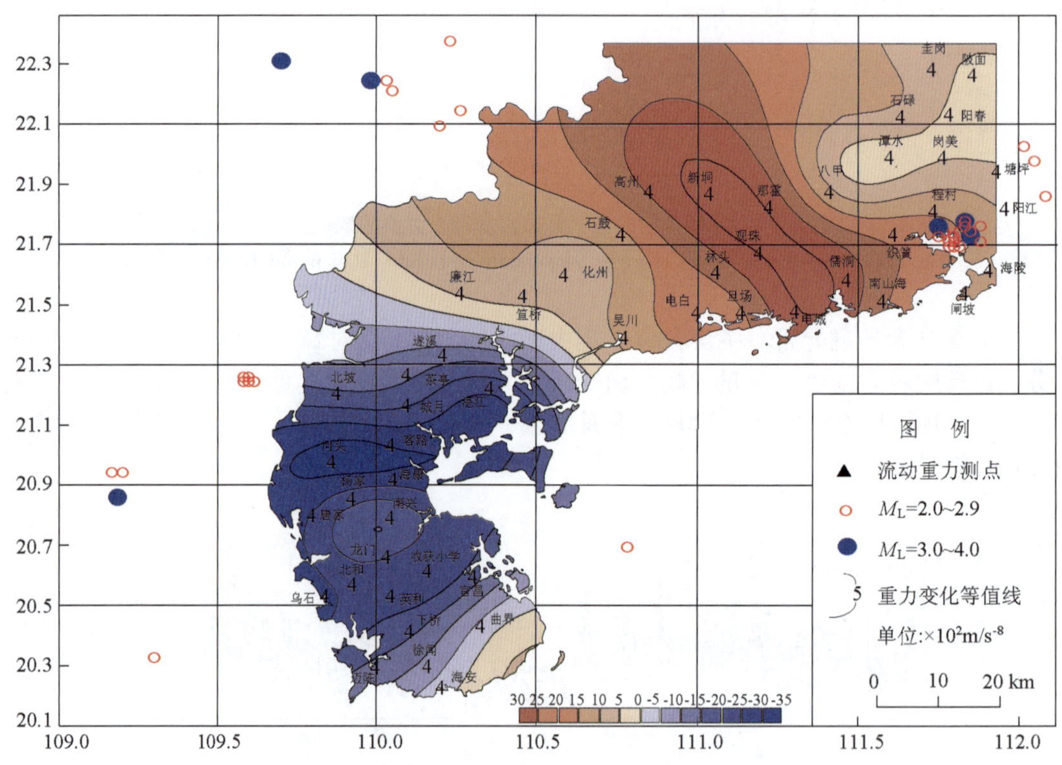

图 17　阳江—雷州半岛测区重力场期变化平面图（2003.9～2004.3）

Fig. 17　The plan of gravity field change in Yangjiang-Leizhou byland detection region

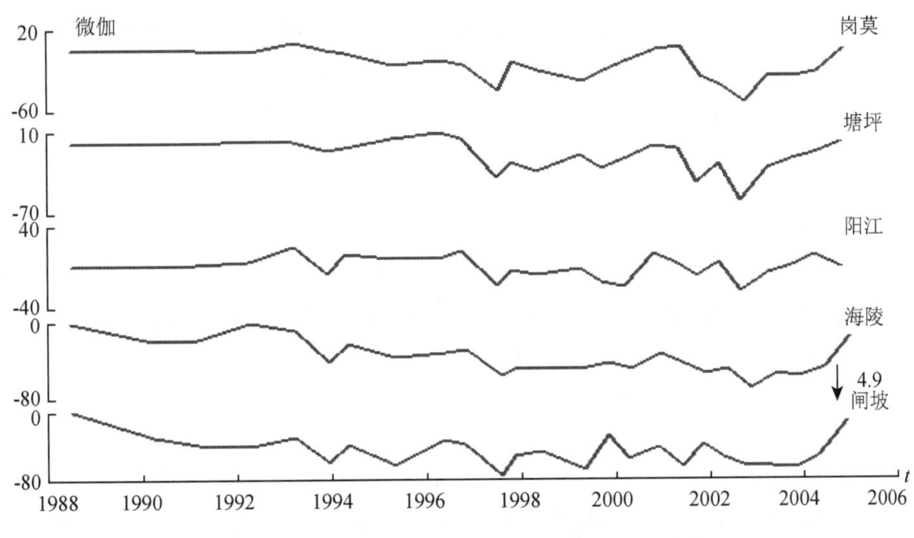

图 18　岗美—阳江—闸坡测线点值时序变化曲线

Fig. 18　Curves of the time sequence change of the detection points values in Gangmei，Yangjiang and Zhapo

七、地震前兆异常特征分析

阳江 4.9 级地震前，地震学异常不显著。仅部分地震活动性参数的时空扫描发现，在震前 1 年出现一些中期异常变化，如响应比、演化指数 *YH* 值、R_m 值、$C(n)$ 值等，但未出现短期和临震异常。

阳江 4.9 级地震震中周围 200km 范围内的 25 台项定点前兆观测项目中，共提取到 6 条异常，均属于临震短期异常，异常项数比为 24%。此外阳江—雷州半岛地区流动重力观测到中期异常。离震中最近的电白，台山水温在 2004 年 9 月出现较为明显的短临异常，信宜 2 号井水氡，1 号井 CO_2 震前十几天临震突跳现象明显。前兆异常变化具有一定的同步性，在一定程度上反映出震中周围地区应力场的不稳定状态。另外，前兆异常显示由中期到短临异常这样一个渐进式的演化过程。

八、震前预测、预防和震后响应

1. 预测情况

2003 年底完成的《2004 年度广东省地震趋势研究》[2] 与分类会商（预测）卡[6] 指出，阳江地区存在发生 4 级左右地震的可能。此次 4.9 级地震发生在预测区内，地点和时间的中期预测正确，但震级偏低。由于震前未出现小震活动明显异常，仅有少部分前兆异常，因此未能作出短期和临震的预测。

2. 震后响应

本次地震发生在深夜。地震发生后，广东省地震局迅速启动《广东省地震局地震应急预案》，向中国地震局，广东省委、省政府报告地震情况。震后 1.5 小时广东省地震局现场工作队出发。当日下午中国地震局专家组也赶赴灾区指导现场灾情调查工作。

预报人员根据阳江地区历史地震活动情况，尤其是历史上该区 4 级以上地震的震后情况，以及当已记录到的 25 次 1 级左右余震，在主震后 1.5 小时做出“近期内发生更大地震的可能性很小，但存在发生非破坏性有感地震可能”的震后趋势判定意见，上报中国地震局和省政府7)。几小时后，又首次在广东地震信息网页上向社会发布8)。结果表明，较明确的震后趋势判断意见和及时的向社会公布对稳定社会起到了重要的作用。

3. 震害分析1)

灾区位于北东东向平冈断裂和北西向洋边海断裂交汇处的洋边海沿岸一带。虽然地震发生在凌晨，造成群众一定程度的恐慌，但没有人员伤亡。阳江地区第四系松散层比较薄，基岩埋藏浅，灾区房屋以框架和砖混结构为主，同时分布数量不少的老旧民房。地震造成 14 个镇的 57 个行政村共 1620 多间房屋遭受不同程度轻微破坏，受影响户数达 21195 户。需要安置人员数为 5 户，共 15 人。经济损失约 2308 万元。

地震后，阳江市委、市政府启动了《破坏性地震应急预案》，主要领导及时赶赴灾区，安抚民心，积极主动采取有力措施，加之广东省地震局给出较明确的震后趋势判定意见，安定了群众情绪，保障了社会安定，没有出现停工停产现象，有效减轻了地震灾害损失。

九、结论与讨论

（1）最初将阳江 4.9 级地震与该区 1986 年 5.0 级和 1987 年 4.8 级地震，均判定为 1969 年 6.4 级主震后的晚期较强余震。理由是：①这 3 次地震与 6.4 级地震的震级差均大于 1 级，符合一般余震的特点；②3 次地震震中分布与主震震中基本一致，位于 6.4 级地震余震区；③3 次地震与 6.4 级主震宏观烈度Ⅴ～Ⅵ区基本重叠，仅范围大小有所不同，说明受同一发震构造控制[3];4,5)。4.9 级地震等震线分布及震源机制表明，平冈断裂是 4.9 级地震的主要控震和发震构造；④4.9 级地震序列与该区已发生的中强地震序列类似，具有余震频度、强度衰减快的特点。但根据近些年国内对历史上的强震区在几十年后重新发生中等地震的一些新的学术认识，我们认为这次 4.9 级地震可能是发生在 1969 年阳江 6.4 级地震余震区内的新的地震活动。

（2）阳江 4.9 级地震前，小震频度和能量未出现明显的异常。定点前兆观测项目提取到的异常，多数为震后总结中发现，且异常量级不大。因此，如何在有中期预测的前提下，在震前发现短临异常，依然是很困难的。

（3）作为一种尝试，广东省地震局首次将震后趋势判断意见在广东地震信息网页上向社会发布。实践证明，及时向社会发布明确的震后趋势意见对稳定社会，安定民心能够起到重要的作用。虽然 4.9 级地震是广东省 10 多年来在陆地发生的最大地震，且全省有感范围较大，但震后当天震区及周边地区就恢复了正常的社会生活秩序，广东省内未出现任何的社会公众恐慌和不安定的现象。

参考文献

[1] 魏柏林，冯绚敏，陈定国等. 东南沿海地震活动特征 [J]. 北京，地震出版社，2001

[2] 康英，杨选，吕金水等. 2004 年 9 月 17 日阳江 4.9 级地震震源参数分析 [J]. 地震，2005，25（3）：109 ~ 114

[3] 叶秀薇，杨马陵，叶东华等. 2004 年 9 月 17 日阳江 4.9 级地震概述 [J]. 华南地震，2005，25（3），69 ~ 77

[4] 刘特培，秦乃岗，陈玉桃. 阳江 M_S4.9 地震活动特征、影响场及应急对策 [J]. 地震地磁观测与研究，26（6），33 ~ 41，2005

参考资料

1）广东省地震局地震应急指挥部办公室，2004 年 9 月 17 日阳江市平冈镇 4.9 级地震应急工作总结，2004

2）广东省地震局，2004 年度广东省地震趋势研究报告，2003

3）广东省地震局，2005 年度广东省地震趋势研究报告，2004

4）魏柏林等，1986 年 1 月 28 日阳江 5.0 级地震调查报告

5）任镇寰等，1987 年 2 月 25 日阳江 5.0 级地震调查和现场震情处理的报告

6）分类会商（预测）卡（B2004-1）

7）震情简报（2004 年第 7 期）

8）阳江地震有关信息的网页截屏

The Yangjiang Earthquake with *M*4.9 on September 17, 2004 in Guangdong Province

Abstract

An earthquake of *M*4.9 occurred in Yangjiang of GuangDong province on September 17th, 2004. The earthquake can be feel north to the Guangzhou and Zhaoqing area, east to the Shenzhen and Macao area, west to the Zhanjiang and Maoming area. The intensity at the epicenter was Ⅵ. The isoseismal was ellipse and the long axis was from NE to NEE. The earthquake aera was about 214.8km^2. The direct economic loss was 2208 million Yuan, but no people were killed in the earthquake.

The earthquake sequence of *M*4.9 in Yangjiang belonged to mainshock-aftershock type and the magnitude of the largest aftershock was *M*3.6. The earthquake sequence was reduced quickly. According to the mechanism of earthquake resource, the aftershock distributing, the intensity of isoseismal and the construct distributing in local area, it can be determined that the Pinggang rupture with NEE direction was the the main rupture surface.

Within the distance of 200km from the epicenter, there were 12 seismic stations with 6 seismic observation stations and 9 seismic precursory observation stations in it. It was included 13 observation items about tilt, radon in water, well water level, water temperature, water quality, overflow gas, geomagnetism etc and the total of the station items was up to 27. There were 9 anomalies in 8 observation items before the earthquake of *M*4.9, a lot of them were short term anomalies.

The earthquake is the largest earthquakes occurred in the land in Guangdong Province for recent 10 years. The Seismological Bureau of Guangdong Province announced the earthquake trend advice to the public for the first time after the earthquake. It was played a positive role for Social and people stability. It was restored to the normal order of social life in the quake zone at the day after the quake.

2004 年 10 月 19 日云南省保山 5.0 级地震

云南省地震局

赵小艳　付　虹　邬成栋

摘　要

2004 年 10 月 19 日云南省保山市发生 5.0 级地震，宏观震中位于隆阳区汉庄镇张家山一带，极震区烈度为Ⅵ度，呈 NW 向椭圆形。地震造成 2 人重伤，13 人轻伤，直接经济损失为 21720 万元。

此次地震序列为双震型，2005 年 1 月 7 日在 5.0 地震左侧发生 4.8 级地震。最大余震 4.2 级，余震分布呈北西向，与相应烈度分布走向一致。节面Ⅱ为主破裂面，主压应力 P 轴方位 NNE 向，地震的发震构造为北北西向的石罗—银川断裂。

震中周围 200km 范围内共有地震台站 23 个，其中测震台 11 个，定点前兆观测台站 21 个。震前共出现 6 个异常项目的 29 项前兆异常。测震学 4 条异常，定点前兆出 22 条异常，宏观异常 3 条。

保山 5.0 级地震发生在云南省地震局 2004 年年中报告提出的 5～6 级地震危险区内，震前根据资料的异常变化，提出下半年保山一带将发生 5 级地震的预测意见，取得了一定的社会效益。

前　言

2004 年 10 月 19 日 06 时 11 分在云南省保山市隆阳区发生 M_S5.0 地震，据云南地震台网测定，微观震中位于北纬北纬 25°06′，东经 99°05′，宏观震中位于隆阳区汉庄镇张家山一带，宏、微观震中距离为 2km。极震区烈度为Ⅵ度。地震造成 2 人重伤，轻伤 13 人，直接经济总损失为 21720 万元。

震区附近历史上多次发生 $M_S \geqslant 5.0$ 级地震，1930 年曾发生过腾冲东北 6.0 级地震。2001 年 4～6 月该区发生了 3 次 5 级地震。

震中周围 200km 范围内共有地震台站 23 个，震前共出现 6 个异常项目的 29 项前兆异常。

2004 年中期会商时根据震兆、前兆的动态变化，提出 2004 年下半年保山一带发生 5 级地震的预测意见，把保山、施甸、腾冲一带作为 5～6 级地震的注意危险区写入年中预测报告。

地震发生后云南省地震局派出工作组赴震区开展震害评估、地震地质考察和余震监视工作。

一、测震台网及地震基本参数

保山 5.0 级地震前，震中周围 200km 范围内共有测震台 14 个，其中 100km 范围内有 5 个，101 ~ 200km 内有 9 个（图 1）。本次地震参数、地震序列采用云南台网资料，震中区测震台网监控能力为 M_L1.0，保山 5.0 级地震基本参数列于表 1 中。

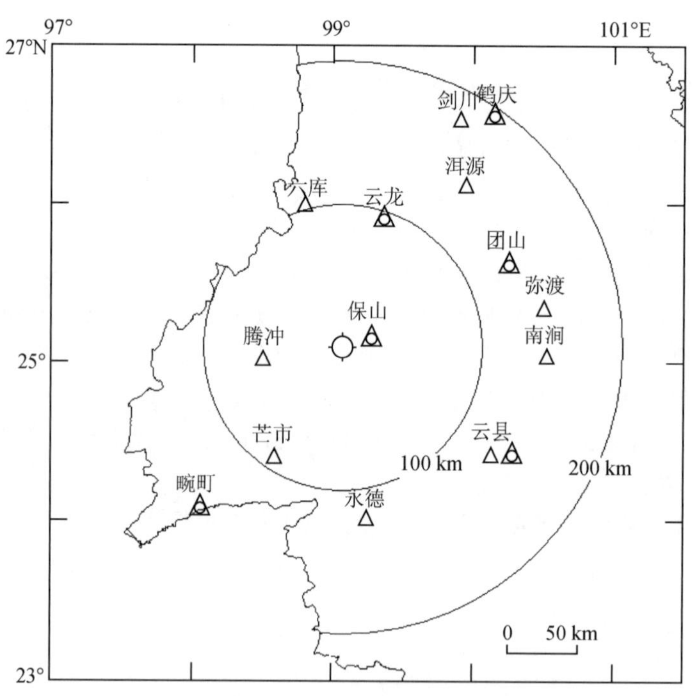

图 1　保山 5.0 级地震震中附近测震台网分布

Fig. 1　Distribution of seismometric stations around the epicentral area of the M_S5.0 Baoshan earthquake

表 1　地震基本参数

Table 1　Basic parameters of the earthquake

编号	发震日期	发震时刻	震中位置		震级	震源深度	震中地名	结果来源
	年 月 日	时 分 秒	φ_N	λ_E	M_S	(km)		
1	2004 10 19	06 11 41	25°06′	99°00′	5.0	6	保山	资料 1)
1	2004 10 19	06 11 41	25°06′	99°05′	5.0	6	保山	资料 2)

二、地震地质背景

2004 年 10 月 19 日保山 5.0 级地震，震区位于保山断块西缘，其西侧为分隔保山断块和腾冲断块的怒江深大断裂。区内断裂构造十分复杂，主要分布有：北西向的蒲缥—施甸断裂、罗明坝—太平断裂、石罗—银川断裂，南北向保山—施甸盆断裂。它们均为第四纪活动断裂，断裂活动强烈，总体走向 320°～345°，断裂面多向南西倾斜，倾角 56°～80°，断层岩显压扭性并具脆性变形特点，均属走滑逆断层。北西向的石罗—银川断裂和蒲缥—施甸断裂、与本次保山 5.0 级地震极震区烈度线的长轴走向基本一致，保山 5.0 地震发震构造与这两组断裂活动关系密切。

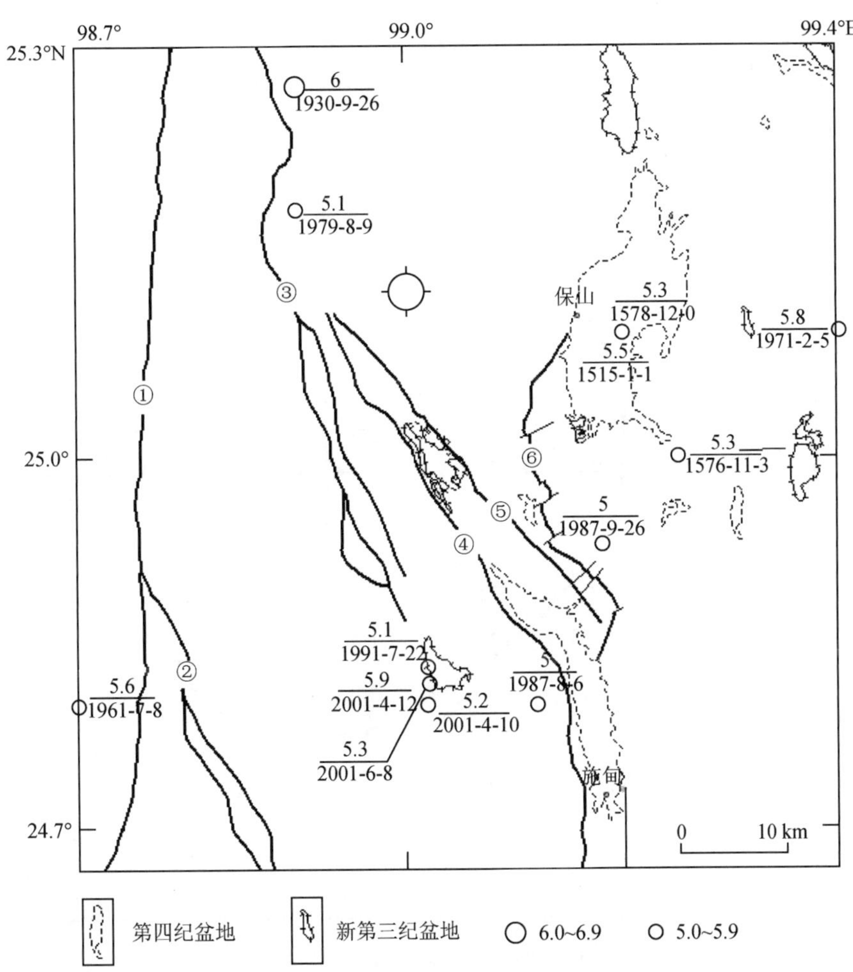

图 2　保山附近地质构造及历史地震震中分布图

Fig. 2　Map of geology structure and distribution of historical epicenter around Baoshan area

①怒江西支断裂；②镇安断裂；③罗明坝—太平断裂；

④蒲缥—施甸断裂；⑤石罗—银川断裂；⑥保山—施甸断裂

震区断陷盆地发育，如保山、施甸盆地均属断陷盆地，保山盆地西缘、施甸盆地两侧均有断裂发育。上述断裂相互交切交汇，形成了本次地震特殊的构造背景，反映了本地区十分复杂、破碎的地质结构，是本地区破坏性地震频繁发生的基本原因。

保山震区西邻腾冲—龙陵地震区，南邻澜沧—耿马地震带，由历史地震资料统计，震区附近（24°40′～25°20′N，东经 98°42′～99°24′E）自有历时地震记录以来，共发生 5 级以上地震 15 次，其中，5.0～5.9 级有 14 次，6.0～6.9 级 1 次，最大地震为 1930 年腾冲东北的 6.0 级地震。由于该地区复杂的地质构造，构成该区多样的地震活动方式，其地震类型有主余型、双震型、震群型等，其中以双震、震群方式活动的地震的比例为 64%，表明该地区地震活动方式以双震、震群为主。本次保山 5.0 级地震也为双震型地震。

三、烈度分布及震害

据现场考察资料，此次地震宏观震中位于隆阳区汉庄镇张家山一带，震区烈度Ⅵ度，等震线形状呈椭圆形，长轴走向为北西向，灾区总面积 443km²（图 3）。

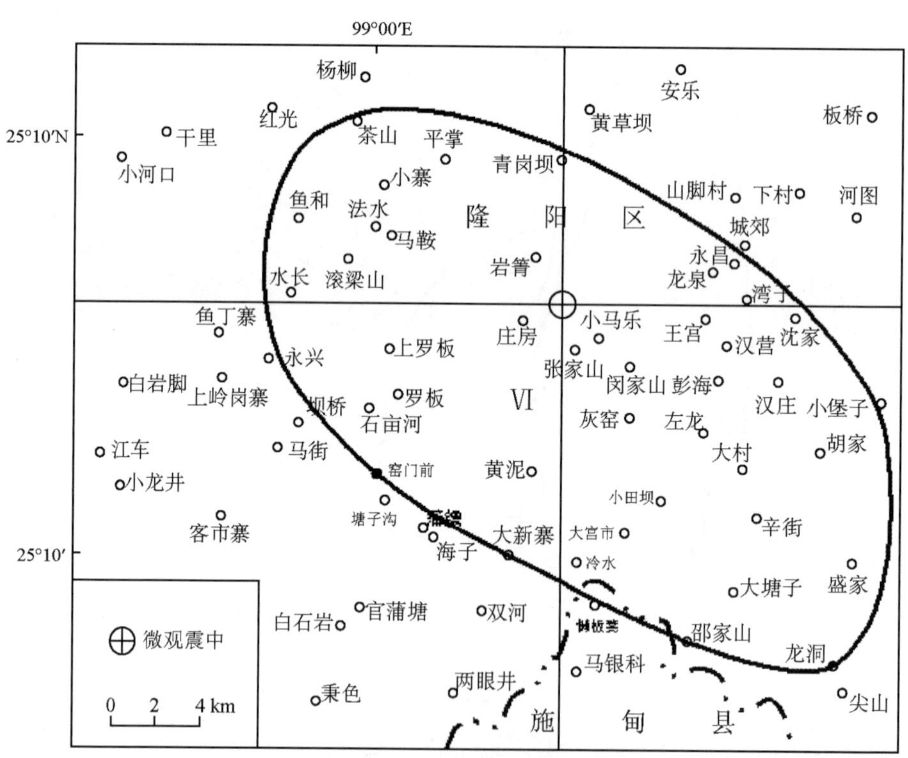

图 3　保山 5.0 级地震等震线图

Fig. 3　Isoseimal map of the M_S5.0 Baoshan earthquake

Ⅵ度区分布在隆阳区内：北自杨柳乡茶山村，南至辛街乡龙洞村，东自汉庄镇小堡子村，西近蒲缥镇永兴村。该烈度区内城乡民房、教育、卫生、工矿企业、机关事业单位的各类房屋建筑、生命线系统包括水利、交通基础设施和构筑物造成了不同程度的破坏。

本次地震震害特征总体可以概括为四方面：

（1）城区震害调查中发现，框架结构房屋填充墙开裂现象较多、局部墙面抹灰层脱落；而多数砖混结构未见破坏。框架结构的破坏比砖混结构稍重。

（2）本次地震灾区人口密度大，数量多以及社会财富高度聚集，因而经济损失较严重。

（3）部分房屋建筑选址不当，坡地建筑不经可行性论证和专门设计，边坡支护质量不高；部分建筑没有经过勘察设计，地基处理不力导致整体倾斜或墙体位错、变形；农村部分房屋年久失修，部分房屋土坯墙浸泡在地面积水中。上述房屋建筑地震破坏严重（有毁坏或局部倒塌现象）是本次地震经济损失较重的另一个重要原因。

（4）教育系统的校舍破坏较重。现场调查发现，与震区其他同类房屋相比，校舍破坏比例较高。

地震造成 2 人重伤，轻伤 13 人，直接经济总损失为 21720 万元。

四、地震序列

据云南地震台网测定，截至 2005 年 2 月 1 日共发生 $M_L \geqslant 1.0$ 级地震 522 次，其中 1 ~ 1.9 级 424 次，2 ~ 2.9 级 75 次，3 ~ 3.9 级 20 次，4 ~ 4.9 级 2 次。

1. 地震序列类型

2004 年 10 月 19 日最大地震能量占整个序列能量的 61%，2005 年 1 月 7 日 4.8 级次大地震占整个序列能量的 31%，最大地震与次大地震震级差 $\Delta M = 5.0 - 4.8 = 0.2$，$\Delta M \leqslant 0.6$，表明 2004 年 10 月 19 日保山 5.0 级地震为双震型地震。

2. 地震序列衰减情况

云南地震台网对保山地震的监控能力达到 $M \geqslant 1.0$ 级。从地震目录和序列 M-t 图来看，余震序列次数多，能量释放充分。从 M-t 看（图 4），5.0 级地震后 3 级以上地震余震次数多，且强度大，10 月 19 日至 11 月 7 日，序列 3 级地震频发，共发生 3 级以上地震 16 次，占整个序列 3 级以上地震的 73%。11 月 7 日后迅速衰减，到 12 月 10 日 3 级地震平静一个多月，其后陆续发生多次 2.5 级以上地震。在 2005 年 1 月 7 日 4.8 级地震前 1 月 3 ~ 7 日，1 级以上地震的强度增强，频度增加，直至发生 4.8 级地震。5.0 级地震序列最大余震为 4.2 级，发生在主震后 12 个小时（表 2）。2005 年 1 月 7 日 4.8 级地震共发生 1 级以上余震 93 次，其中 1 ~ 1.9 级 75 次，2 ~ 2.9 级 15 次，3 ~ 3.9 级 3 次，序列最大余震为 3.4 级，发生在主震后 19 天（表 2）。整个地震序列 b 值为 0.70，震级频度分布均较好地遵从 G-R 关系，h 值为 1.67，p 值为 0.49，序列衰减较慢。

3. 序列空间分布特征

图 9 为云南数字台网记录的保山 5.0 级地震的主震和地震序列震中分布图像。由图可见，5.0 级地震序列分布主体呈北西向，分布在 40km 范围左右，3 级以上余震均分布在主震 10km 范围内。4.8 级地震在 5.0 级地震北西方向 29.2km 处，其余震分布呈北西向。

表 2　保山 5.0 级地震序列目录（$M_L \geq 3.0$ 级）

Table 2　Catalogue of the M_S5.0 Baoshan earthquake sequence（$M_L \geq 3.0$）

编号	发震日期	发震时刻	震中位置		震级		震源深度	震中地名	结果来源
	年 月 日	时 分 秒	φ_N	λ_E	M_L	M_S	（km）		
1	2004 10 19	06 11 40	25°06′	99°05′		5.0	6	保山	资料 1）
2	2004 10 19	06 12 52	25°07′	99°09′	3.1			保山	资料 2）
3	2004 10 19	06 17 51	25°07′	99°09′	3.0			保山	资料 2）
4	2004 10 19	06 18 09	25°03′	99°02′	3.1			保山	资料 2）
5	2004 10 19	06 21 21	25°04′	99°04′	3.1		4	保山	资料 2）
6	2004 10 19	12 52 56	25°03′	99°02′	3.0		7	保山	资料 2）
7	2004 10 19	15 14 32	25°04′	99°02′	3.5		10	保山	资料 2）
8	2004 10 19	15 15 04	25°04′	99°02′	3.8		10	保山	资料 2）
9	2004 10 19	18 40 23	25°05′	99°05′	4.2		10	保山	资料 2）
10	2004 10 19	23 08 07	25°05′	99°02′	3.1		4	保山	资料 2）
11	2004 10 19	23 46 13	25°06′	99°04′	3.0		4	保山	资料 2）
12	2004 10 20	00 21 16	25°07′	99°05′	3.0		3	保山	资料 2）
13	2004 10 24	04 36 03	25°07′	99°05′	3.1		4	保山	资料 2）
14	2004 10 24	05 26 56	25°07′	99°05′	3.5		3	保山	资料 2）
15	2004 10 24	14 05 09	25°04′	99°04′	3.1		10	保山	资料 2）
16	2004 11 01	00 04 29	24°57′	99°21′	3.0		10	保山	资料 2）
17	2004 11 07	09 58 52	25°05′	99°03′	3.3			保山	资料 2）
18	2004 12 10	22 08 47	25°07′	99°04′	3.0		2	保山	资料 2）
19	2005 01 04	17 37 46	24°53′	98°48′	3.5		3	保山	资料 2）
20	2005 01 07	23 50 07	25°10′	98°48′	4.8		5	保山	资料 2）
21	2005 01 08	01 51 28	25°08′	98°49′	3.2		7	保山	资料 2）
22	2005 01 09	23 15 12	25°09′	98°51′	3.2		10	保山	资料 2）
23	2005 01 26	16 50 42	25°04′	99°04′	3.4			保山	资料 2）

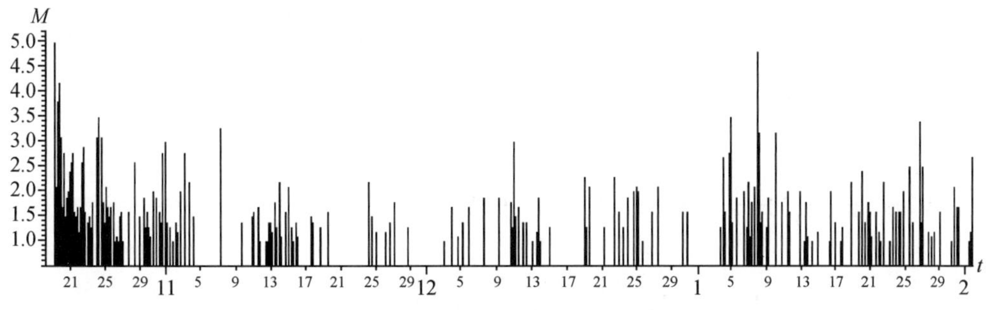

图 4　保山 5.0 级地震序列 M-t 图（$M_L \geqslant 1.0$）

Fig. 4　M-t diagram of the M_S5.0 Baoshan earthquake sequence（$M_L \geqslant 1.0$）

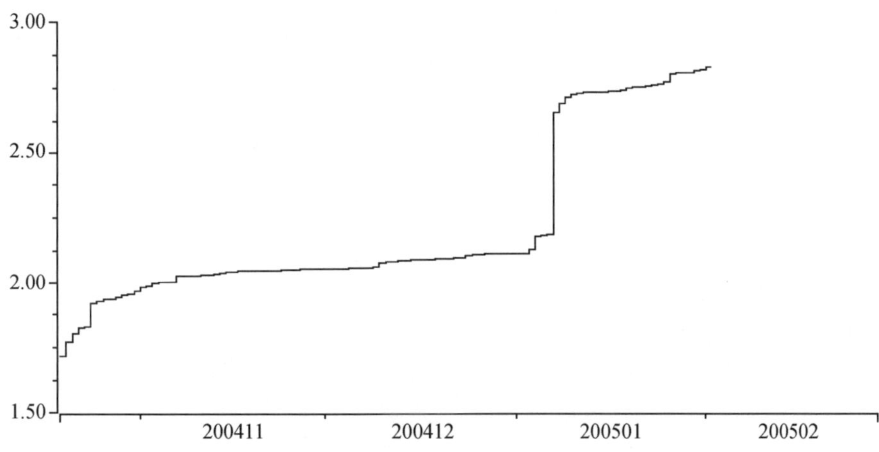

图 5　保山 5.0 级地震序列蠕变曲线图（$M_L \geqslant 1.0$）

Fig. 5　Strain release of the M_S5.0 Baoshan earthquake sequence（$M_L \geqslant 1.0$）

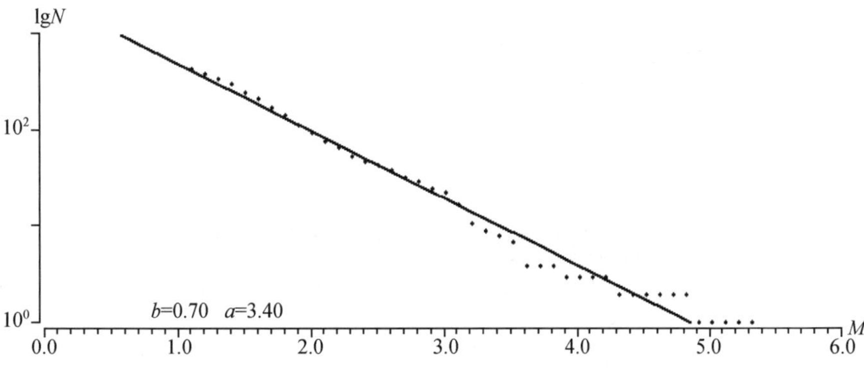

图 6　保山 5.0 级地震序列 b 值图

Fig. 6　b-value diagram of the M_S5.0 Baoshan earthquake sequence

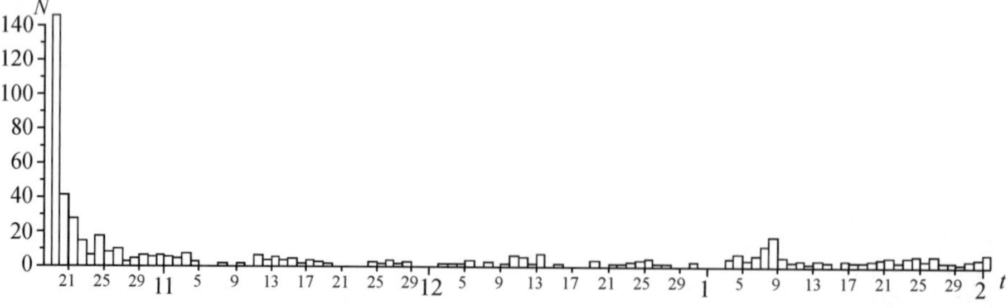

图7　保山5.0级地震序列 N-t 曲线（$M_L \geqslant 1.0$）

Fig. 7　N-t diagram of the M_S5.0 Baoshan earthquake sequence（$M_L \geqslant 1.0$）

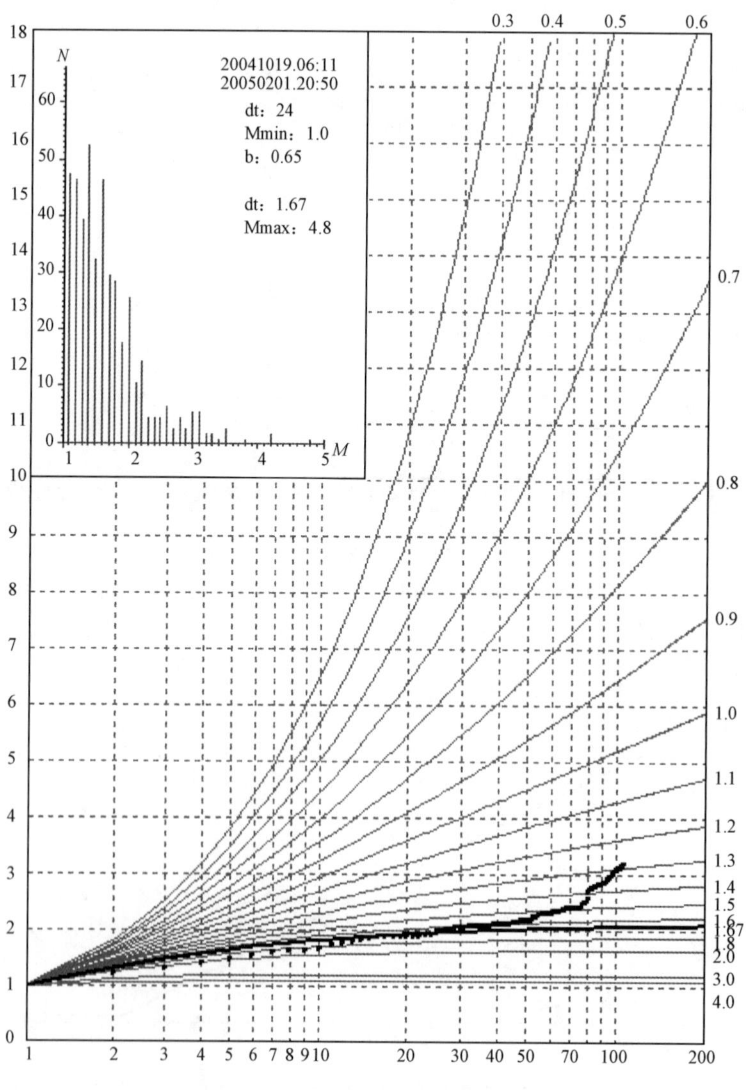

图8　保山5.0级地震序列 h 值图

Fig. 8　h-value diagram of the M_S5.0 Baoshan earthquake sequence

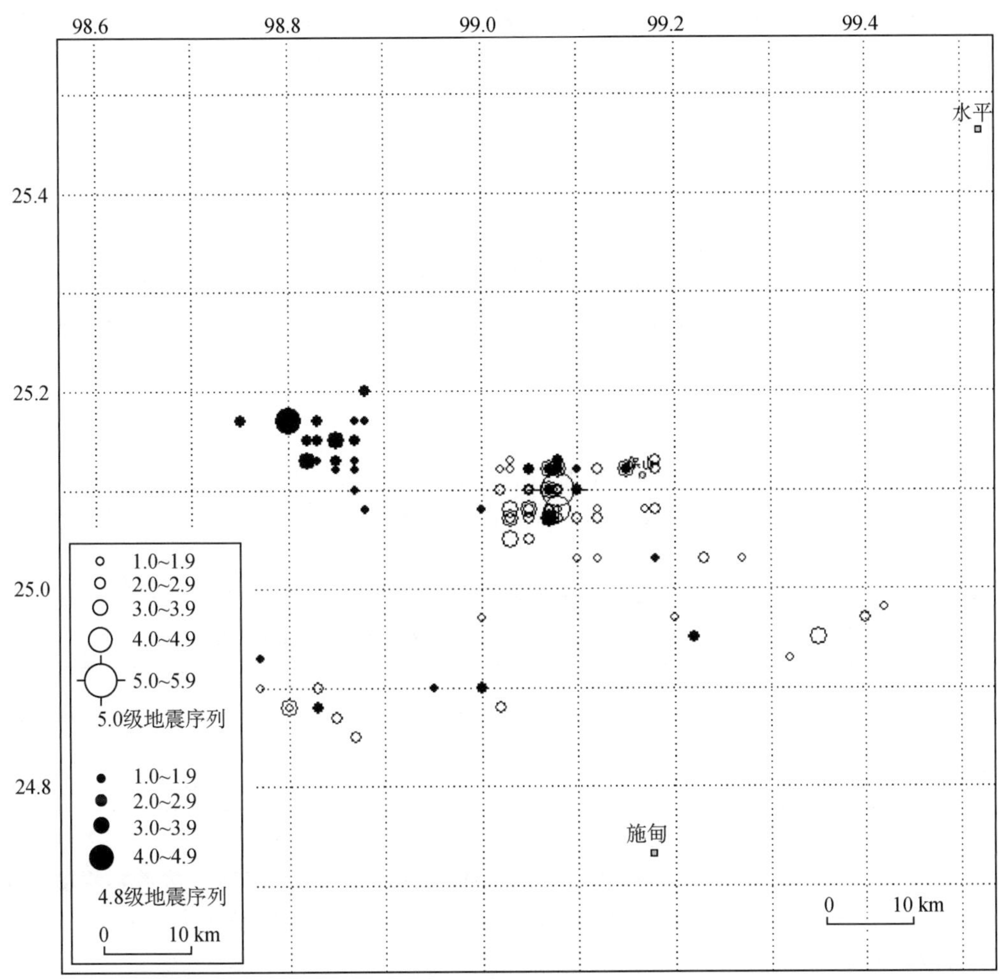

图 9　保山 5.0 级地震序列震中分布图

Fig. 9　Distribution of the the M_S5.0 Baoshan sequence（2004.10.19-2005.2.1）

五、震源机制解及地震主破裂面

利用昆明数字化地震台网记录清晰的 28 个 P 波初动符号，矛盾符号比为 0.12，采用下半球投影作图求解法求解震源机制，得到了震源机制解，其参数列在表 3，图 10 吴尔弗网图解。

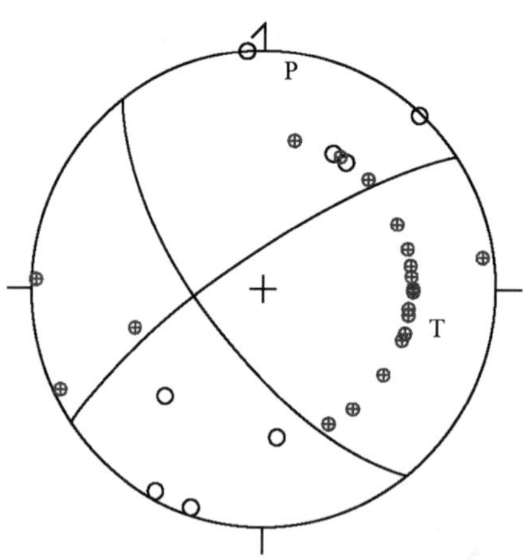

图 10　保山 5.0 级地震震源机制解

Fig. 10　Focal mechanism solution of the M_S5.0 Baoshan earthquake

表 3　5.0 级地震震源机制解

Table 3　Focal mechanism solution of the M_S5.0 Baoshan earthquake

编号	节面Ⅰ			节面Ⅱ			P 轴		T 轴		B 轴		结果来源
	走向	倾向	倾角	走向	倾向	倾角	方位	仰角	方位	仰角	方位	仰角	
1	236	79	22	142	68	168	8	7	101	23	263	66	付虹

结果显示，主震发震应力场是北北东（8°）方向，仰角很小（7°），接近水平的压应力作用。节面Ⅰ走向北东东（236°），倾角陡立（79°），在北北东向接近水平的压应力作用下，具有以左旋走滑为主的错动性质。节面Ⅱ走向北北西（平均 142°），倾角为 68°，在北北东向接近水平的压应力作用下，具有以右旋走滑为主的错动性质。

这次地震的等震线长轴方向为北北西向，5.0、4.8 级地震的余震展布也主要呈北西方向，结合保山地震周边构造来看，节面Ⅱ的走向与北北西向的石罗—银川断裂构造关系较为一致，据此推测，保山 5.0 级地震发生在石罗—银川断裂构造上，是在近水平的压应力作用下产生右旋走滑错动的结果。

六、观测台网及前兆异常

图 1 和图 11 为震中附近地区测震台站和定点前兆观测台站分布图，地震发生在前兆观测台站相对密集地区。

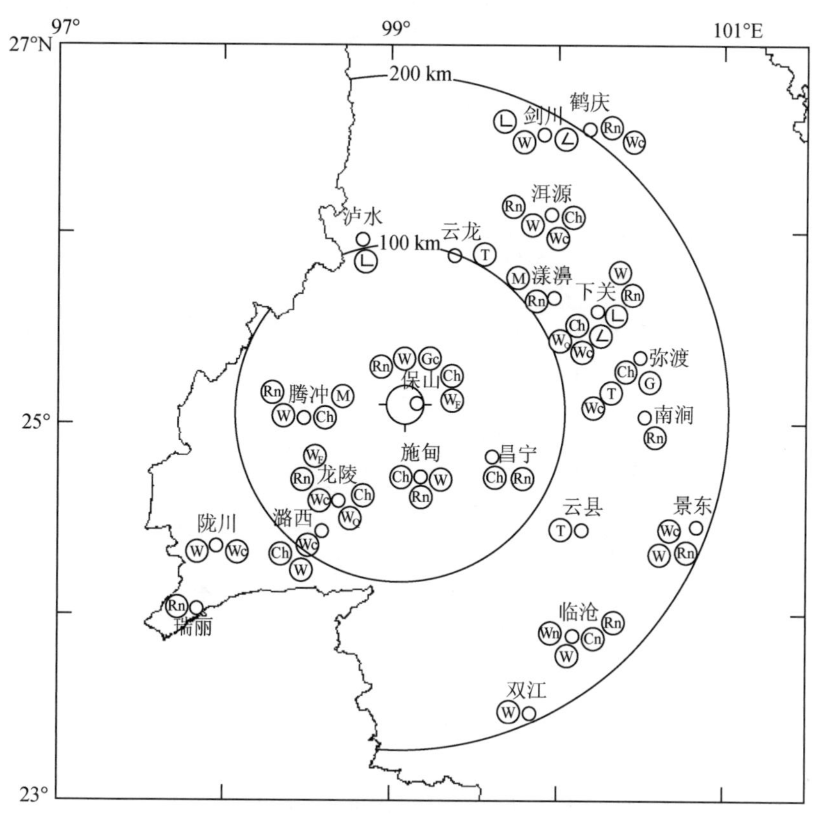

图11　保山5.0级地震前定点前兆观测台站的分布图

Fig. 11　Distribution of the precursor observation stations before the M_S5.0 Baoshan earthquake

保山5.0级地震周围200km范围内共有地震台站23个，其中测震台14个，定点前兆观测台站21个，有水氡、水位、水质、水温、地倾斜、地电、地磁D、地磁Z等10个定点前兆观测项目共98个台项，这些定点前兆观测项目大多数均有5年以上连续可靠的观测资料。在0～100km和101～200km范围内分别有测震台5个和9个，定点前兆观测台站7个和14个。这次地震前共出现6个异常项目（地震频度、b值、水氡、水位、水质、水温），29条前兆异常，其中测震学出现了4条异常，它们是地震频度及b值等；定点前兆出现了22条异常（图12），它们是龙陵水氡、昌宁水氡、保山水氡、保山水位、龙陵流量、腾冲水位、龙陵流量、施甸水位、保山水温、龙陵水温、保山水汞、保山硫酸根离子等。宏观异常3项，分别是洱源九气台温泉水发浑、大理挖色乡水塘水、发浑洱源风羽机井井水发浑。

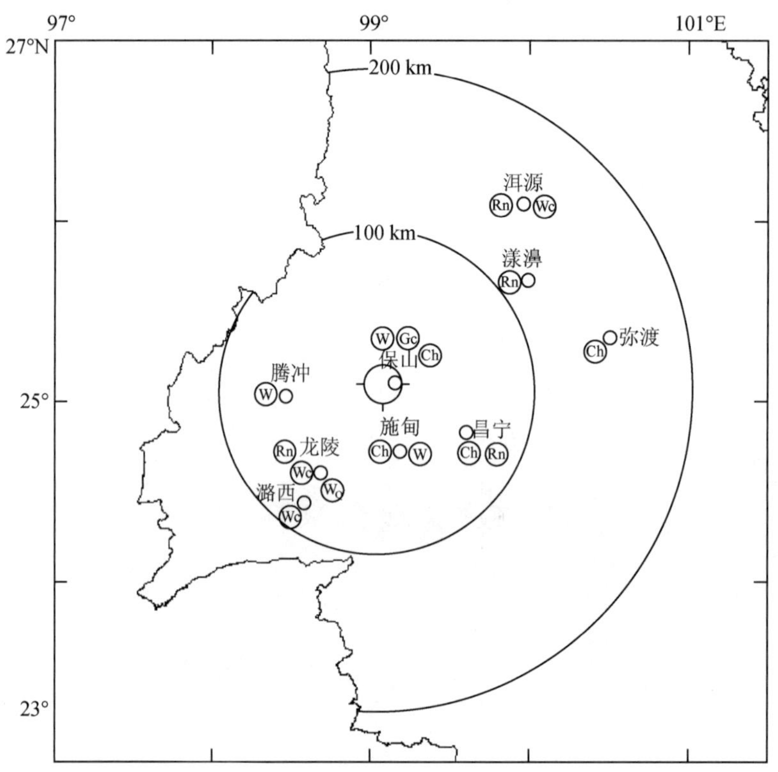

图 12　保山 5.0 级地震定点前兆异常分布图

Fig. 12　Distribution of precursory anormalies of the M_S5.0 Baoshan earthquakeon the fixed observation points

表 5 所列出的 29 条异常，其信度及预测效能简述如下：

（1）异常项目：资料信度高，异常清晰，具有较好的对应率，预测效能好。

（2）定点前兆异常项目：所有观测项目均有 5 年以上连续、完整的资料，震前异常幅度显著，远大于观测误差，并具有较好的地震对应率，预测效能好。

（3）宏观异常：洱源九气台温泉 2004 年 3 月 11 日至 2004 年 3 月 17 日、2004 年 4 月 20 日、2004 年 5 月 9 日出现 3 次水体发浑现象，且每次发浑后洱源及其周边的地区都有 4 级以上地震发生。2004 年 8 月 9 日大理挖色乡水塘水色发浑，经现场落实组综合分析后认为：此次水塘变浑事件属实，察看周边环境可以排除降雨等干扰，可确认为前兆异常。2004 年 6 月 30 日下午 1～7 时，洱源县风羽镇江登村公所机井水发浑，井水能见度小于 2cm，而平时井水清澈见底，该井深 100 多米，自 1978 年建成以来从来没有出现过井水发浑现象，现场落实组综合分析后认为，该井井水发浑现象确为前兆异常。这 3 次宏观异常均表明，表明滇西地区处于应力加载过程中并已产生了深部微破裂。

（4）绝大多数异常均为震前提出，震后总结的异常主要为一些基本形态不清楚的前兆测项，如昌宁电导率、施甸 pH 值、德宏水温、漾濞水氡等，部分异常幅度较大的异常可能事先不能把它当做一个 5 级地震的异常来解释，如龙陵流量。

表 4 异常情况登记表

Table 4 Summmary table of precursory anomalies

序号	观测项目	台站或观测区	分析方法	异常判据及观测误差	震前异常起止时间	震后变化	最大幅度	震中距 km	异常类别	图号	异常特点及备注	震前提出/震后总结
1	地震频度	震中为圆心 50km 范围	$M_L \geqslant 3.0$ 级月频次	$N \geqslant 3$	2004.8～2004.9	处于高值状态	6		S_1	13	震前小地震显示增频异常变化	震前提出
2	地震频度	震中为圆心 50km 范围	$M_L \geqslant 1.0$ 级月频次	$N \geqslant 20$	2004.8～2004.10	处于高值状态	70		S_1	14	震前小地震显著增频	震后总结
3	b 值	云南地区 21°～29°N，97°～106°E	3 个月滑动	$0.78 \geqslant b \geqslant 1.12$	2004.6～2004.7	正常	1.19		S_1	15	震前显示高值异常变化	震前提出
4	震中	云南地区 21～29°N，97～106°E	$M_L \geqslant 3.0$ 级震中	震区 3 级地震活动	2004.7～2004.10	正常			S_2	16	震群震前 3 级地震集中	震前提出
5	水位	保山	日均值	突跳	2004.9.19	处于低值状态	0.25	8	I_1	17	水位上升后大幅下降	震前提出
6	水位	施甸	日均值	破年变	2004.6.23～2004.10.14	逐步恢复年变	60%	43	S_1	18	按年变震前水位应该下降，实际上升破年变	震前提出
7	水位	腾冲	日均值	$\leqslant \sigma$	2004.8.16～2004.12.16	处于低值状态	3.55	62	S_1	19	高值突跳	震前提出
8	流量	龙陵	5 日均值	$\Delta \geqslant 0.001$	2003.12.25～2004.5.20	处于高值状态	0.0023	69	M_2	20	上下波动	震后总结
9	水位	景东	日均值	破年变	2004.4～2004.12	异常	70%	191	M_1	21	该异常对应印尼地震可能性更大	震后总结

续表

序号	观测项目	台站或观测区	分析方法	异常判据及观测误差	震前异常起止时间	震后变化	最大幅度	震中距 km	异常类别	图号	异常特点及备注	震前提出/震后总结
10	水氡	昌宁	旬均值	$\Delta R_n \geqslant 10\%$	2004.4.10 ~ 2004.10.10	正常	17%	61	M_1	22	水氡浓度下降	震前提出
11	水氡	龙陵	五日均值	$R_n \leqslant \sigma$	2004.5.10 ~ 2004.9.10	处于低值状态	56.5	69	S_1	23	水氡浓度大幅度下降	震后总结
12	水氡	漾濞	日均值	$\leqslant 2\sigma$	2004.10.11	正常	44.8ng/l	109	I_2	24	临震大幅突跳异常明显	震后总结
13	水氡	洱源	日均值	突跳	2004.10.13 ~ 2004.10.16	正常	117.5ng/l	146	I_1	25	保山地震后恢复正常，印尼地震前再次出现异常	震前提出
14	水温	保山	日均值	$\Delta \geqslant 0.01$℃	2004.6.28 ~ 2004.10.19	正常	0.0178	8	S_1	26	临震变化显著	震前提出
15	水温	保山市	日均值	突跳	2004.8.22，2004.10.6	仍有突跳	0.11℃	8	S_1	27	大幅突跳印尼地震后仍然存在	震前提出
16	水温	龙陵	日均值	突跳	2004.5.9	正常	1.9℃	69	S_2	28	高值突跳	震前提出
17	水温	德宏	日均值	$\Delta \geqslant 0.01$℃	2003.12.8 ~ 2004.8.15	正常	0.012℃	98	M_1	29	震前 1 年水温大幅波动异常	震后总结
18	水温	洱源	日均值	$\leqslant 2\sigma$	2004.10.13 ~ 2005.1.1	异常	40.20℃	146	I_2	30	该异常虽然在保山地震前出现，但对应印尼地震可能性更大	震前提出
19	水汞	保山	日均值	突跳	2003.12.7 ~ 2004.9.10	正常	1318ng/l	8	M_1	31	持续大幅突跳，临震前幅度更大	震前提出

续表

序号	观测项目	台站或观测区	分析方法	异常判据及观测误差	震前异常起止时间	震后变化	最大幅度	震中距 km	异常类别	图号	异常特点及备注	震前提出/震后总结
20	硫酸根离子	保山	五日均值	$\leqslant\sigma$	2003.11.5 ~ 2004.5.5	正常	6.9ng/l	8	M_2	32	震前 5 个月持续低值异常	震前提出
21	氟离子	保山	日均值	$\geqslant$0.42ng/l	2004.7.4 ~ 2004.10.19	正常	0.445ng/l	8	S_1	33	高值异常变化	震前提出
22	pH 值	施甸	五日均值	$\leqslant\sigma$	2004.1.10 ~ 2004.10.19	趋于平稳	6.668	43	M_2	34	震前 1 年大幅升降变化	震后总结
23	碳酸根离子	施甸	五日均值	$\leqslant\sigma$	2004.5.15 ~ 2004.8.30	正常	367.8ng/l	43	S_2	35	震前半年大幅升降变化	震前提出
24	氟离子	施甸	五日均值	突跳	2004.7.20 ~ 2004.10.19	仍有突跳	0.26ng/l	43	S_1	36	异常直至印尼地震后仍未结束	震前提出
25	电导率	昌宁	日均值	$\leqslant\sigma$ 及突跳	2004.4.25 ~ 2005.1.16	异常	0.066	61	S_1	37	震前半年低值异常，临震出现大幅突跳	震后总结
26	水汞	弥渡	日均值	突跳	2004.10.4	正常	1022ng/l	142	I_1	38	临震突跳幅度大	震前提出
27	宏观	大理挖色乡水塘	水色	发浑	2004.8.9	正常		129	S_2		该水塘不受降雨影响，确为宏观异常	震前提出
28	宏观	洱源九气台温泉	水色	发浑	2004.3.11、4.20、5.9	正常		142	M_1		2004.3 ~ 5 月九气台温泉出现 3 次水体发浑	震前提出
29	宏观	洱源风羽机井	水色	发浑	2004.6.30	正常		141	S_1		水体发浑，能见度低于 2cm	震前提出

注：震前提出/震后总结一栏以年度、周月会商报告中提到的异常为震前提出，反之为震后总结。

表 5　固定前兆观测台（点）与观测项目汇总表

Table 5　Summary sheet of fixed stations and items

序号	台站（点）名称	经纬度		观测项目	资料类别	震中距（km）	备注
		λ（°）	φ（°）				
1	保山	99.16	25.11	测震△	Ⅰ	8	
				水氡	Ⅰ		
				水位	Ⅰ		
				Hg	Ⅰ		
				Ca^{2+}	Ⅱ		
				Mg^{2+}	Ⅱ		
				So_4^{2-}	Ⅱ		
				HCO_3^-	Ⅱ		
				F^-	Ⅱ		
				pH	Ⅱ		
				高精度水温	Ⅱ		
2	施甸	99.17	24.74	水氡	Ⅰ	43	
				水位	Ⅰ		
				HCO_3^-	Ⅱ		
				Ca^{2+}	Ⅱ		
				Mg^{2+}	Ⅱ		
3	昌宁	99.6	24.84	水氡	Ⅰ	61	
				CL^-	Ⅰ		
				pH	Ⅱ		
				电导率	Ⅰ		
4	滕冲	98.49	25.03	测震△	Ⅰ	62	
				水氡	Ⅰ		
				水位	Ⅰ		
				HCO_3^-	Ⅱ		
				Ca^{2+}	Ⅱ		
				Mg^{2+}	Ⅱ		
				地磁（*D*）	Ⅰ		
				地磁（*Z*）	Ⅰ		
				地电阻率	Ⅱ		
				自然电位	Ⅱ		

续表

序号	台站（点）名称	经纬度		观测项目	资料类别	震中距（km）	备注
		λ（°）	φ（°）				
5	龙陵	98.69	24.6	水氡	Ⅰ	69	
				流量	Ⅰ		
				水温	Ⅰ		
				电导率	Ⅱ		
				SO_4^{2-}	Ⅱ		
				HCO^-	Ⅱ		
				F^-	Ⅱ		
6	云龙	99.37	25.90	测震△	Ⅰ	95	
				地倾斜	Ⅱ		
7	潞西	98.58	24.45	测震△	Ⅰ	98	
				水氡	Ⅰ		
				水位	Ⅰ		
				高精度水温	Ⅰ		
				HCO_3^-	Ⅱ		
				F^-	Ⅱ		
				Ca^{2+}	Ⅱ		
				Mg^{2+}	Ⅱ		
8	泸水	98.82	25.99	测震△	Ⅰ	104	
				水氡	Ⅰ		
				水位	Ⅱ		
				高精度水温	Ⅰ		
9	漾濞	99.96	25.66	水氡	Ⅱ	109	
				地磁	Ⅱ		
10	永德	99.25	24.03	测震△	Ⅰ	120	

续表

序号	台站（点）名称	经纬度		观测项目	资料类别	震中距（km）	备注
		λ（°）	φ（°）				
11	下关	100.22	25.60	测震	Ⅰ	128	
				水氡	Ⅰ		
				水位	Ⅰ		
				高精度水温	Ⅰ		
				Hg	Ⅰ		
				CO_2	Ⅱ		
				短水准	Ⅰ		
				短基线	Ⅰ		
12	云县	100.12	24.56	测震△◬	Ⅰ	129	
				地倾斜	Ⅱ		
13	陇川	97.96	24.37	水位	Ⅱ	140	
				水温	Ⅱ		
14	弥渡△	100.49	25.35	测震	Ⅰ	142	
				水氡	Ⅱ		
				Hg	Ⅱ		
				CO_2	Ⅱ		
				高精度水温	Ⅱ		
				地倾斜	Ⅱ		
				重力	Ⅱ		
15	南涧	100.52	25.06	测震△	Ⅱ	144	
				水氡	Ⅱ		
16	洱源	99.95	26.12	测震△	Ⅰ	146	
				水氡	Ⅰ		
				水位	Ⅰ		
				高精度水温	Ⅰ		
				Hg	Ⅰ		
				CO_2	Ⅱ		
17	畹町	98.07	24.09	测震◬	Ⅰ	152	

续表

序号	台站（点）名称	经纬度 λ（°）	经纬度 φ（°）	观测项目	资料类别	震中距（km）	备注
18	临沧	100.08	23.89	水氡	Ⅱ	170	
				水位	Ⅱ		
				高精度水温	Ⅱ		
				F^-	Ⅱ		
19	瑞丽	97.85	24.01	水氡	Ⅱ	174	
20	剑川	99.90	26.53	测震△	Ⅰ	180	
				水位	Ⅱ		
				短基线	Ⅰ		
				短水准	Ⅰ		
21	景东	100.83	24.46	水氡	Ⅱ	191	
				水位	Ⅱ		
				高精度水温	Ⅱ		
22	双江	99.82	23.47	水位	Ⅰ	194	
23	鹤庆	100.17	26.55	测震△◬	Ⅰ	200	
				水氡	Ⅱ		
				二氧化碳	Ⅱ		

分类统计	0 < Δ≤100km	100km < Δ≤200km	总数
测项数 N	17	12	22
台项数 n	47	51	98
测震单项台数 a	0	2	2
形变单项台数 b	1	1	2
电磁单项台数 c	0	0	0
流体单项台数 d	0	3	3
综合台站数 e	6	10	16
综合台中有测震项目的台站数 f	3	6	9
测震台总数 $a+f$	3	8	11
台站总数 $a+b+c+d+e$	7	16	23
备注			

表 6　测震以外固定前兆观测项目与异常统计表

Table 6　Precursory observation items and anomalies statistics

序号	台站（点）名称	经纬度		观测项目	资料类别	震中距（km）	按震中距Δ进行异常同基									
							$0<\Delta\leqslant100$km					$100<\Delta\leqslant200$km				
		λ（°）	φ（°）				L	M	S	I	U	L	M	S	I	U
1	保山	99.16	25.11	水氡	Ⅰ	8	—	—	—	—						
				水位	Ⅰ		—	—	—	√		—	—			
				Hg	Ⅰ		—	√	—	—						
				Ca^{2+}	Ⅱ		—	—	—	—						
				Mg^{2+}	Ⅱ		—	—	—	—						
				SO_4^{2-}	Ⅱ		—	√	—	—						
				HCO_3^-	Ⅱ		—	—	—	—						
				F^-	Ⅱ		—	—	√	—						
				pH	Ⅱ		—	—	—	—						
				高精度水温	Ⅱ		—	—	√	—						
2	施甸	99.17	24.74	水氡	Ⅰ	43	—	—	—	—						
				水位	Ⅰ		—	—	√	—						
				HCO_3^-	Ⅱ		—	—	√	—						
				Ca^{2+}	Ⅱ		—	—	—	—						
				Mg^{2+}	Ⅱ		—	—	—	—						
				F^-			—	—	√	—						
				pH			—	√	—	—						

续表

序号	台站（点）名称	经纬度		观测项目	资料类别	震中距（km）	按震中距 Δ 进行异常同基									
							$0<\Delta\leqslant100$km					$100<\Delta\leqslant200$km				
		λ（°）	φ（°）				L	M	S	I	U	L	M	S	I	U
3	昌宁	99.6	24.84	水氡	Ⅰ	61	—	√	—	—						
				CL^-	Ⅰ		—	—	—	—						
				pH	Ⅱ		—	—	—	—						
				电导率	Ⅰ		—	—	√	—						
4	滕冲	98.49	25.03	水氡	Ⅰ	62	—	—	—	—						
				水位	Ⅰ		—	—	√	—						
				HCO_3^-	Ⅱ		—	—	—	—						
				Ca^{2+}	Ⅱ		—	—	—	—						
				Mg^{2+}	Ⅱ		—	—	—	—						
				地磁 D	Ⅰ		—	—	—	—						
				地磁 Z	Ⅰ		—	—	—	—						
				地电阻率	Ⅱ		—	—	—	—						
				自然电位	Ⅱ		—	—	—	—						

续表

序号	台站（点）名称	经纬度		观测项目	资料类别	震中距（km）	按震中距Δ进行异常同基									
							0＜Δ≤100km					100＜Δ≤200km				
		λ（°）	φ（°）				L	M	S	I	U	L	M	S	I	U
5	龙陵	98.69	24.6	水氡	Ⅰ	69	—	—	√	—						
				流量	Ⅰ		—	√	—	—						
				水温	Ⅰ		—	—	√	—						
				电导率	Ⅱ		—	—	—	—						
				SO_4^{2-}	Ⅱ		—	—	—	—						
				HCO_3^-	Ⅱ		—	—	—	—						
				F^-	Ⅱ		—	—	—	—						
6	云龙	99.37	25.90	地倾斜	Ⅱ	95	—	—	—	—						
				地倾斜	Ⅱ		—	—	—	—						
7	潞西	98.58	24.45	水氡	Ⅰ	98	—	—	—	—						
				水位	Ⅰ		—	—	—	—						
				高精度水温	Ⅰ		—	√	—	—						
				HCO_3^-	Ⅱ		—	—	—	—						
				F^-	Ⅱ		—	—	—	—						
				Ca^{2+}	Ⅱ		—	—	—	—						
				Mg^{2+}	Ⅱ		—	—	—	—						

续表

序号	台站（点）名称	经纬度		观测项目	资料类别	震中距（km）	按震中距Δ进行异常同基									
							0 < Δ ≤ 100km					100 < Δ ≤ 200km				
		λ（°）	φ（°）				L	M	S	I	U	L	M	S	I	U
8	泸水	98.82	25.99	水氡	Ⅰ	104										
				水位	Ⅱ											
				高精度水温	Ⅰ											
9	漾濞	99.96	25.66	水氡	Ⅱ	109									√	
				地磁	Ⅱ											
11	下关	100.22	25.60	水氡	Ⅰ	128										
				水位	Ⅰ											
				高精度水温	Ⅰ											
				Hg	Ⅰ											
				CO_2	Ⅱ											
				短水准	Ⅰ											
				短基线	Ⅰ											
12	云县	100.12	24.56	地倾斜	Ⅱ	129										
13	陇川	97.96	24.37	水位	Ⅱ	140										
				水温	Ⅱ											

续表

序号	台站（点）名称	经纬度		观测项目	资料类别	震中距（km）	按震中距 Δ 进行异常同基									
							0＜Δ≤100km					100＜Δ≤200km				
		λ（°）	φ（°）				L	M	S	I	U	L	M	S	I	U
14	弥渡	100.49	25.35	水氡	Ⅱ	142									√	
				Hg	Ⅱ											
				CO_2	Ⅱ											
				高精度水温	Ⅱ											
				地倾斜	Ⅱ											
				重力	Ⅱ											
15	南涧	100.52	25.06	水氡	Ⅱ	144										
16	洱源	99.95	26.12	水氡	Ⅰ	146									√	
				水位	Ⅰ											
				高精度水温	Ⅰ									√		
				Hg	Ⅰ											
				CO_2	Ⅱ											
18	临沧	100.08	23.89	水氡	Ⅱ	170										
				水位	Ⅱ											
				高精度水温	Ⅱ											
				F^-	Ⅱ											
19	瑞丽	97.85	24.01	水氡	Ⅱ	174										

续表

序号	台站（点）名称	经纬度		观测项目	资料类别	震中距（km）	按震中距 Δ 进行异常同基									
							0 < Δ ≤ 100km					100 < Δ ≤ 200km				
		λ（°）	φ（°）				L	M	S	I	U	L	M	S	I	U
20	剑川	99.90	26.53	水位	Ⅱ	180										
				短基线	Ⅰ											
				短水准	Ⅰ											
21	景东	100.83	24.46	水氡	Ⅱ	191										
				水位	Ⅱ									√		
				高精度水温	Ⅱ											
22	双江	99.82	23.47	水位	Ⅰ	194										
23	鹤庆	100.17	26.55	水氡	Ⅱ	200										
				二氧化碳	Ⅱ											
分类统计	台项	异常台项数					0	6	10	1		0	0	2	3	
		台项总数					47	47	47	47	/	51	51	51	51	/
		异常台项百分比/%					0	13	21	2	/	0	0	4	6	/
	观测台站（点）	异常台站数					0	5	5	1		0	0	2	3	
		台站总数					7	7	7	7	/	14	14	14	14	/
		异常台站百分比/%					0	71	71	14	/	0	0	14	21	/
	测项总数						17					21				
	观测台站总数						7					14				
备注																

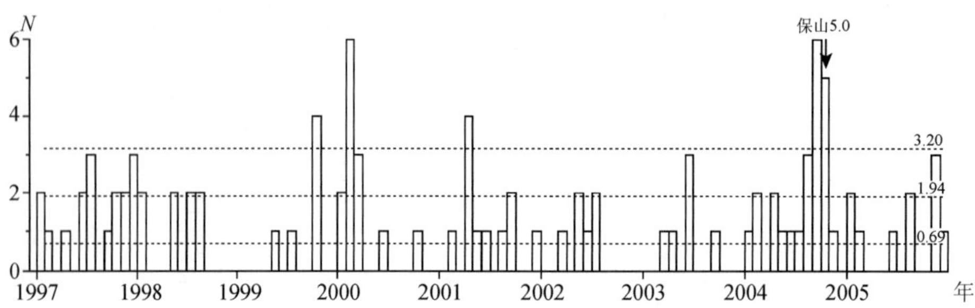

图 13　震中周围 $M \geqslant 3.0$ 级地震震中分布图

Fig. 13　Distribution of $M \geqslant 3.0$ epcenters around the main shock（1997. 1. 1-2005. 12. 31）

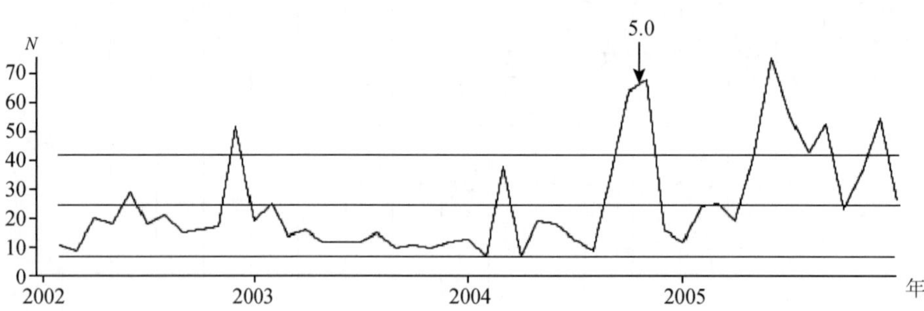

图 14　震中附近 $M \geqslant 1.0$ 级地震月频度曲线图

Fig. 14　The N-t curve of $M \geqslant 1.0$ earthquakes around the mainshock

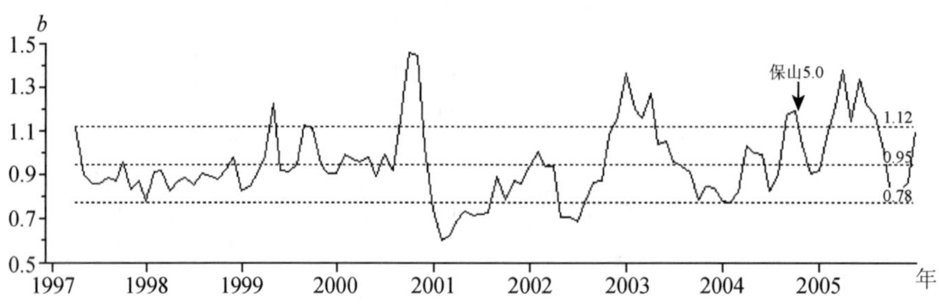

图 15　云南地区 b 值（M2. 5～4. 9）曲线图

Fig. 15　The b-t curve of M2. 5-4. 9 earthquakes in the Yunnan area

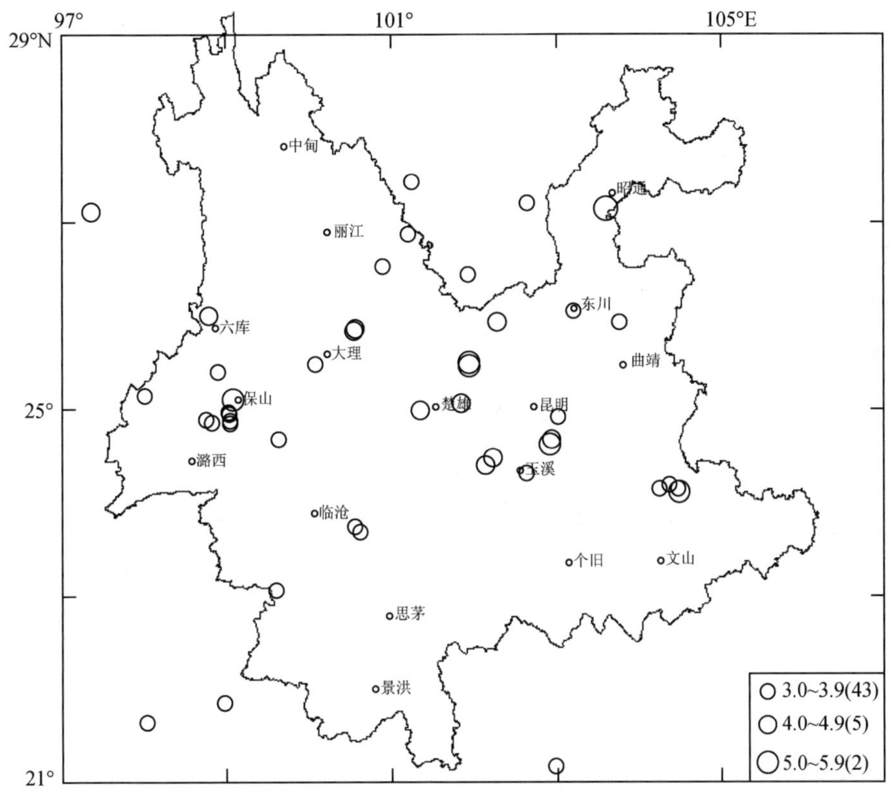

图 16 保山地震震前 3 个月云南地区 3 级地震震中分布图

Fig. 16 Distribution of $M \geqslant 3.0$ earthquakes in Yunnan 3 months before Baoshan $M_S5.0$ happened

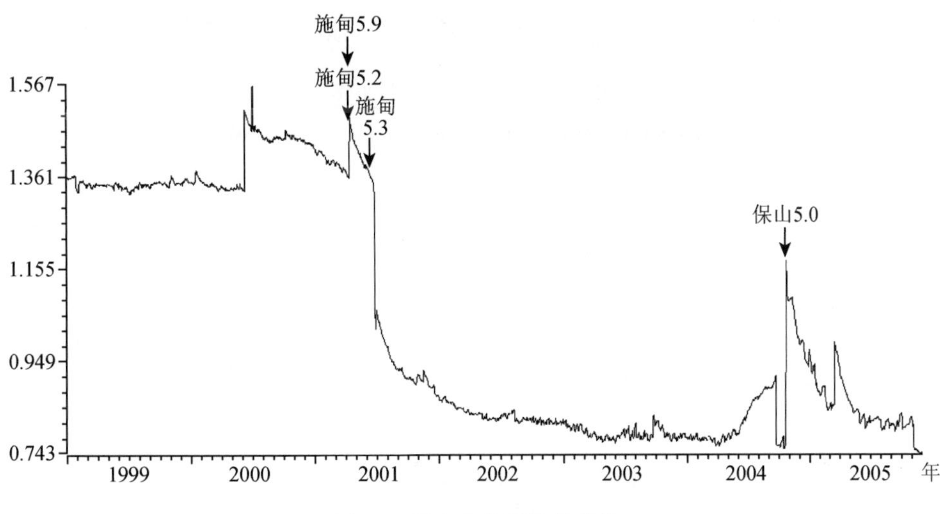

图 17 保山水位日均值曲线

Fig. 17 Daily mean value of water level t in groundwater at Baoshan station

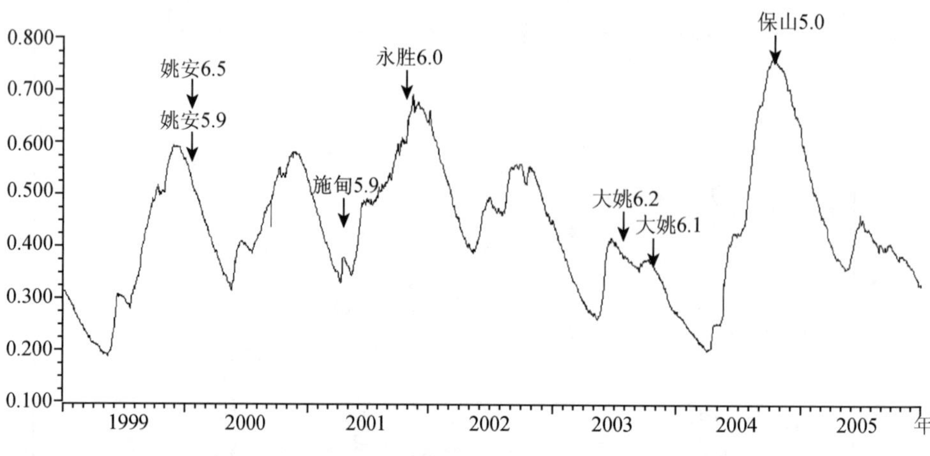

图 18　施甸水位日均值曲线

Fig. 18　Daily mean value of water level t in groundwater at Shidian station

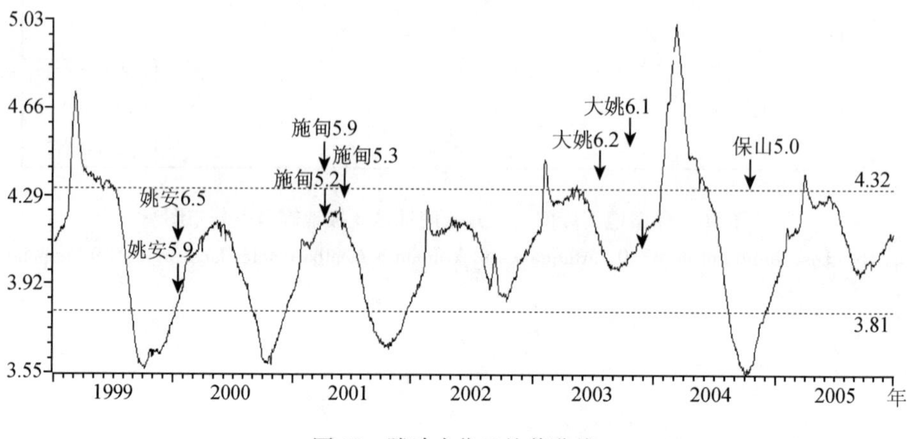

图 19　腾冲水位日均值曲线

Fig. 19　Daily mean value of water level t in groundwater at Tengchongn station

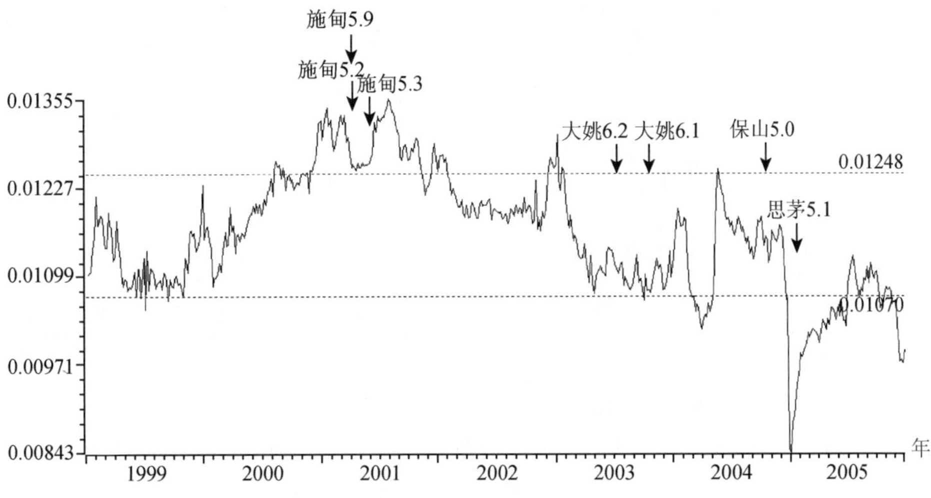

图 20　龙陵流量五日均值曲线

Fig. 20　5-day mean value of water level t in groundwater at Tengchong station

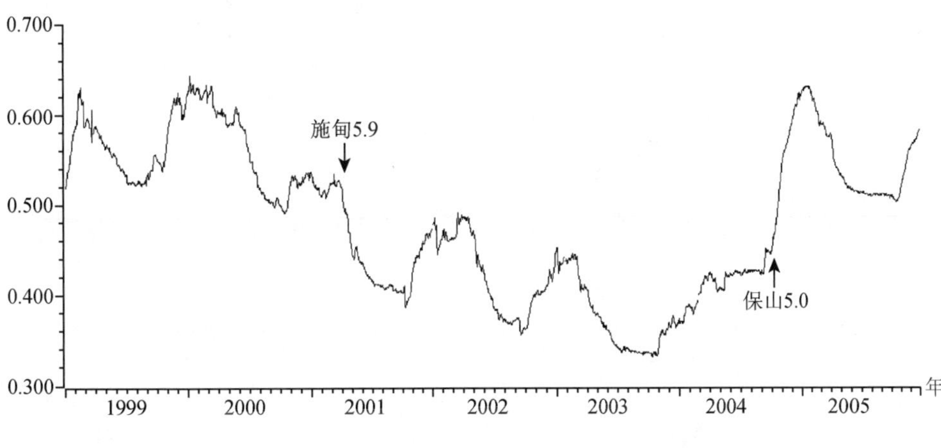

图 21　景东水位日均值曲线

Fig. 21　Daily mean value of water level t in groundwater at Jingdong station

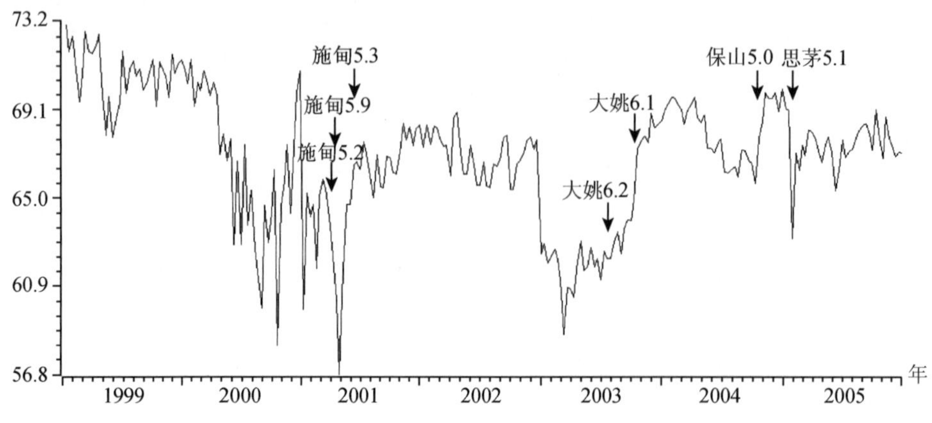

图 22　昌宁水氡旬均值曲线

Fig. 22　10-day mean value of radon content in groundwater ate Changning station

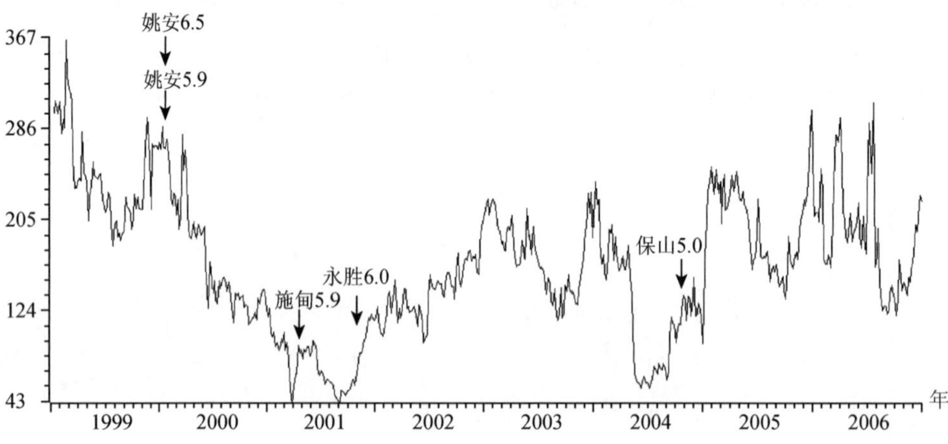

图 23　龙陵水氡五日均值曲线

Fig. 23　5-day mean value of radon content in groundwater ate Longling station

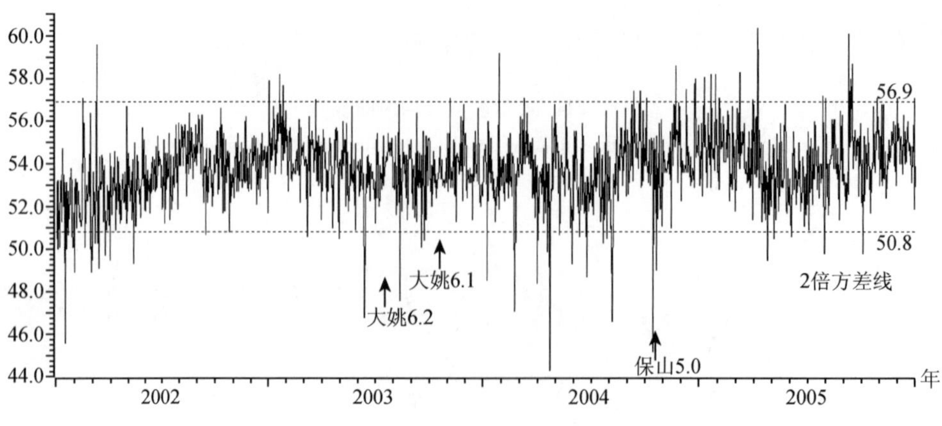

图 24　漾濞水氡日均值曲线

Fig. 24　Daily mean value of radon content in groundwater ate Yangbi station

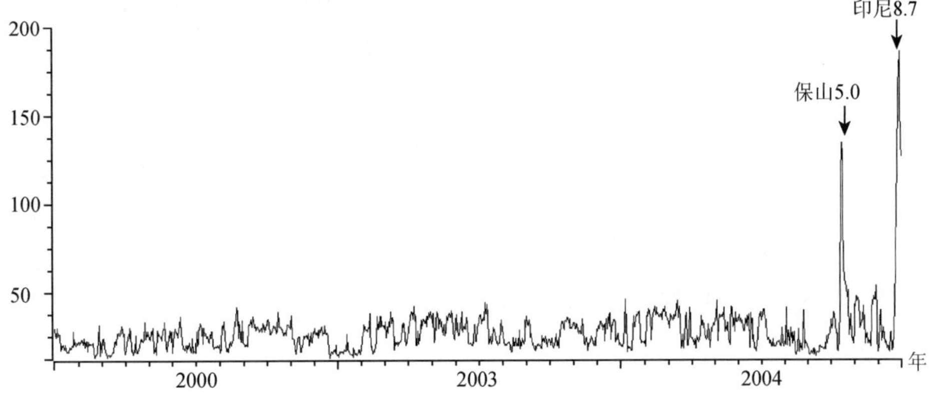

图 25　洱源水氡日均值曲线

Fig. 25　Daily mean value of radon content in groundwater at Eryuan station

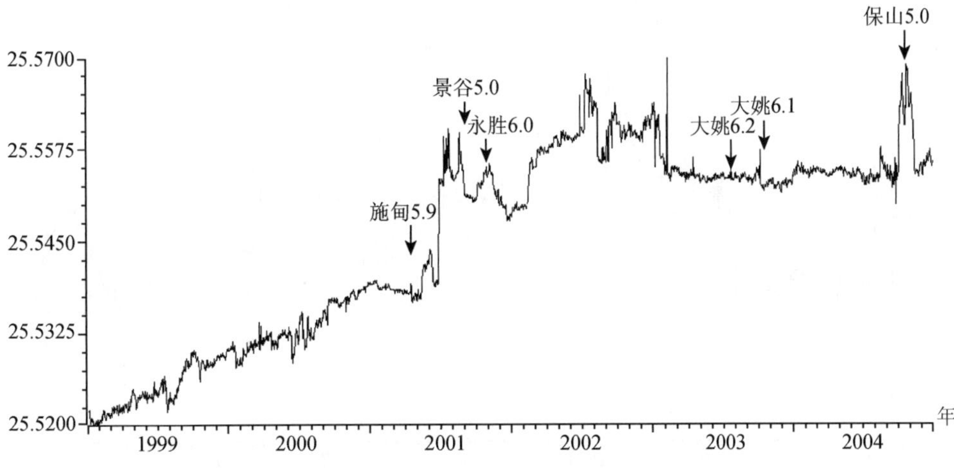

图 26　保山水温日均值曲线

Fig. 26　Daily mean value of water temperature at Baoshan station

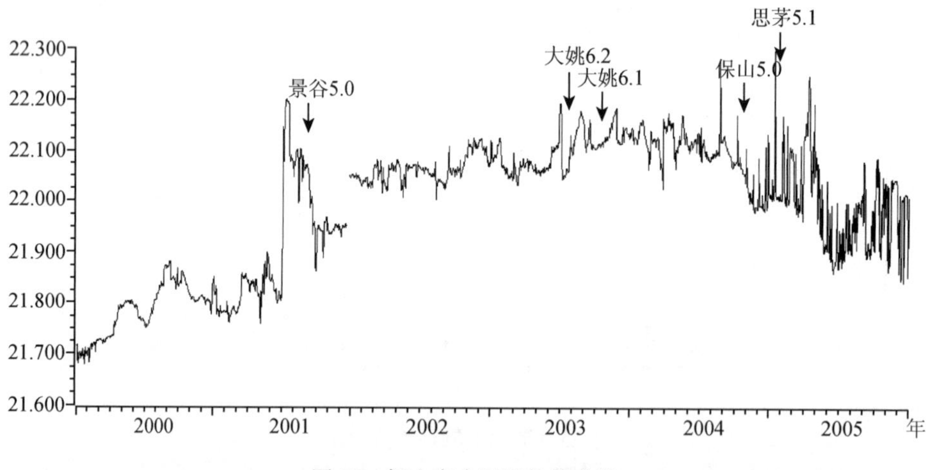

图 27　保山市水温日均值曲线

Fig. 27　Daily mean value of water temperature at Baoshan city station

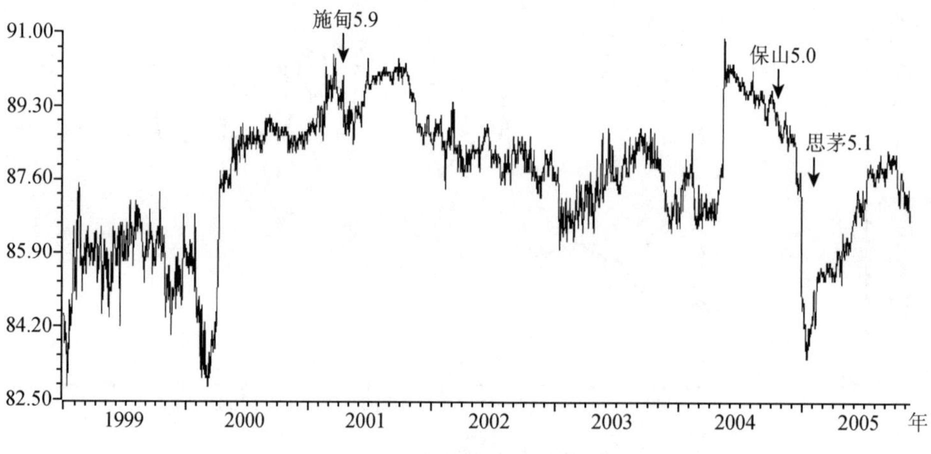

图 28　龙陵水温日均值曲线

Fig. 28　Daily mean value of water temperature at Longling station

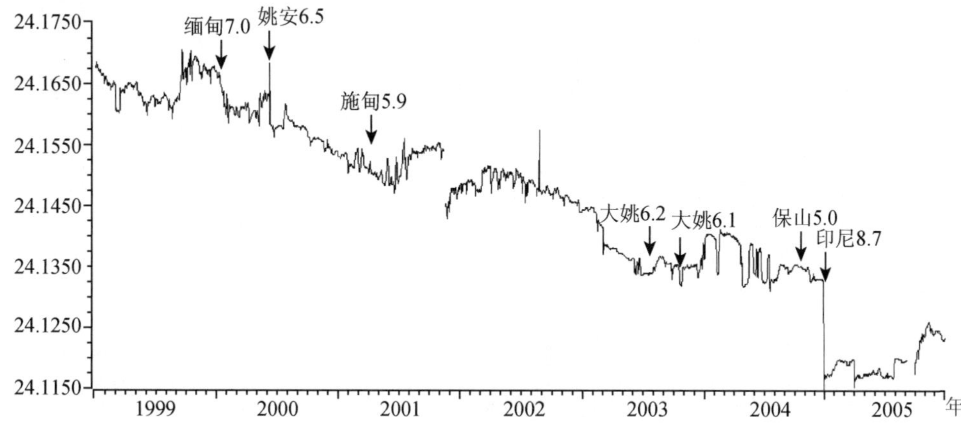

图29 德宏水温日均值曲线

Fig. 29 Daily mean value of water temperature at Dehong station

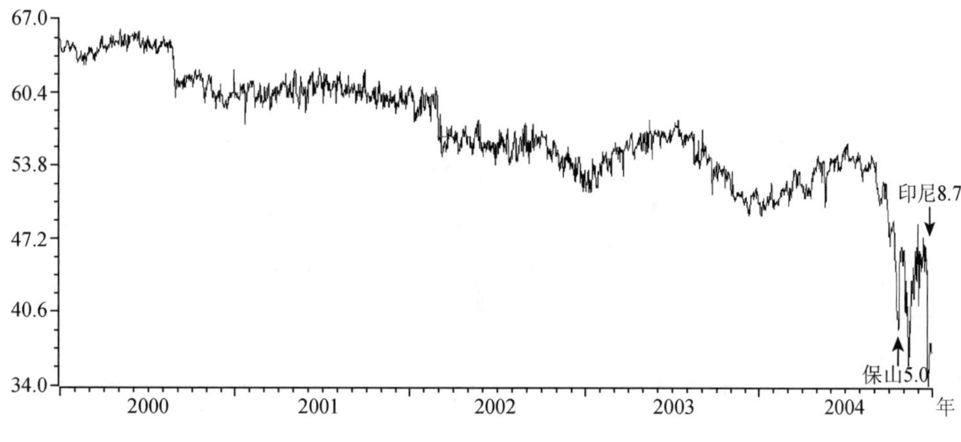

图30 洱源水温日均值曲线

Fig. 30 Daily mean value of water temperature at Eryuan station

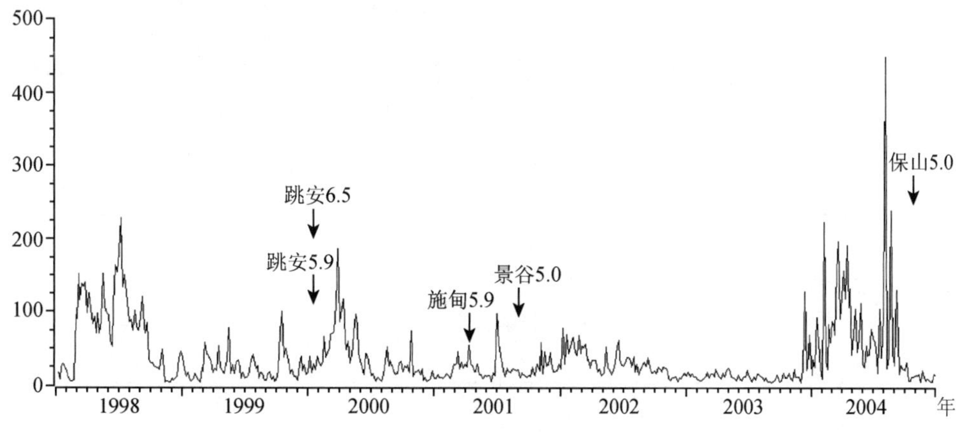

图31 保山水汞五日均值曲线

Fig. 31 5-day mean value of mercury content in groundwater at Baoshan station

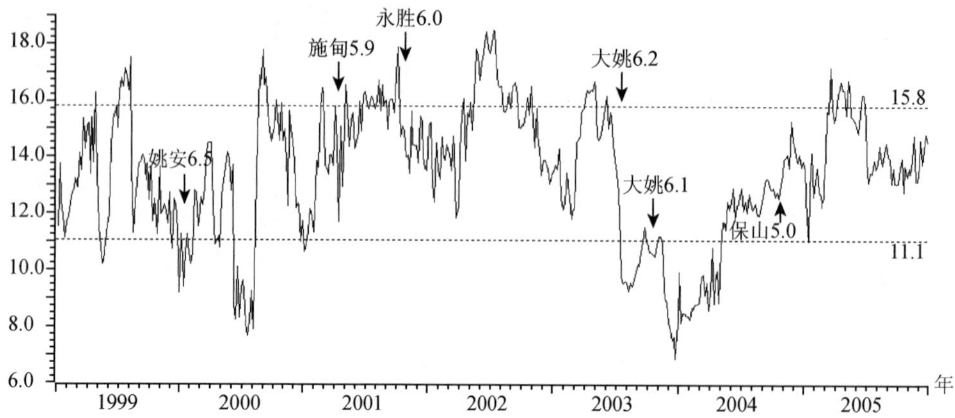

图 32　保山硫酸根离子五日均值曲线

Fig. 32　5-day mean value of SO_4^{2-} content in groundwater at Baoshan station

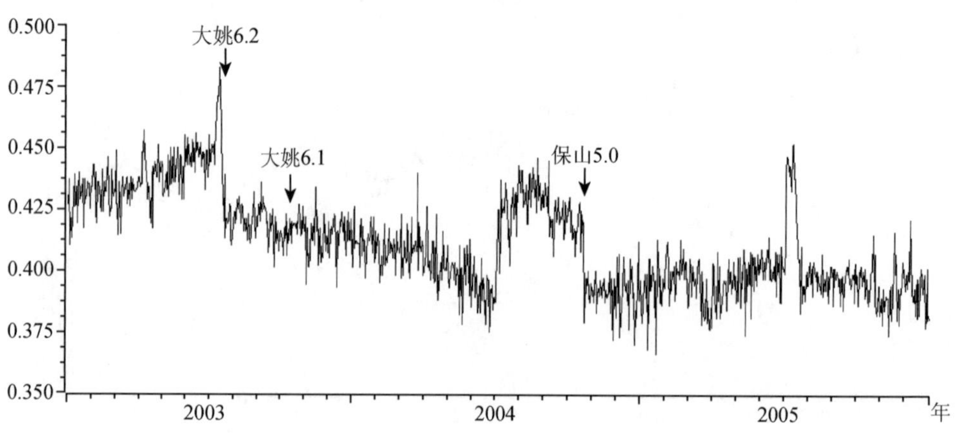

图 33　保山氟离子日均值曲线

Fig. 33　Daily mean value of F^- content in groundwater at Baoshan station

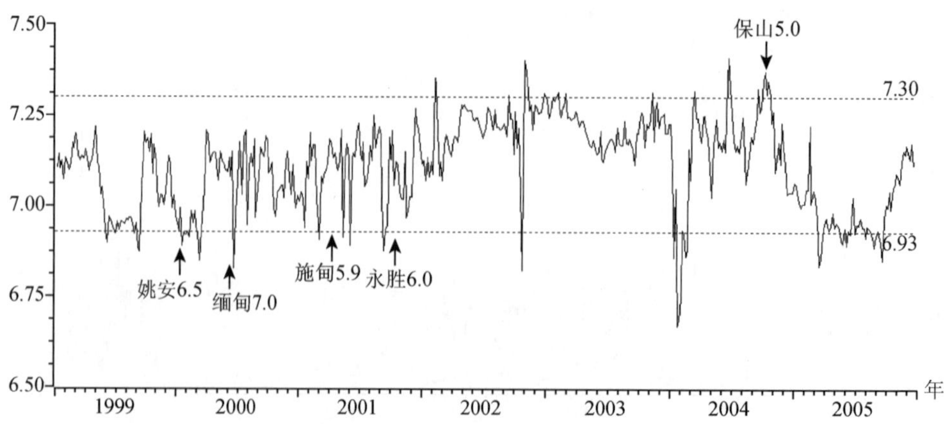

图 34　施甸 pH 值五日均值曲线

Fig. 34　5-day mean value of pH in groundwater at Shidian station

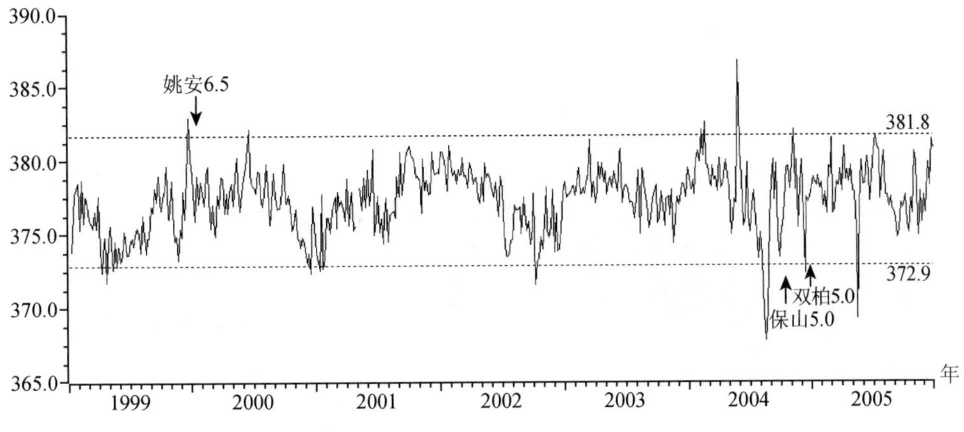

图 35　施甸碳酸根离子五日均值曲线

Fig. 35　5-day mean value of HCO_3^- content in groundwater at Shidian station

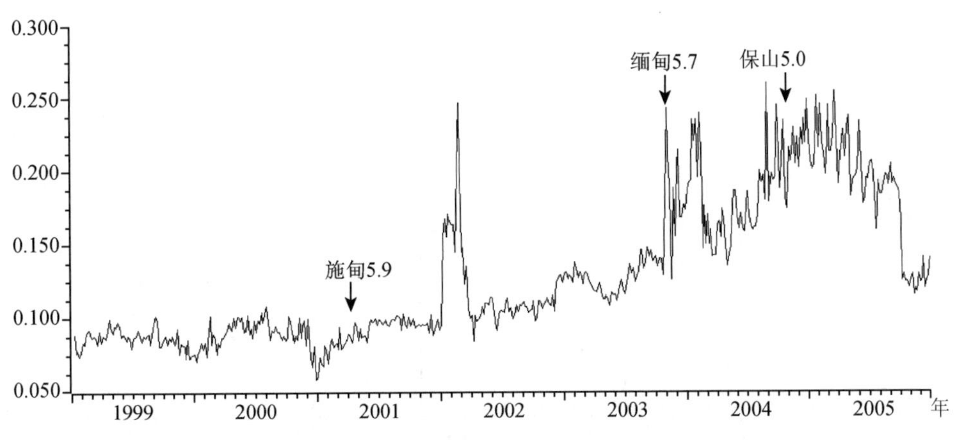

图 36　施甸氟离子五日均值曲线

Fig. 36　5-day mean value of F^- content in groundwater at Shidian station

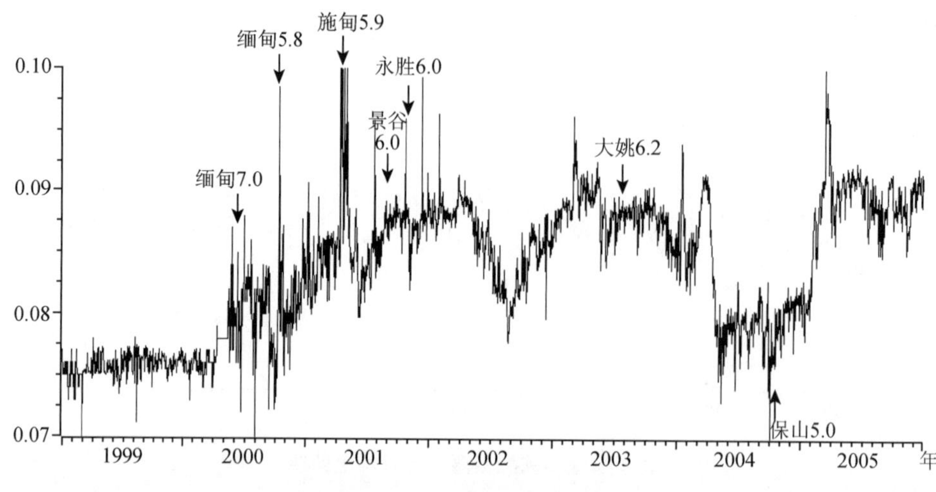

图 37　昌宁电导率日均值曲线

Fig. 37　Daily mean value of geoelectricity at Changning station

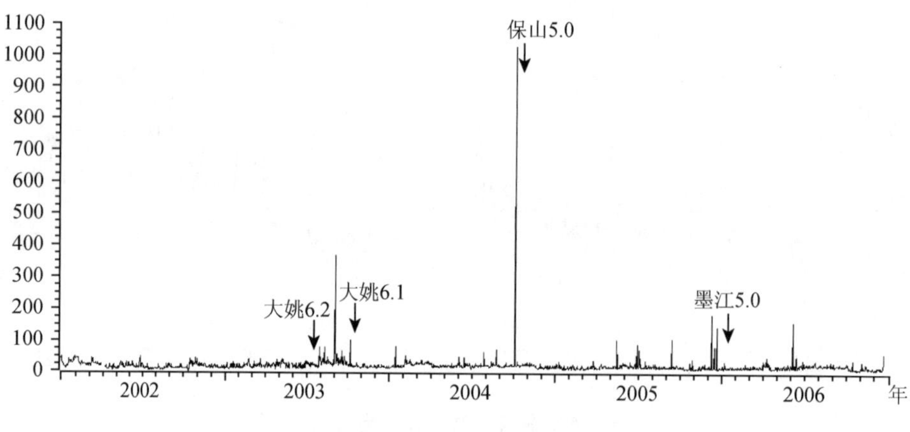

图 38　弥渡水汞日均值曲线

Fig. 38　Daily mean value of mercury content in groundwater at Midu station

七、前兆异常特征

综上所述，保山 5.0 级地震前观测到较丰富的前兆异常现象，异常主要特征如下：

1. 异常数量多

地震前共出现地震学前兆、定点前兆及宏观异常三大类观测项目 29 项异常。其中定点前兆异常尤为显著，震前 11 个台站相应观测到了水氡、水质、水位、水等 4 个观测项目。其中水氡异常 4 项，水位、水温异常均为 5 项，水质异常最多，共 8 项。此外，震前震中附近地震活动异常变化显著。

据文献［2］，云南地区 5 ~5.9 级地震定点前兆异常项目数与异常台项数分别为 3.4 个项目和 4.5 个台项，相比之下，这次地震前兆异常数量明显偏多。

2. 异常持续时间短

震前地震学、定点前兆异常中期异常较少，26 项异常中（不包括宏观异常）仅有 6 项为中期异常，持续时间 5 ~9 个月，主要表现为水氡持续低值异常，水位、水质、水温大幅升降变化。如保山水汞自 2003 年 12 月以来一直持续高值突跳异常，异常持续 9 个月后，在保山 5.0 级地震前一个月结束。

3. 短临突变性异常突出

按时间发展的进程把异常分为临震异常（震前 1 月内）、短期异常（震前 1 ~6 个月）和中期异常（震前 0.5 ~5 年）。则本次保山 5.0 级地震前出现的 26 条异常，临震异常 4 项，短期异常 16 项，中期异常 6 项，短期、临震异常占异常总数的 77%。测震学异常均为短期异常，表明震前地震活动增强现象明显。另外短期异常还有部分水位破年变，水氡持续高/低值异常及水温、水质突跳异常明显。另外震前 1 个月的临震异常，主要表现为水位、水氡、水质的大幅单点突跳。

4. 异常空间集中

保山地震震中 200km 范围内有 8 个台站观测到了显著的异常信度较高的前兆异常，其中有 5 个台站位于震中 100km 范围内，占 200km 范围内总异常前兆台站的 67% （5/8）。其中异常台项为 17 项次，达总前兆异常台项的 77% （17/22），由此表明，本次保山地震前兆异常主要集中分布在震中附近地区，尤其是距离震中较近的保山、施甸、龙陵三个台站观测到了大量的短临异常，对地震地点的预测具有重要的指示意义。

5. 宏观异常

宏观异常出现的时间距离保山地震发震时间分别为 7、2、4 个月，且距离保山地震震中范围也均超过了 100km，对地震时间和地点判断有一定影响。

6. 部分异常可能为印度尼西亚地震的影响

在表 5 中所列的部分异常，有些在保山地震后异常仍然持续，直至 2004 年 12 月苏门答腊 8.7 级地震后异常才趋于结束，这种异常可能是受到苏门答腊地震的影响。如洱源水温异常虽然在保山地震前出现，但保山地震后异常幅度反而更大。另外有部分异常的幅度相当大，是一个 5 级地震所不能解释的，如保山水汞、景东水位、施甸水位等，这种大幅度的异常要用 6 级以上地震地震才能解释，故也不能排除是远场大地震影响所致。

八、震前预测、预防和震后响应

1. 预测情况

这次保山震群前，云南省地震局对该地区的地震危险性作了较好的中、短预测预报工作[1]。

中期预报：2003 年 11 月云南省地震局的《云南省 2004 年地震趋势研究报告》指出：2004 年度云南滇西华坪—剑川—永平—下关—永仁一带有为 6 级左右地震危险。中国地震局于 2003 年 12 月下旬在 2004 年的全国地震趋势会商会上，把滇西华坪—剑川—永平—下关—永仁一带列为 2004 年度全国地震重点危险区。保山 5.0 地震靠近该危险区的南边缘仅 30km。

短期预报：2004 年 6 月云南省地震局的《云南省 2004 年年中地震趋势研究报告》根据保山一带小滇西德地震活动和前兆变化，将保山一带列为 2004 年下半年注意地区。2004 年 8 月鲁甸 5.6 级地震后，保山施甸一带小震频度增强，通过对地震活动和前兆的动态跟踪结果，当时分析认为“在施甸及其邻近地区发生 5.5 级以上地震的可能性不大，但不能排除该区发生 5 级地震的可能”的结论。2004 年 7 月 23 日以云南省地震局《震情反映》（2004 年 7 期）上报中国地震局及云南省委、省政府，明确提出：未来 3 个月省内存在发生 5 ~ 6 级地震的危险，发震危险区仍以滇西地区、玉溪及其周边地区的可能性最大。保山 5.0 地震靠近第一个危险区的南边缘。

本次地震的预测成功，受到云南省委、省政府和中国地震局的表彰和奖励。

2. 震后响应与余震趋势判定

保山 5.0 地震发生后，云南省地震局立即启动地震应急预案，07:20 派出地震现场工作队奔赴震区，13:50 到达现场；中国地震局派出 3 人专家组于次日凌晨 2:30 到达保山。中国地震局、云南省地震局以及保山市地震局的领导和专家组成地震现场工作指挥部，下设秘书、震情监测预报、地震灾评及科考、强震观测、通信及后勤保障等 6 个组，共 33 人开展地震现场工作。10 月 19 ~ 24 日，共计 6 天，12 名灾评人员分乘 9 辆车对云南省保山市隆阳区和施甸县进行灾情调查，行程 1.5 万余千米，共完成 83 个调查点（图 1）的调查工作，其中抽样点 48 个。灾害评估组用 6 天时间完成了 48 个抽样点的野外调查工作，并评定了烈度，编写完成《保山 6.0 级地震灾害损失评估报告》，并上报省人民政府和中国地震局。震情监视组每天跟踪震情形势，召开地震现场紧急会商会，提出明确分析意见及建议，以《震情简报》上报云南省地震局，并通报保山市委、市政府。成功实现了现场强余震的临震预报，得到了当地政府充分肯定和高度评价。

九、总结与讨论

通过本震例研究，我们得出下述重要认识：

（1）2004 年 10 月 19 日保山 5.0 级地震是继 2001 年 4 月 10 日、12 日，6 月 8 日云南省施甸 5.2、5.9、5.3 级地震后保山地区发生的又一次破坏性地震。其震区所处趋于为双震或震群发生率较高的地区，以往该地区双震时间间隔有长有短，如：1976 年龙陵 7.3、7.4 级双震仅间隔 1 小时 37 分，1991 年施甸 5.0、5.1 级双震间隔 21 天，间隔较长。2001 年施甸 5 级震群的间隔最短为 2 天，最长为 57 天，概括了该地区双震间隔长、短分布的特征。本次 2004 年 10 月 19 日保山 5.0 级地震后间隔 79 天后震区西侧 24km 处再次发生 4.8 级地震。本次 2004 年 10 月 19 日保山 5.0 级地震前序列小震活动均出现平静—活动—发震的动态变化，小震活动增强、前兆项数增加、前兆异常幅度增加对地震预报有很好的主导作用。本次中强地震活动的特点可能是该地区中强震群活动特征的典范，对其以后震群序列的中强地震短临预测预报有一定的借鉴[3]。

（2）从保山地震震前所出现的短临地震活动性异常来看，震前小震活动增强、前兆项数增加、前兆异常幅度增加、是保山地震前短临重要标志。

（3）从保山地震前前兆变化来看，地震发生在前兆台网较密集、观测手段较齐全的小滇西地区，前兆异常相对较丰富，在保山地震震中附近的前兆异常开始得较晚，随着地震的临近，异常测项数量及异常范围都不断增加，地下流体中水质异常更为突出；宏观异常全部分布在 100km 范围外，对地震地点判断有一定影响。震前多项目大幅度的突发性异常是本次地震短临阶段前兆异常最显著的特征，表明在前兆台网较密集、观测手段较齐全的地区，可以观测到 5 级强震在孕震短临阶段能观测到信度较高，异常幅度较直观、醒目的前兆异常变化，这对 5 级强震的预测提供了重要经验。地震前突出的前兆异常在空间范围分布上集中于震中区附近，位于与地震震中相邻构造上的观测项目异常尤为显著，这为应用定点前兆异常资料对未来主震发生地点的最后圈定提供了可行的途径。

由于 2004 年 10 月 19 日保山 5.0 级地震距离 2004 年苏门答腊 8.7 级巨震仅 2 个多月，有些在保山地震后异常仍然持续，直至苏门答腊地震后异常才趋于结束，这种异常可能是由于苏门答腊地震的影响。另外有部分异常的幅度相当大，是一个 5 级地震所不能解释的，如保山水汞、景东水位、施甸水位等，这种大幅度的异常要用 6 级以上地震地震才能解释，故也不能排除是远场大地震影响所致。故苏门答腊地震的远场效应，使得在本次震例研究工作中，异常的识别相当困难。

参 考 文 献

[1] 刘翔等，2001 年 4 月 12 日施甸震群特征 [J]，地震研究，2003，9（1）

[2] 罗平等，云南地区地震预报判据和指标研究，地震预报方法实用化研究文集（综合预报专辑），北京：地震出版社，1991

[3] 中国地震局，地震现场工作大纲和技术指南，北京：地震出版社，1998

参 考 资 料

1）国家地震局震害防御司，中国历史强震目录，北京：地震出版社，1995

2）云南省地震局，云南月报目录

3）云南省地震局科技档案室，2004 年 10 月 19 日保山 5.0 级地震灾害损失评估报告，2004.10.26

4）云南省地震局，震情反映，第 7、8、9 号，2004 年

Baoshan Earthquake of M_S 5.0 on October 19, 2004 in Yunnan

Abstract

On Oct. 19, 2004, an earthquake of M_S5.0 occurred in Baoshan district, Yunnan province. The macroscopic epicenter was located in Zhangjia mountain in Hanzhuang country in Longyang district. The intensity in the meizoseismal area was Ⅵ. The shape of isoseismic line was elliptic witgh major axis in NW direction. 2 were seriously injured, and 13 people were slightly injured in this earthquake. The direct economic loss reached 217 million Yuan.

The sequence was of doubleshock type, a M_L4.8 earthquake happened in Jan. 7, 2005, in the left of the M_S5.0 earthquake. The major axis of sequence area is in the NW direction consistend with the meizoseismal area strike. The sode plane Ⅱ of its focal mechanism solution is the main rupture plane with NW direction principal compressive axis. Its seismogenic geological structure was Shiluo-Yinchuan fault in NNW direction.

There were 23 seicmic stations around this earthquake, among them, there were 11 seicmometric stations and 21 precursory observation stations.

There were 29 anomalies with 4 items of seismometric anomalies, 22 precursory observation items, and 3 macroscopic anomalies.

The Baoshan M_S5.0 happened in the hazard zone of the midyear report lodgged by the Earthquake Administration of Yunnan Province. Before the earthquake, according to the anomalies the Baoshan district was considered as an 5-6 earthquake hazard zone which bring some extend of social benefit.

2004 年 12 月 26 日云南省双柏 M_S5.0 地震

云南省地震局

钱晓东　王世芹

摘　要

2004 年 12 月 26 日，云南省双柏县发生 5.0 级地震，宏观震中位于双柏县妥甸镇一带，极震区烈度为Ⅵ度，呈近圆形状，地震造成 1 人死亡、1 人重伤，直接经济损失 4070 万元。

此次地震序列为主震—余震型，最大余震 4.1 级，余震分布在主震西北侧，无明显优势分布，与极震区烈度分布相吻合，节面Ⅰ为主破裂面，主压应力 P 轴方位 NNW 向，地震的发震构造为 NWW 向的王家村断层。

震中周围 200km 范围内共有地震台站 43 个，其中测震台 7 个，定点前兆观测台站 36 个，震前共出现 7 个异常项目 10 条前兆异常，测震学 4 条异常，定点前兆 6 条异常，均为短临异常。

前　言

2004 年 12 月 26 日 15 时 30 分云南省双柏县发生 5.0 级地震，地震发生后云南省地震局派出工作组赴震区开展震害评估、地震地质考察和余震监视工作。据云南省地震台网测定，微观震中位于北纬 24.7°，东经 101.5°。宏观震中位于双柏县妥甸镇马龙—苎麻地一带，极震区烈度为Ⅵ度，微观震中和宏观震中相距 6km。地震造成 1 人死亡、1 人重伤，直接经济损失 4070 万元[1)]。

震区附近历史地震较少，区内最强的地震是 1680 年楚雄 6¾级地震，2001 年楚雄 5.3 级地震与本次地震相距 30km。

震中周围 200km 范围内共有地震台站 43 个，其中测震台 7 个，定点前兆观测台站 36 个，地震前共出现 7 个异常项目的 10 条前兆异常，测震学出现了 4 条异常，定点前兆出现了 6 条异常，均为短临异常。

一、测震台网及地震基本参数

双柏 5.0 级地震前，震中周围 200km 范围内共有有人值守测震台和电信传输测震台 24

个，震中附近地区遥测台和区域台对地震的监控能力分别为 $M_L \geqslant 1.5$ 和 $M_L \geqslant 1.0$ 级。这次地震的基本参数和序列目录采用云南地震台网资料，地震的基本参数见表 1。

图 1 给出了双柏 5.0 震中 200km 范围内的地震台分布，在研究时段内基本可达到 $M_L \geqslant 1.5$ 级地震不遗漏。100km 内有双柏、楚雄、一平浪、易门和安宁 5 个测震台；100～200km 有 19 个测震台。

表 1 列出了不同来源给出的这次地震基本参数，经对比分析，认为云南省地震局正式目录的震中位置更为精确，因此，此次地震基本参数取表 1 中编号 1 结果。

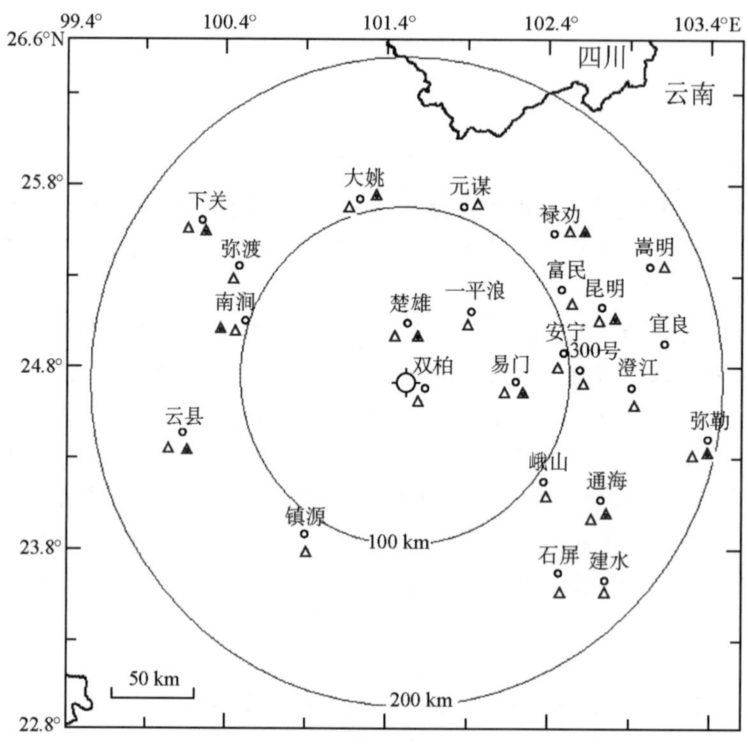

图 1　双柏 5.0 级地震前测震台网分布图

Fig. 1　Distribution of seismometric stations before the M_S5.0 Shuangbai earthquake

表 1　地震基本参数

Table 1　Basic parameters of the earthquake

编号	发震日期	发震时刻	震中位置		震级	震源深度	震中地名	结果来源
	年 月 日	时 分 秒	φ_N	λ_E	M_S	(km)		
1	2004 12 26	15 30 09	24°42′	101°32′	5.0	7	双柏	资料 2)
2	2004 12 26	15 30 09	24°43′	101°32′	5.0	7	双柏	资料 3)
3	2004 12 26	15 30 10	24.45°	101.30°	5.0	15	双柏	中国地震目录

二、地震地质背景

双柏地震震区在地质构造体系中处于云南山字形构造西翼内侧马蹄形盾地北部，西翼受青、藏、滇、缅歹字形构造东支中断影响。这样使西翼构造活动性加强，北西向断裂、褶皱发育，主压应力为南西—北东向；东部为川滇经向构造体系绿汁江断裂带，发育着近南北向褶皱、断裂以及派生入字形断裂；而中、北部为反时针旋卷构造。震区基底由下元古界上昆阳群构成，盖层全由中生界组成，下部为上三迭统，向上则变为侏罗系、白垩系巨厚的陆相红色建造。构造变动主要为燕山运动，共分为三期，早期表现为地壳升降运动；中期东部具褶皱运动性质，西部仍为升降运动；晚期，即四川运动使整个盖层全面发生褶皱、断裂，部分前期构造也有所复活，震区中北部形成 10 余个向斜盆地及断陷盆地，从此奠定了本区的构造雏形。喜山运动以大面积差异性上升为主。第四纪以来，震区地壳显示了间歇性上升，并迄今地壳仍在上升。震区构造的分区性比较明显，震中附近主要有会基关双柏穹隆褶皱区、南华扭动褶皱区和马龙河褶断区[4,5)]。

图 2 给出双柏附近地区主要断裂及历史地震分布。震区断裂并不发育，主要有邑舍河断层、三街断层、小箐河断层、王家村断层。震区属于扬子准地台川滇台背斜的中南段。震区

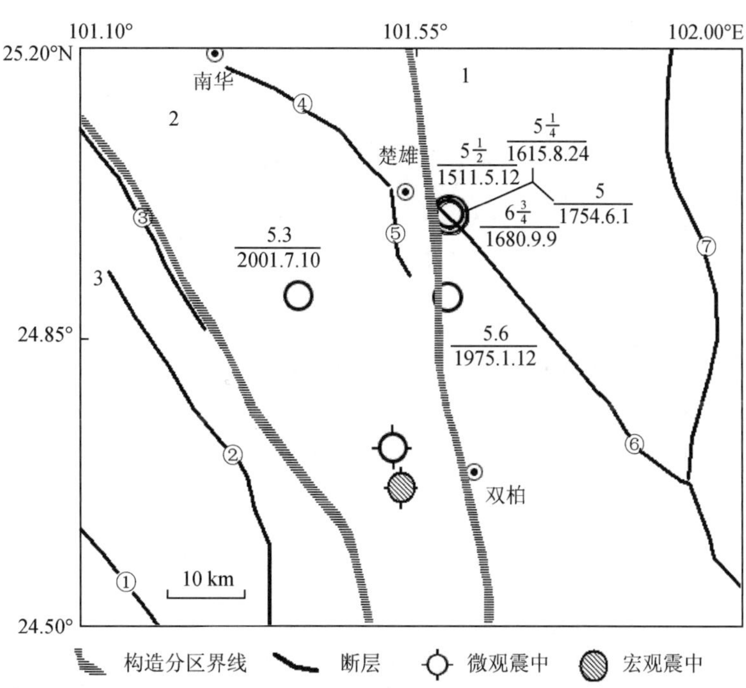

图 2 双柏附近地质构造与历史地震震中分布图

Fig. 2 Map of fault structure and distribution of historical earthquakes around Shuangbai area

①三街断层；②邑舍河断层；③小箐河断层；④南华—楚雄断层；
⑤王家村断层；⑥楚雄—建水断层；⑦元谋断层

1. 会基关双柏穹窿褶断区；2. 南华扭动褶断区；3. 马龙河褶断区

内分布有会基关双柏穹窿褶断区、南华扭动褶皱区和马龙河褶断区。双柏震区位于南华扭动褶皱区东缘，以北北西向为长轴的褶皱和部分倾伏背斜为主，区内分布有山前背斜、阿白厂背斜。区内的王家村断层为南华—楚雄断裂的一部分，走向316°，倾向西，倾角大于60°，西盘上升，为逆断层，全长21km。

震区附近（24°30′～25°12′N，101°6′～102°E）历史地震较少，自公元886年以来共记载到5级以上地震6次，其中5～5.9级5次，6～6.9级1次。区内最强的地震是1680年9月9日楚雄6¾级地震，距离2004年12月26日双柏5.0级地震最近的一次地震是2001年7月10日楚雄5.3级地震，相距30km。

三、烈度分布及震害

经考察[1)]，双柏5.0级地震的宏观震中位于双柏县妥甸镇马龙—苎麻地一带，宏观震中为（24°40′N，101°33′E）。极震区烈度为Ⅵ度（图3）。宏观震中与微观震中相距6km。

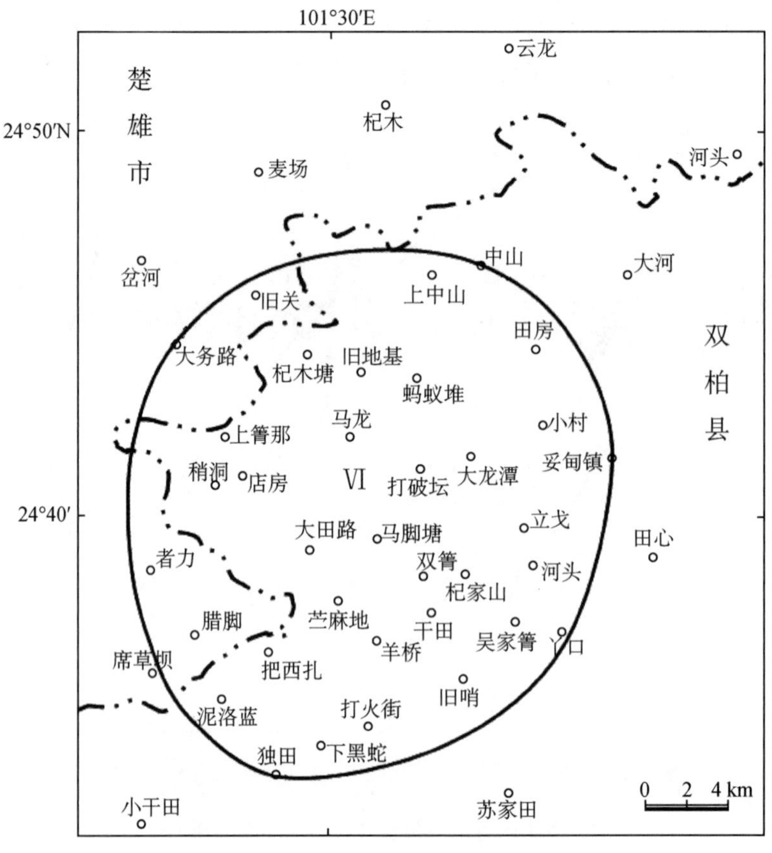

图3 双柏5.0级地震烈度分布图

Fig. 3 Isoseimal map of the 5.0 Shuangbai earthquake

Ⅵ度区的主要特征为：北自双柏县妥甸镇中山村，南至双柏独田乡政府驻地，东自双柏县城所在地妥甸镇，西达楚雄市大地基乡者力村西，面积约 493km^2。等震线形状近圆形。地震造成少数土木结构建筑严重破坏，框架结构建筑仅有轻微破坏，砖混结构建筑少数出现中等或轻微破坏，砖木结构绝大部分出现轻微破坏。

地震造成了震区生命线工程的破坏：供排水系统的破坏，主要表现为供水管道断裂、接头拉脱，毛石砌筑供水管道拉裂、漏水；净化池、蓄水池局部开裂渗水；水厂值班室房屋墙体开裂。交通系统的破坏，主要表现为个别桥梁拱圈纵向开裂，涵洞的涵台、拱圈纵向开裂，路基下沉、边坡崩塌，道班房墙体及围墙开裂等。电力系统的破坏，震区通信系统的线路和房屋遭受一定程度的破坏，线路破坏主要表现为线杆歪斜、移位，明线拉断等；房屋破坏主要表现为变电所、供电所墙体、地板、天花板开裂。水利设施的破坏，水利设施主要有水库、坝塘、引水沟渠、涵洞、供水工程等。此次地震造成了部分水利设施的损坏，主要表现为水库坝体开裂，坝坡局部变形，涵洞洞壁开裂渗水，沟渠局部垮塌，部分人畜饮水工程损坏。

此次地震的震害波及范围广，楚雄市新村镇的民房、校舍、其他公用房屋、生命线工程及水利设施（主要为水窖）也遭受破坏。此次地震灾区涉及楚雄州双柏县和楚雄市，包括 5 个乡（镇），12 个村；受灾人口 35715 人，涉及 9343 户。地震中 1 人死亡，1 人重伤，9343 户、35715 人受灾，造成直接经济损失为 4070 万元。

四、地 震 序 列

双柏地震发生在区域地震活动相对活跃的时期，震前震中区曾出现过小震相对活跃现象。主震后截至 2005 年 2 月 14 日，定出 $M_L \geq 1.0$ 级以上地震震中 128 次，其中 5.0～5.9 级 1 次；4.0～4.9 级 1 次；3.0～3.9 级 1 次；2.0～2.9 级 39 次；1.0～1.9 级 86 次。表 2 给出了 M_L3.0 以上地震序列目录。从表中可以看出，最大强余震发生于主震后约 2 分钟，次大余震发生于主震后 2 天，未发生 2.5～3.7 级余震。

表 2　双柏 M_S5.0 地震序列目录（$M_L \geq 3.0$ 级）

Table 2　Catalogue of the M_S5.0 Shuangbai earthquake sequence ($M_L \geq 3.0$)

编号	发震日期	发震时刻	震中位置		震级	震源深度	震中地名	结果来源
	年 月 日	时 分 秒	φ_N	λ_E	M_L	(km)		
1	2004 12 26	15 30 09	24°42′	101°32′	M_S5.0	7	双柏	资料 2)
2	2004 12 26	15 31 56	24°44′	101°30′	4.0		双柏	资料 2)
3	2004 12 28	14 41 40	24°40′	101°32′	3.8		双柏	资料 2)

从双柏 5.0 级地震余震震中分布图看到（图 4），余震发生于双柏县与楚雄市交界，余震大部分发生于主震西侧，余震空间分布无明显形状。

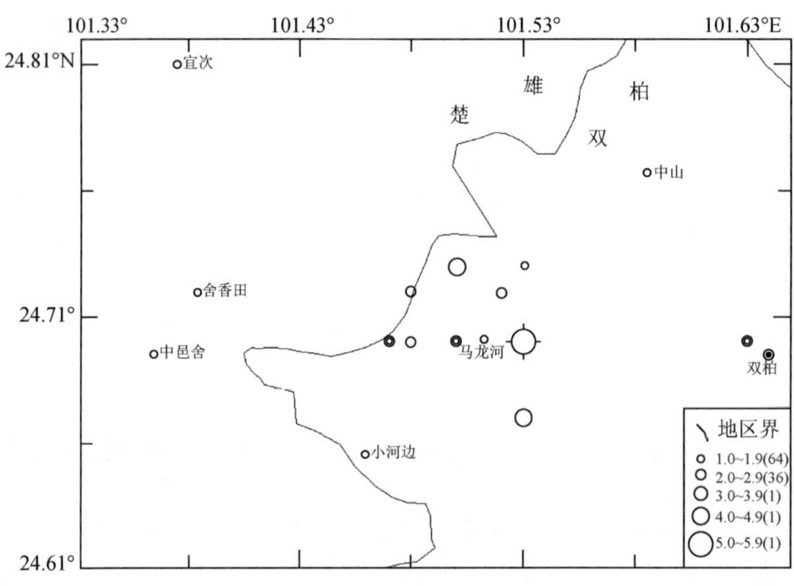

图4　双柏5.0级地震序列震中分布图

Fig. 4　Distribution of the 5.0 Shuangbai earthquake sequence

由余震序列的 M-t 图（图5）看出，余震主要集中发生在主震后前2天内，多数为2.5级以下地震。序列中最大余震为12月26日4.0级强余震，之后，陆续发生一些2级左右小震，12月28日发生一次 M_L3.8 地震，从2005年1月份以后余震发生趋于平稳。

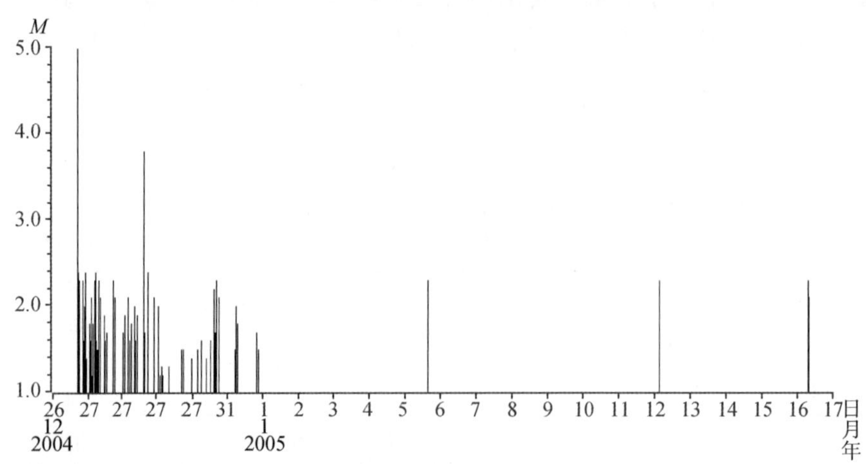

图5　双柏 M_S5.0 地震序列 M-t 图

Fig. 5　M-t diagram of the M_S5.0 Shuangbai earthquake sequence

从地震序列日频次分布（图6）来看，12月26日5.0级主震后，序列衰减系数 p 减小较快，序列逐步衰减，2005年1月9日以后，p 值几乎不随时间的增加而发生太大变化，表明震区余震的发生次数较少。

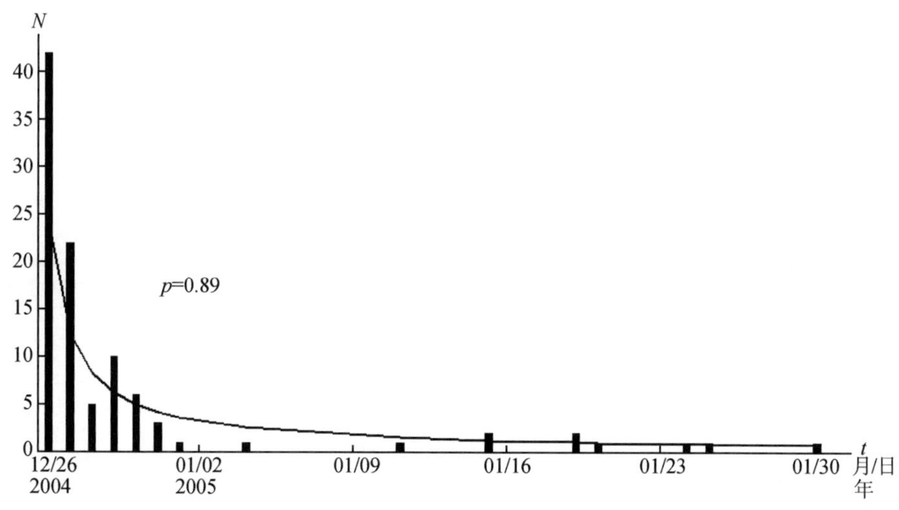

图 6　双柏 M_S5.0 地震序列日频度和 p 值图

Fig. 6　p-value and daily frequency diagram of the M_S5.0 Shuangbai earthquake sequence

从地震序列的 h 图来看（图 7，控制震级为 1.5 级），震后第 2 天 h 值开始增大，虽然震后第 4、5 日 h 值有减小的现象，但仍然小于 1.2，从第 6 日开始，h 值增大明显。

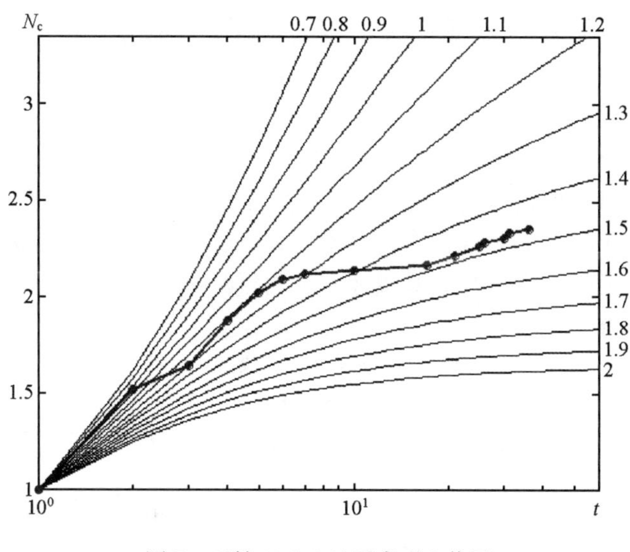

图 7　双柏 M_S5.0 地震序列 h 值图

Fig. 7　h-value diagram of the M_S5.0 Shuangbai earthquake sequence

余震序列 b 值为 0.66（图 8），b 值截距为 4.3。主震释放能量占全序列能量的 95.2%，主震与最大余震震级相差 1.0。根据地震类型的判别指标[1,2]，综合判定双柏 5.0 级地震为主震—余震型。

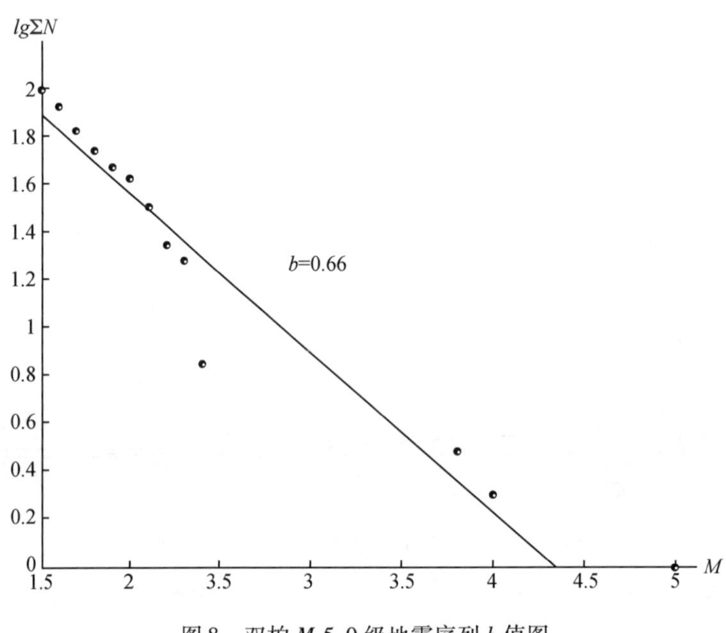

图 8　双柏 M_S5. 0 级地震序列 b 值图

Fig. 8　b-value diagram of the M_S5. 0 Shuangbai earthquake sequence

五、震源机制解及地震主破裂面

据云南传输台网资料，使用初动符号总数为 33 个，运用 P 波初动符号图解断层面法，求得双柏 5. 0 级地震的震源机制解见表 3 及图 9。从图、表可见，节面 I 走向 100°，倾角 83°，滑动角 -171°，P 轴方位 324°，仰角 11°。结合地质构造、烈度等震线和余震活动等资料，分析认为节面 I 和王家村断层的走向基本一致，节面 I 为主震初始破裂面，主破裂面为近北西西向断层面。据此推测双柏 5. 0 级地震是北西西向断裂在近于水平向的主压应力作用下产生右旋走滑的结果。

表 3　双柏 M_S5. 0 地震震源机制解

Table 3　Focal mechanism solution of the M_S5. 0 Shuangbai earthquake

编号	节面 I			节面 II			P 轴		T 轴		B 轴		X 轴		Y 轴		结果来源
	走向	倾角	滑动角	走向	倾角	滑动角	方位	仰角	方位	仰角	方位	仰角	方位	仰角	方位	仰角	
1*	100	83	-171	8	81	-7	324	11	234	1	138	79					付虹

* 矛盾符号比：0. 06。

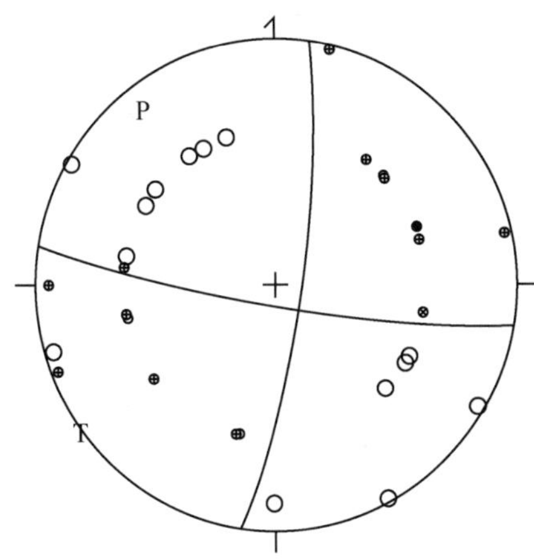

图 9　双柏 $M_S5.0$ 地震震源机制解图

Fig. 9　Focal mechanism solution of the $M_S5.0$ Shuangbai earthquake

六、观测台网及前兆异常

双柏 5.0 级地震发生在云南省前兆台网较密集的地区，距震中周围 200km 范围内分布有 43 个地震台，其中有测震单项观测台 3 个，其他前兆单项观测台 4 个，综合观测台 29 个。有测震、水氡、水位、流量、水温、水汞、地磁 D、地磁 Z、地电阻率、自然电位、地倾斜、地应力、重力、短基线和短水准、电导率等 30 个观测项目，129 个观测台项（图 10），其中 0 ~ 100km 内有 9 个观测台，27 个观测台项；101 ~ 200km 内 34 个观测台，101 个观测台项。定点前兆观测台站观测项目震前都有长期连续可靠的观测资料。

在震中周围 100km 内有前兆观测台站 6 个，21 个台项，未出现异常台项；在 101 ~ 200km 范围内有前兆观测台站 30 个，台项 83 个，出现异常台站和异常台项为 6 个和 6 个，异常台站和异常台项百分比分别为 20% 和 7%，中期异常台站和异常台项百分比分别为 13% 和 5%，短期异常台站和异常台项百分比分别为 7% 和 2%，各类异常的具体情况详见表 4 和图 11。

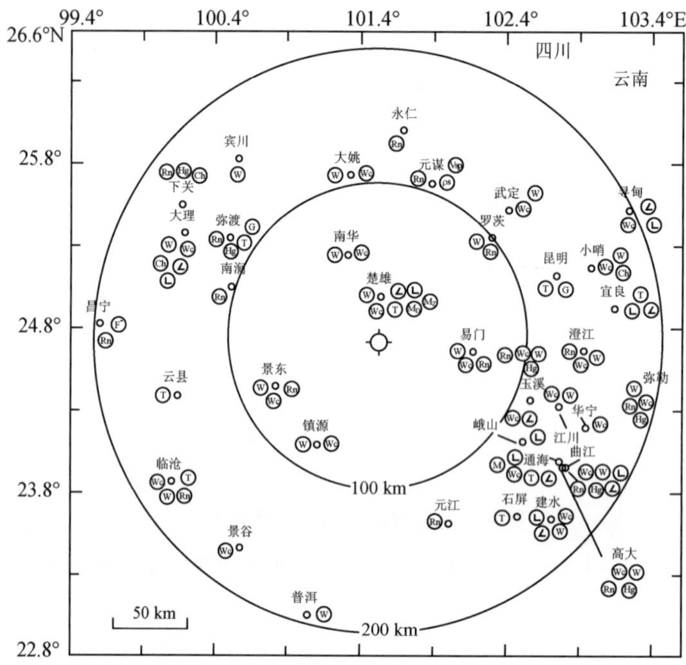

图 10　双柏 M_S5.0 地震前定点前兆观测台站分布图

Fig. 10　Distribution of the precursor monitoring stations before the M_S5.0 Shuangbai earthquake

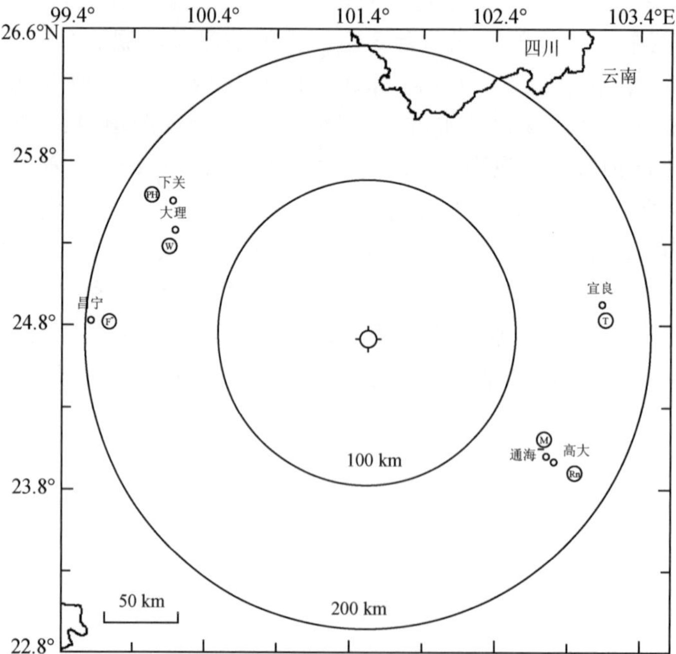

图 11　双柏 M_S5.0 地震定点前兆观测异常分布图

Fig. 11　Distribution of precursory anormalies of the M_S5.0 Shuangbai earthquake on the fixed observation points

表 4　异常情况登记表

Table 4　Summmary table of precursory anomalies

序号	异常项目	台站或观测区	分析方法	异常判据及观测误差	震前异常起止时间	震后变化	最大幅度	震中距（km）	异常类别	图号	异常特点及备注
1	地震频度	震中周围 50km	$M_L \geqslant 2.5$ 级 月频度	$N \geqslant 3$	2004.8	正常	$N=8$		S_2	12	对楚雄一带对应较好
2	地震频度	滇中地区	$M_L \geqslant 3$ 级 1 月为时窗，0.5 月为步长滑动	$N \geqslant 5$	2004.8	正常	$N=5$		S_2	13	对滇中 5 级以上中强地震对应较好
3	归一化环境剪应力τ_{0c}	云南地区 21°～29°N 97°～106°E	$M_L \geqslant 2.5 \sim 4.9$ 级	$\tau_{0c} \geqslant 25.5$MPa	2004.10.20～200412.20	异常	$\tau_{0c}=43$MPa		I_2	14	2 次以上异常高值，其后有 5 级以上地震危险
4	τ_{0c}值	云南地区 21°～29°N 97°～106°E	$M_L \geqslant 2.5 \sim 4.9$ 级 τ_{0c}值空间动态分布	不同时间τ_{0c}高值区空间分布	2004.11.1～2004.12.10	异常消失	$\tau_{0c} \geqslant 25.5$MPa 的区域集中于双柏附近		I_2	15	τ_{0c}值从零星、随机向双柏一带集中、增强
6	地磁 D 分量	通海	旬均值	反向变化	2004.4.30～2005.4.10	持续	持续反向变化	133	M_2	16	震前出现了明显的反向变化
5	水氡	高大	日均值	≥8.3Bg/L	2004.7.6～2004.9.9	正常	10.1Bg/L	147	I_2	17	震前半年高值变化
7	水位	大理	月均值	破年变	2004.5～2005.3	正常	0.59m	154	M_2	18	震前破年变
8	地倾斜	宜良	日均值	出现上升	2004.5.25～2004.8.27	正常	0.606 角秒	166	M_2	19	震前大幅度上升
9	pH 值	下关	五日均值	升速快	2004.7.5～2005.3.5	持续	0.9	166	I_2	20	震前快速增加
10	氟离子	昌宁	五日均值	出现大幅度上升	2004.1.31～2004.5.25	正常	0.0114mg/L	195	M_2	21	震前大幅度上升

七、地震前兆异常特征分析

1. 地震学异常

双柏地震前震中 50km 范围内小震曾出现过增强活动，图 12 给出了震中周围 50km 范围内 M_L≥2. 5 级地震月频度，以 1 个月为时窗、步长。可以看到震前 2 次达到月频度为 3 次的异常控制线，在 2004 年 8 月曾达到 7 次。

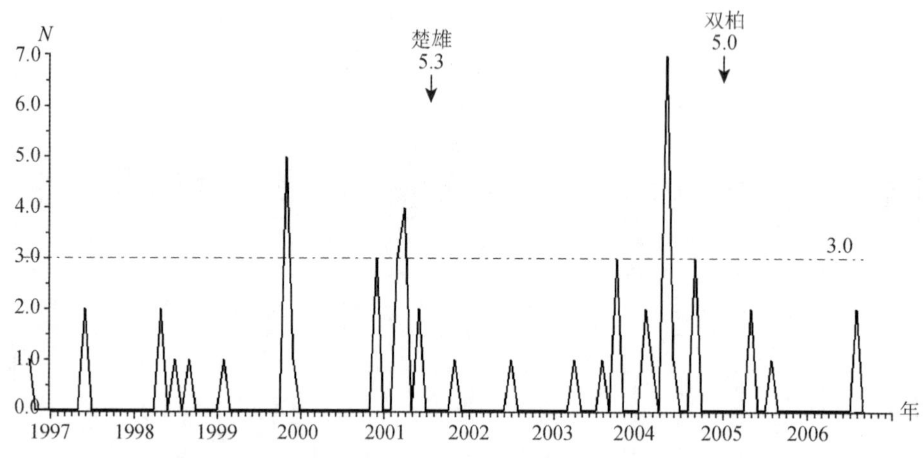

图 12　震中周围 50km 范围内 M_L≥2. 5 级地震月频度曲线

Fig. 12　Curve of monthly frequency of M_L≥2. 5 earthquakes around the epicentral 50 range

震前滇中地区曾出现过地震活动增强现象，图 13 给出滇中地区（北纬 23. 5° ~ 26. 7°，东经 100. 5° ~ 104°）M_L≥3. 0 级地震月频度，以 1 个月为时窗、0. 5 月为步长。可以看到从 1997 年来该区只要月频度达到或超过 5 次，其后都会发生 5 级以上地震，该区 2004 年 8 月出现异常。

双柏地震前环境剪应力时空演化。在时间变化方面[3]（图 14），可以看到，鲁甸、保山地震前均有高值异常出现。保山地震后，从 10 月下旬到 12 月下旬，归一化环境剪应力τ_{0c}值并未恢复到如 2004 年 4 ~ 6 月的正常状态，期间陆续发生了 4 次大于、等于指标线的异常值。如果一次 5 级以上中强地震后，在短时间内再次出现τ_{0c}高值且出现 2 次以上，则其后 3 个月内再次发生中强地震的可能性很大[4]。在空间动态演化方面（图 15），随着时间向双柏 5. 0 地震逼近，尤其是保山地震后云南地区的环境剪应力强度并未逐渐恢复，而是更加增强，高值点持续不断，高值分布区主要集中在滇中的禄劝、元谋、楚雄一带，表明鲁甸、保山地震的发生并还是一次孤立事件，而应当是一次中强地震连发事件。

双柏 5. 0 地震震前出现的地震活动主要特征是，震前震中附近中小地震活动增强，2004 年 8 月 25 ~ 28 日短短 4 天之内发生 7 次 2. 6 ~ 3. 0 级地震，这在 2001 年 3 月楚雄 5. 3 级地震之后从未出现过。从较大区域来看，滇中地区震前出现 3 级地震活动增强也是较为明显的，2004 年 8 月禄丰至楚雄一带出现高达 6 次的 3 级地震集中活动，其后 9 ~ 11 月地震活动水平没有减少，2004 年 10 月还发生元谋 M_L4.6、4.0 和澄江 M_L4.1 中等地震活动。双柏 5.0 地震震前出现的另一个地震活动特征是，震前环境剪应力呈现逐渐增强趋势，在空间分布上表现从随机、分散状态向临震震中附近收缩、集中现象较为突出，笔者曾在 11 月份作出过“未来 2 个月，双柏—楚雄一带有发生 5～6 级地震危险”的正式短临预报意见[6)]。

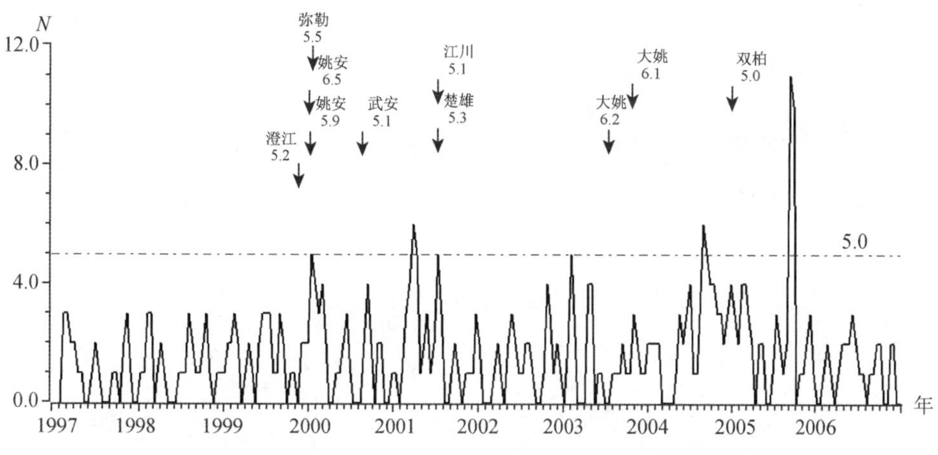

图 13　滇中地区 $M_L \geqslant 3.0$ 级地震月频度曲线

Fig. 13　Curve of monthly frequency of $M_L \geqslant 3.0$ earthquakes in the middle of Yunnan area

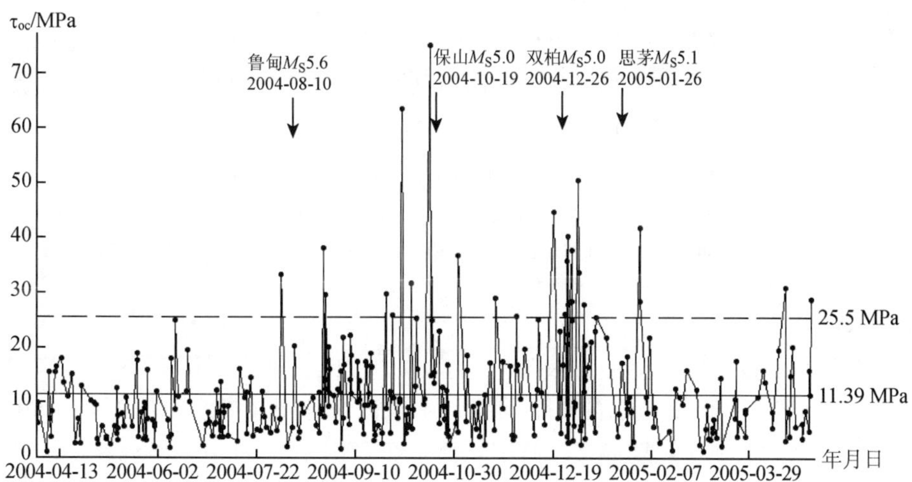

图 14　云南地区 2004 年 4 月至 2005 年 4 月应力参数随时间的变化

Fig. 14　The temporal variation of stress field of Yunnan area during April 2004 ~ April 2005

（图中横实线为均值线，虚线为均值加 1.5 倍方差线，又称异常线）

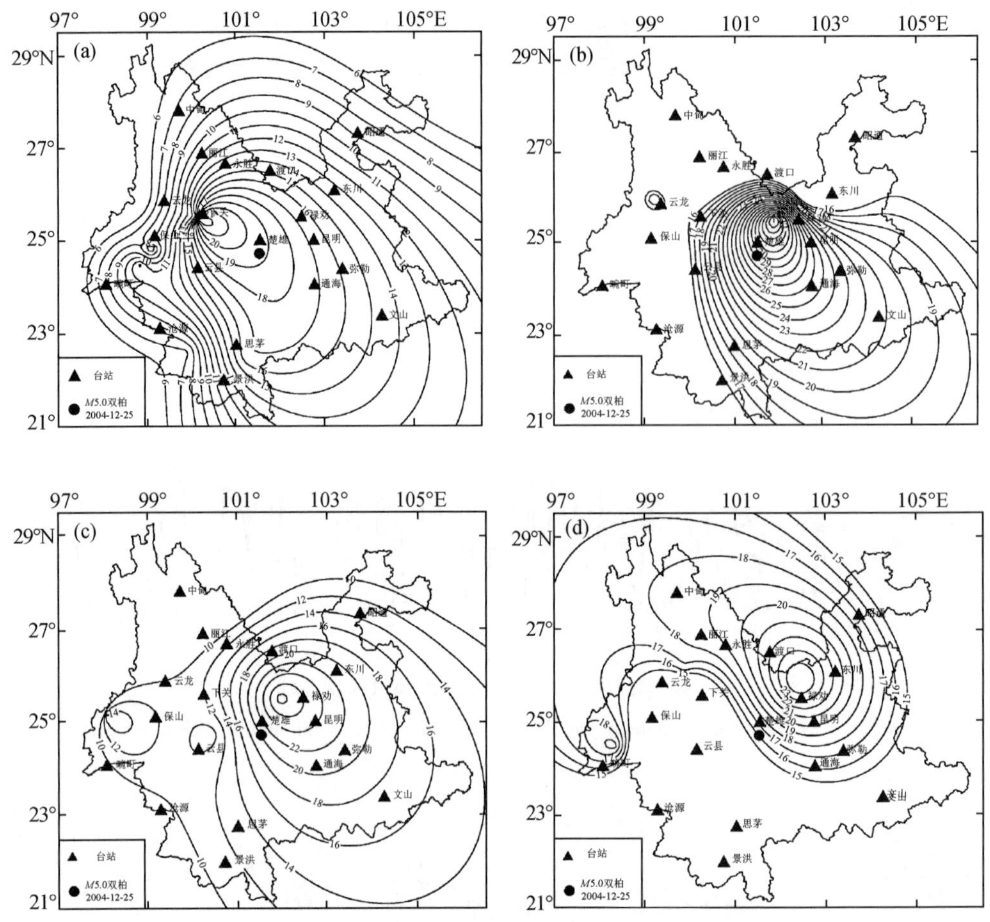

图 15　2004 年 12 月 26 日双柏 M_S5.0 地震前环境剪应力τ_{0c}值空间动态演化

Fig. 15　The dynamic spatial evolution of the ambient shear τ_{0c}

prior to the December 26，2004 Shuangbai earthquake

（a）10 月 20～30 日；（b）11 月 1～18 日；（c）11 月 19～29 日；（d）11 月 30～10 日

2. 前兆异常

双柏 5.0 级地震发生在监测能力较强的云南滇中地区，震区周围 200km 范围定点前兆台站多达 36 个。震区 100km 范围的 6 个前兆台站未记录到明显异常，但 100～200km 范围却有 6 个台站记录到明显中短期异常。昌宁氟离子在 2004 年 1 月最早出现异常，之后 2004 年 4、5 月通海地磁 D 分量、大理水位和宜良地倾斜先后出现中期异常，2004 年 7 月高大水氡和下关 pH 值出现短期异常，尤其是高大水氡在震前短期内出现 2 次高值，较为显著，距离震区 133km，为出现异常台站中较近的台站，可能一定程度反映出震中附近地区介质应力的不稳定。

双柏 5.0 级地震前地磁和形变异常特点是长期平稳变化趋势状态的打破。通海地磁 D 分量，2004 年 4 月 30 日开始打破原来缓慢上升状态，出现连续下降变化，异常一直持续到 2005 年 4 月 10 日（图 16）。宜良地倾斜东西分量，2004 年 5 月 25 日开始出现打破年变向东倾斜，异常持续到 2004 年 8 月 27 日，最大变化幅度为 0.606 角秒，在 2003 年 7 月大姚 6.2 级、2005 年 7 月会泽 5.3 级地震前出现过类似变化（图 19）。水位异常特点是破年变、高水位异常。大理水位 2004 年 5 月后打破正常年变，异常一直持续到 2005 年 3 月，变化最大幅度为 0.59m。在 2003 年 7、10 月大姚 6.2、6.1 级地震前后曾出现过类似异常变化（图 18）。水化学异常特点是短时间出现大幅度变化。高大水氡在 2003 年 7 月大姚 6.2 级地震前出现过增加变化，在 2005 年 8 月文山 5.3 级地震前、2006 年 7 月盐津 5.1 级地震前都出现过较为明显的上升变化异常（图 17）。下关 pH 值 2004 年 7 月 5 日开始上升，一直增加到震后 2005 年 3 月 5 日，增加幅度达 0.9，下关 pH 值在 2003 年大姚 6.2 级和 2006 年 1 月墨江 5.2 级地震前增加现象明显（图 20）。昌宁氟离子 2004 年 1 月 31 日开始大幅度上升，至 2004 年 5 月 25 日异常结束缓慢下降，最大升幅达 0.0114mg/L，在 2003 年 7 月大姚 6.2 级地震前，昌宁氟离子却是出现大幅度下降变化后，在低值缓慢回升过程中发生的地震（图 21）。

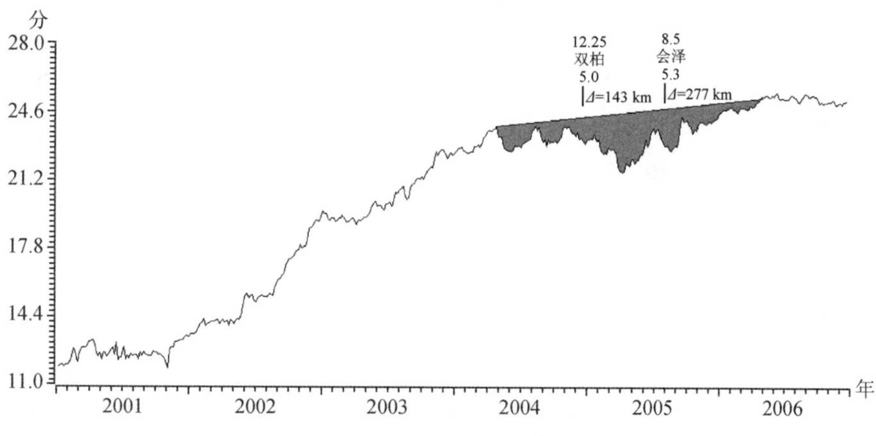

图 16　通海地磁 D 分量日均值曲线

Fig. 16　Curves of daily mean value of geomagnetie at Tonghai station

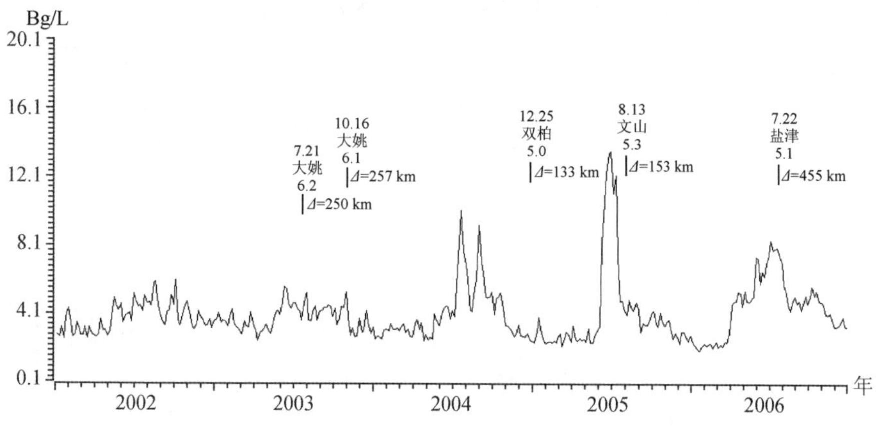

图 17　高大水氡五日均值曲线

Fig. 17　Curves of 5-day mean value of radon content in groundwater at Gaoda station

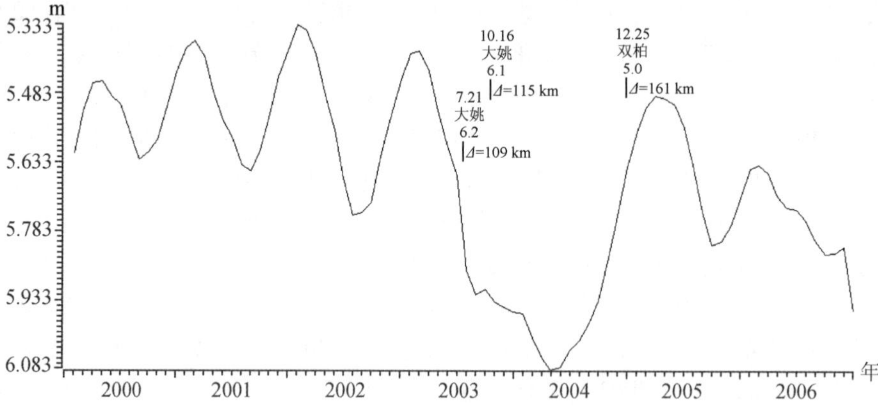

图 18　大理水位月均值曲线

Fig. 18　Curves of monthly mean value of water level at Dali station

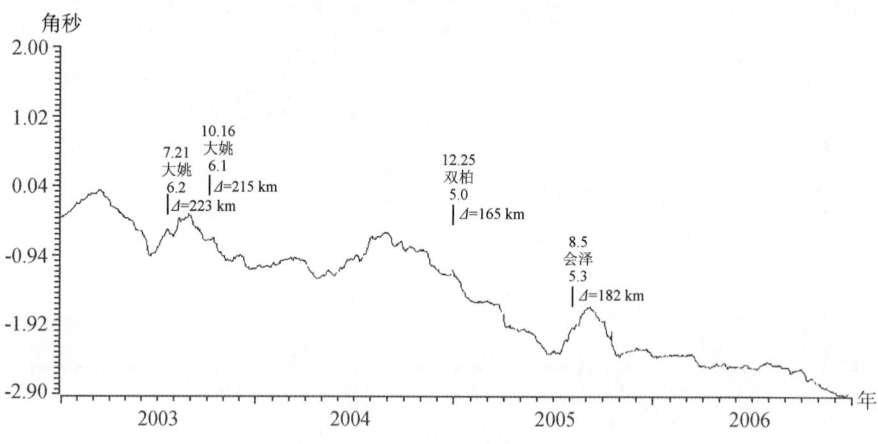

图 19　宜良倾斜日均值

Fig. 19　Curves of daily mean value of tilt at Yiliang station

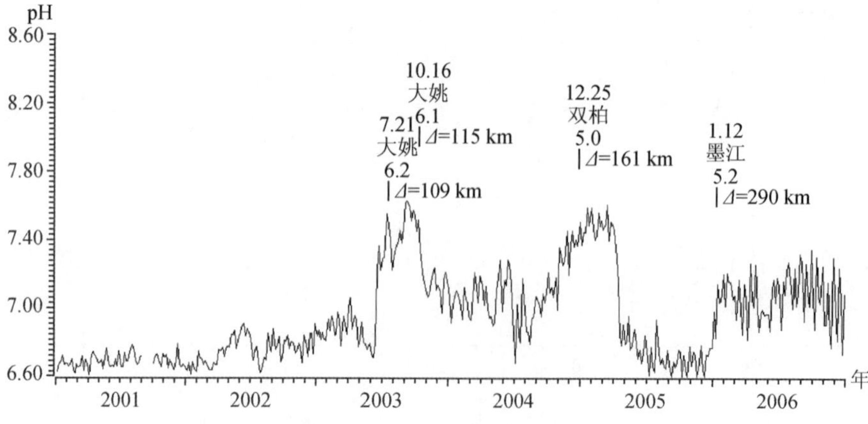

图 20　下关 pH 值五日均值曲线

Fig. 20　Curves of 5-day mean value of pH in groundwater at Xiaguang station

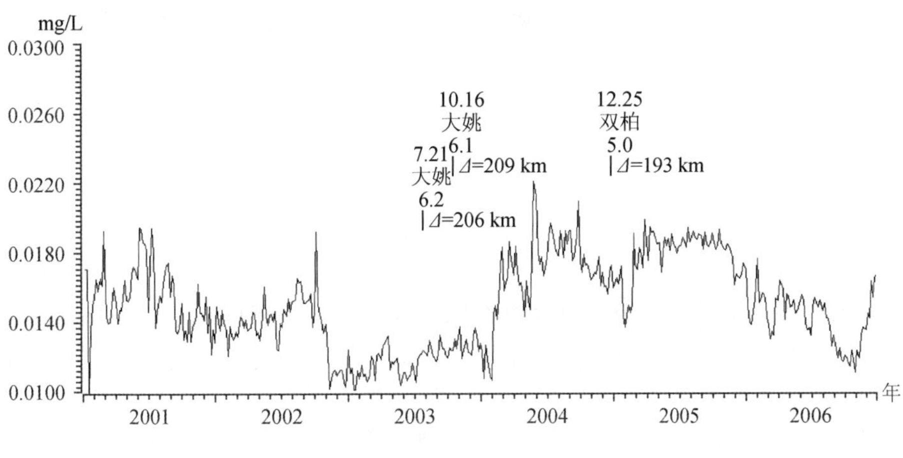

图 21　昌宁氟离子五日均值曲线

Fig. 21　Curves of 5-day mean value of F^+ in groundwater at Changning station

八、震前预测、预防和震后响应

1. 震前预测、预防

2003 年底，云南省地震局对云南省 1 年尺度进行中期预测，对危险区的判定明确提出“滇东双柏—易门—建水—开远—华宁—宜良—寻甸—巧家一带”，将双柏地区列为首位。这次双柏 5.0 级地震发生在危险区内，中期预测效果较好[7)]。

在中期预测的基础上，云南省地震局加强了对危险区的重点短临跟踪。2004 年 7 月底，根据前兆及地震活动情况，云南省地震局分析认为“近期双柏—易门—建水一带，可能发生 5～6 级地震”，以《震情反映》形式报送中国地震局、云南省委、省政府，引起各级主要领导的高度重视[8)]。

2004 年印度尼西亚 8.7 级大地震后，面对当时复杂的地震形势，云南省地震局认为“云南地区强震的危险性进一步加剧”[9)]。在 12 月 29 日会商会上提出“云南地区发生中强地震的危险性在进一步增强，重点危险区为滇西南勐海—思茅—江城—勐腊及其边境一带，可能发生 5～6 级地震”[10)]。2005 年 1 月 26 日发生思茅 5.0 级地震。

2. 震后响应

（1）震后趋势判断。双柏 5.0 地震发生后，中国地震局与云南省地震局专家赶赴地震现场，进行现场地震监测预报工件。现场工作组与云南省地震局分析预报人员密切跟踪地震序列的发展，结合该区历史地震活动情况，于 12 月 29 日对震后趋势进行初步判断，认为该地震属主震—余震型，未来几天震区注意 3 级左右地震的发生。云南省地震局对地震类型和强余震作出了准确的判别和预测。

（2）应急响应。双柏 5.0 地震发生后，云南省地震局立即启动地震应急预案，派出地震现场工作队奔赴震区。2004 年 12 月 26 日 18 时 18 分，第一批现场工作队员到达现场，中国地震局派出4人专家组于次日凌晨3时到达双柏。中国地震局、云南省地震局以及楚雄州地震局的领导和专家组成地震现场工作指挥部，包括秘书、震情分析预报、地震灾评、强震观测、测震、通讯宣传及后勤保障7个组，共39人开展地震现场工作。12月26 ~ 29日，共计4天，8名灾评人员分乘6辆车对楚雄州双柏县和楚雄市进行灾情调查，行程1.1万千米，共完成43个调查点的调查工作，其中抽样点25个。

九、结论与讨论

（1）双柏5.0级地震的发生，尽管存在印度尼西亚8.7级巨震的触发作用，但其自身存在的强震孕育本质，是我们要作深刻认识的。震前1年尺度云南省地震局作出了较好预测，短期5个月尺度也作出了成功的对该区短期地震危险认识，以《震情反映》的形式提出“近期双柏—易门—建水一带可能发生5 ~ 6级地震”的结论，但在短临3个月尺度未能作出有效预测，说明地震预测这个世界性的科学难题还有待地震科学工作者逐步攻克。

（2）对双柏5.0地震后短期内云南复杂的地震形势，云南省地震局作出了正确的判定，根据安达曼弧强烈地震对云南地震活动的影响规律认识以及云南自身存在的地震前兆特征，认为云南地区发生中强地震的危险性在进一步增强，重点危险区为滇西南勐海—思茅—江城—勐腊及其边境一带，可能发生5 ~ 6级地震”，2005年1月26日发生思茅5.0级地震。

（3）对双柏5.0级地震序列的震后趋势判断，云南省地震局也作出了较为成功的预测。双柏地震序列属于余震衰减较快的序列，根据地震序列的特征及该区历史地震序列规律，云南省地震局在震后3天就作出了该序列属于主震—余震型类型、未来注意3级左右地震活动的判断，即不太可能会发生4级左右强烈有感地震的判定，这对当地政府作好抗震救灾工作是极其重要的，事后证明该判断是正确的。

（4）双柏地震的短临跟踪是值得认真总结和思考的。该区中短期出现强震的孕育现象，但短临阶段，比较突出的地震前兆异常出现得较少，尽管出现了一些异常，如环境剪应力参数异常（是一种通过地震波谱分析得到的表征地下介质力学性的参数方法），双柏5.0地震1 ~ 2年前才应用于地震预测预报工作，还处于接受实践检验并完善和提高阶段。

2004年12月26日在苏门答腊发生M_S8.7巨震，地震所在的安达曼弧构造带向北穿过缅甸，与我国南北地震带（滇、川、青、甘）和喜玛拉雅地震带相连，构造上与我国中西部强震活动有关。8.7级巨震后不到20分钟云南宾川发生4.6级地震，其后仅过了12分钟缅甸又发生5.9级地震，本次双柏5.0级地震是在8.7级巨震5个小时后发生的（图22）。巨震后10天的时间内，云南地区还触发了8次震群活动（图23），这些震群主要分布在川滇菱形块体东部和红河断裂带上。8.7巨震后云南地区地震活动显著增强。图24给出云南地区（北纬21° ~ 29°，东经97° ~ 106°）在去除双柏5.0余震后$M_L \geqslant 1.0$级地震日频度，可以看到，云南地区正常情况下小震每日平均发生10次，26日后日频度显著增强，在2004年12月26日至2005年1月5日期间，日频度平均值高达80次，其中2004年12月26、27日分别达到100次，2005年1月4日更是高达125次，2005年3月份以后云南地区地震活动趋于正常活动水平。可见，印度尼西亚苏门答腊M_S8.7巨震对云南地区地震活动的影响是明显的。

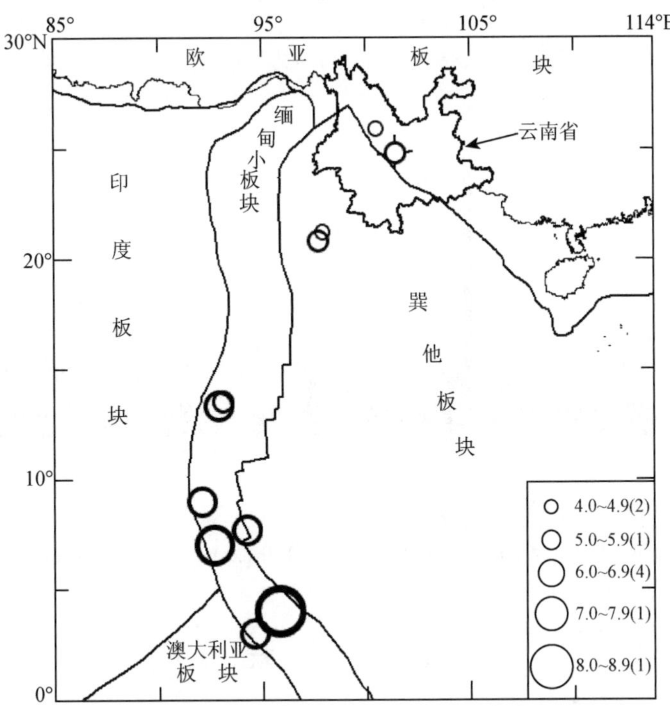

图 22 苏门达腊 8.7 级地震后 1 天内周边地区 $M \geqslant 4.0$ 级地震分布

Fig 22 Distribution of the 8.7 Sumatra earthquake within one day, $M \geqslant 4.0$

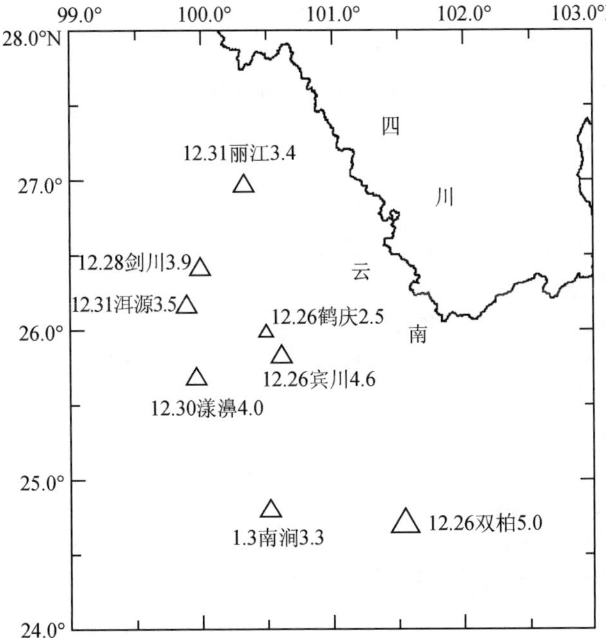

图 23 2004.12.26 ~ 2005.1.4 云南地区的主要震群分布

（图标为震群发生时间，地名和最大地震震级）

Fig 23 Distribution of the earthquake-swarm in Yunnan area during December 26, 2004 ~ January 4, 2005

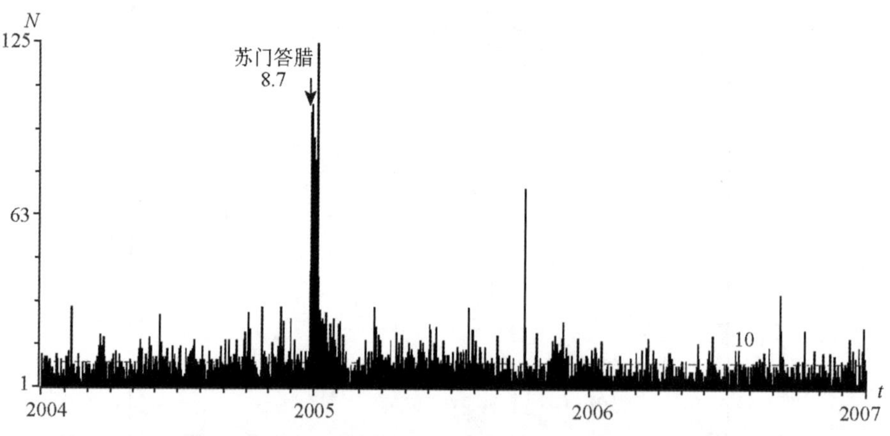

图 24　云南地区 $M_L \geqslant 1.0$ 级地震日频度

Fig. 24　Curve of daily frequency of $M_L \geqslant 1.0$ earthquakes in Yunnan area

参 考 文 献

［1］中国地震局编，1998，地震序列的类型判别和震后趋势估计，地震现场工作大纲和技术指南，北京：地震出版社

［2］吴开统、焦远碧、吕培苓等，1990，地震序列概论，北京：北京大学出版社

［3］钱晓东、邬成栋、秦嘉政，2004，用环境应力参数对云南地区进行地震短临跟踪监视，地震研究，27（1）：1～7

［4］钱晓东、秦嘉政，2006，环境应力参数在 2004 年云南双柏 5.0 地震短期预报中的应用，中国地震，22（2）：161～171

参 考 资 料

1）云南省地震局，双柏 5.0 级地震灾害损失评估报告，2004.12

2）云南省震局，云南省地震正式目录，2004

3）云南地震局，云南省地震速报目录，2004

4）中国人民解放军建字七三〇部队，楚雄幅区域水文地质普查报告，1975.10

5）地质部云南省地质局，楚雄幅地质报告书（上册），1965

6）云南省地震局，地震预报卡，编号 85，2004

7）云南省地震局，云南省 2004 年度地震趋势研究报告，2003

8）云南省地震局，震情反映（200403），2004

9）云南省地震局，震情反映（200411），2004

10）云南省地震局，会商报告（2004）第 77 期

Shuangbai Earthquake of M_S 5.0 on December 26, 2004 in Yunnan

Abstract

On Dec. 26 2004, an earthquake of M_S5.0 occurred in Shuangbai county, Yunnan province. The macroscopic epicenter of the earthquake was located round Tuodian town in Shuangbai county. The intensity in the meizoseismal area was VI. The shape of isoseismic line was nearly circle. One person was killed, one person was seriously wounded. Direct economic loss caused by this earthquake was 407 million Yuan.

The Shuangbai M_S5.0 earthquake sequence was of mainshock aftershock type. The largest aftershock was M_L4.1. Aftershocks were mainly distributed the direction of NW of mainshock, the predominant direction of the aftershock was not obvious. The spatial distribution of the aftershock was consistent with isoseismic line of the meizoseismal area. Nodal plane Ⅰ was the main rupture plane of this earthquake. The zimuth of the principal compressive stress was in NNW direction. The seismogenic structure was the WangJiacun fault.

There were 43 seicmic stations around this earthquake, among them, there were 7 seicmometric stations and 36 precursory observation stations. Before this earthquake there were 10 anomalies in 7 observation items, 4 of them were anomalies in seismic activity and 6 of them were anomalies in precursory observation items, they were imminent anomalies.

报告附件

附表1　固定前兆观测台（点）与观测项目汇总表

序号	台站（点）名称	经纬度（°）		测项	资料类别	震中距Δ（km）	备注
		φ_N	λ_E				
1	双柏	24.68	101.63	测震△		12	
2	楚雄	25.00	101.53	测震△		32	
				水位	Ⅱ		
				水温	Ⅱ		
				水质	Ⅱ		
				地倾斜（摆试）	Ⅱ		
				地倾斜（连通管）	Ⅱ		
				水准	Ⅱ		
				基线	Ⅱ		
				地磁 D	Ⅱ		
				地磁 Z	Ⅱ		
3	一平浪	23.09	101.92	测震△		59	
4	南华	25.25	101.30	水位	Ⅱ	65	
				水温	Ⅱ		
5	易门	24.67	102.17	测震△		66	
				水位	Ⅰ		
				水温	Ⅰ		
				水氡	Ⅰ		
6	景东	24.45	100.82	水位	Ⅰ	78	
				水温	Ⅰ		
				水氡	Ⅰ		
7	镇源	24.10	101.10	测震△		80	
				水位	Ⅱ		
				水温	Ⅱ		
8	安宁	24.87	102.50	测震△		99	
9	罗茨	25.35	102.28	水位	Ⅱ	99	
				水氡	Ⅱ		

续表

序号	台站（点）名称	经纬度（°）		测项	资料类别	震中距Δ（km）	备注
		φ_N	λ_E				
10	南涧	25.05	100.52	测震△		109	
				水氡	Ⅱ		
11	300 号	24.78	102.60	测震△		110	
12	玉溪	24.37	102.55	水位	Ⅰ	111	
				水温	Ⅰ		
				水氡	Ⅰ		
				水汞	Ⅰ		
13	富民	25.22	102.48	测震△		113	
14	元谋	25.68	101.87	测震△		114	
				水氡	Ⅰ		
				电阻率	Ⅰ		
				自然电位	Ⅰ		
15	大姚	25.73	101.32	测震△		116	
				水位	Ⅰ		
				水温	Ⅰ		
16	峨山	24.12	102.50	测震△		120	
				水温	Ⅱ		
				水准	Ⅱ		
				基线	Ⅱ		
17	弥渡	25.35	100.50	测震△		126	
				水氡	Ⅰ		
				水汞	Ⅰ		
				二氧化碳	Ⅰ		
				地倾斜（摆试）	Ⅰ		
				地倾斜（连通管）	Ⅰ		
				重力	Ⅰ		
18	武定	25.52	102.40	水位	Ⅰ	127	
				水温	Ⅰ		
19	禄劝	25.53	102.43	测震△		130	
20	元江	23.62	102.00	水氡	Ⅰ	131	

续表

序号	台站（点）名称	经纬度（°）		测项	资料类别	震中距 Δ（km）	备注
		φ_N	λ_E				
21	昆明	25.13	102.73	测震△		131	
				地倾斜（摆试）	Ⅰ		
				重力	Ⅰ		
22	江川	24.33	102.75	水位	Ⅰ	132	
				水温	Ⅰ		
23	通海	24.00	102.75	测震△		133	
				水温	Ⅰ		
				地倾斜（摆试）	Ⅰ		
				水准	Ⅰ		
				基线	Ⅰ		
				地磁 D	Ⅰ		
				地磁 Z	Ⅰ		
24	澄江	24.67	102.92	测震△		142	
				水位	Ⅱ		
				水温	Ⅱ		
				水氡	Ⅱ		
25	永仁	26.00	101.68	水氡	Ⅱ	144	
26	云县	24.40	100.13	测震△	Ⅱ	145	
				地倾斜（摆试）	Ⅱ		
27	高大	23.99	102.71	水位	Ⅰ	147	
				水温	Ⅰ		
				水氡	Ⅰ		
				水汞	Ⅰ		
28	石屏	23.67	102.47	测震△		151	
				地倾斜（摆试）	Ⅰ		
29	华宁	24.20	102.93	水温	Ⅱ	154	

续表

序号	台站（点）名称	经纬度（°）		测项	资料类别	震中距Δ（km）	备注
		φ_N	λ_E				
30	大理	25.37	100.18	水位	Ⅰ	154	
				水温	Ⅰ		
				水质	Ⅰ		
				水准	Ⅰ		
				基线	Ⅰ		
31	曲江	23.97	102.80	水位	Ⅰ	154	
				水温	Ⅰ		
				水氡	Ⅰ		
				水汞	Ⅰ		
				水准	Ⅰ		
				基线	Ⅰ		
32	小哨	25.17	102.97	水位	Ⅰ	155	
				水温	Ⅰ		
				气体	Ⅰ		
33	宾川	25.83	100.55	水位	Ⅱ	159	
34	宜良	24.92	103.13	测震△		166	
				水准	Ⅱ		
				基线	Ⅱ		
				地倾斜（摆试）	Ⅱ		
35	下关	25.55	100.17	测震△		166	
				水氡	Ⅰ		
				水汞	Ⅰ		
				pH 值	Ⅰ		
36	景谷	23.47	100.58	水温	Ⅰ	168	
37	建水	23.65	102.70	测震△		169	
				水位	Ⅱ		
				水温	Ⅱ		
				水准	Ⅱ		
				基线	Ⅱ		
38	嵩明	25.35	103.03	测震△		170	

续表

序号	台站（点）名称	经纬度（°）		测项	资料类别	震中距Δ（km）	备注
		φ_N	λ_E				
39	临沧	23.87	100.10	水位	Ⅱ	172	
				水温	Ⅱ		
				水氡	Ⅱ		
				地倾斜（摆试）	Ⅱ		
40	普洱	23.07	101.03	水位	Ⅱ	190	
41	昌宁	24.82	099.60	氟离子	Ⅱ	195	
				水氡	Ⅱ		
42	寻甸	25.53	103.25	水温	Ⅱ	195	
				水准	Ⅱ		
				基线	Ⅱ		
43	弥勒	24.40	103.43	测震△		196	
				水位	Ⅰ		
				水温	Ⅰ		
				水氡	Ⅰ		
				水汞	Ⅰ		

分类统计	0＜Δ≤100km	100＜Δ≤200km	200＜Δ≤300km	300＜Δ≤500km	500＜Δ≤800km	总数
测项数 N	11	19				
台项数 n	27	101				
测震单项台数 a	3	4				
形变单项台数 b						
电磁单项台数 c						
流体单项台数 d	1	6				
综合台站数 e	5	24				
综合台中有测震项目的台站数 f	3	14				
测震台总数 $a+f$	6	18				
台站总数 $a+b+c+d+e$	9	34				
备注						

续表

序号	台站（点）名称	经纬度（°）		测项	资料类别	震中距 Δ（km）	备注
		φ_N	λ_E				

注 1：一个台编一个序号，按震中距从近到远排序，研究区域以外的台不编号；分类统计范围根据具体震例确定的研究区域确定，可增列更远范围的栏目（见第 4.2.2 条）。

注 2：地震前兆观测项目定义见 3.1.8 条，简称测项，不同台的相同观测项目算作一个测项，不累计计算；观测台项——以台站为单元计算观测项目，简称台项，不同台的相同项目累计计算；测项数——统计范围内测项的数目；台项数——统计范围内所有台站（点）台项数的累计数。

注 3：综合台站——具有一个以上测项的地震台；测震单项台——仅进行测震单项观测的地震台；其他单项台——仅有测震以外其他单一测项的地震台。

注 4：测震台站分布图上的台站应与本表一致，在同一台站的所有数字和模拟的测震项目应在图上和表中的项目名称后标出（图例见表 1），并计作一个台；地磁台的总强度、垂直分量和偏角各算一个测项，在图和表上应同时给出，同一量的相对与绝对观测算为一个测项；地电阻率和自然电位各作为一个测项；在一个台上同时有摆式地倾斜和连通管观测时，算作两个地倾斜台项，分别用括号注明“摆式”和“连通管”；应变分钻孔应变、体应变和断层应变，在备注栏里注明仪器型号。

附表 2　测震以外固定前兆观测项目与异常统计表

序号	台站（点）名称	测项	资料类别	震中距Δ/km	按震中距Δ范围进行异常统计																								
					$0 < \Delta \leqslant 100$km					$100 < \Delta \leqslant 200$km					$200 < \Delta \leqslant 300$km					$300 < \Delta \leqslant 500$km					$500 < \Delta \leqslant 800$km				
					L	M	S	I	U	L	M	S	I	U	L	M	S	I	U	L	M	S	I	U	L	M	S	I	U
1	楚雄	水位	Ⅱ	32	—	—	—	—																					
		水温	Ⅱ		—	—	—	—																					
		水质	Ⅱ		—	—	—	—																					
		地倾斜（摆式）	Ⅱ		—	—	—	—																					
		地倾斜（连通管）	Ⅱ		—	—	—	—																					
		水准	Ⅱ		—	—	—	—																					
		基线	Ⅱ		—	—	—	—																					
		地磁 *D*	Ⅱ		—	—	—	—																					
		地磁 *Z*	Ⅱ		—	—	—	—																					
2	南华	水位	Ⅱ	65	—	—	—	—																					
		水温	Ⅱ		—	—	—	—																					
3	易门	水位	Ⅰ	66	—	—	—	—																					
		水温	Ⅰ		—	—	—	—																					
		水氡	Ⅰ		—	—	—	—																					

续表

序号	台站（点）名称	测项	资料类别	震中距 Δ/km	按震中距 Δ 范围进行异常统计																								
					0 < Δ≤100km					100 < Δ≤200km					200 < Δ≤300km					300 < Δ≤500km					500 < Δ≤800km				
					L	M	S	I	U	L	M	S	I	U	L	M	S	I	U	L	M	S	I	U	L	M	S	I	U
4	景东	水位	Ⅰ	78	—	—	—	—																					
		水温	Ⅰ		—	—	—	—																					
		水氡	Ⅰ		—	—	—	—																					
5	镇源	水位	Ⅱ	80	—	—	—	—																					
		水温	Ⅱ		—	—	—	—																					
6	罗茨	水位	Ⅱ	99	—	—	—	—																					
		水氡	Ⅱ		—	—	—	—																					
7	南涧	水氡	Ⅱ	109						—	—	—	—																
8	玉溪	水位	Ⅰ	111						—	—	—	—																
		水温	Ⅰ							—	—	—	—																
		水氡	Ⅰ							—	—	—	—																
		水汞	Ⅰ							—	—	—	—																
9	元谋	水氡	Ⅰ	114						—	—	—	—																
		电阻率	Ⅰ							—	—	—	—																
		自然电位	Ⅰ							—	—	—	—																
10	大姚	水位	Ⅰ	116						—	—	—	—																
		水温	Ⅰ							—	—	—	—																

续表

序号	台站（点）名称	测项	资料类别	震中距Δ/km	按震中距Δ范围进行异常统计																								
					0＜Δ≤100km					100＜Δ≤200km					200＜Δ≤300km					300＜Δ≤500km					500＜Δ≤800km				
					L	M	S	I	U	L	M	S	I	U	L	M	S	I	U	L	M	S	I	U	L	M	S	I	U
11	峨山	水温	Ⅱ	120						—	—	—	—																
		水准	Ⅱ							—	—	—	—																
		基线	Ⅱ							—	—	—	—																
12	弥渡	水氡	Ⅰ	126						—	—	—	—																
		水汞	Ⅰ							—	—	—	—																
		二氧化碳	Ⅰ							—	—	—	—																
		地倾斜（摆试）	Ⅰ							—	—	—	—																
		地倾斜（连通管）	Ⅰ							—	—	—	—																
		重力	Ⅰ							—	—	—	—																
13	武定	水位	Ⅰ	127						—	—	—	—																
		水温	Ⅰ							—	—	—	—																
14	元江	水氡	Ⅰ	102						—	—	—	—																
15	昆明	地倾斜（摆试）	Ⅰ	131						—	—	—	—																
		重力	Ⅰ							—	—	—	—																
16	江川	水位	Ⅰ	132						—	—	—	—																
		水温	Ⅰ							—	—	—	—																

续表

序号	台站（点）名称	测项	资料类别	震中距Δ/km	按震中距Δ范围进行异常统计																								
					0 < Δ ≤ 100km					100 < Δ ≤ 200km					200 < Δ ≤ 300km					300 < Δ ≤ 500km					500 < Δ ≤ 800km				
					L	M	S	I	U	L	M	S	I	U	L	M	S	I	U	L	M	S	I	U	L	M	S	I	U
17	高大	水位	Ⅰ	133						—	—	—	—																
		水温	Ⅰ							—	—	—	—																
		水氡	Ⅰ							—	—	—	√																
		水汞	Ⅰ							—	—	—	—																
18	澄江	水位	Ⅱ	142						—	—	—	—																
		水温	Ⅱ							—	—	—	—																
		水氡	Ⅱ							—	—	—	—																
19	永仁	水氡	Ⅱ	144						—	—	—	—																
20	云县	地倾斜（摆试）	Ⅱ	145						—	—	—	—																
21	通海	水温	Ⅰ	147						—	—	—	—																
		地倾斜（摆试）	Ⅰ							—	—	—	—																
		水准	Ⅰ							—	—	—	—																
		基线	Ⅰ							—	—	—	—																
		地磁 *D*	Ⅰ							—	√	—	—																
		地磁 *Z*	Ⅰ							—	—	—	—																

续表

序号	台站（点）名称	测项	资料类别	震中距 Δ/km	按震中距 Δ 范围进行异常统计																								
					0 < Δ ≤ 100km					100 < Δ ≤ 200km					200 < Δ ≤ 300km					300 < Δ ≤ 500km					500 < Δ ≤ 800km				
					L	M	S	I	U	L	M	S	I	U	L	M	S	I	U	L	M	S	I	U	L	M	S	I	U
22	石屏	地倾斜（摆试）	Ⅰ	151						—	—	—	—																
23	华宁	水温	Ⅱ	154						—	—	—	—																
24	大理	水位	Ⅰ	154						—	√	—	—																
		水温	Ⅰ							—	—	—	—																
		水质	Ⅰ							—	—	—	—																
		水准	Ⅰ							—	—	—	—																
		基线	Ⅰ							—	—	—	—																
25	曲江	水位	Ⅰ	154						—	—	—	—																
		水温	Ⅰ							—	—	—	—																
		水氡	Ⅰ							—	—	—	—																
		水汞	Ⅰ							—	—	—	—																
		水准	Ⅰ							—	—	—	—																
		基线	Ⅰ							—	—	—	—																
26	小哨	水位	Ⅰ	155						—	—	—	—																
		水温	Ⅰ							—	—	—	—																
		气体	Ⅰ							—	—	—	—																
27	宾川	水位	Ⅱ	159						—	—	—	—																

续表

序号	台站（点）名称	测项	资料类别	震中距 Δ/km	按震中距 Δ 范围进行异常统计																								
					0 < Δ ≤ 100km					100 < Δ ≤ 200km					200 < Δ ≤ 300km					300 < Δ ≤ 500km					500 < Δ ≤ 800km				
					L	M	S	I	U	L	M	S	I	U	L	M	S	I	U	L	M	S	I	U	L	M	S	I	U
28	宜良	地倾斜（摆试）	Ⅱ	166						—	√	—	—																
		水准	Ⅱ							—	—	—	—																
		基线	Ⅱ							—	—	—	—																
29	下关	水氡	Ⅰ	166						—	—	—	—																
		水汞	Ⅰ							—	—	—	—																
		pH 值	Ⅰ							—	—	—	√																
30	景谷	水温	Ⅰ	168						—	—	—	—																
31	建水	水位	Ⅱ	169						—	—	—	—																
		水温	Ⅱ							—	—	—	—																
		水准	Ⅱ							—	—	—	—																
		基线	Ⅱ							—	—	—	—																
32	临沧	水位	Ⅱ	172						—	—	—	—																
		水温	Ⅱ							—	—	—	—																
		水氡	Ⅱ							—	—	—	—																
		地倾斜（摆试）	Ⅱ							—	—	—	—																
33	普洱	水位	Ⅱ	190						—	—	—	—																

续表

序号	台站（点）名称	测项	资料类别	震中距Δ/km	按震中距Δ范围进行异常统计																								
					0＜Δ≤100km					100＜Δ≤200km					200＜Δ≤300km					300＜Δ≤500km					500＜Δ≤800km				
					L	M	S	I	U	L	M	S	I	U	L	M	S	I	U	L	M	S	I	U	L	M	S	I	U
34	昌宁	氟离子	Ⅱ	195						—	√	—	—																
		水氡	Ⅱ							—	—	—	—																
35	寻甸	水温	Ⅱ	195						—	—	—	—																
		水准	Ⅱ							—	—	—	—																
		基线	Ⅱ							—	—	—	—																
36	弥勒	水位	Ⅰ	196						—	—	—	—																
		水温	Ⅰ							—	—	—	—																
		水氡	Ⅰ							—	—	—	—																
		水汞	Ⅰ							—	—	—	—																
分类统计	台项	异常台项数			0	0	0	0		0	4	0	2																
		台项总数			21	21	21	21		83	83	83	83																
		异常台项百分比（%）			0	0	0	0		0	5	0	2																
	观测台站（点）	异常台站数			0	0	0	0		0	4	0	2																
		台站总数			6	6	6	6		30	30	30	30																
		异常台站百分比（%）			0	0	0	0		0	13	0	7																
	测项总数				10					18																			
	观测台站总数				6					30																			

续表

序号	台站（点）名称	测项	资料类别	震中距 Δ/km	按震中距 Δ 范围进行异常统计																								
					$0<\Delta\leqslant100$km					$100<\Delta\leqslant200$km					$200<\Delta\leqslant300$km					$300<\Delta\leqslant500$km					$500<\Delta\leqslant800$km				
					L	M	S	I	U	L	M	S	I	U	L	M	S	I	U	L	M	S	I	U	L	M	S	I	U
备注																													

注 1：一个台编一个序号，按震中距从近到远排序，研究区域以外的台不编号；异常统计范围根据具体震例确定的研究区域而定，可增列更远范围的栏目（见 4.2.2 条）。

注 2：表 A.5 中“资料类别”为Ⅰ类和Ⅱ类者原则上均应填入此表；但是，虽然“资料类别”为Ⅰ类和Ⅱ类而在此次总结中没有分析使用者不应填入此表。如某一观测项目在某一阶段观测资料质量为Ⅰ、Ⅱ类，某一阶段为Ⅲ类，Ⅲ类阶段的资料不作统计（空格）。

注 3：L、M、S、I 栏中，有异常项目在栏中划“√”，无异常划“—”，目前有争议者划“?”，无资料则空格，后二者不计入台站和台项总数。

注 4：U 类异常，在 U 栏中填异常阶段符号，在相应的异常阶段栏中划“√”。

注 5：表 A.4 中研究区域内固定观测的异常项目、项次、和异常阶段应与本表中的一致；测震以外固定观测项目分布图上的测项应与本表一致。

注 6：异常台项——观测台上出现异常的测项，有几个测项出现异常计为几个异常台项；异常台站——有一个以上测项出现异常的观测台即为异常台站；异常台项数—— 各台异常台项的累计数；异常台站数——异常台站的累计数。

注 7：异常百分比取整数（小数点后 4 舍 5 入）。

2005 年 2 月 15 日新疆维吾尔自治区乌什 6.3 级地震

新疆维吾尔自治区地震局

李莹甄　曲延军

摘　要

2005 年 2 月 15 日，新疆乌什县东北英阿瓦提乡发生 6.3 级地震，震中烈度Ⅶ度。该次地震发生在南天山山前地带的 NE 向迈丹断裂上，震中及附近地区部分房屋倒塌和严重破坏，但无人员伤亡。

6.3 级地震序列为主震余震型，最大余震 5.2 级，余震区长轴 NW 向（精定位结果为 NS 向）。震源机制解主压应力 P 轴方位 160°，节面Ⅰ走向 NE，倾角 75°，与南天山西段区域应力场和迈丹断裂运动属性较为一致，6.3 级地震是在 NNW 向区域应力场作用下，迈丹断裂发生逆冲运动所致。

震中附近地区观测台站不多，300km 范围内共有测震台 6 个，前兆台 9 个。6.3 级地震前共出现 5 个前兆异常项目共计 12 条异常。震前震中周围地区 4 级以上地震频次明显降低，4 级以上地震活动空间分布出现空段，震前 1 个月震中附近出现小震条带。距震中 230km 的巴楚地震台土层应力震前出现明显的短临异常，哈萨克斯坦境内部分流体测点水化学测项出现突出异常。

前　言

据新疆地震台网测定，2005 年 2 月 15 日 7 时 38 分在新疆维吾尔自治区乌什县北偏东约 60km 的英阿瓦提乡农一师 4 团附近发生 6.3 级地震。微观震中北纬 41.72°，东经 79.37°，震中烈度Ⅶ度，加速度 349Gal。新疆乌什县、阿克苏市和阿合奇县等地强烈有感。

地震发生后，新疆地震局立即启动破坏性地震应急预案，及时了解灾情，召开会商会研判震情，会同中国地震局组织地震现场工作队赶赴灾区，开展震情监视、震害损失评估、震后科考等工作。在持续近 1 个月的余震序列跟踪与震情判定中，较好地判定了震后趋势。

6.3 级地震极震区南部为人口密集区，西北大部为无人区。极震区震感强烈，地震没有造成人员伤亡。震害主要表现为土木结构的房屋多数出现裂缝，少数倒毁；砖木房屋墙体出现微小裂缝，个别开裂；砖混房屋基本完好；山谷陡壁局部出现小型滑坡、塌方、崩塌。

这次地震发生在新疆中强地震活动较为频繁的乌什地区。自 1893 年以来震中周围

100km 范围内先后发生过 7 次 6 级地震，这次地震前该区 6 级地震平静 17 年。

在前人对此次地震研究的基础上[1~21]，经过重新整理和研究，认定这次地震前共有 5 个前兆异常项目，共计 12 条异常。

一、测震台网及地震基本参数

图 1 为乌什地震前震中 300km 范围内的地震台站分布，这一范围内总计有 6 个测震台，大部分为模拟记录，有两个台站在 2004 年改造成数字测震台，其中 100km 范围内有乌什和阿克苏测震台，乌什台 2004 年改造成为数字测震台；101～200km 有 2 个测震台，分别是阿合奇台和拜城台；201～300km 也有 2 个测震台，分别是巴楚台和库车台，其中巴楚台 2004 年改造成为数字台。震中附近是新疆地震监测能力相对较弱的地区，研究时段内新疆境内基本只能保证 $M_S \geqslant 2.0$ 级地震不遗漏。

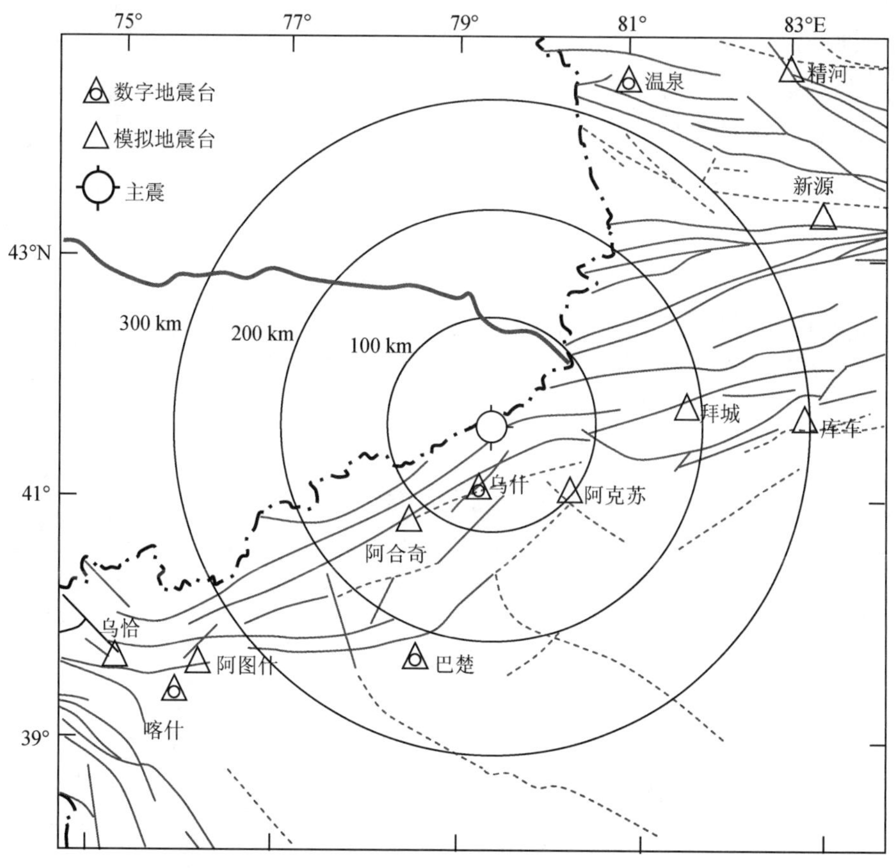

图 1 乌什 6.3 级地震前震中附近测震台站分布图

Fig. 1 Distribution of Seismic Stations Near the Epicenter of Wushi 6.3 Earthquake

表 1 列出不同机构给出的这次地震基本参数。经对比分析，认为新疆维吾尔自治区地震局采用近台交切定位经重新修订的震中位置更为准确，本文采用这一结果。

表 1　乌什 6.3 级地震的基本参数

Table 1　Basic Parameters of Wushi 6.3 Earthquake

编号	发震日期	发震时刻	震中位置（°）		震级	震源深度	震中地名	结果来源
	年 月 日	时 分 秒	φ_N	λ_E	M	(km)		
1	2005 2 15	07 38 09	41.72	79.37	M_S6.3	34	乌什	新疆地震局修订
2	2005 2 15	07 38 11	41.69	79.49	M_S6.0	17	乌什	中国台网中心
5	2005 2 15	07 38 10	41.60	79.30	M_S6.2	33	乌什	国家台网大震速报
6	2005 2 15	07 38 10	41.60	79.20	M_S6.2	33	乌什	国家台网日报目录
7	2005 2 15	07 38 10	41.60	79.20	M_S6.2	33	乌什	全国地震目录
8	2005 2 14（国际标准时）	23 38 09	41.66	79.57	M_S6.2	27		BJI（国际地震中心网站提供）
	2005 2 14（国际标准时）	23 38 07	41.75	79.40	M_w6.2	24		MOS（国际地震中心网站提供）
	2005 2 14（国际标准时）	23 38 08	41.73	79.44	M_w6.0	22		NEIC（国际地震中心网站提供）
	2005 2 14（国际标准时）	23 38 08	41.72	79.27	M_w6.1	23		HRVD（国际地震中心网站提供）
	2005 2 14（国际标准时）	23 38 08	41.67	79.34	M_S6.1	31.4		ISC（国际地震中心网站提供）
	2005 2 14（国际标准时）	23 38 09	41.71	79.44	M_S6.1	25.9		IDE（国际地震中心网站提供）
	2005 2 14（国际标准时）	23 37 49	39.84	81.97	M_b6.0	22		BGS（国际地震中心网站提供）

二、地震地质背景

乌什 6.3 级地震发生在塔里木地块西北缘与天山南缘断陷褶皱带衔接地带，震中位于南天山柯坪断块与库车坳陷的构造结合部位。区域构造走向 NE，大都具有强烈的新构造运动特征[1]（图 2）。

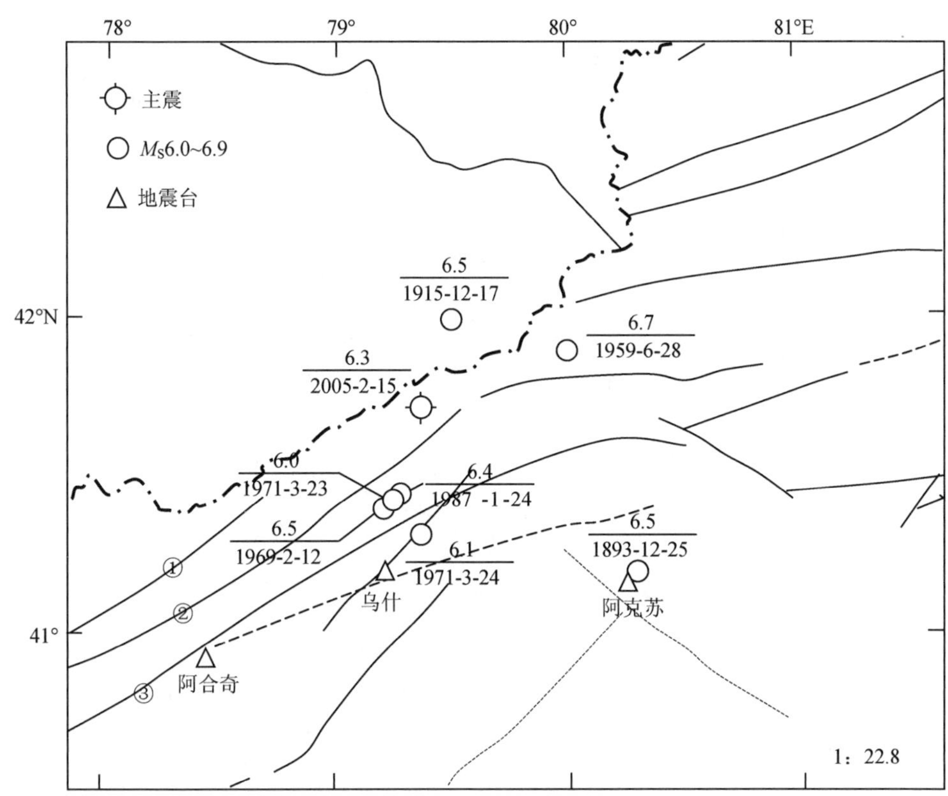

图 2　震中周围地区历史地震分布（1800～2005 年，$M \geqslant 6.0$ 级）与主要活动断层

Fig. 2　Earthquake History and Major Active Faults Surrounding the Epicenter

主要断裂编号：①阔克萨勒断裂；②迈丹断裂；③阿合奇断裂

1. 活动构造

该区主要活动断裂有 NE 向的阔克萨勒断裂、麦丹断裂和阿合奇断裂。阔克萨勒断裂沿阔克萨勒岭南坡延伸，总体走向为 N50°～70°E，断面倾向 NW，倾角 50°～60°，长逾 40km，为左旋逆断层；迈丹断裂走向 NE，倾向 NW，倾角 60°～70°，为左旋逆走滑，在乌什县沙伊拉姆为背斜逆断层带，是塔里木地块与天山山脉的分界断裂；阿合奇断裂长 200km，走向 NEE，倾向 333°，倾角 20°～50°，逆冲性质。

2. 深部构造

由北向南本区重力布格异常值介于 $-250 \times 10^{-5} \sim 200 \times 10^{-5} \mathrm{m/s^2}$ 之间[2]，地壳厚度约为 50km，为重力布格异常和地壳厚度变化的梯度带，是中强地震的多发区[3]。本区 41°N 以南地表 10km 以下为高速区，介质密度较大，完整性较好。41°～42°N 为低速区，介质密度较小且较破碎[4]，此次地震就发生在该低速区内。

3. 形变场

由图 3 新疆地区运动矢量图可以看出，柯坪断裂附近地壳形变差异非常显著，其北部中天山地区表现为较大幅度的南向运动，而南部的乌恰-喀什交汇区则存在相对向北的运动趋势，显示柯坪断裂附近存在最大剪应变高值区[5,6]。

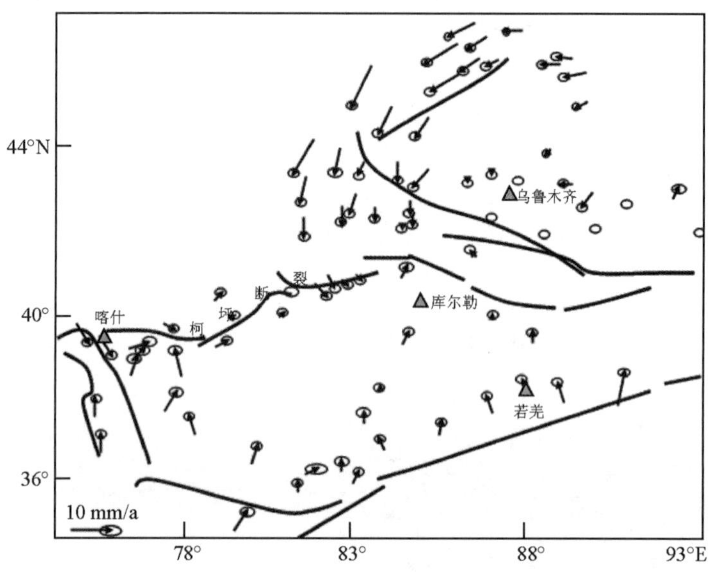

图 3　新疆地区水平运动矢量图 (2001 ~ 2004 年)[6]

Fig. 3　The Direction Field of Horizontal Motion in Xinjiang (2001-2004)

阿克苏以东的塔里木盆地是相对稳定的盆地，而震中区的垂直运动速率为 1 ~ 2mm/a[4]，无论天山山区还是塔里木盆地均以此垂直活动速率在隆起或沉降，隆起走向为 NNE 向，该区复杂的地质构造和强烈的新构造运动有利于发生中强地震。

4. 地震活动性

1893 年以来乌什附近发生过 7 次 6 级以上地震 (图 2)，属新疆中强地震活动较为频繁的地区之一。1988 ~ 2004 年该区没有发生 6 级以上地震，处于平静状态，但在其外围地区强震活跃，先后发生 1997 ~ 1998 年伽师 6 级强震群、1996 ~ 1998 年阿图什两次 6 级地震、2003 年巴楚—伽师 6.8 级地震和昭苏 6.1 级地震。

三、地震影响场和震害1)

1. 地震列度与震害

根据对震区各考察点的调查和烈度综合评定，确定了乌什 6.3 级地震烈度等震线图 (图 4)，包括Ⅵ度和Ⅶ度两个区。由于受自然地理环境和气候条件限制，西部和北部大部分地区缺乏烈度调查的参照点，等震线未能圈闭。本次地震的烈度等震线，大体沿 NE 方向断裂带展布。

(1) Ⅶ度区 (极震区)：包括边防执勤点以南、马场二组以北、闸口以西地区。区内包括英阿瓦提乡的库齐村、贡格拉提村、亚克艾里克村以及牙满苏乡的下牙满苏村、尤里吐孜布拉克村和伊麻木乡马场，还有农一师四团、四团 13 连、14 连等主要人口聚集地，居民点集中分布在Ⅶ度区南部，Ⅶ度区的西北大部分为无人区。Ⅶ度区内人普遍震感强烈，大多数人仓皇出逃；多数土木结构的房屋出现裂缝，少数倒毁，砖木房屋的墙体出现微小裂缝，个别开裂；砖混房屋基本完好；少数不稳定器物翻到；山谷的陡壁局部出现小型滑坡、塌方、崩塌。

（2）Ⅵ度区：包括喀尔吐尔村以南、古鲁克村以北（不包括古鲁克村）、奥特贝希牧场学校以东、四团民族连二排以西。主要包括乌什县城、博孜村、麦盖提农场、亚克瑞克乡的部分村子、亚贝西村以及四团 18 连、民族连和温宿县阿热勒乡结格吉村等主要人口聚集地。Ⅵ度区人普遍震感较强烈，部分人仓皇出逃，多数土木结构的房屋出现细小裂缝，个别严重破坏及倒毁，砖木房屋的墙体出现微小裂缝，个别开裂；砖混房屋基本完好。

（3）Ⅴ度区：主要包括乌什县阿合雅乡的尤喀可库曲麦村、果鲁克村，温宿县阿热勒镇的依尔玛村、皮亚兹其村、阿热勒镇，恰格拉克乡的喀拉萨村、吉格代村、代扑台尔村、恰格拉克以及温宿电厂。Ⅴ度区内，室内的人普遍有感，室外的人多数有感；土木房屋门窗、屋顶颤动作响，灰土掉落，墙皮局部出现微细裂缝；极个别土木房屋的墙皮掉落、残旧木梁断裂。室内悬挂物明显摆动，个别不稳定器物翻到。

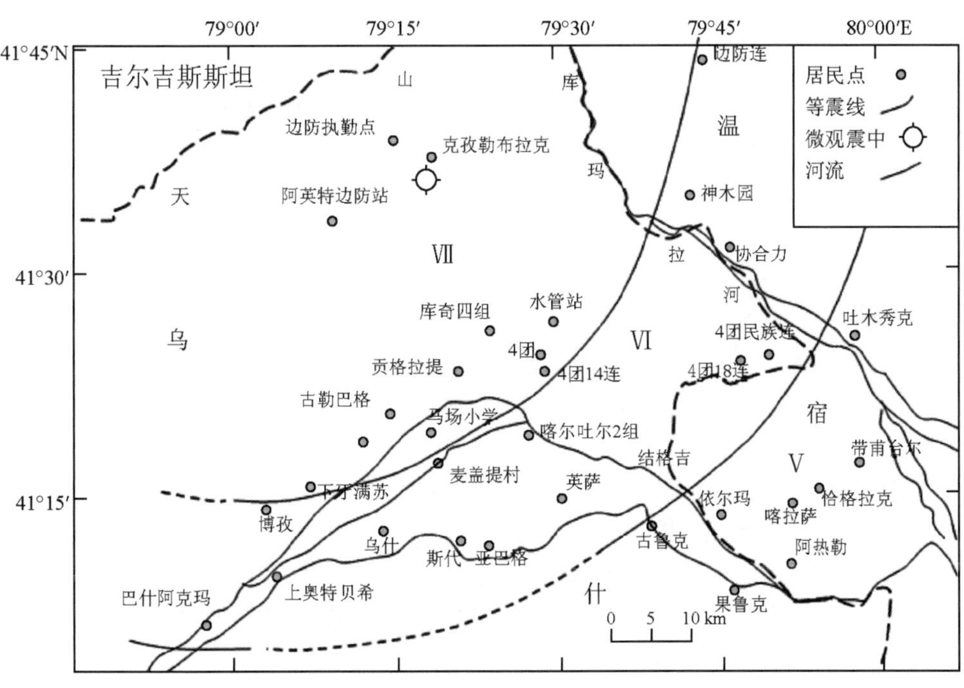

图 4 2005 年 2 月 15 日乌什 6.3 级地震等震线图

Fig. 4 Isoseismic Lines of Wushi 6.3 Earthquake, 15, Feb., 2005

2. 乌什 6.3 级地震的强震加速度记录[7]

设在 4 团和乌什的强震仪记录到了这次地震的加速度时程，记录到的最大加速度为 78Gal。4 团强震台位于Ⅶ度区内，为模拟记录，台址位于居民平房内，地基为黄土。乌什强震台位于Ⅵ度区内，为数字地震记录，地基为基岩。基本数据见表 2。乌什台站的原始地震加速度记录见图 5。

表2　乌什 6.3 级地震的强震加速度记录初步结果

Table 2　Preliminary Result of Acceleration Record of Wushi 6.2 Earthquake

台站	台站位置		震级	仪器型号	影响烈度	震中距（km）	峰值加速度(Gal)／周期		
	东经	北纬					SN	EW	UD
4 团强震台	79°28′18″	41°24′24.4″	6.3	GQⅢ-A	Ⅶ	26	78	52.5	22.4
乌什数字强震台	79°12′15″	41°12′12.7″	6.3	GDQJ-1A	Ⅵ	48	12.3	17.7	9.3
							1.17s	1.26s	1.17s

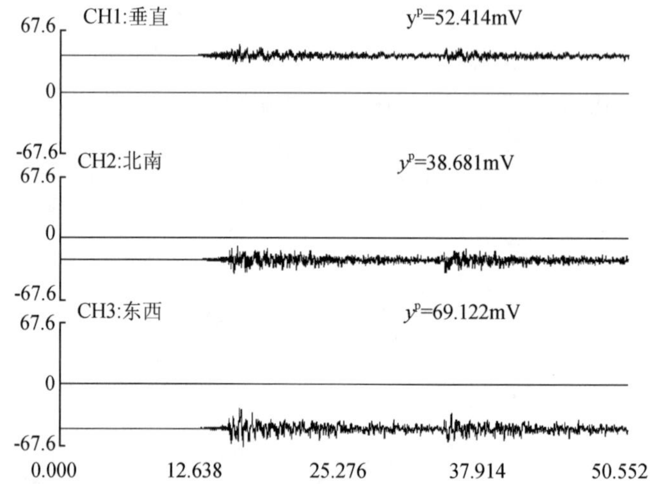

图5　乌什 6.3 级地震的原始数字强震加速度记录（乌什强震台）

Fig. 5　Acceleration Record of Wushi 6.3 Earthquake (Original Data, Wushi Station)

采用中国西部烈度衰减公式[8]计算 4 团的烈度为 $I_a=6.8$、$I_b=6.2$，乌什的烈度为 $I_a=6.1$、$I_b=5.5$，与这次地震烈度符合较好。

4 团及乌什台均位于《中国地震动参数区划图》GB18306-2001 的加速度峰值 0.2g 区内，采用中国西部加速度衰减公式[8]计算出 4 团台的加速度峰值为 87.5Gal，乌什台的加速度峰值为 49.25Gal，震中区加速度值为 349Gal。

四、地 震 序 列

2005 年 1 月 17 日在距离乌什地震震中 35km 的吉尔吉斯斯坦国境内先期发生 5.0 级地震，表明震前震源区附近存在地震活动的增强现象。主震后截至 2005 年 4 月 9 日，新疆地震台网记录 1.0 以上余震 44 次，其中 5.0～5.9 级 1 次，4.0～4.9 级 1 次，3.0～3.9 级 3 次，2.0～2.9 级 10 次，1.0～1.9 级 29 次，最大余震为 5.2。表 2 给出 3.0 级以上地震的序列目录。

由双差定位方法确定乌什地震序列震中分布[9]，并按余震的展布方向由南向北做震源深度剖面图（图6），这次地震以单侧破裂为主，余震大多位于主震以北，破裂延伸方向为近 N 方向。主震发生后前 3 天的余震分布尺度和整个地震序列的余震分布尺度基本相当。

表 3　乌什 6.3 级地震序列目录（$M_S \geqslant 3.0$ 级）

Table 3　Earthquake Sequence List of Wushi 6.3 Earthquake（$M_S \geqslant 3.0$）

编号	发震日期	发震时刻	震中位置		震级	震源深度	震中地名	结果来源
	年 月 日	时 分 秒	φ_N	λ_E	M_S	(km)		
1	2005 2 15	07 38 09	41.72	79.37	6.3	34	乌什	新疆地震局修订
2	2005 2 15	08 06 12	41.69	79.36	3.0	24		
3	2005 2 15	19 16 13	41.76	79.39	5.2	23		
4	2005 2 15	19 33 56	41.71	79.32	4.3	35		
5	2005 2 17	16 36 09	41.67	79.24	3.3	29		
6	2005 2 17	21 32 35	41.69	79.31	3.2	28		

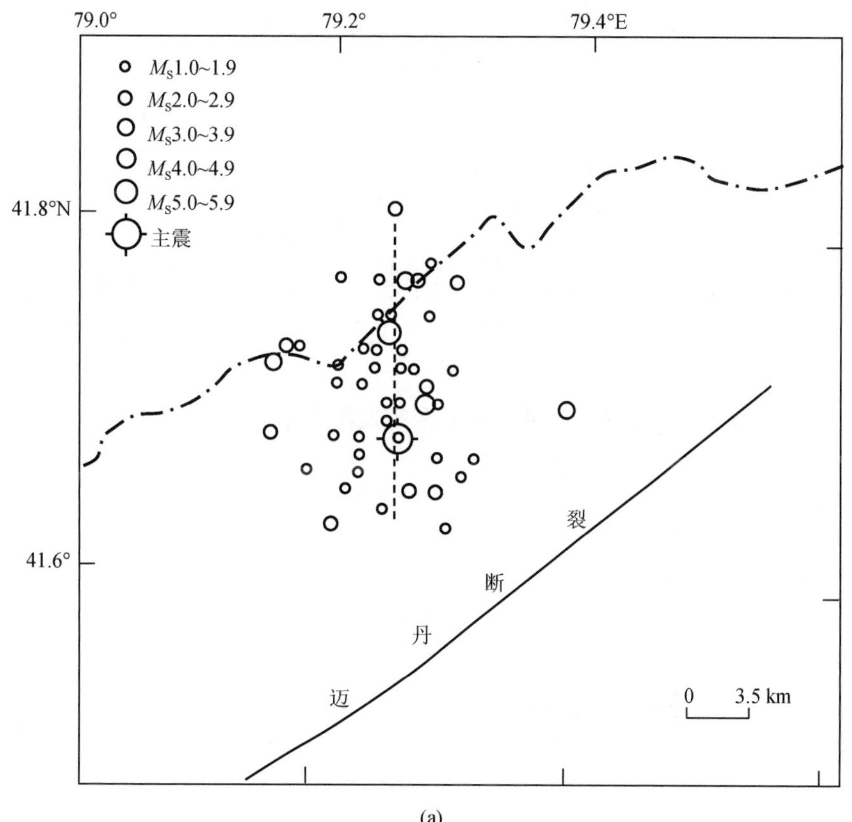

(a)

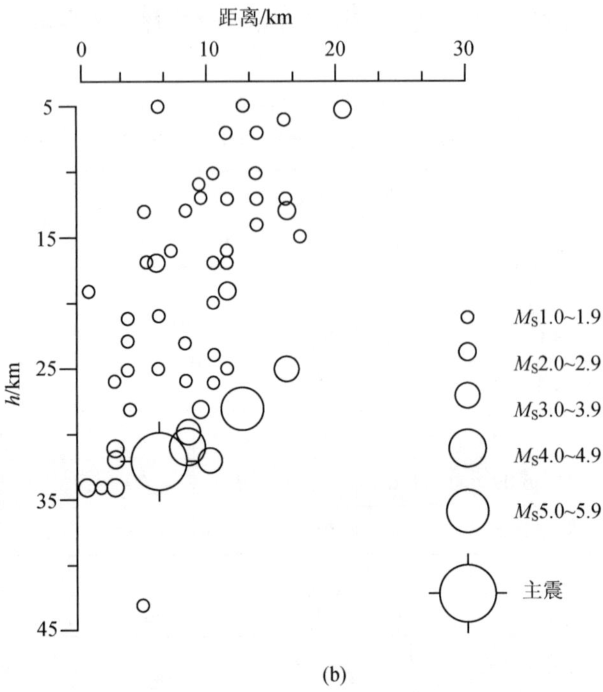

图6　乌什 6.3 级地震序列震中分布（a）与震源深度分布（b）（双差定位结果）

Fig. 6　Epicenter and Focal Depth Distribution of Wushi 6.3 Earthquake (Double Difference Positioning Result)

主震震源深度 34km，余震集中分布在地下 5 ~ 32km，且由南向北逐渐变浅，倾角约 60°。这与迈丹断裂 60° ~70°的倾角基本吻合；但震源深度由南向北逐渐变浅的趋势与迈丹断裂倾向不符。

由余震序列的 *M-t* 图（图7）可见，最大余震发生在主震后 12 小时内，3 级以上余震主

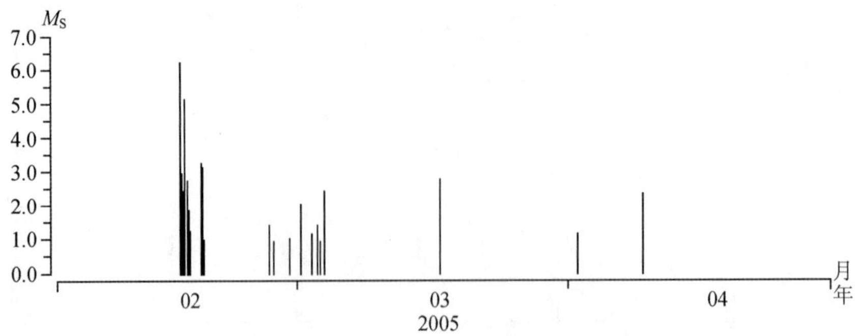

图7　乌什 6.3 级地震序列 *M-t* 图

Fig. 7　*M-t* Plot of Wushi 6.3 Earthquake Sequence

要集中在主震后 3 天。4 月 6 日主震西南 57km 处发生 5.0 地震后，余震序列迅速衰减，至 9 日完全平静。

地震序列频次和能量图显示（图8、图9），主震后3 天余震频次显著高于其他时段，为剩余应力集中释放时段，释放了整个序列 99% 以上的能量，之后序列进入调整期，持续约 15 天，而后迅速衰减至背景活动状态。

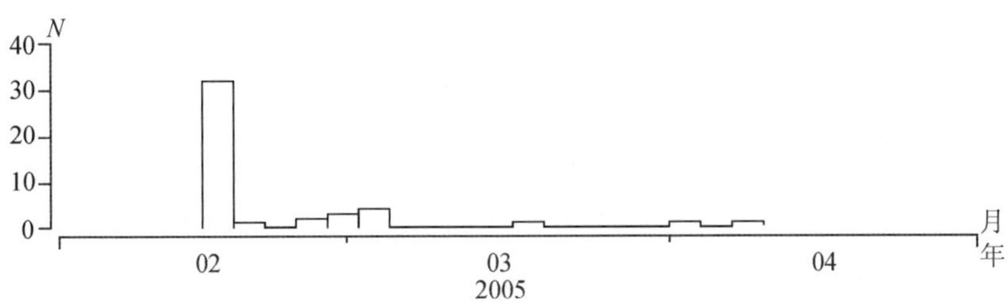

图 8 乌什 6.3 级地震序列频度 N-t 图（3 日为单位）

Fig. 8 N-t Plot of Wushi 6.3 Earthquake Sequence (Dimension of 3 Days)

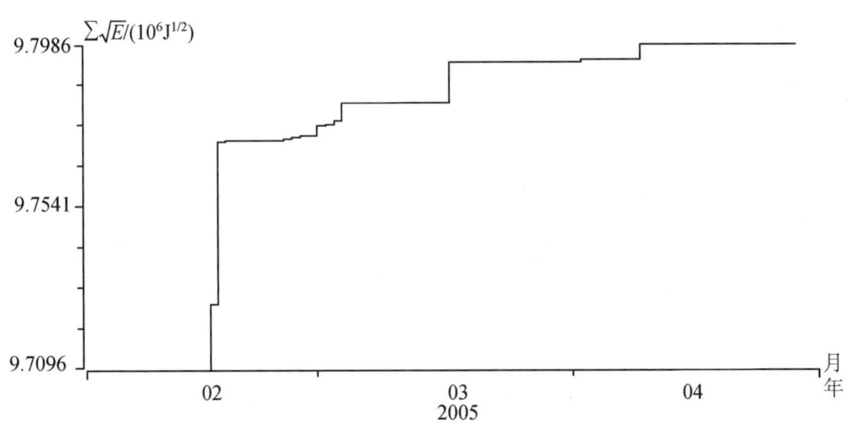

图9 乌什 6.3 级地震序列蠕变图（M_S≥1.0 级）

Fig. 9 Wriggle plot of Wushi 6.3 Earthquake Sequence (M_S≥1.0)

余震序列 b 值为0.38（图10），远低于南天山地震带平均为0.7 的背景值，序列 P 值为 1.06，h 值为 1.0，主震释放能量占全序列的能量的 99.6%，主震与最大余震的震级相差 1.1 级，根据序列类型的判别指标，乌什 6.3 级地震为主—余型序列。

采用乌什台记录的乌什地震 M_S2.5 以上余震波形，利用 S 波段的频谱特性，计算余震序列的应力降[10]（图 11）。主震应力降为 312.76bar，最大余震应力降为 95.87bar，次大余震的应力降为 114.36bar，2.5 级以上地震序列应力降衰减迅速。

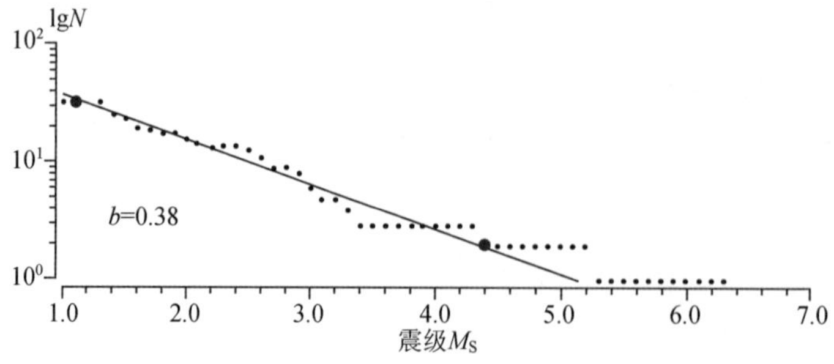

图 10　乌什 6.3 级地震序列频度 lgN-M 图（起始震级 M_S1.3）

Fig. 10　lgN-M plot of Wushi 6.3 Earthquake Sequence（Initial M_S = 1.3）

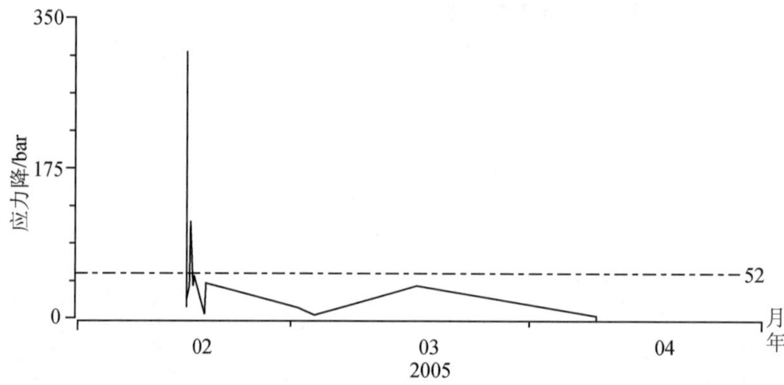

图 11　乌什 6.3 级地震序列应力降图（$M_S \geqslant 2.5$ 级）

Fig. 11　The Decay of Stress of Wushi 6.3 Earthquake Sequence（$M_S \geqslant 2.5$）

五、震源参数和地震破裂面

根据新疆区域台网 30 次清晰的 P 波初动符号，测定了这次 6.3 级地震的震源机制解，矛盾符号比 0.27。另外还收集到中国地震台网和国外一些研究机构做出的震源机制解，与新疆地震局的震源机制解对比显示，除中国地震台网中心的结果外基本一致（表 4、图 12）。由此，以下以新疆地震局结果为主进行论述。

表 4 乌什 6.3 级地震的震源机制解

Table 4 Focal Mechanism Solutions of Wushi 6.3 Earthquake

序号	节面 I			节面 Ⅱ			P 轴		T 轴		N(B)轴		结果来源
	走向	倾角	滑动角	走向	倾角	滑动角	方位	仰角	方位	仰角	方位	仰角	
1	61	71	80	270	22	118	160	30	310	67	64	10	新疆地震局
2	216	77	-91	46	12	-80	124	57	308	33	217	19	中国地震台网中心
3	321	10	152	79	85	81	177	40	340	49	80	9	哈佛大学
4	320	13	148	81	83	79	181	37	339	51	82	11	USGS
5	300	8	133	76	84	85	171	39	340	50	77	5	哈佛大学全球快速 CMT
6	80	75	80	294	18	123	178	29	336	59			NEIC

此次地震断错性质为逆断倾滑型。根据区域断层属性以及极震区烈度和余震分布综合判定，NE 向的节面 I 为断层面，倾角 71°，断面较陡。主压应力 P 轴方位 160°，与区域应力场方向较为一致，P 轴仰角 30°，张应力 T 轴仰角 67°，存在一定的左旋分量。震源机制解与南天山西段区域应力场和迈丹断裂运动属性较为一致，表明此次 6.3 级地震是在 NNW 向区域应力场作用下，迈丹断裂发生逆冲错断所致。对比余震序列震中分布图和震源深度分布

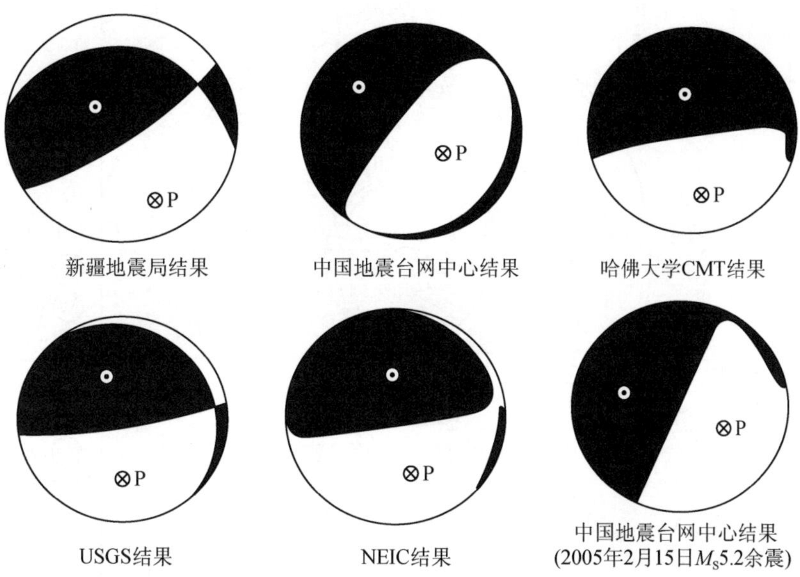

图 12 国内外不同机构给出的乌什 6.3 级及其 5.2 级余震震源机制解

Fig. 12 Focal Mechanism Solutions given by Different Institutions in China and Abroad, Concerning Wushi 6.3 Earthquake and Its 5.2 Aftershock

图6，本次地震属于逆冲型地震，且以单侧破裂为主，由深向浅，沿近N方向破裂，显示出断层下盘向西北方向的下插作用较强。中国地震台网中心给出了5.2级余震的震源机制解，与主震同为倾滑正断性质这与迈丹断裂的运动性质不符，难以解释这次地震的成因。

六、地震前兆观测台网及前兆异常

乌什地震发生在柯坪断块东端，区内台站主要沿南天山构造带分布，震中东南塔里木盆地无台站分布，震中西北有部分哈萨克斯坦国的流体前兆台站。上世纪九十年代起，中哈两国地震科技交流日趋活跃，彼此各有5个流体监测点包括水化学、水温、气体等方面的资料进行互换交流。

由震中附近前兆台站及观测项目分布图可见（图13），震中300km内共有9个前兆观测台，包括地倾斜、钻孔应变、土层应力、断层蠕变、水化学等观测项目，境内绝大部分台站设有测震观测。其中100km范围内有前兆观测台2个，共3个前兆测项；101～200km有前兆观测台2个，2个测项；201～300km有前兆观测台5个，5个测项。由于此次地震前新疆前兆台网正在进行数字化改造，一些模拟前兆观测停测，数字记录又处于试运行阶段，资料不连续。总体上，乌什地震前共出现包括地震学在内的5个异常项目12条异常（表5）。

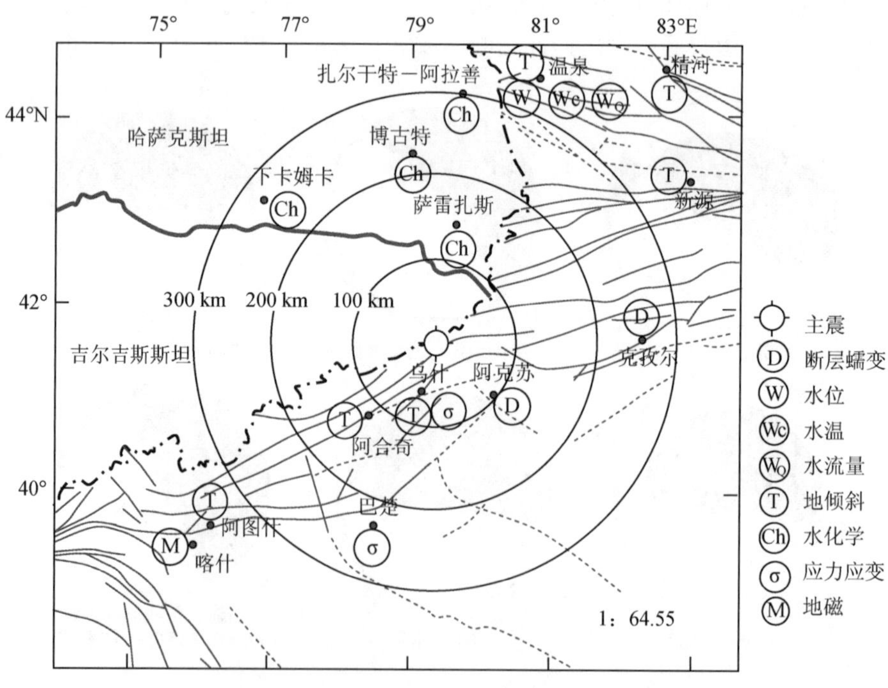

图13　乌什地震前震中附近地震前兆观测台及主要测项分布

Fig. 13　Precursory Observation Stations and Major Item Distriution near the Epicenter of Wushi Earthquake

表 5 地震前兆异常登记表

Table 5 Precursory Anomalies Record

序号	异常项目	台站（点）或观测区	分析方法	异常判据及观测误差	震前异常起止时间	震后变化	最大幅度	震中距（km）	异常类别及可靠性	图号	异常特点及备注
1	地震活动增强平静	震中 140km 范围内	M_S≥4.0 级地震 M-t 图	4 级地震活动 1 年≥3 次；地震平静	2001.9～2005.2	恢复正常		震中周围	M_2	14	震前 27 天震中附近出现 5 级地震
2	4 级地震空段	天山地震带	M_S≥4.0 级地震空间图像	4 级地震明显平静区	2004.10～2005.1	消失	空段长 360km	震中周围	M_1	15	
3	小震条带	震中附近	M_L≥3.0 级地震图像	M_L≥3.0 级集中成带	2005.1.2～2.12	消失	条带长轴 330km		S_1	16	震中位于北西向条带内
4	η 值	以北纬 41.07°、东经 79.52°为中心沿北东向构造长轴 141km、短轴 105km、矩形范围	M_S≥2.0 级时间进程分析（12 月窗长，2 月滑动）	持续 1 年低于 2.09	2003.6～2005.1	恢复	最低值	震中周围	M_1	17	异常幅度较大，地震发生在恢复后
5	D 值	同上	同上	持续 1 年高于 0.158	2004.4～2005.5	未恢复	最高值	震中周围	M_1	18	地震发生在异常过程中
6	地震频度	同上	同上	1 年以上持续低于 26 次	2002.5～2004.5	恢复	最低值	震中周围	M_1	19	地震发生在异常恢复后
7	地震活动度 S 值	同上	同上	持续低于 3.83 年以上	2000.1～2004.12	恢复	最低值	震中周围	M_1	20	异常恢复后发震
8	b 值	同上	同上	持续低于 0.482 半年以上	2003.5～2004.9	出现新异常	最低值	震中周围	M_2	21	异常恢复后发震

续表

序号	异常项目	台站（点）或观测区	分析方法	异常判据及观测误差	震前异常起止时间	震后变化	最大幅度	震中距（km）	异常类别及可靠性	图号	异常特点及备注
9	S波分裂	震中附近	慢S波时间延迟	时间延迟加大	2003.8 ~ 2004.12	恢复背景状态	7ms/km	震中周围	M_1	22a	地震发生前延迟恢复
			快S波偏振方向	偏振方向发生偏转	2003.8 ~ 2004.11	恢复	N30° ~ 80°E为优势方向		M_2	22b	震前1.5年偏离背景应力方向
10	土层应力	巴楚	单分量日速率	在均值附近大幅波动	N59°W：2004.11.23 ~ 2005.2.5	恢复	34/kPa	240km	S_1	23	异常最大幅度时间到发震在1个半月内
					N1°E：2004.12.17 ~ 2005.02.05	恢复	21.5/kPa	240km	S_1	23	
					N16°E：2004.12.15 ~ 2005.02.05	恢复	80.7/kPa	240km	S_1	23	
					垂直向：2005.01.03 ~ 2005.01.30	恢复	153/kPa	240km	I_1	23	
11	水位	博古特井	日均值	年变畸变	2004.09.23 ~ 2005.01.15	恢复	20cm	220km	S_1	24	在异常持续过程中，距震中190km的温泉发生5.1级地震，震后异常没有恢复

续表

序号	异常项目	台站（点）或观测区	分析方法	异常判据及观测误差	震前异常起止时间	震后变化	最大幅度	震中距（km）	异常类别及可靠性	图号	异常特点及备注
12	水化学	下卡姆卡5-T井	钙离子、氯离子日均值	高值突跳	2004.12.11 ~12.16	恢复	钙离子：19.0mg/L 氯离子：36.9mg/L	270km	S_1	25	异常出现64天后发震

1. 地震学异常

1）4级地震增强—平静

由图14，此次地震前在距离震中约150km范围内4级地震存在增强—平静现象。2001年至2003年4级地震平均1年2次左右。1997～1999增强活动之后，2000～2001年4级以上地震出现长时间平静，与此相伴新疆境内也出现大范围4级以上地震平静，2001年11月14日昆仑山口西发生8.1级地震；2004年2～12月，震中附近4级地震再次平静11个月，1个月后发生乌什6.3级地震。

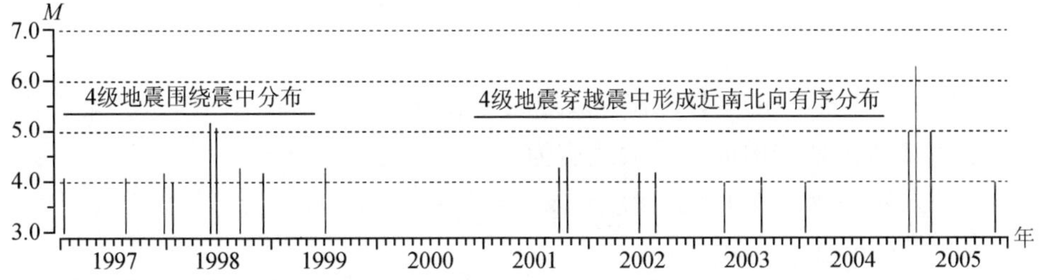

图14　1997～2005年乌什6.3级地震震中140km范围内4级以上地震 M-t 图

Fig. 14　M-t Plot of 4.0 and above Earthquakes, in a radius of 140km of the Epicenter of Wushi 6.3 Earthquake, 1997-2005

从全疆4级地震分布看（图15），2004年10月至2005年1月新疆4级地震活动主要集中在柯坪地块西段和中天山地区，而在乌什地区形成明显的空段。2005年1月19日在未来震中附近的空段内发生一次5.0级地震，紧接着2月15日空段内发生乌什6.3级地震。

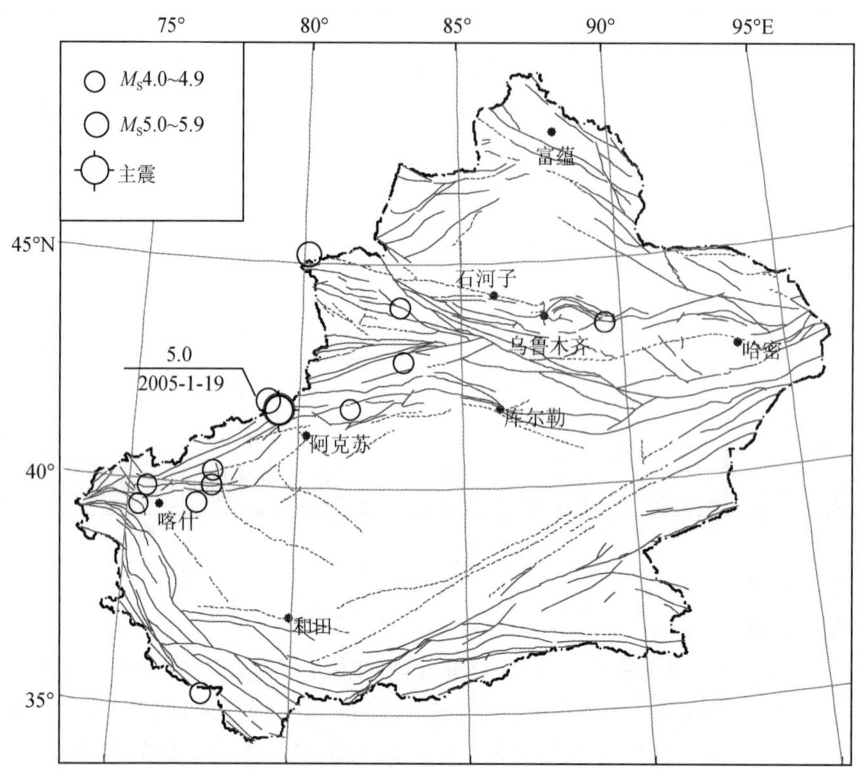

图 15　2004 年 10 月至 2005 年 2 月 14 日新疆 4 级以上地震分布

Fig. 15　M4.0 and above Earthquake Distribution in Xinjiang, during Oct 2004 and 14, Feb, 2005

2）小震条带[11]

2005 年 1 月 2 日至 2 月 12 日，在 4 级地震空段内迅速形成北西向的 M_L3 地震条带（图 16），由 7 个 M_L3 以上地震构成，最大地震为 M_L5.3（正式目录修订为 M_S5.0），该条带与 NE 向迈丹断裂大体正交，乌什 6.3 级地震发生在条带内的空段，与 5.3 级地震相距 35km。

3）地震学参数异常

以北纬 41.07°、东经 79.52°为中心，沿北东向构造选取长轴 140km，短轴 100km 矩形区域内的 M_S2.0 以上地震，以 12 个月为窗长，2 个月为滑动步长，进行乌什地震附近多种地震学参数时间进程扫描分析。

乌什 6.3 级地震前，从 2003 年 6 月起 η 值出现 20 个月的低值异常（$\eta < 2.09$），2005 年 1 月恢复后发震（图 17）。

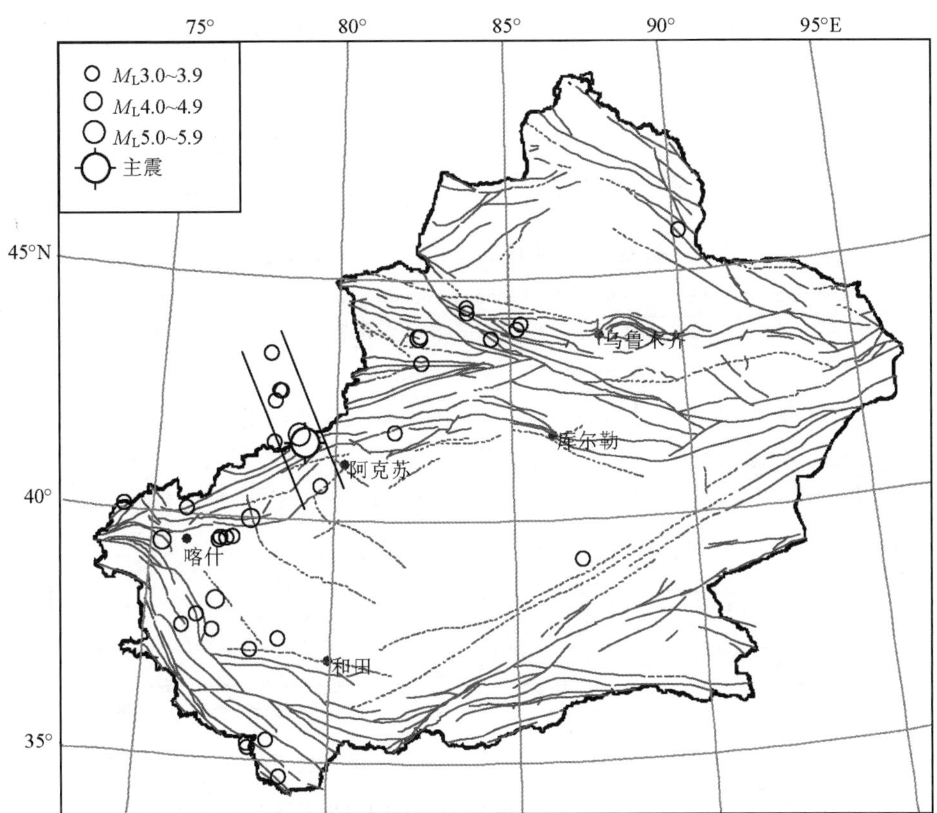

图16 乌什6.3级地震前小震活动图像（据新疆地震快报目录2005.01.02~2005.02.12）

Fig. 16 Minor Earthquake Activities before Wushi 6.3 Earthquake

（According to Letters, 2, Jan, 2005 – 12, Feb, 2005）

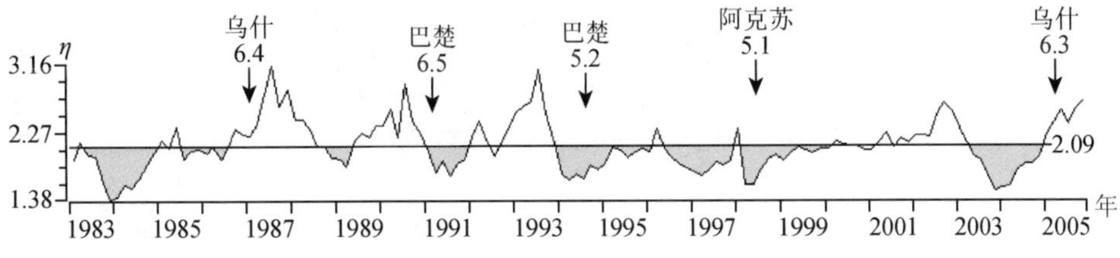

图17 乌什6.3级地震震中周围地区1982年以来 η 值曲线（$\geqslant M_S 2.0$）

Fig. 17 Curve since 1982 in the surrounding area of the Epicenter of Wushi 6.3 Earthquake（$\geqslant M_S 2.0$）

地震危险度 D 值在2004年4月至2005年5月出现高值异常（$D>0.158$），乌什地震发生在高值异常中，震后两个月异常恢复（图18）。

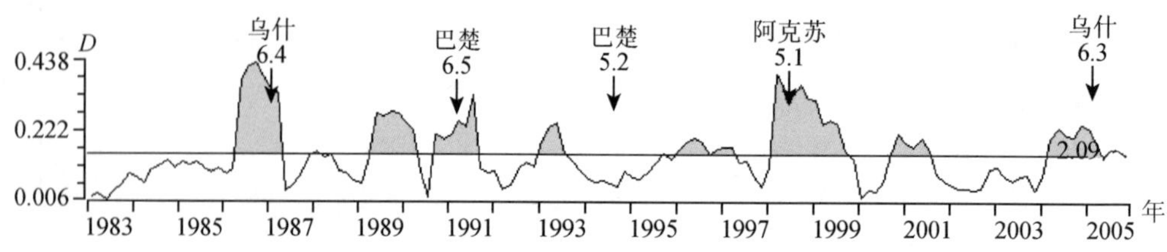

图 18　乌什 6.3 级地震震中周围地区 1982 年以来 D 值曲线（$\geq M_S 2.0$）

Fig. 18　D curve since 1982 in the surrounding area of the Epicenter of Wushi 6.3 Earthquake（$\geq M_S 2.0$）

2 级地震频度在 2002 年 5 月至 2004 年 5 月出现持续 2 年的低值异常（$N<26$），之后地震活动逐渐增强，在最活跃时段发生乌什 6.3 级地震，距离平静异常结束 8 个月（图 19）。

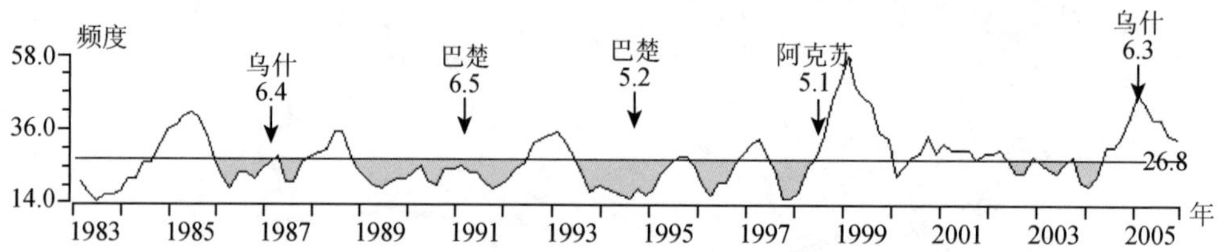

图 19　乌什 6.3 级地震震中周围地区 1982 年以来频度值曲线（$\geq M_S 2.0$）

Fig. 19　Frequency curve since 1982 in the surrounding area of the Epicenter of Wushi 6.3 Earthquake（$\geq M_S 2.0$）

地震活动度 S 值 2000 年 1 月至 2004 年 12 月出现持续 5 年的低值异常（$S<3.83$），震前 1 个月高值回返，震后维持在高值状态（图 20）。

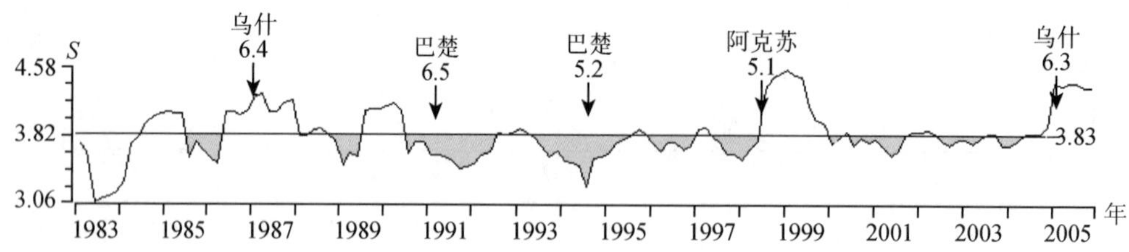

图 20　乌什 6.3 级地震震中周围地区 1982 年以来 S 值曲线（$\geq M_S 2.0$）

Fig. 20　Scurve since 1982 in the surrounding area of the Epicenter of Wushi 6.3 Earthquake（$\geq M_S 2.0$）

b 值在 2003 年 5 月至 2004 年 9 月连续出现低于 0.8 倍均方差的低值异常（$b<0.482$），10 月开始恢复到均值线附近，乌什地震发生在异常恢复后，距离低值异常结束 4 个月（图 21）。

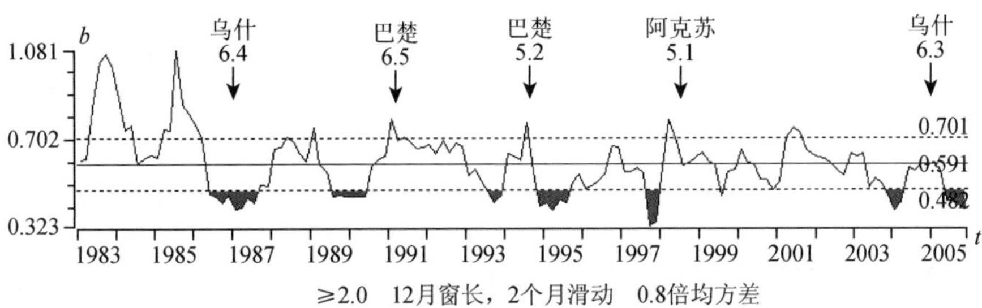

图 21 乌什 6.3 级地震震中周围地区 1982 年以来 b 值曲线

Fig. 21 b curve since 1982 in the surrounding area of the Epicenter of Wushi 6.3 Earthquake（$\geqslant M_S 2.0$）

4）震源区附近 S 波分裂[12]

1984 年 Crampin 提出 EDA 假说[13]，认为地壳岩石中普遍存在着近似直立的定向排列的微裂隙，微裂隙的排列方向与最大水平主压应力方向一致，裂隙密度与应力的大小有关。当 S 波通过 EDA 介质时，平行于裂隙面传播的 S 波比垂直于裂隙面的 S 波的传播速度快，快、慢 S 波的速度差取决于裂隙密度，裂隙密度越大，速度差就越大，慢 S 波的时间延迟也就越大。通过计算快 S 波的偏振方向和慢 S 波的时间延迟，可以得到应力场的最大水平主压应力方向和相对大小。

利用新疆乌什地震台记录的小地震事件对 2005 年 2 月 15 日乌什 6.3 级地震前、后 S 波分裂特征进行了初步的分析。该台使用的地震计为 CTS-1 型甚宽带数字地震仪。地震计坐落在基岩上，记录到的波形数据质量较高。选取的数据是乌什台 1999 年 10 月至 2005 年 12 月记录的地震波形文件。在筛选数据时，考虑到非均匀介质的曲面波前效应，将 S 波窗口从 35°扩展到 45°。同时考虑到为避免地幔介质各向异性的干扰，只挑选震源深度小于 52km（乌什地区地壳厚度）的记录，筛选到 30 条记录。另外，新疆台网比较稀疏，很多小震定位结果没有深度，因此挑选 P 波与 S 波到时差小于 3.0s，P 波水平分量振幅小于垂直分量的地震记录作为补充，得到 20 个符合该条件的记录，使用 SAM 分析方法进行 S 波分裂计算[14]。

慢 S 波时间延迟结果如图 22a 所示，由图 22a 可见，2003 年 8 月以前慢 S 波时间延迟较为分散，变化具有一定的随机性，但 2003 年 8 月之后，慢 S 波时间延迟显示明显的持续增加趋势，持续 l4 个月左右。从 1999 年 10 月 1 日到乌什地震发生，慢 S 波平均时间延迟 4.6234ms/km，标准差为 2.6970ms/km。从地震发生到 2005 年 12 月 31 日，慢 S 波平均时间延迟 3.2325ms/km，标准差为 1.9519ms/km。

与慢 S 波时间延迟分段相对应，2003 年 8 月以前快 S 波偏振方向优势分布在 N30°W 和 N20°E 之间，总体近 NS 向分布；但 2003 年 8 月（6.3 级地震前 1 年半左右）偏振方向出现明显变化，N30°～80°E 区间出现优势集中分布（图 22b）。6.3 级地震之后，偏振方向恢复到较早期的分布状态。

2. 其他前兆观测项目异常

300km 范围 9 个前兆观测台中，乌什台距离 6.3 级地震最近，但震前半年乌什倾斜仪和

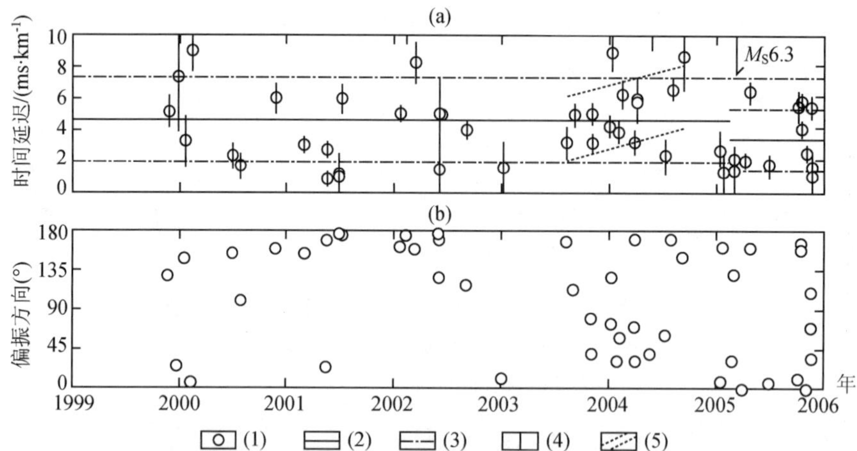

图 22　乌什 6.3 级地震前后震中附近慢 S 波时间延迟（a）和快 S 波偏振方向（b）随时间的变化

Fig. 22　Slow S wave delay (a) and Fast S wave Polarization Direction (b) change over time near the epicenter, pre- and post- the Wushi 6.3 Earthquake

（1）地震事件；（2）时间延迟平均值；（3）标准差；（4）单个地震时间延迟误差；（5）时间延迟增加

形变仪因数字化改造停测而没有资料；阿合奇石英摆地倾斜仪资料干扰较大，震前异常不明显；克孜尔跨断层形变异常不明显。乌什地震前异常较为显著的是巴楚台土层应变和哈萨克斯坦国的下卡姆卡 5-T 井水化学和博古特 2614 井水位变化。

1）巴楚台土层应力[15]

2005 年 2 月 15 日乌什 6.3 级地震前巴楚台土层应力日变化速率出现显著的波动异常。4 个元件的异常特征如表 6 所列。由表 6 和图 23 可看出：①波动异常有比较完整的过程，从

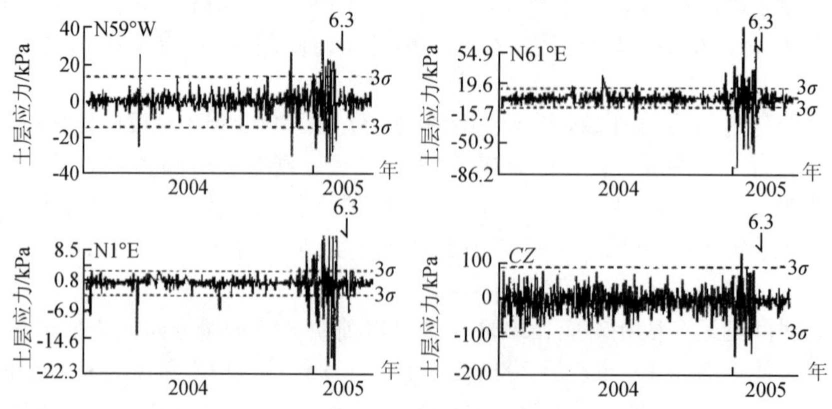

图 23　乌什 6.3 级地震前巴楚应力波动异常图

Fig. 23　Stress Fluctuation Anomalies by Bachu Station before the Wushi 6.3 Earthquake

异常开始到幅度逐步增大达到最大幅度，震后幅度逐步减小至结束；②N61°E、N1°E、N59°W 和垂直分量测值在异常持续期间超过 3 倍均方差的异常分别占 40%、52%、23% 和 14%；③异常持续时间在 2 个月内；④异常开始到发震在 3 个月以内，异常最大幅度时间到发震在 1 个半月以内，异常结束到发震在半个月以内，属于短临异常。

表 6　乌什 6.3 级地震前巴楚应力特征

Table 6　Stress features by Bachu Station before Wushi 6.3 Earthquake

元件方向	均方差	异常出现时间（年-月-日）	异常结束时间（年-月-日）	异常持续时间（天）	异常最大幅度时间（年-月-日）	异常幅度（kPa）	异常开始至发震时间（天）	异常结束至发震的时间（天）	最大幅度至发震的时间（天）
N59°W	4.5	2004-11-23	2005-02-05	75	2005-01-13	34	85	10	33
N1°E	1.5	2004-12-17	2005-02-05	50	2005-01-29	21.5	60	10	17
N61°E	5	2004-12-15	2005-02-05	52	2005-01-03	80.7	62	10	44
垂直	34	2005-01-03	2005-01-30	27	2005-01-03	153	43	12	44

2）博古特 2614 井水位[16]

博古特 2614 号井位于哈萨克斯坦国阿拉木图东，距乌什 6.3 级地震 220km。该井水位存在夏高冬低的年变，年变幅度约为 16cm 左右，其震兆异常特征为年变畸变（图 24）。2002 年下半年水位出现趋势性上升，2004 年 9 月 23 日起该井水位打破正常的年变规律，在年变的下降时段出现上升变化，2005 年 1 月 15 日异常结束，异常持续时间为 114 天，异常

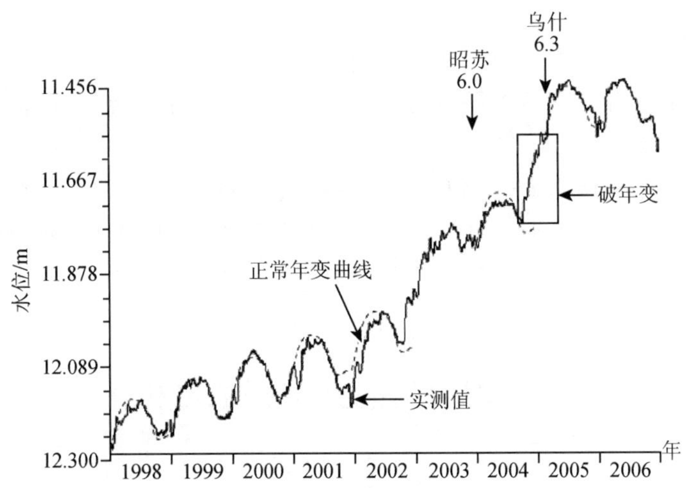

图 24　哈萨克斯坦博古特水位日均值平滑曲线图

Fig. 24　The Smooth Curve of Average Daily Water Level in Bogut, SmootCurves

幅度为 20cm。在异常持续过程中，距离该点 190km 的温泉发生了 5.1 级地震，震后该井水位并没有恢复，而是继续上升。异常结束 31 天后发生乌什 6.3 级地震，震后年变规律恢复，属于中短期异常变化。

3）下卡姆卡 5-T 井钙离子、氯离子[16]

下卡姆卡 5-T 井距中哈边境约 10km，距乌什 6.3 级地震 270km，震前井内钙离子、氯离子出现短期异常（图 25）。2004 年 12 月 11 日钙离子、氯离子出现同步的高值异常，最大幅度分别为 19.0、36.9mg/L，异常持续 5 天，异常出现 64 天后发生乌什地震，属于短期异常。

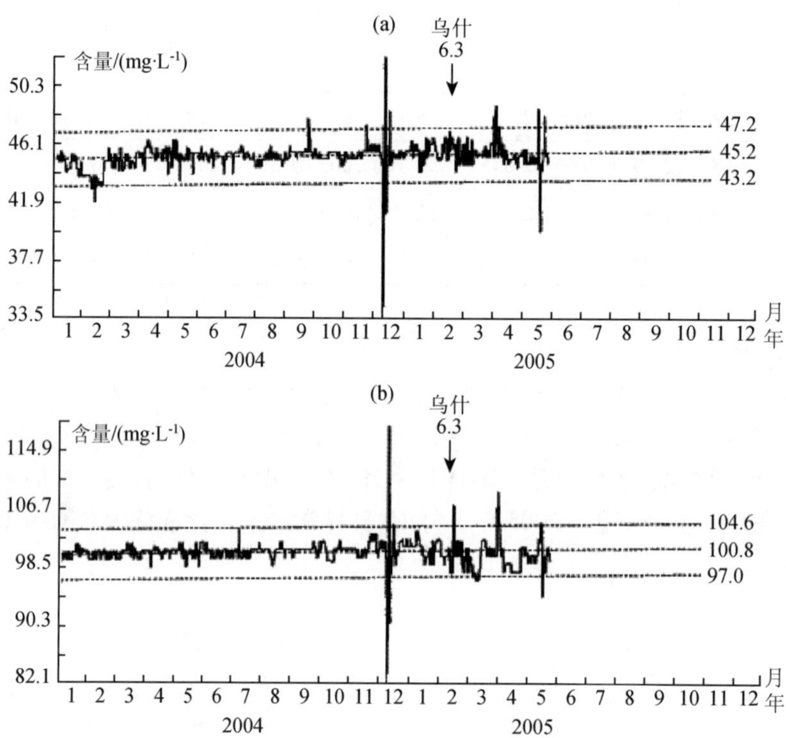

图 25　乌什地震前下卡姆卡 5-T 井钙离子（a）、氯离子（b）日均曲线图

Fig. 25　Daily Average Calcium ion (a) and Chloride ion (b) Amount Curve in Kamu Ka 5-T Well before the Wushi Earthquake

七、地震前兆异常特征分析

1. 地震学异常

乌什地震前震中周围自 1997～2004 年 4 级地震出现“增强—平静—再增强—再平静”的活动过程。震前 5 个月新疆地区 4 级地震集中分布在天山地震带中西部地区，在未来震中附近形成 4 级地震空段，震前 1 个月空段内出现垂直于北东向断裂走向的 M_L3 地震条带，

乌什地震发生在条带的空段内。

部分地震学参数时间扫描在震前出现持续 1 ~2 年的中短期异常，主要包括 η 值、D 值、N 值、S 值和 b 值异常。异常绝大多数在震前恢复，仅 D 值持续到震时，震后恢复。乌什地震后仅 b 值出现新的异常。

由震源附近的 S 波分裂分析，震前 1 年半出现快 S 波偏振方向的明显偏转，转为 EW 向，偏离了区域构造应力场方向；震前 1 年 S 波时间延迟出现明显的持续增加趋势。

2. 其他前兆观测资料

乌什 6.3 级地震发生在新疆地震监测能力较弱的南天山西段，一些台站观测条件较差，资料信度低，又正值前兆仪器数字化改造期间，因而离地震最近的乌什台停记。其他台站受 2003 年巴楚-伽师 6.8 级地震和昭苏 6.1 级地震的影响，乌什地震前中期异常不明显，仅记录到一些短期和短临变化。震前 1 年巴楚土层应力日变速率出现不稳定波动，震前半年哈萨克斯坦国境内的博古特 2614 号井水位出现年变畸变的短期异常变化，下卡姆卡 5-T 井钙离子震前出现同步的高值突跳异常。

除上述前兆异常外，在距乌什地震 390km 的喀什台地磁观测和哈萨克斯坦国卡帕尔-阿拉善 3 号井氦气观测也出现较突出的中短期或短期异常变化[8]。

喀什台距离乌什 6.3 级地震 390km，震前地磁 Z 分量差值在较稳定的上升背景下，出现下降-上升过程（图 26）。对其进行 K-L 最佳直线拟合去除趋势后发现，2004 年 7 月至 2005 年 1 月拟合差值低于均值线（0.0），时间持续大约 7 个月，出现在 2004 年 8 月，之后缓慢回升，回返到均值线后发生乌什 6.3 级地震。喀什台地磁异常属于中短期异常。

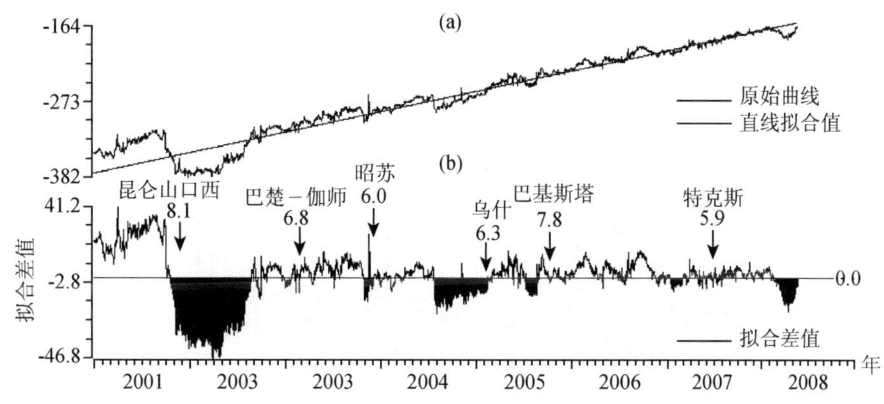

图 26 喀什地磁 Z 分量原始曲线（a）与拟合差值曲线（b）

Fig. 26 The Origin Curve (a) and the Fitting Difference Curve (b) of Kashi Geomagnetic Component Z (by Zhang Yi)

卡帕尔-阿拉善 3 号井位于哈萨克斯坦国阿拉木图东北，距乌什地震 390km，是哈萨克斯坦阿拉木图地震实验场最重要的地下流体监测点之一。该井深 150m，水温背景值为 30℃左右，水中溶解气成分有 N_2（97.84%）、He（0.999%）、Ar（1.16%）、CH_4（0.04%），硫化物总量 255mg/L。

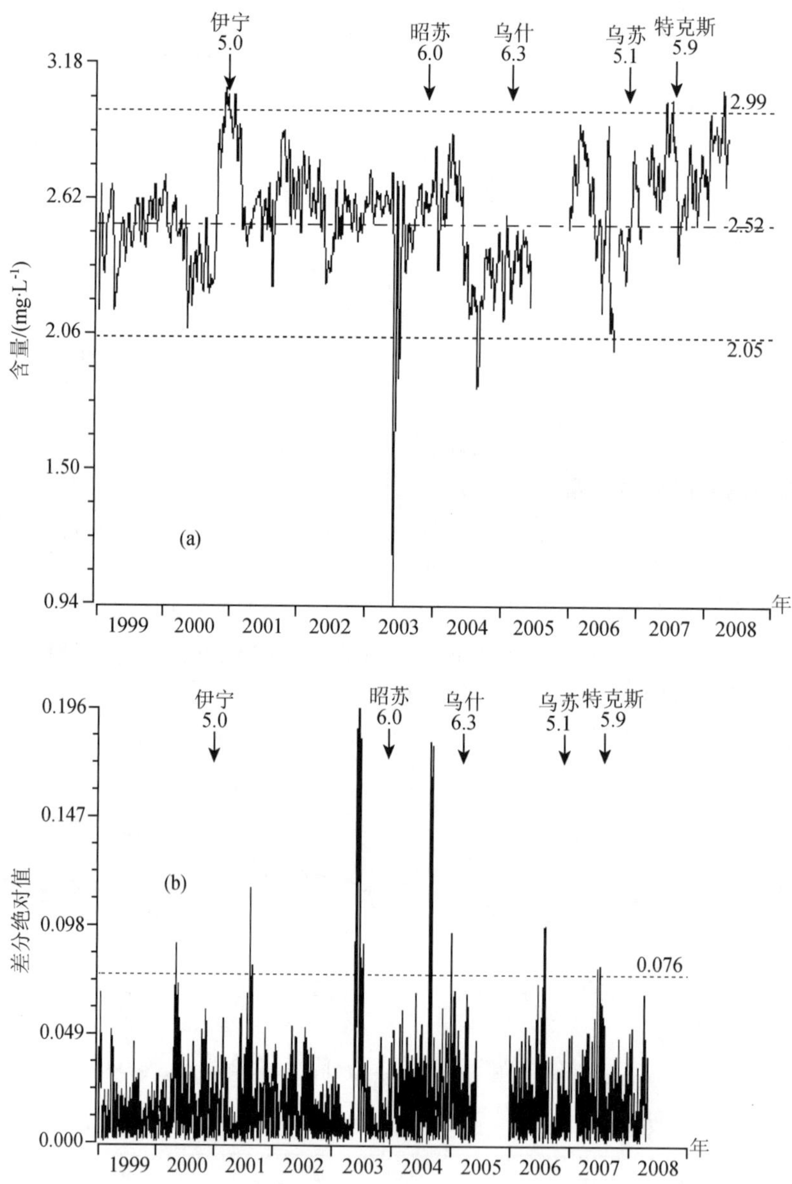

图 27　哈萨克斯坦卡帕尔—阿拉善氦气日均值（a）和差分绝对值（b）曲线图

Fig. 27　The Curve of Daily Average of He in Ka Paer-Alashan, Kazakhstan (a) and Absolute Difference (b)

卡帕尔-阿拉善 4 号井氦气震兆异常为变化幅度超出正常范围，且年变现象消失（图 27）。乌什 6.3 级地震前该井 He 气存在短期异常，2004 年 12 月 28 日该井氦气差分绝对值超出三倍均方差的异常控制线，12 月 30 日恢复，异常持续时间 2 天，异常幅度为 23.37%，异常结束 47 天后发生乌什地震。

从乌什地震前新疆前兆异常的动态演化来看[17]，从 2003 年开始新疆前兆异常比例出现新的增长，期间新疆及边境地区中强震不断。2004 年南天山地震带异常比例继续上升，6.3 级地震震前半年多个前兆台站出现不同程度的短期异常变化，异常先聚集在北天山地区，而后迁移至南天山西段，地震前 3 个月前兆异常数量相对比较多，震前 1 个月有所减少。

八、震前预测预防和震后响应

1. 震前预测情况

乌什-柯坪地区自 1999 ~2004 年一直是新疆 5 ~6 级地震危险区[3~6]，2004 年本地区一些台站进行数字化改造，由于观测条件的改变和限制，新异常判定依据不足，未能将其列入 2005 年度地震危险区。但所开展的新疆三年（2004 ~2006 年）地震大形势研究中，根据地震活动性和部分概率预测方法，将乌什-阿合奇地区列为 6 级地震值得注意区。

中国地震局第一形变监测中心利用新疆地区地倾斜资料和中国地壳运动网络工程 GPS 复测资料，在 1 ~3 年预测研究的基础上，对乌什 6.3 级地震做出了一定程度的年度中期预测[6]。

2. 震后趋势快速判定与震害评估[18]

乌什地震发生后，中国地震局与新疆地震局立即组织专家及有关人员赶赴地震现场，开展地震灾害调查和地震科学考察，并进行现场地震监测预报工作。现场工作组与新疆地震局分析预报人员密切跟踪序列的发展，对乌什台余震序列进行分析，根据震中附近历史地震破裂类型和序列类型、地震空间分布图像和发震构造的对比分析，在主震后 4 个小时内做出快速判定，认为 6.3 级地震较大可能为主—余型地震，短期内震区有可能发生 5 级左右强余震。根据历史地震震害、地震空间位置与震害关系以及地震断错主动盘与震害关系的对比分析，快速估计出 6.3 级地震区的人员伤亡和经济损失较轻。评估意见及时向上级进行了汇报，为顺利开展抗震救灾发挥了积极作用。

3. 震害分析

乌什 6.3 级地震震中位于乌什县英阿瓦提乡及新疆生产建设兵团农二师辖区范围内。乌什县属于国家级贫困县，初步统计英阿瓦提乡约 30 户居民的 50 间房屋倒塌，部分房屋出现裂缝。地震时距震中 40km 的乌什县城震感强烈。据新疆生产建设兵团农二师统计，这次地震造成全师 900 余户 5800 多间房屋受损、6000 户民众住房成为危房，部分中小学校宣布推迟 10 天开学以检修校舍。位于本次地震震中附近的农二师四团损失较重，共有 430 户群众的 2580 间房屋受损，其中 140 户群众的房屋被震裂。该团受灾的 1720 名群众被安全转移和安置，农二师划拨 100 万专项资金帮助四团救灾。

九、结论与讨论

乌什 6.3 级地震发生在塔里木地块西北缘与天山南缘断陷褶皱带衔接地带，震中位于南天山柯坪断块与库车坳陷的构造结合部位，这一区域发育多组平行于造山带、向塔里木地块呈弧形俯冲的断裂，这些断裂有的呈推覆状断达地表，有的呈铲式或反铲式深入基底，有的则沿盖层与基底间的滑脱面发育，深部与根部的山前断裂一起收敛于基底内部。这次地震是在 NNW 向区域应力场作用下 NE 向的迈丹断裂发生的逆冲错断。6.3 级地震为单侧破裂，序列类型为主-余型，震中烈度Ⅶ度，震中区加速度值为 349Gal。

乌什地震前前兆异常以短期异常为主。其中以地震学异常较为突出，主要表现为部分地震学参数时间进程在震前出现持续 1～2 年的中短期异常；震前 1～1.5 年震中附近 S 波分裂出现中短期异常变化；短期异常包括 4 级地震空段和 3 级地震条带（震前已看到这一异常图像，但受到条带内先期发生的 5.0 级地震影响，未能做出进一步的预测）。受到台网改造和观测条件的限制，其他前兆异常不多，主要有巴楚土层应力日变速率短期不稳定波动和境外监测点地下流体和水化学出现的短期、短临异常变化。

文献［6］介绍了利用地倾斜资料和 GPS 复测资料对 2005 年 2 月 15 日乌什 6.3 级地震进行 1～3 年中期预测的依据、方法和过程，认为震前柯坪断裂附近存在最大剪应变高值区，而 2003 和 2004 年新疆地区地倾斜方向线大多交汇于震中附近区域的中天山地区，基于地倾斜方向可能倾向和背向震中的认识，认为柯坪地区存在发生 6 级地震的危险性。

参 考 文 献

［1］张良臣，中国新疆板块构造与动力学特征，新疆第三届天山地质矿产学术讨论会，乌鲁木齐：新疆人民出版社，1～14，1995

［2］殷秀华、史志宏、刘战坡，1°×1°布格重力异常，中国岩石圈地球动力学地图集，北京，地图出版社，1989

［3］肖序常，刘训，高锐等，2004，新疆南部地壳结构和构造演化，北京：商务印书馆

［4］彭树森，1993，大地形变测量所反映的天山最新构造运动，内陆地震，7（2），136～141

［5］王晓强等，2007，利用 GPS 形变资料研究天山及邻近地区地壳水平位移与应变特征，地震学报，29（1），31～37

［6］薄万举、章思亚、刘广余等，2006，新疆乌什 6.2 级地震的中期预测，大地测量与地球动力学，vol. 25（1）：26～30

［7］柔洁、艾买提·乃买提，2005，2005 年 2 月 15 日新疆乌什 6.2 级地震的地震环境及强震加速度记录，国际地震动态，315（32）：19～22

［8］汪素云、俞言祥、高阿甲等，2000，中国高瓦地震的衰减关系的确定，中国地震，20（2），103～104

［9］李志海、王海涛、赵翠萍等，2006，2005 年新疆乌什 6.3 级地震序列研究，大地测量与地球动力学，26（增刊）：97～99

［10］叶建庆、苏有锦、刘学军等，2002，2001 年云南中强地震序列震源参数，地震研究，25（2），115～123

［11］陈荣华、张永仙、薛艳等，2005，2005 年 2 月 15 日新疆乌什 6.2 级地震前的测震学异常，内陆地震，vol. 19（3）：203～205

［12］高歌、王海涛，2006，乌什 6.3 级地震前、后 S 波分裂特征初步研究，内陆地震，vol. 20（2）：139～142

［13］Crampin S.，1984，Effective anisotropie elastic constants for wave propagation through cracked solids［J］. Geophys. J. R. Astron. Soc. 76：135-145。

［14］高原、刘希强、梁维等，2004，剪切波系统分裂方法（SAM）软件系统［J］，中国地震，20（1）：101～107

[15] 蒋靖祥、杜文平、张进等，2007，新疆土层应力地震异常类型初步分析，内陆地震，21（1）：296 ~380

[16] 杨晓芳、高小其、魏若萍，2006，哈萨克斯坦下卡姆卡 5 - T 号井前兆异常特征，内陆地震，20（2）：155 ~ 165

[17] 龙海英、聂晓红、孙甲宁等，2007，新疆 3 次 6 级地震前的短期前兆异常，内陆地震，21（4）：304 ~ 310

[18] 苏乃秦，2006，2005 年 2 月 15 日新疆乌什县 6.3 级地震后震情和灾情快速判定，内陆地震，20（1）：18 ~ 24

[19] 张肇诚主编，1999，中国震例（1986 ~ 1988），北京：地震出版社：142 ~ 155

[20] 董泰，2005，2005 年 2 月 15 日新疆乌什县境内发生 6.2 级地震，国际地震动态，314（2）：48

[21] 史丽艳、朱传庆、杨书江等，2007，新疆乌什地震带断裂研究的综合地球物理方法，西北地震学报，29（2）：156 ~ 160

参 考 资 料

1）新疆地震局，2005 年 2 月 15 日乌什 6.1 级地震现场调查报告，2005 年 2 月

2）新疆地震局，日常地震会商资料汇总——2005 年 1 月月会商报告，2005 年 1 月 26 日

3）新疆地震局，新疆维吾尔自治区 2005 年度地震趋势研究报告，2004 年 11 月

4）新疆地震局，新疆维吾尔自治区 2004 年度地震趋势研究报告，2003 年 11 月

5）新疆地震局，新疆维吾尔自治区 2003 年度地震趋势研究报告，2002 年 11 月

6）新疆地震局，新疆维吾尔自治区 2002 年度地震趋势研究报告，2001 年 11 月

7）中国地震局第一监测中心，2005 年度震情趋势研究报告，2004 年 12 月

The Wushi 6. 3 Earthquake in Xinjiang Uygur Autonomous Region, 15, February, 2005

Abstract

February 15, 2005, a 6. 3 earthquake occurred in Ying'awatixiang, Northeast of Wushi County, Xingjiang, China. With the Epicentral intensity of Ⅶ, the earthquake took place in the southern Tianshan piedmont zone on the Maidan NE fault. Some Buildings in the surrounding areas collapsed and were severely damaged, but fortunately, no people killed or injured.

The 6. 3 earthquake belongs to aftershock earthquake sequence-based type, with the largest aftershock of 5. 2. The aftershock zone axis went in NW (precisely NS), and the decay of aftershocks was normal. Focal mechanisms in the principal stress went in P axis 160 °, section surface Ⅰ to NE, dip 75 °, which was consistent with the stress field of western part of Southern Tianshan, and the motion properties of Maidan fault, indicating that the 6. 3 earthquake was the result of the thrust rupture of Maidan fault, under the effect of regional NNW-trending stress field.

There are a few observation stations near the epicenter region, 6 seismic units and 9 precursors within a radius of 300km. 13 precursory anomalies were found previous of the 6. 3 earthquake, categorized into 5 items. The frequency of 4. 0 and above earthquakes decayed significantly in the surrounding areas before the 6. 3 earthquake. The Surrounding area of the epicenter saw empty segments of 4. 0 and above earthquakes. Minor earthquakes strip appeared one month before the earthquake. The soil stress in Bachu seismic, which is 230km from the epicenter, has experienced short-time and temporary anomalies. Prominent anomalies also occured in the fluid part water chemistry test stations in Kazakhstan.

2005 年 7 月 25 日黑龙江省林甸 5.1 级地震

黑龙江省地震局

赵　谊　高　峰　于　露　马宝君　石　伟

刘长生　王　春　薛佳佳

摘　要

2005 年 7 月 25 日 23 时 43 分 32.57 秒（北京时间），在黑龙江省大庆市林甸县发生 5.1 级地震，震中烈度为Ⅵ度，极震区呈近南北走向的椭圆形。地震造成 1 人死亡，11 人受伤，直接经济损失 2744.68 万元。

林甸县 5.1 级地震属于主震—余震型，最大余震震级 M_L3.7。构造位置位于滨州断裂和德都—大安断裂交汇处。根据地震学研究、宏观地质考察和震源机制推测，北北东向的大安—德都断裂是地震的主破裂面。

震中 200km 以内有地震观测台 11 个，其中哈尔滨台仅有测震观测，齐齐哈尔台和大庆台既有测震观测又有前兆观测，其余 8 个地震台为前兆观测台（包括地磁、地电、水位、水氡、地温和大地微电流等观测项目）。此次地震前共出现 2 个前兆异常项目的 2 条异常，地震学异常 3 条，无宏观异常。

黑龙江省地震局在震前曾一定程度上发现了一些短临异常，并采取在预测区域内布设流动观测设备以及召开震情形势研讨会等一系列措施，但没能做出临震预报。

前　言

2005 年 7 月 25 日 23 时 43 分 32.57 秒（北京时间），在黑龙江省大庆市林甸县发生 5.1 级地震，微观震中为北纬 46.90°，东经 125.00°，宏观震中在林甸县花园乡齐心村，与微观震中基本一致，震中烈度为Ⅵ度。地震受灾范围南起大庆采油六厂（庆新村），北至林甸县黎明乡新民村，东到花园乡卫星村，西到花园乡向阳村之间的区域。齐齐哈尔市、富裕县、依安县、明水县、杜蒙县、安达市、大庆市、林甸县、泰来县、青冈县，吉林省的大安县和镇赉县等地强烈有感。此次地震灾害较为严重，其中受灾最为严重的地区为林甸县花园乡齐心村、黎明乡和大庆油田采油六厂，有 1 人死亡，11 人受伤，直接经济损失 2744.68 万元。

此次地震发生在松辽盆地中部地区，历史上此次地震震中区域未发生过 5 级以上地震，这次地震符合走滑型地震的常规特征。震中200km范围内有11个地震台站，震前出现2个前兆异常项目2条异常，地震学异常3条，无宏观异常。黑龙江省地震分析预报与火山研究中心（简称黑龙江省地震局预报中心，下同）根据地震前的异常情况，在震前采取了向黑龙江省地震局党组汇报震情、在预测区域内布设流动观测设备以及召开震情形势研讨会等一系列措施。

7月25日5.1级地震发生后，黑龙江省地震局立即启动了破坏性地震应急预案，了解灾区情况，按照应急预案要求，地震现场工作队在震后1小时后就出发，凌晨3时赶到震区。中国地震局应急救援司13人组成的中国地震局地震现场工作队也在震后不到12个小时就赶到了灾区，与黑龙江省地震局联合开展了地震科学考察、震情监测、灾害损失评估等工作。

林甸5.1级地震的震例总结是在相关文献、资料整理的基础上，经综合分析研究，本报告得以完成。

本报告中所需要讨论的科学问题是：①信号震的定义和外延，华北地区预报时间尺度和东北地区预报时间的尺度的差异；②松辽盆地中央深断裂大安—德都断裂与嫩江断裂构造活动的相关性问题。

一、测震台网及地震基本参数

图1给出了震中300km范围内的测震台分布，研究时段内可达到$M_S \geqslant 2.5$级地震不遗漏，定位精度为2类。100km范围内有大庆的杏五井、火炬、杏南、庆新、创业、微一井、微四井等七个测震子台组成的大庆测震台网以及碾子山测震台；200～300km有五大连池测震台和宾县测震台。震前的5月30日曾在距离5.1级地震震中150km处的甘南县架设流动地震台，7月3日该台撤消。

表1列出了这次地震不同来源的基本参数，经对比分析，认为黑龙江省地震局的震中位置更为精确，位于滨洲断裂和大安-德都断裂交汇部位。因此，本次地震基本参数采取表1中编号3的结果。

表1　地震基本参数

Table 1　Basic parameters of the *M*5.1 Lindian earthquake

编号	发震日期	发震时刻	震中位置		震级	震源深度	震中	结果来源
	年 月 日	时 分 秒	φ_N	λ_E	M	（km）	地名	
1	2005 7 25	23 43 34	46.90°	125.00°	5.1	15	林甸	中国地震局
2	2005 7 25	23 43 35	46.90°	124.97°	5.3	16		中国地震台网中心
3	2005 7 25	23 43 34	46.98°	125.05°	5.1	5		黑龙江省地震局

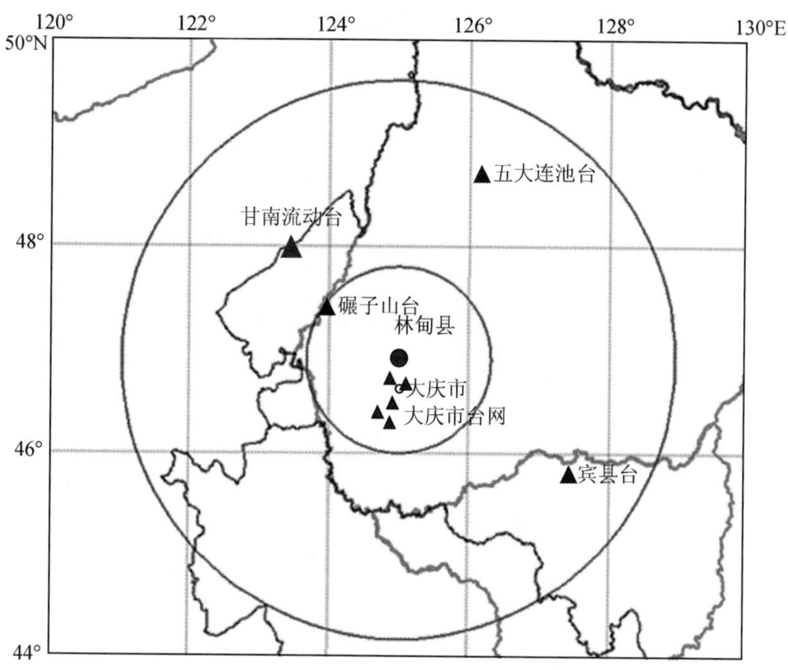

图 1 林甸 $M5.1$ 地震前后震中附近地震台站分布图

Fig. 1 Distribution of earthquake-monitoring stations around the epicentral area before and after the $M5.1$ Lindian earthquake

二、地震地质背景

林甸 5.1 级地震发生在滨洲断裂与大安德都断裂的交汇部位，其附近地区主要断裂分布见图 2 和表 2。本次地震周围主要断裂有大安—德都断裂、滨洲断裂、富裕—明水断裂和嫩江断裂。

表 2 区域主要断裂一览表

Table 2 The regional main fault list

编号	断层名称	长度/km	走向	性质	时代	备注
F1	嫩江断裂	700	NE	正	Q	有 5.0 级左右地震发生
F2	大安—德都断裂	500	NE	逆	N	
F3	第二松花江断裂	500	NW	走滑	Q	1119 年吉林前郭发生 $M_S6\frac{3}{4}$ 地震
F4	富裕—明水断裂	300	NW	走滑	Q2	有中强地震发生
F5	滨州断裂	320	NW	走滑	Q	

续表

编号	断层名称	长度/km	走向	性质	时代	备注
F6	通肯河断裂	100	NNW	逆	Q	
F7	呼兰河断裂	125	NEE	走滑	Q1-Q2	
F8	南北河—勃利断裂	400	NW	走滑	Q	历史上沿断裂有 5 级以上地震
F13	岔林河断裂	250	NW	走滑	Q3	
F10	通河断裂	150	EW	正	Q	
F11	扶余—肇东断裂	250	NE	走滑	Q	沿断裂有小震群

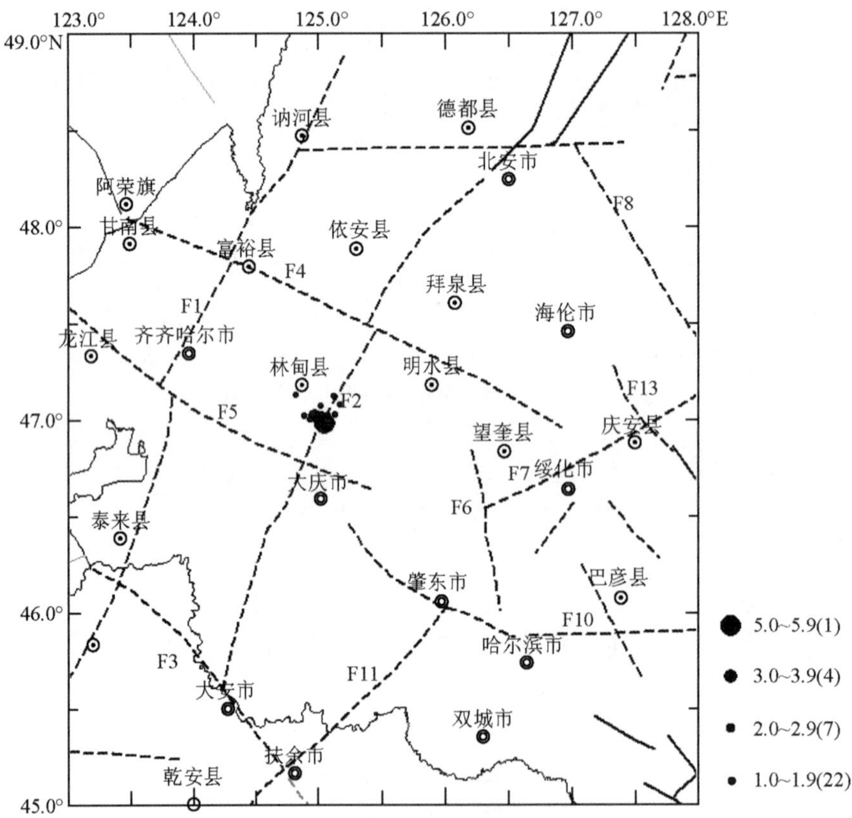

图 2　林甸附近主要断裂

Fig. 2　Near Lindian main fault

大安—德都断裂（F2）发育于松辽盆地内，亦称松辽盆地中央深断裂。南起吉林省大安县经大庆市北至黑龙江省德都县（现今五大连池市）。由于新生代覆盖层巨厚，据地球物理场资料推测，沿断裂重、磁异常密集成带，电测深和人工地震勘探也有明显反映。断裂走向呈北北东 30° ~35°，长约 500km。新构造活动有微弱显示，断裂西部整体抬升缓慢，东侧速率稍大。1986 年在大安—德都断裂北端黑龙江省五大连池火山地区半年内发生 4 次 5 级以上中强地震，2003 ~2005 年在大安—德都断裂的南端吉林省松源地区，每年都出现几十个小震组成的小震群，体现这条断裂的现代地震活动的增强。

滨洲断裂（F5）是一条自哈尔滨至满洲里的隐伏性断裂，在内蒙段扎兰屯—满洲里表现为东西向分布，在黑龙江省境内段表现为北西分布，经哈尔滨、肇东、安达、大庆、延伸至齐齐哈尔南部，与内蒙古的扎兰屯相衔接。据物探资料考证，该断裂在哈尔滨市行政辖区附近尖灭。在黑龙江省境内部分断裂长度为 390km，横切滨东隆起带和松嫩坳陷，对松辽中断（坳）陷带内部构造分区具有控制作用，断裂以张剪性活动为主，活动性不明显。在大庆油田开发过程中，发现这条断裂不仅是电场梯度带，还是地球物理勘探物性的梯度带，但是由于松辽盆地沉积盖层，在松辽盆地内厚度最大为 9000m 左右，因此从构造地质学角度，很难找到地表出露特征。

富裕—明水断裂（F4）为北西走向，长约 300km，是一条规模较大的隐伏断裂。区域地质资料认为该断裂对松嫩中生代断（坳）陷带的形成具有一定的控制作用，是一条具有张扭性质的大型断裂构造。深部构造显示为沿北西向的莫霍面呈斜坡带。对小兴安岭断块与松嫩坳陷的形成与活动具有明显的控制作用。该断裂是 1940 年、1941 年绥化两次 6 级地震的辅助断层，也是 2005 年 2 月 23 日富裕县 M_L4.2 地震发震的辅助断层。

嫩江断裂（F1）带北起嫩江，向西南经齐齐哈尔、泰来进入吉林省境内，南延过白城市，走向北北东，长约 400km，是由多条断裂组成的断裂带，卫星照片显示为多条平行的线状影像带。重磁场均有明显异常显示，根据重磁场确定的断裂位置明显偏东，分析得出断裂倾向东，断裂西侧向上抬升，东侧下降，为断阶式断裂。据断裂带控制沉积分布特点分析，该断裂带形成于中生代，中燕山期活动强烈。该断裂在莫霍面上也有明显显示，是松嫩幔隆区的西部斜坡区与中央幔凹区的分界构造，对松嫩盆地的形成具有明显的控制作用，是一条切割深度达上地幔的岩石圈断裂带。该断裂分三段，嫩江断裂北段、嫩江断裂中段和嫩江断裂南段，三段的地震活动性为北部、中部和南部依次减弱，该断裂最大宽度为 80km，该断裂是 2005 年 2 月 23 日富裕县 M_L4.2 地震的发震断层。

林甸 5.1 级地震主震及余震主要沿大安—德都断裂分布，经过宏观地质考察，主要震害分布范围的长轴方向为近南北向，综合林甸 5.1 级地震主震的震源机制解，确认此次林甸县 5.1 地震的发震断层为大安—德都断裂，辅助断裂为滨洲断裂。

三、地震影响场和震害

根据本次地震的地质和现场灾害考察，破坏较严重的地区位于大庆市林甸县东南的花园乡和大庆油田公司采油六厂，本次地震的宏观震中与微观震中基本一致。Ⅵ和Ⅴ度区等震线长轴呈近南北向（图 3）。

本次地震灾区总面积约为 700km^2。受灾范围主要为南起大庆油田采油六厂（庆新村），北至林甸县黎明乡新民村，东到花园乡卫星村，西到花园乡向阳村之间的区域（图 4），极震区包括林甸县花园乡齐心村七、八屯和大庆采油六厂三矿区。灾区人口共计 62195 人，17770 户，房屋毁坏面积 23105m^2，灾区有两所小学及一所幼儿园遭到显著破坏，直接经济损失总计 2744.68 万元。

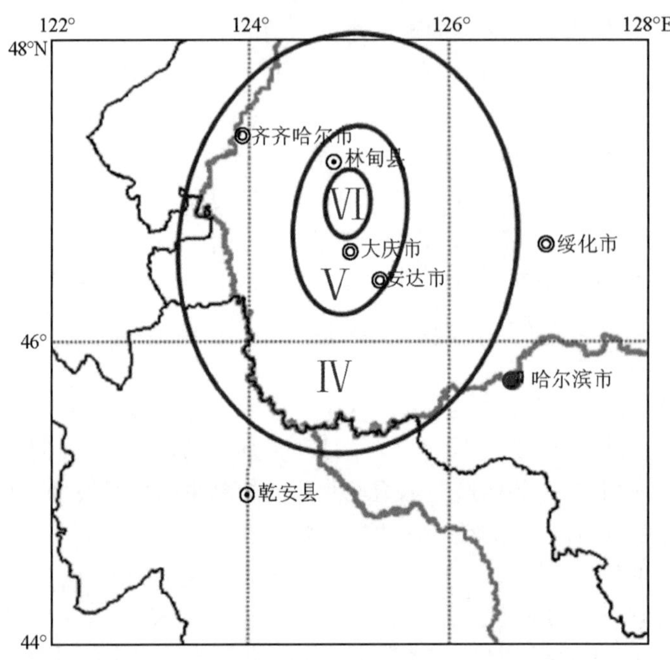

图 3　林甸 5.1 级地震烈度分布图

Fig. 3　Isoseismal map of *M*5.1 Lindian earthquake

Ⅵ度区长轴方向近南北向，长轴长度 70km，短轴长度 40km，总面积约 2200km^2。地震造成房屋严重破坏 1743 间，以花园乡齐心村七、八屯破坏最为严重。震中区南部大庆采油六厂庆新三村多处砖混楼房产生严重裂缝，采油设施受到破坏。

Ⅴ度区长轴方向为近南北向，长轴长度 180km，短轴长度 130km，总面积约 18300km^2，大体范围东至安达，西至林甸县城及红旗镇，南至红岗，北至依安、富裕南部的个别村镇。

四、地 震 序 列

林甸 5.1 级地震发生在区域地震活动相对平静的背景下，无前震活动。主震后截至 2005 年 12 月 14 日，记录到余震 33 次，其中 M_L3.0～3.9 级 4 次；M_L2.0～2.9 级 7 次；M_L1.0～1.9 级 22 次。详见表 3 震级分档统计和表 4 地震目录（据国家科学数据共享中心东北子网地震目录黑龙江省台网目录，下同）。

图 4　林甸 5.1 级地震受灾范围图

Fig. 4　Hit the range of M5.1 Lindian earthquake

表 3　林甸地震序列震级分档统计结果

Table 3　Lindian Earthquake Sequence Classifying Magnitude Statistical Results

震级分档 M_L	1.0 ~ 1.9	2.0 ~ 2.9	3.0 ~ 3.9	4.0 ~ 4.9	5.0 ~ 5.9	合计
次数	22	7	4	0	1	34
占总数比例	0.647	0.205	0.118	0	0.03	1

林甸 5.1 级地震的余震分布在东西 16km、南北 20km，总面积约 260km^2 范围内，主要分布在大安—德都断裂两侧宽度 7km 范围内。震中分布图像显示，余震分布与大安—德都断裂走向基本一致，主震附近余震较为集中，主震的东北方向余震较为稀疏（图 5）。

表 4　林甸 5.1 级地震序列目录

Table 4　Lindian *M*5.1 Earthquake Sequence List

编号	发震日期	发震时刻	震中位置		震级	震源深度
	年 月 日	时 分 秒	φ_N	λ_E	M_L	(km)
1	2005 7 25	23 43 33.9	46°58′	125°03′	5.5	5
2	2005 7 25	23 54 2	47°01′	124°58′	3.1	4
3	2005 7 25	23 57 12.4	46°58′	125°04′	3.7	5
4	2005 7 26	0 33 14.9	47°04′	125°01′	1.7	15
5	2005 7 26	0 50 51.6	46°58′	125°01′	1.5	4
6	2005 7 26	0 54 37.3	47°00′	125°01′	2.3	5
7	2005 7 26	0 57 4.2	47°00′	125°01′	1.6	5
8	2005 7 26	1 1 26.5	46°58′	124°04′	2.9	8
9	2005 7 26	2 39 25.4	47°00′	124°55′	1.7	4
10	2005 7 26	3 24 40.7	47°01′	125°42′	1.8	4
11	2005 7 26	4 17 38	47°01′	125°00′	1.5	4
12	2005 7 26	4 55 52.7	46°58′	124°04′	3.2	5
13	2005 7 26	5 21 52.9	46°58′	125°01′	1.4	5
14	2005 7 26	8 15 17.9	47°01′	125°00′	1.5	5
15	2005 7 26	8 48 10.1	46°58′	125°12′	2.3	13
16	2005 7 26	9 17 30.2	47°07′	125°07′	1.7	5
17	2005 7 26	9 43 08.8	46°58′	125°01′	2.1	5
18	2005 7 26	10 16 10.1	47°01′	125°07′	1.7	4
19	2005 7 27	4 51 7.2	46°58′	125°00′	1.5	5
20	2005 7 30	5 20 9.4	46°46′	124°49′	1.0	8
21	2005 7 30	21 27 21.4	47°00′	125°01′	2.0	11
22	2005 7 31	10 38 25.2	47°01′	125°01′	1.7	5
23	2005 8 3	1 47 38.5	47°48′	125°10′	1.4	5
24	2005 8 4	14 21 35.9	46°58′	125°04′	3.5	5
25	2005 8 7	0 36 43	46°58′	125°04′	2.7	5
26	2005 8 9	3 48 56.4	47°01′	125°04′	2.6	4
27	2005 8 21	13 7 42.5	47°01′	125°04′	1.6	4
28	2005 8 25	11 52 21.5	46°43′	124°54′	1.3	
29	2005 8 25	17 34 37.9	46°43′	124°54′	1.2	

续表

编号	发震日期	发震时刻	震中位置		震级	震源深度
	年 月 日	时 分 秒	φ_N	λ_E	M_L	(km)
30	2005 8 30	12 23 51.2	46°58′	125°03′	1.3	4
31	2005 9 2	17 02 30.8	46°43′	124°54′	1.4	
32	2005 9 15	13 00 34.3	46°43′	124°54′	1.2	
33	2005 11 16	17 21 19.7	46°55′	125°01′	1.7	5
34	2005 12 14	6 00 34.3	46°55′	125°01′	1.5	5

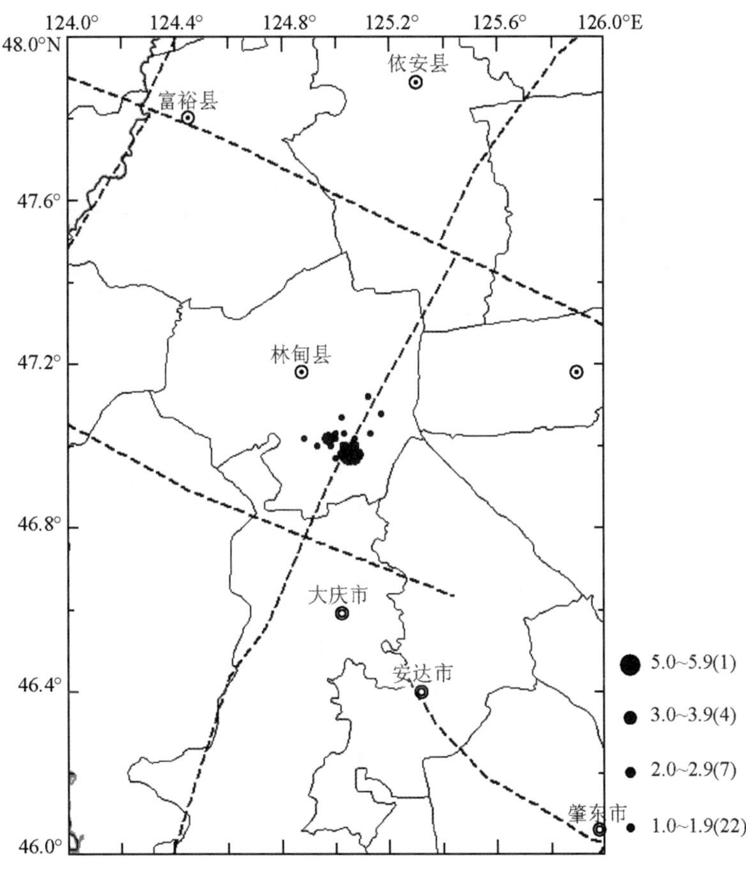

图 5 林甸 M_S5.1 地震序列分布

Fig. 5 Lindian *M*5.1 Earthquake Sequence Distribution

主震后一天，使用 $U\rho K$ 组合判定得到 $U=0.00<0.5$、$\rho=1.21>0.55$、$K=0.006<0.7$，即为非前兆震群。由震级-频度关系分析，震级结构基本合理，b 值 =0.44，以截距法外推最大余震震级为 M_L4.1，与实际最大余震 M_L3.7 基本吻合（图 6）。

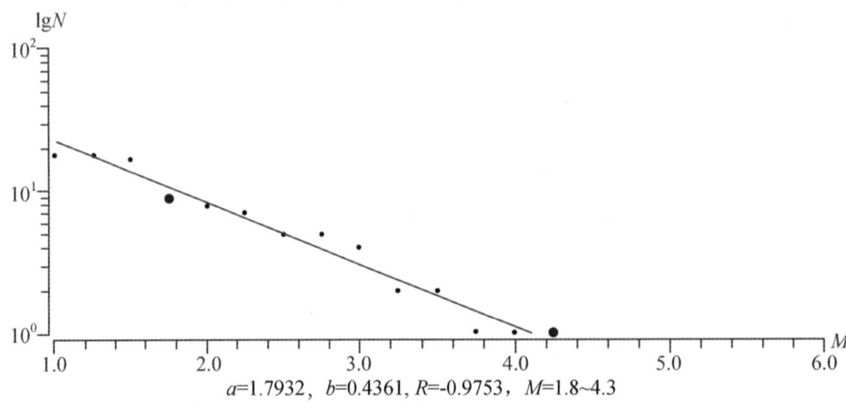

图 6　林甸 5.1 级地震序列 lg*N*-*M* 图

Fig. 6　Lindian *M*5.1 Earthquake Sequence lg*N*-*M* Figure

本次地震序列的 $h=2.0$、$P=1.4063$（图 7）主震能量 2.8183×10^{12}J，占整个地震序列总能量 99%，主震与最大余震级差 $\Delta M=2.0$（以 M_L 计则为 1.8），判定地震序列为主震—余震型。从地震序列震中分布看，地震主要集中在大安—德都断裂两侧，由余震序列的 *M*-*t* 图（图 9）看出，余震主要集中发生在 2005 年 7 月 25 日至 2005 年 8 月 3 日的 10 天内，多数为 2 级地震，最大余震为 7 月 25 日 M_L3.7 余震。2005 年 8 月 3 日之后，M_L2.0 以上地震迅速衰减（图 8），最后一次余震为 2005 年 12 月 14 日 M_L1.5。

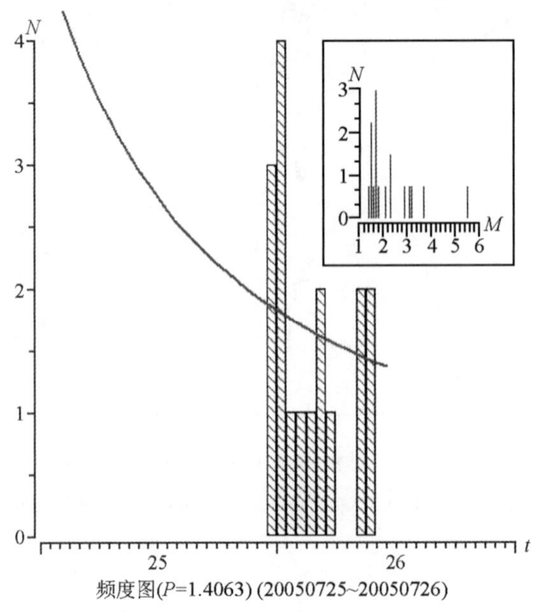

图 7　林甸 5.1 级地震序列 *P* 值图

Fig. 7　Lindian *M*5.1 Earthquake Sequence P Figure

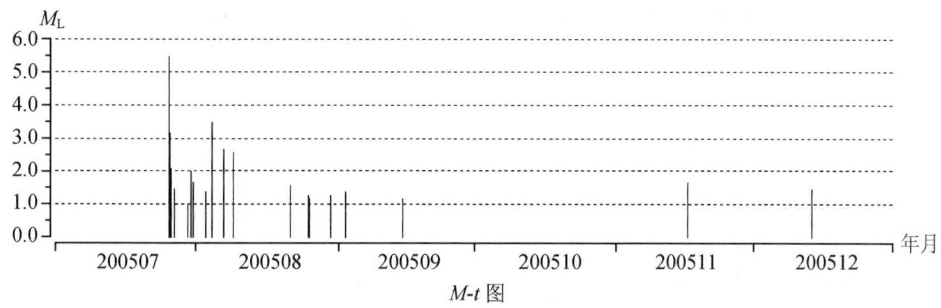

图 8　林甸 M_S5.1 级地震序列 M-t 图

Fig. 8　Lindian M5.1 Earthquake Sequence M-t Figure

五、震源参数和地震破裂面

利用黑龙江省、吉林省和内蒙古自治区地震监测台网 P 波初动较为清晰的记录，分别使用了 19 个台、19 个台和 13 个台的 P 波初动，采用下半球等面积投影法反演了林甸 5.1 级地震和 2 次较大余震的 P 波初动震源机制解（图 9、表 5）。

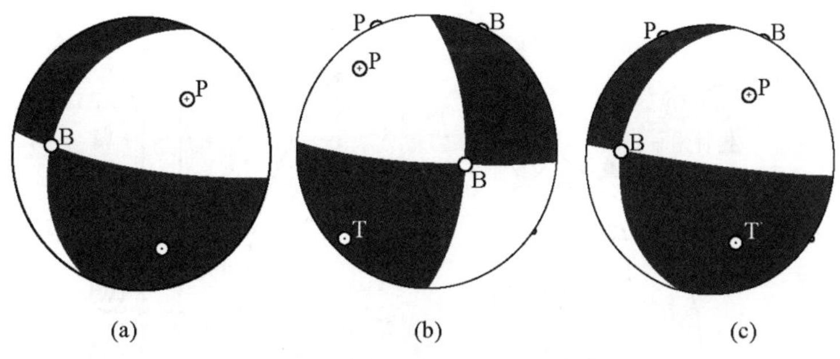

图 9　林甸地震震源机制解

Fig. 9　Lindian earthquake focal mechanism solution

（a）2005 年 7 月 25 日 23 时 43 分 5.1 级；（b）7 月 25 日 23 时 57 分 M_L3.7 级；（c）7 月 26 日 1 时 1 分 M_L2.9 级

本次地震没有明显的地表破裂。从地震构造、余震分布方向、烈度分布走向和房屋破坏分布范围走向，结合震源机制解节界面分析，本次地震活动的破裂面可能是北北东向的大安—德都断裂。其中 5.1 级主震的破裂面表现为右旋—逆冲走滑性质；7 月 25 日 M_L3.7 余震的为左旋—走滑性质；7 月 26 日 M_L2.9 余震的为右旋—逆冲走滑性质。综合推测，林甸 5.1 级主余型地震活动整体上以右旋—逆冲走滑破裂为主，是北东向大安德都断裂与北西向滨州断裂共同作用的结果，其北北东向的大安—德都断裂是主破裂面。

表 5　林甸地震震源机制解

Table 5　Lindian earthquake focal mechanism solution

序号	节面Ⅰ			节面Ⅱ			P 轴		T 轴		结果来源
	走向	倾角	滑动角	走向	倾角	滑动角	方位	仰角	方位	仰角	
a	100	－59	81	355	32	－163	35	61	159	45	黑龙江省地震局
b	181	7	114	25	84	87	118	39	293	51	黑龙江省地震局
c	12	6	－28	80	3	175	323	74	59	68	黑龙江省地震局

六、地震前兆台网及前兆异常

震中附近地区的地震台站及观测项目分布见图 10，其中甘南流动台为林甸地震前增设。在震中 200km 以内共有 11 个台站，其中 3 个台站设有测震观测，其余 8 个台站只有定点前兆观测，包括水位、地电、大地微电流、地磁、水氡和地温等观测项目（表 6）。在 0～100km、100～200km 距离内分别有前兆观测的为地震台站 3 个和 7 个，定点前兆观测项目分别为 3 项和 10 项。

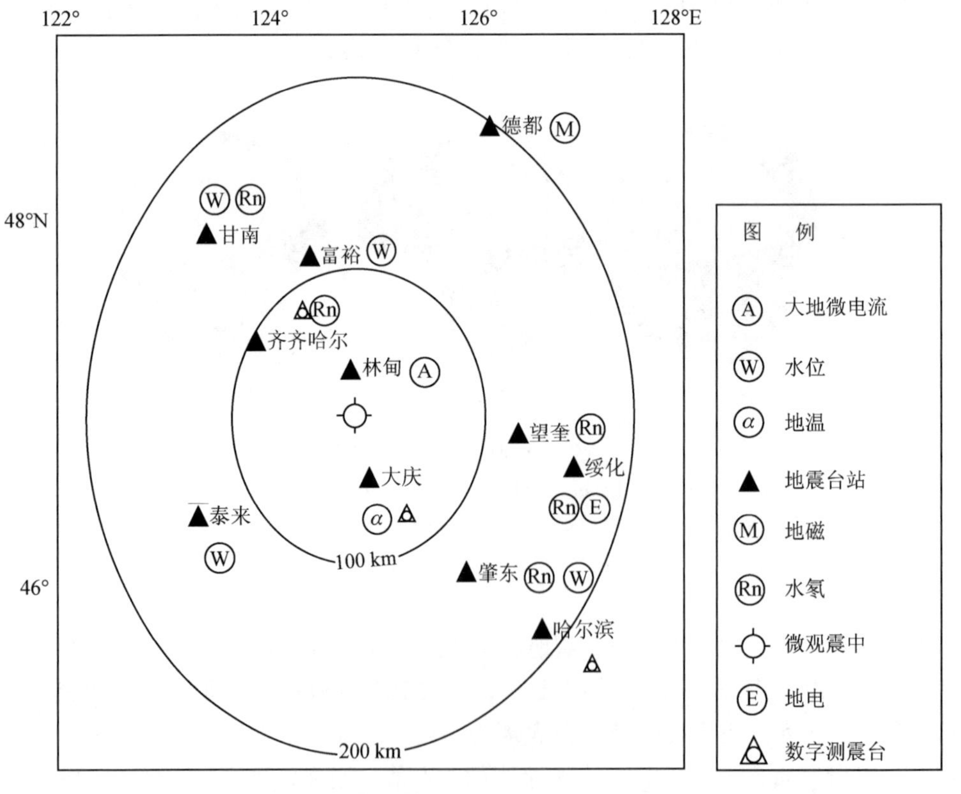

图 10　林甸 5.1 级地震前震中附近地震台站及观测项目分布图

Fig. 10　Distribution of earthquake-monitoring stations around the epicentral area before the M5.1 Lindian earthquake

表 6　林甸地震前测震以外固定前兆观测项目与异常统计表

Table 6　Summmary table of fixed precursory observation projects and anomalies before Lindian M_S5.1 earthquake

序号	台站（点）名称	测项	资料类别	震中距 Δ/km	按震中距 Δ 范围进行异常统计																								
					0 < Δ≤100km					100 < Δ≤200km					200 < Δ≤300km					300 < Δ≤500km					500 < Δ≤800km				
					L	M	S	I	U	L	M	S	I	U	L	M	S	I	U	L	M	S	I	U	L	M	S	I	U
1	林甸	大地微电流	Ⅰ	30	—	—	—	√																					
2	大庆	地温	Ⅱ	56	—	—	—	—																					
3	齐齐哈尔	水氡	Ⅰ	96	—	—	—	—																					
4	富裕	水位	Ⅱ	110						—	—	—	—																
5	望奎	水氡	Ⅰ	130						—	—	—	—																
6	肇东	水氡	Ⅰ	138						—	—	—	—																
		水位	Ⅱ							—	—	—	—																
7	甘南	水氡	Ⅰ	150						—	—	—	—																
		水位	Ⅱ							—	—	—	—																
8	泰来	水位	Ⅱ	152						—	—	—	—																
9	绥化	地电阻率	Ⅰ	162						—	—	—	√																
		水氡	Ⅰ							—	—	—	—																
10	德都	地磁	Ⅰ	198						—	—	—	—																

续表

序号	台站（点）名称	测项	资料类别	震中距 Δ/km	按震中距 Δ 范围进行异常统计																								
					0 < Δ ≤ 100km					100 < Δ ≤ 200km					200 < Δ ≤ 300km					300 < Δ ≤ 500km					500 < Δ ≤ 800km				
					L	M	S	I	U	L	M	S	I	U	L	M	S	I	U	L	M	S	I	U	L	M	S	I	U
分类统计	台项	异常台项数			0	0	0	1		0	0	0	1																
		台项总数			3	3	3	3		10	10	10	10																
		异常台项百分比/%			0	0	0	33		0	0	0	10																
	观测台站（点）	异常台站数			0	0	0	1		0	0	0	1																
		台站总数			3	3	3	3		7	7	7	7																
		异常台站百分比/%			0	0	0	33		0	0	0	14																
	测项总数（13）				3					10																			
	观测台站总数（10）				3					7																			
备注																													

1. 地震学异常

2003～2005 年大嫩江断裂带北段出现一个北东向地震条带，2005 年 2 月 23 日在富裕县发生 M_L4.2 地震，该地区 M_L4.0 地震平静 1 年，并且其附近形成地震条带（图 11），根据王俊国信号震的判定条件，认为此次地震为信号震。第五活跃期以来，松辽盆地震内地震活动较弱，1999～2004 年在松辽盆地内形成 M_L3.0 地震空区（图 12）。

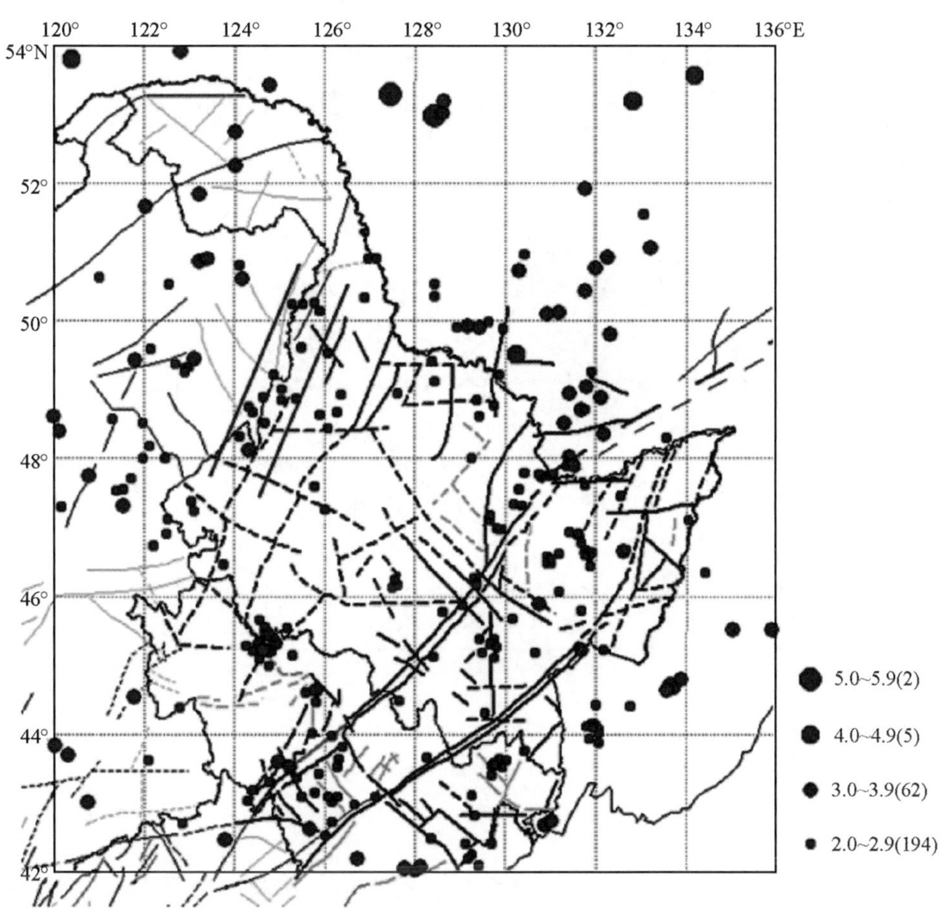

图 11　嫩江断裂带条带（2003 年 1 月至 2004 年 12 月）

Fig. 11　Nenjiang Fault seismic belt

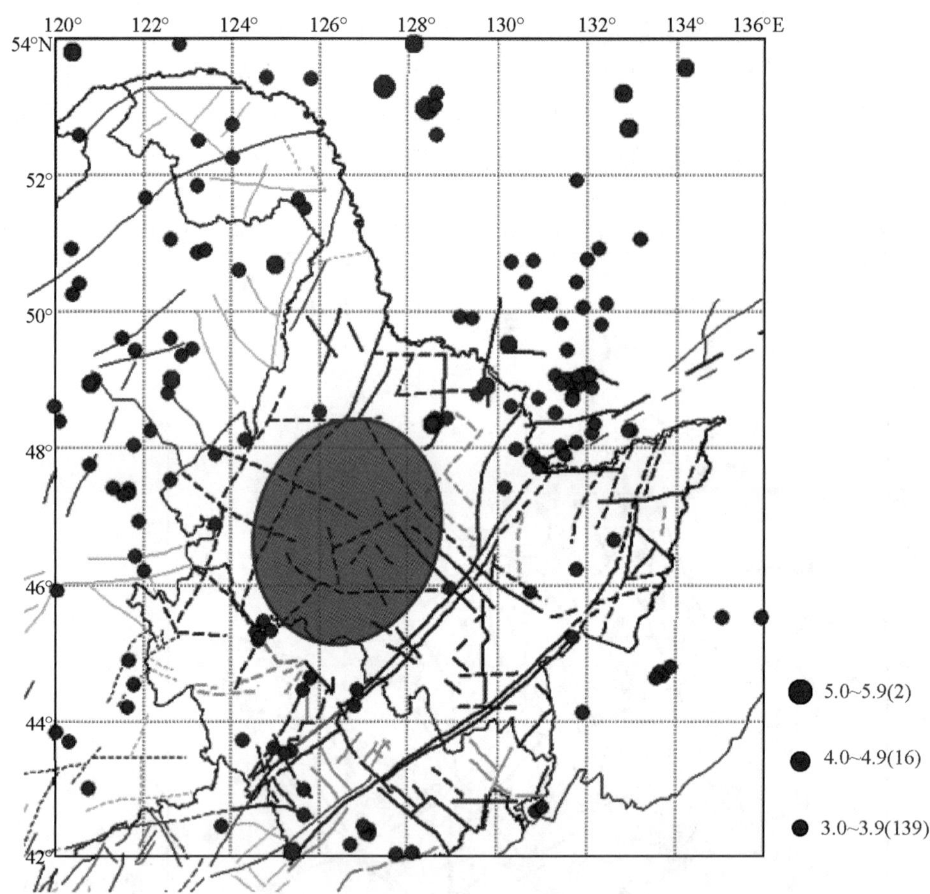

图 12 松辽盆地地震空区（1999 年 1 月至 2004 年 12 月，≥M_L3.0）

Fig. 12 Songliao basin area earthquake empty

2. 前兆异常

震中 200km 范围内的 13 个前兆观测项目中，有 2 项前兆观测项目震前存在临震异常（表 7、图 13）。

1）林甸大地微电流异常

林甸井地处大安—德都断裂和滨州断裂的交汇部位，地理坐标为 47°17′N，124°87′E，位于林甸地震震中北东方向约 30km。林甸井大地微电流从 2004 年 8 月 4 日开始观测，初始阶段数据不是很稳定，从 2005 年 1 月 28 日数据开始稳定，背景值为 0.0464nA，2005 年 5 月 30 日出现波动，达观测以来的最高值 1.001nA，但只持续了 2 天，数据开始恢复，2005 年 6 月 14 日数据日趋上升，自 2005 年 7 月 18 日 0.254nA 数据突升值 8.0759nA，比正常值增大了 31 倍，异常持续 4 天，2005 年 7 月 22 日数据恢复正常，背景值为 0.179nA，2005 年 7 月 25 日发生林甸发震，地震发生后数据正常，偶有干扰引起的小幅波动。

表 7　地震前兆异常登记表

Table 7　Summary table of earthquake precursory anomalies

序号	异常项目	台站（点）或观测区	分析方法	异常判据及观测误差	震前异常起止时间	震后变化	最大幅度	震中距（km）	异常类别及可靠性	图号	异常特点及备注	震前提出/震后总结
1	地震条带	大兴安岭隆起带东侧	$M_L \geqslant 2.0$ 级	在大兴安岭隆起带形成 $M_L 2.0$ 地震条带	2003.1 ~ 2004.12		2 年	70	I_L	11	2003 年 1 月到 2004 年 12 月在大兴安岭隆起带形成 $M_L 2.0$ 地震条带	震前提出
2	地震空区	松辽盆地内部	$M_L \geqslant 3.0$ 级	$M_L 3.0$ 地震异常平静	1999.1 ~ 2004.12		6 年		I_L	12	1999 年 1 月到 2004 年 12 月在松辽盆地内形成 $M_L 3.0$ 地震空区	震前提出
3	信号震	松辽盆地内部		2005 年 2 月 23 日富裕发生 $M_L 4.2$ 地震	2005.2 ~ 2005.7			90	I_M		2005 年 2 月 23 日富裕发生 $M_L 4.2$ 地震为信号震	震前提出
4	大地微电流	林甸	日均值	大幅上升	2005.7.18 ~ 2005.7.22	恢复	增大 31 倍	30	I_I	14	2005 年 7 月 18 日至 2005 年 7 月 22 日出现的震前临震异常	震前提出
5	地电阻率	绥化	日均值	走平	2005.6.23 ~ 2005.7.26	恢复		162	I_I	15	2005 年 6 月 23 日至 2005 年 7 月 26 日出现的震前临震异常	震前提出

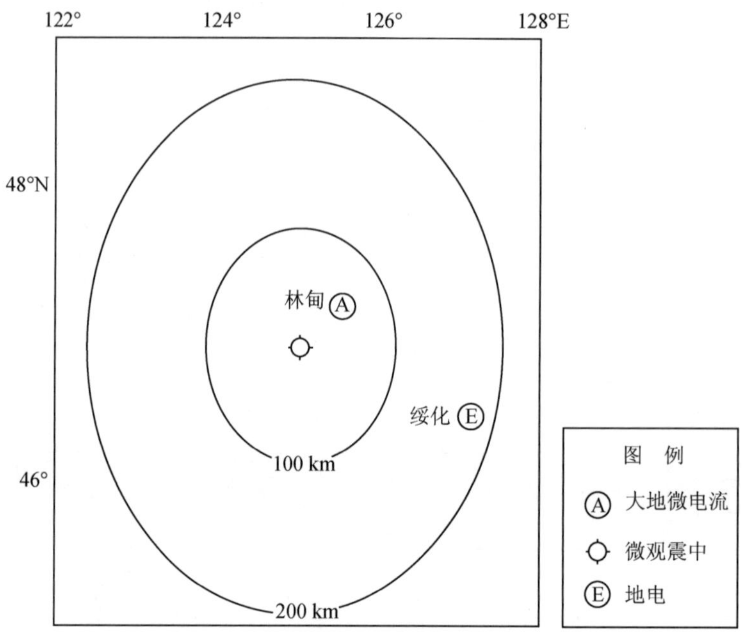

图 13 林甸 5.1 级地震前异常分布图

Fig. 13 Distribution of earthquake precursory anomalies before the *M*5.1 Lindian earthquake

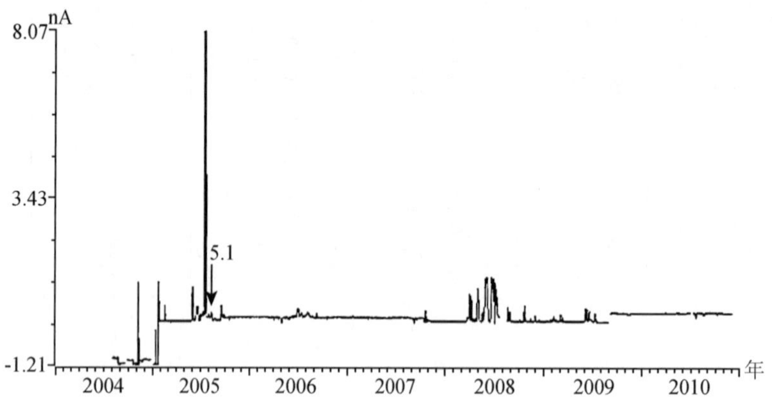

图 14 林甸大地微电流日均值曲线

Fig. 14 Curve of daily mean value of Micro-current at Lindian well

2）绥化地电阻率异常

绥化台位于林甸地震震中东偏南方向 162km，台址位置为东经 126°57′50″，北纬 46°37′30″。该台所处构造部位属于新华夏系第二沉降带，测区内分布有走向约 N18°E 的绥化—逊克主干断裂带，还有 NE 向、NW 向断裂与之斜交或斜切。测区第四系岩性为黄土状亚黏土及砂砾石，厚约 80m 左右，其下部白垩系主要为砂砾岩，泥岩，厚度可达 2000m，该台布设有 N45°E 和 N45°W 两个测道。绥化电阻率观测点距离震中约 162km，该测点的正常年变规律为，每年的2、3 月为低值，7、8 月份为高值，但在2005 年6 月末，也就是林甸地震前一个月，南北（NS）、东西（EW）向数据均出现了走平的异常现象，直至林甸 5.1 级地震后，两个分向的数据开始恢复正常上升的状态。

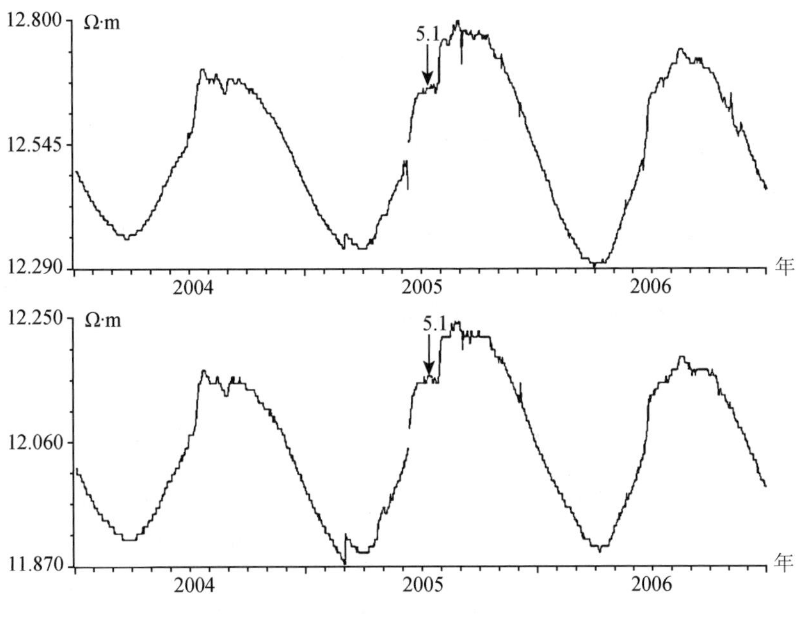

图 15 绥化地电阻率日均值曲线图

Fig. 15 The daily mean value of the ground resistivity at Suihua seismic station

七、地震前兆异常特征分析

测震异常：2003 ~2005 年大嫩江断裂带北段出现一个北东向地震条带，2005 年 2 月 23 日在富裕—明水断裂附近富裕县发生 M_L4. 2 地震，该地区 M_L4. 0 级地震平静 1 年，并且其附近有地震条带，根据信号震的判定条件[1,2]，判断此次地震为信号震。同时东北地区的第五活跃期以来，松辽盆地震内地震活动较弱，1999 ~2004 年在松辽盆地内形成 M_L3. 0 地震空区。以此判断以富裕县 M_L4. 2 地震震中为圆心，150km 为半径，与嫩江断裂和大安-德都断裂的两个交汇点及附近，即北部齐齐哈尔地区的甘南县，南部的大庆市两个地区存在 M_L5. 7（M_L4. 2 +1. 5）地震背景。

其他前兆观测资料异常：在 200km 范围内固定前兆台站主要分布在震中北部和东部地区，13 个测项仅有 2 个测项出现异常，占 15. 4%，且均为临震异常。即林甸井大地微电流异常，表现为震前大幅度上升；绥化地电阻率异常，主要表现为短期走平形态。

八、震前预测、预防和震后响应

1. 预测与预防

黑龙江省地震局在“2005年度黑龙江省地震趋势研究报告”中指出，“2005年度黑龙江省及其邻近地区存在5级以上地震的可能，地震活动的主体地区可能是嫩江断裂西侧齐齐哈尔至博克图地区，仍要注意的地区是绥化—松源附近地区”（实际的林甸5.1级地震震中距离预测区域约70km）。

2005年2月23日，黑龙江省富裕县发生M_L4.2地震，黑龙江省地震局预报中心在3月21日召开“富裕M_L4.2地震是否是该地区未来中强地震的信号震”的专题研讨会，根据地震条带、信号镇以及地震空区，判定富裕M_L4.2地震震中100km范围内，大庆地区，甘南地区存在发生M_L5.7（M_L4.2+1.5）级地震背景。3月22日向局党组汇报，指出“2005年2月23日（正月十五）富裕M_L4.2地震是该地区未来中强地震的信号震，下一步将向南迁，震级随之增大，将达到5级。”（图16）

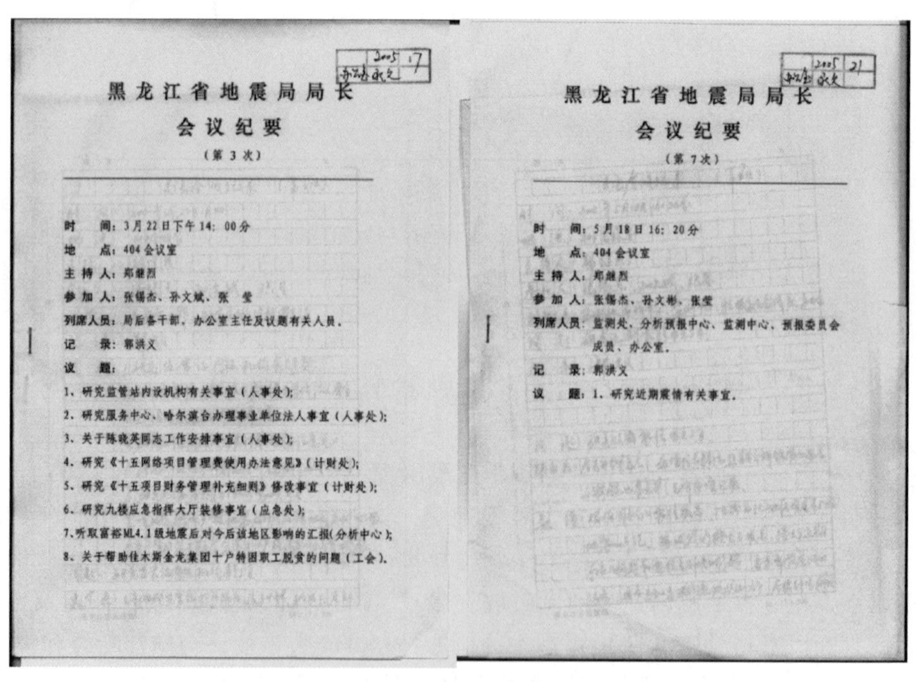

黑龙江省地震局局长
会议纪要
（第3次）

时　　间：3月22日下午14：00分
地　　点：404会议室
主 持 人：郑继烈
参 加 人：张锡杰、孙文斌、张　莹
列席人员：局后备干部，办公室主任及议题有关人员。
记　　录：郭洪文
议　　题：
1、研究监管站内设机构有关事宜（人事处）；
2、研究服务中心、哈尔滨台办理事业单位法人事宜（人事处）；
3、关于陈晓英同志工作安排事宜（人事处）；
4、研究《十五网络项目管理费使用办法意见》（计财处）；
5、研究《十五项目财务管理补充细则》修改事宜（计财处）；
6、研究九楼应急指挥大厅装修事宜（应急处）；
7、听取富裕ML4.1级地震后对今后该地区影响的汇报（分析中心）；
8、关于帮助佳木斯金龙集团十户特困职工脱贫的问题（工会）。

黑龙江省地震局局长
会议纪要
（第7次）

时　　间：5月18日16：20分
地　　点：404会议室
主 持 人：郑继烈
参 加 人：张锡杰、孙文彬、张莹
列席人员：监测处、分析预报中心、监测中心、预报委员会成员、办公室。
记　　录：郭洪文
议　　题：1、研究近期震情有关事宜。

图16　黑龙江省地震局党组会议纪要

Fig. 16　The minutes of Communist organization in Seimological Bureau of Heilongjiang Province

2005年5月18日黑龙江省地震局预报中心向局党组汇报，指出“根据5月13日中国地震局2005年4月29日第四期《震情监视报告》的月会商意见，‘吉林抚松4级地震发生在长白山火山震区西北50km处。1970年以来震中100km范围内仅发生4次4级地震。前三次抚松4级地震后，华北、东北地区在其后40天内都发生过M_L5地震。因此，5月份华北、东北地区有发生 5 级左右地震的可能。’提出加强年度趋势会商会提出的两个重点危险区的流动观测，加强群测群防，向黑龙江省政府通报震情。”

2005 年 5 月 29 日至 6 月 3 日黑龙江省地震局对嫩江断裂宏观异常点进行落实情况检查及召开嫩江断裂构造片区震情研讨会，并在 5 月 30 日至 7 月 3 日在嫩江断裂构造带的甘南县架设流动地震台进行地震流动观测。

2. 震后响应

7 月 25 日 5.1 级地震发生后，黑龙江省地震局立即启动了破坏性地震应急预案，了解灾区情况，按照应急预案要求，现场工作队在地震发生 1 小时后出发，凌晨 3 时赶到震区。由中国地震局应急救援司 13 人组成的中国地震局地震现场工作队，在震后不到 12 个小时也赶到了灾区，开展地震灾害损失评估、震情监测、科学考察等工作。

林甸 5.1 级地震发生在区域地震活动相对平静的背景下，无前震活动。主震后截至 2005 年 12 月 14 日，记录到余震 33 次，其中 $M_L3.0\sim3.9$ 级 4 次；$M_L2.0\sim2.9$ 级 7 次；$M_L1.0\sim1.9$ 级 22 次。由余震序列的 M-t 图看出，余震主要集中发生在前 10 天内，多数为 2 级地震。序列中最大余震为 7 月 25 日 $M_L3.7$ 余震，之后，$M_L2.0$ 以上地震迅速衰减，最后一次 $M_L2.0$ 余震发生在 12 月 14 日，因此我们判定林甸 $M_S5.1$ 地震为主震—余震型，判断该区域未来发生同等级别地震的可能性不大。

九、结论与讨论

（1）大兴安岭隆起带地震序列类型基本一致。2005 年林甸 5.1 级地震属于主余震序列，能量释放与发生在大兴安岭隆起带西侧 2004 年内蒙东乌珠 5.9 级地震序列相似（共发生余震 41 次，最大余震 $M_L3.5$ 地震），主震释放能量占全序列总能量的 99%，体现了大兴安岭隆起带的构造活动特征。

（2）这次地震活动体现了松辽盆地中央深断裂大安—德都断裂与嫩江断裂构造活动的相关性问题，表明了松辽盆地构造活动的整体性。富裕 $M_L4.2$ 地震所处构造位置为嫩江断裂（大兴安岭隆起带中央断裂）与富裕—明水断裂的交汇处，而林甸 5.1 级地震所处的构造位置在滨州断裂和德都—大安断裂交汇处，可见 2005 年大兴安岭隆起带的构造活动主要体现在它的东侧的大兴安岭隆起带中央断裂即嫩江断裂与富裕—明水断裂，松辽盆地中央深断裂即大安—德都断裂和北西向滨洲断裂围成的菱形块体之中（以下称菱形块体）。

（3）根据以往研究：华北地区信号震的判定条件为[2]：①信号震一般发生在局部的 $M_L\geq4.0$ 级地震平静区内，平静时间大约 2～3 年。②信号震发生在中小地震活动条带上或附近，条带持续时间一般在 1.5～2 年。③信号震发生在中小地震活动空区内部或附近，空区持续时间与未来强震震级大小有关，一般 1.5 年以上。

根据黑龙江省地震实践的验证：①1999 年到 2004 年在松辽盆地内菱形块体之中形成 $M_L3.0$ 地震空区，平静时间达到 5 年之久。我们判定这种空区为一种孕震空区，而不是背景空区。②富裕 $M_L4.2$ 地震发生在以 $M_L2.0$ 地震为震级下限形成的条带附近，而这一地震条带是从 2003 年 1 月到 2004 年 12 月，持续近两年时间的地震条带。③富裕 $M_L4.2$ 地震就发生在中小地震活动空区内部。

经验表明：在松辽盆地内部很少有大于 M_L4.0 小于 M_L5.0 的孤立地震。一般情况下，这一孤立出现的地震都具有某种构造指示意义。震例表明：华北地区信号震一般发生在强震前 2 年之内，多数发生在 15 个月之内，预测强震震级一般为 $M_S \geq 6.0$ 级，而林甸 5.1 级地震则发生在富裕 M_L4.2 信号震发震后的 6 个月之内，震级也未超过 M_S6.0，所以，华北地区预报时间尺度和东北地区预报时间尺度之间是存在差异的。

（4）本次地震之所以能够在震前有所察觉，主要原因是对富裕 M_L4.2 信号震的正确判断以及对前兆异常现象的正确认识。

参 考 文 献

[1] 陆远忠、沈建文、宋俊高，地震空区与逼近地震［J］，地震学报，1982，Vol.4，No.4，327～336

[2] 王俊国、王林瑛、吴晓芝、陈佩燕、白彤霞、何巧云，华北地区强震前的信号震及预测意义［J］，地震，2004，Vol.24，No.3，51～60

参 考 资 料

1）黑龙江省地震局、中国地震局地震现场联合工作队，2005 年 7 月 25 日大庆市林甸县 M_S5.1 地震灾害损失评估报告，2005 年 8 月 2 日

2）黑龙江省地质矿产局，1993，黑龙江省区域地质志［M］，北京：地质出版社

3）黑龙江省地震局，黑龙江省 2005 年度地震趋势预测研究报告，2004 年 11 月

4）黑龙江省地震局，黑龙江省 2006 年度地震趋势预测研究报告，2005 年 11 月

Lindian Earthquake of M_S 5.1 on July 25, 2005 in Heilongjiang Province

Abstract

July 25, 2005, 23: 43 32.57 (Beijing time), an earthquake of M_S5.1 occurred in Lindian, Daqing City, Heilongjiang Province, with an epicentral intensity of VI degree and the extreme seismic zone was nearly oval in a N-S direction. The earthquake caused 1 person died and 11 injured, the direct economic loss amounted to 27.4468 million Yuan.

The Lindian M_S5.1 earthquake belongs to the main-aftershock type, the largest aftershock was a M_L3.7. The geotectonic location of the earthquake is in the intersection area of Binzhou fault and Dedu-Daan fault. Based on seismic studies, macro-geological study and focal mechanisms speculated that, Dedu-Daan fault in an N-N-E direction was the main rupture plane of the earthquake.

There are 11 seismic stations within the ranges of 200km distant to the epicenter, Harbin seismic station is just seismic observation projects, both seismic observations in of Qiqihar and Daqing seismic have precursory observation project, the rest 8 seismic are precursory observatory stations (including geomagnetism, geoelectricity, water level, water radon, ground temperature, earth micro-current and other observation projects). There were two precursory anomalies before the earthquake, three seismology exceptions, and no macroscopic abnormalities.

Seismological Bureau of Heilongjiang Province found some short-term anomalies before the earthquake to a certain extent, and taken a series of measures, including laying mobile observation equipment in the forecast area and convening the seminar about the earthquake situation and the situation, but failed to make the earthquake prediction in the short.

2005 年 8 月 5 日云南省会泽 5.3 级地震

云南省地震局

刘丽芳　王世芹　付　虹

摘　要

2005 年 8 月 5 日云南省会泽县发生 5.3 级地震，宏观震中位于会泽县娜姑、老厂之间，极震区烈度Ⅵ度，呈北东向椭圆形。地震造成 19 人受伤，直接经济损失 10440 万元。

此次地震序列为前震—主震—余震型，最大前震 4.7 级发生在主震前 29 分钟，最大余震 4.0 级，余震分布在主震西侧，优势方向不明显。综合地质构造、烈度等震线和余震活动分析认为，节面Ⅰ为主破裂面，主压应力 P 轴方位为 NW 向，本次地震的发震构造可能与深部北东向隐伏断裂有关，是在近水平向的主压应力作用下，产生右旋走滑破裂活动的结果。

震中附近 200km 范围内共有地震台 37 个，其中测震台 20 个，定点前兆观测台 19 个，震前共出现 7 个异常项目 13 条异常，主要为短临异常。

本文对会泽地震前兆进行了识别，对异常特征及有关重要现象进行了讨论，对震后小江断裂带附近的地震活动趋势进行了探讨。

前　言

2005 年 8 月 5 日 22 时 14 分在云南省会泽县发生 5.3 级地震，据云南地震台网测定微观震中 24°33′N、103°09′E，宏观震中 24°33′4N、103°08′E，微、宏观震中相距约 3km，极震区烈度为Ⅵ度。

震区附近历史上多次发生 $M \geqslant 5.0$ 级地震，最强的地震是 1733 年 8 月 2 日东川紫牛坡 7¾级地震[2]。

震中附近 200km 范围内共有地震台 37 个，震前共出现 7 个异常项目 13 条异常，主要为短临异常。

会泽 5.3 级地震发生在云南省地震局 2005 年度地震危险区预测区内，但预测震级偏高。地震发生后，云南省地震局立即启动地震应急预案，赶赴震区开展震情监视、现场考察和震害评估工作[1];1)。

一、测震台网及地震基本参数

会泽 5.3 级地震前，震中周围 200km 范围内共有有人值守测震台和电信传输测震台 20 个，其中 100km 范围内有 4 个，101 ~200km 范围内有 16 个（图 1）。震中附近地区遥测台和区域台对地震的监控能力分别为 $M_L \geqslant 1.0$ 和 $M_L \geqslant 1.7$ 级。这次地震的震中位置采用云南地震台网测定结果，震级采用中国地震台网测定结果，地震序列目录采用云南地震台网资料，地震的基本参数见表 1。

表 1　地震基本参数

Table 1　Basic parameters of the earthquake

编号	发震日期	发震时刻	震中位置		震级		震源深度	震中地名	结果来源
	年 月 日	时 分 秒	φ_N	λ_E	M_S	M_b	(km)		
1	2005 08 05	22 14 43.1	26°33′	103°09′	5.4		21	会泽	资料 2）
2	2005 08 05	22 14 43.1	26°33′	103°09′	5.4		21	会泽	资料 3）
3	2005 08 05	22 14 43.3	26.6°	103.1°	5.3	10		云南会泽与四川会东间	中国地震台网速报
4	2005 08 05	22 14 43.3	26°35′	103°03′	5.3				四川地震台网月报
5	2005 08 05	22 14 48.0	26.57°	103.04°	4.8	5.2	42.4	云南	哈佛大学

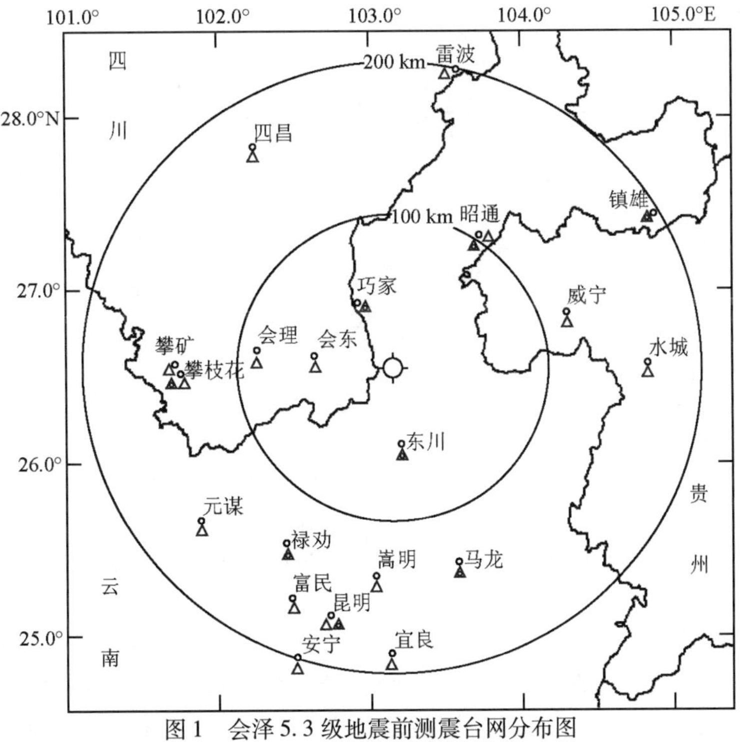

图 1　会泽 5.3 级地震前测震台网分布图

Fig. 1　Distribution of seismometric stations before the M_S5.3 Huize earthquake

二、地震地质背景

会泽震区主要由次级滇中块体和滇东块体构成，二者以则木河、小江断裂为界，会泽地震位于滇东块体的西缘及滇中块体东边界的转折部位。区内断块的新构造运动呈掀斜式隆升，地壳在新生代的间歇性抬升形成了层状地貌，断块间的差异抬升致使夷平面呈台阶状发育。区内第四纪中晚期活动断裂发育，主要有南北、北西—北北西、北东向三组断裂，前两组断裂活动性较大，运动方式主要为左旋走滑，其中小江及则木河全新世断裂的地震活动性最大；第三组断裂为中—晚更新世活动断裂，运动方式主要为右旋走滑，地震的发生主要与这三组断裂的活动有关。

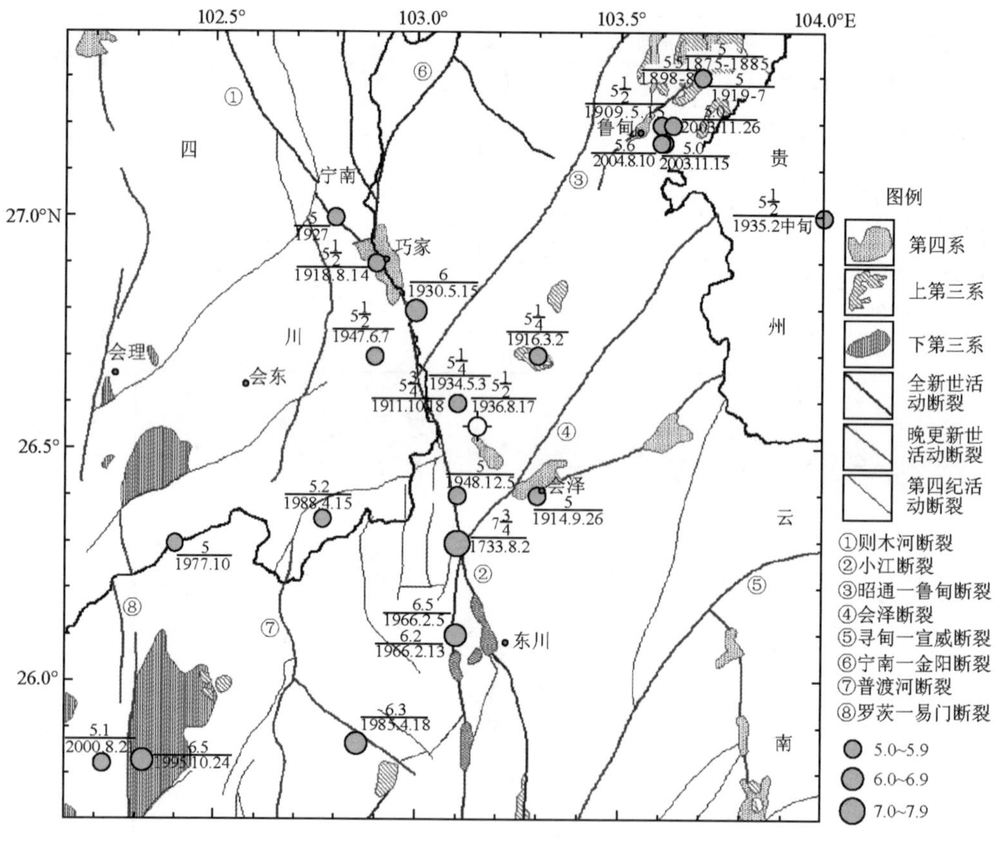

图 2　会泽震区附近地质构造与历史地震震中分布图

Fig. 2　Map of fault structure and distribution of historical earthquakes around Huize area

会泽震区位于东川—建水地幔坳陷区的中心部位，地壳厚度约 44～45 km，该区中地壳下部，有一厚 9～14 km、顶部埋深 23～27 km、波速 516～517 km／s 的低速层（推测是一个温度很高的软弱构造层位）[3]。小江断裂及罗茨—易门断裂在地震测深剖面上显示应属切割莫霍面的超壳深断裂。会泽地震震中区的地表没有断裂，但其深部存在一较明显的北东向舌形重力异常，因此不排除深部（上地壳底部）存在隐伏断裂的可能性。元谋、鲁甸、彝良一线存在一条较明显的北东向线状重力异常带，推测其深部可能存在一条较大规模的北东向隐伏断裂。研究区中部有一北东向巨型磁异常密集带，该带西南边界为寻甸、威宁一线，西北以武定、彝良一线为界。据地质背景、岩浆活动带和地震密集带推测，震区及其邻近地区的深部（中上地壳）可能存在北东向隐伏断裂。

会泽震区的主要活动断裂有小江断裂、会泽断裂和昭通鲁甸断裂。会泽断裂西起小江断裂，往东到会泽分为南、北两支。南支断裂经中路卡、者海、矿山镇延入贵州境内；北支断裂经桃园、下寨延入贵州境内，云南省内长约100 km，总体走向北东，产状320°∠20°。断裂第四纪活动较强，对会泽、者海第四纪盆地的形成、发育具有明显控制作用。地貌上有断崖、沟槽、断错水系及温泉出露。新生代以来以右旋走滑活动为主，断错的最新地层是中更新统。沿断裂有中强地震记录。会泽断裂与五星背斜、背罗箐断裂等构成包括断裂、褶皱的北东向五星构造带，此次会泽5.3级地震的微观震中就位于该构造带上。

震区附近（25.70°~27.40°N，101.10°~104.00°E）自公元624年以来共记载到5级以上地震28次，其中5~5.9级22次，6~6.9级5次，7~7.9级1次。区内最强的地震是1733年8月2日东川紫牛坡7¾级地震[2]。6级以上的强震均发生在震区南北两侧的小江断裂带，如1733年东川7¾级地震，1930年巧家6级地震，1966年东川6.5、6.2级地震。

三、地震影响场和震害

经考察[1];1)，会泽5.3级地震的宏观震中位于娜姑、老厂之间，极震区烈度为Ⅵ度，灾区涉及云南和四川两省。

Ⅵ度区：等震线形状呈椭圆形，走向北东，沿60°方向展布。北自昭通市巧家县马树镇政府驻地，南到昆明市东川区拖布卡镇的蒋家湾村，东自会泽县大桥乡政府驻地以东，西至四川省境内，南至新山乡的大荞地，南西至雪山乡的朱家村。总面积约1200km^2，其中云南境内约占880 km^2，四川境内面积约353 km^2。

在以礼河河流切割较深而高差悬殊大的西侧山梁上的炉房和金沙江河谷阶地老河床上的蒙姑出现了两个Ⅶ度异常点。在场地有膨胀土软弱层，座落于盆地内的大桥、磨盘卡、杨梅山、马树等村庄和位于桃园断层附近边坡上的竹箐、五星、披嘎等村庄震害指数较高，震害明显加重。在评估区以外，昭通市巧家县的崇溪乡的部分民房、校舍、水池、小水窖也遭受了不同程度的破坏。

本次地震云南灾区主要涉及曲靖市会泽县的娜姑、老厂、五星、大桥、昭通市巧家县的蒙姑、炉房、马树和昆明市东川区的拖布卡等8个乡（镇），75个行政村。地震造成村镇民用房屋破坏，倒塌民房总计110895m^2，生命线工程及水利等基础设施损坏；多处出现滑坡、崩塌。地震中无人死亡，19人轻伤，50992户、198747人受灾，云南震区失去居所人数5948人，直接经济总损失为10440万元。四川灾区主要涉及会东县松坪、嘎吉、鲁吉、淌塘、大桥、野租6个区的28个乡，个别民房出现了倒塌等破坏现象，生命线工程、电站、水利等基础设施受损，受灾人口110097人，涉及27301户[4]。

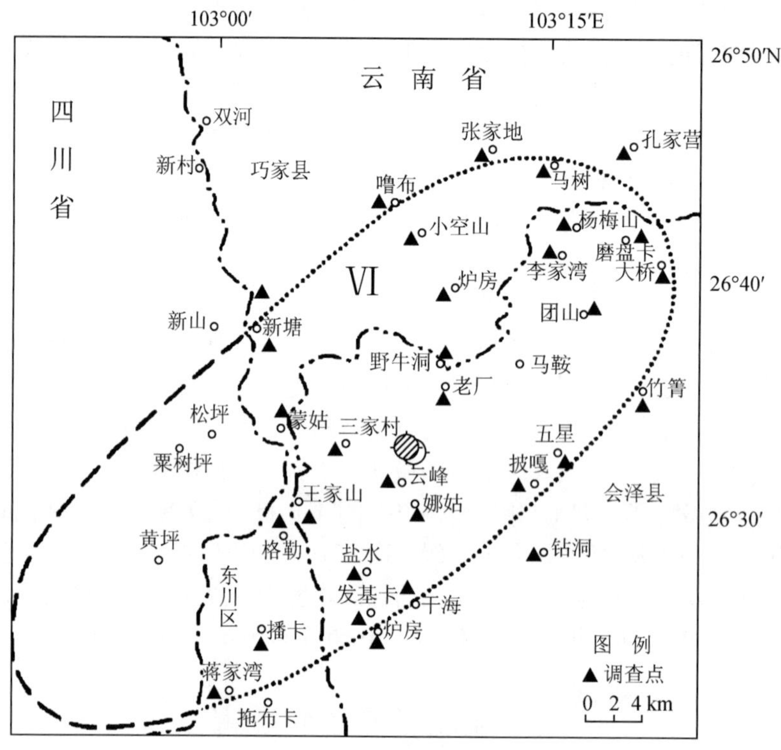

图 3　会泽 5.3 级地震烈度分布图

Fig. 3　Isoseimal map of the M_S5.3 Huize earthquake

2005 年 8 月 5 日会泽 5.3 级地震后，云南省地震局相关人员携带 5 台 K2 型数字强震仪，连夜赶往会泽震区，并于 8 月 6 日完成了以礼河四级电站、五星乡、娜姑乡云峰村、老厂乡、会泽县城数字强震仪的布设。截至 2005 年 8 月 9 日，共获取余震记录 12 条，其中最大加速度值达 40Gal。以礼河四级电站、五星乡卫生院和会泽县地震局三台 K2 型数字强震仪一直观测到 2005 年 10 月下旬[4)]。表 2 和图 4 给出了会泽 5.3 级地震强震动观测台站参数和空间分布。

表 2　会泽 5.3 级地震强震动观测台站参数

Table 2　Parameters of strong motion observation stations for the M_S5.3 Huize earthquake

观测地点	仪器号	触发阀值	场地	经度	纬度	海拔/m
以礼河四级电站	2283	(10) Gal	土层	26.5363°	103.0649°	810
云峰村村委会	2292	(8) Gal	土层	26.5367°	103.1325°	2115
老厂乡派出所	2277	(8) Gal	土层	26.5970°	103.1687°	
五星乡卫生院	2300	(8) Gal	土层	26.5511°	103.2605°	
会泽县地震局	2287	(8) Gal	土层	26.4137°	103.2992°	

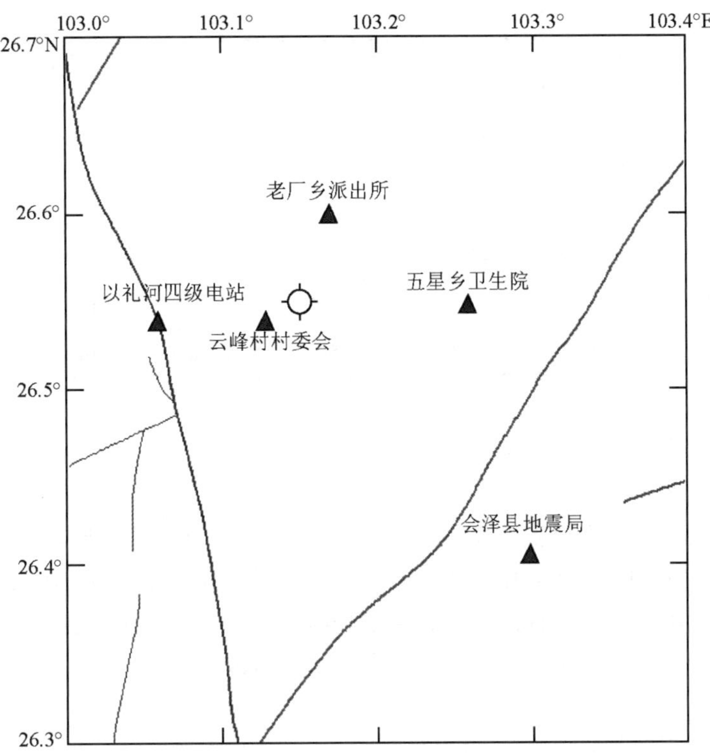

图 4　会泽 5.3 级地震强震动观测台站分布图

Fig. 4　Distribution of strong motion observation stations of the M_S5.3 Huize earthquake

四、地 震 序 列

会泽 5.3 级地震前发生了 M_S4.7 和 M_L3.6 两次前震，截至 2005 年 11 月 5 日，该序列共发生 1.0 级以上地震 72 次，其中 1.0～1.9 级 21 次，2.0～2.9 级 45 次；3.0～3.9 级 3 次；4.0～4.9 级 2 次；5.0～5.9 级 1 次，最大前震为 M_S4.7，发生在主震前 29 分钟，最大余震为 M_L4.0，发生在主震后 8 小时。序列主震与最大前震的震级差为 0.7 级，与最大余震的震级差为 1.7 级。

该序列 $M_L \geq 3.0$ 级地震目录见表 3。

表 3　会泽 5.3 级地震序列目录（$M_L \geq 3.0$ 级）

Table 3　Catalogue of the Huize M_S5.3 earthquake sequence（$M_L \geq 3.0$）

编号	发震日期	发震时刻	震中位置		震级	震源深度	震中地名	资料来源
	年 月 日	时 分 秒	φ_N	λ_E		(km)		
1	2005 08 05	21 45 12.0	25°33′	103°08′	M_S4.7	21	会泽	资料 2）

续表

编号	发震日期	发震时刻	震中位置		震级	震源深度 (km)	震中地名	资料来源
	年 月 日	时 分 秒	φ_N	λ_E				
2	2005 08 05	21 53 38.1	26°33′	103°09′	M_L 3.6	14	会泽	资料 2)
3	2005 08 05	22 14 43.1	26°33′	103°09′	M_S5.4	21	会泽	资料 2)
4	2005 08 06	07 40 56.3	26°33′	103°09′	M_L 4.0	16	会泽	资料 2)
5	2005 08 06	14 39 17.7	26°33′	103°00′	M_L 3.0	11	会东	资料 2)
6	2005 08 20	23 34 59.7	26°12′	103°01′	M_L 3.4	10	东川	资料 2)

根据云南地震台网资料分析会泽地震序列有以下特征:

1. 序列空间分布

图 5 给出了会泽 5.3 级地震序列震中分布。8 月 5 ~6 日,序列地震集中发生在主震附近,8 月 8 日开始序列地震活动出现明显的扩散,分别在小江断裂上的巧家和东川活动。总体来看,序列地震活动与地震构造相关,主要分布在北北西向小江断裂及其与北东向昭通鲁甸断裂、会泽断裂和包括断裂、褶皱的北东向五星构造带的交汇处。

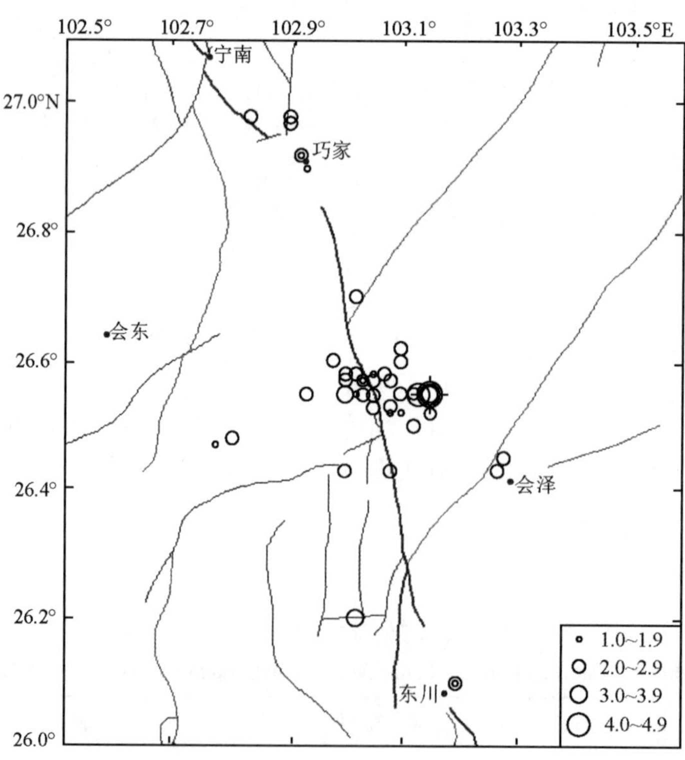

图 5 会泽 5.3 级地震序列震中分布图

Fig. 5 Distribution of the M_S5.3 Huize earthquake sequence

2. 序列时间分布

图6、图7 分别给出了会泽 5.3 级地震序列 $M_L \geqslant 1.0$ 级地震 M-t 图和日频度衰减曲线。由图可见，主震发生的前 5 天内，序列地震活动强度和频度都较高，其中 8 月 5 ~6 日共发生 1 级以上地震 22 次，序列前震和最大强余震均发生在这一时段；8 月 7 ~9 日共发生了 11 次 1 级以上地震，但无 3 级以上地震发生，总体上强度较弱。之后，仅于 8 月 20 日在主震以南的小江断裂带附近发生了 $M_L3.4$ 级地震，序列活动表现出了明显的衰减特征。

会泽 5.3 级地震序列余震震源深度展布较深（图 8），其中 $h \geqslant 30$km 的地震占全序列的 5%，$15 \leqslant h < 30$km 的地震占全序列的 29%，1996 ~2005 年会泽地震震中周围 50km 范围内 $M \geqslant 2.0$ 级地震震源深度均小于 20km[5]。显然会泽 5.3 级地震后的微破裂相对地震活动处于正常背景下的微破裂深度要深，但余震的破裂深度在主震后前 10 天较深，之后逐渐变向浅部破裂。

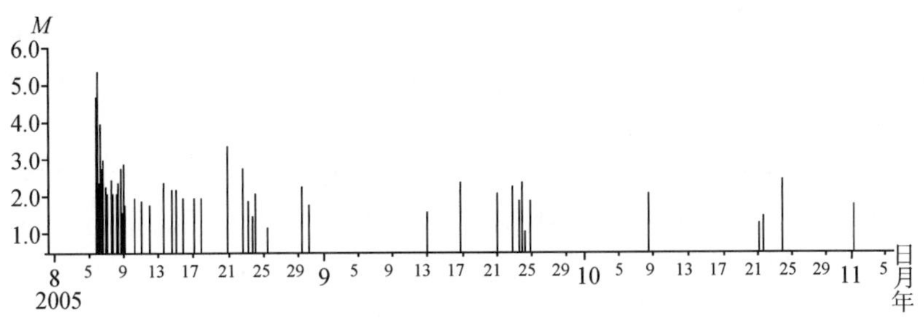

图6 会泽 5.3 级地震序列 M-t 图

Fig. 6 M-t diagram of the $M_S5.3$ Huize earthquake sequence

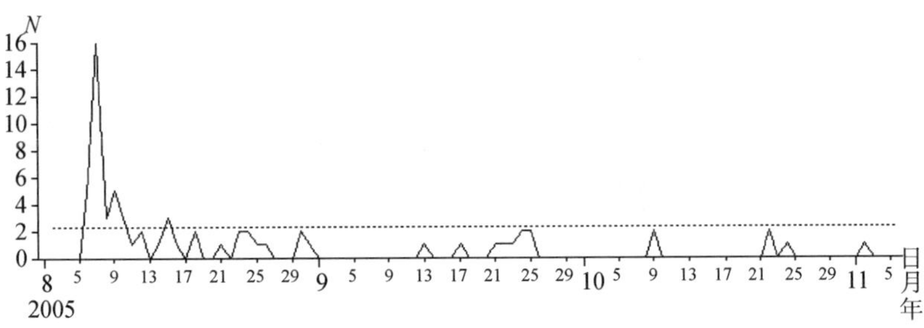

图7 会泽 5.3 级地震序列日频度曲线

Fig. 7 Daily frequency of the $M_S5.3$ Huize earthquake sequence

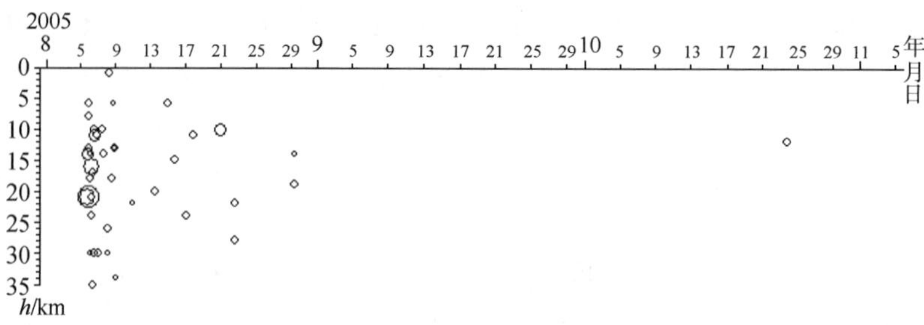

图8　会泽5.3级地震序列震源深度随时间分布

Fig. 8　Time distribution of focal depth of the M_S5.3 Huize earthquake sequence

3. *b* 值、*p* 值、*h* 值和应变释放

根据云南省地震台网的监控能力，分析认为该序列 $M_L \geqslant 2.0$ 级地震资料完整，故序列参数的起算震级取 M_L2.0。会泽 M_S5.3 主震前发生了 M_S4.7 和 M_L3.6 地震，与主震时间间隔分别为29分钟和20分钟，距离接近（表2），为该序列的前震。若剔除这两次前震，计算得到主震发生后序列 b 值为0.79（图9），这一结果与李忠华等[6]由云南地区15次主震余震型地震序列的 b 值平均值为 0.75 ± 0.18 的结果一致；$p = 0.32$（图10）；$h = 1.5$（图11），按刘正荣等[7]的研究结果，序列 h 值大于1.0，是主震—余震型；图12给出了序列蠕变曲线，并计算得到主震能量占序列总能量98.9%，如果计算中考虑序列的两次前震，得到主震占序列总能量的90.7%，与吴开统等[8]给出的主震—余震型的判定指标一致。综上分析认为，会泽5.3级地震序列类型为前震—主震—余震型。

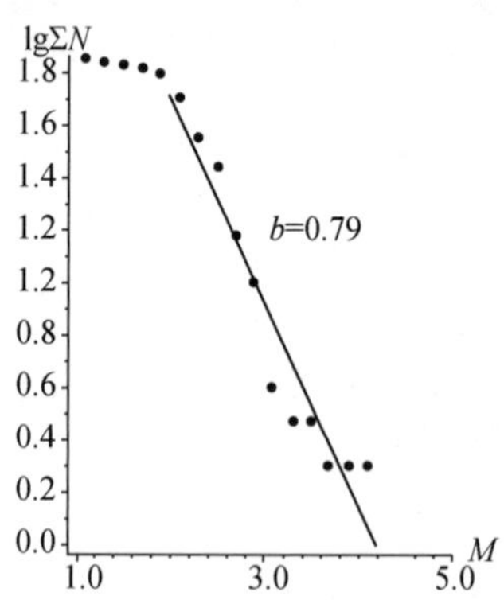

图9　会泽5.3级地震序列 b 值图

Fig. 9　*b*-value diagram of the M_S5.3 Huize earthquake sequence

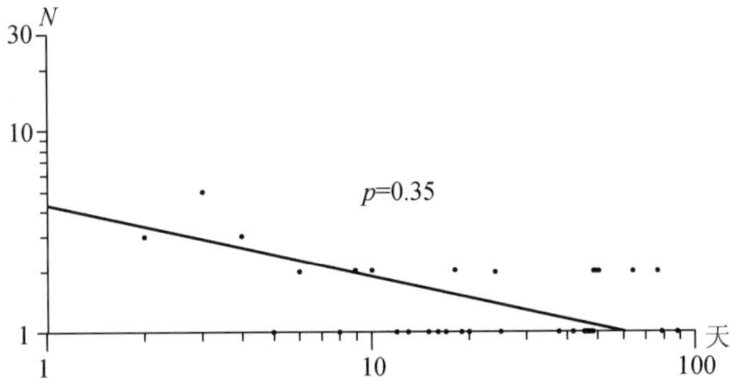

图10　会泽5.3级地震序列p值图

Fig. 10　p-value diagram of the M_S5.3 Huize earthquake sequence

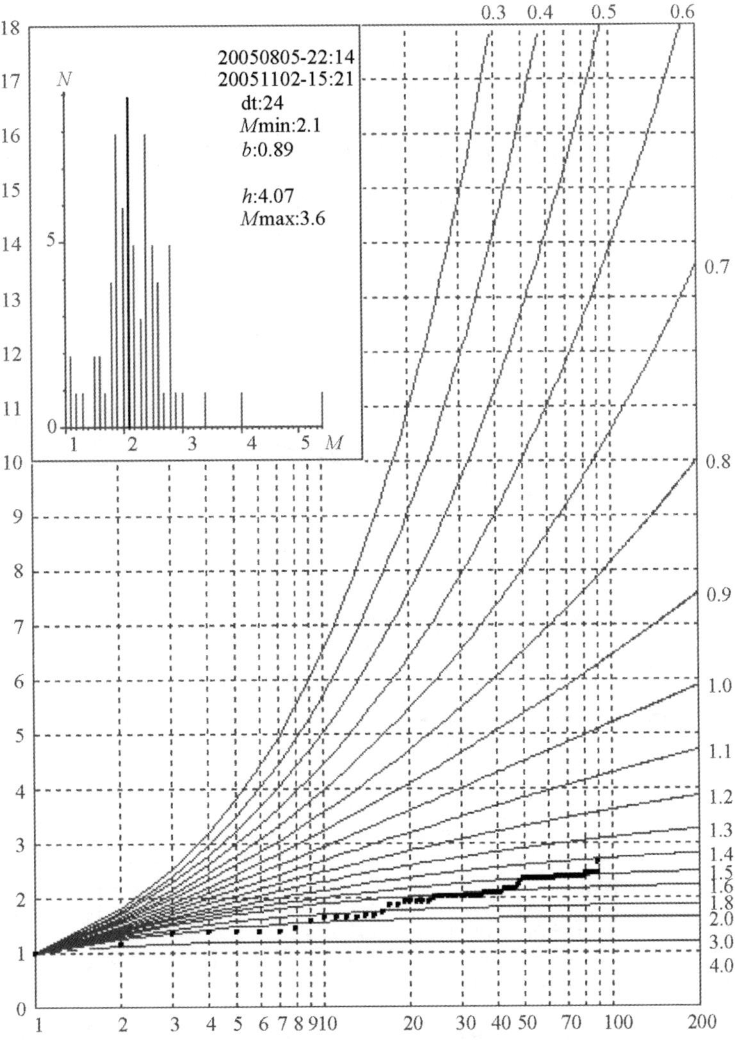

图11　会泽5.3级地震序列h值图

Fig. 11　h-value diagram of the M_S5.3 Huize earthquake sequence

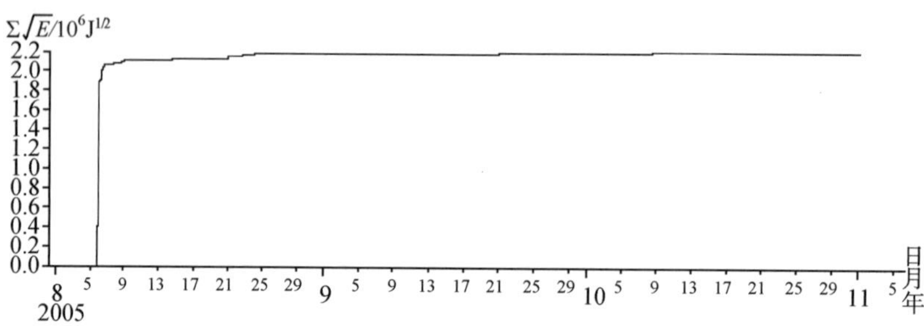

图 12　会泽 5.3 级地震序列蠕变曲线

Fig. 12　Strain release of the M_S5.3 Huize earthquake sequence

五、震源参数及地震破裂面

利用云南区域数字地震台网记录的波形数据资料，采用 P 波初动求解震源机制解的方法，求得会泽 5.3 级、4.7 级地震的震源机制解（图 13，表 4、表 5）。

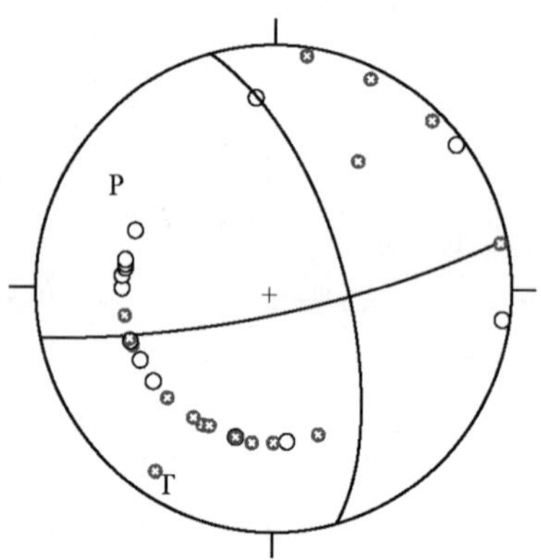

图 13　会泽 5.3 级地震震源机制解图

Fig. 13　Focal mechanism solution of theM_S5.3 Huize earthquake

表 4　会泽 5.3 级地震矩张量解

Table 4　the CMT of the M_S5.3 Huize earthquake

编号	节面 I			节面 II			矩张量						地震矩 M_0 (dyn · cm)	矩震级 M_w	结果来源
	走向	倾角	滑动角	走向	倾角	滑动角	M_{rr}	M_{tt}	M_{pp}	M_{rt}	M_{rp}	M_{tp}			
1	252	73	-178	161	88	-17	-0.325	0.658	-0.334	-0.196	0.191	-0.687	8.82×10^{17}	5.2	Harvard

表 5　会泽 5.3 级地震震源机制解

Table 5　Focal mechanism solution of the M_S5.3 Huize earthquake

序号	节面Ⅰ			节面Ⅱ			P 轴		T 轴		N 轴		结果来源
	走向	倾角	滑动角	走向	倾角	滑动角	方位	仰角	方位	仰角	方位	仰角	
1	78	81	-155	344	65	-10	304	24	209	11	97	64	王绍晋*
2	252	73	-178	161	88	-17	115	13	208	11	335	73	Harvard
3	50	84		142	66		278	12	184	21	36	65	牟雅元等[6]

* 初动矛盾符号比为 0.17

震源机制解表明，节面Ⅰ走向 78°，倾角 81°，滑动角 -178°，是一个近垂直的节面，与北东向深部隐伏断裂的右旋走滑破裂活动特性一致[3]，也有研究认为与四川岩坝（淌塘）断层的性质基本一致[9]，表明这个节面很可能就是破裂面。节面Ⅱ走向 344°，倾角 65°，滑动角 -10°，与北北西向小江断裂带北段的左旋走滑破裂活动性质一致，表明如果这个节面是破裂面，也有可能是这个断层发生了错动[3],[9],[10]。P 轴方位为 304°，仰角为 24°，P 轴产生的压应力方向与区域应力场方向基本一致。烈度等震线和序列空间分布，极震区烈度长轴呈北东向分布。会泽 5.3 级地震序列的发生可能与小江断裂与五星断裂褶皱带结合部位的破裂活动有关。

结合地质构造、烈度等震线和余震活动等资料，综合分析认为节面Ⅰ最有可能为主震初始破裂面。会泽 5.3 级地震的发生可能与深部北东向隐伏断裂有关，是在近水平向的主压应力作用下，产生右旋走滑破裂活动的结果。

六、地震前兆观测台网及前兆异常

会泽 5.3 级地震发生在云南省前兆台网较稀疏的地区，距震中周围 200km 范围内分布有 20 个地震台，其中有测震单项观测台 8 个，其他前兆单项观测台 4 个，综合观测台 18 个。有测震、水氡、水位、水温、水氡、地磁 D、地磁 Z、地电阻率、自然电位、地倾斜、地应力、重力、短基线、短水准、水化学等 15 个观测项目，42 个观测台项（图 14），其中 0～100km 内有 7 个观测台，13 个观测台项；101～200km 内 23 个观测台，31 个观测台项。定点前兆观测台站观测项目震前都有长期连续可靠的观测资料。

会泽 5.3 级地震前共出现了 7 个异常项目 13 项次异常，地震活动性出现了 4 项次异常，包括地震频度、地震 b 值、破裂时间空间异常。定点前兆观测出现了 7 个项目 9 项次异常（图 15），包括水氡、水位、水温、地倾斜、地磁 D 分量、自然电位、电阻率。

在震中周围 100km 内有前兆观测台站 5 个，12 个台项，出现异常台站 2 个，异常台项 2 项，异常台站和异常台项百分比分别为 40% 和 17%；在 101～200km 范围内有前兆观测台站 14 个，台项 30 个，出现异常台站和异常台项为 5 个和 7 项，异常台站和异常台项百分比分别为 36% 和 23%。各类异常的具体情况详见表 6 和图 16 至图 28。

表 6 异常情况登记表

Table 6 Summmary table of precursory anomalies

序号	异常项目	台站或观测区	分析方法	异常判据及观测误差	震前异常起止时间	震后变化	最大幅度	震中距Δ (km)	异常类别及可靠性	图号	异常特点及备注
1	地震平静	云南省	$M \geqslant 4.0$ 级地震，时间间隔	$dT \geqslant$ 60 天	2005.4 ~ 2005.8	正常	$dT=$ 107 天		S_1	16	是震前密切跟踪的一条重要指标
2	地震 b 值	云南地区 20° ~ 30°N, 96° ~ 107°E	$M_L 2.5 \sim 4.9$ 地震，3 月时窗 1 月步长滑动 b 值	$b \geqslant 1.1$ 或 $b \leqslant 0.8$	2005.3 ~ 2005.8	正常	$b=$ 1.21		S_1	17	云南短临跟踪工作中常用的指标，震前出现明显的高 b 值异常
3	地震频度	震中周围 O (26.55°N, 103.15°E) $R=100$km	$M_L \geqslant 2.0$ 级地震，月频度	$N \geqslant 6$	2004.10 ~ 2005.8	持续	$N=9$		M_1	18	震中附近地区的中小地震活动增强非常显著
4	破裂时间法	云南地区 21° ~ 29°N, 97° ~ 106°E	用搜索半径内的 $M_L \geqslant$ 2.5 级地震前兆事件构建能量加速释放曲线进行主震位置搜索	$C>1.0$	2004.10 ~ 2005.8	正常	$C=10$		M_1	19	会泽地震前出现明显的 NSR 高值区，震前 2 个月秦嘉政等[8]采用该方法得到了为次地震的中短期预测结果
5	水氡	会泽	月均值	大幅升降变化	2005.4.5 ~ 6.30	持续	1.2 Bq/l	5	S_2	20	震前出现趋势下降，为震前发现的异常

续表

序号	异常项目	台站或观测区	分析方法	异常判据及观测误差	震前异常起止时间	震后变化	最大幅度	震中距Δ (km)	异常类别及可靠性	图号	异常特点及备注
6	水氡	元谋	日均值	≥45.9 Bq/l	2005.6.19 ~6.22	正常	47.9	160	S_2	21	震前出现明显的上升变化
7	地磁 *D* 分量	巧家	日均值	加速上升	2005.3.8 ~6.6	正常	6.9 分	47	S_2		震前加速上升，并在高值波动，为震前发现的异常
8	自然电位 EW	元谋	日均值	加速上升	2005.7.28 ~8.13	正常	61.6 mV	160	I_2		震前加速上升，显示临震异常，为震前发现的异常
9	电阻率 EW	元谋	日均值	大幅变化	2005.7.5 ~7.11	正常	0.5 Ω·m	160	I_2		震前变化幅度显著，为震前发现的异常
10	水温	沾益	日均值	≥0.005 ℃	2005.5.21 ~7.24	正常	0.0072℃	102	S_2	25	震前出现大幅度波动变化，为震前发现的异常
11	水温	曲靖	日均值	≥0.002 °C	2005.6.8 ~6.12	正常	0.004 ℃	135	S_2	26	震前出现明显的降温异常，为震前发现的异常
12	水位	鱼洞	日均值	≥ 0.143m	2005.7.4 ~7.16	正常	0.42m	103	I2	27	震前出现快速上升变化，为震前发现的异常

续表

序号	异常项目	台站或观测区	分析方法	异常判据及观测误差	震前异常起止时间	震后变化	最大幅度	震中距Δ（km）	异常类别及可靠性	图号	异常特点及备注
13	地倾斜	攀枝花	五日均值	趋势改变	2005.3.25～9.10	正常	10749毫秒	140	S_2	28	出现显著的趋势上升变化

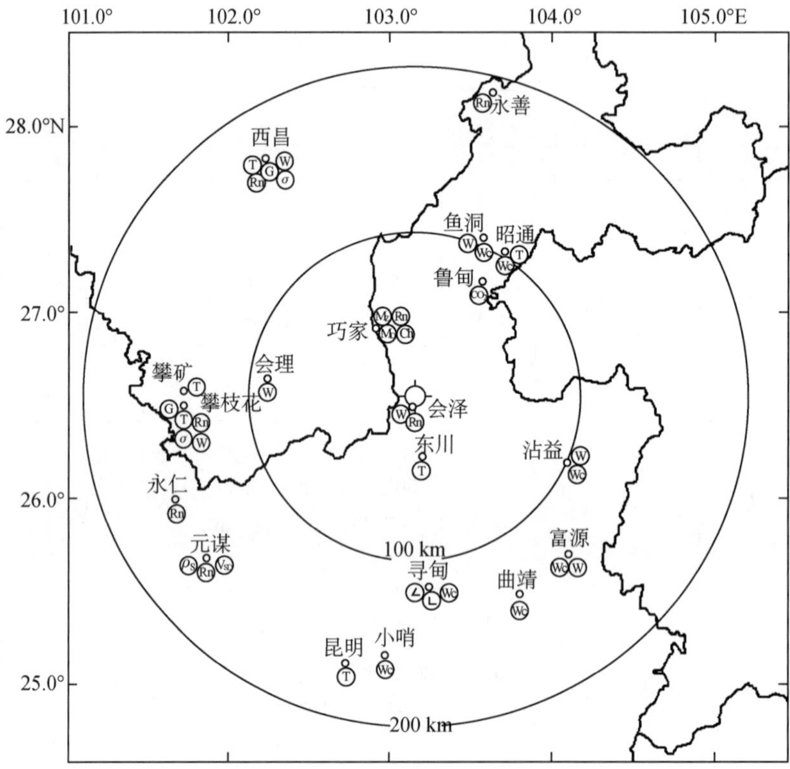

图 14　会泽 5.3 级地震前定点前兆观测台站分布图

Fig. 14　Distribution of the precursor monitoring stations before the M_S5.3 Huize earthquake

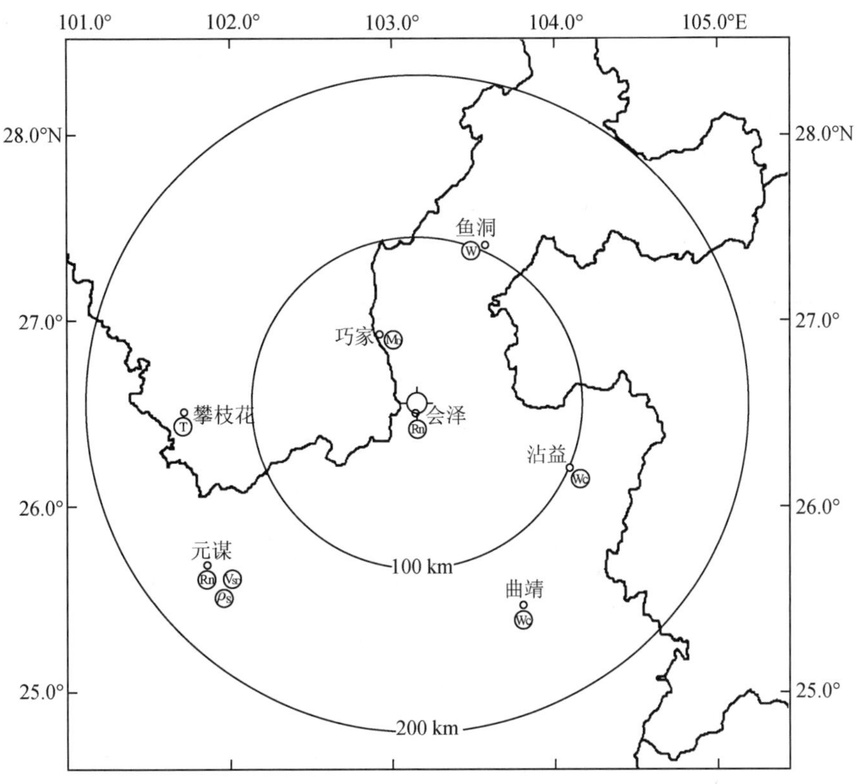

图15　会泽5.3级地震定点前兆观测异常分布图

Fig. 15　Distribution of precursory anormalies of the M_S5.3 Huize earthquake on the fixed observation points

会泽地震前，云南地震台网已经具备了较高的地震监控能力，所以地震活动性的资料信度高。我们在这里给出的异常主要是在日常震情跟踪视与预报工作中常用的，在震前发现的异常。会泽5.3级地震前在云南地震活动水平总体并不高，虽然云南省 $M \geq 4.0$ 级地震时间间隔达到60天的异常指标后，我们便作为一条重要的异常指标进行跟踪监视，云南地区 M_L2.5~4.9地震月 b 值高值异常也在日常地震会商中提出，但在准确把握会泽地震的时间和地点上还是有一定的偏差。

表6中给出的前兆异常均有连续完整的观测资料，资料信度高，异常均有较好的地震对应率。对会泽5.3级地震前的前兆变化进行了认真分析和再认识，给出了表6中的异常。表6和图20~26中列出的异常，大多数是地震前发现和重点跟踪监视的，如曲靖水温、沾益水温、鱼洞水位、会泽水氡、巧家地磁 D 分量、元谋自然电位、元谋电阻率等，从源兆和场兆的角度分析，表中列出的异常与会泽5.3级地震后第8天发生的文山5.3级地震无关。

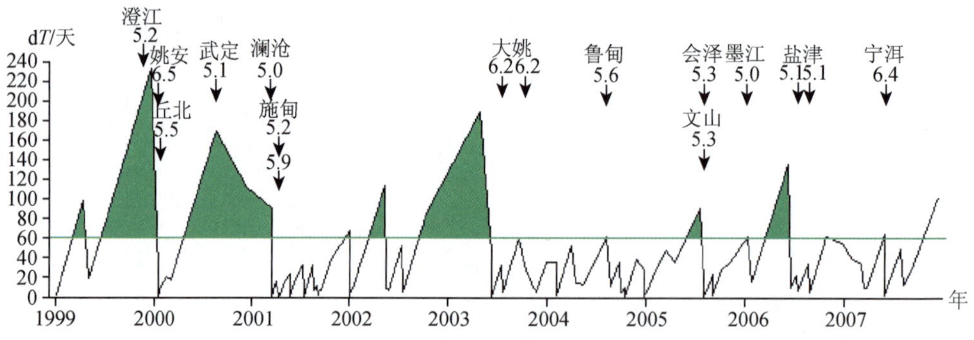

图 16　云南省 $M \geqslant 4.0$ 级地震 dT-t 曲线

Fig. 16　time interval curve for the $M \geqslant 4.0$ earthquakes in Yunnan Province

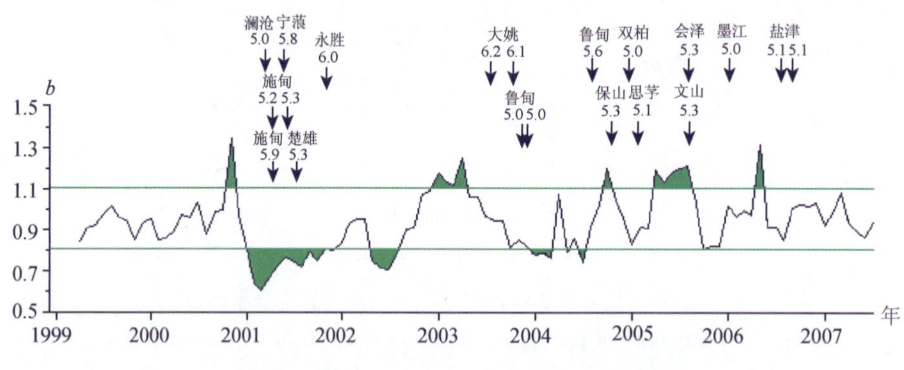

图 17　云南地区 M_L2.5～4.9 地震 b 值曲线

Fig. 17　Curve of b value of M_L2.5-4.9 earthquakes in Yunnan area

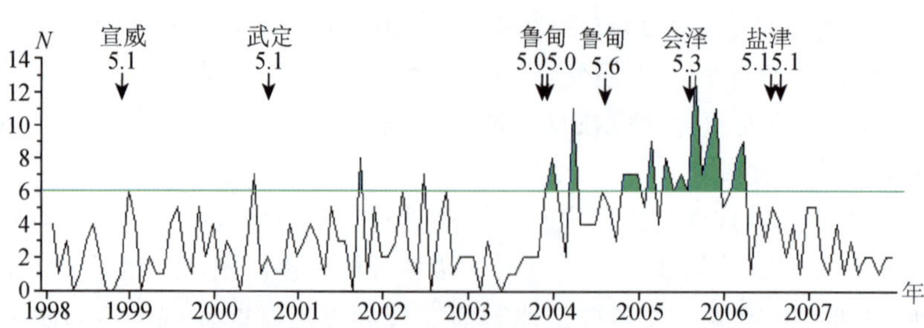

图 18　震中周围 100km 范围内 $M_L \geqslant 2.0$ 地震月频度曲线

Fig. 18　Curve of monthly frequency of $M_L \geqslant 2.0$ earthquakes about 100km around the epicentral erea

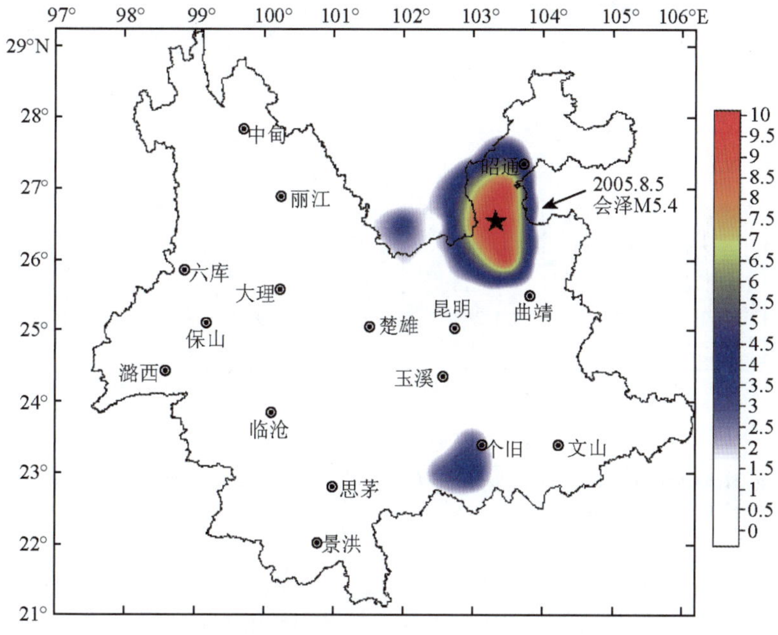

图19 云南地区破裂时间法NSR异常空间分布图

Fig. 19 Spatial distribution of NSR anomalies using time-rupture methods in Yunnan area

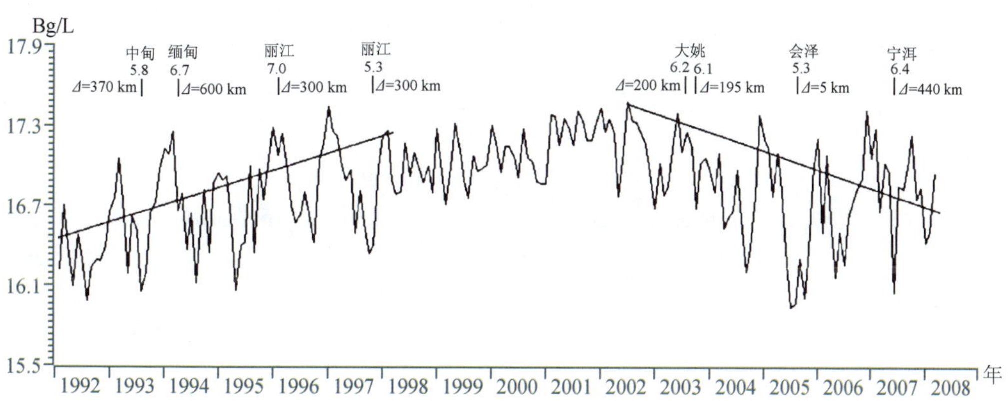

图20 会泽水氡月均值曲线

Fig. 20 Curve of monthly mean value of radon content in groundwater at Huize station

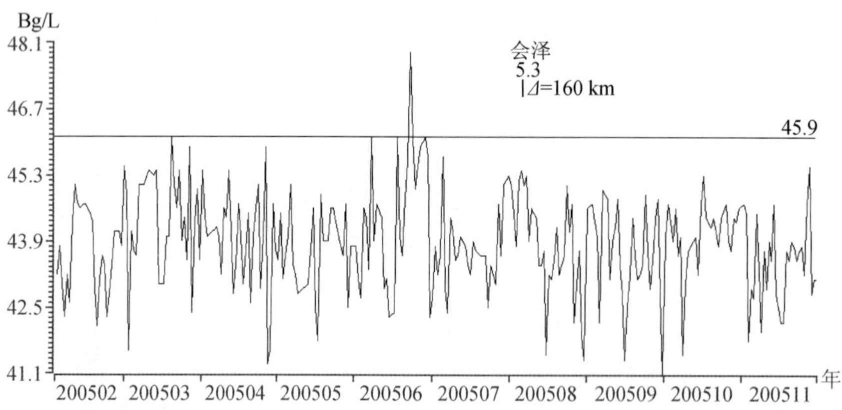

图 21　巧家地磁 *D* 分量日均值曲线

Fig. 21　Curve of daily mean value of *D* variation of geomagnetism at Qiaojia station

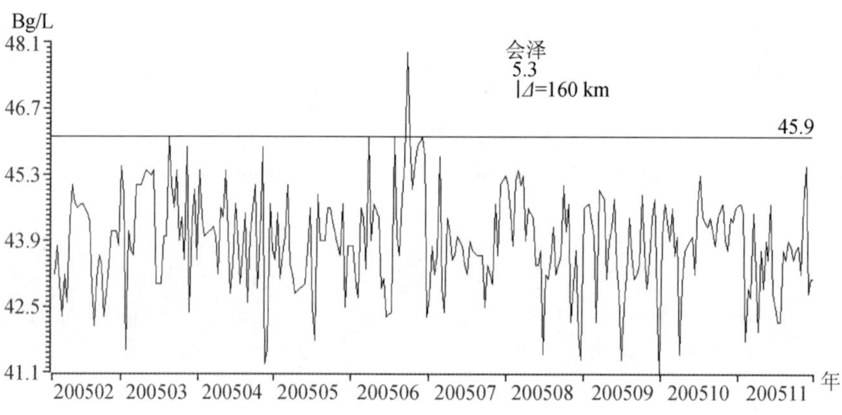

图 22　元谋水氡日均值曲线

Fig. 22　Curve of daily mean value of radon content in groundwater at Yuanmou station

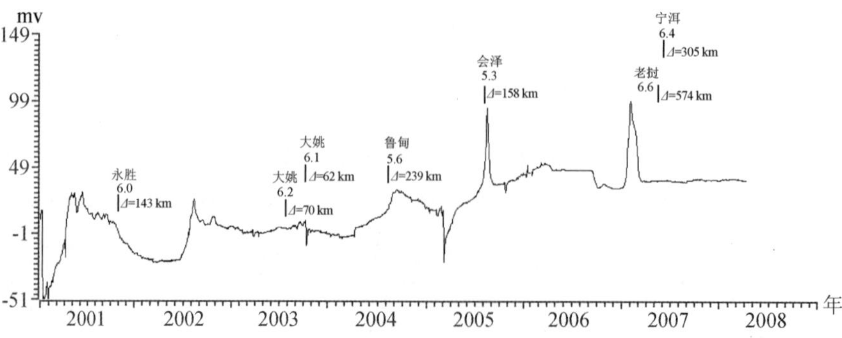

图 23　元谋自然电位 EW 向日均值曲线

Fig. 23　Curve of daily mean value of EW-ward spontaneous potential at Yuanmou station

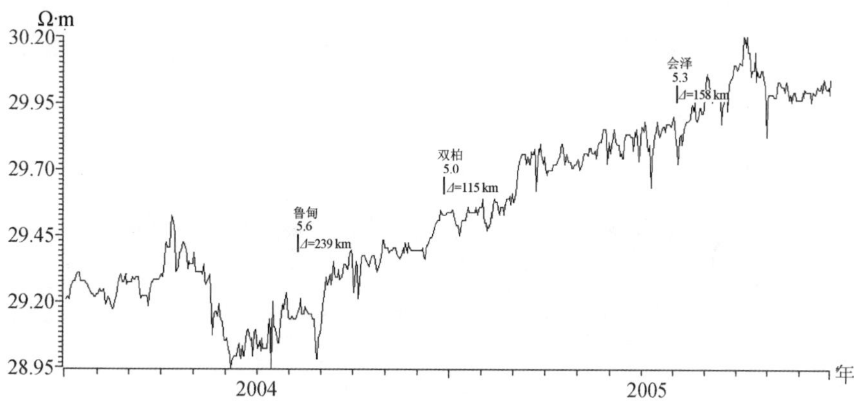

图 24　元谋电阻率 EW 向日均值曲线

Fig. 24　Curve of daily mean value of EW-ward apparent resistivity at Yuanmou station

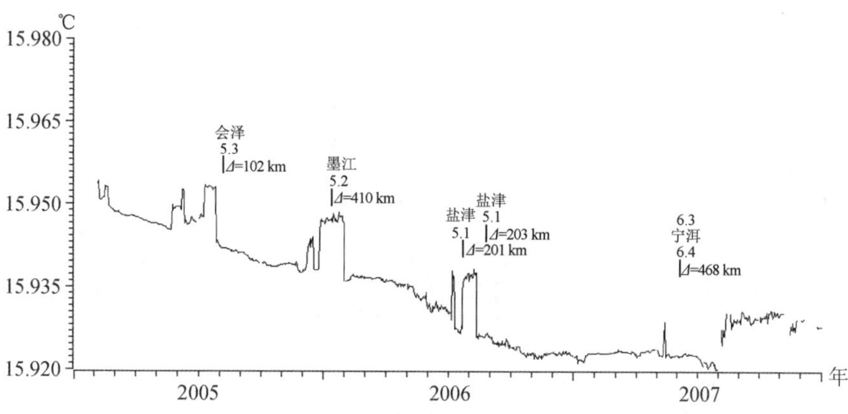

图 25　沾益水温日均值曲线

Fig. 25　Curve of daily mean value of water temperature at Zhanyi station

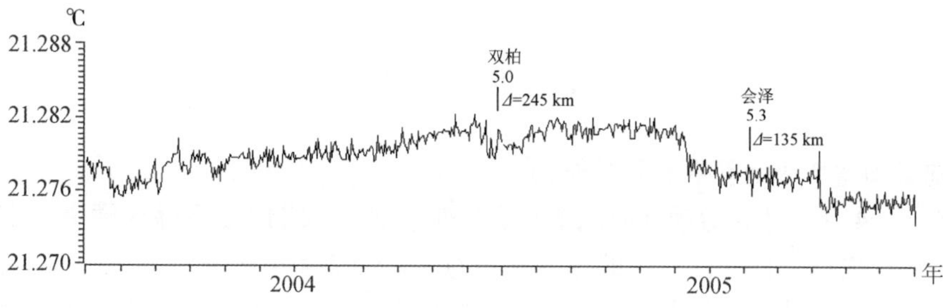

图 26　曲靖水温日均值曲线

Fig. 26　Curve of daily mean value of water temperature at Qijing station

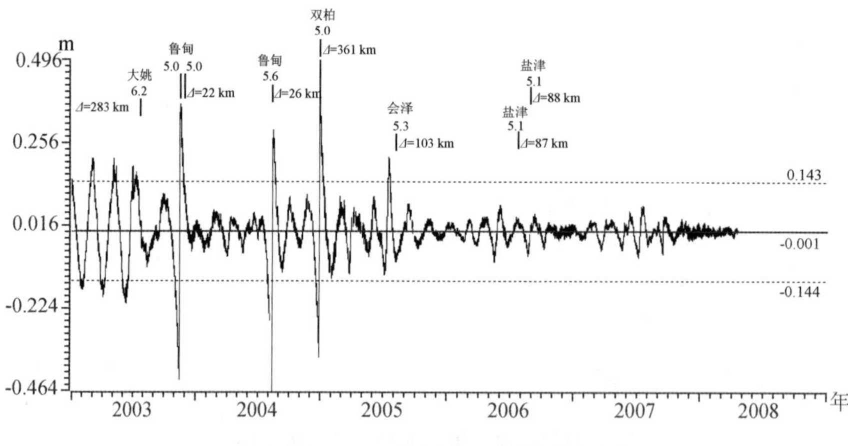

图 27　鱼洞水位日均值曲线

Fig. 27　Curve of daily mean value of water level at Yudong station

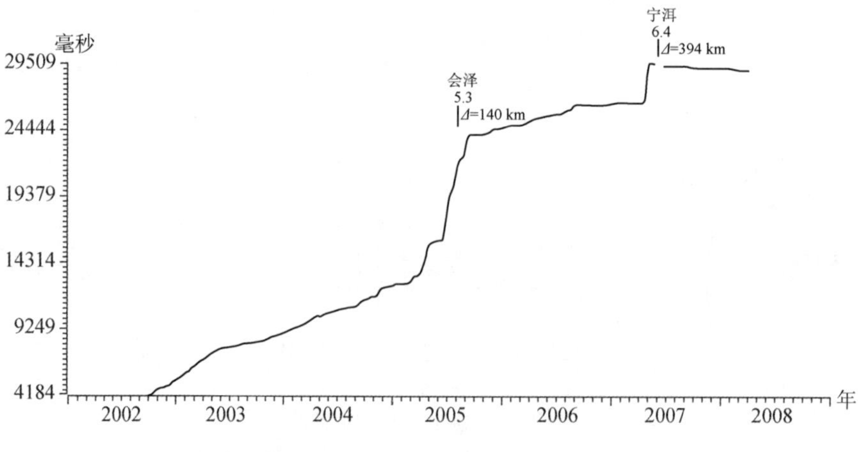

图 28　攀枝花地倾斜五日均值曲线

Fig. 28　Curve of five-day mean value of tilt at Panzhihua station

七、地震前兆异常特征分析

1. 震前云南省 4 级地震平静异常显著

云南省 $M \geqslant 4.0$ 级地震活动在时间上具有群集特征，4 级地震平静异常是云南中强震前重要的震兆特征之一，且具有较好的短期预报意义。将 $M \geqslant 4.0$ 级地震平静 60 天作为异常指标，预报未来 3 个月内云南省内 5 级以上地震及邻近地区 6 级以上地震的对应率达 80%，通过 R 值检验，因该短期预报指标可操作性强，在近年云南省地震预测预报实际工作中得到了应用[11]。会泽 5.3 级地震前 2005 年 1 月 28 日永胜 4.8 级地震至 2006 年 6 月 14 日耿马

4.2 级地震，平静了 61 天；2005 年 4 月 19 日云龙 4.6 级地震至 2007 年 6 月 1 日武定 4.2 级地震，平静了 64 天，连续两次的平静表明云南省及邻近地区有发生中强地震的危险。从另一层面讲，4 级地震的平静表明会泽地震前云南地区地震活动主要表现为相对平静。

2. 震中附近地震中小地震活动中短期内增频异常显著

改进型破裂时间法进行动态模拟主震前前兆地震事件的能量加速释放过程，进行中短期预测，是一种新的预测方法。对主震发生前出现地震活动增强的现象的震例具有较好的预测效果，秦嘉政等[9]利用云南地区 $M_L \geqslant 2.0$ 级地震对会泽 5.3 级地震采用主震位置搜索技术预测的 NSR 异常，与实际地震发生符合得非常好，并且在震前两个月给出了这样的中短期预测结果。同样地，从地震频度时间进程曲线看，震中周围 100km 范围内自 2004 年 10 月开始出现了明显的 $M_L \geqslant 2.0$ 级地震活动增频，类似的异常在 2003 年 11 月 15 日鲁甸 5.0 级、11 月 26 日鲁甸 5.0 级地震和 2004 年 8 月 10 日鲁甸 5.6 级地震前出现过，表明这一地区发震的可能性较大。会泽地震后，该区域中小地震活动水平仍然较高，2006 年 7 月 22 日盐津 5.1 级，8 月 25 日盐津 5.1 级地震发生后活动明显减弱。

震中周围地区中小地震活动的增频，表明区域介质处于非稳定状态，有中强地震在孕育之中，再配合 4 级地震平静、地震 b 值异常等短期指标，对中强地震趋势的把握具有很好的指导和实践意义。

3. 震前前兆异常数量增多，短临异常突出

会泽 5.3 级地震前，短临阶段前兆异常数量明显增多，异常台项数呈明显上升趋势，在震前发现了显著的地下流体和地磁异常变化，并重点跟踪监视[12]，如元谋自然电位 EW 向 7 月 28 日至 8 月 13 日加速突升，元谋电阻率 EW 向 2005 年 7 月 5～11 日下降 $0.5\Omega \cdot m$，有临震异常显示。通过再认识与研究，提出了 9 项次前兆异常，其中曲靖水温、沾益水温、鱼洞水位、会泽水氡、巧家地磁 D 分量、元谋自然电位、元谋电阻率为震前发现并跟踪的异常，共 7 项次。这 9 项次异常以短临异常为主，其中短期异常（S）6 项次，临震异常（I）3 项次（元谋自然电位、元谋电阻率、鱼洞水位）。

4. 近源前兆异常特征

从 9 项次异常的空间分布来看，100km 范围内有 2 项次，分别为会泽水氡和巧家地磁 D 分量，100～200km 范围内的有 7 项次。会泽水氡观测点距会泽震中仅 5km，观测资料连续可靠，2005 年 4 月 5 日至 6 月 30 日出现明显的趋势下降异常变化，异常形态明显，持续时间长；巧家地磁 D 分量 2005 年 3～6 月破年变加速上升，6 月转折，在高值波动。也就是说，距震中较近的会泽水氡和巧家地磁 D 分量在震前都出现了持续时间较长，幅度显著的异常形态，这一现象可考虑对未来中强地震的空间指示意义。

八、震前预测、预报和震后响应

1. 震前预测

较准确的中期预测预报：《云南省 2004 年度地震趋势研究报告》中指出[5)]：2005 年度滇东易门—石屏—建水—华宁—宜良—寻甸—巧家一带存在发生 6.5 级左右地震的危险。结果会泽地震就发生在这一危险区内，预测时间和地点均正确，但预测震级偏高。

2. 震后响应

地震发生后，云南省地震局立即启动地震应急预案，派出地震现场工作队奔赴震区。8月6日凌晨4：00第一批现场工作队员到达现场；中国地震局派出3人专家组于8月6日16：00到达会泽。中国地震局、云南省地震局、曲靖市地震局的领导和专家以及云南省地震灾害紧急救援队官兵组成地震现场工作指挥部，下设秘书、震情监测预报、地震灾评及科考、强震观测、通信、宣传及后勤保障等7个组，共53人开展地震现场工作[1];1)。

3. 序列判定

对会泽5.3级地震序列，云南省地震局后方工作组在规定的时间内给出了趋势判定意见，序列类型判定了主震—余震型，结果证明序列判断是正确的。8月6～9日，地震现场震情监视工作组在现场工作期间，向当地政府和有关部门发出《震情简报》，对序列做出较准确的震后趋势判定。

4. 灾害评估

8月6～10日，10名灾评人员分乘6辆车对会泽县、巧家县和昆明市东川区的部分乡（镇）进行灾情调查，共完成41个居民点的调查工作，其中抽样点23个，此外，还调查了19所学校、16个卫生院所及计生站、20项生命线工程及水利设施，行程约1.2万余公里。8月11日，省地震灾害评定委员会召集民政、地震、建设、抗震办、国土资源、财政、卫生、教育、电信等云南省厅局级的评委和领导共21人对会泽5.3级地震灾害损失进行了评审[1];1)。

九、结论与讨论

（1）会泽震区位于滇东块体的西缘及滇中块体东边界的转折部位，地质构造复杂，褶皱、断裂发育，会泽地震发生在北北西向断裂与北东向断裂褶皱带的交汇部位，震中靠近小江断裂带。但震区深部在刚性上地壳下有一埋深23～27km的低速软流层[2]，这种构造有利应变能的积累和释放，再结合震源机制解、烈度等震线和余震活动空间分布认为，会泽5.3级地震的发生可能与深部北东向隐伏断裂有关，是在近水平向的主压应力作用下，产生右旋走滑破裂的结果。

（2）会泽5.3级地震宏观震中位于娜姑、老厂之间，极震区烈度为Ⅵ度，灾区涉及云南和四川两省。震区复杂的地质构造和破碎的地层，使个别居民点震害明显加重。但主震前发生了29分钟发生了M_S4.7和M_L3.5地震，震感强烈，起到了预警作用，震区居民先后逃到户外避震，尽管地震造成了民房倒塌和两个Ⅶ度烈度点，但是没有造成人员重伤和死亡。

（3）会泽5.3级地震序列类型为前震—主震—余震型，序列震源深度较深，可能有利于上方更浅部岩石形成较大的破裂。用最小二乘法计算会泽地震序列b值，若剔除M_S4.7和M_L3.6这两次前震，计算得到主震发生后序列b值为0.79，这一结果与李忠华等[6]由云南地区15次主震余震型地震序列的b值平均值为0.75±0.18的结果一致。从主震后趋势判定的角度来看，求得的b值对判定地震序列类型来说结果表明是正确的。若不剔除M_S4.7和M_L3.6两次前震，求得整个序列b值为0.57，低于1996～2005年震中周围100范围$M_L \geqslant$ 2.0级地震背景b值（为0.74）[5]，云南中强地震序列低b值异常通常为震源及附近地区地下岩石处于不稳定状态的前兆，对后续地震危险程度具有一定的指示意义[6]，2003 年 11 月 16 日鲁甸 5.0 级、2003 年 11 月 26 日鲁甸 5.0 级、2004 年 8 月 10 日鲁甸 5.6 级和 2005 年 8 月 5 日会泽 5.3 级地震序列均为低 b 值，表明震中周围地壳岩石积累的参量没有完全释放，事实上，2006 年在该区附近又先后发生了 7 月 22 日盐津 5.1 级和 8 月 25 日盐津 5.1 级地震。

（4）会泽地震前最突出的短期地震活动异常为云南省 4 级地震平静、云南地区地震高 b 值。中期异常为震中周围中小地震活动增频。前兆异常短临阶段地震异常台项数呈明显上升趋势，在震前发现了显著的地下流体和地磁异常变化，并重点跟踪监视，9 项次中短期异常（S）6 项次，临震异常（I）3 项次，100km 范围内有 2 项次，100 ~ 200km 范围内的有 7 项次。距震中较近的会泽水氡和巧家地磁 D 分量在震前都出现了持续时间较长，幅度显著的异常形态，这一现象可考虑对未来中强地震的空间指示意义。事实上，在会泽地震前云南省 4 级地震平静和前兆异常数量的增加，表明云南存在发生中强地震的危险性，已引起短临跟踪工作的高度警惕与重视。但是震前仅在震中附近地区表现为中小地震活动趋势增强，但震兆异常特征并不明显，使得我们对会泽地震发生的时间和地点不能做出较好的预测和判定。

参 考 文 献

[1] 非明伦，卢永坤，冉华等. 2005. 云南会泽 5.3 级地震灾害现场调查与烈度分布. 地震研究，28（4）：415 ~ 418

[2] 四川云南西藏地震简目编辑组. 1988. 西南地震简目. 成都：四川科学技术出版社

[3] 谢英情，付正新，非明伦等. 2005. 云南会泽 5.3 级地震构造背景及其序列活动特征. 地震研究，28（4）：408 ~ 414

[4] 龙德雄，吴今生，王松等. 2006. 2005 年 8 月 5 日云南省会泽—四川省会东 5.3 级地震四川震区地震宏观烈度考察. 四川地震，118（1）：

[5] 刘翔，王绍晋，钱晓东等. 2006. 2005 年会泽 5.3 级地震与小江断裂地震活动关系研究. 地震研究，29（4）：332 ~ 337

[6] 李忠华，苏有锦，蔡明军等. 2000. 云南地区地震序列的 P 值和 b 值变化特征. 地震，20（4）：74 ~ 77

[7] 刘正荣等. 1986. 地震频度衰减与地震预报. 地震研究，9（1）

[8] 吴开统，焦远碧，吕培苓等. 1990. 地震序列概论，北京：北京大学出版社

[9] 牟雅元，龙思胜、龙德雄. 2006. 2005 年 8 月 5 日四川会东—云南会泽间 5.3 级地震序列初步研究. 四川地震，119（2）：13 ~ 16

[10] 秦嘉政，钱晓东. 2005. 云南会泽和文山 5.3 级地震研究. 地震研究，28（4）：403 ~ 407

[11] 和宏伟，秦嘉政，石绍先等. 2002. 云南中强震前 4 级地震活动平静特征研究. 地震研究，增刊（B），55 ~ 63

[12] 李树华，张立，王世芹等. 2006. 2005 年云南会泽、文山 5.3 级地震短临预报及再认识. 地震研究，29（3）：225 ~ 229

参 考 资 料

1）云南省地震局，2005 年 8 月 5 日会泽—会东 5.3 级地震灾害损失评估报告，2005.8

2）云南省地震局，云南省地震目录（1997 ~ 2001 年）（铅印本），2005

3）云南地震遥测中心，云南地震速报目录，2005

4) 云南省地震局，二〇〇五年八月五日会泽地震强震观测，2005
5) 云南省地震局，云南省 2005 年度地震趋势研究报告，2004

The M_S 5.3 Huize Earthquake on August 5, 2005 in Yunnan Province

Abstract

On August 5, 2005, an earthquake of $M_S = 5.3$ occurred in Huize county in Yunnan province. The macroscopic epicenter was located between Nagu and Laochang village in Huize. The idensity in the magistoseismic area was Ⅵ. The meizoseismal area was elliptic with major axis in NE direction. The disaster caused 19 persons injured. . The total economical loss caused by this earthquake was about 104.40 million yuan.

The earthquake sequence was of foreshock-mainshock-aftershock type. Its large foreshock was of M_S4.7 about 29 minutes before the mainshock. Its large aftershock was of M_L4.0. The aftershocks mainly occurred in the western of mainshock and had no dominant direction. Synthetically analyzing the fault structure, Isoseimal, and aftershock distribution, the nodal plane Ⅰ was the main rupture plane of this earthquake, and the azimuth of the P axis was NW ward. It is guessed that the seismogenic structure was the deep and insidious fault in NE ward, and the earthquake was the result of the right-lateral strike-slip with near horizontal principal press stress.

There were 37 seismic stations within the distance of 200km from epicenter, 20 of them were seismometric stations and 19 of them were precursory observation stations. Before this event there were 13 anomalies in 7 observation iteM_S, which were mainly short-term.

The precursor anomalies of the M_S5.3 Huize earthquake are judged. The characteristic of anomalies and other important phenomena are analyzed and discussed. Furthmore, the seismic tendency around the Xiaojiang fault after the Huize earthquake was discussed in this paper.

报告附件

附表 1 固定前兆观测台（点）与观测项目汇总表

序号	台站（点）名称	经纬度（°）		测项	资料类别	震中距 Δ/km	备注
		ϕ_N	λ_E				
1	会泽	26.50	103.15	水位	Ⅱ	5	
				水氡	Ⅱ		
2	东川	26.11	103.20	测震◬	Ⅰ	49	
				地倾斜	Ⅱ		
3	巧家	26.92	102.92	测震◬	Ⅰ	47	
				地磁 *D*	Ⅱ		
				地磁 *Z*	Ⅱ		
				水氡	Ⅱ		
				HCO^{3-}	Ⅱ		
				F^-	Ⅱ		
				Ca^{2+}	Ⅱ		
				Mg^{2+}	Ⅱ		
4	会东	26.63	102.63	测震△	Ⅰ	52	
5	鲁甸	27.17	103.57	CO_2	Ⅱ	90	
6	会理	26.65	102.25	测震△	Ⅰ	90	
				水位	Ⅱ		
7	宣威	26.20	104.10	水温	Ⅱ	102	
				水位	Ⅱ		
8	鱼洞	27.40	103.57	水温	Ⅱ	103	
				水位	Ⅱ		
9	昭通	27.32	103.72	测震◬	Ⅰ	104	
				地倾斜	Ⅱ		
				水温	Ⅱ		
10	寻甸	25.53	103.25	水温	Ⅱ	114	
				短水准	Ⅱ		
				短基线	Ⅱ		
11	威宁	26.87	104.30	测震△	Ⅰ	120	
12	禄劝	25.54	102.45	测震◬	Ⅰ	132	
				水氡	Ⅱ		
13	马龙	25.43	103.58	测震◬	Ⅰ	132	
14	嵩明	25.35	103.03	测震△	Ⅰ	134	
				F^-	Ⅱ		
15	富源	25.70	104.10	水温	Ⅱ	134	
				水位	Ⅱ		
16	曲靖	25.48	103.80	水温	Ⅱ	135	
17	攀枝花	26.50	101.73	测震◬	Ⅰ	140	
				地倾斜	Ⅱ		
				重力	Ⅱ		

续表

序号	台站（点）名称	经纬度° ϕ_N	经纬度° λ_E	测项	资料类别	震中距 Δ/km	备注
17	攀枝花	26.50	101.73	地应力	Ⅱ	140	
				水位	Ⅱ		
				水氡	Ⅱ		
18	攀矿	26.58	101.03	测震△	Ⅰ	142	
				地倾斜	Ⅱ		
19	小哨	25.17	102.97	水温	Ⅱ	154	
20	元谋	25.68	101.87	测震△	Ⅰ	160	
				水氡	Ⅱ		
				视电阻率	Ⅱ		
				自然电位	Ⅱ		
21	富民	25.22	102.48	测震△	Ⅰ	163	
22	昆明	25.13	102.73	测震◬	Ⅰ	163	
				重力	Ⅱ		
23	水城	26.58	104.83	测震△	Ⅰ	167	
24	西昌	27.83	102.23	测震△	Ⅰ	169	
				地倾斜	Ⅱ		
				水位	Ⅱ		
				水氡	Ⅱ		
				重力	Ⅱ		
				地应力	Ⅱ		
25	宜良	24.93	103.12	测震△	Ⅰ	184	
				水氡	Ⅱ		
26	永善	28.18	103.63	水氡	Ⅱ	187	
27	镇雄	27.44	104.87	测震◬	Ⅰ	197	
28	安宁	24.88	102.50	测震△	Ⅰ	198	
29	雷波	28.27	103.57	测震△	Ⅰ	200	

分类统计	0<Δ≤100km	100<Δ≤200km	200<Δ≤300km	300<Δ≤500km	500<Δ≤800km	总数
测项数 N	11	12				
台项数 n	13	31				
测震单项台数 a	1	7				
形变单项台数 b	0	0				
电磁单项台数 c	0	0				
流体单项台数 d	1	3				

续表

序号	台站（点）名称	经纬度°		测项	资料类别	震中距 Δ/km	备注
		ϕ_N	λ_E				

综合台站数 e	5	13				
综合台中有测震项目的台站数 f	3	9				
测震台总数 $a+f$	4	16				
台站总数 $a+b+c+d+e$	7	23				

备注	

附表 2　测震以外固定前兆观测项目与异常统计表

序号	台站(点)名称	测项	资料类别	震中距 Δ/km	按震中距 Δ 范围进行异常统计																								
					0＜Δ≤100km					100＜Δ≤200 km					200＜Δ≤300 km					300＜Δ≤500km					500＜Δ≤800km				
					L	M	S	I	U	L	M	S	I	U	L	M	S	I	U	L	M	S	I	U	L	M	S	I	U
1	会泽	水位	Ⅱ	5	—	—	—	—																					
		水氡	Ⅱ		—	—	√	—																					
2	东川	地倾斜	Ⅱ	36	—	—	—	—																					
3	巧家	地磁 *D*	Ⅱ	47	—	—	√	—																					
		地磁 *Z*	Ⅱ		—	—	—	—																					
		水氡	Ⅱ		—	—	—	—																					
		HCO^{3-}	Ⅱ		—	—	—	—																					
		F^{-}	Ⅱ		—	—	—	—																					
		Ca^{2+}	Ⅱ		—	—	—	—																					
		Mg^{2+}	Ⅱ		—	—	—	—																					
4	鲁甸	CO_2	Ⅱ	90	—	—	—	—																					
5	会理	水位	Ⅱ	90	—	—	—	—																					
6	沾益	水温	Ⅱ	102						—	—	√	—																
		水位	Ⅱ							—	—	—	—																
7	鱼洞	水温	Ⅱ	103						—	—	—	—																
		水位	Ⅱ							—	—	—	√																
8	昭通	地倾斜	Ⅱ	104						—	—	—	—																
		水温	Ⅱ							—	—	—	—																

续表

序号	台站（点）名称	测项	资料类别	震中距 Δ/km	按震中距 Δ 范围进行异常统计																								
					0 < Δ ≤ 100km					100 < Δ ≤ 200 km					200 < Δ ≤ 300 km					300 < Δ ≤ 500km					500 < Δ ≤ 800km				
					L	M	S	I	U	L	M	S	I	U	L	M	S	I	U	L	M	S	I	U	L	M	S	I	U
9	昭通	水温	Ⅱ	114						—	—	—	—																
		短水准	Ⅱ							—	—	—	—																
		短基线	Ⅱ							—	—	—	—																
10	富源	水温	Ⅱ	134	—					—	—	—	—																
		水位	Ⅱ							—	—	—	—																
11	曲靖	水温	Ⅱ	135						—	—	√	—																
12	攀枝花	地倾斜	Ⅱ	140						—	—	√	—																
		重力	Ⅱ							—	—	—	—																
		地应力	Ⅱ							—	—	—	—																
		水位	Ⅱ							—	—	—	—																
		水氡	Ⅱ							—	—	—	—																
13	攀矿	地倾斜	Ⅱ	142						—	—	—	—																
14	小哨	水温	Ⅱ	154						—	—	—	—																
15	永仁	水氡	Ⅱ	159						—	—	—	—																
16	元谋	水氡	Ⅱ	160						—	—	√	—																
		视电阻率	Ⅱ							—	—	—	√																
		自然电位	Ⅱ							—	—	—	√																
17	昆明	重力	Ⅱ	163						—	—	—	—																

续表

序号	台站（点）名称	测项	资料类别	震中距Δ/km	按震中距Δ范围进行异常统计																								
					0＜Δ≤100km					100＜Δ≤200 km					200＜Δ≤300 km					300＜Δ≤500km					500＜Δ≤800km				
					L	M	S	I	U	L	M	S	I	U	L	M	S	I	U	L	M	S	I	U	L	M	S	I	U
18	西昌	地倾斜	Ⅱ	169						—	—	—	—																
		重力	Ⅱ							—	—	—	—																
		地应力	Ⅱ							—	—	—	—																
		水位	Ⅱ							—	—	—	—																
		水氡	Ⅱ							—	—	—	—																
19	永善	水氡	Ⅱ	187						—	—	—	—																
分类统计	台项	异常台项数			0	0	2	0		0	0	4	3																
		台项总数			12	12	12	12		30	30	30	30																
		异常台项百分比/%			0	0	17	0		0	0	13	10																
	观测台站（点）	异常台站数			0	0	2	0		0	0	4	2																
		台站总数			5	5	5	5		14	14	14	14																
		异常台站百人分比/%			0	0	40	0		0	0	29	14																
	测项总数				10					11																			
	观测台站总数				5					14																			
备注																													

2005 年 8 月 5 日云南省文山 5.3 级地震

云南省地震局

邬成栋　付　虹

摘　要

2005 年 8 月 13 日云南文山发生了 5.3 级地震，宏观震中位于文山县红甸、马塘之间，极震区烈度为Ⅵ度，呈 NW 向的椭圆形。地震造成 2 人重伤，27 人轻伤，直接经济损失 9220 万元。

该地震为主震—余震型，余震序列较为丰富，最大余震 3.8 级。震源机制解中节面Ⅰ走向 100°，主压应力轴方位 NNW 向，据地质考察、等震线长轴方向和余震分布综合分析，节面Ⅰ为主破裂面。5.3 级地震可能是在接近水平的主压应力作用下，产生右旋错动的结果。发震构造可能是文山麻栗坡断裂带西支文山西山断裂。

震中 200km 范围内共有 20 个地震台站，其中测震观测台 9 个，定点前兆观测台站 20 个。震前共出现 6 个异常项目 10 项次前兆异常。其中地震学异常 4 项次，定点前兆异常 6 项次。以短期异常为主。

本次地震震后趋势判定正确，对稳定人心、安定社会起到了较好的作用。

前　言

据云南测震台网测定，2005 年 8 月 13 日 12 时 58 分云南文山西北发生 5.3 级地震，微观震中为北纬 23°36′，东经 104°04′，宏观震中位于文山县红甸、马塘之间，极震区烈度为Ⅵ度。地震主要涉及文山州文山县、砚山县的 10 个乡（镇），44 个行政村；受灾人口 127624 人。地震造成 2 人重伤，27 人轻伤，直接经济损失 9220 万元。

震中附近 200km 范围内共有 20 个地震观测台，震前共出现 6 个异常项目，10 项次前兆异常。其中地震学 3 个异常项目 4 项次异常，定点前兆 3 个异常项目 6 项次异常。

震区在云南地区属于相对少震区，5.3 级地震附近历史上发生过多次 5 级中强震，最大地震为 5.5 级。

震前对本次地震未能做出预测。震后云南省地震局立即派出工作组赶赴现场，进行地震灾害损失评估和震情动态跟踪监视等工作。针对震后传言，及时准确对震后趋势作出了判定，对稳定人心、安定社会起到了较好的作用。

一、测震台网及地震基本参数

文山 5.3 级地震震中周围 200km 范围内有 9 个测震台（包括数字记录和模拟记录），其中 100km 范围内有 2 个，101 ~ 200km 内有 7 个（图 1）。云南省测震台网对这次地震发震地区的 M_L2.0 以上的地震基本能控制，本次地震的基本参数如表 1 所示。

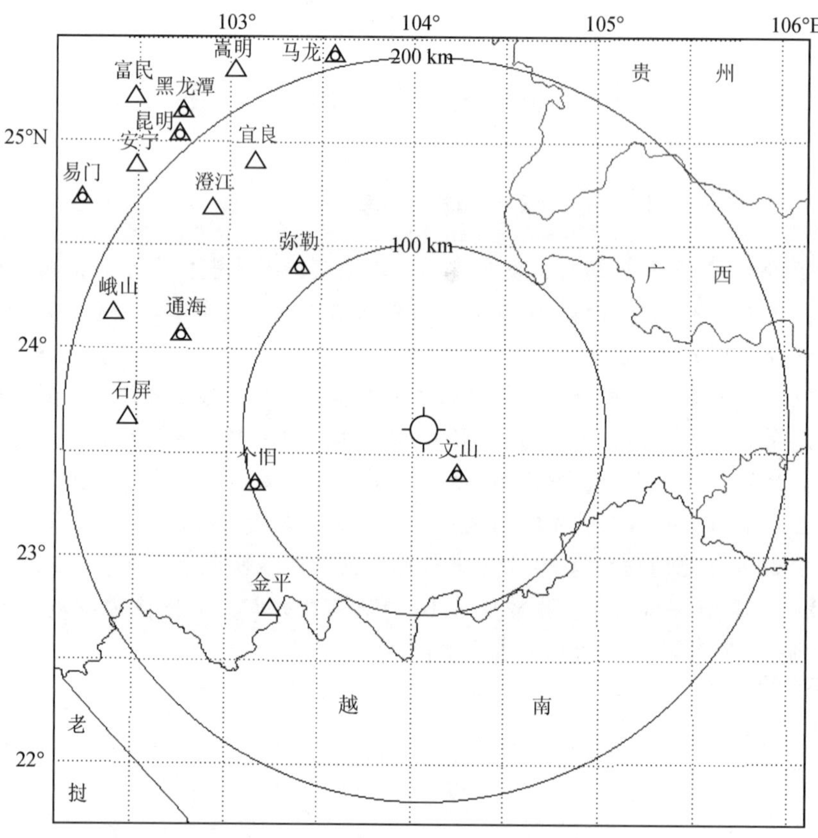

图 1　文山 5.3 级地震前震中附近测震台站分布图

Fig. 1 Distribution of seismometric stations around the epicentral area before the M_S5.3 Wenshan earthquake

表 1　地震基本参数

Table 1　Basic parameters of the earthquake

编号	发震日期	发震时刻	震中位置		震级				震源深度/km	震中地名	结果来源
	年 月 日	时 分 秒	φ_N	λ_E	$M(M_S)$	M_L	M_b	M_w			
1	2005 08 13	12 58 43.0	23.62	104.07	5.3	4.7			3	文山	云南台网
2	2005 08 13	12 58 44.3	23.5	104.1	5.3				15	文山	中国局
3	2005 08 13	12 58 43.9	23.6	104.13	5.2	5.1	4.7		11	文山	BJI

续表

编号	发震日期	发震时刻	震中位置		震级				震源深度/km	震中地名	结果来源
	年 月 日	时 分 秒	φ_N	λ_E	M（M_S）	M_L	M_b	M_w			
4	2005 08 13	12 58 44.85	23.627	104.103	4.5		4.8		10	文山	NEIC
5	2005 08 13	12 58 42.58	23.4532	104.0976	4.5		4.7		10	文山	ISC

二、地震地质背景

震区位于滇东南褶皱带二级构造单元所属的文山—富宁断褶束与丘北—广南断褶束交界区三级构造单元的中部。文（山）麻（栗坡）断裂为震区主要地震构造。该断裂带由文山西山断裂和文山东山断裂两支组成，北西起自砚山县平远街盆地北西，向东南经文山盆地，至麻栗坡北东的迷洒附近延入越南，云南省内长约 100km。总体走向北西，倾向南东。

本次地震可能与断裂带西支文山西山断裂活动有关，该断裂对文山盆地发育有明显控制作用，全新世以来断裂活动不明显，为晚更新世活动断裂。历史上发生过 1885 年 12 月 22 日丘北 5¼级地震。

震区在云南地区属于相对少震区，据该区最早地震记录公元 1885 年至今统计，本次地震 100km 范围内共发生 $M_S \geqslant 5$ 级地震 9 次，无 6 级以上地震发生。最大地震为 5.5 级。

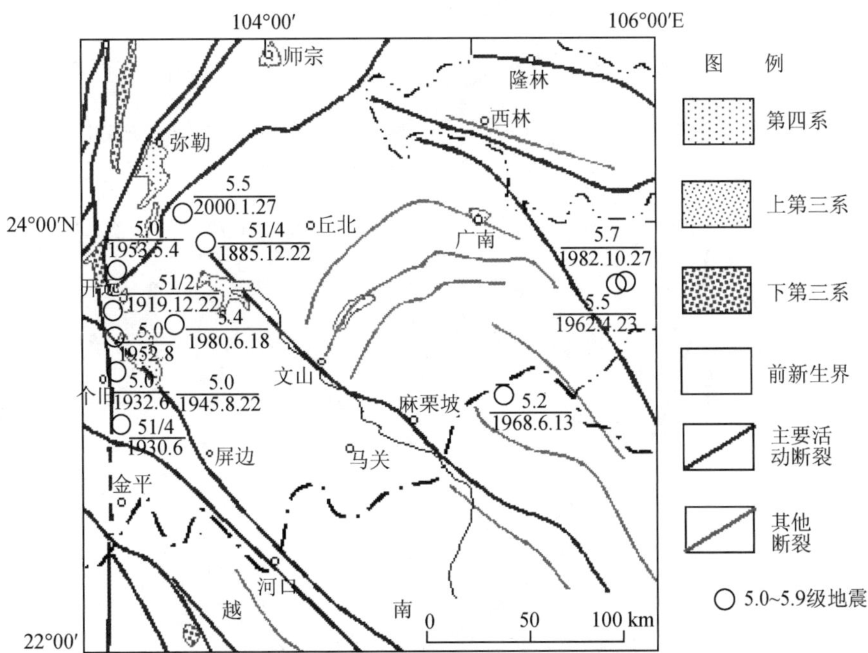

图 2 文山 5.3 级地震震区地质构造及历史地震分布图

Fig. 2 Map of geological structure and distribution of historical earthquakes around the M_S5.3 Wenshan epicentral area

表2　文山 5.3 级地震震中附近历史地震目录（M_S≥5.0 级）

Table 2　Catalogue of historical earthquakes around the M5.3 Wenshan epicentral area（M_S≥5.0 级）

编号	发震日期	震中位置（°）		震级 M_S	震中地名	结果来源
	年 月 日	φ_N	λ_E			
1	1885 12 22	23.9	103.7	5¼	丘北西南	文献［1］
2	1919 12 22	23.7	103.3		开远	文献［1］
3	1930 06	23.4	103.2		个旧南	文献［1］
4	1932 06	23.4	103.2	5.0	个旧南	文献［1］
5	1945 08 22	23.4	103.2	5.0	个旧南	文献［1］
6	1952 08	23.6	103.2	5.0	开远个旧间	文献［1］
7	1953 05 04	24.2	103.2	5.0	弥勒西南	文献［1］
8	1980 06 18	23.5	103.9	5.4	蒙自东北	文献［1］
9	2000 01 27	24.2	103.6	5.5	丘北弥勒间	资料1）

三、地震影响场和震害

据震区的考察资料，文山 5.3 级地震的宏观震中位于文山县红甸、马塘之间，极震区烈度为Ⅵ度，等震线形状呈椭圆形，长轴走向为北西向。长轴 43.2km，短轴 26.2km，灾区面积为 890km^2。

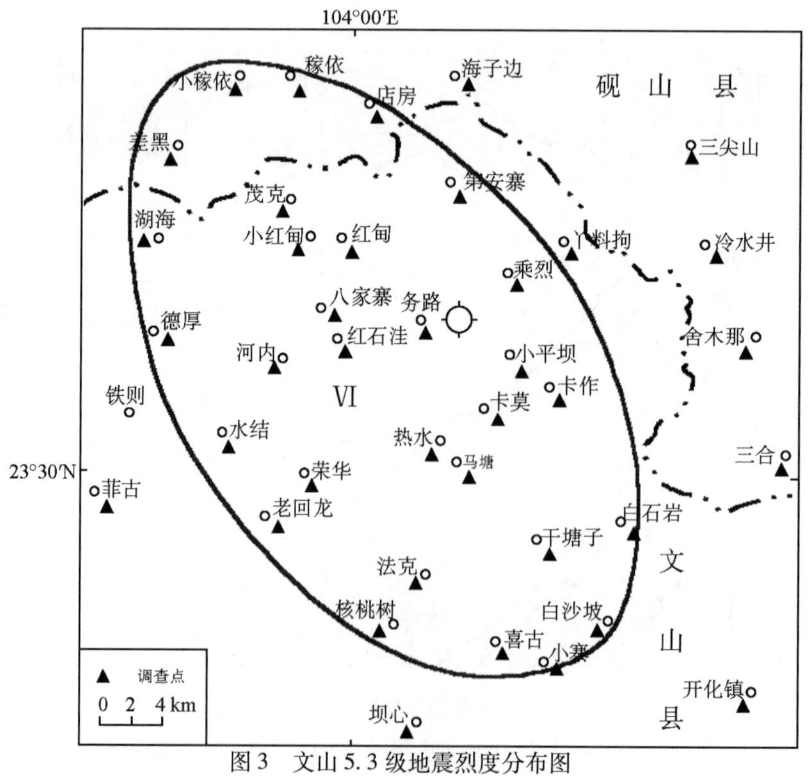

图3　文山 5.3 级地震烈度分布图

Fig. 3　Intensity distribution of the M_S5.3 Wenshan earthquake

Ⅵ度区范围（极震区）：北自砚山县稼依镇的小稼依村，南到文山县开化镇的白沙坡村，东自文山县马塘镇白石岩村，西至文山县德厚镇政府驻地一带。Ⅵ度区面积约 890km^2。

本次地震主要涉及文山州文山县的红甸乡、马塘镇、秉烈乡、德厚镇、老回龙镇、坝心乡、喜古乡、开化镇及砚山县的稼依镇、平远镇等 10 个乡（镇），44 个行政村；受灾人口 127624 人，涉及 27137 户。地震中无人死亡，2 人重伤，27 人轻伤。

地震造成村镇民用房屋破坏，震后个别房屋有倒毁现象，少数房屋墙体开裂，部分房屋墙体见裂纹或裂缝，灰皮或灰块掉落。生命线工程及水利、通信、交通等部分基础设施受到损坏。地震给当地烤烟经济造成一定损失，主要表现为烤烟房墙体开裂失效，倒塌或者局部倒塌，烟叶得不到及时烘烤。文山 5.3 级地震造成的直接经济总损失为 9220 万元。

四、地 震 序 列

据云南地震台网测定，2005 年 8 月 13 日文山 5.3 级地震后至 2005 年 11 月 28 日共发生余震 701 次，其中 M_L0.0 ~ 0.9 地震 18 次、M_L1.0 ~ 1.9 地震 541 次、M_L2.0 ~ 2.9 地震 126 次、M_L3.0 ~ 3.9 地震 16 次，最大余震为 10 月 16 日发生的 M_L3.8 地震。表 3 给出了 M_L 3.5 以上地震序列目录。

表 3　文山 5.3 级地震序列目录（$M_L \geqslant 3.5$ 级）

Table 3　Catalogue of the M_S5.3 Wenshan earthquake sequence（$M_L \geqslant 3.5$）

编号	发震日期	发震时刻	震中位置		震级		震源深度（km）	震中地名	结果来源
	年 月 日	时 分 秒	φ_N	λ_E	M_L	M_S			
1	2005 08 13	12 58 43.0	23°37′	104°04′		5.3	3	文山	资料 1）
2	2005 08 17	01 59 25.6	23°33′	104°03′	3.5		25	文山	资料 1）
3	2005 08 17	02 29 24.7	23°36′	104°06′	3.5		3	文山	资料 1）
4	2005 08 20	13 27 33.1	23°36′	104°08′	3.5		6	文山	资料 1）
5	2005 10 16	17 19 24.2	23°36′	104°07′	3.8			文山	资料 1）
6	2005 10 16	17 19 33.5	23°33′	104°09′	3.7		1	文山	资料 1）

1. 地震序列类型

该序列主震能量与序列总能量比 $E_1/\sum E = 0.97$，次大地震为 10 月 16 日发生的 M_L3.8 地震，与主震的震级差 $\Delta M = 1.5$，据此判定该序列为主震—余震型。

2. 地震序列衰减情况

虽然云南省测震台网仅能基本控制这次地震发震地区的 M_L2.0 以上的地震，但文山地震台距离本次地震震中仅约 30km，可以给出较好的单台序列。

本次地震无明显前震，余震序列较为丰富，余震活动以 2～3 级地震为主。序列 M-t 图显示 3 级以上地震主要发生在震后 10 天内，震后一个月左右序列有一次较强起伏，发生了最大强余震 3.8 级地震（图 4）。日频度曲线及应变释放曲线均显示 5.3 级地震后余震迅速衰减（图 5、图 6）。

序列峰值震级是 1.3，取 1.3 级以上的地震计算系列参数。序列的 b 值为 0.95，h 值为 1.4 大于 1，p 值为 0.84，都显示了该序列的地震类型为主震—余震型，衰减正常（图 7 至图 9）。

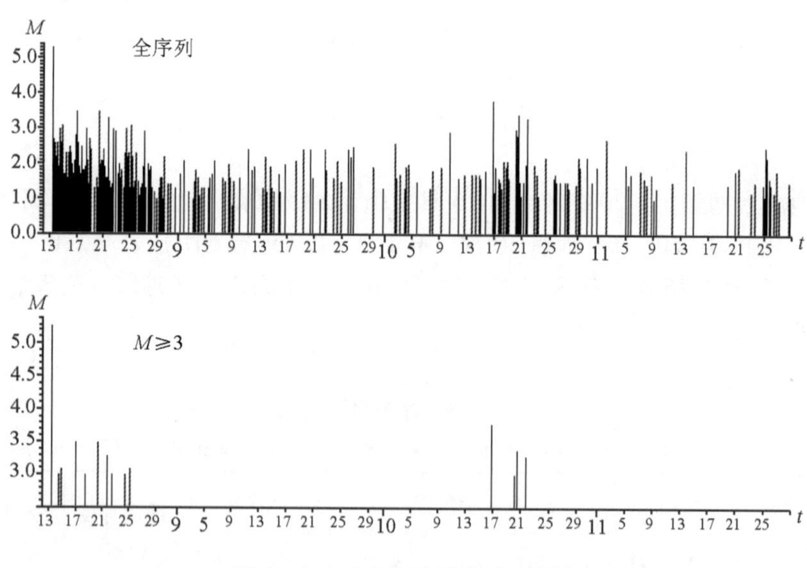

图 4　文山 5.3 级地震序列 M-t 图

Fig. 4　M-t diagram of the M_S5.3 Wenshan earthquake sequence

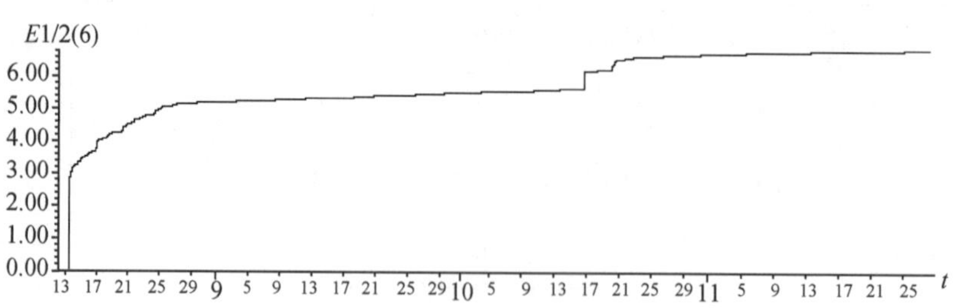

图 5　文山 5.3 级地震序列应变释放曲线

Fig. 5　Curve of strain release of the M_S5.3 Wenshan earthquake sequence

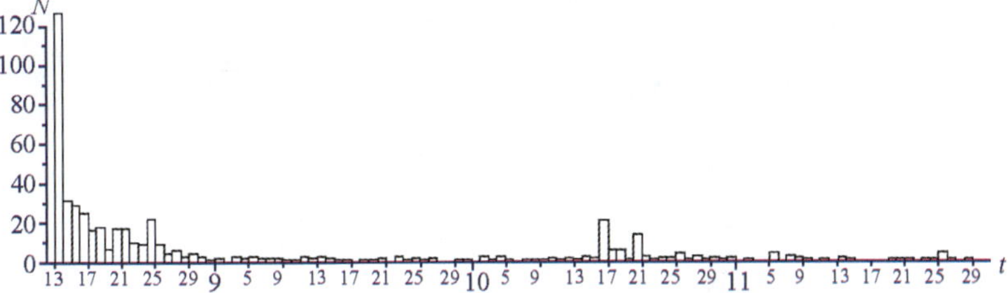

图 6　文山 5.3 级地震序列日频度曲线

Fig. 6　Curve of daily frequency of the M_S5.3 Wenshan earthquake sequence

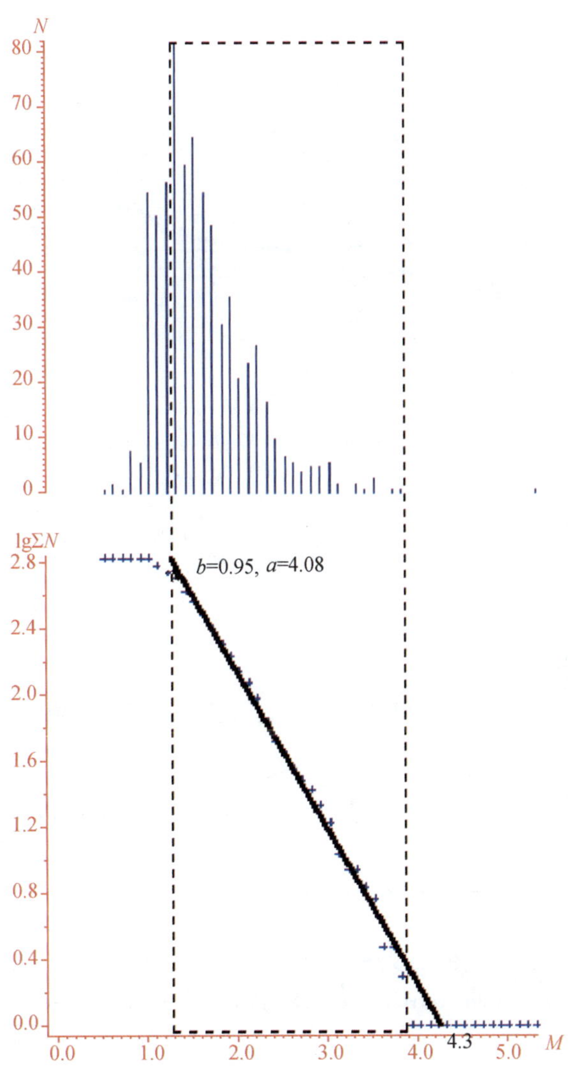

图 7　文山 5.3 级地震序列 b 值曲线

Fig. 7　b-value curve of the M_S5.3 Wenshan earthquake sequence

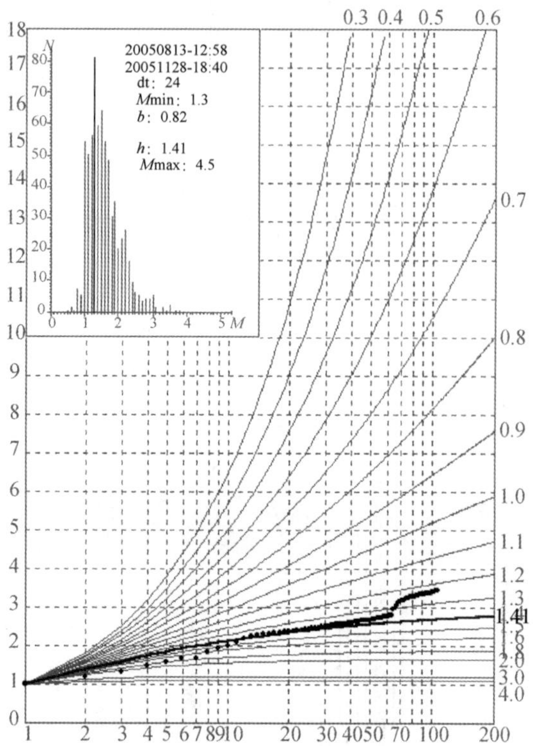

图 8　文山 5.3 级地震序列 h 值曲线

Fig. 8　h-value curve of the $M_S5.3$ Wenshan earthquake sequence

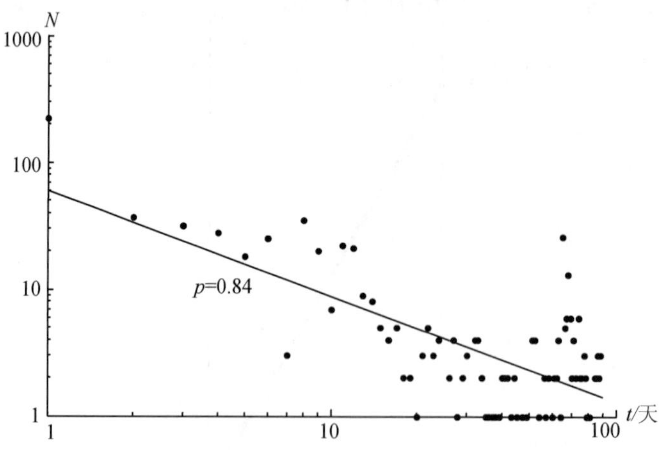

图 9　文山 5.3 级地震序列 p 值曲线

Fig. 9　p-value curve of the $M_S5.3$ Wenshan earthquake sequence

3. 余震空间分布

整个序列分布为北北东向(图 10),长轴约为 45km。但较强的 $M \geqslant 3$ 级余震集中在一个较小的范围内(图 11),其优势分布近似于北西向。

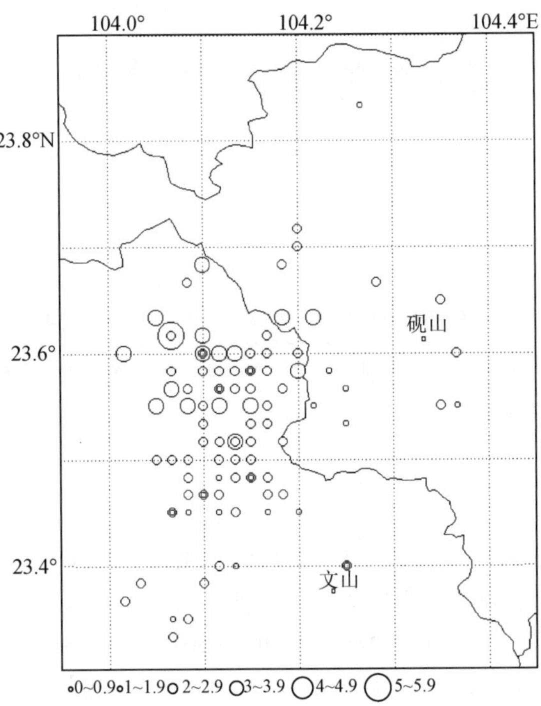

图 10 文山 5.3 级地震序列震中分布图

Fig. 10 Epicentral distribution of the M_S5.3 Wenshan earthquake sequence

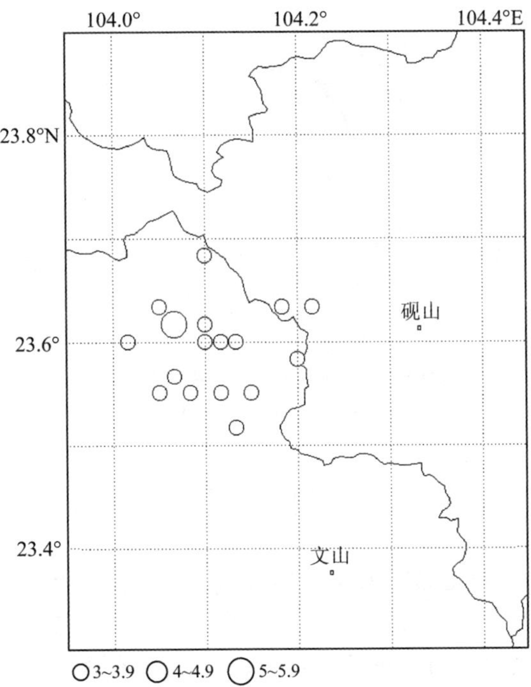

图 11 序列 $M \geqslant 3$ 级地震分布

Fig. 11 Epicentral distribution of $M \geqslant 3$ earthquake of sequence

五、震源参数和地震破裂面

根据昆明台网记录，用P波初动符号图解断层面法求得文山5.3级地震的震源机制解见表4和图12。

表4　文山5.3级地震震源机制解

Table 4　Focal mechanism solutions of the M_S5.3 Wenshan earthquakes

编号	节面Ⅰ			节面Ⅱ			P轴		T轴		N(B)轴		X轴		Y轴		结果来源
	走向	倾角	滑动角	走向	倾角	滑动角	方位	仰角	方位	仰角	方位	仰角	方位	仰角	方位	仰角	
1	100	85	-162	8	72	-5	325	16	233	9	114	71					文献[2]

由上述结果知，5.3级地震节面Ⅰ走向100°，倾角85°，滑动角－162°；节面Ⅱ走向8°，倾角72°，滑动角－162°。这次地震的等震线长轴方向为北西向与节面Ⅰ较接近。文山震区的余震分布总体呈北北东向与节面Ⅱ较吻合，但较强余震优势分布为北西向又与节面Ⅰ较为吻合。前已述震区主要地震构造为文麻断裂，断裂总体走向北西，倾向南东。本次地震可能与断裂带西支文山西山断裂活动有关，因此走向100°的节面Ⅰ是5.3级地震的破裂面的可能性较大。5.3级地震可能是在接近水平的主压应力作用下，产生右旋错动的结果。

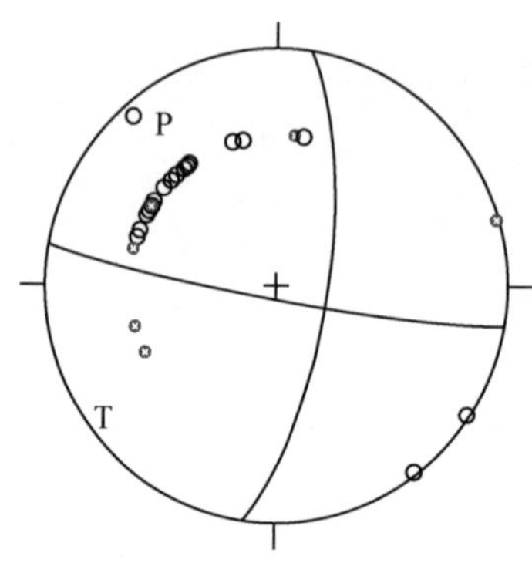

图12　文山5.3级地震震源机制解

Fig. 12　Focal mechanism solutions of the M_S5.3Wenshan earthquakes

六、地震前兆观测台网及前兆异常

文山5.3级地震震中200km范围内共有20个地震台站，其中测震台站9个，其他前兆观测台站20个（图1、图13），有测震、水氡、水位、水温、地倾斜、地磁、跨断层短水准、短基线等11个观测项目。其中0～100km和101～200km分别有地震台3个和17个，测震台2个和7个，前兆台3个和17个，测震学以外的观测项目4个和10个，台项6个和54个。这些前兆定点台站，除个别数字化台站资料不连续外，绝大多数震前都有连续观测的资料。

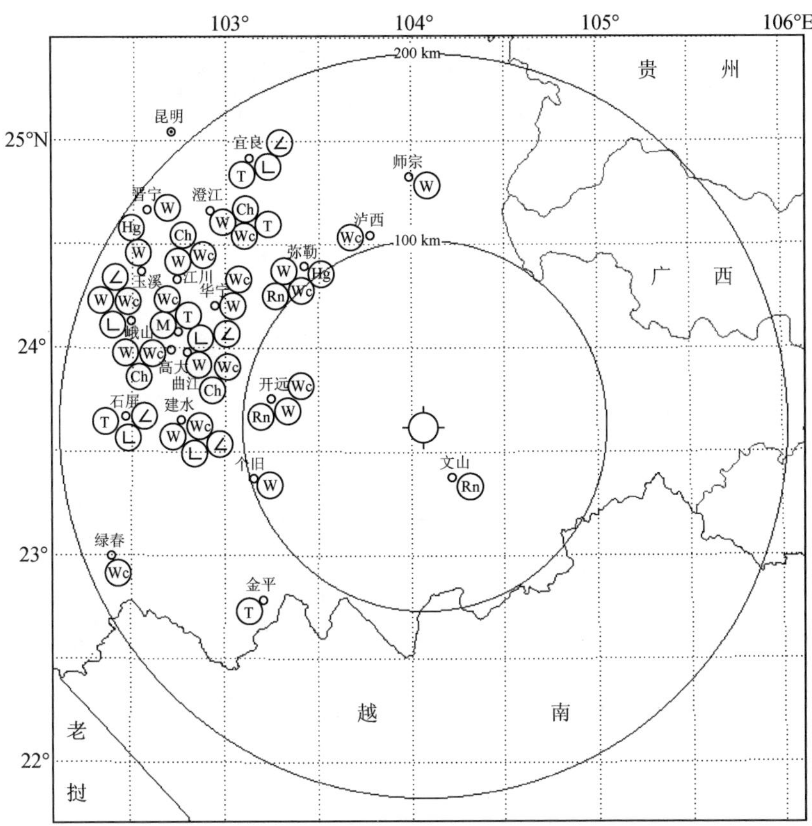

图 13 文山 5.3 级地震定点前兆观测台站的分布图

Fig. 13 Distribution of the precursory monitoring stations before the M_S5.3 Wenshan earthquake

此次地震前共出现 6 个项目，10 项次异常（表 5）。其中 3 个项目 4 项次地震学异常，3 个项目 6 项次前兆观测异常（图 14）。各类异常的具体情况详见表 5 和图 15 至图 24。

表 5 地震前兆异常登记表

Table 5 Summary table of precursory anomalies

序号	异常项目	台站（点）或观测区	分析方法	异常判据及观测误差	震前异常起止时间	震后变化	最大幅度	震中矩 Δ/km	异常类别及可靠性	图号	异常特点及备注
1	地震频度	文山 5.3 级地震震中 100km 范围内	$M \geqslant 3$ 级地震月频度	显著增加	2004.8 ~ 2005.7	正常	5 次		M_1	15	近场区 3 级地震的显著增强是云南地区较好的预报指标
2		文山 5.3 级地震震中 50km 范围内	$M \geqslant 2.5$ 级地震 6 月累计 1 月滑动频度	显著增加	2005.1 ~ 2005.7	正常	17 次		M_1	16	该异常说明震源区震前有微破裂出现

续表

序号	异常项目	台站（点）或观测区	分析方法	异常判据及观测误差	震前异常起止时间	震后变化	最大幅度	震中矩 Δ/km	异常类别及可靠性	图号	异常特点及备注
3	地震平静	云南省内	地震间隔时间	$\Delta_t \geqslant 90$ 天	2005.4.19 ~8.5	正常	108 天		S_1	17	短期阶段云南省内出现4级地震平静
4	b 值	23° ~ 29°N 102° ~ 106°E	2.5 ~4.9 级地震6 月累计1 月滑动	$b \geqslant 1.2$	2005.5 ~ 2005.7	正常	1.3		S_1	18	该异常多次对应滇东及附近地区地震
5	氡	文山	日均值	连续上升	2005.6.10 ~8.20	正常	7.3 Bq/L	32	S_2	19	震前提出异常，水氡测值增大
6		高大	日均值	高值异常	2005.6.9 ~7.12	正常	15.7 Bq/L	144	S_2	20	震前提出异常，水氡测值增大
7		江川	日均值	高值异常	2005.5.30 ~8.30	正常	94.8 Bq/L	156	S_2	21	震前提出异常，水氡测值增大
8	水位	师宗	日均值	<3.5m	2005.6.29 ~7.27	正常	0.5m	135	S_2	22	震后总结，下降幅度大，震后恢复
9		江川	日均值	加速下降	2005.5.14 ~7.18	正常	0.14m	156	S_2	23	震前提出异常，转折回升后发震
10	水温	澄江	日均值	加速下降	2005.4.25 ~7.15	正常	0.5	165	S_2	24	震后总结，下降幅度大，转折回升后发震

文山附近属云南地震监测能力相对较弱的地区，在震中周围 100km 内仅有前兆观测台站 3 个台项 5 个，出现异常台站 1 个异常台项 1 个，异常台站和异常台项百分比分别为 33% 和 20%；在 101 ~200km 范围内有前兆台站 17 个台项 50 个，出现异常台站 4 个异常台项 5 个，异常台站和异常台项百分比分别为 24% 和 10%。

表 5 中给出的地震学异常，基本上是在日常监视预报中使用的常规方法得到的异常，近场区地震频度显著增加，省内 4 级地震平静，是云南地区比较好的预报指标。

表 5 中给出的前兆观测异常，大多数的异常判别标准是经验的，部分异常是震前就看到并给出的，因此对指导未来的地震预报有一定的参考价值。另据文献 [3]，2005 年 5 ~7 月云南地区连续出现磁暴和地磁低点位移，全省气压于 2005 年 7 月 12 日大幅上升达预报指标。磁暴、地磁低点位移和环境因子的同步变化，是预测时间的较好指标。

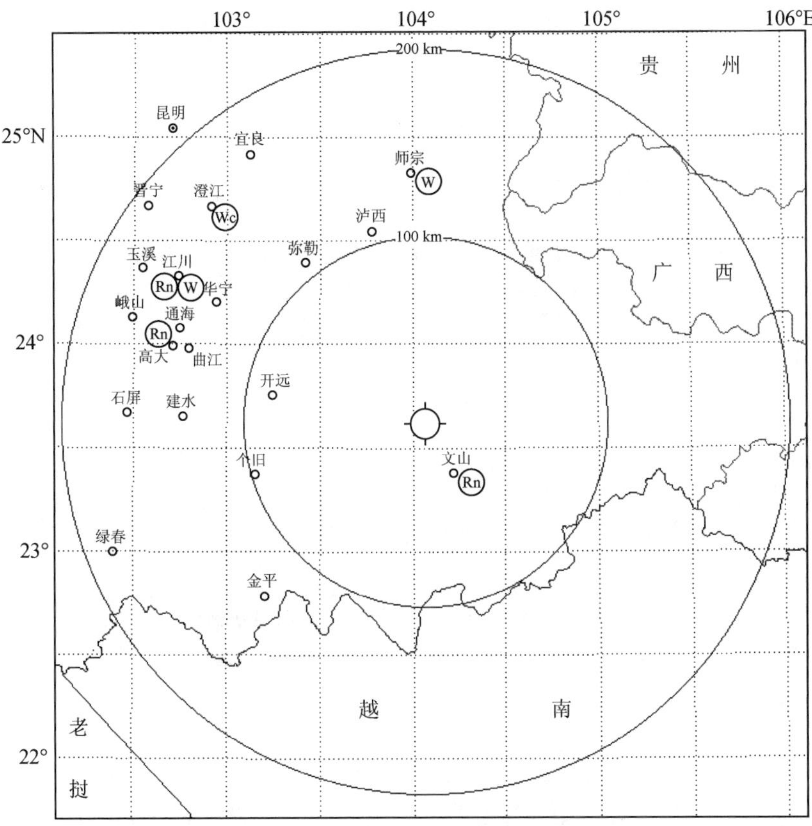

图14　文山5.3级地震定点前兆异常分布图

Fig. 14　Distribution of precursory anomalies on the fixed observation points before the M_S5.3 Wenshan earthquake

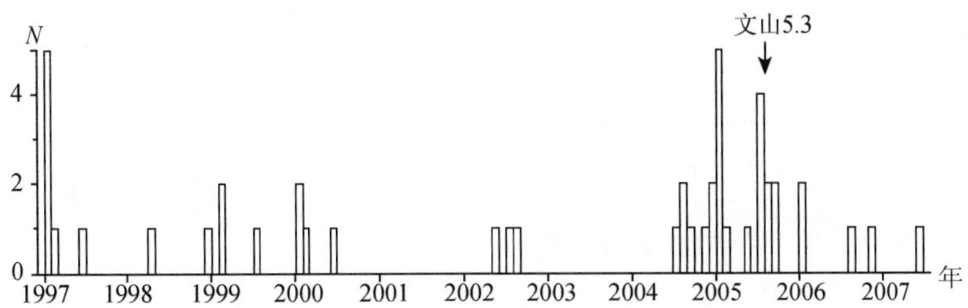

图15　文山5.3级地震震中100km范围内$M \geqslant 3$级地震月频度图

Fig. 15　Monthly frequency of $M \geqslant 3$ earthquakes in 100km away from the epicenter of Wenshan 5.3 earthquake

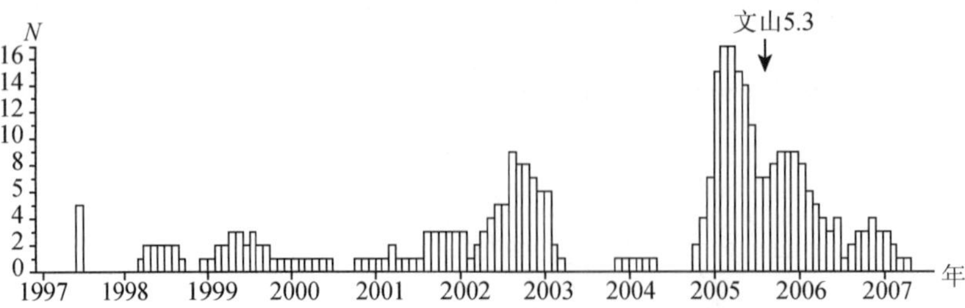

图 16　文山 5.3 级地震震中 50km 范围内 $M \geqslant 2.5$ 级地震频度图

Fig. 16　Frequency of $M \geqslant 2.5$ earthquakes in 50km away from the epicenter of Wenshan 5.3 earthquake

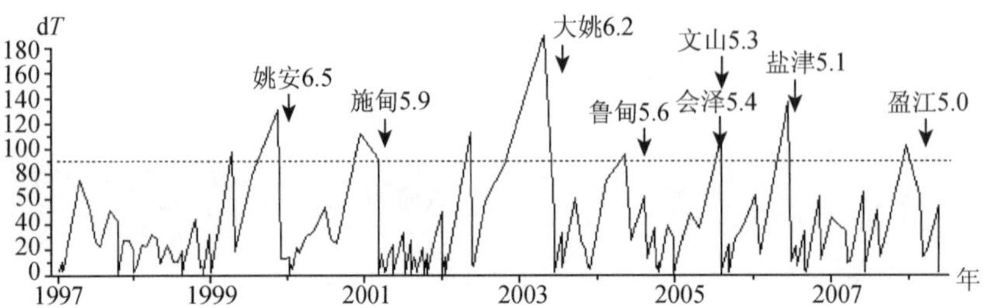

图 17　云南省内 $M \geqslant 4$ 级地震时间间隔曲线

Fig. 17　Interval curve of $M \geqslant 4$ earthquake in Yunnan Province

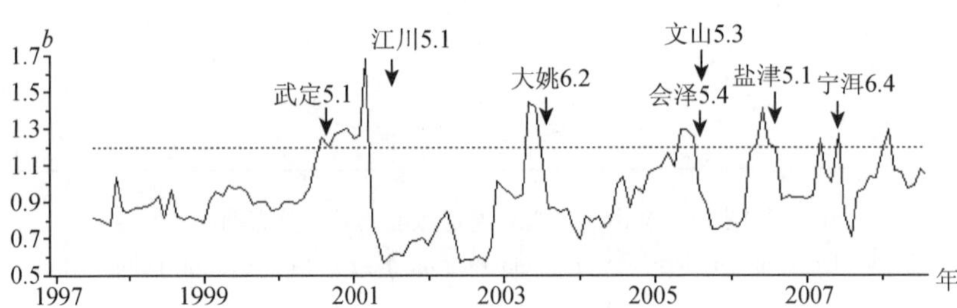

图 18　滇东地区 2.5～4.9 级地震 b 值

Fig. 18　b-value of $M2.5 \sim 4.9$ earthquakes in eastern Yunnan

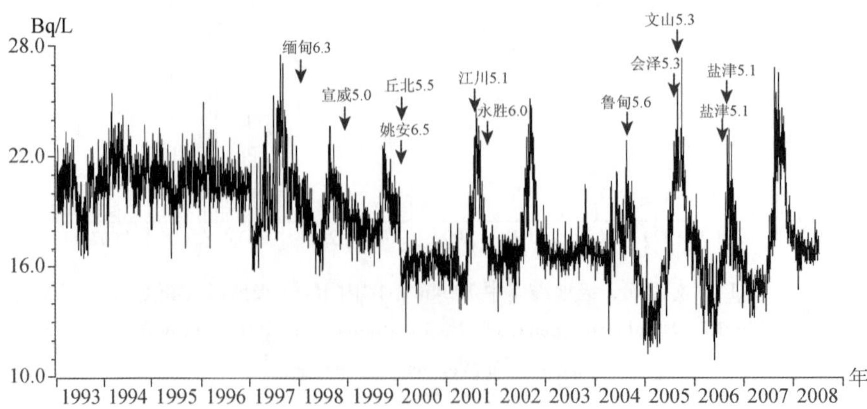

图 19　文山水氡日均值曲线

Fig. 19　Daily average value curve of Wenshan radon content in groundwater

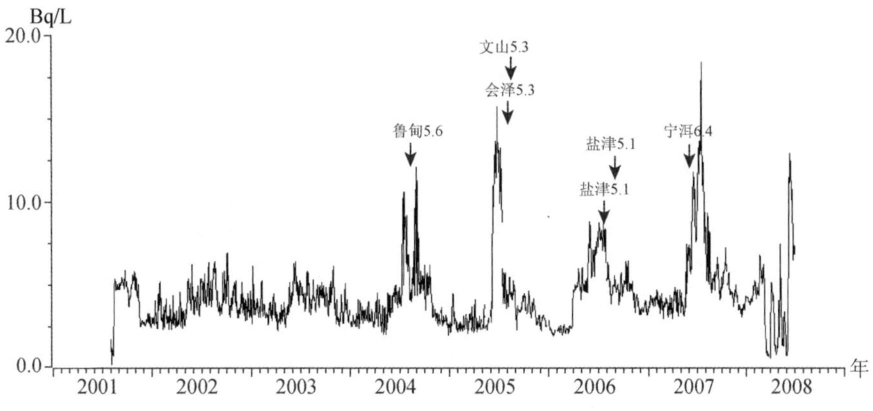

图 20 高大气氡日均值曲线

Fig. 20 Daily average value curve of Gaoda radon content in air

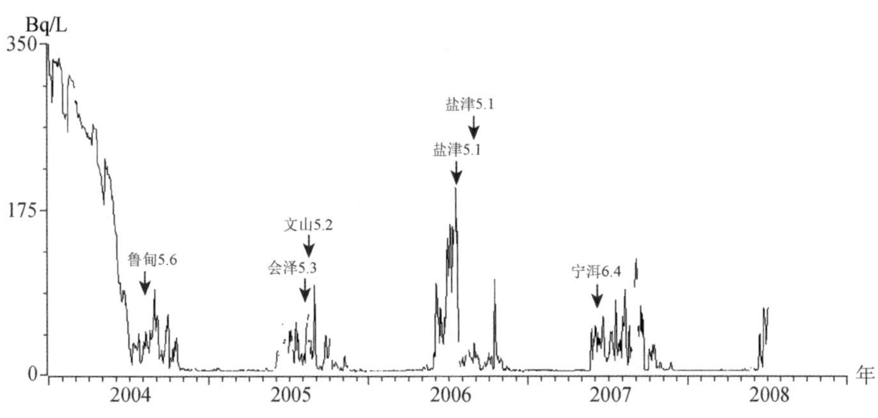

图 21 江川气氡日均值曲线

Fig. 21 Daily average value curve of Jiangchuan radon content in air

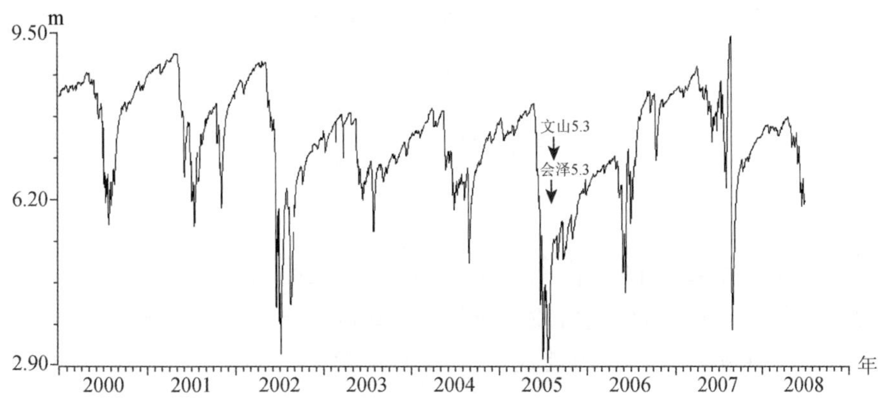

图 22 师宗水位日均值曲线

Fig. 22 Daily average value curve of Shizong water level

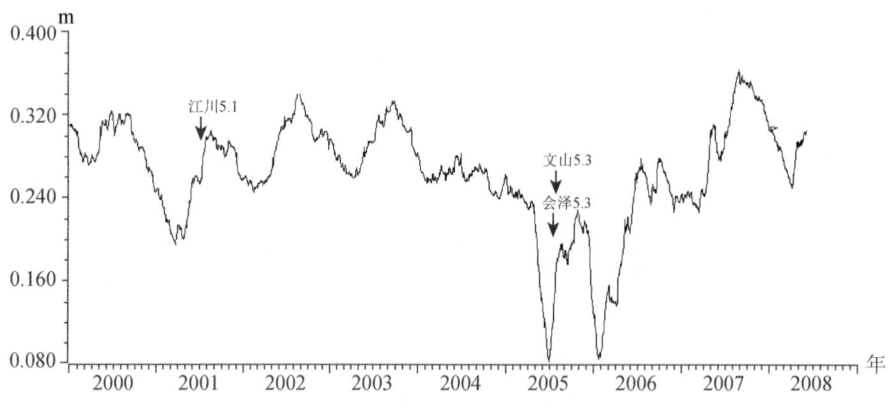

图 23　江川水位日均值曲线

Fig. 23　Daily average value curve of Jiangchuan water level

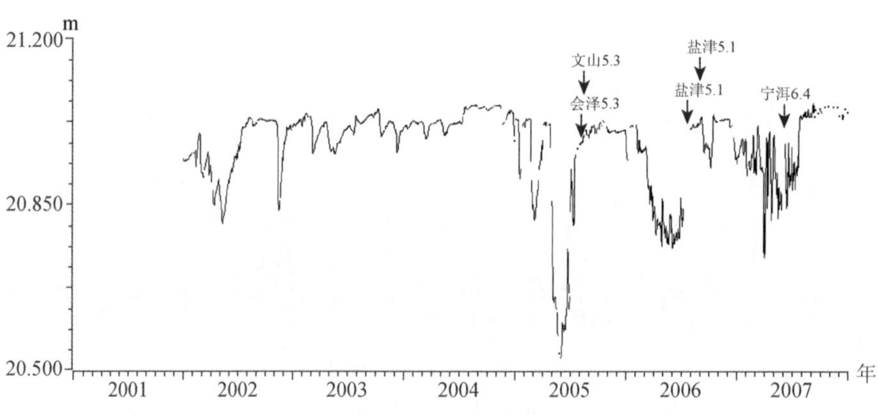

图 24　澄江水温日均值曲线

Fig. 24　Daily average value curve of Chengjiang water temperature

七、地震前兆异常特征分析

1. 地震学异常

强震前中小地震频度增加在云南地区具有普适性，且地震活动增强的地区距离主震的距离并不是太远。本次地震孕育中期阶段震源区和近场区小震活动明显增加。2004 年 8 月震中附近的丘北发生 4.0 和 3.7 级地震地震活动开始增强，2005 年 1 月震中西南发生 3 级震群。

短期阶段省内出现了 4 级地震平静超过 90 天的异常，从 1997 年云南进入新的 $M \geqslant 6.8$ 级强震平静后，省内 4 级地震平静大于 90 天后基本上都有 5 级以上地震发生，与平静相应的，滇东地区 b 值也出现高值异常。同时 3 级以上地震活动水平也较低。平静是个能量积累的过程，也是云南比较好的预报指标。

进入临震阶段，在经过一段时间的平静后，文山及附近地区 7 月份 3 级以上地震再次出现活跃，最大强度达到 4.0 级地震。

2. 前兆异常

前兆异常以短期异常为主，没有半年以上的中期及长期异常。主要是云南的地震较多，同一观测曲线异常有时对应多次地震，不可能区分这些异常只与哪一次地震有关。文山地震震前出现水氡、水位和水温 3 个测项 6 条异常，异常数量不多。

水氡异常最为突出，文山、高大、江川水氡异常均在震前给出，其中距离震中最近的文山水氡在 6 月份突升以前，自 2004 年 10 月以来的下降也与正常动态不同，其异常的持续时间要相对长些。可能说明距震中较近的观测点异常出现得要早些，持续时间也相对较长。

各项异常集中出现在震前的 5 ~7 月份，异常趋于结束或转折后发震。磁暴、地磁低点位移和环境因子也在 7 月出现同步变化，异常在时间上的集中对强震的时间预测有重要的指示作用。

3. 前兆异常特征

本次地震异常数量总体偏少。文山附近地震台站数量较少，属云南地震监测能力相对较弱的地区，因而震前异常数量较少，从统计比例上，异常台站比和台项比也较低。各学科异常项次总计 10 项，其中地震学异常 4 项，定点前兆异常 6 项。

异常时间分布显示，最早出现的异常为地震活动性异常，以趋势性中短期异常为主，持续时间最长的为一年。定点前兆异常均为短期异常。不过，本次地震前一周，滇东北 8 月 5 日还发生过会泽 5.3 级地震，对异常判断带来了困难。

异常空间分布显示，除震中小震活动增频和文山水氡异常外，其他异常主要分布在震中西北面。从震中距离分析，定点前兆异常在 100km 范围内也仅有文山水氡 1 项，其他均分布在 101 ~200km 范围。无论从异常的方位分布还是按震中距离分布看，两者都与台网测项的分布明显有关。

八、震前预测、预防和震后响应

1. 震前预测情况

文山是云南的少震地区，有记载以来，文山麻栗坡断裂也仅发生过一次 5 级以上地震。对这一地区仅是从地质角度指出该地区断裂第四纪以来具有明显的活动，而全新世以来断裂活动不明显，但现今仍有活动存在中强地震的危险性。

从一年尺度的中短期预测上看，虽然文山地区小震活动有明显增强，但更多的是将其视为窗口地震，对文山本地将发生破坏性地震认识不足，及至临震均未能给出预测意见。

2. 震后响应与余震趋势判定

地震发生后，云南省地震局立即启动地震应急预案，派出地震现场工作队奔赴震区。8 月 13 日下午 17：00 第一批现场工作队员到达现场；中国地震局派出 4 人专家组于次日凌晨 03：30 到达文山，广东省地震局，广西地震局也派出相关人员参与本次地震现场工作。现场工作队开展了灾害损失评估、地震监测、震情监视预报、灾情传送等工作。经过4天的艰苦努力于2005年8月17日圆满完成了各项任务。

文山5.3级地震发生在距文山州政府所在的文山县城30余公里的地方，震时城区强烈有感。由于文山是云南的少震区，当地群众对地震了解、认识较少，有人传闻当天晚上有7级大地震。谣言通过现代化的通讯手段，以手机短信的形式一传十，十传百……很快传遍了县城，城区上万群众聚集在几个大的广场上不回家。接到文山州政府的报告后，云南省地震局采取了积极应对措施。根据野外实地考察，认为文山5.3级地震与文麻断裂带活动有关，据断裂两侧的地貌特征（如河流阶地发育）分析，断裂全新世以来活动不明显，不具备发生大震的构造条件；同时根据历史地震序列类比分析，及时判断该地震序列为主震—余震型，3～4级余震会持续活动一段时间。把明确的意见用广播电视的形式播出后，群众迅速疏散回家，对稳定人心、安定社会起到了较好的作用，得到了文山州委、州政府的好评。

九、结论与讨论

通过本震例的研究，得到以下认识：

（1）1995年以来的近8年云南地区的5级以上地震集中在滇西北地区发生，自2003年11月鲁甸5.1、5.0地震后，地震活动格局有了明显变化，其后2004年8月、2005年8月滇东北鲁甸再次发生5级地震，本次文山地震也发生在构造相对稳定的滇东地区，其后几年内5级地震从盐津、鲁甸、会泽、双柏到墨江、思茅形成了新的地震活动条带，地震活动的这种高度有序的图像，可能是应力场增强、大震危险性逼近的信号。

（2）中期阶段近场区附近地震活动增强，是云南地区较好的预报指标。短期阶段省内4级地震平静，前兆异常集中出现，临震前异常结束或转折，对预测时间提供了有益的信息。另外水氡群体异常突出，距离震中最近的文山水氡异常的持续时间要相对长些，对地点预报有一定的指示意义。从本次震例的异常来看，中期阶段地震学异常可以为地点判定提供主要依据，短期阶段前兆异常对时间的指示优于地震学指标，二者结合可能是提高地震预报准确率的途径之一。

（3）本次地震震前认识到的前兆异常与震后总结存在一定的差异。其原因一方面震后以已发生地震为目标，以震中为参考点，选取一定范围内的台站去寻找前兆异常，部分震前看来不太确定的前兆异常与本次地震的对应也较好；另一方面本次地震前云南前兆台网于2001年开始实行数字化改造，至本次地震前仅有4年观测资料，资料积累的时间尚短，加上部分数字化观测仪器稳定性较差，影响了震前对这些测项的分析与认识。

（4）虽然近场区附近2～3级地震活动增强为未来地震震中提供了较好的地点指示，但文山在云南属于少震地区，这一地区的小震活动过去更多的认为其具有“窗口”效用，并不能准确的判定该地区有发生5级以上破坏性地震的危险。加上前兆观测相对薄弱，从前兆分布上看，震前也很难判定文山会发生地震。因此，对于少震地区如何判定其强震孕育已进入短临阶段并做出短临预测仍是十分困难的。这些初步的认识，还有待进一步深入探讨。

参考文献

[1] 云南省地震局等，西南地震简目，成都：四川科技出版社，1988
[2] 秦嘉政、钱晓东，云南会泽和文山 5.3 级地震研究，地震研究，28（4），403 ~ 407，2005
[3] 李树华、张立、王世芹等，2005 年云南会泽、文山 5.3 级地震短临预报及再认识，地震研究，29（1），43 ~ 49，2006
[4] 陈棋福主编，中国震例（1995 ~ 1996），北京：地震出版社，2002
[5] 陈棋福主编，中国震例（1997 ~ 1999），北京：地震出版社，2002

参考资料

1）云南省地震局，云南地震目录（区域台网）
2）云南省地震局，2005 年 8 月 13 日文山 5.3 级地震灾害损失评估报告，2005.8
3）云南省地震局，云南省 2005 年度地震趋势研究报告，2004.11
4）中国地震局，震例总结规范，2007.6

Wenshan Earthquake with M_S 5.3 on August 13, 2005 in Yunnan Province

Abstract

An earthquake with *M*5.3 occurred in Wenshan County, Yunnan Province on August 13, 2005, the macro-epicenter of this earthquake located between Hongdian and Matang. The epicenter intensity was Ⅵ degree, the meizoseismal area was elliptic with major axis in NW direction. The earthquake caused 2 people injured heavily and 27 people injured lightly, direct economic loss reaches 92.20 million RMB.

The earthquake is main shock- aftershock type. The number of earthquakes is very abundant in the sequence. The maximal aftershock with *M*3.8 happened. The focal mechanism solution of the earthquake shows the nodal plane Ⅰ with strike is 100°, the principal compressive stress was in NNW direction. Synthetically analysed result of geological examination, long axis direction of isoseismic line and distribution of aftershocks, we presumed it is rupture plane of the *M*5.3 earthquake. The earthquake resulted from right lateral strike slip of fault under the nearly horizontal principal compressive stress. The seismogenic structure possibly is the west section of the Wenshan-malipo fault, Wenshan-xishan fault.

There are 20 stations 200km away from epicenter; among of them there are 9 seismometric stations and 20 precursor stations. 10 abnormities exist before Wenshan 5.3 earthquake. Seismic abnormity is 4 iteM_S. Precursor abnormity is 6 iteM_S. The precursor abnormity mainly is short-term abnormity.

The post seismic tendency of the earthquake was rightly determined. It has a good effect for stabilize society.

报告附件

附表1　固定前兆观测台（点）与观测项目汇总表

序号	台站（点）名称	经纬度（°）		测项	资料类别	震中距Δ（km）	备注
		φ_N	λ_E				
1	文山	23.38	104.24	测震△̂		32	
				水氡	Ⅱ		
2	开远	23.75	103.25	水位	Ⅱ	85	
				水温	Ⅱ		
				气氡	Ⅱ		
				He	Ⅲ		
3	个旧	23.37	103.15	测震△△̂		98	
				水位	Ⅱ		
4	泸西	24.55	103.78	水温	Ⅱ	108	
5	弥勒	24.40	103.43	测震△̂		108	
				水位	Ⅱ		
				水温	Ⅱ		
				气氡	Ⅱ		
				气汞	Ⅱ		
				He	Ⅲ		
6	金平	22.78	103.22	测震△		128	
				地倾斜（摆式）	Ⅰ		
7	华宁	24.20	102.95	水位	Ⅱ	131	
				水温	Ⅱ		
8	建水	23.65	102.77	短水准	Ⅰ	133	
				短基线	Ⅰ		
				水位	Ⅱ		
				水温	Ⅱ		
9	师宗	24.83	103.99	水位	Ⅱ	135	
10	曲江	23.98	102.80	水位	Ⅱ	135	
				水温	Ⅱ		
				水汞	Ⅱ		
				水氡	Ⅱ		

续表

序号	台站（点）名称	经纬度（°） φ_N	经纬度（°） λ_E	测项	资料类别	震中距 Δ（km）	备注
11	通海	24.07	102.75	测震△		143	
				地倾斜（摆式）	Ⅰ		
				短水准	Ⅰ		
				短基线	Ⅰ		
				地磁 Z	Ⅱ		
				地磁 D	Ⅱ		
				水温	Ⅱ		
12	高大	23.99	102.71	水位	Ⅱ	144	
				水温	Ⅱ		
				气氡	Ⅱ		
				气汞	Ⅱ		
13	江川	24.33	102.75	水位	Ⅱ	156	
				水温	Ⅱ		
				气氡	Ⅱ		
				气汞	Ⅱ		
				He	Ⅲ		
14	石屏	23.67	102.47	测震△		163	
				地倾斜（摆式）	Ⅰ		
				短水准	Ⅰ		
				短基线	Ⅰ		
15	澄江	24.67	102.92	测震△		165	
				地倾斜（摆式）	Ⅰ		
				水位	Ⅱ		
				水温	Ⅱ		
				气氡	Ⅱ		
				气汞	Ⅱ		

续表

序号	台站（点）名称	经纬度（°）		测项	资料类别	震中距Δ（km）	备注
		φ_N	λ_E				
16	峨山	24.13	102.50	测震△		169	
				短水准	Ⅰ		
				短基线	Ⅰ		
				水位	Ⅱ		
				水温	Ⅱ		
				地倾斜（摆式）	Ⅲ		
17	宜良	24.92	103.13	测震△		173	
				地倾斜（摆式）	Ⅰ		
				短水准	Ⅰ		
				短基线	Ⅰ		
18	玉溪	24.37	102.55	水位	Ⅱ	176	
				气汞	Ⅱ		
				水温	Ⅲ		
19	绿春	23.00	102.40	水温	Ⅱ	184	
20	晋宁	24.67	102.58	地倾斜（摆式）	Ⅰ	191	

分类统计	$0<\Delta\leqslant100$km	$100<\Delta\leqslant200$km	总数
测项数 N	5	11	11
台项数 n	8	61	69
测震单项台数 a	0	0	0
形变单项台数 b	0	1	1
电磁单项台数 c	0	0	0
流体单项台数 d	0	3	3
综合台站数 e	3	13	16
综合台中有测震项目的台站数 f	2	7	9
测震台总数 $a+f$	2	7	9
台站总数 $a+b+c+d+e$	3	17	20
备注			

附表 2　测震以外固定前兆观测项目与异常统计表

序号	台站（点）名称	测项	资料类别	震中距Δ（km）	按震中距Δ范围进行异常统计									
					0 < Δ ≤ 100km					100 < Δ ≤ 200km				
					L	M	S	I	U	L	M	S	I	U
1	文山	水氡	Ⅱ	32	—	—	√	—						
2	开远	水位	Ⅱ	85	—	—	—	—						
		水温	Ⅱ		—	—	—	—						
		气氡	Ⅱ		—	—	—	—						
3	个旧	水位	Ⅱ	98	—	—	—	—						
4	泸西	水温	Ⅱ	108						—	—	—	—	
5	弥勒	水位	Ⅱ	108						—	—	—	—	
		水温	Ⅱ							—	—	—	—	
		气氡	Ⅱ							—	—	—	—	
		气汞	Ⅱ							—	—	—	—	
6	金平	地倾斜	Ⅰ	128										
7	华宁	水位	Ⅱ	131						—	—	—	—	
		水温	Ⅱ							—	—	—	—	
8	建水	短水准	Ⅰ	133						—	—	—	—	
		短基线	Ⅰ							—	—	—	—	
		水位	Ⅱ							—	—	—	—	
		水温	Ⅱ							—	—	—	—	
9	师宗	水位	Ⅱ	135						—	—	√	—	
10	曲江	水位	Ⅱ	135						—	—	—	—	
		水温	Ⅱ							—	—	—	—	
		水汞	Ⅱ							—	—	—	—	
		水氡	Ⅱ							—	—	—	—	
11	通海	地倾斜	Ⅰ	143						—	—	—	—	
		短水准	Ⅰ							—	—	—	—	
		短基线	Ⅱ							—	—	—	—	
		地磁 Z	Ⅱ							—	—	—	—	
		地磁 D	Ⅱ							—	—	—	—	
		水温	Ⅱ							—	—	—	—	

续表

序号	台站（点）名称	测项	资料类别	震中距Δ（km）	按震中距Δ范围进行异常统计									
					0＜Δ≤100km					100＜Δ≤200km				
					L	M	S	I	U	L	M	S	I	U
12	高大	水位	Ⅱ	144						—	—	—	—	
		水温	Ⅱ							—	—	—	—	
		气氡	Ⅱ							—	—	√	—	
		气汞	Ⅱ							—	—	—	—	
13	江川	水位	Ⅱ	156						—	—	√	—	
		水温	Ⅱ							—	—	—	—	
		气氡	Ⅱ							—	—	√	—	
		气汞	Ⅱ							—	—	—	—	
14	石屏	地倾斜	Ⅰ	163						—	—	—	—	
		短水准	Ⅰ							—	—	—	—	
		短基线	Ⅱ							—	—	—	—	
15	澄江	水位	Ⅱ	165						—	—	—	—	
		水温	Ⅱ							—	—	√	—	
		气氡	Ⅱ							—	—	√	—	
		气汞	Ⅱ							—	—	—	—	
		地倾斜	Ⅰ							—	—	—	—	
16	峨山	短水准	Ⅰ	169						—	—	—	—	
		短基线	Ⅱ							—	—	√	—	
		水位	Ⅱ							—	—	—	—	
		水温	Ⅱ							—	—	—	—	
17	宜良	地倾斜	Ⅰ	173						—	—	—	—	
		短水准	Ⅰ							—	—	—	—	
		短基线	Ⅱ							—	—	—	—	
18	玉溪	水位	Ⅱ	176						—	—	—	—	
		气汞	Ⅱ							—	—	—	—	
19	绿春	水温	Ⅱ	184						—	—	—	—	
20	晋宁	地倾斜	Ⅰ	191						—	—	—	—	

续表

序号	台站（点）名称	测项	资料类别	震中距Δ（km）	按震中距Δ范围进行异常统计									
					0＜Δ≤100km					100＜Δ≤200km				
					L	M	S	I	U	L	M	S	I	U
分类统计	台项	异常台项数			0	0	1	0		0	0	5	0	
		台项总数			5	5	5	5		50	50	50	50	
		异常台项百分比/%			0	0	20	0		0	0	10	0	
	观测台站（点）	异常台站数			0	0	1	0		0	0	4	0	
		台站总数			3	3	3	3		17	17	17	17	
		异常台站百分比/%			0	0	30	0		0	0	24	0	
	测项总数（55）				5					50				
	观测台站总数（20）				3					17				
备注														

2005 年 11 月 26 日江西省九江—瑞昌 5.7 级地震

江西省地震局

高建华　吕　坚　杨雪超　汤兰荣

摘　要

2005 年 11 月 26 日江西省九江县与瑞昌市交界地区发生 5.7 级地震，宏观震中位于九江县港口乡，震中烈度为Ⅶ度，极震区呈北东向的椭圆形，地震造成 13 人死亡，重伤 82 人，轻伤 683 人，直接经济损失 20.38 亿元。

此次地震序列为主震—余震型，最大余震 4.8 级。余震分布优势长轴走向为北西向，与极震区长轴走向并不一致，节面Ⅰ为主破裂面，主压应力 P 轴方位 NWW 向。5.7 级的地震发震构造与广济—襄樊断裂有关，4.8 级地震的发震构造为 NE 向的丁家山—桂林桥—武宁断裂。

在震中 200km 内共有地震观测台 13 个，其中 11 个台设有测震观测，9 个台站有其他前兆观测。地震前共出现 10 条异常，地震学异常明显，共 5 条，定点前兆观测异常 4 条，宏观异常 1 条。

前　言

2005 年 11 月 26 日 08 时 49 分，江西省九江县与瑞昌市交界地区发生 5.7 级地震。微观震中为 29°41′N，115°44′E，宏观震中在九江县港口乡（29°42′N，115°42′E）。震中烈度为Ⅶ度，地震波及湖北、安徽、江苏、上海、浙江、湖南、福建等省市，有感范围较大。地震造成 13 人死亡，重伤 82 人，轻伤 683 人，直接经济损失 20.38 亿元[1)]。

5.7 级地震发生的南华活动地块区北部是华南腹地地震活动较为活跃的地区，历史上多次发生中强地震，200km 范围内共有地震台站 13 个，震前出现 5 个异常项目 10 条异常，宏观异常 1 条。

一、测震台网及地震基本参数

图 1 给出了震中 200km 范围内的地震台站分布，其中江西省九江台、南昌台、修水台，湖北省武汉台、黄梅台、南川台、麻城台和安徽省安庆台、佛子岭、石家河、豹子崖等 11 个台有测震观测，在研究时段内基本可达到 $M \geqslant 2.5$ 级地震不遗漏。100km 内仅有九江和黄

梅测震台，其中九江测震台是观测范围为 20s-20Hz 的数字化记录，而黄梅测震台为 DD-2 仪模拟记录，总体而言区域地震监测能力相对较弱。

表 1 列出了不同来源给出的这次地震基本参数，经对比分析，认为文献［1］经重新修订后的震中位置更为精确，因此，本次地震基本参数取表 1 中编号 4 的结果。

表 1　地震基本参数

Table 1　Basic parameters of the earthquake

编号	发震日期	发震时刻	震中位置		震级 M	震源深度	震中	结果来源
	年 月 日	时 分 秒	φ_N	λ_E		(km)	地名	
1	2005 11 26	08 49 37	29.69°	115.74°	5.7	11	九江—瑞昌	文献［1］
2	2005 11 26	08 49 39	29.7°	115.7°	5.7	15	九江—瑞昌	中国地震局
3	2005 11 26	08 49 37	29.72°	115.72°	5.5	12	九江—瑞昌	江西省地震局
4	2005 11 26	08 49 38	29.66°	115.72°	5.7	10	九江—瑞昌	USGS

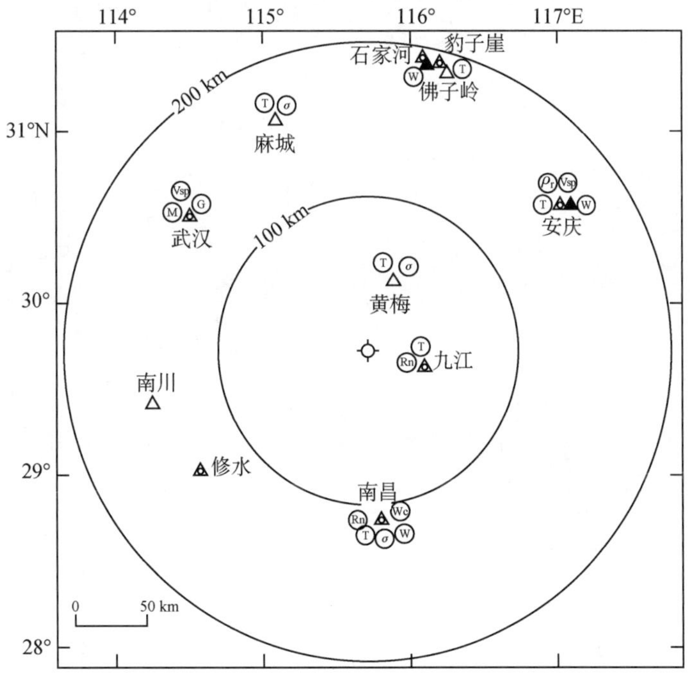

图 1　九江—瑞昌 5.7 级地震前震中附近地震台站及观测项目分布图

Fig. 1　Distribution of seismometrci stations and observations project around the epicentral area before the M_S 5.7 Jiujiang-Ruichang earthquake

二、地震地质背景

2005 年 11 月 26 日九江—瑞昌 5.7 级地震发生在南华活动地块区北部的扬子准地台与秦岭—大别褶皱带的接壤部位[2]，其震中附近区域的主要断裂分布见图 2。

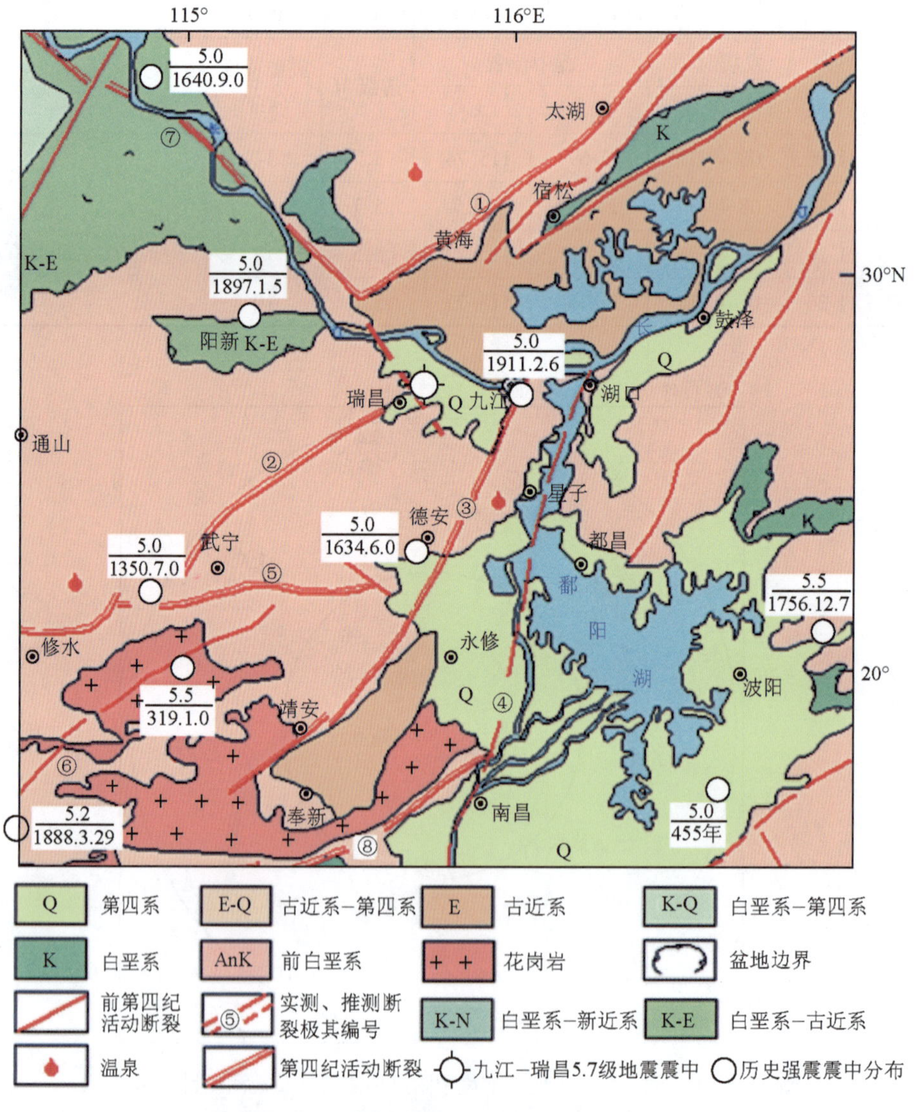

图 2　区域地质构造和历史中强地震分布图

Fig. 2　Map of geological structure and distribution of historical earthquakes

主要断裂名称：①庐江—广济断裂　②修水—武宁—瑞昌断裂　③九江—靖安断裂　④新干—湖口断裂　⑤渣津—柘林断裂　⑥铜鼓—武宁断裂　⑦襄樊—广济断裂　⑧宜丰—景德镇断裂

在1°×1°的布格重力异常图上，九江—瑞昌5.7级地震发生在淮阳断块隆起和九岭断块隆起包夹长江中下游断块拗陷形成的北东向和北西向的重力梯度带的汇聚部位，布格重力异常为-25×10^{-5}m/s^2[3,4]。根据横穿大别造山带的人工地震测深及与之重合的重力剖面研究结果，震中附近的地壳厚度约35km[5,6]。

自1959～1990年观测的高精度垂直形变资料反映本区域以1mm/a左右的速度在下沉，下沉等值线优势走向为北西向。据全球定位系统（GPS）观测，震中邻近地区的水平运动方向基本一致为N45°～50°E，运动速率基本一致为8～10mm/a，说明区内各点之间的相对差异运动微弱，呈整体性运动特征[7]。

九江—瑞昌5.7级地震发生在华南地震区长江中游地震带的东南段，微观震中则位于新构造运动时期形成的瑞昌盆地内，震中附近区域主要发育有北东向的庐江—广济断裂带与武宁—瑞昌断裂带、北北东走向的九江—靖安断裂带和北西向的广济—襄樊断裂带等，历史上发生了多次中强破坏性地震，呈现北西向和北东向带状分布的格局。距离九江—瑞昌5.7级地震最近的历史中强地震为1911年2月6日江西九江市5级地震，二者相距约29km。自1970年有完整的小震仪器记录以来，距离九江—瑞昌5.7级地震震中50 km范围内发生了1972年9月12日湖北阳新M_L4.5、1995年4月15日江西瑞昌M_L4.9和2004年1月26日江西德安M_L4.1级3次中等地震，是华南腹地地震活动较为活跃的地区，本次九江—瑞昌5.7级地震也是邻近区域近百年来最显著的地震之一。

三、地震影响场和震害1,2)

九江—瑞昌5.7级地震受灾范围波及江西省九江县、瑞昌市以及湖北省黄梅县、武穴市。江西南昌市、九江市、彭泽县、星子县、德安县、永修县、武宁县、修水县以及湖北省通山县、阳新县、安徽省宿松县、望江县等地震感强烈，共造成13人死亡，82人重伤，683人轻伤，是1806年江西会昌发生6级地震以来，江西境内震级最大、死亡人数最多、损失最大、灾害最严重地震。

根据地震现场科学考察报告，灾区地震烈度分为Ⅶ度区和Ⅵ度区；震中区烈度达Ⅶ度，等值线呈椭圆形分布，长轴方向为北北东向（图3）。

Ⅶ度区主要包括九江县的城门乡、新合乡、新塘乡、港口镇乡、狮子镇、瑞昌市市区以及瑞昌市航海仪器厂以东地区，还包括长江北岸的小池镇少部分地区，长轴约24km，北东东方向，短轴约15km，面积260km^{2}左右。区内土木结构房屋部分倒塌或毁坏，砖木结构房屋也有少部分严重破坏或倒塌；城市砖混结构房屋轻微或中等破坏，少数严重破坏，基本没有倒塌，农村砖混结构房屋多数达中等以上破坏，少数局部倒塌；框架结构非承重墙严重裂缝，柱子、楼板等主体结构基本完好，局部轻微破坏。区内有小范围砂土液化和喷砂冒水现象，其中赛湖农厂二分厂十三连棉花地里喷砂冒水高达几米，持续时间较长，导致地里淤积大量黄泥砂。在Ⅶ度区内出现数十处地陷，地陷深3～9m，规模大小不一，大的直径10多米，小的直径两三米，其中赛湖农厂二分厂十三连棉花地里的三个大陷坑连成一片，长达98m。

Ⅵ度区包括九江县、九江市周岭以西地区、瑞昌市花园以东地区、黄梅县坝口—陈杨武一带以南地区和武穴市、阳新县、德安县部分地区，长轴约61km，走向北东，短轴约

45km，面积 1800km²。区内土木民宅及牲畜棚圈破坏较严重，其中九江市新港镇太平桥村有几处局部倒塌；砖木房屋大部分有轻微裂缝，少数中度破坏，没有倒塌；农村砖混房屋 15% 达中度破坏，50% 轻微破坏，个别严重破坏，约 30% 基本完好。Ⅵ度区内也有宏观破坏现象，九江市新港镇太平桥村有一处出现大规模崩塌，瑞昌市高丰镇永丰村出现地陷，其他地方也有小规模塌陷。受震区地形地貌和场地条件的影响，Ⅵ度区向东侧靠庐山方向呈收敛局态势，向西南西北方向辐射面积较大。

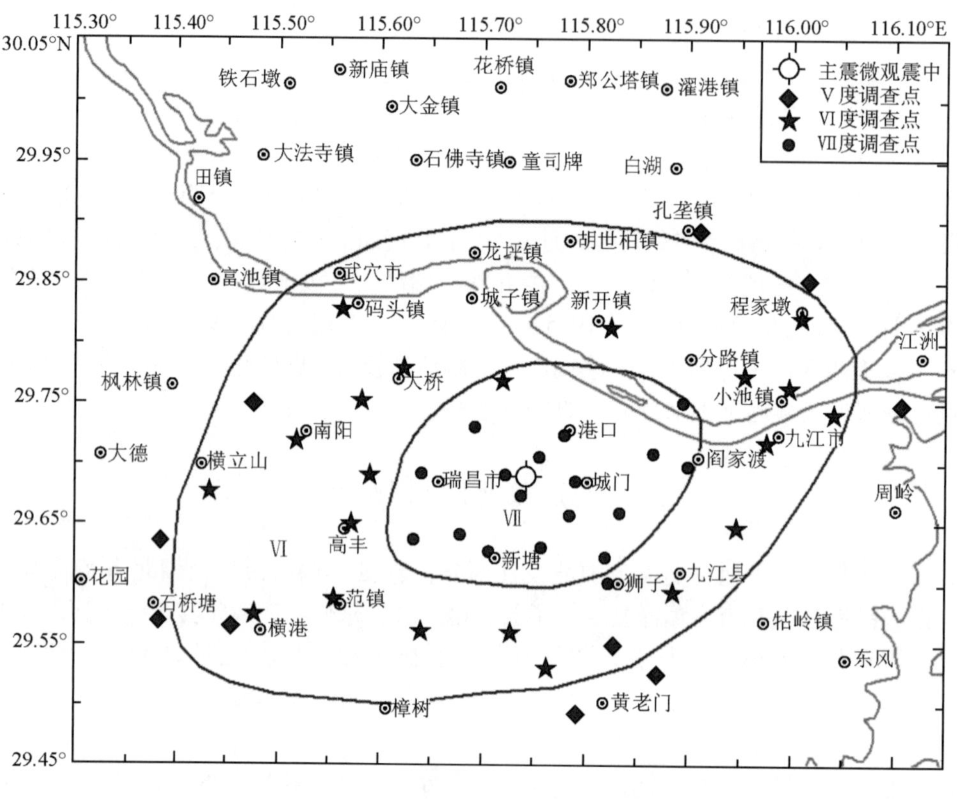

图 3　九江—瑞昌 5.7 级地震烈度分布图

Fig. 3　Intensity distribution of the M_S 5.7 Jiujiang-Ruichang earthquake

四、地 震 序 列

九江—瑞昌 5.7 级地震震前未记录到小震活动。本次地震序列持续时间较长，2005 年 11 月 26 日主震后，至 2007 年初，震源区仍可见零星 $M_L \geqslant 2.0$ 级小震活动。截至 2006 年 9 月 21 日，共记录到 $M_L \geqslant 2.0$ 级 103 次，详见表 2 震级分档统计和表 3 地震目录（据江西台网目录，下同）。

表 2 九江—瑞昌地震序列震级分档统计结果

Table 2 Magnitude interval statistics of the M_S 5.7 Jiujiang-Ruichang earthquake sequence

震级分档 M_L	0.1~0.9	1.0~1.9	2.0~2.9	3.0~3.9	4.0~4.9	5.0~5.9	6.0~6.9	合计
次数	1581	643	87	12	2	1	1	2327
占总数比例	0.6794	0.2763	0.0374	0.0052	0.0009	0.0004	0.0004	1.00

表 3 九江—瑞昌地震序列目录（$M_L \geqslant 3.0$ 级）

Table 3 Catalogue of the M_S 5.7 Jiujiang-Ruichang earthquake sequence（$M_L \geqslant 3.0$）

编号	发震日期	发震时刻	震中位置		震级 M_L	震源深度（km）
	年 月 日	时 分 秒	φ_N	λ_E		
1	2005 11 26	8 49 37	29°41′	115°44′	6.0	11
2	2005 11 26	9 8 9	29°41′	115°44′	3.0	
3	2005 11 26	9 10 36	29°42′	115°42′	3.1	
4	2005 11 26	9 25 23	29°40′	115°44′	4.5	10
5	2005 11 26	12 55 39	29°40′	115°44′	5.4	10
6	2005 11 26	16 21 59	29°42′	115°42′	3.1	
7	2005 11 27	8 48 8	29°42′	115°42′	3.3	
8	2005 11 27	19 35 6	29°42′	115°42′	3.0	
9	2005 11 27	23 44 10	29°42′	115°42′	3.0	
10	2005 12 3	7 46 36	29°42′	115°45′	4.0	16
11	2005 12 4	2 27 50	29°41′	115°47′	3.0	17
12	2005 12 19	23 38 17	29°42′	115°42′	3.4	
13	2006 4 15	16 49 32	29°41′	115°42′	3.2	12
14	2006 4 25	6 41 27	29°39′	115°46′	3.5	
15	2006 4 29	3 28 41	29°43′	115°47′	3.0	
16	2006 6 16	6 9 1	29°40′	115°44′	3.5	

序列余震活动主要集中在东西向 16km、南北向 10km、总面积约 110km^2 的范围内，包括全部 M_L3 级以上余震。震中分布图像总体似纺棰型，向东侧收敛、向西侧发散，而从主震与几次强余震位置看，似乎呈北东与北西共轭分布，但不太明显（图 4）。

序列 M-t 图显示：余震主要集中在主震后 3~8 天内，共发生 11 次 M_L3 以上余震，占整个序列的 73%，最大余震为 M_L5.4（M_S4.8），发生在主震后 4 小时。之后余震频度与强度均明显下降，最后一次 M_L3 级以上地震发生在 2006 年 6 月 16 日（图 5）。

序列频度衰减具有明显的阶段性，大体可分为 5 个阶段，首先是快速衰减阶段，由主震当日最高 206 次（可能存在震后初期的小震缺失），迅速下降至 94 次，持续 4 天；随后再度上升，至12月3日7时46分发生M_L4.0余震后下降，最高日频度122次，持续时间4天；第三阶段日频度为50次左右，仅及前两阶段一半弱，持续时间仍为4天，最大震级M_L3.0；第四阶段为12月8～19日，日频度平均在25次，期间发生一次M_L3.4；12日20日开始的第五阶段日频度基本在10次以下，未发生M_L3级以上地震，表现为低频小震活动，但持续时间很长，直至2007年仍偶见小震活动，并在2006年4～6月份出现一次增强，发生2次M_L3.5余震（图6）。

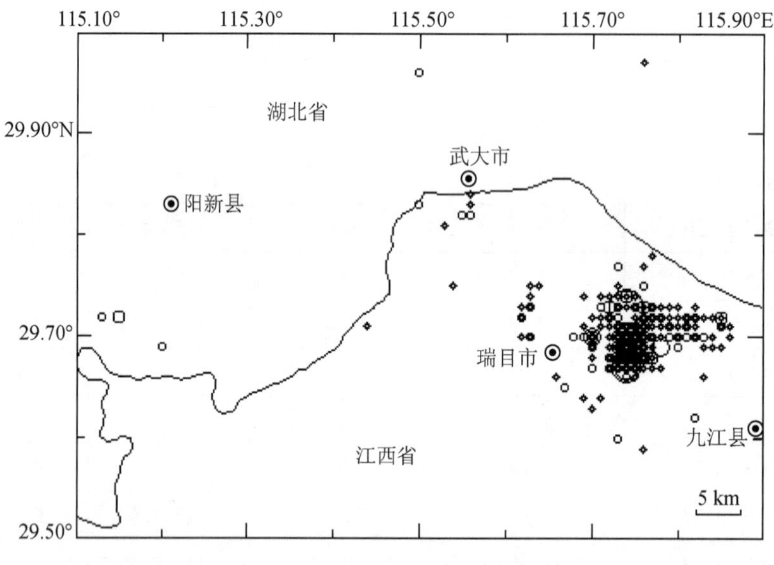

图4 九江—瑞昌地震序列分布

Fig. 4 Distribution of the M_S 5.7 Jiujiang-Ruichang earthquake sequence

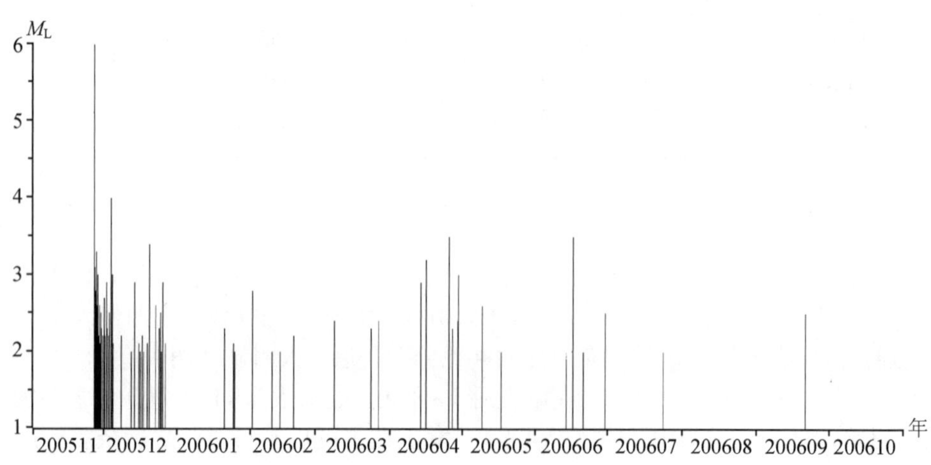

图5 九江—瑞昌地震序列 M-t 图（$M_L \geqslant 2.0$级）

Fig. 5 M-t digaram of the M_S 5.7 Jiujiang-Ruichang earthquake sequence

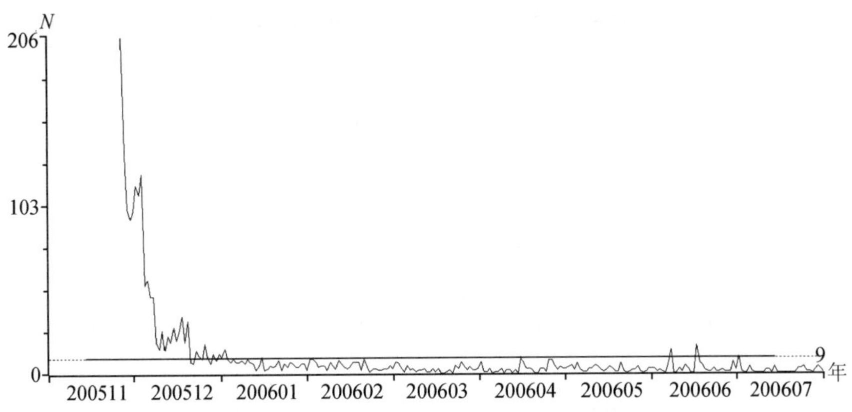

图 6　九江—瑞昌地震序列频次分布图

Fig. 6　Frequency distribution diagram of the M_S 5.7 Jiujiang-Ruichang earthquake sequence

由震级-频度关系分析，震级结构基本合理，震级频度关系式为 $\lg N=3.25-0.64M$，$R=-0.9775$，拟合最大余震震级为 $M_L5.5$，与实际最大余震 $M_L5.4$ 较为吻合（图7）。

主震后一天，使用 $U\rho K$ 组合判定得到 $U=0.20<0.5$、$\rho=1.02>0.55$、$K=0.23<0.7$，即为非前兆震群；序列 $h=1.05$、$P=1.09$，主震能量 2.2387×10^{13}J，占整个序列总能量的93.58%，主震与最大余震级差 $\Delta M=0.8$（以 M_L 计则为0.7），$2.4\geqslant\Delta M\geqslant0.7$，判定为主震—余震型。

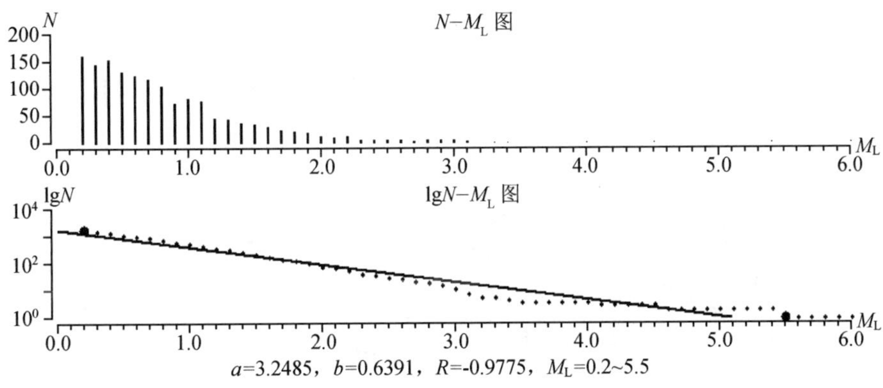

图 7　九江—瑞昌地震序列 b 值拟合

Fig. 7 b-value curve of the M_S5.7 Jiujiang-Ruichang earthquake sequence

五、震源参数和地震破裂面

在诸多九江—瑞昌 5.7 级地震的震源机制解研究结果中，文献［8］和文献［1］的研究结果与哈佛大学的研究结果较为一致，而且三者使用的资料和方法各不相同，结果可靠。

表 4　2005 年 11 月 26 日 5.7 级地震震源机制解

Table 4　Focal mechanism solutions of the M_S 5.7 Jiujiang-Ruichang earthquake sequence

序号	节面 I			节面 B			P 轴		T 轴		资料来源
	走向	倾角	滑动角	走向	倾角	滑动角	方位	仰角	方位	仰角	
1	324°	55°	18°	223°	75°	144°	277°	13°	178°	36°	文献［8］
2	325°	61°	20°	224°	72°	149°	277°	8°	182°	33°	Harward
3	334°	64°	16°	237°	76°	153°	287°	8°	193°	29°	文献［1］
4	84°	89°	-98°	346°	8°	－8°	182°	43°	346°	45°	CSN
5	5°	23°	74°	202°	68°	97°	287°	23°	124°	66°	USGS

由于九江—瑞昌 5.7 级的微观震中位于长江南岸的瑞昌盆地内，地震造成的破坏明显受到局部地质和水文条件的影响，且地震震中被两条 NE 向断裂和两条 NW 向隐伏构造线所包围（图 8），发震构造的认定具有一定困难。对于本次地震的发震构造，现场科考[1,2]将其确定为 NE 向的丁家山—桂林桥—武宁断裂（图 8 中的断裂②）北段，主要依据为地震烈度等震线呈 NE 向长轴分布和该断裂为瑞昌盆地北西缘控制断裂，地表表现较清楚。但本研究认为判定该断裂为 5.7 级主震的发震构造依据不足，值得商榷，主要理由如下：

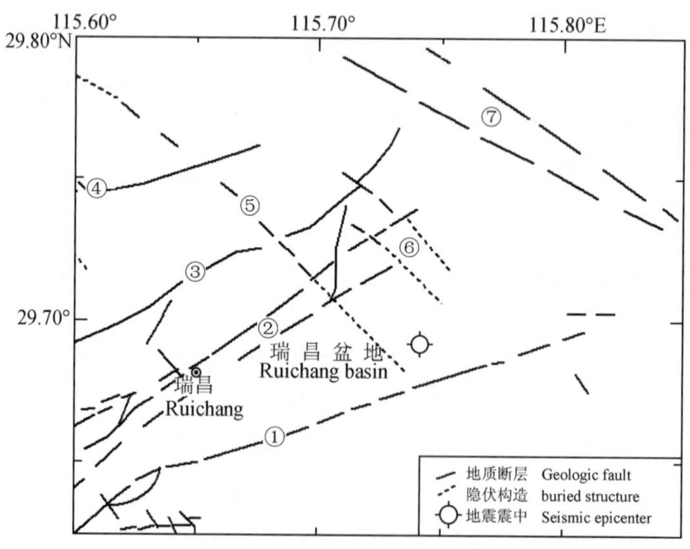

图 8　震区地质构造与精定位主震震中

Fig. 8　Map of geological structure and distribution of the main earthquake after precise relocation

①刘家—范家铺—城门山断裂；②丁家山—桂林桥—武宁断裂；③望夫山—大浪断裂；④武山—南阳断裂；⑤洋鸡山—武山—通江岭断裂；⑥丁家山—狮子岛断裂；⑦长江断裂

（1）现有的资料[3~5]和地震现场考察结果均显示，该断裂在赛湖—马家垅—丁家山（主震震中以北）一带，表现为梯状正断层，走向约 NE50°，倾向 SE，倾角 80°左右。根据该断裂的走向和倾向，如果其为发震构造，主震震源机制解应有一组节面的走向在 0°～90°间，这与主震的震源机制解结果不符。

（2）如果该断裂为唯一的发震构造，根据其走向和倾向，地震序列总体上应该呈现 NE 向优势分布，震源深度由 NW 至 SE 逐渐变深，这一推测结果与文献［1］的地震序列精确定位结果不太一致（图 9）。精确定位后本次地震序列 $M_L \geqslant 1.0$ 级地震呈现 NNW 向优势分布，余震区分布长轴大约 5km，短轴大约 3km。如果考虑 $M_L \geqslant 1.5$ 级地震，则似乎存在 NW 向（剖面 $B—A$）和 NE 向（剖面 $C—D$）的两组共轭分布。图 10 显示，地震序列在震源深度上的变化也较为复杂，大致以震源深度 11.5km 处为分界：沿剖面 $A—B$（与主震震源机制解的节面 I 走向大致垂直），上部地震震源深度总体上由 NW 至 SE 逐渐变浅，下部地震震源深度总体上由 NW 至 SE 逐渐变深；沿剖面 $C—D$（与主震震源机制解的节面 Ⅱ 走向大致垂直），上部地震震源深度总体上由 SW 至 NE 逐渐变深，下部地震震源深度总体上由 SW 至 NE 逐渐变浅。

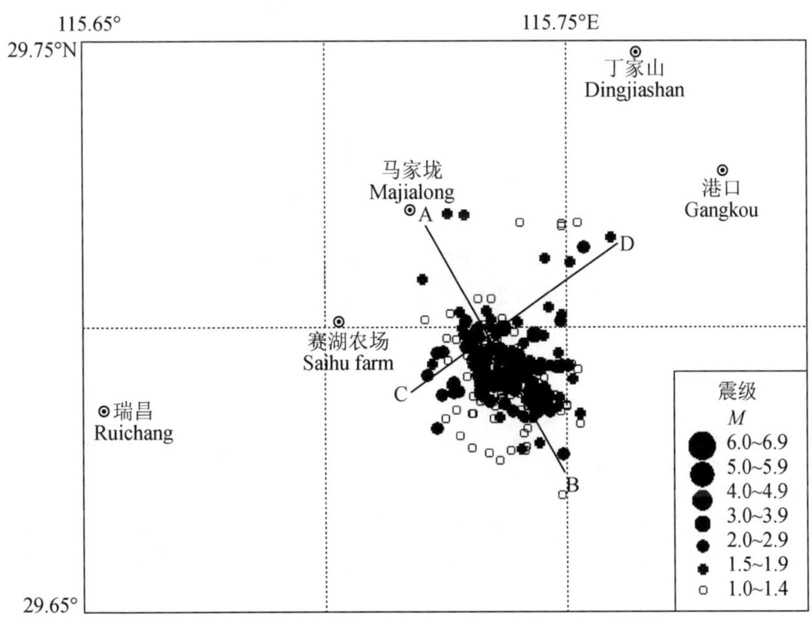

图 9 精定位后的地震震中分布

Fig. 9 Epicenter distribution of seismic sequence after precise relocation

（3）文献［9］采用初动和振幅比方法，研究得到 112 次 $M_L \geqslant 1.5$ 级余震的震源机制解存在 NEE 和 NNW 向两组优势走向分布，倾角以 50°～60°居多（图 11）。考虑到 5.7 级主震和 4.8 级强余震的震源机制解也不完全一致，而且这两次地震在震中空间分布和震源深度分布上也有所区别。从图 9 和图 10 可以看出，5.7（M_L6.0）级主震发生在 $M_L \geqslant 1.5$ 级地震序列的 SE 端，震源深度小于 11.5km 的上部；而 4.8（M_L5.3）级强余震发生在 $M_L \geqslant 1.5$ 级地震序列 NW 向分布（剖面 $B—A$）和 NE 向分布（剖面 $C—D$）的共轭部位，震源深度大于 11.5km 的下部。上述现象表明在 5.7 级主震发生后，余震活动从 SE 往 NW、从浅部往深部发展，在破裂过程中可能遇到障碍体，触发了 4.8 级强余震，因此这两次地震的发震构造可能并不唯一，需进一步讨论。在与发震构造可能有关的 4 条断裂或隐伏构造线中（图 8），断裂①为刘家—范家铺—城门山断裂北段，该断裂为瑞昌盆地南东缘控制断裂，在震中一带走向 NE，倾向 SE，倾角 70°～85°，根据其走向和倾向，发震破裂时地震位置应该处于该断裂以南而不在瑞昌盆地内，显然与实际情况不符，该断裂不太可能为本次地震的发震构造；

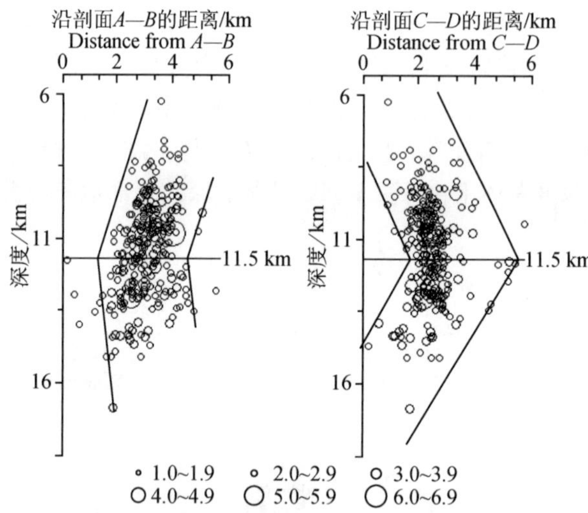

图 10　沿不同剖面的震源深度图

Fig. 10　Profile of focal depth along different cross-section

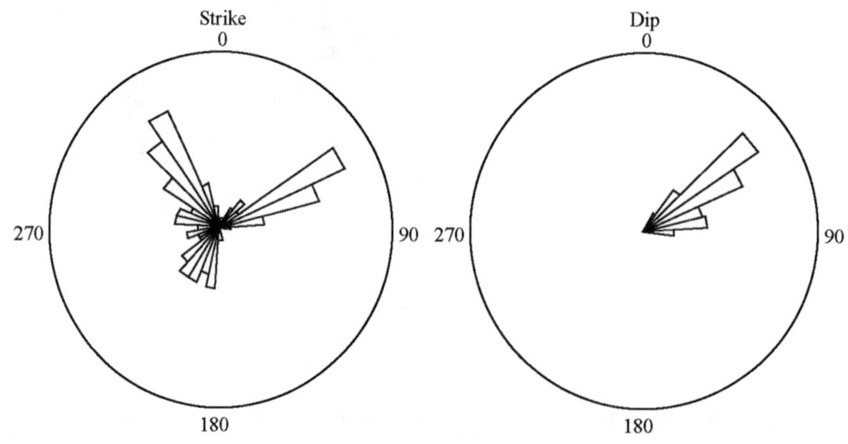

图 11　$M_L \geqslant 1.5$ 级余震序列震源机制解的走向和倾角分布玫瑰图

Fig. 11　The rose map of $M_L \geqslant 1.5$ aftershock focal mechanisms' strike and dip distribution

断裂⑥为丁家山—狮子岛断裂，属性目前未见详细资料阐述，如果其倾向 NE，则走向 NW，发震破裂时地震位置应该处于该断裂以东而不在瑞昌盆地内；如果其倾向 SW，则走向 SE，发震破裂时 5.7 级主震和 4.8 级强余震的震源机制解中应有一组节面的走向在 90°～180°间，上述两种推测结果都与实际情况不一致，该断裂也不太可能为本次地震的发震构造。在排除了上述两条断裂后，研究认为 NE 向的丁家山—桂林桥—武宁断裂（图 8 中的断裂②）北段应该是 4.8 级强余震的发震构造，因为该断裂的属性与 4.8 级强余震的震源机制解节面 I 的性质吻合，发震破裂时其余震可能出现沿剖面 C—D（与震源机制解节面 I 走向大体一致）的一组优势分布，震源深度沿剖面 A—B（与震源机制解的节面 I 走向大致垂直）总体上可能由 NW 至 SE 逐渐变深，这一推测结果与本研究结果是较为一致的。对于 5.7 级主震的发震构造，我们认为图 8 中的断裂⑤即洋鸡山—武山—通江岭 NW 向断裂不可忽视，因为本次地震序列整体上表现为 NNW—NW 向优势分布，且震源分布图像和主震震源机制解的节面 Ⅱ 性质均显示可能存在一条倾向 NE、走向 NW 的断层活动，引发了 5.7 级主震，而这条 NW 向断裂位于地震震中以西，可能向东南延伸并隐伏在瑞昌盆地中，在区域大比例尺的航磁图（图 13）与重力梯度图（图 12）上均有显示，很可能为本次地震序列主震的发震构造。

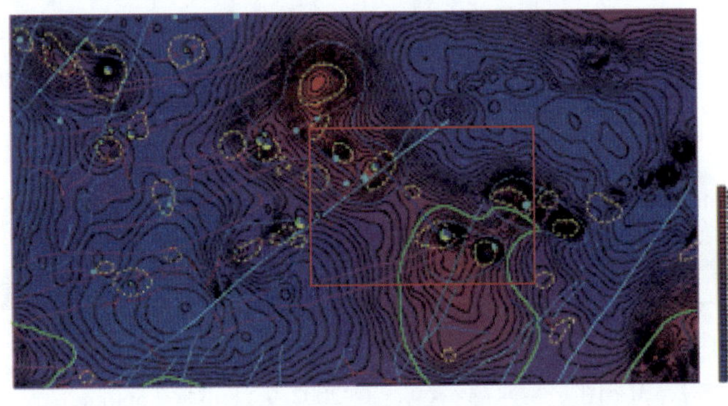

图 12　区域航磁图

Fig. 12　Map of the regional aeromagnetic

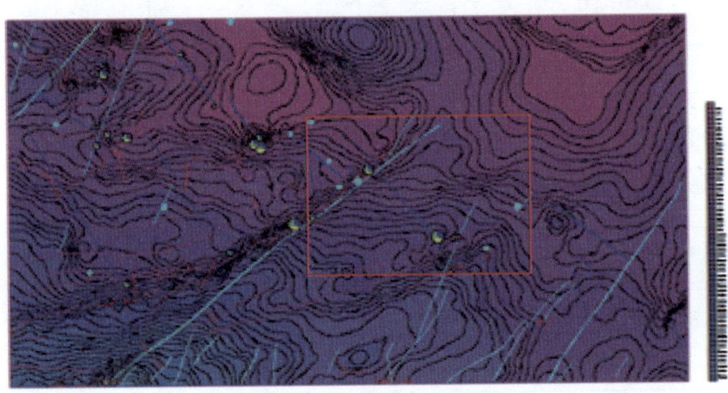

图 13　区域重力梯度图

Fig. 13　Map of the regional gravity gradient

六、地震前兆台网及前兆异常

震中附近地区的地震台站及观测项目分布见图1。江西、湖北、安徽三省交界地区前兆观测台站较稀疏且不均匀，九江—瑞昌5.7级地震震中以南地区仅有一个台站有综合前兆测项观测。

在震中200km以内共有地震台站13个，其中11个台设有测震观测，9个台站有其他前兆观测，包括地倾斜、洞体应变、定点重力、地电阻率、大地电场、地磁F、地磁Z、地磁H、水位、水氡和水温等共25个台项。在0～100km、100～200km距离内分别有地震台站2个和11个，其中定点前兆观测台站2个和7个，台项分别为5项和20项。

此次地震前共出现10条异常，其中地震活动性异常4项5条；定点前兆观测异常3项4条；宏观异常1条（表5）。

表5　地震前兆异常登记表

Table 5　Summary of precursory anomalies

序号	异常项目	台站（点）或观测区	分析方法	异常判据及观测误差	震前异常起止时间	震后变化	最大幅度	震中距 Δ/km	异常类型及可靠性	图号	异常特点及备注
1	地震丛集活动	震中200 km范围内	$M_L \geqslant 3.0$级地震M-t图	$M_L \geqslant 3.0$级地震活动持续3年	2002.10～2005.10	转入平静	持续3年	震中	M_1	14	
2	地震活动平静	28°～31°N, 113°～118°E	空间分布 M-t图	$M_L \geqslant 2.5$级地平静、空区	1999.12～2002.9	恢复正常	空区长560km		M_1	15，16	
3	地震活动增强	28°～31°N, 113°～118°E	空间分布、M-t图、应变释放	空间集中、$M_L \geqslant 3.0$级地震活动丛集、应变释放加速	2002.10～2005.10	恢复正常	M_L = 4.1级		M_1	15，16	
4	条带	湖北西部—江西九江	$M_L \geqslant 3.0$级地震M-t图、空间分布	$M_L \geqslant 3.0$级地震成带分布	2002.10～2004.3	恢复正常	条带长1000km		M_1	17，18	
5	条带	江西九江—黄海北部	$M_L \geqslant 4.0$级地震M-t图、空间分布	$M_L \geqslant 4.0$级地震成带分布	2004.11～2005.7	恢复正常	条带长1000km		M_1	17，19，20	

续表

序号	异常项目	台站（点）或观测区	分析方法	异常判据及观测误差		震前异常起止时间	震后变化	最大幅度	震中距 Δ/km	异常类型及可靠性	图号	异常特点及备注
6	地倾斜（石英摆）	九江	日均值	打破年变形态		1996～			38	L_2	21	南北向年变形态消失，东西向年变幅度减小、方向变化
7	地倾斜（水管）	黄梅	单分量整点值	速率增大，反向		2005.9～	基本恢复	0.23″	49	S_1	22	两分量速率出现同步变化，速率明显增大
8	洞体应变（伸缩）	黄梅	单分量整点值	速率明显增大		2005.10.21～	异常继续	5.3×10^{-7}	49	S_1	23	速率增大，震前反向变化
9	水位	皖33井	日均值	大幅上升		2005.7.22～	没有恢复		190	S_2	24	异常幅度大
10	10	宏观异常	震中附近	动物习性观察	鸡	震前2天				I_1		乱飞乱叫，不进圈
					猪							乱叫，“飞”出圈

1. 地震学异常

在震中区域200 km范围内，1970年以来M_L3以上地震有成丛（成对）发生的特点。2002年10月开始的丛集$M_L3\sim4$地震相对活跃，至5.7级地震前，持续了近3年时间，先后发生2次M_L4地震（图14）。在上述3级以上地震平静—增强的丛集活动背景上，区域地震活动也经历了一个空间图像围空—集中的演化过程，1999年12月至2002年2月，赣鄂皖

交界地区 M_L2.5 以上地震平静，形成北东东向的 M_L2.5 以上地震空区；2002 年 10 月至 2005 年 10 月，M_L3 以上地震的丛集相对集中在上述空区，并表现为加速活动（图 15、图 16）。2002 年 10 月开始，中国大陆中东部地区 M_L3～4 地震活动形成地震条带在九江及其邻区交汇（图 15）。一是 2002 年 10 月至 2004 年 3 月出现的湖北西部—江西九江 $M_L \geq 3.0$ 级地震条带，这条北西向地震条带一直延伸到甘青川交界地区（图 17、图 18）；二是 2002 年以来华东南地区还出现了一条以郯庐断裂带为核心，从江西九江—黄海北部的北东向 $M_L \geq$ 4.0 级地震巨型活动带（图 17、图 19），2004 年 11 月至 2005 年 7 月仅 9 个月间发生了 7 次 $M_L \geq 4.0$ 级地震，其中南端连发 3 次 M_L4 级以上地震，表现出明显的加速活动过程（图 20），显示了该部位应力集中的特征。

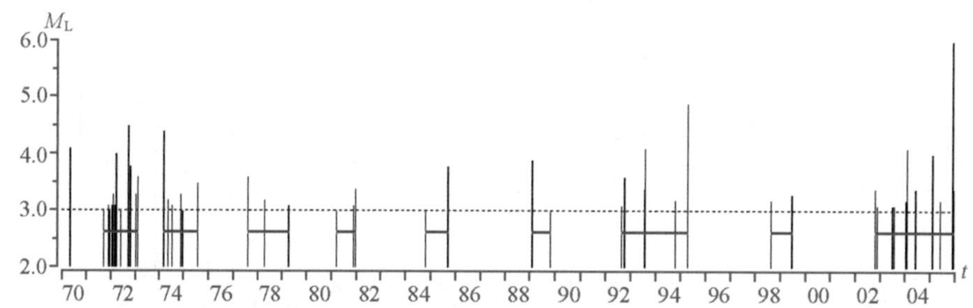

图 14　九江—瑞昌 5.7 级地震震中 180km 范围 1970～2005 年 $M_L \geq 3.0$ 级地震 M-t 图

Fig. 14　M-t of $M_L \geq 3.0$ earthquakes surrounding the epicenter region within 180km of the M_S 5.7 Jiujiang-Ruichang earthquake sequence from 1970 to 2005

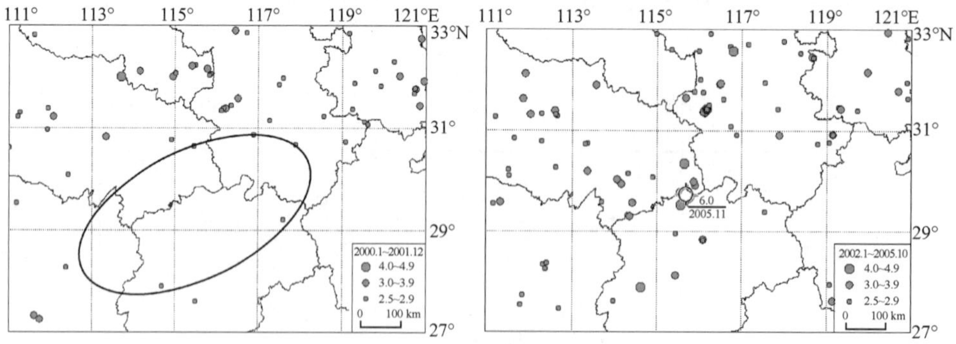

图 15　2000～2005 年 10 月地震平静—增强空间图像

Fig. 15　Space image of earthquake quiet-enhanced from 2000 to October 2005

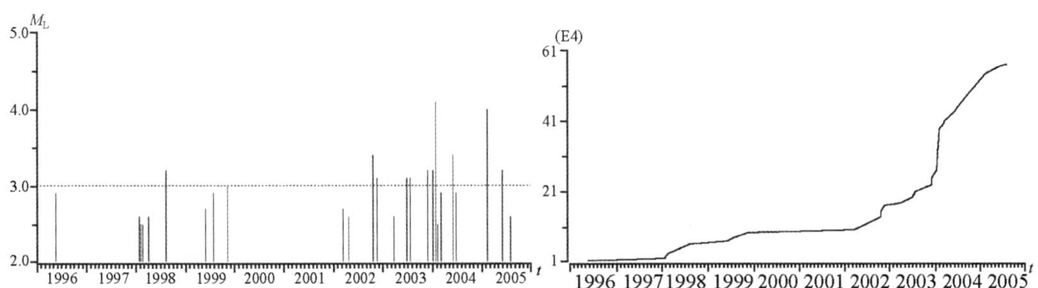

图 16　2000～2005 年 10 月地震平静—增强时间图像

Fig. 16　Time image of earthquake quiet-enhanced from 2000 to October 2005

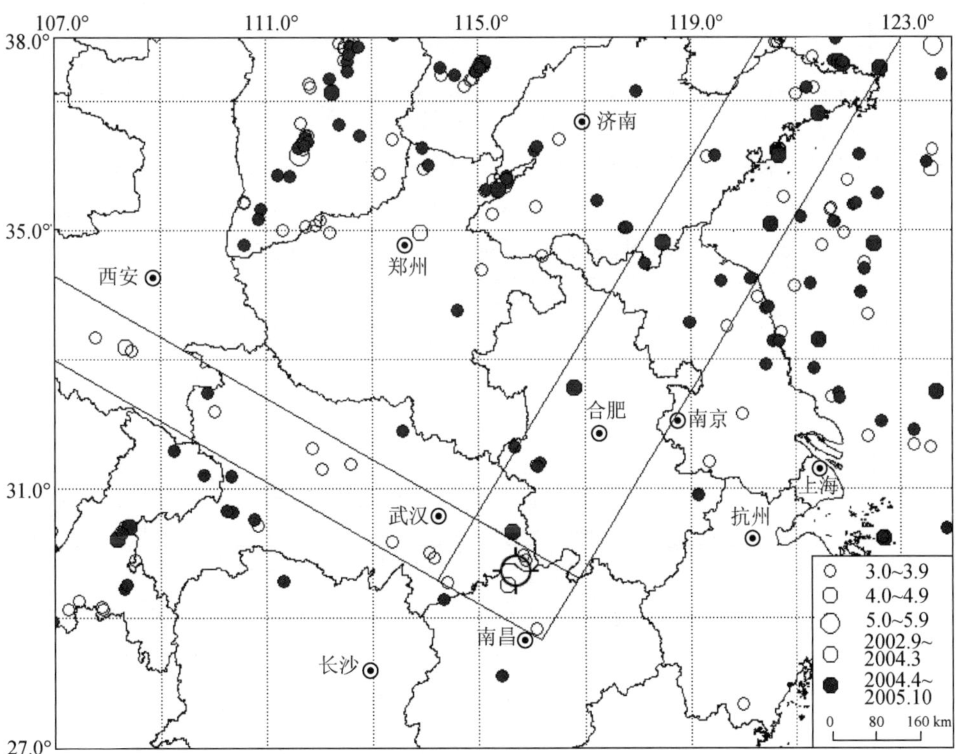

图 17　九江—瑞昌 5.7 级地震前中国东中部地区 $M_L \geq 3.0$ 级地震震中分布

Fig. 17　Distribution of $M_L \geq 3.0$ earthquakes in eastern and central regions of China before the M_S 5.7 Jiujiang-Ruichang earthquake

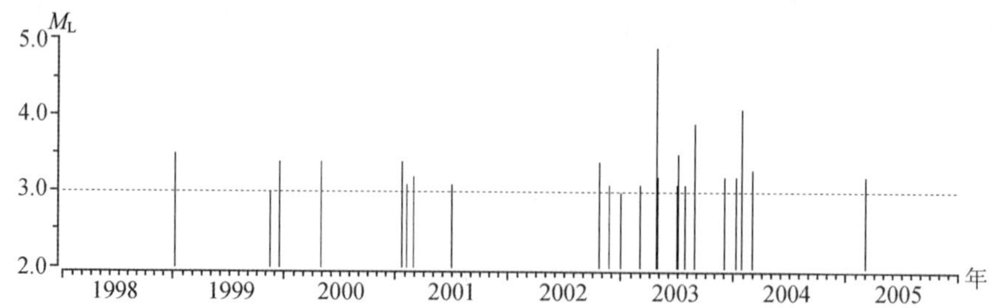

图 18　九江—瑞昌 5.7 级地震前北西向条带 1998～2005 年 10 月 M_L≥3.0 级地震 M-t 图

Fig. 18　M-t of north west M_L≥3.0 earthquakes strip before the M_S 5.7 Jiujiang-Ruichang earthquake sequence from 1998 to October 2005

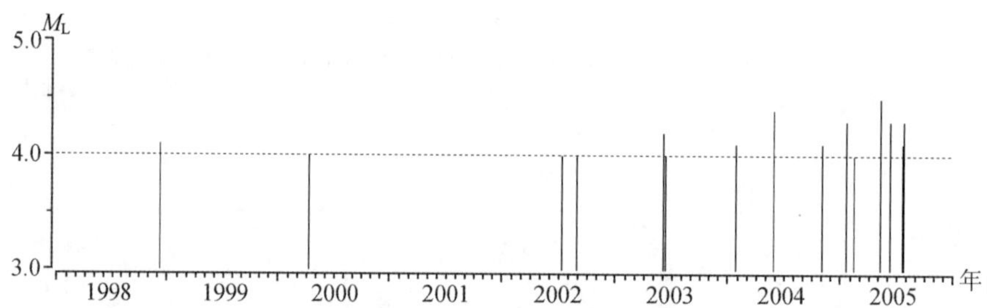

图 19　九江—瑞昌 5.7 级地震前北东向条带 1998～2005 10 月年 M_L≥4.0 级地震 M-t 图

Fig. 19　M-t of north east M_L≥4.0 earthquakes strip before the M_S 5.7 Jiujiang-Ruichang earthquake sequence from 1998 to October 2005

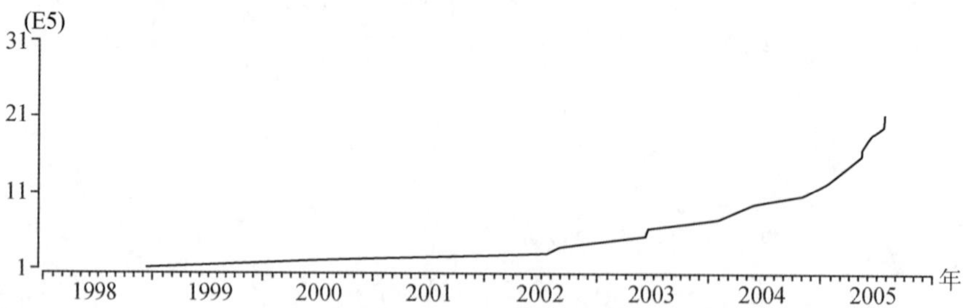

图 20　九江—瑞昌 5.7 级地震前北东向条带 1998～2005 年 10 月年应变释放曲线

Fig. 20　Strain release of north east strip before the M_S 5.7 Jiujiang-Ruichang earthquake sequence from 1998 to October 2005

2. 其他前兆观测项目异常

在 200km 范围内的 25 个前兆观测台项中，有 4 台项前兆观测项目（洞体应变、地倾斜、水位）震前存在不同程度的异常。

长期异常：九江台石英丝水平摆倾斜仪在 1995 年瑞昌 M_L4.9 地震后，两个分量都出现破年变异常，东西向运动方向也发生转折（图 21），异常持续到 2003 年 12 月，2004 年之后由于“十五”改造停止观测。

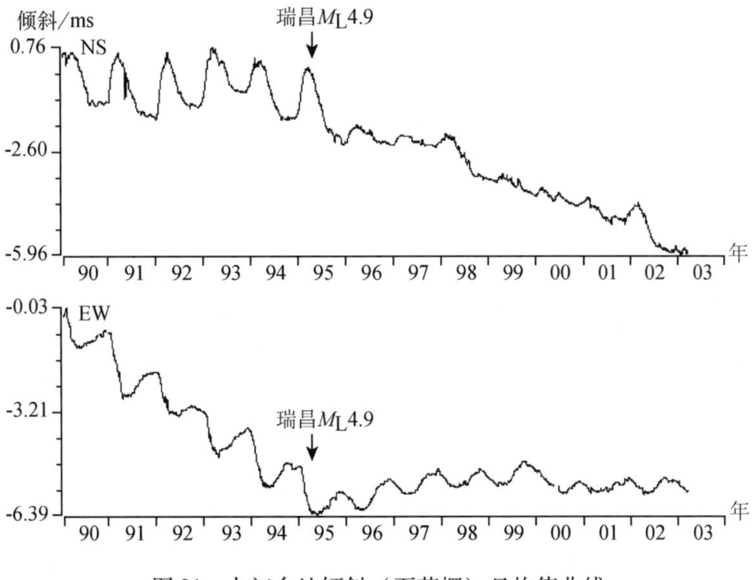

图 21　九江台地倾斜（石英摆）日均值曲线

Fig. 21　Curve of daily mean value of quartzose tilt at Jiujiang station

短期异常：黄梅台水管倾斜仪 EW、NS 向分别于 2005 年 9、10 月开始向东、北倾变，显示出短期异常。EW 向 11 月 7 日突然转向西倾 0.23″，恢复到正常时发震，显示出临震突变异常（图 22）。伸缩仪两分向从 2005 年 10 月 21 日起同时出现突发性的下降：NS、EW 分量下降量级分别达 5.3×10^{-7}、4.5×10^{-7}（图 23）。皖 33 井位水位自 2005 年 7 月 22 日开始大幅度上升，之后起伏变化，直至溢出井口（图 24）。

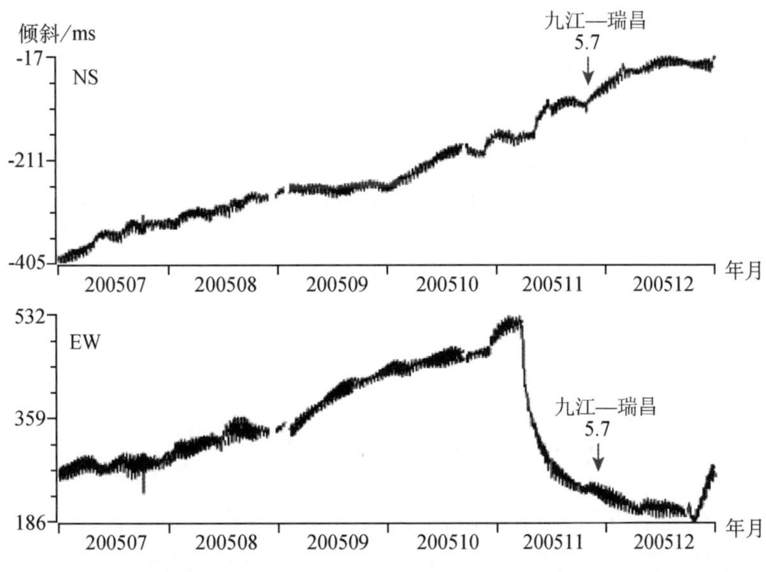

图 22　黄梅台地倾斜（水管）整点值曲线变化图

Fig. 22　Curve of the whole point of value of tube tilt at Huangmei station

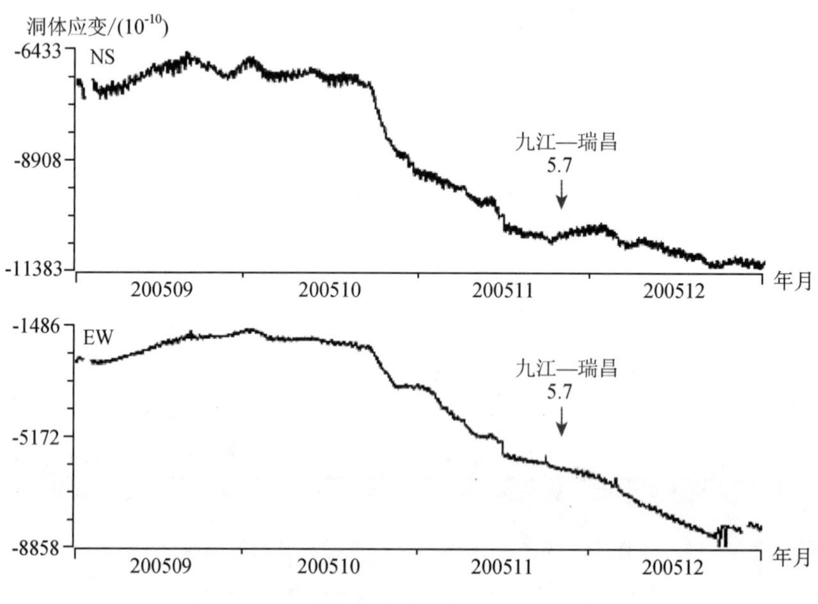

图 23　黄梅台洞体应变单分量整点值曲线变化图

Fig. 23　Curve of the whole point of value of cavity strain at Huangmei station

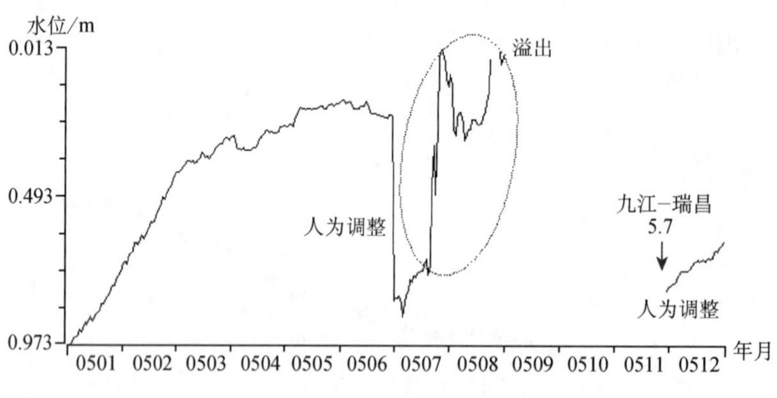

图 24　霍山皖 33 井水位日均值曲线变化图

Fig. 24　Curve of daily mean value of water temperature at Huoshanwan station

七、地震前兆异常特征分析

地震学异常：在震中 200 km 范围内，近 30 多年 M_L3 以上地震有成丛（成对）活动的特点，地震丛集持续时间较长可作为一种地震活动异常表现；震前 5 年地震学时空异常基本特征表现为地震平静（空区）→增强（空间集中）；震前 2～3 年表现出明显的 M_L3 以上地震相对活跃、空间集中成带和应变释放加速等异常；异常以中长期为主。

其他前兆观测资料异常：在 200km 范围内前兆台站主要分布在震中以北的地区，25 个台项仅 4 台项前兆观测资料出现异常，其中以地壳形变异常为主，地下流体异常有少量异常，其他学科未发现明显前兆异常。异常以短期为主。离震中最近的黄梅台个别地壳形变测项有异常显示，其中地倾斜在震前 20 天出现较为明显的临震突变异常，一定程度上反映出震中周围地区应力场的不稳定状态。宏观异常较少，震前 2 天在震中附近有少量动物的行为异常。

八、震前预测、预防和震后响应

1. 预测情况

在震前一周召开的江西省 2006 年度地震趋势会商会上，根据存在的地震活动性异常判定赣北及其邻区未来 1 ~3 年可能发生 5 级左右地震，但在年度和短临尺度未对此次地震作出有效预测。当时的预测依据与判定意见为：从 1981 年开始，3 级以上地震存在明显的成丛活动，目前所处的第 6 个 3 级以上地震丛活动已经接近 3 年，发生了江西德安 M_L4.1 和湖北蕲春 M_L4.0 两次显著地震，继续持续活动的可能性不大，2006 年度可能处在丛间间歇时段。前兆方面，目前我省中北部各台项观测资料分析未出现明显异常，综合分析认为 2006 年度赣北及其邻区发生 5 级以上地震的可能性不大。

2. 震后趋势判定

现场工作组与江西省地震局分析预报人员密切跟踪序列的发展，对地震类型和强余震趋势作出了准确的判定。江西省地震局在主震发生后 1 小时，作出了此次地震为主—余型地震的初判意见。现场工作组于 26 日晚间会商判定此次地震为主—余型地震，在原震区发生超过主震震级地震的可能性较小，但余震活动将持续一段时间。据此趋势判定，江西省地震应急指挥部确定了灾民安置等方面的应对决策。先后适时向社会发布了地震趋势判定意见和“震区 20km 以外的地区，除房屋严重受损外，露宿在外的居民可以搬到室内居住。”等安定民心的意见和公告，震后第三天全面复课，维护了社会稳定。江西省地震局对地震类型和强余震作出了准确的判定和预测。

3. 震后响应

地震发生后，江西省地震局立即启动省局地震应急预案，迅速派出多支现场工作队伍，并报请省政府启动《江西省地震应急预案》Ⅱ级响应。中国地震局也迅即派出了中国地震局应急工作组，协调组织了中国地震局搜救中心、工程力学研究所，以及江西、福建、湖北、山东、江苏、湖南、安徽、上海等省（市）地震局科技人员开展现场震情监测、地震分析预报、地震灾害损失评估和科学考察、建筑物安全鉴定、防震减灾知识宣传等工作。在灾区现场参加应急工作的地震系统工作人员达 128 人，发挥了地震系统统一领导、分级负责、协同配合、靠前指挥的应急工作优势。

九、结论与讨论

（1）赣北及其邻区近 20 多年来的地震活动具有明显的增强趋势、表现出时间成丛、重复的空间围空—集中和应变释放逐渐加速等有序性，反映了九江—瑞昌 5.7 级地震的孕震过程。九江—瑞昌 5.7 级地震发生前，存在区域性中长期异常背景，孕震区存在地震活动丛集—逐步增强、空间围空—集中、应变释放加速等中长期异常。九江—瑞昌 5.7 级地震前的地震丛集持续时间及其中等地震频度都要超过瑞昌 M_L4.9 地震，处于明显的加速释放阶段，具有一定的前兆意义。

（2）九江—瑞昌 5.7 级地震是近几年中国大陆东部震灾损失最重的一次地震。此次地震震害较重的原因，主要有以下几点：

一是灾区抗震设防标准低。此次地震震中处于瑞昌市城区边缘，老城区及城乡结合部位的房屋大多是老旧建筑，普遍缺乏抗震能力。农村地区的房屋结构不合理，砌筑质量差，普遍不设防。二是灾区地基地质条件差，软土、地表水系和水体发育（沿赛湖和长江两岸），广泛分布古湖泊等软土地基，一定程度上加重了地震灾害。三是江西省及其周边地区历史地震频率不高，震级不大，广大群众对地震的认知程度和警惕性不够。一些民众慌忙逃生，没有注意躲避塌落的墙体砖块，是造成人员死亡和重伤的直接原因。

（3）目前所有的资料尚不能提供有力的证据来解释和阐明本次地震的发震构造。根据地震学研究结果推测 5.7 级主震与瑞昌盆地内隐伏的 1 条北西向断裂活动有关，该隐伏断裂可能为广济—襄樊断裂往东南的延伸部分；4.8 级强余震的发震构造为 NE 向的丁家山—桂林桥—武宁断裂北段。

参 考 文 献

[1] 吕坚，倪四道，沈小七等. 2007. 九江—瑞昌地震的精确定位及其发震构造初探. 中国地震，23（2）：166～174

[2] 张培震，邓起东，张国民等. 2003. 中国大陆强震活动与活动地块. 中国科学（D 辑），33（4）：12～20

[3] 崔学军. 2002. 赣江断裂特征及其与郯庐断裂接合关系. 博士学位论文，1～82，武汉：中国地质大学

[4] 朱介寿，蔡学林，曹家敏等. 2005. 中国华南及东海地区岩石圈三维结构及演化. 北京：地质出版社

[5] 王椿镛，张先康，丁志峰等. 1997. 大别造山带上部地壳结构的有限差分层析成像. 地球物理学报，40（4）：495～502

[6] 王椿镛，张先康，陈步云等. 1997. 大别造山带地壳结构研究. 中国科学（D 辑），27（3）：221～226

[7] 张静华，李延兴，郭良迁等. 2005. 华南块体的现今构造运动与内部形变. 大地测量与地球动力学，25（3）：57～62

[8] 吕坚，郑勇，倪四道等. 2008. 2005 年 11 月 26 日九江—瑞昌 M_S5.7、M_S4.8 地震的震源机制解与发震构造研究. 地球物理学报，51（1）：158～164

[9] 曾文敬，赵爱平，汤兰荣等. 2009. 九江—瑞昌 5.7 级地震余震震源机制解. 大地测量与地球动力学，29（4）：42～47

参 考 资 料

1）江西省地震局，2005 年 11 月 26 日九江—瑞昌 5.7 级地震灾害损失评估报告，2005. 12. 7

2）江西省地震局，九江—瑞昌 5.7 级地震现场科学考察报告，2005. 12

3）地质部江西省地质局区域地质测量队，中华人民共和国地质图（1:200000 瑞昌幅），1966

4）地矿部江西地质矿产勘查开发局赣西北大队区调分队，中华人民共和国地质图说明书（1:50000 万，瑞

昌县幅），1987

5）江西省地质矿产局，中华人民共和国地质图（1∶50000 瑞昌县幅），1987

6）江西省地震局，2006 年度江西地震趋势研究报告，2005. 11

The Jiujiang-Ruichang M_S 5.7 Earthquake on November 26, 2005 in Jiangxi Province

Abstract

An earthquake of M_S 5.7 occurred at the border area of Jiujiang and Ruichang in Jiangxi provine, The macroscopic epicenter was near gangkou of Jiujiang. The intensity at the epicenter was Ⅶ, and The extreme seismic District was the NWW elliptical, The earthquake caused about 13 people dead, 82 people Seriously injured, 683 peolle injured, and the direct economic loss was 2038 million yuan.

The earthquake sequence was of the mainshock-aftershock type. The largest aftershock was M_S 4.8. The aftershocks advantage of the long axis toward the northwest, the trend is not consistent with the long axis of the meizoseismal area, and Nodal plane I was main rupture surface, the principal compressive stress P axis azimuth was NWW direction. The 5.7 earthquake seismogenic structure Had a relationship with Guangji-Xiangfan fracture, and the 4.8 earthquake seismogenic structure had a relationship with the NE to Dingjiashan-Guilin ridge the-wuning fracture.

Within the distance of 200km from the epicenter there were 13 seismic stations, 11 of them were comprehensive monitoring stations, 9 of them were seismometric station. Before this event there appeared 10 anomalies in 5 items, in which 4 anomalies from precursory observations, 1 anomalie from Macroscipic.

报告附件

附表1　固定前兆观测台（点）与观测项目汇总表

序号	台站（点）名称	经纬度（°）		测项	资料类别	震中距 Δ/km	备注
		φ_N	λ_E				
1	九江	29.65	116.10	测震		38	地倾斜2003年3月停记，水氡2003年12月停记
				地倾斜（摆式）	Ⅰ		
				水氡	Ⅱ		
2	黄梅	30.13	115.90	测震		49	
				地倾斜（连通管）	Ⅱ		
				洞体应变	Ⅱ		
				地倾斜（摆式）	Ⅱ		
3	南昌	28.76	115.80	测震		106	
				地倾斜（连通管）	Ⅱ		
				洞体应变	Ⅱ		
				水位	Ⅱ		
				地热	Ⅰ		
				水氡	Ⅱ		
4	修水	29.04	114.57	测震		135	
5	南川	29.41	114.25	测震		146	
6	武汉	30.51	114.51	测震		146	
				地磁 F	Ⅰ		
				地磁 H	Ⅰ		
				地磁 Z	Ⅰ		
				大地电场	Ⅱ		
				重力	Ⅱ		
7	安庆	30.58	117.02	测震		158	
				地倾斜（连通管）	Ⅱ		
				地电阻率	Ⅰ		
				自然电位	Ⅱ		
8	安庆皖23井	30.58	117.02	测震		161	
				水位	Ⅱ		

续表

序号	台站（点）名称	经纬度（°）		测项	资料类别	震中距 Δ/km	备注
		φ_N	λ_E				
9	麻城	31.13	115.16	测震		165	因“十五”改造影响2005年9月前兆仪器全部停止观测
				地倾斜（连通管）	Ⅰ		
				地倾斜（摆式）	Ⅱ		
				洞体应变	Ⅰ		
10	佛子岭	31.35	116.27	测震		188	
				地倾斜（连通管）	Ⅰ		
				地倾斜（摆式）	Ⅰ		
11	霍山皖33井	31.40	116.10	水位	Ⅱ	190	
12	豹子崖	31.40	116.22	测震		192	
13	石家河	31.44	116.09	测震		194	

分类统计	0<Δ≤100km	100km<Δ≤200km		总数
测项数 N	4	12		16
台项数 n	7	30		37
测震单项台数 a	0	4		4
形变单项台数 b	0	0		0
电磁单项台数 c	0	0		0
流体单项台数 d	0	2		2
综合台站数 e	2	5		7
综合台中有测震项目的台站数 f	2	5		7
测震台总数 $a+f$	2	9		11
台站总数 $a+b+c+d+e$	2	11		13
备注				

附表 2　测震以外固定前兆观测项目与异常统计表

序号	台站（点）名称	测项	资料类别	震中距 Δ/km	按震中距 Δ 范围进行异常统计									
					0 < Δ≤100km					100 < Δ≤200km				
					L	M	S	I	U	L	M	S	I	U
1	九江	地倾斜（摆式）	Ⅰ	38	√	—	—	—						
		水氡	Ⅱ											
2	黄梅	地倾斜（连通管）	Ⅱ	49	—	—	√	—						
		洞体应变	Ⅱ		—	—	√	—						
		地倾斜（摆式）	Ⅱ		—	—	—	—						
3	南昌	地倾斜（连通管）	Ⅱ	106						—	—	—	—	
		洞体应变	Ⅱ							—	—	—	—	
		水位	Ⅱ							—	—	—	—	
		地热	Ⅰ							—	√	—	—	
		水氡	Ⅱ							—	—	—	—	
4	武汉	地磁 F	Ⅰ	146						—	—	—	—	
		地磁 H	Ⅰ							—	—	—	—	
		地磁 Z	Ⅰ							—	—	—	—	
		大地电场	Ⅱ							—	—	—	—	
		重力	Ⅱ							—	—	—	—	
5	安庆	地倾斜（连通管）	Ⅱ	158						—	—	—	—	
		地电阻率	Ⅰ							—	—	—	—	
		自然电位	Ⅱ							—	—	—	—	
6	安庆皖 23 井	水位	Ⅱ	161										
7	麻城	地倾斜（连通管）	Ⅱ	165						—	—	—	—	
		地倾斜（摆式）	Ⅱ							—	—	—	—	
		洞体应变	Ⅰ							—	—	—	—	
8	佛子岭	地倾斜（连通管）	Ⅰ	178						—	—	—	—	
		地倾斜（摆式）	Ⅰ							—	—	—	—	
9	霍山皖 33 井	水位	Ⅱ	190						—	—	√	—	

续表

序号	台站（点）名称	测项	资料类别	震中距Δ/km	按震中距Δ范围进行异常统计									
					0 < Δ≤100km					100 < Δ≤200km				
					L	M	S	I	U	L	M	S	I	U
分类统计	台项	异常台项数			1	0	2	0		0	1	1	0	
		台项总数			4	4	4	4	/	20	20	20	20	/
		异常台项百分比/%			25	0	50	0	/	0	5	5	0	/
	观测台站（点）	异常台站数			1	1	1	0		0	1	2	0	
		台站总数			2	2	2	2	/	7	7	7	7	/
		异常台站百分比/%			50	50	50	0	/	0	14	29	0	/
	测项总数（16）				4					12				
	观测台站总数（9）				2					7				

2006 年 3 月 31 日吉林省前郭尔罗斯—乾安 5.0 级地震

吉林省地震局

盘晓东　李　克

摘　要

2006 年 3 月 31 日在吉林省前郭尔罗斯—乾安交界处发生 5.0 级地震。宏观震中位于松原市前郭尔罗斯县查干花镇老营吐、腰英吐村一带，震中烈度达Ⅵ⁺度，极震区长轴呈北西方向展布。这次地震发生在东北断块区松辽断陷带中央坳陷区内，地震造成 2 人轻伤，直接经济损失 1.1 亿元。

本次地震的余震活动比较丰富，从 3 月 31 日至 4 月 14 日止，共记录余震序列 112 次，其中 0 ~ 0.9 级 70 次，1 ~ 1.9 级 27 次，2 ~ 2.9 级 10 次，3 ~ 3.9 级 5 次。余震分布初期为 NW 方向，后期为 NS 方向。最大地震为 4 月 8 日 01 时 26 分的 3.7 级地震。震源机制解的节面Ⅰ走向 51°，P 轴走向 10°，主破裂方向为北西向，可能与北西向查干泡—道字井断裂有关。

2006 年初召开的全省地震会商会上提出了明确的中期预报意见，对应较好。认为 2006 年吉林省存在发生 $M_S \geqslant 5.0$ 级地震的可能，重点监视区为第二松花江断裂的松原及邻近地区。在震中 200 km 以内有 13 个地震台，100 km 内有 2 个地震台。5.0 级地震前出现 9 项测震学异常和前兆异常。震后趋势判定正确，取得了一定的社会效益。

前　言

2006 年 3 月 31 日 20 时 23 分，吉林省前郭尔罗斯—乾安（以下简称前郭—乾安）交界处发生 5.0 级地震。吉林省地震台网测定的微观震中为 44°36′N、124°03′E，宏观震中为 44°40′N、124°07′E，震中烈度达Ⅵ⁺度。这次地震有感范围较大，地震震感波及全省大部分地区、辽宁省和黑龙江省部分地区也有一定程度的震感。地震造成 2 人轻伤，直接经济损失 1.1 亿元。

5.0 级地震发生在松辽断陷带中央坳陷区内的中小地震活动区，在震中东北约 70km 处，曾发生过 1119 年前郭尔罗斯 6¾级地震。震中附近地震观测台站较少，200km 范围内共有地震台 9 个。震前出现 6 项前兆手段异常。

前郭—乾安 5.0 级地震发生在区域地震活动相对平静的背景下，并发生在中期预测和年度确定的地震危险区内，且在 2006 年初召开的全省地震会商会上提出了明确的中期预报意见，向吉林省政府提出了本省的年度中期预测意见。国务院，吉林省委、省政府高度重视，吉林省抗震救灾指挥部立即成立地震应急工作领导小组，按吉林省地震应急预案开展应急指挥工作，吉林省地震局也成立了地震应急工作队与中国地震局地震应急工作组一起成立前方地震应急指挥部，开展地震应急指挥及地震灾害损失评估等工作。

本报告是在整理近些年来的测震及前兆资料基础上，加以分析研究而完成的。

一、测震台网及地震基本参数

1. 测震台网

图 1 给出了震中 200 km 范围内的测震台站分布，所有 6 个地震台，在研究时段内基本可达到 $M_S \geqslant 2.0$ 级地震不遗漏。100 km 内仅有松原台；100 ~ 200 km 省内分别有净月台、四平台、白城台、榆树台和阿古拉台等四个测震台。

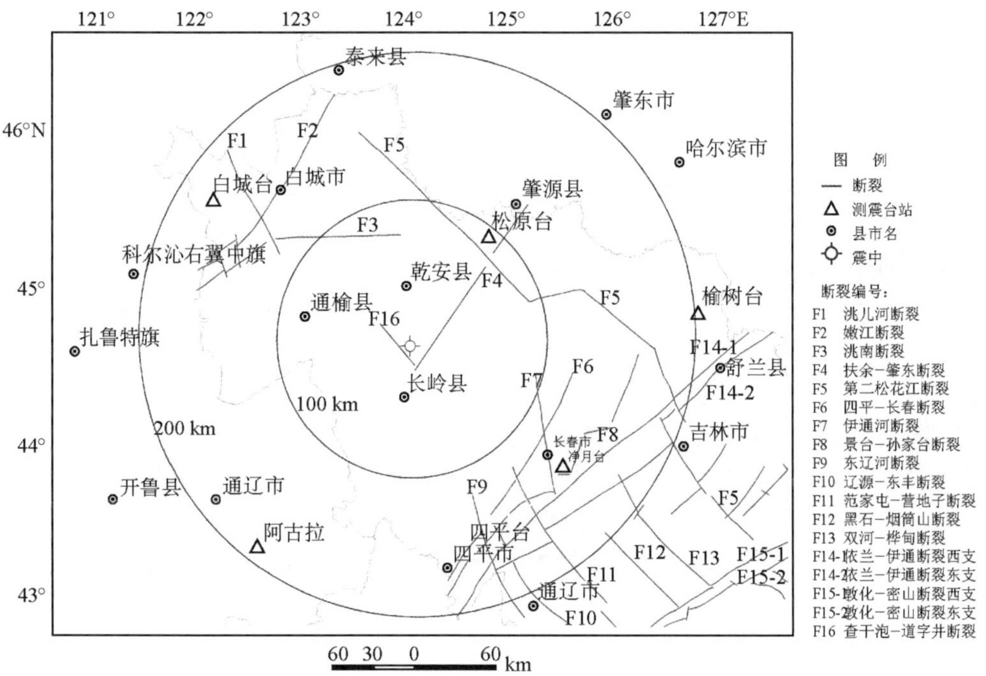

图 1 吉林省前郭—乾安 5.0 级地震前震中附近地震台站分布图（截至 2006 年 12 月）

Fig. 1 Distribution of earthquake-monitoring stations around the epicentral area before the *M*5.0 Qianguo-Qian'an Earthquake

2. 地震基本参数

吉林省地震台网测定结果为：

时间：2006 年 03 月 31 日 20 时 23 分

地点：吉林省省前郭县查干花镇（北纬 44.60°，东经 124.05°）

震级：M 5.0

地震基本参数见表 1。表 1 列出了吉林省地震台网、国家地震台网和美国地质调查局给出的这次地震基本参数，经对比分析，认为吉林省地震台网测定的结果更为精确，因此，此次地震基本参数取表 1 中编号 1 结果。

表 1　前郭—乾安 *M*5.0 地震基本参数

Table 1　Basic Parameters of The *M*5.0 Qianguo-Qian'an Earthquake

编号	发震日期	发震时刻	震中位置		震级 M	震源深度 (km)	震中地名	结果来源
	年 月 日	时 分 秒	φ_N	λ_E				
1	2006 3 11	20 23 14.0	44°36′	124°03′	5.0	10	前郭	吉林省地震台网
2	2006 3 11	20 23 15.7	44°42′	124°06′	5.0	15	前郭	国家地震台网
3	2006 3 11	20 23 17.9	44°37′	124°07′	4.9	10	前郭	美国地调查局

二、地震地质背景

本次地震发生在东北断块区松辽断陷带中央坳陷区内。地貌上为广阔的平原区。松嫩中断（坳）陷带是一个大型的中、新生代内陆断（坳）陷盆地。其附近地区主要断裂分布及历史地震分布见图 2。

坳陷区的边界断裂发育，西有嫩江断裂，东界为依兰—伊通断裂，南界有赤峰—开原断裂，北界有讷莫尔河断裂。基底由古生代变质岩系组成，中、新生代形成大型断陷—拗陷型盆地，两翼不对称，西翼陡、东翼缓，沉降幅度西深东浅。区内的松辽断块拗陷可进一步分成：北部拗陷、南部拗陷和松辽分水岭隆起。

震中区断（坳）陷盆地基底之上的中生代沉积层厚约 4000m。上部第四系沉积层在震中区厚度约 80m。深部地质构造单元位于松辽平原上地幔隆起区北北东向中央隆起带的东部斜坡带上。地震位于扶余—肇东北东向断裂和查干泡—道字井北西向断裂的交汇部位附近。

区域范围内有 NE、NW 等十几条断裂构造，但与此次地震有关的地震构造主要有两条断裂，即扶余—肇东断裂和查干泡—道字井北西向断裂。

扶余—肇东断裂：这是一条位于松辽盆地中部的北东向断裂。断裂北起肇东，经扶余至怀德杨大城子一带，为松辽断块沉降带内次一级构造单元的分界线，西侧为中央凹陷带，东侧为东南隆起区。断裂大致位于基底等深线的陡变带上，控制了西侧晚白垩世、第三纪沉积物的分布。根据中石化地震勘探反射资料，肇东—扶余断裂对基底埋深有明显的控制作用，断裂西侧为坳陷区，东侧为相对隆起区，断裂大致位于基底等深线的陡变带上。该断裂使其两侧地貌形态明显不同，沿第二松花江南展布的 NW 向地貌陡坎，与沿断裂展布的 NE 向套呼太—哈达山陡坎汇于哈达山。断层西北侧出现 NE 向谷地，白垩系基岩不再出露，落差数十米。断裂在哈达山陡崖白垩系中有断层出露，走向北东 45°，近直立，取断层泥，采用热释光法测年，年龄为 57.20 万 ±4.40 万年，即最新活动时代为中更新世早期。肇东附近，断裂西北盘基底埋深最厚达 5600m，而东南盘最厚为 4600m，两盘相差 1000m，反映断裂前第四纪活动明显。该断裂在近场区内呈隐伏状，未见任何断错地貌，没有发现晚更新世断错地表活动的迹象，在卫星影像上断裂迹象不明显。第四系等厚线展布方向与断裂走向也不太一致，至少说明晚更新世以来断裂活动已不明显，应属于早、中更新世活动断裂。该断裂与北西向第二松花江断裂带交汇处发生 1119 年前郭 6¾级地震，在松原附近 2003 年小震活动频繁。

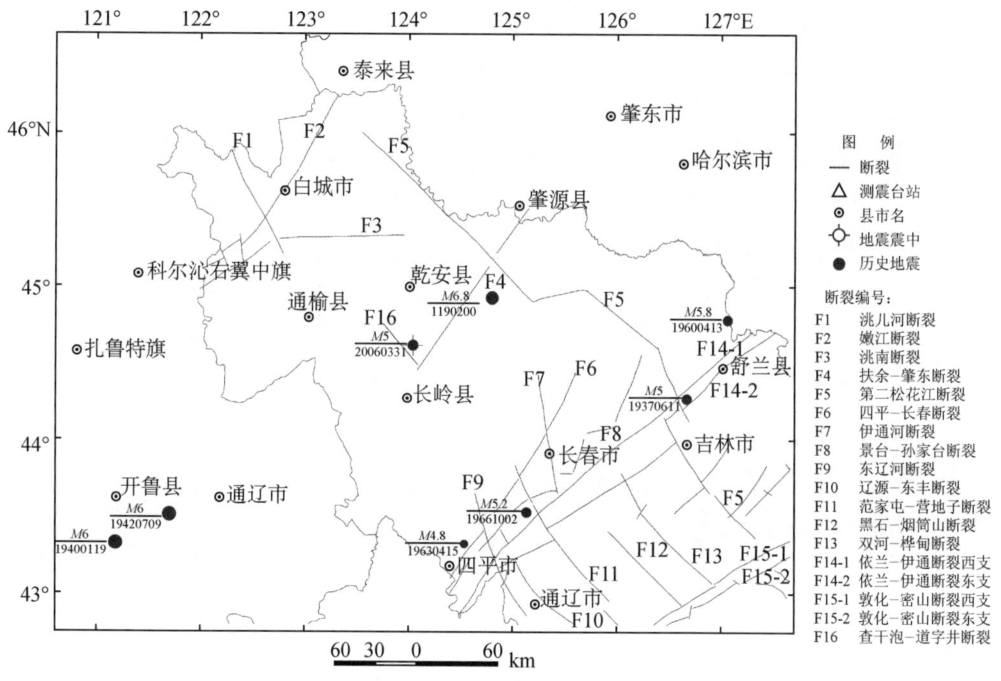

图 2　研究区域主要断裂及历史地震震中分布图

Fig. 2　Major faults and distribution of historical earthquakes around research area

查干泡—道字井北西向断裂：该断裂位走向 NW，倾向 SW，属于正断层。该断层属于基底断裂。根据中石化公司地震反射剖面资料，该断层断错了 T_2 和 T_1 反射层，消失在了新近系中。该断裂在白垩纪有过活动，到白垩纪末开始衰亡。根据野外调查该断裂在地貌上也没有任何显示。综上，该断裂应为前第四纪断裂。

三、地震影响场和震害

据现场实地考察及调查资料，本次地震宏观震中为 44°40′N、124°07′E，震中烈度达Ⅵ⁺度。等震线呈北西向近椭圆形分布（图 3）。

本次地震宏观震中位于松原市前郭县查干花镇老营吐、腰英吐村一带。宏观震中与测定震中位置基本一致，震害波及乾安县安子镇部分村屯。南北长约50多公里，东西宽约30多公里的区域，受到不同程度的破坏，范围约1565km^2。调查中发现，震中区房屋质量差别很大。震中区查干花镇老营吐、腰英吐村一带的房屋多为土房或土木结构平房，多数房屋抗震性能较差；少部分房屋为单层砖房。

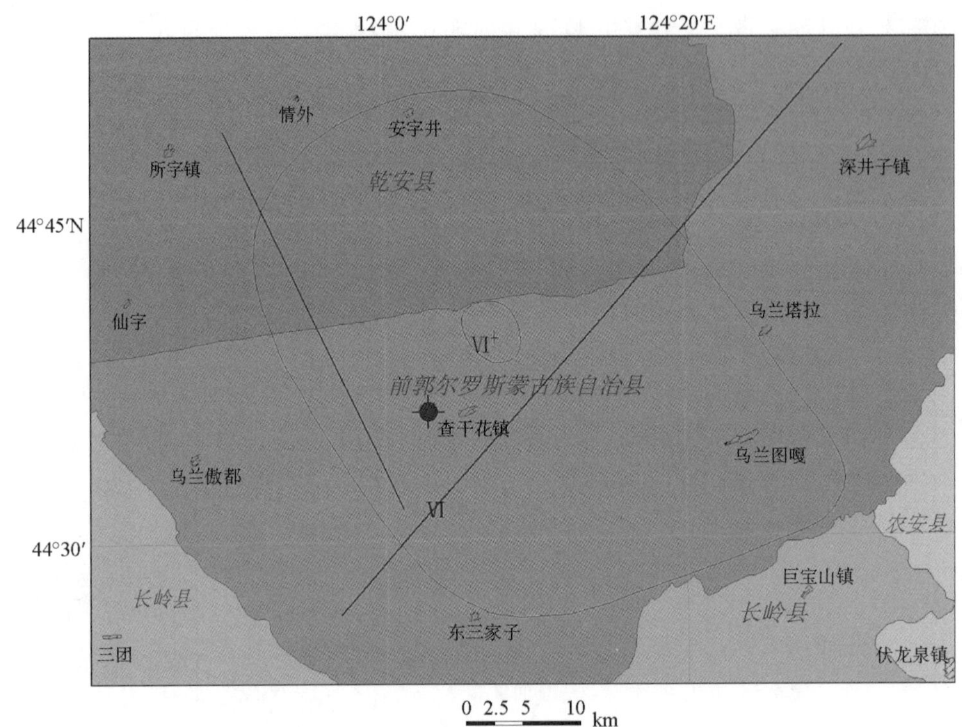

图3　吉林省前郭—乾安5.0级地震等烈度线图

Fig. 3　Isoseismal map of the *M*5.0 Qianguo-Qian'an earthquake

Ⅵ度强区：极震区内以查干花镇老营吐、腰英吐村房屋损坏最为严重，腰英吐小学承重纵强横向发生宽约5cm裂缝，屋顶烟囱倒落，另有多处砖瓦掉落；另有个别村民房屋门、窗洞口发生较大裂缝，墙体连接处出现2～5cm裂缝。地震未造成人畜死亡，2人受轻伤，本区范围如图3所示。

Ⅵ度区：区内居民普遍强烈有感，门窗响声较大，并拌有玻璃损坏。多处土房和砖房门窗洞口、墙体连接处出现新鲜2～5mm裂缝，多为原有裂缝加剧现象。大体范围为北起安子镇南至东三家子、乌兰图嘎以南地区；东起乌兰塔拉西至后蒙户屯、董家窝堡一带。

灾害特点：

(1) 震中区破坏较多和较严重的是砂土泥浆砌筑的砖木平房。这些砖木结构房屋墙体砌砖黏合力差，基本无抗震措施，震害较严重。但破坏范围不大。

(2) 地震有感范围大，黑龙江的大庆市、哈尔滨市，内蒙古通辽市、呼和浩特市，辽宁彰武等地也有不同程度震感。

极震区地震烈度达Ⅵ度强。地面未产生地震地质灾害，仅在腰英吐村西侧取土坑内见原有温度变化产生的裂隙加宽的现象。地震破坏分布范围呈为椭园形，长轴方向呈北西向。结合震中区的区域地质构造特征、主要断裂走向与震源机制解结果，对发震构造的分析认为，本次地震位于查干泡—道字井北西向断裂与北东向扶余—肇东断裂的交汇部位，可能主要受查干泡—道字井北西向断裂影响。

本次地震波及范围较大，前郭、乾安两县的安字镇、查干花镇等9乡（镇）5场受灾较重，受灾面积约1565km^2。受影响的人口较多，据了解波及较重的村庄约200个，受影响的人口约为30万人，直接经济损失为1.1亿元。

四、地震序列

本次地震的余震活动比较丰富，从3月31日至4月14日止，共记录余震序列112次，见表2。其中0~0.9级70次，1~1.9级27次，2~2.9级10次，3~3.9级5次。余震分布初期为NW方向，后期为NS方向。2.5级以上余震目录见表3，最大地震为4月8日01时26分的3.7级地震。余震分布近似为南北方向（图4）。余震的时间序列呈起伏地衰减，丛集间时间间隔逐渐增大，4月5、8日分别有一个小的丛集过程。根据现有的地震序列资料，计算了U值、K值、F值、h值、b值、p值等相关的震群判定参数，并提出了后续震情的判定意见。除余震序列中的h值（$h<0.6$）较小、b值（0.45）较低外，其他结果显示序列为主震—余震型（图5、图6）。

表2 前郭—乾安5.0级地震序列目录

Table 2 Catalogue of seismic sequence in the *M*5.0 Qianguo-Qian'an earthquake

序号	发震日期	发震时刻	震中位置		震级	震源深度	震中	结果来源
	年 月 日	时 分 秒	φ_N	λ_E	M_L	(km)	地名	
1	2006 03 31	20 23 14.0	44°36′	124°3′	5.2	10	前郭	吉林省地震局
2	2006 03 31	21 29 40.0	44°36′	124°0′	2.4		前郭	吉林省地震局
3	2006 03 31	23 07 49.0	44°36′	124°3′	2.3		前郭	吉林省地震局
4	2006 04 01	10 53 30.0	44°36′	124°6′	3.2		前郭	吉林省地震局
5	2006 04 01	20 00 00.0	44°40′	124°7′	1.3		前郭	吉林省地震局
6	2006 04 01	22 12 20.0	44°39′	124°6′	1.9		前郭	吉林省地震局
7	2006 04 01	04 15 40.0	44°40′	124°6′	1.7		前郭	吉林省地震局
8	2006 04 01	05 05 00.0	44°40′	124°7′	1		前郭	吉林省地震局
9	2006 04 01	06 05 00.0	44°40′	124°7′	0.8		前郭	吉林省地震局
10	2006 04 01	07 51 42.0	44°43′	124°6′	1.8		前郭	吉林省地震局

续表

序号	发震日期	发震时刻	震中位置		震级 M_L	震源深度（km）	震中地名	结果来源
	年 月 日	时 分 秒	φ_N	λ_E				
11	2006 04 01	09 51 12.0	44°43′	124°6′	1.6		前郭	吉林省地震局
12	2006 04 01	12 32 00.0	44°43′	124°7′	0.6		前郭	吉林省地震局
13	2006 04 01	12 46 00.0	44°40′	124°7′	0.5		前郭	吉林省地震局
14	2006 04 01	14 35 00.0	44°40′	124°7′	0.4		前郭	吉林省地震局
15	2006 04 01	20 24 20.0	44°36′	124°0′	2.3		前郭	吉林省地震局
16	2006 04 01	21 35 00.0	44°40′	124°7′	0.3		前郭	吉林省地震局
17	2006 04 01	22 13 00.0	44°40′	124°8′	1.6		前郭	吉林省地震局
18	2006 04 01	22 20 00.0	44°40′	124°8′	0.6		前郭	吉林省地震局
19	2006 04 01	22 31 00.0	44°41′	124°8′	1.2		前郭	吉林省地震局
20	2006 04 02	01 32 52.0	44°40′	124°9′	1.4		前郭	吉林省地震局
21	2006 04 02	03 50 00.0	44°41′	124°8′	0.7		前郭	吉林省地震局
22	2006 04 02	04 00 30.0	44°40′	124°7′	1.6		前郭	吉林省地震局
23	2006 04 02	04 03 00.0	44°41′	124°8′	0.9		前郭	吉林省地震局
24	2006 04 02	04 04 27.0	44°39′	124°7′	1.5		前郭	吉林省地震局
25	2006 04 02	04 52 00.0	44°41′	124°8′	0		前郭	吉林省地震局
26	2006 04 02	06 52 00.0	44°41′	124°8′	0.1		前郭	吉林省地震局
27	2006 04 02	08 00 39.0	44°40′	124°7′	2		前郭	吉林省地震局
28	2006 04 02	09 28 18.0	44°40′	124°7′	1.6		前郭	吉林省地震局
29	2006 04 02	11 48 28.0	44°40′	124°9′	1.9		前郭	吉林省地震局
30	2006 04 02	18 12 40.0	44°36′	124°0′	3		前郭	吉林省地震局
31	2006 04 02	20 03 44.0	44°39′	124°7′	2.6		前郭	吉林省地震局
32	2006 04 02	20 04 17.0	44°36′	124°3′	3.3		前郭	吉林省地震局
33	2006 04 02	20 24 42.0	44°40′	124°7′	1.6		前郭	吉林省地震局
34	2006 04 02	20 37 00.0	44°40′	124°8′	1.6		前郭	吉林省地震局
35	2006 04 02	21 57 00.0	44°41′	124°8′	1.5		前郭	吉林省地震局
36	2006 04 03	02 00 00.0	44°41′	124°8′	1.4		前郭	吉林省地震局
37	2006 04 03	02 10 00.0	44°41′	124°8′	1		前郭	吉林省地震局
38	2006 04 03	04 42 00.0	44°40′	124°7′	1.2		前郭	吉林省地震局
39	2006 04 03	09 14 49.0	44°55′	124°20′	2.2		前郭	吉林省地震局
40	2006 04 03	14 13 00.0	44°43′	124°7′	1.3		前郭	吉林省地震局

续表

序号	发震日期	发震时刻	震中位置		震级	震源深度	震中	结果来源
	年 月 日	时 分 秒	φ_N	λ_E	M_L	(km)	地名	
41	2006 04 03	22 34 00.0	44°40′	124°7′	0.6		前郭	吉林省地震局
42	2006 04 03	23 44 00.0	44°42′	124°7′	0.3		前郭	吉林省地震局
43	2006 04 04	02 06 00.0	44°39′	124°7′	0.5		前郭	吉林省地震局
44	2006 04 04	10 60 00.0	44°52′	124°7′	0.5		前郭	吉林省地震局
45	2006 04 04	11 30 00.0	44°40′	124°7′	0.1		前郭	吉林省地震局
46	2006 04 04	13 04 10.0	44°40′	124°7′	0.5		前郭	吉林省地震局
47	2006 04 04	13 04 30.0	44°38′	124°7′	0.3		前郭	吉林省地震局
48	2006 04 04	13 33 00.0	44°38′	124°7′	0.6		前郭	吉林省地震局
49	2006 04 04	14 04 00.0	44°40′	124°7′	0.7		前郭	吉林省地震局
50	2006 04 04	14 26 00.0	44°40′	124°7′	0.9		前郭	吉林省地震局
51	2006 04 04	22 24 00.0	44°40′	124°7′	0.1		前郭	吉林省地震局
52	2006 04 05	00 03 00.0	44°41′	124°7′	0.4		前郭	吉林省地震局
53	2006 04 05	00 23 00.0	44°42′	124°7′	0		前郭	吉林省地震局
54	2006 04 05	01 19 00.0	44°42′	124°7′	0.1		前郭	吉林省地震局
55	2006 04 05	01 43 00.0	44°40′	124°7′	0.9		前郭	吉林省地震局
56	2006 04 05	02 03 00.0	44°40′	124°7′	0.9		前郭	吉林省地震局
57	2006 04 05	02 11 00.0	44°43′	124°7′	0.5		前郭	吉林省地震局
58	2006 04 05	03 48 00.0	44°43′	124°7′	0.3		前郭	吉林省地震局
59	2006 04 05	03 48 00.0	44°40′	124°7′	1		前郭	吉林省地震局
60	2006 04 05	04 14 00.0	44°39′	124°7′	0.6		前郭	吉林省地震局
61	2006 04 05	04 18 00.0	44°42′	124°7′	0.2		前郭	吉林省地震局
62	2006 04 05	15 29 00.0	44°39′	124°7′	1.4		前郭	吉林省地震局
63	2006 04 05	15 53 00.0	44°39′	124°7′	0.4		前郭	吉林省地震局
64	2006 04 05	20 31 00.0	44°41′	124°7′	2		前郭	吉林省地震局
65	2006 04 05	20 32 00.0	44°41′	124°7′	1		前郭	吉林省地震局
66	2006 04 05	40 42 00.0	44°42′	124°7′	0		前郭	吉林省地震局
67	2006 04 05	21 25 00.0	44°44′	124°7′	0.2		前郭	吉林省地震局
68	2006 04 05	22 03 00.0	44°42′	124°7′	0.2		前郭	吉林省地震局
69	2006 04 05	22 09 00.0	44°42′	124°7′	0.2		前郭	吉林省地震局
70	2006 04 05	22 16 00.0	44°42′	124°7′	0.1		前郭	吉林省地震局

续表

序号	发震日期	发震时刻	震中位置		震级 M_L	震源深度 (km)	震中地名	结果来源
	年 月 日	时 分 秒	φ_N	λ_E				
71	2006 04 05	22 16 00.0	44°39′	124°7′	0.6		前郭	吉林省地震局
72	2006 04 05	22 46 00.0	44°43′	124°7′	1.3		前郭	吉林省地震局
73	2006 04 06	00 54 00.0	44°42′	124°7′	0.1		前郭	吉林省地震局
74	2006 04 06	03 29 00.0	44°41′	124°7′	0.2		前郭	吉林省地震局
75	2006 04 06	06 06 42.0	44°42′	124°7′	0.3		前郭	吉林省地震局
76	2006 04 06	13 21 00.0	44°42′	124°7′	0		前郭	吉林省地震局
77	2006 04 06	20 21 00.0	44°43′	124°7′	0		前郭	吉林省地震局
78	2006 04 06	20 22 00.0	44°42′	124°7′	0		前郭	吉林省地震局
79	2006 04 06	20 29 00.0	44°39′	124°7′	0.8		前郭	吉林省地震局
80	2006 04 06	20 32 00.0	44°40′	124°7′	0.3		前郭	吉林省地震局
81	2006 04 06	20 55 00.0	44°39′	124°7′	0.8		前郭	吉林省地震局
82	2006 04 06	20 56 00.0	44°39′	124°7′	0.8		前郭	吉林省地震局
83	2006 04 06	21 49 00.0	44°39′	124°7′	0.8		前郭	吉林省地震局
84	2006 04 06	22 54 00.0	44°40′	124°7′	0.1		前郭	吉林省地震局
85	2006 04 06	23 31 00.0	44°39′	124°7′	0.5		前郭	吉林省地震局
86	2006 04 06	23 33 00.0	44°39′	124°7′	0.5		前郭	吉林省地震局
87	2006 04 07	02 40 00.0	44°40′	124°8′	0		前郭	吉林省地震局
88	2006 04 07	16 06 00.0	44°39′	124°7′	0.7		前郭	吉林省地震局
89	2006 04 07	21 10 01.0	44°42′	124°8′	0		前郭	吉林省地震局
90	2006 04 07	21 39 41.0	44°43′	124°5′	0.4		前郭	吉林省地震局
91	2006 04 08	01 26 00.0	44°39′	124°7′	3.7		前郭	吉林省地震局
92	2006 04 08	01 44 00.0	44°38′	124°7′	2.4		前郭	吉林省地震局
93	2006 04 08	03 20 00.0	44°43′	124°7′	1		前郭	吉林省地震局
94	2006 04 08	08 12 00.0	44°43′	124°7′	1.6		前郭	吉林省地震局
95	2006 04 08	08 12 40.0	44°36′	124°6′	2.3		前郭	吉林省地震局
96	2006 04 08	12 49 00.0	44°43′	124°7′	1.2		前郭	吉林省地震局
97	2006 04 09	22 30 37.0	44°43′	124°8′	0.4		前郭	吉林省地震局
98	2006 04 10	04 51 22.0	44°43′	124°8′	0.3		前郭	吉林省地震局
99	2006 04 10	12 08 38.0	44°43′	124°8′	0		前郭	吉林省地震局
100	2006 04 10	12 16 51.0	44°43′	124°8′	0		前郭	吉林省地震局

续表

序号	发震日期	发震时刻	震中位置		震级	震源深度	震中	结果来源
	年 月 日	时 分 秒	φ_N	λ_E	M_L	(km)	地名	
101	2006 04 10	12 28 48.0	44°43′	124°8′	0		前郭	吉林省地震局
102	2006 04 11	12 26 51.0	44°43′	124°7′	0.7		前郭	吉林省地震局
103	2006 04 12	01 21 21.0	44°43′	124°8′	0.2		前郭	吉林省地震局
104	2006 04 12	21 31 45.0	45°22′	124°38′	3.3		前郭	吉林省地震局
105	2006 04 13	07 11 00.0	44°39′	124°7′	0.1		前郭	吉林省地震局
106	2006 04 13	18 41 52.0	44°30′	124°4′	2.2		前郭	吉林省地震局
107	2006 04 14	04 52 50.0	44°30′	124°4′	0.9		前郭	吉林省地震局
108	2006 04 14	20 33 00.0	44°24′	124°4′	1		前郭	吉林省地震局
109	2006 04 14	23 13 10.0	44°24′	124°4′	0		前郭	吉林省地震局
110	2006 04 14	22 08 54.0	44°24′	124°4′	0		前郭	吉林省地震局
111	2006 04 14	23 10 00.0	44°40′	124°7′	0		前郭	吉林省地震局
112	2006 04 14	13 01 00.0	44°39′	124°7′	0.2		前郭	吉林省地震局

表 3 M_L2.5 以上地震序列目录

Table 3 Catalogue of seismic sequence with $M_L \geqslant 2.5$

序号	发震日期	发震时刻	震中位置		震级	震源深度	震中	结果来源
	年 月 日	时 分 秒	φ_N	λ_E	M_L	(km)	地名	
1	2006 03 31	20 23 14.0	44°36′	124°3′	5.2	10	前郭	吉林省地震局
2	2006 04 01	10 53 30.0	44°36′	124°6′	3.2		前郭	吉林省地震局
3	2006 04 02	18 12 40.0	44°36′	124°0′	3		前郭	吉林省地震局
4	2006 04 02	20 03 44.0	44°39′	124°7′	2.6		前郭	吉林省地震局
5	2006 04 02	20 04 17.0	44°36′	124°3′	3.3		前郭	吉林省地震局
6	2006 04 08	01 26 00.0	44°39′	124°7′	3.7		前郭	吉林省地震局
7	2006 04 12	21 31 45.0	45°22′	124°38′	3.3		前郭	吉林省地震局

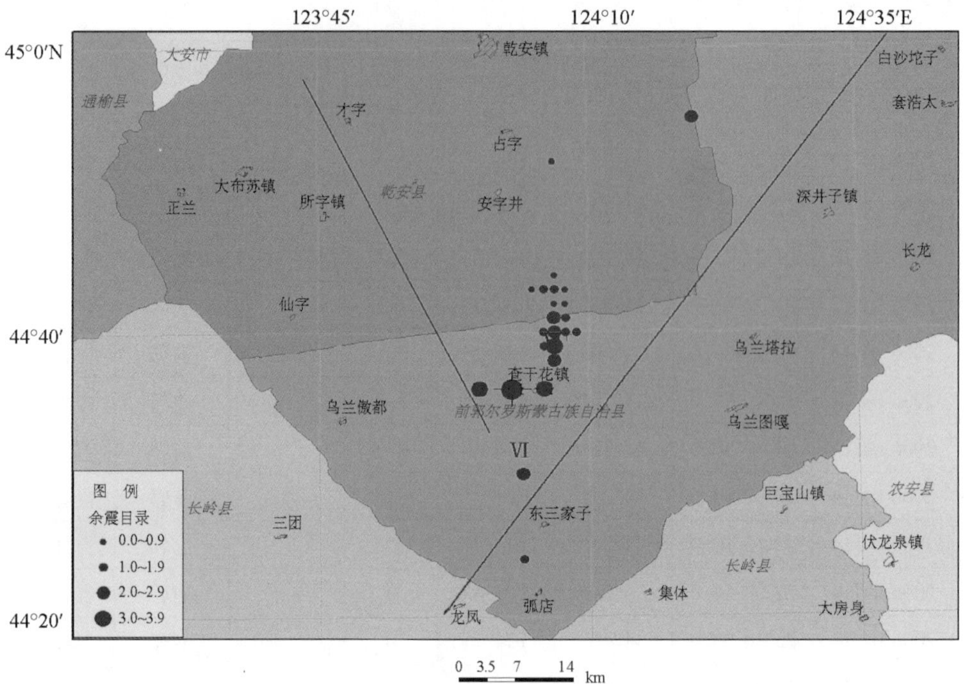

图4　前郭—乾安5.0级地震序列分布

Fig. 4　Distribution of seismic sequence of the *M*5.0 Qianguo-Qian'an earthquake

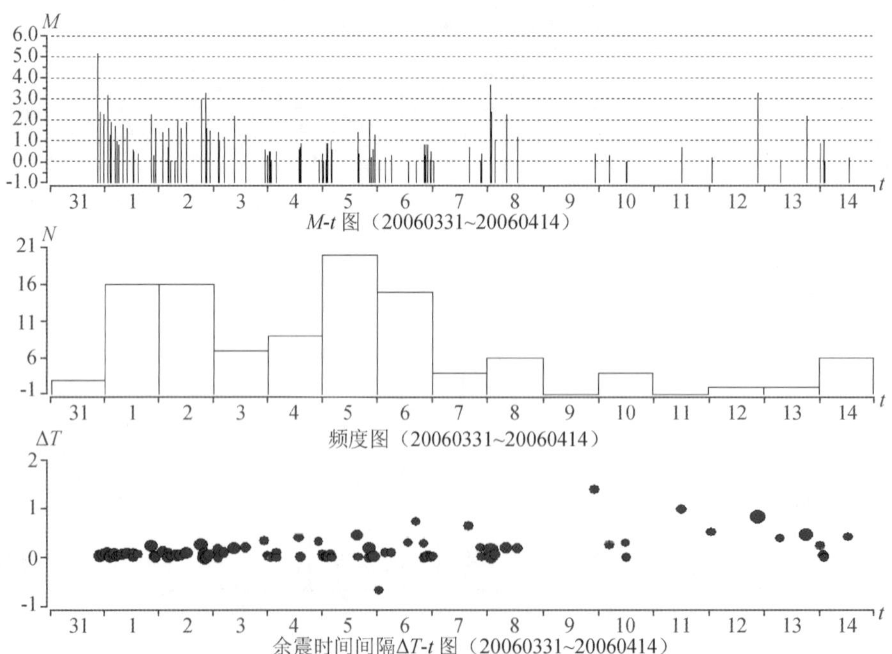

图5　乾安—前郭5.0级地震序列 *M-t*、频度、余震时间间隔图

Fig. 5　*M-t*、frequency and time interval between earthquakes diagram of the *M*5.0 Qianguo-Qian'an earthquake sequence

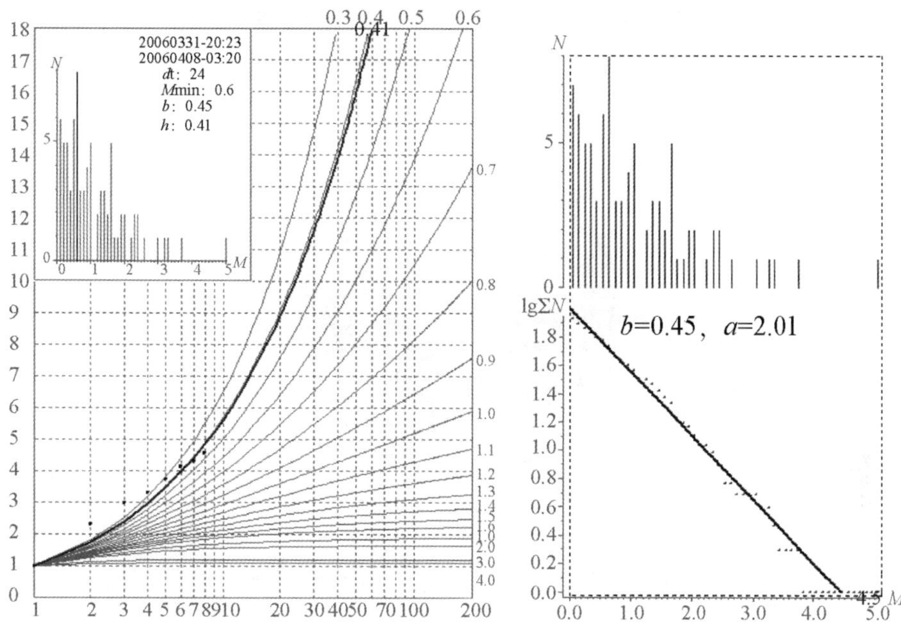

图6　余震序列参数计算结果

Fig. 6　Calculations of the parameters for aftershock sequence

五、震源参数和地震破裂面

利用吉林省地震台网及邻区地震台网P波初动资料，给出5.0级地震及主要余震的震源机制解如下（图7、表4）：

表4　震源机制结果表

Table 4　Focal mechanism solutions of the *M*5.0 Qianguo-Qian'an earthquake

序号	时间 年 月 日	震级 M_L	节面A			节面B			P轴		T轴		N轴	
			走向	倾向	滑动角	走向	倾向	滑动角	方位	倾角	方位	倾角	方位	倾角
1	2006-03-31	5.2	51°	69°	-30°	153°	62°	-156°	10°	36°	103°	4°	199°	54°
2	2006-04-01	3.2	61°	76°	-19°	156°	72°	-165°	18°	24°	109°	3°	205°	66°
3	2006-04-02	3.3	248°	43°	-162°	351°	78°	-49°	222°	42°	111°	22°	1°	40°
4	2006-04-08	3.7	309°	62°	-113°	171°	36°	-54°	178°	65°	55°	14°	320°	20°

利用省台网和邻近台站数字化资料，给出了此次地震相关的震源参数：

$M_L = 5.2$ $f_0 = 2.793899\mathrm{Hz}$ $M_0 = 1.25653\mathrm{e}+015\mathrm{N}\cdot\mathrm{m}$

应力降 = 5.413421bar 破裂半径 = 466.544776m

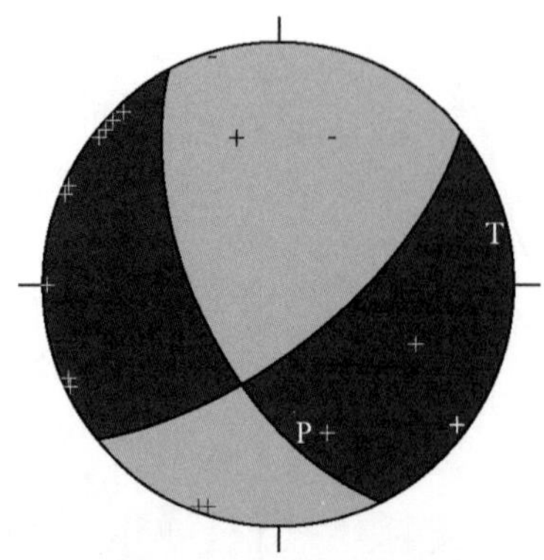

图7 5.0 级地震震源机制解

Fig. 7 Focal mechanism solutions of the $M5.0$ Qianguo-Qian'an earthquake

从图 7 和表 4 中可以看出，2006 年 3 月 31 日 5.2 级地震的主破裂面应为 B 节面，其走向为 153°，倾向 62°，滑动角为 -156°，P 轴方位 10°，倾角 36°，T 轴方位为 103°，倾角 4°。2006 年 4 月 8 日发生的最大余震的发生在余震条带的南部，从该震的震源机制解分析，该震微破裂为走滑兼正断层性质。节面与余震条带走向基本一致。

表 4 中包括主震在内的 4 次地震的震源性质均为走滑兼正断层性质。4 次地震的主压应力轴方向比较一致，均接近南北方向。主震的极震烈度长轴走向为北西方向[2]，余震为南北向条带，判断发震断层为隐伏的北西方向的断层。震源机制结果显示，在近南北的主压应力作用下主震及余震断层为走滑兼正断层分量[3]。这表明地震断层除存在剪切作用外，还兼有张性特征，可能与北西向查干泡—道字井断裂有关。

六、地震前兆观测台网及前兆异常

震中附近地震台站分布见图 8。地震发生在松辽断陷坳陷带内，区内台站主要沿北东向断裂构造及其与北西向断裂构造交汇部位分布，震中区附近 60km 范围内震前无台站分布。

在震中 200 km 以内共有所有 10 个地震台，分别有测震及前兆观测，在研究时段内基本可达到 $M_S \geqslant 2.0$ 级地震不遗漏。100 km 内仅有松原台及前郭台；100 ~ 200 km 省内分别有净月台、四平台、白城台、榆树四个测震台和前郭台、四平台、合隆三个前兆台；邻省有泰来台、肇东台、通辽台等三个测震及前兆台。只有四个台站在地震前存在测项异常，如图 8，其中大庆台位于震中 200km 之外。此次地震前共出现 8 项前兆手段异常（表 5）。

表5　地震前兆异常登记表

Table 5　Summary table of earthquake precursory anomalies

序号	异常项目	台站（点）或观测区	分析方法	异常判据及观测误差	震前异常起止时间	震后变化	最大幅度	震中距 Δ/km	异常类型及可靠性	图号	异常特点及备注
1	地震活动增强	东北地区	$M_S \geqslant 3.5$ 级地震 M-t 图	中等地震活动增强	2005～2006	恢复正常		震中周围	M_1	10	震前1年中等地震活动增强
2	地震条带	东北西部	图像法		2004～2006	恢复正常		条带内			中期异常
3	地温	大庆台	日均值分析	日均值降幅超出正常幅度	2006.3.8～3.10	恢复	0.003℃	200			短临异常
4	3	地震空区	吉林省西部	图像法		2005～2006	恢复正常		空区内		中期异常
5	NE向自然电位	白城台	日均值分析	加速上升后黑白加速下降	2006.2.14～2.3.1	恢复		150			短临异常
6	钙离子氯离子镁离子碳酸氢根离子	前郭台	日均值分析	同步下降	2006.1～	震后仍未恢复、呈趋势异常	1%	70			短临异常显著
7	水汞	白城台	日均值分析	两次出现超二倍均方差异常	2005.12～2006.2	震后恢复	10%	150			短临异常
8	水温	四平台	日均值分析	加速上升	2006.1～5	震后恢复		164			短临异常
9	水位	四平	日均值分析	破年变形态	2005.10～2006.4	震后恢复		164			中短期异常

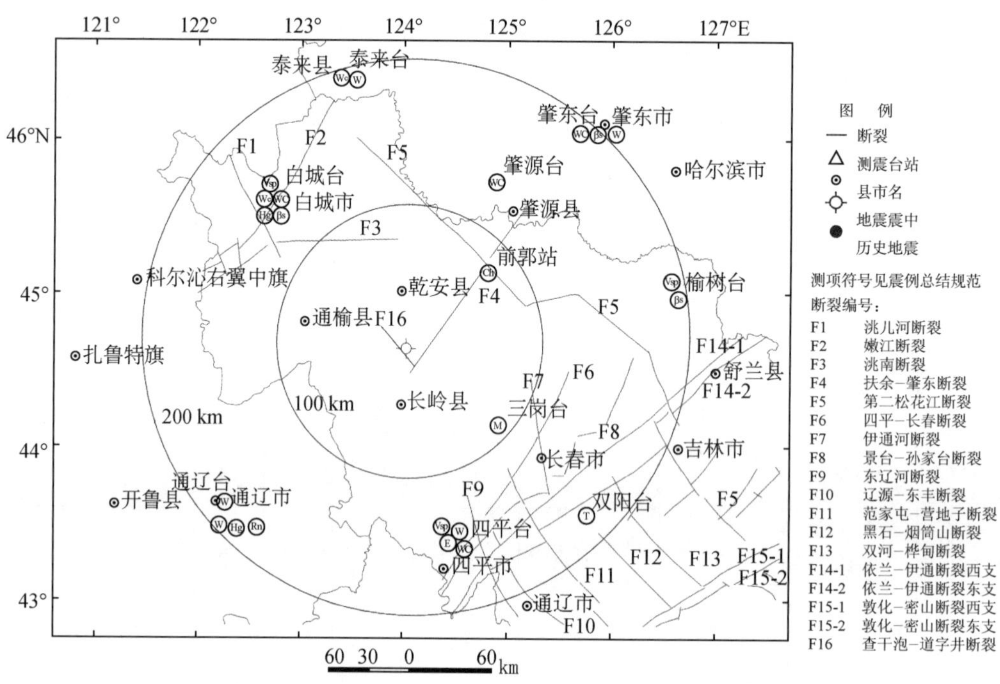

图 8　前兆台站分布图

Fig. 8　Distribution of earthquake-monitoring stations

1. 地震学异常

测震学异常：在东北中强地震前 1 年左右，4 级左右的中等地震频次明显增加，见图 9。

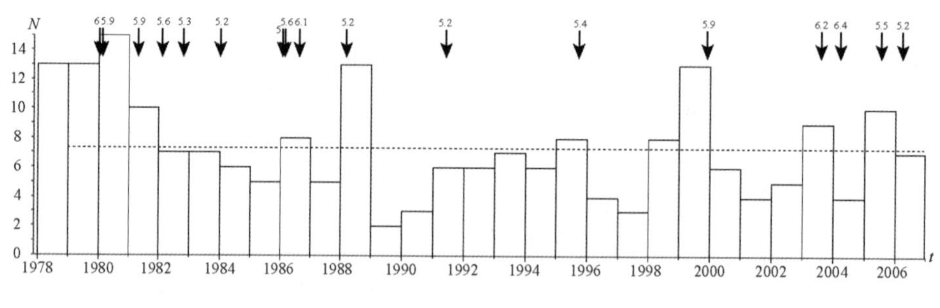

图 9　东北地区 $M_S \geqslant 4.5$ 级地震频度图

Fig. 9　Seismic frequency with $M_S \geqslant 4.5$ in Northeast China

研究表明，东北地区几次中强震在地震发生前，中等地震的活动明显增强。1980 年 3 月 2 日博克图 5.0 级地震之前，在震中附近连续发生多次 4 级以上地震，最后在地震活动的密集区发生主震；1986 年德都 5.6 级地震之前，先在伊舒带北段发生一组 3.5 以上地震，然后迁移到勃利—神树断裂东侧，最后在其西端与孙吴断裂交汇处发震；2003 年巴林左旗 5.9 级地震发生前，东北地区 3.5 级以上地震活跃，并在震前 1 年形成了地震空区，而未来的主震就发生在空区的空段；2005 年林甸地震前，出现了北西向的地震条带，主震发生在条带的空段上；2006 乾安—前郭地震前，出现北西和近东西的两条条带，而主震发生两个条带的交汇部位（图 10）。

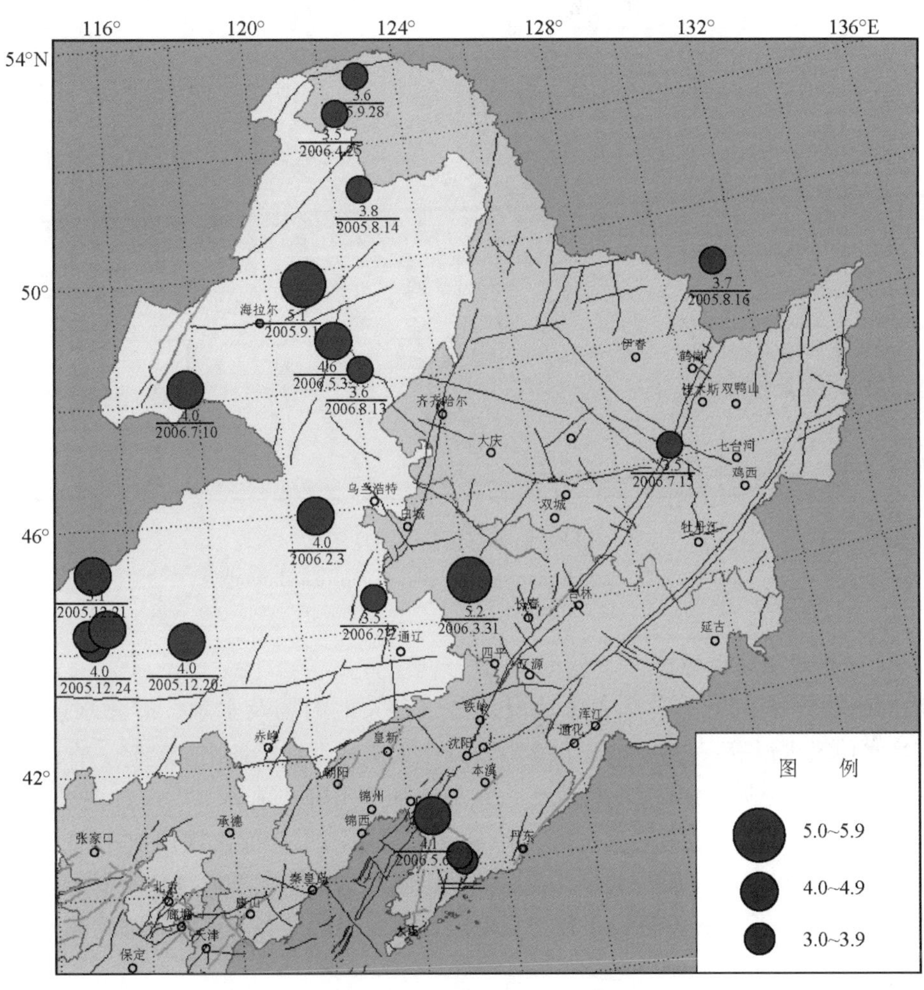

图 10 东北地区中强地震前 $M_L \geqslant 3.5$ 级震中分布图

Fig. 10 Distribution of epicenter with $M_L \geqslant 3.5$ befor medium-strong earthquakes in Northeast China

从省内的地震活动性分析，2006 年 2 月开始，该区地震活动性明显增强，先后发生了 2 月 2 日 2.7 级、2 月 27 日 3.4 级和 3 月 9 日 2.8 级地震。在近 2 个月时间内在该区连续发生 3 次 2.5 以上地震，地震活动性为东北地区同期之最强；特别是 2 月 27 日发生在前郭乌兰图嘎的 3.4 级地震距这次 5.0 级地震震中仅 25km，完全可以视为这次地震的信号震。

吉林省地震局分析预报中心测震组在 2 月 26 日月会商会上，根据省内近期地震活动，在该区曾划分出一个地震空区，而这次地震就发生在空区边缘附近（图 11）。

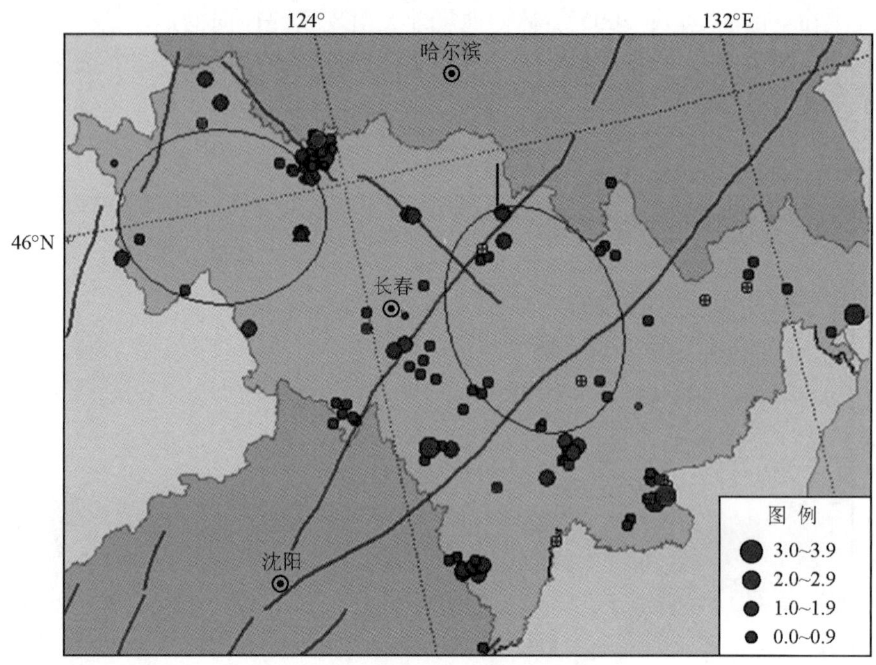

图 11　地震空区分布示意图

Fig. 11　Sketch map of the seismic gap distribution

2. 其他前兆观测项目异常

在震前出现异常观测手段有大庆地温、白城台大地电场及水汞、前郭水质（四种离子）、四平水位及水温等[1]（图 12 至图 17）。其中大庆地温在震前出现陡降，在低值区持续数天后发震；白城大地电场年变形态改变，在震前加速下降；前郭水质中各离子（氯、钙、碳酸氢根和镁离子）均出现不同程度的下降异常；白城水汞震前出现两次小幅超二倍均方差低值异常；四平水温震前加速上升，震后恢复；四平水位则在震前打破年变形态。由于这次地震震级不太高，上述异常多发生在松辽断陷盆地及周边，多表现为短临前兆异常变化。

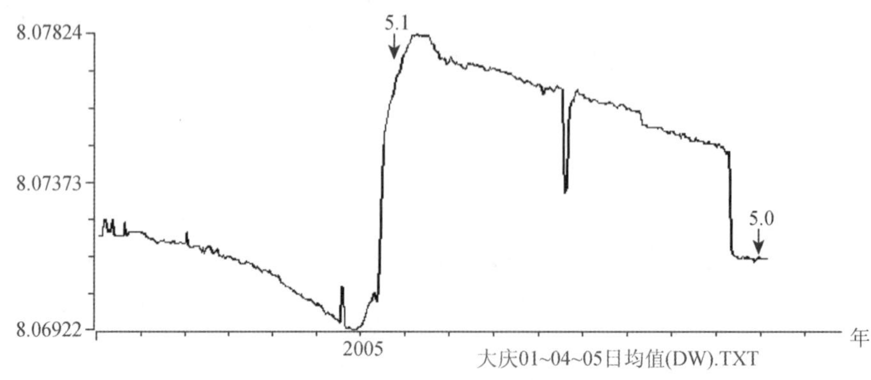

图 12　大庆地温异常图

Fig. 12　Curve of geothermal anomaly of Daqing station

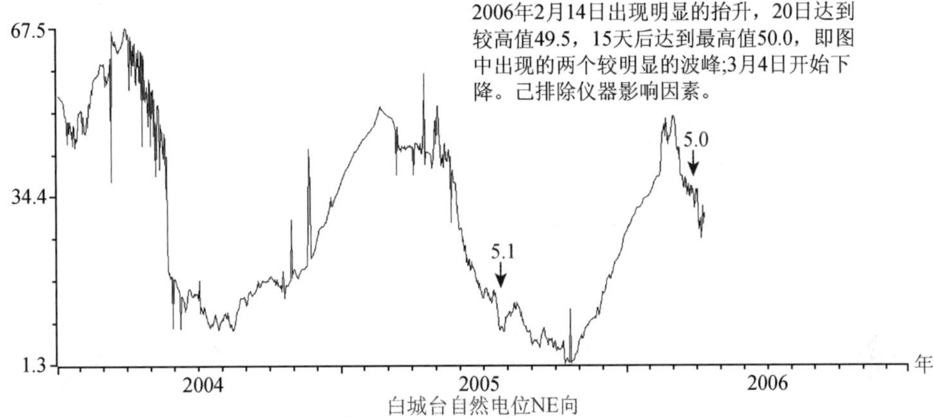

图 13　白城大地电场异常曲线图

Fig. 13　Curve of geoelectric field anomaly of Baicheng station

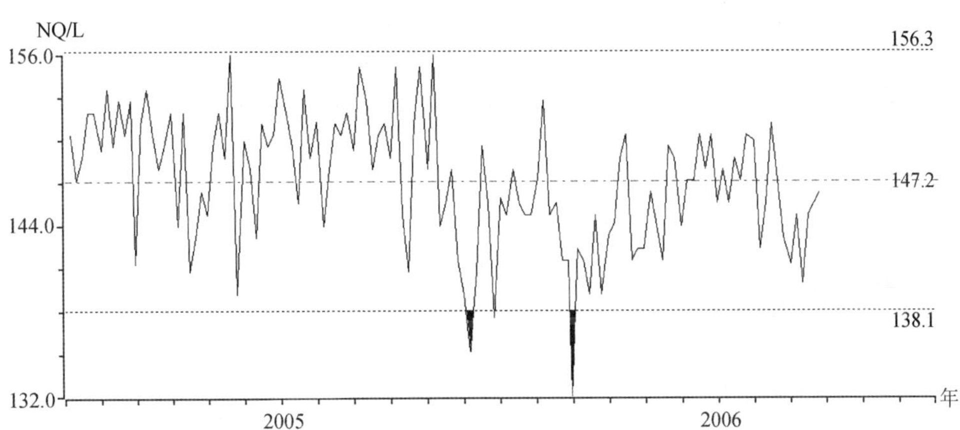

图 14　白城台水汞异常曲线图

Fig. 14　Curve of water mercury anomaly of Baicheng station

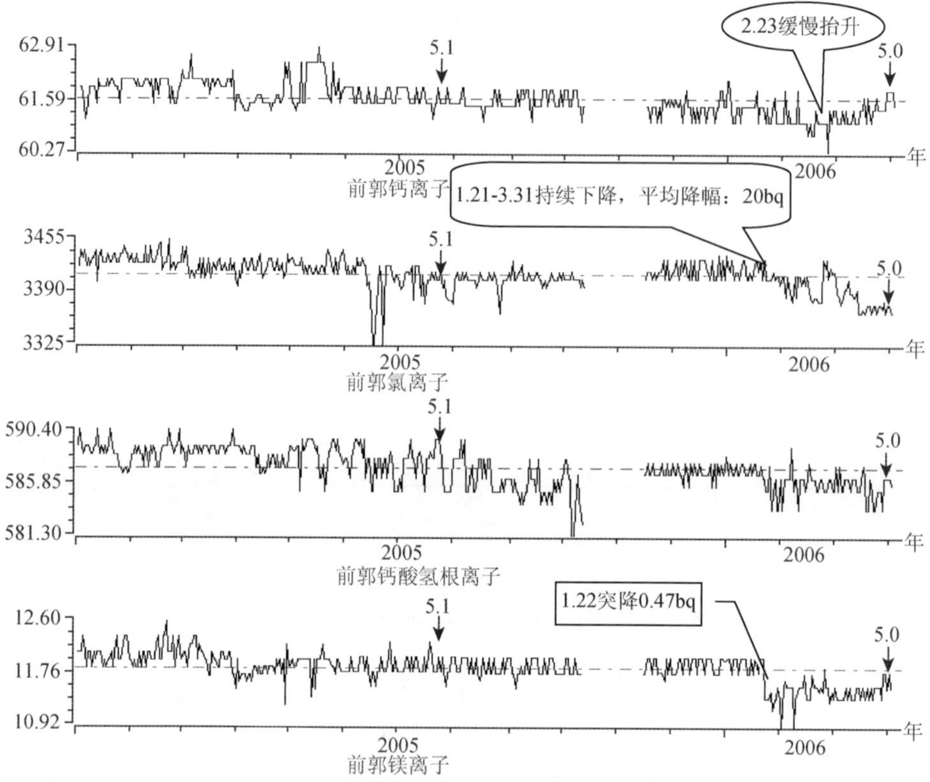

图 15　前郭水质观测异常曲线图

Fig. 15　Curve of water quality observation anomaly of Qianguo station

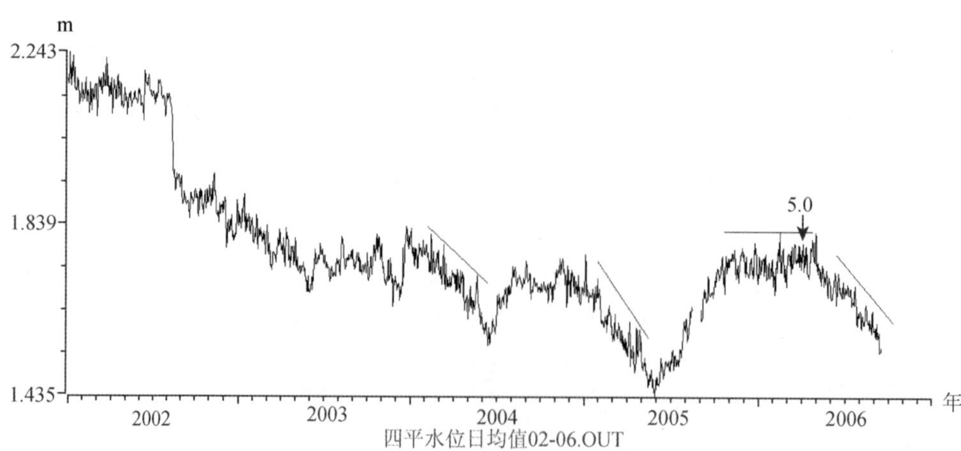

图 16　四平水位异常曲线图

Fig. 16　Curve of water level anomaly of Siping station

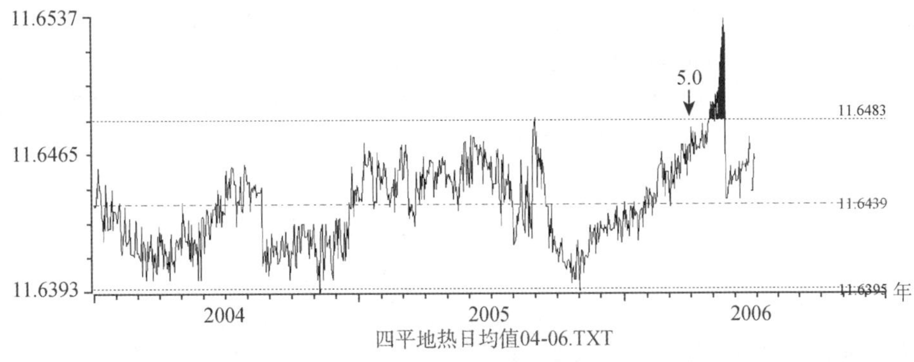

图17 四平水温异常曲线图

Fig. 17 Curve of water temperature anomaly of Siping station

七、地震前兆异常特征分析

1. 地震学异常

松辽盆地及周边近60年来，共出现了4次成对发生的中强地震（图9），分别为1941～1942、1980～1981、1986年的通辽、绥化、博克图和德都地震，显示了很强的原地重复性。自2002年起，在松辽盆地的西缘和中部先后发生了4次中强地震，且其发生位置靠近前4次发生双震型地点所围限的区域内，因此未来在这个区域再发生中强地震的可能性较大，并且在东北中强地震前1年左右，4级左右的中等地震频次明显增加，2006乾安—前郭地震前，出现北西和近东西的两条条带，而主震发生两个条带的交汇部位。2006年2月开始，该区地震活动性明显增强，先后发生了2月2日2.7级、2月27日3.4级和3月9日2.8级地震。在近2个月时间内在该区连续发生3次2.5以上地震，地震活动性为东北地区同期之最强；测震学指标出现了不同程度的异常。

2. 其他前兆观测资料异常

此次地震前出现明显的地震前兆异常，根据资料分析总结，主要有大庆地温、白城台大地电场及水汞、前郭水质（四种离子）、四平水位及水温等项异常，这些异常多发生在松辽断陷盆地及周边，多表现为短临前兆异常变化，异常的空间分布与区域构造背景相一致，异常的时间上则表现出短临前兆异常变化，但表现形式有所不同。

八、震前预测、预防和震后响应

1. 预测情况

在2005年底吉林省召开的2006年地震趋势会商会上，根据近年来东北中强地震活动的时空演化图像和地震迁移、时序间隔等研究，认为在今后的一二年内，松辽盆地周边和内部发生中强地震的可能性很大。2003年吉林省松原市发生了显著震群事件，共发生地震79次，其中3级以上地震5次。该地曾于1119年发生过6¾级破坏性地震，这次震群的发生表明松辽盆地及周边构造应力场活动趋于加强。2005 年 7 月 25 日黑龙江省林甸 5.1 级地震后，该地即发生了 $M_L3.2$ 响应地震，显示了同一控震构造间的呼应关系。

根据东北及吉林省地震学指标的时空扫描及前兆观测中存在的趋势性异常，经综合分析，认为 2006 年吉林省存在发生 5 级以上地震的可能性，并确定了松原地区为吉林省 2006 年地震重点监视区，提出了明确的中期地震预报意见。

2006 年 2 月开始，该区地震活动性明显增强，先后发生了 2 月 2 日 2.7 级、2 月 27 日 3.4 级和 3 月 9 日 2.8 级地震。在近 2 个月时间内在该区连续发生 3 次 2.5 以上地震，地震活动性为东北地区同期之最强；特别是 2 月 27 日发生在前郭县乌兰图嘎的 3.4 级地震距这次 5.0 级地震震中仅 25km，完全可以视为这次地震的信号震。吉林省局分析预报中心在 2 月 26 日月会商会上，根据省内近期地震活动，在该区曾划分出一个地震空区，而这次地震就发生在空区边缘附近。在震前松原市及邻区前兆台站也出现了异常观测手段，如大庆地温在震前出现陡降；白城大地电场年变形态改变，在震前加速下降；前郭水质中的离子中的钙、碳酸氢根和镁离子均出现不同程度的下降异常等等。但上述异常多以年变形态发生变化为主，均未显示出较好的短期－临震变化特征，虽然未能实现临震预测，但仍可以从中得到不少的经验和教训。

此次地震前，也较好地采取了监测预防措施。2006 年 1 月，吉林省地震局以文件方式，向吉林省政府及有关单位通报了 2006 年度吉林省地震趋势预测意见，并建议有关部门在重点监视区加强预防工作。2006 年 1 月 25～29 日，松嫩平原地震联防会议在松原市召开，重点研究了松原市及周边的震情，并就进一步加强监视和数据共享等工作进行了部署。2006 年 3 月 18 日省地震局领导去松原检查地震工作，特别就松原市地震形势及加强监视工作与松原市有关领导进行了交谈，对下一步地震预防工作提出了指导性意见。

2. 震后趋势判定

地震发生后，吉林省地震局迅速进行地震定位，并上报中国地震局和吉林省政府。吉林省地震局随即启动了破坏性地震应急预案，组成现场地震工作队赶赴地震现场。中国地震局应急救援专家组于 2006 年 4 月 1 日到达地震现场，指导地震现场工作。地震现场工作队在极震区附近架设了由老英吐、腰英吐、二龙山、后英、乾安、长岭 6 个台站组成流动监测台网开展地震应急观测工作。

现场工作队地震分析预报人员密切跟踪地震序列的发展，对此次地震的余震序列进行分析，发现余震的时间序列呈起伏地衰减，从集间时间间隔逐渐增大，初步判定该地震属主震-余震型，并预测近期震区内再发生 5 级以上地震的可能性较小。

3. 震害分析

地震发生后，吉林省地震局立即派出地震现场工作队，同随后到达的中国地震局应急救援组、松原市地震局工作组、黑龙江省地震局和大庆地震台网的工作人员联合开展了现场监测及震害评估工作。并在当地政府的大力支持下，在较短的时间内完成了这次地震的震害评估工作，共完成 51 个抽样调查点的现场工作，评定了地震烈度并初步考察了发震构造。

此次地震所造成的房屋破坏主要集中在老英吐村、腰英吐村等一带呈北西向 30 多公里的区域（地震烈度Ⅵ度强），地震烈度Ⅵ度破坏区域，约 1565km²，呈长轴为北西向的近椭圆形区域。从此次地震发生的震灾情况分析，对比国内外其他地震灾害的震例，发现此次地震的有感范围很大，而破坏程度相对较轻。此次地震震中地点位于远离城市的偏远农村，灾区受破坏结构类型单一，几乎 99% 以上的受损工程结构全部为单层砖木结构和单层木架土坯墙结构房屋。从结构破坏特点上，本次地震竖向地震作用明显，脉冲作用突出，导致了土坯房及老旧房屋破坏加重。

九、结论与讨论

（1）此次地震发生在东北断块区松辽断（坳）陷带南部长岭凹陷内。主震发生前两个月，震中区地震活动性明显增强，以发生 2～3 级左右的小震为主要特征，在空间上主要分布在北东向、北西向断裂及其附近。从区域上，东北地区中强地震前 1 年左右，4 级左右的中等地震频次明显增加，2006 乾安—前郭地震前，出现北西和近东西的两条条带，而主震发生两个条带的交汇部位。测震学指标出现了一定程度的异常表现，前兆手段也在震前出现了 9 项异常，主要为流体及地电手段，这些异常多发生在松辽断陷盆地及周边，多表现为短临前兆异常变化，异常的空间分布与区域构造背景相一致，异常的时间上则表现出短临前兆异常变化，但表现形式有所不同。异常的幅度并不大，可能与此次地震的震级不高有关，这些异常的时空分布特征表现松辽盆地及周边区域震前区域应力场处于一种不稳定状态。说明了，地震的孕育和发生，与区域构造背景及区域应力场变化有着密切的联系，为今后开展地震成因机制研究提供了一定的参考。

（2）此次地震发生在吉林省地震局中期预测和 2006 年度地震危险区内，地震强度预测基本准确。这表明东北地区大形势的研究和判断是卓有在效的。2006 年以来，省地震局对松原地区的震情跟踪监视工作一直都未放松过，多次与当地政府沟通，加强地震监视工作。2006 年 3 月 26 日东北三省地震研讨会在松原市召开，地震专家在会上就地震形势进行了广泛的讨论，要求加强松原地区的地震监视工作，受到了松原市政府的高度重视。自 2006 年年初以来前郭水化学测项、白城水汞、四平水位水温等出现了明显的异常，吉林省地震局对这些前兆异常进行过认真的会商，并形成会商意见上报中国地震局。因此，应当指出，吉林省地震局在此次地震发生前对其进行过成功的中期预报。但短临异常的出现与未来发震的判定工作是比较困难的，因此短临预报存在着较大的困难和局限性。

参 考 文 献

［1］张洪艳，卢燕红，康建红等．吉林前郭—乾安 M_S5.0 地震前地下流体异常特征分析［J］．东北地震研究，2007，23（3）：38～44

［2］盘晓东，王军亮等．乾安—前郭 5.0 级地震的构造背景分析［J］．东北地震研究，2007，23（1）：8～14

［3］吕政，张京辉等．2006 年 3 月 31 日吉林省前郭—乾安 M_S 4.8 地震序列［J］．国际地震动态，2006，334（10）：27～32

参 考 资 料

1）吉林省地震局，吉林省 2006 年度地震趋势研究报告，2005，打印稿

2）吉林省地震局，吉林前郭—乾安 5.0 级地震总结报告，2006，打印稿

Earthquake of *M*5. 0 on March 31, 2006 in the Joint Region between Qianguo and Qian'an of Jilin Province

Abstract

The M_S5. 0 earthquake occurred in the joint between Qianguo and Qian'an of Jilin Province on March 31, 2006. The micro-epicenter located in Laoyingtu and Yaoyingtu village, Chaganhua town, Qianguo country, Song yuan. the epicenter intensity reached to Ⅵ$^+$, the long axis direction of Meizoseismal Area outspreaded along the north-west. This earthquake happened at the central depression area which located in Songliao fault zone in Fault-block area of northeast China, two people were injured and the direct economic losses reached to Rmb110mn in this earthquake.

The earthquake abounded with aftershocks, about 112 aftershocks were totally recorded between March 31 and April 14, with 0-0. 9 magnitude 70 times, 1-1. 9 magnitude 27 times, 2-2. 9 magnitude 10 times, 3-3. 9 magnitude 5 times. The early distribution of aftershocks was NW direction, later for NS direction. The biggest aftershock (*M*3. 7) happened on 01: 26 April 8. The the quarter Ⅰ direction of the focal mechanism solution to 51° face, P axis to 10 °, the main direction of rupture was NW, it maybe associated with the N-W Chaganpao-daozijing fault.

The mid-term forecast advice was proposed exactly that $M_S \geqslant 5.0$ earthquake will be happened in Jilin province during the earthquake situation consultation meeting in the early 2006. The key monitoring area was near the SongYuan and adjacent area of the second Songhua river fault. In the epicenter of 200 km have 13 seismic stations, 2 stations in 100km. Nine abnormal precursory anomalies and learn ride appeared before M_S5. 0 earthquake. Giving an accurate judgment of the earthquake activity trend prediction, that obtained good social benefit.

报告附件

附表 1 固定前兆观测台（点）与观测项目汇总表

序号	台站（点）名称	经纬度（°）		测项	资料类别	震中距Δ（km）	备注
		φ_N	λ_E				
1	前郭	45.1	124.8	钙离子	Ⅱ	81	
				氯离子	Ⅱ		
				镁离子	Ⅱ		
				碳酸氢根离子	Ⅱ		
2	三岗	44.1	124.9	地磁 D	Ⅰ	89	
				地磁 Z	Ⅰ		
3	松原	45.3	124.8	测震△		94	
4	肇源	45.7	124.9	水温	Ⅱ	126	
5	净月	43.801	125.448	测震△		142	
6	白城	45.6	122.8	地电阻率	Ⅱ	154	
				大地电场	Ⅱ		
				水汞	Ⅱ		
				水温	Ⅱ		
				水位	Ⅱ		
		45.553	122.203	测震△		157	
7	四平	43.3	124.534	测震△		169	
				地电阻率	Ⅱ		
				大地电场	Ⅱ		
				水温	Ⅰ		
				水位	Ⅰ		
8	双阳	43.5	125.7	垂直摆倾斜	Ⅱ	81	
				石英伸缩	Ⅱ		
				水管倾斜	Ⅱ		
9	阿古拉	43.303	122.627	测震△		185	
10	通辽	43.6	122.3	水位	Ⅱ	192	
				水温	Ⅱ		
				气氡	Ⅱ		
				气汞	Ⅱ		

续表

序号	台站（点）名称	经纬度（°）		测项	资料类别	震中距Δ（km）	备注
		φ_N	λ_E				
11	榆树	44.763	126.749	测震△		200	
		44.9	126.6	地电阻率	Ⅰ	196	
				大地电场	Ⅱ		
12	肇东	46.0	125.9	水温	Ⅱ		
				水位	Ⅱ	198	
				地电阻率	Ⅱ		
13	泰来	46.4	123.4	水温	Ⅱ	199	
				水位	Ⅱ		

分类统计	0＜Δ≤100km	100＜Δ≤200km		总数
测项数 N	6	10		16
台项数 n	3	10		13
测震单项台数 a	1	2		3
形变单项台数 b	0	1		1
电磁单项台数 c	2	1		3
流体单项台数 d	1	3		4
综合台站数 e	0	3		3
综合台中有测震项目的台站数 f	0	3		3
测震台总数 $a+f$	1	5		6
台站总数 $a+b+c+d+e$	3	10		13
备注				

附表2 测震以外固定前兆观测项目与异常统计表

Table. 2 Statistical chart of fixed earthquake precursor observation projects and anomaly except seismometry

序号	台站（点）名称	测项	资料类别	震中距Δ（km）	按震中距Δ范围进行异常统计														
					0<Δ≤100km					100<Δ≤200km					200<Δ≤300km				
					L	M	S	I	U	L	M	S	I	U	L	M	S	I	U
1	前郭	钙离子	Ⅱ	81	—	—	√	—											
		氯离子	Ⅱ		—	—	√	—											
		镁离子	Ⅱ		—	—	√	—											
		碳酸氢根离子	Ⅱ		—	—	√	—											
2	白城	地电阻率	Ⅱ	154						—	√	√	—						
		水汞	Ⅱ							—	—	√	—						
3	四平	水位	Ⅰ	169						—	√	—	—						
		水温	Ⅰ							—	—	√	—						
分类统计	台项	异常台项数			0	0	4	0		0	2	3	0						
		台项总数			4	4	4	4		4	4	4	2	/					
		异常台项百分比/%			0	0	100	0		0	50	75	0	/					
	观测台站（点）	异常台站数			0	0	1	0		0	2	2	0						
		台站总数			1	1	1	1		2	2	2	2	/					
		异常台站百人分比/%			0	0	1	0		0	100	100	0	/					
	测项总数				4					4									
	观测台站总数				2					2									
备注																			

2006 年 6 月 21 日甘肃省文县 5.0 级地震

甘肃省地震局

代　炜　郑卫平　刘小凤

摘　要

2006 年 6 月 21 日甘肃省陇南市南部文县发生 5.0 级地震。宏观震中位于文县临江乡至桥头乡一带，震中烈度Ⅵ度强，局部显现Ⅶ度烈度异常区。地震造成造成 1 人死亡，19 人受伤。直接经济损失 7335.1 万元。

此次地震序列为主震—余震型，最大余震 3.6 级，余震成北西向条带状分布，与极震区长轴走向呈一定的夹角，节面Ⅰ为主破裂面，主压应力 P 轴方位为 NE60°，该地震的震源机制以走滑为主，发震构造为石坊—临江断裂。

震中 200km 范围内共有 6 个地震观测台，均有测震观测项目和定点前兆观测项目。震前共出现 4 个异常项目 6 条异常。

甘肃省地震局对文县 5.0 级地震做出了较好的中期预测，该地震发生在 2006 年甘东南地震重点危险区边缘。震前未提出短临预测意见，震后趋势判定基本正确。

前　言

2006 年 6 月 21 日 00 是 52 分 55.0 秒，甘肃省陇南市南部文县发生 5.0 级地震。甘肃省地震台网测定的微观震中为 N33°03′，E104°56′，宏观震中位于文县临江乡至桥头乡一带。震中烈度达到Ⅵ度强，局部地方存在Ⅶ度烈度异常区。这次地震的有感范围较大，甘肃省文县五库乡、山仓乡、桥头乡、临江镇，陇南市城区震感强烈；陕西省宝鸡市凤县和汉中市宁强县、略阳县有震感。地震造成 1 人死亡，19 人受伤，其中 6 人重伤，直接经济损失 7335.1 万元。

文县 5.0 级地震发生的甘肃东南部地区，地震活动水平高。在震中西北方向 45km 处曾发生 1879 年武都 8 级地震。震中 200km 范围内共有 6 个地震台，地震前共出现 4 个异常异项目，6 条异常，其中测震学异常 1 项次，前兆测项水氡、应变和地电阻率共 5 项次异常。震前甘肃省地震局根据地震活动性及武都应变、武都水氡等趋势性异常作了中期预测，未提出短临预测意见，震后趋势判定基本正确。

地震发生后，由中国地震局、甘肃省地震局、陕西省地震局和陇南市地震局组成现场联合工作队对该地震进行了考察，并进行了灾害损失评估。

本研究报告是在有关文献和资料的基础上，经过重新整理和分析研究而完成的。

一、测震台网及地震基本参数

震中 200km 范围内有 6 个地震台分布（图 1）。100km 内分别是武都台和文县台；100 ~ 200km 分别是天水台、成县台、岷县台，迭部台。震中位于南北地震带的中北段地区，地震监测能力较强，在研究时段内基本可达到 $M \geqslant 2.0$ 级地震不遗漏。

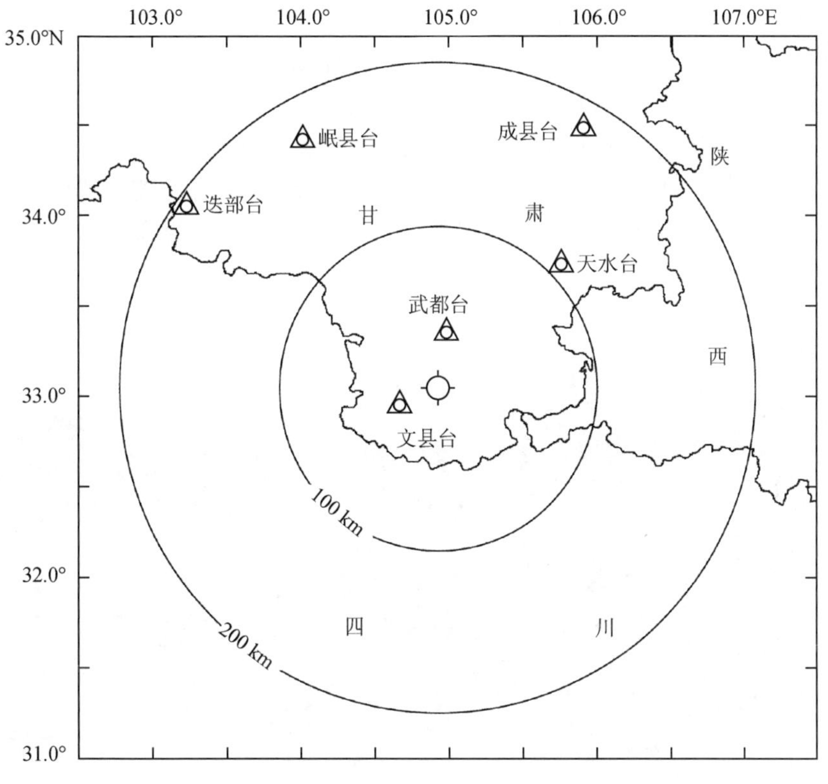

图 1 文县 5.0 级地震前震中附近地震台站分布图

Fig. 1 Distribution of earthquake-monitoring stations around the epicentral area before the M_S 5.0 Wenxian earthquake

表 1 列出了不同来源给出的地震基本参数，经过分析认为，甘肃地震局给出的震中位置更为精确。因此，此次地震基本参数采用表 1 中编号 1 的结果。

表1　文县5.0级地震基本参数

Table 1　Basic parameters of the M_S 5.0 Wenxian earthquake

编号	发震日期	发震时刻	震中位置		震级				震源深度	震中地名	结果来源
	年 月 日	时 分 秒	φ_N	λ_E	M_S	M_L	M_b	M_w	(km)		
1	2006 6 21	00 52 55	33°03′	104°56′		5.4			15	文县	甘肃省地震局
2	2006 6 21	00 52 57	33.1	105	5				33	甘肃武都、文县间	中国地震台网中心

二、地震地质背景

文县5.0级地震发生在南北地震带中段，地处秦岭褶皱系与松潘—甘孜褶皱系交界处，在区域上受青藏地块的挤压，是中强以上地震多发区。该地震发生在近东西至北东向的石坊—临江断裂带上，断裂倾角较陡，为60°～80°，主体断面北倾。卫星影像上表现为十分清晰的线性构造，线性构造线两侧冲沟及山脊被错的地貌特征，表明该断裂将多个近南北向的山梁及梁间小溪左旋错动，并在多个山梁的东西山坡形成坡中沟，错动量最大的地区是杨家沟一带，水平位移量估计在10m左右（冯希杰等，2005）。自有地震记载以来，震中50km范围内发生5级以上地震7次，最大地震为1879年武都8级，距离文县5.0级地震30km（表2）。自1880年以后，在长达126年的时间内未发生过5级以上地震，属于中强以上地震活动水平低的地区。

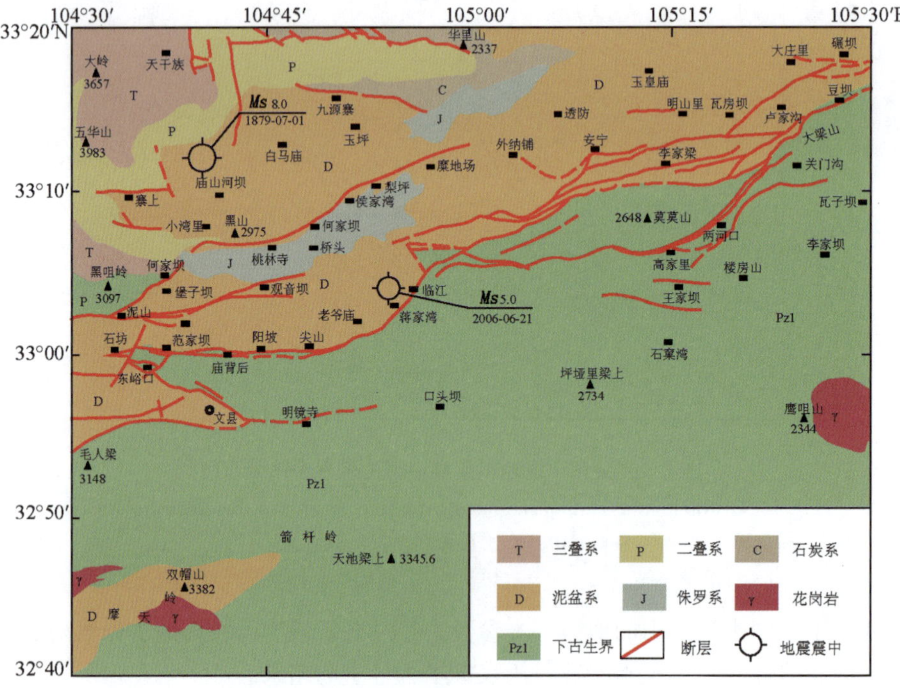

图2　5.0级地震震中区及周围地质构造图

Fig 2　Geological structure map of epicentral region, M_S 5.0, 2006, Wenxian

表 2　震中区历史地震参数

Table 2　earthquake parameters of the M_S 5.0 Wenxian

序号	发震时间	参考位置		震级 M_S	参考地名	距离（km）
		北纬（°）	东经（°）			
1	1581 年 7 月	32.9	104.6	5½	甘肃文县	43
2	1634 年冬	33.2	104.8	5½	甘肃文县	22
3	1677 年 9	33.4	104.9	5½	甘肃武都	35
4	1822 年 4 年 24 月	33	104.6	5½	甘肃文县	39
5	1879 年 6 月 29 日	33.2	105	5¾	甘肃武都附近	11
6	1879 年 7 月 1 日	33.2	104.7	8	甘肃武都南	30
7	1880 年 6 月 22 日	32.9	104.6	5½	甘肃文县	43

三、地震影响场和震害

据考察资料，宏观震中位于文县临江乡至桥头乡一带。震中烈度Ⅵ度，局部地区存在Ⅶ度异常，等震线呈东西向椭圆形分布（图 3）。地震造成造成 1 人死亡，19 人受伤，其中 6 人重伤，直接经济损失 7335.1 万元。

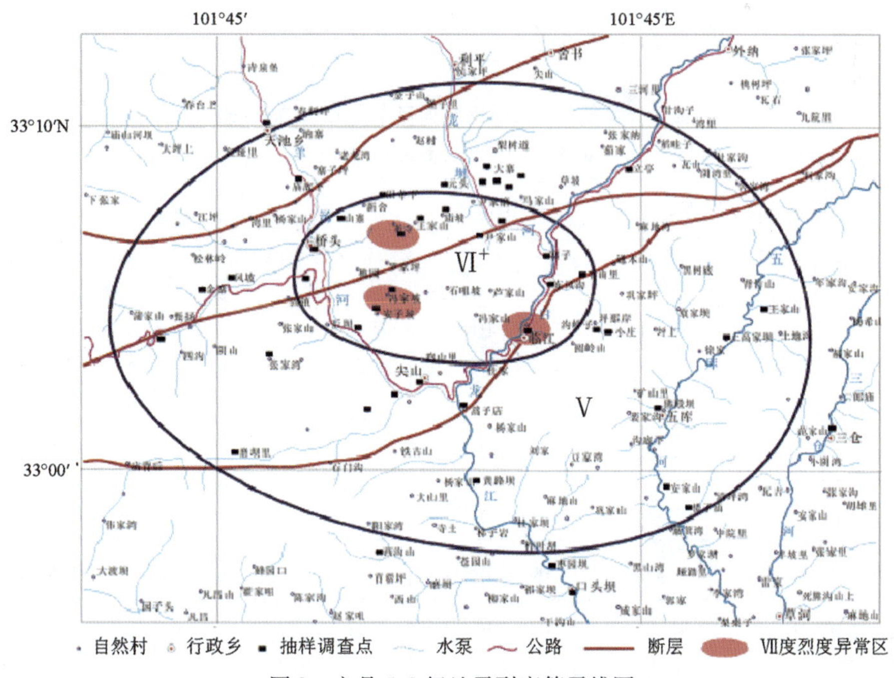

图 3　文县 5.0 级地震烈度等震线图

Fig. 3　Isoseismal map of the M_S 5.0 Wenxian earthquake

Ⅶ度异常点：新寺村，冯家坡，安子坡，临江镇欧家坝和铧厂等村（社）属于烈度异常点。新寺村25户村民的房屋严重破坏。临江镇欧家坝、铧厂2个村（社）20余户村民大部分房屋遭受破坏，多为墙体开裂、屋顶掉瓦，部分梁柱扭转。临江镇中学1栋三层教学楼多处裂缝，部分校舍屋顶坍塌。

Ⅵ度区：本次地震的极震区，包括桥头乡以东，临江镇以西，梨坪乡一带。呈近东西走向的椭圆形。区内砖混结构房的破坏主要表现为墙体裂缝，土木结构房的破坏表现为大梁与墙体脱位，山墙与后墙粘结部位开裂也较普遍，土坯房的破坏主要表现为倒塌和墙体开裂。其中尹家磨小学4间土木结构房屋成危房，造成学校停课；畦次坝村小学陈旧校舍墙体裂缝较为普遍，教室黑板裂缝严重，影响使用；河口村小学4间土木房屋结构房屋基本完好；木元村小学砖混结构校舍12间，其中5间出现轻微裂缝，其余完好；尖山村小学土木结构教室3间，房顶瓦片摇落。

区内地形地貌较复杂路段，路基滑塌、开裂下沉严重，部分道路中断。道路上方山体滑塌，滑塌体体积几立方米至几千立方米不等；滚石较多，其中滚石最大的直径约有3m，滑塌体与滚石阻塞公路十分严重。

四、地 震 序 列

文县5.0级地震发生在区域地震活动相对平静的背景下，震前没有明显的前震活动。主震后截至2006年10月底，共发生余震246次，其中单台（文县台）地震169次，可定位地震76次，其中1.0 ~ 1.9级50次，2.0 ~ 2.9级13次，3.0 ~ 3.9级5次，4.0 ~ 4.9级1次。表3给出了$M_L \geqslant 3.0$级的地震序列目录。

表3　文县5.0级地震序列目录（$M_L > 3.0$级）

Table. 3　Catalogue of the M_S 5.0 Wenxian earthquake sequence（$M_L > 3.0$）

编号	发震日期	发震时刻	震中位置		震级	震源深度	震中	结果来源
	年 月 日	时 分 秒	φ_N	λ_E	M_L	(km)	地名	
1	2006 06 21	00 52 54	33°04′	104°54′	5.4	15	文县	甘肃地震台网
2	2006 06 21	01 39 39	33°03′	104°55′	3.4	12	文县	甘肃地震台网
3	2006 06 21	02 21 36	33°01′	104°57′	3.3	18	文县	甘肃地震台网
4	2006 06 21	16 56 20	33°01′	104°56′	4	18	文县	甘肃地震台网
5	2006 07 30	14 47 20	33°06′	104°56′	3.3	16	文县	甘肃地震台网
6	2006 07 31	10 54 59	33°04′	104°56′	3.3	16	文县	甘肃地震台网
7	2006 08 01	04 59 13	33°04′	104°58′	3	10	文县	甘肃地震台网

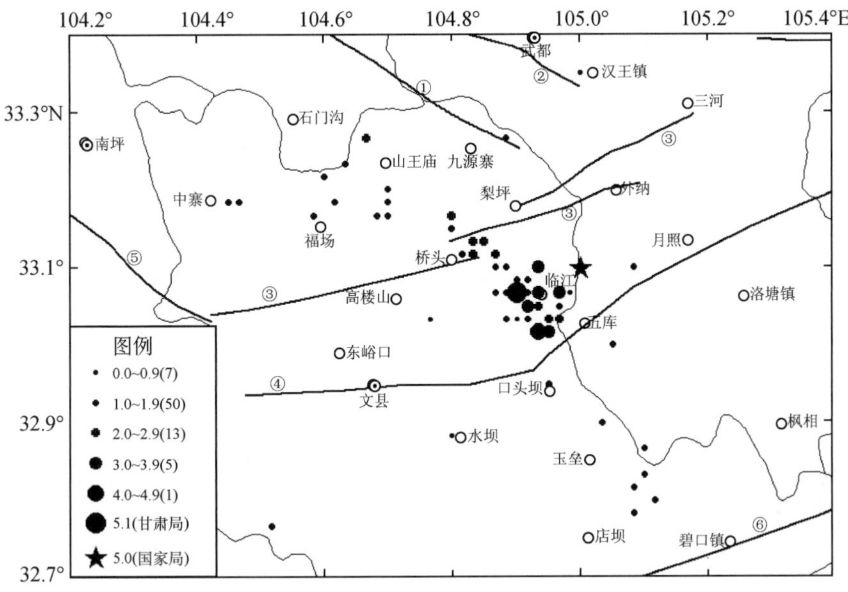

图4　文县5.0级地震序列分布

Fig. 4　Distribution of the M_S 5.0 Wenxian earthquake sequence

图4给出了文县5.0级地震序列的震中分布图。图中五星为中国地震台网中心定位的地震震中，最大的实心圆是甘肃地震台网中心定位的地震震中，二者相距9km。余震主要分布在主震的两侧，呈北西方向展布。较大余震分布在主震的东南侧，较小的余震分布在主震的西北侧。因此可以判定该地震震源破裂为单侧破裂。余震由东向西眼主震破裂方向强度逐步衰减。在余震优势分布方向没有早期断层和此次地震形成的地面破裂带。

从序列 M-t 图（图5）可以看出，余震主要集中在震后5天内，序列中 M_L2.0以下地震比较多。最大余震发生在主震发生16小时后，震级 M_L4.1。

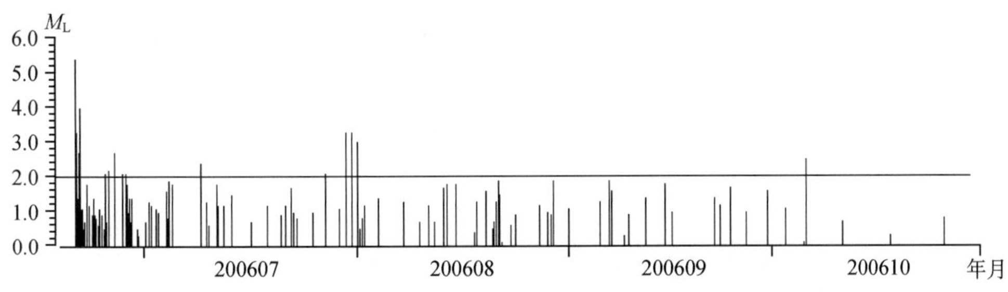

图5　文县 M_S5.0级地震序列的 M-t 图

Fig. 5　The diagram of the M_S 5.0 Wenxian earthquake sequence

该序列日频度曲线（图6）表明序列能量释放很快。主震释放能量占全序列能量的

96%，主震和最大余震震级之差为1.4，序列 b 值0.56，序列的 h 值为1.2（图7），表明该序列为主震—余震型地震序列。

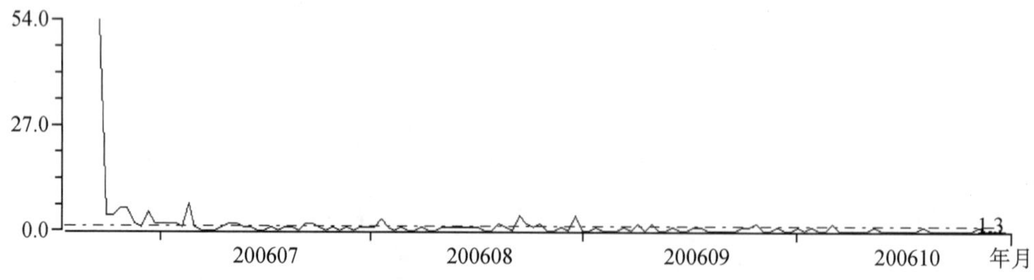

图6　文县5.0级地震日频次曲线

Fig. 6　Earthquake frequency of the M_S 5.0 Wenxian sequence

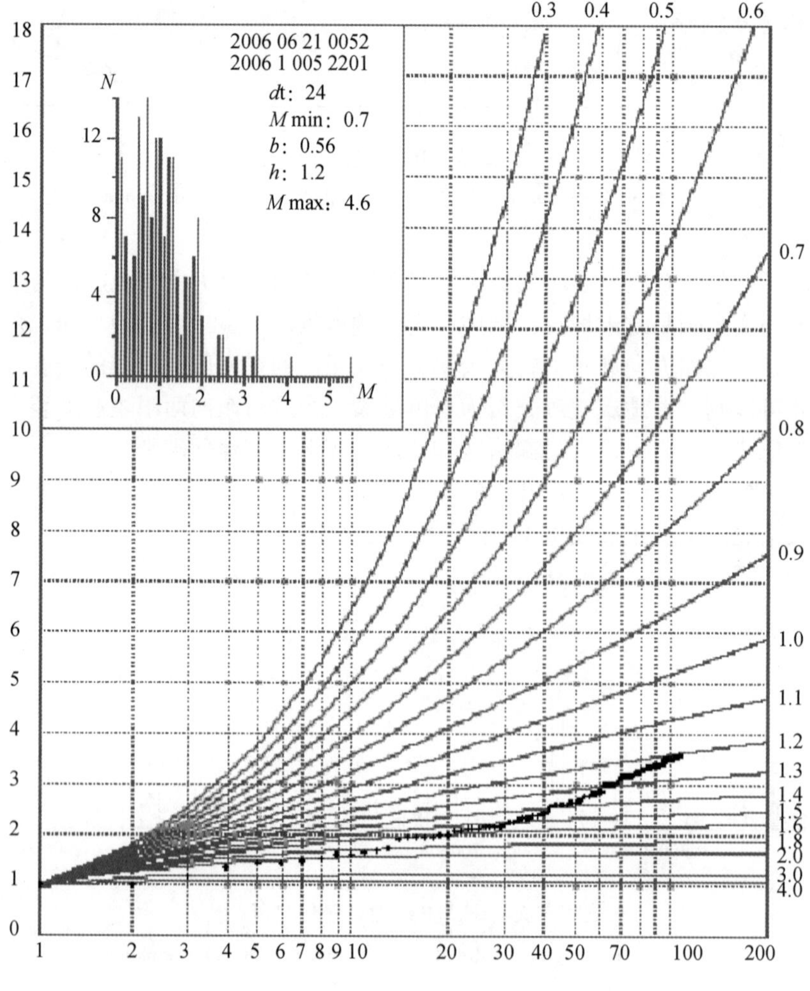

图7　文县5.0级地震序列的 h 值曲线

Fig. 7　h-t diagram of the M_S 5.0 Wenxian earthquake sequence

五、震源参数和地震破裂面

收集甘肃地震台网清晰的 P 波初动 18 次、四川地震台网、陕西地震台网的相关波形数据，用 P、SV、SH 波的初动和振幅比的方法（Snoke，1980）和初动方法分别求得该地震震源机制（图 8、表 4）。

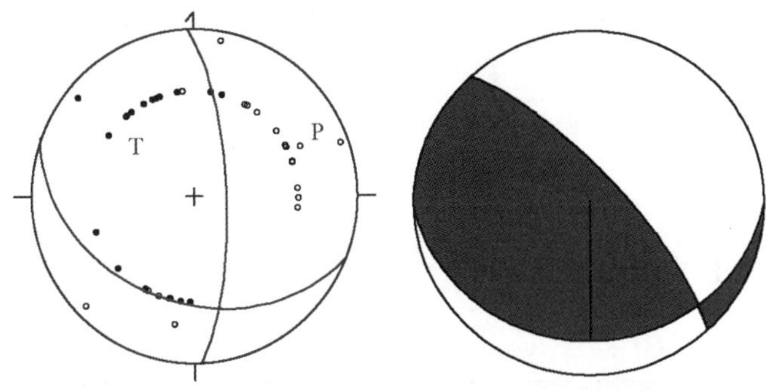

图 8　由初动方法和 Snoke 方法获得的文县 5.0 级地震震源机制解

Fig. 8　Focal mechanism solutions of the M_S 5.0 Wenxian earthquake with P and snoke

表 4　文县 5.0 级地震震源机制解

Tabal. 4　Focal mechanism solutions of the M_S 5.0 Wenxian earthquake

节面 I			节面 II			P 轴		B 轴		T 轴		Mod	方法
strike	dip	rake	strike	dip	rake	dip	angle	dip	angle	dip	angle		
76.77	25	11	337.19	85	115	46	36	155	25	272	44		Snoke 方法
111	36	28	358	74	123	63	22	305	50	168	32	0.11	P 波初动法

本次地震为一次中等强度地震，没有形成地表破裂带，根据航片解译结果，震源机制结果、地震序列精确定位结果以及现场地震灾害调查结果尝试综合判定本次地震的发震构造：

（1）航片解译结果表明该地区存在多条活动强烈的近东西向至北北东向的左旋错动的断裂，文县 5.0 级地震的发震构造很有可能就是其中的一条。现今地壳应力状态表明该地区的主压应力方向为北北西—南南东，应力值为 40 ~ 100kg/cm^2（曾秋生等，1989）。石坊—临江断裂在尖山以东转为北东向，并向南弧形突出，转折部位容易造成应力的积累与应变的集中，NE60°主压应力作用下，石坊—尖山段容易造成走向滑动，而在尖山—临江段容易形成逆冲，等震线长轴方向东西，长短轴差异不大。

（2）震源机制表明该地震为一倾滑型地震，节面Ⅰ走向 111°，节面Ⅱ走向 358°，节面Ⅰ的走向和石坊—临江断裂以及石鸡坝—观音坝断裂走向相近。若以节面Ⅰ作为主断面，文县 5.0 级地震为逆冲兼左旋走滑型，和已有断层的运动特征相一致，若以节面Ⅱ为主断面，文县 5.0 级地震为逆冲兼右旋走滑型，与已有断层的性质相反，且该地区并不存在近南北向的断裂。震源机制反演结果表明该地震的主压应力方向为NE63°，这一结果与该地区区域构造应力不一致，因此，文县5.0级地震不是区域构造应力作用下的产物。同时，该地震余震分布优势方向为北北西，但总体还是比较集中，符合倾滑型地震余震分布的特点。

（3）现场地震灾害调查结果表明文县5.0级地震可能是东西向断裂活动的结果。根据以上结果可以排除石鸡坝—观音坝断裂作为5.0级地震发震构造的可能。完全可以确定石坊—临江断裂为该地震的发震构造。

六、地震前兆观测台网及前兆异常

文县5.0级地震震中200km范围内共有6个地震台，其中6个测震观测单项台，2个综合观测台，2个前兆单项观测台；总计有测震、水氡、流量、电阻率、形变、地磁、应力和应变等7个定点观测项目（图9）。地震前共出现4个异常项目，6条异常，其中测震学异常1条，有水氡、应变和地电阻率共5条异常。

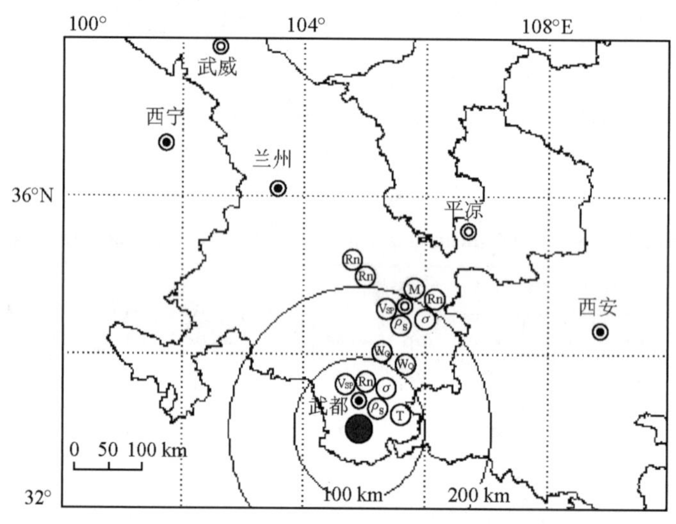

图9　文县5.0级地震前定点前兆观测台站分布图

Fig. 9　Distribution of precursory monitoring stations before the M_S5.0 Wenxian earthquake

1. 地震学异常

M_L3.0地震活动异常。利用全国小震目录，对以震中为中心10°×7°范围内，震前一年甘东南地区发生的M_L3.0地震，以1°×1°的窗长和步长进行空间频度扫描，发现文县5.0级地震地区的M_L3.0地震频度明显高于其他地区（图10）。M_L3.0地震增强活动过程中存在地震连发现象，地震连发事件发生在2005年9月3日的M_L3.5、M_L2.3、M_L3.8地震。距本次地震发生9个月时间。震前一年的增强活动区域在震前存在平静现象，地震平静自2006年1月2日开始至震前。因此文县5.0级地震之前一年M_L3.0地震活动存在增强活动，伴随地震连发现象，增强活动结束平静一段时间后发震。

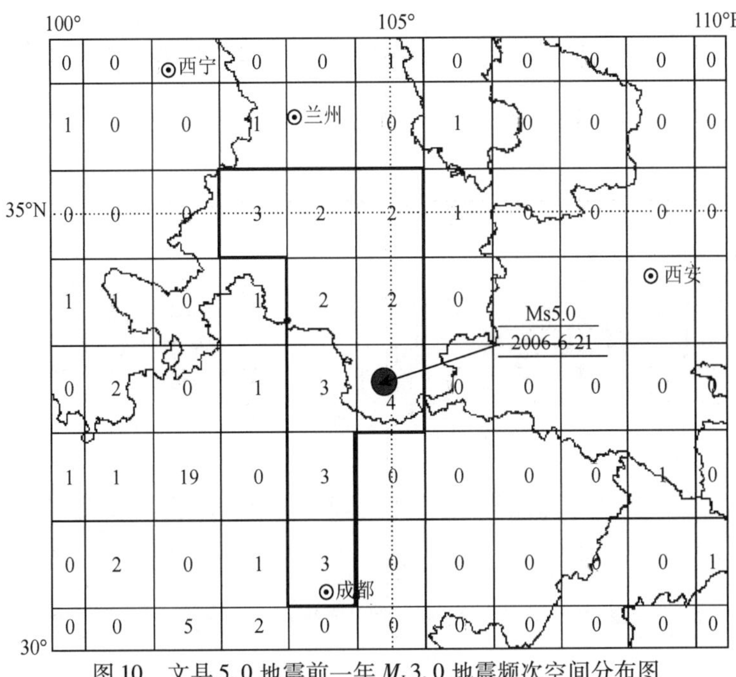

图 10　文县 5.0 地震前一年 M_L3.0 地震频次空间分布图

Fig. 10　Space and frequence distribution of earthquakes M_L 3.0 before one year of M_S5.0 Wenxian

2. 其他前兆观测项目异常

在 200km 范围内，存在水氡、应变和地电阻率共 5 项次异常。异常主要表现为武都电阻率日均值破年变异常（图 11）；武都水氡日值破年变异常（图 12）；武都应变日均值趋势改变异常（图 13）；西和水氡日值破年变异常（图 14）；天水水氡日值破年变异常（图 15）等。这些异常都表现为中期尺度的异常。

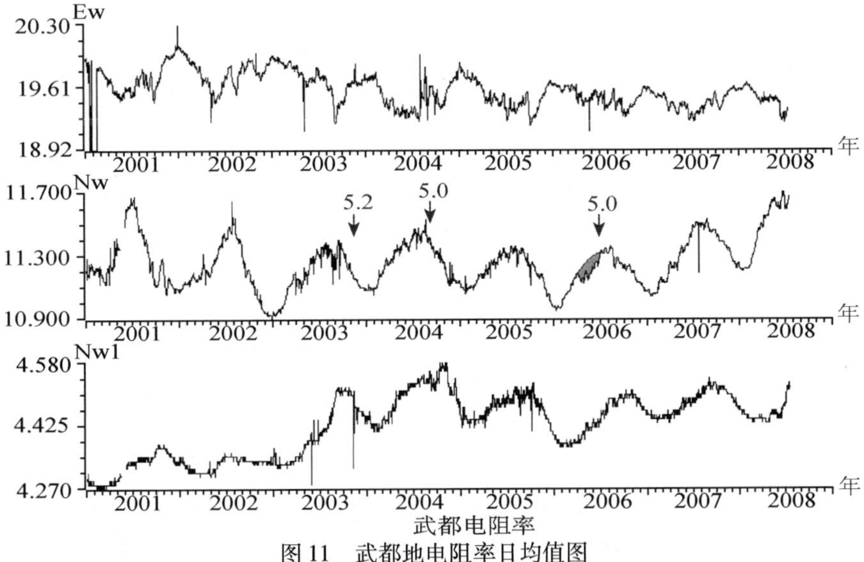

图 11　武都地电阻率日均值图

Fig. 11　Curve of daily mean value of apparent resistivity at Wudu station

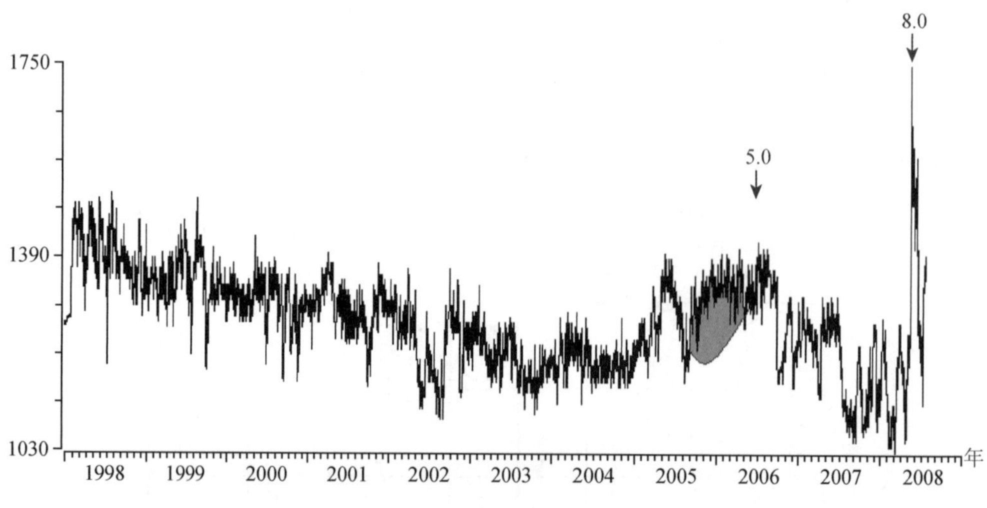

图 12　武都水氡日均值图

Fig. 12　Daily value curve of radon content in groundwater at Wudu station

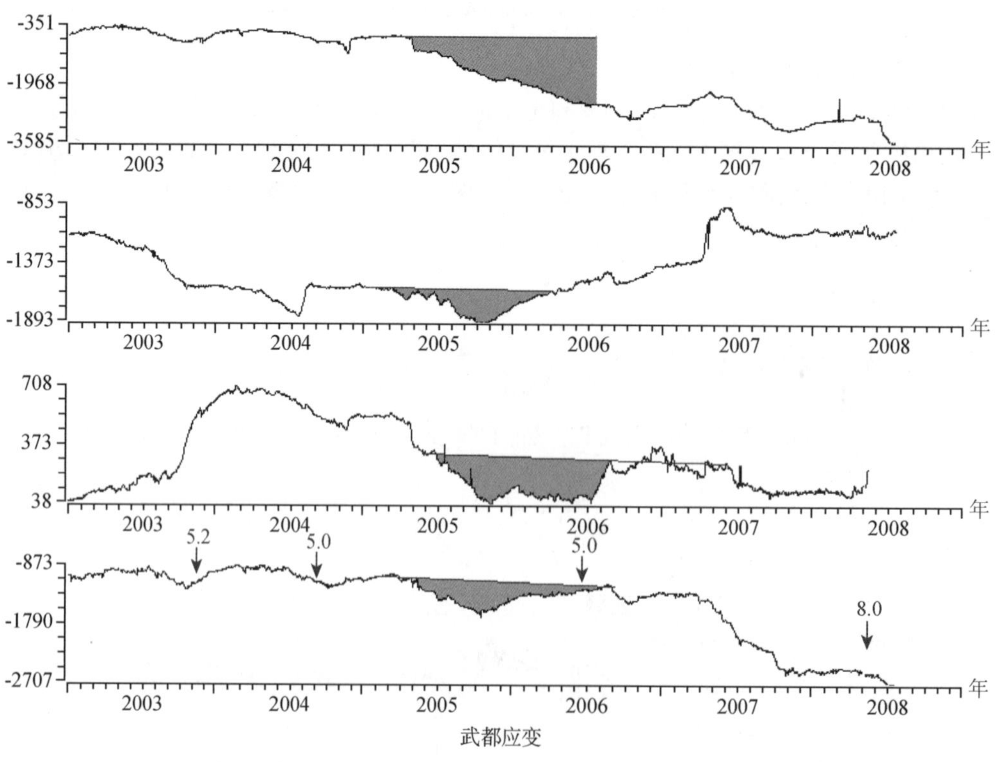

图 13　武都应变日均值曲线

Fig. 13　Curve of daily mean value of borehole strain at Wudu station

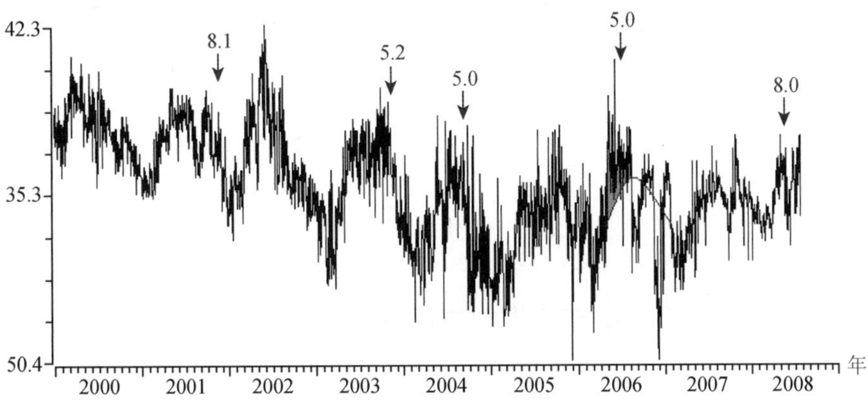

图14　西和水氡日均值图

Fig. 14　Daily value curve of radon content in groundwater at Xihe station

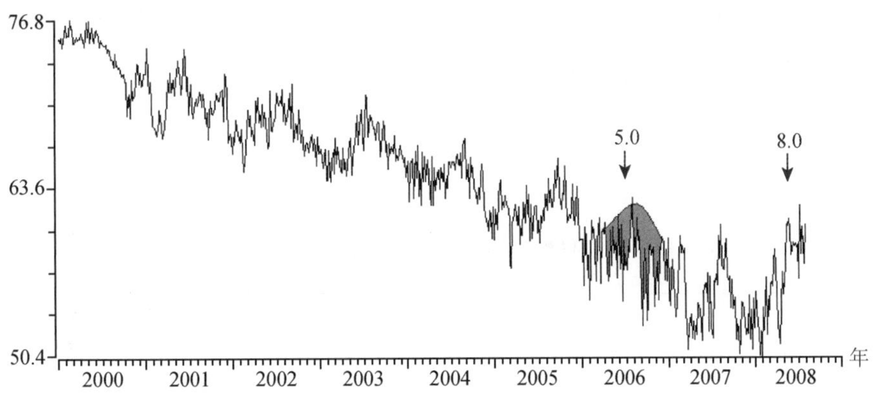

图15　天水水氡日均值图

Fig. 15　Daily value curve of radon content in groundwater at Tianshui station

表5　地震前兆异常登记表

Table 5　Summary table of earthquake precursory anomalies

序号	异常项目	台站（点）或观测区	分析方法	异常判据及观测误差	震前异常起止时间	震后变化	最大幅度	震中距 Δ/km	异常类别及可靠性	图号	异常特点	备注
1	M_L3.0地震活动异常	震中200km以内	时间、空间扫描	震前地震活动增强与平静	2005.6～2006.6	恢复正常		震中周围			先增强，后平静	震后发现

续表

序号	异常项目	台站（点）或观测区	分析方法	异常判据及观测误差	震前异常起止时间	震后变化	最大幅度	震中距Δ/km	异常类别及可靠性	图号	异常特点	备注
2	电阻率	武都	日均值	破年变	NW：2006.04～	正常	0.90%	80	S_1	1	破年变低值异常	震前发现
3	水氡	武都	日值	破年变	2005.10～2006.05	正常	140Bq/L	80	M_1	2	破年变高值异常，震前有预测	震前发现
4	应变	武都	日均值	趋势改变	EW：2005.05～	持续	1.8×10^{-7}	80	M_1	3	出现各向同性、压性异常，震前有预测	震前发现
					NE：2005.03～	转为舒张	3.1×10^{-8}					
					NS：2005.06～	正常	2.9×10^{-8}					
					NW：2005.05～	正常	5.5×10^{-8}					
5	水氡	西和	日值	破年变	2006.04～		6Bq/L	110	S_2	4	破年变高值异常	震前发现
6	水氡	天水	日值	破年变	2006.03～	持续	6.5Bq/L	200	S_1	5	破年变低值异常，震前有预测	震前发现

七、地震前兆异常特征分析

文县5.0级地震前的主要异常特征如下：

1. 空间特征

2006年6月21日文县5.0级地震发生在甘肃省南部甘川交界地区，200km范围内定点前兆观测台站不多，受地理环境的限制，大多分布在震中的北东方向。震中200km范围内定点前兆有5个测项、3个台项的异常，0～100km和101～200km范围内各有1个和2个台项的异常，最近的为武都地电、水氡和应变，距震中约80km；200km内最远的为天水水氡，距震中200km。

2. 时间特征

文县5.0级地震前，地震活动性以中期异常为主要特征，6项9条异常中，有7条为半年以上的中期异常，占地震活动性异常项次的78%。而200km范围内的定点前兆5个测项的异常中，半年以上的中期异常为 2 个（武都水氡和应变见图 12、13）台项，占异常测项总数的 40%；1～6 个月的短期异常为 3 个测项（图 11、图 14、图 15），占异常总数的 60%；没有 1 个月以内的临震异常。可见定点前兆以短期异常为主要特征。

3. 数量特征

文县 5.0 级地震前 200km 范围内共 4 个前兆观测台站 6 个观测项目，有 3 个台站的 3 个测项、8 个测道出现前兆异常，其中 100km 范围内共 1 个前兆观测台站，有 1 个台站的 3 个测项、6 个测道出现前兆异常，分别占异常项目和测道总次数的 100% 和 75%；101～200km 范围内共 3 个前兆观测台站，有 2 个台站的 1 个测项、2 个测道出现前兆异常，分别占异常项目和测道总次数的 67% 和 25%；可见 100km 范围内异常测项比和测道比均较高。

八、震前预测、预防和震后响应

1. 震前预测

在 2006 年度甘肃省地震趋势研究报告中，研究确定了三个地震危险区，其中祁连山中东段和甘东南甘青川交界地震重点危险区存在发生 6 级左右地震的可能，祁连山西段注意监视区存在发生 5～6 级地震的可能。甘东南地震重点危险区包括漳县、礼县、舟曲、迭部、碌曲、夏河、岷县一带，文县 5.0 级地震发生在甘东南地震重点地震危险区东部边界附近，空间距离约 90km。

1）中期震情跟踪

2005 年 8 月发生的 3 级以上地震在空间上集中在甘东南地区。8 月 12 日卓尼 3.0 级和 8 月 22 日通渭 2.6 级地震打破了甘东南地区 1～7 月未发生 3 级以上地震的平静状态。结合甘东南地区 1 月 5 日马尔康 4.6 级地震，1 月 19 日宕昌 3.4 级地震，4 月 9 日宕昌 4.5 级地震等中等地震活跃的实际情况，8 月 30 日月会商结论为："下周（月）甘东南地区和祁连山中东段存在发生 5 级左右地震的可能，密切跟踪祁连山西段的震情发展变化"。9 月 3 日 19 时 08 分、20 时 24 分文县 2.9 级、3.2 级地震发生在宕昌—武都—松潘小震活动水平较高地区，经临时会商认为："短期内甘东南地区存在发生 5 级左右地震的可能，仍需密切各种资料的发展变化"。针对 2006 年 1 月 2 日发生在迭部白龙江断裂东端 4.0 级地震召开的临时会商会，对近一段时间甘东南地区的地震活动性进行了梳理，发现两个重要的现象，一是自 2005 年 11 月开始甘青川交界地区 4 级地震活动增强明显。二是甘东南—甘青川交界地区出现了多次中等地震成组活动现象。1 月 16 日对甘东南地区震情重新研判，认为甘肃及边邻地区存在发生 5～6 级地震的背景。2006 年 2 月甘肃及边邻地区地震活动低水平活动，一方面多次出现间隔数小时的小震成组活动现象，另一方面 3 级以上地震活动水平偏低，一度排除了省内发生 5.5 级以上地震的可能。2 月 22 日月会商结论是："下周（月）甘肃省内发生 5.5 级以上地震的可能性较小，注意跟踪各种资料的发展变化"。

2）短期阶段震情跟踪

2006 年 3 月 27 日宕昌 4.5 级地震发生在临潭—宕昌断裂带东端，与 2005 年 4 月 9 日宕昌 4.5 级地震构成较大地震的成组活动。该地震的位置与 2005 年 11 月 2 日以来沿北西向白龙江断裂形成 2 级地震平静相关，认为短期内应注意甘青川交界地区震情发展与变化。

2006年3月29日月会商分析认为："下周（月）甘肃省内发生6级以上地震的可能性较小，注意甘东南和祁连山中东段地区震情发展和变化。" 4月26日月会商分析认为："下周（月）省内发生6级以上地震的可能性较小，但存在发生5级左右地震的背景条件，重点跟踪甘东南和祁连山中东段地区"。5月30日月会商分析认为："下周（月）甘肃省内存在发生5级左右地震的背景条件，重点注意祁连山中东段和甘东南—甘青川交界地区"。可见，短期阶段还是提出了比较明确的预测意见。

3）短临阶段震情跟踪

文县5.0级地震前没有做出明确的短临预测意见，也没有发现较突出的异常现象。纵观文县5.0级地震前的震情跟踪过程，发现在5.0级地震前的中期阶段，中等地震活动明显增强，到短期阶段，小地震成组地震活动明显增强，而在短临阶段，震中所在区域的地震活动处于弱活动状态。各阶段前兆异常也以趋势异常为主。

2. 震后趋势判定

文县5.0级地震后，甘肃省地震局立即组织地震预报评审委员会专家和全体分析预报人员召开了紧急会商会，对地震类型和震后趋势进行了分析判定。6月21日会商结论认为，"原震区24小时内发生更大地震的可能性不大，短期内甘东南地区不排除5～6级地震发生的可能"。地震序列分析结果为"截至6月22日11点00分，全序列共发生余震50次，其中3.0～3.9级地震2次、2.0～2.9级地震3次、1.0～1.9级地震10次、0.1～0.9级地震25次，最大余震为3.3级。余震呈NW向分布，长轴约为9.5km。根据现有序列分析认为属于主震—余震型的可能性较大"。震后趋势及序列类型的判定基本正确。

3. 震后响应

文县5.0级地震后，中国地震局、甘肃省地震局、陕西省地震局和陇南市地震局现场联合工作队在震区约800多平方公里的范围内，完成29个抽样点、152个调查点的调查，调查建筑物、工程结构、地表破坏500余处，累计行程4000余公里，取得了地震灾害损失评估的第一手资料，为得到科学、准确的评估结果提供了保障。

文县5.0级地震发生后，甘肃省地震局立即启动地震应急预案，派出地震现场工作队和现场强震观测组赶赴震区进行现场调查和强化地震监测工作。6月24日下发了甘震发66号文"甘肃省地震局关于加强震情跟踪工作的通知"，采取了一系列行之有效的跟踪措施。一是加大了资料的分析处理力度，坚持每天跟踪分析资料，保证重大宏、微观异常的及时发现，强化了各类异常的跟踪和分析；二是严格执行定期会商和会商意见按时上报制度，随时召开震情紧急会商会，及时做出震后趋势判定意见；三是加大了异常落实力度，对突出的前兆变化及时开展各种形式的落实工作，并于6月28日至7月1日在甘东南地区开展了现场异常落实；四是强化了文县5.0级地震序列的跟踪与分析，震后分别于12小时、24小时、48小时，3天、一周召开临时和周月会商会，及时将序列发展情况上报，并对序列和震后趋势进行了动态跟踪与判定；五是加强对数字化前兆、数字化测震资料的应用力度，希望通过数字化资料的应用，尽可能在中等—中强地震前发现一些新的短临信息，进一步提高了短临预报的科学含量和能力。在文县地震后的震情短临跟踪过程中，强化分析处理的资料包括电磁学科31个台项98个测项、地下流体24个台项60个测项、形变学科2个台项6个测项的资料；六是加强了与中国地震局、边邻省区和地方地震局和中心台震情信息的沟通，加大了对四川、青海、宁夏、二测中心的资料的应用力度。

4. 地震灾害特点

（1）震中及附近地区民房绝大多数是土木结构房屋，抗震性能较差，虽然震后屋架完全倒塌很少，但墙体开裂、屋顶掉瓦、房脊震落和部分柱子发生移位，扭转等震害十分普遍。

（2）由于震区地质条件复杂，多数村子座落在半山坡和河谷地区，地震引发多处滑坡、山石崩塌等地质灾害，损坏公路，砸坏通信和电力设施，导致交通运输、通信和供电中断，加重了地震灾害。

九、结论与讨论

（1）2006 年 6 月 21 日文县 5.0 级地震发生在构造环境复杂地区。文县 5.0 级地震之前已有 100 多年没有发生中强地震，该地震的发生说明该地区地震活动开始增强。文县 5.0 级地震没有形成地表破裂带，给研究发震构造带来一定的困难，1:20 万地质图显示该区域存在多条断裂，遥感资料解译结果表明石坊—临江断裂为较活动断裂。两种方法所得震源机制结果表明该地震为左旋走滑兼有逆冲型地震，主压应力方向为 NE60°。双差法定位结果也支持该地震为走滑兼有逆冲型地震，余震的分布与断裂的逆冲有关。结合多种结果联合分析认为该地震的发震构造为石坊—临江断裂，主压构造应力方向为 NE60°。文县 5.0 级地震发生在甘东南南部石坊—临江断裂上，在同一构造应力作用下做左旋走滑运动和逆冲运动的结果。

（2）地震发生在 M_L3 地震集中活动，又出现异常平静的背景下。研究区域范围内其他前兆观测资料存在水氡、应变和地电阻率共 5 项次异常。异常主要表现为武都水氡日值破年变异常；武都电阻率日均值破年变异常；武都应变日均值趋势改变异常；西和水氡日值破年变异常，天水水氡日值破年变异常等。这些异常都表现为中期尺度的异常。

（3）此次地震没有发生在 2006 年度地震危险区内。近几年甘东南地区的年度地震危险区，因为考虑到西秦岭北缘断裂以及宕昌—碌曲等断裂的地震危险性更大，所以年度地震危险区的范围偏西，直接导致该地震的预测失败。地震震情跟踪隐约在遵从“小震闹，大震到”的思路，中小地震活动增强，认为地震发生的危险性更高，曾一度指出地震发生的地点，并将地震发生的时间提至短期，还先后排除了 5.5 级、6 级地震。但随着地震活动的减弱和迁移，认为地震发生的可能性不大或者仅有背景等，对该地区地震的孕育发生规律还需要进一步的跟踪研究。地震发生后，甘肃省地震局立即启动地震应急预案，开展地震应急。指示陇南市和文县地震局先期组织地震现场工作队赶赴震区进行现场调查；组织召开紧急会商会，同时派出地震现场工作队和现场强震观测组赶赴震区进行现场调查和强化地震监测工作。甘肃省地震局预报中心在序列判定、余震预测方面做了大量的工作，判定该地震序列为主震—余震型地震序列，最大余震在 4 级左右，为震后救灾作出了贡献，取得了一定的社会效益和减灾实效。

参 考 文 献

冯希杰，董星宏，刘春等 . 2005. 范家坝—临江断裂活动与1879年甘肃武都南8辑地震的讨论 . 地震地质，27（1），155～163

刘杰，郑斯华，康英等 . 2004. 利用P波和S波初动和振幅比计算中小地震的震源机制解 . 地震，24（1），19～26

马占虎，周志宇，高晓明等 . 2007. 2006年甘肃文县5.0级地震灾害损失评估 . 西北地震学报，29（3）：256～263

卢海峰，马保起，刘光勋 . 2006. 甘肃文县北部北东东向断裂带新构造活动特征 . 地震研究，29（2）：143～146

王周元，范世宏，姬凤英等 . 1984. 甘肃地区地壳速度的非均匀分布 . 西北地震学报，18（2）：18～25

曾秋生 . 1989. 中国现今地壳应力状态，地质力学学报 . 中国地质力学学院，地质力学研究所所刊，第12号，197～207

地质部陕西省地质局区域地质测量队 . 1967. 1∶20万地质图 . 碧口幅 . 北京：地质出版社

地质部陕西省地质局区域地质测量队 . 1970. 1∶20万地质图 . 文县幅 . 北京：地质出版社

代炜，陈立琼，严武建 . 2006. 2006年文县5.0级地震序列特征和震前 M_L3.0地震活动特征 . 西北地震学报，29（3），293～295

代炜，张辉，冯建刚等 . 2009. 2006年文县5.0级地震的发震构造研究 . 地震地质，31（3），424～432

参 考 资 料

甘肃省地震局，甘肃地质目录库（2006）

甘肃省地震局，2006年度甘肃地震趋势研究报告卷宗，2005.11

中国地震局兰州地震研究所，论著编号：LC2009029

中国地震局兰州地震研究所地震地质队，文县地震地质工作小结，1975

中国地震局兰州地震研究所地震地质队，陕甘宁青大震调查，1984

Wenxian Earthquake of M_S 5.0 on June 21, 2006 in Gansu Province

Abstract

A earthquake of M_S 5.0 occurred at the Linjiang town Wenxian county in Gansu province on June 21, 2006. Its macroseismic epicenter was located in the zone form Linjing town to Qiaotou town. The intensity at the epicenter was Ⅵ. A people was killed, 19 people were injured and the direct economic loss was 73.35152 million yuan.

The earthquake sequences belonged to mainshock-aftershock type and the magnitude of the largest aftershock was M_L 3.6. Aftershocks destributed NW zone and exist a angle with direction of the major axis of the meizoseismal area. Triggering seismic fault is the Shifang-Linjiang fault, based on the stress of role tectonic stress NE60, the fault caused this earthquake. Focal mechanism derived from the two methods results showed that the earthquake is left-lateral slip.

Within the distance of 200km from the epicenter, there were 16 stations, 6 seismometric stations, 2 composite stations, and 2 single precursory stations among the 16 stanions. Totle 7 reguler seismic observation project such as seismometric observations, water radon, flow measure, resistivity, deformation, geo magnetic, stress and strain. There were 6 anomaly observation iteM_S before the earthquake.

2006 年 7 月 22 日、8 月 25 日云南省盐津 5.1、5.1 级地震

云南省地震局

张　立　赵小艳

摘　要

2006 年 7 月 22 日、8 月 25 日，云南省昭通市盐津县相继发生了 5.1、5.1 级地震，宏观震中位于盐津县中和、豆沙至大关县吉利一带，极震区烈度Ⅶ度，等震线形状呈北东向椭圆形。地震造成 24 人死亡，28 人重伤，158 人轻伤，直接经济损失 44170 万元。

此次地震序列为震群型。最大余震 3.5 级，余震分布呈北北西向分布。7 月 22 日盐津 5.1 级地震后，8 月 25、29 日相继发生 5.1 级、4.7 级地震，3 次 $M_S \geqslant 4.7$ 级地震震源机制解基本一致，节面Ⅰ为主破裂面，主压应力 P 轴方位近南北向，北西向马边—盐津断裂为发震构造。

盐津地震周围 200km 范围内共有地震台站 33 个，其中测震观测台 9 个，定点前兆观测台站 24 个。震前共出现 11 个异常项目，21 条前兆异常，多为短临异常。

前　言

2006 年 7 月 22 日 09 时 10 分、2006 年 8 月 25 日 13 时 51 分、2006 年 8 月 29 日 09 时 14 分云南省盐津县先后发生 5.1 级、5.1 级和 4.7 级地震，据云南地震台网测定，微观震中分别为 28°01′N、104°08′E，28°02′N、104°06′E 和 27°59′N、104°09′E，7 月 22 日 5.1 级地震宏观震中位于盐津县豆沙至大关县吉利一带，震区烈度Ⅵ度，有个别Ⅶ度破坏点[1)]；8 月 25 日 5.1 级和 29 日 4.7 级地震宏观震中位于盐津县中和、豆沙至吉利一带，极震区烈度为Ⅶ度，有个别Ⅷ度破坏点[2)]。地震造成 24 人死亡，28 人重伤，153 个轻伤，失踪 1 人，直接经济损失 44170 万元人民币[1,2)]。

这三次地震是继 2005 年 8 月 5 日会泽 5.3 地震后发生在滇东北地区的又一组中强地震；历史上震中区附近曾发生过多次 $M_S \geqslant 5$ 级地震，地震最大强度为 1974 年大关 7.1 级地震，震中 300km 范围内共有地震台 33 个，震前出现了 11 个异常项目的 21 条前兆异常。

盐津地震前，云南省地震局对该地区的地震危险性作了较好的短期预测预报[3)]。

地震发生后云南省地震局 2 次派出工作组赴震区开展震害评估、地震地质考察和余震监

视工作。

一、测震台网及地震基本参数

盐津地震前，震中周围 200km 内共有测震台 9 个（遥测台 6 个），其中 100km 内有 4 个，101～200km 内有 5 个（图 1）。本次地震参数、地震序列采用云南台网资料，震中区测震台网监控能力为 M_L1.1，盐津地震基本参数列于表 1 中。

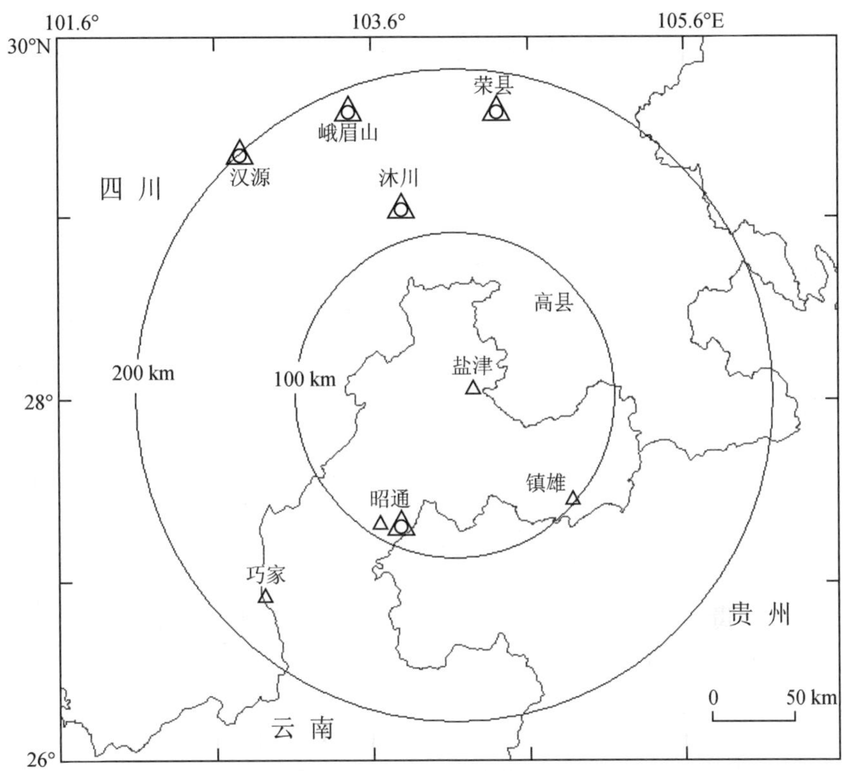

图 1　盐津地震前震中附近测震台网分布图

Fig. 1　Distribution of seismometric stations around the epicentral area of Yanjin earthquake

表 1　地震基本参数

Table 1　Basic parameters of the earthquake

编号	发震日期	发震时刻	震中位置		震级				震源深度	震中	结果
	年 月 日	时 分 秒	φ_N	λ_E	M_S	M_L	M_b	M_w	(km)	地名	来源
1	2006 07 22	09 10 22.0	28°01′	104°08′	5.1				9	盐津	资料 6)
2	2006 08 25	13 51 40.7	28°02′	104°06′	5.1				7	盐津	
3	2006 08 29	09 14 17.1	27°59′	104°09′		4.7			8	盐津	资料 7)

2006年7月22日09时10分云南盐津县发生5.1级地震后，数字流动台网在震中区架设了豆沙、吉利、柿子、万古4个临时观测站，结合盐津地震台，形成一个以5个观测台为主的地震现场监测台网，截止到7月26日台网撤离，台网共记录到盐津余震1037次，震级范围为－1.8～3.5级。2006年8月25日13时51分云南盐津县再次发生5.1级地震后，数字流动台网在震中区架设了豆沙、吉利和柿子3个临时观测站，结合盐津地震台，形成一个以4个观测台为主的地震现场监测台网，截至8月31日台网撤离，台网共记录到余震1339次，震级范围为－2.2～4.9级。表2列出了两次地震现场台网临时台站基本参数。台网分布见图2[1]。

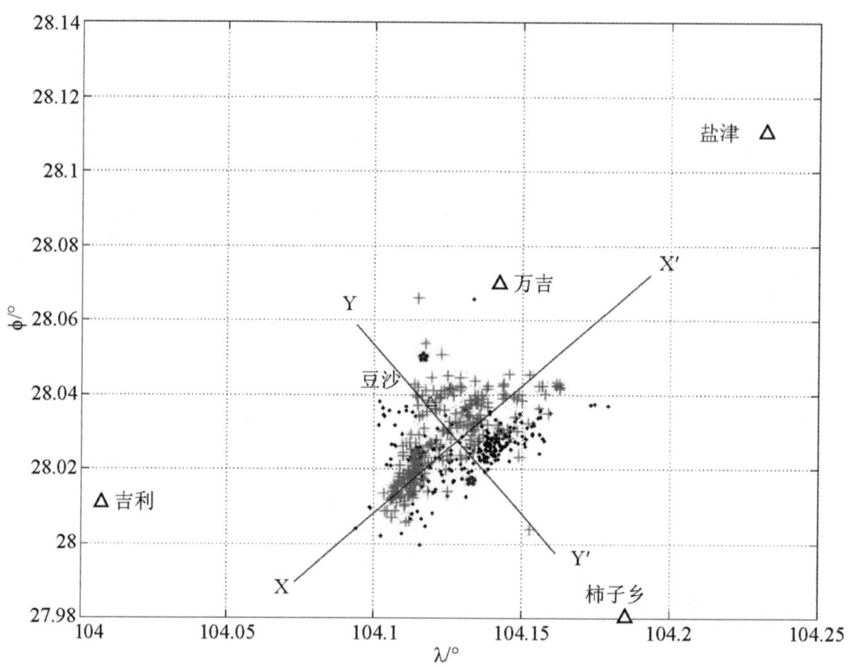

图2　盐津地震序列及地震现场流动测震台站分布（据叶建庆）

Fig. 2　Distribution of the Yanjin earthquake sequence and seismometric running stations

（三角形表示台站，黑色圆点为7月22日地震序列分布，绿色十字为8月25日地震序列分布，红色五星为两次5.1级地震）

表2　地震现场流动台网台站基本参数

Table 2　Basic parameters of the seismometric running stations

编号	台名	东经（°）	北纬（°）	高层（m）	台基	响应灵敏度 Count/（μ m/s）			现场台网监测时间
						垂直向	东西向	南北向	
1	豆沙	104.1191	28.0379	566.52	土层	291	291	291	7月22～26日
2	盐津	104.2315	28.1116	475.22	砂岩	291	291	291	7月23～26日
3	柿子	104.1846	27.9803	463.93	石灰岩	291	291	291	7月23～26日
4	吉利	104.0067	28.0109	556.67	土层	291	291	291	7月23～26日

续表

编号	台名	东经（°）	北纬（°）	高层（m）	台基	响应灵敏度 Count/（μ m/s）			现场台网监测时间
						垂直向	东西向	南北向	
5	万古	104.1427	28.0699	694.21	土层	291	291	291	7 月 23～26 日
6	豆沙	104.1191	28.0379	566.52	土层	291	291	291	8 月 25～31 日
7	盐津	104.2315	28.1116	475.22	砂岩	291	291	291	8 月 25～31 日
8	柿子	104.1846	27.9803	463.93	石灰岩	291	291	291	8 月 25～31 日
9	吉利	104.0067	28.0109	556.67	土层	291	291	291	8 月 25～31 日

二、地震地质背景

震区地处扬子准地台滇东台褶带滇东北台褶束，区内地质构造复杂，发育有北东向、北西向和近东西向三组区域褶皱、断裂构造，主要断裂有北东向的五莲峰—化蓥山断裂和北西向马边—盐津断裂[1)]。

历史上震中区附近（27°00′～28°30′N，103°00′～104°48′E）共记有 22 次 $M \geqslant 5.0$ 级地震，其中 5.0～5.9 级地震 20 次，6.0～6.9 级地震 1 次，7 级以上地震 1 次。震区附近发生过 1974 年大关 7.1 级地震。盐津三次地震位于马边—盐津断裂东西端的东侧，震中位于北东向大田坝向斜的西部，震中附近地表未见明显的断裂构造（图 3）。

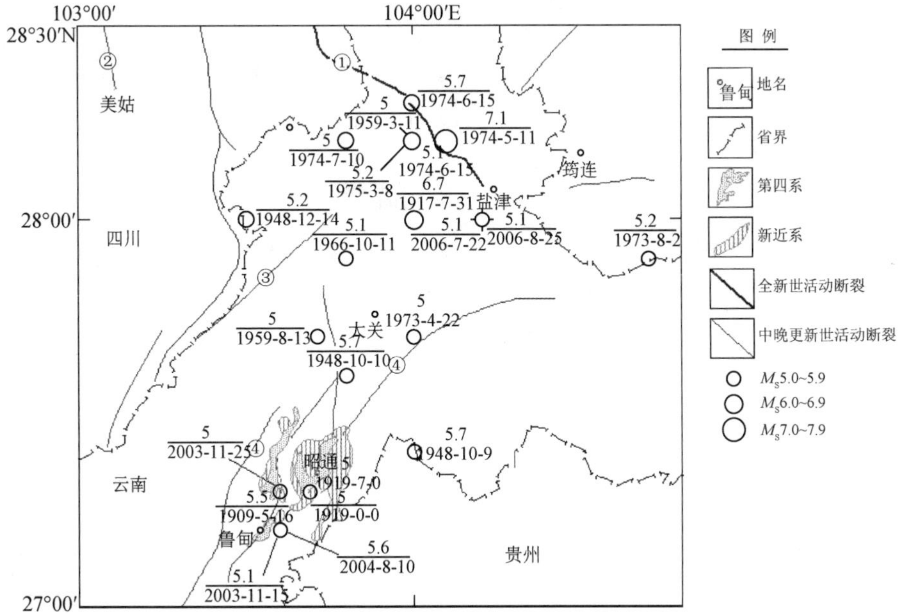

图 3 震中附近地质构造及历史地震震中分布图（据谢英情）

Fig. 3 Map of geological structure and distribution of historical earthquakes around the epicenter

断裂名称：①荥经－马边－盐津断裂 ②西河－美姑断裂 ③五莲峰－华蓥山断裂 ④昭鲁断裂

三、地震影响场和震害[1,2)]

7 月 22 日 5.1 级地震的烈度分布如图 4a。宏观震中位于盐津县豆沙至大关县吉利一带，震区烈度Ⅵ度，有个别Ⅶ破坏点，等震线形状呈椭圆形，长轴走向为北东向。总面积为 890km^2。

Ⅵ度区范围：北自盐津县普洱镇桐子村，南到大关县天星镇南甸村，东自盐津县盐井镇黎山村，西近大关县高桥乡政府驻地。Ⅵ度区面积约 890km^2。其中，大关县吉利至豆沙一带的房屋有倒塌现象，滚石、崩塌、滑坡现象常见。

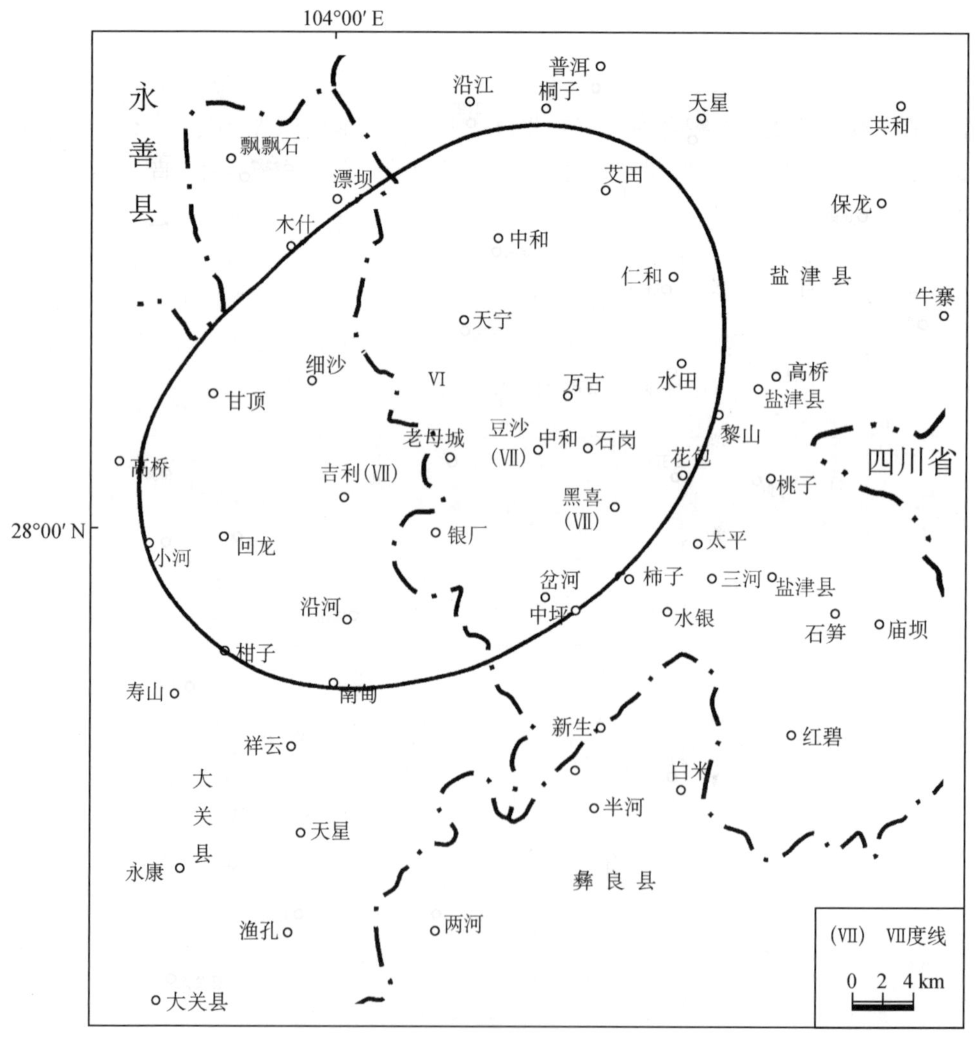

(a)

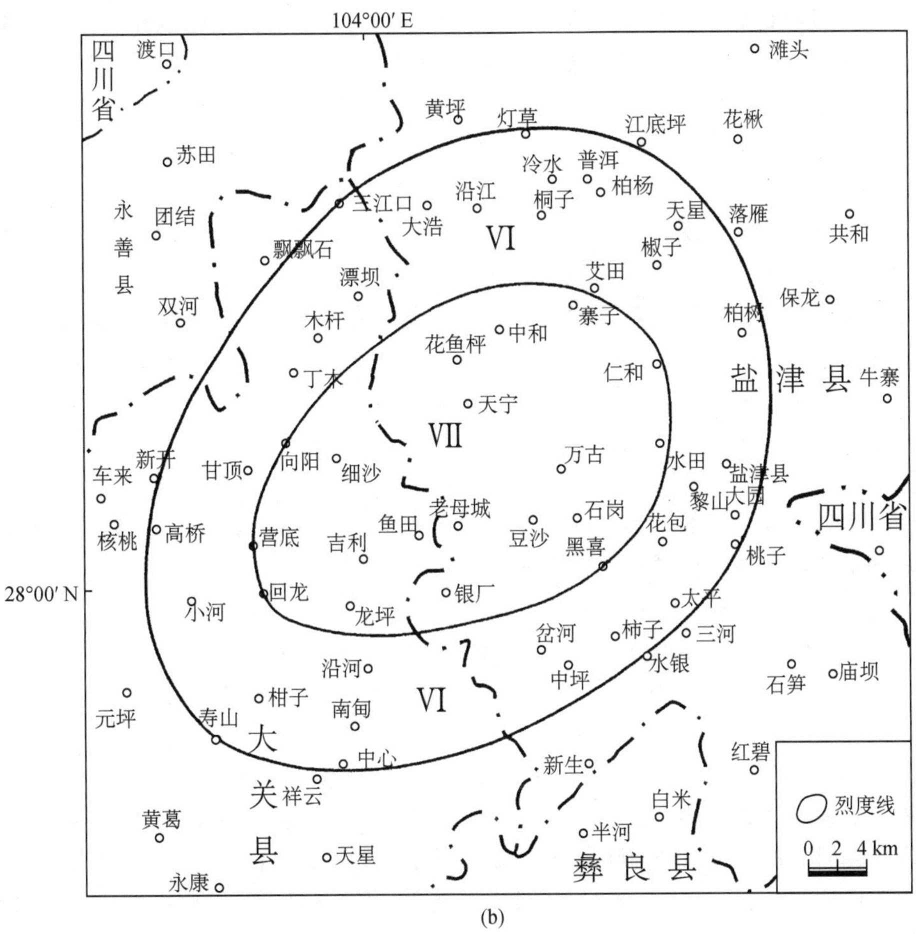

图 4 盐津地震烈度分布图

Fig. 4 Isoseimal map of the Yanjin earthquakes

(a) 7 月 22 日盐津 5.1 级地震烈度分布图；(b) 盐津 3 次 $M \geq 4.7$ 级地震综合烈度分布图

本次地震灾区主要涉及盐津县的豆沙、柿子、盐井、中和、普洱和大关县的吉利、天星、木杆、高桥和寿山等 10 个乡（镇），48 个村民委员会；地震造成村镇民用房屋破坏，生命线工程及水利等基础设施损坏。地震中有 22 人死亡，13 人重伤，101 个轻伤，失踪 1 人，20606 人无家可归，35428 户、151168 人受灾。地震造成直接经济总损失达 23900 万元人民币。

8 月 25 日 5.1 级和 8 月 29 日 4.7 级地震是继 7 月 22 日 5.1 级地震之后再次发生的两次地震，震害叠加，宏观震中位于盐津县中和、豆沙至大关县吉利一带，极震区烈度为Ⅶ度，有个别Ⅷ度破坏点。

地震造成村镇民用房屋破坏，生命线工程及水利等基础设施损坏。地震中有 2 人死亡，15 人重伤，52 个轻伤，18509 人无家可归，61170 户、262538 人受灾。地震造成直接经济总损失达 20270 万元人民币。

三次地震震中相近、间隔时间不久，灾区重叠、震害叠加，现场调查很难区分前发地震和两次续发地震的震害。盐津三次地震的综合烈度如图 4b。宏观震中位于盐津县中和、豆沙至大前县吉利一带，极震区烈度Ⅶ度，等震线形状呈椭圆形，长轴走向为北东向。Ⅵ度以上区域总面积为 1350km^2。

Ⅶ度区北自盐津县中和镇寨子村，南至大关县吉利镇龙坪村，东起盐津县盐井镇水思村，西到大关县吉利镇营底村，面积约 501km^2。其中，盐津县中和、豆沙至大关县吉利一带的房屋有倒塌现象，滚石、崩塌现象常见。

Ⅵ度区范围：北自盐津县普洱镇灯草村，南到大关县天星镇中心村以南，东自盐津县盐井镇柏树村以东，西至大关县高桥乡政府驻地以西。Ⅵ度区面积约 849km^2。

三次地震灾区主要涉及盐津县的中和、普洱、豆沙、盐井、柿子和大关县的吉利、天星、木杆、高桥和寿山等 10 个乡（镇），74 个村民委员会；受灾人口 262538 人，涉及 61170 户。

评估区以外，大关县的翠华、悦乐、彝良县的两河、钟鸣、永善县的团结和绥江县的会仪等乡镇的部分民房也遭受了不同程度的破坏。

三次地震震害主要特征为：

（1）三次地震灾区范围广，达 1350km^2；震害重，造成较大范围的Ⅶ度破坏区（501km^2），甚至有个别Ⅷ度破坏点。主要原因大致如下：①震源深度浅、地表振动强烈。两次 5.1 级地震的震源深度分别是 9 和 7km，强震组布置在豆沙镇政府驻地的强震仪获得 4.7 级地震的地震记录：南北方向峰值加速度 494.2Gal，东西方向峰值加速度 389.3Gal、垂直方向峰值加速度 282.1Gal。②震害叠加。两次 5.1 级地震微观震中相距仅 4km，灾区重叠；发震间隔时间短，仅一个月零 3 天，震害叠加；8 月 25 日 5.1 级地震以后发生多次强余震，也造成明显的震害叠加，8 月 27 日 3.3 级和 8 月 29 日 4.7 级地震中有民房倒塌现象；强余震造成的零星倒塌现象广泛分布在灾区；8 月 27 日 2.8 级余震中有一个死亡，8 月 29 日 4.7 级地震造成 27 人受伤，其中重伤 8 人。震害叠加是灾区地震灾害重的重要原因之一。震害叠加使前震灾区的外延区域出现明显震害，灾区范围扩大。

（2）震区地处滇东北高原乌蒙山区，地形起伏，盐津县最高海拔 2263m，最低 300m，相对高差近 2000m。震区沟壑纵横，山高谷深，山体破碎，震区滚石、崩塌和滑坡现象普遍，地震地质灾害比较严重。滚石、崩塌和滑坡体造成部分人员伤亡和房屋破坏；公路破坏，交通堵塞甚至中断；电力、通信中断。

（3）震区绝大多数土（石）木结构房屋是抗震能力极差的土搁梁房屋，没有木构架，主要由夯土墙或者堆石墙体承重，部分砖混结构房屋为土石墙承重，地震中极易倒塌或严重破坏。灾区房屋建筑抗震性能差。

（4）震区多数水库为 20 世纪 50、60 年代建造，个别为 40 年代建造，后经多次排险加固，但病险难以根除。这次地震造成水库坝体开裂或使已有裂缝加宽变长变深，已有的病险进一步加剧，出现险情。

（5）震区多灾并发。在一个月余的时间内连续发生三次破坏性地震和多次强余震，震害叠加，造成人员伤亡、房倒屋塌，引发滚石、崩塌和滑坡等地质灾害；灾区多次遭受大风和暴雨袭击，造成大片农作物和林木倒折，引发局部滑坡，加重灾情。震区各县区属国家级的扶贫县，2005 年盐津、大关县农民人均收入分别是 1037 元、1022 元（全国农民人均收入 3200 元），当地自然条件较差，经济发展缓慢，财政收支倒挂、入不敷出，十分困难，抗震自救能力很弱。

7 月 22 日 5. 1 级地震后，当日深夜 22：30，在盐津县公安指挥中心的教室内及柿子乡乡政府架设了 2 台最新进口的美国 K2 型数字强震仪进行强震应急观测（图 5）。从 22 日 22：30 到 25 日下午 14：30 为止，2 台强震仪器共记录到地震事件 114 条[4)]。

8 月 25 日 5. 1 级地后，于 26 日 01 时和 11 时，在豆沙乡政府和万古村村委会架设了 2 台最新进口的美国 K2 型数字强震仪进行强震应急观测（图 5）。截至 8 月 30 日，2 台强震仪共记录到 83 条强震记录，其中最大余震为 4. 7 级。对于 4. 7 级余震，豆沙强震记录的峰值为：东西向 389. 3Gal，南北向 494. 2Gal，垂直向 282. 1Gal；万古强震记录的峰值为：东西向 188. 3Gal，南北向 220. 1Gal，垂直向 149. 0Gal。豆沙 4. 7 级余震的强震记录中，（S—P）约为 0. 95s[5)]。

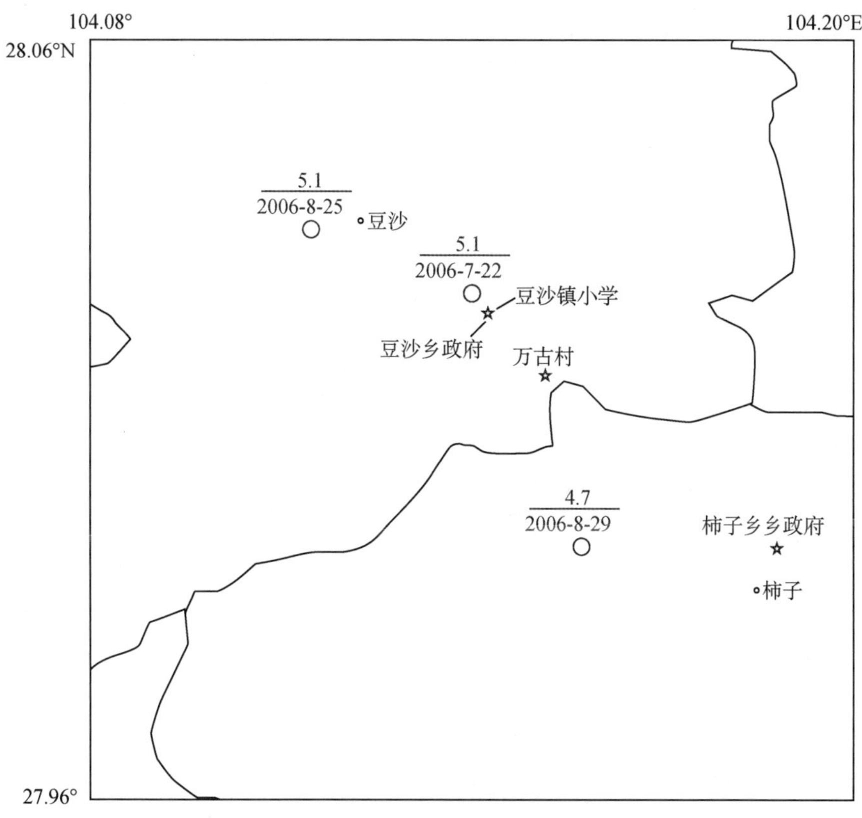

图 5 盐津地震流动强震观测点分布图

Fig. 5 Distribution of the Yanjin earthquake strong shock running stations

四、地震序列

据昆明区域数字地震台网记录，2006 年 7 月 22 日至 10 月 7 日，共记录到 $M_L \geqslant 1.0$ 级 147 次，其中 1.0～1.9 级 110 次，2.0～2.9 级 29 次，3.0～3.9 级 5 次，4.0～4.9 级 1 次，5.0～5.9 级 2 次。序列最大余震为 4.7 级，发生在第一次 5.1 级地震 37 天后、第二次 5.1 级地震 3 天后（表 3）。

表 3　盐津 5.1、5.1 级地震序列目录（$M_L \geqslant 2.5$ 级）

Table3　Catalogue of the M_S 5.1 and M_S 5.1 Yanjin earthquake sequence（$M_L \geqslant 2.5$）

编号	发震日期	发震时刻	震中位置		震级				震源深度	震中	结果
	年 月 日	时 分 秒	φ_N	λ_E	M_S	M_L	M_b	M_w	h/km	地名	来源
1	2006 07 22	09 10 22	28°01′	104°08′	5.1				9	盐津	资料 4）
2	2006 07 22	23 16 28	28°00′	104°25′		2.5			6	盐津	
3	2006 07 24	00 03 00	28°03′	104°05′		2.8			1	盐津	
4	2006 08 09	11 04 13	27°56′	104°13′		2.7			5	彝良	
5	2006 08 15	12 59 40	27°52′	104°16′		2.9			0	盐津	
6	2006 08 23	12 06 21	28°02′	104°05′		2.5			7	盐津	
7	2006 08 25	08 57 31	28°02′	104°05′		3.2			1	盐津	
8	2006 08 25	13 51 40	28°02′	104°06′	5.1				7	盐津	
9	2006 08 25	23 49 11	28°01′	104°09′		2.6			0	盐津	
10	2006 08 26	02 00 18	28°01′	104°09′		2.6			9	盐津	
11	2006 08 26	02 36 47	28°01′	104°09′		3.3			0	盐津	
12	2006 08 27	08 54 01	28°00′	104°06′		2.8			4	盐津	
13	2006 08 29	09 14 17	27°59′	104°08′		4.7			8	盐津	
14	2006 08 29	12 19 34	27°59′	104°09′		2.6			9	盐津	
15	2006 08 29	12 43 05	28°00′	104°08′		2.8			8	盐津	
16	2006 08 29	13 15 29	27°59′	104°09′		2.5			0	盐津	
17	2006 08 29	13 35 43	28°00′	104°08′		3.1			2	盐津	
18	2006 08 31	08 26 11	28°01′	104°06′		3.5			0	盐津	
19	2006 10 02	14 15 58	28°01′	104°06′		3.2			2	盐津	

本次地震序列的 2 次最大地震和次大地震震级差为 $\Delta M = 5.1 - 4.7 = 0.4$，$\Delta M \leqslant 0.6$，地震序列为震群型。7 月 22 日 M_S5.1 地震、8 月 25 日 M_S5.1 地震的能量均占整个序列总能量

的 44.14%。这 2 次事件占了整个序列能量释放的 88.28%，2006 年 8 月 29 日 M_L 4.7 地震占整个序列能量释放的 11.08%（图 10）。

由图 8 可见，第一个 5.1 级地震之后共发生地震 37 次，其中 1.0 ~ 1.9 级 26 次，2.0 ~ 2.9 级 10 次，3.0 ~ 3.9 级 1 次，序列小震次数明显偏少，序列缺 3 ~ 4 地震。从图 6a 可以看出，第一个 5.1 级地震序列震级频度分布线性关系离散，b = 0.58，h 值也有明显异常，取最小震级为 1.1，h = 0.36。从盐津地震序列 1 级以上地震 M-t 图也可以看出，7 月 24 日发生一次 2.8 级地震后，直至 8 月 1 日 1.5 级地震，序列出现了 8 天的平静异常，紧接着余震却出现增强活动过程。因此，从整个序列的衰减情况并结合参数计算来看，7 月 22 日盐津 5.1 级地震序列衰减明显异常，指示震中及附近可能有更大或相当地震发生。

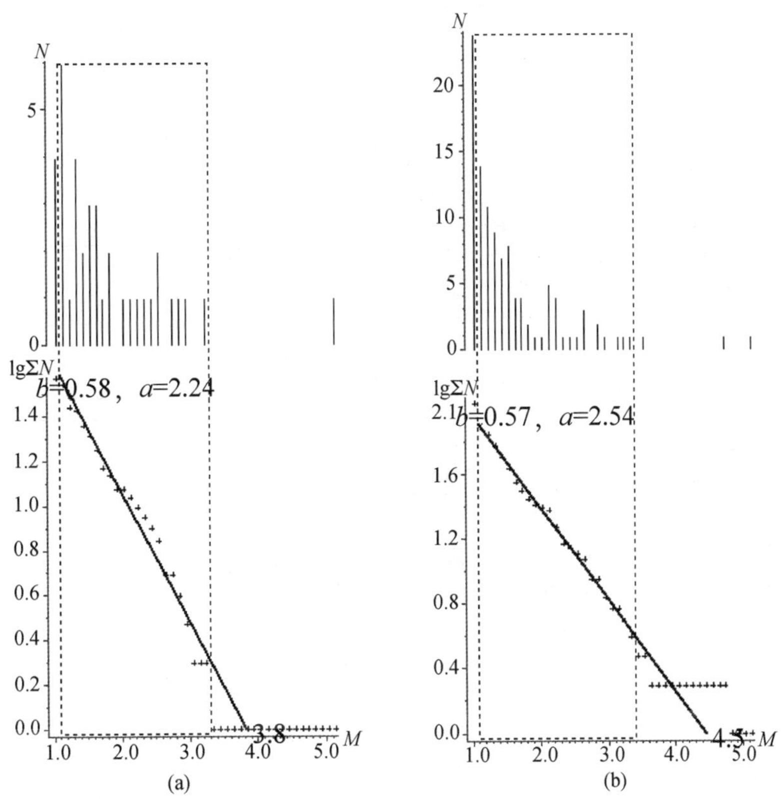

图 6　盐津地震序列 b 值图

Fig. 6　b-value diagram of the Yanjin earthquake sequence

（a）盐津第一个 5.1 级地震序列 b 值图（$M_L \geqslant 1.1$ 级）；

（b）盐津第二个 5.1 级地震序列 b 值图（$M_L \geqslant 1.1$ 级）

8 月 25 日 5.1 级地震之后共发生地震 108 次，其中 1.0 ~ 1.9 级 84 次，2.0 ~ 2.9 级 19 次，3.0 ~ 3.9 级 4 次，4.0 ~ 4.9 级 1 次，相比 7 月 22 日 5.1 级地震，小震次数明显增多。8 月 25 日 5.1 级地震序列震级频度分布均较好的遵从 G-R 关系，其序列 b 值为 0.57。由图 7b 还可以得到第二个 5.1 级地震序列 h = 1.2。从盐津地震序列 1 级以上地震 M-t 图及蠕变曲线也可以看出，虽然序列强度及频度随时间的变化有一定的起伏，但从总的来说，序列余震次数多，能量释放充分。8 月 25 日 5.1 级地震后，序列震级频度分布线性关系较好，h 值恢复正常，随时间增长，序列逐渐趋于结束。

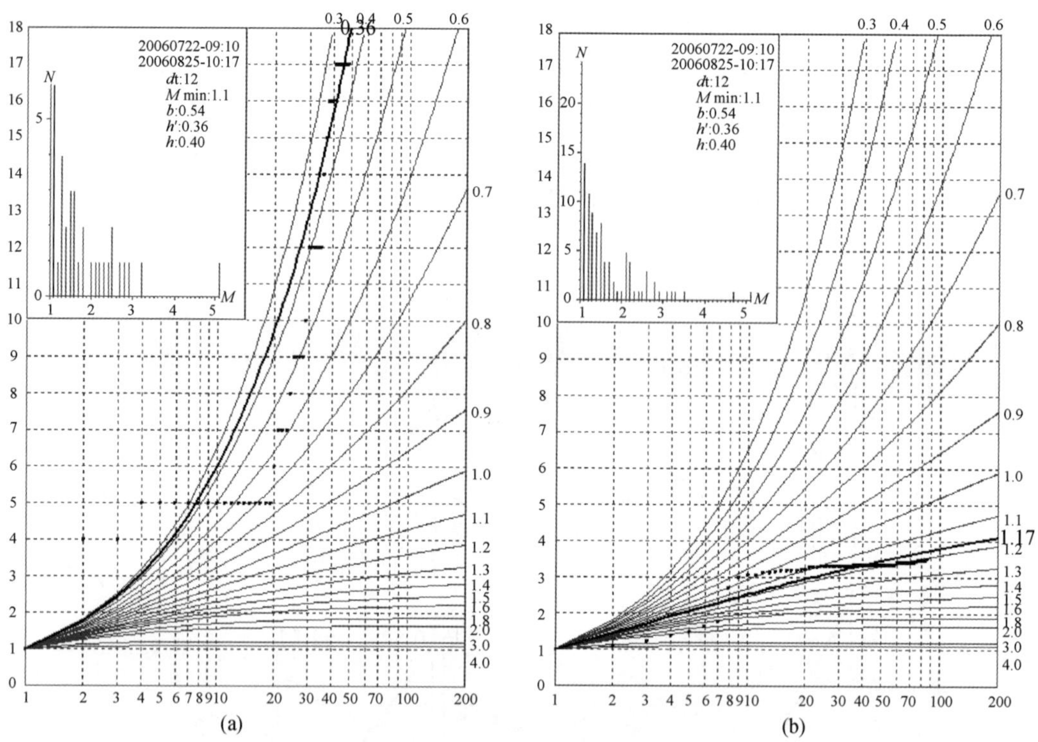

图 7　盐津地震序列 h 值图

Fig. 7　h-value diagram of the Yanjin earthquake sequence

(a) 盐津第一个 5.1 级地震序列 h 值图（$M_L \geqslant 1.1$ 级）；

(b) 盐津第二个 5.1 级地震序列 h 值图（$M_L \geqslant 1.1$ 级）

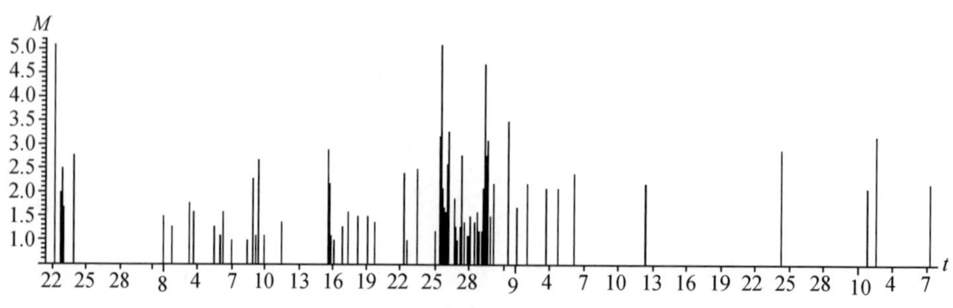

图 8　盐津地震序列 M-t 图

Fig. 8　M-t diagram of the Yanjin earthquake sequence（$M_L \geqslant 1.0$）

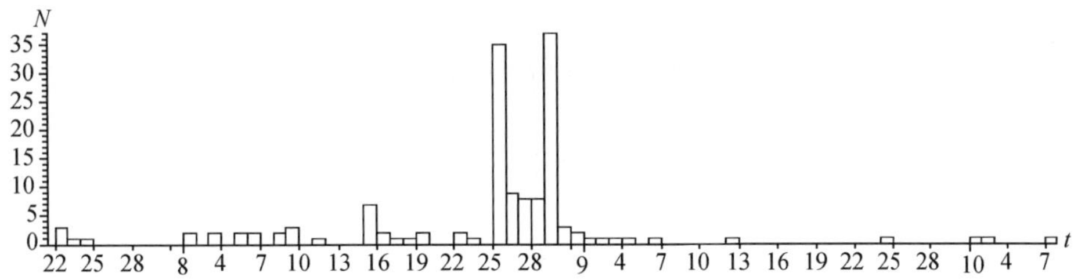

图 9 盐津地震序列 N-t 图

Fig. 9 Curve of daily frequency of the Yanjin earthquake sequence

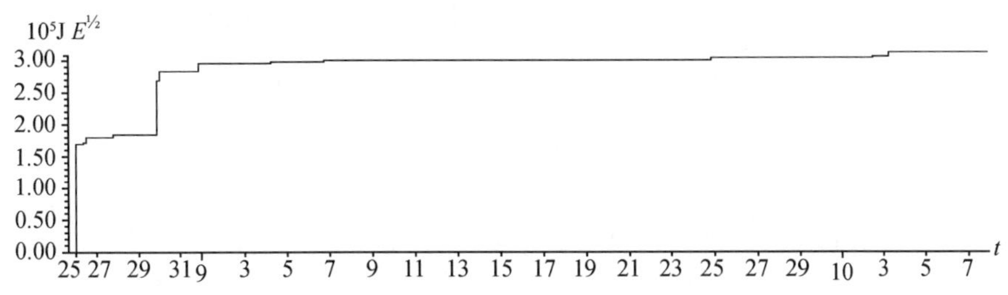

图 10 盐津地震序列蠕变曲线图

Fig. 10 Curve of strain release of the Yanjin earthquake sequence

由地震序列震中分布图（图 11）可见，2 次盐津地震序列余震震中空间分布形成北西向，其动态变化过程为：7 月 22 日 5.1 级地震序列，呈北西向分布；8 月 25 日 5.1 级序列震中分布优势不明显，但总体呈北北西向分布。

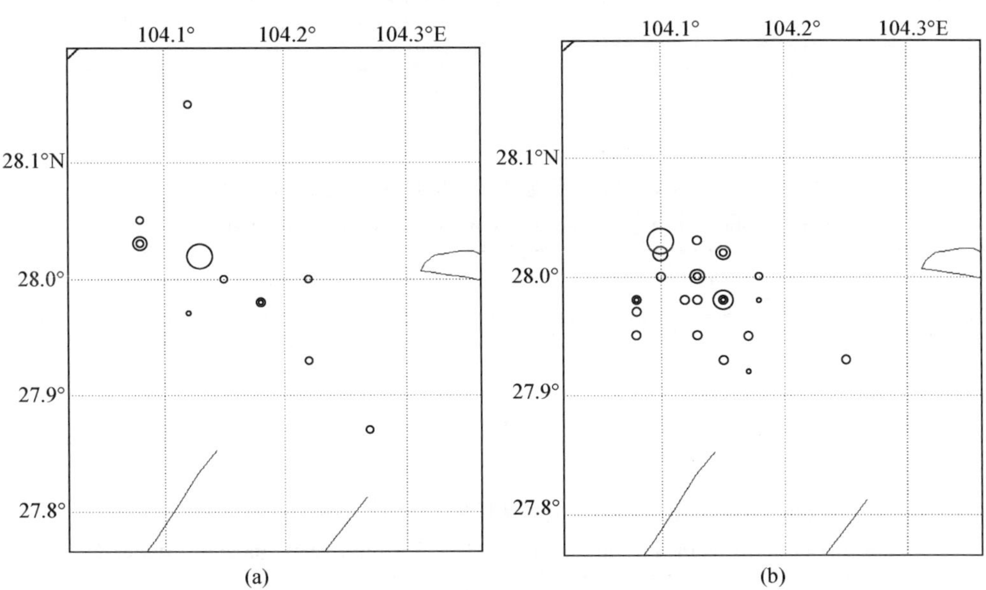

图 11 盐津地震序列余震震中分布图

Fig. 11 Epicentral distribution of the Yanjin earthquake sequence

（a）盐津第一个 5.1 级地震序列余震震中分布图；（b）盐津第二个 5.1 级地震序列余震震中分布图

五、震源机制解及地震破裂面

应用P波初动符号图解断层面法，获得盐津5.1、5.1、4.7级地震震源机制解见表4至表6和图12。

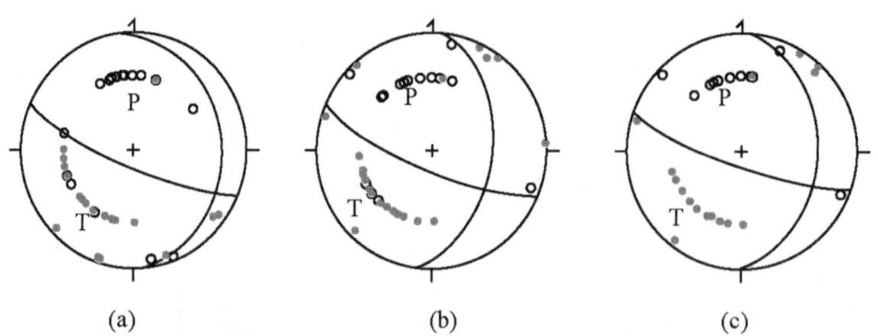

图12　盐津5.1、5.1、4.7级地震震源机制解（据付虹）

Fig. 12　Focal mechanism solution of the M_S 5.1，5.1，4.7 Yanjin earthquake

（a）7月22日5.1级；（b）8月25日5.1级；（c）8月29日4.7级

表4　盐津7月22日5.1级地震震源机制解（茅盾符号比0.220）

Table 4　Focal mechanism solution of the M_S5.1 Yanjin earthquake on July 22

编号	节面Ⅰ			节面Ⅱ			P轴		T轴		B轴		X轴		Y轴		结果来源
	走向	倾角	滑动角	走向	倾角	滑动角	方位	仰角	方位	仰角	方位	仰角	方位	仰角	方位	仰角	
1	113	78	－110	352	23	－32	359	53	219	30	117	19					付虹
2	146	67	242	347	59	27	296	5	201	40	31	49					HRV

表5　盐津8月25日5.1级地震震源机制解（茅盾符号比0.135）

Table 5　Focal mechanism solution of the M_S5.1 Yanjin earthquake on August 25

编号	节面Ⅰ			节面Ⅱ			P轴		T轴		B轴		X轴		Y轴		结果来源
	走向	倾角	滑动角	走向	倾角	滑动角	方位	仰角	方位	仰角	方位	仰角	方位	仰角	方位	仰角	
1	112	76	－133	7	45	－21	341	43	232	19	125	41					付虹
2	130	57	232	45	50	355	295	4	198	198	27	32					HRV

表6　盐津8月29日4.7级地震震源机制解（茅盾符号比0.091）

Table 6　Focal mechanism solution of the M_S 4.7 Yanjin earthquake on August 29

编号	节面Ⅰ			节面Ⅱ			P轴		T轴		B轴		X轴		Y轴		结果来源
	走向	倾角	滑动角	走向	倾角	滑动角	方位	仰角	方位	仰角	方位	仰角	方位	仰角	方位	仰角	
1	109	79	－123	3	35	－20	345	46	225	26	117	32					付虹

由图 12 和表 4 至表 6 可见，3 个地震的节面 I 方位 113°、112°、109°，走向 NW，倾向 NW，主压应力 P 轴方位 359°、341°、345°，为 NNW 向。7 月 22 日 5.1 级 8 月 29 日 4.7 级为正断为主兼有走滑型地震，8 月 25 日 5.1 级为走滑型地震。

极震区马边—盐津断裂为 NW 走向，它与节面 I 走向基本一致，余震分布优势方向呈 NW 向。综上所述，认为节面 I 是该次地震的主破裂面，其走向与极震区的马边—盐津断裂基本一致，盐津地震是在 NNW 向的主压应力作用下，NW 向马边—盐津断裂发生左旋走滑错动的结果。

六、地震前兆观测台网及前兆异常

图 1 和图 13 为震中附近地区测震台站和定点前兆观测台站分布图，地震发生在前兆观测台站相对密集地区。

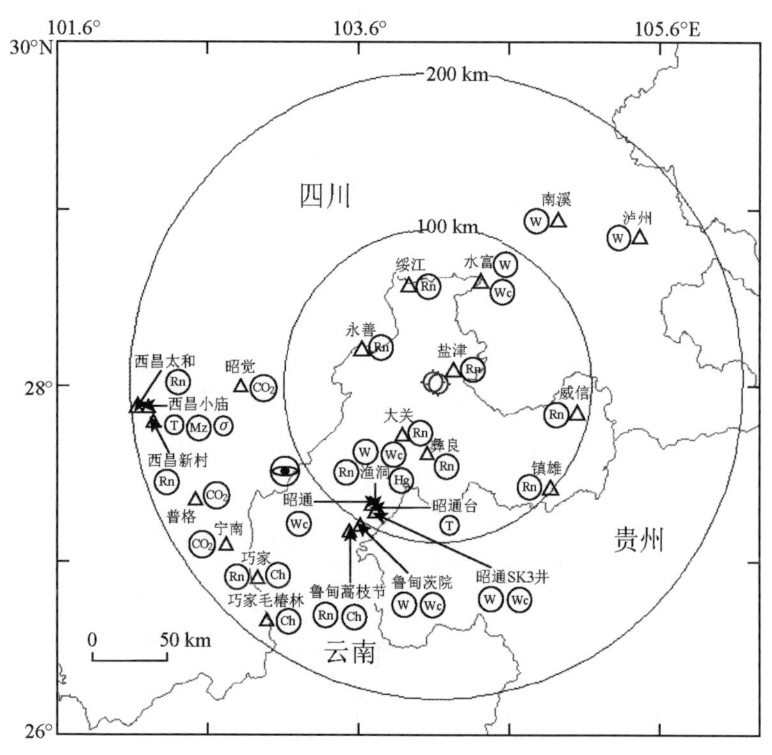

图 13 盐津地震前定点前兆观测台站分布图

Fig. 13 Distribution of the precursor monitoring stations before Yanjin earthquakes

盐津地震周围 200km 范围内共有地震台站 33 个，其中测震观测台 9 个，定点前兆观测台站 24 个，有水位、水温、水化含量、断层气二氧化碳、形变等共 34 个台项，这些前兆观测项目大多数均有 3 年以上连续可靠的观测资料。在 0 ~ 100km 和 101 ~ 200km 范围内分别有测震台 4 个和 5 个，定点前兆观测台站 12 个和 12 个，测震学以外的观测项目 17 个和 17

个。这次地震前共出现 11 个异常项目，21 条前兆异常，其中测震学出现了 3 项异常，它们是地震频度、地震空区和 *b* 值等；定点前兆出现了 8 项异常（图 14），它们是水位、水温、水汞、pH 值、碳酸氢根、二氧化碳、形变和宏观等。

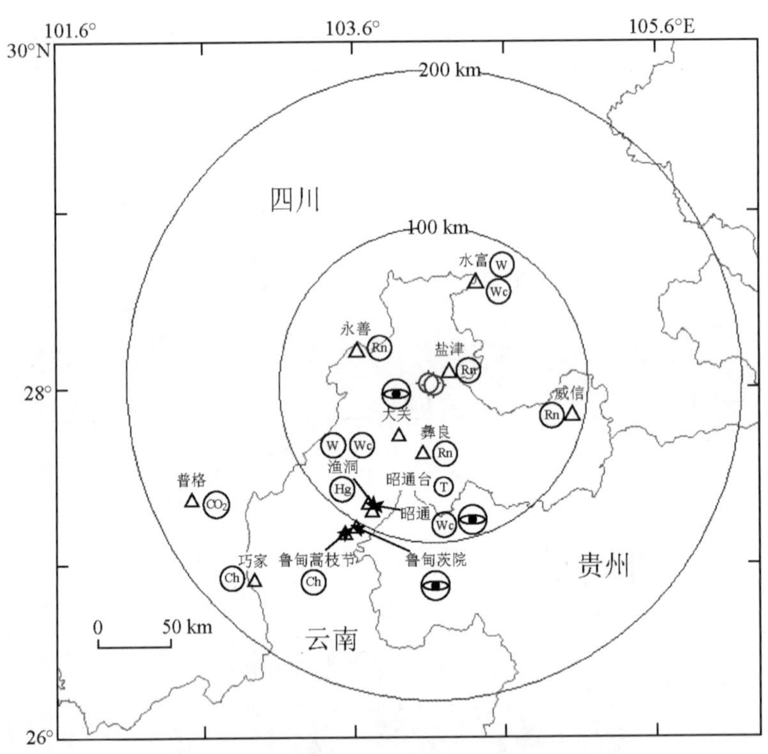

图 14　盐津地震定点前兆异常分布图

Fig. 14　Distribution of precursory anomalies on the fixed observation points before Yanjin earthquakes

各类异常的具体情况详见表 7 和图 15 至图 23。

表 7　异常情况登记表

Table 7　Summary table of precursory anomalies

序号	异常项目	台站（点）或观测区	分析方法	异常判据及观测误差	震前异常起止时间	震后变化	最大幅度	震中距 Δ/km	异常类别及可靠性	图号	异常特点及备注	震前提出/震后总结
1	地震频度	滇东北 26.5°～30.0°N，102°～106°E	$M_L \geqslant 3.0$ 级月频度	≥3	2006.4、6～	2006.10 恢复	8		S_1	15	震前显示增频异常变化	震前提出

续表

序号	异常项目	台站（点）或观测区	分析方法	异常判据及观测误差	震前异常起止时间	震后变化	最大幅度	震中距 Δ/km	异常类别及可靠性	图号	异常特点及备注	震前提出/震后总结
2	地震空区	震中周围 26°~30°N，101.8~105.8°E	$M_L \geq 3.0$ 震中分布图	内外频度比	2005.8~2006.4	空区消失			$L_{1\,I}$	16	地震前出现 $M_L \geq 3.0$ 级地震空区，主震发生在空区内	震前提出
3	*b* 值	滇东北（26.5°~29.0°，102.5°~105.5°）	M_L：2.0~4.9 6 个月时窗 1 个月滑动	>1 倍方差	2006.3~7	正常	1.17		$S_{1\,II}$	17	震前 *b* 值显示高值异常变化	震前提出
4	水位	渔洞	整点值	破年变	2006.5.15~5.29，2006.8.18~9.7	异常持续至 12.27 结束	290mm 200mm	93	S_1 I_1	18	大幅度上升	震前提出
5	水温	渔洞	整点值线性拟合差	突升	2006.6.22.8~9 时	异常持续至 9 月 6 日 13 时结束	0.002 ℃	93	S_1	18	突升	震前提出
6	气汞	渔洞	整点值	高值	2006.4.1~	9.11 结束	86.0 ng/L	93	S1	18	高值持续	震前提出
7	水位	水富	日均值	破年变	2006.2.21~23	正常	8.2m	69	S_1	19	大幅度下降	震前提出
8	水温	水富	日差分	破正常动态	2006.6.15~7.22	正常	0.0113~0.0163 ℃	69	S_1	19	大幅度波动	震前提出
9	水氡	永善	日均值	破正常动态	2006.7.1~27	正常	1.7 Bq/L	60	I_1	20	大幅度波动	震前提出
10	水氡	盐津	日均值	破正常动态	2006.4.18~7.4	正常	9.0 Bq/L	9	S_1	20	大幅度升降	震前提出

续表

序号	异常项目	台站（点）或观测区	分析方法	异常判据及观测误差	震前异常起止时间	震后变化	最大幅度	震中距 Δ/km	异常类别及可靠性	图号	异常特点及备注	震前提出/震后总结
11	水氡	彝良	5 日均值	破正常动态	2006. 3. 5～	10. 25 结束	7. 9 Bq/L	49	S_1	20	持续上升	震前提出
12	水氡	威信	日均值	破正常动态	2006. 6. 29	正常	14. 74 Bq/L	85	I_1	20	突降	震前提出
13	水温	昭通	日均值	突升	2006. 7. 7～11	正常	0. 0006 ℃	91	I_1	21	突升	震前提出
14	pH	鲁甸	日均值	破正常动态	2006. 6. 25～8. 2	正常	3. 03	114	I_1	21	大幅度波动	震前提出
15	巧家碳酸氢根	巧家	日均值	破年变	2006. 5. 22～7. 4	正常	24. 0 mg/L	175	S_1	21	快速上升	震前提出
16	二氧化碳	普格	日均值	破年变	2006. 5. 16～7. 30	正常	72 mg/d	179	S_1	22	高值	震前提出
17	地倾斜	昭通	EW 向日均值	破年变	2004. 7. 13～	2007. 6. 19	2. 47 角秒	93	M_1	23	持续东倾	震前提出
18	宏观	昭通	机井水	冒泡、变色	2006. 4. 14	正常		91	S_1		地下水翻花冒泡、乳白色	震前提出
19	宏观	大关	冷泉水	变色	2006. 5. 10	正常		45	S_1		水中有白色产状物质	震前提出
20	宏观	大关	泉水	变色	2006. 8. 13	正常		45	I_1		水为纯白色	震前提出
21	宏观	鲁甸	井水	变色	2006. 6. 20, 2006. 8. 15～16	正常			S_1 I_1		饮用水发浑	震前提出

将表 7 所列异常项目、历史对应率及其预测效能简述如下：

（1）地震学项目，资料均起始 1997 年，历史上多次对应研究区内及邻区 $M_S \geqslant 5$ 级地震，预测效能较好。

（2）定点前兆异常项目：渔洞、盐津、彝良、鲁甸和普格观测点测项有连续 2 ~5 年资料，其余观测点测项有 10 年连续资料，上述各类前兆项目多对应研究区及邻区 $M_S \geqslant 5$ 级地震，有一定预测效能。

（3）宏观异常：昭通、大关、鲁甸县井水、泉水有不同程度的冒泡、浑浊现象，2003 年鲁甸 5.1 级地震前也出现过类似现象。经现场落实组综合分析后认为：此次水变浑事件属实，确属地震宏观异常，反映滇东北地区处于加载过程中并已产生了深部微破裂。

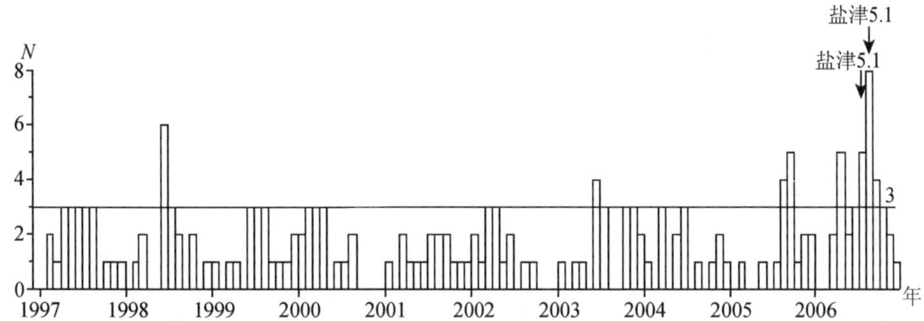

图 15 滇东北（26.5° ~30.0°，102° ~106°）$M_L \geqslant 3.0$ 级地震月频度曲线图（1997 ~2006 年）

Fig. 15 Curve of monthly frequency for the $M_L \geqslant 3.0$ earthquakes in the northeast of Yunnan province

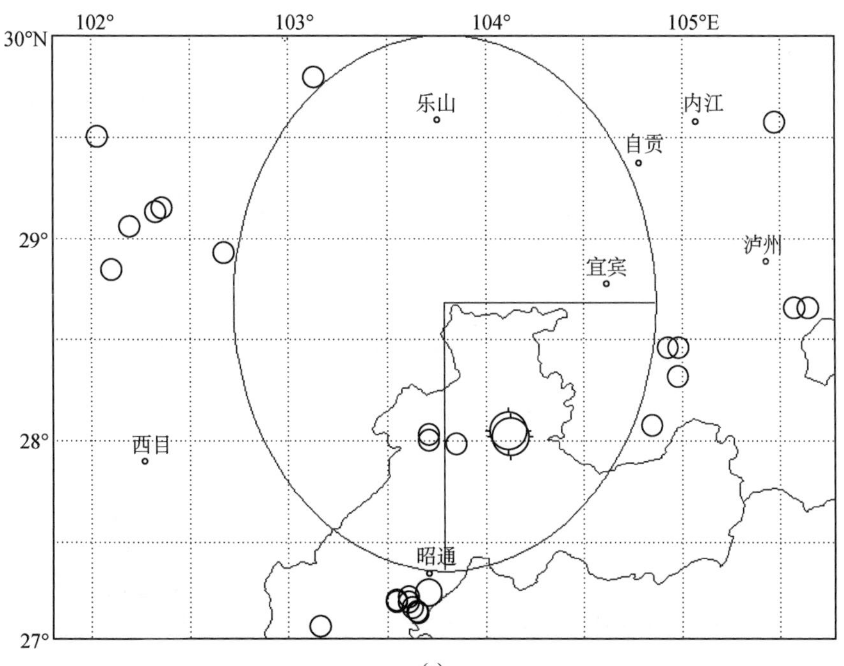

(a)

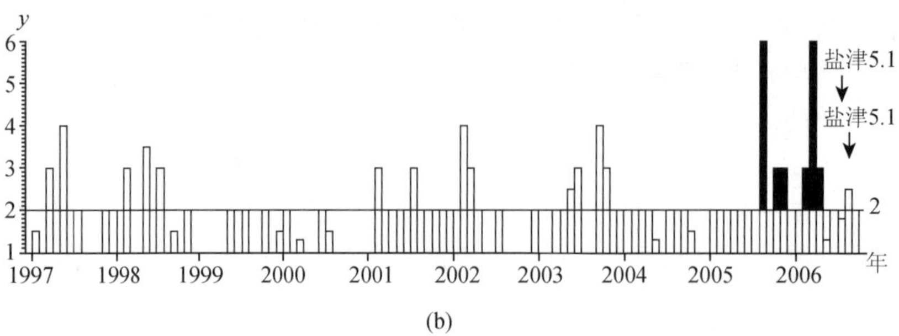

(b)

图 16　震中周围（26～30°N，101.8°～105.8°E）$M_L \geqslant 3.0$ 级地震空区

Fig. 16　Vacancy area for the $M_L \geqslant 3.0$ earthquakes around the epicenter

（a）震中周围 $M_L \geqslant 3.0$ 级地震震中分布图（2005.9～2006.7）;

（b）空区内外 $M_L \geqslant 3.0$ 级地震频度比（1997.1～2006.12）

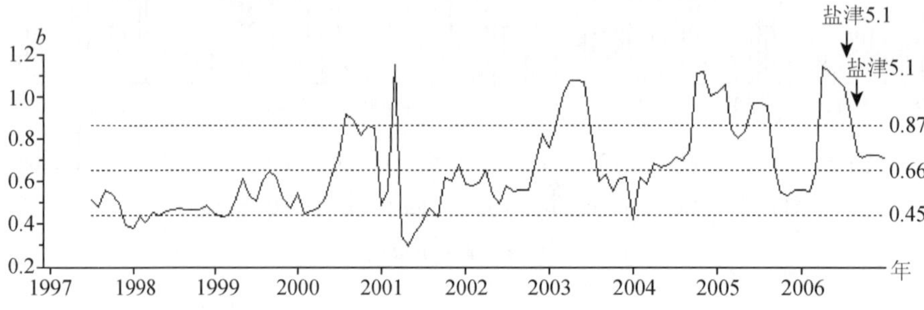

图 17　滇东北（26.5°～29.0°，102.5°～105.5°）b 值（M_L2.0～4.9）曲线图

Fig. 17　Monthly b-value Curve for the M_L2.0-4.9 earthquakes in the northeast of Yunnan province

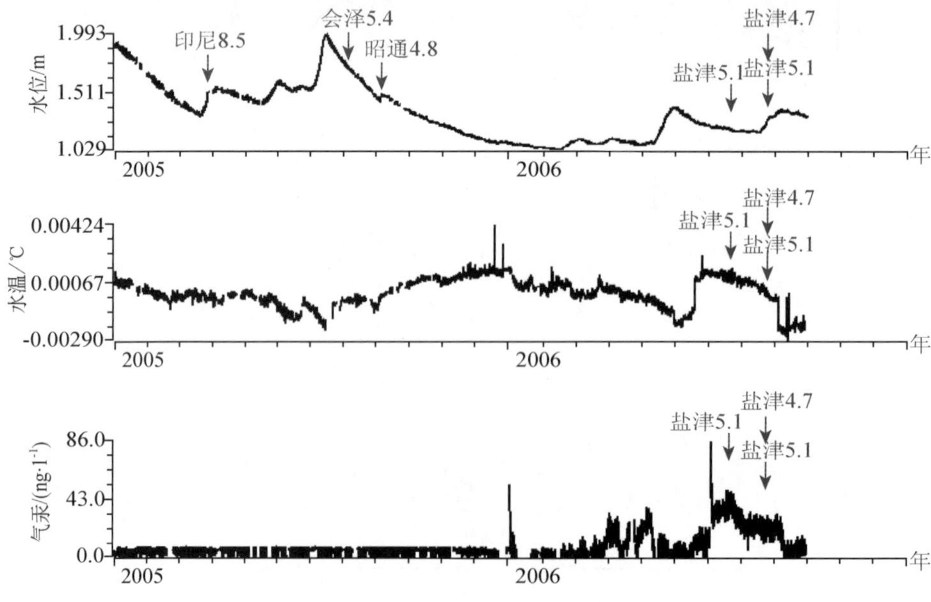

图 18　昭通渔洞数字化观测井水位、水温、气汞异常

Fig. 18　Anomalies of water level, water temperature, Hg content in groundwater at Zaotong Yudong station

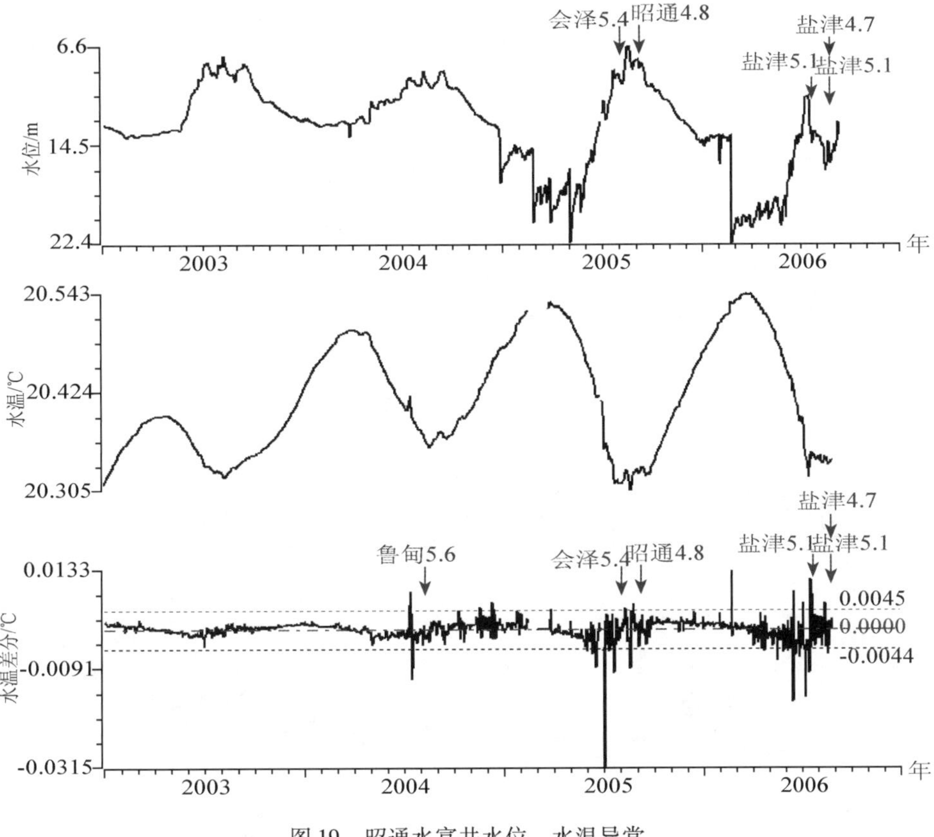

图 19 昭通水富井水位、水温异常

Fig. 19 Anomalies of water level, water temperature at Zaotong Shuifu station

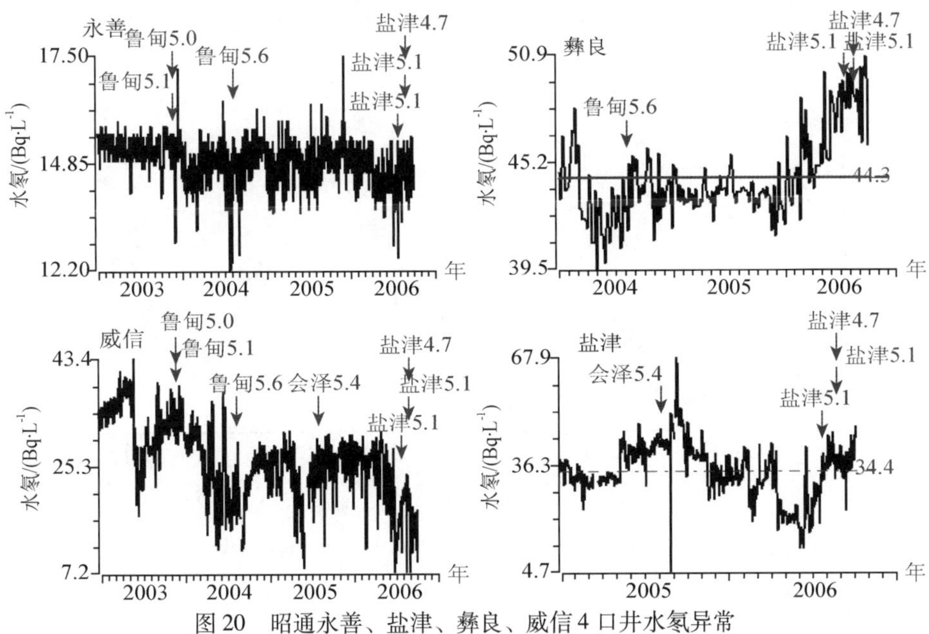

图 20 昭通永善、盐津、彝良、威信 4 口井水氡异常

Fig. 20 Anomalies of Rn content in groundwater at Zaotong Yongshan, Yanjin, Yilang and Weixin station

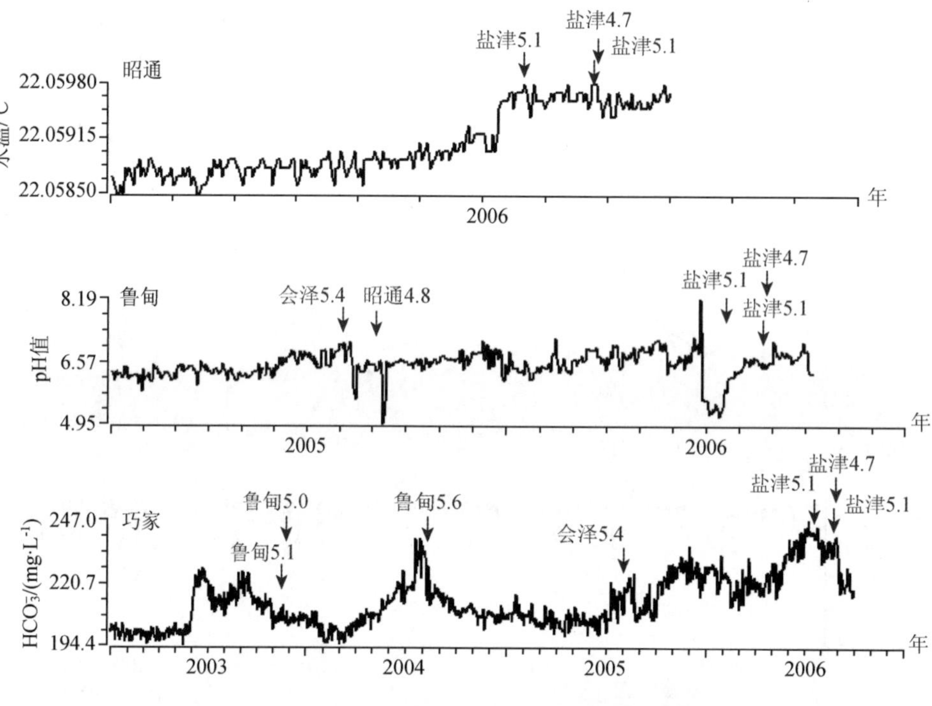

图 21　昭通水温、鲁甸 pH 值、巧家碳酸氢根异常

Fig. 21　Anomalies of water temperature at Zaotong station, pH value at Ludian station and HCO_3^- content in groundwater at Qiaojia station

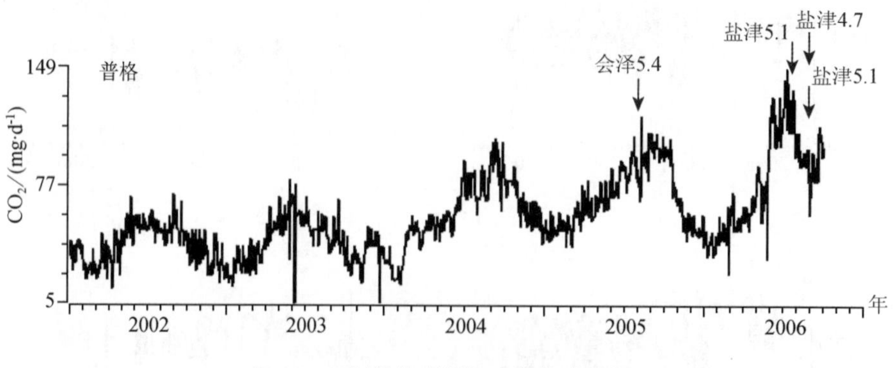

图 22　四川普格断层气二氧化碳异常

Fig. 22　Anomalies of CO_2 content at Sichuan Puge station

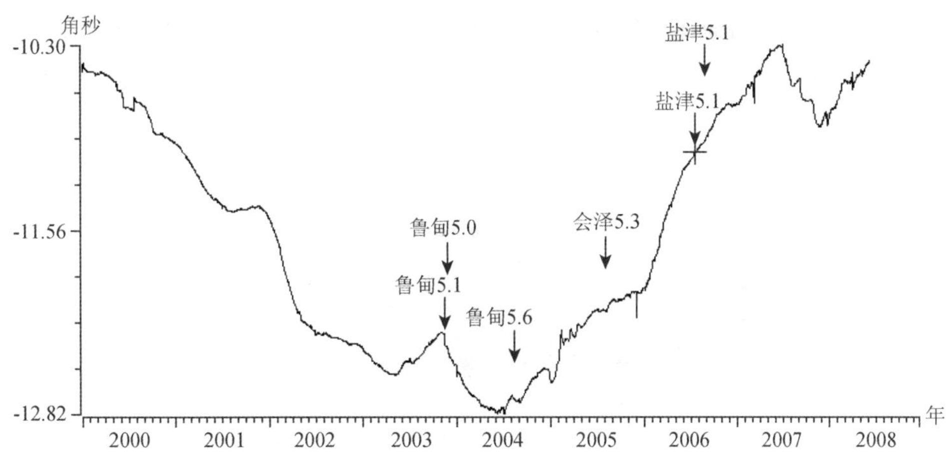

图 23 昭通倾斜石英摆东西向日均值曲线

Fig. 23 Daily mean value curves of tilt at Zaotong station

七、地震前兆异常特征分析

综上所述，盐津地震前兆异常主要特征如下：

1. 震兆异常特征

震前震中周围地震空区为中期异常，滇东北地震频度和 b 值为短期异常，没有明显临震异常显示，且于震前 11 ~4 个月出现，震后均恢复正常。

2. 定点前兆异常待征

前兆异常除昭通倾斜为中期异常外，均为流体短期、临震异常。流体异常的时空分布特征分析如下：

1）空间分布特征

由表 7、图 14 可见，盐津地震前地下流体的前兆异常集中在昭通地区，在距震中 8 ~175km 范围内流体观测台站均有异常显示，异常测项有水位、水温、水氡、汞、碳酸氢根、pH 值及地倾斜。昭通外围地区的四川普格断层气二氧化碳观测点，距震中 179km，其震前异常较为明显。

定点前兆异常分布表 8 可见，异常主要集中分布于距震中 100km 范围内。

表 8 异常分布表

Table 8 Spatial distribution of precursory anomalies

范围（km）	观测数量		异常数量		异常数量/观测数量/%	
	台站	台项	台站	台项	台站	台项
1 ~100	12	17	8	11	66.7%	64.7%
100 ~200	12	17	3	3	25%	17.6%
合计	24	34	11	14	45.87%	41.2%

从图 24 中可以看出，尽管盐津地震震级不大，但从异常空间演化来分析，震前 5 个月表现为从震中向外围扩散的态势，在距地震发生 2 个月左右，异常出现向震中收缩的情况，在距发震 1 个月前，异常基本围绕在震中 100km 左右的范围内。

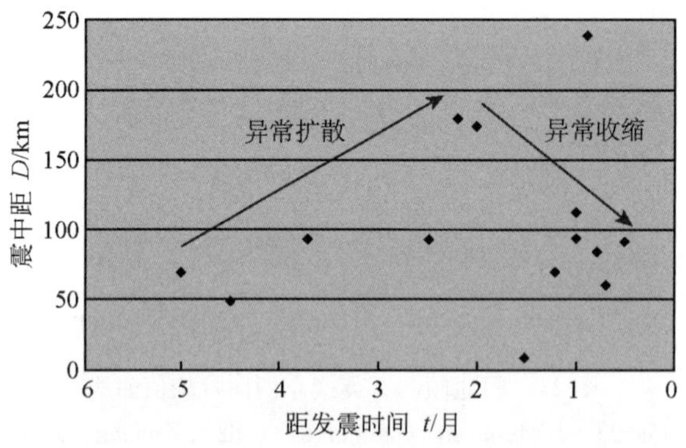

图 24　盐津地震流体前兆时空演化过程图

Fig. 24　Temporal and Spatial evolutive course of precursory anomalies of the Yanjin earthquare in liquid

2）时间进程特征[2]

从前兆月异常频次分析（图 25），前兆异常数量在盐津地震前 5 个月开始逐月增多，6 月份达最高值，7 月 22 日盐津 5.1 级地震后（8 月）大部分异常恢复，但还有 5 项异常仍持续，至 8 月 25 日盐津 5.1 级地震后所有异常才结束，说明未结束的持续性前兆异常与震区后续地震密切相关[2]。前兆异常的出现在时间进程上具有明显的阶段性，前兆异常数量越临近地震发生短期异常的数量越多，这对预报发震时间是很有意义的。

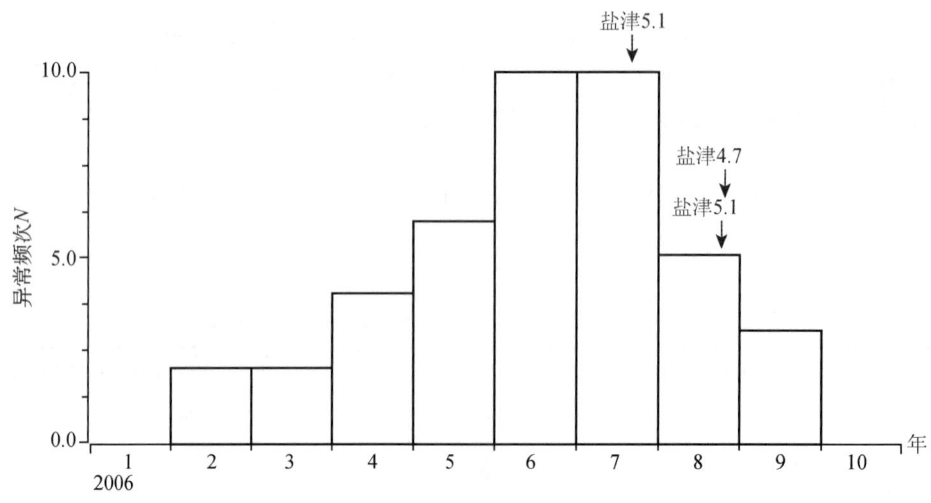

图 25　盐津地震前兆异常月频次图

Fig. 25　Monthly frequency of precursory anomalies of the Yanjin earthquakes

3. 宏观异常特征[2]

盐津地震前宏观异常为地下水异常，属构造活动宏观异常，异常出现时间至发震时间相隔 10 天至 3 个月，属短临异常，且距震中 45 ~ 114km，短临异常主要分布于地震以南，形成一宏观异常密集区。从宏观异常出现的时间及空间分布分析，反映了滇东北地区构造活动加剧，在这样的背景下，滇东北的中强地震活动有可能会增强，宏观异常表现出对未来发震地点、发震时间的预测预报意义。

八、震前预测、预防和震后响应

1. 震前预测、预防

1）预报卡

云南省地震局 2006 年填报了 A 类卡一份。内容为 2006 年 7 月 6 日至 9 月 30 日云南有发生 5 ~6 级地震的危险，主要危险区为：①小江带大关、巧家、会泽至石屏、建水一带；2. 滇西北东带中甸、永平至腾冲一带。在预报期间云南省发生了 7 月 22 日盐津 5.1 级和 8 月 25 日盐津 5.1 级地震，预测时间、地点和震级三要素均正确。

2）震情反映

2006 年 1 ~5 月，云南地区地震活动处于相对较低的活动水平，6 月份在滇西北东带、小江断裂带和滇西南地区出现了明显的增强活动，且出现了较多的前兆和宏观异常，7 月 12 日以《震情反映》（200604）上报中国地震局、云南省委、省政府，明确指出："在未来 3 个月内，云南地区存在发生 5 ~6 级地震的危险，主要危险区为滇东的大关—巧家至石屏—建水一带，滇西的中甸—宁蒗至永平—腾冲一带以及滇西南的思茅—勐腊地区"。采取的措施有①密切跟踪云南震兆、前兆宏微观异常的动态变化，适时进行震情的进一步分析预测；加强震情监视，有突出变化和短临震情分析意见及时上报。②我局通知有关州市地震部门，采取必要的内部强化震情监视及应震、应急准备措施。③加强震情值班。各级地震部门严格执行 24 小时值班制度和重大震情报告制度。并建议有关州市政府主管领导要适时召开抗震救灾指挥部工作会议，听取地震部门汇报震情，研究防范措施，相关领导要熟悉地震应急预案，强化震情观念，做好相关备震工作。按"内紧外松"的原则，确保社会稳定。结果发生了 2006 年 7 月 22 日和 8 月 25 日盐津 2 次 5.1 级地震。

2006 年 7 月 22 日盐津 5.1 级地震后，以《震情反映》（200608）上报中国地震局、云南省委、省政府，指出："盐津 5.1 级地震发生后，表明云南地区中强地震活动在今后一段时间仍有增强趋势，目前云南地区中小地震活动及地震前兆异常仍在继续发展，因此，对 2006 年度提出的地震重点监视防御区应继续加强短临跟踪工作。"虽然我们对 7 月 22 日的地震类型判断错误，8 月 25 日盐津又发生 5.1 级地震，但我们对地震趋势把握基本是正确的。

2. 序列判断

7 月 22 日发生了盐津 5.1 级地震，在规定时间内给出了该地震序列的趋势判断意见，8 月 25 日盐津 5.1 级地震后，认为近期震区及其附近有继续发生 5 级左右地震的可能，8 月 27 日在地震现场干部会上将该预报意见告知当地政府，8 月 29 日又发生了 4.7 级地震（中国地震局定 4.9 级），大关县政府的办公大楼，在前 2 次 5.1 级地震时未造成明显破坏，4.7 级地震后墙体出现明显的开裂现象，因此当地政府认为我们对序列的判断是准确的。在 8 月 25 日盐津地震现场，针对大地震的谣传，明确指出震区附近大震活动的能量积累不够，目前不可能发生 6 级以上的地震，对稳定社会取到了积极的作用。

3. 震后响应

7 月 22 日盐津地震发生后，云南省地震局立即启动地震应急预案，派出了由 26 人组成的现场工作组，对这次地震进行地震灾害损失评估、强震观测和震情监视。

灾害评估组对豆沙镇石门村的灾害损失情况进行调查；测震组在豆沙镇、县政府等地完成 2 台测震仪架设工作；强震组分别在豆沙镇和柿子乡架设强震仪。云南省地震灾害紧急救援队出动 5 辆车 23 人，前往地震灾区进行地震紧急救援工作。由中国地震局苗崇刚副司长率队的中国地震局现场工作组及时投入地震灾害调查工作，并协调、指导地方各级政府开展地震现场抗震救灾工作。

震后先后共组织各类救灾人员近千人深入灾区，疏散安置灾民，运送救治伤员，维护灾区交通、治安秩序。云南省民政厅、昭通市政府组织 700 顶帐篷、1000 床棉被、500 床毛毯紧急运往灾区，安置无家可归灾民，并发放救灾资金 71 万元，粮食 9.5 万公斤及其他相关生活物资。基本确保了广大灾民有临时住所、有被盖、有饭吃、有水喝、有伤有病能医治，确保无次生灾害发生。

8 月 25 日盐津地震发生后，云南省地震局立即启动破坏性地震应急预案，成立前、后方指挥部，召开紧急会商会，判定地震趋势。成立了由胡永龙副局长为指挥长的 25 人现场工作队，在震后 50 分钟后分批出发，赶赴地震灾区。开展震害调查、损失评估和震情监视及趋势分析工作，并派出应急队伍赶赴地震灾区进行地震紧急救援工作，同时，省地震局领导及相关部门领导立即召开地震应急工作会议，部署震后应急救援和宣传工作。

云南省地震局现场工作队下设秘书组、震情监视预报组、测震组、强震组、灾评组、后勤组、通讯组 7 个工作组。测震组在大关县吉利镇，盐津县豆沙镇、柿子乡架设 3 台测震仪，灾评组将分为 8 个小组（大关 3 组，盐津 5 组）展开灾害调查工作。

据云南遥测台网测定，2006 年 8 月 29 日 09 时 14 分，在昭通地区盐津县境内（北纬 28°00″，东经 104°11″，震源深度 8km）发生 4.7 级地震。强余震发生后，省地震局胡永龙副局长立即召开紧急会议，部署了各项工作：立即通报了各级政府；部署收集此次强余震地震灾情；立即了解我局现场工作队在开展现场工作是否有突发情况发生；要求盐津和大关两县抓紧收集此次强余震后新的灾害损失情况报现场指挥部。

九、结论与讨论

（1）盐津地震序列为震群型。7 月 22 日盐津 5.1 级后，从序列的衰减情况并结合参数计算来看，序列衰减明显异常，指示震中及附近地区有更大或相当地震发生的危险。8 月 25 日 5.1 级地震之后虽然序列强度及频度随时间的变化有一定的起伏，但与前一地震相比，序列余震次数多、能量释放充分、震级频度分布线性关系较好、h 值恢复正常，随时间增长，序列逐渐趋于结束。

（2）临近盐津地震发生的短期阶段，流体前兆异常数量有增多的趋势。在时间进程上，前兆异常数量越临近地震发生短期异常的数量越多。从异常点的时空演化特征来看，距震中 180km 以内的观测井在震前异常显著，异常演化具有扩散、收缩现象。

（3）短临阶段在近震区附近出现宏观异常，异常点分布在距震中 120km 范围内，且形成一宏观异常密集区。盐津地震前云南地区除昭通地区出现宏观异常外，还有玉溪地区及大理地区也出现了地下水宏观异常，但这两个地区的流体观测异常特征不明显。由此可见，在地震孕育的短临阶段，地下流体微观异常与宏观异常的配套综合分析是提高预测能力的重要依据。

参 考 文 献

[1] 叶建庆．陈慧．刘学军等．2006 年云南盐津 5.1 级双震的监测与研究．地震研究，2008，31（2）：134 ~ 141

[2] 张立．申玻．官致君等．盐津地震流体前兆异常分析．地震，2007，27（2）：121 ~ 129

参 考 资 料

1）云南省地震局，2006 年 7 月 22 日盐津 5.1 级地震灾害直接损失评估报告，2006 年 7 月

2）云南省地震局，2006 年 8 月 25、29 日盐津 5.1、4.7 级地震灾害直接损失评估报告，2006 年 9 月

3）云南省地震局，云南省 2007 年度地震趋势研究的报告，2006 年 11 月

4）云南省地震局，2006 年 7 月 22 日盐津 5.1 级地震强震动应急观测报告，2006 年 7 月

5）云南省地震局，2006 年 8 月 25 日盐津 5.1 级地震现场强震动应急观测工作报告，2006 年 9 月

6）云南省地震局，云南地震目录（2006）

7）云南省地震局，云南地震速报目录（2006）

Yanjin Earthquake of M_S 5.1 on July 22, M_S 5.1 on August 25, 2006 in Yunnan Province

Abstract

On Jul. 22, Aug. 25 and 29, 2006, three earthquakes with M_S5.1, M_S5.1 occurred in Yanjin county, Zhaotong city, Yunnan Province. The macroseismic epicenters of the 3 earthquakes were located in Zhonghe, Dousha in Yanjin town and Jili in Daguan town. The intensity in the meizoseismal area is Ⅶ The shape of the isoseismal area is elliptic with major axis in NE direction. 24 people were killed, 28 people were seriously injured, 158 people were slightly injured. The total economic loss caused by these earthquakes was 442 million Yuan RMB.

The event was shock swarm type. The biggest aftershock was M_L3.5. The major axis of aftershock areas is NNW direction. After the Jul. 22 Yanjin M_S5.1, another M_S 5.1 and M_L4.7 were happened on Aug. 25, 29 respectively. The mechanism results of the 3 earthquakes are basic uniform. Nodal plane I were the main rupture plane, strike of principal compressive P axis were nearly SN direction. Their seismogenic geological structure was the Mabian-Yanjin fault in NW direction.

There were 33 seicmic stations around this earthquake, among them, there were 9 seicmometric stations and 24 precursory observation stations. There were 11 anomalies iteM_S with 21 precursory anomalies and most of them are short-imminent term anomalies.

报 告 附 件

附件一：

附表 1　固定前兆观测台（点）与观测项目汇总表

序号	台站（点）名称	经纬度（°）		测项	资料类别	震中距 Δ（km）	备注
		φ_N	λ_E				
1	盐津	28.106	104.232	测震△		9	
				水氡	Ⅱ		
2	大关	27.737	103.889	水氡	Ⅱ	43	
3	彝良	27.634	104.059	水氡		49	
				CO_2	Ⅱ		
4	永善	28.226	103.629	水氡	Ⅱ	60	
5	水富	28.618	104.408	水位	Ⅱ	69	
				水温	Ⅱ		
				倾斜仪（摆式）	Ⅱ		
6	溪江	28.599	103.95	水氡	Ⅱ	71	
7	威信	27.856	105.038	水氡	Ⅱ	85	
8	昭通	27.35	103.69	水温	Ⅱ	91	
9	渔洞	27.35	103.717	水位	Ⅰ	93	
				水氡	Ⅱ		
				气氡	Ⅱ		
				气汞	Ⅰ		
10	昭通台	27.32	103.72	测震△		93	
				测震◬			
				倾斜仪（摆式）	Ⅱ		
11	昭通 SK3 井	27.289	103.711	水位	Ⅱ	93	
				水温	Ⅱ		
12	镇雄	27.44	104.87	测震△		98	
				水氡	Ⅱ		
13	高县	28.64	104.74	测震◬		100	
14	鲁甸茨院	27.210	103.610	水位	Ⅱ	106	
				水温	Ⅱ		

续表

序号	台站（点）名称	经纬度（°）		测项	资料类别	震中距Δ（km）	备注
		φ_N	λ_E				
15	鲁甸蒿枝节	27.182	103.55	水氡	Ⅱ	114	
				F^-	Ⅱ		
				HCO_3^-	Ⅱ		
				pH 值	Ⅱ		
				电导率	Ⅱ		
16	沐川	29.05	103.79	测震△		119	
17	南溪	28.98	104.92	水位	Ⅱ	129	
18	昭觉	28.02	102.83	CO_2	Ⅱ	135	
19	泸州	28.88	105.48	水位	Ⅱ	159	
20	巧家	26.92	102.94	测震△	Ⅱ	175	
				水氡	Ⅱ		
				F^-	Ⅱ		
				HCO_3^-	Ⅱ		
				Ca^{2+}	Ⅱ		
				Mg^{2+}	Ⅱ		
				CO_2	Ⅱ		
21	宁南	27.10	102.72	CO_2	Ⅱ	178	
22	普格	27.37	102.53	CO_2	Ⅱ	179	
23	荣县	29.57	104.39	测震△		179	
24	峨眉山	29.58	103.45	测震△		183	
25	巧家毛椿林	26.68	103.0	F^-	Ⅱ	190	
				HCO_3^-	Ⅱ		
				Ca^{2+}	Ⅱ		
				Mg^{2+}	Ⅱ		
26	汉源	29.34	102.77	测震△		191	
27	西昌新村	27.83	102.25	水氡	Ⅱ	192	
28	西昌小庙	27.90	102.22	倾斜仪（摆式）	Ⅱ	195	
				地磁重直分量	Ⅱ		
				钻孔应变	Ⅱ		

续表

序号	台站（点）名称	经纬度（°）		测项	资料类别	震中距 Δ（km）	备注
		φ_N	λ_E				
29	西昌太和	27.9	102.13	水氡	Ⅱ	198	

分类统计	0 < Δ ≤100km	100 < Δ ≤200km	200 < Δ ≤300km	300 < Δ ≤500km	500 < Δ ≤800km	总数
测项数 N	7	10				17
台项数 n	19	18				37
测震单项台数 a	1	4				5
形变单项台数 b	1	0				1
电磁单项台数 c	0	0				0
流体单项台数 d	7	7				14
综合台站数 e	4	5				9
综合台中有测震项目的台站数 f	0	1				1
测震台总数 $a+f$	1	5				6
台站总数 $a+b+c+d+e$	13	16				29
备注						

附表 2　测震以外固定前兆观测项目与异常统计表

序号	台站（点）名称	测项	资料类别	震中距 Δ/km	按震中距 Δ 范围进行异常统计																								
					0 < Δ ≤ 100km					100 < Δ ≤ 200 km					200 < Δ ≤ 300 km					300 < Δ ≤ 500km					500 < Δ ≤ 800km				
					L	M	S	I	U	L	M	S	I	U	L	M	S	I	U	L	M	S	I	U	L	M	S	I	U
1	盐津	水氡	Ⅱ	9	—	—	√	—																					
2	大关	水氡	Ⅱ	43	—	—	—	—																					
3	彝良	水氡	Ⅱ	49	—	—	√	—																					
		CO_2	Ⅱ		—	—	—	—																					
4	永善	水氡	Ⅱ	60	—	—	—	√																					
5	水富	水位	Ⅱ	69	—	—	√	—																					
		水温	Ⅱ		—	—	√	—																					
		倾斜仪（摆式）	Ⅱ		—	—	—	—																					
6	溪江	水氡	Ⅱ	71	—	—	—	—																					
7	威信	水氡	Ⅱ	85	—	—	—	√																					
8	昭通	水温	Ⅱ	91	—	—	—	√																					
9	渔洞	水位	Ⅰ	93	—	—	√	—																					
		水温	Ⅱ		—	—	√	—																					
		气氡	Ⅱ		—	—	—	—																					
		气汞	Ⅰ		—	—	√	—																					
10	昭通台	倾斜仪（摆式）	Ⅱ	93	—	√	—	—																					

续表

序号	台站（点）名称	测项	资料类别	震中距Δ/km	按震中距Δ范围进行异常统计																								
					0<Δ≤100km					100<Δ≤200 km					200<Δ≤300 km					300<Δ≤500km					500<Δ≤800km				
					L	M	S	I	U	L	M	S	I	U	L	M	S	I	U	L	M	S	I	U	L	M	S	I	U
11	昭通SK3井	水位	Ⅱ	93	—	—	—	—																					
		水温	Ⅱ		—	—	—	—																					
12	镇雄	水氡	Ⅱ	98	—	—	—	—																					
13	鲁甸茨院	水位	Ⅱ	106						—	—	—	—																
		水温	Ⅱ							—	—	—	—																
14	鲁甸蒿枝节	水氡	Ⅱ	114						—	—	—	—																
		F^-	Ⅱ							—	—	—	—																
		HCO_3^-	Ⅱ							—	—	—	—																
		pH值	Ⅱ							—	—	—	√																
		电导率	Ⅱ							—	—	—	—																
15	南溪	水位	Ⅱ	129						—	—	—	—																
16	昭觉	CO_2	Ⅱ	135						—	—	—	—																
17	泸州	水位	Ⅱ	159						—	—	—	—																

续表

序号	台站（点）名称	测项	资料类别	震中距 Δ/km	按震中距 Δ 范围进行异常统计																								
					$0<\Delta\leqslant100$km					$100<\Delta\leqslant200$ km					$200<\Delta\leqslant300$ km					$300<\Delta\leqslant500$km					$500<\Delta\leqslant800$km				
					L	M	S	I	U	L	M	S	I	U	L	M	S	I	U	L	M	S	I	U	L	M	S	I	U
18	巧家	水氡	Ⅱ	175						—	—	—	—																
		F^-	Ⅱ							—	—	—	—																
		HCO_3^-	Ⅱ							—	—	√	—																
		Ca^{2+}	Ⅱ							—	—	—	—																
		Mg^{2+}	Ⅱ							—	—	—	—																
		CO_2	Ⅱ							—	—	—	—																
19	宁南	CO_2	Ⅱ	178						—	—	—	—																
20	普格	CO_2	Ⅱ	179						—	—	√	—																
21	巧家毛椿林	F^-	Ⅱ	190						—	—	—	—																
		HCO_3^-	Ⅱ							—	—	—	—																
		Ca^{2+}	Ⅱ							—	—	—	—																
		Mg^{2+}	Ⅱ							—	—	—	—																
22	西昌新村	水氡	Ⅱ	192						—	—	—	—																

续表

序号	台站（点）名称	测项	资料类别	震中距 Δ/km	按震中距 Δ 范围进行异常统计																								
					0 < Δ ≤ 100km					100 < Δ ≤ 200 km					200 < Δ ≤ 300 km					300 < Δ ≤ 500km					500 < Δ ≤ 800km				
					L	M	S	I	U	L	M	S	I	U	L	M	S	I	U	L	M	S	I	U	L	M	S	I	U
23	西昌小庙	倾斜仪（摆式）	Ⅱ	195						—	—	—	—																
		地磁垂直分量	Ⅱ							—	—	—	—																
		钻孔应变	Ⅱ							—	—	—	—																
24	西昌太和	水氡	Ⅱ	198						—	—	—	—																
分类统计	台项	异常台项数			0	1	7	3	0	0	0	2	1	0															
		台项总数			19	19	19	19	19	27	27	27	27	27															
		异常台项百分比/%			0	5	37	16	0	0	0	7	4	0															
	观测台站（点）	异常台站数			0	1	4	3	0	0	0	2	1	0															
		台站总数			12	12	12	12	12	12	12	12	12	12															
		异常台站百人分比/%			0	8	33	25	0	0	0	17	8	0															
	测项总数				7					13																			
	观测台站总数				12					12																			
备注																													

附件二：

附表 2－1　地震参数表

序号	数据项中文名	填　入　项	备注
1	地震编号		××××（不填）
2	主震编号		×（不填）
3	发震日期	2006 年 7 月 22 日 2006 年 8 月 25 日	年　月　日
4	发震时刻	09 时 10 分 22 秒 13 时 51 分 40.7 秒	时　分　秒
5	震中纬度	28.02°N 28.03°N	×××.××（°）
6	震中经度	104.13°E 104.10°E	××.××××（°）
7	震源深度 h	9 7	×××（km）
8	地震震级 M^*（面波震级 M_S）	5.1 5.1	×.×
9	近震震级 M_L		×.×
10	体波震级 M_b		×.×
11	矩震级 M_w		×.×
12	震中烈度	Ⅵ Ⅶ	大写罗马数字表示
13	宏观震中纬度		×××.××（°）
14	宏观震中经度		××.××××（°）
15	宏观震源深度		×××（km）
16	地名编号	657500	邮政编码
17	地名	云南省盐津县	省县（市）15 个汉字

注：地震震级 M 指 GB 17740—1999 规定的地震震级

附表 2－2　地震构造背景

序号	数据项中文名	填　入　项	备注
1	地震编号		××××（不填）
2	主震编号		×（不填）

续表

序号	数据项中文名	填　入　项	备注
3	大地构造单元编号		D×××（不填）
4	地壳构造	震区地处扬子准地台滇东台褶带滇东北台褶束	最多 45 个汉字
5	深部构造	有北东向、北西向和近东西向三组区域褶皱、断裂构造	最多 30 个汉字
6	区域形变场	区域主压应力场，呈 NNW 向	最多 40 个汉字
7	震中附近地质构造	主要断裂有北东向的五莲峰—化蓥山断裂和北西向马边—盐津断裂	最多 40 个汉字
8	主震断层活动性	活动强烈	最多 30 个汉字
9	孕震构造		最多 10 个汉字
10	发震构迁	NW 向马边—盐津断裂	最多 15 个汉字
11	发震断层		最多 15 个汉字

附表 2-3　震害统计数

序号	数据项中文名	填　入　项	备注
1	地震编号		××××（不填）
2	主震编号		×（不填）
3	宏观震中	7 月 21 日 5.1 级在盐津县豆沙至大关县吉利一带。 2006 年 8 月 25 日 5.1 级和 8 月 29 日 4.7 级在盐津县中和、豆沙至大关县吉利一带	最多 15 个汉字，写到最小的代表地名
4	等震线形态	7 月 21 日 5.1 级极震区烈度为Ⅵ度，呈椭圆形，长轴走向为北东向。 2006 年 8 月 25 日 5.1 级和 8 月 29 日 4.7 级极震区烈度为Ⅶ度，呈椭圆形，长轴走向为北东向	最多 20 个汉字，形态并注明长轴方向
5	极震区面积	7 月 21 日 5.1 级：890km² 2006 年 8 月 25 日 5.1 级和 8 月 29 日 4.7 级：1350km²	×××（km²）
6	极震区长轴（直径）	/	××.×（km）

续表

序号	数据项中文名	填　入　项	备注
7	极震区短轴（直径）	/	××.×（km）
8	有感面积	/	×××××（km）
9	有感范围	/	最多30个汉字，注明各边界的代表地名
10	主震地烈缝	/	最多30个汉字，方向、长度、宽度及力学性质等
11	其他地表破坏	/	最多30个汉字，有无滑坡、陷坑、喷砂冒水等其它地表破坏现象
12	其他重大破坏	/	最多30个汉字
13	伤亡总人数	7月21日5.1级：22人死亡，13人重伤，101个轻伤，失踪1人。2006年8月25日5.1级和8月29日4.7级：2人死亡，15人重伤，52个轻伤	单位：人
14	死亡总人数	7月21日5.1级：22人 2006年8月25日5.1级和8月29日4.7级：2人	单位：人
15	房屋破坏总间数	/	单位：间
16	房屋破坏面积	/	m^2
17	直接经济损失	/	单位：万元
18	间接经济损失	/	单位：万元
19	经济损失总额	7月21日5.1级：23900万元 2006年8月25日5.1级和8月29日4.7级：20270万元	单位：万元

附表2-4　地震序列

序号	数据项中文名	填　入　项	备注
1	地震编号		××××（不填）
2	主震编号		×（不填）
3	前震类别	/	×（1. 有前震序列；2. 单个前震；3. 无前震）
4	前震开始时间	/	年　月　日

续表

序号	数据项中文名	填　入　项	备注
5	最大前震震级	/	×.×
6	次大前震震级	/	×.×
7	前震序列 b 值	/	×.×
8	前震序列 h 值	/	×.×
9	前震与主震的最近距离	/	×××（km）
10	最大余震发生时间*	2008 年 8 月 29 日	年　月　日
11	最大余震震级	4.7	×.×
12	次大余震震级	3.5	×.×
13	余震序列 b 值	7 月 22 日 5.1 级：0.58 8 月 25 日 5.1 级：0.57	×.××
14	余震序列 h 值	7 月 22 日 5.1 级：0.36 8 月 25 日 5.1 级：1.20	×.××
15	余震序列 p 值	/	×.××（可分段给出）
16	$E_{主}/E_{总}$	0.8828	×.××
17	地震序列号		××××（不填）
18	地震序列类型	震群型	×（1. 主震余震型；2. 震群型；3. 弧立型）
19	余震分布形态	余震活动总体呈北西向	最多 25 个汉字，说明余震分布的形态及优势分布方向等

注：如多个相同，取第一个。

附表 2－5　震源机制解

序号	数据项中文名	填　入　项	备注
1	地震编号		××××（不填）
2	主震编号		×（不填）
3	节面Ⅰ：走向	113.0 112.0	×××.×（°）
4	节面Ⅰ：倾角	78.0 76.0	××.×（°）
5	节面Ⅰ：滑动角	－110.0 －133.0	××.×（°）

续表

序号	数据项中文名	填　入　项	备注
6	节面Ⅱ：走向	352.0 7.0	×××.×（°）
7	节面Ⅱ：倾角	23.0 45.0	××.×（°）
8	节面Ⅱ：滑动角	-32.0 -21.0	××.×（°）
9	P轴：方位	359.0 341.0	×××.×（°）
10	P轴：仰角	53.0 43.0	××.×（°）
11	T轴：方位	219.0 232.0	×××.×（°）
12	T轴：仰角	30.0 19.0	××.×（°）
13	N（B）轴：方位	117.0 125.0	×××.×（°）
14	N（B）轴：仰角	19.0 41.0	××.×（°）
15	X轴：方位		×××.×（°）
16	X轴：仰角		××.×（°）
17	Y轴：仰角		××.×（°）
18	Y轴：仰角		××.×（°）
19	使用初动符号总数	37 33	
20	矛盾符号比	0.220 0.135	0.×××
21	方法	P波初动符号图解断层面法	最多16个汉字
22	资料来源	云南省地震预报研究中心	最多11个汉字，填入文献号或测定人所在单位
23	测定人	付虹	最多10个汉字

续表

序号	数据项中文名	填　入　项	备注
24	主破裂面	北西向断层面	最多 5 个汉字
25	断层错动方式	左旋走滑为主	最多 10 个汉字

附表 2－6　地震矩张量解

序号	数据项中文名	填　入　项	备注
1	地震编号		××××（不填）
2	主震编号		×（不填）
3	节面Ⅰ：走向	146 130	×××. ×（°）
4	节面Ⅰ：倾角	67 57	××. ×（°）
5	节面Ⅰ：滑动角	242 232	××. ×（°）
6	节面Ⅱ：走向	347 45	×××. ×（°）
7	节面Ⅱ：倾角	59 50	××. ×（°）
8	节面Ⅱ：滑动角	27 355	××. ×（°）
9	矩张量：M_{xx}	1. 650 2. 900	×. $\times_{e+}$××（N · m）
10	矩张量：M_{yy}	0. 974 0. 787	×. $\times_{e+}$××（N · m）
11	矩张量：M_{zz}	－2. 620 －3. 680	×. $\times_{e+}$××（N · m）
12	矩张量：M_{xy}	－0. 914 －1. 260	×. $\times_{e+}$××（N · m）
13	矩张量：M_{yz}	－0. 066 －0. 042	×. $\times_{e+}$××（N · m）
14	矩张量：M_{zx}	－2. 120 －2. 490	×. $\times_{e+}$××（N · m）

续表

序号	数据项中文名	填　入　项	备注
15	地震矩：M_0		×. ×$_{e+}$×× （N·m）
16	地震级：M_w	4.9 5.0	×. ×
17	使用地震台站数		
18	使用地震波形数		
19	分析方法	矩心矩张量法	最多16个汉字
20	资料来源	哈佛大学震源机制中心	最多11个汉字，填入文献号或测定人所在单位
21	分析波段	长周期体波	
22	测定人		最多10个汉字
23	主破裂面	北西向断层面	最多5个汉字
24	断层错动方式	左旋走滑为主	最多10个汉字

附表2－7　历史地震情况记录

序号	数据项中文名	填　入　项	备注
1	地震编号		××××（不填）
2	主震编号		×（不填）
3	历史地震统计范围	（27°00′～28°30′）N （103°00′～104°48′）E	最多30个汉字，尽可能给出经纬度
4	历史地震统计开始时间	1900年	年
5	历史地震统计结束时间	2006年	年
6	历史地震最大震级	7.1	×. ××
7	历史最大震级	1974年5月11日凌晨3时25分，在昭通地区大关，永善两县交界处的团结，木杆乡一带（北纬28.1°/东经103.9°发生7.1级地震）	最多30个汉字
8	历史地震最高烈度	Ⅸ	×××
9	历史最高列度地震	1974年大关7.1级地震震中烈度达Ⅸ度	最多30个汉字
10	M_S≥6历史地震次数	2	××（次）
11	M_S≥5.0历史地震次数	22	××（次）

附表 2－8　预测预报情况登记表

序号	数据项中文名	填　入　项	备注
1	地震编号	/	××××（不填）
2	主震编号	/	×（不填）
3	预报单位名称	云南省地震预报研究中心	最多 11 个汉字
4	预报人	云南省地震预报研究中心	最多 10 个汉字
5	预报依据	4 级地震平静异常、3 级以上地震出现明显增频现象、滇东北地区水位、水氡、水温前兆异常	最多 50 个汉字
6	预报时间开始	2006 年 7 月 6 日	年　月　日
7	预报时间结束	2006 年 9 月 30 日	年　月　日
8	预报最大震级	6.00	×.×
9	预报最小震级	5.00	×.×
10	预报地点	1. 小江带大关、巧家、会泽至石屏、建水一带； 2. 滇西北东带中甸、永平至腾冲一带	最多 20 个汉字
11	预报区南界纬度	/	××.×××（°）
12	预报区北界纬度	/	××.×××（°）
13	预报区东界纬度	/	××.×××（°）
14	预报区西界纬度	/	××.×××（°）
15	提出预报时间	/	年　月　日
16	预报郊果	在预报期间云南省发生了 7 月 22 日盐津 5.1 级和 8 月 25 日盐津 5.1 级地震，预测时间、地点和震级三要素均正确	最多 15 个汉字
17	证明附录名称	预报卡	最多 40 个汉字

附件三：

（1）预报卡：

地震预报卡片

类别：B　编号：01　填报人（或单位）：云南省地震预报研究中心

填报时间：2006 年 07 月 05 日

签发人（或单位盖章）：

※ ※※※※※※※※※※※※※※※※※※※※※※※※※※※※※※

预报内容：

1.时间：2006 年 07 月 06 日至 2006 年 09 月 30 日

2.震级（Ms）：5～6 级

3.地域：

范围：1.小江带大关、巧家、会泽至石屏、建水一带；2.滇西北东带中甸、永平至腾冲一带；滇西南地区值得注意。

参考点经纬度：＿＿＿＿N，＿＿＿＿E

以参考点为中心的封闭区域最大半径（距）：＿＿＿＿公里

4. 地震类型 ① 地震（震群）名：＿＿＿＿＿＿

②地震类型：＿＿＿＿＿＿

填卡须知

1. 预报意见分为 A、B、C、D 四类，一张卡片只能填报其中一类，
   A 类：中期预报。时间 3 个月至 1 年。
   B 类：短临预报，时间 3 个月以内（含 3 个月）
   C 类：地震趋势估计，类型判定，余震预报，震群判定。
   D 类：安全预报。
2. 三要素填报规定：
   ① 时间分为：
   中期：一年，半年及 3 个月至 1 年内的其他时段。
   短临：3 个月，2 个月，1 个月，20 天，15 天，10 天，一周
   ② 震级分为：5.0-5.9，5.5-6.4，6.0-6.9，6.5-7.5，≥5 五档。
   ③ 地域最大半径（距）：

	5.0-5.9	5.5-6.4	6.0-6.9	6.5-7.5	≥7
中期：	≤100	≤125	≤125	≤150	≤150
短临：	≤75	≤100	≤100	≤125	≤125

预报地域可为任意形状的封闭形，参考点为其几何中心点。

3. 本卡片必须严格按规定填报，否则无效。
4. 本卡寄送云南省地震预报研究中心。
5. 原省局印发的各种预报卡一律作废。
6. 本卡可复印使用。

以下由接收部门填写：

收卡人：＿＿＿＿收到卡片时间：＿＿＿＿

评审意见：

评审人（或单位盖章）

评审时间：　　年　　月　　日

（2）震情反映：

	2006	
办公室	Y	

机密★三个月

震 情 反 映

（2006 4）

云 南 省 地 震 局　　　　二〇〇六年七月十二日

云南近期震情形势分析

一、震情形势分析

自2006年1月以来，云南地区地震活动在前5个月内均处于相对较低的活动水平，共发生3级以上地震63次，月均13次。但6月以来，一个多月已发生3级以上地震28次，主要发震地区为滇西北东带、滇东的小江断裂带以及滇西南地区，省内地震活动水平明显增强。

云南省6级地震已平静1000天，接近云南历史上6级地震发生间隔的极限值。5级地震平静也远超过预测指标，从年初的墨江5.0级地震至今已180天没有5级以上地震发生，预示着省内发生中强地震的危险性较大。

另外，近期在滇西、滇东以及滇西南地区都出现了较多的地震前兆异常和一些地震宏观异常。

经我局综合会商，分析地震活动异常和地震前兆异常指标结果认为：

在未来3个月内，云南地区存在发生5－6级地震的危险，主要危险地区为滇东的大关—巧家至石屏—建水一带，滇西的中甸—宁蒗至永平—腾冲一带以及滇西南的思茅—勐腊地区。

二、采取的措施

（一）密切跟踪云南震兆、前兆宏微观异常的动态变化，适时进行震情的进一步分析预测；加强震情监视，有突出变化和短临震情分析意见及时上报。

（二）我局通知有关州市地震部门，采取必要的内部强化震情监视及应震、应急准备措施。

（三）加强震情值班。各级地震部门严格执行24小时值班制度和重大震情报告制度。

三、建议

（一）有关州市政府主管领导要适时召开抗震救灾指挥部工作会议，听取地震部门汇报震情，研究防范措施，相关领导要熟悉地震应急预案，强化震情观念，做好相关备震工作。

（二）按“内紧外松”的原则、确保社会稳定。

机密★1个月

震 情 反 映

(2006 08)

云 南 省 地 震 局　　　　二〇〇六年七月二十二日

云南近期震情形势分析

2006年7月22日9时10分，在云南滇东北盐津县(北纬: 28° 0′，东经: 104° 12′)发生5.1级地震。震源深度9km，距盐津县城约14Km，距昆明市340Km。该次地震发生在云南省地震局7月12日上报省政府的《震情反映》(2006 4)预报区范围内。地震发生后，云南省地震局立即召开了震情紧急会商会，给出了近期云南震情趋势预测意见:

一、近期云南震情趋势发展

盐津5.1级地震发生后，表明云南地区中强地震活动在今后一段时期仍有增强趋势，目前云南地区中小地震活动及地震前兆异常仍在持续发展。因此，对2006年度提出的地震重点监视防御区应继续加强短临跟踪工作。

二、盐津 5.1 级地震震区的震情分析

历史上滇东北地区是我省中强地震频繁发生的地区之一，该地区的地震类型比较复杂，有双震型、孤立型和主震—余震型。根据1900年以来在盐津 5.1 级震中周围 50Km 范围的中强地震活动统计，所发生的 5 次中强地震以主震—余震型为主。截止现在为止，盐津 5.1 级地震后无余震发生，近几天震区注意 3—4 级地震的活动。

报：中国地震局。省委、省政府，白恩培书记、徐荣凯省长、杨应楠秘书长、孔垂柱副省长，黄毅秘书长、何兴泽副秘书长。

发：有关州、市地震局(防震减灾局)，局有关单位。

2006 年 11 月 23 日新疆维吾尔自治区乌苏 5.1 级地震

新疆维吾尔自治区地震局

聂晓红　曲延军

摘　要

2006 年 11 月 23 日新疆乌苏县发生 5.1 级地震，震中位置为 44°15′N，83°27′E。由于地震发生于山区，无法判定极震区方向，震中烈度为Ⅴ度。此次地震未造成人员伤亡，直接经济损失 126.39 万元。

此次地震类型为孤立型，最大余震 2.8 级。2.0 级以上余震分布在主震南侧。震源断错为右旋走滑逆断性质，节面Ⅰ为主破裂面，主压应力 P 轴方位北西方向。发震构造为博罗克努—阿其克库都克断裂。

地震发生在新疆监测能力较好的北天山地区，200km 范围内共有 11 个地震台，其中测震单台 2 个，地温观测点 1 个，综合观测台（点）8 个。震前震中周围出现 10 项测震学异常，4 项地倾斜异常，2 项水化学异常，其中短临异常 2 项。

前　言

2006 年 11 月 23 日 19 时 04 分，新疆乌苏县境内发生 5.1 级地震。新疆地震局台网测定的微观震中为北纬 44.2°，东经 83.5°，由于极震区位于南部山区且交通不便、居民点少，因而无法确定宏观震中，震中烈度为Ⅴ度。本次地震有感范围 200km，东到石河子市，南到新源县，西至伊宁县。本次地震无人员伤亡，造成直接经济损失为 126.39 万元。

此次地震序列为孤立型，最大余震 2.8 级，余震活动衰减迅速。震源断错性质为右旋走滑逆断，震源机制解中走向 108°的节面Ⅰ为此次地震的破裂面，P 轴方位北西。博罗克努—阿其克库都克断裂带为该地震的发震构造。

震中附近属新疆监测能力相对较强的地区，200km 范围内有测震台站 7 个，其他前兆台站 8 个。震前共出现 12 个异常项目的 16 条前兆异常，震前异常多为地震活动异常，形变异常多为趋势异常，仅地下流体出现 2 项短临异常。

地震发生后，新疆地震局立即派出地震现场工作组和地震前兆跟踪组赴震区开展现场调查、震害评估和震情跟踪监视工作。

该地震发生在年度确定的地震危险区内，但由于前兆异常不显著，没有短临预报意见。

在相关文献、资料整理的基础上，经综合分析研究，本报告得以完成。

一、测震台网及地震基本参数

图1给出震中200km范围内的地震台分布，共计11个地震台（点），其中7个有测震观测，在研究时段内基本可达到1.0级以上地震不遗漏。100km内有精河、新源和乌苏测震台，100～200km分别有克拉玛依、柳树沟、石场和察布查尔测震台。

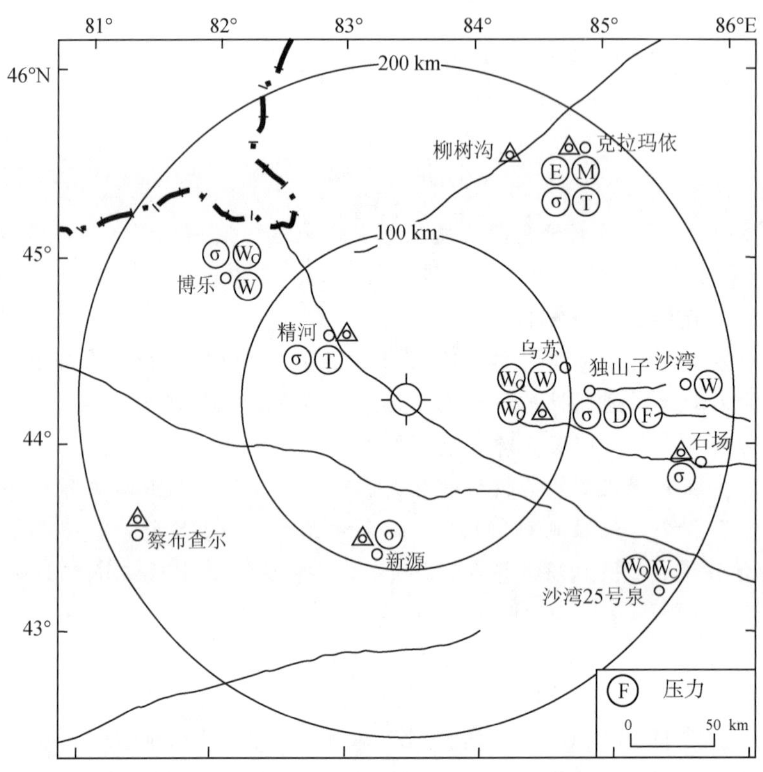

图1　乌苏5.1级地震前震中附近地震台站及观测项目分布图

Fig. 1　Distribution of earthquake-monitoring stations around the epicentral area before the *M*5.1 Wusu

表1列出不同来源给出的这次地震基本参数，经对比分析，认为新疆地震局经重新修订后的震中位置较为准确，因此本报告地震基本参数取表1中编号1结果。

表1　乌苏5.1级地震基本参数

Table 1　Basic parameters of the *M*5.1 Wusu earthquake

编号	发震日期	发震时刻	震中位置		震级	震源深度	震中地名	结果来源
	年 月 日	时 分 秒	φ_N	λ_E	M	(km)		
1	2006 11 23	19 04 42.5	44°15′	83°27′	5.1	36	乌苏	新疆地震局[1]

续表

编号	发震日期	发震时刻	震中位置		震级	震源深度	震中地名	结果来源
	年 月 日	时 分 秒	φ_N	λ_E	M	(km)		
2	2006 11 23	19 04 42.7	44°12′	83°30′	5.1	15	乌苏、精河间	国家台网中心
3	2006 11 23	19 04 47.4	44°16′	83°28′	4.7	30	乌苏	Harvard
4	2006 11 23	19 04 44.6	44°16′	83°32′	5.3b	21	乌苏	Harvard
5	2006 11 23	19 04 44.7	44°14′	83°29′	5.4b	22	乌苏	IRIS

二、地震地质背景[1]

此次地震震中位于北天山北部博罗克努山脉。在大地构造上，震中位于博罗克努地槽褶皱带，南部为伊犁地块；在地貌上，北天山博罗克努山脉山区河谷深切，地形陡峭，冲沟、阶地发育；在地层岩性上，博罗克努山区主要为古生代石炭系中统地层，主要为一套浅变质砂岩、凝灰岩等，分布少量第三系砂岩、泥岩，南部有华力西期花岗岩。

震中所处构造为博罗克努复背斜，为北西向构造，构造较为复杂（图2）。博罗克努—

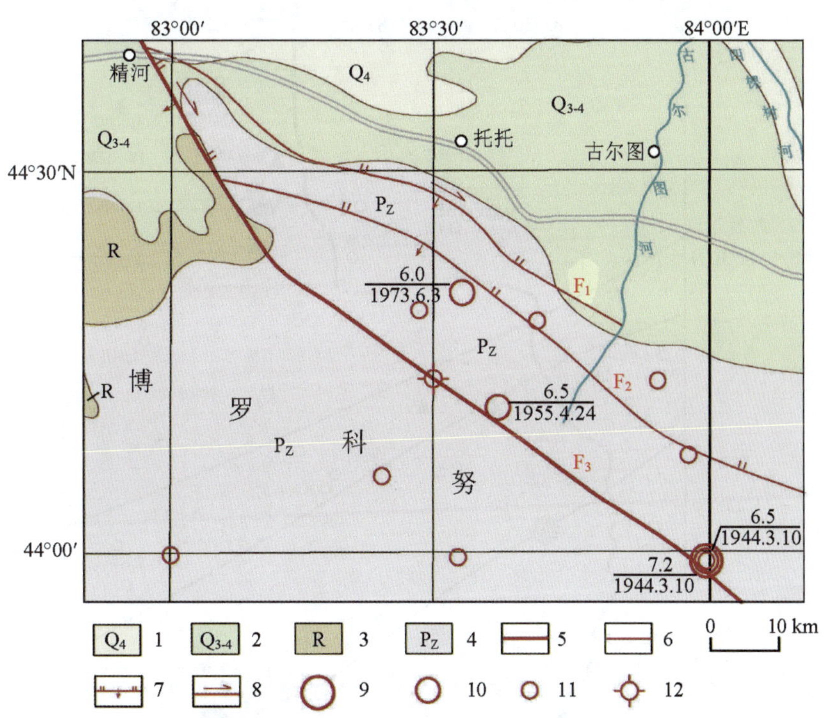

图2　乌苏附近地区主要断裂及历史地震震中分布图

Fig. 2　Major faults and distribution of historical earthquakes around Wusu area

F1 四棵树河—古尔图断裂；F2 亚玛特断裂；F3 博罗克努—阿其克库都克断裂

1. 全新统；2. 晚更新统—全新统；3. 第三系；4. 古生界；5. 全新世活动断裂；6. 更新世活动断裂；7. 逆断层；8. 走滑断层；9. M=7.0～7.9；10. M=6.0～6.9；11. M=5.0～5.9；12. 2006年11月23日5.1级地震

阿其克库都克断裂带是划分准噶尔—北天山褶皱系与天山褶皱系的分界断裂，该断裂沿NW—EW—NE向延伸，总体呈向南突出的弧形，全长1400多公里，断层面总体南倾，倾角50° ~ 80°。断裂带上有几十米到数公里宽的挤压破碎带，地表与钻孔中均可见到大量的糜棱岩、碎裂岩、擦痕及强烈柔皱的各种片理化岩石，以及退色变质现象。断裂在地貌上形成南高北低的断层阶地，航、卫片上有明显的线性影像。沿断裂有狭长凹地、干谷、盐沼地等一系列长形断陷盆地分布，其中沉积有厚达上千米以上的中、新生代地层，同时还分布有大小不等的长条状古生代基性—酸性侵入岩，断裂西段是现代热泉异常带，有40℃温泉多处。该断裂活动具长期性与多期性，由其所控制的地层看，早古生代以前即已形成，之后多次活动，在精河东南和艾比湖南该断裂切割全新世地层，表明该断裂是全新世活动断裂。该断裂曾于1944年3月10日发生过7.2级地震，此外该断裂上还发生过1955年4月24日6.5级地震，沿断裂至今仍有5级左右地震不断发生。综合分析认为博罗克努—阿其克库都克断裂带是2006年11月23日乌苏5.1级地震的发震构造。

三、地震影响场和震害[1]

由于极震区位于南部山区交通不便，居民点少，5.1级地震又不大可能造成地表破裂现

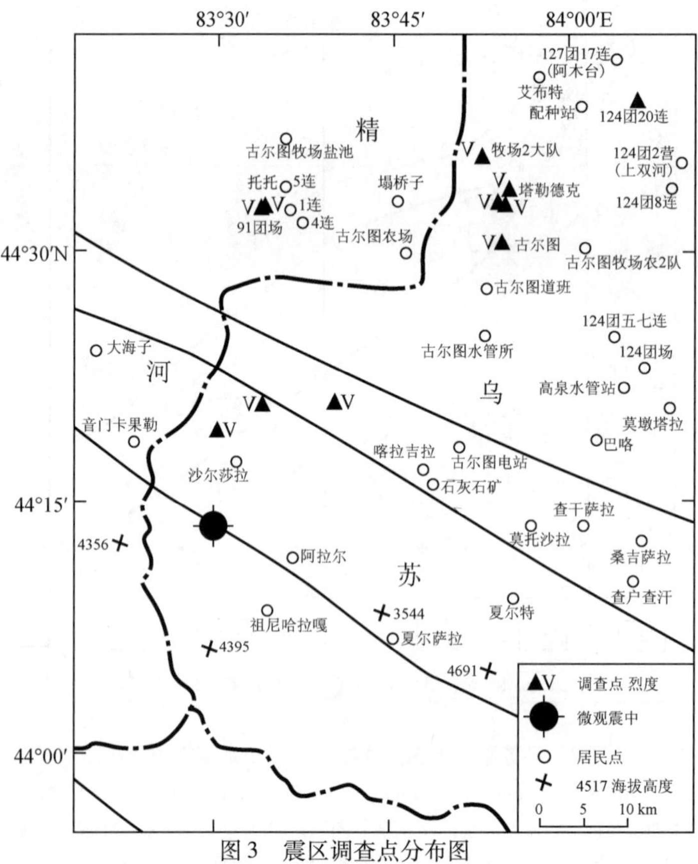

图3　震区调查点分布图

Fig. 3　Distribution map of investigation in epicenter region

象，因此无法确定宏观震中。根据现场实地考察及调查资料，震中烈度可能为Ⅴ度，地震有感范围为 200km，东到石河子市，南到新源县，西至伊宁县。

Ⅴ度区：调查点分布较散，包括乌苏市鼠疫防疫站、乌苏林场管理站、精河县托托乡、乌苏市古尔图镇等（图 3）。

乌苏市鼠疫防疫站距微观震中 10km，地震时听见轰隆隆的地声，水平震感强烈。沿途山区路边有少量的崩塌落石，1 ~ 2m³ 大小，岩块直径最大 50cm。

乌苏林场管理站听见轰隆隆的地声，水平震感强烈。附近玉滚格勒河Ⅱ级堆积阶地前缘陡直，距河水面高度达 30m，在靠近山嘴处，沿阶地前缘发育 30 ~ 40m 长滑塌。

精河县托托乡部分砖混结构教室和土木住宅发育有斜向八字形裂缝和竖向裂缝。水平震感强烈。

乌苏市古尔图镇场部居民水平震感强烈。部分老旧土木结构房屋和砖木结构平房发育有竖向裂缝和斜向裂缝，门梁角有竖向裂缝，部分房屋墙角出现下沉。

本次地震多数房屋未遭到破坏，无人员伤亡，造成总经济损失为 126.39 万元，其中乌苏市直接经济损失 79.42 万元；精河县直接经济损失 46.97 万元，属一般破坏性地震。

四、地 震 序 列

2006 年 11 月 23 日乌苏 5.1 级地震后，截至 2007 年 2 月 24 日共定出 1.0 级以上地震 19 次，其中 5.0 ~ 5.9 级 1 次；2.0 ~ 2.9 级 4 次；1.0 ~ 1.9 级 14 次。表 2 给出新疆地震局地震台网定位的 2.0 以上地震序列目录，主震后 3 个月内最大余震为 2.8 地震，与主震震级差 2.3 级，主震能量占总序列的 99.95%，由此认为该地震序列类型为孤立型。

表 2　乌苏 5.1 级地震序列目录（$M_S \geqslant 2.0$ 级）

Table 2　Catalogue of the *M*5.1 Wusu earthquake sequence（$M_S \geqslant 2.0$）

编号	发震日期	发震时刻	震中位置		震级 M_S	震源深度	震中	结果
	年 月 日	时 分 秒	φ_N	λ_E		（km）	地名	来源
1	2006 11 23	19 04 42.5	44°15′	83°27′	5.1	36	乌苏	资料 1）
2	2006 11 24	01 43 56.2	44°14′	83°22′	2.3	15	精河	
3	2006 12 04	04 23 07.0	44°01′	83°43′	2.8	24	乌苏	
4	2007 01 30	06 15 27.3	44°00′	83°09′	2.0	19	尼勒克	
5	2007 02 07	04 26 08.0	44°03′	83°09′	2.5	10	尼勒克	

图 4 给出震后 3 个月震中附近 1.0 级以上余震分布，可见余震主要沿北西向的博罗克努—阿其克库都克断裂分布。分析余震序列 *M-t* 图（图 5），余震在主震后 15 日内较为集中，之后迅速衰减。

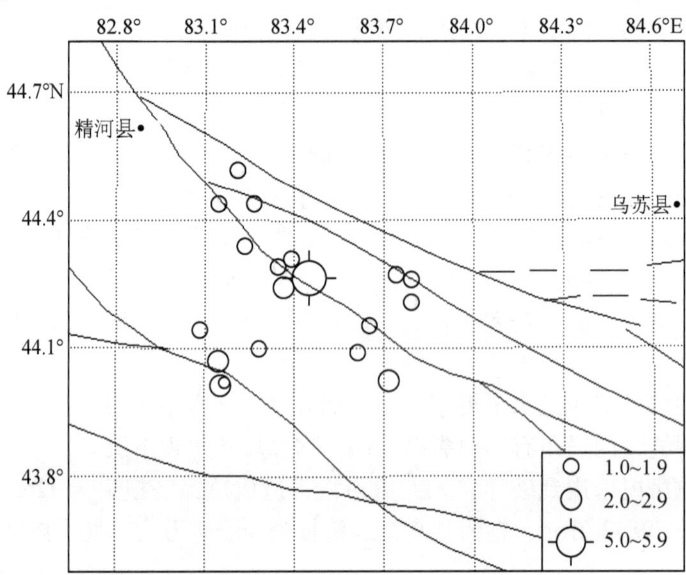

图4　乌苏 5.1 级地震 M_S≥1.0 级余震震中分布图

Fig. 4　Epicentral distribution of the aftershocks of the M_S 5.1 Wusu earthquake

对余震序列进行5日窗长、2日步长的时间扫描，给出5.1级地震序列的频度（图6）和应变曲线（图7），可见频次衰减较快，但2007年1月底开始频度略有回返；主震后未出现较大能量的释放。利用最小二乘法计算该序列 b 值为0.72（图8），与该区背景值0.65基本相当。

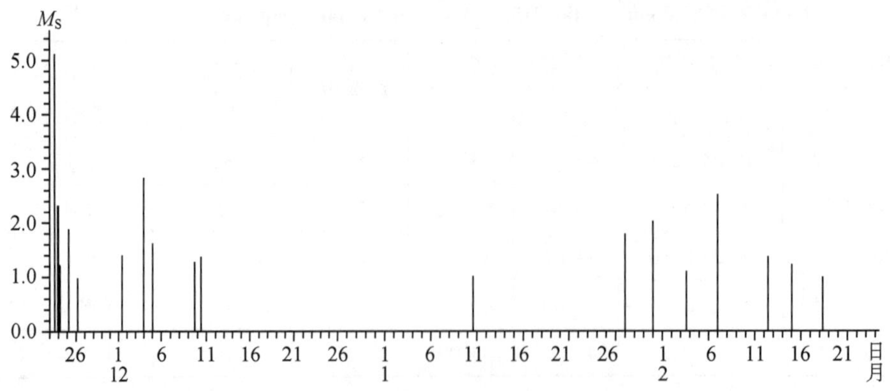

图5　乌苏 5.1 级地震序列 M-t 图

Fig. 5　M-t diagram of the M_S5.1 Wusu earthquake sequence

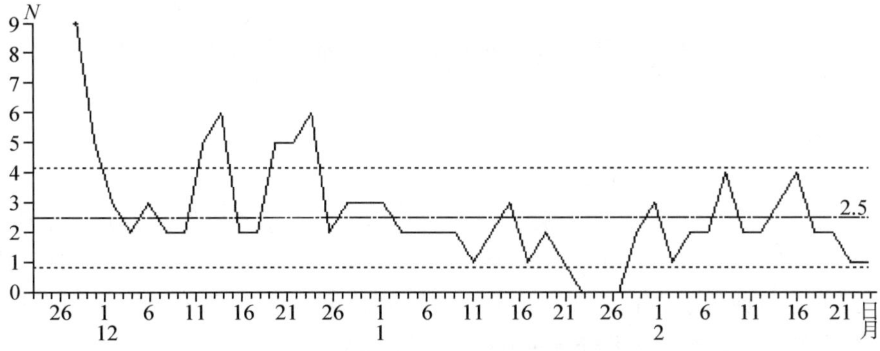

图 6　乌苏 5.1 级地震序列日频度图

Fig. 6　The daily frequency distribution of the M_S5.1 Wusu earthquake sequence

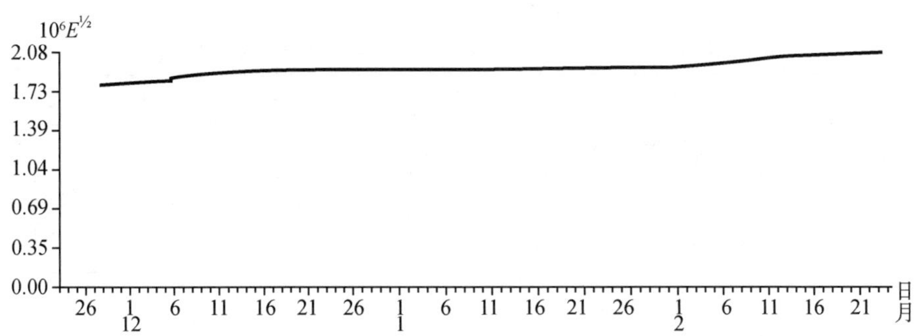

图 7　乌苏 5.1 级地震序列应变释放曲线

Fig. 7　Strain release curve for the M_S5.1 Wusu earthquake sequence

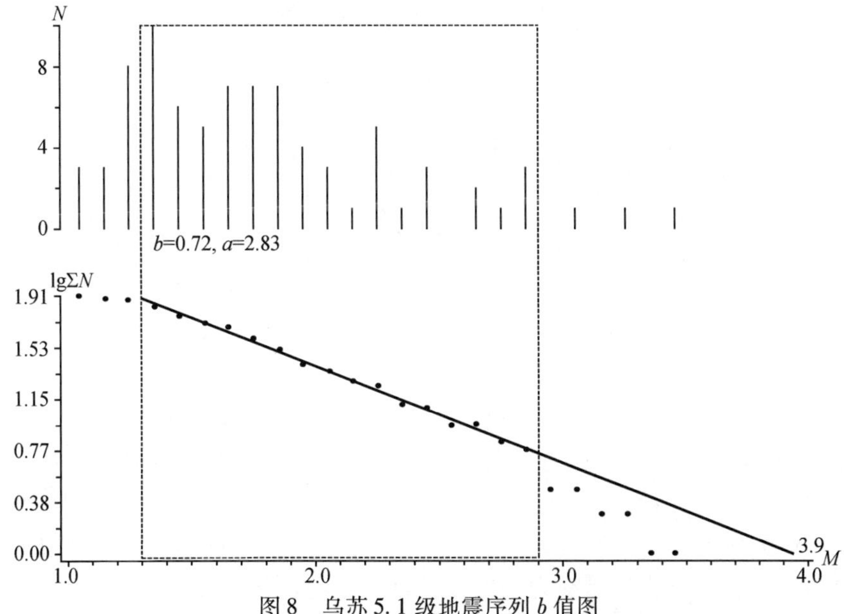

图 8　乌苏 5.1 级地震序列 b 值图

Fig. 8　b-value diagram of the M_S5.1 Wusu earthquake sequence

五、震源机制解和地震破裂面[2]

表3和图9给出乌苏5.1级地震的震源机制解结果。利用了新疆境内台网有清楚初动的27个台站资料，矛盾符号比为0.074。

表3显示此次地震震源断错为右旋走滑逆断性质。根据博罗克努—阿其克库都克断裂的走向（NW向）和余震分布认为，走向为北西的节面I是此次地震的破裂面，其方位为108°，倾角78°，节面近乎直立，5.1级地震破裂滑动角154°。主压应力P轴方位为158°，倾角9°，接近水平，主张应力T轴方位64°，倾角26°，表明5.1级地震为主要受NW向水平作用力产生右旋走滑错动的结果。

表3　乌苏5.1级地震的震源机制解

Table 3　Focal mechanism solution of the M_S 5.6 Wusu earthquake

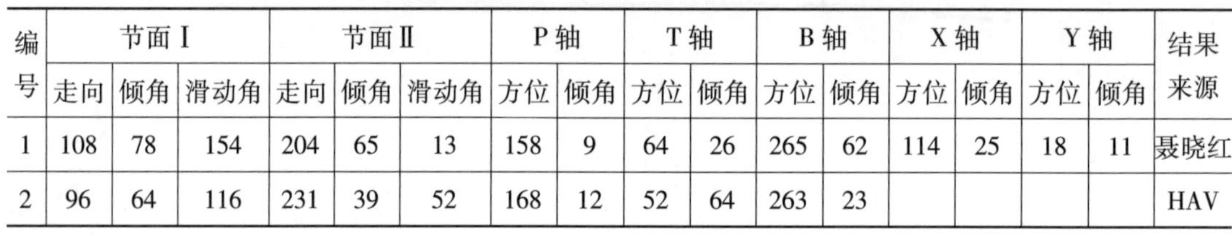

编号	节面I			节面II			P轴		T轴		B轴		X轴		Y轴		结果来源
	走向	倾角	滑动角	走向	倾角	滑动角	方位	倾角	方位	倾角	方位	倾角	方位	倾角	方位	倾角	
1	108	78	154	204	65	13	158	9	64	26	265	62	114	25	18	11	聂晓红
2	96	64	116	231	39	52	168	12	52	64	263	23					HAV

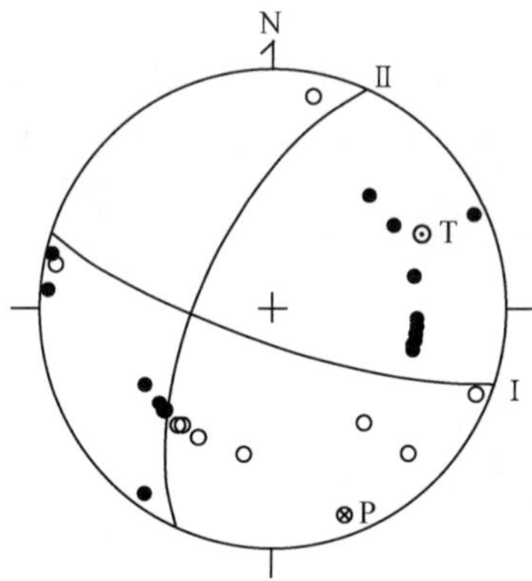

图9　乌苏5.1级地震震源机制解

Fig. 9　Focal mechanism solution of the M_S 5.6 Wusu earthquake

北天山西段主要构造为NW走向，根据新疆历史中强地震震源机制解结果显示，该区域多数地震震源机制解主压应力P轴方位大致垂直构造走向，以NS—NNE向为主，震源断错多数为倾滑型错动。与此前几次地震不同的是，此次地震断错性质为走滑。根据《新疆2006年度地震趋势研究报告》中GPS观测结果，在这次地震震源区周围主压应力方向近NNW向，可能反映出震前大区域应力场处在一种调整和变化过程中之，应力场的变化会产生局部构造破裂特征的差异[1]。

六、观测台网及前兆异常

震中附近地区地震台站及观测项目分布见图 1。地震发生在北天山西段，属新疆监测能力较强地区，区内台站分布相对均匀。震中 200km 范围内有测震台站 7 个，均为数字地震台；有定点前兆测项的地震台 6 个（包括地倾斜、钻孔应变、地磁和地电等观测项目），有地下流体测项的 4 个（包括水位、水温、流量和压力）。

在 0～100km 范围内有地震台站 3 个，均为数字地震记录和前兆观测并存的综合性台站；101～200km 范围内地震台站 8 个，其中测震台站 2 个，地温台站 1 个，综合台站 5 个。此次地震前共出现 12 个异常项目 16 条异常（表 4）。

表 4　地震异常情况登记表

Table 4　Abnormity situation

序号	观测项目	台站或观测区	分析方法	异常判据及分析方法	震前异常起止时间	震后变化	最大幅度	震中距（km）	异常类别	图号	异常特点	备注
1	AC	43.14°～44.93°N 82.58°～86.07°E	时间扫描	高于 1.80	2004.08～2005.10	正常	2.40	200	M_I	12	高值异常	
2	A（b）		时间扫描	低于 5.55	2005.04～2006.08	正常	5.22	200	M_I	12	低值异常	
3	η 值		时间扫描	$\eta \leqslant 2.2602$	2003.08～2006.10	正常	1.86	200	M_I	12	低值异常	
4	b 值		时间扫描	$b \geqslant 0.70$	2004.02～2006.09	正常	0.84	200	M_I	12	高值异常	
5	空区	42.74°～44.06°N 82.34°～84.97°E 43.05°～44.54°N 81.86°～84.36°E	$M_S \geqslant$ 3.0 级空间分布	空区内外频次比	2004.09～2005.09 2005.10～2006.07	空区瓦解	围空 13 个月长轴 216km 围空 10 个月长轴 204km		M_I	10		震前
6	应力降	43.30°～44.30°N 83.00°～85.00°E	D =（$\Delta\sigma$ − 本底值）/本底值	$D > 0$	2005.10～2006.09	恢复	$D = 4.7$		M_I	11	D 值越大，地震危险性越大	震前

续表

序号	观测项目	台站或观测区	分析方法	异常判据及分析方法	震前异常起止时间	震后变化	最大幅度	震中距(km)	异常类别	图号	异常特点	备注
7	地震窗	温泉窗 石场窗 乌苏窗	单台小震频次	超过历史警戒线	2006.02～2006.10	恢复		200 180 100	M_I		超越历史警戒线	震前
8	初动半周期	精河台	时间进程	低于均值	2006.01～2006.08	恢复		57	M_I	13	低值异常持续8个月，震前回返	震后总结
9	地震波振幅比	新源	时间进程	低于均值	2006.08～2006.12	恢复		100	S_I	14	震前异常恢复，震后一个月异常恢复	震后总结
10	震群	沙湾	空间分布	小震频次	2006.02～2006.05	恢复	138次	166	S_I		同一位置发生2次小震群活动	震前
11	地倾斜	精河石英摆	矢量图	年变幅增大，年变反向	2005.03～震前	持续	0.56″	60	M_I	15	年变幅明显增大	震后总结
		精河水管仪	单分量多项式拟合时序曲线	加速	2003.10～震前	持续	1.1″	60	M_I	16	NS向加速N倾，EW向加速E倾	震后总结
		石场石英摆	多项式拟合曲线	破年变速率加快	2005.05～??	持续		180	M_I	18	2005年初EW向加速E倾，震前1年结束；2004年6月NS向加速N倾，2005年转为S倾，2006年5月再次出现加速N倾	震前

续表

序号	观测项目	台站或观测区	分析方法	异常判据及分析方法	震前异常起止时间	震后变化	最大幅度	震中距 (km)	异常类别	图号	异常特点	备注
11	地倾斜	新源竖值摆	单分量时序曲线	速率加速方向不稳定	2005.07 ~ 2006.06	恢复	0.4″	100	M_I	17	2005 年 7 月 NS 向加速 N 倾，2006 年 6 月中旬恢复 S 倾	震前
12	地下流体和水化学	25 号泉	水温	突降—突升	2006.11.07 ~ 2006.11.09 2006.11.18 ~ 2006.11.23	恢复	突降近 25℃ 突升近 2℃	190	I_I	19	2006 年 11 月 7 日水温迅速下降，9 日恢复，15 日水温突然上升，17 日后逐步下降	震后总结
		26 号泉	流量	溢出冒泡	2006.09 ~ 2006.10 2006.10 ~震前	恢复持续		160	M_I I_I	20	流量增大有气体溢出	震后总结

1. 地震活动性异常

1）地震空区[3~6]

“十五”课题研究中，对 1970 年以来北天山 5 级以上地震前围空现象进行了全时空扫描，结果表明，北天山 13 次 5 级地震前 9 次形成围空，围空时间为 0.5 ~ 1.5 年。2006 年 11 月 23 日乌苏 5.1 级地震前，在震源区附近的北天山西段地区连续形成 2 个 3 级地震围空图像（图 10），即 2004 年 9 月至 2005 年 9 月在乌苏西南地区形成 3 级地震围空，并在 2005 年 10 月 11 日被打破；紧接着 2005 年 10 月至 2006 年 7 月在该空区西侧再次形成 3 级地震围空，该围空内部 2006 年 11 月 1 日和 2 日连续发生 4.1 级和 3.9 级地震，空区被打破。2006 年 11 月 23 日乌苏 5.1 级地震即发生在第二次形成的 3 级地震围空内部。震后 1 年北天山西段的 3 级地震分布没有再形成新的空区。

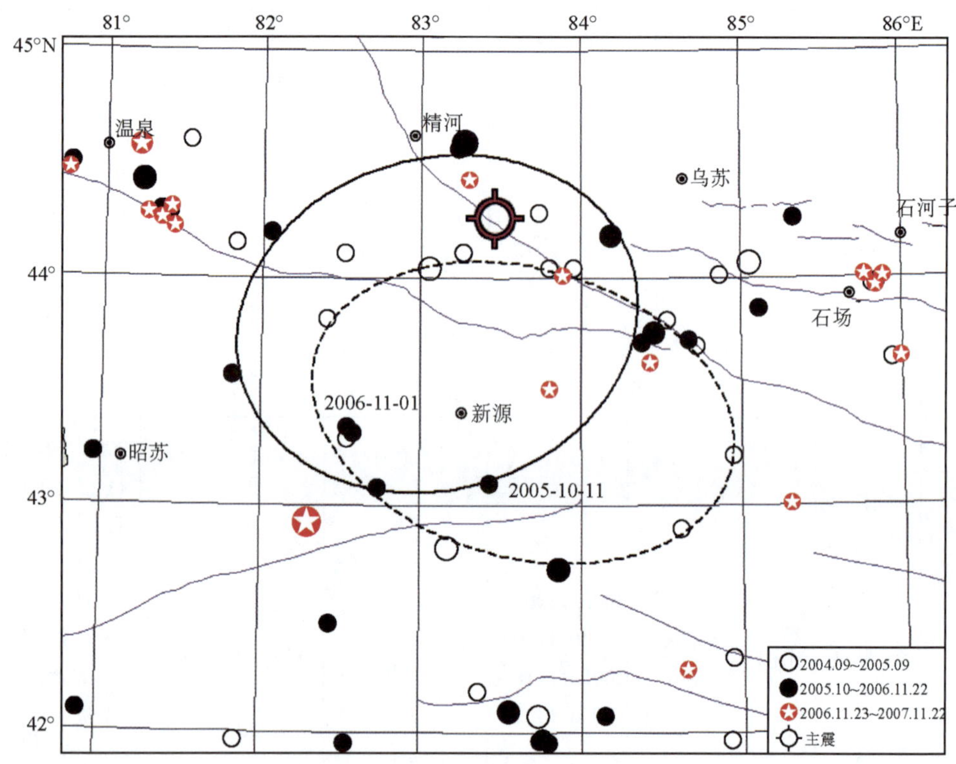

图 10　乌苏地震前 3 级地震围空

Fig. 10　Seismic gap of *M*3 earthquake before Wusu earthquake

2）*地震窗*[3～6]

地震窗方法是以台站为中心，选取 S—P≤10s 的 M_L1.0 以上小地震月频次，其超过历史警戒线为异常。新疆地震窗口网由 14 个地震窗组成，是新疆地区日常地震监测预报的重要方法之一。北天山地区有 4 个地震窗。乌苏 5.1 级地震前 9 个月距震中 100km 的乌苏窗出现异常；震前半年距震中 180km 的石场窗异常；震前 3 个月（2006 年 9 月）距震中 200km 的温泉地震窗异常。其中温泉地震窗 1993 年以来异常出现后 4 个月内对应周围 5.1 级以上地震的对应率为 56%，曾对库车 1999 年 5.7 级和 2003 年昭苏 6.0 级等地震有较好的反映。乌苏 5.1 级地震前地震窗的异常显示，随着地震的临近，异常从近距离向远距离发展。

3）*震群活动*[3,4,6]

乌苏 5.1 级地震发生前，沙湾附近出现 2 次小震群活动。一次是 2 月 25 日至 3 月 6 日，共发生 M_L1.0 以上小震 88 次，包括 2 次 3 级地震，最大震级 M_L3.8。另一次是 5 月 14～24 日，共发生地震 32 次，最大震级 M_L2.4。这两次震群空间位置基本一致。沙湾距离乌苏地震震中 166km，沙湾小震群发生在天山北缘一套中新生代地层复式褶皱前沿与准噶尔地块相衔接的边缘地带，震群发生区域及邻区地质构造复杂，新构造运动强烈，介质相对破碎，历史震群活动较为频繁。而乌苏地震结束后，11 月 24～26 日沙湾震群又一次活跃，共发生地震 14 次，最大震级 M_L2.4。

4）应力降[3~5]

运用新疆数字化地震波资料，从应力降时空演化特征中寻找中强震前后震源区应力状态和孕震状态变化信息，研究中在一定程度上扣除震级大小对应力降的影响。定义某区域应力降相对于长期平均值（即本底值）的增长幅度为：$D=(\Delta\sigma-\text{本底值})/\text{本底值}$。若 $D\leqslant0$，表示该区的应力降值处于长期本底值之下；若 $D>0$，表示该区的应力降值比长期本底值高。D 值越大，则该区的地震危险性越大。

夏爱国等[7]对 2001 年以来新疆天山中东段 152 次 $M_S\geqslant2.5$ 级地震应力降的研究表明，中强地震前该地区 D 值有明显的增高趋势。研究时段内，研究区内仅发生了 2003 年石河子 5.4 级地震，该地震前 D 值为 4.5。2006 年 11 月 23 日乌苏 5.1 级地震前，震源区附近地区的 D 值达到 4.7（图 11），与 2003 年 2 月 14 日石河子 5.4 级地震前 D 值（4.5）的水平基本相当[2]。

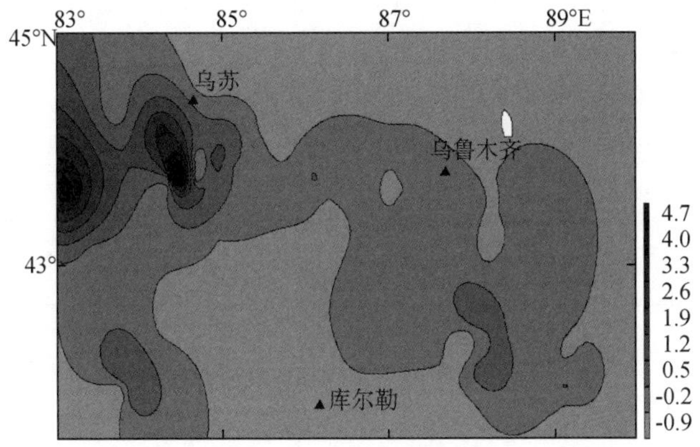

图 11　乌苏 5.1 级地震前应力降值增长幅度 D 值分布

Fig. 11　Distribution of stress drop D value before Wusu $M5.1$ earthquake

5）地震学参数时间扫描[3~5]

选取震中 200km 范围内，1980 年以来 $M_S\geqslant2.0$ 级地震，以 12 个月为窗长、2 个月为滑动步长，对该区域进行地震活动性时间扫描，图 12 给出震前出现异常的各项参数，分别是 AC 值、η 值、$A(b)$ 值和 b 值。表 5 给出乌苏 5.1 级地震前震中附近区域出现的异常。表中以异常持续半年以上为确定异常，异常结束后对应的目标地震发震时间最长不超过 1.5 年。由图中可以看出，AC 值异常开始到结束，持续 1 年 2 个月，从异常结束到发震 13 个月；η 值异常持续时间最长，达 3 年，震前 1 个月恢复；b 值异常持续时间 2 年半，震前 2 个月异常恢复；$A(b)$ 值异常持续时间 1 年 4 个月，震前 3 个月异常恢复。

表 5　北天山西段地震活动性参数异常时间统计表

Fig. 5　Statistics table of abnormal time of seismic activity parameters in northern Tian Shan

地震活动性参数	异常起止始时间	异常对应比例	地震对应率
AC	2004.08～2005.10	3/5	3/5
b	2004.02～2006.09	4/8	4/5

续表

地震活动性参数	异常起止始时间	异常对应比例	地震对应率
η	2003.08 ~2006.10	4/8	4/5
$A(b)$	2005.04 ~2006.08	5/7	4/5

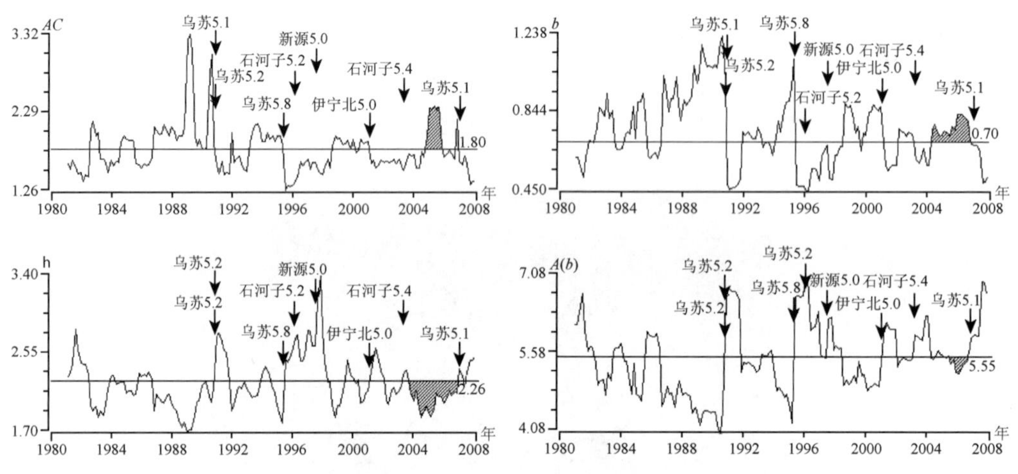

图 12　地震活动性参数时间扫描

Fig. 12　Time curve of parameters of seismic activity

6）初动半周期[6]

精河地震台处于博阿断裂、准噶尔南缘断裂和科古琴三大断裂带交汇区，海拔高度 387 m，台基岩性为黑灰色千枚岩河凝灰岩，地质年代为古生代，所使用的地震计型号为 DD-1，采样率为 50Hz，仪器工作状态良好。该地震台距离乌苏 5.1 级地震仅 57km。挑选精河台 S－P≤8s，P 波有清晰初动的 M_L≤4.5 级地震进行初动半周期分析。结果显示（图 13），2006 年 1 ~7 月精河台初动半周期出现较一致的低值过程，持续 7 个月，5 月中旬至 7 月呈现回返趋势，8 月后异常基本结束。

7）振幅比[6]

新源地震台位于天山深部腹地，其东部为南北天山交汇处，周围有多条地震断裂带，呈现多个方向展布，海拔 1045m，台基岩性为石炭纪安山岩，2006 年 5 月架设 CTS-1E 型宽频带数字地震仪。该台距离乌苏 5.1 级地震约 100km。对其周围不同方向地震进行振幅比时间进程分析。

图 14 给出新源台其中一个分区（乌苏地震在该分区内）的振幅比时间进程曲线。图中显示，该台振幅比在 2006 年 8 月至 2006 年 12 月出现了 4 个月的低值过程，震后异常持续了 1 个月后恢复。

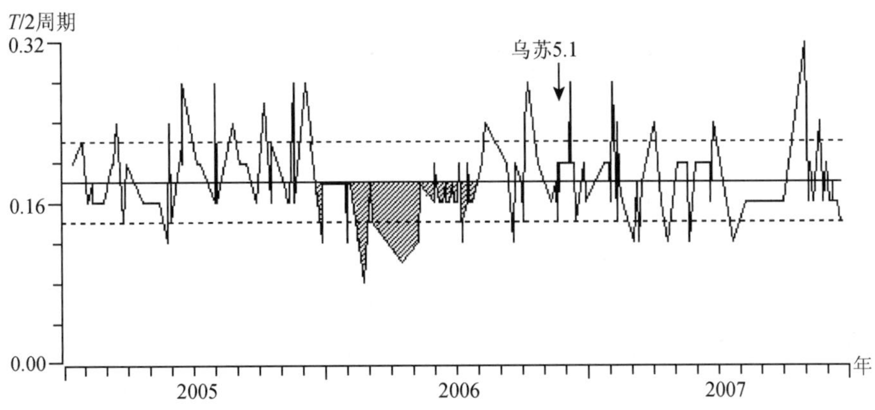

图 13 精河台初动半周期时间分布

Fig. 13 Distribution of time of half period of first arrive for Jinghe station

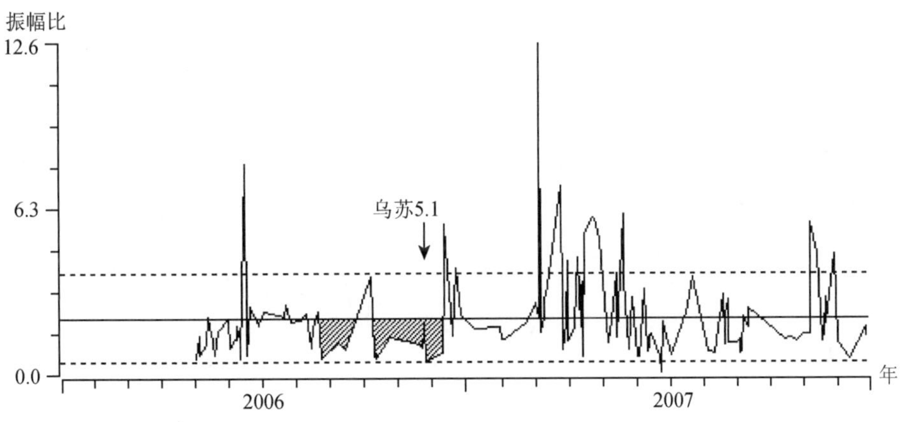

图 14 新源台振幅比分区时间进程曲线

Fig. 14 Time curve of ratio of amplitude for subarea of Xinyuan station

2. 前兆观测项目异常[8]

1）地倾斜

精河台石英摆倾斜仪自 1990 年以后开始连续观测，日变形态正常、年变清晰。2005 年 8 月初 NS 向出现趋势性加速 N 倾，在 2005 年 10 月 8 日巴基斯坦 7.8 级强震后速率减缓，并转向 S 倾。2006 年 2 月初再次转为趋势性 N 倾变化，至 9 月初快速转向 S 倾。乌苏 5.1 级地震后，该资料异常仍然持续（图 15），2007 年距离该台 200km 的特克斯发生 5.9 级地震。

精河台水管倾斜仪从 2001 年开始连续观测，年变形态清晰。2005 年 4 月资料 NS 向出现加速 N 倾，同年 8 月 EW 向也开始出现加速 E 倾的情况，乌苏 5.1 级地震后，该资料异常仍然持续（图 16），2007 年距离该台 200km 的特克斯发生 5.9 级地震。

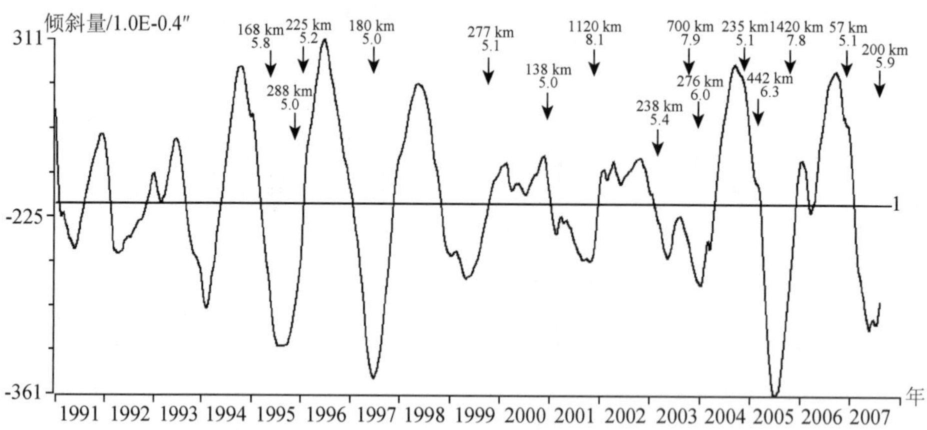

图 15　精河石英倾斜仪 NS 向去年变、去趋势后的时序曲线

Fig. 15　Time curve of the datum of quartz NS direction clinometer with no annual change and no trend in Jinghe station

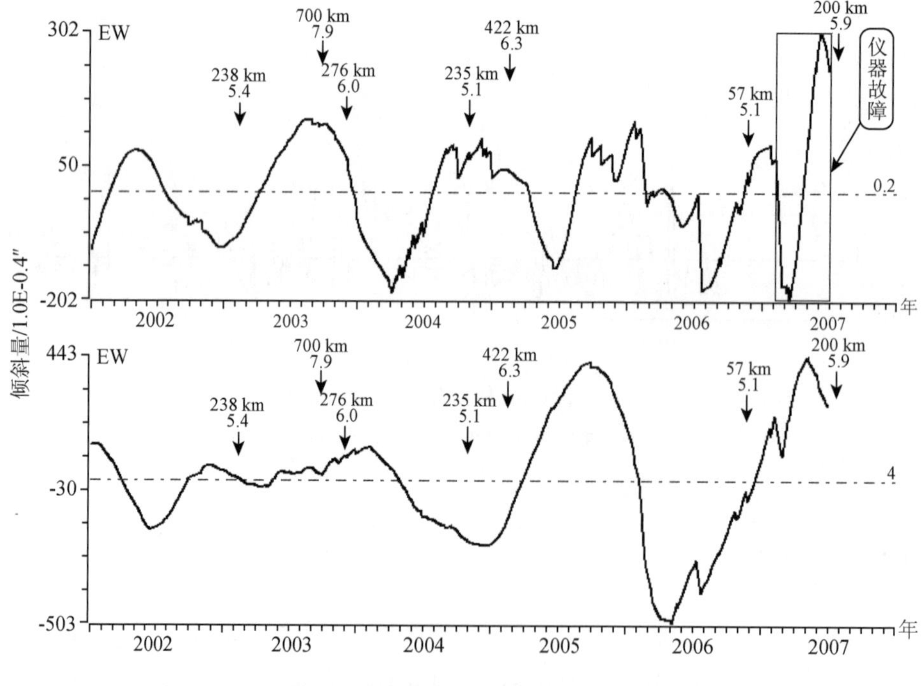

图 16　精河水管倾斜仪去年变、去趋势时序曲线

Fig. 16　Time curve of the datum of hose clinometer with no annual change and no trend in Jinghe station

新源台竖直摆倾斜仪从 1992 年开始观测，其总体趋势为 NS 向 S 倾、EW 向 E 倾。2005 年 7 月开始，NS 向出现加速 N 倾变化，在 N 倾变化过程中伴随有明显的速率和方向的不稳定异常变化，2006 年 6 月初资料恢复原有趋势（图 17）。

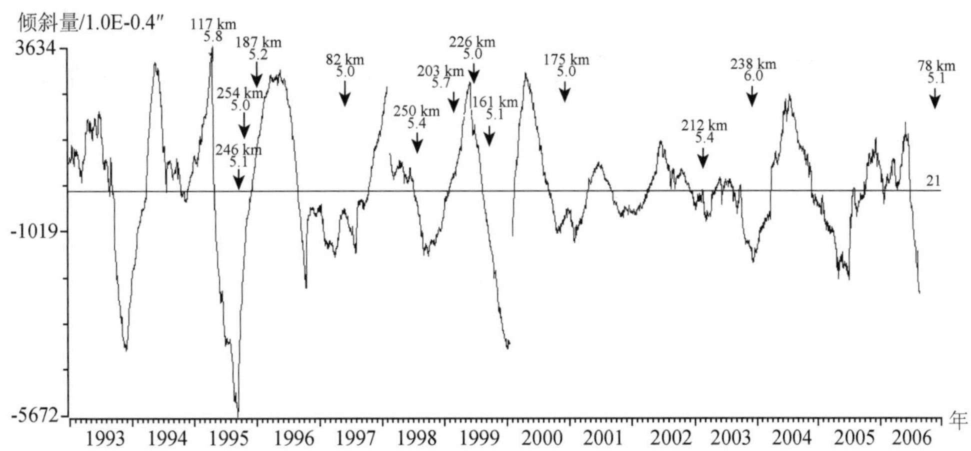

图 17　新源竖直摆倾斜仪 NS 向去趋势时序曲线

Fig. 17　Time curve of the datum of Vertical pendulum NS direction tiltmeter with no trend in Xinyuan station

石场台石英摆倾斜仪 1991 年开始进行观测，年变清晰。2005 年初该资料 EW 向出现加速 E 倾的变化，1 年之后转为 W 倾，该异常结束后 1 年发生乌苏 5.1 级地震。而 NS 向在 2004 年 6 月出现加速 N 倾变化，持续 1 年后转为 S 倾变化。2006 年 5 月资料再次出现加速 N 倾的变化，在乌苏 5.1 级地震前速率逐渐减缓（图 18）。

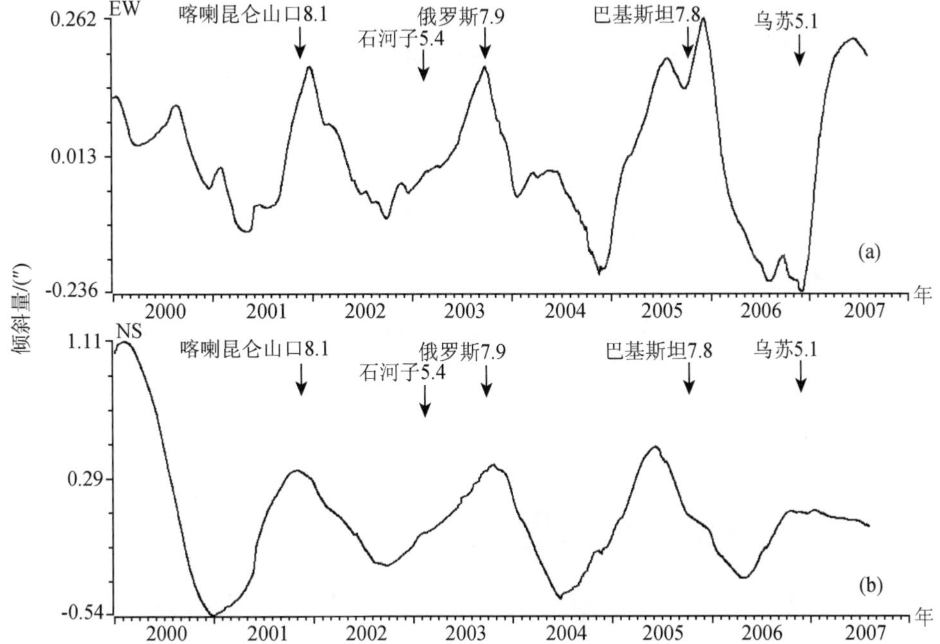

图 18　石场台地倾斜 EW 向（a）、NS 向（b）消除年变及趋势后的时序曲线

Fig. 18　Time curve of ground tiltmeter with no annual and no trend in Shichang station

2）地下流体

2006年11月23日乌苏5.1级地震前，沙湾25号泉（图19）和乌苏东南26号泥火山泉动水位出现短临异常（图20）。

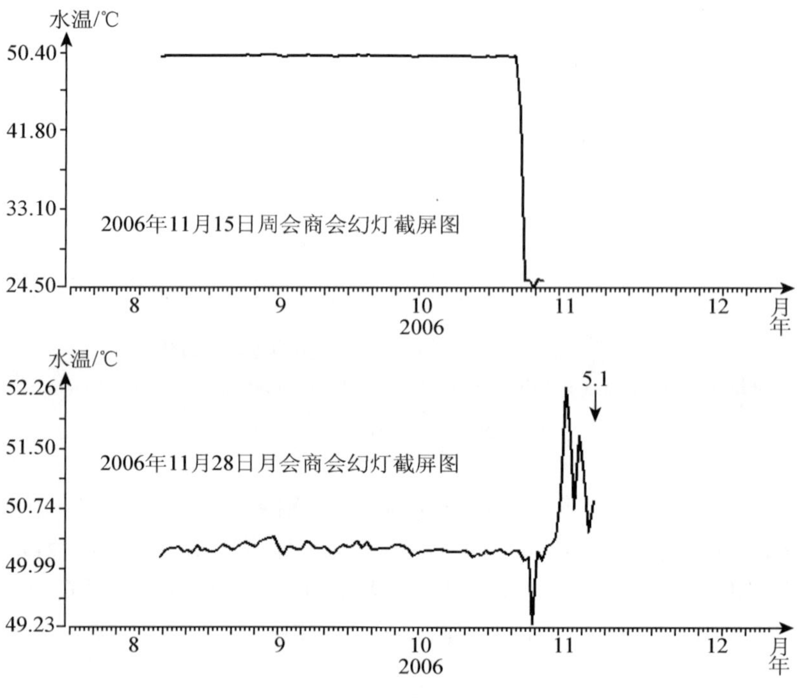

图19　25号泉动水温

Fig. 19　Dynamic water temperature of No. 25 spring

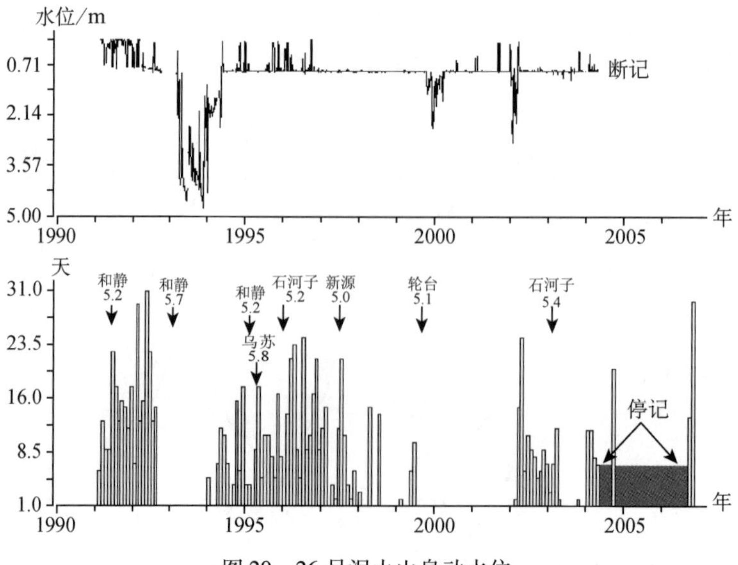

图20　26号泥火山泉动水位

Fig. 20　Dynamic water level of No. 26 Mud volcano spring

沙湾 25 号泉始建于 1988 年，2006 年 8 月 21 日停测，同日于该测点旁的鸡蛋泉开始新的记录，2006 年 11 月 7~9 日鸡蛋泉水温突然从 50.4℃降至 24.5℃，11 月 10 日恢复正常，11 月 15 日鸡蛋泉水温再次出现突然上升变化，20 日达到最高值 52.26℃。在 2006 年 11 月 15 日周会商会中相关人员提到该项资料出现的异常变化，但由于鸡蛋泉测点仅记录 2 个多月，资料稳定性无法判定，因此在预测过程中未对该资料进行评述，而震后总结认为，该异常可能是一项较好的短临异常。

乌苏东南 26 号泥火山泉距震中 160km，其动水位自 2006 年 9 月 26 日开始出现流量增大现象，且涌流液体中含有大量气体；9 月 30 日，因气体含量猛增导致水位记录图纸上出现许多幅度达 2~3cm 的“脉冲峰”[7]；10 月 5 日开始，水已涌到井口，水位记录浮子被抛出井管。此时模拟水位仪已经不能准确记录其真实流量，只能通过水位记录纸上脉冲曲线的幅度来反映井管内上涌泥浆量及气泡大小。2006 年 10 月 26 日后，“脉冲峰”幅度有所减小，但是这种流量增大并含有大量气体现象，截至 11 月 20 日之后仍在持续。

七、前兆异常特征分析

乌苏 5.1 级地震前共出现异常项目 12 项，异常项次 16 次。其中地震活动性异常 10 项，包括空区，震情窗、震群、应力降 D 值、AC 值、b 值、η 值、$A(b)$ 值、初动半周期和地震波振幅比值；地倾斜观测 1 项（包括精河石英摆、精河水管仪、新源竖直摆和石场石英摆）；地下流体观测异常测项 1 项（包括 25 号泉水温和 26 号泉流量），主要为流量、气体。异常特征主要表现在中短期异常较多，共 11 项，而短临异常仅 1 项。

乌苏 5.1 级地震 200km 范围内出现的地震活动性异常 10 项次，占总异常项次的比例为 10/16（62.5%），定点前兆观测异常出现 4 个项次，占总异常项次的比例为 4/16（25%），地下流体观测异常项次 2 次。地震活动性异常所占比例相对较大，异常多为中期异常，短期异常表现为部分单台地震波参数。定点前兆观测项目异常相对较少，但异常项次较多，主要表现为地倾斜的速率不稳、年变畸变等。地下流体在该区域覆盖相对较为密集，但震前仅出现 2 个测项的异常，但临震异常显著。

乌苏 5.1 级地震后，地震活动性异常均已恢复，前兆异常中除新源台异常在震前就已恢复外，其他前兆异常在震后仍然持续。实际上，在乌苏 5.1 级地震西南 180km 处，相隔 8 个月，于 2007 年 7 月 20 日发生了特克斯 5.9 级地震。这可能是前兆异常持续的原因。

八、应急响应和抗震设防工作

1. 预报情况

2006 年 11 月 23 日 19 时乌苏 5.1 级地震发生在 2006 年度新疆地震局划定的乌苏—精河 5~6 级地震危险区内，同时也落在震前划定的 2007 年度北天山乌苏—精河 5~6 级地震危险区内（2006 年 11 月 20 日完成研究报告，主要结论报中国地震台网中心；附件 2、附件 3）。表明该地震前有较好的年度预测。

2006 年 7 月 26 日预报中心填报了 1 张短临预报卡（附件 4），该预报意见的主要依据包括新疆5、6级地震平静，南天山东段的增强—平静过程，北天山5个月的4级地震平静被打破、应力降高于历史背景水平，库尔勒台3套倾斜资料、库车台地倾斜资料和库尔勒跨断层资料出现短期异常。该预报意见的定点前兆依据主要来源于南天山东段，但在预报时结合了地震活动性资料综合划定了危险区域，2006年11月23日乌苏5.1级地震距离预报区域边缘30km，该预报意见对此次地震具有一定程度的预报（附件4）。

2. 震后趋势判定

乌苏5.1级地震发生后，新疆地震局立即启动地震应急预案，安排各部门开展应急工作，电话通知当地地震机构立即了解灾情。分别派出地震现场工作组和由预报中心、地下水研究中心组成地震前兆跟踪组赴震区开展现场调查、震害评估和前兆震情跟踪工作。

震后现场工作组和新疆地震局预报中心人员密切跟踪序列的发展，对乌苏5.1级地震序列进行分析，综合该区历史地震序列类型和序列发展情况，认为该地震可能属孤立型。

3. 震害分析[1]

由于本次地震震中位于北天山高山区，震中区人烟稀少，也无重要的工程设施，因此无人员伤亡，造成总经济损失为126.39万元，其中乌苏市直接经济损失79.42万元，精河县直接经济损失46.97万元，属一般破坏性地震。

该地震距离震中最近的居民点为精河县托托乡（35km）和乌苏市古尔图镇（50km），仅有少量的老旧土木结构房屋产生轻微破坏，个别达到中等破坏，而这些房屋之前许多已是危房；个别没有采取抗震措施的砖木、砖混结构的房屋也产生了轻微破坏和中等破坏，其他大多数房屋完好。本次地震震中地区抗震安居房完好无损，受到损害的基本为当地基本无抗震设防的老旧房屋。

九、结论和讨论

2005年8月26日墨玉5.5级地震后，新疆境内5级地震活动平静达11个多月。2006年8月6日新疆边境中、塔、阿交界地区发生5.5级地震，9月12日在新疆境内和田西南发生5.8级地震。乌苏5.1级地震正是发生在新疆地区5级地震长达11个月的平静被打破之后。而和田5.8级地震前新疆3级、4级地震较为集中分布于西昆仑地震带，北天山地区3级以上地震异常平静。2006年10月后，北天山地区出现3、4级地震集中活跃，在半个月内连续发生4次4左右地震，中小地震活动表现出短期增强特点。

乌苏5.1级地震震前地震活动性异常较为显著，但多为中期异常。其他前兆异常基本为地倾斜的中短期异常，表现形式为矢量方向不稳，速率加快。地下流体异常仅2项，异常表现为水温的急剧下降和上升、流量增大、涌流液体中含有大量气体，是该地震前出现的仅有的两项短临异常。近场前兆异常多为中短期异常，虽然有短临异常表现，但由于异常比例较低且资料观测时间（附件4）短，因此无法为震前形势的判定提供依据。

此次地震发生在新疆维吾尔自治区地震局2006年度和2007年度北天山5～6级地震危险区内，地震三要素的预测较为准确（新疆地震局2006年度和2007年度分别在精河—石场和精河—新源—乌苏地区划定了5～6级地震危险区，2006年11月23乌苏5.1级地震就发生在这2个危险区内，三要素准确。那个时期危险区震级偏低三要素都准确吗？危险区震级肯定不是 5 级!)。该地震前新疆地震活动在前期显著平静背景下，短期内从境外至境内地震活动状态发生了明显转折变化。2006 年 7 月底新疆地震局预报中心根据前期地震活动资料的显著变化和部分定点前兆资料的变化，填报了 3 个月的短期预报意见。其后通过历史中强地震活动统计结果，并结合当时地震增强活动特点和新疆强震活动的动力条件等综合分析认为，“年底前或稍长时间，新疆有发生 6 级左右地震的可能，优势发震区域是南天山西段与西昆仑交汇地区，其次为天山中部地区”。9 月 20 日新疆地震局对震情提出明确的判定意见，并上报自治区人民政府和中国地震局[9]。乌苏 5.1 级地震虽然没有发生在预测区域内，地震强度也明显偏低，但也表明了新疆地震局在对近期震情发展的把握中有所查觉。

参 考 文 献

[1] 新疆地震局地震现场工作组，2006 年 11 月 23 日乌苏 5.1 级地震现场调查报告，2006

[2] 高国英、李莹甄、聂晓红等，2007，2006 年 11 月 23 日乌苏 5.1 级地震后北天山地震形势分析，内陆地震，(21) 2，113 ~ 118

[3] 新疆维吾尔自治区 2006 年度地震趋势研究报告，2005

[4] 2006 年中新疆地震趋势会商会部分震情研究报告汇编，2006

[5] 新疆维吾尔自治区 2007 年度地震趋势研究报告，2006

[6] 李莹甄、夏爱国、龙海英等，2007，2006 年 11 月 23 日新疆乌苏 5.1 级地震及部分地震学前兆异常特征，地震，(27) 4，121 ~ 128

[7] 夏爱国、赵翠萍，2006，天山中东段地区震源参数的初步应用研究，内陆地震，(20) 3，245 ~ 251

[8] 孙甲宁、温和平、杨晓芳，2008，2006 年乌苏 M_S5.1 地震前后北天山地倾斜资料异常分析，西北地震学报，(30) 1，66 ~ 70

[9] 高国英、李莹甄、聂晓红等，2007，2006 年 11 月 23 日乌苏 5.1 级地震后北天山地震形势分析，内陆地震，(21) 2，113 ~ 118

The Wusu, Xinjiang Uygur Autonomous Region, Earthquake of M_S 5.1 of November 23, 2006

Abstract

An earthquake of M_S5. 1 occurred in Wusu country on November 23, 2006 in Xinjiang Uygur Autonomous Region, The micro-epicenter is 44°15′N, 83°37′E, because the earthquake occurred in the mountain region, the meizoseismic area can not decided, the epicenter intensity maybe Ⅴ. No person was injured, the economic loss was 1. 2639 millions yuan RMB.

The earthquake sequence is isolation type, the most aftershock is M2. 8. the $M \geqslant 2.0$ aftershock distribute in the south side of main earthquake. The epicenter type is right slip reverse fault, the plane Ⅰ is the main rupture plane, the main stress P axis is northwest direction. Boluokenu-Aqikekudu fault is the seismic fault.

The earthquake occurred in the northern Tian Shan region with better monitoring ability region, there are two seismic stations, one ground temperature station, eight synthesis observation stations in its 200km region. There is ten abnormities of seismology, one abnormity of ground tiltmeters, one abnormity of water chemistry; but only one short-term abnormity.

报 告 附 件

附件一：2006 年 11 月 22 日周会商震情监视报告

秘密★三个月

震情监视报告

<table>
<tr><td>单位</td><td colspan="2">新疆地震局预报中心</td><td>会商会类型</td><td>周会商会</td></tr>
<tr><td rowspan="2">期数</td><td colspan="2">（2006）第 55 期</td><td>会商会地点</td><td>局会商室</td></tr>
<tr><td colspan="2">（总　字）第 393 期</td><td>会商会时间</td><td>2006 年 11 月 22 日 12 时 00 分</td></tr>
<tr><td>主持人</td><td colspan="2">王季达</td><td>发送时间</td><td>11 月 22 日 14 时 10 分</td></tr>
<tr><td>签发人</td><td colspan="2">王海涛</td><td>收到时间</td><td>月　日　时</td></tr>
<tr><td colspan="2">Apnet</td><td>AP65</td><td>发送人</td><td>杨欣</td></tr>
</table>

1. 本周地震活动及震情实况

2006 年 11 月 13 ~ 19 日，全疆共交出地震 145 次，其中 M_L1.0 ~ 1.9 级 79 次；2.0 ~ 2.9 级 56 次；3.0 ~ 3.9 级 9 次；4.0 ~ 4.9 级 1 次。最大地震为 2006 年 11 月 18 日伽师 M_L4.3。另外，边境地区共交出地震 21 次。分区统计情况如下：

乌鲁木齐地区 20 次（M_{Lmax} = 3.6）；北天山西段 19 次（M_{Lmax} = 3.0）；南天山东段 18 次（$M_{L,max}$ = 3.0）；柯坪块区 47 次（$M_{L,max}$ = 4.3）；乌恰地区 13 次（$M_{L,max}$ = 2.7）；西昆仑地区 6 次（$M_{L,max}$ = 3.3）；富蕴地区 1 次（M_{Lmax} = 1.9）；其他地区 21 次（M_{Lmax} = 3.6）。

本周全疆地震活动水平略高于上周。小地震相对集中分布在柯坪、乌鲁木齐地区和北天山西段。截至 11 月 20 日，共记录中塔交界 5.6 级地震余震 1994 次，序列 h 值为 1.1。

2. 前兆异常情况

沙湾 26 号泥火山泉水流量仍处于喷涌状态；11 月上旬阿勒泰地倾斜两分向同步出现的大速率短期不稳定变化可能与气温变化有关；库车台倾斜仪矢量方向仍处于持续不稳定状态；库尔勒台钻孔应变 NS 向继续处于张性异常上升状态，现异常量级已达 5398×10^{-8}。全疆其他前兆测项未出现明显的异常变化。

3. 综合分析

本周全疆地震活动水平仍然偏低。除北天山西段个别地下流体测项和南天山东段部分定点形变观测资料有异常显示外，全疆其他前兆资料未有明显短临异常变化。

4. 地震趋势预测

（1）未来一周新疆地震活动水平将与上周持平。

（2）乌鲁木齐地区未来一周发生 M_S5.0 以上地震的可能性不大。

（3）密切跟踪南北天山地震重点监视区内的震情变化。

附件二：2006 年度新疆地震趋势研究报告附表和危险区图

附表 2－1　2006 年度地震危险区预测及异常项目统计表

危险区	预测时段	危险区范围	预测震级	概率	主要异常项目表
喀什—乌恰地区	2005. 12～2006. 12	38. 2°～40. 7° 73. 5°～76. 8°	6～7 级	0. 78	地震空区、震群、地震窗；b 值、A 值空间扫描；缺震、地震频度、η 值、地震活动度 S 值时间扫描；阿合奇、阿图什台地倾斜
库车—拜城地区	2005. 12～2006. 12	41. 4°～42. 8° 80. 8°～84. 1°	6 级左右	0. 80	地震增强；地震窗；震群；地震频度时间扫描；b 值、A 值空间扫描；拜城、库车台地震尾波持续时间比；库尔勒台地倾叙、钻孔应变；阿克苏台断层测量
乌苏—精河地区	2005. 12～2006. 12	43. 2°～44. 5° 82. 6°～85. 6°	5～6 级	0. 75	空区、地震活动增强、尾波 Q 值、应力降；地震频度、A 值空间扫描；地震活动 S 值时间扫描；精河台倾斜仪、水管仪；石场、新源台地倾斜、独山子、呼图壁台断层测量；33 号井水流量、水温；4 号井水温、水位；4 号、10 号泉氡气、二氧化碳

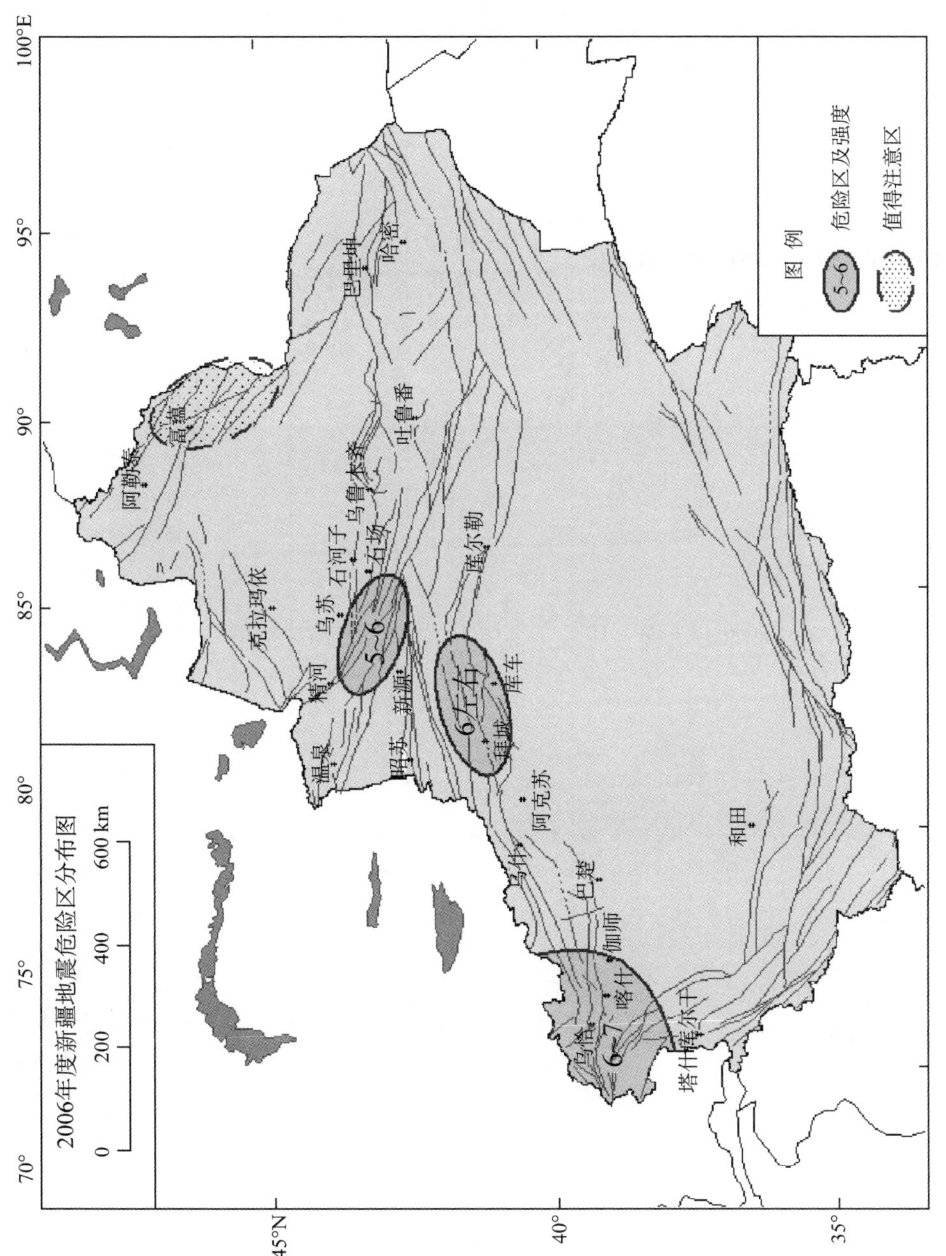

附图2-1　2006年度新疆地震危险区分布图

附件三：2006 年度新疆地震趋势研究报告附表和危险区图

附表 3－1　2007 年度地震危险区预测及异常项目统计表

危险区	预测时段	危险区范围	预测震级	概率	主要异常项目表
乌恰—塔什库尔干地区	2006. 12～2007. 12	37. 5°～40. 7° 73. 5°～76. 6°	6. 5 级左右	0. 78	地震地区、地震平静、地震条带、震群、地震窗；*b* 值、*A* 值空间扫描；缺震、地震频度空间扫描、*η* 值、地震活动度 *S* 值时间扫描、*A*（*b*）值、*GL* 值
库车—拜城地区	2006. 12～2007. 12	41. 2°～42. 7° 81. 4°～84. 7°	6 级左右	0. 80	地震增强、地震条带；地震窗；震群；地震频度时间扫描；*b* 值、*A* 值空间扫描；地震活动度 *S* 值时间扫描、*A*（*b*）、*GL* 值、*Rm* 值；库尔勒台地倾斜、钻孔应变；库车台地倾斜、克孜尔断层测量
乌苏—精河—新源地区	2006. 12～2007. 12	43. 0°～44. 5° 82. 5°～85. 3°	5～6 级	0. 75	地震空区；地震窗；应力降；地震频度、*b* 值、*η* 值、*Rm* 值、*A* 值空间扫描；精河台倾斜仪、水管仪；石场、新源、雅山台地倾斜、呼图壁台断层测量；33 号井水流量、水温；26 号泥火山泉水流量

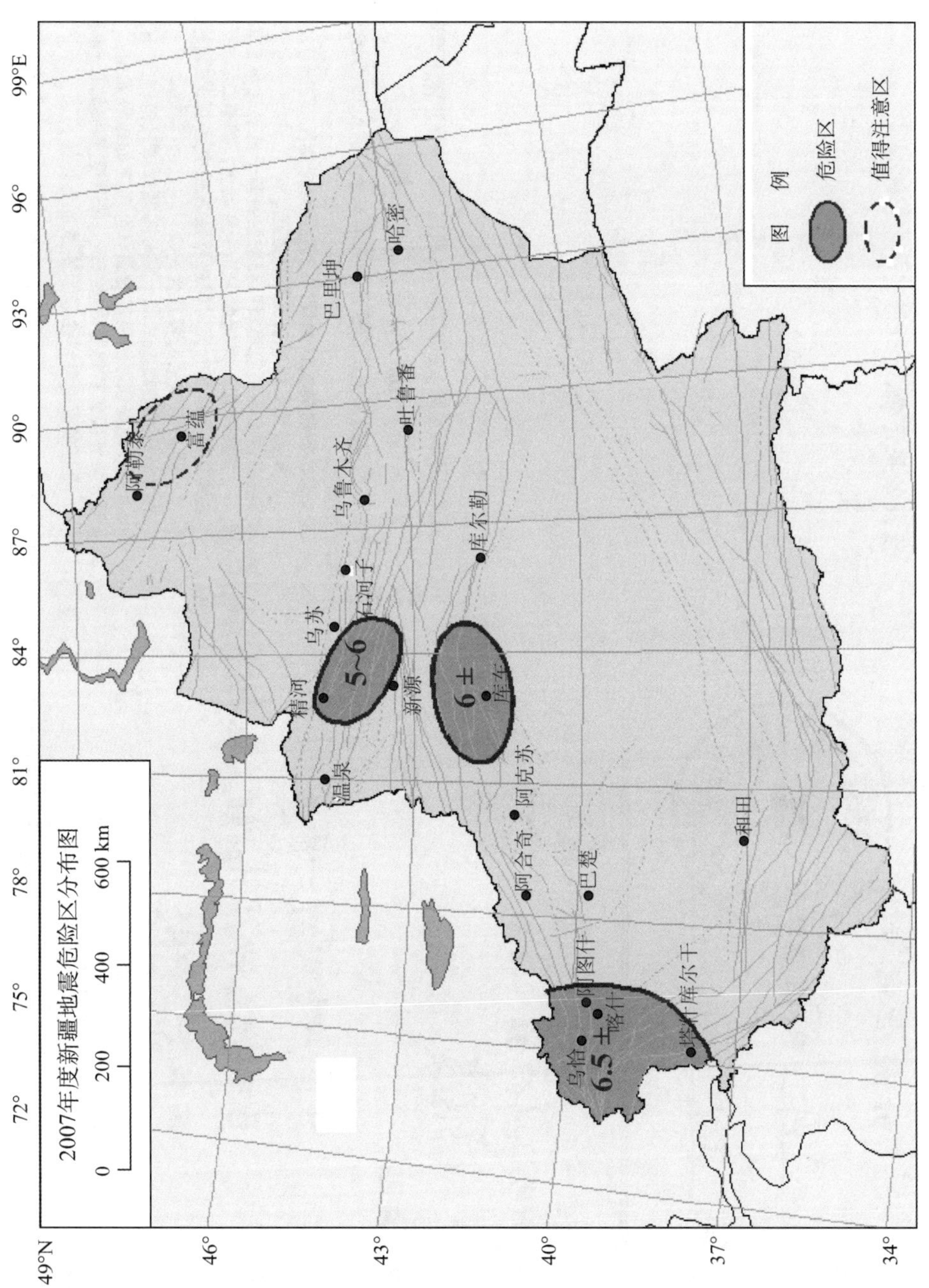

附图3-1　2007年度新疆地震危险区分布图

附件四：

分类会商卡片

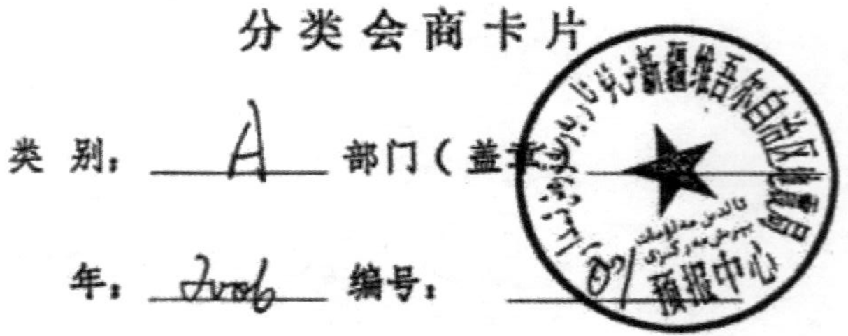

类　别：A　部门（盖章）：

年：2006　编号：01

会商时间：2006.07.26　地点：

主 持 人：刘国英

预报意见：

时　间：2006.08.01—2006.11.01

区　域：以44°N 42.7°，E84.0°为中心半径150km范围

震　级：5—6级

上报时间和部门：2006年7月26日
新疆地震局预报中心

预报效果评价：

说　　明

1. 类别：指分析预报工作管理条例第21条中的A、B、C、D四类。
2. 预报意见：
   A类必须明确填报地震可能发生的时间段（三个月内），哪些地县区域以及震级范围。
   B类必须明确填报地震可能发生的若干地县区域范围和震级范围，时间至少几个月以上，甚至二、三年。
   C类只指出未来一定时间内可能发生中强以上或较大地震。区域范围可较大。
   D类必须明确指出未来一定时间某区域范围内没有破坏性地震或可能引起较强社会反应的地震发生。
3. 预报效果评价包括地震活动实况、三要素预报的正确程度、预报依据的科学性以及决策能力评估。

预报根据

※※※※※※※※※※※※※※※※※※

（附 图）

1. 目前，新疆6级地震平静17个月，5级地震平静11个月。震例统计显示，当新疆境内5级地震平静达到11个月之后，新疆及周边地区发生6级以上地震的可能性很大。
2. 自2005年开始，南天山中东段就是新疆中级地震的集中活动区域，目前地震平静区域地震活动已出现结束一平静过程。
3. 7月21日精河4.1级地震打破北天山地区中级地震平静5个月左右。4.1级、3.7级（余震）地震的应力降值高于区域历史背景值1倍以上。
4. 南天山东段的库尔勒台3套定点形变资料近期都有不同期异常变化。库车台地倾斜仪在恢复正常过程中，再次出现异常变化。
5. 库尔勒台钻孔应变资料近期有明显异常变化。